ROM(s)

DeGarmo's

MATERIALS AND PROCESSES IN MANUFACTURING

TENTH EDITION

THE WILEY BICENTENNIAL—KNOWLEDGE FOR GENERATIONS

*E*ach generation has its unique needs and aspirations. When Charles Wiley first opened his small printing shop in lower Manhattan in 1807, it was a generation of boundless potential searching for an identity. And we were there, helping to define a new American literary tradition. Over half a century later, in the midst of the Second Industrial Revolution, it was a generation focused on building the future. Once again, we were there, supplying the critical scientific, technical, and engineering knowledge that helped frame the world. Throughout the 20th Century, and into the new millennium, nations began to reach out beyond their own borders and a new international community was born. Wiley was there, expanding its operations around the world to enable a global exchange of ideas, opinions, and know-how.

For 200 years, Wiley has been an integral part of each generation's journey, enabling the flow of information and understanding necessary to meet their needs and fulfill their aspirations. Today, bold new technologies are changing the way we live and learn. Wiley will be there, providing you the must-have knowledge you need to imagine new worlds, new possibilities, and new opportunities.

Generations come and go, but you can always count on Wiley to provide you the knowledge you need, when and where you need it!

WILLIAM J. PESCE
PRESIDENT AND CHIEF EXECUTIVE OFFICER

PETER BOOTH WILEY
CHAIRMAN OF THE BOARD

DeGarmo's
MATERIALS AND PROCESSES IN MANUFACTURING

TENTH EDITION

J T. Black
Auburn University-Emeritus

Ronald A. Kohser
University of Missouri–Rolla

John Wiley & Sons, Inc

Cover photos:

The sketchbooks of Leonardo da Vinci (1452-1519) contain two drawings that are of particular interest to the authors of this text. One is a crude sketch of an underwater device, or submarine, with the elongated sausage shape characteristic of many later day successes. The other, reproduced on the cover of this edition, is a "flying-machine," that bears an uncanny resemblance to a modern-day helicopter. Unlike many of Leonardo's creations, he apparently made no attempt to further refine the concepts, since there was never a subsequent sketch of either.

Was this man really such a genius? We have no way of knowing, but he may have realized that the construction materials of his day were totally inadequate for either task. One would not want to build a submarine or helicopter from wood, stone or leather. Today's submarines are constructed from corrosion-resistant, high-strength metals that are also selected for their ability to be fabricated by welding. Aerospace materials must offer high-strength and light-weight, along with fatigue- and fracture-resistance. The rotor arms of modern helicopters are now being made from fiber-reinforced composite materials. The components of the engine and drive assembly have some of the most demanding requirements of modern engineering.

The materials and processes presented in this book are the tools that enable ideas to be converted into reality. The myriad of manufactured items, and the range of uses and applications, demonstrates the success of those materials and processes. Like Leonardo, however, today's designers continue to push the limits–lighter, stronger, more corrosion resistant, closer to net-shape, more economical. New materials will certainly be developed, and new processes will expand our capabilities. It is the goal of this text to present the capabilities and limitations of current technology with a look toward future advances that hopefully will enable today's dreams to become tomorrow's reality.

EXECUTIVE EDITOR	Joseph Hayton
EXECUTIVE PUBLISHER	Don Fowley
ASSOCIATE PUBLISHER	Daniel Sayre
EDITORIAL ASSISTANT	Sandra Kim
MARKETING MANAGER	Phyllis Cerys
SENIOR PRODUCTION EDITOR	Sandra Dumas
CREATIVE DIRECTOR	Harry Nolan
DESIGNER	Hope Miller
MEDIA EDITORS	Steve Chasey/Lauren Sapira
PRODUCTION MANAGEMENT SERVICES	Ingrao Associates/Suzanne Ingrao
COVER PHOTO CREDITS:	
TOP PHOTO	© Kean Collection/Getty Images
BOTTOM PHOTO	© Michael Melford/The Image Bank/Getty Images

This book was set in 10/12 Times Ten by Preparé and printed and bound by *Courier/Westford*. The cover was printed by *Courier/Westford*.

This book is printed on acid free paper.

ISBN   978-0470-05512-0

Printed in the United States of America

10 9 8 7 6 5 4 3

In the world of manufacturing, significant changes and trends are having a profound impact on our everyday lives. Whether we like it or not, we all live in a technological society, a world of manufactured goods. Every day we come in contact with hundreds of manufactured items, from the bedroom to the kitchen, to the workplace, we use appliances, phones, cars, trains, and planes, TVs, VCRs, DVD's, furniture, clothing, and so on. These goods are manufactured in factories all over the world using manufacturing processes. What are the trends in the manufacturing world, and how do they impact manufacturing processes?

■ TRENDS IN MANUFACTURING

First, manufacturing has become a global activity with U.S. companies sending work to other countries (China, Taiwan, Mexico) to take advantage of low-cost labor, while many foreign companies are building plants in the United States, to be nearer their marketplace, The automobile manufacturers and their suppliers use just about every process described in this book and some that we do not describe, often because they are closely held secrets.

Second, many manufacturing companies are redesigning their factories (their manufacturing systems) becoming lean producers. and learning how to make goods better (higher quality), cheaper, faster in a flexible way (i.e., they are more responsive to the customers). Almost every plant that you can visit these days is doing something to make itself leaner. Many of them have adopted some version of the Toyota Production System. More importantly, these manufacturing factories are designed with the internal customer (the workforce) in mind, so things like ergonomics and safety are key design requirements. So while this book is all about materials and processes for making the products, the design of the factory cannot be ignored when it comes to making the external customer happy with the product and the internal customer satisfied with the employer.

Third, the number and variety of products and the materials from which they are made continue to proliferate, while production quantities have become smaller. Existing processes must be modified to be more flexible, and new processes must be developed.

Fourth, consumers want better quality and reliability, so the methods, processes, and people responsible for the quality must be continually improved. The trend toward zero defects and continuous improvement requires continuous improvements of the manufacturing system.

Finally the new product development effort to reduce the *time-to-market* for new products is continuing. Many companies are taking wholistic or system wide perspectives, including concurrent engineering efforts to bring product design and manufacturing closer to the customer. There are two key aspects here. First, products are designed to be easier to manufacture and assemble (*called design for manufacture/assembly*). Second, the manufacturing system design is flexible (able to accept new products), so the company can be competitive in the global marketplace.

Basically, manufacturing is a *value-adding* activity, where the conversion of materials into products adds value to the original material. Thus, the objective of a company engaged in manufacturing is to add value and do so in the most efficient manner, using the least amount of time, material, money, space, and labor. To minimize waste and maximize efficiency, the processes and operations need to be properly selected and arranged to permit smooth and controlled flow of material through the factory and provide for product variety. Meeting these goals requires a well-designed and efficient manufacturing system.

■ PURPOSE OF THE BOOK

The purpose of this book is to give design and manufacturing engineers and technicians basic information on materials, manufacturing processes and systems. The materials section focuses on properties and behavior. Thus, aspects of smelting and refining (or other material production processes) are presented only as they affect manufacturing and manufactured products. In terms of the processes used to manufacture items (converting materials into products), this text seeks to provide a descriptive introduction to a wide variety of options, emphasizing how each process works and its relative advantages and limitations. Our goal is to present this material in a way that can be understood by individuals seeing it for the very first time. This is not a graduate text where the objective is to thoroughly understand and optimize manufacturing processes. Mathematical models and analytical equations are used only when they enhance the basic understanding of the material. So, while the text is an introductory text, we do attempt to incorporate new and emerging technologies like a welding process that is being adapted to alter and improve material properties and performance without creating a joint.

The book also serves to introduce the *language of manufacturing*. Just as there is a big difference between a gun hand and a hand gun, there is a big difference between an engine lathe and a lathe engine. Everyday English words (words like *climb, bloom, allowance, chuck, coin, head,* and *ironing*) have entirely different meanings on the factory floor, a place where misunderstandings can be very costly. Pity the engineer who has to go on the plant floor not knowing an engine lathe from a milling machine or what a press brake can do. This engineer quickly loses all credibility with the people who make the products (and pay the engineers' salaries). However, the modern manufacturing engineer must be able to deal with real workplace problem-solving techniques like Taguchi methods and six sigma and developing manufacturing cells to make product families. This requires redesign of all the elements of the manufacturing systems—the machine tools and manufacturing processes, the workholding devices, the material handling equipment, and the retraining of the people who work in the system.

■ HISTORY OF THE TEXT

In 1957, E. Paul DeGarmo was a mechanical engineering professor at the University of California, Berkley when he wrote the first edition of Materials and Processes in Manufacturing. The book quickly became the emulated standard for introductory texts in manufacturing. Second, third, and fourth editions followed in 1962, 1969, and 1974. DeGarmo had begun teaching at Berkeley in 1937, after earning his M.S. in mechanical engineering from California Institute of Technology. He worked as a factory control engineer at Firestone Tire and Rubber Company while attending Caltech. DeGarmo was a founder of the Department of Industrial Engineering (now Industrial Engineering and Operations Research) and served as its chair from 1956–1960. He was also assistant dean of the College of Engineering for three years while continuing his teaching responsibilities.

He retired from active teaching in 1971 and he continued his research, writing, and consulting for many years. In 1977, after the publication of the fourth edition of *Materials and Processes in Manufacturing*, he received a letter from Ron Kohser, then an assistant professor at Missouri-Rolla who had many suggestions regarding the materials chapters. DeGarmo asked Kohser to rewrite those chapters for the fifth edition, which Ron did. After the fifth edition DeGarmo decided he was really going to retire and after a national search, recruited J T. Black, then a Professor at Ohio State, to co-author the book. For the sixth edition, seventh edition, eighth and ninth editions (published in 1984, 1988, and 1997, respectively, by Macmillan, Prentice Hall and 1999 and 2003 by John Wiley & Sons), Ron Kohser and J T. Black have shared the responsibility for the text. The chapters on engineering materials, casting, forming, powder metallurgy, joining and non-destructive testing have been written or revised by Ron Kohser. J T. Black has assumed the responsibility for the introduction and chapters on material removal, metrology, surface finishing, quality control and manufacturing systems design.

DeGarmo died in 2000, three weeks short of his 93rd birthday. His wife Mary died in 1995; he is survived by his sons, David and Richard, and many grandchildren. For this 10th edition, we honor our mentor E. Paul DeGarmo with a change in the title to include his name. We are forever indebted to Paul for selecting us to carry on the tradition of his book on its' fiftieth anniversary!

■ 50TH ANNIVERSARY EDITION!

Any long-term user of this book will note a significant change in its title—from *Materials and Processes in Manufacturing* by DeGarmo, Black, and Kohser to *DeGarmo's Materials and Processes in Manufacturing* by Black and Kohser. Paul DeGarmo initiated this text in 1957 and nurtured it through a number of editions. Even after his retirement, through his death in 2000, Paul maintained an active interest and involvement. In recognition, the 9th edition, published in 2003, carried his name as a posthumous coauthor. For 50 years, this text has been known by many as simply "DeGarmo," and it is this identity that we wish to continue by moving his name to become a preface to the former title.

In 1957 Dr. DeGarmo observed that engineering education had begun to place more emphasis on the underlying sciences at the expense of hands on experience. Most of his students were coming to college with little familiarity with materials, machine tools, and manufacturing methods that their predecessors had acquired through the old "shop" classes. If these engineers and technicians were to successfully convert their ideas into reality, they needed a foundation in materials and processes, with emphasis on their opportunities and their limitations. He sought to provide a text that could be used in either a one- or two-semester course designed to meet these objectives. The materials sections were written with an emphasis on use and application. Processes and machine tools were described in terms of what they could do, how they do it, and their relative advantages and limitations, including economic considerations. Recognizing that many students would be encountering the material for the first time, clear description was accompanied by numerous visual illustrations.

Paul's efforts were well received, and the book quickly became the standard text in many schools and curricula. As materials and processes evolved, advances were incorporated into subsequent editions. Computer usage, quality control, and automation were added to the text, along with other topics, so that it continued to provide state-of-the-art instruction in both materials and processes. As competing books entered the market, one was forced to note that their subject material and organization tended to mimic the DeGarrno text.

Professors Black and Kohser are proud to continue Paul's legacy. It is fitting that this 10th edition will be published in 2007, 50 years following the initial efforts of Professor DeGarmo. It is further fitting that his name continue to appear on this 50th anniversary edition and any subsequent editions.

■ THE 10TH EDITION

E. Paul DeGarmo wanted a book that explained to engineers how the things they designed are made. *DeGarmo's Materials and Processes in Manufacturing* is still written providing a broad, basic introduction to the fundamentals of manufacturing. The book begins with a survey of engineering materials, the "stuff" that manufacturing begins with, and seeks to provide the basic information that can be used to match the properties of a material to the service requirements of a component. A variety of engineering materials are presented, along with their properties and means of modifying them. The materials section can be used in curricula that lack preparatory courses in metallurgy, materials science, or strength of materials, or where the student has not yet been exposed to those topics. In addition, various chapters in this section can be used as supplements to a basic materials course, providing additional information on topics such as heat treatment, plastics, composites, and material selection.

Following the materials chapters, measurement and nondestructive testing are introduced with a manufacturing perspective. Then chapters on casting, forming, powder metallurgy, material removal, and joining are all developed as families of manufacturing processes.

Each section begins with a presentation of the fundamentals on which those processes are based. This is followed by a discussion of the various process alternatives, which can be selected to operate individually or be combined into an integrated system.

In the last two chapters there is some in depth material on surface engineering and quality control. Engineers need to know how to determine process capability and if they get involved in six sigma projects, to know what sigma really measures. There is also introductory material on surface integrity, since so many processes produce the finished surface and residual stresses in the components.

■ WHAT'S NEW IN 10e:

- New chapter on measurement, inspection and testing
- New chapter on *electronic processes*
- New examples of basic calculations in machining chapters
- NC chapter reorganized with more examples
- Reclassification of metal deformation processes into bulk and sheet
- Expanded coverage of new and emerging technology, such as friction-stir welding
- Expanded coverage of polymers; ceramic materials and composites, and the processes that are unique to those materials.

Throughout the book, case studies have been designed to make students aware of the great importance of properly coordinating design, material selection, and manufacturing to produce a satisfactory and reliable product.

The text is intended for use by engineering (mechanical, manufacturing, and industrial) and engineering technology students, in both two- and four-year undergraduate degree programs. In addition, the book is also used by engineers and technologists in other disciplines concerned with design and manufacturing (such as aerospace and electronics). Factory personnel will find this book to be a valuable reference that concisely presents the various production alternatives and the advantages and limitations of each. Additional or more in-depth information on specific materials or processes can be found in the various references posted on the internet along with chapters on rapid prototyping, automation and enterprise systems.

■ SUPPLEMENTS

For instructors adopting the text for use in their course, an *instructor solutions manual* is available through the book website: www.wiley.com/college/degarmo. Also available on the website is a set of *powerpoint lecture slides* created by Philip Appel at Gonzaga University.

Three additional chapters, as identified in the table of contents, are available on the book website. The registration card attached on the inside front cover provides information on how to access and download this material. If the registration card is missing, access can be purchased directly on the website www.wiley.com/college/degarmo, by clicking on "student companion site" and then an the links to the chapter titles.

■ ACKNOWLEDGMENTS

The authors wish to acknowledge the multitude of assistance, information, and illustrations that have been provided by a variety of industries, professional organizations, and trade associations. The text has become known for the large number of clear and helpful photos and illustrations that have been graciously provided by a variety of sources. In some cases, equipment is photographed or depicted without safety guards, so as to show important details, and personnel are not wearing certain items of safety apparel that would be worn during normal operation.

Over the many editions, there have been hundreds of reviewers, faculty, and students who have made suggestions and corrections to the text. We continue to be grateful for the time and interest that they have put into this book. In this edition we benefited from the comments of the following reviewers: J. Don Book, Pittsburg State University;

Jan Brink, Midwestern State University; Rene A.Chapelle, University of Houston; Joe Chow, Florida International University; Kurt Colvin, California Polytechnic State University, Pomona; Subi Dinda, Oakland University; Roman Dubrovsky, New Jersey Institute of Technology; Richard B. Griffin, Texas A&M University–Main; Rodney G. Handy, Purdue University; T. Kesavadas, State University of New York, Buffalo; John Lee, San Jose State University; H. Joel Lenoir, Western Kentucky University; Steven Y.Liang, Georgia Institute of Technology;Victor Okhuysen, California Polytechnic State University, Pomona; Lewis N. Payton, Auburn University; Zhijian Pei, Kansas State University; William Schoech, Valparaiso University; Mala M.Sharma, Bucknell University; Bharat S.Thakkar, Illinois Institute of Technology; and Alan Zoyhofski, Rochester Institute of Technology.

The authors would also like to acknowledge the contributions of Dr. Elliot Stern for the dynamics of machining section in Chapter 20. Dr. Brian Paul for his work on the rapid prototyping and electronics chapters, and Dr. Barney Klamecki for his help with the 9th edition.

As always, our wives have played a major role in preparing the manuscript. Carol Black and Barb Kohser have endured being "textbook widows" during the time when the last four editions were written. Not only did they provide loving support, but Carol also provided hours of expert proofreading, typing, and editing as the manuscript was prepared.

Finally special thanks to our acquisitions editor, Joseph P. Hayton, for putting up with two procrastinating professors, who tried both his patience and his abilities as he coordinated all the various activities required to produce this text as scheduled. We also thank Suzanne Ingrao and Sandra Dumas for all their help in bringing the 10th edition to reality.

■ ABOUT THE AUTHORS

J T. Black received his Ph.D. from Mechanical and Industrial Engineering, University of Illinois, Urbana in 1969, an M.S. in Industrial Engineering from West Virginia University in 1963 and his B.S. in Industrial Engineering, Lehigh University in 1960. J T. is Professor Emeritus from Industrial and Systems Engineering in the Samuel Ginn College of Engineering at Auburn University. He was the Chairman and a Professor of Industrial and Systems Engineering at The University of Alabama-Huntsville He also taught at The Ohio State University, the University of Rhode Island, the University of Vermont, the University of Illinois and West Virginia University. J T. is a Fellow in the American Society of Mechanical Engineers, the Institute of Industrial Engineering and the Society of Manufacturing Engineers. J loves to write music (mostly down home country) and poetry. Co-authoring with Ron Kohser makes this book a success, just as picking his doubles partner in tennis has given him the #1 doubles ranking for 65 year olds in the State of Alabama.

Ron Kohser received his Ph.D. from the Lehigh University Institute for Metal Forming in 1975. Ron is currently in his 32nd year on the faculty of the University of Missouri-Rolla, where he is a Professor of Metallurgical Engineering and Dean's Teaching Scholar. While maintaining a full commitment to classroom instruction, he has served as department chair and Associate Dean for Undergraduate Instruction. He currently teaches courses in Metallurgy for Engineers, Introduction to Manufacturing Processes, and Material Selection, Fabrication and Failure Analysis. In addition to the academic responsibilities, Ron and his wife Barb operate *A Miner Indulgence*, a bed-and-breakfast in Rolla, Missouri.

CONTENTS

Contents **xiii**

Chapter 37   Manufacturing Automation (web-based chapter)

(www.wiley.com/college/DeGarmo)

Chapter 38   The Enterprise (web-based chapter)

(www.wiley.com/college/DeGarmo)

Chapter 39   Rapid Prototyping, Tooling And Fabrication (web-based chapter)

(www.wiley.com/college/DeGarmo)

INTRODUCTION TO DEGARMO'S MATERIALS AND PROCESSES IN MANUFACTURING

■ 1.1 MATERIALS, MANUFACTURING, AND THE STANDARD OF LIVING

Manufacturing is critical to a country's economic welfare and standard of living because the standard of living in any society is determined, primarily, by the *goods* and *services* that are available to its people. Manufacturing companies contribute about 20% of the GNP, employ about 18% of the workforce, and account for 40% of the exports of the United States. In most cases, materials are utilized in the form of manufactured goods. Manufacturing and assembly represent the organized activities that convert raw materials into salable goods. The manufactured goods are typically divided into two classes: producer goods and consumer goods. *Producer goods* are those goods manufactured for other companies to use to manufacture either producer or consumer goods. *Consumer goods* are those purchased directly by the consumer or the general public. For example, someone has to build the machine tool (a lathe) that produces (using machining processes) the large rolls that are sold to the rolling mill factory to be used to roll the sheets of steel that are then formed (using dies) to become the body panels of your car. Similarly, many service industries depend heavily on the use of manufactured products, just as the agricultural industry is heavily dependent on the use of large farming machines for efficient production.

Converting materials from one form to another adds value to them. The more efficiently materials can be produced and converted into the desired products that function with the prescribed quality, the greater will be the companies' productivity and the better will be the standard of living of the employees.

The history of man has been linked to his ability to work with materials, beginning with the Stone Age and ranging through the eras of copper and bronze, the Iron Age, and recently the age of steel. While ferrous materials still dominate the manufacturing world, we are entering the age of tailor-made plastics, composite materials, and exotic alloys.

A good example of this progression is shown in Figure 1-1. The goal of the manufacturer of any product or service is to continually improve. For a given product or service, this improvement process usually follows an S-shaped curve, as shown in Figure 1-1(a), often called a product life-cycle curve. After the initial invention/creation, a period of rapid growth in performance occurs, with relatively few resources required. However, each improvement becomes progressively more difficult. For a delta gain,

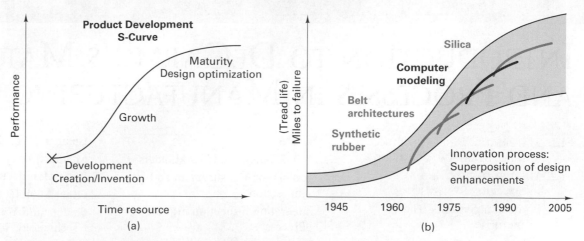

FIGURE 1-1 (a) A product development curve usually has an "S"-shape. (b) Example of the S-curve for the pneumatic radial tire. *(Courtesy of: Bart Thomas, Michelin).*

more money and time and ingenuity are required. Finally, the product or service enters the maturity phase, during which additional performance gains become very costly.

For example, in the automobile tire industry, Figure 1-1b shows the evolution of radial tire performance from its birth in 1946 to the present. Growth in performance is actually the superposition of many different improvements in material, processes, and design.

These innovations, known as *sustaining technology*, serve to continually bring more value to the consumer of existing products and services. In general, sustaining manufacturing technology is the backbone of American industry and the ever-increasing productivity metric.

Although materials are no longer used only in their natural state, there is obviously an absolute limit to the amounts of many materials available here on earth. Therefore, as the variety of man-made materials continues to increase, resources must be used efficiently and recycled whenever possible. Of course, recycling only postpones the exhaustion date.

Like materials, processes have also proliferated greatly in the last 40 years, with new processes being developed to handle the new materials more efficiently and with less waste. A good example is the laser, invented around 1960, which now finds many uses in manufacturing, measurement, inspection, heat treating, welding, and more. New developments in manufacturing technology often account for improvements in productivity. Even when the technology is proprietary, the competition often gains access to it, usually quite quickly.

Starting with the product design, materials, labor, and equipment are interactive factors in manufacturing that must be combined properly (integrated) to achieve low cost, superior quality, and on-time delivery. Typically, as shown in Figure 1-2, 40% of the

FIGURE 1-2 Manufacturing cost is the largest part of the selling price, usually around 40%. The largest part of the manufacturing cost is materials, usually 50%.

selling price of a product is *manufacturing cost*. Since the selling price is determined by the customer, maintaining profit often depends on reducing manufacturing cost. The internal customers who really make the product, called direct labor, are usually the targets of automation, but typically they account for only about 10% of the manufacturing cost even though they are the main element in increasing productivity. In Chapter 39, a manufacturing strategy is presented that attacks the materials cost, indirect costs, and general administration costs, in addition to labor costs. The materials costs include the cost of storing and handling the materials within the plant. The strategy is called *lean production*.

Referring again to the total expenses shown in Figure 1-2 (selling price less profit), about 68% of dollars are spent on people, the breakdown being about 15% for engineers; 25% for marketing, sales, and general management people; 5% for direct labor, and 10% for indirect labor. The average labor cost in manufacturing in the United States was around $15 per hour for hourly workers in 2000. Reductions in direct labor will have only marginal effects on the total people costs. The optimal combination of factors for producing a small quantity of a given product may be very inefficient for a larger quantity of the same product. Consequently, a systems approach, taking all the factors into account, must be used. *This requires a sound and broad understanding on the part of the decision makers on the value of materials, processes, and equipment to the company, accompanied by an understanding of the manufacturing systems.* Materials and processes in manufacturing systems are what this book is all about.

■ 1.2 MANUFACTURING AND PRODUCTION SYSTEMS

Manufacturing is the economic term for making goods and services available to satisfy human wants. Manufacturing implies creating value by applying useful mental or physical labor. The *manufacturing processes* are collected together to form a *manufacturing system* (MS). The manufacturing system is a complex arrangement of physical elements characterized by measurable parameters (Figure 1-3). The manufacturing system takes inputs and produces products for the external customer.

The entire company is often referred to as the enterprise or, in this textbook, the production system. The production system includes the manufacturing system, as shown in Figure 1-4, and services it. In this book, a *production system* will refer to the total company and will include within it the *manufacturing system (SPSs)*. The production system includes the manufacturing system plus all the other functional areas of the plant for information, design, analysis, and control. These subsystems are connected by various means to each other to produce either goods or services or both.

Goods refer to material things. *Services* are nonmaterial things that we buy to satisfy our wants, needs, or desires. *Service production systems (SPSs)* include transportation, banking, finance, savings and loan, insurance, utilities, health care, education, communication, entertainment, sporting events, and so forth. They are useful labors that do not directly produce a product. Manufacturing has the responsibility for designing processes (sequences of operations and processes) and systems to create (make or

FIGURE 1-3
The manufacturing system converts inputs to outputs using processes to add value to the goods for the external customer.

FIGURE 1-4 The functions and systems of the production system, which includes (and services) the manufacturing system. The functional departments are connected by formal and informal information systems designed to service the manufacturing system that produces the goods.

manufacture) the product as designed. The system must exhibit flexibility to meet customer demand (volumes and mixes of products) as well as changes in product design.

As shown in Table 1-1, production terms have a definite rank of importance, somewhat like rank in the army. Confusing *system* with *section* is similar to mistaking a colonel for a corporal. In either case, knowledge of rank is necessary. The terms tend to overlap because of the inconsistencies of popular usage.

An obvious problem exists here in the terminology of manufacturing and production. The same term can refer to different things. For example, *drill* can refer to the machine tool that does these kinds of operations; the operation itself, which can be done on many different kinds of machines; or the cutting tool, which exists in many different forms. It is therefore important to use modifiers whenever possible: "Use the *radial* drill *press* to drill a hole with a 1-in.-diameter spade drill." The emphasis of this book will be

TABLE 1-1   Production Terms for Manufacturing Production Systems

Term	Meaning	Examples
Production system The enterprise	All aspects of workers, machines, and information, considered collectively, needed to manufacture parts or products; integration of all units of the system is critical.	Company that makes engines, assembly plant, glassmaking factory, foundry; sometimes called the enterprise or the business.
Manufacturing system (sequence of operations, collection of processes)	The collection of manufacturing processes and operations resulting in specific end products; an arrangement or layout of many processes, materials-handling equipment, and operators.	Rolling steel plates, manufacturing of automobiles, series of connected operations or processes, a job shop, a flow shop, a continuous process.
Machine or machine tool or manufacturing process	A specific piece of equipment designed to accomplish specific processes, often called a *machine tool;* machine tools linked together to make a manufacturing system.	Spot welding, milling machine, lathe, drill press, forge, drop hammer, die caster, punch press, grinder, etc.
Job (sometimes called a *station;* a collection of tasks)	A collection of operations done on machines or a collection of tasks performed by one worker at one location on the assembly line.	Operation of machines, inspection, final assembly; e.g., forklift driver has the job of moving materials.
Operation (sometimes called a *process*)	A specific action or treatment, often done on a machine, the collection of which makes up the job of a worker.	Drill, ream, bend, solder, turn, face, mill extrude, inspect, load.
Tools or tooling	Refers to the implements used to hold, cut, shape, or deform the work materials; called *cutting tools* if referring to machining; can refer to *jigs* and *fixtures* in workholding and *punches* and *dies* in metal forming.	Grinding wheel, drill bit, end milling cutter, die, mold, clamp, three-jaw chuck, fixture.

directed toward the understanding of the processes, machines, and tools required for manufacturing and how they interact with the materials being processed. In the last section of the book, an introduction to systems aspects is presented.

PRODUCTION SYSTEM—THE ENTERPRISE

The highest-ranking term in the hierarchy is *production system.* A production system includes people, money, equipment, materials and supplies, markets, management, and the manufacturing system. In fact, all aspects of commerce (manufacturing, sales, advertising, profit, and distribution) are involved. Table 1-2 provides a partial list of production systems.

TABLE 1-2   Partial List of Production Systems for Producer and Consumer Goods

Aerospace and airplanes	Foods (canned, dairy, meats, etc.)
Appliances	Footwear
Automotive (cars, trucks, vans, wagons, etc.)	Furniture
Beverages	Glass
Building supplies (hardware)	Hospital suppliers
Cement and asphalt	Leather and fur goods
Ceramics	Machines
Chemicals and allied industries	Marine engineering
Clothing (garments)	Metals (steel, aluminum, etc.)
Construction	Natural resources (oil, coal, forest, pulp and paper)
Construction materials (brick, block, panels)	Publishing and printing (books, CDs, newspapers)
Drugs, soaps, cosmetics	Restaurants
Electrical and microelectronics	Retail (food, department stores, etc.)
Energy (power, gas, electric)	Ship building
Engineering	Textiles
Equipment and machinery (agricultural, construction	Tire and rubber
and electrical products, electronics, household products,	Tobacco
industrial machine tools, office equipment, computers,	Transportation vehicles (railroad, airline, truck, bus)
power generators)	Vehicles (bikes, cycles, ATVs, snowmobiles)

TABLE 1-3   Types of Service Industries
Advertising and marketing
Communication (telephone, computer networks)
Education
Entertainment (radio, TV, movies, plays)
Equipment and furniture rental
Financial (banks, investment companies, loan companies)
Health care
Insurance
Transportation and car rental
Travel (hotel, motel, cruise lines)

Much of the information given *for manufacturing production systems* (MPSs) is relevant to the *service production system*. Many MPSs require an SPS for proper product sales. This is particularly true in industries, such as the food (restaurant) industry, in which customer service is as important as quality and on-time delivery. Table 1-3 provides a short list of service industries.

MANUFACTURING SYSTEMS

A collection of operations and processes used to obtain a desired product(s) or component(s) is called a *manufacturing system*. The manufacturing system is therefore the design or *arrangement of the manufacturing processes*. Control of a system applies to overall control of the whole, not merely of the individual processes or equipment. The entire manufacturing system must be controlled in order to schedule and control production, inventory levels, product quality, output rates, and so forth.

MANUFACTURING PROCESSES

A *manufacturing process* converts unfinished materials to finished products, often using machines or machine tools. For example, injection molding, die casting, progressive stamping, milling, arc welding, painting, assembling, testing, pasteurizing, homogenizing, and annealing are commonly called *processes* or *manufacturing processes*. The term *process* often implies a sequence of steps, processes, or operations for production of goods and services, as shown in Figure 1-5, which shows the processes to manufacture an Olympic-type medal.

A *machine tool* is an assembly of related mechanisms on a frame or bed that together produce a desired result. Generally, motors, controls, and auxiliary devices are included. Cutting tools and workholding devices are considered separately.

A machine tool may do a single process (e.g., cutoff saw) or multiple processes, or it may manufacture an entire component. Machine sizes vary from a tabletop drill press to a 1000-ton forging press.

JOB AND STATION

In the classical manufacturing system, a *job* is the total of the work or duties a worker performs. A station is a location or area where a production worker performs tasks or his job.

A job is a group of related operations and tasks performed at one station or series of stations in cells. For example, the job at a final assembly station may consist of four tasks:

1. Attach carburetor.
2. Connect gas line.
3. Connect vacuum line.
4. Connect accelerator rod.

The job of a turret lathe (a semiautomatic machine) operator may include the following operations and tasks: load, start, index and stop, unload, inspect. The operator's job may also include setting up the machine (i.e., getting ready for manufacturing). Other machine operations include drilling, reaming, facing, turning, chamfering, and knurling. The operator can run more than one machine or service at more than one station.

The terms *job* and *station* have been carried over to unmanned machines. A job is a group of related operations generally performed at one station, and a *station* is a position or location in a machine (or process) where specific operations are performed. A simple machine may have only one station. Complex machines can be composed of many stations. The job at a station often includes many simultaneous operations, such as "drill all five holes" by multiple spindle drills. In the planning of a job, a process plan is often developed (by the engineer) to describe how a component is made using a

How an olympic medal is made using the CAD/CAM process

1) An oversized 3D plaster model is made from the artist's conceptual drawings.

Artist's model of medal

Laser scan model

Model of medal

2) The model is scanned with a laser to produce a digital computer called a Computer-aided design (CAD).

CAD

Computer

CAM

CNC machine tool

Top

Die set

Cavity in die forms medal

Bottom

3) The computer has software to produce a program to drive numerical control machine to cut a die set.

4) Blanks are cut from bronze medal sheet stock using an abrasive water jet under 2-axis CNC control.

High-pressure water inlet

Abrasive cutting head

Sheet stock (bronze)

Blank

Air supply port valve

Abrasive metering system

Abrasive feed line

5) The blanks are heated and placed between the top die and bottom die. Very high pressure is applied by a press at very slow rates. The blank plastically deforms into the medal. This press is called hot isostatic pressing.

Blank

Formed medal

Additional finishing steps in the process include chemical etching; gold or silver plating; packaging

FIGURE 1-5 The manufacturing process for making Olympic medals has many steps or operations, beginning with design and including die making.

	Simplified Sequence of Operations
Raw material bar stock cylinder with flat ends	Cut bar stock to length; centerdrill ends.
Multiple cylinders made by turning (see Figure 1-12)	Turn (on a lathe); rough turn and finish turn.
External cylindrical / Three external cylinders and four flats	Turn (on a lathe) the smaller external cylindrical surfaces.
Three cylinders and six flats	Mill the flat on the right end. Mill the slot on the left end.
Four internal holes	Drill four holes on left end. Tap (internal threads) holes.

FIGURE 1-6   The component called a pinion shaft is manufactured by a "sequence of operations" to produce various geometric surfaces. The engineer figures out the sequence and selects the tooling to perform the steps.

sequence of operation. So, for example, the part shown in Figure 1-6 is produced by a set of machining operations. This information can be embedded in a CNC program, as shown in Figure 1-5.

OPERATION

An *operation* is a distinct action performed to produce a desired result or effect. Typical manual machine operations are loading and unloading. Operations can be divided into suboperational elements. For example, loading is made up of picking up a part, placing part in jig, closing jig. However, suboperational elements will not be discussed here.

Operations categorized by function are:

1. *Materials handling and transporting*: change in position of the product
2. *Processing*: change in volume and quality, including assembly and disassembly; can include packaging
3. *Packaging*: special processing; may be temporary or permanent for shipping
4. *Inspecting and testing*: comparison to the standard or check of process behavior
5. *Storing*: time lapses without further operations

These basic operations may occur more than once in some processes, or they may sometimes be omitted. *Remember, it is the manufacturing processes that change the value and quality of the materials*. Defective processes produce poor quality or scrap. Other operations may be necessary but do not, in general, add value, whereas operations performed by machines that do material processing usually do add value.

TREATMENTS

Treatments operate continuously on the workpiece. They usually alter or modify the product-in-process without tool contact. Heat treating, curing, galvanizing, plating, finishing, (chemical) cleaning, and painting are examples of treatments. Treatments usually add value to the part.

These processes are difficult to include in cells because they often have long cycle times, are hazardous to the workers' health, or are unpleasant to be around because of high heat or chemicals. They are often done in large tanks or furnaces or rooms. The cycle time for these processes may *dictate* the cycle times for the entire system. These operations also tend to be material specific. Many manufactured products are given decorative and protective surface treatments that control the finished appearance. A customer may not buy a new vehicle because it has a visible defect in the chrome bumper, although this defect will not alter the operation of the car.

TOOLS, TOOLING, AND WORKHOLDERS

The lowest mechanism in the production term rank is the *tool*. Tools are used to hold, cut, shape, or form the unfinished product. Common hand tools include the saw, hammer, screwdriver, chisel, punch, sandpaper, drill, clamp, file, torch, and grindstone.

Basically, machines are mechanized versions of such hand tools and are called cutting tools. Some examples of tools for cutting are drill bits, reamers, single-point turning tools, milling cutters, saw blades, broaches, and grinding wheels. Noncutting tools for forming include extrusion dies, punches, and molds.

Tools also include workholders, jigs, and fixtures. These tools and cutting tools are generally referred to as the *tooling*, which usually must be considered (purchased) separate from machine tools. Cutting tools wear and fail and must be periodically replaced before parts are ruined. The workholding devices must be able to locate and secure the workpieces during processing in a repeatable, mistake-proof way.

TOOLING FOR MEASUREMENT AND INSPECTION

Measuring tools and instruments are also important for manufacturing. Common examples of measuring tools are rulers, calipers, micrometers, and gages. Precision devices that use laser optics or vision systems coupled with sophisticated electronics are becoming commonplace. Vision systems and coordinate measuring machines are becoming critical elements for achieving superior quality.

INTEGRATING INSPECTION INTO THE PROCESS

The integration of the *inspection* process into the manufacturing process or the manufacturing system is a critical step toward building products of superior quality. An example will help. Compare an electric typewriter with a computer that does word processing. The electric typewriter is flexible. It types whatever words are wanted in whatever order. It can type in Pica, Elite, or Orator, but the font (disk or ball that has the appropriate type size on it) has to be changed according to the size and face of type wanted. The computer can do all of this but can also, through its software, do italics, darken the words, vary the spacing to justify the right margin, plus many other functions. It checks immediately for incorrect spelling and other defects like repeated words. The software system provides a signal to the hardware to flash the word so that the operator will know something is wrong and can make an immediate correction. If the system were designed to prevent the typist from typing repeated words, then this would be a *poka-yoke*, a defect prevention. Defect prevention is better than immediate defect detection and correction. Ultimately, the system should be able to forecast the probability of a defect, correcting the problem at the source. This means that the typist would have to be removed from the process loop, perhaps by having the system type out what it is told (convert oral to written directly). Poka-yoke devices and source inspection techniques are keys to designing manufacturing systems that produce superior-quality products at low cost.

PRODUCTS AND FABRICATIONS

In manufacturing, material things (goods) are made to satisfy human wants. *Products* result from manufacture. Manufacture also includes conversion processes such as refining, smelting, and mining.

Products can be manufactured by fabricating or by processing. *Fabricating* is the manufacture of a product from pieces such as parts, components, or assemblies. Individual products or parts can also be fabricated. Separable discrete items such as tires, nails, spoons, screws, refrigerators, or hinges are fabricated.

Processing is also used to refer to the manufacture of a product by continuous means, or by a continuous series of operations, for a specific purpose. Continuous items such as steel strip, beverages, breakfast foods, tubing, chemicals, and petroleum are "processed." Many processed products are marketed as discrete items, such as bottles of beer, bolts of cloth, spools of wire, and sacks of flour.

Separable discrete products, both piece parts and assemblies, are fabricated in a *plant, factory,* or *mill,* for instance, a textile or rolling mill. Products that *flow* (liquids, gases, grains, or powders) are processed in a *plant* or *refinery.* The *continuous-process industries* such as petroleum and chemical plants are sometimes called processing industries or flow industries.

To a lesser extent, the terms *fabricating industries* and *manufacturing industries* are used when referring to fabricators or manufacturers of large products composed of many parts, such as a car, a plane, or a tractor. Manufacturing often includes continuous-process treatments such as electroplating, heating, demagnetizing, and extrusion forming.

Construction or building is making goods by means other than manufacturing or processing in factories. Construction is a form of project manufacturing of useful goods like houses, highways, and buildings. The public may not consider construction as manufacturing because the work is not usually done in a plant or factory, but it can be. There is a company in Delaware that can build a custom house of any design in its factory, truck it to the building site, and assemble it on a foundation in two or three weeks.

Agriculture, fisheries, and *commercial fishing* produce real goods from useful labor. Lumbering is similar to both agriculture and mining in some respects, and mining should be considered processing. Processes that convert the raw materials from agriculture, fishing, lumbering, and mining into other usable and consumable products are also forms of manufacturing.

WORKPIECE AND ITS CONFIGURATION

In the manufacturing of goods, the primary objective is to produce a component having a desired geometry, size, and finish. Every component has a shape that is bounded by various types of surfaces of certain sizes that are spaced and arranged relative to each other. Consequently, a component is manufactured by producing the surfaces that bound the shape. Surfaces may be:

1. Plane or flat
2. Cylindrical (external or internal)
3. Conical (external or internal)
4. Irregular (curved or warped)

Figure 1-6 illustrates how a shape can be analyzed and broken up into these basic bounding surfaces. Parts are manufactured by using a set or sequence of processes that will either (1) remove portions of a rough block of material (bar stock, casting, forging) so as to produce and leave the desired bounding surface, or (2) cause material to form into a stable configuration that has the required bounding surfaces (casting, forging). Consequently, in designing an object, the designer specifies the shape, size, and arrangement of the bounding surface. The part design must be analyzed to determine what materials will provide the desired properties, including mating to other components, and what processes can best be employed to obtain the end product at the most reasonable cost. This is often the job of the manufacturing engineer.

ROLES OF ENGINEERS IN MANUFACTURING

Many engineers have as their function the designing of products. The products are brought into reality through the processing or fabrication of materials. In this capacity designers are a key factor in the material selection and manufacturing procedure. A *design engineer*, better than any other person, should know what the design is to accomplish, what assumptions can be made about service loads and requirements, what service environment the product must withstand, and what appearance the final product is to have. To meet these requirements, the material(s) to be used must be selected and specified. In most cases, to utilize the material and to enable the product to have the desired form, the designer knows that certain *manufacturing processes* will have to be employed. In many instances, the selection of a specific material may dictate what processing must be used. On the other hand, when certain processes must be used, the design may have to be modified in order for the process to be utilized effectively and economically. Certain dimensional sizes can dictate the processing, and some processes require certain sizes for the parts going into them. In converting the design into reality, many decisions must be made. In most instances, they can be made most effectively at the design stage. It is thus apparent that design engineers are a vital factor in the manufacturing process, and it is indeed a blessing to the company if they can *design for manufacturing*, that is, design the product so that it can be manufactured and/or *assembled* economically (i.e., at low unit cost). Design for manufacturing uses the knowledge of manufacturing processes, and so the design and manufacturing engineers should work together to integrate design and manufacturing activities.

Manufacturing engineers select and coordinate specific processes and equipment to be used, or supervise and manage their use. Some design special tooling is used so that standard machines can be utilized in producing specific products. These engineers must have a broad knowledge of manufacturing processes and material behavior so that desired operations can be done effectively and efficiently without overloading or damaging machines and without adversely affecting the materials being processed. Although it is not obvious, the most hostile environment the material may ever encounter in its lifetime is the processing environment.

Industrial or *manufacturing engineers* are responsible for manufacturing systems design (or layout) of factories. They must take into account the interrelationships of the design and the properties of the materials that the machines are going to process as well as the interreaction of the materials and processes. The choice of machines and equipment used in manufacturing and their arrangement in the factory are also design tasks.

Materials engineers devote their major efforts to developing new and better materials. They, too, must be concerned with how these materials can be processed and with the effects that the processing will have on the properties of the materials. Although their roles may be quite different, it is apparent that a large proportion of engineers must concern themselves with the interrelationships of materials and manufacturing processes.

As an example of the close interrelationship of design, materials selection, and the selection and use of manufacturing processes, consider the common desk stapler. Suppose that this item is sold at the retail store for $20. The wholesale outlet sold the stapler for $16 and the manufacturer probably received about $10 for it. Staplers typically consist of 10 to 12 parts and some rivets and pins. Thus the manufacturer had to produce and assemble the 10 parts for about $1 per part. Only by giving a great deal of attention to design, selection of materials, selection of processes, selection of equipment used for manufacturing (tooling), and utilization of personnel could such a result be achieved.

The stapler is a relatively simple product, yet the problems involved in its manufacture are typical of those that manufacturing industries must deal with. The elements of design, materials, and processes are all closely related, each having its effect on the performance of the device and the other elements. For example, suppose the designer calls for the component that holds the staples to be a metal part. Will it be a machined part rather than a formed part? Entirely different processes and materials need to be specified depending on the choice. Or, if a part is to be changed from metal to plastic, then a whole new set of fundamentally different materials and processes would need to come into play. Such changes would also have a significant impact on cost.

CHANGING WORLD COMPETITION

In recent years, major changes in the world of goods manufacturing have taken place. Three of these are:

1. Worldwide competition for global products and their manufacture
2. High-tech manufacturing or advanced technology
3. New manufacturing systems designs, strategies, and management

Worldwide (global) competition is a fact of manufacturing life, and it will get stronger in the future. The goods you buy today may have been made anywhere in the world. The second aspect, advanced manufacturing technology, usually refers to new machine tools or processes with computer-aided manufacturing. Producing machine tools is a small industry with enormous leverage. Improved processes lead to better components and more durable goods. However, the new technology is often purchased from companies that have developed the technology, so this approach is important but may not provide a unique competitive advantage if your competitors can also buy the technology, provided that they have the capital. Some companies develop their own unique process technology and try to keep it proprietary as long as they can. A good example of unique process technology is the numerical control machine tool, shown in Figure 1-5 and discussed in Chapter 27. Computer-controlled machines are now common to the factory floor.

The third change and perhaps the real key to success in manufacturing is to build a manufacturing system that can deliver, on time to the customer, super-quality goods at the lowest possible cost in a flexible way. This change reflects an effort to improve markedly the methodology by which goods are produced rather than simply upgrading the manufacturing process technology.

Manufacturing system design is discussed extensively in the last section of the book, and we recommend that students examine this material closely after they have gained a working knowledge of materials and processes. The next section provides a brief introduction to manufacturing system designs.

MANUFACTURING SYSTEM DESIGNS

Five manufacturing system designs can be identified: the job shop, the flow shop, the linked-cell shop, the project shop, and the continuous process. The latter system primarily deals with liquids and/or gases (such as an oil refinery) rather than solids or discrete parts.

The most common of these layouts is the *job shop*, characterized by large varieties of components, general-purpose machines, and a functional layout (Figure 1-7). This means that machines are collected by function (all lathes together, all broaches together, all milling machines together) and the parts are routed around the shop in small lots to the various machines. The inset shows the multiple paths through the shop and a detail on one of the seven broaching machine tools. The material is moved from machine to machine in carts or containers and is called the *lot or batch*.

Flow shops are characterized by larger volumes of the same part or assembly, special-purpose machines and equipment, less variety, and more mechanization. Flow shop layouts are typically either continuous or interrupted and can be for manufacturing or assembly, as shown in Figure 1-8. If *continuous*, a production line is built that basically runs one large-volume complex item in great quantity and nothing else. The common light bulb is made this way. A transfer line producing an engine block is another typical

FIGURE 1-7   This rack bar machining area is functionally designed so it operates like a job shop, with lathes, broaches, and grinders lined up.

FIGURE 1-8   The moving assembly line for cars is an example of the flow shop.

BEFORE: Layout with conveyor–Subassembly with two conveyors

AFTER: U-Shape layout–Conveyors removed

Eliminate conveyor

Standing, walking workers

8 Seated Workers

Measurable Parameters

	Before:	After:
Output	700	1056 units/shift
In-process inventory	750	8 units
Personnel	10	8 operators
Daily output per person	70	132 units
Cycle time	0.60 minute	0.43 minute

Directly link processes to eliminate in-process inventory

FIGURE 1-9   The traditional subassembly lines can be redesigned into U-shaped cells as part of the conversion of mass production to lean production.

example. If *interrupted*, the line manufactures large lots but is periodically "changed over" to run a similar but different component.

The *linked-cell* manufacturing system (L-CMS) is composed of manufacturing and subassembly cells (Figure 1-9) connected to final assembly (linked) using a unique form of inventory and information control called *kanban*. The L-CMS is used in lean production systems where manufacturing processes and subassemblies are restructured into U-shaped cells so they can operate on a one-piece-flow basis, like final assembly.

The *project shop* is characterized by the immobility of the item being manufactured. In the construction industry, bridges and roads are good examples. In the manufacture of goods, large airplanes, ships, large machine tools, and locomotives are manufactured in project shops. It is necessary that the workers, machines, and materials come to the site. The number of end items is not very large, and therefore the lot sizes of the components going into the end item are not large. Thus the job shop usually supplies parts and subassemblies to the project shop in small lots.

Continuous processes are used to manufacture liquids, oils, gases, and powders. These manufacturing systems are usually large plants producing goods for other producers or mass-producing canned or bottled goods for consumers. The manufacturing engineer in these factories is often a chemical engineer.

Naturally, there are many hybrid forms of these manufacturing systems, but the job shop is the most common system. Because of its design, the job shop has been shown to be the least cost-efficient of all the systems. Component parts in a typical job shop spend only 5% of their time in machines and the rest of the time waiting or being moved from one functional area to the next. Once the part is on the machine, it is actually being processed (i.e., having value added to it by the changing of its shape) only about 30% to 40% of the time. The rest of the time parts are being loaded, unloaded, inspected, and so on. The advent of *numerical control* machines increased the percentage of time that the machine is making chips because tool movements are programmed and the machines can automatically change tools or load or unload parts.

However, there are a number of trends that are forcing manufacturing management to consider means by which the job shop system itself can be redesigned to improve its overall efficiency. These trends have forced manufacturing companies to convert their batch-oriented job shops into linked-cell manufacturing systems, with the manufacturing and subassembly cells structured around specific products. Another way to identify families of products with a similar set of manufacturing processes is called group technology.

Group technology (GT) can be used to restructure the factory floor. GT is a concept whereby similar parts are grouped together into part families. Parts of similar size and shape can often be processed through a *similar set of processes*. A part family based on manufacturing would have the same set or sequences of manufacturing processes. The set of processes is called a cell. Thus, with GT, job shops can be restructured into cells, each cell specializing in a particular family of parts. The parts are handled less, machine setup time is shorter, in-process inventory is lower, and the time needed for parts to get through the manufacturing system (called the throughput time) is greatly reduced.

BASIC MANUFACTURING PROCESSES

It is the manufacturing processes that create or add value to a product. The manufacturing processes can be classified as:

- Casting, foundry, or molding processes
- Forming or metalworking processes
- Machining (material removal) processes
- Joining and assembly
- Surface treatments (finishing)
- Rapid prototyping
- Heat treating
- Other

These classifications are not mutually exclusive. For example, some finishing processes involve a small amount of metal removal or metal forming. A laser can be used either for joining or for metal removal or heat treating. Occasionally, we have a process such as shearing, which is really metal cutting but is viewed as a (sheet) metalforming process. Assembly may involve processes other than joining. The categories of process types are far from perfect.

Casting and *molding* processes are widely used to produce parts that often require other follow-on processes, such as machining. Casting uses molten metal to fill a cavity. The metal retains the desired shape of the mold cavity after solidification. An important advantage of casting and molding is that, in a single step, materials can be converted from a crude form into a desired shape. In most cases, a secondary advantage is that excess or scrap material can easily be recycled. Figure 1-10 illustrates schematically some of the basic steps in the lost-wax casting process, one of many processes used in the foundry industry.

Casting processes are commonly classified into two types: permanent mold (a mold can be used repeatedly) or nonpermanent mold (a new mold must be prepared for each casting made). Molding processes for plastics and composites are included in the chapters on forming processes.

Forming and *shearing* operations typically utilize material (metal or plastics) that has been previously cast or molded. In many cases the materials pass through a series of forming or shearing operations, so the form of the material for a specific operation may be the result of all the prior operations. The basic purpose of forming and shearing is to modify the shape and size and/or physical properties of the material.

Metalforming and *shearing operations* are done both "hot" and "cold," a reference to the temperature of the material at the time it is being processed with respect to the

The Lost-Wax Casting Process

The most common casting method through the ages was flit *cire-perdue* or lost-wax process. Although expensive and time consuming, the lost-way method allows the artist to accurately reproduce the delicate nuances of the original model.

Models depicting the casting of Rodin's *Sorrow* (1889).

1

The artist creates a sculpted model, generally made of plaster, clay, marble, stone, or wood.

2

The surface of the model is coated with a protective substance. Then the model is put into a bed of very fine elastic material held in place by a rigid outer mold. When the model is removed, its impression remains.

3

Fireproof clay is carefully put into the impression. making a sharply defined duplicate of the artist's original model.

4

The surface of this second clay model is slightly scraped away. When this second model is returned to the mold, there is a gap between the model and the mold. This gap is where the wax will be poured. The final bronze will be of the same thickness as the gap that is created by the scraping.

5

After closing the mold around the clay model, hot wax is poured into the gap between the model and the mold. This stage is crucial in producing a perfect reproduction of the initial sculpture. The result is a clay model covered with wax, which is then hand-finished to fidelity, incorporating the artist's signature, cast number and a foundry seal.

6

A network of wax pipes, called sprues and gates. is attached to the wax model. These pipes will allow the wax to escape as it melts. The pipes will also spread the molten metal evenly throughout the mold and will let air escape as the metal is poured in.

7

A finely granulated ceramic is applied to the surface of the model and its pipes until it becomes thick and coarse. The result, now called an "investment mold," is then dried and heated causing the wax to melt and flow out of the mold. leaving a space between the fire resistant clay model and the investment mold. Accordingly, this method is called the "lost-wax process."

8

The investment mold is then heated to a high temperature (over 1,000 °F). Except for a place to pour in the liquid bronze at the top, the mold is covered with a layer of cladding (a protective metal coating), which must be completely dry before bronze pouring begins.

9

Molten bronze (over 2,000 °F) is then poured into the investment mold filling the space left by the "lost" wax. When everything has cooled, the cladding and investment mold are broken and the metal appears. The bronze sculpture and its sprues and gates are an exact reproduction of the wax in step six.

10

The network of sprues and gates is then removed and the surface of the bronze is often chiseled and filed so that no trace of them can be seen. This process of hand-finishing the bronze to perfection is called "chasing." Any remains of the fireproof clay model left inside the bronze are also removed now.

11

When the chasing is finished, hot or cold oxides are applied to the surface of the bronze, creating a thin layer of corrosion. This layer -slightly brown, green, or blue in color-is called the "patina:" The patina protects and enlivens the surface of the bronze.

FIGURE 1-10   A manufacturing process represents a sequence of steps. Here are the steps in the lost-wax casting process.

Metalforming Process for Automobile Fender

FIGURE 1-11   The forming process used to make a fender for a car.

temperature at which this material can recrystallize (i.e., grow new grain structure). Figure 1-11 shows the process by which the fender of a car is made using a series of metalforming processes.

Machining or *metal removal processes* refer to the removal of certain selected areas from a part in order to obtain a desired shape or finish. Chips are formed by interaction of a cutting tool with the material being machined. Figure 1-12 shows a chip being formed by a single-point cutting tool in a machine tool called a lathe. The manufacturing engineer may be called upon to specify the cutting parameters such as cutting speed, feed, or depth of cut (DOC). The engineer may also have to select the cutting tools for the job.

Cutting tools used to perform the basic turning on the lathe are shown in Figure 1-12. The cutting tools are mounted in machine tools, which provide the required movements of the tool with respect to the work (or vice versa) to accomplish the process desired. In recent years many new machining processes have been developed.

The seven basic machining processes are *shaping, drilling, turning, milling, sawing, broaching*, and *abrasive machining*. With the exception of shaping, each of the basic processes has a chapter dedicated to it. Historically, eight basic types of machine tools

The Machining Process
(turning on a lathe)

Cutting tool   Workpiece

The workpiece is mounted in a machine tool (lathe) with a cutting tool.

Workpiece

Original diameter — Cutting speed *V* — Final diameter

Depth of cut

Feed (inch/rev)   Tool   Chip

The workpiece is rotated while the tool is fed at some feed rate (inches per revolution). The desired cutting speed *V* determines the rpm of the workpiece. This process is called turning.

The cutting tool interacts with the workpiece to form a chip by a shearing process. The tool shown here is an indexable carbide insert tool with a chip-breaking groove.

FIGURE 1-12 Single-point metalcutting process (turning) produces a chip while creating a new surface on the workpiece.

have been developed to accomplish the basic processes. These machine tools are called shapers (and planers), drill presses, lathes, boring machines, milling machines, saws, broaches, and grinders. Most of these machine tools are capable of performing more than one of the basic machining processes. Shortly after numerical control was invented, *machining centers* were developed that could combine many of the basic processes, plus other related processes, into a single machine tool with a single workpiece setup. Included with the machining processes are processes wherein metal is removed by chemical, electrical, electrochemical, or thermal sources. Generally speaking, these nontraditional processes have evolved to fill a specific need when conventional processes were too expensive or too slow when machining very hard materials. One of the first uses of a laser was to machine holes in ultra-high-strength metals. Lasers are being used today to drill tiny holes in turbine blades for jet engines. Because of its ability to produce components with great precision and accuracy, metalcutting, using machine tools, is recognized as having great value-adding capability.

In recent years a new family of processes has emerged called *rapid prototyping*. These additive-type processes produce first, or prototype, components directly from the software using specialized machines driven by computer-aided design packages. The prototypes can be field tested and modifications to the design quickly implemented. Early versions of these machines produced only nonmetalic components, but modern machines can make metal parts. In contrast, the machining processes are recognized as having great *value-adding capability*, that is, the ability to produce components with great precision and accuracy.

Perhaps the largest collection of processes, in terms of both diversity and quantity, are the *joining processes*, which include the following:

1. Mechanical fastening
2. Soldering and brazing
3. Welding
4. Press, shrink, or snap fittings
5. Adhesive bonding
6. Assembly processes

Many of these processes are often found in the assembly area of the plant. Figure 1-13 provides one example where all but welding are used in the sequence of operations to produce a computer.

At the lowest level, microelectronic fabrication methods produce entire *integrated circuits* (ICs) of solid-state (no moving parts) components, with wiring and connections, on a single piece of semiconductor material, usually single-crystalline silicon. Arrays of ICs are produced on thin, round disks of semiconductor material called *wafers*. Once the semiconductor on the wafer has been fabricated, the finished wafer is cut up into individual ICs, or *chips*. Next, at level 2, these chips are individually housed with connectors or leads making up "dies" that are placed into "packages" using adhesives. The packages provide protection from the elements and a connection between the die and another subassembly called the printed circuit board (PCB). At level 3, IC packages, along with other discrete components (e.g., resistors, capacitors, etc.), are soldered onto PCBs and then assembled with even larger circuits on PCBs. This is sometimes referred to as *electronic assembly*. Electronic packages at this level are called *cards* or *printed wiring assemblies* (PWAs). Next, series of cards are combined on a *backpanel* PCB, also known as a *motherboard* or simply a *board*. This level of packaging is sometimes referred to as *card-on-board* packaging. Ultimately, card-on-board assemblies are put into

FIGURE 1-13 How an electronic product is made.

housings using mechanical fasteners and snap fitting and finally integrated with power supplies and other electronic peripherals through the use of cables to produce final commercial products.

Finishing processes are yet another class of processes typically employed for cleaning, removing burrs left by machining, or providing protective and/or decorative surfaces on workpieces. Surface treatments include chemical and mechanical cleaning, deburring, painting, plating, buffing, galvanizing, and anodizing.

Heat treatment is the heating and cooling of a metal for the specific purpose of altering its metallurgical and mechanical properties. Because changing and controlling these properties is so important in the processing and performance of metals, heat treatment is a very important manufacturing process. Each type of metal reacts differently to heat treatment. Consequently, a designer should know not only how a selected metal can be altered by heat treatment but, equally important, *how a selected metal will react, favorably or unfavorably, to any heating or cooling that may be incidental to the manufacturing processes*.

OTHER MANUFACTURING OPERATIONS

In addition to the processes already described, there are some other fundamental manufacturing operations that we must consider. *Inspection* determines whether the desired objectives stated by the designer in the specifications have been achieved. This activity provides feedback to design and manufacturing with regard to the process behavior. Essential to this inspection function are measurement activities.

In *testing*, a product is tried by actual function or operation or by subjection to external effects. Although a test is a form of inspection, it is often not viewed that way. In manufacturing, parts and materials are inspected for conformance to the dimensional and physical specifications, while testing may simulate the environmental or usage demands to be made on a product after it is placed in service. Complex processes may require many tests and inspections. Testing includes life-cycle tests, destructive tests, nondestructive testing to check for processing defects, wind-tunnel tests, road tests, and overload tests.

Transportation of goods in the factory is often referred to as *material handling* or *conveyance* of the goods and refers to the transporting of unfinished goods (work-in-process) in the plant and supplies to and from, between, and during manufacturing operations. Loading, positioning, and unloading are also material-handling operations. Transportation, by truck or train, is material handling between factories. Proper manufacturing system design and mechanization can reduce material handling in countless ways.

Automatic material handling is a critical part of continuous automatic manufacturing. The word *automation* is derived from automatic material handling. Material handling, a fundamental operation done by people and by conveyors and loaders, often includes positioning the workpiece within the machine by indexing, shuttle bars, slides, and clamps. In recent years, wire-guided automated guided vehicles (AGVs) and automatic storage and retrieval systems (AS/RSs) have been developed in an attempt to replace forklift trucks on the factory floor. Another form of material handling, the mechanized removal of waste (chips, trimming, and cutoffs), can be more difficult than handling the product. Chip removal must be done before a tangle of scrap chips damages tooling or creates defective workpieces.

Most texts on manufacturing processes do not mention *packaging*, yet the packaging is often the first thing the customer sees. Also, packaging often maintains the product's quality between completion and use. (Packaging is also used in electronics manufacturing to refer to placing microelectronic chips in containers for mounting on circuit boards.) Packaging can also prepare the product for delivery to the user. It varies from filling ampules with antibiotics to steel-strapping aluminum ingots into palletized loads. A product may require several packaging operations. For example, Hershey Kisses are (1) individually wrapped in foil, (2) placed in bags, (3) put into boxes, and (4) placed in shipping cartons.

Weighing, filling, sealing, and labeling are packaging operations that are highly automated in many industries. When possible, the cartons or wrappings are formed from material on rolls in the packaging machine. Packaging is a specialty combining elements of product design (styling), material handling, and quality control. Some packages cost more than their contents, for example, cosmetics and razor blades.

During *storage*, nothing happens intentionally to the product or part except the passage of time. Part or product deterioration on the shelf is called *shelf life*, meaning that items can rust, age, rot, spoil, embrittle, corrode, creep, and otherwise change in state or structure, while supposedly nothing is happening to them. Storage is detrimental, wasting the company's time and money. The best strategy is to keep the product moving with as little storage as possible. Storage during processing must be *eliminated*, not automated or computerized. Companies should avoid investing heavily in large automated systems that do not alter the bottom line. Have the outputs improved with respect to the inputs, or has storage simply increased the costs (indirectly) without improving either the quality or the throughput time?

By not storing a product, the company avoids having to (1) remember where the product is stored, (2) retrieve it, (3) worry about its deteriorating, or (4) pay storage costs. Storage is the biggest waste of all and should be eliminated at every opportunity.

UNDERSTAND YOUR PROCESS TECHNOLOGY

Understanding the process technology of the company is very important for everyone in the company. Manufacturing technology affects the design of the product and the manufacturing system, the way in which the manufacturing system can be controlled, the types of people employed, and the materials that can be processed. Table 1-4 outlines the factors that characterize a process technology. Take a process you are familiar with and think about these factors. One valid criticism of American companies is that their managers seem to have an aversion to understanding their companies' manufacturing technologies. Failure to understand the company business (i.e., its fundamental process technology) can lead to the failure of the company.

The way to overcome technological aversion is to run the process and study the technology. Only someone who has run a drill press can understand the sensitive relationship between feed rate and drill torque and thrust. All processes have these "know-how" features. Those who run the processes must be part of the decision making for the factory. The CEO who takes a vacation working on the plant floor and learning the processes will be well on the way to being the head of a successful company.

PRODUCT LIFE CYCLE AND LIFE-CYCLE COST

Manufacturing systems are dynamic and change with time. There is a general, traditional relationship between a *product's life cycle* and the kind of manufacturing system used to make the product. Figure 1-14 simplifies the life cycle into these steps:

1. *Startup*. New product or new company, low volume, small company.

2. *Rapid growth*. Products become standardized and volume increases rapidly. Company's ability to meet demand stresses its capacity.

3. *Maturation*. Standard designs emerge. Process development is very important.

TABLE 1-4   Characterizing a Process Technology

Mechanics (statics and dynamics of the process)
  How does the process work?
  What are the process mechanics (statics, dynamics, friction)?
  What physically happens, and what makes it happen? (Understand the physics.)

Economics or costs
  What are the tooling costs, the engineering costs?
  Which costs are short term, which long term?
  What are the setup costs?

Time spans
  How long does it take to set up the process initially?
  What is the throughput time?
  How can these times be shortened?
  How long does it take to run a part once it is set up (cycle time)?
  What process parameters affect the cycle time?

Constraints
  What are the process limits?
  What cannot be done?
  What constrains this process (sizes, speeds, forces, volumes, power, cost)?
  What is very hard to do within an acceptable time/cost frame?

Uncertainties and process reliability
  What can go wrong?
  How can this machine fail?
  What do people worry about with this process?
  Is this a reliable, stable process?

Skills
  What operator skills are critical?
  What is not done automatically?
  How long does it take to learn to do this process?

Flexibility
  Can this process be adapted easily for new parts of a new design or material?
  How does the process react to changes in part design and demand?
  What changes are easy to do?

Process capability
  What are the accuracy and precision of the process?
  What tolerances does the process meet? (What is the process capability?)
  How repeatable are those tolerances?

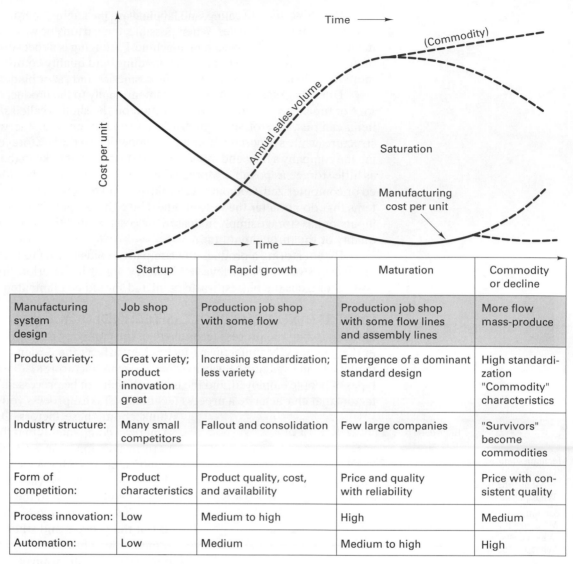

	Startup	Rapid growth	Maturation	Commodity or decline
Manufacturing system design	Job shop	Production job shop with some flow	Production job shop with some flow lines and assembly lines	More flow mass-produce
Product variety:	Great variety; product innovation great	Increasing standardization; less variety	Emergence of a dominant standard design	High standardization "Commodity" characteristics
Industry structure:	Many small competitors	Fallout and consolidation	Few large companies	"Survivors" become commodities
Form of competition:	Product characteristics	Product quality, cost, and availability	Price and quality with reliability	Price with consistent quality
Process innovation:	Low	Medium to high	High	Medium
Automation:	Low	Medium	Medium to high	High

FIGURE 1-14 Product life-cycle costs change with the classic manufacturing system designs.

4. *Commodity.* Long-life, standard-of-the-industry type of product or

5. *Decline.* Product is slowly replaced by improved products.

The maturation of a product in the marketplace generally leads to fewer competitors, with competition based more on price and on-time delivery than on unique product features. As the competitive focus shifts during the different stages of the product life cycle, the requirements placed on manufacturing—cost, quality, flexibility, and delivery dependability—also change. The stage of the product life cycle affects the product design stability, the length of the product development cycle, the frequency of engineering change orders, and the commonality of components, all of which have implications for manufacturing process technology.

During the design phase of the product, much of the cost of manufacturing and assembly is determined. Assembly of the product is inherently integrative as it focuses on pairs and groups of parts.

It is crucial to achieve this integration during the design phase because about 70% of the life-cycle cost of a product is determined when it is designed. Design choices determine materials, fabrication methods, assembly methods, and, to a lesser degree, material-handling options, inspection techniques, and other aspects of the production system. Manufacturing engineers and internal customers can influence only a small part of the overall cost if they are presented with a finished design that does not reflect their concerns. Therefore all aspects of production should be included if product designs are to result in real functional integration.

Life-cycle costs include the costs of all the materials, manufacture, use, repair, and disposal of a product. Early design decisions determine about 60% of the cost, and all activities up to the start of full-scale development determine about 75%. Later decisions can make only minor changes to the ultimate total unless the design of the manufacturing system is changed.

In short, the concept of product life-cycle provides a framework for thinking about the product's evolution through time and the kind of market segments that are likely to develop at various times. Analysis of life-cycle costs shows that the design of the manufacturing system determines the cost per unit, which generally decreases over time with process improvements and increased volumes. For additional discussion on reliability and maintainability of manufacturing equipment, see SAE publication M-110.2.

The linked-cell manufacturing system design discussed in Chapter 39 enables companies to decrease cost per unit significantly while maintaining flexibility and making smooth transitions from low-volume to high-volume manufacturing.

Low-cost manufacturing does not just happen. There is a close, interdependent relationship among the design of a product, the selection of materials, the selection of processes and equipment, the design of the processes, and tooling selection and design. Each of these steps must be carefully considered, planned, and coordinated before manufacturing starts. This lead time, particularly for complicated products, may take months, or even years, and the expenditure of large amounts of money may be involved. Typically, the lead time for a completely new model of an automobile or a modern aircraft may be two to five years.

Some of the steps involved in getting the product from the original idea stage to daily manufacturing are discussed in Chapter 9 in more detail. The steps are closely related to each other. For example, the design of the tooling is dependent on the design of the parts to be produced. It is often possible to simplify the tooling if certain changes are made in the design of the parts or the design of the manufacturing systems. Similarly, the material selection will affect the design of the tooling or the processes selected. Can the design be altered so that it can be produced with tooling already on hand and thus avoid the purchase of new equipment? Close coordination of all the various phases of design and manufacture is essential if economy is to result.

With the advent of computers and computer-controlled machines, the integration of the design function and the manufacturing function through the computer is a reality. This is usually called CAD/CAM (computed-aided design/computer-aided manufacturing). The key is a common database from which detailed drawings can be made for the designer and the manufacturer and from which programs can be generated to make all the tooling. In addition, extensive computer-aided testing and inspection (CATI) of the manufactured parts is taking place. There is no doubt that this trend will continue at ever-accelerating rates as computers become cheaper and smarter, but at this time, the computers necessary to accomplish complete computer-integrated manufacturing (CIM) are expensive and the software very complex. Implementing CIM requires a lot of manpower as well.

MANUFACTURING SYSTEM DESIGN

When designing a manufacturing system, two customers must be taken into consideration: the external customer who buys the product and the internal customer who makes the product. The external customer is likely to be global and demand greater variety with superior quality and reliability. The internal customer is often empowered to make critical decisions about how to make the products. The Toyota Motor Company is making vehicles in 25 countries. Their truck plant in Indiana has the capacity to make 150,000 vehicles per year (creating 2300 new jobs), using the Toyota Production System (TPS). An appreciation of the complexity of the manufacturing system design problem is shown in Figure 1-15, where the choices in system design are reflected against the number of different products, or parts being made in the system, often called variety. Clearly, there are many choices regarding which method (or system) to use to make the goods.

FIGURE 1-15 Different manufacturing system designs produce goods at different production rates.

This part variety-production rate matrix shows examples of particular manufacturing system designs. This matrix was developed by Black based on real factory data. Notice there is a large amount of overlap in the middle of the matrix, so the manufacturing engineer has many choices regarding which method or system to use to make the goods. This book will show the connection between the process and the manufacturing system used to produce the products, turning raw materials into finished goods.

A manufacturer never really knows how large or diverse a market will be. If a diverse and specialized market emerges, a company with a focused flow-line system may be too inflexible to meet the varying demand. If a large but homogeneous market develops, a manufacturer with a flexible system may find production costs too high and the flexibility unexploitable. Another general relationship between manufacturing system designs and production volumes is shown in Figure 1-16.

NEW MANUFACTURING SYSTEMS

The manufacturing process technology described herein is available worldwide. Many countries have about the same level of process development when it comes to manufacturing technology. Much of the technology existing in the world today was developed in the United States, Germany, France, and Japan. More recently Taiwan, Korea, and China have been making great inroads into American markets, particularly in the automotive and electronics industries. What many people have failed to recognize is that many companies have developed and promoted a totally different kind of manufacturing production system. This new system, called the Toyota Production System, will take its place with the American Armory System and the Ford System for mass production. This new system, developed by the Toyota Motor Company, is known worldwide as the lean manufacturing system.

Many American companies have successfully adopted some version of the Toyota system. The experience of dozens of these companies is amalgamated into 10 steps, which, if followed, can make any company a factory with a future.

For the lean production to work, 100%-good units flow rhythmically to subsequent processes without interruption. In order to accomplish this, an integrated quality control (IQC) program has to be developed. The responsibility for quality has been given to manufacturing. All the employees are inspectors and are empowered to make it right the first time. There is a companywide attitude toward constant quality improvement. Make quality easy to see, stop the line when something goes wrong, and inspect things 100% if necessary to prevent defects from occurring. The results of this system are astonishing in terms of quality, low cost, and on-time delivery of goods to the customer.

The most important factor in economical and successful manufacturing is the manner in which the resources—labor, materials, and capital—are organized and managed so as to provide effective coordination, responsibility, and control. Part of the success of

The figure shows in a general way the relationship between manufacturing systems and production volumes. The upper left represents systems with low flexibility but high efficiency compared to the lower right, where volumes are low and so is efficiency. Where a particular company lies in this matrix is determined by many forces, not all of which are controllable. The job of manufacturing and industrial engineers is to design and implement a system which can achieve low unit cost, superior quality, with on-time delivery in a flexible way.

FIGURE 1-16   This figure shows in a general way the relationship between manufacturing systems and production volumes.

lean production can be attributed to a different management approach. This approach is characterized by a holistic attitude toward people.

The real secret of successful manufacturing lies in designing a manufacturing system in which everyone who works in the system understands how the system works, how goods are controlled, with the decision making placed at the correct level. The engineers also must possess a broad fundamental knowledge of design, metallurgy, processing, economics, accounting, and human relations. In the manufacturing game, low-cost mass production is the result of teamwork within an integrated manufacturing/production system. This is the key to producing superior quality at less cost with on-time delivery.

■ Key Words

assembling	heat treatment	machine tool	product life cycle
casting	inspection	machining	production system
consumer goods	job	manufacturing cost	project shop
continuous process	job shop	manufacturing engineer	shearing
design engineer	joining	manufacturing process	station
fabricating	lean production	manufacturing system	tooling
filing jig	lean shop	molding	tools
flow shop	lost-wax casting	numerical control	treatments
forming	linked-cell manufacturing	operation	welding
group technology	system	producer goods	

■ Review Questions

1. What role does manufacturing play relative to the standard of living of a country?
2. Aren't all goods really consumer goods, depending on how you define the customer? Discuss.
3. Give examples of a job shop, flow shop, and project shop.
4. How does a system differ from a process? From a machine tool? From a job? From an operation?
5. Is a cutting tool the same thing as a machine tool?
6. What are the major classifications of basic manufacturing processes?

7. Could casting be used to produce a complex-shaped part to be made from a hard-to-machine metal? How else could the part be made?
8. In the lost-wax casting process, what happens to the foam?
9. In making a gold medal, what do we mean by a "relief image" cut into the die?
10. How is a railroad station like a station on an assembly line?
11. Since no work is being done on a part when it is in storage, it does not cost you anything. True or false? Explain.
12. What forming processes are used to make a paper clip?

13. We can analogize your university to a manufacturing system that produces graduates. Assuming that it takes four years to get a college degree and that each course really adds value to the student's knowledge base, what percentage of the four years is "value adding" (percentage of time in class plus two hours of preparation for each hour in class)?
14. It is acknowledged that chip-type machining is basically an inefficient process. Yet it is probably used more than any other to produce desired shapes. Why?
15. Compare Figure 1-1 and Figure 1-14. What are the stages of the product life cycle for an audiocassette tape?
16. In a modern safety razor with three or four blades that sells for $1, what do you think the cost of the blades might be?
17. List three purposes of packaging operations.
18. Assembly is defined as "the putting together of all the different parts to make a complete machine." Think of (and describe) an assembly process. Is making a club sandwich an assembly process? What about carving a turkey? Is this an assembly process?
19. What are the physical elements in a manufacturing system?
20. In the production system, who usually figures out how to make the product?
21. In Figure 1-7, what do the lines connecting the processes represent?
22. Characterize the process of squeezing toothpaste from a tube (extrusion of toothpaste) using Table 1-4 as a guideline. See the index for help on extrusion.
23. What are the major process steps in the assembly of an automobile?
24. What difficulties would result if production planning and scheduling were omitted from the procedure outlined in Chapter 9 for making a product in a job shop?

25. It has been said that low-cost products are more likely to be more carefully designed than high-priced items. Do you think this is true? Why or why not?
26. Proprietary processes are closely held or guarded company secrets. The chemical makeup of a lubricant for an extrusion process is a good example. Give another example of a proprietary process.
27. If the rolls for the cold-rolling mill that produces the sheet metal used in your car cost $300,000 to $400,000, how is it that your car can still cost less than $20,000?
28. Make a list of service production systems, giving an example of each.
29. What is the fundamental difference between an SPS and an MPS?
30. In the process of buying a calf, raising it to a cow, and disassembling it into "cuts" of meat for sale, where is the "value added"?
31. What kind of process is powder metallurgy: casting or forming?
32. In view of Figure 1-2, who really determines the selling price per unit?
33. What costs make up manufacturing cost (sometimes called factory cost)?
34. What are major phases of a product life cycle?
35. How many different manufacturing systems might be used to make a component with annual projected sales of 16,000 parts per year with 10 to 12 different models (varieties)?
36. In general, as the annual volume for a product increases, the unit cost decreases. Explain.

■ Problems

1. The Toyota truck plant in Indiana produces 150,000 trucks per year. The plant runs one eight-hour shift, 300 days per year, and makes 500 trucks per day. About 1300 people work on the final assembly line. Each car has about 20 labor hours per car in it.
   a. Assuming the truck sells for $16,000 and workers earn $30 per hour in wages and benefits, what percentage of the cost of the truck is in direct labor?
   b. What is the production rate of the final assembly line?
2. Suppose you wanted to redesign a stapler to have fewer components. (You should be able to find a stapler at a local discount store.) How much did it cost? How many parts does it

have? Make up a "new parts" list and indicate which parts would have to be redesigned and which parts would be eliminated. Estimate the manufacturing cost of the stapler assuming that manufacturing costs are 40% of the selling price. What are the disadvantages of your new stapler design versus the old stapler?
3. A company is considering making automobile bumpers from aluminum instead of from steel. List some of the factors it would have to consider in arriving at its decision.
4. Many companies are critically examining the relationship of product design to manufacturing and assembly. Why do they call this concurrent engineering?

www.wiley.com/college/degarmo

*C*hapter 1 CASE STUDY

Famous Manufacturing Engineers

Manufacturing engineering is that engineering function charged with the responsibility of interpreting product design in terms of manufacturing requirements and process capability. Specifically: the manufacturing engineer may:

- Determine how the product is to be made in terms of specific manufacturing processes.
- Design workholding and work transporting tooling or containers.
- Select the tools (including the tool materials) that will machine or form the work materials.
- Select, design, and specify devices and instruments which inspect that which has been manufactured to determine its quality.
- Design and evaluate the performance of the manufacturing system.
- Perform all these functions (and many more) related to the actual making of the product at the most reasonable cost per unit without sacrifice of the functional requirements or the users' service life.

There's no great glory in being a great manufacturing engineer (MfE). If you want to be a manufacturing engineer, you had better be ready to got your hands dirty.

Of course, there are exceptions. There have been some very famous manufacturing engineers.

For example:

- John Wilkinson of Bersham England built a boring mill in 1775 to bore the cast iron cylinders for James Watt's steam engine. How good was this machine?
- Eli Whitney was said to have invented the cotton gin, a machine to separate seeds from cotton. His machine was patented but was so simple, anyone could make one. He was credited with "interchangeability" – but we know Thomas Jefferson observed interchangeability in France in 1785 and probably the French gunsmith LeBlanc is the real inventor here. Jefferson tried to bring the idea to America and Whitney certainly did. He took 10 muskets to congress, disassembled them, and scattered the pieces. Interchangeable parts permitted them to be reassembled. He was given a contract for 2,000 guns to be made in 2 years. But what is the rest of his story?
- Joe Brown started a business in Rhode Island in 1833 making lathes and small tools as well as timepieces (watchmaker). Lucian Sharp joined the company in 1848 and developed a pocket sheet metal gage in 1877 a 1 inch micrometer and in 1862 the universal milling machine.
- Sam Colt at age 16 he salled to Calcutta on the Brig "Curve". He whittled a wood model of a revolver on this

voyage. He saved his money and had models of a gun built in Hartford by Anson Chase for which he got a patent. He set up a factory in New Jersey – but he could not sell his guns to the army – too complicated. He sold to the Texas Rangers and the Florida Frontiersmen but he had to close the plant. In 1846 the Mexican war broke out. General Zachary Taylor and Captain Sam Walters wanted to buy guns. Colt had none but accepted orders for 1000 guns and constructed a model (Walker Colt) and arranged to have them made at Whitney's (now 40 years old) plant in Whitneyville. Here he learned about mass production methods. In 1848 he rented a plant in Hartford and the Colt legend spread. In 1853 he had built one of the worlds largest arms plant in Connecticut which had 1400 machine tools. Colt helped start the careers of

- E K Root – mechanic and superintendent – paying him a salary of $25,000 in the 1800's. Abolished hand work – jigs and fixtures.
- Francis Pratt and Amos Whitney – famous machine tool builders.
- William Gleason – gear manufacturer
- E. P. Bullard – invented the Mult–An–Matic Multiple spindle machine which cut the time to make a flywheel from 18 minutes to slightly over 1 minute. Sold this to Ford.
- Christopher Sponer.
- E. J. Kingsbury invented a drilling machine to drill holes through toy wheel hubs that had a spring loaded cam which enables the head to sense the condition of the casting and modify feed rate automatically.

Now here are some more names from the past of famous and not so famous manufacturing, mechanical, and industrial engineers. Relate them to the development of manufacturing processes or manufacturing system designs.

- Eli Whitney
- Henry Ford
- Charles Sorenson
- Sam Colt
- John Parsons
- Eiji Toyoda
- Elisha Root
- John Hall
- Thomas Blanchard
- Fred Taylor
- Talichi Ohno
- Ambrose Swasey

CHAPTER 2

PROPERTIES OF MATERIALS

■ 2.1 INTRODUCTION

Structure — Properties

Processing — Performance

FIGURE 2-1
The manufacturing relationships among structure, properties, processing, and performance.

Manufacturing has been accurately defined as the activities that are performed in the conversion of "stuff" into "things." Successful products begin with appropriate materials. You wouldn't build an airplane out of lead or an automobile out of concrete—you need to start with the right stuff. But "stuff" rarely comes in the right shape, size, and quantity for the desired use. Parts and components must be produced by subjecting engineering materials to one or more processes (often a series of operations) that alter their shape, their properties, or both. Much of a manufacturing education relates to an understanding of (1) the *structure* of materials, (2) the *properties* of materials, (3) the *processing* of materials, and (4) the *performance* of materials, and the interrelations between these four factors, as illustrated in Figure 2-1.

This chapter will begin to address the properties of engineering materials. Chapters 3 and 4 will discuss the subject of "structure" and begin to provide the whys behind various properties. Chapter 5 introduces the possibility of modifying structure to produce desired properties. Most engineering materials do not have a single set of properties but rather offer a range or spectrum of possibilities. Taking advantage of this range, we might want to intentionally make a material weak and ductile for easy shaping (forming loads are low and tool life is extended) and then, once the shape has been produced, make the material strong for enhanced performance in use.

When selecting a material for a product or application, it is important to ensure that its properties will be adequate for the anticipated operating conditions. The various requirements of each part or component must first be estimated or determined. These requirements typically include mechanical characteristics (strength, rigidity, resistance to fracture, the ability to withstand vibrations or impacts) and physical characteristics (weight, electrical properties, appearance) as well as features relating to the service environment (ability to operate under extremes of temperature or to resist corrosion). Candidate materials must possess the desired properties within their range of possibilities.

To help evaluate the properties of engineering materials, a variety of standard tests have been developed, and data from these tests have been tabulated and made readily available. Proper use of this data often requires sound engineering judgment. It is important to consider which of the evaluated properties are significant, under what conditions the test values were determined, and what restrictions or limitations should be placed on their use. Only by being familiar with the various test procedures, their capabilities, and their limitations can one determine if the resulting data are applicable to a particular problem.

METALLIC AND NONMETALLIC MATERIALS

While engineering materials are often grouped as metals, ceramics, polymers, and composites, a simpler distinction might be to separate them into metallic and nonmetallic. The common *metallic* materials include iron, copper, aluminum, magnesium, nickel, titanium, lead, tin, and zinc as well as the alloys of these metals, such as steel, brass, and bronze. They possess the metallic properties of luster, high thermal conductivity, and high electrical conductivity; they are relatively ductile; and some have good magnetic properties. Some common *nonmetals* are wood, brick, concrete, glass, rubber, and plastics. Their properties vary widely, but they generally tend to be weaker, less ductile, and less dense than the metals, and to have poor electrical and thermal conductivities.

Although metals have traditionally been the more important of the two groups, the nonmetallic materials have become increasingly important in modern manufacturing. Advanced ceramics, composite materials, and engineered plastics have emerged in a number of applications. In many cases, metals and nonmetals are viewed as competing materials, with selection being based on how well each is capable of providing the required properties. Where both perform adequately, total cost often becomes the deciding factor, where total cost includes both the cost of the material and the cost of fabricating the desired component. Factors such as product lifetime, environmental impact, energy requirements, and recyclability are also considered.

PHYSICAL AND MECHANICAL PROPERTIES

A common means of distinguishing one material from another is through their *physical properties*. These include such features as density (weight); melting point; optical properties (transparency, opaqueness, or color); the thermal properties of specific heat, coefficient of thermal expansion, and thermal conductivity; electrical conductivity; and magnetic properties. In some cases, physical properties are of prime importance when selecting a material, and several will be discussed in more detail near the end of this chapter.

More often, however, material selection is dominated by the properties that describe how a material responds to applied loads or forces. These *mechanical properties* are usually determined by subjecting prepared specimens to standard test conditions. When using test results, however, it is important to remember that they apply only to the specific conditions that were employed. The actual service conditions of engineered products rarely duplicate the conditions of laboratory testing, so considerable caution should be exercised when applying test results.

STRESS AND STRAIN

When a force or load is applied to a material, it deforms or distorts (becomes *strained*), and internal reactive forces (*stresses*) are transmitted through the solid. For example, if a weight, W, is suspended from a bar of uniform cross section and length L, as in Figure 2-2, the bar will elongate by an amount ΔL. For a given weight, the magnitude of the *elongation*, ΔL, depends on the original length of the bar. The amount of elongation per unit length, expressed as $e = \Delta L/L$, is called the *unit strain*. Although the ratio is that of a length to another length and is therefore dimensionless, strain is usually expressed in terms of millimeters per meter, inches per inch, or simply as a percentage.

Application of the force also produces reactive stresses, which serve to transmit the load through the bar and on to its supports. *Stress* is defined as the force or load being transmitted divided by the cross-sectional area transmitting the load. Thus, in Figure 2-2, the stress is $S = W/A$, where A is the cross-sectional area of the supporting bar. Stress is normally expressed in megapascals (in SI units, where a pascal is 1 newton per square meter) or pounds per square inch (in the English system).

In Figure 2-2, the weight tends to stretch or lengthen the bar, so the strain is known as a *tensile strain* and the stress as a *tensile stress*. Other types of loadings produce other types of stresses and strains (Figure 2-3). Compressive forces tend to shorten the material and produce *compressive stresses and strains*. *Shear stresses and strains* result when two forces acting on a body are offset with respect to one another.

FIGURE 2-2 Tension loading and the resultant elongation.

FIGURE 2-3 Examples of tension, compression, and shear loading, and their response.

■ 2.2 STATIC PROPERTIES

When the forces that are applied to a material are constant, or nearly so, they are said to be *static*. Since static loadings are observed in many applications, it is important to characterize the behavior of materials under these conditions. For design engineers, the strength of a material may be of primary concern, along with the amount of elastic stretching or deflection that may be experienced while under load. Manufacturing engineers, looking to shape products, may be more concerned with the ability to mechanically deform the material without fracture.

As a result, a number of standardized tests have been developed to evaluate the *static properties* of engineering materials. Test results can be used to determine if a given material or batch of material has the necessary properties to meet specified requirements. Other tests provide the materials characterization base used for material selection. In all cases, it is important to determine that the service conditions are indeed similar to those of testing. Even when the service conditions differ, the results of standard tests may be helpful in qualitatively rating and comparing various materials.

TENSILE TEST

The most common of the static tests is the *uniaxial tensile test*. The test begins with the preparation of a standard specimen with prescribed geometry, like the round and flat specimens described in Figure 2-4. The standard specimens ensure meaningful and re-

FIGURE 2-4 Two common types of standard tensile test specimens: (a) round; (b) flat. Dimensions are in inches, with millimeters in parentheses.

(a)   (b)

FIGURE 2-5   (a) Hydraulic universal (tension and compression) testing machine; (b) schematic of the load frame showing how upward motion of the darkened yoke can produce tension or compression with respect to the stationary (white) crosspiece. *(Courtesy of Satec Systems, Inc., Grove City, PA.)*

producible results, and are designed to produce uniform uniaxial tension in the central portion of the specimen while ensuring reduced stresses in the enlarged ends or shoulders that are gripped.

Strength Properties.   The standard specimen is then loaded in tension in a testing machine like the one shown in Figure 2-5. A force or load, W, is applied and measured by the testing machine, while the elongation or stretch (ΔL) of a specified length (*gage length*) is simultaneously monitored. A plot of the coordinated load–elongation data produces a curve similar to that of Figure 2-6. Since the loads will differ for different-sized specimens and the amount of elongation will vary with different gage lengths, it is important to remove these geometric or size effects if we are to produce data that are characteristic of a given material, not a particular specimen. If the load is divided by the

FIGURE 2-6   Engineering stress–strain diagram for a low-carbon steel.

original cross-sectional area, A_o, and the elongation is divided by the *original* gage length, L_o, the size effects are eliminated and the resulting plot becomes known as an *engineering stress–engineering strain curve* (see Figure 2-6). This is simply a load–elongation plot with the scales of both axes modified to remove the effects of specimen size.

In Figure 2-6 it can be noted that the initial response is linear. Up to a certain point, the stress and strain are directly proportional to one another. The stress at which this proportionality ceases is known as the *proportional limit*. Below this value, the material obeys *Hooke's law*, which states that the strain is directly proportional to the stress. The proportionality constant, or ratio of stress to strain, is known as *Young's modulus* or the *modulus of elasticity*. This is an inherent property of a given material[1] and is of considerable engineering importance. As a measure of *stiffness*, it indicates the ability of a material to resist deflection or stretching when loaded and is commonly designated by the symbol E.

Up to a certain stress, if the load is removed, the specimen will return to its original length. The response is elastic or recoverable, like the stretching and relaxation of a rubber band. The uppermost stress for which this behavior is observed is known as the *elastic limit*. For most materials the elastic limit and proportional limit are almost identical, with the elastic limit being slightly higher. Neither quantity should be assigned great engineering significance, however, because the determined values are often dependent on the sensitivity and precision of the test equipment.

The amount of energy that a material can absorb while in the elastic range is called the *resilience*. The area under a load–elongation curve is the product of a force and a distance, and is therefore a measure of the energy absorbed by the specimen. If the area is determined up to the elastic limit, the absorbed energy will be elastic (or potential) energy and is regained when the specimen is unloaded. If we perform the same calculation on an engineering stress–engineering strain diagram, the area beneath the elastic region corresponds to an energy per unit volume and is known as the *modulus of resilience*.

Elongation beyond the elastic limit becomes unrecoverable and is known as *plastic deformation*. When the load is removed, only the elastic stretching will be recovered, and the specimen will retain a permanent change in shape. For most components, the onset of plastic flow represents failure, since the part dimensions will now be outside of allowable tolerances. In manufacturing processes where plastic deformation is used to produce a desired shape, the applied stresses must be sufficiently above the elastic limit to induce the required amount of plastic flow. Permanent deformation, therefore, may be either desirable or undesirable, and it is important to determine the conditions where elastic behavior transitions to plastic flow.

Whenever the elastic limit is exceeded, increases in strain no longer require proportionate increases in stress. For some materials, a stress value may be reached where additional strain occurs without any further increase in stress. This stress is known as the *yield point*, or *yield-point stress*. For low-carbon steels, with curves like that in Figure 2-6, two distinct points are significant. The highest stress preceding extensive strain is known as the *upper yield point*, and the lower, relatively constant, "run-out" value is known as the *lower yield point*. The lower value is the one that usually appears in tabulated data.

Most materials, however, do not have a well-defined yield point and exhibit stress–strain curves more like that shown in Figure 2-7. For these materials, the elastic-to-plastic transition is not distinct, and detection of plastic deformation would be dependent upon machine sensitivity. To solve this dilemma, we elect to define a useful and easily determined property known as the *offset yield strength*. Offset yield strength does *not* describe the onset of plastic deformation but instead defines the stress required to produce a given, but tolerable, amount of permanent strain. By setting this strain, or "offset," to 0.2% (a common value), we can determine the stress required to plastically

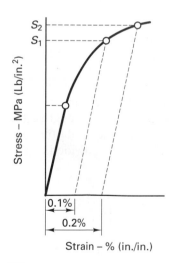

FIGURE 2-7 Stress–strain diagram for a material not having a well-defined yield point, showing the offset method for determining yield strength. S_1 is the 0.1% offset yield strength; S_2 is the 0.2% offset yield strength.

---

[1] The modulus of elasticity is determined by the binding forces between the atoms. Since these forces cannot be changed, the elastic modulus is characteristic of a specific material and is not alterable by the structure modifications that can be induced by processing.

deform a 1-inch length to a final length of 1.002 inches (a 0.2% strain). If the applied stresses are then kept below the 0.2% offset yield strength of the material, the user can be guaranteed that any resulting plastic deformation will be less than 0.2% of the original dimension. While 0.2% is a common offset for many mechanical products, applications that cannot tolerate that amount of deformation may specify offset values of 0.1% or even 0.02%.

Offset yield strength is determined by drawing a line parallel to the elastic line, but displaced by the offset strain, and reporting the stress where the constructed line intersects the actual stress–strain curve. Figure 2-7 shows the determination of both 0.1% offset and 0.2% offset values, S_1 and S_2, respectively. The intersection values are reproducible and independent of equipment sensitivity. Offset yield values are meaningless unless they are reported in conjunction with the amount of offset strain used in their determination. The 0.2% value is most common and is generally assumed unless another number is specified.

As shown in Figure 2-6, the load (or engineering stress) required to produce additional plastic deformation continues to increase. The load that a material or specimen can bear (load-bearing ability) can be computed by multiplying the material strength times its cross-sectional area. During tensile deformation, the specimen is getting longer. The cross-sectional area is decreasing, but the load-bearing ability of the specimen continues to increase! For this to occur, the material must be getting stronger. The mechanism for this phenomenon will be discussed in Chapter 3, where we will learn that the strength of a metal continues to increase with increased deformation.

During the plastic deformation portion of a tensile test, the weakest location of the specimen undergoes deformation and becomes stronger. Since it is no longer the weakest location, another location assumes that status and deforms. As a consequence, the specimen strengthens uniformly and maintains its original cylindrical or rectangular geometry. As plastic deformation progresses, however, the additional increments of strength decrease in magnitude, and a point is reached where the decrease in area cancels or dominates the increase in strength. When this occurs, the load-bearing ability peaks, and the force required to continue straining the specimen begins to decrease, as seen in Figure 2-6. The stress at which the load-bearing ability peaks is known as the *ultimate strength*, *tensile strength*, or *ultimate tensile strength* of the material. The weakest location in the test specimen at that time continues to be the weakest location by virtue of the decrease in area, and further deformation becomes localized. This localized reduction in cross-sectional area, known as *necking*, is shown in Figure 2-8.

If the straining is continued, necking becomes intensified and the tensile specimen will ultimately fracture. The stress at which fracture occurs is known as the *breaking strength* or *fracture strength*. For ductile materials, necking precedes fracture, and the breaking strength is less than the ultimate tensile strength. For a brittle material, fracture usually terminates the stress–strain curve before necking, and possibly before the onset of plastic flow.

Ductility and Brittleness.   When evaluating the suitability of a material for certain manufacturing processes or its appropriateness for a given application, the amount of plasticity that precedes fracture, or the *ductility*, can often be a significant property. For metal deformation processes, the greater the ductility, the more a material can be deformed without fracture. Ductility also plays a key role in toughness, a property that will be described shortly.

FIGURE 2-8   A standard 0.505-in.-diameter tensile specimen showing a necked region developed prior to failure.

FIGURE 2-9 Final elongation in various segments of a tensile test specimen: (a) original geometry; (b) shape after fracture.

One of the simplest ways to evaluate ductility is to determine the *percent elongation* of a tensile test specimen at the time of fracture. As shown in Figure 2-9, ductile materials do not elongate uniformly when loaded beyond necking. If the percent change of the entire 8-in. gage length is computed, the elongation is 31%. However, if only the center 2-in. segment is considered, the elongation of that portion is 60%. A valid comparison of material behavior, therefore, requires similar specimens with the same standard gage length.

In many cases, material "failure" is defined as the onset of localized deformation or necking. Consider a sheet of metal being formed into an automobile body panel. If we are to ensure uniform strength and corrosion resistance in the final panel, the operation must be performed in such a way as to maintain uniform sheet thickness. For this application, a more meaningful measure of material ductility would be the *uniform elongation* or the *percent elongation prior to the onset of necking*. This value can be determined by constructing a line parallel to the elastic portion of the diagram, passing through the point of highest force or stress. The intercept where the line crosses the strain axis denotes the available uniform elongation. Since the additional deformation that occurs after necking is not considered, uniform elongation is always less than the total elongation at fracture (the generally reported elongation value).

Another measure of ductility is the *percent reduction in area* that occurs in the necked region of the specimen. This can be computed as

$$\text{R.A.} = \frac{A_o - A_f}{A_o} \times 100\%$$

where A_o is the original cross-sectional area and A_f is the smallest area in the necked region. Percent reduction in area, therefore, can range from 0% (for a brittle glass specimen that breaks with no change in area) to 100% (for extremely plastic soft bubble gum that pinches down to a point before fracture).

When materials fail with little or no ductility, they are said to be *brittle*. Brittleness, however, is simply the lack of ductility and should not be confused with a lack of strength. Strong materials can be brittle, and brittle materials can be strong.

Toughness. *Toughness*, or *modulus of toughness*, is the work per unit volume required to fracture a material. The tensile test can provide one measure of this property, since toughness corresponds to the total area under the stress–strain curve from test initiation to fracture, and thereby encompasses both strength and ductility. Caution should be exercised when using toughness data, however, because the work or energy needed to fracture can vary markedly with different conditions of testing. Variations in the temperature or the speed of loading can significantly alter both the stress–strain curve and the toughness.

In most cases, toughness is associated with impact or shock loadings, and the values obtained from high-speed (dynamic) impact tests often fail to correlate with those obtained from the relatively slow-speed (static) tensile test.

True Stress–True Strain Curves.   The stress–strain curve in Figure 2-6 is a plot of *engineering stress, S,* versus *engineering strain, e,* where S is computed as the applied load divided by the original cross-sectional area and e is the elongation, ΔL, divided by the original gage length, L_o. As the test progresses, the cross section of the test specimen changes continually, first in a uniform manner and then nonuniformly after necking begins. The actual stress should be computed based on the instantaneous cross-sectional area, not the original. Since the area is decreasing, the actual or true stress will be greater than the engineering stress plotted in Figure 2-6. *True stress, σ,* can be computed by taking simultaneous readings of the load and the minimum specimen diameter. The actual area can then be computed, and true stress can be determined as

$$\sigma = W/A$$

The determination of *true strain* is a bit more complex. In place of the change in length divided by the original length that was used to compute engineering strain, true strain is defined as the summation of the incremental strains that occur throughout the test. For a specimen that has been stretched from length L_o to length L, the *true, natural,* or *logarithmic strain* would be:

$$\varepsilon = \int_{L_o}^{L} \frac{d\ell}{\ell} = \ln \frac{L}{L_o} = 2 \ln \frac{D_o}{D}$$

The last equality makes use of the relationship for cylindrical specimens

$$\frac{L}{L_o} = \frac{A_o}{A} = \frac{D_o^2}{D^2}$$

that applies only up to the onset of necking.

Figure 2-10 depicts the type of curve that results when the data from a uniaxial tensile test are converted to the form of true stress versus true strain. Since the true stress is a measure of the material strength at any point during the test, it will continue to rise even after necking. Data beyond the onset of necking should be used with extreme caution, since the geometry of the neck transforms the stress state from uniaxial tension (stretching in one direction with compensating contractions in the other two) to triaxial tension, in which the material is stretched or restrained in all three directions. Because of the triaxial tension, voids or cracks (Figure 2-11) tend to form in the necked region and serve as a precursor to final fracture. Measurements of the external diameter no longer reflect the true load-bearing area, and the data are further distorted.

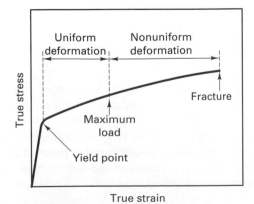

FIGURE 2-10   True stress–true strain curve for an engineering metal.

FIGURE 2-11   Section of a tensile test specimen stopped just prior to failure, showing a crack already started in the necked region. *(Courtesy of E. R. Parker.)*

FIGURE 2-12   Stress–strain diagram obtained by unloading and reloading a specimen.

FIGURE 2-13   True stress–true strain curves for metals with large and small strain hardening. Metals with larger n values experience larger amounts of strengthening for a given strain.

Strain Hardening and the Strain-Hardening Exponent.   Figure 2-12 is a true stress–true strain diagram, which has been modified to show how a ductile metal (such as steel) will behave when subjected to slow loading and unloading. Loading and unloading within the elastic region will result in simply cycling up and down the linear portion of the curve between points O and A. However, if the initial loading is carried through point B (in the plastic region), unloading will follow the path BeC, which is approximately parallel to the line OA, and the specimen will exhibit a permanent elongation of the amount OC. Upon reloading from point C, elastic behavior is again observed as the stress follows the line CfD, a slightly different path from that of unloading. Point D is now the yield point or yield stress for the material in its partially deformed state. A comparison of points A and D reveals that plastic deformation has made the material stronger. If the test were again interrupted at point E, we would find a new, even higher-yield stress. Thus, within the region of plastic deformation, each of the points along the true stress–true strain curve represents the yield stress for the corresponding value of strain.

When metals are plastically deformed, they become harder and stronger, a phenomenon known as *strain hardening*. If a stress is capable of producing plastic deformation, an even greater stress will be required to continue the flow. In Chapter 3 we will discuss the atomic-scale features that are responsible for this phenomenon.

Various materials strain-harden at different rates; that is, for a given amount of deformation different materials will exhibit different increases in strength. One method of describing this behavior is to mathematically fit the plastic region of the true stress–true strain curve to the equation

$$\sigma = K\,\varepsilon^n$$

and determine the best-fit value of n, the *strain-hardening exponent*.[2] As shown in Figure 2-13, a material with a high value of n will have a significant increase in material strength with a small amount of deformation. A material with a small n value will show little change in strength with plastic deformation.

Damping Capacity.   In Figure 2-12 the unloading and reloading of the specimen follow slightly different paths. The area between the two curves is proportional to the amount of energy that is converted from mechanical form to heat and is therefore absorbed by the material. When this area is large, the material is said to exhibit good *damping capacity* and is able to absorb mechanical vibrations or damp them out quickly. This is an important property in applications such as crankshafts and machinery bases. Gray cast iron is used in many applications because of its high damping capacity. Materials with low damping capacity, such as brass and steel, readily transmit both sound and vibrations.

Rate Considerations.   The rate or speed at which a tensile test is conducted can have a significant effect on the various properties. *Strain rate* sensitivity varies widely for engineering materials. Plastics and polymers are very sensitive to testing speed. Steels are also sensitive, but aluminum is rather insensitive. Those materials that are sensitive to speed variations exhibit higher strengths and lower ductility when speed is increased. It is important to recognize that standard testing selects a standard speed, which may or may not correlate with the conditions of product application.

COMPRESSION TESTS

When a material is subjected to compressive loadings, the relationships between stress and strain are similar to those for a tension test. Up to a certain value of stress, the material behaves elastically. Beyond this value, plastic flow occurs. In general, however, a compression test is more difficult to conduct than a standard tensile test. Test specimens must have larger cross-sectional areas to resist bending or buckling. As deformation proceeds, the material strengthens by strain hardening and the cross section of the specimen increases, combining to produce a substantial increase in required load. Friction between

---

[2] Taking the logarithm of both sides of the equation yields $\log \sigma = \log K + n \log \varepsilon$. This is the same form as the equation $y = mx + b$, the equation of a straight line with slope m and intercept b. Therefore, if the true stress–true strain data were plotted on a log-log scale with stress on the y-axis, the slope of the data in the plastic region would be n.

the testing machine surfaces and the ends of the test specimen will alter the results if not properly considered. The type of service for which the material is intended, however, should be the primary factor in determining whether the testing should be performed in tension or compression.

HARDNESS TESTING

The wear resistance and strength of a material can also be evaluated by assessing its "hardness." *Hardness* is actually a hard-to-define property of engineering materials, and a number of different tests have been developed using various phenomena. The most common of the hardness tests are based on resistance to permanent deformation in the form of penetration or indentation. Other tests evaluate resistance to scratching, wear resistance, resistance to cutting or drilling, or elastic rebound (energy absorption under impact loading). Since these phenomena are not the same, the results of the various tests often do not correlate with one another. Caution should be exercised to ensure that the selected test clearly evaluates the phenomena of interest.

Brinell Hardness Test.   The *Brinell hardness test* was one of the earliest accepted methods of measuring hardness. A tungsten carbide or hardened steel ball 10 mm in diameter is pressed into the flat surface of a material by a standard load of 500 or 3000 kg, and the load is maintained for 10 to 15 seconds to permit the full amount of plastic deformation to occur. The load and ball are then removed, and the diameter of the resulting spherical indentation (usually in the range of 2 to 5 mm) is measured using a special grid or traveling microscope. The *Brinell hardness number* (BHN) is equal to the load divided by the surface area of the spherical indentation when the units are expressed as kilograms per square millimeter.

In actual practice, the Brinell hardness number is determined from tables that correlate the Brinell number with the diameter of the indentation produced under the various loads. Figure 2-14 shows a typical Brinell tester, along with a schematic of the testing procedure. Portable testers are available for use on pieces that are too large to be brought to a benchtop machine.

The Brinell test measures hardness over a relatively large area and is somewhat indifferent to small-scale variations in the material structure. It is relatively simple and

FIGURE 2-14   (a) Brinell hardness tester; (b) Brinell test sequence showing loading and measurement of the indentation under magnification with a scale calibrated in millimeters. [(a) Courtesy of Wilson Instruments Division, Instron Corp., Norwood, MA]

(a)

Force (kgf)

10-mm (0.4-in.) ball

(b)

easy to conduct, and is used extensively on irons and steels. On the negative side, however, the Brinell test has the following limitations:

1. It cannot be used on very hard or very soft materials.

2. The results may not be valid for thin specimens. It is best if the thickness of material is at least 10 times the depth of the indentation. Some standards specify the minimum hardnesses for which the tests on thin specimens will be considered valid.

3. The test is not valid for case-hardened surfaces.

4. The test must be conducted far enough from the edge of the material so that no edge bulging occurs.

5. The substantial indentation may be objectionable on finished parts.

6. The edge or rim of the indentation may not be clearly defined or may be difficult to see.

The Rockwell Test. The widely used *Rockwell hardness test* is similar to the Brinell test, with the hardness value again being determined through an indentation produced under a static load. Figure 2-15a shows the key features of the Rockwell test. A small indenter, either a small-diameter steel ball or a diamond-tipped cone called a *brale*, is first seated firmly against the material by the application of a 10-kg "minor" load. This causes a slight elastic penetration into the surface and removes the effects of any surface irregularities. The indicator on the screen of the tester, like the one shown in Figure 2-15b, is then set to zero, and a "major" load of 60, 100, or 150 kg is applied to the indenter to produce a deeper penetration (i.e., plastic deformation). When the indicating pointer has come to rest, the major load is removed. With the minor load still applied, the tester now indicates the Rockwell hardness number on either a dial gage or digital display. This number is really an indication of the *depth* of the plastic or permanent penetration that was produced by the major load, with each unit representing a penetration depth of 2 μm.

Different combinations of major loads and indenters are designated by letters and are used for materials with various levels of strength. Table 2-1 provides a partial listing of the Rockwell scales and typical materials for which they are used. Because of the different scales, a Rockwell hardness number must be accompanied by the letter corresponding to the particular combination of load and indenter used in its determination. The notation R_C60 (or Rockwell C 60), for example, indicates that a 120° diamond-tipped brale indenter was used in combination with a major load of 150 kg, and

FIGURE 2-15 (a) Operating principle of the Rockwell hardness tester; (b) typical Rockwell hardness tester with digital readout. *[(a) Courtesy of Wilson Instruments Division, Instron Corp., Norwood, MA; (b) courtesy of MTI Corporation, Aurora, IL.]*

TABLE 2-1	Some Common Rockwell Hardness Tests		
Scale Symbol	Penetrator	Load (kg)	Typical Materials
A	Brale	60	Cemented carbides, thin steel, shallow case-hardened steel
B	$\frac{1}{16}$-in. ball	100	Copper alloys, soft steels, aluminum alloys, malleable iron
C	Brale	150	Steel, hard cast irons, titanium, deep case-hardened steel
D	Brale	100	Thin steel, medium case-hardened steel
E	$\frac{1}{8}$-in. ball	100	Cast iron, aluminum, magnesium
F	$\frac{1}{16}$-in. ball	60	Annealed coppers, thin soft sheet metals
G	$\frac{1}{16}$-in. ball	150	Hard copper alloys, malleable irons
H	$\frac{1}{8}$-in ball	60	Aluminum, zinc, lead

a reading of 60 was obtained. The B and C scales are used more extensively than the others, with B being common for copper and aluminum and C for steels.[3]

Rockwell tests should not be conducted on thin materials (typically less than 1.5 mm or 1/16 in.), on rough surfaces, or on materials that are not homogeneous, such as gray cast iron. Because of the small size of the indentation, variations in roughness, composition, or structure can greatly influence test results. For thin materials, or where a very shallow indentation is desired (as in the evaluation of surface-hardening treatments such as nitriding or carburizing), the *Rockwell superficial hardness test* is preferred. Operating on the same Rockwell principle, this test employs smaller major and minor loads (15 or 45 kg and 3 kg, respectively) and uses a more sensitive depth-measuring device.

In comparison with the Brinell test, the Rockwell test offers the attractive advantage of direct readings in a single step. Because it requires little (if any) surface preparation and can be conducted quite rapidly (up to 300 tests per hour or 5 per minute), it is often used for quality control purposes, such as determining if an incoming product meets specification, assuring that a heat treatment was performed properly, or simply monitoring the properties of products at various stages of manufacture. It has the additional advantage of producing a small indentation that can be easily concealed on the finished product or easily removed in a later operation.

Vickers Hardness Test.   The *Vickers hardness test* is also similar to the Brinell test but uses a 136° square-based diamond pyramid as the indenter and loads between 1 and 120 kg. Like the Brinell value, the Vickers hardness number is also defined as load divided by the surface area of the indentation expressed in units of kilograms per square millimeter. The advantages of the Vickers approach include increased accuracy in determining the diagonal of a square impression as opposed to the diameter of a circle and the assurance that even light loads will produce some plastic deformation. The use of diamond as the indenter material enables the test to evaluate any material and effectively places the hardness of all materials on a single scale.

Like the other indentation or penetration methods, the Vickers test has a number of attractive features: (1) it is simple to conduct, (2) little time is involved, (3) little surface preparation is required, (4) the marks are quite small and are easily hidden or removed, (5) the test can be done on location, (6) it is relatively inexpensive, and (7) it provides results that can be used to evaluate material strength or assess product quality.

Microhardness Tests.   Various *microhardness tests* have been developed for applications where it is necessary to determine the hardness of a very precise area of material or where the material or modified surface layer is exceptionally thin. These tests might be more appropriately termed *microindentation* hardness tests, since it is the size of the indentation that is extremely small, not the measured value of hardness. Special machines, such as the one shown in Figure 2-16,

FIGURE 2-16   Microhardness tester. *(Courtesy of LECO Corporation. St. Joseph, MI)*

---

[3] The Rockwell C number is computed as 100 − (depth of penetration in μm/2 μm), while the Rockwell B number is 130 − (depth of penetration in μm/2 μm).

(a)   (b)

FIGURE 2-17   (a) Comparison of the diamond-tipped indenters used in the Vickers and Knoop hardness tests. (b) Series of Knoop hardness indentations progressing left-to-right across a surface-hardened steel specimen (hardened surface to unhardened core). *(Courtesy Buehler Ltd., Lake Bluff, IL.)*

FIGURE 2-18   Durometer hardness tester. *(Courtesy of Newage Testing Instruments, Southampton, PA.)*

have been constructed for this type of testing. The location for the test is selected under high magnification. A small diamond penetrator is then loaded with a predetermined load ranging from 25 to 3600 g. In the *Knoop test*, an elongated diamond-shaped indenter (long diagonal seven times the short diagonal) is used and the length of the indentation is measured with the aid of a microscope. Figure 2-17 compares the indenters for the Vickers and Knoop tests, and shows a series of Knoop indentations progressing left-to-right across a surface-hardened steel specimen, from the hardened surface to the unhardened core. The hardness value, known as the *Knoop hardness number*, is again obtained by dividing the load in kilograms by the projected area of the indentation, expressed in square millimeters. A light-load Vickers test can also be used to determine microhardness.

Other Hardness Determinations.   When testing soft, elastic materials, such as rubbers and nonrigid plastics, a *durometer* can be used. This instrument, shown in Figure 2-18, measures the resistance of a material to elastic penetration by a spring-loaded conical steel indenter. No permanent deformation occurs. A similar test, used to evaluate the strength of molding sands used in the foundry industry, will be described in Chapter 14.

In the *scleroscope* test, hardness is measured by the rebound of a small diamond-tipped "hammer" that is dropped from a fixed height onto the surface of the material to be tested. This test evaluates the resilience of a material, and the surface on which the test is conducted must have a fairly high polish to yield good results. Because the test is based on resilience, scleroscope hardness numbers should only be used to compare similar materials. A comparison between steel and rubber, for example, would not be valid.

Another definition of hardness is the ability of a material to resist being scratched. A crude but useful test that employs this principle is the *file test*, where one determines if a material can be cut by a simple metalworking file. The test can be either a pass–fail test using a single file or a semiquantitative evaluation using a series of files that have been pretreated to various levels of known hardness.

Relationships among the Various Hardness Tests.   Since the various hardness tests often evaluate different material phenomena, there are no simple relationships between the different types of hardness numbers. Approximate relationships have been developed, however, by testing the same material on a variety of devices. Table 2-2 presents a correlation of hardness values for plain carbon and low-alloy steels. It may be noted that for Rockwell C numbers above 20, the Brinell values are approximately 10 times the Rockwell number. Also, for Brinell values below 320, the Vickers and Brinell values agree quite closely. Since the relationships among the various tests will differ with material, mechanical processing, and heat treatment, correlations such as Table 2-2 should be used with caution.

TABLE 2-2 Hardness Conversion Table for Steels

Brinell Number	Vickers Number	Rockwell Number		Scleroscope Number	Tensile Strength	
		C	B		ksi	MPa
	940	68		97	368	2537
757[a]	860	66		92	352	2427
722[a]	800	64		88	337	2324
686[a]	745	62		84	324	2234
660[a]	700	60		81	311	2144
615[a]	655	58		78	298	2055
559[a]	595	55		73	276	1903
500	545	52		69	256	1765
475	510	50		67	247	1703
452	485	48		65	238	1641
431	459	46		62	212	1462
410	435	44		58	204	1407
390	412	42		56	196	1351
370	392	40		53	189	1303
350	370	38	110	51	176	1213
341	350	36	109	48	165	1138
321	327	34	108	45	155	1069
302	305	32	107	43	146	1007
285	287	30	105	40	138	951
277	279	28	104	39	34	924
262	263	26	103	37	128	883
248	248	24	102	36	122	841
228	240	20	98	34	116	800
210	222	17	96	32	107	738
202	213	14	94	30	99	683
192	202	12	92	29	95	655
183	192	9	90	28	91	627
174	182	7	88	26	87	600
166	175	4	86	25	83	572
159[a]	167	2	84	24	80	552
153	162		82	23	76	524
148[a]	156		80	22	74	510
140	148		78	22	71	490
135	142		76	21	68	469
131	137		74	20	66	455
126	132		72	20	64	441
121	121		70		62	427
112	114		66		58	

[a] Tungsten, carbide ball; others, standard ball.

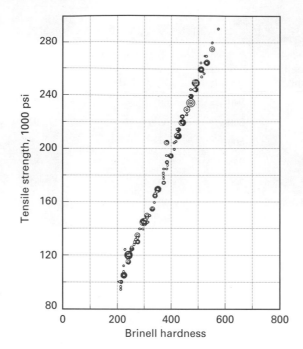

FIGURE 2-19 Relationship of hardness and tensile strength for a group of standard alloy steels. *(Courtesy of ASM International, Materials Park, OH.)*

Relationship of Hardness to Tensile Strength. Table 2-2 and Figure 2-19 show a definite relationship between tensile strength and hardness. For plain carbon and low-alloy steels, the tensile strength (in pounds per square inch) can be estimated by multiplying the Brinell hardness number by 500. In this way, an inexpensive and quick hardness test can be used to provide a close approximation of the tensile strength of the steel. For other materials, however, the relationship is different and may even exhibit too much variation to be dependable. The multiplying factor for age-hardened aluminum is about 600, while for soft brass it is around 800.

■ 2.3 DYNAMIC PROPERTIES

In many engineering applications, products or components are subjected to various types of dynamic loading. These may include (1) sudden impacts or loads that vary rapidly in magnitude, (2) repeated cycles of loading and unloading, or (3) frequent changes in the mode of loading, such as from tension to compression. To handle these conditions, we must be able to characterize the mechanical properties of engineering materials under dynamic loadings.

Most dynamic tests subject standard specimens to a well-controlled set of test conditions. The conditions of actual application, however, rarely duplicate the controlled conditions of a standardized test. While identical tests on different materials can indeed provide a comparison of material behavior, the assumption that similar results can be expected for similar conditions may not always be true. Since dynamic conditions can vary greatly, the quantitative results of standardized tests should be used with extreme caution, and one should always be aware of the test limitations.

IMPACT TEST

Several tests have been developed to evaluate the *toughness* or fracture resistance of a material when it is subjected to a rapidly applied load, or impact. Of the tests that have become common, two basic types have emerged: (1) bending impacts, which include the standard Charpy and Izod tests, and (2) tension impacts.

The bending impact tests utilize specimens that are supported as beams. In the *Charpy test*, shown schematically in Figure 2-20, the standard specimen is a square bar containing a V-, keyhole-, or U-shaped notch. The test specimen is positioned horizontally, supported on the ends, and an impact is applied to the center, behind the notch, to complete a three-point bending. The *Izod test* specimen, while somewhat similar in size

FIGURE 2-20   (a) Standard Charpy impact specimens. Illustrated are keyhole and U notches; dimensions are in millimeters with inches in parentheses. (b) Standard V-notch specimen showing the three-point bending type of impact loading.

FIGURE 2-21   (a) Izod impact specimen; (b) cantilever mode of loading in the Izod test.

and appearance, is supported vertically as a cantilever beam and is impacted on the unsupported end, striking from the side of the notch (Figure 2-21). Impact testers, like the one shown in Figure 2-22, supply a predetermined impact energy in the form of a swinging pendulum. After breaking or deforming the specimen, the pendulum continues its upward swing with an energy equal to its original minus that absorbed by the impacted specimen. The loss of energy is measured by the angle that the pendulum attains during its upward swing.

The test specimens for bending impacts must be prepared with geometric precision to ensure consistent and reproducible results. Notch profile is extremely critical, for the test measures the energy required to both initiate and propagate a fracture. The effect of notch profile is shown dramatically in Figure 2-23. Here two specimens have been made from the same piece of steel with the same reduced cross-sectional area. The one with the keyhole notch fractures and absorbs only 43 ft-lb of energy, whereas the unnotched specimen resists fracture and absorbs 65 ft-lb during the impact.

FIGURE 2-22   Impact testing machine. *(Courtesy of Tinius Olsen Inc., Horsham, PA.)*

FIGURE 2-23   Notched and unnotched impact specimens before and after testing. Both specimens had the same cross-sectional area, but the notched specimen fractures while the other doesn't.

Caution should also be placed on the use of impact data for design purposes. The test results apply only to standard specimens containing a standard notch. Moreover, the tests evaluate material behavior under very specific conditions. Changes in the form of the notch, minor variations in the overall specimen geometry, or faster or slower rates of loading (speed of the pendulum) can all produce significant changes in the results. Under conditions of sharp notches, wide specimens, and rapid loading, many ductile materials lose their energy-absorbing capability and fail in a brittle manner. [For example, the standard impact test should not be used to evaluate materials for bullet-proof armor, since the velocities of loading are extremely different.]

The results of standard tests, however, can be quite valuable in assessing a material's sensitivity to notches and the multiaxial stresses that exist around a notch. Materials whose properties vary with notch geometry are termed *notch-sensitive*. Good surface finish and the absence of scratches, gouges, and defects in workmanship will be key to satisfactory performance. Materials that are *notch-insensitive* can often be used with as-cast or rough-machined surfaces with no risk of premature failure.

Impact testing can also be performed at a variety of temperatures. As will be seen later in this chapter, the evaluation of how fracture resistance changes with temperature can be crucial to success when selecting engineering materials for low-temperature service.

The *tensile impact test*, illustrated schematically in Figure 2-24, eliminates the use of a notched specimen and thereby avoids many of the objections inherent in the Charpy and Izod tests. Turned specimens are subjected to uniaxial impact loadings applied through drop weights, modified pendulums, or variable-speed flywheels.

FIGURE 2-24   Tensile impact test.

FATIGUE AND THE ENDURANCE LIMIT

Materials can also fail by fracture if they are subjected to repeated applications of stress, even though the peak stresses have magnitudes less than the ultimate tensile strength and usually less than the yield strength. This phenomenon, known as *fatigue*, can result from either the cyclic repetition of a particular loading cycle or entirely random variations in stress. Almost 90% of all metallic fractures are in some degree attributed to fatigue.

For experimental simplicity, a periodic, sinusoidal loading is often utilized, and conditions of equal-magnitude tension–compression reversals provide further simplification. These conditions can be achieved by placing a cylindrical specimen in a rotating drive and hanging a weight so as to produce elastic bending along the axis, as shown in Figure 2-25. As a result of the elastic bending, material at the bottom of the specimen is stretched, or loaded in tension, while material on the top surface is compressed. As the specimen turns, the surface of the specimen experiences a sinusoidal application of tension and compression with each rotation.

By conducting multiple tests, subjecting identical specimens to different levels of maximum loading, and recording the number of cycles necessary to achieve fracture, curves such as that in Figure 2-26 can be produced. These curves are known as *stress versus number of cycles*, or *S–N, curves*. If the material being evaluated in Figure 2-26 were subjected to a standard tensile test, it would require a stress in excess of 480 MPa (70,000 psi) to induce failure. Under cyclic loading with a peak stress of only 380 MPa

FIGURE 2-25   Schematic diagram of a Moore rotating-beam fatigue machine. *(Adapted from Hayden et al., "The Structure and Properties of Materials", Vol 3, p. 15, Wiley, 1965.)*

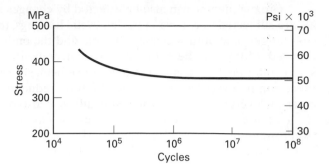

FIGURE 2-26 Typical $S-N$ curve for steel showing an endurance limit. Specific numbers will vary with the type of steel and treatment.

(55,000 psi), the specimen will fail after about 100,000 cycles. If the peak stress were further reduced to 350 MPa (51,000 psi), the fatigue lifetime would be extended by an order of magnitude to approximately 1,000,000 cycles. With a further reduction to any value below 340 MPa (49,000 psi), the specimen would not fail by fatigue, regardless of the number of stress application cycles.

The stress below which the material will not fail regardless of the number of load cycles is known as the *endurance limit* or *endurance strength*, and may be an important criterion in many designs. Above this value, any point on the curve is the *fatigue strength*, the maximum stress that can be sustained for a specified number of loading cycles.

A different number of loading cycles is generally required to determine the endurance limit for different materials. For steels, 10 million cycles are usually sufficient. For several of the nonferrous metals, 500 million cycles may be required. For aluminum, the curve continues to drop such that, if aluminum has an endurance limit, it is at such a low value that a cheaper and much weaker material could be used. In essence, if aluminum is used under realistic stresses and cyclic loading, it will fail by fatigue after a finite lifetime.

The fatigue resistance of an actual product is sensitive to a number of additional factors. One of the most important of these is the presence of stress raisers (or stress concentrators), such as sharp corners, small surface cracks, machining marks, or surface gouges. Data for the $S-N$ curves are obtained from polished-surface, "flaw-free" specimens, and the reported lifetime is the cumulative number of cycles required to initiate a fatigue crack and then grow or propagate it to failure. If a part already contains a surface crack or flaw, the number of cycles required for crack initiation can be reduced significantly. In addition, the stress concentrator magnifies the stress experienced at the tip of the crack, accelerating the rate of subsequent crack growth. Great care should be taken to eliminate stress raisers and surface flaws on parts that will be subjected to cyclic loadings. Proper design and good manufacturing practices are often more important than material selection and heat treatment.

Operating temperature can also affect the fatigue performance of a material. Figure 2-27 shows $S-N$ curves for Inconel 625 (a high-temperature Ni–Cr–Fe alloy) determined over a range of temperatures. As temperature is increased, the fatigue strength drops significantly. Since most test data are generated at room temperature, caution should be exercised when the product application involves elevated service temperatures.

FIGURE 2-27 Fatigue strength of Inconel alloy 625 at various temperatures. *(Courtesy of Huntington Alloy Products Division, The International Nickel Company, Inc., Toronto, Canada.)*

Fatigue lifetime can also be affected by changes in the environment. When metals are subjected to corrosion during cyclic loadings, the condition is known as corrosion fatigue, and both specimen lifetime and the endurance limit can be significantly reduced. Moreover, the nature of the environmental attack need not be severe. For some materials, tests conducted in air have been shown to have shorter lifetimes than those run in a vacuum, and further lifetime reductions have been observed with increasing levels of humidity. The test results can also be dependent on the frequency of the loading cycles. For slower frequencies, the environment has a longer time to act between loadings. At high frequencies, the environmental effects may be somewhat masked. The application of test data to actual products, therefore, requires considerable caution.

Residual stresses can also alter fatigue behavior. If the specimen surface is in a state of compression, such as that produced from shot peening, carburizing, or burnishing, it is more difficult to initiate a fatigue crack, and lifetime is extended. Conversely, processes that produce residual tension on the surface, such as welding or machining, can significantly reduce the fatigue lifetime of a product.

If the magnitude of the load varies during service, the fatigue response can be extremely complex. For example, consider the wing of a commercial airplane. As the wing vibrates during flight, the wing–fuselage joint is subjected to a large number of low–stress loadings. The large number of these load applications may be far less damaging, however, than a few high-stress loadings, like those that occur when the plane contacts the runway during landing. From a different perspective, however, the heavy loads may be sufficient to stretch and blunt a sharp fatigue crack, requiring many additional small-load cycles to "reinitiate" it. Evaluating how materials respond to complex patterns of loading is an area of great importance to design engineers.

Since reliable fatigue data may take a considerable time to generate, we may prefer to estimate fatigue behavior from properties that can be determined more quickly. Table 2-3 shows the approximate ratio of the endurance limit to the ultimate tensile strength for several engineering metals. For many steels the endurance limit can be approximated by 0.5 times the ultimate tensile strength as determined by a standard tensile test. For the nonferrous metals, however, the ratio is significantly lower.

TABLE 2-3   Ratio of Endurance Limit to Tensile Strength for Various Materials

Material	Ratio
Aluminum	0.38
Beryllium copper (heat-treated)	0.29
Copper, hard	0.33
Magnesium	0.38
Steel	
AISI 1035	0.46
Screw stock	0.44
AISI 4140 normalized	0.54
Wrought iron	0.63

FATIGUE FAILURES

Components that fail as a result of repeated or cyclic loadings are commonly called *fatigue failures*. These fractures form a major part of a larger group known as progressive fractures. Consider the fracture surface shown in Figure 2-28. The two arrows identify the points of fracture initiation, which often correspond to discontinuities in the form of surface cracks, sharp corners, machining marks, or even "metallurgical notches," such as an abrupt change in metal structure. With each repeated application of load, the stress at the tip of the crack exceeds the strength of the material, and the crack grows a very small amount. Crack growth continues with each successive application of load until the remaining cross section is no longer sufficient to withstand the peak stresses. Sudden overload fracture then occurs through the remainder of the material. The overall fracture surface tends to exhibit two distinct regions: a smooth, relatively flat region where the crack was propagating by cyclic fatigue, and a coarse, ragged region, corresponding to the ductile overload tearing.

The smooth areas of the fracture often contain a series of parallel ridges radiating outward from the origin of the crack. These ridges may not be visible under normal examination, however. They may be extremely fine; they may have been obliterated by a rubbing action during the compressive stage of repeated loading; or they may be very few in number if the failure occurred after only a few cycles of loading ("low-cycle fatigue"). Electron microscopy may be required to reveal the ridges, or *fatigue striations*,

FIGURE 2-28 Progressive fracture of an axle within a ball-bearing ring, starting at two points (arrows).

that are characteristic of fatigue failure. Figure 2-29 shows an example of these markings at high magnification.

For some fatigue failures, the overload area may exhibit a crystalline appearance, and the failure is sometimes attributed to the metal having "crystallized." As will be noted in Chapter 3, engineering metals are almost always crystalline materials. The final overload fracture simply propagated along the intercrystalline surfaces (grain boundaries) and revealed the already-existing crystalline nature of the material. The conclusion that the material crystallized is totally erroneous, and the term is a definite misnomer.

Another common error is to classify all progressive-type failures as fatigue failures. Other progressive failure mechanisms, such as creep failure and stress–corrosion cracking, will also produce the characteristic two-region fracture. In addition, the same mechanism can produce fractures with different appearances depending on the magnitude of the load, type of loading (torsion, bending, or tension), temperature, and operating environment. Correct interpretation of a metal failure generally requires far more information than that acquired by a visual examination of the fracture surface.

A final misconception regarding fatigue failures is to assume that the failure is time dependent. The failure of materials under repeated loads below their static strength is primarily a function of the magnitude and number of loading cycles. If the frequency of loading is increased, the time to failure should decrease proportionately. If the time does not change, the failure is dominated by one or more environmental factors, and fatigue is a secondary component.

FIGURE 2-29 Fatigue fracture of AISI type 304 stainless steel viewed in a scanning electron microscope at 810X. Well-defined striations are visible. *(From "Interpretation of SEM Fractographs," Metals Handbook, Vol. 9, 8th ed., ASM International, Materials Park, OH, 1970, p. 70.)*

■ 2.4 TEMPERATURE EFFECTS (BOTH HIGH AND LOW)

The test data used in design and engineering decisions should always be obtained under conditions that simulate those of actual service. A number of engineered structures, such as aircraft, space vehicles, gas turbines, and nuclear power plants, are required to operate under temperatures as low as $-130°C$ ($-200°F$) or as high as $1250°C$ ($2300°F$). To cover these extremes, the designer must consider both the short- and long-range effects of temperature on the mechanical and physical properties of the material being considered. From a manufacturing viewpoint, the effects of temperature are equally important. Numerous manufacturing processes involve heat, and the elevated temperature and processing may alter the material properties in both favorable and unfavorable ways. A material can often be processed successfully, or economically, only because heating or cooling can be used to change its properties.

Elevated temperatures can be quite useful in modifying the strength and ductility of a material. Figure 2-30 summarizes the results of tensile tests conducted over a wide range of temperatures using a medium-carbon steel. Similar effects are presented for magnesium in Figure 2-31. As expected, an increase in temperature will typically induce a decrease in strength and hardness and an increase in elongation. For manufacturing operations such as metalforming, heating to elevated temperature may be extremely attractive because the material is now both weaker and more ductile.

FIGURE 2-30 The effects of temperature on the tensile properties of a medium-carbon steel.

FIGURE 2-31 The effects of temperature on the tensile properties of magnesium.

Figure 2-32 shows the combined effects of temperature and strain rate (speed of testing) on the ultimate tensile strength of copper. For a given temperature, the *rate of deformation* can also have a strong influence on mechanical properties. Room-temperature standard-rate tensile test data will be of little value if the application involves a material being hot-rolled at speeds of 1300 m/min (5000 ft/min).

The effect of temperature on impact properties became the subject of intense study in the 1940s when the increased use of welded-steel construction led to catastrophic failures of ships and other structures while operating in cold environments. Welding produces a monolithic (single-piece) product where cracks can propagate through a joint and continue on to other sections of the structure! Figure 2-33 shows the effect of decreasing temperature on the impact properties of two low-carbon steels. Although similar in form, the two curves are significantly different. The steel indicated by the solid line becomes brittle (requires very little energy to fracture) at temperatures below −4°C (25°F) while the other steel retains good fracture resistance down to −26°C (−15°F). The temperature at which the response goes from high energy absorption to low energy absorption is known as the *ductile-to-brittle transition temperature*. While all steels tend to exhibit this transition, the temperature at which it occurs varies with carbon content and alloy. Special caution should be taken, therefore, when selecting steels for low-temperature applications.

FIGURE 2-32 The effects of temperature and strain rate on the tensile strength of copper. *(From A. Nadai and M. J. Manjoine,* Journal of Applied Mechanics, *Vol. 8, 1941, p. A82, courtesy of ASME.)*

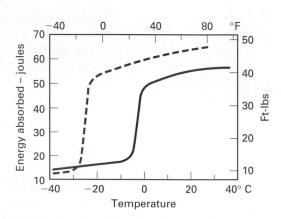

FIGURE 2-33 The effect of temperature on the impact properties of two low-carbon steels.

FIGURE 2-34 Longitudinal and transverse notch toughness impact data: steel from the *Titanic* versus modern steel plate, with both longitudinal and transverse specimens. *(Courtesy I&SM, September 1999, p. 33, Iron and Steel Society, Warrendale, PA.)*

Figure 2-34 shows the ductile-to-brittle transition temperature for steel salvaged from the *Titanic* compared to currently used ship plate material. While both are quality materials for their era, the *Titanic* steel has a much higher transition temperature and is generally more brittle. Recalling that the water temperature at the time the *Titanic* struck the iceberg was −2°C, the results show that the steel would have been quite brittle. Two curves are provided for each material, reflecting specimens in different orientation with respect to the direction of product rolling. Here we see that processing features can further affect the properties and performance of a material.

CREEP

Long-term exposure to elevated temperatures can also lead to failure by a phenomenon known as *creep*. If a tensile-type specimen is subjected to a constant load at elevated temperature, it will elongate continuously until rupture occurs, even though the applied stress is below the yield strength of the material at the temperature of testing. While the rate of elongation is often quite small, creep can be an important consideration when designing equipment such as steam or gas turbines, power plant boilers, and other devices that operate under loads or pressures for long periods of time at high temperature.

If a test specimen is subjected to conditions of fixed load and fixed elevated temperature, an elongation-versus-time plot can be generated, similar to the one shown in Figure 2-35. The curve contains three distinct stages: a short-lived initial stage, a rather long second stage where the elongation rate is somewhat linear, and a short-lived third stage leading to fracture. Two significant pieces of engineering data are obtained from this curve: the rate of elongation in the second stage, or *creep rate*, and the total elapsed

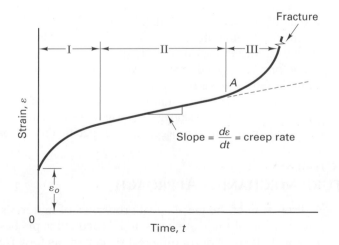

FIGURE 2-35 Creep curve for a single specimen at a fixed temperature, showing the three stages of creep and reported creep rate. Note the nonzero strain at time zero due to the initial application of the load.

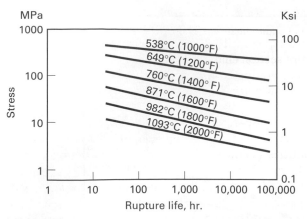

FIGURE 2-36   Stress–rupture diagram of solution-annealed Incoloy alloy 800 (Fe–Ni–Cr alloy). *(Courtesy of Huntington Alloy Products Division, The International Nickel Company, Inc., Toronto, Canada.)*

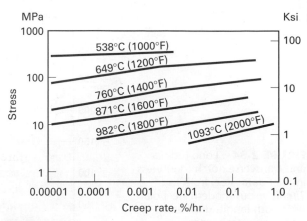

FIGURE 2-37   Creep-rate properties of solution-annealed Incoloy alloy 800. *(Courtesy of Huntington Alloy Products Division, The International Nickel Company, Inc., Toronto, Canada.)*

time to rupture. These results are unique to the material being tested and the specific conditions of the test. Tests conducted at higher temperatures or with higher applied loads would exhibit higher creep rates and shorter rupture times.

When creep behavior is a concern, multiple tests are conducted over a range of temperatures and stresses, and the rupture time data are collected into a single *stress–rupture diagram*, like the one shown in Figure 2-36. This simple engineering tool provides an overall picture of material performance at elevated temperature. In a similar manner, creep-rate data can also be plotted to show the effects of temperature and stress. Figure 2-37 presents a creep-rate diagram for a high-temperature nickel-based alloy.

■ 2.5 MACHINABILITY, FORMABILITY, AND WELDABILITY

While it is common to assume that the various "-ability" terms also refer to specific material properties, they actually refer to the way a material responds to specific processing techniques. As a result, they can be quite nebulous. *Machinability*, for example, depends not only on the material being machined but also on the specific machining process; the conditions of that process, such as cutting speed; and the aspects of that process that are of greatest interest. Machinability ratings are generally based on relative tool life. In certain applications, however, we may be more interested in how easy a metal is to cut, or how it performs under high-speed machining, and less interested in the tool life or the resulting surface finish. For other applications, surface finish or the formation of fine chips may be the most desirable feature. As a result, the term *machinability* may mean different things to different people, and it frequently involves multiple properties of a material interacting with the conditions of a process.

In a similar manner, *malleability*, *workability*, and *formability* all refer to a material's suitability for plastic deformation processing. Since a material often behaves differently at different temperatures, a material with good "hot formability" may have poor deformation characteristics at room temperature. Furthermore, materials that flow nicely at low deformation speeds may behave in a brittle manner when loaded at rapid rates. Formability, therefore, needs to be evaluated for a specific combination of material, process, and process conditions. The results cannot be extrapolated or transferred to other processes or process conditions. Likewise, the *weldability* of a material may also depend on the specific welding or joining process and the specific process parameters.

■ 2.6 FRACTURE TOUGHNESS AND THE FRACTURE MECHANICS APPROACH

A discussion of the mechanical properties of materials would not be complete without mention of the many tests and design concepts based on the fracture mechanics approach. Instead of treating test specimens as flaw-free materials, fracture mechanics begins with the premise that *all materials contain flaws or defects of some given size*. These may be *material defects*, such as pores, cracks, or inclusions; *manufacturing defects*, in the form of machining marks, arc strikes, or contact damage to external surfaces; or *design defects*, such as abrupt section changes, excessively small fillet radii, and holes. When the specimen is subjected to loads, the applied stresses are amplified or intensified in the vicinity of these defects, potentially causing accelerated failure or failure under unexpected conditions.

Fracture mechanics seeks to identify the conditions under which a defect will grow or propagate to failure and, if possible, the rate of crack or defect growth. The methods concentrate on three principal quantities: (1) the size of the largest or most critical flaw, usually denoted as a; (2) the applied stress, denoted by σ; and (3) the fracture toughness, a quantity that describes the resistance of a material to fracture or crack growth, which is usually denoted by K with subscripts to signify the conditions of testing. Equations have been developed that relate these three quantities (at the onset of crack growth or propagation) for various specimen geometries, flaw locations, and flaw orientations. If nondestructive testing or quality control methods have been applied, the size of the largest flaw that could go undetected is often known. By mathematically placing this worst possible flaw in the worst possible location and orientation, and coupling this with the largest applied stress for that location, a designer can determine the value of fracture toughness necessary to prevent that flaw from propagating during service. Specifying any two of the three parameters allows the computation of the third. If the material and stress conditions were defined, the size of the maximum permissible flaw could be computed. Inspection conditions could then be selected to ensure that flaws greater than this magnitude are cause for product rejection. Finally, if a component is found to have a significant flaw and the material is known, the maximum operating stress can be determined that will ensure no further growth of that flaw.

In the past, detection of a flaw or defect was usually cause for rejection of the part (detection = rejection). With enhanced methods and sensitivities of inspection, almost every product can now be shown to contain flaws. Fracture mechanics comes to the rescue. According to the philosophy of fracture mechanics, each of the flaws or defects in a material can be either *dormant* or *dynamic*. Dormant defects are those whose size remains unchanged through the lifetime of the part and are indeed permissible. A major goal of fracture mechanics, therefore, is to define the distinction between dormant and dynamic for the specific conditions of material, part geometry, and applied loading.[4] Alternative efforts to prevent material fracture generally involve overdesign, excessive inspection, or the use of premium-quality materials—all of which increase cost and possibly compromise performance.

Fracture mechanics can also be applied to fatigue, which has already been cited as causing as much as 90% of all dynamic failures. The standard method of fatigue testing applies cyclic loads to polished, flaw-free specimens, and the reported lifetime consists of both crack initiation and crack propagation. In contrast, fracture mechanics focuses on the growth of an already-existing flaw. Figure 2-38 shows the *crack growth rate* (change in size per loading cycle denoted as *da/dN*) plotted as a function of the fracture mechanics parameter, ΔK (where ΔK increases with an increase in either the flaw size and/or the magnitude of applied stress). Since the fracture mechanics approach begins with an existing flaw, it provides a far more realistic guarantee of minimum service life.

Fracture mechanics is a truly integrated blend of design (applied stresses), inspection (flaw-size determination), and materials (fracture toughness). The approach has proven valuable in many areas where fractures could be catastrophic.

---

[4] The basic equation of fracture mechanics assumes the form of $K \geq \alpha\,\sigma\,\sqrt{\pi a}$, where K is the *fracture toughness* of the material (a material property), σ is the maximum applied tensile stress, a is the size of the largest or most critical flaw, and α is a dimensionless factor that considers the flaw location and flaw shape. The left side of the equation considers the material and the right side describes the usage condition (a combination of flaw and loading). The relationship is usually described as a greater than or equal. When the material number is greater than the usage condition, the flaw is dormant. When equality is achieved, the flaw becomes dynamic, and crack growth or fracture occurs.

FIGURE 2-38   Plot of the fatigue crack growth rate vs. ΔK for a typical steel—the fracture mechanics approach. Similar-shaped curves are obtained for most engineering metals. *(Courtesy of ASM International.)*

■ 2.7 PHYSICAL PROPERTIES

For certain applications, the *physical properties* of an engineering material may be even more important than the mechanical ones. These include the thermal, electrical, magnetic, and optical characteristics.

We have already seen several ways in which the mechanical properties of materials change with variations in temperature. In addition to these effects, there are some truly *thermal properties* that should be considered. The *heat capacity* or *specific heat* of a material is the amount of energy that must be added to or removed from a given mass of material to produce a 1° change in temperature. This property is extremely important in processes such as casting, where heat must be extracted rapidly to promote solidification, or heat treatment, where large quantities of material are heated and cooled. *Thermal conductivity* measures the rate at which heat can be transported through a material. While this may be tabulated separately in reference texts, it is helpful to remember that for metals, thermal conductivity is directly proportional to electrical conductivity. Metals such as copper, gold, and aluminum that possess good electrical conductivity are also good transporters of thermal energy. *Thermal expansion* is another important thermal property. Most materials expand upon heating and contract upon cooling, but the amount of expansion or contraction will vary with the material. For components that are machined at room temperature but put in service at elevated temperatures, or castings that solidify at elevated temperatures and then cool, the manufactured dimensions must be adjusted to compensate for the subsequent changes.

Electrical conductivity and *electrical resistivity* may also be important design considerations. These properties will vary not only with the material but also with the temperature and the way the material has been processed.

From the standpoint of *magnetic response*, materials are often classified as diamagnetic, paramagnetic, ferromagnetic, antiferromagnetic, and ferrimagnetic. These terms refer to the way in which the material responds to an applied magnetic field. Material properties, such as saturation strength, remanence, and magnetic hardness or softness, describe the strength, duration, and nature of this response.

Still other physical properties that may assume importance include *weight* or *density*, *melting* and *boiling points*, and the various *optical properties*, such as the ability to transmit, absorb, or reflect light or other electromagnetic radiation.

■ 2.8 TESTING STANDARDS AND CONCERNS

When evaluating the mechanical and physical properties of materials, it is important that testing be conducted in a standardized and reproducible manner. ASTM International, formerly the American Society of Testing and Materials, maintains and updates many testing standards, and it is important to become familiar with their contents. For example, ASTM specification E370 describes the "Standard Test Methods and Definitions for Mechanical Testing of Steel Products." Tensile testing is described in specifications E8 and E83, impact testing in E23, creep in E139, and penetration hardness in E10. Other specifications describe fracture mechanics testing as well as the procedures to evaluate corrosion resistance, compressive strength, shear strength, torsional properties, and corrosion–fatigue.

In addition, it is important to note not only the material being tested but also the location from which the specimen was taken and its orientation. Rolled sheet, rolled plate, and rolled bars, for example, will have different properties when tested parallel to the direction of rolling (longitudinal) and perpendicular to the rolling direction (transverse). This variation of properties with direction, known as *anisotropy*, may be crucial to the success or failure of a product.

■ Key Words

ASTM	elongation	necking	stress
anisotropy	endurance limit	nonmetal	stress–rupture diagram
brale	engineering strain	notch-sensitive	structure
breaking strength	engineering stress	notch-insensitive	tensile strength
Brinell hardness number	fatigue	offset yield strength	tensile impact test
Brinell hardness test	fatigue strength	percent elongation	tensile test
brittle	fatigue striations	percent reduction in area	tension
Charpy test	formability	performance	thermal conductivity
compression	fracture toughness	physical properties	thermal expansion
crack growth rate (da/dN)	gage length	plastic deformation	time to rupture (rupture time)
creep	hardness	processing properties	toughness
creep rate	heat capacity	proportional limit	transition temperature
damping	Hooke's law	resilience	true strain
density	impact test	Rockwell hardness test	true stress
design defects	Izod test	*S–N* curve	ultimate tensile strength
dormant flaw	Knoop hardness	sclerosope	uniaxial tensile test
ductile-to-brittle transition temperature	machinability	shear	uniform elongation
ductility	manufacturing defects	specific heat	Vickers hardness test
durometer	material defects	static properties	weldability
dynamic flaw	mechanical properties	stiffness	yield point
elastic limit	metal	strain	Young's modulus
electrical conductivity	microhardness	strain hardening	
electrical resistivity	modulus of elasticity	strain–hardening exponent	
	modulus of resilience	strain rate	

■ Review Questions

1. A knowledge of what four aspects is critical to the successful application of a material in an engineering design?
2. What are some properties commonly associated with metallic materials?
3. What are some of the more common nonmetallic engineering materials?
4. What are some of the important physical properties of materials?
5. Why should caution be exercised when applying the results from any of the standard mechanical property tests?
6. What are the standard units used to report stress and strain in the English system? In the metric or SI system?
7. What are static properties?
8. What is the most common static test related to mechanical properties?
9. Why might Young's modulus or stiffness be an important material property?
10. What are some of the tensile test properties that are used to describe or define the elastic-to-plastic transition in a material?
11. Why is it important to specify the "offset" when providing yield strength data?
12. What are two tensile test properties that can be used to describe the ductility of a material?
13. Is a brittle material a weak material? What does "brittleness" mean?
14. What is the toughness of a material?
15. What is the difference between true stress and engineering stress? True strain and engineering strain?
16. What is strain hardening or work hardening? How might this phenomenon be measured or reported? How might it be used in manufacturing?
17. How might tensile test data be misleading for a "strain rate sensitive" material?
18. What are some of the different material characteristics or responses that have been associated with the term *hardness*?
19. What are the similarities and differences between the Brinell and Rockwell hardness tests?
20. Why are there different Rockwell hardness scales?
21. When might a microhardness test be preferred over the more standard Brinell or Rockwell tests?
22. Why might the various types of hardness tests fail to agree with one another?
23. What is the relationship between penetration hardness and the ultimate tensile strength for steel?
24. Describe several types of dynamic loading.
25. Why should the results of standardized dynamic tests be applied with considerable caution?
26. What are the two most common types of bending impact tests? How are the specimens supported and loaded in each?
27. What aspects or features can significantly alter impact data?
28. What is "notch-sensitivity" and how might it be important in the performance of a product?
29. What is the endurance limit? What occurs when stresses are above it? Below it?
30. Are the stresses applied during a fatigue test above or below the yield strength (as determined in a tensile test)?
31. What features may significantly alter the fatigue lifetime or fatigue behavior of a material?
32. What relationship can be used to estimate the endurance limit of a steel?
33. What material, design, or manufacturing features can contribute to the initiation of a fatigue crack?
34. What are fatigue striations and why do they form?
35. Why is it important for a designer or engineer to know a material's properties at all possible temperatures of operation?
36. Why should one use caution when using steel at low temperature?
37. How might we evaluate the long-term effect of elevated temperature on an engineering material?
38. What is a stress–rupture diagram, and how is one developed?
39. Why are terms such as *machinability, formability,* and *weldability* considered to be poorly defined and therefore quite nebulous?
40. What is the basic premise of the fracture mechanics approach to testing and design?
41. What three principal quantities does fracture mechanics attempt to relate?
42. What are the three most common thermal properties of a material, and what do they measure?
43. Describe an engineering application where the density of the selected material would be an important material consideration.
44. Why is it important that property testing be performed in a standardized and reproducible manner?
45. Why is it important to consider the orientation of a test specimen with respect to the overall piece of material?

■ Problems

1. Select a product or component for which physical properties are more important than mechanical properties.
   a. Describe the product or component and its function.
   b. What are the most important properties or characteristics?
   c. What are the secondary properties or characteristics that would also be desirable?
2. Repeat Problem 1 for a product or component whose dominant required properties are of a static mechanical nature.
3. Repeat Problem 1 for a product or component whose dominant requirements are dynamic mechanical properties.
4. One of the important considerations when selecting a material for an application is to determine the highest and lowest operating temperatures along with the companion properties that must be present at each extreme. The ductile-to-brittle transition temperature, discussed in Section 2.4, has been an important factor in a number of failures. An article that summarized the features of 56 catastrophic brittle fractures that made headline news between 1888 and 1956 noted that low temperatures were present in nearly every case. The water temperature at the time of the sinking of the *Titanic* was above

the freezing point for salt water but below the transition point for the steel used in construction of the hull of the ship.

   a. Which of the common engineering materials exhibits a ductile-to-brittle transition?

   b. For plain carbon and low-alloy steels, what is a typical value (or range of values) for the transition temperature?

   c. What type of material would you recommend for construction of a small vessel to transport liquid nitrogen within a building or laboratory?

   d. Figure 2-34 summarizes the results of impact testing performed on hull plate from the *Titanic* and similar material produced for modern steel-hulled ships. Why should there be a difference between specimens cut longitudinally (along the rolling direction) and transversely (across the rolling direction)? What advances in steel making have led to the significant improvement in low-temperature impact properties?

5. Several of the property tests described in this chapter produce results that are quite sensitive to the presence or absence of notches or other flaws. The fracture mechanics approach to materials testing incorporates flaws into the tests and evaluates their performance. The review article mentioned in Problem 4 cites the key role of a flaw or defect in nearly all of the headline-news fractures.

   a. What are some of the various "flaws or defects" that might be present in a product? Consider flaws that might be present in the starting material, flaws that might be introduced during manufacture, and flaws that might occur due to shipping, handling, use, maintenance, or repair.

   b. What particular properties might be most sensitive to flaws or defects?

   c. Discuss the relationship of flaws to the various types of loading (tension vs. compression, torsion, shear).

   d. Fracture mechanics considers both surface and interior flaws and assigns terms such as *crack initiator*, *crack propagator*, and *crack arrestor*. Briefly discuss why location and orientation may be as important as the physical size of a flaw.

www.wiley.com/college/degarmo

Chapter 2 CASE STUDY

Separation of Mixed Materials

Because of the amount of handling that occurs during material production, within warehouses, and during manufacturing operations, along with the handling of loading, unloading, and shipping, material mix-ups and mixed materials are not an uncommon occurrence. Mixed materials also occur when industrial scrap is collected or when discarded products are used as raw materials through recycling. Assume that you have equipment to perform each of the tests described in this chapter (as well as access to the full spectrum of household and department store items and even a small machine shop). For each of the following material combinations, determine a procedure that would permit separation of the mixed materials. Use standard datasource references to help identify distinguishable properties.

1. Steel and aluminum cans that have been submitted for recycling
2. Stainless steel sheets of Type 430 ferritic stainless and Type 316 austenitic stainless.
3. 6061-T6 aluminum and AZ91 magnesium that have become mixed in a batch of machine shop scrap.
4. Transparent bottles of polyethylene and polypropylene (both thermoplastic polymers) that have been collected for recycling.
5. Hot-rolled bars of AISI 1008 and 1040 steel.
6. Hot-rolled bars of AISI 1040 (plain-carbon) steel and 4140 steel (a molybdenum-containing alloy)

CHAPTER 3

NATURE OF METALS AND ALLOYS

■ 3.1 STRUCTURE–PROPERTY–PROCESSING–PERFORMANCE RELATIONSHIPS

As discussed in Chapter 2, the success of many engineering activities depends on the selection of engineering materials whose properties match the requirements of the application. Primitive cultures were often limited to the naturally occurring materials in their environment. As civilization developed, the spectrum of engineering materials expanded. Materials could be processed and their properties altered and possibly enhanced. The alloying or heat treatment of metals and the firing of ceramics are examples of techniques that can substantially alter the properties of a material. Fewer compromises were required, and enhanced design possibilities emerged. Products, in turn, became more sophisticated. While the early successes in altering materials were largely the result of trial and error, we now recognize that the *properties* and *performance* of a material are a direct result of its *structure* and *processing*. If we want to change the properties, we will most likely have to induce changes in the material structure.

Since all materials are composed of the same basic components—particles that include *protons, neutrons,* and *electrons*—it is amazing that so many different materials exist with such widely varying properties. This variation can be explained, however, by the many possible combinations these units can assume in a macroscopic assembly. The subatomic particles combine in different arrangements to form the various elemental *atoms*, each having a nucleus of protons and neutrons surrounded by the proper number of electrons to maintain charge neutrality. The specific arrangement of the electrons surrounding the nucleus affects the electrical, magnetic, thermal, and optical properties as well as the way the atoms bond to one another. Atomic bonding then produces a higher level of structure, which may be in the form of a *molecule, crystal*, or *amorphous aggregate*. This structure, along with the imperfections that may be present, has a profound effect on the mechanical properties. The size, shape, and arrangement of multiple crystals, or the mixture of two or more different structures within a material, produces a higher level of structure, known as *microstructure*. Variations in microstructure further affect the material properties.

Because of the ability to control structures through processing and the ability to develop new structures through techniques such as composite materials, engineers now have at their disposal a wide variety of materials with an almost unlimited range of properties. The specific properties of these materials depend on all levels of structure, from subatomic to macroscopic (Figure 3-1). This chapter will attempt to develop an understanding of the basic structure of engineering materials and how changes in that structure affect their properties and performance.

FIGURE 3-1 General relationship between structural level and the various types of engineering properties.

■ 3.2 THE STRUCTURE OF ATOMS

Experiments have revealed that atoms consist of a relatively dense nucleus composed of positively charged protons and neutral particles of nearly identical mass, known as neutrons. Surrounding the nucleus are the negatively charged electrons, which appear in numbers equal to the protons, so as to maintain a neutral charge balance. Distinct groupings of these basic particles produce the known elements, ranging from the relatively simple hydrogen atom to the unstable transuranium atoms over 250 times as heavy. Except for density and specific heat, however, the weight of atoms has very little influence on their engineering properties.

The light electrons that surround the nucleus play an extremely significant role in determining material properties. These electrons are arranged in a characteristic structure consisting of shells and subshells, each of which can contain only a limited number of electrons. The first shell, nearest the nucleus, can contain only two. The second shell can contain eight, and the third, 32. Each shell and subshell is most stable when it is completely filled. For atoms containing electrons in the third shell and beyond, however, relative stability is achieved with eight electrons in the outermost layer or subshell.

If an atom has slightly less than the number of outer-layer electrons required for stability, it will readily accept an electron from another source. It will then have one electron more than the number of protons and becomes a negatively charged atom, or *negative ion*. Depending on the number of additional electrons, ions can have negative charges of 1, 2, 3, or more. Conversely, if an atom has a slight excess of electrons beyond the number required for stability (such as sodium, with one electron in the third shell), it will readily give up the excess electron and become a *positive ion*. The remaining electrons become more strongly bound, so further removal of electrons becomes progressively more difficult.

The number of electrons surrounding the nucleus of a neutral atom is called the *atomic number*. More important, however, are those electrons in the outermost shell or subshell, which are known as *valence electrons*. These are influential in determining chemical properties, electrical conductivity, some mechanical properties, the nature of interatomic bonding, atom size, and optical characteristics. Elements with similar electron configurations in their outer shells tend to have similar properties.

■ 3.3 ATOMIC BONDING

Atoms are rarely found as free and independent units; they are usually linked or bonded to other atoms in some manner as a result of interatomic attraction. The electron structure of the atoms plays the dominant role in determining the nature of the bond.

Three types of *primary bonds* are generally recognized, the simplest of which is the *ionic bond*. If more than one type of atom is present, the outermost electrons can break free from atoms with excesses in their valence shell, transforming them into positive ions. The electrons then transfer to atoms with deficiencies in their outer shell, converting them into negative ions. The positive and negative ions have an electrostatic attraction for each other, resulting in a strong bonding force. Figure 3-2 presents a crude schematic of the ionic bonding process for sodium and chlorine. Ionized atoms do not usually unite in

FIGURE 3-2 Ionization of sodium and chlorine, producing stable outer shells by electron transfer.

FIGURE 3-3 Three-dimensional structure of the sodium chloride crystal. Note how the various ions are surrounded by ions of the opposite charge.

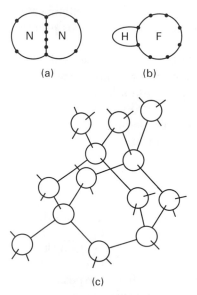

FIGURE 3-4 Formation of a chlorine molecule by the electron sharing of a covalent bond.

FIGURE 3-5 Examples of covalent bonding in (a) nitrogen molecule, (b) HF, and (c) diamond.

simple pairs, however. All positively charged atoms attract all negatively charged atoms. Therefore, each sodium ion will attempt to surround itself with negative chlorine ions, and each chlorine ion will attempt to surround itself with positive sodium ions. Since the attraction is equal in all directions, the result will be a three-dimensional structure, like the one shown in Figure 3-3. Since charge neutrality must be maintained within the structure, equal numbers of positive and negative charges must be present in each neighborhood. General characteristics of materials joined by ionic bonds include moderate to high strength, high hardness, brittleness, high melting point, and low electrical conductivity (since electrons are captive to atoms, charge transport requires atom—or ion—movement).

A second type of primary bond is the *covalent* type. Here the atoms in the assembly find it impossible to produce completed shells by electron transfer but achieve the same goal through electron sharing. Adjacent atoms share outer-shell electrons so that each achieves a stable electron configuration. The shared (negatively charged) electrons locate between the positive nuclei, forming a positive–negative–positive bonding link. Figure 3-4 illustrates this type of bond for a pair of chlorine atoms, each of which contains seven electrons in the valence shell. The result is a stable two-atom molecule, Cl_2. Stable molecules can also form from the sharing of more than one electron from each atom, as in the case of nitrogen (Figure 3-5a). The atoms in the assembly need not be identical (as in HF, Figure 3-5b), the sharing does not have to be equal, and a single atom can share electrons with more than one other atom. For atoms such as carbon and silicon, with four electrons in the valence shell, one atom may share its valence electrons with each of four neighboring atoms. The resulting structure is a three-dimensional network of bonded atoms, like the one shown in Figure 3-5c. Like the ionic bond, the covalent bond tends to produce materials with high strength and high melting point. Since atom movement within the three-dimensional structure (plastic deformation) requires the breaking of discrete bonds, covalent materials are characteristically brittle. Electrical conductivity depends on bond strength, ranging from conductive tin (weak covalent bonding), through semiconductive silicon and germanium, to insulating diamond (carbon). Ionic or covalent bonds are commonly found in ceramic and polymeric materials.

A third type of primary bond is possible when a complete outer shell cannot be formed by either electron transfer or electron sharing. This bond is known as the *metallic bond* (Figure 3-6). If each of the atoms in an aggregate contains only a few valence electrons (one, two, or three), these electrons can be easily removed to produce "stable" ions. The positive ions (nucleus and inner, nonvalence electrons) then arrange in a three-dimensional periodic array and are surrounded by wandering, universally shared valence electrons, sometimes referred to as an electron cloud or electron gas. These highly mobile, free electrons account for the high electrical and thermal conductivity values as well as the opaque (nontransparent) characteristic observed in metals (the free electrons are able to absorb the various discrete energies of light radiation). They also provide the "cement" required for the positive–negative–positive attractions that result in bonding. Bond strength, and therefore material strength and melting temperature, varies over a wide range. More significant, however, is the observation that the positive ions can now move within the structure without the breaking of discrete bonds. Materials bonded by metallic bonds can be deformed by atom-movement mechanisms and produce an altered shape that is every bit as strong as the original. This phenomenon is the basis of metal plasticity, enabling the wide variety of forming processes used in the fabrication of metal products.

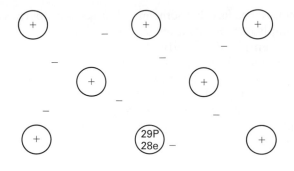

FIGURE 3-6   Schematic of the metallic bond showing the positive ions and free electrons for copper. Each positive-charged ion contains a nucleus with 29 protons and stable electron shells and subshells containing the remaining 28 electrons.

■ 3.4 SECONDARY BONDS

Weak or secondary bonds, known as *van der Waals forces*, can form between molecules that possess a nonsymmetrical distribution of electrical charge. Some molecules, such as hydrogen fluoride and water,[1] can be viewed as electric dipoles. Certain portions of the molecule tend to be more positive or negative than others (an effect referred to as *polarization*). The negative part of one molecule tends to attract the positive region of another, forming a weak bond. Van der Waals forces contribute to the mechanical properties of a number of molecular polymers, such as polyethylene and polyvinyl chloride (PVC).

■ 3.5 ATOM ARRANGEMENTS IN MATERIALS

As atoms bond together to form aggregates, we find that the particular arrangement of the atoms has a significant effect on the material properties. Depending on the manner of atomic grouping, materials are classified as having *molecular structures, crystal structures,* or *amorphous structures*.

Molecular structures have a distinct number of atoms that are held together by primary bonds. There is only a weak attraction, however, between a given molecule and other similar groupings. Typical examples of molecules include O_2, H_2O, and C_2H_4 (ethylene). Each molecule is free to act more or less independently, so these materials exhibit relatively low melting and boiling points. Molecular materials tend to be weak, since the molecules can move easily with respect to one another. Upon changes of state from solid to liquid or liquid to gas, the molecules remain as distinct entities.

Solid metals and most minerals have a crystalline structure. Here the atoms are arranged in a three-dimensional geometric array known as a *lattice*. Lattices are describable through a unit building block, or *unit cell*, that is essentially repeated throughout space. Crystalline structures will be discussed more fully in Section 3.6.

In an amorphous structure, such as glass, the atoms have a certain degree of local order (arrangement with respect to neighboring atoms), but when viewed as an aggregate, they lack the periodically ordered arrangement that is characteristic of a crystalline solid.

■ 3.6 CRYSTAL STRUCTURES OF METALS

From a manufacturing viewpoint, metals are an extremely important class of materials. They are frequently the materials being processed and often form both the tool and the machinery performing the processing. More than 50 of the known chemical elements are classified as metals, and about 40 have commercial importance. They are characterized by the metallic bond and possess the distinguishing characteristics of strength, good electrical and thermal conductivity, luster, the ability to be plastically deformed to a fair degree without fracturing, and a relatively high specific gravity (density) compared to

---

[1] The H_2O molecule can be viewed as a 109° boomerang or elbow with oxygen in the middle and the two hydrogens on the extending arms. The eight valence electrons (six from oxygen and two from hydrogen) associate with oxygen, giving it a negative charge. The hydrogen arms are positive. Therefore, when two or more water molecules are present, the positive hydrogen locations of one molecule are attracted to the oxygen location of adjacent molecules.

nonmetals. The fact that some metals possess properties different from the general pattern simply expands their engineering utility.

When metals solidify, the atoms assume a crystalline structure; that is, they arrange themselves in a geometric lattice. Many metals exist in only one lattice form. Some, however, can exist in the solid state in two or more lattice forms, with the particular form depending on the conditions of temperature and pressure. These metals are said to be *allotropic* or *polymorphic* (poly means "more than one"; morph means "structure"), and the change from one lattice form to another is called an *allotropic transformation*. The most notable example of such a metal is iron, where the allotropic change makes possible heat-treating procedures that yield a wide range of final properties. It is largely because of its allotropy that iron has become the basis of our most important alloys.

There are 14 basic types of crystal structures or lattices. Fortunately, however, nearly all of the commercially important metals solidify into one of three lattice types: body-centered cubic, face-centered cubic, or hexagonal close-packed. Table 3-1 lists the room-temperature structure for a number of common metals. Figure 3-7 compares the structures to one another as well as to the easily visualized, but rarely observed, simple cubic structure.

To begin our study of crystals, consider the *simple cubic* structure illustrated in Figure 3-7a. This crystal can be constructed by placing single atoms on all corners of a cube and then linking identical cube units together. Assuming that the atoms are rigid spheres with atomic radii touching one another, computation reveals that only 52% of available space is occupied. Each atom is in direct contact with only six neighbors (plus and minus directions along the *x*, *y*, and *z* axes). Both of these observations are unfavorable to the metallic bond, where atoms desire the greatest number of nearest neighbors and high-efficiency packing.

TABLE 3-1   Types of Lattices for Common Metals at Room Temperature

Metal	Lattice Type
Aluminum	Face-centered cubic
Copper	Face-centered cubic
Gold	Face-centered cubic
Iron	Body-centered cubic
Lead	Face-centered cubic
Magnesium	Hexagonal
Silver	Face-centered cubic
Tin	Body-centered tetragonal
Titanium	Hexagonal

	Lattice structure	Unit cell schematic	Ping-Pong ball model	Number of nearest neighbors	Packing efficiency	Typical metals
a	Simple cubic			6	52%	None
b	Body-centered cubic			8	68%	Fe, Cr, Mn, Cb, W, Ta, Ti, V, Na, K
c	Face-centered cubic			12	74%	Fe, Al, Cu, Ni, Ca, Au, Ag, Pb, Pt
d	Hexagonal close-packed			12	74%	Be, Cd, Mg, Zn, Zr

FIGURE 3-7   Comparison of crystal structures: simple cubic, body-centered cubic, face-centered cubic, and hexagonal close-packed.

The largest region of unoccupied space is in the geometric center of the cube, where a sphere of 0.732 times the atom diameter could be inserted.[2] If the cube is expanded to permit the insertion of an entire atom, the *body-centered-cubic (BCC)* structure results (Figure 3-7b). Each atom now has eight nearest neighbors, and 68% of the space is occupied. This structure is more favorable to metals and is observed in room-temperature iron, chromium, manganese, and the other metals listed in Figure 3-7b. Compared to materials with other structures, body-centered-cubic metals tend to be high strength.

In seeking efficient packing and a large number of adjacent neighbors, consider maximizing the number of spheres in a single layer and then stacking those layers. The layer of maximized packing is known as a *close-packed plane* and exhibits the hexagonal symmetry shown in Figure 3-8. The next layer is positioned with spheres occupying either the "point-up" or "point-down" recesses in the original layer. Depending on the sequence in which the various layers are stacked, two distinctly different structures can be produced. Both have twelve nearest neighbors (six within the original plane and three from each of the layers above and below) and a 74% efficiency of occupying space.

If the layers are stacked in sets of three (original location, point-up recess of the original layer, and point-down recess of the original layer), the resulting structure can also be viewed as an expanded cube with an atom inserted in the center of each of the six cube faces. This is the *face-centered-cubic (FCC)* structure shown in Figure 3-7c. It is the preferred structure for many of the engineering metals and tends to provide the exceptionally high ductility (ability to be plastically deformed without fracture) that is characteristic of aluminum, copper, silver, gold, and elevated-temperature iron.

A stacking sequence of any two alternating layers results in a structure known as *hexagonal close-packed (HCP)*, where the individual close-packed planes can be clearly identified (Figure 3-7d). Metals having this structure, such as magnesium and zinc, tend to have poor ductility, fail in a brittle manner, and often require special processing procedures.

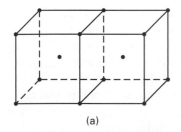

FIGURE 3-8 Close-packed atomic plane showing three directions of atom touching or close packing.

■ 3.7 DEVELOPMENT OF A GRAIN STRUCTURE

When a metal solidifies, a small particle of solid forms from the liquid with a lattice structure characteristic of the given material. This particle then acts like a seed or nucleus and grows as other atoms attach themselves. The basic crystalline unit, or unit cell, is repeated, as illustrated in the examples of Figure 3-9.

In actual solidification, many nuclei form independently throughout the liquid and have random orientations with respect to one another. Each then grows until it begins to interfere with its neighbors. Since the adjacent lattice structures have different alignments or orientations, growth cannot produce a single continuous structure, and a polycrystalline solid is produced. Figure 3-10 provides a two-dimensional illustration of this phenomenon. The small, continuous volumes of solid are known as crystals or *grains*, and the surfaces that divide them (i.e., the surfaces of crystalline discontinuity) are known as *grain boundaries*. The process of solidification is one of crystal *nucleation and growth*.

(a)

(b)

FIGURE 3-9 Growth of crystals to produce an extended lattice: (a) line schematic; (b) Ping-Pong ball model.

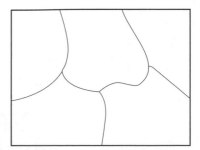

FIGURE 3-10 Schematic representation of the growth of crystals to produce a polycrystalline material.

---

[2] The diagonal of a cube is equal to $\sqrt{3}$ times the length of the cube edge, and the cube edge is here equal to two atomic radii or one atomic diameter. Thus the diagonal is equal to 1.732 times the atom diameter and is made up of an atomic radius, open space, and another atomic radius. Since two radii equal one diameter, the open space must be equal in size to 0.732 times the atomic diameter.

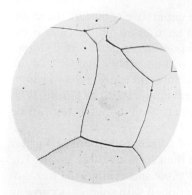

FIGURE 3-11
Photomicrograph of alpha ferrite (essentially pure iron) showing grains and grain boundaries; *(Courtesy of United States Steel Corp., Pittsburgh, PA.)*

Grains are the smallest unit of structure in a metal that can be observed with an ordinary light microscope. If a piece of metal is polished to mirror finish with a series of abrasives and then exposed to an attacking chemical for a short time (etched), the grain structure can be revealed. The atoms along the grain boundaries are more loosely bonded and tend to react with the chemical more readily than those that are part of the grain interior. When viewed under reflected light, the attacked boundaries scatter light and appear dark compared to the relatively unaffected (still flat) grains (Figure 3-11). In some cases, the individual grains may be large enough to be seen by the unaided eye, as with some galvanized steels, but usually magnification is required.

The number and size of the grains in a metal vary with the rate of nucleation and the rate of growth. The greater the nucleation rate, the smaller the resulting grains. Conversely, the greater the rate of growth, the larger the grains. Because the resulting *grain structure* will influence certain mechanical and physical properties, it is an important property for an engineer to control and specify. One means of specification is through the *ASTM grain size number*, defined in ASTM specification E112 as

$$N = 2^{n-1}$$

where N is the number of grains per square inch visible in a prepared specimen at 100X magnification, and n is the ASTM grain size number. Low ASTM numbers mean a few massive grains, while high numbers refer to materials with many small grains.

■ 3.8 ELASTIC DEFORMATION

The mechanical properties of a material are highly dependent on its crystal structure. An understanding of mechanical behavior, therefore, begins with an understanding of the way crystals react to mechanical loads. Most studies begin with carefully prepared single crystals. Through them, we learn that the mechanical behavior is dependent on (1) the type of lattice, (2) the interatomic forces (i.e., bond strength), (3) the spacing between adjacent planes of atoms, and (4) the density of the atoms on the various planes.

If the applied loads are relatively low, the crystals respond by simply stretching or compressing the distance between atoms (Figure 3-12). The basic lattice unit does not change, and all of the atoms remain in their original positions relative to one another. The applied load serves only to alter the force balance of the atomic bonds, and the atoms assume new equilibrium positions with the applied load as an additional component of force. If the load is removed, the atoms return to their original positions and the crystal resumes its original size and shape. The mechanical response is *elastic* in nature, and the amount of stretch or compression is directly proportional to the applied load or stress.

Elongation or compression in the direction of loading results in an opposite change of dimensions at right angles to that direction. The ratio of lateral contraction to axial tensile strain is known as *Poisson's ratio*. This value is always less than 0.5 and is usually about 0.3.

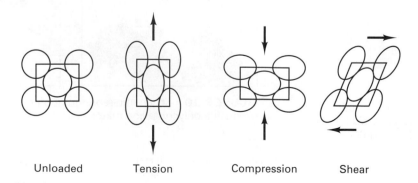

FIGURE 3-12   Distortion of a crystal lattice in response to various elastic loadings.

Unloaded   Tension   Compression   Shear

■ 3.9 PLASTIC DEFORMATION

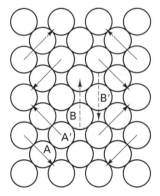

As the magnitude of applied load becomes greater, distortion (or elastic strain) continues to increase, and a point is reached where the atoms either (1) break bonds to produce a fracture or (2) slide over one another in a way that would reduce the load. For metallic materials, the second phenomenon generally requires lower loads and occurs preferentially. The atomic planes shear over one another to produce a net displacement or permanent shift of atom positions, known as *plastic deformation*. Conceptually, this is similar to the distortion of a deck of playing cards when one card slides over another. The actual mechanism, however, is really a progressive one rather than one in which all of the atoms in a plane shift simultaneously. More significantly, however, the result is a permanent change in shape that occurs without a concurrent deterioration in properties.

Recalling that a crystal structure is a regular and periodic arrangement of atoms in space, we see that it becomes possible to link the atoms into flat planes in an almost infinite number of ways. Planes having different orientations with respect to the surfaces of the unit cell will have different atomic densities and different spacing between adjacent, parallel planes, as shown in Figure 3-13. Given the choice of all possibilities, plastic deformation tends to occur along planes having the highest atomic density and greatest separation. The rationale for this can be seen in the simplified two-dimensional array of Figure 3-14. Planes A and A′ have higher density and greater separation than planes B and B′. In visualizing relative motion, we see that the atoms of B and B′ would interfere significantly with one another, whereas planes A and A′ do not experience this difficulty.

Although Figure 3-14 represents the planes of sliding as lines, crystal structures are actually three-dimensional. Within the preferred planes are also preferred directions. If sliding occurs in a direction that corresponds to one of the close-packed directions (shown as dark lines in Figure 3-8), atoms can simply follow one another rather than each having to negotiate its own path. Plastic deformation, therefore, tends to occur by the preferential sliding of maximum-density planes (close-packed planes if present) in directions of closest packing. The specific combination of plane and direction is called a *slip system*, and the resulting shear deformation or sliding is known as *slip*.

The ability to deform a given metal depends on the ease of shearing one atomic plane over an adjacent one and the orientation of the plane with respect to the applied load. Consider, for example, the deck of playing cards. The deck will not "deform" when laid flat on the table and pressed from the top or when stacked on edge and pressed uniformly. The cards will slide over one another, however, if the deck is skewed with respect to the applied load so as to induce a shear stress along the plane of sliding.

With this understanding, consider the deformation properties of the three most common crystal structures:

1. *Body-centered cubic.* In the BCC structure, there are no close-packed planes. Slip occurs on the most favorable alternatives, which are those planes with the greatest interplanar spacing (six of which are illustrated in Figure 3-15). Within these planes, slip occurs along the directions of closest packing, which are the cube diagonals. If each specific combination of plane and direction is considered as a separate slip system, we find that the BCC materials contain 48 attractive ways to slip (plastically deform). The probability that one or more of these systems will be oriented in a favorable manner is great, but the force required to produce deformation is extremely large since there are no

FIGURE 3-13 Schematic diagram showing crystalline planes with different atomic densities and interplanar spacings.

FIGURE 3-14 Simple schematic illustrating the lower deformation resistance of planes with higher atomic density and larger interplanar spacing.

FIGURE 3-15 Slip planes within the BCC, FCC, and HCP crystal structures.

BCC FCC HCP

close-packed planes. Materials with this structure generally possess high strength with moderate ductility. (Refer to the typical BCC metals in Figure 3-7.)

2. *Face-centered cubic.*   In the FCC structure, each unit cell contains four close-packed planes, as illustrated in Figure 3-15. Each of those planes contains three close-packed directions, or face diagonals, giving 12 possible means of slip. Again, the probability that one or more of these will be favorably oriented is great, and this time, the force required to induce slip is quite low. Metals with the FCC structure are relatively weak and possess excellent ductility, as can be confirmed by a check of the metals listed in Figure 3-7.

3. *Hexagonal close-packed.* The hexagonal lattice also contains close-packed planes, but only one such plane exists within the lattice. Although this plane contains three close-packed directions and the force required to produce slip is again rather low, the probability of favorable orientation to the applied load is small (especially if one considers a polycrystalline aggregate). As a result, metals with the HCP structure tend to have low ductility and are often classified as brittle.

■ 3.10 DISLOCATION THEORY OF SLIPPAGE

A theoretical calculation of the strength of metals based on the sliding of entire atomic planes over one another predicts yield strengths on the order of 3 million pounds per square inch, or 20,000 MPa. The observed strengths in actual testing are typically 100 to 150 times less than this value. Extremely small laboratory-grown crystals, however, have been shown to exhibit the full theoretical strength.

An explanation can be provided by the fact that plastic deformation does not occur by all of the atoms in one plane slipping simultaneously over all the atoms of an adjacent plane. Instead, deformation is the result of the progressive slippage of a localized disruption known as a *dislocation.* Consider a simple analogy. A carpet has been rolled onto a floor, and we now want to move it a short distance in a given direction. One approach would be to pull on one end and try to "shear the carpet across the floor," simultaneously overcoming the frictional resistance of the entire area of contact. This would require a large force acting over a small distance. An alternative approach might be to form a wrinkle at one end of the carpet and walk the wrinkle across the floor to produce a net shift in the carpet as a whole—a low-force-over-large-distance approach to the same task. In the region of the wrinkle, there is an excess of carpet with respect to the floor beneath it, and the movement of this excess is relatively easy.

Electron microscopes have revealed that metal crystals do not have all of their atoms in perfect arrangement but rather contain a variety of localized imperfections. Two such imperfections are the *edge dislocation* and *screw dislocation* (Figure 3-16). Edge dislocations are the edges of extra half-planes of atoms. Screw dislocations correspond to partial tearing of the crystal plane. In each case the dislocation is a disruption to the regular, periodic arrangement of atoms and can be moved about with a rather low applied force. It is the motion of these atomic-scale dislocations under applied load that is responsible for the observed macroscopic plastic deformation.

All engineering metals contain dislocations, usually in abundant quantities. The ease of deformation depends on the ease of inducing dislocation movement. Barriers to dislocation motion tend to increase the overall strength of a metal. These barriers take the form of other crystal imperfections and may be of the point type (missing atoms or *vacancies,* extra atoms or *interstitials,* or *substitution atoms* of a different variety, as may occur in an alloy), line type (another *dislocation*), or surface type (*crystal grain boundary* or *free surface*). To increase the strength of a material, we can either remove all defects to create a perfect crystal (extremely difficult) or work to impede the movement of existing dislocations by adding other crystalline defects.

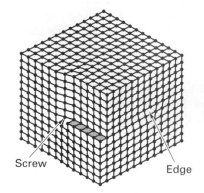

FIGURE 3-16 Schematic representation of screw and edge dislocations.

Screw   Edge

■ 3.11 STRAIN HARDENING OR WORK HARDENING

As noted in our discussion of the tensile test in Chapter 2, most metals become stronger when they are plastically deformed, a phenomenon known as *strain hardening* or *work hardening.* Understanding of this phenomenon can now come from our knowledge of dislocations and a further extension of the carpet analogy. Suppose that this time our goal is to move the carpet diagonally. The best way would be to move a wrinkle in one direction, and then move a second one perpendicular to the first. But suppose that both wrinkles were started simultaneously. We would find that wrinkle 1 would impede the motion of wrinkle 2, and vice versa. In essence, the feature that makes deformation easy can also serve to impede the motion of other, similar dislocations.

In metals, plastic deformation occurs through dislocation movement. As dislocations move, they are more likely to encounter and interact with other dislocations or crystalline defects, thereby producing resistance to further motion. In addition, mechanisms exist that markedly increase the number of dislocations in a metal during deformation (usually by several orders of magnitude), thereby enhancing the probability of interaction.

The effects of strain hardening become attractive when one considers that mechanical working (metalforming) is frequently used in the production of metal products. Since strength can be increased substantially during deformation, a strain-hardened (deformed), inexpensive metal can often be substituted for a more costly, stronger one that is machined to shape. As the product shape is being formed, the material is simultaneously becoming stronger.

Experimental evidence has confirmed the dislocation and slippage theory of deformation. A transmission electron microscope can reveal images of the individual dislocations in a thin metal section, and studies confirm both the increase in number and the interaction during deformation. Macroscopic observations also lend support. When a load is applied to a single metal crystal, deformation begins on the slip system that is most favorably oriented. The net result is often an observable slip and rotation, like that of a skewed deck of cards (Figure 3-17). Dislocation motion becomes more difficult as strain hardening produces increased resistance and rotation makes the slip system orientation less favorable. Further deformation may then occur on alternative systems that now offer less resistance, a phenomenon known as *cross slip*.

FIGURE 3-17 Schematic representation of slip and crystal rotation resulting from deformation.

■ 3.12 PLASTIC DEFORMATION IN POLYCRYSTALLINE METALS

Commercial metals are not single crystals but usually take the form of polycrystalline aggregates. Within each crystal, deformation proceeds in the manner previously described. Since the various grains have different orientations, an applied load will produce different deformations within each of the crystals. This can be seen in Figure 3-18, where a metal has been polished and then deformed. The relief of the polished surface reveals the different slip planes for each of the grains.

One should note that the slip lines do not cross from one grain to another. The grain boundaries act as barriers to the dislocation motion (i.e., the defect is confined to the crystal in which it occurs). As a result, metals with a finer grain structure—more grains per unit area—tend to exhibit greater strength and hardness, coupled with increased impact resistance. This near-universal enhancement of properties is an attractive motivation for grain size control during processing.

FIGURE 3-18 Slip lines in a polycrystalline material. *(From Richard Hertzberg,* Deformation and Fracture Mechanics of Engineering Materials; *courtesy of John Wiley & Sons, Inc.)*

■ 3.13 GRAIN SHAPE AND ANISOTROPIC PROPERTIES

When a metal is deformed, the grains tend to elongate in the direction of metal flow (Figure 3-19). Accompanying the nonsymmetric structure are nonsymmetric or directionally varying properties. Mechanical properties (such as strength and ductility), as well as physical properties (such as electrical and magnetic characteristics), may all exhibit directional differences. Properties that vary with direction are said to be *anisotropic*. Properties that are uniform in all directions are *isotropic*.

The directional variation of properties can be harmful or beneficial, and therefore such variation assumes importance to both the part designer and the part manufacturer. If the metal flow is controlled in processes such as forging, enhanced strength or fracture resistance can be imparted to certain locations. Caution should be exercised, however, since an improvement in one direction is generally accompanied by a decline in another. Moreover, directional variation in properties may create problems during further processing operations, such as the further forming of rolled metal sheets.

FIGURE 3-19 Deformed grains in a cold-worked 1008 steel after 50% reduction by rolling. *(From* Metals Handbook, *8th ed. ASM International, Materials Park, OH, 1972.)*

■ 3.14 FRACTURE OF METALS

If too much plastic deformation is attempted, the metal may respond by fracture. When plastic deformation precedes the break, the fracture is known as a *ductile fracture*. Fractures can also occur before the onset of plastic deformation. These sudden, catastrophic failures, known as *brittle fractures*, are more common in metals having the BCC or HCP crystal structures. Whether the fracture is ductile or brittle, however, often depends on the specific conditions of material, temperature, state of stress, and rate of loading.

■ 3.15 COLD WORKING, RECRYSTALLIZATION, AND HOT WORKING

During plastic deformation, a portion of the deformation energy is stored within the material in the form of additional dislocations and increased grain boundary surface area.[3] If a deformed polycrystalline metal is subsequently heated to a high enough temperature, the material will seek to lower its energy. New crystals nucleate and grow to consume and replace the original structure (Figure 3-20). This process of reducing the internal energy through the formation of new crystals is known as *recrystallization*. The temperature at which recrystallization occurs is different for each metal and also varies with the amount of prior deformation. The greater the amount of prior deformation,

---

[3] A sphere has the least amount of surface area of any shape to contain a given volume of material. When the shape becomes altered from that of a sphere, the surface area must increase. Consider a round balloon filled with air. If the balloon is stretched or flattened into another shape, the rubber balloon is stretched further. When the applied load is removed, the balloon snaps back to its original shape, the one involving the least surface energy. Metals behave in an analogous manner. During deformation, the distortion of the crystals increases the energy of the material. Given the opportunity, the material will try to lower its energy by returning to spherical grains.

FIGURE 3-20   Recrystallization of 70–30 cartridge brass: (a) cold-worked 33%; (b) heated at 580°C (1075°F) for three seconds, (c) four seconds, and (d) eight seconds; 45X. *(Courtesy of J.E. Burke, General Electric Company, Fairfield, CT.)*

the more stored energy and the lower the recrystallization temperature. There is a lower limit, however, below which recrystallization will not take place in a reasonable amount of time. Table 3-2 gives the lowest practical recrystallization temperatures for several materials. This temperature can often be estimated by taking 0.4 times the melting point of the metal when the melting point is expressed as an absolute temperature (Kelvin or Rankine). This is also the temperature at which atomic diffusion (atom movement within the solid) becomes significant, indicating that diffusion is an important mechanism in recrystallization.

When metals are plastically deformed at temperatures below their recrystallization temperature, the process is called *cold working*. The metal strengthens by strain hardening, and the resultant structure consists of distorted grains. As deformation continues, the metal decreases in ductility and may ultimately fracture. It is a common practice, therefore, to recrystallize the material after a certain amount of cold work. Through this *recrystallization anneal*, the structure is replaced by one of new crystals that have never experienced deformation. All strain hardening is lost, but ductility is restored, and the material is now capable of further deformation without the danger of fracture.

If the temperature of deformation is sufficiently above the recrystallization temperature, the deformation process becomes *hot working*. Recrystallization begins as soon as sufficient driving energy is created, (i.e., deformation and recrystallization take place simultaneously), and extremely large deformations are now possible. Since a recrystallized grain structure is constantly forming, the final product will not exhibit strain hardening.

Recrystallization can also be used to control or improve the grain structure of a material. A coarse grain structure can be converted to a more attractive fine grain structure through recrystallization. The material must first be plastically deformed to store sufficient energy to provide the driving force. Subsequent control of the recrystallization process then establishes the final grain size.

TABLE 3-2	The Lowest Recrystallization Temperature of Common Metals

Metal	Temperature [°F(°C)]
Aluminum	300 (150)
Copper	390 (200)
Gold	390 (200)
Iron	840 (450)
Lead	Below room temperature
Magnesium	300 (150)
Nickel	1100 (590)
Silver	390 (200)
Tin	Below room temperature
Zinc	Room temperature

■ 3.16 GRAIN GROWTH

Recrystallization is a continuous process in which a material seeks to lower its overall energy. Ideally, recrystallization will result in a structure of uniform crystals with a comparatively small grain size. If a metal is held at or above its recrystallization temperature for any appreciable time, however, the grains in the recrystallized structure will continue to increase in size. In effect, some of the grains become larger at the expense of their smaller neighbors as the material seeks to further lower its energy by decreasing the amount of grain boundary surface area. Since engineering properties tend to diminish as the size of the grains increase, control of recrystallization is of prime importance. A deformed material should be held at elevated temperature just long enough to complete the recrystallization process. The temperature should then be decreased to stop the process and avoid the property changes that accompany *grain growth*.

■ 3.17 ALLOYS AND ALLOY TYPES

Our discussion thus far has been directed toward the nature and behavior of pure metals. For most manufacturing applications, however, metals are not used in their pure form. Instead, engineering metals tend to be *alloys*, materials composed of two or more different elements, and they tend to exhibit their own characteristic properties.

There are three ways in which a metal might respond to the addition of another element. The first, and probably the simplest, response occurs when the *two materials are insoluble in one another in the solid state*. In this case the base metal and the alloying addition each maintains its individual identity, structure, and properties. The alloy in effect becomes a composite structure, consisting of two types of building blocks in an intimate mechanical mixture.

The second possibility occurs when the *two elements exhibit some degree of solubility in the solid state*. The two materials form a *solid solution*, where the alloy element dissolves in the base metal. The solutions can be (1) *substitutional* or (2) *interstitial*. In the substitutional solution, atoms of the alloy element occupy lattice sites normally filled by atoms of the base metal. In an interstitial solution, the alloy element atoms squeeze into the open spaces between the atoms of the base metal lattice.

A third possibility exists where the *elements combine to form intermetallic compounds*. In this case, the atoms of the alloying element interact with the atoms of the base metal in definite proportions and in definite geometric relationships. The bonding is primarily of the nonmetallic variety (i.e., ionic or covalent), and the lattice structures are often quite complex. Because of the type of bonding, intermetallic compounds tend to be hard, but brittle, high-strength materials.

Even though alloys are composed of more than one type of atom, their structure is still one of crystalline lattices and grains. Their behavior in response to applied loadings is similar to that of pure metals, with some features reflecting the increased level of structural complexity. Dislocation movement can be further impeded by the presence of unlike atoms. If neighboring grains have different chemistries and/or structures, they may respond differently to the same type and magnitude of load.

■ 3.18 ATOMIC STRUCTURE AND ELECTRICAL PROPERTIES

In addition to mechanical properties, the structure of a material also influences its physical properties, such as its electrical behavior. *Electrical conductivity* refers to the net movement of charge through a material. In metals, the charge carriers are the valence electrons. The more perfect the atomic arrangement, the greater the freedom of electron movement and the higher the electrical conductivity. Lattice imperfections or irregularities provide impediments to electron transport, and lower conductivity.

The electrical resistance of a metal, therefore, depends largely on two factors: (1) lattice imperfections and (2) temperature. Vacant atomic sites, interstitial atoms, substitutional atoms, dislocations, and grain boundaries all act as disruptions to the regularity of a crystalline lattice. Thermal energy causes the atoms to vibrate about their equilibrium position. These vibrations cause the atoms to be out of position, which further interferes with electron travel. For a metal, electrical conductivity will decrease with an increase in temperature. As the temperature drops, the number and type of crystalline imperfections become more of a factor. The best metallic conductors, therefore, are pure metals with large grain size, at low temperature.

The electrical conductivity of a metal is due to the movement of the free electrons in the metallic bond. For covalently bonded materials, however, bonds must be broken to provide the electrons for charge transport. Therefore, the electrical properties of these materials is a function of bond strength. Diamond, for instance, has strong bonds and is a strong insulator. Silicon and germanium have weaker bonds that are more easily broken by thermal energy. These materials are known as *intrinsic semiconductors*, since moderate amounts of thermal energy enable them to conduct small amounts of electricity. Continuing down Group IV of the periodic table of elements, we find that tin has such weak bonding that a high number of bonds are broken at room temperature, and the electrical behavior resembles that of a metal.

The electrical conductivity of intrinsic semiconductors can be substantially improved by a process known as *doping*. Silicon and germanium each have four valence electrons and form four covalent bonds. If one of the bonding atoms were replaced with an atom containing five valence electrons, such as phosphorus or arsenic, four covalent bonds would form, leaving an additional valence electron that is not involved in the bonding process. This extra electron would be free to move about and provide additional conductivity. Materials doped in this manner are known as *n-type extrinsic semiconductors*.

A similar effect can be created by substituting an atom with only three valence electrons, such as aluminum. An electron will be missing from one of the bonds, creating an *electron hole*. When a voltage is applied, a nearby electron can jump into this hole, creating a hole in the location that it vacated. Movement of electron holes is equivalent to a countermovement of electrons and thus provides additional conductivity. Materials containing dopants with three valence electrons are known as *p-type semiconductors*. The ability to control the electrical conductivity of semiconductor material is the functional basis of solid-state electronics and circuitry.

In ionically bonded materials, all electrons are captive to atoms (ions). Charge transport, therefore, requires the movement of entire atoms, not electrons. Consider a large block of salt (sodium chloride). It is a good electrical insulator until it becomes wet, whereupon the ions are free to move in the liquid solution and conductivity is observed.

■ Key Words

allotropic	edge dislocation	ionic bond	screw dislocation
alloy	elastic deformation	isotropic	secondary bonds
amorphous structure	electrical conductivity	lattice	simple cubic
anisotropic	extrinsic semiconductor	metallic bond	slip
ASTM grain size number	face-centered cubic	microstructure	slip system
body-centered cubic	grain	molecular structure	solid solution
brittle fracture	grain boundary	negative ion	strain hardening
close-packed planes	grain growth	nucleation and growth	structure
cold work	grain size	plastic deformation	substitutional atom
covalent bond	hexagonal close-packed	Poisson's ratio	unit cell
cross slip	hot work	polarization	vacancy
crystal structure	intermetallic compound	polymorphic	valence electrons
dislocation	interstitial	positive ion	van der Waals forces
doping	intrinsic semiconductor	primary bond	work hardening
ductile fracture	ion	recrystallization	

■ Review Questions

1. Why might an engineer be concerned with controlling or altering the structure of a material?
2. What are the next levels of structure that are greater than the atom?
3. What is meant by the term *microstructure*?
4. What is an ion and what are the two varieties?
5. What properties or characteristics of a material are influenced by the valence electrons?
6. What are the three types of primary bonds, and what types of atoms do they unite?
7. What are some general characteristics of ionically bonded materials?
8. What are some general properties and characteristics of covalently bonded materials?
9. What are some unique property features of materials bonded by metallic bonds?
10. For what common engineering materials are van der Waals forces important?
11. What is the difference between a crystalline material and one with an amorphous structure?
12. What is a lattice? A unit cell?
13. What are some of the general characteristics of metallic materials?
14. What is an allotropic material?
15. Why is the simple cubic crystal structure not observed in the engineering metals?
16. What are the three most common crystal structures found in metals?
17. What is the efficiency of filling space with spheres in the simple cubic structure? Body-centered-cubic structure? Face-centered-cubic structure? Hexagonal-close-packed structure?
18. What is the dominant characteristic of body-centered-cubic metals? Face-centered-cubic metals? Hexagonal-close-packed metals?
19. What is a grain boundary?
20. What is the most common means of describing or quantifying the grain size of a solid metal?
21. What is implied by a low ASTM grain size number? A large ASTM grain size number?
22. How does a metallic crystal respond to low applied loads?
23. What is plastic deformation?
24. What is a slip system in a material? What types of planes and directions tend to be preferred?
25. What structural features account for each of the dominant properties cited in Question 18?
26. What is a dislocation? What role do dislocations play in determining the mechanical properties of a metal?
27. What are some of the common barriers to dislocation movement that can be used to strengthen metals?
28. What are the three major types of point defects in crystalline materials?
29. What is the mechanism (or mechanisms) responsible for the observed deformation strengthening or strain hardening of a metal?
30. Why is a fine grain size often desired in an engineering metal?
31. What is an anisotropic property? Why might anisotropy be a concern?
32. What is the difference between brittle fracture and ductile fracture?
33. How does a metal increase its internal energy during plastic deformation?
34. In what ways can recrystallization be used to enable large amounts of deformation without fear of fracture?
35. What is the major distinguishing feature between hot and cold working?
36. Why is grain growth usually undesirable?
37. What types of structures can be produced when an alloy element is added to a base metal?
38. As a result of ionic or covalent bonding, what types of mechanical properties are characteristic of intermetallic compounds?
39. How is electrical charge transported in a metal (electrical conductivity)?
40. What features in a metal structure tend to impede or reduce electrical conductivity?
41. What is the difference between an intrinsic semiconductor and an extrinsic semiconductor?

■ Problems

1. It is not uncommon for subsequent processing to expose manufactured products to extremely elevated temperatures. Zinc coatings can be applied by immersion into a bath of molten zinc (hot-dip galvanizing). Welding actually melts and resolidifies the crystalline metals. Brazing deposits molten filler metal. How might each of the structural features listed below, and their associated properties, be altered by an exposure to elevated temperature?
   a. A recrystallized polycrystalline metal
   b. A cold-worked metal
   c. A solid-solution alloy such as brass, where zinc atoms dissolve and disperse throughout copper

2. Polyethylene consists of fibrous molecules of covalently bonded atoms tangled and interacting like the fibers of a cotton ball. Weaker van der Waals forces act between the molecules with a strength that is inversely related to separation distance.
   a. What properties of polyethylene can be attributed to the covalent bonding?
   b. What properties are most likely the result of the weaker van der Waals forces?
   c. If we pull on the ends of a cotton ball, the cotton fibers go from a random arrangement to an array of somewhat aligned fibers. Assuming we get a similar response from deformed polyethylene, how might properties change? Why?

www.wiley.com/college/degarmo

EQUILIBRIUM PHASE DIAGRAMS AND THE IRON–CARBON SYSTEM

■ 4.1 INTRODUCTION

As our study of engineering materials becomes more focused on specific metals and alloys, it is increasingly important that we acquire an understanding of their natural characteristics and properties. What is the basic structure of the material? Is the material uniform throughout, or is it a mixture of two or more distinct components? If there are multiple components, how much of each is present, and what are the different chemistries? Is there a component that may impart undesired properties or characteristics? What will happen if temperature is increased or decreased, pressure is changed, or chemistry is varied? The answers to these and other important questions can be obtained through the use of *equilibrium phase diagrams*.

■ 4.2 PHASES

Before we move to a discussion of equilibrium phase diagrams, it is important that we first develop a working definition of the term *phase*. As a starting definition, a phase is simply a form of material possessing a characteristic structure and characteristic properties. Uniformity of chemistry, structure, and properties is assumed throughout a phase. More rigorously, a phase has *a definable structure, a uniform and identifiable chemistry* (also known as *composition*), and distinct *boundaries* or *interfaces* that separate it from other different phases.

A phase can be continuous (like the air in a room) or discontinuous (like grains of salt in a shaker). A phase can be solid, liquid, or gas. In addition, a phase can be a pure substance or a solution, provided that the structure and composition are uniform throughout. Alcohol and water mix in all proportions and will therefore form a single phase when combined. There are no boundaries across which structure and/or chemistry changes. Oil and water, on the other hand, tend to form isolated regions with distinct boundaries and must be regarded as two distinct phases. Ice cubes in water are another two-phase system, since there are two distinct structures with interfaces between them.

■ 4.3 EQUILIBRIUM PHASE DIAGRAMS

An *equilibrium phase diagram* is a graphic mapping of the natural tendencies of a material or a material system, assuming that equilibrium has been attained for all possible conditions. There are three primary variables to be considered: *temperature, pressure,*

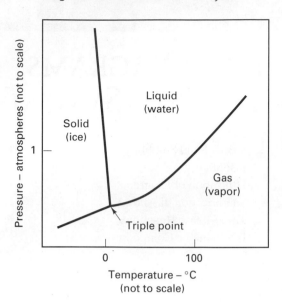

FIGURE 4-1 Pressure–temperature equilibrium phase diagram for water.

and *composition*. The simplest phase diagram is a pressure–temperature (*P–T*) diagram for a fixed-composition material. Areas of the diagram are assigned to the various phases, with the boundaries indicating the equilibrium conditions of transition.

As an introduction, consider the pressure–temperature diagram for water, presented as Figure 4-1. With the composition fixed as H_2O, the diagram maps the stable form of water for various conditions of temperature and pressure. If the pressure is held constant and temperature is varied, the region boundaries denote the melting and boiling points. For example, at 1 atmosphere pressure, the diagram shows that water melts at 0°C and boils at 100°C. Still other uses are possible. Locate a temperature where the stable phase is liquid at atmospheric pressure. Maintaining the pressure at 1 atmosphere, drop the temperature until the material goes from liquid to solid (i.e., ice). Now, maintain that new temperature and begin to decrease the pressure. A transition will be encountered where solid goes directly to gas without melting (sublimation). The combined process just described, known as *freeze drying*, is employed in the manufacture of numerous dehydrated products. With an appropriate phase diagram, process conditions can be determined that might reduce the amount of required cooling and the magnitude of pressure drop required for sublimation. A process operating about the triple point would be most efficient.

TEMPERATURE–COMPOSITION DIAGRAMS

While the *P–T* diagram for water is an excellent introduction to phase diagrams, *P–T* phase diagrams are rarely used for engineering applications. Most engineering processes are conducted at atmospheric pressure, and variations are more likely to occur in temperature and composition. The most useful mapping, therefore, is usually a *temperature–composition phase diagram* at 1 atmosphere pressure. For the remainder of the chapter, this will be the form of phase diagram that will be considered.

For mapping purposes, temperature is placed on the vertical axis and composition on the horizontal. Figure 4-2 shows the axes for mapping the A–B system, where the left-hand vertical corresponds to pure material A, and the percentage of B (usually expressed in weight percent) increases as we move toward pure material B at the right side of the diagram. The temperature range often includes only solids and liquids, since few processes involve engineering materials in the gaseous state. Experimental investigations that provide the details of the diagram take the form of either vertical or horizontal scans that seek to locate the transitions between phases.

COOLING CURVES

Considerable information can be obtained from vertical scans through the diagram where a fixed-composition material is heated and slowly cooled. If the cooling history

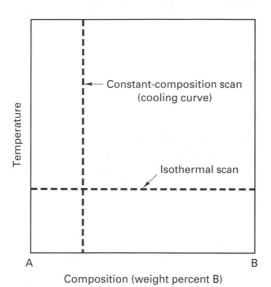

FIGURE 4-2 Mapping axes for a temperature–composition equilibrium phase diagram.

is plotted in the form of a temperature-versus-time plot, known as a *cooling curve*, the transitions in structure will appear as characteristic points, such as slope changes or isothermal (constant-temperature) holds.

Consider the system composed of sodium chloride (common table salt) and water. Five different cooling curves are presented in Figure 4-3. Curve (a) is for pure water being cooled from the liquid state. A decreasing-temperature line is observed for the liquid where the removal of heat produces a concurrent drop in temperature. When the freezing point of 0°C is reached (point *a*), the material begins to change state and releases heat energy as part of the liquid-to-solid transition. Heat is being continuously extracted from the system, but since its source is now the change in state, there is no companion decrease in temperature. An isothermal or constant-temperature hold (*a*–*b*) is observed until the solidification is complete. From this point, as heat extraction continues, the newly formed solid experiences a steady drop in temperature. This type of curve is characteristic of pure metals and other substances with a distinct melting point.

Curve (b) in Figure 4-3 presents the cooling curve for a solution of 10% salt in water. The liquid region undergoes continuous cooling down to point *c*, where the slope abruptly decreases. At this temperature, small particles of ice (i.e., solid) begin to form

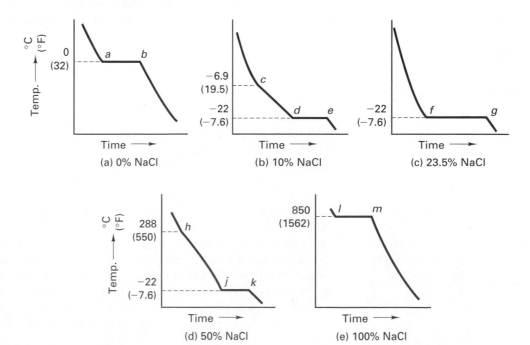

FIGURE 4-3 Cooling curves for five different solutions of salt and water: (a) 0% NaCl; (b) 10% NaCl; (c) 23.5% NaCl; (d) 50% NaCl; (e) 100% NaCl.

FIGURE 4-4 Partial equilibrium diagram for NaCl and H_2O derived from cooling-curve information.

and the reduced slope is attributed to the energy released in this transition. The formation of these ice particles leaves the remaining solution richer in salt and imparts a lower freezing temperature. Further cooling results in the formation of additional solid, which continues to enrich the solution and further lowers the freezing point of the remaining liquid. Instead of possessing a distinct melting point or freezing point, this material is said to have a *freezing range*. When the temperature of point *d* is reached, the remaining liquid undergoes an abrupt reaction and solidifies into an intimate mixture of solid salt and solid water (discussed later), and an isothermal hold is observed. Further extraction of heat produces a drop in the temperature of the fully solidified material.

For a solution of 23.5% salt in water, a distinct freezing point is again observed, as shown in curve (c). Compositions with richer salt concentration, such as curve (d), show phenomena similar to those in curve (b), but with salt being the first solid to form from the liquid. Finally, the curve for pure salt, curve (e), exhibits behavior similar to that of pure water.

If the observed transition points are now transferred to a temperature–composition diagram, such as Figure 4-4, we have the beginnings of a map that summarizes the behavior of the system. Line *a–c–f–h–l*, denoting the lowest temperature at which the material is totally liquid, is known as the *liquidus* line. Line *d–f–j* denotes a particular three-phase reaction and will be discussed later. Between the lines, two phases coexist, one being a liquid and the other a solid. The equilibrium phase diagram, therefore, can be viewed as a collective presentation of cooling-curve data for an entire range of alloy compositions.

Our cooling-curve studies have provided some key information regarding the salt–water system, including some insight into the use of salt on highways in the winter. With the addition of salt, the freezing point of water can be lowered from 0°C (32°F) to as low as −22°C (−7.6°F).

SOLUBILITY STUDIES

The observant reader will note that the ends of the diagram still remain undetermined. Both pure materials have a distinct melting point, below which they appear as a pure solid. Can ice retain some salt in a single-phase solid solution? Can solid salt hold some water and remain a single phase? If so, how much, and does the amount vary with temperature? Completion of the diagram, therefore, requires several horizontal scans to determine any *solubility limits* and their possible variation with temperature.

These isothermal (constant-temperature) scans usually require the preparation of specimens over a range of composition and their subsequent examination by X-ray techniques, microscopy, or other methods to determine whether the structure and chemistry are uniform or indicate a two-phase mixture. As we move away from a pure material, we often encounter a single-phase solid solution, in which a small amount of one component is dissolved and dispersed throughout the other. If there is a limit to this solubility, there will be a line in the phase diagram, known as a *solvus* line, denoting the conditions where the single-phase solid solution becomes a two-phase mixture. Figure 4-5 presents the equilibrium phase diagram for the lead–tin system, using the conventional notation in

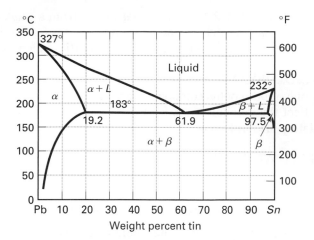

FIGURE 4-5 Lead–tin equilibrium phase diagram.

which Greek letters are used to denote the various single-phase solids. The upper portion of the diagram closely resembles the salt–water diagram, but the partial solubility of one material in the other can be observed on both ends of the diagram.[1]

COMPLETE SOLUBILITY IN BOTH LIQUID AND SOLID STATES

Having developed the basic concepts of equilibrium phase diagrams, we now consider a series of examples in which solubility changes. If two materials are completely soluble in each other in both the liquid and solid states, a rather simple diagram results, like the copper–nickel diagram of Figure 4-6. The upper line is the *liquidus* line, the lowest temperature for which the material is 100% liquid. Above the liquidus, the two materials form a uniform-chemistry liquid solution. The lower line, denoting the highest temperature at which the material is completely solid, is known as a *solidus* line. Below the solidus, the materials form a solid-state solution in which the two types of atoms are uniformly distributed throughout a single crystalline lattice. Between the liquidus and solidus is a *freezing range*, a two-phase region where liquid and solid solutions coexist.

PARTIAL SOLID SOLUBILITY

Many materials do not exhibit complete solubility in the solid state. Each is often soluble in the other up to a certain limit or saturation point, which varies with temperature. Such a diagram has already been observed for the lead–tin system in Figure 4-5.

At the point of maximum solubility, 183°C, lead can hold up to 19.2 wt% tin in a single-phase solution and tin can hold up to 2.5% lead within its structure and still be a single phase. If the temperature is decreased, however, the amount of *solute* that can be held in solution decreases in a continuous manner. If a saturated solution of tin in lead

FIGURE 4-6 Copper–nickel equilibrium phase diagram, showing complete solubility in both liquid and solid states.

---

[1] Lead–tin solders have had a long history in joining electronic components. With the miniaturization of components and the evolution of the circuit board to multitudes of circuits on single chips, exposure to the potentially damaging temperatures of the soldering operation became an increasing concern. Figure 4-5 reveals why 60–40 solder (60 wt% tin) became the primary joining material in the lead–tin system. Of all possible alloys, it has the lowest (all-liquid) melting temperature.

is cooled from 183°C, the material will go from a single-phase solution to a two-phase mixture as a tin-rich second phase precipitates from solution. This change in structure can be used to alter and control the properties in a number of engineering alloys.

INSOLUBILITY

If one or both of the components are totally insoluble in the other, the diagrams will also reflect this phenomenon. Figure 4-7 illustrates the case where component A is completely insoluble in component B in both the liquid and solid states.

UTILIZATION OF DIAGRAMS

Before moving to more complex diagrams, let us first return to a simple phase diagram, such as the one in Figure 4-8, and develop several useful tools. For each point of temperature and composition, we would like to obtain three pieces of information:

1. *The phases present.* The stable phases can be determined by simply locating the point of consideration on the temperature–composition mapping and identifying the region of the diagram in which the point appears.

2. *The composition of each phase.* If the point lies in a single-phase region, there is only one component present, and the composition (or chemistry) of the phase is simply the composition of the alloy being considered. If the point lies in a two-phase region, a *tie-line* must be constructed. A tie-line is simply an isothermal (constant-temperature) line drawn through the point of consideration, terminating at the boundaries of the single-phase regions on either side. The compositions where the tie-line intersects the neighboring single-phase regions will be the compositions of those respective phases in the two-phase mixture. For example, consider point *a* in Figure 4-8. The tie-line for this temperature runs from S_2 to L_2. The tie-line intersects the solid-phase region at point S_2. Therefore, the solid in the two-phase mixture at point *a* has the composition of point S_2. Since the other end of the tie-line intersects the liquid region at L_2, the liquid phase that is present at point *a* will have the composition of point L_2.

3. *The amount of each phase present.* If the point lies in a single-phase region, all of the material, or 100%, must be of that phase. If the point lies in a two-phase region, the relative amounts of the two components can be determined by a *lever-law* calculation using the previously drawn tie-line. Consider the cooling of alloy X in Figure 4-8 in a manner sufficiently slow so as to preserve equilibrium. At temperatures above t_1 the material is in a single-phase liquid state. Temperature t_1, therefore, is the lowest temperature at which the alloy is 100% liquid. If we draw a tie-line at this temperature, it runs from S_1 to L_1 and lies entirely to the left of composition X. At temperature t_3, the alloy is

FIGURE 4-7 Equilibrium diagram of two materials that are completely insoluble in each other in both the liquid and solid states.

FIGURE 4-8 Equilibrium diagram showing the changes that occur during the cooling of alloy X.

completely solid, and the tie-line lies completely to the right of composition X. As the alloy cools from temperature t_1 to temperature t_3, the amount of solid goes from zero to 100% while the segment of the tie-line that lies to the right of composition X also goes from zero to 100%. Similarly, the amount of liquid goes from 100% to zero as the segment of the tie-line lying to the left of composition X undergoes a similar change. Extrapolating these observations to intermediate temperatures, such as temperature t_2, we predict that the fraction of the tie-line that lies to the left of point a corresponds to the fraction of the material that is liquid. This fraction can be computed as:

$$\frac{a - S_2}{L_2 - S_2} \times 100\%$$

where the values of a, S_2, and L_2 are their composition values in weight percent B. In a similar manner, the fraction of solid corresponds to the fraction of the tie-line that lies to the right of point a. (*Note*: These mathematical relations could be rigorously derived from the conservation of either A or B atoms, as the material divides into the two different compositions of S_2 and L_2.) Since the calculations consider the tie-line as a lever with the phases at each end and the fulcrum at the composition line, they are called *lever-law* calculations.

Equilibrium phase diagrams can also be used to provide an overall picture of an alloy system or to identify the transition points for phase changes in a given alloy. For example, the temperature required to redissolve a second phase or melt an alloy can be easily determined. The various changes that will occur during the slow heating or slow cooling of a material can now be predicted. In fact, most of the questions posed at the beginning of this chapter can be answered.

SOLIDIFICATION OF ALLOY X

Let us now apply the tools that we have just developed, tie-lines and lever laws, to follow the solidification of alloy X in Figure 4-8. At temperature t_1, the first minute amount of solid forms with the chemistry of point S_1. As the temperature drops, more solid forms, but the chemistries of both the solid and liquid phases shift to follow the tie-line endpoints. The chemistry of the liquid follows the liquidus line, and the chemistry of the solid follows the solidus. Finally, at temperature t_3, solidification is complete, and the composition of the single-phase solid is now that of alloy X, as required.

The composition of the first solid to form is different from that of the final solid. If the cooling is sufficiently slow, such that equilibrium is maintained or approximated, the composition of the entire mass of solid changes during cooling and follows the endpoint of the tie-line. These chemistry changes are made possible by diffusion, the process in which atoms migrate through the crystal lattice given sufficient time at elevated temperature. If the cooling rate is too rapid, however, the temperature may drop before sufficient diffusion occurs. The resultant material will have a nonuniform chemistry. The initial solid that formed will retain a chemistry that is different from the solid regions that form later. When these nonequilibrium variations occur on a microscopic level, the resultant structure is referred to as being *cored*. Variation on a larger scale is called *macrosegregation*.

THREE-PHASE REACTIONS

Several of the phase diagrams that were presented earlier contain a feature in which phase regions are separated by a horizontal (or constant-temperature) line. These lines are further characterized by either a V intersecting from above or an inverted-V intersecting from below. The intersection of the V and the line denotes the location of a *three-phase equilibrium reaction*.

One common type of three-phase reaction, known as a *eutectic*, has already been observed in Figures 4-4 and 4-5. It is possible to understand these reactions through use of the tie-line and lever-law concepts that have been developed. Refer to the lead–tin diagram of Figure 4-5 and consider any alloy containing between 19.2 and 97.5 wt% tin at a temperature just above the 183°C horizontal line. Tie-line and lever-law computations reveal that the material contains either a lead-rich or tin-rich solid and remaining liquid. At this temperature, any liquid that is present will have a composition of 61.9 wt% tin, regardless of the overall composition of the alloy. If we now focus on this liquid and allow it to cool to just below 183°C, a transition occurs in which the liquid of composition 61.9% tin transforms to a mixture of lead-rich solid with 19.2% tin and tin-rich solid containing 97.5% tin. The three-phase reaction that occurs upon cooling through 183°C can be written as:

$$\text{liquid}_{61.9\% \text{ Sn}} \xrightarrow{183°C} \alpha_{19.2\% \text{ Sn}} + \beta_{97.5\% \text{ Sn}}$$

Note the similarity to the very simple chemical reaction in which water dissociates, or separates, into hydrogen and oxygen: $H_2O \longrightarrow H_2 + \frac{1}{2}O_2$. Since the two solids in the lead–tin eutectic reaction have chemistries on either side of the original liquid, a similar separation must have occurred. Any chemical separation requires atom movement, but the distances involved in a eutectic reaction cannot be great. The resulting structure, known as *eutectic structure*, will be an intimate mixture of the two single-phase solids, with a multitude of interphase boundaries.

For a given reaction, the eutectic structure always forms from the same chemistry at the same temperature and therefore has its own characteristic set of physical and mechanical properties. Alloys with the eutectic composition have the lowest melting point of all neighboring alloys and generally possess relatively high strength. For these reasons, they are often used as casting alloys or as filler material in soldering or brazing operations.

The eutectic reaction can be written in the general form of:

$$\text{liquid} \longrightarrow \text{solid}_1 + \text{solid}_2$$

Figure 4-9 summarizes the various forms of three-phase reactions that may occur in engineering systems, along with the generic description of the reaction shown below the figures.[2] These include the *peritectic*, *monotectic*, and *syntectic* reactions, where the suffix *-ic* denotes that at least one of the three phases in the reaction is a liquid. If the same prefix appears with an *-oid* suffix, the reaction is of a similar form but all phases involved are solids. Two such reactions are the *eutectoid* and the *peritectoid*. The separation eutectoid produces an extremely fine two-phase mixture, and the combination peritectoid reaction is very sluggish since all of the chemistry changes must occur within (usually crystalline) solids.

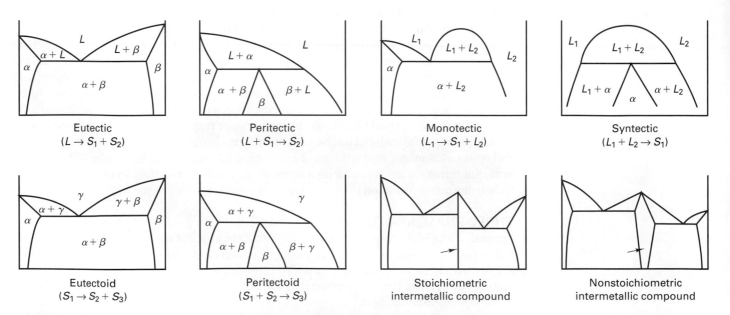

FIGURE 4-9   Schematic summary of three-phase reactions and intermetallic compounds.

---

[2] To determine the specific form of a three-phase reaction, locate its horizontal line and the V intersecting from either above or below the line. Go above the point of the V and write the phases that are present. Then go below and identify the equilibrium phases. Write the reaction as the phases above the line transform to those below. Apply this method to the diagrams in Figure 4-9 to identify the specific reactions, and compare them to their generic forms presented below the figures, remembering that Greek letters denote single-phase solids.

INTERMETALLIC COMPOUNDS

A final phase diagram feature occurs in alloy systems where the bonding attraction between the component materials is strong enough to form stable compounds. These compounds are single-phase solids and tend to break the diagram into recognizable subareas. If components A and B form a compound, and the compound cannot tolerate any deviation from its fixed atomic ratio, the product is known as a *stoichiometric intermetallic compound* and it appears as a single vertical line in the diagram. (*Note:* This will be seen for the Fe_3C iron carbide at 6.67 wt.% carbon in the upcoming iron–carbon equilibrium diagram.) If some degree of chemical deviation is tolerable, the vertical line expands into a single-phase region, and the compound is known as a *nonstoichiometric intermetallic compound*. Figure 4-9 shows schematic representations of both stoichiometric and nonstoichiometric compounds.

Intermetallic compounds appear as single phases in the middle of equilibrium diagrams, with locations consistent with whole-number atomic ratios, such as AB, A_2B, AB_2, A_3B, AB_3, and so on.[3] In general, they tend to be hard, brittle materials, since these properties are a consequence of their ionic or covalent bonding. If they are present in large quantities or lie along grain boundaries in the form of a continuous film, the overall alloy can be extremely brittle. If the same compound is dispersed throughout the alloy in the form of small discrete particles, the result can be a considerable strengthening of the base metal.

COMPLEX DIAGRAMS

The equilibrium diagrams for actual alloy systems may be one of the basic types just discussed or some combination of them. In some cases the diagrams appear to be quite complex and formidable. However, by focusing on a particular composition and analyzing specific points using the tie-line and lever-law concepts, even the most complex diagram can be interpreted and understood. If the properties of the various components are known, phase diagrams can be used to predict the behavior of resultant structures.

■ 4.4 IRON–CARBON EQUILIBRIUM DIAGRAM

Steel, composed primarily of iron and carbon, is clearly the most important of the engineering metals. For this reason, the iron–carbon equilibrium diagram assumes special importance. The diagram most frequently encountered, however, is not the full iron–carbon diagram but the iron–iron carbide diagram shown in Figure 4-10. Here, a

FIGURE 4-10   The iron–carbon equilibrium phase diagram. Single phases are α, ferrite; γ, austenite; δ, δ–ferrite; Fe_3C, cementite.

[3] The use of "weight percent" along the horizontal axis tends to mask the whole-number atomic ratio of intermetallic compounds. Many equilibrium phase diagrams now include a second horizontal scale to reflect "atomic percent." Intermetallic compounds then appear at atomic percents of 25, 33, 50, 67, 75, and similar values that reflect whole-number atomic ratios.

stoichiometric intermetallic compound, Fe_3C, is used to terminate the carbon range at 6.67 wt% carbon. The names of key phases and structures, and the specific notations used on the diagram, have evolved historically and will be used in their generally accepted form.

There are four single-phase solids within the diagram. Three of these occur in pure iron, and the fourth is the iron carbide intermetallic that forms at 6.67% carbon. Upon cooling, pure iron solidifies into a body-centered-cubic solid that is stable down to 1394°C (2541°F). Known as *delta-ferrite*, this phase is present only at extremely elevated temperatures and has little engineering importance. From 1394° to 912°C (2541° to 1674°F) pure iron assumes a face-centered-cubic structure known as *austenite* in honor of the famed metallurgist Roberts-Austen of England. Designated by the Greek letter γ, austenite exhibits the high formability that is characteristic of the face-centered-cubic structure and is capable of dissolving over 2% carbon in single-phase solid solution. Hot forming of steel takes advantage of the low strength, high ductility, and chemical uniformity of austenite. Most of the heat treatments of steel begin by forming the high-temperature austenite structure. Alpha-ferrite, or more commonly just *ferrite*, is the stable form of iron at temperatures below 912°C (1674°C). This body-centered-cubic structure can hold only 0.02 wt% carbon in solid solution and forces the creation of a two-phase mixture in most steels. Upon further cooling to 770°C (1418°F), iron undergoes a transition from nonmagnetic to magnetic. The temperature of this transition is known as the Curie point, but because it is not associated with any change in phase (but is an atomic-level transition), it does not appear on the equilibrium phase diagram.

The fourth single phase is the stoichiometric intermetallic compound Fe_3C, which goes by the name *cementite*, or iron–carbide. Like most intermetallics, it is quite hard and brittle, and care should be exercised in controlling the structures in which it occurs. Alloys with excessive amounts of cementite, or cementite in undesirable form, tend to have brittle characteristics. Because cementite dissociates prior to melting, its exact melting point is unknown, and the liquidus line remains undetermined in the high-carbon region of the diagram.

Three distinct three-phase reactions can also be identified. At 1495°C (2723°F), a *peritectic* reaction occurs for alloys with a low weight percentage of carbon. Because of its high temperature and the extensive single-phase austenite region immediately below it, the peritectic reaction rarely assumes any engineering significance. A *eutectic* is observed at 1148°C (2098°F), with the eutectic composition of 4.3% carbon. All alloys containing more than 2.11% carbon will experience the eutectic reaction and are classified by the general term *cast irons*. The final three-phase reaction is a *eutectoid* at 727°C (1341°F) with a eutectoid composition of 0.77 wt% carbon. Alloys with less than 2.11% carbon miss the eutectic reaction and form a two-phase mixture when they cool through the eutectoid. These alloys are known as *steels*. The point of maximum solubility of carbon in iron, 2.11 wt%, therefore, forms an arbitrary separation between steels and cast irons.

■ 4.5 STEELS AND THE SIMPLIFIED IRON–CARBON DIAGRAM

If we focus on the materials normally known as steel, the phase diagram of Figure 4-10 can be simplified considerably. Those portions near the delta phase (or peritectic) region are of little significance, and the higher-carbon region of the eutectic reaction only applies to cast irons. By deleting these segments and focusing on the eutectoid reaction, we can use the simplified diagram of Figure 4-11 to provide an understanding of the properties and processing of steel.

Rather than beginning with liquid, our considerations generally begin with high-temperature, face-centered-cubic, single-phase austenite. The key transition will be the conversion of austenite to the two-phase ferrite plus carbide mixture as the temperature drops. Control of this reaction, which arises as a result of the drastically different carbon solubilities of the face-centered and body-centered structures, enables a wide range of properties to be achieved through heat treatment.

To begin to understand these processes, consider a steel of the eutectoid composition, 0.77% carbon, being slow cooled along line x–x' in Figure 4-11. At the upper

FIGURE 4-11   Simplified iron–carbon phase diagram with labeled regions. Figure 4-10 shows the more standard Greek letter notation.

FIGURE 4-12   Pearlite; 1000X. *(Courtesy of United States Steel Corp., Pittsburgh, PA.)*

temperatures, only austenite is present, with the 0.77% carbon being dissolved in solid solution within the face-centered structure. When the steel cools through 727°C (1341°F), several changes occur simultaneously. The iron wants to change crystal structure from the face-centered-cubic austenite to the body-centered-cubic ferrite, but the ferrite can only contain 0.02% carbon in solid solution. The excess carbon is rejected and forms the carbon-rich intermetallic known as cementite. The net reaction at the eutectoid, therefore, is:

$$\text{austenite}_{0.77\% \text{ C; FCC}} \longrightarrow \text{ferrite}_{0.02\% \text{ C; BCC}} + \text{cementite}_{6.67\% \text{ C}}$$

Since the chemical separation occurs entirely within crystalline solids, the resultant structure is a fine mixture of ferrite and cementite. Specimens prepared by polishing and etching in a weak solution of nitric acid and alcohol reveal a lamellar structure composed of alternating layers or plates, as shown in Figure 4-12. Since it always forms from a fixed composition at a fixed temperature, this structure has its own set of characteristic properties (even though it is composed of two distinct phases) and goes by the name *pearlite* because of its metallic luster and resemblance to mother-of-pearl when viewed at low magnification.

Steels having less than the eutectoid amount of carbon (less than 0.77%) are called *hypoeutectoid steels* (*hypo* means "less than"). Consider the cooling of a typical hypoeutectoid alloy along line *y–y′* in Figure 4-11. At high temperatures the material is entirely austenite. Upon cooling, however, it enters a region where the stable phases are ferrite and austenite. Tie-line and lever-law calculations show that the low-carbon ferrite nucleates and grows, leaving the remaining austenite richer in carbon. At 727°C (1341°F), the remaining austenite will have assumed the eutectoid composition (0.77% carbon), and further cooling transforms it to pearlite. The resulting structure, therefore, is a mixture of *primary* or *proeutectoid ferrite* (ferrite that forms before the eutectoid reaction) and regions of pearlite as shown in Figure 4-13.

Hypereutectoid steels (*hyper* means "greater than") are those that contain more than the eutectoid amount of carbon. When such a steel cools, as along line *z–z′* in Figure 4-11, the process is similar to the hypoeutectoid case, except that the primary or proeutectoid phase is now cementite instead of ferrite. As the carbon-rich phase nucleates and grows, the remaining austenite decreases in carbon content, again reaching the

FIGURE 4-13   Photomicrograph of a hypoeutectoid steel showing regions of primary ferrite (white) and pearlite; 500X. *(Courtesy of United States Steel Corp., Pittsburgh, PA.)*

FIGURE 4-14 Photomicrograph of a hypereutectoid steel showing primary cementite along grain boundaries; 500X. *(Courtesy of United States Steel Corp., Pittsburgh, PA.)*

eutectoid composition at 727°C (1341°F). As before, this austenite transforms to pearlite upon slow cooling through the eutectoid temperature. Figure 4-14 is a photomicrograph of the resulting structure, which consists of primary cementite and pearlite. In this case the continuous network of primary cementite (an intermetallic) will cause the material to be extremely brittle.

It should be noted that the transitions just described are for equilibrium conditions, which can be approximated by slow cooling. Upon slow heating, the transitions will occur in the reverse manner. When the alloys are cooled rapidly, however, entirely different results may be obtained, since sufficient time may not be provided for the normal phase reactions to occur. In these cases, the equilibrium phase diagram is no longer a valid tool for engineering analysis. Since the rapid-cool processes are important in the heat treatment of steels and other metals, their characteristics will be discussed in Chapter 5, and new tools will be introduced to aid our understanding. Steels and other ferrous metals, including stainless steels and tool steels, will be further developed in Chapter 6.

■ 4.6 CAST IRONS

Iron–carbon alloys with more than 2.11% carbon experience the eutectic reaction during cooling and are known as *cast irons*. The term *cast iron* applies to an entire family of metals with a wide variety of properties. Being relatively inexpensive, with good fluidity and rather low liquidus temperatures, they are readily cast and occupy an important place in engineering applications.

Most commercial cast irons also contain a significant amount of silicon. A typical cast iron contains 2.0 to 4.0% carbon, 0.5 to 3.0% silicon, less than 1.0% manganese, and less than 0.2% sulfur. Silicon produces several major effects. First, it partially substitutes for carbon, so that use of the equilibrium phase diagram requires replacing the weight percent carbon scale with a *carbon equivalent*. Several formulations exist to compute the carbon equivalent, with the simplest being the weight percent carbon plus one-third the weight percent silicon:

$$\text{carbon equivalent (CE)} = (\text{wt \% carbon}) + \tfrac{1}{3}(\text{wt \% silicon})$$

The high silicon enhances the oxidation and corrosion resistance of cast irons by promoting the formation of a tightly adhering surface oxide. Silicon also tends to promote the formation of graphite as the carbon-rich single phase instead of the Fe_3C intermetallic. The eutectic reaction now has two distinct possibilities, as indicated in the modified phase diagram of Figure 4-15:

$$\text{liquid} \longrightarrow \text{austenite} + Fe_3C$$
$$\text{liquid} \longrightarrow \text{austenite} + \text{graphite}$$

The final microstructure of cast iron, therefore, has two possible extremes: (1) all of the carbon-rich phase being intermetallic Fe_3C and (2) all of the carbon-rich phase being *graphite*. In practice, both of these extremes can be approached by controlling the chemistry and other process variables. Graphite formation is promoted by slow cooling, high carbon and silicon contents, heavy or thick section sizes, *inoculation* practices, and the presence of sulfur, phosphorus, aluminum, magnesium, antimony, tin, copper, nickel, and cobalt. Cementite (Fe_3C) formation is favored by fast cooling, low carbon and silicon levels, thin sections, and alloy additions of titanium, vanadium, zirconium, chromium, manganese, and molybdenum.

TYPES OF CAST IRON

Various types of cast iron can be produced, depending on the chemical composition, cooling rate, and the type and amount of inoculants that are used. (Inoculants and

FIGURE 4-15 An iron–carbon diagram showing two possible high-carbon phases. Solid lines denote the iron–graphite system; dashed lines denote iron–cementite (or iron–carbide).

FIGURE 4-16
Photomicrograph of typical gray cast iron; 1000X. *(Courtesy of Bethlehem Steel Corporation, Bethlehem, PA.)*

inoculation practice will be discussed shortly.) *Gray cast iron*, the least expensive and most common variety, is characterized by those features that promote the formation of graphite. Typical compositions range from 2.5 to 4.0% carbon, 1.0 to 3.0% silicon, and 0.4 to 1.0% manganese. The microstructure consists of three-dimensional graphite flakes (which form during the eutectic reaction) dispersed in a matrix of ferrite, pearlite, or other iron-based structure that forms from the cooling of austenite. Figure 4-16 presents a typical section through gray cast iron, showing the graphite flakes dispersed throughout the metal matrix. Because the graphite flakes have no appreciable strength, they act essentially as voids in the structure. The pointed edges of the flakes act as preexisting notches or crack initiation sites, giving the material a characteristic brittle nature. Since a large portion of any fracture follows the graphite flakes, the freshly exposed fracture surfaces have a characteristic gray appearance, and a graphite smudge can usually be obtained if one rubs a finger across the fracture. On a more positive note, the formation of the lower-density graphite reduces the amount of shrinkage that occurs when the liquid goes to solid, making possible the production of more complex iron castings.

The size, shape, and distribution of the graphite flakes have a considerable effect on the overall properties of gray cast iron. When maximum strength is desired, small, uniformly distributed flakes with a minimum amount of intersection are preferred. A more effective means of controlling strength, however, is through control of the metal matrix structure, which is in turn controlled by the carbon and silicon contents and the cooling rate of the casting. Gray cast iron is normally sold by *class*, with the class number corresponding to the minimum tensile strength in thousands of pounds per square inch. Class 20 iron (minimum tensile strength of 20,000 psi) consists of high-carbon-equivalent metal with a ferrite matrix. Higher strengths, up to class 40, can be obtained with lower carbon equivalents and a pearlite matrix. To go above class 40, alloying is required to provide solid solution strengthening, and heat-treatment practices must be performed to modify the matrix. Gray cast irons can be obtained up through class 80, but regardless of strength the presence of the graphite flakes results in extremely low ductility.

Gray cast irons offer excellent compressive strength (compressive forces do not promote crack propagation, so compressive strength is typically 3–4 times tensile strength), excellent machinability (graphite acts to break up the chips and lubricate contact surfaces), good resistance to adhesive wear and galling (graphite flakes self-lubricate), and outstanding sound- and vibration-damping characteristics (graphite flakes absorb transmitted energy). Table 4-1 compares the relative damping capacities of various engineering metals, and clearly shows the unique characteristic of the high-carbon-equivalent gray cast

TABLE 4-1	Relative Damping Capacity of Various Metals
Material	Damping Capacity[a]
Gray iron (high carbon equivalent)	100–500
Gray iron (low carbon equivalent)	20–100
Ductile iron	5–20
Malleable iron	8–15
White iron	2–4
Steel	4
Aluminum	0.4

[a] Natural log of the ratio of successive amplitudes

irons (20–25 times better than steel and 250 times better than aluminum!). High silicon contents promote good corrosion resistance and the enhanced fluidity desired for casting operations. For these reasons, coupled with low cost, high thermal conductivity, low rate of thermal expansion, good stiffness, resistance to thermal fatigue, and 100% recyclability, gray cast iron is specified for a number of applications, including automotive engine blocks, heads, and cylinder liners; transmission housings; machine tool bases; and large equipment parts that are subjected to compressive loads and vibrations.

White cast iron has all of its excess carbon in the form of iron carbide and receives its name from the white surface that appears when the material is fractured. Features promoting its formation are those that favor cementite over graphite: a low carbon equivalent (1.8 to 3.6% carbon, 0.5 to 1.9% silicon, and 0.25 to 0.8% manganese) and rapid cooling.

Because the large amount of iron carbide dominates the microstructure, white cast iron is very hard and brittle, and finds applications where high abrasion resistance is the dominant requirement. For these uses it is also common to pursue the hard, wear-resistant *martensite* structure as the metal matrix. (*Note:* This structure will be described in Chapter 5.) In this way, both the metal matrix and the high-carbon second phase contribute to the wear-resistant characteristics of the material.

White cast iron surfaces can also be formed over a base of another material. For example, mill rolls that require extreme wear resistance may have a white cast iron surface over a steel interior. Accelerated cooling rates produced by tapered sections or metal chill bars placed in the molding sand can be used to produce white iron surfaces at selected locations of a gray iron casting. Where regions of white and gray cast iron occur in the same component, there is generally a transition region comprised of both white and gray irons, known as the *mottled zone*.

When white cast iron is exposed to an extended heat treatment at temperatures in the range of 900°C (1650°F), the cementite will dissociate into its component elements, and some or all of the carbon will be converted into irregularly-shaped nodules of graphite (also referred to as clump or popcorn graphite). The product, known as *malleable cast iron*, has significantly greater ductility than that of gray cast iron because the more favorable graphite shape removes the internal notches. The rapid cooling required to produce the starting white iron structure restricts the size and thickness of malleable iron products such that most weigh less than 5 kg (10 lb).

Various types of malleable iron can be produced, depending on the type of heat treatment that is employed. If the white iron is heated and held for a prolonged time just below the melting point, the carbon in the cementite converts to graphite (first-stage graphitization). Subsequent slow cooling through the eutectoid reaction causes the carbon-containing austenite to transform to ferrite and more graphite (second-stage graphitization). The resulting product, known as *ferritic malleable cast iron*, has a structure of irregular particles of graphite dispersed in a ferrite matrix (Figure 4-17). Typical properties would be: 10% elongation; 35-ksi (240-MPa) yield strength; 50-ksi (345-MPa) tensile strength; and excellent impact strength, corrosion resistance, and machinability. The heat-treatment times, however, are quite lengthy, often involving over 100 hours at elevated temperature.

If the material is cooled more rapidly through the eutectoid transformation, the carbon in the austenite does not form additional graphite but is retained in a pearlite or

FIGURE 4-17
Photomicrograph of malleable iron showing the irregular graphite spheroids, here in a ferrite matrix. *(Courtesy of Iron Castings Society, Rocky River, OH.)*

FIGURE 4-18   Ductile cast iron with (a) ferrite matrix and (b) pearlite matrix; 500X. Note the spheroidal shape of the graphite nodule in each photo.

martensite matrix. The resulting *pearlitic malleable cast iron* is characterized by higher strength and lower ductility than its ferritic counterpart. Properties range from 1 to 4% elongation, 45- to 85-ksi (310- to 590-MPa) yield strength, and 65- to 105-ksi (450- to 725-MPa) tensile strength, with reduced machinability compared to the ferritic material.

The modified graphite structure of malleable iron provided quite an improvement in properties compared to gray cast iron, but it would be even more attractive if it could be obtained directly upon solidification rather than through a prolonged heat treatment at highly elevated temperature. If a high-carbon-equivalent cast iron is sufficiently low in sulfur (either by original chemistry or by desulfurization), the addition of certain materials can promote graphite formation and change the morphology (shape) of the graphite product. If ferrosilicon is injected into the melt (*inoculation*), it will promote the formation of graphite. If magnesium (in the form of an MgFeSi or MgNi alloy) is also added just prior to solidification, the graphite will form as smooth-surface spheres. The latter addition is known as a *nodulizer*, and the product becomes *ductile* or *nodular cast iron*. Subsequent control of cooling can produce a variety of matrix structures, with ferrite or pearlite being the most common (Figure 4-18). By controlling the matrix structure, properties can be produced that span a wide range from 2 to 18% elongation, 40- to 90-ksi (275- to 620-MPa) yield strength, and 60- to 120-ksi (415- to 825-MPa) tensile strength. The combination of good ductility, high strength, toughness, wear resistance, machinability, low-melting-point castability, and up to a 10% weight reduction compared to steel makes ductile iron an attractive engineering material. High silicon–molybdenum ductile irons offer excellent high-temperature strength and good corrosion resistance. Unfortunately, the costs of a nodulizer, higher-grade melting stock, better furnaces, and the improved process control required for its manufacture combine to place it among the most expensive of the cast irons.

Austempered ductile iron (ADI), ductile iron that has undergone a special austempering heat treatment to modify and enhance its properties,[4] has emerged as a significant engineering material. It combines the ability to cast intricate shapes with strength, fatigue, and wear-resistance properties that are similar to those of heat-treated steel. Compared to conventional as-cast ductile iron, it offers nearly double the strength at the same level of ductility. Compared to steel, it also offers an 8 to 10% reduction in density (so strength-to-weight ratio is excellent) and enhanced damping capability, both due to the graphite nodules, but generally poorer machinability and with about a 20% lower elastic modulus. Table 4-2 compares some typical mechanical properties of malleable and ductile irons with the five grades of austempered ductile cast iron that are specified in ASTM Standard A-897.

Compacted graphite cast iron (CGI) is also attracting considerable attention. Produced by a method similar to that used to make ductile iron (an Mg–Ce–Ti addition is made), compacted graphite iron is characterized by a graphite structure that is intermediate to the flake graphite of gray iron and the nodular graphite of ductile iron, and it tends to possess some of the desirable properties and characteristics of each. Table 4-3 shows how the properties of compacted graphite iron bridge the gap between gray and ductile. Strength, stiffness, and ductility are greater than those of gray iron, while castability, machinability, thermal conductivity, and damping capacity all exceed those of ductile. Impact and fatigue properties are good.

ASTM Specification A842 identifies five grades of CGI—250, 300, 350, 400, and 450—where the numbers correspond to tensile strength in megapascals. Areas of application tend to be those where the mechanical properties of gray iron are insufficient and those of ductile iron, along with its higher cost, are considered to be overkill. More specific, compacted graphite iron is attractive when the desired properties include high strength, castability, machinability, thermal conductivity, and thermal shock resistance.

---

[4] The austempering process begins by heating the metal to a temperature between 1500° and 1750°F (815° to 955°C) and holding for sufficient time to saturate the austenite with carbon. The metal is then rapidly cooled to an austempering temperature between 450° and 750°F (230°–400°C), where it is held until all crystal structure changes have completed, and then cooled to room temperature. High austempering temperatures give good toughness and fatigue properties, while lower austempering temperatures give better strength and wear resistance.

TABLE 4-2	Typical Mechanical Properties of Malleable, Ductile, and Austempered Ductile Cast Irons					
Class or Grade	Minimum Yield Strength		Minimum Tensile Strength		Minimum Percentage Elongation	Brinell Hardness Number
	ksi	MPa	ksi	MPa		
Malleable Iron[a]						
M3210	32	224	50	345	10	156 max
M4504	45	310	65	448	4	163–217
M5003	50	345	75	517	3	187–241
M5503	55	379	75	517	3	187–241
M7002	70	483	90	621	2	229–269
M8501	85	586	105	724	1	269–302
Ductile Iron[b]						
60–40–18	40	276	60	414	18	149–187
65–45–12	45	310	65	448	12	170–207
80–50–06	55	379	80	552	6	187–248
100–70–03	70	483	100	689	3	217–269
120–90–02	90	621	120	827	2	240–300
Austempered Ductile Iron[c]						
1	80	550	125	850	10	269–321
2	100	700	150	1050	7	302–363
3	125	850	175	1200	4	341–444
4	155	1100	200	1400	1	388–477
5	185	1300	230	1600	–	444–555

[a] ASTM Specification A602 (Also SAE J 158).
[b] ASTM Specification A536.
[c] ASTM Specification A897.

TABLE 4-3	Typical Properties of Pearlitic Gray, Compacted Graphite, and Ductile Cast Irons		
Property	Gray	CGI	Ductile
Tensile strength (MPa)	250	450	750
Elastic modulus (Gpa)	105	145	160
Elongation (%)	0	1.5	5
Thermal conductivity (w/mk)	48	37	28
Relative damping capacity (Gray = 1)	1	0.35	0.22

THE ROLE OF PROCESSING ON PROPERTIES

While typical properties have been presented for the various types of cast iron, it should be noted that the properties of all metals are influenced by how they are processed. For cast materials, properties will vary with the manner of solidification and cooling. Because cast components often have complex geometries, the cooling rate may vary from location to location, with companion variation in properties. To assure compliance with industry specifications, standard geometry test bars are often cast along with manufactured products so the material can be evaluated and properties ensured independent of product geometry.

■ Key Words

austempered ductile iron
austenite
carbon equivalent
cast iron
cementite
class
compacted graphite
complete solubility
composition
cooling curve
cored structure
ductile cast iron
equilibrium

eutectic
eutectic structure
eutectoid
ferrite
freezing range
graphite
gray cast iron
hypereutectoid
hypoeutectoid
inoculation
interfaces
intermetallic compound
lever law

liquidus
macrosegregation
malleable cast iron
martensite
monotectic
mottled zone
nodular cast iron
nodulizer
nonstoichiometric
pearlite
peritectic
peritectoid
phase

phase diagram
primary phase
solidus
solubility limit
solute
solvus
steel
stoichiometric
syntectic
three-phase reaction
tie-line
white cast iron

■ Review Questions

1. What are some features that are useful in defining a phase?
2. Supplement the examples provided in the text with another example of a single phase that is each of the following: continuous, discontinuous, gaseous, and a liquid solution.
3. What is an equilibrium phase diagram?
4. What three primary variables are generally considered in equilibrium phase diagrams?
5. Why is a pressure–temperature phase diagram not that useful for most engineering applications?
6. What is a cooling curve?
7. What features in a cooling curve indicate some form of change in a material's structure?
8. What is a solubility limit, and how might it be determined?
9. In general, how does the solubility of one material in another change as temperature is increased?
10. Describe the conditions of complete solubility, partial solubility, and insolubility.
11. What types of changes occur upon cooling through a liquidus line? A solidus line? A solvus line?
12. What three pieces of information can be obtained for each point in an equilibrium phase diagram?
13. What is a tie-line? For what types of phase diagram regions would it be useful?
14. What points on a tie-line are used to determine the chemistry (or composition) of the component phases?
15. What tool can be used to compute the relative amounts of the component phases in a two-phase mixture? How does this tool work?
16. What is a cored structure? Under what conditions is it produced?
17. What features in a phase diagram can be used to identify three-phase reactions?
18. What is the general form of a eutectic reaction?
19. What is the general form of the eutectic structure?
20. Why are alloys of eutectic composition attractive for casting and as filler metals in soldering and brazing?
21. What is a stoichiometric intermetallic compound, and how would it appear in a temperature–composition phase diagram? How would a nonstoichiometric intermetallic compound appear?
22. What type of mechanical properties would be expected for intermetallic compounds?
23. In what form(s) might intermetallic compounds be undesirable in an engineering material? In what form(s) might they be attractive?

24. What are the four single phases in the iron–iron carbide diagram? Provide both the phase diagram notation and the assigned name.
25. What feature in the iron–carbon diagram is used to distinguish between cast irons and steels?
26. What features of austenite make it attractive for forming operations? What features make it attractive as a starting structure for many heat treatments?
27. Which of the three-phase reactions in the iron–carbon diagram is most important in understanding the behavior of steels? Write this reaction in terms of the interacting phases and their composition.
28. Describe the relative ability of iron to dissolve carbon in solution when in the form of austenite (the elevated temperature phase) and when in the form of ferrite at room temperature.
29. What is pearlite? Describe its structure.
30. What is a hypoeutectoid steel, and what structure will it assume upon slow cooling? What is a hypereutectoid steel and how will its structure differ from that of a hypoeutectoid?
31. In addition to iron and carbon, what other element is present in rather large amounts in cast iron?
32. What is a carbon equivalent and how is it computed?
33. What are the two possible high-carbon phases in cast irons? What features tend to favor the formation of each?
34. Describe the microstructure of gray cast iron.
35. Which of the structural units is generally altered to increase the strength of a gray cast iron?
36. What are some of the attractive engineering properties of gray cast iron?
37. What are some of the key limitations to the engineering use of gray cast iron?
38. What is the dominant mechanical property of white cast iron?
39. What structural feature is responsible for the increased ductility and fracture resistance of malleable cast iron?
40. How is malleable cast iron produced?
41. What is unique about the graphite that forms in ductile cast iron?
42. What requirements of ductile iron manufacture are responsible for its increased cost over materials such as gray cast iron?
43. What are some of the attractive features of austempered ductile cast iron?
44. Compacted graphite iron has a structure and properties intermediate to what two other types of cast irons?

■ Problems

1. Obtain a binary (two-component) phase diagram for a system not discussed in this chapter. Identify each:
   a. Single Phase
   b. Three-phase reaction
   c. Intermetallic compound
2. Identify at least one easily identified product or component that is currently being produced from each of the following types of cast irons:
   a. Gray cast iron
   b. White cast iron
   c. Malleable cast iron
   d. Ductile cast iron
   e. Compacted graphite cast iron
3. Find an example where one of the types of cast iron has been used in place of a previous material. What feature or features might have prompted the substitution?

www.wiley.com/college/degarmo

*C*hapter 4 CASE STUDY

The Blacksmith Anvils

As an officer in the Western-America Blacksmith Association, you have determined that a number of your members would like to have a modern equivalent of an 1870-vintage blacksmith anvil. Your objective is to replicate the design but utilize the advantageous features of today's engineering materials. You hope ultimately to identify a producer who will make a limited number of these items for sale and distribution through your monthly magazine. The proposed design is a large forging anvil that has a total length of 20 inches. The top surfaces must be resistant to wear, deformation, and chipping. Estimated mechanical properties call for a yield strength in excess of 70 ksi, an elongation greater than 2%, and a Brinell hardness of 200 or more on the top surface. You feel confident that you will be able to secure a minimum of 500 orders.

1. Discuss the various properties that this part must possess to adequately perform its intended task.
2. Discuss the various concerns that would influence the proposed method of fabricating the anvils.
3. Assuming that the anvils will be made from some form of ferrous metal, consider the properties of the various types of cast irons and steels with regard to this application. Which material would you recommend? Why?
4. How would you propose that a production run of 500 replica anvils be produced?

CHAPTER 5

HEAT TREATMENT

■ 5.1 INTRODUCTION

In the previous chapters, you have been introduced to the interrelationship among the structure, properties, processing, and performance of engineering materials. Chapters 3 and 4 considered aspects of structure, while Chapter 2 focused on properties. In this chapter, we begin to expand on and incorporate processing so that the structure and companion properties can be manipulated and controlled.

Many engineering materials can be characterized not by a single set of properties but by an entire spectrum of possibilities that can be selected and varied at will. *Heat treatment* is the term used to describe *the controlled heating and cooling of materials for the purpose of altering their structures and properties*. The same material can be made weak and ductile for ease in manufacture, and then retreated to provide high strength and good fracture resistance for use and application. Because both physical and mechanical properties (such as strength, toughness, machinability, wear resistance, and corrosion resistance) can be altered by heat treatment and these changes can be induced with no concurrent change in product shape, heat treatment is one of the most important and widely used manufacturing processes.

Technically, the term *heat treatment* applies only to processes where the heating and cooling are performed for the specific purpose of altering properties, but heating and cooling often occur as incidental phases of other manufacturing processes, such as hot forming or welding. The material properties will be altered, however, just as though an intentional heat treatment had been performed, and the results can be either beneficial or harmful. For this reason, both the individual who selects material and the person who specifies its processing must be fully aware of the possible changes that can occur during heating or cooling activities. Heat treatment should be fully integrated with other manufacturing processes if effective results are to be obtained. To provide a basic understanding, this chapter will present both the theory of heat-treatment and a survey of the more common heat-treatment processes. Since more than 90% of all heat treatment is performed on steel and other ferrous metals, these materials will receive the bulk of our attention.

■ 5.2 PROCESSING HEAT TREATMENTS

The term *heat treatment* is often associated with those thermal processes that increase the strength of a material, but the broader definition permits inclusion of another set of processes that we will call *processing heat treatments*. These are often performed as a means of preparing the material for fabrication. Specific objectives may be the improvement of machining characteristics, the reduction of forming forces, or the restoration of ductility to enable further processing.

EQUILIBRIUM DIAGRAMS AS AIDS

Most of the processing heat treatments involve rather slow cooling or extended times at elevated temperatures. These conditions tend to approximate equilibrium, and the resulting structures, therefore, can be reasonably predicted through the use of an *equilibrium phase diagram* (presented in Chapter 4). These diagrams can be used to determine the temperatures that must be attained to produce a desired starting structure, and to describe the changes that will then occur upon subsequent cooling. It should be noted, however, that these diagrams are for true equilibrium conditions, and any departure from equilibrium may lead to substantially different results.

PROCESSING HEAT TREATMENTS FOR STEEL

Because many of the processing heat treatments are applied to plain-carbon and low-alloy steels, they will be presented here with the simplified iron–carbon equilibrium diagram of Figure 4-11 serving as a reference guide. Figure 5-1 shows this diagram with the key transition lines labeled in standard notation. The eutectoid line is designated by the symbol A_1, and A_3 designates the boundary between austenite and ferrite + austenite.[1] The transition from austenite to austenite + cementite is designated as the A_{cm} line.

A number of process heat-treating operations have been classified under the general term of *annealing*. These may be employed to reduce strength or hardness, remove residual stresses, improve toughness, restore ductility, refine grain size, reduce segregation, or alter the electrical or magnetic properties of the material. By producing a certain desired structure, characteristics can be imparted that will be favorable to subsequent operations (such as machining or forming) or applications. Because of the variety of anneals, it is important to designate the specific treatment, which is usually indicated by a preceding adjective. The specific temperatures, cooling rate, and details of the process will depend on the material being treated and the objectives of the treatment.

In the process of *full annealing*, hypoeutectoid steels (less than 0.77% carbon) are heated to 30° to 60°C (50° to 100°F) above the A_3 temperature, held for sufficient time to convert the structure to homogeneous single-phase austenite of uniform composition and temperature, and then slowly cooled at a controlled rate through the A_1 temperature. Cooling is usually done in the furnace by decreasing the temperature by 10° to 30°C (20° to 50°F) per hour to at least 30°C (50°F) below the A_1 temperature.

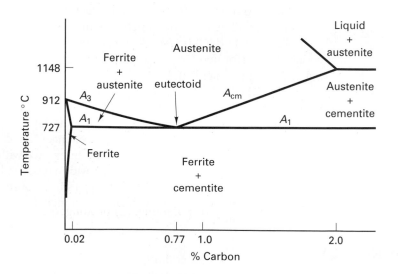

FIGURE 5-1 Simplified iron–carbon phase diagram for steels with transition lines labeled in standard notation as A_1, A_3, and A_{cm}.

---

[1] Historically, an A_2 line once appeared between the A_1 and A_3. This line designated the magnetic property change known as the Curie point. Since this transition was later shown to be an atomic change, not a change in phase, the line was deleted from the equilibrium phase diagram without a companion relabeling.

At this point all structural changes are complete, and the metal can be removed from the furnace and air cooled to room temperature. The resulting structure is one of coarse pearlite (widely spaced layers or lamellae) with excess ferrite in amounts predicted by the equilibrium phase diagram. In this condition, the steel is quite soft and ductile.

The procedure to full-anneal a hypereutectoid alloy (greater than 0.77% carbon) is basically the same, except that the original heating is only into the austenite plus cementite region (30° to 60°C above the A_1). If the material is slow cooled from the all-austenite region, a continuous network of cementite may form on the grain boundaries and make the entire material brittle. When properly annealed, a hypereutectoid steel will have a structure of coarse pearlite with excess cementite in dispersed spheroidal form.

While full anneals produce the softest and weakest properties, they are quite time consuming, and considerable amounts of energy must be spent to maintain the elevated temperatures required during soaking and furnace cooling. When maximum softness and ductility are not required and cost savings are desirable, *normalizing* may be specified. In this process, the steel is heated to 60°C (100°F) above the A_3 (hypoeutectoid) or A_{cm} (hypereutectoid) temperature, held at this temperature to produce uniform austenite, and then removed from the furnace and allowed to cool in still air. The resultant structures and properties will depend on the subsequent cooling rate. Wide variations are possible, depending on the size and geometry of the product, but fine pearlite with excess ferrite or cementite is generally produced.

One should note a key difference between full annealing and normalizing. In the full anneal, the furnace imposes identical cooling conditions at all locations within the metal, which results in identical structures and properties. With normalizing, the cooling will be different at different locations. Properties will vary between surface and interior, and different thickness regions will also have different properties. When subsequent processing involves a substantial amount of machining that may be automated, the added cost of a full anneal may be justified, since it produces a product with uniform machining characteristics at all locations.

If cold working has severely strain-hardened a metal, it is often desirable to restore the ductility, either for service or to permit further processing without danger of fracture. This is often achieved through the *recrystallization* process described in Chapter 3. When the material is a low-carbon steel (<0.25% carbon), the specific procedure is known as a *process anneal*. The steel is heated to a temperature slightly below the A_1, held long enough to induce recrystallization of the dominant ferrite phase, and then cooled at a desired rate (usually in still air). Since the entire process is performed at temperatures within the same phase region, the process simply induces a change in phase morphology (size, shape, and distribution). The material is not heated to as high a temperature as in the full-anneal or normalizing process, so a process anneal is somewhat cheaper and tends to produce less scaling.

A *stress-relief anneal* may be employed to reduce the *residual stresses* in large steel castings, welded assemblies, and cold-formed products. Parts are heated to temperatures below the A_1 (between 550° and 650°C or 1000° and 1200°F), held for a period of time, and then slow cooled to prevent the creation of additional stresses. Times and temperatures vary with the condition of the component, but the basic microstructure and associated mechanical properties generally remain unchanged.

When high-carbon steels (>0.60% carbon) are to undergo extensive machining or cold forming, a process known as *spheroidization* is often employed. Here the objective is to produce a structure in which all of the cementite is in the form of small spheroids or globules dispersed throughout a ferrite matrix. This can be accomplished by a variety of techniques, including (1) prolonged heating at a temperature just below the A_1 followed by relatively slow cooling, (2) prolonged cycling between temperatures slightly above and slightly below the A_1, or (3) in the case of tool or high-alloy steels, heating to 750° to 800°C (1400° to 1500°F) or higher and holding at this temperature for several hours, followed by slow cooling.

Although the selection of a processing heat treatment often depends on the desired objectives, steel composition strongly influences the choice. Process anneals are restricted to low-carbon steels, and spheroidization is a treatment for high-carbon material.

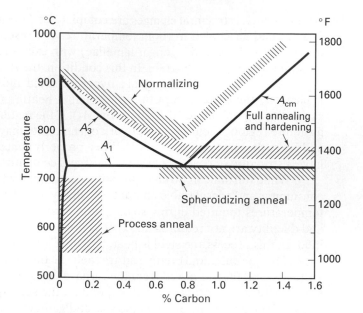

FIGURE 5-2 Graphical summary of the process heat treatments for steels on an equilibrium diagram.

Normalizing and full annealing can be applied to all carbon contents, but even here, preferences are noted. Since different cooling rates do not produce a wide variation of properties in low-carbon steels, the air cool of a normalizing treatment often produces acceptable uniformity. For higher carbon contents, such as the 0.4 to 0.6% range, different cooling rates can produce wider property variations, and the uniform furnace cooling of a full anneal is often preferred. Figure 5-2 provides a graphical summary of the process heat treatments.

HEAT TREATMENTS FOR NONFERROUS METALS

Most of the nonferrous metals do not have the significant phase transitions observed in the iron–carbon system, and for them, the process heat treatments do not play such a significant role. Aside from the strengthening treatment of precipitation hardening, which is discussed later, the nonferrous metals are usually heat treated for three purposes: (1) to produce a uniform, homogeneous structure, (2) to provide stress relief, or (3) to bring about recrystallization. Castings that have been cooled too rapidly can possess a segregated solidification structure known as coring (discussed more fully in Chapter 4). *Homogenization* can be achieved by heating to moderate temperatures and then holding for a sufficient time to allow thorough diffusion to take place. Similarly, heating for several hours at relatively low temperatures can reduce the internal stresses that are often produced by forming, welding, or brazing. *Recrystallization* (discussed in Chapter 3) is a function of the particular metal, the amount of prior straining, and the desired recrystallization time. In general, the more a metal has been strained, the lower the recrystallization temperature or the shorter the time. Without prior straining, however, recrystallization will not occur and heating will only produce undesirable grain growth.

■ 5.3 HEAT TREATMENTS USED TO INCREASE STRENGTH

Six major mechanisms are available to increase the strength of metals:

1. Solid-solution strengthening

2. Strain hardening

3. Grain size refinement

4. Precipitation hardening

5. Dispersion hardening

6. Phase transformations

All of these can be induced or altered by heat treatment, but all may not be applicable to a specific metal or alloy.

In *solid-solution strengthening*, a base metal dissolves other atoms, either as *substitutional solutions*, where the new atoms occupy sites in the host crystal lattice, or as *interstitial solutions*, where the new atoms squeeze into "holes" between the atoms of the base lattice. The amount of strengthening depends on the amount of dissolved solute and the size difference of the atoms involved. Since distortion of the host structure makes dislocation movement more difficult, the greater the size difference, the more effective the addition.

Strain hardening (discussed in Chapter 3) produces an increase in strength by means of plastic deformation under cold-working conditions.

Because grain boundaries act as barriers to dislocation motion, a metal with small grains tends to be stronger than the same metal with larger grains. Thus *grain size refinement* can be used to increase strength, except at elevated temperatures, where grain growth can occur and grain boundary diffusion contributes to creep and failure. It is important to note that grain size refinement is one of the few processes that can improve strength without a companion loss of ductility and toughness.

In *precipitation hardening*, or *age hardening*, strength is obtained from a nonequilibrium structure that is produced by a three-step heat treatment. Details of this method will be provided in Section 5.4.

Strength obtained by dispersing second-phase particles throughout a base material is known as *dispersion hardening*. To be effective, the dispersed particles should be stronger than the matrix, adding strength through both their reinforcing action and the additional interfacial surfaces that present barriers to dislocation movement.

Phase transformation strengthening involves those alloys that can be heated to form a single phase at elevated temperature and subsequently transform to one or more low-temperature phases upon cooling. When this feature is used to increase strength, the cooling is usually rapid and the phases that are produced are usually of a nonequilibrium nature.

■ 5.4 STRENGTHENING HEAT TREATMENTS FOR NONFERROUS METALS

All six of the mechanisms just described can be used to increase the strength of nonferrous metals. Solid-solution strengthening can impart strength to single-phase materials. Strain hardening can be quite useful if sufficient ductility is present. Alloys containing eutectic structure exhibit considerable dispersion hardening. Among all of the possibilities, however, the most effective strengthening mechanism for the nonferrous metals tends to be precipitation hardening.

PRECIPITATION OR AGE HARDENING

To be a candidate for precipitation hardening, an alloy system must exhibit solubility that decreases with decreasing temperature, such as the aluminum-rich portion of the aluminum–copper system shown in Figure 5-3 and enlarged in Figure 5-4. Consider the alloy with 4% copper, and use the phase diagram to determine its equilibrium structure. Liquid metal solidifies into a single-phase solid (α phase). At 1000°F, the full 4% of copper is dissolved and distributed throughout the alpha crystals. As the temperature drops, the maximum solubility of copper in aluminum decreases from 5.65% at 1018°F to less than 0.2% at room temperature. Upon cooling through the solvus (or solubility limit) line at 930°F, the 4% copper alloy enters a two-phase region, and copper-rich theta-phase precipitates form and grow. (*Note*: Theta-phase is actually a hard, brittle intermetallic compound with the chemical formula of $CuAl_2$.) The equilibrium structure, therefore, would be an aluminum-rich alpha-phase structure with coarse theta-phase precipitates, generally lying along alpha-phase grain boundaries where the nucleation of second-phase particles can benefit from the existing interfacial surface.

Whenever two or more phases are present, the material exhibits dispersion strengthening. Dislocations are confined to their own crystal and cannot cross interfacial

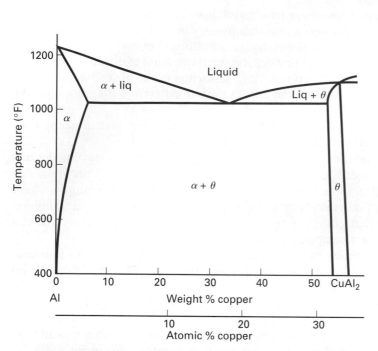

FIGURE 5-3 High-aluminum section of the aluminum–copper equilibrium phase diagram.

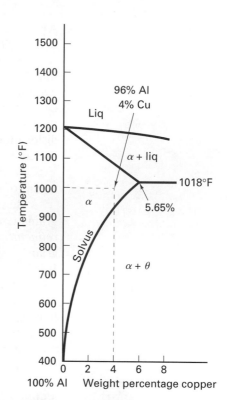

FIGURE 5-4 Enlargement of the solvus-line region of the aluminum–copper equilibrium diagram of Figure 5-3.

boundaries. Therefore, each interface between alpha-phase and the theta-phase precipitate is a strengthening boundary. Take a particle of theta precipitate and cut it into two halves. Forming the two half-size precipitates has just added two additional interfaces, corresponding to both sides of the cut. If the particle were to be further cut into quarters, eighths, and sixteenths, we would expect strength to increase as we continually add interfacial surface. Ideally, we would like to have millions of ultra-small particles dispersed throughout the alpha-phase structure. When we try to form this more desirable nonequilibrium configuration, however, we gain an unexpected benefit that adds significant strength. This new nonequilibrium treatment is known as *age hardening* or *precipitation hardening*.

The process of precipitation hardening is actually a three-step sequence. The first step, known as *solution treatment*, erases the room-temperature structure and redissolves any existing precipitate. The metal is heated to a temperature above the solvus and held in the single-phase region for sufficient time to redissolve the second phase and uniformly distribute the solute atoms (in this case, copper).

If the alloy were slow cooled, the second-phase precipitate would nucleate and the material would revert back to a structure similar to equilibrium. To prevent this from happening, age-hardening alloys are *quenched* from their solution treatment temperature. The rapid-cool quenching, usually in water, suppresses diffusion, trapping atoms in place. The result is a room-temperature *supersaturated* solid solution. In the alloy discussed above, the alpha phase would now be holding 4% copper in solution at room temperature—far in excess of its equilibrium maximum of <0.2%. In this nonequilibrium quenched condition, the material is often soft and can be easily straightened, formed, or machined.

If the supersaturated material were now reheated to a temperature where atom movement (diffusion) could occur, the alloy would attempt to form its equilibrium structure. If the reheating temperature remained within the two-phase region, the excess solute atoms would precipitate out of the supersaturated matrix. This stage of the process, known as *aging*, is actually a continuous transition. Solute atoms begin to cluster at

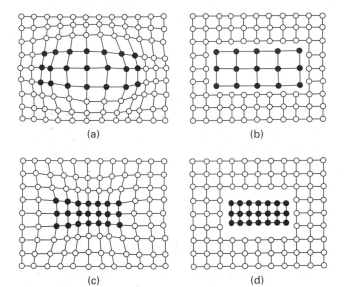

FIGURE 5-5 Two-dimensional illustrations depicting (a) a coherent precipitate cluster where the precipitate atoms are larger than those in the host structure, and (b) its companion overaged or discrete second-phase precipitate particle. Parts (c) and (d) show equivalent sketches where the precipitate atoms are smaller than the host.

locations within the parent crystal, still occupying atom sites within the original lattice. Various transitions may then occur, leading ultimately to the formation of distinct second-phase particles with their own characteristic chemistry and crystal structure.

A key concept in the aging sequence is that of *coherency* or crystalline continuity. If the clustered solute atoms continue to occupy lattice sites within the parent structure, the crystal planes remain continuous in all directions, and the clusters of solute atoms (which are of different size and possibly different valence from the host material) tend to distort or strain the adjacent lattice for a sizable distance in all directions, as illustrated in Figures 5-5a and 5-5c. For this reason, each small cluster appears to be much larger with respect to its ability to interfere with dislocation motion (i.e., impart strength). When the clusters reach a certain size, however, the associated strain becomes so great that the clusters can lower their energy by breaking free from the parent structure to form distinct second-phase particles with their own crystal structure and well-defined interphase boundaries, as shown in Figures 5-5b and 5-5d. Coherency is lost and the strengthening reverts to *dispersion hardening*, where dislocation interference is limited to the actual size of the particle. Strength and hardness decrease, and the material is said to be *overaged*.

Figure 5-6 presents a family of aging curves for the 4% copper–96% aluminum alloy. For higher aging temperatures, the peak properties are achieved in a shorter time,

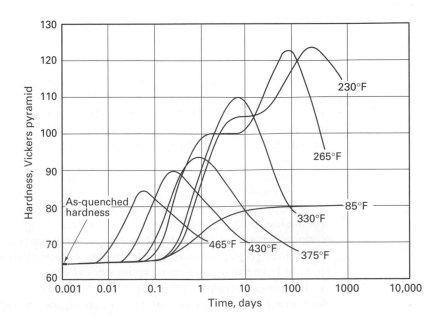

FIGURE 5-6 Aging curves for the Al–4%Cu alloy at various temperatures showing peak strengths and times of attainment. *(Adapted from* Journal of the Institute for Metals, *Vol. 79, p. 321, 1951.)*

but the peak hardness (or strength) is not as great as can be achieved with the finer precipitates and larger amounts that form at lower aging temperatures. Selection of the aging conditions (temperature and time) is a decision that is made on the basis of desired strength, available equipment, and production constraints.

The aging step can be used to divide precipitation-hardening materials into two types: (1) *naturally aging* materials, where room temperature is sufficient to move the unstable supersaturated solution toward the stable two-phase structure, and (2) *artificially aging* materials, where elevated temperatures are required to provide the necessary diffusion. With natural aging materials, such as aluminum alloy rivets, some form of refrigeration may be required to retain the after-quench condition of softness. Upon removal from the refrigeration, the rivets are easily headed but progress to full strength after several days at room temperature.

Since artificial aging requires elevated temperature to provide diffusion, the aging process can be stopped at any time by simply dropping the temperature (quenching). Diffusion is halted, and the current structure and properties are "locked-in," provided that the material is not subsequently exposed to elevated temperatures that would reactivate diffusion. When diffusion is possible, the material will always attempt to revert to its equilibrium structure! According to Figure 5-6, if the 4% copper alloy were aged for one day at 375°F and then quenched to prevent overaging, the metal would attain a hardness of 94 Vickers (and the associated strength) and retain these properties throughout its useful lifetime provided subsequent diffusion did not occur. If a higher strength is required, a lower temperature and longer time could be selected.

Precipitation hardening is an extremely effective strengthening mechanism and is responsible for the attractive engineering properties of many aluminum, copper, magnesium, and titanium alloys. In many cases, the strength can more than double that observed upon conventional cooling. While other strengthening methods are traditionally used with steels and cast irons, those methods have been combined with age hardening to produce some of the highest-strength ferrous alloys, such as the maraging steels and precipitation hardenable stainless.

■ 5.5 STRENGTHENING HEAT TREATMENTS FOR STEEL

Iron-based metals have been heat treated for centuries, and today over 90% of all heat-treatment operations are performed on steel. The striking changes that resulted from plunging red-hot metal into cold water or some other quenching medium were awe-inspiring to the ancients. Those who performed these acts in the making of swords and armor were looked upon as possessing unusual powers, and much superstition arose regarding the process. Because quality was directly related to the act of quenching, great importance was placed on the quenching medium that was used. Urine, for example, was found to be a superior quenching medium, and that from a red-haired boy was deemed particularly effective, as was that from a 3-year-old goat fed only ferns.

ISOTHERMAL TRANSFORMATION DIAGRAM

It has only been within the last century that the art of heat treating has begun to turn into a science. One of the major barriers to understanding was the fact that the strengthening treatments were nonequilibrium in nature. Minor variations in cooling often produced major variations in structure and properties.

A useful aid to understanding nonequilibrium heat treatment processes is the *isothermal-transformation* (I-T) or *time-temperature-transformation* (T-T-T) *diagram*. The information in this diagram is obtained by heating thin specimens of a particular steel to produce elevated-temperature uniform-chemistry austenite, "instantaneously" quenching to a temperature where austenite is no longer the stable phase, holding for variable periods of time at this new temperature, and observing the resultant structures via metallographic photomicrographs (i.e., optical microscope examination).

For simplicity, consider a carbon steel of eutectoid composition (0.77% carbon) and its T-T-T diagram shown as Figure 5-7. Above the A_1 temperature of 1341°F (727°C), austenite is the stable phase and will persist regardless of the time. Below this temper-

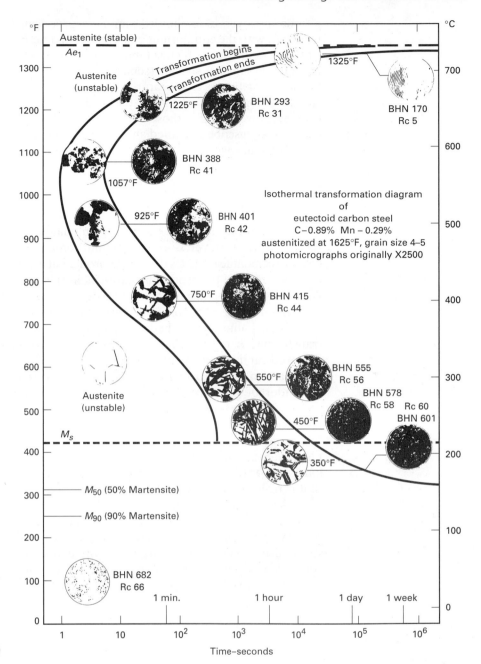

FIGURE 5-7 Isothermal-transformation diagram (T-T-T diagram) for eutectoid composition steel. Structures resulting from transformation at various temperatures are shown as insets. *(Courtesy of United States Steel Corp., Pittsburgh, PA.)*

ature, the face-centered austenite would like to transform to body-centered ferrite and carbon-rich cementite. Two factors control the rate of transition: (1) the motivation or driving force for the change and (2) the ability to form the desired products (i.e., the ability to rearrange the atoms through diffusion). The region below 1341°F in Figure 5-7 can be interpreted as follows. Zero time corresponds to a sample "instantaneously" quenched to its new, lower temperature. The structure is usually unstable austenite. As time passes (moving horizontally across the diagram), a line is encountered representing the start of transformation and a second line indicating completion of the phase change. At elevated temperatures (just below 1341°F), atom movement within the solid (diffusion) is rapid, but the rather sluggish driving force dominates the kinetics. At a low temperature, the driving force is high but diffusion is quite limited. The kinetics of phase transformation are most rapid at a compromise intermediate temperature, resulting in the characteristic C-curve shape. The portion of the C that extends farthest to the left is known as the *nose* of the T-T-T or I-T diagram.

If the transformation occurs between the A_1 temperature and the nose of the curve, the departure from equilibrium is not very great. The austenite transforms into alternating layers of ferrite and cementite, producing the *pearlite* structure that was introduced with the equilibrium phase diagram description in Chapter 4. Since the diffusion rate is greater at higher temperatures, pearlite produced under those conditions has a larger lamellar spacing (separation distance between similar layers). The pearlite formed near the A_1 temperature is known as *coarse pearlite*, while the closer-spaced structures formed near the nose are called *fine pearlite*. Since the resulting structures and properties are similar to those of the near-equilibrium process heat treatments, the procedure just described is called an *isothermal anneal*.

If the austenite is quenched to a temperature between the nose and the temperature designated as M_s, a different structure is produced. These transformation conditions are a significant departure from equilibrium, and the amount of diffusion required to form the continuous layers within pearlite is no longer available. The metal still has the goal of changing crystal structure from face-centered austenite to body-centered ferrite, with the excess carbon being accommodated in the form of cementite. The resulting structure, however, does not contain cementite layers but rather a dispersion of discrete cementite particles dispersed throughout a matrix of ferrite. Electron microscopy may be required to resolve the carbides in this structure, which is known as *bainite*. Because of the fine dispersion of carbide, it is stronger than fine pearlite, and ductility is retained because the soft ferrite is the continuous matrix.

If austenite is quenched to a temperature below the M_s line, a different type of transformation occurs. The steel still wants to change its crystal structure from face-centered cubic to body-centered cubic, but it can no longer expel the amount of carbon necessary to form ferrite. Responding to the severe nonequilibrium conditions, it simply undergoes an abrupt change in crystal structure with no significant movement of carbon. The excess carbon becomes trapped, distorting the structure into a body-centered tetragonal crystal lattice (distorted body-centered cubic), with the amount of distortion being proportional to the amount of excess carbon. The new structure, shown in Figure 5-8, is known as *martensite*, and, with sufficient carbon, it is exceptionally strong, hard, and brittle. The highly distorted lattice effectively blocks the dislocation motion that is necessary for metal deformation.

As shown in Figure 5-9, the hardness and strength of steel with the martensitic structure are strong functions of the carbon content. Below 0.10% carbon, martensite is not very strong. Since no diffusion occurs during the transformation, higher-carbon steels form higher-carbon martensite, with an increase in strength and hardness and a concurrent decrease in toughness and ductility. From 0.3 to 0.7% carbon, strength and hardness increase rapidly. Above 0.7% carbon, however, the rise is far less dramatic and

FIGURE 5-8 Photomicrograph of martensite; 1000X. *(Courtesy of United States Steel Corp., Pittsburgh, PA.)*

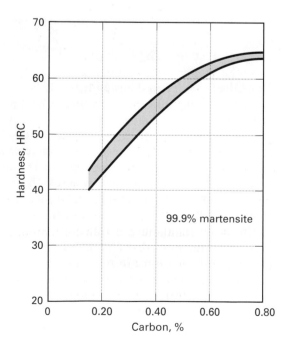

FIGURE 5-9 Effect of carbon on the hardness of martensite.

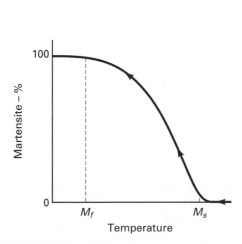

FIGURE 5-10   Schematic representation depicting the amount of martensite formed upon quenching to various temperatures from M_s through M_f.

FIGURE 5-11   Variation of M_s and M_f temperatures with carbon content. Note that for high-carbon steels, completion of the martensite transformation requires cooling to below room temperature.

may actually be a decline, a feature related to the presence of retained austenite (to be described below).

Unlike the other structure transformations, the *amount* of martensite that forms is not a function of time, but rather depends only on the lowest temperature that is encountered during the quench. This feature is shown in Figure 5-10, where the amount of martensite is recorded as a function of temperature. Returning to the C curve of Figure 5-7, there is a temperature designated as M_{50}, where the structure is 50% martensite and 50% untransformed austenite. At the lower M_{90} temperature, the structure has become 90% martensite. If no further cooling were to occur, the untransformed austenite could remain within the structure. This *retained austenite* can cause loss of strength or hardness, dimensional instability, and cracking or brittleness. Since many quenches are to room temperature, retained austenite becomes a significant problem when the martensite finish, or 100% martensite, temperature lies below room temperature. Figure 5-11 presents the martensite start and martensite finish temperatures for a range of carbon contents. Higher carbon contents, as well as most alloy additions, decrease all martensite-related temperatures, and materials with these chemistries may require refrigeration or a quench in dry ice or liquid nitrogen to produce full hardness.

It is important to note that all of the transformations that occur below the A_1 temperature are one-way transitions (austenite to something). The steel is simply seeking to change its crystal structure, and the various products are the result of this change. It is impossible, therefore, to convert one transformation product to another without first reheating to above the A_1 temperature to again form the face-centered-cubic austenite.

T-T-T diagrams can be quite useful in determining the kinetics of transformation and the nature of the products. The left-hand curve shows the elapsed time (at constant temperature) before the transformation begins, and the right-hand curve shows the time required to complete the transformation. If hypo- or hypereutectoid steels were considered, additional regions would have to be added to the diagram to incorporate the primary equilibrium phases that form below the A_3 or A_{cm} temperatures. These regions would not extend below the nose, however, since the nonequilibrium bainite and martensite structures can exist with variable amounts of carbon, unlike the near-equilibrium pearlite. Figure 5-12 shows the T-T-T curve for a 0.5%-carbon hypoeutectoid steel, showing the additional region for the primary ferrite.

FIGURE 5-12 Isothermal-transformation diagram for a hypoeutectoid steel (1050) showing the additional region for primary ferrite.

TEMPERING OF MARTENSITE

Despite its great strength, medium- or high-carbon martensite in its as-quenched form lacks sufficient toughness and ductility to be a useful engineering structure. A subsequent heating, known as *tempering*, is usually required to impart the necessary ductility and fracture resistance, and relax undesirable residual stresses. As with most property-changing processes, there is a concurrent drop in other features, most notably strength and hardness.

Martensite is a supersaturated solid solution of carbon in alpha-ferrite and, therefore, is a metastable structure. When heated into the range of 100° to 700°C (200° to 1300°F), the excess carbon atoms are rejected from solution, and the structure moves toward a mixture of the stable phases of ferrite and cementite. This decomposition of martensite into ferrite and cementite is a time- and temperature-dependent, diffusion-controlled phenomenon with a continuous spectrum of intermediate and transitory structures.

Table 5-1 presents a chart-type comparison of the previously discussed precipitation-hardening process and the austenitize-quench-and-temper sequence. Both are non-equilibrium heat treatments that involve three distinct stages. In both, the first step is an elevated temperature soaking designed to erase the prior structure, redissolving material to produce a uniform-chemistry, single-phase starting condition. Both treatments follow this soak with a rapid-cool quench. In precipitation hardening, the purpose of the quench is to prevent nucleation of the second phase, thereby producing a supersaturated solid solution. This material is usually soft, weak, and ductile, with good toughness. Subsequent aging (reheating within the temperatures of the stable two-phase region) allows the material to move toward the formation of the stable two-phase structure and sacrifices toughness and ductility for an increase in strength. When the proper balance is achieved, the temperature is dropped, diffusion ceases, and the current structure and properties are preserved, provided that the material is never subsequently exposed to any elevated temperature that would reactivate diffusion and permit the structure to move further toward equilibrium.

For steels, the quench induces a phase transformation as the material changes from the face-centered-cubic austenite to the distorted body-centered structure known as martensite. The quench product is again a supersaturated, single-phase solid solution, this time of carbon in iron, but the associated properties are the reverse of precipitation hardening. Martensite is strong and hard but relatively brittle. When the material is

TABLE 5-1	Comparison of Age Hardening with the Quench-and-Temper Process		
Heat Treatment	Step 1	Step 2	Step 3
Age hardening	*Solution treatment.* Heat into the stable single-phase region (above the solvus) and hold to form a uniform-chemistry single-phase solid solution.	*Quench.* Rapid cool to form a nonequilibrium supersaturated single-phase solid solution (crystal structure remains unchanged, material is soft and ductile).	*Age.* A controlled reheat in the stable two-phase region (below the solvus). The material moves toward the formation of the stable two-phase structure, becoming stronger and harder. The properties can be "frozen in" by dropping the temperature to stop further diffusion.
Quench and temper for steel	*Austenitize.* Heat into the stable single-phase region (above the A_3 or A_{cm}) and hold to form a uniform-chemistry single-phase solid solution (austenite).	*Quench.* Rapid cool to form a nonequilibrium supersaturated single-phase solid solution (crystal structure changes to body-centered martensite, which is hard but brittle).	*Temper.* A controlled reheat in the stable two-phase region (below the A_1). The material moves toward the formation of the stable two-phase structure, becoming weaker but tougher. The properties can be "frozen in" by dropping the temperature to stop further diffusion.

tempered (reheated to a temperature within the stable two-phase region), strength and hardness are sacrificed for an increase in ductility and toughness. Figure 5-13 shows the final properties of a steel that has been tempered at a variety of temperatures. During tempering, diffusion enables movement *toward* the stable two-phase structure, and a drop in temperature can again halt diffusion and lock in properties. By quenching steel to form martensite and then tempering it at various temperatures, an infinite range of structures and corresponding properties can be produced. This procedure is known as the *quench-and-temper process* and the product, which offers an outstanding combination of strength and toughness, is called *tempered martensite.*

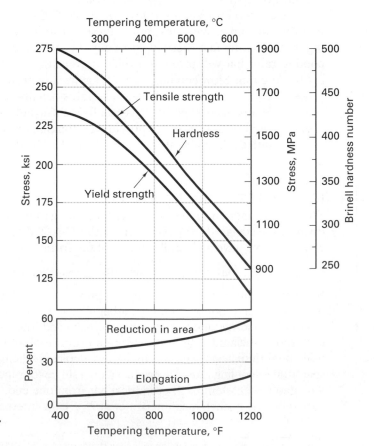

FIGURE 5-13 Properties of an AISI 4140 steel that has been austenitized, oil-quenched, and tempered at various temperatures. (Adapted from *Engineering Properties of Steel, ASM International, Materials Park, OH., 1982.*)

FIGURE 5-14 C-C-T diagram for a eutectoid composition steel (bold), with several superimposed cooling curves and the resultant structures. The lighter curves are the T-T-T transitions for the same steel. *(Courtesy of United States Steel Corp., Pittsburgh, PA.)*

CONTINUOUS COOLING TRANSFORMATIONS

While the T-T-T diagrams provide considerable information about the structures obtained through nonequilibrium thermal processing, the assumptions of instantaneous cooling followed by constant temperature transformation rarely match reality. Actual parts generally experience continuous cooling from elevated temperature, and a diagram showing the results of this type of cooling at various rates would be far more useful. What would be the result if the temperature were to be decreased at a rate of 500°F per second, 50°F per second, or 5°F per second?

A *continuous-cooling-transformation* (C-C-T) *diagram*, like the one shown in Figure 5-14, can provide answers to these questions and numerous others. If the cooling is sufficiently fast (dashed curve A), the structure will be martensite. The slowest cooling rate that will produce a fully martensitic structure is referred to as the *critical cooling rate*. Slow cooling (curve D) generally produces coarse pearlite along with a possible primary phase. Intermediate rates usually result in mixed structures, since the time at any one temperature is usually insufficient to complete the transformation. If each structure is regarded as providing a companion set of properties, the wide range of possibilities obtainable through the controlled heating and cooling of steel becomes even more evident.

JOMINY TEST FOR HARDENABILITY

The C-C-T diagram shows that different cooling rates produce different structures with different associated properties. Since the C-C-T diagram will change with material chemistry, we have a general relation of the form:

$$\text{material} + \text{cooling rate} \longrightarrow \text{structure} \longrightarrow \text{properties}$$

For a given material, its C-C-T diagram provides the link between cooling rate and structure. Engineers who focus on use and application, however, are often more interested in material properties and how they may be achieved. The *Jominy end-quench hardenability test*[2] and associated diagrams provide a useful tool and expand our understanding of nonequilibrium heat treatment. In this test, depicted schematically in Figure 5-15, an entire spectrum of cooling rates are produced on a single four-inch-long specimen by quenching a heated (i.e., austenitized) cylindrical bar from one end. The quench is standardized by specifying the quench medium (water at 75°F), the internal nozzle diameter ($\frac{1}{2}$ inch), the

FIGURE 5-15 Schematic diagram of the Jominy hardenability test.

Air-cooled end

Jig

1″ dia.

Test bar

4″

1/2″

Water quench

1/2″

[2] This test is described in detail in the following standards: ASTM A255, SAE J406, DIN 50191, and ISO 642.

FIGURE 5-16   Typical hardness distribution along a Jominy test specimen.

water pressure (that producing a $2\frac{1}{2}$-inch vertical fountain), and the gap between the nozzle and the specimen ($\frac{1}{2}$ inch). Since none of the water contacts the side of the specimen, all cooling is directional along the axis of the bar. One end sees rapid cooling, while the other is essentially air-cooled. Since the thermal conductivity of steel does not change over the normal ranges of carbon and alloy additions, a characteristic cooling rate can be assigned to each location along the length of the standard-geometry specimen.

After the test bar has cooled to room temperature, a flat region is ground along opposite sides and Rockwell C hardness readings are taken every $\frac{1}{16}$ inch along the bar. The resulting data are then plotted as shown in Figure 5-16. The hardness values are correlated with position, and since the cooling rate is known for each location within the bar, the hardnesses are indirectly correlated with the cooling rate that produced them. Since hardness is also an indicator of strength, we have experimentally linked cooling rate to resultant strength in a simple and efficient manner.

For any given material, application of the test assumes that equivalent cooling conditions will produce equivalent results. If the cooling rate is known for a specific location within a part (from experimentation or theory), the properties at that location can be predicted to be those at the Jominy test bar location with the same cooling rate. Conversely, if specific properties are required, the necessary cooling rate can be determined. Should the cooling rates be restricted by either geometry or processing limitations, various materials can be compared and a satisfactory alloy selected. Figure 5-17 shows the Jominy curves for several engineering steels. Since differences in chemical composition can exist between heats of the same grade of steel, hardenability data are

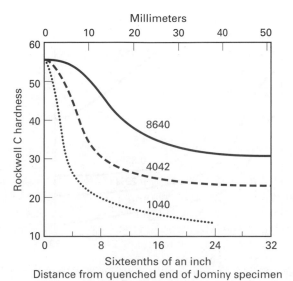

FIGURE 5-17   Jominy hardness curves for engineering steels with the same carbon content but varying types and amounts of alloy elements.

often presented in the form of bands, where the upper curve corresponds to the maximum expected hardness and the lower curve to the minimum. The data for actual heats should then fall between these two extremes.

HARDENABILITY CONSIDERATIONS

Several key effects must be considered if we are to understand the heat treatment of steel: (1) the effect of carbon content, (2) the effect of alloy additions, and (3) the effect of various quenching conditions. The first two relate to the material being treated and the third to the heat-treatment process.

Hardness is a mechanical property related to strength and is a strong function of the carbon content of a steel and the particular microstructure. With different heat treatments, the same steel can have different hardness values. *Hardenability* is a measure of the depth to which full hardness can be obtained under a normal hardening cycle and is related primarily to the amounts and types of alloying elements. Hardenability, therefore, is a material property dependent upon chemical composition. In Figure 5-17, all of the steels have the same carbon content, but they differ in the type and amounts of alloy elements. The maximum hardness is the same in all cases, but the depth of hardening (or the way hardness varies with distance from the quenched end) varies considerably. Figure 5-18 shows Jominy test results for steels containing the same alloying elements but variable amounts of carbon. Note the change in peak hardness as carbon content is increased.

The results of a heat-treat operation depend on both the hardenability of the metal and the rate of heat extraction. The primary reason for adding alloy elements to commercial steels is to increase their hardenability, not to improve their strength. Steels with greater hardenability can achieve a desired level of strength or hardness with slower rates of cooling and, for this reason, can be completely hardened in thicker sections. Slower cooling also serves to reduce the amount of quench-induced distortion and the likelihood of *quench cracking*.

An accurate determination of need is required if steels are to be selected for specific applications. Strength tends to be associated with carbon content, and a general rule is to select the lowest possible level that will meet the specifications. Because heat can be extracted only from the surface of a metal, the size of the piece and the depth of required hardening set the conditions for hardenability and quench. For a given quench condition, different alloys will produce different results. Because alloy additions increase the cost of a material, it is best to select only what is required to ensure compliance with specifications. Money is often wasted by specifying an alloy steel

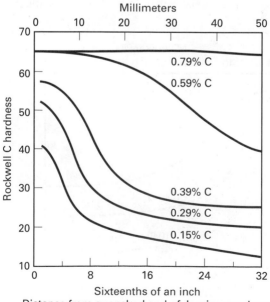

FIGURE 5-18 Jominy hardness curves for engineering steels with identical alloy conditions but variable carbon content.

for an application where a plain-carbon steel, or a steel with lower alloy content (less costly), would be satisfactory. When greater depth of hardness is required, another alternative is to modify the quench conditions so that a faster cooling rate is achieved. Quench changes may be limited, however, by cracking or warping problems as well as other considerations relating to the size, shape, complexity, and desired precision of the part being treated.

QUENCH MEDIA

Quenchants are selected to provide the cooling rates required to produce the desired structure and properties in the size and shape part being treated. Quench media vary in their effectiveness, and one can best understand the variation by considering the three stages of quenching. Let's begin with a piece of hot metal being inserted into a tank of liquid quenchant. If the temperature of the metal is above the boiling point of the quenchant, the liquid adjacent to the metal will vaporize and form a thin gaseous layer between the metal and the liquid. Cooling is slow through this *vapor jacket* (first stage) since the gas has an insulating effect, and heat transfer is largely through radiation. Bubbles soon nucleate, however, and break the jacket. New liquid contacts the hot metal, vaporizes (removing its heat of vaporization from the metal), forms another bubble, and the process continues. Because large quantities of heat are required to vaporize a liquid, this *second stage of quenching* (or nucleate boiling phase) produces rapid rates of cooling down to the boiling point of the quenchant. At this point, vaporization can no longer occur. Heat transfer must now take place by conduction across the solid–liquid interface, aided by convection or stirring within the liquid. In this *third stage of quenching*, the slower cooling by conduction and convection continues from the boiling point to room temperature. Breakdown of the vapor jacket, bubble removal, and convection cooling can all be aided by moving the metal through the liquid or flowing the liquid over the metal surface. Various liquids offer different heats of vaporization, viscosities, and boiling points. Quenches can be further tailored by varying the temperature of the liquid and the degree of flow or agitation.

Water is a fairly effective quenching medium because of its high heat of vaporization and the fact that the second stage of quenching extends down to 100°C (212°F), usually well into the temperatures for martensite formation. Water is also cheap, readily available, easily stored, nontoxic, nonflammable, smokeless, and easy to filter and pump. Agitation is usually recommended with a water quench, however, since the clinging tendency of the bubbles may cause soft spots on the metal. Other problems associated with a water quench include its oxidizing nature (i.e., its corrosiveness) and the tendency to produce excessive distortion and possible cracking.

While *brine* (salt water) has a similar heat of vaporization and boiling point to water, it produces more rapid cooling because the salt nucleates bubbles, forcing a quick transition through the vapor jacket stage. Unfortunately, the salt in a brine quench also tends to accelerate corrosion problems unless all residues are completely removed by a subsequent rinse. Different types of salts can be used, including sodium or potassium hydroxide, and various degrees of agitation or spraying can be used to adjust the effectiveness of the quench. (*Note:* Because of all of the dissolved salts, the urines cited in quenching folklore are actually quite similar to the brines of today.)

If a slower cooling rate is desired, *oil* quenches are often utilized. Various oils are available that have high flash points and different degrees of quenching effectiveness. Since the boiling points can be quite high, the transition to third-stage cooling usually precedes the martensite start temperature. The slower cooling through the M_s-to-M_f martensite transformation leads to a milder temperature gradient within the piece, reduced distortion, and reduced likelihood of cracking. Heating the oil actually increases its cooling ability, since the reduced viscosity assists bubble formation and removal. Problems associated with oil quenchants include water contamination, smoke, fumes, spill and disposal problems, and fire hazard. In addition, quench oils tend to be somewhat expensive.

Quite often, there is a need for a quenchant that will cool more rapidly than the oils but slower than water or brine. To fill this gap, a number of water-based *polymer quench* solutions (also called *synthetic quenchants*) have been developed. Tailored quenchants can be produced by varying the concentrations of the components (such as liquid organic polymers, corrosion inhibitors, and water) and adjusting the operating temperature and amount of agitation. The polymer quenchants provide extremely uniform and reproducible results, are less corrosive than water and brine, and are less of a fire hazard than oils (no fires, fumes, smoke, or need for air pollution control apparatus). Distortion and cracking are less of a problem since the boiling point can be adjusted to be above the martensite start temperature. In addition, the polymer-rich film that forms initially on the hot metal part serves to modify the cooling rate.

If slow cooling is required, molten salt baths can be employed to provide a medium where the quench goes directly to the third stage of cooling. Still slower cooling can be obtained by cooling in still air, burying the hot material in sand, or a variety of other methods.

High-pressure gas quenching uses a stream of flowing gas to extract heat, and the cooling rates can be adjusted by controlling the gas velocity and pressure. Results are comparable to oil quenching, with far fewer environmental and safety concerns. From an environmental perspective, *vegetable oils* may also be an attractive quenchant. They are biodegradable, offer low toxicity, and are a renewable resource.

DESIGN CONCERNS, RESIDUAL STRESSES, DISTORTION, AND CRACKING

Product design and material selection play important roles in the satisfactory and economical heat treatment of parts. Proper consideration of these factors usually leads to simpler, more economical, and more reliable products. Failure to relate design and materials to heat-treatment procedures usually produces disappointing or variable results and may lead to a variety of service failures.

From the viewpoint of heat treatment, undesirable design features include (1) nonuniform sections or thicknesses, (2) sharp interior corners, and (3) sharp exterior corners. Since these features often find their way into the design of parts, the designer should be aware of their effect on heat treatment. Undesirable results may include nonuniform structure and properties, undesirable residual stresses, cracking, warping, and dimensional changes.

Heat can only be extracted from a piece through its exposed surfaces. Therefore, if the piece to be hardened has a nonuniform cross section, any thin region will cool rapidly and may fully harden, while thick regions may harden only on the surface, if at all. The shape that might be closest to ideal from the viewpoint of quenching would be a doughnut. The uniform cross section with high exposed surface area and absence of sharp corners would be quite attractive. Since most shapes are designed to perform a function, however, compromises are usually necessary.

Residual stresses are the often-complex stresses that are present within a body, independent of any applied load. They can be induced in a number of ways, but the complex dimensional changes that can occur during heat treatment are a primary cause. Thermal expansion during heating and thermal contraction during cooling are well-understood phenomena, but when these occur in a nonuniform manner, the results can be extremely complex. In addition, the various phases and structures that can exist within a material usually possess different densities. Volume expansions or contractions accompany any phase transformation. For example, when austenite transforms to martensite, there is a volume expansion of up to 4%. Transformations to ferrite, pearlite, or other room-temperature structures also involve volume expansions but of a smaller magnitude.

If all of the temperature changes occurred uniformly throughout a part, all of the associated dimensional changes would occur simultaneously and the resultant product would be free of residual stresses. Most of the parts being heat treated, however, experience nonuniform temperatures during the cooling or quenching operation. Consider a block of hot aluminum being cooled by water sprays from top and bottom. For simplicity, let us model the block as a three-layer sandwich, as in the top sequence of Figure 5-19. At the start of the quench, all layers are uniformly hot. As the water spray begins, the surface layers cool and contract, but the center layer does not experience the quench and remains hot. Since the part is actually one piece, however, the various

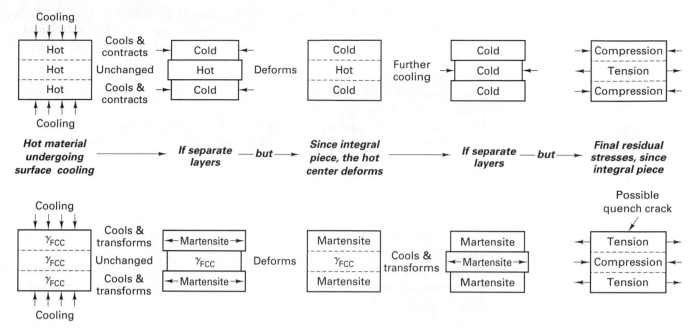

FIGURE 5-19   Three-layer model of a plate undergoing cooling. The upper sequence depicts a material such as aluminum that contracts upon cooling while the bottom sequence depicts steel, which expands during the cooling-induced phase transformation.

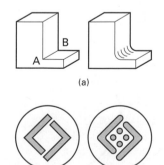

FIGURE 5-20   (a) Shape containing nonuniform sections and a sharp interior corner that may crack during quenching. This is improved by using a large radius to join the sections. (b) Original design containing sharp corner holes, which can be further modified to produce more uniform sections.

layers must accommodate each other. The contracting surface layers exert compressive forces on the hot, weak interior, causing it to also contract,—but by plastic deformation, not thermal contraction. As time passes, the interior now cools and wants to contract but finds itself sandwiched between the cold, strong surface layers. It pulls on the surface layers, placing them in compression, while the surface layers restrict its movement, creating tension in the interior. While the net force is zero (since there is no applied load), counterbalancing tension and compression stresses exist within the product.

Let us now change the material to steel and repeat the sequence. When heated, all three layers are hot, face-centered-cubic austenite, as shown in the bottom sequence of Figure 5-19. Upon quenching, the surface layers transform to martensite (the structure changes to body-centered tetragonal) and *expand*! The expanding surfaces deform the soft, weak, (and still hot) austenite center, which then cools and wants to expand as it undergoes the crystal structure change to martensite or another body-centered, room-temperature product. The hard, strong surface layers hold it back, placing the center in compression, while the expanding center tries to stretch the surface layers, producing surface tension. If the tension at the surface becomes great enough, cracking can result, a phenomenon known as *quench cracking*. One should note that for the rapid-quench conditions just described, aluminum will never quench crack, because the surface is in compression, but the steel might, since the residual stresses are reversed. If the cooling conditions were not symmetrical, there might be more contraction or expansion on one side, and the block might warp. With more complex shapes and the accompanying nonuniform cooling, the residual stresses induced by heat treatment can be extremely complex.

Various techniques can be employed to minimize or reduce the problems associated with residual stresses. If product cross sections can be made more uniform, temperature differences can be minimized and will not be concentrated at any specific location. If this is not possible, slower cooling may be recommended, coupled with a material that will provide the desired properties after a slower oil or air quench. Since materials with greater hardenability are more expensive, design alternatives should generally precede quench modifications.

When temperature differences and the resultant residual stresses become severe or localized, cracking or distortion problems can be expected. Figure 5-20a shows an example where a sharp interior corner has been placed at a change in cross section. Upon quenching, stresses will concentrate along line A–B, and a crack is almost certain to result. If changes in cross section are required, they should be gradual, as in the redesigned version of Figure 5-20a. Generous fillets at interior corners, radiused exterior corners, and smooth transitions all reduce problems. A material with greater hardenability and a less severe quench would also help. Figure 5-20b shows the cross section of a blanking die that consistently cracked during hardening. Rounding the sharp corners and adding additional holes to provide a more uniform cross section during quenching eliminated the problem.

One of the ominous features of poor product and process design is the fact that the residual stresses may not produce immediate failure but may contribute to failure at a later time. Applied stresses add to the residual stresses already present within the part. Therefore, it is possible for applied stresses that are well within the "safe" designed limit to couple with residual stresses and produce a value sufficient to induce failure. Residual stresses can also accelerate corrosion reactions. Dimensional changes or warping can occur when subsequent machining or grinding operations upset the equilibrium balance of the residual stresses. After the removal of some material, the remaining piece adjusts its shape to produce a new (sum of forces equal zero) equilibrium balance. When we consider all of the possible difficulties, it is apparent that considerable time and money can be saved if good design, material selection, and heat-treatment practices are employed.

TECHNIQUES TO REDUCE CRACKING AND DISTORTION

The steel segment of Figure 5-19 has already introduced the phenomena of quench cracking, and Figure 5-21a further illustrates its cause using a T-T-T diagram (a misuse of the diagram, since continuous cooling is employed, but helpful for visualization). The surface and center of a quenched product generally have different cooling rates. As a result, when the surface has cooled and is transforming to martensite with its companion expansion, the center is still hot and remains soft, untransformed austenite. At a later time, the center cools to the martensite transformation, and it expands. The result is significant tension in the cold, hard surface and possible cracking.

Figure 5-21b depicts two variations of rapid quenching that have been developed to produce strong structures while reducing the likelihood of cracking. A rapid cool must still be employed to prevent transformation to the softer, weaker pearlitic structure, but instead of quenching through the martensite transformation, the component is rapidly quenched into a liquid medium that is now several degrees above the martensite start (M_s) temperature, such as hot oil or molten salt. Holding for a period of time in this bath allows the entire piece to return to a nearly uniform temperature.

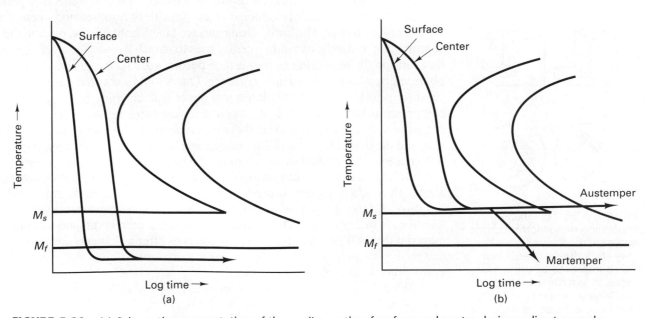

FIGURE 5-21 (a) Schematic representation of the cooling paths of surface and center during a direct quench; (b) The modified cooling paths experienced during the austempering and martempering processes.

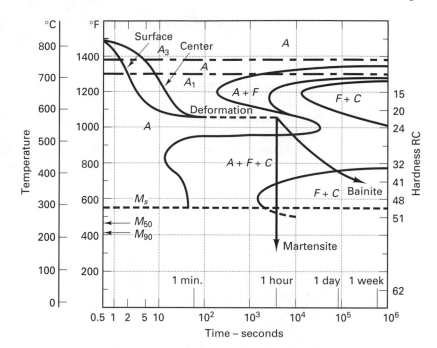

FIGURE 5-22 T-T-T diagram for 4340 steel, showing the "bay," along with a schematic of the ausforming process. *(Courtesy of United States Steel Corp., Pittsburgh, PA.)*

If the material is simply held at this temperature for sufficient time, the austenite will transform to bainite, which usually has sufficient toughness that a subsequent temper is not required. This process is known as *austempering*. If the material is brought to a uniform temperature and then slow cooled through the martensite transformation, the process is known as *martempering*, or *marquenching*. The resulting structure is martensite, which must be tempered the same as the martensite that forms directly upon quenching. In both of these processes, all transformations (and related volume expansions) occur at the same time, thereby eliminating the residual stresses and tendency to crack.

AUSFORMING

A process that is often confused with austempering is *ausforming*. Certain alloys tend to retard the pearlite transformation more than the bainite reaction and produce a T-T-T curve of the shape shown in Figure 5-22. If this material is heated to form austenite and then quenched to the temperature of the "bay" between the pearlite and bainite reactions, it can retain its austenite structure for a useful period of time. Deformation can be performed on an austenite structure at a temperature where it technically should not exist. Benefits include the increased ductility of the face-centered-cubic crystal structure, the finer grain size that forms upon recrystallization at the lower temperature, and the possibility of some degree of strain hardening. Following the deformation, the metal can be slowly cooled to produce bainite or rapidly quenched to martensite, which must then be tempered. The resulting product has exceptional strength and ductility, coupled with good toughness, creep resistance, and fatigue life—properties that are superior to those produced if the deformation and transformation processes are conducted in their normal separated sequence. Ausforming is an example of a growing class of *thermomechanical processes* in which deformation and heat treatment are intimately combined.

■ 5.6 SURFACE HARDENING OF STEEL

Many products require different properties at different locations. Quite frequently, this variation takes the form of a hard, wear-resistant surface coupled with a tough, fracture-resistant core. The methods developed to produce the varied properties can be classified into three basic groups: (1) selective heating of the surface, (2) altered surface chemistry, and (3) deposition of an additional surface layer. The first two approaches will be discussed in the next sections, while platings and coatings will be described in Chapter 36.

SELECTIVE HEATING TECHNIQUES

If a steel has sufficient carbon to attain the desired surface hardness, generally greater than 0.3%, the different properties can often be obtained simply by varying the thermal histories of the various regions. Core properties are set by a bulk treatment, with the surface properties being established by a subsequent surface treatment. Maximum hardness depends on the carbon content of the material, while the depth of that hardness depends on both the depth of heating and the material's hardenability. The various methods generally differ in the way the surface is brought to elevated temperature.

Flame hardening uses an oxy-acetylene flame to raise the surface temperature high enough to reform austenite. The surface is then water quenched[3] to produce martensite and tempered to the desired level of toughness. Heat input is quite rapid, leaving the interior at low temperature and free from any significant change. Considerable flexibility is provided since the rate and depth of heating can be easily varied. Depth of hardening can range from thin skins to over 6 mm. ($\frac{1}{4}$ inch). Flame hardening is often used on large objects, since alternative methods tend to be limited by both size and shape. Equipment varies from crude handheld torches to fully automated and computerized units.

When using *induction hardening*, the steel part is placed inside a conductor coil, which is then energized with alternating current. The changing magnetic field induces surface currents in the steel, which heat by electrical resistance. The heating rates can be extremely rapid, and energy efficiency is high. Rapid cooling is then provided by either an immersion quench or water spray using a quench ring that follows the induction coil.

Induction heating is particularly well suited to surface hardening since the rate and depth of heating can be controlled directly through the amperage and frequency of the generator. Induction hardening is ideal for round bars and cylindrical parts but can also be adapted to more complex geometries. The process offers high quality, good reproducibility, and the possibility of automation. Figure 5-23 shows a partial cross section of an induction-hardened gear, where hardening has been applied to those areas expected to see high wear. Distortion during hardening is negligible since the dark areas remain cool and rigid throughout the entire process.

Laser-beam hardening has been used to produce hardened surfaces on a wide variety of geometries. An absorptive coating such as zinc or manganese phosphate is often applied to the steel to improve the efficiency of converting light energy into heat. The surface is then scanned with the laser, where beam size, beam intensity, and scanning speed have been selected to obtain the desired amount of heat input and depth of heating. Because of the localized heating of the beam, it is possible for the process to be *autoquenching* (cooled simply by conductive transfer into the underlying metal), but a water or oil quench can also be used. Through laser-beam hardening, a 0.4% carbon steel can attain surface hardnesses as high as Rockwell C 65. Typical depth of hardening is between 0.5 and 1 mm (0.02 to 0.04 inches). The process operates at high speeds, produces little distortion, induces residual compressive stresses on the surface, and can be used to harden selected surface areas while leaving the remaining surfaces unaffected. Computer software and automation can be used to control the process parameters, and conventional mirrors and optics can be used to shape and manipulate the beam.

Electron-beam hardening is similar to laser-beam hardening. Here the heat source is a beam of high-energy electrons rather than a beam of light, with the charged particles being focused and directed by electromagnetic controls. Like laser-beam treating, the process can be readily automated, and production equipment can perform a variety of operations with efficiencies often greater than 90%. Electrons cannot travel in air, however, so the entire operation must be performed in a hard vacuum, which is the major limitation of this process. More information on laser- and electron-beam techniques, as well as other means of heating material, is provided in Chapters 31 through 34.

FIGURE 5-23 Section of gear teeth showing induction-hardened surfaces. *(Courtesy of Ajax Tocco Magnathermic Corp., Warren, OH.)*

---

[3] There is no real danger of surface cracking during the water quench. When the surface is reaustenitized, the soft austenite adjusts to the colder, stronger underlying material. Upon quenching, the surface austenite expands during transformation. The interior is still cold and restrains the expansion, producing a surface in compression with no tendency toward cracking. This is just the opposite of the conditions that occur during the through-hardening of a furnace-soaked workpiece!

Still other surface-heating techniques employ immersion in a pool of molten lead or molten salt (*lead pot* or *salt bath* heating). These processes are attractive for treating complex-shaped products and hardening relatively inaccessible surfaces.

TECHNIQUES INVOLVING ALTERED SURFACE CHEMISTRY

If the steels contain insufficient carbon to achieve the desired surface properties, or the difference in surface and interior properties is too great for a single chemistry material, an alternative approach is to alter the surface chemistry. The most common technique within this category, *carburizing*, involves the diffusion of carbon into the elevated-temperature, face-centered-cubic, austenite structure, at temperatures between 800° and 1050°C (1450° and 1950°F). When sufficient carbon has diffused to the desired depth, the parts are then thermally processed. Direct quenching from the carburization treatment is the simplest alternative, and the different carbon contents and cooling rates can often produce the desired variation in properties. Alternative processes include a slow cool from the high-temperature carburizing treatment, followed by a lower-temperature reaustenitizing and quenching, or a duplex process involving a bulk treatment and a separate surface heat treatment. These latter processes are more involved and more costly but produce improved product properties. The carbon content of the surface usually varies from 0.7 to 1.2% depending on the material, the process, and the desired results. Case depth may range from a few thousandths of an inch to over $\frac{3}{8}$ inch.

The various carburizing processes differ in the source of the carbon. The most common is *gas carburizing*, where a hot, carbon-containing gas surrounds the parts. The process is fast, is easily controlled, and produces accurate and uniformly modified surfaces. In *pack-carburizing*, the steel components are surrounded by a high-carbon solid material (such as carbon powder, charcoal, or cast iron turnings) and heated in a furnace. The hot carburizing compound produces CO gas, which reacts with the metal, releasing carbon, which is readily absorbed by the hot austenite. A molten bath supplies the carbon in *liquid carburizing*. At one time, liquid-carburizing baths contained cyanide, which supplied both carbon and nitrogen to the surface. Safety and environmental concerns now dictate the use of noncyanide liquid compounds that are generally used to produce thin cases on small parts.

Nitriding hardens the surface by producing alloy nitrides in special steels that contain nitride-forming elements like aluminum, chromium, molybdenum, or vanadium. The parts are first heat treated and tempered at 525° to 675°C (1000° to 1250°F). After cleaning and removal of any decarburized surface material, they are heated in an atmosphere containing dissociated ammonia (nitrogen and hydrogen) for 10 to 40 hours at 500° to 625°C (950° to 1150°F). Since the temperatures are below the A_1 temperature, the nitrogen is diffusing into ferrite, not austenite, and subsequent cooling will not induce a phase transformation. The diffused nitrogen forms alloy nitrides, hardening the metal to a depth of about 0.65 mm (0.025 in.). Extremely hard cases are formed and distortion is low. No subsequent thermal processing is required. In fact, subsequent heating should be avoided because the thermal expansions and contractions will crack the hard nitrided case. Finish grinding should also be avoided because the nitrided layer is exceptionally thin.

Ionitriding is a plasma process that has emerged as an attractive alternative to the conventional method of nitriding. Parts to be treated are placed in an evacuated "furnace" and a direct current potential of 500 to 1000 volts is applied between the parts and the furnace walls. Low-pressure nitrogen gas is introduced into the chamber and becomes ionized. The ions are accelerated toward the negatively charged product surface, where they impact and generate sufficient heat to promote inward diffusion. This is the only heat associated with the process; the "furnace" acts only as a vacuum container and electrode. Advantages of the process include shorter cycle times, reduced consumption of gases, significantly reduced energy costs, and reduced space requirements. Product quality is improved over that of conventional nitriding, and the process is applicable to a wider range of materials. *Ion carburizing* is a parallel process in which low-pressure methane is substituted for the low-pressure nitrogen, and carbon diffuses into the surface.

Carbon and nitrogen can be added simultaneously. If ammonia is added to a standard carburizing atmosphere, and the steel is heated to a temperature where the structure is austenite, the process becomes one of *carbonitriding*. The temperature is usually lower than for standard carburizing, and the treatment time is somewhat shorter. If CO_2 is added to the ammonia of nitriding and the process is carried out at a temperature below the A_1, the process is known as *nitrocarburizing*. The resulting surface resists scuffing, and fatigue resistance is improved.

Ion plating and ion implantation are other technologies that permit modification of the surface chemistry. In Chapter 35 we will expand on the surface treatment of materials and compare the processes discussed in the preceding sections to various platings, coatings, and other techniques.

■ 5.7 FURNACES

FURNACE TYPES AND FURNACE ATMOSPHERES

To facilitate production heat treatment, many styles of furnaces have been developed in a wide range of sizes, each having its characteristic advantages and disadvantages. These furnaces are generally classified as batch or continuous type. *Batch furnaces*, in which the workpiece remains stationary throughout its treatment, are preferred for large parts or small lots of a particular part or grade of steel. *Continuous furnaces* move the components through the heat-treatment operation at rates selected to be compatible with the other manufacturing operations. Continuous furnaces are used for large production runs where the same or similar parts undergo the same thermal processing. The workpieces are moved through the furnace by some type of transfer mechanism (conveyor belt, walking beam, pusher, roller, or monorail) and often fall into a quench tank to complete the treatment. By incorporating various zones, complex cycles of heating, holding, and quenching/cooling can be conducted in an exact and repeatable manner with low labor cost.

Horizontal batch furnaces, often called *box furnaces* because of their overall shape, generally use gas or electricity as their source of heat. As shown in Figure 5-24, a door is provided on one end to allow the work to be inserted and removed. When large or very long workpieces are to be heated, a *car-bottom box furnace* may be employed, like the one shown in Figure 5-25. Here the work is loaded onto a refractory-topped flatcar, which can be rolled into and out of the furnace on railway rails.

In a *bell furnace*, the heating elements are contained within a bottomless "bell" that is lowered over the work. An airtight inner shell is often placed over the workpieces to

FIGURE 5-24   Box-type electric heat-treating furnace. *(Courtesy of Lindberg, a Division of TPS/SPX, Watsontown, PA.)*

FIGURE 5-25   Car-bottom box furnace. *(Courtesy of Sola/Hevi-Duty, Rosemont, IL.)*

contain a protective atmosphere during the heating and cooling operations. After the work is heated, the furnace unit can be lifted off and transferred to another batch, while the inner shell maintains a protective atmosphere during cooling. If extremely slow cooling is desired, an insulated cover can be placed over the heated shell. An interesting modification of the bell design is the *elevator furnace*, where the bell remains stationary and the workpieces are raised into it on a movable platform that then forms the bottom of the furnace. By placing a quench tank below the furnace, this design enables the workpieces to first be raised into the furnace and then lowered into the quench tank. It is extremely attractive for applications where the work must be quenched as soon as possible after being removed from the heat.

When long, slender parts are positioned horizontally, there is little resistance to sagging or warping. For these types of workpieces, a *vertical pit furnace* is preferred. These furnaces are usually cylindrical chambers sunk into the floor with a door on top that can be swung aside to allow suspended workpieces to be lowered into the furnace. They can also be used to heat large quantities of small parts by loading them into wire-mesh baskets that are then stacked within the column.

While all of the furnaces that have been described can heat in air, most commercial furnaces can also employ *artificial gas atmospheres*. These are selected to prevent scaling or tarnishing, to prevent decarburization, or even to provide carbon or nitrogen for surface modification. Many of the artificial atmospheres are generated from either the combustion or decomposition of natural gas, but nitrogen-based atmospheres frequently offer reduced cost, energy savings, increased safety, and environmental attractiveness. Other common atmospheres include argon, dissociated ammonia, dry hydrogen, helium, steam, and vacuum.

The heating rates of gas atmosphere furnaces can be significantly increased by incorporating the *fluidized-bed* concept. These furnaces consist of a bed of dry, inert particles, such as aluminum oxide (a ceramic), which are heated and fluidized (suspended) in a stream of upward-flowing gas. Products introduced into the bed become engulfed in the particles, which then radiate uniform heat. Temperature and atmosphere can be altered quickly, and high heat-transfer rates, high thermal efficiency, and low fuel consumption have been observed. Since atmosphere changes can be performed in minutes, a single furnace can be used for nitriding, stress relieving, carburizing, carbonitriding, annealing, and hardening.

When a liquid heating medium is preferred, *salt bath furnaces* are a popular choice. Electrically conductive salt can be heated by passing a current between two electrodes suspended in the bath. The electrical currents also cause the bath to circulate and thereby maintain uniform temperature. Nonconductive salts can be heated by some form of immersion heater, or the containment vessels can be externally fired. In these furnaces, the molten salt not only serves as a uniform source of heat but also can be selected to prevent scaling or decarburization. A *lead pot* is a similar device, where molten lead replaces salt as the heat-transfer medium.

Electrical induction heating is another popular means of heating conductive materials, such as metal. Small parts can be through-heated and hardened. Long products can be heated and quenched in a continuous manner by passing them through a stationary heating coil or by having a moving coil traverse a stationary part. Localized or selective heating can also be performed at rapid production rates. Flexibility is another attractive feature, since a standard induction unit can be adapted to a wide variety of products simply by changing the induction coil and adjusting the equipment settings.

FURNACE CONTROL

All heat-treatment operations should be conducted with rigid control if the desired results are to be obtained in a consistent fashion. Most furnaces are equipped with one or more temperature sensors, which can be coupled to a controller or computer to regulate the temperature and the rate of heating or cooling. It should be remembered, however, that it is the temperature of the workpiece, not the temperature of the furnace, that controls the result, and it is this temperature that should be monitored.

■ 5.8 HEAT TREATMENT AND ENERGY

Because of the elevated temperatures and the time required at those temperatures, heat treatments can consume considerable amounts of energy. However, if one considers the broader picture, heat treatment may actually prove to be an energy conservation measure. The manufacture of higher-quality, more durable products can often eliminate the need for frequent replacements. Higher strengths may also permit the use of less material in the manufacture of a product, thereby saving additional energy.

Further savings can often be obtained by integrating the manufacturing operations. For example, a direct quench and temper from hot forging may be used to replace the conventional sequence of forge, conventional air cool, reheat, quench, and temper. One should note, however, that the integrated procedure quenches from the conditions of forging, which generally have greater variability in temperature, uniformity of that temperature, and austenite grain size. If these variations are too great, the additional energy for the reheat and soak may be well justified.

Heat treatment, a business worth $15–20 billion a year in the United States, impacts nearly every industrial market sector. It is both capital intensive (specialized and dedicated equipment) and energy intensive. Industry goals currently include reducing energy consumption, reducing processing times, reducing emissions, increasing furnace life, improving heat transfer during heating and cooling, reducing distortion, and improving uniformity of structure and properties, both within a given part and throughout an entire production quantity.

■ Key Words

A_1	dispersion hardening	laser-beam hardening	recrystallization
A_3	electron-beam hardening	martempering	residual stresses
A_{cm}	equilibrium phase diagram	martensite	retained austenite
age hardening	flame hardening	natural aging	solid-solution strengthening
aging	fluidized-bed furnaces	nitriding	solution treatment
annealing	full anneal	normalize	spheroidization
artificial aging	grain size refinement	overaged	strain hardening
artificial gas atmospheres	hardenability	pearlite	stress-relief anneal
ausforming	hardness	phase transformation	substitutional solution
austempering	heat treatment	strengthening	surface hardening
autoquenching	homogenization	polymer quench	synthetic quenchant
bainite	induction hardening	precipitation hardening	T-T-T diagram
batch furnaces	interstitial solution	process anneal	tempered martensite
C-C-T diagram	ionitriding	processing heat treatments	tempering
carburizing	isothermal anneal	quench and temper	thermomechanical processing
coherency	isothermal-transformation	quench cracking	vapor jacket
continuous furnaces	diagram	quenchant	
critical cooling rate	Jominy test	quenching	

■ Review Questions

1. What is heat treatment?
2. What types of properties can be altered through heat treatment?
3. Why should people performing hot forming or welding be aware of the effects of heat treatment?
4. What is the broad goal of the processing heat treatments? Cite some of the specific objectives that may be sought.
5. Why might equilibrium phase diagrams be useful aids in designing and understanding the processing heat treatments?
6. What are the A_1, A_3, and A_{cm} lines?
7. What are some possible objectives of annealing operations?
8. While full anneals often produce the softest and most ductile structures, what may be some of the objections or undesirable features of these treatments?
9. Why are the hypereutectoid steels not furnace cooled from the all-austenite region?
10. What is the major process difference between full annealing and normalizing?
11. While normalizing is less expensive than a full anneal, some manufacturers cite cost saving through the use of a full anneal. How is this achieved?
12. What are some of the process heat treatments that can be performed without reaustenitizing the material (heating above the A_1 temperature)?
13. What types of steel would be candidates for a process anneal? Spheroidization?

14. How might steel composition influence the selection of a processing heat treatment?
15. Other than increasing strength, for what three purposes are nonferrous metals often heat treated?
16. What are the six major mechanisms that can be used to increase the strength of a metal?
17. What is the most effective strengthening mechanism for the nonferrous metals?
18. What are the three steps in an age-hardening treatment?
19. What is the difference between a coherent precipitate and a distinct second-phase particle? Why does coherency offer significant strengthening?
20. What is overaging?
21. What is the difference between natural and artificial aging? Which offers more flexibility? Over which does the engineer have more control?
22. Why is it more difficult to understand the nonequilibrium strengthening treatments?
23. What types of heating and cooling conditions are imposed in an I-T or T-T-T diagram? Are they realistic for the processing of commercial items?
24. What are the stable equilibrium phases for steels at temperatures below the A_1 temperature?
25. What are some nonequilibrium structures that appear in the T-T-T diagram for a eutectoid composition steel?
26. Which steel structure is produced by a diffusionless phase change?
27. What is the major factor that influences the strength and hardness of martensite?
28. Most structure changes proceed to completion over time. The martensite transformation is different. What must be done to produce more martensite in a partially transformed structure?
29. Why is retained austenite an undesirable structure in heat-treated steels?
30. What types of steels are more prone to retained austenite?
31. Why are martensitic structures usually tempered before being put into use? What properties increase during tempering? Which ones decrease?
32. In what ways is the quench-and-temper heat treatment similar to age hardening? How are the property changes different in the two processes?
33. What is a C-C-T diagram? Why is it more useful than a T-T-T diagram?
34. What two features combine to determine the structure and properties of a heat-treated steel?
35. How do the various locations of a Jominy test specimen correlate with cooling rate?
36. What conditions are used to standardize the quench in the Jominy test?
37. What is the assumption that allows the data from a Jominy test to be used to predict the properties of various locations on a manufactured product?
38. What is hardenability? What capabilities are provided by high-hardenability materials?
39. What are the three stages of liquid quenching?
40. What are some of the major advantages and disadvantages of a water quench?
41. Why is an oil quench less likely to produce quench cracks than water or brine?
42. What are some of the attractive qualities of a polymer or synthetic quench?
43. What are some undesirable design features that may be present in parts that are to be heat treated?
44. Why would the residual stresses in steel be different from the residual stresses in an identically processed aluminum part?
45. What are some of the potentially undesirable effects of residual stresses?
46. What causes quench cracking to occur when steel is rapidly cooled?
47. Describe several techniques that utilize simultaneous transformation to reduce residual stresses in steel products.
48. What is thermomechanical processing?
49. What are some of the methods that can be used to selectively alter the surface properties of metal parts?
50. What are some of the attractive features of surface hardening with a laser beam?
51. How can a laser beam be manipulated and focused? How are these operations performed with an electron beam?
52. What is carburizing?
53. Why does a carburized part have to be further heat treated after the carbon is diffused into the surface? What are the various options?
54. In what ways might ionitriding be more attractive than conventional nitriding or carburizing?
55. For what type of products or product mixes might a batch furnace be preferred to a continuous furnace?
56. What are some possible functions of artificial atmospheres in a heat-treating furnace?
57. How are parts heated in a fluidized-bed furnace? What are some of the attractive features?
58. In what ways might the heat treatment of metals actually be an energy conservation measure?

■ Problems

1. A number of heat treatments have been devised to harden the surfaces of steel and other engineering metals. Consider the following processes:
   a. Flame hardening
   b. Induction hardening
   c. Laser-beam hardening
   d. Carburizing
   e. Nitriding
   f. Ionitriding
   For each of these processes, provide information relating to:
   (1) A basic description of how the process works
   (2) Typical materials on which the process is performed
   (3) Type of equipment required
   (4) Typical times, temperatures, and atmospheres required
   (5) Typical depth of hardening and reasonable limits
   (6) Hardness achievable
   (7) Subsequent treatments or processes that might be required
   (8) Information relating to distortion and/or stresses
   (9) Ability to use the process to harden selective areas
2. Investigate the nine areas in Problem 1 for one of the lesser known surface-modification treatments, such as ion implantation, boriding, chromizing, or other similar techniques.
3. This chapter presented four processing-type heat treatments whose primary objective is to soften, weaken, enhance

ductility, or promote machinability. Consider each of the following processes as they are applied to steels:

a. Full annealing
b. Normalizing
c. Process annealing
d. Spheroidizing

Provide information relating to:

1. A basic description of how the process works and what its primary objectives are
2. Typical materials on which the process is performed
3. Type of equipment used
4. Typical times, temperatures, and atmospheres required
5. Recommended rates of heating and cooling
6. Typical properties achieved

4. A number of different quenchants were discussed in the chapter, including brine, water, oil, synthetic polymer mixes, and even high-pressure gas flow. Select two of these and investigate the environmental concerns that may accompany their use.

5. Traditional manufacturing generally separates mechanical processing (such as forging, extrusion, presswork, or machining) and thermal processing (heat treatment), and applies them as sequential operations. Ausforming was presented as an example of a thermomechanical process where the mechanical and thermal processes are performed concurrently. When this is done, the resulting structures and properties are often quite different from the traditional. Identify another thermomechanical process and discuss its use and attributes.

6. It has been noted that *hot oil* is often a more effective quench than *cold oil*. Can you explain this apparent contradiction?

www.wiley.com/college/degarmo

Chapter 5 CASE STUDY

A Carpenter's Claw Hammer

Carpenter claw hammers are actually a rather sophisticated metallurgical product, since the loadings differ for the various locations. The claw sees static bending, while the eye sustains impacts, and the striking face sees impact contact with potentially hard surfaces.

As one might expect, the optimum properties and microstructures vary with location. While hammer handles have been made from a variety of materials, including heat-treated tubular 4140 steel, our problem will focus on the head.

The following information was obtained from the American National Standards Institute (ANSI) Standard B173.1, "American National Standard Safety Requirements for Nail Hammers." This is a voluntary specification (recommendation only) developed as a "guide to aid the manufacturer, the consumer, and the general public."

According to the ANSI specification:

"Hammerheads shall be forged in one piece from special quality hot rolled carbon steel bars." While the specification allows for steels ranging from 1045 to 1088, two major manufacturers of high-quality tools have used 1078 steel as their material of choice, so we will go with their selection.

The hammer striking face "shall be hardened and tempered to a Rockwell hardness of not less than C 40 or more than C 60, and the steel directly behind the striking face shall be a toughened supporting core gradually decreasing in hardness."

"Hammer claws shall be hardened to a Rockwell hardness of not less than C 40 or more than C 55 for a minimum length of $\frac{3}{4}$-inch from the tip end; the remaining length to the base of the V-slot shall be of the same through hardness, or shall contain a toughened core gradually decreasing in hardness to the core center."

While there is no specification for the eye region, many manufacturers prefer for this area to have the greatest toughness (i.e., even softer still—as low as R_C 25!).

In essence, we are looking at a single piece of heat-treated steel that preferably exhibits different properties at different locations. For example, one top-quality hammer has a striking face of R_C 55 to 58, coupled with a claw of R_C 46 to 48. Another top-quality hammer has a striking face hardness of R_C 50 to 58, claw tip hardness of R_C 47

to 55, and a hardness in the crotch of the V of R_C 44 to 52. The rim of the striking face is softened to a lower hardness (R_C 41 to 48) to prevent chipping—a characteristic feature of this particular manufacturer.

Fixing our material as the above-used 1078 hot-rolled steel bar, and using forging as our shaping process:

1. What problems might be expected if the material on the striking face were too hard? Too soft? Consider each with respect to possible liability.
2. Describe some heat-treatment processes or sequences that could be used to produce a quality product like those described above.
3. Discuss the methods of heating, cooling or quenching, target temperatures, and so on that you are proposing to accomplish this task.
4. Finally, how might you duplicate the rim softening being achieved by the cited manufacturer?
5. Inexpensive hammers frequently use a single material and single heat treatment, rendering the properties similar for all locations. What are the major compromises? If these hammers were to be used by a professional carpenter, how might they be deficient?

CHAPTER 6

FERROUS METALS AND ALLOYS

■ 6.1 INTRODUCTION TO HISTORY-DEPENDENT MATERIALS

Engineering materials are available with a wide range of useful properties and characteristics. Some of these are inherent to the particular material, but many others can be varied by controlling the manner of production and the details of processing. Metals are classic examples of such "history-dependent" materials. Their final properties are clearly affected by their past processing history. The particular details of the smelting and refining process control the resulting purity and the type and nature of any influential contaminants. The solidification process imparts structural features that may be transmitted to the final product. Preliminary operations such as the rolling of sheet or plate often impart directional variations to properties, and their impact should be considered during subsequent processing and use. Thus, while it is easy to take the attitude that "metals come from warehouses," it is important to recognize that aspects of prior processing can significantly influence further operations as well as the final properties of the product. The breadth of this book does not permit full coverage of the processes and methods involved in the production of engineering metals, but certain aspects will be presented because of their role in affecting subsequent performance.

■ 6.2 FERROUS METALS

In this chapter we will introduce the major *ferrous* (iron-based) *metals* and *alloys*, summarized in Figure 6-1. These materials made possible the Industrial Revolution and have been the backbone of modern civilization. We see them everywhere in our lives—in the cars we drive, the homes in which we live, the cans we open, and the appliances that enhance our standard of living. Numerous varieties have been developed over the years to meet the specific needs of various industries. The developments and improvements have continued, with recent decades seeing the introduction of a number of new varieties and even classes of ferrous metals. According to the American Iron and Steel Institute, over 50% of the steels made today did not exist 10 years ago and over 70% of the steel used in automotive production meets this criteria. The newer steels are stronger than ever, easier to shape, and more corrosion resistant.

In addition, all steel is recyclable, and this recycling does not involve any loss in material quality. In fact, more steel is recycled each year than all other materials combined, including aluminum, glass, and paper. Because steel is magnetic, it is easily separated and recovered from demolished buildings, junked automobiles, and discarded

FIGURE 6-1 Classification of common ferrous metals and alloys.

appliances. In 2003, nearly 70 million tons of steel were recycled in the United States, for an overall recycling rate of nearly 71%. The recycling rate for steel cans was 60.2%, 89.7% for large appliances, and 102.9% for automobiles. That's correct—more steel was recovered from scrap cars than was used in the production of new vehicles! Each ton of recycled steel saves over 4000 pounds of raw materials and 74% of the energy required to make a ton of new steel.

■ 6.3 IRON

For centuries, *iron* has been the most important of the engineering metals. While iron is the fourth most plentiful element in the earth's crust, it is rarely found in the metallic state. Instead, it occurs in a variety of mineral compounds, known as ores, the most attractive of which are iron oxides coupled with companion impurities. To produce metallic iron, the ores are processed in a manner that breaks the iron–oxygen bonds (chemical reducing reactions). Ore, limestone, coke (carbon), and air are continuously introduced into specifically designed furnaces and molten metal is periodically withdrawn.

Within the furnace, other oxides (which were impurities in the original ore) will also be reduced. All of the phosphorus and most of the manganese will enter the molten iron. Oxides of silicon and sulfur compounds are partially reduced, and these elements also become part of the resulting metal. Other contaminant elements, such as calcium, magnesium, and aluminum, are collected in the limestone-based slag and are largely removed from the system. The resulting *pig iron* tends to have roughly the following composition:

Carbon	3.0–4.5%
Manganese	0.15–2.5%
Phosphorus	0.1–2.0%
Silicon	1.0–3.0%
Sulfur	0.05–0.1%

A small portion of this iron is cast directly into final shape and is classified as cast iron. Most commercial cast irons, however, are produced by recycling scrap iron and steel, with the possible addition of some newly produced pig iron. The metallurgical properties of cast iron have been presented in Chapter 4, and its melting and utilization in the casting process will be developed in Chapters 11 through 13. Most pig iron, however, is further processed into steel.

■ 6.4 STEEL

Steel is an extremely useful engineering material. It offers strength, rigidity, and durability. From a manufacturing perspective, its formability, joinability, and paintability, as well as repairability, are all attractive. For the past 20 years, steel has accounted for about 55% of the weight of a typical passenger car and is expected to continue at this level. While the automotive and construction industries are major consumers of steel, the material is also used extensively in containers, appliances, and machinery as well as the infrastructure of such industries as oil and gas.

The manufacture of *steel* is essentially an oxidation process that decreases the amount of carbon, silicon, manganese, phosphorus, and sulfur in a molten mixture of pig iron and/or steel scrap. In 1856, the Kelly–Bessemer process opened up the industry by enabling the manufacture of commercial quantities of steel. The open-hearth process surpassed the Bessemer process in tonnage produced in 1908 and was producing over 90% of all steel in 1960. Currently, most of our commercial steels are produced by a variety of oxygen and electric-arc furnaces.

In many of the current processes, air or oxygen passes over or through the molten metal to drive a variety of exothermic refining reactions. Carbon oxidizes to form gaseous CO or CO_2, which then exits the melt. Other elements, such as silicon and phosphorus, are similarly oxidized and, being lighter than the metal, rise to be collected in a removable slag. At the same time, oxygen and other elements from the reaction gases dissolve in the molten metal and may later become a cause for concern.

SOLIDIFICATION CONCERNS

Regardless of the method by which the steel is made, it must undergo a change from liquid to solid before it can become a usable product. The liquid can be converted directly into finish-shape steel castings or solidified into a form suitable for further processing. In most cases, some form of continuous casting produces the feedstock material for subsequent forging or rolling operations.

Prior to solidification, we want to remove as much contamination as possible. The molten metal is poured from the steelmaking furnaces into containment vessels, known as *ladles*. Historically, the ladles simply served as transfer and pouring containers, but they have recently emerged as the site for additional processing. *Ladle metallurgy* refers to a variety of processes designed to provide final purification and to fine-tune both the chemistry and temperature of the melt. Alloy additions can be made; carbon can be further reduced; dissolved gases can be reduced or removed; and steps can be taken to control subsequent grain size, limit inclusion content, reduce sulfur, and control the shape of any included sulfides. Stirring, degassing, reheating, and the injection of powdered alloys or cored wire can all be performed to increase the cleanliness of the steel and provide for tighter control of the chemistry and properties.

The processed liquid is then poured from these ladles into molds or some form of *continuous caster*, usually through a bottom-pouring process such as the one shown schematically in Figure 6-2. By extracting the metal from the bottom of the ladle, slag and floating matter are not transferred, and a cleaner product results. Figure 6-3a illus-

Refractory sleeves

Lever for pouring

Graphite stopper

Graphite pouring hole

FIGURE 6-2   Diagram of a bottom-pouring ladle.

FIGURE 6-3   (a) Schematic representation of the continuous casting process for producing billets, slabs, and bars. (b) Simultaneous continuous casting of multiple strands. *(a) (Courtesy of Materials Engineering, Penton Publishing, New York, NY.)*

trates a typical continuous caster, in which molten metal flows from a ladle, through a tundish, and into a bottomless, water-cooled mold, usually made of copper. Cooling is controlled so that the outside has solidified before the metal exits the mold. Direct water sprays further cool the emerging metal to complete the solidification. The solid metal can then be cut to desired length, or, since the cast solid is still hot, it can be bent and fed horizontally through a short reheat furnace or directly to a rolling operation. If the size and shape of the mold are varied, products can be cast with a variety of cross sections with names such as slab, bloom, billet, and strand. Figure 6-3b depicts the simultaneous casting of multiple strands. Compared to the casting of discrete ingots, continuous casting offers significant reduction in cost, energy, and scrap. In addition, the products have improved surfaces, more uniform chemical composition, and fewer oxide inclusions.

DEOXIDATION AND DEGASSIFICATION

As a result of the steelmaking process, large amounts of oxygen can be dissolved in the molten metal. During the subsequent cooling and solidification, the solubility levels decrease significantly, as shown in Figure 6-4, and the oxygen and other gases are rejected. The rejected oxygen frequently links with carbon to produce carbon monoxide

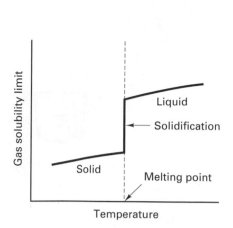

FIGURE 6-4 Solubility of gas in a metal as a function of temperature showing significant decrease upon solidification.

FIGURE 6-5 Method of degassing steel by pouring through a vacuum.

gas, which may escape through the liquid or become trapped to produce a porous solid. The bubble-induced porosity may take various forms ranging from small, dispersed voids to large blowholes. While these pores can often be welded shut during subsequent hot forming, some may not be fully closed, and others may not weld upon closure. Cracks and internal voids can persist into a finished product.

Porosity problems can often be avoided by either removing the oxygen prior to solidification or by making sure it does not reemerge as a gas. Aluminum, ferromanganese, or ferrosilicon can be added to molten steel to provide a material whose affinity for oxygen is higher than that of carbon. The rejected oxygen then reacts with these *deoxidizer* additions to produce solid metal oxides that are either removed from the molten metal or become dispersed throughout the structure.

While deoxidizer additions can effectively tie up dissolved oxygen, small amounts of other gases, such as hydrogen and nitrogen, can also have deleterious effects on the performance of steels. This is particularly important for alloy steels because the solubility of these gases tends to be increased by alloy additions, such as vanadium, niobium, and chromium. Alternative degassing processes have been devised that reduce the amounts of all dissolved gases. Figure 6-5 illustrates one form of *vacuum degassing*, in which an ingot mold is placed in an evacuated chamber, and a stream of molten metal passes through a vacuum during pouring. Because a large amount of exposed surface is created during the pouring operation, the vacuum is able to extract most of the dissolved gas.

An alternative to vacuum degassing is the *consumable-electrode remelting* process, where an already-solidified metal electrode replaces the ladle of molten metal. As the electrode is progressively remelted, molten droplets pass through a vacuum, and the extremely high surface area again provides an effective means of gas removal. If the melting is done by an electric arc, the process is known as *vacuum arc remelting* (VAR). If induction heating is used, the process becomes *vacuum induction melting* (VIM). Both are highly effective in removing dissolved gases, but they fail to remove any nonmetallic impurities that may be present in the metal.

The *electroslag remelting* process (ESR), shown in Figure 6-6, can be used to produce extremely clean, gas-free metal. A solid electrode is again melted and recast using an electric current, but the entire remelting is conducted under a blanket of molten flux. Nonmetallic impurities float and are collected in the flux, leaving a newly solidified metal structure with greatly improved quality. No vacuum is required, since the molten material is confined beneath the flux and the progressive freezing permits easy escape for the rejected gas. This process is simply a large-scale version of the electroslag welding process that will be discussed in Chapter 34.

(a)

Electrode clamp

Stinger

Electrode

Transformer

Cooling water in

Movable mold

Slag pool

Molten metal pool
Ingot
Cooling water out

Starting plate

Water cooling

(b)

FIGURE 6-6   (a) Production of an ingot by the electroslag remelting process; (b) schematic representation of this process showing the starting electrode, melting arc, and resolidified ingot. *(Courtesy of Carpenter Technology Corporation, Reading, PA.)*

PLAIN-CARBON STEEL

While theoretically an alloy of only iron and carbon, commercial steel actually contains manganese, phosphorus, sulfur, and silicon in significant and detectable amounts. When these four additional elements are present in their normal percentages and no minimum amount is specified for any other constituent, the product is referred to as *plain-carbon steel*. Strength is primarily a function of carbon content, increasing with increasing carbon, as shown in Table 6-1. Unfortunately, the ductility, toughness, and weldability of plain-carbon steels decrease as the carbon content is increased, and hardenability is quite low. In addition, the properties of ordinary carbon steels are impaired by both high and low temperatures (loss of strength and embrittlement, respectively), and they are subject to corrosion in most environments.

Plain-carbon steels are generally classed into three subgroups based on their carbon content. *Low-carbon steels* have less than 0.20% carbon and possess good formability (can be strengthened by cold work) and weldability. Their structures are usually ferrite and pearlite, and the material is generally used as it comes from the hot-forming or cold-forming processes, or in the as-welded condition. *Medium-carbon steels* have between 0.20 and 0.50% carbon, and they can be quenched to form martensite or bainite if the section size is small and a severe water or brine quench is used. The best balance of properties is obtained at these carbon levels, where the high toughness and ductility of the low-carbon material is in good compromise with the strength and hardness that come with higher carbon contents. These steels are extremely popular and find numerous mechanical applications. *High-carbon steels* have more than 0.50% carbon. Toughness and formability are quite low, but hardness and wear resistance are high. Severe quenches can form martensite, but hardenability is still poor. Quench cracking is often a problem when the material is pushed to its limit. Figure 6-7 depicts the characteristic properties of low-, medium-, and high-carbon steels using a

TABLE 6-1	Effect of Carbon on the Strength of Annealed Pain-Carbon Steels[a]		
		Minimum Tensile Strength	
Type of Steel	Carbon Content	Mpa	ksi
1020	0.20%	414	60
1030	0.30%	448	65
1040	0.40%	517	75
1050	0.50%	621	90

[a] Data are from ASTM Specification A732.

FIGURE 6-7 A comparison of low-carbon, medium-carbon, and high-carbon steels in terms of their relative balance of properties. (a) Low-carbon steel has excellent ductility and fracture resistance, but lower strength; (b) medium-carbon steel has balanced properties; (c) high-carbon steel has high strength and hardness at the expense of ductility and fracture resistance.

balance of properties that shows the offsetting characteristics of "strength and hardness" and "ductility and toughness."

Compared to other engineering materials, the carbon steels offer high strength and high stiffness, coupled with reasonable toughness. Unfortunately, they also rust easily and generally require some form of surface protection, such as paint, galvanizing, or other coating. The plain-carbon steels are generally the lowest-cost steel material and should be given first consideration for many applications. Their limitations, however, may become restrictive. When improved performance is required, these steels can often be upgraded by the addition of one or more alloying elements.

ALLOY STEELS

The differentiation between plain-carbon and alloy steel is often somewhat arbitrary. Both contain carbon, manganese, and usually silicon. Copper and boron are possible additions to both classes. Steels containing more than 1.65% manganese, 0.60% silicon, or 0.60% copper are usually designated as *alloy steels*. Also, a steel is considered to be an alloy steel if a definite or minimum amount of other alloying element is specified. The most common alloy elements are chromium, nickel, molybdenum, vanadium, tungsten, cobalt, boron, and copper, as well as manganese, silicon, phosphorus, and sulfur in amounts greater than are normally present. If the steel contains less than 8% of total alloy addition, it is considered to be a *low-alloy steel*. Steels with more than 8% alloying elements are *high-alloy steels*.

In general, alloying elements are added to steels in small percentages (usually less than 5%) to improve strength or hardenability, or in much larger amounts (often up to 20%) to produce special properties such as corrosion resistance or stability at high or low temperatures. Additions of manganese, silicon, or aluminum may be made during the steelmaking process to remove dissolved oxygen from the melt. Manganese, silicon, nickel, and copper add strength by forming solid solutions in ferrite. Chromium, vanadium, molybdenum, tungsten, and other elements increase strength by forming dispersed second-phase carbides. Nickel and copper can be added in small amounts to improve corrosion resistance. Nickel has been shown to impart increased toughness and impact resistance, and molybdenum helps resist embrittlement. Zirconium, cerium, and calcium can also promote increased toughness by controlling the shape of inclusions. Machinability can be enhanced through the formation of manganese sulfides or by additions of lead, bismuth, selenium, or tellurium. Still other additions can be used to provide ferrite or austenite grain size control.

Selection of an alloy steel still begins with identifying the proper carbon content. Table 6-2 shows the effect of carbon on the strength of quenched-and-tempered alloy steels. The strength values are significantly higher than those of Table 6-1, reflecting the difference between the annealed and quenched-and-tempered microstructures. The 4130 steel has about 1.2% total alloying elements, 4330 has 3.0%, and 8630 has about 1.3%, yet all have the same quenched-and-tempered tensile strength. Strength and hardness depend primarily on carbon content. The primary

TABLE 6-2	Effect of Carbon on the Strength of Quenched-and-Tempered Alloy Steels[a]		

Type of Steel	Carbon Content	Minimum Tensile Strength	
		Mpa	ksi
4130	0.30%	1030	150
4330	0.30%	1030	150
8630	0.30%	1030	150
4140	0.40%	1241	180
4340	0.40%	1241	180

[a] Data from ASTM Specification A732.

TABLE 6-3 Principal Effects of Major Alloying Elements in Steel

Element	Percentage	Primary Function
Aluminum	0.95–1.30	Alloying element in nitriding steels
Bismuth	—	Improves machinability
Boron	0.001–0.003	Powerful hardenability agent
Chromium	0.5–2	Increase of hardenability
	4–18	Corrosion resistance
Copper	0.1–0.4	Corrosion resistance
Lead	—	Improved machinability
Manganese	0.25–0.40	Combines with sulfur to prevent brittleness
	>1	Increases hardenability by lowering transformation points and causing transformations to be sluggish
Molybdenum	0.2–5	Stable carbides; inhibits grain growth
Nickel	2–5	Toughener
	12–20	Corrosion resistance
Silicon	0.2–0.7	Increases strength
	2	Spring steels
	Higher percentages	Improves magnetic properties
Sulfur	0.08–0.15	Free-machining properties
Titanium	—	Fixes carbon in inert particles
		Reduces martensitic hardness in chromium steels
Tungsten	—	Hardness at high temperatures
Vanadium	0.15	Stable carbides; increases strength while retaining ductility, Promotes fine grain structure

role of an alloy addition is usually to increase *hardenability*, but other effects are also possible, such as modified toughness or machinability. The most common hardenability-enhancing elements (in order of decreasing effectiveness) are manganese, molybdenum, chromium, silicon, and nickel. Boron is an extremely powerful hardenability agent. Only a few thousandths of a percent are sufficient to produce a significant effect in low-carbon steels, but the results diminish rapidly with increasing carbon content. Since no carbide formation or ferrite strengthening accompanies the addition, improved machinability and cold-forming characteristics may favor the use of boron in place of other hardenability additions. Small amounts of vanadium can also be quite effective, but the response drops off as the quantity is increased.

Table 6-3 summarizes the primary effects of the common alloying elements in steel. A working knowledge of this information may be useful in selecting an alloy steel to meet a given set of requirements. Alloying elements are often used in combination, however, resulting in the immense variety of alloy steels that are commercially available. To provide some degree of simplification, a classification system has been developed and has achieved general acceptance in a variety of industries.

AISI–SAE CLASSIFICATION SYSTEM
The most common classification scheme for alloy steels is the *AISI–SAE identification system*. This system, which classifies alloys by chemistry, was started by the Society of Automotive Engineers (SAE) to provide some standardization for the steels used in the automotive industry. It was later adopted and expanded by the American Iron and Steel Institute (AISI) and has been incorporated into the Universal Numbering System that was developed to include all engineering metals. Both plain-carbon and low-alloy steels are identified by a four-digit number, where the first number indicates the major alloying elements and the second number designates a subgrouping within the major alloy system. These first two digits can be interpreted by looking them up on a list, such as the one presented in Table 6-4. The last two digits of the number indicate the approximate amount of carbon, expressed as "points," where one point is equal to 0.01%. Thus, a 1080 steel would be a plain-carbon steel with 0.80% carbon. Similarly, a 4340 steel

TABLE 6-4	AISI–SAE Standard Steel Designations and Associated Chemistries						

AISI Number	Type	Alloying Elements (%)					
		Mn	Ni	Cr	Mo	V	Other
1xxx	Carbon steels						
10xx	Plain carbon						
11xx	Free cutting (S)						
12xx	Free cutting (S) and (P)						
15xx	High manganese						
13xx	High manganese	1.60–1.90					
2xxx	Nickel steels		3.5–5.0				
3xxx	Nockel–chromium		1.0–3.5	0.5–1.75			
4xxx	Molybdenum						
40xx	Mo				0.15–0.30		
41xx	Mo, Cr			0.40–1.10	0.08–0.35		
43xx	Mo, Cr, Ni		1.65–2.00	0.40–0.90	0.20–0.30		
44xx	Mo				0.35–0.60		
46xx	Mo, Ni (low)		0.70–2.00		0.15–0.30		
47xx	Mo, Cr, Ni		0.90–1.20	0.35–0.55	0.15–0.40		
48xx	Mo, Ni (high)		3.25–3.75		0.20–0.30		
5xxx	Chromium						
50xx				0.20–0.60			
51xx				0.70–1.15			
6xxx	Chromum–vanadium						
61xx				0.50–1.10		0.10–0.15	
8xxx	Ni, Cr, Mo						
81xx			0.20–0.40	0.30–0.55	0.08–0.15		
86xx			0.40–0.70	0.40–0.60	0.15–0.25		
87xx			0.40–0.70	0.40–0.60	0.20–0.30		
88xx			0.40–0.70	0.40–0.60	0.30–0.40		
9xxx	Other						
92xx	High silicon						1.20–2.20Si
93xx	Ni, Cr, Mo		3.00–3.50	1.00–1.40	0.08–0.15		
94xx	Ni, Cr, Mo		0.30–0.60	0.30–0.50	0.08–0.15		

would be a Mo–Cr–Ni alloy with 0.40% carbon. Because of the double-digit groupings, these steels are identified as a "ten eighty" and a "forty-three forty."

Letters may also be incorporated into the designation. The letter *B* between the second and third digits indicates that the base metal has been supplemented by the addition of boron. Similarly, an *L* in this position indicates a lead addition for enhanced machinability. A letter prefix may also be employed to designate the process used to produce the steel, such as *E* for electric furnace.

When hardenability is a major requirement, one might consider the H grades of AISI steels, designated by an *H* suffix attached to the standard designation. The chemistry specifications are somewhat less stringent, but the steel must now meet a hardenability standard. The hardness values obtained for each point of a Jominy test specimen (see Chapter 5) must lie within a predetermined band for that particular type of steel.

Other designation organizations, such as the American Society for Testing and Materials (ASTM) and the U.S. government ("Military" specifications and federal), have specification systems based more on specific applications. Acceptance into a given classification is generally determined by physical or mechanical properties rather than the chemistry of the metal. ASTM designations are often used when specifying structural steels.

SELECTING ALLOY STEELS

From the previous discussion it is apparent that two or more alloying elements can often produce similar effects. Thus, when properly heat treated, steels with substantially different chemical compositions can possess almost identical mechanical properties. Figure 6-8 clearly demonstrates this fact, which becomes particularly important when one realizes that some alloying elements can be very costly and others may be in short supply due to emergencies or political constraints. Overspecification has often been employed to guarantee success despite sloppy manufacturing and heat-treatment practice. The correct steel, however, is usually the least expensive one that can be consistently processed to achieve the desired properties. This usually involves taking advantage of the effects provided by all of the alloy elements.

When selecting alloy steels, it is also important to consider both use and fabrication. For one product, it might be permissible to increase the carbon content to obtain greater strength. For another application, such as one involving assembly by welding, it might be best to keep the carbon content low and use a balanced amount of alloy elements, obtaining the desired strength while minimizing the risk of weld cracking. Steel selection involves defining the required properties, determining the best microstructure to provide those properties (strength can be achieved through alloying, cold work, and heat treatment, as well as combinations thereof), determining the method of part or product manufacture (casting, machining, metalforming, etc.), and selecting the steel with the best carbon content and hardenability characteristics to achieve those goals.

HIGH-STRENGTH LOW-ALLOY STRUCTURAL STEELS

Among the general categories of alloy steels are: (1) the *constructional alloys*, where the desired properties are typically developed by a separate thermal treatment and the specific alloy elements tend to be selected for their effect on hardenability, and (2) the *high-strength low-alloy* (HSLA) or microalloyed types, which rely largely on chemical composition to develop the desired properties in the as-rolled or normalized condition. The constructional alloys are usually purchased by AISI–SAE identification, which

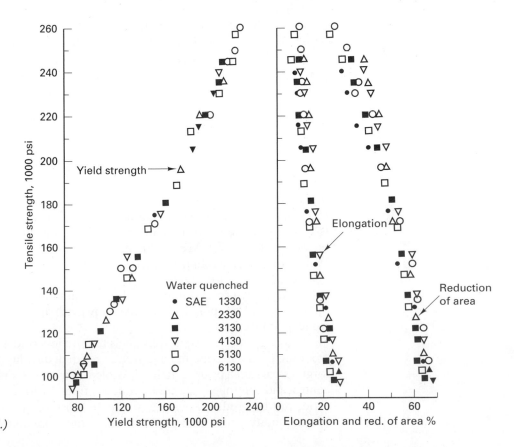

FIGURE 6-8 Relationships between the mechanical properties of a variety of properly heat-treated AISI–SAE alloy steels. *(Courtesy of ASM International, Materials Park, OH.)*

TABLE 6-5 Typical Compositions and Strength Properties of Several Groups of High-Strength Low-Alloy Structural Steels

Group	C	Mn	Si	Cb	V	Yield ksi	Yield MPa	Tensile ksi	Tensile MPa	Elongation in 2 in. (%)
Columbium or vanadium	0.20	1.25	0.30	0.01	0.01	55	379	70	483	20
Low manganese–vanadium	0.10	0.50	0.10		0.02	40	276	60	414	35
Manganese–copper	0.25	1.20	0.30			50	345	75	517	20
Manganese–vanadium–copper	0.22	1.25	0.30		0.02	50	345	70	483	22

[a] All have 0.04% P, 0.05% S, and 0.20% Cu.

effectively specifies chemistry. The HSLA materials generally focus on product (size and shape) and desired properties. When steels are specified by mechanical properties, the supplier or producer is free to adjust the chemistry (within limits), and substantial cost savings may result. To assure success, however, it is important that all of the necessary properties be specified.

The HSLA materials provide increased strength-to-weight compared to conventional carbon steels for only a modest increase in cost. They are available in a variety of forms, including sheet, strip, plate, structural shapes, and bars. The dominant property requirements generally are high yield strength, good weldability, and acceptable corrosion resistance. Ductility and hardenability may be somewhat limited, however. The increase in strength, and the resistance to martensite formation in a weld zone, is obtained by controlling the amounts of carbon, manganese, and silicon, with the addition of small amounts of niobium, vanadium, titanium, or other alloys. About 0.2% copper can be added to improve corrosion resistance.

Because of their higher yield strength, weight savings of 20 to 30% can often be achieved with no sacrifice to strength or safety. Rolled and welded HSLA steels are being used in automobiles, trains, bridges, and buildings. Because of their low alloy content and high-volume application, their cost is often little more than that of the ordinary plain-carbon steels. Table 6-5 presents the chemistries and properties of several of the more common types.

MICROALLOYED STEELS IN MANUFACTURED PRODUCTS

In terms of both cost and performance, *microalloyed steels* occupy a position between carbon steels and the alloy grades, and they are being used increasingly as substitutes for heat-treated steels in the manufacture of small- to medium-sized discrete parts. These low- and medium-carbon steels contain small amounts (0.05 to 0.15%) of alloying elements, such as niobium, vanadium, titanium, molybdenum, zirconium, boron, rare earth elements, or combinations thereof. The primary effect of the alloy addition is to provide grain refinement and/or precipitation strengthening. Yield strengths between 500 and 750 MPa (70 and 110 ksi) can be obtained without heat treatment. Weldability can be retained or even improved if the carbon content is simultaneously decreased. In essence, these steels offer maximum strength with minimum carbon, while simultaneously preserving weldability, machinability, and formability. Compared to a quenched-and-tempered alternative, however, ductility and toughness are generally somewhat inferior.

Cold-formed microalloyed steels require less cold work to achieve a desired level of strength, so they tend to have greater residual ductility. Hot-formed products, such as forgings, can often be used in the air-cooled condition. By means of accurate temperature control and controlled-rate cooling directly from the forming operation, mechanical properties can be produced that approximate those of quenched-and-tempered material. Machinability can be enhanced because of the more uniform hardness and the fact that the ferrite–pearlite structure of the microalloyed steel is often more machinable than the ferrite–carbide structure of the quenched-and-tempered variety. Fatigue life and wear resistance can also be superior to those of the heat-treated counterparts.

In applications where the properties are adequate, microalloyed steels can often provide attractive cost savings. Energy savings can be substantial, straightening or stress relieving after heat treatment is no longer necessary, and quench cracking is not a problem. Due to the increase in material strength, the size and weight of finished products can often be reduced. As a result, the cost of a finished forging could be reduced by 5 to 25%.

If these materials are to attain their optimum properties, certain precautions must be observed. During the elevated-temperature segments of processing, the material must be heated high enough to place all of the alloys into solution. After forming, the products should be rapidly air cooled to 540° to 600°C (1000° to 1100°F) before dropping into collector boxes. In addition, microalloyed steels tend to through-harden upon air cooling, so products fail to exhibit the lower-strength, higher-toughness interiors that are typical of the quenched-and-tempered materials.

BAKE-HARDENABLE STEEL SHEET

Bake-hardenable steel has assumed a significant role in automotive sheet applications. These low-carbon steels are processed in such a way that they are resistant to aging during normal storage but begin to age during sheet metal forming. A subsequent exposure to heat during the paint-baking operation completes the aging process and adds an additional 35 to 70 MPa (5 to 10 ksi), raising the final yield strength to approximately 275 MPa (40 ksi). Since the increase in strength occurs after the forming operation, the material offers good formability coupled with improved dent resistance in the final product. In addition, it allows weight savings to be achieved without compromising the attractive features of steel sheet, which include spot weldability, good crash energy absorption, low cost, and full recyclability.

ADVANCED HIGH-STRENGTH STEELS (AHSS)

Since 2000, there have been significant developments in automotive materials, with large amounts of low-carbon and HSLA steels likely to be replaced by the advanced high-strength steels (AHSS). The AHSS steels are primarily ferrite-phase, soft steels with varying amounts of martensite, bainite, and retained austenite, which offer high strength with enhanced ductility. While previous high-strength grades, such as HSLA, suffered from reduced formability, the AHSS materials enable the stamping or hydroforming of more complex parts. Parts can often be integrated into single pieces, eliminating the cost and time associated with assembly, and the higher strength provides improved fatigue and crash performance, along with the possibility of weight reduction.

Dual-phase steels form when we quench material from a temperature that is above the A_1 but below the A_3, where the structure consists of ferrite and high-carbon austenite. During the quench, the ferrite remains unaffected, while the high-carbon austenite transforms to high-carbon martensite. A low- or medium-carbon steel now has a mixed microstructure of weak, ductile ferrite combined with high-strength, high-hardness, high-carbon martensite. The dual-phase structure offers strengths that are comparable to HSLA materials, coupled with improved forming characteristics and no loss in weldability. The high work-hardening rates and excellent elongation lead to a high ultimate tensile strength coupled with a low initial yield strength. The high strain rate sensitivity means that the faster the steel is crushed, the more energy it absorbs—a feature that further enhances the crash resistance of automotive structures.

While the dual-phase steels have structures of ferrite and martensite, *transformation-induced plasticity* (TRIP) steels contain ferrite with a combination of martensite, bainite, and retained austenite. Since the retained austenite transforms progressively to martensite as the steel is deformed, the high work-hardening rates of the TRIP steels persist to high strains, offering significant advantages in operations such as stretch forming and deep drawing. Alternatively, the transformation to martensite and high rate of work hardening can also be used to provide excellent energy absorption during crash deformation.

FIGURE 6-9 Relative strength and formability (elongation) of conventional, high-strength low-alloy, and advanced high-strength steels. B.H. = bake hardenable; DP = dual phase; and Mart = martensitic.

Complex-phase (CP) steels and martensitic (Mart) steels offer even higher strengths with useful capacity for deformation and energy absorption. The CP steels have a microstructure of ferrite and a higher volume fraction of hard phases (martensite and bainite), strengthened further by a fine precipitate of niobium, titanium, or vanadium carbides or nitrides. The Mart steels are almost entirely martensite and can have tensile strengths up to 1700 MPa (245 ksi).

Figure 6-9 shows the relative strengths and formability (elongation) of the conventional steels (including mild steels and bake-hardenable steels), HSLA steels, and the newer AHSS materials. The carbon–manganese steels would bridge the gap between the mild and bake-hardenable and the HSLA steels. Also included in this figure are some useful distinctions between low-strength steels (ultimate tensile strength below 270 MPa or 40 ksi), high-strength steel, and ultra-high-strength steels (ultimate tensile strength above 700 MPa or 100 ksi).

FREE-MACHINING STEELS

The increased use of high-speed, automated machining has spurred the use and development of several varieties of *free-machining steels*. These steels machine readily and form small chips when cut. The smaller chips reduce the length of contact between the chip and cutting tool, thereby reducing the associated friction and heat, as well as required power and wear on the cutting tool. The formation of small chips also reduces the likelihood of chip entanglement in the machine and makes chip removal much easier. On the negative side, free-machining steels often carry a cost premium of 15 to 20% over conventional alloys, but this increase may be easily recovered through higher machining speeds, larger depths of cut, and extended tool life.

Free-machining steels are basically carbon steels that have been modified by an addition of sulfur, lead, bismuth, selenium, tellurium, or phosphorus plus sulfur to enhance machinability. Sulfur combines with manganese to form soft manganese sulfide inclusions. These, in turn, serve as chip-breaking discontinuities within the structure. The inclusions also provide a built-in lubricant that prevents formation of a built-up edge on the cutting tool and imparts an improved cutting geometry (see Chapter 21). In leaded materials, the insoluble lead particles work in much the same way.

The bismuth free-machining steels are an attractive alternative to the previous varieties. Bismuth is more environmentally acceptable (compared to lead), has a reduced tendency to form stringers, and can be more uniformly dispersed since its density is a better match to that of iron. Machinability is improved because the heat generated by cutting is sufficient to form a thin film of liquid bismuth that lasts for only fractions of a microsecond. Tool life is noticeably extended and the machined product is still weldable.

The use of free-machining steels is not without compromise, however. Ductility and impact properties are somewhat reduced compared to the unmodified steels. Copper-based braze joints tend to embrittle when used to join bismuth free-machining steels, and the machining additions reduce the strength of shrink-fit assemblies. If these compromises are objectionable, other methods may be used to enhance machinability. For example, the machinability of steels can be improved by cold working the metal. As the strength and hardness of the metal increase, the metal loses ductility, and subsequent machining produces chips that tear away more readily and fracture into smaller segments.

PRECOATED STEEL SHEET

Traditional sheet metal fabrication involves the shaping of components from bare steel, followed by the finishing (or coating) of these products on a piece-by-piece basis. In this sequence, it is not uncommon for the finishing processes to be the most expensive and time-consuming stages of manufacture, since it involves handling, manipulation, and possible curing or drying, as well as adherence to the various EPA (environmental) and OSHA (safety and health) requirements.

An alternative to this procedure is to purchase mill-coated steel sheet, where the steel supplier applies the coating when the material is still in the form of a long, continuous strip. Cleaning, pretreatment, coating, and curing can all be performed in a continuous manner, producing a coating that is uniform in thickness and offers improved adhesion. Numerous coatings can be specified, including the entire spectrum of dipped and plated metals (including aluminum, zinc, and chromium), vinyls, paints, primers, and other polymers or organics. Many of these coatings are specially formulated to endure the rigors of subsequent forming and bending. The continuous sheets can also be printed, striped, or embossed to provide a number of visual effects. Extra caution must be exercised during handling and fabrication to prevent damage to the coating, but the additional effort and expense are often less than the cost of finishing individual pieces.

STEELS FOR ELECTRICAL AND MAGNETIC APPLICATIONS

Soft magnetic materials can be magnetized by relatively low-strength magnetic fields but lose almost all of their magnetism when the applied field is removed. They are widely used in products such as solenoids, transformers, motors, and generators. The most common soft magnetic materials are high-purity iron, low-carbon steels, iron–silicon electrical steels, amorphous ferromagnetic alloys, iron–nickel alloys, and soft ferrites (ceramic material).

In recent years, the *amorphous metals* have shown attractive electrical and magnetic properties. Since the material has no crystal structure, grains, or grain boundaries, (1) the magnetic domains can move freely in response to magnetic fields, (2) the properties are the same in all directions, and (3) corrosion resistance is improved. The high magnetic strength and low hysteresis losses offer the possibility of smaller, lighter-weight magnets. When used to replace silicon steel in power transformer cores, this material has the potential of reducing core losses by as much as 50%.

To exhibit permanent magnetism, materials must remain magnetized when removed from the applied field. While most permanent magnets are ceramic materials or complex metal alloys, cobalt alloy steels (containing up to 36% cobalt) may be specified for electrical equipment where high magnetic densities are required.

MARAGING STEELS

When superhigh strength is required from a steel, the *maraging* grades become a very attractive option. These alloys contain between 15 and 25% nickel, plus significant amounts of cobalt, molybdenum, and titanium, all added to a very low-carbon steel. They can be hot worked at elevated temperatures, machined, or cold worked in the air-cooled condition, and then aged to yield strengths in excess of 1725 MPa (250 ksi), with good residual elongation.

Maraging alloys are very useful in applications where ultra-high strength and good toughness are important. They can be welded, provided the welding is followed by the full solution and aging treatment. As might be expected from the large amount of alloy additions (over 30%) and multistep thermal processing, maraging steels are quite expensive and should be specified only when their outstanding properties are absolutely required.

STEELS FOR HIGH-TEMPERATURE SERVICE

As a general rule of thumb, plain-carbon steels should not be used at temperatures in excess of about 250°C (500°F). Conventional alloy steels extend this upper limit to around 350°C (650°F). Continued developments in areas such as missiles and jet aircraft, however, have increased the demand for metals that offer good strength characteristics, corrosion resistance, and creep resistance at operating temperatures in excess of 550°C (1000°F).

The high-temperature ferrous alloys tend to be low-carbon materials with less than 0.1% carbon. At their peak operating temperatures, 1000-hour rupture stresses tend to be quite low, often in the neighborhood of 50 MPa (7 ksi). While iron is also a major component of other high-temperature alloys, when the amounts fall below 50%, the metal is not generally classified as a ferrous material. High strength at high temperature usually requires the more expensive nonferrous materials that will be discussed in Chapter 7.

■ 6.5 STAINLESS STEELS

Low-carbon steel with the addition of 4 to 6% chromium acquires good resistance to many of the corrosive media encountered in the chemical industry. This behavior is attributed to the formation of a strongly adherent iron chromium oxide on the surface. If more improved corrosion resistance and outstanding appearance are required, materials should be specified that use a superior oxide that forms when the amount of chromium in solution (excluding chromium carbides and other forms where the chromium is no longer available to react with oxygen) exceeds 12%. When damaged, this tough, adherent, corrosion-resistant oxide (which is only 1–2 nanometers thick) actually heals itself, provided oxygen is present, even in very small amounts. Materials that form this superior protective oxide are known as the *true stainless steels*.

Several classification schemes have been devised to categorize these alloys. The American Iron and Steel Institute groups the metals by chemistry and assigns a three-digit number that identifies the basic family and the particular alloy within that family. In this text, however, we will group these alloys into microstructural families, since it is the basic structure that controls the engineering properties of the metal. Table 6-6 presents the AISI designation scheme for stainless steels and correlates it with the microstructural families.

Chromium is a ferrite stabilizer; that is, the addition of chromium tends to increase the temperature range over which ferrite is the stable structure. With sufficient chromium and a low level of carbon, a corrosion-resistant iron alloy can be produced that is ferrite at all temperatures below solidification. These alloys are known as the *ferritic stainless steels*. They possess rather limited ductility and poor toughness but are readily weldable. No martensite can form in the welds because there is no possibility of forming the face-centered-cubic (FCC) austenite structure that can then transform during cooling. These alloys cannot be heat treated, and poor ductility limits the amount of strengthening by cold work. The primary source of strength is the body-centered-cubic (BCC) crystal structure combined with the effects of solid solution strengthening. Characteristic of BCC metals, the ferritic stainless steels exhibit a ductile-to-brittle transition as the temperature is reduced. The ferritic alloys are the cheapest type of stainless steel, however, and, as such, they should be given first consideration when a stainless alloy is required.

If increased strength is needed, the *martensitic stainless steels* should be considered. For these alloys, carbon is added and the chromium content is reduced to a level where the material can be austenite (FCC) at high temperature and ferrite (BCC) at low. The carbon can be dissolved in the face-centered-cubic austenite, which can then be quenched

TABLE 6-6 AISI Designation Scheme for Stainless Steels

Series	Alloys	Structure
200	Chromium, nickel, manganese, or nitrogen	Austenitic
300	Chromium and nickel	Austenitic
400	Chromium and possibly carbon	Ferritic or martensitic
500	Low chromium (<12%) and possibly carbon	Martensitic

to trap it in a body-centered martensitic structure. The carbon contents can be varied up to 1.2% to provide a range of strengths and hardnesses. Caution should be taken, however, to ensure more than 12% chromium remains in solution. Slow cools may allow the carbon and chromium to react and form chromium carbides. When this occurs, the chromium is not available to react with oxygen and form the protective oxide. As a result, the martensitic stainless steels may only exhibit good corrosion resistance when in the martensitic condition (when the chromium is trapped in atomic solution) and may be susceptible to red rust when annealed or normalized for machining or fabrication. The martensitic stainless steels cost about $1\frac{1}{2}$ times as much as the ferritic alloys, with part of the increase being due to the additional heat treatment, which generally consists of an austenitization, quench, stress relief, and temper. They are less corrosion resistant than the other varieties and tend to be used when strength and hardness are the dominant requirements.

Nickel is an austenite stabilizer; with sufficient amounts of both chromium and nickel (and low carbon), it is possible to produce a stainless steel in which austenite is the stable structure from elevated to cryogenic temperatures. Known as *austenitic stainless steels*, these alloys may cost two to three times as much as the ferritic variety, but here the added expense is attributed to the cost of the nickel and chromium alloys. Manganese and nitrogen are also austenite stabilizers and may be substituted for some of the nickel to produce a lower-cost, somewhat lower-quality austenitic stainless steel (the AISI 200-series).

Austenitic stainless steels are easily identified by their nonmagnetic characteristic (the ferritic and martensitic stainlesses are attracted to a magnet). They are highly resistant to corrosion in almost all media (except hydrochloric acid and other halide acids and salts) and may be polished to a mirror finish, thereby combining attractive appearance and corrosion resistance. Formability is outstanding (characteristic of the FCC crystal structure), and these steels strengthen significantly when cold worked. The following table shows the response of the popular 304 alloy (also known as 18-8 because of the composition of 18% chromium and 8% nickel) to a small amount of cold work:

	Water Quench	Cold Rolled 15%
Yield strength [MPa (ksi)]	260 (38)	805 (117)
Tensile strength [MPa (ksi)]	620 (90)	965 (140)
Elongation in 2 in. (%)	68	11

The austenitic stainless steels offer the best combination of corrosion resistance and toughness of the stainless varieties. Since they are also some of the most costly, they should not be specified where the less expensive ferritic or martensitic alloys would be adequate or where a true stainless steel is not required. Figure 6-10 lists some of the popular alloys from each of the three major structural classifications and schematically denotes some of their key properties. Table 6-7 shows the basic types and the primary mechanism of strengthening.

A fourth and special class of stainless steels is the *precipitation-hardening* variety. These alloys are basically martensitic or austenitic types, modified by the addition of alloying elements such as aluminum that permit the precipitation of hard intermetallic compounds at the temperatures used to temper martensite. With the addition of age hardening, these materials are capable of attaining high-strength properties such as a 1790-MPa (260-ksi) yield strength, 1825-MPa (265-ksi) tensile strength, and a 2% elongation. Since the additional alloys and extra processing make the precipitation-hardening alloys some of the most expensive stainless steels, they should be used only when their high-strength feature is absolutely required.

While the four structures described above constitute the bulk of stainless steels, there are also some additional variants. *Duplex stainless steels* contain between 18 and 25% chromium, 4 to 7% nickel, and up to 4% molybdenum; they can be water quenched

TABLE 6-7 Primary Strengthening Mechanism for the Various Types of Stainless Steel

Type of Stainless Steel	Primary Strengthening Mechanism
Ferritic	Solid-solution strengthening
Martensitic	Phase transformation strengthening (martensite)
Austenitic	Cold work (deformation strengthening)

	AISI type	Usage
Martensitic (hardenable by heat treatment)	410 420 440C	General purpose
		Hardenable by heat treatment
Ferritic (more corrosion resistant than martensitic, but not hardenable by heat treatment)	405 430 446	
		Hardenable by cold working
Austenitic (best corrosion resistance, but hardenable only by cold working)	201 202 301 302 302B 304L 310 316 321	For elevated-temperature service
		Modified for welding
		Superior corrosion resistance

FIGURE 6-10 Popular alloys and key properties for different types of stainless steels.

from a hot-working temperature to produce a microstructure that is approximately half ferrite and half austenite. This mixed structure offers a higher yield strength and greater resistance to stress corrosion cracking and pitting corrosion than either the full-austenitic or full-ferritic grades.

Since stainless steels are difficult to machine because of their work-hardening properties and their tendency to seize during cutting, special *free-machining alloys* have been produced within each family. Additions of sulfur or selenium can raise machinability to approximately that of a medium-carbon steel.

Cast stainless steels have structures and properties that are similar to the wrought grades but are specified by the designations of the Alloy Casting Institute. The C series, used primarily to impart corrosion resistance, are used in valves, pumps, and fittings. The H grades (heat-resistant), designed to provide useful properties at elevated temperature, are used for furnace parts and turbine components.

Several potential problems are unique to the family of stainless steels. Since the protective oxide provides the excellent corrosion resistance, this feature can be lost whenever the amount of chromium in solution drops below 12%. A localized depletion of chromium can occur when elevated temperatures allow chromium carbides to form along grain boundaries (*sensitization*). To prevent their formation, one can keep the carbon content of stainless steels as low as possible, usually below 0.10%. Another method is to tie up existing carbon with small amounts of stabilizing elements, such as titanium or niobium, that have a stronger affinity for carbon than does chromium. Rapidly cooling these metals through the carbide-forming range of 480° to 820°C (900° to 1500°F) also works to prevent carbide formation.

Another problem with high-chromium stainless steels is an embrittlement that can occur after long times at elevated temperatures. This is attributed to the formation of a brittle compound that forms at elevated temperature and coats grain boundaries. Known as *sigma phase*, this material then provides a brittle crack path through the metal. Stainless steels used in high-temperature service should be checked periodically to detect and monitor sigma-phase formation.

■ 6.6 TOOL STEELS

Tool steels are high-carbon, high-strength, ferrous alloys that have been modified by alloy additions to provide a desired balance of strength, toughness, and wear resistance when properly heat treated. Several classification systems have been developed, some using chemistry as a basis and others employing hardening method or major mechanical property. The AISI system uses a letter designation to identify basic features such as quenching method, primary application, special alloy or characteristic, or specific industry

TABLE 6-8	Basic Types of Tool Steel and Corresponding AISI Grades	
Type	AISI Grade	Significant Characteristic
1. Water-Hardening	W	
2. Cold-work	O	Oil-hardening
	A	Air-hardening medium alloy
	D	High-carbon–high-chromium
3. Shock-resisting	S	
4. High-speed	T	Tungsten alloy
	M	Molybdenum alloy
5. Hot-work	H	H1–H19: chromium alloy
		H20–H39: tungsten alloy
		H40–H59: molybdenum alloy
6. Plastic-mold	P	
7. Special-purpose	L	Low alloy
	F	Carbon–tungsten

involved. Table 6-8 lists seven basic families of tool steels, the corresponding AISI letter grades, and the associated feature or characteristic. Individual alloys within the letter grades are then listed numerically to produce a letter–number identification system.

Water-hardening tool steels (W grade) are essentially high-carbon plain-carbon steels. They are the least expensive variety and are used for a wide range of parts that are usually quite small and not subject to severe usage or elevated temperature. Because strength and hardness are functions of the carbon content, a wide range of properties can be achieved through composition variation. Hardenability is low, so these steels must be quenched in water to attain high hardness. They can be used only for relatively thin sections if the full depth of hardness is desired. They are also rather brittle, particularly at higher hardness.

Typical uses of the various plain-carbon steels are as follows:

0.60–0.75% carbon: machine parts, chisels, setscrews, and similar products where medium hardness is required, coupled with good toughness and shock resistance

0.75–0.90% carbon: forging dies, hammers, and sledges

0.90–1.10% carbon: general-purpose tooling applications that require a good balance of wear resistance and toughness, such as drills, cutters, shear blades, and other heavy-duty cutting edges

1.10–1.30% carbon: small drills, lathe tools, razor blades, and other light-duty applications in which extreme hardness is required without great toughness

In applications where improved toughness is required, small amounts of manganese, silicon, and molybdenum are often added. Vanadium additions of about 0.20% are used to form strong, stable carbides that retain fine grain size during heat treatment. One of the main weaknesses of the plain-carbon tool steels is their loss of hardness at elevated temperature, which can occur with prolonged exposure to temperatures over 150°C (300°F).

When larger parts must be hardened or distortion must be minimized, the *cold-work tool steels* are usually recommended. The alloy additions and higher hardenability of the *oil-* or *air-hardening grades* (O and A designations, respectively) enable hardening by less severe quenches. Tighter dimensional tolerances can be maintained during heat treatment, and the cracking tendency is reduced. The *high-chromium tool steels,* designated by the letter *D,* contain between 10 and 18% chromium, and are also air-hardening and offer outstanding deep-hardening wear resistance. Blanking, stamping, and cold-forming dies, punches, and other tools for large production runs are all common applications for this class. Because these steels do not have the alloy content necessary to resist softening at elevated temperatures, they should not be used for applications that involve prolonged service at temperatures in excess of 250°C (500°F).

Shock-resisting tool steels (S designation) offer the high toughness needed for impact applications. Low carbon content (approximately 0.5% carbon) is usually specified to assure the necessary toughness, with carbide-forming alloys providing the necessary abrasion resistance, hardenability, and hot-work characteristics. Applications include parts for pneumatic tooling, chisels, punches, and shear blades.

High-speed tool steels are used for cutting tools and other applications where strength and hardness must be retained at temperatures up to or exceeding red-heat (about 760°C or 1400°F). One popular member of the tungsten high-speed tool steels (T designation) is the T1 alloy, which contains 0.7% carbon, 18% tungsten, 4% chromium, and 1% vanadium. It offers a balanced combination of shock resistance and abrasion resistance and is used for a wide variety of cutting applications. The molybdenum high-speed steels (M designation) were developed to reduce the amount of tungsten and chromium required to produce the high-speed properties.

Hot-work tool steels (H designation) were developed to provide strength and hardness during prolonged exposure to elevated temperature. All employ substantial additions of carbide-forming alloys. H1 to H19 are chromium-based alloys with about 5.0% chromium; H20 to H39 are tungsten-based types with 9 to 18% tungsten coupled with 3 to 4% chromium; and H20 to H59 are molybdenum-based. The chromium types tend to be less expensive than the tungsten or molybdenum alloys.

Other types of tool steels include (1) the *plastic mold steels* (P designation), designed to meet the requirements of zinc die casting and plastic injection molding dies; (2) the *low-alloy special-purpose tool steels* (L designation), such as the L6 extreme toughness variety; and (3) the *carbon–tungsten type* of special-purpose tool steels (F designation), which are water hardening but substantially more wear-resistant than the plain-carbon tool steels.

Most tool steels are wrought materials, but some are designed specifically for fabrication by casting. Powder metallurgy processing has also been used to produce special compositions that are difficult or impossible to produce by wrought or cast methods or provide key structural enhancements. By subjecting the water-atomized powders to hot-isostatic pressing (HIP), 100%-dense billets can be produced with fine grain size and small, uniformly distributed carbide particles. These materials offer superior wear resistance compared to conventional tool steels, combined with useful levels of toughness.

■ 6.7 ALLOY CAST STEELS AND IRONS

The effects of alloying elements are the same regardless of the process used to produce the final shape. When the desired shape is to be made by casting, some alloys can be used to enhance process-specific features, such as fluidity and as-solidified properties. If a ferrous casting alloy contains less than about 2.0% carbon, it is considered to be a *cast steel*. Alloys with more than 2% carbon are *cast irons*.

Most cast irons are used in the as-cast condition, with the only heat treatment being a stress relief or annealing. For these applications, the alloy elements are selected for their ability to alter properties by (1) affecting the formation of graphite or cementite, (2) modifying the morphology of the carbon-rich phase, (3) strengthening the matrix material, or (4) enhancing wear resistance through the formation of alloy carbides. Nickel, for example, promotes graphite formation and tends to promote finer graphite structures. Chromium retards graphite formation and stabilizes cementite. These alloys are frequently used together in a ratio of two or three parts of nickel to one part of chromium. Between 0.5 and 1.0% molybdenum is often added to gray cast iron to impart additional strength, form alloy carbides, and help to control the size of the graphite flakes.

High-alloy cast irons have been designed to provide enhanced corrosion resistance and/or good elevated-temperature service. Within this family, the austenitic gray cast irons, which contain about 14% nickel, 5% copper, and 2.5% chromium, offer good corrosion resistance to many acids and alkalis at temperatures up to about 800°C (1500°F). Alloy cast irons and cast steels are usually specified by their ASTM designation numbers, which relate the materials to their mechanical properties and intended service applications. The Society of Automotive Engineers also has specifications for cast steels used in the automotive industry.

Cast steels are generally used whenever a cast iron is not adequate for the application. Compared to cast irons, the cast steels offer enhanced stiffness, toughness, and ductility over a wide range of operating temperatures and can be readily welded. They are usually heat treated to produce a final quenched-and-tempered structure, and the alloy additions are selected to provide the desired hardenability and balance of properties. The enhanced properties come with a price, however, since the cast steels have a higher melting point (more energy to melt and higher cost refractories are necessary), less fluidity (leading to increased probability of incomplete die or mold filling), and increased shrinkage (since graphite is not formed during solidification). The diverse applications take advantage of the material's structural strength and its ability to contain pressure, resist impacts, withstand elevated temperatures, and resist wear.

■ Key Words

advanced high-strength steel (AHSS)
air-hardenable tool steel
AISI–SAE designation
alloy steel
amorphous metals
austenitic stainless steel
bake-hardenable steel
cast iron
cast steel
cold-work tool steel
continuous casting
degassification

deoxidation
dual-phase steels
duplex stainless steel
electroslag remelting
ferritic stainless steel
ferrous metals
free-machining steel
high-carbon steel
high-speed tool steel
high-strength low-alloy steel (HSLA)
hot-work tool steel
iron

ladle metallurgy
ladles
low-carbon steel
maraging steel
martensitic stainless steel
medium-carbon steel
microalloyed steel
oil-hardenable tool steel
pig iron
plain-carbon steel
precipitation-hardenable stainless steel
precoated steel

sensitization
shock-resisting tool steel
sigma phase
solidification shrinkage
stainless steel
steel
tool steel
TRIP steels (transformation-induced plasticity)
vacuum arc remelting
vacuum degassing
vacuum induction melting
water-hardenable tool steel

■ Review Questions

1. Why might it be important to know the prior processing history of an engineering material?
2. What is a ferrous material?
3. How does the recycling of steel compare with the recycling of aluminum?
4. What properties or characteristics have made steel such an attractive engineering material?
5. When iron ore is reduced to metallic iron, what other elements are generally present in the metal?
6. How does steel differ from pig iron?
7. What are some of the modification processes that can be performed on a steel during ladle metallurgy operations?
8. What is the advantage of pouring molten metal from the bottom of a ladle?
9. What are some of the attractive economic and processing advantages of continuous casting?
10. What are some of the techniques used to reduce the amount of dissolved oxygen in molten steel?
11. How might other gases, such as nitrogen and hydrogen, be reduced?
12. What are some of the attractive features of electroslag remelting?
13. What is plain-carbon steel?
14. What is considered a low-carbon steel? Medium-carbon? High-carbon?
15. What properties account for the high-volume use of medium-carbon steels?
16. Why should plain-carbon steels be given first consideration for applications requiring steel?
17. What are some of the common alloy elements added to steel?
18. For what different reasons might alloying elements be added to steel?
19. What are some of the alloy elements that tend to form stable carbides within a steel?
20. What alloys are particularly effective in increasing the hardenability of steel?
21. What is the significance of the last two digits in a typical four-digit AISI–SAE steel designation?
22. How are letters incorporated into the AISI–SAE designation system for steel, and what do some of the more common ones mean?
23. What is an H-grade steel, and when should it be considered in a material specification?
24. Why should the proposed fabrication processes enter into consideration when selecting a steel?
25. How are the final properties usually obtained in the constructional alloy steels? In the HSLA steels?
26. How are HSLA steels specified?
27. What are microalloyed steels?
28. What are some of the potential benefits that may be obtained through the use of microalloyed steels?
29. What is the primary attraction of the bake-hardenable steels?
30. What are advanced high-strength steels (AHSS)?
31. What are the two phases that are present in dual-phase steels?
32. What is the "transformation" that occurs during the deformation of the "transformation-induced plasticity (TRIP) steels?
33. What are some of the various alloy additions that have been used to improve the machinability of steels?
34. What are some of the compromises associated with the use of free-machining steels?
35. What factors might be used to justify the added expense of precoated steel sheet?
36. Why have the amorphous metals attracted attention as potential materials for magnetic applications?

37. What are maraging steels, and for what conditions might they be required?
38. What are the typical elevated temperature limits of plain-carbon and alloy steels?
39. What is a stainless steel?
40. What feature is responsible for the observed corrosion resistance of stainless steels?
41. Why should ferritic stainless steels be given first consideration when selecting a stainless steel?
42. Which of the major types of stainless steel is likely to contain significant amounts of carbon? Why?
43. Under what conditions might a martensitic stainless steel "rust" when exposed to a hostile environment?
44. What are some of the unique properties of austenitic stainless steels?

45. How can an austenitic stainless steel be easily identified?
46. What two structures are present in a duplex stainless steel?
47. What is sensitization of a stainless steel, and how can it be prevented?
48. What is a tool steel?
49. How does the AISI–SAE designation system for tool steels differ from that for plain-carbon and alloy steels?
50. For what types of applications might an air-hardenable tool steel be attractive?
51. What alloying elements are used to produce the hot-worked tool steels?
52. What are some of the reasons that alloy additions are made to cast irons that will be used in their as-cast condition?
53. When should a cast steel be used instead of a cast iron?

www.wiley.com/college/degarmo

Chapter 6 CASE STUDY

Interior Tub of a Top-Loading Washing Machine

The interior tub of a washing machine is the container that holds the clothes during the washing and rinsing cycles, but it also contains the perforations that permit removal of the water by draining and spinning. The component will see mechanical loadings from the weight of the clothes and water, and also the dynamic action of spinning water-laden fabrics. There will be exposure to a wide range of water quality, as well as the full spectrum of soaps, detergents, bleaches, and other laundry additives. The surfaces should also be resistant to the impact and abrasion of buttons, zippers, and snaps.

This part has traditionally been manufactured by the sequential deep drawing, perforating, and trimming of metal sheet, followed by some form of surface-coating treatment. For a long time, the standard material was "enameling iron"—a steel sheet with less than 0.03% carbon that was then coated with a fired porcelain enamel. Due to the difficulties of producing ultra-low-carbon material in today's steelmaking operations, enameling iron became increasingly scarce, and manufacturers were forced to substitute the lowest-carbon, most readily available material, namely 1008 steel. This substitution further required modification of the enameling process to prevent defects and blistering from CO evolution.

Your employer is presently manufacturing these tubs from 1008 steel sheet with a subsequent coating of fired porcelain enamel. Your marketing staff, however, reports that consumers tend to view a stainless steel tub to be of higher quality. As a result, your supervisor has asked you to evaluate the merits of converting to this material. You must first familiarize yourself with the current product (the base material, the forming process, and the porcelain enameling),

and then determine what might be involved in converting to stainless steel. Consider the following specific questions:

1. What are the obvious pros and cons of the present product and process? Where would you expect most problems to occur in the current manufacturing process? Which aspects of fabrication are likely to be the most costly?
2. What would be the pros and cons of converting to stainless steel? In what ways would the product be superior? Are there any assets or liabilities associated with product fabrication from stainless material?
3. Which stainless steel would you recommend? Begin by considering the basic types (ferritic, austenitic, and martensitic) and then refine your selection to a specific alloy if possible. Discuss the rationale for your selection.
4. Since deep drawing is a metal deformation process, we could use cold working (strain hardening) as a strengthening mechanism. Would you find this to be attractive, or would you prefer to use a recrystallization anneal after drawing and prior to use? Why? If you elect to use cold work, might you want to at least perform a stress-relief heat treatment prior to use? Could this be done and still preserve the deformation strengthening? In deep drawing, the deformation is not uniform (increasing as we move up the sidewalls of the container), and the bottom of the tub simply retains the properties of the starting sheet. In order to ensure a minimum amount of strength at all locations, it may be desirable to begin the drawing with a partially cold-rolled sheet. Do you find this suggestion to be desirable? Why or why not?
5. After drawing and perforating, the residual drawing lubricant is removed from the part. Would any additional surface treatment be required? What would be your recommendation?

NONFERROUS METALS AND ALLOYS

■ 7.1 INTRODUCTION

Nonferrous metals and alloys have assumed increasingly important roles in modern technology. Because of their number and the fact that their properties vary widely, they provide an almost limitless range of properties for the design engineer. While they tend to be more costly than iron or steel, these metals often possess certain properties or combinations of properties that are not available in the ferrous metals, such as:

1. Resistance to corrosion
2. Ease of fabrication
3. High electrical and thermal conductivity
4. Light weight
5. Strength at elevated temperatures
6. Color

Nearly all the nonferrous alloys possess at least two of the qualities listed above, and some possess nearly all. For many applications, specific combinations of these properties are highly desirable. Each year, the average American requires about 65 pounds of aluminum, 21 pounds of copper, 12 pounds of lead, 11 pounds of zinc, and 25 pounds of various other nonferrous metals. Figure 7-1 classifies some of the nonferrous metals by advantageous engineering properties, and Table 7-1 shows the increasing role of the nonferrous metals in a typical family vehicle.

As a whole, the strength of the nonferrous alloys is generally inferior to that of steel. Also, the modulus of elasticity is usually lower, a fact that places them at a distinct disadvantage when stiffness is a required characteristic. Ease of fabrication is often attractive. Those alloys with low melting points are easy to cast in sand molds, permanent molds, or dies. Many alloys have high ductility coupled with low yield points, the ideal combination for cold working. Good machinability is also characteristic of many nonferrous

TABLE 7-1	The Material Content of a Typical Family Vehicle (in pounds)		
Material	1978	1990	2002
Steel	2103	1682.5	1757
Stainless steel	26	34	56.5
Cast iron	512	454	328
Plastics	180	229	255
Aluminum	112.5	158.5	279.5
Copper	37	48.5	50
Zinc	31	18.5	8.5
Magnesium	1	3	9.5
Powder metal	15.5	24	40.5
Other materials	551.5	488.5	573
Total	3569.5	3140.5	3357.5

FIGURE 7-1   Some common nonferrous metals and alloys, classified by attractive engineering property.

alloys. The savings obtained through ease of fabrication can often overcome the higher cost of the nonferrous material and justify its use in place of steel. Weldability is the one fabrication area where the nonferrous alloys tend to be somewhat inferior to steel. With modern joining techniques, however, it is generally possible to produce satisfactory weldments in all of the nonferrous metals.

■ 7.2 COPPER AND COPPER ALLOYS

GENERAL PROPERTIES AND CHARACTERISTICS

Copper has been an important engineering metal for over 6000 years. As a pure metal, it has been the backbone of the electrical industry. It is also the base metal of a number of alloys, generically known as brasses and bronzes. Compared to other engineering materials, copper and copper alloys offer three important properties: (1) *high electrical and thermal conductivity*, (2) *useful strength with high ductility*, and (3) *corrosion resistance* to a wide range of media. Because of its excellent conductivity, about one-third of all copper produced is used in some form of electrical application, such as the commutators shown in Figure 7-2. Other large areas of use include plumbing, heating, and air conditioning.

Pure copper in its annealed state has a tensile strength of only about 200 MPa (30 ksi), with an elongation of nearly 60%. Through cold working, the tensile strength can be more than doubled to over 450 MPa (65 ksi), with a decrease in elongation to about 5%. Because of its relatively low strength and high ductility, copper is a very desirable metal for applications where extensive forming is required. Since the recrystallization temperature for copper is less than 260°C (500°F), the hardening effects of cold working can also be easily removed. Copper and copper alloys lend themselves nicely to the whole spectrum of fabrication processes, including casting, machining, joining, and surface finishing by either plating or polishing.

Unfortunately, copper is *heavier than iron*. While strength can be quite high, the strength-to-weight ratio for copper alloys is usually less than that for the weaker aluminum and magnesium materials. In addition, problems can occur when copper is used at elevated temperature. Copper alloys tend to soften when heated above 220°C (400°F), and if copper is stressed for a long period of time at high temperature, intercrystalline failure can occur at about half of its normal room-temperature strength. While offering good resistance to adhesive wear, copper and copper alloys have poor abrasive wear characteristics.

The low-temperature properties of copper are quite attractive, however. Strength tends to increase as temperatures drop, and the material does not embrittle, retaining attractive ductility even under cryogenic conditions. Conductivity also tends to increase with a drop in temperature.

Copper and copper alloys respond well to strengthening methods, with the strongest alloy being 15 to 20 times stronger than the weakest. Because of the wide range of properties, the material can often be tailored to the specific needs of a design. Elastic stiffness is between 50 and 60% of steel. Additional features include being

FIGURE 7-2   Copper and copper alloys are used for a variety of electrical applications, such as these electrical commutators. *(Courtesy of The Electric Materials Company, North East, PA.)*

nonmagnetic, nonpyrophoric (slivers or particles do not burn in air — i.e., nonsparking), and nonbiofouling (inhibits marine organism growth), as well as offering a wide spectrum of colors, including yellow, red, brown, and silver.

COMMERCIALLY PURE COPPER
Refined copper containing between 0.02 and 0.05% oxygen is called *electrolytic tough-pitch* (ETP) copper. It is often used as a base for copper alloys and may be used for electrical applications, such as wire and cable, when the highest conductivity is not required. For superior conductivity, additional refining can reduce the oxygen content and produce *oxygen-free high-conductivity* (OFHC) copper. The better grades of conductor copper now have a conductivity rating of about 102% IACS, reflecting metallurgical improvements made since 1913, when the International Annealed Copper Standard (IACS) was established and the conductivity of pure copper was set at 100% IACS.

COPPER-BASED ALLOYS
As a pure metal, copper is not used extensively in manufactured products, except in electrical applications, and even here alloy additions of silver, arsenic, cadmium, and zirconium are used to enhance various properties without significantly impairing conductivity. More often, copper is the base metal for an alloy, where it imparts its good ductility, corrosion resistance, and electrical and thermal conductivity. A full spectrum of mechanical properties is available, ranging from pure copper, which is soft and ductile, through alloys whose properties can rival those of quenched-and-tempered steel.

Copper-based alloys are commonly designated using a system of numbers standardized by the Copper Development Association (CDA). Table 7-2 presents a breakdown of this system, which has been further adopted by the American Society for Testing and Materials (ASTM), the Society of Automotive Engineers (SAE), and the U.S. government. Alloys numbered from 100 to 199 are mostly copper with less than 2% alloy addition. Numbers 200 to 799 are *wrought*[1] alloys, and the 800 and 900 series are *casting alloys*. When converted to the Unified Numbering System for metals and alloys, the three-digit numbers are converted to five digits by placing two zeros at the end, and the letter *C* is used as a prefix to denote the copper base.

COPPER–ZINC ALLOYS
Zinc is by far the most popular alloying addition, and the resulting alloys are generally known as some form of *brass*. If the zinc content is less than 36%, the brass is a single-phase solid solution. Since this structure is identified as the alpha phase, these alloys are

TABLE 7-2 Designation System for Copper and Copper Alloys (Copper Development Association System)

	Wrought Alloys		Cast Alloys
100–155	Commercial coppers	833–838	Red brasses and leaded red brasses
162–199	High-copper alloys	842–848	Semired brasses and leaded semired brasses
200–299	Copper–zinc alloys (brasses)	852–858	Yellow brasses and leaded yellow brasses
300–399	Copper–zinc–lead alloys (leaded brasses)	861–868	Manganese and leaded manganese bronzes
400–499	Copper–zinc–tin alloys (tin brasses)	872–879	Silicon bronzes and silicon brasses
500–529	Copper–tin alloys (phosphor bronzes)	902–917	Tin bronzes
532–548	Copper–tin–lead alloys (leaded phosphor bronzes)	922–929	Leaded tin bronzes
600–642	Copper–aluminum alloys (aluminum bronzes)	932–945	High-leaded tin bronzes
647–661	Copper–silicon alloys (silicon bronzes)	947–949	Nickel–tin bronzes
667–699	Miscellaneous copper–zinc alloys	952–958	Aluminum bronzes
700–725	Copper–nickel alloys	962–966	Copper nickels
732–799	Copper–nickel–zinc alloys (nickel silvers)	973–978	Leaded nickel bronzes

[1] The term *wrought* means "shaped or fabricated in the solid state." Key properties for wrought material generally relate to ductility. *Cast* alloys are shaped as a liquid, where the attractive features include low melting point, high fluidity, and good as-solidified strength.

often called *alpha brasses*. They are quite ductile and formable, with both strength and ductility increasing with the zinc content up to about 36%. The alpha brasses can be strengthened significantly by cold working and are commercially available in various degrees of cold-worked strength and hardness. Cartridge brass, the 70% copper–30% zinc alloy, offers the best overall combination of strength and ductility. As its name implies, it has become a popular material for sheet-forming operations like deep drawing.

With more than 36% zinc, the copper–zinc alloys enter a two-phase region involving a brittle, zinc-rich phase, and ductility drops markedly. While cold-working properties are rather poor for these high-zinc brasses, deformation can be performed easily at elevated temperature.

Many applications of these alloys result from the high electrical and thermal conductivity coupled with useful engineering strength. The wide range of colors (red, orange, yellow, silver, and white), enhanced by further variations that can be produced through the addition of a third alloy element, account for a number of decorative uses. Since the plating characteristics are excellent, the material is also a frequently used base for decorative chrome or similar coatings. Another attractive property of alpha brass is its ability to have rubber vulcanized to it without any special treatment except thorough cleaning. As a result, brass is widely used in mechanical rubber goods.

Most brasses have good corrosion resistance. In the range of 0 to 40% zinc, the addition of a small amount of tin imparts improved resistance to seawater corrosion. Cartridge brass with tin becomes admiralty brass, and the 40% zinc Muntz metal with a tin addition is called naval brass. Brasses with 20 to 36% zinc, however, are subject to a selective corrosion, known as *dezincification*, when exposed to acidic or salt solutions. Brasses with more than 15% zinc often experience *season cracking* or *stress–corrosion cracking*. Both stress and exposure to corrosive media are required for this failure to occur (but residual stresses and atmospheric moisture may be sufficient!). As a result, cold-worked brass is usually stress relieved (to remove the residual stresses) before being placed in service.

When high machinability is required, as with automatic screw-machine stock, 2 to 3% lead can be added to the brass to ensure the formation of free-breaking chips. Brass casting alloys are quite popular for use in plumbing fixtures and fittings, low-pressure valves, and a variety of decorative hardware. They have good fluidity during pouring and attractive low melting points. An alloy containing between 50 and 55% copper and the remainder zinc is often used as a filler metal in brazing. It is an effective material for joining steel, cast iron, brasses, and copper, producing joints that are nearly as strong as those obtained by welding.

Table 7-3 lists some of the more common copper–zinc alloys and their composition, properties, and typical uses.

COPPER–TIN ALLOYS

Since tin is more costly than zinc, alloys of copper and tin, commonly called *tin bronzes*, are usually specified when they offer some form of special property or characteristic. The term *bronze* is often confusing, however, since it can be used to designate any copper alloy where the major alloy addition is not zinc or nickel. To provide clarification, the major alloy addition is usually included in the designation name.

The tin bronzes usually contain less than 12% tin. (Strength continues to increase as tin is added up to about 20%, but the high-tin alloys tend to be brittle.) Tin bronzes offer good strength, toughness, wear resistance, and corrosion resistance. They are often used for bearings, gears, and fittings that are subjected to heavy compressive loads. When the copper–tin alloys are used for bearing applications, up to 10% lead is frequently added.

The most popular wrought alloy is phosphor bronze, which usually contains from 1 to 11% tin. Alloy 521 (CDA), with 8% tin, is typical of this class. Hard sheet has a tensile strength of 760 MPa (110 ksi) and an elongation of 3%. Soft sheet has a tensile strength of 380 MPa (55 ksi) and 65% elongation. The material is often specified for pump parts, gears, springs, and bearings.

TABLE 7-3 Composition, Properties, and Uses of Some Common Copper–Zinc Alloys

CDA Number	Common Name	Composition(%)					Condition	Tensile Strength		Elongation in 2 in. (%)	Typical Uses
		Cu	Zn	Sn	Pb	Mn		ksi	MPa		
220	Commercial bronze	90	10				Soft sheet	38	262	45	Screen wire, hardware, screws. jewelry
							Hard sheet	64	441	4	
240	Low brass	80	20				Spring	73	503	3	Drawing, architectural work, ornamental
							Annealed sheet	47	324	47	
							Hard	75	517	7	
260	Cartridge brass	70	30				Spring	91	627	3	Munitions, hardware, musical instruments, tubing
							Annealed sheet	53	365	54	
							Hard	76	524	7	
270	Yellow brass	65	35				Spring	92	634	3	Cold forming, radiator cores, springs, screws
							Annealed sheet	46	317	64	
							Hard	76	524	7	
280	Muntz metal	60	40				Hot-rolled	54	372	45	Architectural work; condenser tube
							Cold-rolled	80	551	5	
443–445	Admiralty metal	71	28	1			Soft	45	310	60	Condenser tube (salt water), heat exchangers
							Hard	95	655	5	
360	Free-cutting brass	61.5	35.3		3		Soft	47	324	60	Screw-machine parts
							Hard	62	427	20	
675	Manganese bronze	58.5	39	1		0.1	Soft	65	448	33	Clutch disks, pump rods, valve stems, high-strength propellers
							Bars, half hard	84	579	19	

Alloy 905 is a bronze casting alloy containing 10% tin and 2% zinc. In the as-cast condition, the tensile strength is about 310 MPa (45 ksi), with an elongation of 45%. It has very good resistance to seawater corrosion and is used on ships for pipe fittings, gears, pump parts, bushings, and bearings.

Bronzes can also be made by mixing powders of copper and tin, followed by low-density powder metallurgy processing (described in Chapter 19). The porous product can be used as a filter for high-temperature or corrosive media, or it can be infiltrated with oil to produce self-lubricating bearings.

COPPER–NICKEL ALLOYS

Copper and nickel exhibit complete solubility (as shown previously in Figure 4-6), and a wide range of useful alloys have been developed. Key features include high thermal conductivity, high-temperature strength, and corrosion resistance to a range of materials, including seawater. These properties, coupled with a high resistance to stress–corrosion cracking, make the copper–nickel alloys a good choice for heat exchangers, cookware, desalination apparatus, and a wide variety of coinage. *Cupronickels* contain 2 to 30% nickel. *Nickel silvers* contain no silver, but 10 to 30% nickel and at least 5% zinc. The bright silvery luster makes them attractive for ornamental applications, and they are also used for musical instruments. An alloy with 45% nickel is known as *constantan*, and the 67%-nickel material is called *Monel*. Monel will be discussed later in the chapter as a nickel alloy.

OTHER COPPER-BASED ALLOYS

The copper alloys discussed previously acquire their strength primarily through solid-solution strengthening and cold work. Within the copper-alloy family, alloys containing aluminum, silicon, or beryllium can be strengthened by precipitation hardening.

Aluminum–bronze alloys are best known for their combination of high strength and excellent corrosion resistance, and they are often considered to be cost-effective alternatives to stainless steel and nickel-based alloys. The wrought alloys can be strengthened by solid-solution strengthening, cold work, and the precipitation of iron- or nickel-rich phases. With less than 8% aluminum, the alloys are very ductile. When aluminum exceeds 9%, however, the ductility drops and the hardness approaches that of steel. Still higher aluminum contents result in brittle, but wear-resistant, materials. By varying the aluminum content and heat treatment, the tensile strength can range from about 415 to 1000 MPa (60 to 145 ksi). Typical applications include marine hardware, power shafts, sleeve bearings, and pump and valve components for handling seawater, sour mine water, and various industrial fluids. Cast alloys are available for applications where casting is the preferred means of manufacture. Since aluminum bronze exhibits large amounts of solidification shrinkage, castings made of this material should be designed with this in mind.

Silicon–bronzes contain up to 4% silicon and 1.5% zinc (higher zinc contents may be used when the material is to be cast). Strength, formability, machinability, and corrosion resistance are all quite good. Tensile strengths range from a soft condition of about 380 MPa (55 ksi) through a maximum that approaches 900 MPa (130 ksi). Uses include boiler, tank, and stove applications, which require a combination of weldability, high strength, and corrosion resistance.

Copper–beryllium alloys, which ordinarily contain less than 2% beryllium, can be age hardened to produce the highest strengths of the copper-based metals but are quite expensive to use. When annealed, the material has a yield strength of 170 MPa (25 ksi), tensile strength of 480 MPa (70 ksi), and an elongation of 50%. After heat treatment, these properties can rise to 1100 MPa (160 ksi), 1250 MPa (180 ksi), and 5%, respectively. Cold work coupled with age hardening can produce even stronger material. The modulus of elasticity is about 125,000 MPa (8×10^6 psi), and the endurance limit is around 275 MPa (40 ksi). These properties make the material an excellent choice for electrical contact springs, but cost limits application to small components requiring long life and high reliability. Other applications, such as spark-resistant safety tools and spot-welding electrodes, utilize the unique combination of properties: (1) the material has the strength of heat-treated steel, but is also (2) nonsparking, nonmagnetic, and electrically and thermally conductive. Concerns over the toxicity of beryllium have created a demand for substitute alloys with similar properties, but no clear alternative has emerged.

LEAD-FREE CASTING ALLOYS

For many years, lead has been a common alloy additive to cast copper alloys. It helped to fill and seal the microporosity that forms during solidification, thereby providing the pressure tightness required for use with pressurized gases and fluids. The lead also acted as a lubricant and chip-breaker, enhancing the machinability and machined surface finish. Many plumbing components have been made from leaded red and semi-red brass casting alloys.

With increased concern about lead in drinking water and the introduction of environmental regulations, efforts were made to develop lead-free copper-based casting alloys. Among the most common are the EnviroBrass alloys, which use *bismuth* and *selenium* as substitutes for lead. Bismuth is not known to be toxic for humans and has been used in a popular remedy for an upset stomach. Selenium is an essential nutrient for humans. While somewhat lower in ductility, the new alloys have been shown to have mechanical properties, machinability and platability that are quite similar to the traditional leaded materials.

■ 7.3 ALUMINUM AND ALUMINUM ALLOYS

GENERAL PROPERTIES AND CHARACTERISTICS

Although *aluminum* has only been a commercial metal for about 120 years, it now ranks second to steel in both worldwide quantity and expenditure, and it is clearly the most important of the nonferrous metals. It has achieved importance in virtually all segments of the economy, with principal uses in transportation, containers and packaging, building construction, electrical applications, consumer durables, and mechanical equipment. We are all familiar with uses such as aluminum cookware, window frames, aluminum siding, and the ever-present aluminum beverage can.

A number of unique and attractive properties account for the engineering significance of aluminum. These include its workability, light weight, corrosion resistance, good electrical and thermal conductivity, optical reflectivity, and a nearly limitless array of available finishes. Aluminum has a specific gravity of 2.7 compared to 7.85 for steel, making aluminum about one-third the weight of steel for an equivalent volume. Cost comparisons are often made on the basis of cost per pound, where aluminum is at a distinct disadvantage, being four to five times more expensive than carbon steel. There are a number of applications, however, where a more appropriate comparison would be based on cost per unit volume. A pound of aluminum produces three times as many same-size parts as a pound of steel, so the cost difference becomes markedly less.

Aluminum can be recycled repeatedly with no loss in quality, and recycling saves 95% of the energy required to produce aluminum from ore. Since the 1980s, the overall reclamation rate for aluminum has been over 50%. The aluminum can is the most recycled beverage container in North America, and over 85% of all aluminum used in cars is recovered at the end of their useful life.

A serious weakness of aluminum from an engineering viewpoint is its relatively low modulus of elasticity, which is also about one-third that of steel. Under identical loadings, an aluminum component will deflect three times as much as a steel component of the same design. Since the modulus of elasticity cannot be significantly altered by alloying or heat treatment, it is usually necessary to provide stiffness and buckling resistance through design features such as ribs or corrugations. These can be incorporated with relative ease, however, because aluminum adapts easily to the full spectrum of fabrication processes.

COMMERCIALLY PURE ALUMINUM

In its pure state, aluminum is soft, ductile, and not very strong. In the annealed condition, pure aluminum has only about one-fifth the strength of hot-rolled structural steel. Commercially pure aluminum, therefore, is used primarily for its physical rather than its mechanical properties.

Electrical-conductor-grade aluminum is used in large quantities and has replaced copper in many applications, such as electrical transmission lines. Commonly designated by the letters *EC*, this grade contains a minimum of 99.45% aluminum and has an electrical conductivity that is 62% that of copper for the same-size wire and 200% that of copper on an equal-weight basis.

ALUMINUMS FOR MECHANICAL APPLICATIONS

For nonelectrical applications, most aluminum is used in the form of alloys. These have much greater strength than pure aluminum yet retain the advantages of light weight, good conductivity, and corrosion resistance. While usually weaker than steel, some alloys are now available that have tensile properties (except for ductility) that are comparable to those of the high-strength low-alloy (HSLA) structural grades. Since alloys can be as much as 30 times stronger than pure aluminum, designers can frequently optimize their design and then tailor the material to their specific requirements. Some alloys are specifically designed for casting, while others are intended for the manufacture of wrought products.

On a strength-to-weight basis, most of the aluminum alloys are superior to steel and other structural metals, but wear, creep, and fatigue properties are generally rather poor. Aluminum alloys have a finite fatigue life at all reasonable values of applied stress. In addition, aluminum alloys rapidly lose their strength and dimensions change by creep when temperature is increased. As a result, most aluminum alloys should not be considered for applications involving service temperatures much above 150°C(300°F). At subzero temperatures, however, aluminum is actually stronger than at room temperature with no loss in ductility. Both the adhesive and the abrasive varieties of wear can be extremely damaging to aluminum alloys.

The selection of steel or aluminum for any given component is often a matter of cost, but considerations of light weight, corrosion resistance, low maintenance expense, and high thermal or electrical conductivity may be sufficient to justify the added cost of aluminum. With the drive for lighter, more fuel-efficient vehicles, the fraction of aluminum targeted for transportation applications rose from 19.4% in 1992 to 31.8% in 2002. The use of aluminum doubled in cars and tripled in sport-utility vehicles (SUVs) and light trucks. Aluminum is being used in body panels, engine blocks, manifolds, transmission housings, and wheels. An aluminum space frame, such as the ones shown in Figure 7-3 for the 2005 Ford GT and the 2006 Corvette Z06, can reduce the overall weight of the structure, enhance recyclability, and reduce the number of parts required for the primary body structure. The all-aluminum space frame of the 2006 Z06 Corvette resulted in a 30% reduction in weight from the all-steel design of the previous model. In 2001, aluminum passed plastics as a percentage of automotive material content and is now second only to steel and iron. The average North American automobile now contains over 125 kg (280 pounds) of aluminum.

(a)

(b)

FIGURE 7-3 (a) The space frame chassis for the 2005 Ford GT is comprised of 35 aluminum extrusions, 7 complex castings, 2 semisolid castings, and various aluminum panels, some superplastically formed. (b) The aluminum frame of the 2006 Corvette Z06 yielded a 30% weight savings compared to the previous steel design. *[(a) Courtesy Ford Motor Company, Dearborn, MI; and HydroAluminum of North America, Linthicum, MD). (b) (Courtesy of General Motors, Detroit, MI.)]*

CORROSION RESISTANCE OF ALUMINUM AND ITS ALLOYS

Pure aluminum is very reactive and forms a tight, adherent oxide coating on the surface as soon as it is exposed to air. This oxide is resistant to many corrosive media and serves as a corrosion-resistant barrier to protect the underlying metal. Like stainless steels, the corrosion resistance of aluminum is actually a property of the oxide, not the metal itself. Since the oxide formation is somewhat retarded when alloys are added, aluminum alloys do not have quite the corrosion resistance of pure aluminum.

The oxide coating also causes difficulty when welding. To produce consistent-quality resistance welds, it is usually necessary to remove the tenacious oxide immediately before welding. For fusion welding, special fluxes or protective inert gas atmospheres must be used to prevent material oxidation. While welding aluminum may be more difficult than steel, suitable techniques have been developed to permit the production of high-quality, cost-effective welds with most of the welding processes.

CLASSIFICATION SYSTEM

Aluminum alloys can be divided into two major groups based on the method of fabrication. *Wrought alloys* are those that are shaped as solids and are therefore designed to have attractive forming characteristics, such as low yield strength, high ductility, good fracture resistance, and good strain hardening. *Casting alloys* achieve their shape as they solidify in molds or dies. Attractive features for the casting alloys include low melting point, high fluidity, and attractive as-solidified structures and properties. Clearly, these properties are distinctly different, and the alloys that have been designed to meet them are also different. As a result, separate classification systems exist for the wrought and cast aluminum alloys.

WROUGHT ALUMINUM ALLOYS

The wrought aluminum alloys are generally identified using the standard four-digit designation system for aluminums. The first digit indicates the major alloy element or elements as described below:

Major Alloying Element	
Aluminum, 99.00% and greater	1xxx
Copper	2xxx
Manganese	3xxx
Silicon	4xxx
Magnesium	5xxx
Magnesium and silicon	6xxx
Zinc	7xxx
Other element	8xxx

The second digit is usually zero. Nonzero numbers are used to indicate some form of modification or improvement to the original alloy. The last two digits simply indicate the particular alloy within the family. For example, 2024 simply means alloy number 24 within the 2xxx, or aluminum–copper, system. For the 1xxx series, the last three digits are used to denote the purity of the aluminum.

The four digits of a wrought aluminum designation identify the chemistry of the alloy. Additional information about the alloy condition is then provided through a *temper designation*, in the form of a letter or letter–number suffix using the following system:

-F: as fabricated

-H: strain-hardened

    -H1: strain-hardened by working to desired dimensions; a second digit, 1 through 9, indicates the degree of hardening, 8 being commercially full-hard and 9 extra-hard

    -H2: strain-hardened by cold working, followed by partial annealing

    -H3: strain-hardened and stabilized

-O: annealed

-T: thermally treated (heat treated)

-T1: cooled from hot working and naturally aged

-T2: cooled from hot working, cold-worked, and naturally aged

-T3: solution-heat-treated, cold-worked, and naturally aged

-T4: solution-heat-treated and naturally aged

-T5: cooled from hot working and artificially aged

-T6: solution-heat-treated and artificially aged

-T7: solution-heat-treated and stabilized

-T8: solution-heat-treated, cold-worked, and artificially aged

-T9: solution-heat-treated, artificially aged, and cold-worked

-T10: cooled from hot working, cold-worked, and artificially aged

-W: solution-heat-treated only

The various wrought alloys are often divided into two basic types: those that achieve strength by solid-solution strengthening and cold working, and those that can be strengthened by heat treatment (age hardening). Table 7-4 lists some of the common wrought aluminum alloys in each family. It can be noted that the work-hardenable alloys (those that cannot be age hardened) are primarily those in the 1xxx (pure aluminum), 3xxx (aluminum– manganese), and 5xxx (aluminum–magnesium) series. A comparison of the annealed (*O* suffix) and cold-worked (*H* suffix) conditions reveals the amount of strengthening achievable through strain hardening.

The precipitation-hardenable alloys are found primarily in the 2xxx, 6xxx, and 7xxx series. By comparing the properties in the heat-treated condition to those of the strain-hardened alloys, we see that heat treatment offers significantly higher strength. Alloy 2017, the original *duralumin*, is probably the oldest age-hardenable aluminum alloy. The 2024 alloy is stronger and has seen considerable use in aircraft applications. An attractive feature of the 2xxx series is the fact that ductility does not significantly decrease during the strengthening heat treatment. Within the 7xxx series are some newer alloys with strengths that approach or exceed those of the high-strength structural steels. Ductility, however, is generally low, and fabrication is more difficult than for the 2xxx-type alloys. Nevertheless, the 7xxx series alloys have also found wide use in aircraft applications. To maintain properties, age-hardened alloys should not be used at temperatures over 175°C (350°F). Welding should be performed with considerable caution since the exposure to elevated temperature will significantly diminish the strengthening achieved through either cold working or age hardening.

Because of their two-phase structure, the heat-treatable alloys tend to have poorer corrosion resistance than either pure aluminum or the single-phase work-hardenable alloys. When both high strength and superior corrosion resistance are desired, wrought aluminum is often produced as *Alclad* material. A thin layer of corrosion-resistant aluminum is bonded to one or both surfaces of a high-strength alloy during rolling, and the material is further processed as a composite.

Because only moderate temperatures are required to lower the strength of aluminum alloys, extrusions and forgings are relatively easy to produce and are manufactured in large quantities. Deep drawing and other sheet-metal-forming operations can also be carried out quite easily. In general, the high ductility and low yield strength of the aluminum alloys make them appropriate for almost all forming operations. Good dimensional tolerances and fairly intricate shapes can be produced with relative ease.

The machinability of aluminum-based alloys, however, can vary greatly, and special tools and techniques may be desirable if large amounts of machining are required. Free-machining alloys, such as 2011, have been developed for screw-machine work. These special alloys can be machined at very high speeds and have replaced brass screw-machine stock in many applications.

TABLE 7-4 Composition and Properties of Some Wrought Aluminum Alloys in Various Conditions

Designation[a]	Cu	Si	Mn	Mg	Others	Form Tested	Tensile Strength ksi	MPa	Yield Strength[b] ksi	MPa	Elongation in 2 in. (%)	Brinell Hardness	Uses and Characteristics
	\multicolumn{5}{} Composition (%) Aluminum = Balance											
Work-Hardening Alloys—Not Heat-Treatable													
1100–0	0.12				99 Al	$\frac{1}{16}$-in. sheet	13	90	5	34	35	23	Commercial Al: good forming properties
1100–H14						$\frac{1}{16}$-in. sheet	16	110	14	97	9	32	Good corrosion resistance, low yield strength
1100–H18						$\frac{1}{16}$-in. sheet	24	165	21	145	5	44	Cooking utensils; sheet and tubing
3003–0	0.12		1.2			$\frac{1}{16}$-in. sheet	16	110	6	41	30	28	Similar to 1100
3003–H14						$\frac{1}{16}$-in. sheet	22	152	21	145	8	40	Slightly stronger and less ductile
3003–H18						$\frac{1}{16}$-in. sheet	29	200	27	186	4	55	Cooking utensils; sheet-metal work
5052–0				2.5		$\frac{1}{16}$-in. sheet	28	193	13	90	25	45	Strongest work-hardening alloy
5052–H32					0.25 Cr	$\frac{1}{16}$-in. sheet	33	228	28	193	12	60	Highly yield strength and fatigue limit
5052–H36						$\frac{1}{16}$-in. sheet	40	276	35	241	8	73	Highly stressed sheet-metal products
Precipitation-Hardening Alloys—Heat-Treatable													
2017–0	4.0	0.5	0.7	0.6		$\frac{1}{16}$-in. sheet	26	179	10	69	20	45	Duralumin, original strong alloy
2017–T4						$\frac{1}{16}$-in. sheet	62	428	40	276	20	105	Hardened by quenching and aging
2024–0	4.4		0.6	1.5		$\frac{1}{16}$-in. sheet	27	186	11	76	20	42	Stronger than 2017
2024–T4						$\frac{1}{16}$-in. sheet	64	441	45	290	19	120	Used widely in aircraft construction
2014–0	4.4	0.8	0.8	0.5		$\frac{1}{2}$-in. extruded shapes	27	186	14	97	12	45	Strong alloy for extruded shapes
2014–T6						Forgings	65	448	55	379	10	125	Strong forging alloy
2014–T6						$\frac{1}{16}$-in. sheet	70	483	60	413	8		Higher yield strength than Alclad 2024
Alclad 2014–T6	4.5	1.0	0.8	0.4	0.3 Cr	$\frac{1}{16}$-in. sheet	63	434	56	386	7	60	Clad with heat-treatable alloy[c]
7075–0	1.6		0.2	2.5	5.6 Zn	$\frac{1}{16}$-in. sheet	33	228	15	103	17		Alloy of highest strength
7075–T6						$\frac{1}{16}$-in. sheet	76	524	67	462	11	150	Lower ductility than 2024
Alclad 7075–T6						$\frac{1}{16}$-in. sheet	76	524	67	462	11		Strongest Alclad product
7075–T6						$\frac{1}{2}$-in. extruded shapes	80	552	70	483	6		Strongest alloy for extrusions
6061–T6	0.28	0.6		1.0	0.20 Cr	$\frac{1}{2}$-in. extruded shapes	42	290	40	276	12	95	Strong, corrosion resistant
6063–T6		0.4		0.7		extruded	35	241	31	214	12	80	Good forming properties and corrosion resistance
6151–T6		0.9		0.6	0.25 Cr	Forgings	48	331	43	297	17	90	For intricate forgings
2025–T6	4.5	0.8	0.8			Forgings	55	379	30	207	18	100	Good forgeability, lower cost
2018–T6	4			0.7	2 Ni	Forgings	55	379	40	276	10	100	Strong at elevated temperatures; forged pistons
4032–T6	0.9	12.2		1.1	0.9 Ni	Forgings	55	379	46	317	9	115	Forged aircraft pistons
2011–T3	5.5			(0.5 Bi)	0.5 Pb	$\frac{1}{2}$-in. rod	55	379	43	297	15	95	Free cutting, screw-machine products

[a] O, annealed; T, quenched and aged; H, cold-rolled to hard temper.

[b] Yield strength taken at 0.2% permanent set.

[c] Cladding alloy: 1.0 Mg, 0.7 Si, 0.5 Mn.

149

Color anodizing offers an inexpensive and attractive means of surface finishing. A thick aluminum oxide is produced on the surface. Colored dye is then placed on the porous surface and is sealed by immersion into hot water. The result is the colored metallic finish commonly observed on products such as bicycle frames and softball bats.

ALUMINUM CASTING ALLOYS

Although its low melting temperature tends to make it suitable for casting, pure aluminum is seldom cast. Its high shrinkage upon solidification (about 7%) and susceptibility to hot cracking cause considerable difficulty, and scrap is high. By adding small amounts of alloying elements, however, very suitable casting characteristics can be obtained and strength can be increased. Aluminum alloys are cast in considerable quantity by a variety of processes. Many of the most popular alloys contain enough silicon to produce the eutectic reaction, which is characterized by a low melting point and high as-cast strength. Silicon also improves the fluidity of the metal, making it easier to produce complex shapes or thin sections, but high silicon also produces an abrasive, difficult-to-cut material. Copper, zinc, and magnesium are other popular alloy additions that permit the formation of age-hardening precipitates.

Table 7-5 lists some of the commercial aluminum casting alloys and uses the three-digit designation system of the Aluminum Association to designate alloy chemistry. The first digit indicates the alloy group as follows:

Major Alloying Element	
Aluminum, 99.00% and greater	1xx.x
Copper	2xx.x
Silicon with Cu and/or Mg	3xx.x
Silicon	4xx.x
Magnesium	5xx.x
Zinc	7xx.x
Tin	8xx.x
Other elements	9xx.x

The second and third digits identify the particular alloy or aluminum purity, and the last digit, separated by a decimal point, indicates the product form (e.g., casting or ingot). A letter before the numerical designation indicates a modification of the original alloy, such as a small variation in the amount of an alloying element or impurity.

Aluminum casting alloys have been designed for both properties and process. When the strength requirements are low, as-cast properties are usually adequate. High-strength castings usually require the use of alloys that can subsequently be heat treated. Sand casting has the fewest process restrictions. The aluminum alloys used for permanent mold casting are designed to have lower coefficients of thermal expansion (or contraction) because the molds offer restraint to the dimensional changes that occur upon cooling. Die-casting alloys require high degrees of fluidity because they are often cast in thin sections. Most of the die-casting alloys are also designed to produce high "as-cast" strength without heat treatment, using the rapid cooling conditions of the die-casting process to promote a fine grain size and fine eutectic structure. Tensile strengths of the aluminum permanent-mold and die-casting alloys can be in excess of 275 MPa (40 ksi).

ALUMINUM–LITHIUM ALLOYS

Lithium is the lightest of all metallic elements, and in the search for aluminum alloys with higher strength, greater stiffness, and lighter weight, aluminum–lithium alloys have emerged. Each percent of lithium reduces the overall weight by 3% and increases stiffness by 6%. The initially developed alloys offered 8 to 10% lower density, 15 to 20% greater stiffness, strengths comparable to those of existing alloys, and good resistance to fatigue crack propagation. Unfortunately, fracture toughness, ductility, and stress–corrosion resistance were poorer than for conventional alloys. The current-generation

TABLE 7-5 Composition, Properties, and Uses of Some Aluminum Casting Alloys

Alloy Designation[a]	Process[b]	Composition (%) (Major Alloys > 1%) Cu	Si	Mg	Zn	Fe	Other	Temper	Tensile Strength ksi[c]	MPa	Elongation in 2 in. (%)	Uses and Chracteristics
208	S	4.0	3.0		1.0	1.2		F	19	1.31	1.5	General-purposes and castings, can be heat treated
242	S, P	4.0		1.6		1.0	2.0 Ni	T61	40	276	–	Withstands elevated temperatures
295	S	4.5	1.0			1.0		T6	32	221	3.0	Structural castings, heat-treatable
296	P	4.5	2.5			1.2		T6	35	241	2.0	Permanent-mold version of 295
308	P	4.5	5.5		1.0	1.0		F	24	166	–	General-purpose permanent mold
319	S, P	3.5	6.0		1.0	1.0		T6	31	214	1.5	Superior casting characteristics
354	P	1.8	9.0					–	–	–	–	High-strength, aircraft
355	S, P	1.3	5.0					T6	32	221	2.0	High strength and pressure tightness
C355	S, P	1.3	5.0					T61	40	276	3.0	Stronger and more ductile than 355
356	S, P		7.0					T6	30	207	3.0	Excellent castability and impact strength
A356	S, P		7.0					T61	37	255	5.0	Stronger and more ductile than 356
357	S, P		7.0					T6	45	310	3.0	High strength-to-weight castings
359	S, P		9.0					–	–	–	–	High-strength aircraft usage
360	D		9.5			2.0		F	44[d]	303	2.5[d]	Good corrosion resistance and strength
A360	D		9.5			2.0		F	46[d]	317	3.5[d]	Similar to 360
380	D	3.5	8.5		3.0	2.0		F	46[d]	317	2.5[d]	High strength and hardness
A380	D	3.5	8.5		3.0	1.3		F	47[d]	324	3.5[d]	Similar to 380
383	D	1.5	10.5		3.0	1.3		F	45[d]	310	3.5[d]	High strength and hardness
384	D	3.75	11.3		1.0	1.3		F	48[d]	331	2.5	High strength and hardness
413	D	1.0	12.0			2.0		F	43[d]	297	2.5[d]	General-purpose, good castability
A413	D	1.0	12.0			1.3		F	42[d]	290	3.5[d]	Similar to 413
443	D	5.25				2.0		F	33[d]	228	9.0[d]	General-purpose, good castability
B443	S, P	5.25				2.0		F	17	117	3.0	General-purpose casting alloy
514	S			4.0				F	22	152	6.0	High corrosion resistance
518	D			8.0		1.8		F	45[d]	310	5.0[d]	Good corrosion resistance, strength, and toughness
520	S			10.0				T4	42	290	12.0	High strength with good ductility
535	S			6.9				F	35	241	9.0	Good corrosion resistance and machinability
712	S				5.8			F	34	234	4.0	Good properties without heat treatment
713	S, P				7.5	1.1		F	32	221	3.0	Similar to 712
771	S				7.0			T6	42	290	5.0	Aircraft and computer components
850	S, P	1.0					6.3 Sn, 1.0 Ni	T5	16	110	5.0	Bearing alloy

[a] Aluminum Association.   [b] Sand-cast; P, permanent-mold-cast; D, die cast.   [c] Minimum figures unless noted.   [d] Typical values.

alloys are aluminum–copper–lithium, with about 4% copper and no more than 2% lithium. The weight benefits are still sufficient to warrant use in a number of aerospace applications, and the fact that they can be fabricated by conventional processes make them attractive alternatives to the advanced composites.

Since aluminum alloys can comprise as much as 80% of the weight of commercial aircraft, even small percentage reductions can be significant. Improved strength and stiffness can further facilitate weight reduction. Fuel savings over the life of the airplane would more than compensate for any additional manufacturing expense. As an example of potential, the weight of the external liquid-hydrogen tank on the U.S. space shuttle booster rocket was reduced by approximately 3400 kg (7500 lb) by conversion to an aluminum–lithium alloy.

ALUMINUM FOAM

A material known as "stabilized aluminum foam" can be made by mixing ceramic particles with molten aluminum and blowing gas into the mixture. The bubbles remain through solidification, yielding a structure that resembles metallic Styrofoam. Originally developed around 2000 for automotive, aerospace, and military applications, the material has found additional uses in architecture and design. Strength-to-weight is outstanding, and the material offers excellent energy absorption. The fuel cells of race cars have been shrouded with aluminum foam, and foam fill has been inserted between the front of cars and the driver compartment. Tubular structures can be filled with foam to increase strength, absorb energy, and provide resistance to crushing. Still other applications capitalize on the excellent thermal insulation, vibration damping, and sound absorption that results from the numerous trapped air pockets.

■ 7.4 MAGNESIUM AND MAGNESIUM ALLOYS

GENERAL PROPERTIES AND CHARACTERISTICS

Magnesium is the lightest of the commercially important metals, having a specific gravity of about 1.74 (two-thirds that of aluminum, one-fourth that of steel, and only slightly higher than fiber-reinforced plastics). Like aluminum, magnesium is relatively weak in the pure state and for engineering purposes is almost always used as an alloy. Even in alloy form, however, the metal is characterized by poor wear, creep, and fatigue properties. It has the highest thermal expansion of all engineering metals. Strength drops rapidly when the temperature exceeds 100°C (200°F), so magnesium should not be considered for elevated-temperature service. Its modulus of elasticity is even less than that of aluminum, being between one-fourth and one-fifth that of steel. Thick sections are required to provide adequate stiffness, but the alloy is so light that it is often possible to use thicker sections for the required rigidity and still have a lighter structure than can be obtained with any other metal. Cost per unit volume is low, so the use of thick sections is generally not prohibitive. Moreover, since a large portion of magnesium components are cast, the thicker sections actually become a desirable feature. Ductility is frequently low, a characteristic of the hexagonal-close-packed (HCP) crystal structure, but some alloys have values exceeding 10%.

On the more positive side, magnesium alloys have a relatively high strength-to-weight ratio, with some commercial alloys attaining strengths as high as 380 MPa (55 ksi). High energy absorption means good damping of noise and vibration, as well as impact and dent resistance. While many magnesium alloys require enamel or lacquer finishes to impart adequate corrosion resistance, this property has been improved markedly with the development of higher-purity alloys. In the absence of unfavorable galvanic couples, these materials have excellent corrosion resistance and are finding applications in a wide range of markets, including automotive, aerospace, power tools, sporting goods, and electronic products (where they offer a combination of electromagnetic shielding, light weight, and durability exceeding that of plastics and alternative metals). While aluminum alloys are often used for the load-bearing members of mechanical structures, magnesium alloys are best suited for those applications where lightness is the primary consideration and strength is a secondary requirement.

MAGNESIUM ALLOYS AND THEIR FABRICATION

A designation system for magnesium alloys has been developed by the ASTM, identifying both chemical composition and temper, and is presented in specification B93. Two prefix letters designate the two largest alloying metals in order of decreasing amount, using the following format:

A	aluminum	F	iron	M	manganese	R	chromium
B	bismuth	H	thorium	N	nickel	S	silicon
C	copper	K	zirconium	P	lead	T	tin
D	cadmium	L	beryllium	Q	silver	Z	zinc
E	rare earth						

Aluminum is the most common alloying element and, along with zinc, zirconium, and thorium, promotes precipitation hardening. Manganese improves corrosion resistance, and tin improves castability. The two letters are then followed by two or three numbers and a possible suffix letter. The numbers correspond to the rounded-off whole-number percentages of the two main alloy elements and are arranged in the same order as the letters. Thus the AZ91 alloy would contain approximately 9% aluminum and 1% zinc. A suffix letter is used to denote variations of the same base alloy, such as AZ91A. The temper-designation suffix is quite similar to that used with the aluminum alloys. Table 7-6 lists some of the more common magnesium alloys together with their properties and uses.

Sand, permanent-mold, die, semisolid, and investment casting are all well developed for magnesium alloys and take advantage of the low melting points and high fluidity. Die casting is clearly the most popular manufacturing process for magnesium, accounting for 70% of all castings. Although the magnesium alloys typically cost about twice as much as aluminum, the hot-chamber die-casting process used with magnesium is easier, more economical, and 40 to 50% faster than the cold-chamber process generally required for aluminum. Wall thickness, draft angle, and dimensional tolerances are all lower than for both aluminum die castings and thermoplastic moldings. Die life is significantly greater than that observed with aluminum. As a result, magnesium die castings compete well with aluminum[2] and often replace plastic injection-molded components when improved stiffness or dimensional stability, or the benefits of electrical or thermal conductivity, are required.

Forming behavior is poor at room temperature, but most conventional processes can be performed when the material is heated to temperatures between 250° and 500°C (480° and 775°F). Since these temperatures are easily attained and generally do not require a protective atmosphere, many formed and drawn magnesium products are manufactured. Magnesium extrusions and sheet metal products have properties similar to the more common wrought aluminum alloys. While slightly heavier than plastics, they offer an order of magnitude or greater improvement in stiffness or rigidity.

The machinability of magnesium alloys is the best of any commercial metal and, in many applications, the savings in machining costs, achieved through deeper cuts, higher cutting speeds, and longer tool life, more than compensate for the increased cost of the material. It is necessary, however, to keep the tools sharp and provide adequate cooling for the chips.

Magnesium alloys can be spot welded almost as easily as aluminum, but scratch brushing or chemical cleaning is necessary before forming the weld. Fusion welding is best performed with processes using an inert shielding atmosphere of argon or helium gas.

While heat treatments can be used to increase strength, the added increment achieved by age hardening is far less than observed with aluminum. In fact, the strongest magnesium alloy is only about three times stronger than the weakest. Because of this, designs must be made to accommodate the material, rather than the material being tailored to the design.

Considerable misinformation exists regarding the fire hazards when processing or using magnesium alloys. It is true that magnesium alloys are highly combustible

---

[2] The most common magnesium die-casting alloy, AZ91, has the same yield strength and ductility as the most common die-cast aluminum, alloy 380.

TABLE 7-6   Composition, Properties, and Characteristics of Common Magnesium Alloys

Alloy	Temper	Composition (%)						Tensile Strength[a]		Yield Strength[a]		Elongation in 2 in. (%)	Uses and Characteristics
		Al	Rare Earths	Mn	Th	Zn	Zr	ksi	MPa	ksi	MPa		
AM60A	F	6.0		0.13				30	207	17	117	6	Die castings
AM100A	T4	10.0		0.1				34	234	10	69	6	Sand and permanent-mold castings
AZ31B	F	3.0				1.0		32	221	15	103	6	Sheet, plate, extrusions, forgings
AZ61A	F	6.5				1.0		36	248	16	110	7	Sheet, plate, extrusions forgings
AZ63A	T5	6.0				3.0		34	234	11	76	7	Sand and permanent-mold castings
AZ80A	T5	8.5				0.5		34	234	22	152	2	High-strength forgings, extrusions
AZ81A	T4	7.6				0.7		34	234	11	76	7	Sand and permanent-mold castings
AZ91A	F	9.0				0.7		34	234	23	159	3	Die castings
AZ92A	T4	9.0				2.0		34	234	11	76	6	High-strength sand and permanent-mold castings
EZ33A	T5		3.2			2.6	0.7	20	138	14	97	2	Sand and permanent-mold castings
HK31A	H24				3.2		0.7	33	228	24	166	4	Sheet and plates; castings in T6 temper
HM21A	T5			0.8	2.0			33	228	25	172	3	High-temperature (800°F) sheets, plates, forgings
HZ32A	T5				3.2	2.1		27	186	13	90	4	Sand and permanent-mold castings
ZH62A	T5				1.8	5.7	0.7	35	241	22	152	5	Sand and permanent-mold castings
ZK51A	T5					4.6	0.7	34	234	20	138	5	Sand and permanent-mold castings
ZK60A	T5					5.5	0.45	38	262	20	138	7	Extrusions, forgings

[a] Properties are minimums for the designated temper.

when in a finely divided form, such as powder or fine chips, and this hazard should never be ignored. In the form of sheet, bar, extruded product, or finished castings, however, magnesium alloys rarely present a fire hazard. When the metal is heated above 700°C (950°F), a noncombustible, oxygen-free atmosphere is recommended to suppress burning, which will initiate around 600°C (1100°F). Casting operations often require additional precautions due to the reactivity of magnesium with sand and water.

■ 7.5 ZINC-BASED ALLOYS

Over 50% of all metallic *zinc* is used in the *galvanizing* of iron and steel. In this process the iron-based material is coated with a layer of zinc by one of a variety of processes that include direct immersion in a bath of molten metal (hot dipping) and electrolytic plating. The resultant coating provides excellent corrosion resistance, even when the surface is badly scratched or marred. Moreover, the corrosion resistance will persist until all of the sacrificial zinc has been depleted.

Zinc is also used as the base metal for a variety of die-casting alloys. For this purpose, zinc offers low cost, a low melting point (only 380°C or 715°F), and the attractive

TABLE 7-7 Composition and Properties of Some Zinc Die-Casting Alloys

Alloy	#3 SAE 903 ASTM AG40A	#5 SAE 925 ASTM AC41A	#7 ASTM AG408	ZA-8 S[a]	ZA-8 P	ZA-8 D	ZA-12 S	ZA-12 P	ZA-12 D	ZA-27 S	ZA-27 P	ZA-27 D
Composition[b]												
Aluminum	3.5–4.3	3.5–4.3	3.5–4.3	8.0–8.8			10.5–11.5			25.0–28.0		
Copper	0.25 max	0.75–1.25	0.25 max	0.8–1.3			0.5–1.2			2.0–2.5		
Zinc	balance	balance	balance	balance			balance			balance		
Properties												
Density (g/cc)	6.6	6.6	6.6	6.3			6.0			5.0		
Yield strength (MPa)	221	228	221	200	206	290	214	269	317	372		379
(ksi)	32	33	32	29	30	42	31	39	46	54		55
Tensile strength (MPa)	283	328	283	263	255	374	317	345	400	441		421
(ksi)	41	48	41	38	37	54	46	50	58	64		61
Elongation (% in 2 in.)	10	7	13	2	2	10	3	3	7	6		3
Impact strength (J)	58	65	58	20		42	25		29	47		5
Modulus of elasticity (GPa)	85.5	85.5	85.5	85.5			82.7			77.9		
Machinability[c]	E	E	E	E			VG			G		

[a]S, sand-cast; P, permanent-mold cast; D, die-cast.
[b]Also contains small amounts of Fe, Pb, Cd, Sn, and Ni.
[c]E, excellent; VG, very good; G, good.

property of not adversely affecting steel dies when in contact with molten metal. Unfortunately, pure zinc is almost as heavy as steel and is also rather weak and brittle. Therefore, when alloys are designed for die casting, the alloy elements are usually selected for their ability to increase strength and toughness in the as-cast condition while retaining the low melting point.

The composition and properties of common zinc die-casting alloys are presented in Table 7-7. Alloy AG40A (also known as alloy 903 or Zamak 3) is widely used because of its excellent dimensional stability, and alloy AC41A (also known as alloy 925 or Zamak 5) offers higher strength and better corrosion resistance. As a whole, the zinc die-casting alloys offer a reasonably high strength and impact resistance, along with the ability to be cast to close dimensional limits with extremely thin sections. The dimensions are quite stable, and the products can be finish machined at a minimum of cost. Resistance to surface corrosion is adequate for a number of applications, and the material can be surface finished by a variety of means that include polishing, plating, painting, anodizing, or a chromate conversion coating. Energy costs are low (low melting temperature), tool life is excellent, and the zinc alloys can be efficiently recycled. While the rigidity is low compared to that of other metals, it is far superior to engineering plastics, and zinc die castings often compete with plastic injection moldings.

The attractiveness of zinc die casting has been further enhanced by the zinc–aluminum casting alloys (ZA-8, ZA-12, and ZA-27, with 8, 12, and 27% aluminum, respectively). Initially developed for sand, permanent-mold, and graphite-mold casting, these alloys can also be die cast to achieve higher strength (up to 60 ksi or 415 MPa), hardness (up to 120 BHN), creep resistance and wear resistance, and lighter weight than is possible with any of the conventional alloys. Because of their lower melting and casting costs, these materials are becoming attractive alternatives to the conventional aluminum, brass, and bronze casting alloys, as well as cast iron.

■ 7.6 TITANIUM AND TITANIUM ALLOYS

Titanium is a strong, lightweight, corrosion-resistant metal that has been of commercial importance since about 1950. Because its properties are generally between those of steel and aluminum, its importance has been increasing rapidly. The yield strength of commercially pure titanium is about 210 MPa (30 ksi), but this can be raised to 1300 MPa (190 ksi) or higher through alloying and heat treatment, a strength comparable to that of many heat-treated alloy steels. Density, on the other hand, is only 56% that of steel (making strength-to-weight quite attractive), and the modulus of elasticity ratio is also about one-half. Good mechanical properties are retained up to temperatures of 535°C (1000°F), so the metal is often considered to be a high-temperature engineering material. On the negative side, titanium and its alloys suffer from high cost, fabrication difficulties, a high energy content (they require about 10 times as much energy to produce as steel), and a high reactivity at elevated temperatures (above 535°C).

Titanium alloys are designated by major alloy and amount (see ASTM specification B-265), and are generally grouped into three classes based on their microstructural features. These classes are known as alpha-, beta-, and alpha-beta-titanium alloys, the terms denoting the stable phase or phases at room temperature. Alloying elements can be used to stabilize the hexagonal-close-packed alpha phase or the body-centered-cubic beta phase, and heat treatments can be applied to manipulate structure and improve properties. Fabrication can be by casting (generally investment or graphite mold), forging, rolling, extrusion, or welding, provided that special process modifications and controls are implemented. Advanced processing methods include powder metallurgy, mechanical alloying, rapid-solidification processing (RSP), superplastic forming, diffusion bonding, and hot-isostatic pressing (HIP).

While titanium is an abundant metal, it is difficult to extract from ore, difficult to process, and difficult to fabricate. These difficulties make it significantly more expensive than either steel or aluminum, so its uses relate primarily to its light weight, high strength-to-weight ratio, good stiffness, good fatigue strength and fracture toughness, excellent corrosion resistance (the result of a thin, tenacious oxide coating), and the retention of mechanical properties at elevated temperatures. Aluminum, magnesium, and beryllium are the only base metals that are lighter than titanium, and none of these come close in either mechanical performance or elevated-temperature properties. Aerospace applications tend to dominate, with titanium comprising up to 40% of the structural weight of high-performance military fighters. Titanium and titanium alloys are also used in such diverse areas as chemical- and electrochemical-processing equipment, food-processing equipment, heat exchangers, marine implements, medical implants, high-performance bicycle and automotive components, and sporting goods. They are often used in place of steel where weight savings are desired and to replace aluminums where high-temperature performance is necessary. Some bonding applications utilize the unique property that titanium wets glass and some ceramics. The titanium–6% aluminum–4% vanadium alloy is the most popular titanium alloy, accounting for nearly 50% of all titanium usage worldwide. Figure 7-4 shows the elevated temperature strength retention of several titanium alloys.

FIGURE 7-4 Strength retention at elevated temperature for various titanium alloys.

■ 7.7 Nickel-Based Alloys

Nickel-based alloys are most noted for their outstanding strength and corrosion resistance, particularly at high temperatures, and are available in a wide range of wrought and cast grades. Wrought alloys are generally known by tradenames, such as Monel, Hastelloy, Inconel, Incoloy, and others. Cast alloys are generally identified by Alloy Casting Institute or ASTM designations. General characteristics include good formability (face-centered-cubic crystal structure), good creep resistance, and the retention of strength and ductility at cold or even cryogenic temperatures.

Monel metal, an alloy containing about 67% nickel and 30% copper, has been used for years in the chemical- and food-processing industries because of its outstanding corrosion characteristics. In fact, Monel probably has better corrosion resistance to more media than any other commercial alloy. It is particularly resistant to salt water, sulfuric acid, and even high-velocity, high-temperature steam. For the latter reason, Monel has been used for steam turbine blades. It can be polished to have an excellent appearance, similar to that of stainless steel, and is often used in ornamental trim and household ware. In its most common form, Monel has a tensile strength ranging from 500 to 1200 MPa (70 to 170 ksi), with a companion elongation ranging between 2 and 50%.

Nickel-based alloys have also been used for electrical resistors and heating elements. These materials are primarily nickel–chromium alloys and are known by the trade name *Nichrome*. They have excellent resistance to oxidation while retaining useful strength at red heats. *Invar*, an alloy of nickel and 36% iron, has a near-zero thermal expansion and is used where dimensions cannot change with a change in temperature.

Other nickel-based alloys have been designed to provide good mechanical properties at extremely high temperatures and are generally classified as *superalloys*. These alloys will be discussed along with other, similar materials in the following section.

■ 7.8 Superalloys and Other Metals Designed for High-Temperature Service

Titanium and titanium alloys have already been cited as being useful in providing strength at elevated temperatures, but the maximum temperature for these materials is approximately 535°C (1000°C). Jet engine, gas-turbine, rocket, and nuclear applications often require materials that possess high strength, creep resistance, oxidation and corrosion resistance, and fatigue resistance at temperatures up to and in excess of 1100°C (2000°C). Other application areas include heat exchangers, chemical reaction vessels, and furnace components.

One class of materials offering these properties is the *superalloys*, first developed in the 1940s for use in the elevated-temperature areas of turbojet aircraft. These alloys are based on *nickel, iron and nickel,* or *cobalt* and have the ability to retain most of their strength even after long exposures to extremely high temperatures. Strength comes from solid-solution strengthening, precipitation hardening, and dispersed alloy carbides or oxides. The nickel-based alloys tend to have higher strengths at room temperature, with yield strengths up to 1200 MPa (175 ksi) and ultimate tensile strengths as high as 1450 MPa (210 ksi). The 1000-hour rupture strengths of the nickel-based alloys at 815°C (1500°F) are also higher than those of the cobalt-based material. Unfortunately, the density of all superalloy metals is significantly greater than that of iron, so their use is often at the expense of additional weight.

Most of the superalloys are difficult to form or machine, so methods such as electrodischarge, electrochemical, or ultrasonic machining are often used, or the products are made to final shape as investment castings. Powder metallurgy techniques are also used extensively. Because of their ingredients, all of the alloys are quite expensive, and this limits their use to small or critical parts where the cost is not the determining factor.

A number of engineering applications require materials whose temperature limits exceed those of the superalloys. Figure 7-5 shows the high-temperature exhaust of a jet engine. One reference estimates that the exhaust of future jet engines will reach temperatures in excess of 1425°C (2600°F). Rocket nozzles go well beyond this point. Materials such as TD-nickel (a powder metallurgy nickel alloy containing 2% dispersed thorium oxide) can operate

FIGURE 7-5 Superalloys and refractory metals are needed to withstand the high temperatures of jet engine exhaust. *(Courtesy of Northrop Grumman Corporation, Los Angeles, CA.)*

°F

— 3000 Refractory metals

— 2200 Superalloys

— 1000 Titanium

— 650 Alloy steels
— 500 Carbon steels

— 350 Aluminum

FIGURE 7-6 Temperature scale indicating the upper limit to useful mechanical properties for various engineering metals.

at service temperatures somewhat above 1100°C (2000°F). Going to higher temperatures, we look to the *refractory metals,* which include *niobium, molybdenum, tantalum, rhenium,* and *tungsten.* All have melting points near or in excess of 2500°C (4500°F). They retain a significant fraction of their strength at elevated temperature and can be used at temperatures as high as 1650°C (3000°F) provided that protective ceramic coatings effectively isolate them from gases in their operating environment. Coating technology is quite challenging, however, since the ceramic coatings must (1) have a high melting point, (2) not react with the metal they are protecting, (3) provide a diffusion barrier to oxygen and other gases, and (4) have thermal-expansion characteristics that match the underlying metal. While the refractory metals could be used at higher temperatures, the uppermost temperature is currently being set by limitations and restrictions imposed by the coating.

Table 7-8 presents key properties for several refractory metals. Unfortunately, all are heavier than steel, and several are significantly heavier. In fact, tungsten, with a density about 1.7 times that of lead, is often used in counterbalances, compact flywheels, and weights, with applications as diverse as military projectiles, gyratory compasses, and golf clubs.

Other materials and technologies that offer promise for high-temperature service include intermetallic compounds, engineered ceramics, and advanced coating systems. The *intermetallic compounds* provide properties that are between those of metals and ceramics, and they are excellent candidates for high-temperature applications. They are hard, stiff, creep resistant, and oxidation resistant, with good high-temperature strength that often increases with temperature. The titanium and nickel aluminides offer the additional benefit of being significantly lighter than the superalloys. Unfortunately, the intermetallics are also characterized by poor ductility, poor fracture toughness, and poor fatigue resistance. They are difficult to fabricate using traditional techniques, such as forming and welding. On a positive note, research and development efforts have begun to overcome some of these limitations, and the intermetallics are now appearing in commercial products.

Figure 7-6 compares the upper limit for useful mechanical properties for a variety of engineering metals.

■ 7.9 LEAD AND TIN, AND THEIR ALLOYS

The dominant properties of *lead* and lead alloys are high density coupled with strength and stiffness values that are among the lowest of the engineering metals. The principal uses of lead as a pure metal include storage batteries, cable cladding, and radiation-

TABLE 7-8	Properties of Some Refractory Meals						
		Room Temperature				Elevated Temperature [1832°F (1000°C)]	
Metal	Melting Temperature [°F(°C)]	Density (g/cm³)	Yield Strength (ksi)	Tensile Strength (ksi)	Elongation (%)	Yield Strength (ksi)	Tensile Strength (ksi)
Molybdenum	4730 (2610)	10.22	80	120	10	30	50
Niobium	4480 (2470)	8.57	20	45	25	8	17
Tantalum	5430 (3000)	16.6	35	50	35	24	27
Tungsten	6170 (3410)	19.25	220	300	3	15	66

absorbing or sound- and vibration-damping shields. Lead-acid batteries are clearly the dominant product, and over 60% of U.S. lead consumption is generated from battery recycling. Other applications utilize the properties of good corrosion resistance, low melting point, and the ease of casting or forming. As a pure metal, *tin* is used primarily as a corrosion-resistant coating on steel.

In the form of alloys, lead and tin are almost always used together. Bearing material and *solder* are the two most important uses. One of the oldest and best bearing materials is an alloy of 84% tin, 8% copper, and 8% antimony, known as genuine or tin *babbitt*. Because of the high cost of tin, however, lead babbitt, composed of 85% lead, 5% tin, 10% antimony, and 0.5% copper, is a more widely used bearing material. The tin and antimony combine to form hard particles within the softer lead matrix. The shaft rides on the harder particles with low friction, while the softer matrix acts as a cushion that can distort sufficiently to compensate for misalignment and assure a proper fit between the two surfaces. For slow speeds and moderate loads, the lead-based babbitts have proven to be quite adequate.

Soft solders are basically lead–tin alloys with a chemical composition near the eutectic value of 61.9% tin (see Figure 4-5). While the eutectic alloy has the lowest melting temperature, the high cost of tin has forced many users to specify solders with a lower-than-optimum tin content. A variety of compositions are available, each with its own characteristic melting range. Environmental concerns and recent legislation have prompted a move toward lead-free solders for applications involving water supply and distribution. Additional information on solders and soldering is provided in Chapter 34.

■ 7.10 SOME LESSER KNOWN METALS AND ALLOYS

Several of the lesser known metals have achieved importance as a result of their somewhat unique physical and mechanical properties. *Beryllium* combines a density less than aluminum with a stiffness greater than steel and is transparent to X-rays. *Hafnium, thorium,* and *beryllium* are used in nuclear reactors because of their low neutron-absorption characteristics. Depleted *uranium,* because of its very high density (19.1 g/cm^3), is useful in special applications where maximum weight must be put into a limited space, such as counterweights or flywheels. *Cobalt,* in addition to its use as a base metal for superalloys, is used as a binder in various powder-based components and sintered carbides, where it provides good high-temperature strength. *Zirconium* is used for its outstanding corrosion resistance to most acids, chlorides, and organic acids. It offers high strength, good weldability and fatigue resistance, and attractive neutron-absorption characteristics. *Rare earth metals* have been incorporated into magnets that offer increased strength compared to the standard ferrite variety. Neodymium–iron–boron and samarium–cobalt are two common varieties.

While the precious metals (*gold,silver,* and the platinum group metals—*platinum, palladium, rhodium, ruthenium, iridium,* and *osmium*) may seem unlikely as engineering materials, they offer outstanding corrosion resistance and electrical conductivity, often under extreme conditions of temperature and environment.

■ 7.11 METALLIC GLASSES

Metallic glasses, or amorphous metals, have existed in the form of thin ribbons and fine powders since the 1960s. By cooling liquid metal at a rate that exceeds 10^5 to 10^6 C/second, a rigid solid is produced that lacks crystalline structure. Since the structure also lacks the crystalline "defects" of grain boundaries and dislocations, the materials exhibit extraordinary mechanical properties (high strength, large elastic strain, good toughness, and wear resistance), unusual magnetic behavior, and high corrosion resistance.

Recent developments have enabled the production of amorphous metal with cooling rates of only 1 to 100 °C/second. Known as *bulk metallic glass (BMG),* complex-shaped parts of this material with thicknesses up to several centimeters can now be produced by conventional casting methods, such as die casting. Because the material goes from liquid to glass, not liquid to crystalline solid, precision products can be made with a total shrinkage that is often less than 0.5%. Pellets or powders of bulk metallic glass can also be produced, and since many of the alloys have low melting temperatures, products can be made by reheating to a soft condition and forming by processes that are conventionally used to shape thermoplastic polymers (compression molding, extrusion, blow molding, and injection molding). Applications have just begun to emerge in areas as diverse as load-bearing structures, electronic casings, replacement joints, and sporting goods. In addition, metallic glasses have also been developed that retain their glassy structure at temperatures as high as 870°C (1600°F).

■ 7.12 GRAPHITE

While technically not a metal, *graphite* is an engineering material with considerable potential. It offers properties of both a metal and nonmetal, including good thermal and electrical conductivity, inertness, the ability to withstand high temperature, and lubricity. In addition, it possesses the unique property of increasing in strength as the temperature is elevated. Polycrystralline graphites can have mechanical strengths up to 70 MPa (10 ksi) at room temperature, which double when the temperature reaches 2500°C (4500°F).

Large quantities of graphite are used as electrodes in arc furnaces, but other uses are developing rapidly. The addition of small amounts of borides, carbides, nitrides, and silicides greatly lowers the oxidation rate at elevated temperatures and improves the mechanical strength. This makes the material highly suitable for use as rocket-nozzle inserts and as permanent molds for casting various metals, where it costs less than tool steel, requires no heat treating, and has a lower coefficient of thermal expansion. It can be machined quite readily to excellent surface finishes. Graphite fibers have also found extensive use in composite materials. This application will be discussed in Chapter 8.

■ Key Words

Alclad	cobalt	molybdenum	stress–corrosion cracking
aluminum	copper	Monel	superalloys
amorphous metal	dezincification	nickel	tantalum
babbitt	galvanizing	niobium	temper designation
beryllium	graphite	nonferrous	tin
bismuth	intermetallic compound	refractory metals	titanium
brass	lead	rhenium	tungsten
bronze	magnesium	selenium	wrought
cast	metallic glass	solder	zinc

■ Review Questions

1. What types of properties do nonferrous metals possess that may not be available in the ferrous metals?
2. In what respects are the nonferrous metals generally inferior to steel?
3. For what type of fabrication processes might the low-melting-point alloys be attractive?
4. What are the three properties of copper and copper alloys that account for many of their uses and applications?
5. What properties make copper attractive for cold-working processes?
6. What are some of the limiting properties of copper that might restrict its area of application?
7. Why does the copper designation system separate wrought and cast alloys? What properties are attractive for each group?
8. What are some of the attractive engineering properties that account for the wide use of the copper–zinc alpha brasses?
9. Why might cold-worked brass require a stress relief prior to being placed in service?
10. Why might the term *bronze* be potentially confusing when used in reference to a copper-based alloy?

11. What are some attractive engineering properties of copper–nickel alloys?
12. Describe the somewhat unique property combination that exists in heat-treated copper–beryllium alloys. What has limited its use in recent years?
13. What alloys have been used to replace lead in copper casting alloys being targeted to drinking water applications?
14. What are some of the attractive engineering properties of aluminum and aluminum alloys?
15. How does aluminum compare to steel in terms of weight? Discuss the merits of comparing cost per unit weight versus cost per unit volume.
16. What is the primary benefit of aluminum recycling compared to making new aluminum from ore?
17. How does aluminum compare to copper in terms of electrical conductivity?
18. What features might limit the mechanical uses and applications of aluminum and aluminum alloys?
19. What features make aluminum attractive for transportation applications?

20. How is the corrosion-resistance mechanism observed in aluminum and aluminum alloys similar to that observed in stainless steels?
21. How are the wrought alloys distinguished from the cast alloys in the aluminum designation system? Why would these two groups of metals have distinctly different properties?
22. What feature in the wrought aluminum designation scheme is used to denote the condition or structure of a given alloy?
23. What is the primary strengthening mechanism in the high-strength "aircraft-quality" aluminum alloys?
24. What unique combination of properties is offered by the composite Alclad materials?
25. What surface finishing technique is used in the production of numerous metallic colored aluminum products.
26. What specific material properties might make an aluminum casting alloy attractive for permanent mold casting? For die casting?
27. What features have limited the success and expansion of the aluminum–lithium alloys?
28. What are some possible applications of aluminum foam?
29. What are some attractive and restrictive properties of magnesium and magnesium alloys?
30. Describe the designation system applied to magnesium alloys.
31. In what way can ductility be imparted to magnesium alloys so that they can be formed by conventional processes?
32. Under what conditions should magnesium be considered to be a flammable or explosive material?

33. What is the primary application of pure zinc? Of the zinc-based engineering alloys?
34. What are some of the attractive features of the zinc–aluminum casting alloys?
35. What are some of the attractive engineering properties of titanium and titanium alloys?
36. What feature is used to provide the metallurgical classification of titanium alloys?
37. What temperature is generally considered to be the upper limit for which titanium alloys retain their useful engineering properties?
38. What conditions favor the selection and use of nickel-based alloys?
39. What property of Monel alloys dominates most of the applications?
40. What metals or combinations of metals form the bases of the superalloys?
41. What class of metals or alloys must be used when the operating temperatures exceed the limits of the superalloys?
42. Which metals are classified as refractory metals?
43. What are some general characteristics of intermetallic compounds?
44. What is the dominant product for which lead is used?
45. What features makes beryllium a unique lightweight metal?
46. What are some of the attractive properties of metallic glasses?
47. What unique property of graphite makes it attractive for elevated-temperature applications?

www.wiley.com/college/degarmo

Chapter 7   CASE STUDY

Nonsparking Wrench

The Ivanwold/Kendoric tool-manufacturing company is considering an expansion of its line of conventional hand tools to include safety tools capable of being used in areas such as gas leaks where the potential of explosion or fire exists. Conventional irons and steels are pyrophoric (i.e., small slivers or fragments can burn in air, forming sparks if dropped or impacted on a hard surface).

You are asked to evaluate potential materials and processes that might be used to manufacture a nonsparking pipe wrench. This product is to be produced in the same shape and range of sizes as conventional pipe wrenches and needs to possess all of the same characteristic properties (strength in the handle, hardness in the teeth, fracture resistance, corrosion resistance, etc.). In addition, the new safety wrench must be nonsparking (or nonpyrophoric).

Your initial review of the nonferrous metals reveals that aluminum is nonpyrophoric but lacks the strength and wear resistance needed in the teeth and jaw region of the wrench. Copper is also nonpyrphoric but is heavier than

steel, and this may be unattractive for the larger wrenches. Copper–2% beryllium can be age hardened to provide the strength and hardness properties equivalent to the steel that is currently being used for the jaws of the wrench, but the cost of this material is also quite high. Titanium is difficult to fabricate and may not possess the needed hardness and wear resistance. Mixed materials may create an unattractive galvanic corrosion cell. Both forging and casting appear to be viable means of forming the desired shape. You want to produce a quality product but also wish to make the wrench in the most economical manner possible so that the new line of safety tools is attractive to potential customers.

Suggest some alternative manufacturing systems (materials coupled with companion methods of fabrication) that could be used to produce the desired wrench. What might be the advantages and disadvantages of each? Which of your alternatives would you recommend to your supervisor?

CHAPTER 8

NONMETALLIC MATERIALS: PLASTICS, ELASTOMERS, CERAMICS, AND COMPOSITES

■ 8.1 INTRODUCTION

Because of their wide range of attractive properties, the nonmetallic materials have always played a significant role in manufacturing. Wood has been a key engineering material down through the centuries, and artisans have learned to select and use the various types and grades to manufacture a broad spectrum of quality products. Stone and rock continue to be key construction materials, and clay products can be traced to antiquity. Even leather has been a construction material and was used for fenders in early automobiles.

More recently, however, the family of *nonmetallic materials* has expanded from the natural materials just described and now includes an extensive list of plastics (polymers), elastomers, ceramics, and composites. Most of these are manufactured materials, so a wide variety of properties and characteristics can be obtained. New variations are being created on a continuous basis, and their uses and applications are expanding rapidly. Many observers now refer to a materials revolution as these new materials compete with and complement steel, aluminum, and the other more traditional engineering metals. New products have emerged, utilizing the new properties, and existing products are continually being reevaluated for the possibility of material substitution. As the design requirements of products continue to push the limits of traditional materials, the role of the manufactured nonmetallic materials will no doubt continue to expand.

Because of the breadth and number of nonmetallic materials, we will not attempt to provide information about all of them. Instead, the emphasis will be on the basic nature and properties of the various families so that the reader will be able to determine if they may be reasonable candidates for specific products and applications. For detailed information about specific materials within these families, more extensive and dedicated texts, handbooks, and compilations should be consulted.

■ 8.2 PLASTICS

It is difficult to provide a precise definition of the term *plastics*. From a technical viewpoint, the term is applied to engineered materials characterized by large molecules that are built up by the joining of smaller molecules. On a more practical level, these materials are natural or synthetic resins, or their compounds, that can be molded, extruded, cast, or used as thin films or coatings. They offer low density, low tooling costs, good resistance to corrosion and chemicals, cost reduction, and design versatility. From a chemical viewpoint, most are organic substances containing hydrogen, oxygen, carbon, and nitrogen.

In less than a century, we have gone from a world without plastic to a world where its use and applications are limitless. The United States currently produces more plastic than steel, aluminum, and copper combined. Plastics are used to save lives in applications such as artificial organs, shatter-proof glass, and bullet-proof vests. They reduce the weight of cars, provide thermal insulation to our homes, and encapsulate our medicines. They form the base material in products as diverse as shower curtains, contact lenses, and clothing, and compose some of the primary components in televisions, computers, cell phones, and furniture. Even the Statue of Liberty has a plastic coating to protect it from corrosion.

FIGURE 8-1   The linking of carbon and hydrogen to form methane and ethane molecules. Each dash represents a shared electron pair or covalent bond.

MOLECULAR STRUCTURE OF PLASTICS

To understand the properties of plastics, it is important to first understand their molecular structure. For simplicity, let's begin with the paraffin-type hydrocarbons, in which carbon and hydrogen combine in the relationship C_nH_{2n+2}. Theoretically, the atoms can link together indefinitely to form very large molecules, extending the series depicted in Figure 8-1. The bonds between the various atoms are all pairs of shared electrons (covalent bonds). Bonding within the molecule, therefore, is quite strong, but the attractive forces between adjacent molecules are much weaker. Because there is no provision for additional atoms to be added to the chain, these molecules are said to be *saturated*.

Carbon and hydrogen can also form molecules where the carbon atoms are held together by double or triple covalent bonds. Ethylene and acetylene are common examples (Figure 8-2). Because these molecules do not have the maximum number of hydrogen atoms, they are said to be *unsaturated* and are important in the polymerization process, where small molecules link to form large ones with the same constituent atoms.

In all of the described molecules, four electron pairs surround each carbon atom and one electron pair is shared with each hydrogen atom. Other atoms or structures can be substituted for carbon and hydrogen. Chlorine, fluorine, or even a benzene ring can take the place of hydrogen. Oxygen, silicon, sulfur, or nitrogen can take the place of carbon. Because of these substitutions, a wide range of organic compounds can be created.

Ethylene   Acetylene

FIGURE 8-2   Double and triple covalent bonds exist between the carbon atoms in unsaturated ethylene and acetylene molecules.

ISOMERS

The same kind and number of atoms can also unite in different structural arrangements, known as *isomers*, and these ultimately behave as different compounds with different engineering properties. Figure 8-3 shows an example of this feature, involving propyl and isopropyl alcohol. Isomers can be considered analogous to allotropism or polymorphism in crystalline materials, where the same material possesses different properties because of different crystal structures.

Propyl Alcohol   Isopropyl Alcohol

FIGURE 8-3   Linking of eight hydrogen, one oxygen, and three carbon atoms to form two isomers: propyl alcohol and isopropyl alcohol. Note the different locations of the —OH attachment.

FIGURE 8-4 Addition polymerization—the linking of monomers; in this case, identical ethylene molecules.

FORMING MOLECULES BY POLYMERIZATION

The polymerization process, or linking of molecules, occurs by either an *addition* or *condensation* mechanism. Figure 8-4. illustrates polymerization by addition, where a number of basic units (*monomers*) link together to form a large molecule (*polymer*) in which there is a repeated unit (*mer*). Activators or catalysts, such as benzoyl peroxide, initiate and terminate the chain. Thus, the amount of activator relative to the amount of monomer determines the average molecular weight (or average length) of the polymer chain. The average number of mers in the polymer, known as the *degree of polymerization*, ranges from 75 to 750 for most commercial plastics. Chain length controls many of the properties of a plastic. Increasing the chain length tends to increase toughness, creep resistance, melting temperature, melt viscosity, and difficulty in processing.

Copolymers are a special category of polymer where two different types of mers are combined into the same addition chain. The formation of copolymers (Figure 8-5), analogous to alloys in metals, greatly expands the possibilities of creating new types of plastics with improved physical and mechanical properties. *Terpolymers* further extend the possibilities by combining three different monomers.

FIGURE 8-5 Addition polymerization with two kinds of mers—here, the copolymerization of butadiene and styrene.

In contrast to polymerization by addition, where all of the original atoms appear in the product molecule, *condensation polymerization* occurs when reactive molecules combine with one another to produce a polymer plus small, by-product molecules, such as water. Heat, pressure, and catalysts are often required to drive the reaction. Figure 8-6 illustrates the reaction between phenol and formaldehyde to form Bakelite, first performed in 1910. The structure of condensation polymers can be either linear chains or a three-dimensional framework in which all atoms are linked by strong, primary bonds.

FIGURE 8-6 The formation of phenol–formaldehyde (Bakelite) by condensation polymerization. Note the H_2O or water by-product.

THERMOSETTING AND THERMOPLASTIC MATERIALS

The terms *thermosetting* and *thermoplastic* refer to the material's response to elevated temperature. Addition polymers (or linear condensation polymers) can be viewed as long chains of bonded carbon atoms with attached pendants of hydrogen, fluorine, chlorine, or benzene rings. All of the bonds within the molecules are strong covalent bonds. The attraction between neighboring molecules is through the much weaker van der Waals forces. For these materials, the intermolecular forces strongly influence the mechanical and physical properties. In general, the linear polymers tend to be flexible and tough. Because the intermolecular bonds are weakened by elevated temperature, plastics of this type soften with increasing temperature and the individual molecules can slide over each other in a molding process. When the material is cooled, it becomes harder and stronger. The softening and hardening of these thermoplastic or heat-softening materials can be repeated as often as desired, and no chemical change is involved.

Because thermoplastic materials contain molecules of different lengths, they do not have a definite melting temperature but, instead, soften over a range of temperatures. Above the temperature required for melting, the material can be poured and cast, or formed by injection molding. When cooled to a temperature where it is fully solid, the material can retain its amorphous structure, but with companion properties that are somewhat rubbery. The application of a force produces both elastic and plastic deformation. Large amounts of permanent deformation are available and make this range attractive for molding and extrusion. At still lower temperatures, the bonds become stronger and the polymer is stiffer and somewhat leathery. Many commercial polymers, such as polyethylene, have useful strength in this condition. When further cooled below the glass transition temperature, however, the linear polymer retains its amorphous structure but becomes hard, brittle, and glasslike.

Many thermoplastics can partially *crystallize*[1] when cooled below the melting temperature. This should not be confused with the crystal structures discussed previously in this text. When polymers "crystallize," the chains closely align over appreciable distances, with a companion increase in density. In addition, the polymer becomes stiffer, harder, less ductile, and more resistant to solvents and heat. The ability of a polymer to crystallize depends on the complexity of its molecules, the degree of polymerization (length of the chains), the cooling rate, and the amount of deformation during cooling.

The mechanical behavior of an amorphous (noncrystallized) thermoplastic polymer can be modeled by a common cotton ball. The individual molecules are bonded within by strong covalent bonds and are analogous to the individual fibers of cotton. The bonding forces between molecules are much weaker and are similar to the friction forces between the strands of cotton. When pulled or stretched, plastic deformation occurs by slippage between adjacent fibers or molecular chains. Methods to increase the strength of thermoplastics, therefore, focus on restricting intermolecular slippage. Longer chains have less freedom of movement and are therefore stronger. Connecting adjacent chains to one another with primary bond cross-links, as with the sulfur links when vulcanizing rubber, can also impede deformation. Since the strength of the secondary bonds is inversely related to the separation distance between the molecules, processes such as deformation or crystallization can be used to produce a tight parallel alignment of adjacent molecules and a concurrent increase in strength, stiffness, and density. Polymers with larger side structures, such as chlorine atoms or benzene rings, may be stronger or weaker than those with just hydrogen, depending on whether the dominant effect is the impediment to slippage or the increased separation distance. Branched polymers, where the chains divide in a Y with primary bonds linking all segments of the chain, are often weaker since branching reduces the density and close packing of the chains. Physical, mechanical, and electrical properties all vary with the above changes in structure.

---

[1] It should be noted that the term *crystallize*, when applied to polymers, has a different meaning than when applied to metals and ceramics. Metals and ceramics are crystalline materials, meaning that the atoms occupy sites in a regular, periodic array, known as a lattice. In polymers, it is not the atoms that become aligned, but the molecules. Since van der Waals bonding has a bond strength that is inversely related to the separation distance, the parallel alignment of the crystallized state is a lower-energy configuration and is promoted by slow cooling and equilibrium-type processing conditions.

The four most common thermoplastic polymers are: polyethylene (PE), polypropylene (PP), polystyrene (PS), and polyvinyl chloride (PVC).

In contrast to the thermoplastic polymers, *thermosetting plastics* usually have a highly cross-linked or three-dimensional framework structure in which all atoms are connected by strong, covalent bonds. These materials are generally produced by condensation polymerization where elevated temperature promotes an irreversible reaction, hence the term *thermosetting*. Once set, subsequent heating will not produce the softening observed with the thermoplastics. Instead, thermosetting materials maintain their mechanical properties up to the temperature at which they char or burn. Since deformation requires the breaking of primary bonds, the thermosetting polymers are significantly stronger and more rigid than the thermoplastics. They can resist higher temperatures and have greater dimensional stability, but they also have lower ductility and poorer impact properties.

As a helpful analogy, thermoplastic polymers are a lot like candle wax. They can be softened or melted by heat, and then cooled to assume a solid shape. Thermosets are more like egg whites or bread dough. Heating changes their structure and properties in an irreversible fashion.

Although classification of a polymer as thermosetting or thermoplastic provides insight as to properties and performance, it also has a strong effect on fabrication. For example, thermoplastics can be easily molded. After the hot, soft material has been formed to the desired shape, however, the mold must be cooled so that the plastic will harden and be able to retain its shape upon removal. The repetitive heating and cooling cycles affect mold life, and the time required for the thermal cycles influences productivity. When a part is produced from thermosetting materials, the mold can remain at a constant temperature throughout the entire process, but the setting or curing of the resins now determines the time in the mold. Since the material hardens as a result of the reaction and has strength and rigidity even when hot, product removal can be performed without cooling the mold.

PROPERTIES AND APPLICATIONS

Because there are so many varieties of plastics and new ones are being developed almost continuously, it is helpful to have knowledge of both the general properties of plastics and the unique or specific properties of the various families. General properties of plastics include:

1. *Light weight.* Most plastics have specific gravities between 1.1 and 1.6, compared with about 1.75 for magnesium (the lightest engineering metal).
2. *Corrosion resistance.* Many plastics perform well in hostile, corrosive, or chemical environments. Some are notably resistant to acid corrosion.
3. *Electrical resistance.* Plastics are widely used as insulating materials.
4. *Low thermal conductivity.* Plastics are relatively good thermal insulators.
5. *Variety of optical properties.* Many plastics have an almost unlimited color range, and the color goes throughout, not just on the surface. Both transparent and opaque materials are available.
6. *Formability or ease of fabrication.* Objects can frequently be produced from plastics in a single operation. Raw material can be converted to final shape through such processes as casting, extrusion, and molding. Relatively low temperatures are required for the forming of plastics.
7. *Surface finish.* The same processes that produce the shape also produce excellent surface finish. Additional surface finishing may not be required.
8. *Comparatively low cost.* The low cost of plastics generally applies to both the material itself and the manufacturing process. Plastics frequently offer reduced tool costs and high rates of production.
9. *Low energy content.*

While the attractive features of plastics tend to be in the area of physical properties, the inferior features generally relate to mechanical strength. Plastics can be flexible or rigid, but none of the plastics possess strength properties that approach those of the engineering metals unless they are reinforced in the form of a composite. Their low density allows them to compete effectively on a strength-to-weight (or specific strength) basis, however. Many have low impact strength, although several (such as ABS, high-density polyethylene, and polycarbonate) are exceptions to this rule. Aluminum is nearly 10 times more rigid than a high-rigidity plastic, and steel is 30 times more rigid.

The dimensional stability of plastics tends to be greatly inferior to that of metals, and the coefficient of thermal expansion is rather high. Thermoplastics are quite sensitive to heat, and their strength often drops rapidly as temperatures increase above normal environmental conditions. Thermosetting materials offer good strength retention at elevated temperature but have an upper limit of about 250°C (500°F). Low-temperature properties are generally inferior to those of other materials. While the corrosion resistance of plastics is generally good, they often absorb moisture, and this, in turn, decreases strength. Some thermoplastics can exhibit a 50% drop in tensile strength as the humidity increases from 0 to 100%. Radiation, both ultraviolet and particulate, can markedly alter the properties. Many plastics used in an outdoor environment have ultimately failed due to the cumulative effect of ultraviolet radiation. Plastics are also difficult to repair if broken.

Table 8-1 summarizes the properties of a number of common plastics. By considering the information in this table along with the preceding discussion of general properties, it becomes apparent that plastics are best used in applications that require materials with low to moderate strength, light weight, low electrical and/or thermal conductivity, a wide range of available colors, and ease of fabrication into finished products. No other family of materials can offer this combination of properties. Because of their light weight, attractive appearance, and ease of fabrication, plastics have been selected for many packaging and container applications. This classification includes such items as household appliance housings, clock cases, and exteriors of electronic products, where the primary role is to contain the interior mechanisms. Applications such as insulation on electrical wires and handles for hot articles capitalize on the low electrical and thermal conductivities. Soft, pliable, foamed plastics are used extensively as cushioning material. Rigid foams are used inside sheet metal structures to provide compressive strength. Nylon has been used for gears, acrylic for lenses, and polycarbonate for safety helmets and unbreakable windows.

There are many applications where only one or two of the properties of plastics are sufficient to justify their use. When special characteristics are desired that are not normally found in the commercial plastics, composite materials can often be designed that use a polymeric matrix. For example, high directional strength may be achieved by incorporating a fabric or fiber reinforcement within a plastic resin. These materials will be discussed in some detail later in this chapter.

COMMON TYPES OR FAMILIES OF PLASTICS
The following is a brief descriptive summary of the types of plastics listed in Table 8-1.

THERMOPLASTICS

ABS: contains acrylonitrile, butadiene, and styrene; low weight, good strength, and very tough; resists heat, weather, and chemicals quite well; dimensionally stable but flammable

Acrylics: highest optical clarity, transmitting over 90% of light; common trade names include Lucite and Plexiglas; high-impact, flexural, tensile, and dielectric strengths; available in a wide range of colors; resist weathering

Cellulose acetate: wide range of colors; good insulating qualities; easily molded; high moisture absorption in most grades

Cellulose acetate butyrate: higher impact strength and moisture resistance than cellulose acetate; will withstand rougher usage

Ethyl cellulose: high electrical resistance and impact strength; retains toughness at low temperatures

TABLE 8-1 Properties and Major Characteristics of Common Types of Plastics

Material	Specific Gravity	Tensile Strength (1000 lb/in.²)	Impact Strength Izod (ft-lb/in. of Notch)	Top Working Temperature [°F(°C)]	Dielectric Strength[b] (V/mil)	24-Hour Water Absorption (%)	Weatherability	Colorability	Optical Clarity	Chemical Resistance	Injection Molding	Extrusions	Formable Sheet	Film	Fiber	Compression or Transfer Moldings	Castings	Reinforced Plastics	Moldings	Industrial Thermosetting Laminates	Foam
Thermoplastics																					
ABS material	1.02–1.06	4–8	1.3–10.0		300–400	0.2–0.3	0	X		0	•	•	•								
Acetal	1.4	10	1.5	250(121)	1200	0.22		X		0	•	•									
Acrylics	1.12–1.19	5.5–10	0.2–2.3	200(93)	400–530	0.2–0.4	X	X	X	0	•	•	•	•	•			•			
Cellulose acetate	1.25–1.50	3–8	0.75–4.0	260(127)	300–600	2.0–6.0		X	X		•	•	•	•	•						
Cellulose acetate butyrate	1.18–1.24	2–6	0.6–3.2	130(54)	250–350	1.8–2.1	X	X	X		•	•	•								
Cellulose propionate	1.19–1.24	1–5	0.8–9	140(60)	300	1.8–2.1			X		•	•									
Chlorinated polyether	1.4	6	3.3	300(149)	400	0.01		X		X	•	•									
Ethyl cellulose	1.16	3–6	1.8–4.0	150(66)	350	1.6–2.2					•	•	•	•							
TFE–fluorocarbon	2.1–2.3	1.5–3	2.5–4.0	500(260)	450	0	X	X		X	•	•	•	•	•	•					
CFE–fluorocarbon	2.1–2.15	4.5–6	3.5–3.6	390(199)	550	0	X			X	•	•	•	•	•						
Nylon	1.1–1.2	8–10	2	250(121)	385–470	0.4–5.5	X	0	0	0	•	•			•						
Polycarbonate	1.2	9.5	14	250(121)	400	0.15		0	0	X	•	•									
Polyethylene	0.96	4	10	200(93)	440	0.003		0	X	X	•	•		•							
Polypropylene	0.9–1.27	3.4–5.3	1.02	230(110)	520–800	0.03		0	X	X	•	•			•						•
Polystyrene	1.05–1.15	5–9	0.3–0.6	190(8)	400–600	<0.2		X	X	0	•	•	•	•							
Modified polystyrene	1.0–1.1	2.5–6	0.25–11.0	212(100)	300–600	0.03–02		X	X	X	•	•									
Vinyl	1.16–1.55	1–5.9	0.25–2.0	220(104)	25–500	0.2–1		X	X	X	•	•	•	•	•						•
Thermosetting plastics																					
Epoxy	1.1–1.7	4–13	0.4–1.5	325(163)	500	0.1–0.5	X	X		X						•	•	•		•	•
Melamine	1.76–1.98	5–8		350(177)	460	0.1		X		0							•				
Phenolic	1.2–1.45	5–9	0.25–5	300(149)	100–500	0.2–0.6				0						•	•	•		•	•
Polyester (other than molding compounds)	1.06–1.46	4–10	0.18–0.4	300(149)	340–570	0.5		X	X	0				•	•		•	•			
Polyester (alkyd, DAP)	1.6–1.75	3.2–8	3.6–8		250–350	0.16–0.67		X	X							•	•	•	•	•	
Silicone	2.0	3–5	0.2–3.0	550(288)	250–350	0.4–0.5									•	•	•			•	
Urea	1.41–1.80	4–8.5	0.2–0.5	185(85)	300–600	1–3	X	X	0								•			•	•

a X denotes a principal reason for its use; 0 indicates a secondary reason.

b Short-time ASTM test.

Fluorocarbons: inert to most chemicals; high temperature resistance; very low coefficients of friction (Teflon); used for nonlubricated bearings and nonstick coatings for cooking utensils and electrical irons

Nylon (polyamides): low coefficient of friction; good strength, abrasion resistance, and toughness; excellent dimensional stability; good heat resistance; used for small gears and bearings, zip fasteners, and as monofilaments for textiles, fishing line, and ropes

Polycarbonates: high strength and outstanding toughness; good dimensional stability; transparent or easily colored

Polyethylenes: the most common polymer; inexpensive, tough, good chemical resistance to acids, bases, and salts; high electrical resistance; low strength; easy to shape and join; reasonably clear in thin-film form; subject to weathering via ultraviolet light; flammable; used for grocery bags, milk jugs and other food containers, tubes, pipes, sheeting, and electrical wire insulation. Variations include: low-density polyethylene (LDPE—floats in water), high-density polyethylene (HDPE), ultra-high-molecular-weight polyethylene (UHMW)

PMMA (polymethyl methacrylate): hard, brittle (at room temperature), transparent or easily colored; used for items like tool handles and the interior windows of airplanes

Polypropylene: inexpensive; stronger, stiffer, and better heat resistance than polyethylene; transparent; reasonable toughness; used for beverage containers, luggage, pipes, and ropes

Polystyrenes: high dimensional stability and stiffness with low water absorption; best all-around dielectric; clear, hard, and brittle at room temperature; often used for rigid packaging; can be foamed to produce expanded polystyrene (trade name of Styrofoam); burns readily; softens at about 95°C

Polyvinyl chloride (PVC): general-purpose thermoplastic; good resistance to ultraviolet light (good for outside applications); easily molded or extruded; always used with fillers, plasticizers, and pigments; uses include gas and water pipes as well as window frames

Vinyls: wide range of types, from thin, rubbery films to rigid forms; tear resistant; good aging properties; good dimensional stability and water resistance in rigid forms; used for floor and wall covering, upholstery fabrics, and lightweight water hose; common trade names include Saran and Tygon

THERMOSETS

Epoxies: good strength, toughness, elasticity, chemical resistance, moisture resistance, and dimensional stability; easily compounded to cure at room temperature; used as adhesives, bonding agents, coatings, and in fiber laminates

Melamines: excellent resistance to heat, water, and many chemicals; full range of translucent and opaque colors; excellent electric arc resistance; tableware (but stained by coffee); used extensively in treating paper and cloth to impart water-repellent properties

Phenolics: oldest of the plastics but still widely used; hard, strong, low cost, and easily molded, but rather brittle; resistant to heat and moisture; dimensionally stable; opaque, but with a wide color range; wide variety of forms: sheet, rod, tube, and laminate; trade names include Bakelite.

Polyesters (can be thermoplastic or thermoset): strong and good resistance to environmental influences; uses include boat and car bodies, pipes, vents and ducts, textiles, adhesives, coatings, and laminates

Silicones: heat and weather resistant; low moisture absorption; chemically inert; high dielectric properties; excellent sealants

Urea–formaldehyde: properties similar to those of phenolics but available in lighter colors; useful in containers and housings, but not outdoors; used in lighting fixtures because of translucence in thin sections; as a foam, may be used as household insulation

ADDITIVE AGENTS IN PLASTICS

For most uses, additional materials are incorporated into plastics to (1) impart or improve properties, (2) reduce cost, (3) improve moldability, and/or (4) impart color. These *additive constituents* are usually classified as *fillers and reinforcements*, *plasticizers*, *lubricants*, *coloring agents*, *stabilizers*, *antioxidants*, and *flame retardants*.

Ordinarily, *fillers* comprise a large percentage of the total volume of a molded plastic product. Their primary roles are to improve strength, stiffness, or toughness; reduce shrinkage; reduce weight; or simply serve as an extender, providing cost-saving bulk (often at the expense of reduced moldability). To a large degree, they determine the general properties of a molded plastic. Selection tends to favor materials that are much less expensive than the plastic resin. Some of the most common fillers and their properties are:

1. *Wood flour* (fine sawdust): a general-purpose filler; low cost with fair strength; good moldability
2. *Cloth fibers*: improved impact strength; fair moldability
3. *Macerated cloth*: high impact strength; limited moldability
4. *Glass fibers*: high strength; dimensional stability; translucence
5. *Mica*: excellent electrical properties and low moisture absorption
6. *Calcium carbonate, silica, talc, and clay*: serve primarily as extenders

When fillers are used with a plastic resin, the resin acts as a binder, surrounding the filler material and holding the mass together. The surface of a molded part, therefore, will be almost pure resin with no exposed filler. Cutting or scratching through the shiny surface will expose the less attractive filler.

Coloring agents may be either *dyes*, which are soluble in the resins, or insoluble *pigments*, which impart color simply by their presence. In general, dyes are used for transparent plastics and pigments for the opaque ones. Optical brighteners can also be used to enhance appearance. Carbon black can provide both a black color and electrical conductivity.

Plasticizers can be added in small amounts to reduce viscosity and improve the flow of the plastic during molding or to increase the flexibility of thermoplastic products by reducing the intermolecular contact and strength of the secondary bonds between the polymer chains. When used for molding purposes, the amount of plasticizer is governed by the intricacy of the mold. In general, it should be kept to a minimum because it is likely to affect the stability of the finished product through a gradual aging loss. When used for flexibility, plasticizers should be selected with minimum volatility, so as to impart the desired property for as long as possible.

Lubricants such as waxes, stearates, and soaps can be added to improve the moldability of plastics and to facilitate removal of parts from the mold. They are also used to keep thin polymer sheets from sticking to each other when stacked or rolled. Only a minimum amount should be used, however, because the lubricants adversely affect most engineering properties.

Heat, light (especially ultraviolet), and oxidation tend to degrade polymers. Stabilizers and antioxidants can be added to retard these effects. Flame retardants can be added when nonflammability is important. Antistatic agents allow for the migration of electrical charge and may be incorporated into plastics used for applications such as electronics packaging. Antimicrobial additives can provide long-term protection from both fungus (such as mildew) and bacteria. Fibers can be incorporated to increase strength and stiffness, and metal flakes, fibers, or powders can modify electrical and magnetic properties. Table 8-2 summarizes the purposes of the various additives.

ORIENTED PLASTICS

Because the intermolecular bond strength increases with reduced separation distance, any processing that aligns the molecules parallel to the applied load can be used to give the long-chain thermoplastics high strength in a given direction.[2] This orientation process

---

[2] The effects of orienting can be observed in the common disposable thin-walled plastic drinking cup. Start at the top lip. Place a sharp bend in the lip and then tear down the side wall. The material tears easily. Move around the lip about 1/2 inch and make another side-wall tear—also easy. Now try to tear across the strip that you have created. This tear is much more difficult, since you are tearing across molecules that have been oriented vertically along the cup walls by the cup-forming operation.

TABLE 8-2	Additive Agents in Plastics and Their Purpose
Type	**Purpose**
Fillers	Enhance mechanical properties, reduce shrinkage, reduce weight, or provide bulk
Plasticizer	Increase flexibility, improve flow during molding, reduce elastic modulus
Lubricant	Improve moldability and extraction from molds
Coloring agents (dyes and pigments)	Impart color
Stabilizers	Retard degradation due to heat or light
Antioxidants	Retard degradation due to oxidation
Flame retardants	Reduce flammability

can be accomplished by a forming process, such as stretching, rolling, or extrusion. The material is usually heated prior to the orienting process to aid in overcoming the intermolecular forces and is cooled immediately afterward to "freeze" the molecules in the desired orientation.

Orienting may increase the tensile strength by more than 50%, but a 25% increase is more typical. In addition, the elongation may be increased by several hundred percent. If the oriented plastics are reheated, they tend to deform back toward their original shape, a phenomenon known as *viscoelastic memory*. The various shrink-wrap materials are examples of this effect.

ENGINEERING PLASTICS

The standard polymers tend to be lightweight, corrosion-resistant materials with low strength and low stiffness. They are relatively inexpensive and are readily formed into a wide range of useful shapes, but they are not suitable for use at elevated temperatures.

In contrast, a group of plastics has been developed with improved thermal properties (up to 350°C, or 650°F), enhanced impact and stress resistance, high rigidity, superior electrical characteristics, excellent processing properties, and little dimensional change with varying temperature and humidity. These true engineering plastics include the polyamides, polyacetals, polyacrylates, polycarbonates, modified polyphenylene oxides, polybutylene terephtalates, polyketones, polysulfones, polyetherimides, and liquid crystal polymers. While stabilizers, fibrous reinforcements, and particulate fillers can upgrade the conventional plastics, there is usually an accompanying reduction in other properties. The engineering plastics offer a more balanced set of properties. They are usually produced in small quantities, however, and are often quite expensive.

Materials producers have also developed electroconductive polymers with tailored electrical and electronic properties and high–crystalline polymers with properties comparable to some metals.

PLASTICS AS ADHESIVES

Polymeric adhesives are used in many industrial applications. They are quite attractive for the bonding of dissimilar materials, such as metals to nonmetals, and have even been used to replace welding or riveting. A wide range of mechanical properties are available through variations in composition and additives, and a variety of curing mechanisms can be used. Examples can be found from the thermoplastics (hot-melt glues), thermosets (two-part epoxies), and even elastomers (silicone adhesives). The seven most common structural adhesives are epoxies, urethanes, cyanoacrylates, acrylics, anaerobics, hot melts, and silicones. Selection usually involves consideration of the manufacturing conditions, the substrates to be bonded, the end-use environment, and cost. The various features of adhesive bonding are discussed in greater detail in Chapter 35.

PLASTICS FOR TOOLING

Polymers can also provide inexpensive tooling for applications where pressures, temperatures, and wear requirements are not extreme. Because of their wide range of properties, their ease of conversion into desired shapes, and their excellent properties when loaded in compression, plastics have been widely used in applications such as jigs, fixtures, and a wide variety of forming-die components. Both thermoplastic and thermoset polymers (particularly the cold-setting types) have been used. By using plastics in these applications, costs can be reduced and smaller quantities of products can be economically justified. In addition, the tooling can often be produced in a much shorter time, enabling quicker production.

FOAMED PLASTICS

A number of polymeric materials can be produced in the form of foams that incorporate arrays of gaseous voids in their structure. These materials are extremely versatile, with properties ranging from soft and flexible to hard and rigid. The softer foams are generally used for cushioning in upholstery and automobile seats, and in various applications such as vibration absorbers. Semirigid foams find use in floatation devices, refrigerator insulation, disposable food trays and containers, building insulation panels, and sound attenuation. Rigid foams have been used as construction materials for boats, airplane components, electronic encapsulation, and furniture.

Foamed materials can be made by a wide variety of processes; they can either be made as discrete products or used as a "foamed-in-place" material. In addition to the sound-and vibration-attenuation properties mentioned above, foams offer light weight and the possibility of improved stiffness and reduced cost (less material to make the part).

POLYMER COATINGS

Polymer coatings are used extensively to enhance appearance, but they have also assumed a significant role in providing corrosion protection. The tough, thick coatings must adhere to the substrate; not chip or peel; and resist exposure to heat, moisture, salt, and chemicals. Polymer coatings have been replacing chrome and cadmium due to environmental concerns relating to the heavy metals. In addition, polymers provide better resistance to the effects of acid rain.

PLASTICS VERSUS OTHER MATERIALS

Polymeric materials have successfully competed with traditional materials in a number of areas. Plastics have replaced glass in containers and other transparent products. PVC pipe and fittings compete with copper and brass in many plumbing applications. Plastics have even replaced ceramics in areas as diverse as sewer pipe and lavatory facilities.

While plastics and metals are often viewed as competing materials, their engineering properties are really quite different. Many of the attractive features of plastics have already been discussed. In addition to these, we can add (1) the ability to be fabricated with lower tooling costs; (2) the ability to be molded at the same rate as product assembly, thereby reducing inventory; (3) a possible reduction in assembly operations and easier assembly through snap fits, friction welds, or the use of self-tapping fasteners; (4) the ability to reuse manufacturing scrap; and (5) reduced finishing costs.

Metals, on the other hand, are often cheaper and offer faster fabrication speeds and greater impact resistance. They are considerably stronger and more rigid and can withstand traditional paint cure temperatures. In addition, resistance to flames, acids, and various solvents is significantly better. Table 8-3 compares the mechanical properties of selected polymers to annealed, commercially pure aluminum and annealed 1040 steel. Note the mechanical superiority of the metals, even though they are being presented in their weakest condition. Table 8-4 compares the cost per pound and elastic modulus of several engineering plastics with values for steel and aluminum. When the size of the part is fixed, cost per cubic inch becomes a more valid comparison, and the figures show plastics to be quite competitive because of their low density.

TABLE 8-3	Property Comparison of Metals and Polymers			
Material	Condition	TS (ksi)	E (10^8 psi)	Elongation
Polyethylene	Branched	2	0.025	90–650
Polyethylene	Crystallized	4	0.100	50–800
Polyvinylchloride	Cl-sides	8	0.375	2–40
Polystyrene	Benzene-sides	7	0.500	1–3
Bakelite	Framework	7	1.0	1
Aluminum	Annealed	13	10.0	15–30
1040 steel	Annealed	75	30.0	30

TABLE 8-4	Comparison of Materials (Modulus and Cost)[a]		
Material	Modulus ($\times 10^6$ psi)	$/pound	$/in.^{3}
Aluminum	10.0	1.20	0.122
Steel	30.0	0.30–0.40	0.075–0.10
Nylon	0.1	1.80	0.129
ABS	0.3	1.00	0.034
High-density polyethylene	0.1	0.80	0.030
Polycarbonate	0.35	2.25	0.097
Polypropylene	0.2	0.80	0.025
Polystyrene	0.3	0.90	0.027
Epoxy (bisphenol)	0.45	1.15	0.036

[a] Cost figures are 2006 values and are clearly subject to change.

The automotive industry is a good indication of the expanding use of plastics. Polymeric materials now account for over 250 pounds of a typical vehicle, compared to only 25 pounds in 1960, 105 in 1970, 195 in 1980, and 229 in 1990. In addition to the traditional application areas of dashboards, interiors, body panels, and trim, plastics are now being used for bumpers, intake manifolds, valve covers, fuel tanks, and fuel lines and fittings. If we include clips and fasteners, there are now over 1000 plastic parts in a typical automobile.

RECYCLING OF PLASTICS

Because of the wide variety of types and compositions, all with similar physical properties, the recycling of mixed plastics is far more difficult than the recycling of mixed metals. These materials must be sorted not only on the basis of resin type, but also by type of filler and color.

If the various types of resins can be identified and kept separate, many of the thermoplastic materials can be readily recycled into useful products. Packaging is the largest single market for plastics, and there is currently a well-established network to collect and recycle PET (the polyester used in soft-drink bottles) and high-density polyethylene (the plastic used in milk, juice, and water jugs). The properties generally deteriorate with recycling, however, so applications must often be downgraded with reuse. PET is being recycled into new bottles, fiber-fill insulation, and carpeting. Recycled polyethylene is used for new containers, plastic bags, and recycling bins. Polystyrene has been recycled into cafeteria trays and videocassette cases. Plastic "lumber" offers weather and insect resistance and a reduction in required maintenance (but at higher cost than traditional wood).

When thermoplastics and thermosets are mixed in varying amounts, the material is often regarded more as an alternative fuel (competing with coal and oil) than as a resource for recycling into quality products. On an equivalent-weight basis, polystyrene and polyethylene have heat contents greater than fuel oil and far in excess of paper and wood. As a recycling alternative, decomposition processes can be used to break polymers down into useful building blocks. Hydrolysis (exposure to high-pressure steam) and pyrolysis (heating in the absence of oxygen) methods can be used to convert plastics into simple petrochemical materials, but even these processes require some control of the input material. As a result, only about one-third of all plastic now finds a second life.

■ 8.3 ELASTOMERS

The term *elastomer,* a contraction of the words *elastic polymer,* refers to a special class of linear polymers that display an exceptionally large amount of elastic deformation when a force is applied. Many can be stretched to several times their original length. Upon release of the force, the deformation can be completely recovered as the material quickly returns to its original shape. In addition, the cycle can be repeated numerous times with identical results, as with the stretching of a rubber band.

The elastic properties of most engineering materials are the result of a change in the distance between adjacent atoms (i.e., bond length) when loads are applied. Hooke's law is commonly obeyed, where twice the force produces twice the stretch. When the applied load is removed, the interatomic forces return all of the atoms to their original position and the elastic deformation is recovered completely.

In the elastomeric polymers, the linear chain-type molecules are twisted or curled, much like a coil spring. When a force is applied, the polymer stretches by uncoiling. When the load is removed, the molecules recoil as the bond angles return to their original, unloaded values, and the material returns to its original size and shape. The relationship between force and stretch, however, does not follow Hooke's law.

In reality, the behavior of elastomers is a bit more complex. While the chains indeed uncoil when placed under load, they can also slide with respect to one another to produce a small degree of viscous deformation. When the load is removed, the molecules return to their coiled shape, but the viscous deformation is not recovered and there is some permanent change in shape.

By linking the coiled molecules to one another by strong covalent bonds, a process known as *cross-linking*, it is possible to restrict the viscous deformation while retaining the large elastic response. The elasticity or rigidity of the product can be determined by controlling the number of cross-links. Small amounts of cross-linking leave the elastomer soft and flexible, as in a rubber band. Additional cross-linking further restricts the uncoiling, and the material becomes harder, stiffer, and more brittle, like the rubber used in bowling balls. Since the cross-linked bonds can only be destroyed by extremely high temperatures, the engineering elastomers can be tailored to possess a wide range of stable properties and stress–strain characteristics.

If placed under constant strain, however, even highly cross-linked material will exhibit some viscous flow over time. Consider a rubber band stretched between two nails. While the dimensions remain fixed, the force or stress being applied to the nails will continually decrease. This phenomenon is known as *stress relaxation*. The rate of this relaxation depends on the material, the force, and the temperature.

RUBBER

Natural rubber, the oldest commercial elastomer, is made from latex, a secretion from the inner bark of a tropical tree. In its crude form it is an excellent adhesive, and many cements can be made by dissolving it in suitable solvents. Its use as an engineering material dates from 1839, when Charles Goodyear discovered that it could be vulcanized (cross-linked) by the addition of about 30% sulfur followed by heating to a suitable temperature. The cross-linking restricts the movement of the molecular chains and imparts strength. Subsequent research found that the properties could be further improved by various additives (such as carbon black), which act as stiffeners, tougheners, and antioxidants. Accelerators have been found that speed up the vulcanization process. These have enabled a reduction in the amount of sulfur, such that most rubber compounds now contain less than 3% sulfur. Softeners can be added to facilitate processing, and fillers can be used to add bulk.

Rubber can now be compounded to provide a wide range of characteristics, ranging from soft and gummy to extremely hard. When additional strength is required, textile cords or fabrics can be coated with rubber. The fibers carry the load, and the rubber serves as a matrix to join the cords while isolating them from one another to prevent chafing. For severe service, steel wires can be used as the load-bearing medium. Vehicle tires and heavy-duty conveyor belts are examples of this technology.

Natural rubber compounds are outstanding for their flexibility; good electrical insulation; low internal friction; and resistance to most inorganic acids, salts, and alkalis. However, they have poor resistance to petroleum products, such as oil, gasoline, and naphtha. In addition, they lose their strength at elevated temperatures, so it is advisable that they not be used at temperatures above 80°C (175°F). Unless they are specially compounded, they also deteriorate fairly rapidly in direct sunlight.

ARTIFICIAL ELASTOMERS

In an attempt to overcome some of these limitations, as well as the uncertainty in the supply and price of natural rubber, a number of synthetic or artificial elastomers have been developed and have come to assume great commercial importance. While some are a bit inferior to natural rubber, others offer distinctly different and, frequently, superior properties. Polyisoprene is the synthetic that is closest to duplicating natural rubber. Styrene–butadiene is an oil-derivative, high-volume substitute for natural rubber that has become the standard material for passenger-car tires. For this material, some form of reinforcement is generally required to provide the desired tensile strength, tear resistance, and durability. Neoprenes have properties similar to natural rubber, with better resistance to oils, ozone, oxidation, and flame. They are used for a wide range of applications, including automotive hoses and belts, footwear, tires, mounting cushions, and seals. A number of other artificial elastomers are available and are identified by both chemical and commercial trade names.

Silicone rubbers look and feel like organic rubber but are based on a linear chain of silicon and oxygen atoms (not carbon). Various mixes and blends offer retention of physical properties at elevated temperatures [as hot as 230°C (450°F)]; flexibility at low temperatures [as low as −100°C (−150°F)]; resistance to acids, bases, and other aqueous and organic fluids; resistance to flex fatigue; ability to absorb energy and provide damping; good weatherability; ozone resistance; and availability in a variety of different hardnesses.

Elastomers can also be classified as thermosets or thermoplastics. The thermoset materials are formed during the irreversible vulcanization (cross-linking) process, which may be somewhat time consuming. Thermoplastic elastomers eliminate the vulcanization cycle and can be processed into products by all of the conventional thermoplastic polymer processes (injection molding, extrusion, blow molding, thermoforming, and others). They soften at elevated temperatures, which the thermosets easily withstand, but offer good low-temperature flexibility, scrap recyclability, availability in a variety of colors, and high gripping friction. Unfortunately, many are more costly than the conventional rubber materials.

SELECTION OF AN ELASTOMER

Elastomeric materials can now be selected and used for a wide range of engineering applications, where they impart properties that include shock absorption, noise and vibration control, sealing, corrosion protection, abrasion protection, friction modification, electrical and thermal insulation, waterproofing, and load bearing. Selection of an elastomer for a specific application requires consideration of many factors, including the mechanical and physical service requirements, the operating environment (including temperature), the desired lifetime, the ability to manufacture the product, and cost. There are a number of families, and within each family there exists a wide range of available properties. Moreover, almost any physical or mechanical property can be altered through additives, which can also be used to enhance processing or reduce cost, and modifications of the processing parameters.

Table 8-5 lists some of the more common artificial elastomers, along with natural rubber for comparison, and gives their properties and some typical uses.

ELASTOMERS FOR TOOLING APPLICATIONS

When an elastomer is confined, it acts like a fluid, transmitting force uniformly in all directions. For this reason, elastomers can be substituted for one-half of a die set in sheet-metal-forming operations. Elastomers are also used to perform bulging and to form reentrant sections that would be impossible to form with rigid dies except through the use of costly multipiece tooling. The engineering elastomers have become increasingly popular as tool materials because they can be compounded to range from very soft to very hard; hold up well under compressive loading; are impervious to oils, solvents, and other similar fluids; and can be made into a desired shape quickly and economically. In addition, the elastomeric tooling will not mark or damage highly polished or prepainted surfaces. The urethanes are currently the most popular elastomer for tooling applications.

■ 8.4 CERAMICS

The first materials used by humans were natural materials such as wood and stone. The discovery that certain clays could be mixed, shaped, and hardened by firing led to what was probably the first man-made material. Traditional ceramic products, such as bricks and pottery, have continued to be key materials throughout history. More recently, *ceramic materials* have assumed important roles in a number of engineering applications. Most of these utilize their outstanding physical properties, including the ability to withstand high temperatures, provide a wide variety of electrical and magnetic properties, and resist wear. In general, ceramics are hard, brittle, high-melting-point materials with low electrical and thermal conductivity, low thermal expansion, good chemical and thermal stability, good creep resistance, high elastic modulus, and high compressive

TABLE 8-5 Properties and Uses of Common Elastomers

Elastomer	Specific Gravity	Durometer Hardness	Tensile Strength (psi) Pure Gum	Tensile Strength (psi) Black	Elongation (%) Pure Gum	Elongation (%) Black	Service Temperature [F(C)] Min.	Service Temperature [F(C)] Max.	Resistance to:[a] Oil	Resistance to:[a] Water Swell	Resistance to:[a] Tear	Typical Application
Natural rubber	0.93	20–100	2500	4000	75	650	−65(54)	180(82)	P	G	G	Tires, gaskets, hose
Polyacrylate	1.10	40–100	350	2500	600	400	0(−18)	300(149)	G	P	F	Oil hose, O-rings
EDPM (ethylene propylene)	0.85	30–100	1	3		500	−40(−40)	300(149)	P	G	G	Electric insulation, footwear, hose, belts
Chlorosulfonated polyethylene	1.10	50–90	4	2		400	−65(−54)	250(121)	G	E	G	Tank lining, chemical hose; shoes, soles and heels
Polychloroprene (neoprene)	1.23	20–90	3500	4000	800	550	−50(−46)	225(107)	G	G	G	Wire insulation, belts, hose, gaskets, seals, linings
Polybutadiene	1.93	30–100	1000	3000	800	550	−80(−62)	212(100)	P	P	G	Tires, soles and heels, gaskets, seals
Polyisoprene	0.94	20–100	3000	4000	600	600	−65(−54)	180(82)	P	G	G	Same as natural rubber
Polysulfide	1.34	20–80	350	1000	600	400	−65(−54)	180(82)	E	G	G	Seals, gaskets, diaphragms, valve disks
SBR (styrene-butadiene)	0.94	40–100	2			1200	−65(−54)	225(107)	P	G	G	Molded mechanical goods, disposable pharmaceutical items
Silicone	1.1	25–90		1200		450	−120(−84)	450(232)	F	E	P	Electric insulation, seals, gaskets, O-rings
Epichlorohydrin	1.27	40–90		2		325	−50(−46)	250(121)	G	G	G	Diaphragms, seals, molded goods, low-temperature parts
Urethane	0.85	62–95	5000		700		−54(−65)	212(100)	E	F	E	Caster wheels, heels, foam padding
Fluoroelastomers	1.65	60–90	1	3		400	−40(−40)	450(232)	E	E	F	O-rings, seals, gaskets, roll coverings

[a] P, poor; F, fair; G, good; E, excellent.

strengths that are retained at elevated temperature. A family of "structural ceramics" has also emerged, and these materials now provide enhanced mechanical properties that make them attractive for many load-bearing applications.

Glass and glass products now account for about half of the ceramic materials market. Advanced ceramic materials (including the structural ceramics, electrical and magnetic ceramics, and fiber-optic material) compose another 20%. Whiteware and porcelain enameled products (such as household appliances) account for about 10% each, while refractories and structural clay products make up most of the difference.

NATURE AND STRUCTURE OF CERAMICS

Ceramic materials are compounds of metallic and nonmetallic elements (often in the form of oxides, carbides, and nitrides) and exist in a wide variety of compositions and forms. Most have crystalline structures, but unlike metals, the bonding electrons are generally captive in strong ionic or covalent bonds. The absence of free electrons makes the ceramic materials poor electrical conductors and results in many being transparent in thin sections. Because of the strength of the primary bonds, most ceramics have high melting temperatures, high rigidity, and high compressive strength.

The crystal structures of ceramic materials can be quite different from those observed in metals. In many ceramics, atoms of significantly different size must be accommodated within the same structure, and the interstitial sites, therefore, become extremely important. Charge neutrality must be maintained throughout ionic structures. Covalent materials must have structures with a limited number of nearest neighbors, set by the number of shared-electron bonds. These features often dictate a less efficient packing, and hence lower densities, than those observed for metallic materials. As with metals, the same chemistry material can often exist in more than one structural arrangement (polymorphism). Silica (SiO_2), for example, can exist in three forms—quartz, tridymite, and crystobalite—depending on the conditions of temperature and pressure.

Ceramic materials can also exist in the form of chains, similar to the linear molecules in plastics. Like the polymeric materials having this structure, the bonds between the chains are not as strong as those within the chains. Consequently, when forces are applied, cleavage or shear can occur between the chains. In other ceramics, the atoms bond in the form of sheets, producing layered structures. Relatively weak bonds exist between the sheets, and these interfacial surfaces become the preferred sites for fracture. Mica is a good example of such a material. A noncrystalline structure is also possible in solid ceramics. This *amorphous* condition is referred to as the *glassy state*, and the materials are known as *glasses*.

Elevated temperatures can be used to decrease the viscosity of glass, allowing the atoms to move as groups and the material to be shaped and formed. When the temperature is dropped, the material again becomes hard and rigid. The crystalline ceramics do not soften, but they can creep at elevated temperature by means of grain boundary sliding. Therefore, when ceramic materials are produced for elevated-temperature service, large grain size is generally desired.

CERAMICS ARE BRITTLE BUT CAN BE TOUGH

Both crystalline and noncrystalline ceramics tend to be brittle. The glass materials have a three-dimensional network of strong primary bonds that impart brittleness. The crystalline materials do contain dislocations, but for ceramic materials, brittle fracture tends to occur at stresses lower than those required to induce plastic deformation.

There is little that can be done to alter the brittle nature of ceramic materials. However, the energy required to induce brittle fracture (the material toughness) can often be increased. Tempered glass uses rapid cooling of the surfaces to induce residual surface compression. The surfaces cool, contract, and harden. As the center then cools and tries to contract, it compresses or squeezes the surface. Since fractures initiate on the surface, the applied stresses must first cancel the residual compression before they can become tensile. Cermet materials surround particles of brittle ceramic with a continuous matrix of tough, fracture-resistant metal. Ceramic–ceramic composites use weak interfaces that separate or delaminate to become crack arrestors or crack diverters, allowing the structure to continue carrying the load.

Stabilization involves compounding or alloying to eliminate crystal structure changes and the dimensional expansions or contractions that accompany them. Nonuniform heating or cooling can now occur without the stresses that induce fracture. Transformation toughening stops the progress of a crack by crystal structure changes that occur when volume expansion is permitted. Fine grain size, high purity, and high density can be promoted by enhanced processing, and these all act to improve toughness.

CLAY AND WHITEWARE PRODUCTS

Many ceramic products are still based on *clay*, to which various amounts of quartz and feldspar and other materials are added. Selected proportions are mixed with water, shaped, dried, and fired to produce the structural clay products of brick, roof and structural tiles, drainage pipe, and sewer pipe, as well as the *whiteware* products of sanitary ware (toilets, sinks, and bathtubs), dinnerware, china, decorative floor and wall tile, pottery, and other artware.

REFRACTORY MATERIALS

Refractory materials are ceramics that have been designed to provide acceptable mechanical or chemical properties at high operating temperatures. They may take the form of bricks and shaped products, bulk materials (often used as coatings), and insulating ceramic fibers. Most are based on stable oxide compounds, where the coarse oxide particles are bonded by finer refractory material. Various carbides, nitrides, and borides can also be used in refractory applications.

Refractory ceramics fall into three distinct classes: *acidic, basic,* and *neutral.* Common acidic refractories are based on silica (SiO_2) and alumina (Al_2O_3) and can be compounded to provide high-temperature resistance along with high hardness and good mechanical properties. The insulating tiles on the U.S. space shuttle were made from machinable silica ceramic. Magnesium oxide (MgO) is the core material for most basic refractories. These are generally more expensive than the acidic materials but are often required in metal-processing applications to provide compatibility with the metal. Neutral refractories, containing chromite (Cr_2O_3), are often used to separate the acidic and basic materials since they tend to attack one another. The combination is often attractive when a basic refractory is necessary on the surface for chemical reasons, and the cheaper acidic material is used beneath to provide strength and insulation. Figure 8-7 shows a variety of high-strength alumina components.

ABRASIVES

Because of their high hardness, ceramic materials, such as silicon carbide and aluminum oxide (alumina), are often used for abrasive applications, such as grinding. Materials such as manufactured diamond and cubic boron nitride have such phenomenal properties that they are often termed *superabrasives.* Materials used for abrasive applications are discussed in greater detail in Chapter 29.

FIGURE 8-7 A variety of high-strength alumina (acid refractory) components, including a filter for molten metal. *(Courtesy of Wesgo Division, GTE, Hayward, CA.)*

CERAMICS FOR ELECTRICAL AND MAGNETIC APPLICATIONS

Ceramic materials also offer a variety of useful electrical and magnetic properties. Some ceramics, such as silicon carbide, are used as resistors and heating elements for electric furnaces. Others have semiconducting properties and are used for thermistors and rectifiers. Dielectric, piezoelectric, and ferroelectric behavior can also be utilized in many applications. Barium titanate, for example, is used in capacitors and transducers. High-density clay-based ceramics and aluminum oxide make excellent high-voltage insulators. The magnetic ferrites have been used in a number of magnetic applications. Considerable attention has also been directed toward the "high-temperature"ceramic superconductors.

GLASSES

When some molten ceramics are cooled at a rate that exceeds a critical value, the material solidifies into a hard, rigid, noncrystalline (i.e., amorphous) solid, known as a *glass*. Most commercial glasses are based on silica (SiO_2), lime ($CaCO_3$), and sodium carbonate ($NaCO_3$), with additives to alter the structure or reduce the melting point. Various chemistries can be used to optimize optical properties, thermal stability, and resistance to thermal shock.

Glass is soft and moldable when hot, making shaping rather straightforward. When cool and solid, glass is strong in compression but brittle and weak in tension. In addition, most glasses exhibit excellent resistance to weathering and attack by most chemicals. Traditional applications include automotive and window glass, bottles and other containers, light bulbs, mirrors, lenses, and fiberglass insulation. There is also a wide variety of specialty applications, including glass fiber for fiber-optic communications, glass fiber to reinforce composites, cookware, TV tubes and monitors, and a variety of medical and biological products. Glass and other ceramic fibers have been used for filtration, where they provide a chemical inertness and the possibility to withstand elevated temperature.

GLASS CERAMICS

These materials are first shaped as a glass and then heat treated to promote partial devitrification or crystallization of the material, resulting in a structure that contains large amounts of crystalline material within an amorphous base. Since they were initially formed as a glass, glass ceramics do not have the strength-limiting or fracture-inducing porosity that is characteristic of the conventional sintered ceramics. Strength is greater than with the traditional glasses, and the crystalline phase helps to retard creep at high temperatures. Since the thermal expansion coefficient is near zero, the material has good resistance to thermal shock. The white Pyroceram (trade name) material commonly found in Corningware is a common example of a glass ceramic.

CERMETS

Cermets are combinations of metals and ceramics (usually oxides, carbides, nitrides, or carbonitrides) united into a single product by the procedures of powder metallurgy. This usually involves pressing mixed powders at pressures ranging from 70 to 280 MPa (10 to 40 ksi) followed by sintering in a controlled-atmosphere furnace at about 1650°C (3000°F). Cermets combine the high hardness and refractory characteristics of ceramics with the toughness and thermal shock resistance of metals. They are used as crucibles, jet engine nozzles, and aircraft brakes, as well as in other applications requiring hardness, strength, and toughness at elevated temperature. Cemented tungsten carbide (tungsten carbide particles cemented in a cobalt binder) has been used in dies and cutting tools for quite some time. The more advanced cermets now enable higher cutting speeds than those achievable with high-speed tool steel, tungsten carbide, or the coated carbides. See Chapter 22.

CEMENTS

Various ceramic materials can harden by chemical reaction, enabling their use as a binder that does not require firing or sintering. Sodium silicate hardens in the presence of carbon dioxide and is used to produce sand cores in metal casting. Plaster of paris and portland cement both harden by hydration reactions.

CERAMIC COATINGS

A wide spectrum of enamels, glazes, and other ceramic coatings have been developed to decorate, seal, and protect substrate materials. Porcelain enamel can be applied to carbon steel in the perforated tubs of washing machines, where the material must withstand the scratching of zippers, buttons, and snaps along with the full spectrum of laundry products. Chemical reaction vessels are often glass lined.

CERAMICS FOR MECHANICAL APPLICATIONS: THE STRUCTURAL AND ADVANCED CERAMICS

Because of the strong ionic or covalent bonding and high shear resistance, ceramic materials tend to have low ductility and high compressive strength. Theoretically, ceramics could also have high tensile strengths. However, because of their high melting points and lack of ductility, most ceramics are processed in the solid state, where products are made from powdered material. After various means of compaction, voids remain between the powder particles, and a portion of these persists through the sintering process. Contamination can also occur on particle surfaces and then become part of the internal structure of the product. As a result, full theoretical density is extremely difficult to achieve, and small cracks, pores, and impurity inclusions tend to be an integral part of most ceramic materials. These act as mechanical stress concentrators. As loads are applied, the effect of these flaws cannot be reduced through plastic flow, and the result is generally a brittle fracture. Applying the principles of fracture mechanics,[3] we find that ceramics are sensitive to very small flaws. Tensile failures typically occur at stress values between 20 and 210 MPa (3 and 30 ksi), more than an order of magnitude less than the corresponding strength in compression.

Since the number, size, shape, and location of the flaws are likely to differ from part to part, ceramic parts produced from identical material by identical methods often fail at very different applied loads. As a result, the mechanical properties of ceramic products tend to follow a statistical spread that is much less predictable than for metals. This feature tends to limit the use of ceramics in critical high-strength applications.

If the various flaws and defects could be eliminated or reduced to very small size, high and consistent tensile strengths could be obtained. Hardness, wear resistance, and strength at elevated temperatures would be attractive properties, along with light weight (specific gravities of 2.3 to 3.85), high stiffness, dimensional stability, low thermal conductivity, corrosion resistance, and chemical inertness. Reliability might be low, however, and failure would still occur by brittle fracture. Because of the poor thermal conductivity, thermal shock may be a problem. The cost of these "flaw-free" or "restricted flaw" materials would be rather high. Joining to other engineering materials and machining would be extremely difficult, so products would have to be fabricated through the use of net-shape processing.

Advanced, structural, or engineering ceramics is an emerging technology with a broad base of current and potential applications. The base materials currently include silicon nitride, silicon carbide, partially stabilized zirconia, transformation-toughened zirconia, alumina, sialons, boron carbide, boron nitride, titanium diboride, and ceramic composites (such as ceramic fibers in a glass, glass–ceramic, or ceramic matrix). The materials and products are characterized by high strength, high fracture toughness, fine grain size, and little or no porosity. Applications include a wide variety of wear-

---

[3] According to the principles of fracture mechanics, fracture will occur in a brittle material when the fracture toughness, K, is equal to a product involving a dimensionless geometric factor, α, the applied stress, σ, and the square root of the number π (3.14) times the size of the most critical flaw, a.

$$K = \alpha \sigma (\pi a)^{1/2}$$

When the right–hand side is less than the value of K, the material bears the load without breaking. Fracture occurs when the combination of applied stress and flaw size equals the critical value K. Since K is a material property, any attempt to increase the load or stress a material can withstand must be achieved by a companion reduction in flaw size.

FIGURE 8-8 Gas-turbine rotors made of silicon nitride. The lightweight material (one-half the weight of stainless steel) offers strength at elevated temperature as well as excellent resistance to corrosion and thermal shock. *(Courtesy of Wesgo Division, GTE, Hayward, CA.)*

resistant parts (including bearings, seals, valves, and dies), cutting tools, punches, dies, and engine components, as well as use in heat exchangers, gas turbines, and furnaces. Porous products have been used as substrate material for catalytic converters and as filters for streams of molten metal. Biocompatible ceramics have been used as substitutes for joints and bones and as dental implants.

Alumina (or aluminum oxide) ceramics are the most common for industrial applications. They are relatively inexpensive and offer high hardness and abrasion resistance, low density, and high electrical resistivity. Alumina is strong in compression and retains useful properties at temperatures as high as 1900°C (3500°F), but it is limited by low toughness, low tensile strength, and susceptibility to thermal shock and attack by highly corrosive media. Due to its high melting point, it is generally processed in a powder form.

Silicon carbide and silicon nitride offer excellent strength and wear resistance with moderate toughness. They work well in high-stress, high-temperature applications, such as turbine blades, and may well replace nickel- or cobalt-based superalloys. Figure 8-8 shows gas-turbine rotors made from injection-molded silicon nitride. They are designed to operate at 1250°C (2300°F), where the material retains over half of its room-temperature strength and does not require external cooling. Figure 8-9 shows some additional silicon nitride products.

Sialon (a *si*licon–*al*uminum–*o*xygen–*n*itrogen structural ceramic) is really a solid solution of alumina and silicon nitride, and it bridges the gap between them. More aluminum oxide enhances hardness, while more silicon nitride improves toughness. The resulting material is stronger than steel, extremely hard, and as light as aluminum. It has good resistance to corrosion, wear, and thermal shock; is an electrical insulator; and retains good tensile and compressive strength up to 1400°C (2550°F). It has excellent dimensional stability, with a coefficient of thermal expansion that is only one-third that of steel and one-tenth that of plastic. When overloaded, however, it exhibits the ceramic property of failure by brittle fracture.

Zirconia is inert to most metals and retains strength to temperatures well over 2200°C (4000°F). Partially stabilized zirconia combines the zirconia characteristics of resistance to thermal shock, wear, and corrosion; low thermal conductivity; and low friction coefficient with the enhanced strength and toughness brought about by doping the material with oxides of calcium, yttrium, or magnesium. Transformation-toughened zirconia has even greater toughness as a result of dispersed second phases throughout the ceramic matrix. When a crack approaches the metastable phase, it transforms to a more stable structure, increasing in volume to compress and stop the crack.

The high cost of the structural ceramics continues to be a barrier to their widespread acceptance. High-grade ceramics are currently several times more expensive than their metal counterparts. Even factoring in enhanced lifetime and improved performance, there is still a need to reduce cost. Work continues, however, toward the development of a low-cost, high-strength, high-toughness ceramic with a useful temperature range. Parallel efforts are under way to ensure flaw detection in the range of 10 to 50 mm. If these efforts are successful, ceramics could compete where tool steels,

FIGURE 8-9 A variety of components manufactured from silicon nitride, including an exhaust valve and turbine blade. *(Courtesy of Wesgo Division, GTE, Hayward, CA.)*

TABLE 8-6	Properties of Some Structural Ceramics				
Material	Density (g/cm^3)	Tensile Strength (ksi)	Compressive Strength (ksi)	Modulus of Elasticity (10^6 psi)	Fracture Toughness (ksi $\sqrt{in}$.)
Al$_2$O$_3$	3.98	30	400	56	5
Sialon	3.25	60	500	45	9
SiC	3.1	25	560	60	4
ZrO$_2$ (partially stabilized)	5.8	65	270	30	10
ZrO$_2$ (transformation toughened)	5.8	50	250	29	11
Si$_3$N$_4$ (hot pressed)	3.2	80	500	45	5

powdered metals, coated materials, and tungsten carbide are now being used. Potential applications include engines, turbochargers, gas turbines, bearings, pump and valve seals, and other products that operate under high-temperature, high-stress environments.

A ceramic automobile engine has been discussed for a number of years. By allowing higher operating temperatures, engine efficiency could be increased. Sliding friction would be reduced and there would be no need for cooling. The radiator, water pump, coolant, fan belt, and water lines could all be eliminated. The net result would be up to a 30% reduction in fuel consumption. Unfortunately, this is still a dream because of the inability to produce large, complex-shaped products with few small-sized flaws.

Table 8-6 provides the mechanical properties of some of today's structural ceramics.

ADVANCED CERAMICS AS CUTTING TOOLS

Their high hardness, retention of hardness at elevated temperature, and low reactivity with metals make ceramic materials attractive for cutting applications, and cutting tools have improved significantly through advances in ceramic technology. Silicon carbide is a common abrasive in many grinding wheels. Cobalt-bonded tungsten carbide has been a popular alternative to high-speed tool steels for many tool and die applications. Many carbide tools are now enhanced by a variety of vapor-deposited ceramic coatings. Thin layers of titanium carbide, titanium nitride, and aluminum oxide can inhibit reactions between the metal being cut and the binder phase of the carbide. This results in a significant reduction in friction and wear that enables faster rates of cutting. Silicon nitride, boron carbide, cubic boron nitride, and polycrystalline diamond cutting tools now offer even greater tool life, higher cutting speeds, and reduced machine downtime. With advanced tool materials, cutting speeds can be increased from 60 to 1500 m/min (200 to 5000 ft/min). The use of these ultra-high-speed materials, however, requires companion developments in the machine tools themselves. High-speed spindles must be perfectly balanced, and workholding devices must withstand high centrifugal forces. Chip-removal methods must be able to remove the chips as fast as they are formed.

As environmental regulations become more stringent, dry machining may be pursued as a means of reducing or eliminating coolant- and lubricant-disposal problems. Ceramic materials are currently the best materials for dry operations. Ceramic tools have also been used in the direct machining of materials that once required grinding, a process sometimes called *hard machining*. Figure 8-10 shows the combination of toughness and hardness for a variety of cutting-tool materials.

■ 8.5 COMPOSITE MATERIALS

A *composite material* is a nonuniform solid consisting of two or more different materials that are mechanically or metallurgically bonded together. Each of the various components retains its identity in the composite and maintains its characteristic structure and properties. There are recognizable interfaces between the materials. The composite material, however, generally possesses characteristic properties (or combinations of

FIGURE 8-10   Graphical mapping of the combined toughness and hardness for a variety of cutting-tool materials. Note the superior hardness of the ceramic materials.

properties), such as stiffness, strength, weight, high-temperature performance, corrosion resistance, hardness, and conductivity, which are not possible with the individual components by themselves. Analysis of these properties shows that they depend on (1) the properties of the individual components; (2) the relative amounts of the components; (3) the size, shape, and distribution of the discontinuous components; (4) the orientation of the various components; and (5) the degree of bonding between the components. The materials involved can be organics, metals, or ceramics. Hence a wide range of freedom exists, and composite materials can often be designed to meet a desired set of engineering properties and characteristics.

There are many types of composite materials and several methods of classifying them. One method is based on geometry and consists of three distinct families: laminar or layered composites, particulate composites, and fiber-reinforced composites.

LAMINAR OR LAYERED COMPOSITES

Laminar composites have distinct layers of material bonded together in some manner and include thin coatings, thicker protective surfaces, claddings, bimetallics, laminates, sandwiches, and others. They are used to impart properties such as reduced cost, enhanced corrosion resistance or wear resistance, electrical insulation or conductivity, unique expansion characteristics, lighter weight, improved strength, or altered appearance.

Plywood is probably the most common engineering material in this category and is an example of a laminate material. Layers of wood veneer are adhesively bonded with their grain orientations at various angles to one another. Strength and fracture resistance are improved, properties are somewhat uniform within the plane of the sheet, swelling and shrinkage tendencies are minimized, and large pieces are available at reasonable cost. Safety glass is another laminate in which a layer of polymeric adhesive is placed between two pieces of glass and serves to retain the fragments when the glass is broken. *A*ramid–*a*luminum–*l*aminates (Arall) consist of thin sheets of aluminum bonded with woven adhesive-impregnated aramid fibers. The combination offers light weight coupled with high fracture, impact, and fatigue resistance.

Laminated plastics are made from layers of reinforcing material that have been impregnated with thermosetting resins, bonded together, and cured under heat and pressure. They can be produced as sheets, or rolled around a mandrel to produce a tube, or rolled tightly to form a rod. Various resins have been used with reinforcements of paper, cotton or nylon fabric, asbestos, or glass fiber (usually in woven form). Common applications include a variety of decorative items, such as Formica countertops, imitation hardwood flooring, and furniture. When combined with a metal layer on one or both surfaces, the material is used for printed circuit boards.

FIGURE 8-11 Schematic of a bimetallic strip where material A has the greater coefficient of thermal expansion. Note the response to cold and hot temperatures.

Bimetallic strip is a laminate of two metals with significantly different coefficients of thermal expansion. As Figure 8-11 illustrates, changes in temperature now produce flexing or curvature in the product. The unique property of shape varying with temperature is often employed in thermostat and other heat-sensing applications. Still other laminar composites are designed to provide enhanced surface characteristics while retaining a low-cost, high-strength, or lightweight core. Many clad materials fit this description. Alclad metal, for example, consists of high strength, age-hardenable aluminum with an exterior cladding of one of the more corrosion-resistant, single-phase, non-heat-treatable aluminum alloys. Stainless steel has been applied to cheaper, less corrosion-resistant, substrates. U.S. coinage is a laminate, designed to conserve the more costly, high-nickel-content material while providing a lustrous, corrosion-resistant surface. Other laminates have surface layers that have been selected primarily for enhanced wear resistance, improved appearance, or electrical conductivity.

Sandwich material is a laminar structure composed of a thick, low-density core placed between thin, high-density surfaces. Corrugated cardboard is an example of a sandwich structure. Other engineering sandwiches incorporate cores of a polymer foam or honeycomb structure to produce a lightweight, high-strength, high-rigidity composite.

It should be noted that the properties of laminar composites are always anisotropic—that is, they are not the same in all directions. Because of the variation in structure, properties will always be different in the direction perpendicular to the layers.

PARTICULATE COMPOSITES

Particulate composites consist of discrete particles of one material surrounded by a matrix of another material. Concrete is a classic example, consisting of sand and gravel particles surrounded by hydrated cement. Asphalt consists of similar aggregate in a matrix of bitumin, a thermoplastic polymer. In both of these examples, the particles are rather coarse. Other particulate composites involve extremely fine particles and include many of the multicomponent powder metallurgy products, specifically those where the dispersed particles do not diffuse into the matrix material.

Dispersion-strengthened materials are particulate composites where a small amount of hard, brittle, small-sized particles (typically, oxides or carbides) are dispersed throughout a softer, more ductile metal matrix. Since the dispersed material is not soluble in the matrix, it does not redissolve, overage, or overtemper when the material is heated. Pronounced strengthening can be induced, which decreases only gradually as temperature is increased. Creep resistance, therefore, is improved significantly. Examples of dispersion-strengthened materials include sintered aluminum powder (SAP), which consists of an aluminum matrix strengthened by up to 14% aluminum oxide, and thoria-dispersed (or TD) nickel, a nickel alloy containing 1 to 2 wt% thoria (ThO_2). Because of the metal–ceramic mix and the desire to distribute materials of differing density, the dispersion-strengthened composites are generally produced by powder metallurgy techniques.

Other types of particulate composites, known as *true particulate composites*, contain large amounts of coarse particles. They are usually designed to produce some desired combination of properties rather than increased strength. Cemented carbides, for example, consist of hard ceramic particles, such as tungsten carbide, tantalum carbide, or titanium carbide, embedded in a metal matrix, which is usually cobalt. Although the hard, stiff carbide could withstand the high temperatures and pressure of cutting, it is extremely brittle. Toughness is imparted by combining the carbide particles with cobalt powder, pressing the material into the desired shape, heating to melt the cobalt, and then resolidifying the compacted material. Varying levels of toughness can be imparted by varying the amount of cobalt in the composite.

Grinding and cutting wheels are often formed by bonding abrasives, such as alumina (Al_2O_3), silicon carbide (SiC), cubic boron nitride (CBN), or diamond, in a matrix of glass or polymeric material. As the hard particles wear, they fracture or pull out of the matrix, exposing fresh, new cutting edges. By combining tungsten powder and powdered silver or copper, electrical contacts can be produced that offer both high conductivity and resistance to wear and arc erosion. Foundry molds and cores are often made from sand (particles) and an organic or inorganic binder (matrix).

Metal–matrix composites of the particulate type have been made by introducing a variety of ceramic or glass particles into aluminum or magnesium matrices. Particulate-toughened ceramics using zirconia and alumina matrices are being used as bearings, bushings, valve seats, die inserts, and cutting-tool inserts. Many plastics could be considered to be particulate composites because the additive fillers and extenders are actually dispersed particles. Designation as a particulate composite, however, is usually reserved for polymers where the particles are added for the primary purpose of property modification. One such example is the combination of granite particles in an epoxy matrix that is currently being used in some machine tool bases. This unique material offers high strength and a vibration-damping capacity that exceeds that of gray cast iron.

Because of their unique geometry, the properties of particulate composites are usually *isotropic*, that is, uniform in all directions. This may be particularly important in engineering applications.

FIBER-REINFORCED COMPOSITES

The most popular type of composite material is the *fiber-reinforced composite* geometry, where continuous or discontinuous thin fibers of one material are embedded in a matrix of another. The objective is usually to enhance strength, stiffness, fatigue resistance, or strength-to-weight ratio by incorporating strong, stiff, but possibly brittle, fibers in a softer, more ductile matrix. The matrix supports and transmits forces to the fibers, protects them from environments and handling, and provides ductility and toughness, while the fibers carry most of the load and impart enhanced stiffness. Wood and bamboo are two naturally occurring fiber composites, consisting of cellulose fibers in a lignin matrix. Bricks of straw and mud may well have been the first human-made material of this variety, dating back to near 800 B.C. Automobile tires now use fibers of nylon, rayon, aramid (Kevlar), or steel in various numbers and orientations to reinforce the rubber and provide added strength and durability. Steel-reinforced concrete is actually a double composite, consisting of a particulate matrix reinforced with steel fibers.

Glass-fiber-reinforced resins, the first of the modern fibrous composites, were developed shortly after World War II in an attempt to produce lightweight materials with high strength and high stiffness. Glass fibers about 10 μm in diameter are bonded in a variety of polymers, generally epoxy or polyester resins. Between 30 and 60% by volume is made up of fibers of either E-type borosilicate glass (tensile strength of 500 ksi and elastic modulus of 10.5×10^6 psi) or the stronger, stiffer, high-performance S-type magnesia–alumina–silicate glass (with tensile strength of 670 ksi and elastic modulus of 12.4×10^6 psi).[4]

---

[4] It is important to note that a fiber of material tends to be stronger than the same material in bulk form. The size of any flaw is limited to the diameter of the fiber, and the complete failure of a given fiber does not propagate through the assembly, as would occur in an identical bulk material.

Glass fibers are still the most widely used reinforcement, primarily because of their lower cost and adequate properties for many applications. Current uses of glass-fiber-reinforced plastics include sporting goods, boat hulls, and bathtubs. Limitations of the glass-fiber material are generally related to strength and stiffness. Alternative fibers have been developed for applications requiring enhanced properties. Boron–tungsten fibers (boron deposited on a tungsten core) offer an elastic modulus of 55×10^6 psi with tensile strengths in excess of 400 ksi. Silicon carbide filaments (SiC on tungsten) have an even higher modulus of elasticity.

Graphite (or carbon) and aramid (DuPont tradename of Kevlar) are other popular reinforcing fibers. Graphite fibers can be either the PAN type, produced by the thermal pyrolysis of synthetic organic fibers, primarily polyacrylonitrile, or pitch type, made from petroleum pitch. They have low density and a range of high tensile strengths (600 to 750 ksi) and high elastic moduli (40 to 65×10^6 psi). Graphite's negative thermal-expansion coefficient can also be used to offset the positive values of most matrix materials, leading to composites with low or zero thermal expansion. Kevlar is an organic aramid fiber with a tensile strength up to 650 ksi, elastic modulus of 27×10^6 psi, a density approximately one-half that of aluminum, and good toughness. In addition, it is flame retardant and transparent to radio signals, making it attractive for a number of military and aerospace applications where the service temperature is not excessive.

Ceramic fibers, metal wires, and specially grown whiskers have also been used as reinforcing fibers for high-strength, high-temperature applications. Metal fibers can also be used to provide electrical conductivity or shielding from electromagnetic interference to a lightweight polymeric matrix. With the demand for less expensive, environmentally friendly materials, the natural fibers have also assumed an engineering material role. Cotton, hemp, flax, jute, coir (coconut husk), and sisal have found use in various composites. Thermoplastic fibers, such as nylon and polyester, have been used to enhance the toughness and impact strength of the brittle thermoset resins.

Table 8-7 lists some of the key engineering properties for several of the common reinforcing fibers. Since the objectives are often high strength coupled with light weight, or high stiffness coupled with light weight, properties are often reported as *specific strength* and *specific stiffness*, where the strength or stiffness values are divided by density.

The orientation of the fibers within the composite is often key to properties and performance. Sheet–molding compound, bulk-molding compound, and fiberglass generally contain short, randomly oriented fibers. Long, unidirectional fibers can be used to produce highly directional properties, with the fiber directions being tailored to the direction of loading. Woven fabrics or tapes can be produced and then layered in various orientations to produce a plywood-like product. The layered materials can then be stitched together to add a third dimension to the weave, and complex three-dimensional shapes can be woven from fibers and later injected with a matrix material.

The properties of fiber-reinforced composites depend strongly on several characteristics: (1) the properties of the fiber material; (2) the volume fraction of fibers; (3) the *aspect ratio* of the fibers, that is, the length-to-diameter ratio; (4) the orientation of the fibers; (5) the degree of bonding between the fiber and the matrix; and (6) the properties of the matrix. While more fibers tend to provide greater strength and stiffness, the volume fraction of fibers generally cannot exceed 80% to allow for a continuous matrix. Long, thin fibers (higher aspect ratio) provide greater strength, and a strong bond is usually desired between the fiber and matrix.

TABLE 8-7 Properties and Characteristics of Some Common Reinforcing Fibers

Fiber Material	Specific Strength[a] (10^6 in.)	Specific Stiffness[b] (10^6 in.)	Density (lb/in.^{3})	Melting Temperature[c] (°F)
Al_2O_3 whiskers	21.0	434	0.142	3600
Boron	4.7	647	0.085	3690
Ceramic fiber (mullite)	1.1	200	0.110	5430
E-type glass	5.6	114	0.092	<3140
High-strength graphite	7.4	742	0.054	6690
High-modulus graphite	5.0	1430	0.054	6690
Kevlar	10.1	347	0.052	—
SiC whiskers	26.2	608	0.114	4890

[a] Strength divided by density.
[b] Elastic modulus divided by density.
[c] Or maximum temperature of use.

The matrix materials should be strong, tough, and ductile so that they can transmit the loads to the fibers and prevent cracks from propagating through the composite. In addition, the matrix material is often responsible for providing the electrical properties, chemical behavior, and elevated-temperature stability. For polymer-matrix composites, both thermosetting and thermoplastic resins have been used. The thermosets provide high strength and high stiffness, and the low-viscosity, uncured resins readily impregnate the fibers. Popular thermosets include epoxies, polyesters, bismaleimides, and polyimides. From a manufacturing viewpoint, it may be easier and faster to heat and cool a thermoplastic than to cure a thermoset. Moreover, the thermoplastics are tougher and more tolerant to damage. Polyethylene, polystyrene, and nylon are traditional thermoplastic matrix materials. Improved high-temperature and chemical-resistant properties can be achieved with the thermoplastic polyimides, polyphenylene sulfide (PPS), polyether ether ketone (PEEK), and the liquid-crystal polymers. When reinforced with high-strength, high-modulus fibers, these materials can show dramatic improvements in strength, stiffness, toughness, and dimensional stability.

ADVANCED FIBER-REINFORCED COMPOSITES

Advanced composites are materials that have been developed for applications requiring exceptional combinations of strength, stiffness, and light weight. Fiber content generally exceeds 50% (by weight), and the modulus of elasticity is typically greater than 16×10^6 psi. Superior creep and fatigue resistance, low thermal expansion, low friction and wear, vibration-damping characteristics, and environmental stability are other properties that may also be required in these materials.

There are four basic types of advanced composites where the matrix material is matched to the fiber and the conditions of application:

1. The advanced *organic or resin-matrix composites* frequently use high-strength, high-modulus fibers of graphite, aramid (Kevlar), or boron. Properties can be put in desired locations or orientations at about one half the weight of aluminum (or one-sixth that of steel). Thermal expansion can be designed to be low or even negative. Unfortunately, these materials have a maximum service temperature of about 315°C (600°F) because the polymer matrix loses strength when heated. Table 8-8 compares the properties of some of the common resin-matrix composites with those of several of the lightweight or low-thermal-expansion metals. Typical applications include sporting equipment (tennis rackets, skis, golf clubs, and fishing poles), lightweight armor plate, and a myriad of low-temperature aerospace components.

TABLE 8-8 Properties of Several Fiber-Reinforced Composites (in the Fiber Direction) Compared to Lightweight or Low-Thermal-Expansion Metals

Material	Specific Strength[a] (10^6 in.)	Specific Stiffness[b] (10^6 in.)	Density (lb/in.^{3})	Thermal Expansion Coefficient [in./(in.-°F)]	Thermal Conductivity [Btu/(hr-ft-°F)]
Boron–epoxy	3.3	457	0.07	2.2	1.1
Glass–epoxy (woven cloth)	0.7	45	0.065	6	0.1
Graphite–epoxy: high modulus (unidirectional)	2.1	700	0.063	−0.5	75
Graphite–epoxy: high strength (unidirectional)	5.4	400	0.056	−0.3	3
Kevlar–epoxy (woven cloth)	1	80	0.5	1	0.5
Aluminum	0.7	100	0.10	13	100
Beryllium	1.1	700	0.07	7.5	120
Invar[c]	0.2	70	0.29	1	6
Titanium	0.8	100	0.16	5	4

[a] Strength divided by density.
[b] Elastic modulus divided by density.
[c] A low-expansion metal containing 36% Ni and 64% Fe.

2. *Metal-matrix composites* (MMCs) can be used for operating temperatures up to 1250°C (2300°F), where the conditions require high strength, high stiffness, good electrical and/or thermal conductivity, exceptional wear resistance, and good ductility and toughness. The ductile matrix material can be aluminum, copper, magnesium, titanium, nickel, superalloy, or even intermetallic compound, and the reinforcing fibers may be graphite, boron carbide, alumina, or silicon carbide. Fine whiskers (tiny needle-like single crystals of 1 to 10 μm in diameter) of sapphire, silicon carbide, and silicon nitride have also been used as the reinforcement, as well as wires of titanium, tungsten, molybdenum, beryllium, and stainless steel. The reinforcing fibers may be either continuous or discontinuous and typically comprise between 10 and 60% of the composite by volume. Compared to the engineering metals, these composites offer higher stiffness and strength (especially at elevated temperatures); a lower coefficient of thermal expansion; better elevated-temperature properties; and enhanced resistance to fatigue, abrasion, and wear. Compared to the organic matrix composites, they offer higher heat resistance as well as improved electrical and thermal conductivity. They are nonflammable, do not absorb water or gases, and are corrosion resistant to fuels and solvents. Unfortunately, these materials are quite expensive. The vastly different thermal expansions of the components may lead to debonding, and the assemblies may be prone to degradation through interdiffusion or galvanic corrosion. Graphite-reinforced aluminum can be designed to have near-zero thermal expansion in the fiber direction. Aluminum-oxide-reinforced aluminum has been used in automotive connecting rods to provide stiffness and fatigue resistance with lighter weight. Aluminum reinforced with silicon carbide has been fabricated into automotive drive shafts, cylinder liners, and brake drums as well as aircraft wing panels, all offering significant weight savings. Fiber-reinforced superalloys may well become a preferred material for applications such as turbine blades.

3. *Carbon–carbon composites* (graphite fibers in a graphite or carbon matrix) offer the possibility of a heat-resistant material that could operate at temperatures above 2000°C (3600°F), along with a strength that is 20 times that of conventional graphite, a density that is 30% lighter (1.38 g/cm^3), and a low coefficient of thermal expansion. Not only does this material withstand high temperatures, it actually gets stronger when heated. Companion properties include good toughness, good thermal and electrical conductivity, and resistance to corrosion and abrasion. For temperatures over 540°C (1000°F), however, the composite requires some form of coating to protect it from oxidizing. Various coatings can be used for different temperature ranges. Current applications include the nose cone and leading edge of the space shuttle, aircraft and racing car disc brakes, automotive clutches, aerospace turbines and jet engine components, rocket nozzles, and surgical implants.

4. *Ceramic-matrix composites* (CMCs) offer light weight, high-temperature strength and stiffness, and good dimensional and environmental stability. The matrix provides high temperature resistance. Glass matrices can operate at temperatures as high as 1500°C (2700°F). The crystalline ceramics, usually based on alumina, silicon carbide, silicon nitride, boron nitride, titanium diboride, or zirconia, can be used at even higher temperatures. The fibers add directional strength, increase fracture toughness, improve thermal shock resistance, and can be incorporated in unwoven, woven, knitted, and braided form. Typical reinforcements include carbon fiber, glass fiber, fibers of the various matrix materials, and ceramic whiskers. Composites with discontinuous fibers tend to be used primarily for wear applications, such as cutting tools, forming dies, and automotive parts such as valve guides. Other applications include lightweight armor plate and radomes. Continuous-fiber ceramic composites are used for applications involving the combination of high temperatures and high stresses, and have been shown to fail in a noncatastrophic manner. Application examples include gas-turbine components, high-pressure heat exchangers, and high-temperature filters. Unfortunately, the cost of ceramic–ceramic composites ranges from high to extremely high, so applications are restricted to those where the benefits are quite attractive.

HYBRID COMPOSITES

Hybrid composites involve two or more different types of fibers in a common matrix. The particular combination of fibers is usually selected to balance strength and stiffness, provide dimensional stability, reduce cost, reduce weight, or improve fatigue and fracture resistance. Types of hybrid composites include (1) interply (alternating layers of fibers), (2) intraply (mixed strands in the same layer), (3) interply–intraply, (4) selected placement (where the more costly material is used only where needed), and (5) interply knitting (where plies of one fiber are stitched together with fibers of another type).

DESIGN AND FABRICATION

The design of composite materials involves the selection of the component materials; the determination of the relative amounts of each component; the determination of size, shape, distribution, and orientation of the components; and the selection of an appropriate fabrication method. Many of the possible fabrication methods have been specifically developed for use with composite materials. For example, fibrous composites can be manufactured into useful shapes through compression molding, filament winding, pultrusion (where bundles of coated fibers are drawn through a heated die), cloth lamination, and autoclave curing (where pressure and elevated temperature are applied simultaneously). A variety of fiber-containing thermoset resins premixed with fillers and additives (*bulk-molding compounds*) can be shaped and cured by compression, transfer, and injection molding to produce three-dimensional fiber-reinforced products for numerous applications. Sheets of glass-fiber-reinforced thermoset resin, again with fillers and additives (*sheet-molding compound*), can be press formed to provide lightweight, corrosion-resistant products that are similar to those made from sheet metal. The reinforcing fibers can be short and random, directionally oriented, or fully continuous in a specified direction. With a wide spectrum of materials, geometries, and processes, it is now possible to tailor a composite material product for a specific application. As one example, consider the cargo beds for pickup trucks, where composite products offer reduced weight coupled with resistance to dents, scratches, and corrosion.

A significant portion of Chapter 15 is devoted to a more complete description of the fabrication methods that have been developed for composite materials.

ASSETS AND LIMITATIONS

Figure 8-12 graphically presents the strength-to-weight ratios of various aerospace materials as a function of temperature. The superiority of the various advanced composites over the conventional aerospace metals is clearly evident. The weight of a graphite–epoxy composite I-beam is less than one-fifth that of steel, one-third that of titanium, and one-half that of aluminum. Its ultimate tensile strength equals or exceeds that of the other three materials, and it possesses an almost infinite fatigue life. The greatest limitations of this and other composites are their relative brittleness and the high cost of both materials and fabrication.

While there has been considerable advancement in the field, manufacturing with composites can still be quite labor intensive, and there is a persistent lack of trained designers, established design guidelines and data, information about fabrication costs, and reliable methods of quality control and inspection. It is often difficult to predict the interfacial bond strength, the strength of the composite and its response to impacts, and the probable modes of failure. Defects can involve delaminations, voids, missing layers, contamination, fiber breakage, and (hard-to-detect) improperly cured resin. There is often concern about heat resistance. Many composites with polymeric matrices are sensitive to moisture, acids, chlorides, organic solvents, oils, and ultraviolet radiation, and they tend to cure forever, causing continually changing properties. In addition, most composites have limited ability to be repaired if damaged, preventive maintenance procedures are not well established, and recycling is often extremely difficult. Assembly operations with composites generally require the use of industrial adhesives.

On the positive side, the availability of a corrosion-resistant material with strength and stiffness greater than those of steel at only one-fifth the weight may be sufficient to justify some engineering compromises. Reinforcement fibers can be oriented in

FIGURE 8-12 The strength-to-weight ratio of various aerospace materials as a function of temperature. Note the superiority of the various fiber-reinforced composites. *(Adapted with permission of DuPont Company, Wilmington, DE.)*

the direction of maximum stiffness and strength. In addition, products can often be designed to significantly reduce the number of parts, number of fasteners, assembly time, and cost.

AREAS OF APPLICATION

Many composite materials are stronger than steel, lighter than aluminum, and stiffer than titanium. They can also possess low thermal conductivity, good heat resistance, good fatigue life, low corrosion rates, and adequate wear resistance. For these reasons they have become well established in several areas.

Aerospace applications frequently require light weight, high strength, stiffness, and fatigue resistance. As a result, composites may well account for a considerable fraction of the weight of a current airplane design. Figure 8-13 shows a schematic of the F-22 Raptor fighter airplane. Traditional materials, such as aluminum and steel, make up only about 20% of the F-22 structure by weight. Its higher speed, longer range, greater agility, and reduced detectability are made possible through the use of 42% titanium and 24% composite material. Boeing's new 787, a 200-seat intercontinental commercial airliner, will have a majority of its primary structure, including wings and fuselage, made of polymer-matrix (carbon–epoxy) composites. A titanium–graphite composite will also be used in the wings of this aircraft, which will use 15 to 20% less fuel than current wide-body planes. The Airbus Industries' new A380 wide-body plane will also utilize a high proportion of composite materials (about 16% by weight) and will mark the introduction of a new composite material, known as glass-reinforced aluminum, a laminate composite of alternating layers of aluminum and glass prepreg. The new material, which enables a 25% reduction in the weight of fuselage skin, is more fatigue resistant than aluminum and less expensive than a full composite.

Sports are highly competitive, and fractions of a second or tenths of a millimeter often decide victories. As a result, both professionals and amateurs are willing to invest in athletic equipment that will improve performance. The materials of choice have evolved from naturally occurring wood, twine, gut, and rubber to a wide variety of high-technology metals, polymers, ceramics, and composites. Golf club shafts, baseball bats, fishing

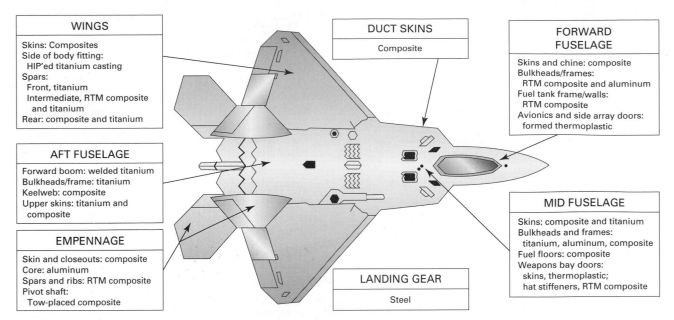

WINGS

Skins: Composites
Side of body fitting:
  HIP'ed titanium casting
Spars:
  Front, titanium
  Intermediate, RTM composite
    and titanium
Rear: composite and titanium

AFT FUSELAGE

Forward boom: welded titanium
Bulkheads/frame: titanium
Keelweb: composite
Upper skins: titanium and
  composite

EMPENNAGE

Skin and closeouts: composite
Core: aluminum
Spars and ribs: RTM composite
Pivot shaft:
  Tow-placed composite

DUCT SKINS

Composite

FORWARD FUSELAGE

Skins and chine: composite
Bulkheads/frames:
  RTM composite and aluminum
Fuel tank frame/walls:
  RTM composite
Avionics and side array doors:
  formed thermoplastic

MID FUSELAGE

Skins: composite and titanium
Bulkheads and frames:
  titanium, aluminum, composite
Fuel floors: composite
Weapons bay doors:
  skins, thermoplastic;
  hat stiffeners, RTM composite

LANDING GEAR

Steel

FIGURE 8-13   Schematic diagram showing the materials used in the various sections of the F-22 Raptor fighter airplane. Traditional materials, such as aluminum and steel, comprise only 20% by weight. Titanium accounts for 42%, and 24% is composite material. The plane is capable of flying at Mach 2. (*Note:* RTM is resin-transfer molding.) (*Reprinted with permission of ASM International, Metals Park, OH.*)

rods, archery bows, tennis rackets, bicycle frames, skis, and snowboards are now available in a wide variety of fibrous composites. Figure 8-14 shows several of these applications.

In addition to body panels, automotive uses of composite materials include drive shafts, springs, and bumpers. Weight savings compared to existing parts is generally 20 to 25%. Truck manufacturers now use fiber-reinforced composites for cab shells and bodies, oil pans, fan shrouds, instrument panels, and engine covers.

Other applications include such diverse products as boat hulls, bathroom shower and tub structures, chairs, architectural panels, agricultural tanks and containers, pipes and vessels for the chemical industry, and external housings for a variety of consumer and industrial products.

FIGURE 8-14   Composite materials are often used in sporting goods to improve performance through light weight, high stiffness, and high strength, and also to provide attractive styling. (Left) A composite material snowboard; (right) composites being used in a fishing rod, water ski, and tennis racquet.

■ Key Words

abrasive	composite	hybrid composite	rubber
addition polymerization	condensation polymerization	isomer	saturated monomer
additive agents	copolymer	isotropic	sheet-molding compound
advanced ceramic	cross-linking	laminar composite	specific stiffness
advanced composite	crystallized polymer	mer	specific strength
amorphous	degree of polymerization	metal–matrix composite	stress relaxation
anisotropic	dispersion-strengthened	nonmetallic materials	superabrasive
aspect ratio	material	oriented plastics	terpolymer
bulk-molding compound	elastomer	particulate composite	thermoplastic
carbon–carbon composite	fiber-reinforced composite	plastic	thermosetting
ceramic	fillers	plasticizer	unsaturated monomer
ceramic–matrix composite	foamed plastic	polymer	viscoelastic memory
cermet	glass	refractory material	whiteware
clay			

■ Review Questions

1. What are some naturally occurring nonmetallic materials that have been used for engineering applications?
2. What are some material families that would be classified under the general term *nonmetallic engineering materials*?
3. How might plastics be defined from the viewpoints of chemistry, structure, fabrication, and processing?
4. What is the primary type of atomic bonding within polymers?
5. What is the difference between a saturated and an unsaturated molecule?
6. What is an isomer?
7. Describe and differentiate the two means of forming polymers: addition polymerization and condensation polymerization.
8. What is degree of polymerization?
9. Describe and differentiate thermoplastic and thermosetting plastics.
10. Describe the mechanism by which thermoplastic polymers soften under heat and deform under pressure.
11. What does it mean when a polymer "crystallizes"?
12. What are some of the ways that a thermoplastic polymer can be made stronger?
13. What are the four most common thermoplastic polymers?
14. Why are thermosetting polymers characteristically brittle?
15. How do thermosetting polymers respond to subsequent heating?
16. Describe how thermoplastic or thermosetting characteristics affect productivity during the fabrication of a molded part.
17. What are some attractive engineering properties of polymeric materials?
18. What are some limiting properties of plastics, and in what general area do they fall?
19. What are some environmental conditions that might adversely affect the engineering properties of plastics?
20. What are some reasons that additive agents are incorporated into plastics?
21. What are some functions of a filler material in a polymer?
22. What are some of the more common filler materials used in plastics?
23. What is the difference between a dye and a pigment?
24. What is the role of a stabilizer or antioxidant?
25. What is an oriented plastic, and what is the primary engineering benefit?
26. What are some properties and characteristics of the "engineering plastics"?
27. Describe the use of plastic materials as adhesives. In tooling applications.
28. Describe some of the applications for foamed plastics.
29. What manufacturing features can enhance the attractiveness of plastics as a product material?
30. Which type of plastic is most easily recycled?
31. Why is the recycling of mixed plastics more difficult than the recycling of mixed metals?
32. What is the unique mechanical property of elastomeric materials, and what structural feature is responsible for it?
33. How can cross-linking be used to control the engineering properties of elastomers?
34. What are some of the materials that can be added to natural rubber, and for what purpose?
35. What are some of the attractive features of the silicone rubbers?
36. What is the most common use of an elastomer in a tooling application?
37. What are some outstanding physical properties of ceramic materials?
38. Why are the crystal structures of ceramics frequently more complex than those observed for metals?
39. What is the common name given for ceramic material in the noncrystalline, or amorphous, state?
40. What are some of the ways that toughness can be imparted to ceramic materials?
41. What is the dominant property of refractory ceramics?
42. What is the dominant property of ceramic abrasives?
43. How are glass products formed or shaped?
44. What are some of the specialty applications of glass?
45. What are cermets, and what properties or combination of properties do they offer?
46. Why do most ceramic materials fail to possess their theoretically high tensile strength?
47. Why do the mechanical properties of ceramics generally show a wider statistical spread than the same properties of metals?
48. If all significant flaws or defects could be eliminated from the structural ceramics, what properties might be present and what features might still limit their possible applications?
49. What are some specific materials that are classified as structural ceramics?
50. What are some attractive and limiting properties of sialon (one of the structural ceramics)?

51. What are some ceramic materials that are currently being used for cutting-tool applications, and what features or properties make them attractive?
52. What is a composite material?
53. What are the basic features of a composite material that influence and determine its properties?
54. What are the three primary geometries of composite materials?
55. What feature in a bimetallic strip makes its shape sensitive to temperature?
56. What is the attractive aspect of the strength that is induced by the particles in a dispersion-strengthened particulate composite material?
58. Which of the three primary composite geometries is most likely to possess isotropic properties?
59. What is the primary role of the matrix in a fiber-reinforced composite? Of the fibers?
60. What are some of the more popular fiber materials used in fiber-reinforced composite materials?
61. What is specific strength? Specific stiffness?
62. What are some possible fiber orientations or arrangements in a fiber-reinforced composite material?
63. What are some features that influence the properties of fiber-reinforced composites?
64. What are "advanced composites"?
65. In what ways are metal–matrix composites superior to straight engineering metals? To organic–matrix composites?
66. What features might be imparted by the fibers in a ceramic–matrix composite?
67. What are hybrid composites?
68. What is bulk-molding compound and how is it used? Sheet-molding compound?
69. What are some major limitations to the extensive use of composite materials in engineering applications?
70. What are some properties of composite materials that make them attractive for aerospace applications?

■ Problems

1. a. One of Leonardo da Vinci's sketchbooks contains a crude sketch of an underwater boat (or submarine). Leonardo did not attempt to develop or refine this sketch further, possibly because he recognized that the engineering materials of his day (wood, stone, and leather) were inadequate for the task. What properties would be required for the body of a submersible vehicle? What materials might you consider?
   b. Another of Leonardo's sketches bears a crude resemblance to a helicopter—a flying machine. What properties would be desirable in a material that would be used for this type of application?
   c. Try to identify a possible engineering product that would require a material with properties that do not exist among today's engineering materials. For your application, what are the demanding features or requirements? If a material were to be developed for this application, from what family or group do you think it would emerge? Why?
2. Select a product (or component of a product) that can reasonably be made from materials from two or more of the basic materials families (metals, polymers, ceramics, and composites).
   a. Describe briefly the function of the product or component.
   b. What properties would be required for this product or component to perform its function?
   c. What two materials groups might provide reasonable candidates for your product?
   d. Select a candidate material from the first of your two families and describe its characteristics. In what ways does it meet your requirements? How might it fall short of the needs?
   e. Repeat part d for a candidate material from the second material family.
   f. Compare the two materials to each other. Which of the two would you prefer? Why?
3. Coatings have been applied to cutting-tool materials since the early 1970s. Desirable properties of these coatings include high-temperature stability, chemical stability, low coefficient of friction, high hardness for edge retention, and good resistance to abrasive wear. Consider the ceramic material coatings of titanium carbide (TiC), titanium nitride (TiN), and aluminum oxide (Al_2O_3), and compare them with respect to the conditions required for deposition and the performance of the resulting coatings.
4. Ceramic engines continue to constitute an area of considerable interest and are frequently discussed in the popular literature. If perfected, they would allow higher operating temperatures with a companion increase in engine efficiency. In addition, they would lower sliding friction and permit the elimination of radiators, fan belts, cooling system pumps, coolant lines, and coolant. The net result would be reduced weight and a more compact design. Estimated fuel savings could amount to 30% or more.
   a. What are the primary limitations to the successful manufacture of such a product?
   b. What types of ceramic materials would you consider to be appropriate?
   c. What methods of fabrication could produce a product of the required size and shape?
   d. What types of special material properties or special processing might be required?
5. Material recyclability has become an important requirement in many manufactured products.
   a. Consider each of the four major materials groups (metals, polymers, ceramics, and composites) and evaluate each for recyclability. What properties or characteristics tend to limit or restrict recyclability?
   b. Which materials within each group are currently being recycled in large or reasonable quantities?
   c. Europe has recently legislated extensive recycling of automobiles and electronic products. How might this legislation change the material makeup of these products?
   d. Consider a typical family automobile and discuss how the factors of (1) recyclability, (2) fuel economy, and (3) energy required to produce materials and convert them to products might favor distinctly different engineering materials.

Chapter 8   CASE STUDY

Two-Wheel Dolly Handles

The items illustrated in the figure are the handle grips for an industrial-quality, pneumatic-tire, two-wheel dolly. They are designed to be bolted onto box-channel tubular sections using four bolt-holes, which are sized to accommodate $\frac{3}{8}$-inch-diameter bolts. The major service requirements are strength, durability, fracture resistance, reasonable appearance, and possibly light weight.

Your employer is currently marketing such a dolly using handles that are made as permanent-mold aluminum castings. The firm is in the process of updating its line and is reevaluating the design and manufacture of each of its products. The dolly is part of your assignment, and you have been asked specifically to determine whether the handles should be replaced by an alternative material, such as a polymer or low-cost composite.

Investigate the properties and cost* of alternative materials, including means of fabricating the desired shape, and make your recommendation. Since the existing design was for cast metal, you might want to make minor modifications. Make sure the alternative materials possess adequate properties in the bolt-hole region, and if not, recommend some form of reinforcement.

_____

* Since the size of the part will remain relatively unchanged, material costs should be compared on the basis of \$/in³ or \$/cm³ and not \$/lb or \$/kg.

MATERIAL SELECTION

■ 9.1 INTRODUCTION

The objective of manufacturing operations is to make products or components that adequately perform their intended task. Meeting this objective implies the manufacture of components from selected engineering materials, with the required geometrical shape and precision and with companion material structures and properties that are optimized for the service environment. The ideal product is one that will just meet all requirements. Anything better will usually incur added cost through higher-grade materials, enhanced processing, or improved properties that may not be necessary. Anything worse will likely cause product failure, dissatisfied customers, and the possibility of unemployment.

It was not that long ago that each of the materials groups had its own well-defined uses and markets. Metals were specified when strength, toughness, and durability were the primary requirements. Ceramics were generally limited to low-value applications where heat or chemical resistance was required and any loadings were compressive. Glass was used for its optical transparency, and plastics were relegated to low-value applications where low cost and light weight were attractive features and performance properties were secondary.

Such clear delineations no longer exist. Many of the metal alloys in use today did not exist as little as 30 years ago, and the common alloys that have been in use for a century or more have been much improved due to advances in metallurgy and production processes. New on the scene are amorphous metals, dispersion-strengthened alloys produced by powder metallurgy, mechanical alloyed products, and directionally solidified materials. Ceramics, polymers, and composites are now available with specific properties that often transcend the traditional limits and boundaries. Advanced structural materials offer higher strength and stiffness; strength at elevated temperature; light weight; and resistance to corrosion, creep, and fatigue. Other materials have enhanced thermal, electrical, optical, magnetic, and chemical properties.

To the inexperienced individual, "wood is wood," but to the carpenter or craftsman, oak is best for one application, while maple excels for another, and yellow pine is preferred for a third. The ninth edition of "Woldman's Engineering Alloys"[1] includes over 56,000 metal alloys, and that doesn't consider polymers, ceramics, or composites. Even if we eliminate the obsolete and obscure, we are still left with tens of thousands of options from which to select the "right" or "best" material for the task at hand.

Unfortunately, the availability of so many alternatives has often led to poor materials selection. Money can be wasted in the unnecessary specification of an expensive alloy or one that is difficult to fabricate. At other times, these materials may be absolutely necessary, and selection of a cheaper alloy would mean certain failure. It is the responsibility of the design and manufacturing engineer, therefore, to be

---

[1]*Woldman's Engineering Alloys*, 9th edition, edited by J. Frick, ASM International, Metals Park, OH, 2000.

knowledgeable in the area of engineering materials and to be able to make the best selection among the numerous alternatives.

In addition, it is also important that the material selection process be one of constant reevaluation. New materials are continually being developed, others may no longer be available, and prices are always subject to change. Concerns regarding environmental pollution, recycling, and worker health and safety may impose new constraints. Desires for weight reduction, energy savings, or improved corrosion resistance may well motivate a change in engineering material. Pressures from domestic and foreign competition, increased demand for quality and serviceability, or negative customer feedback can all prompt a reevaluation. Finally, the proliferation of product liability actions, many of which are the result of improper material use, has further emphasized the need for constant reevaluation of the engineering materials in a product.

The automotive industry alone consumes approximately 60 million metric tons of engineering materials worldwide every year—primarily steel, cast iron, aluminum, copper, glass, lead, polymers, rubber, and zinc. In recent years, the drive toward lighter, more fuel-efficient vehicles has led to an increase in the use of the lightweight metals and high-strength steels, as well as plastics and composites.

A million metric tons of engineering materials go into aerospace applications every year. The principal materials tend to be aluminum, magnesium, titanium, superalloys, polymers, rubber, steel, metal-matrix composites, and polymer-matrix composites. Competition is intense, and materials substitutions are frequent. The use of advanced composite materials in aircraft construction has risen from less than 2% in 1970 to the point where they now account for one-quarter of the weight of the U.S. Air Force's Advanced Tactical Fighter and will soon appear in the main fuselage of commercial planes. Titanium is used extensively for applications that include the exterior skins surrounding the engines, as well as the engine frames. The cutaway section of the Rolls Royce jet engine in Figure 9-1a reveals the myriad of components—each with its own characteristic shape, precision, stresses, and operating temperatures—that require a variety of engineering materials. Figure 9-1b shows an actual engine in a manner that reveals both its size and complexity. The intake fan diameter is nearly 3 meters in diameter (9 ft, 8 in.).

The earliest two-wheeled bicycle frames were constructed of wood, with various methods and materials employed at the joints. Then, for nearly a century, the requirements of yield strength, stiffness, and acceptable weight were met by steel tubing, either low-carbon plain-carbon or thinner-walled, higher-strength chrome-moly steel tubing, with either a brazed or welded assembly.

(a)

FIGURE 9-1 (a) Cutaway drawing showing the internal design of a Rolls-Royce jet engine. Notice the number and intricacy of the components. Ambient air enters the front, and hot exhaust exits the back. A wide range of materials and processes will be used in its construction. (b) A full-size Rolls-Royce engine, showing the size and complexity of the product. (*Courtesy of Rolls-Royce Corporation, London, England.*)

(b)

FIGURE 9-2   (a) A traditional two-wheel bicycle frame (1970s vintage) made from joined segments of metal tubing; (b) a top-of-the-line (Tour de France or triathlon-type) bicycle with one-piece frame, made from fiber-reinforced polymer–matrix composite. *(Courtesy of Trek Bicycle Corporation, Waterloo, WI.)*

In the 1970s a full circle occurred. Where a pair of bicycle builders (the Wright brothers) pioneered aerospace, the aerospace industry returned to revolutionize bicycles. Lightweight frames were constructed from the aerospace materials of high-strength aluminum, titanium, graphite-reinforced polymer, and even beryllium. Wall thickness and cross-section profiles were often modified to provide strength and rigidity. Materials paralleled function as bicycles specialized into road bikes, high-durability mountain bikes, and ultralight racing bikes. Further building on the aerospace experience, the century-old tubular frame has recently been surpassed by one-piece monocoque frames of either die-cast magnesium or continually wound carbon-fiber epoxy tapes with or without selective metal reinforcements. One top-of-the-line carbon-fiber frame now weighs only 2.5 pounds! Figure 9-2 compares a traditional tubular frame with one of the newer designs.

Window frames were once made almost exclusively from wood. While wood remains a competitive material, a trip to any building supply will reveal a selection that includes anodized aluminum in a range of colors, as well as frames made from colored vinyl and other polymers. Each has its companion advantages and limitations. Auto bodies were fabricated from steel sheet and assembled by resistance spot welding. Designers now select from steel, aluminum, and polymeric sheet-molding compounds and may use adhesive bonding to produce the joints.

The vacuum cleaner assembly shown in Figure 9-3, while not a current model, is typical of many engineering products, where a variety of materials are used for the various components. Table 9-1 lists the material changes that were recommended in just one past revision of the appliance. The materials for 12 components were changed completely, and that for a thirteenth was modified. Eleven different reasons were given for the changes. An increased emphasis on lighter weight has brought about even further changes in both design and materials.

The list of available engineering materials now includes metals and alloys, ceramics, plastics, elastomers, glasses, concrete, composite materials, and others. It is not surprising, therefore, that a single person might have difficulty making the necessary decisions concerning the materials in even a simple manufactured product. More frequently, the design engineer or design team will work in conjunction with various materials specialists to select the materials that will be needed to convert today's designs into tomorrow's reality.

■ 9.2 MATERIAL SELECTION AND MANUFACTURING PROCESSES

The interdependence between materials and their processing must also be recognized. New processes frequently accompany new materials, and their implementation can often cut production costs and improve product quality. A change in material may well require a change in the manufacturing process. Conversely, improvements in

- Aluminum die casting
- Aluminum die casting
- Nickel-plated steel stamping
- Molded natural rubber
- Injection-molded ABS
- Zinc-coated steel stamping
- Extruded PVC
- ABS with soft PVC tire
- Zinc-coated steel stamping
- Zinc-coated steel stamping
- Steel stamping
- Aluminum die casting
- PVC foot pad, case-hardened and bright nickel-plated SAE 1010 steel lever
- Black oxide-finished and lacquered SAE 1113 steel
- Medium-density polyethylene

FIGURE 9-3 Materials used in various parts of a vacuum cleaner assembly. *(Courtesy of Advanced Materials and Processes, ASM International, Metals Park, OH.)*

TABLE 9-1 Examples of Material Selection and Substitution in the Redesign of a Vacuum Cleaner

Part	Former Material	New Material	Benefits
Bottom plate	Assembly of steel stampings	One-piece aluminum die casting	More convenient servicing
Wheels (carrier and caster)	Molded phenolic	Molded medium-density polyethylene	Reduced noise
Wheel mounting	Screw-machine parts	Preassembled with a cold-headed steel shaft	Simplified replacement, more economical
Agitator brush	Horsehair bristles in a die-cast zinc or aluminum brush back	Nylon bristles stapled to a polyethylene brush back	Nylon bristles last seven times longer and are now cheaper than horsehair
Switch toggle	Bakelite molding	Molded ABS	Breakage eliminated
Handle tube	AISI 1010 lock-seam tubing	Electric seam-welded tubing	Less expensive, better dimensional control
Handle bail	Steel stamping	Die-cast aluminum	Better appearance, allowed lower profile for cleaning under furniture
Motor hood	Molded cellulose acetate (replaced Bakelite)	Molded ABS	Reasonable cost, equal impact strength, much improved heat and moisture resistance: eliminated warpage problems
Extension-tube spring latch	Nickel-plated spring steel, extruded PVC cover	Molded acetal resin	More economical
Crevice tool	Wrapped fiber paper	Molded polyethylene	More flexibility
Rug nozzle	Molded ABS	High-impact styrene	Reduced costs
Hose	PVC-coated wire with a single-ply PVC extruded covering	PVC-coated wire with a two-ply PVC extruded covering separated by a nylon reinforcement	More durability, lower cost
Bellows, cleaning-tool nozzles, cord insulation, bumper strips	Rubber	PVC	More economical, better aging and color, less marking

Source: Metal Progress, by permission.

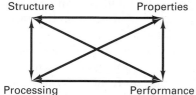

FIGURE 9-4 Schematic showing the interrelation among material, properties, processing, and performance.

processes may enable a reevaluation of the materials being processed. Improper processing of a well-chosen material can definitely result in a defective product. If satisfactory products are to be made, considerable care must be exercised in selecting *both* the *engineering materials* and the *manufacturing processes* used to produce the product.

Most textbooks on materials and manufacturing processes spend considerable time discussing the interrelationships between the structure and properties of engineering materials, the processes used to produce a product, and the subsequent performance. As Figure 9-4 attempts to depict, each of these aspects is directly related to all of the others. An engineering material may possess different properties depending upon its structure. Processing of that material can alter the structure, which in turn will alter the properties. Altered properties certainly alter performance. The objective of manufacturing, therefore, is to devise an optimized system of material and processes to produce the desired product.

■ 9.3 THE DESIGN PROCESS

The first step in the manufacturing process is *design*—the determining in rather precise detail what it is that we want to produce and, for each component of the product or assembly, what properties it must possess, what to make it out of, how to make it, how many to make, and what conditions it will see during use.

Design usually takes place in several distinct stages: (1) conceptual, (2) functional, and (3) production. During the *conceptual-design* stage, the designer is concerned primarily with the functions that the product is to fulfill. Several concepts are often considered, and a determination is made that the concept is either not practical, or is sound and should be developed further. Here the only concern about materials is that materials exist that could provide the desired properties. If such materials are not available, consideration is given to whether there is a reasonable prospect that new ones could be developed within the limitations of cost and time.

At the *functional- or engineering-design* stage, a workable design is developed, including a detailed plan for manufacturing. Geometric features are determined and dimensions are specified, along with allowable tolerances. Specific materials are selected for each component. Consideration is given to appearance, cost, reliability, producibility, and serviceability, in addition to the various functional factors. It is important to have a complete understanding of the functions and performance requirements of each component and to perform a thorough materials analysis, selection, and specification. If these decisions are deferred, they may end up being made by individuals who are less knowledgeable about all of the functional aspects of the product.

Often, a *prototype* or working model is constructed to permit a full evaluation of the product. It is possible that the prototype evaluation will show that some changes have to be made in either the design or material before the product can be advanced to production. This should not be taken, however, as an excuse for not doing a thorough job. It is strongly recommended that all prototypes be built with the same materials that will be used in production and, where possible, with the same manufacturing techniques. It is of little value to have a perfectly functioning prototype that cannot be manufactured economically in the desired volume or one that is substantially different from what the production units will be like.[2]

---

[2]Because of the prohibitive cost of a dedicated die or pattern, as might be required for forging or casting, one-of-a-kind or limited-quantity prototype parts are often made by machining or one of the newer rapid-prototype techniques. If the objective is simply to verify dimensional fit and interaction, the prototype material may be selected for compatibility with the prototype process. If performance is to be verified, however, it is best to use the proper material and process. Machining, for example, simply cuts through the material structure imparted in the manufacture of the starting bar or plate. Casting erases all prior structure during melting and establishes a new structure during solidification. Metalforming processes reorient the starting structure by plastic flow. The altered features caused by these processes may lead to altered performance.

In the *production-design* stage, we look to full production and determine if the proposed solution is compatible with production speeds and quantities. Can the parts be processed economically, and will they be of the desired quality?

As actual manufacturing begins, changes in both the materials and processes may be suggested. In most cases, however, changes made after the tooling and machinery have been placed in production tend to be quite costly. Good up-front material selection and thorough product evaluation can do much to eliminate the need for change.

As production continues, the availability of new materials and new processes may well present possibilities for cost reduction or improved performance. Before adopting new materials, however, the candidates should be evaluated very carefully to ensure that all of their characteristics related to both processing and performance are well established. Remember that it is indeed rare that as much is known about the properties and reliability of a new material as an established one. Numerous product failures and product liability cases have resulted from new materials being substituted before their long-term properties were fully known.

■ 9.4 PROCEDURES FOR MATERIAL SELECTION

The selection of an appropriate material and its subsequent conversion into a useful product with desired shape and properties can be a rather complex process. Nearly every engineered item goes through a sequence of activities that includes: design → material selection → process selection → production → evaluation → and possible redesign or modification. Numerous engineering decisions must be made along the way.

Several methods have been developed for approaching a design and selection problem. The *case-history method* is one of the simplest. Begin by evaluating what has been done in the past (engineering material and method of manufacture) or what a competitor is currently doing. This can yield important information that will serve as a starting base. Then, either duplicate or modify the details of that solution. The basic assumption of this approach is that similar requirements can be met with similar solutions.

The case-history approach is quite useful, and many manufacturers continually examine and evaluate their competitors' products for just this purpose. The real issue here, however, is "how similar is similar." A minor variation in service requirement, such as a different operating temperature or a new corrosive environment, may be sufficient to justify a totally different material and manufacturing method. In addition, this approach tends to preclude the use of new materials, new technology, and any manufacturing advances that may have occurred since the formulation of the original solution. It is equally unwise, however, to totally ignore the benefits and insights that can be gained through past experience.

Other design and selection activities occur during the *modification of an existing product*, generally in an effort to reduce cost, improve quality, or overcome a problem or defect that has been encountered. A customer may have requested a product like the current one but capable of operating at higher temperatures, or in an acidic environment, or at higher pressure. Efforts here generally begin with an evaluation of the current product and its present method of manufacture. The most frequent pitfall, however, is to overlook one of the original design requirements and recommend a change that in some way compromises the total performance of the product. Examples of such oversights, where materials have been changed to meet a specific objective, are provided in Section 9.8.

The safest and most comprehensive approach to part manufacture is to follow the full sequence of design, material selection, and process selection, considering all aspects and all alternatives. This is the approach one would take in the *development of an entirely new product*.

Before any decisions are made, take the time to fully define the needs of the product. What exactly is the "target" that we wish to hit? We must develop a clear picture of all of the characteristics necessary for this part to adequately perform its intended function and do so with no prior biases about material or method of fabrication. These requirements will fall into three major areas: (1) shape or geometry considerations, (2) property requirements, and (3) manufacturing concerns. By first formulating these requirements, we will be in a better position to evaluate candidate materials and companion methods of fabrication.

GEOMETRIC CONSIDERATIONS

A dimensioned sketch can answer many of the questions about the size, shape, and complexity of a part, and these *geometric* or *shape considerations* will have a strong influence on decisions relating to the proposed method or methods of fabrication.[3] While many features of part geometry are somewhat obvious, geometric considerations are often more complex than first imagined. Typical questions might include:

1. What is the relative size of the component?
2. How complex is its shape? Are there any axes or planes of symmetry? Are there any uniform cross sections? Could the component be divided into several simpler shapes that might be easier to manufacture?
3. How many dimensions must be specified?
4. How precise must these dimensions be? Are all precise? How many are restrictive, and which ones?
5. How does this component interact geometrically with other components? Are there any restrictions imposed by the interaction?
6. What are the surface-finish requirements? Must all surfaces be finished? Which ones do not?
7. How much can each dimension change by wear or corrosion and the part still function adequately?
8. Could a minor change in part geometry increase the ease of manufacture or improve the performance (fracture resistance, fatigue resistance, etc.) of the part?

Producing the right shape is only part of the desired objective. If the part is to perform adequately, it must also possess the necessary *mechanical and physical properties*, as well as the ability to endure anticipated environments for a specified period of time. *Environmental considerations* should include all aspects of shipping, storage, and use! Some key questions include those listed in the following three sections.

MECHANICAL PROPERTIES

1. How much static strength is required?
2. If the part is accidentally overloaded, is it permissible to have a sudden brittle fracture, or is plastic deformation and distortion a desirable precursor to failure?
3. How much can the material bend, stretch, twist, or compress under load and still function properly?
4. Are any impact loadings anticipated? If so, of what type, magnitude, and velocity?
5. Can you envision vibrations or cyclic loadings? If so, of what type, magnitude, and frequency?
6. Is wear resistance desired? Where? How much? How deep?
7. Will all of the above requirements be needed over the entire range of operating temperature? If not, which properties are needed at the lowest extreme? At the highest extreme?

---

[3]Die casting, for example, can be used to produce parts ranging from less than an ounce to more than 100 pounds, but the ideal wall thickness should be less than 5/16 inch. Permanent mold casting can produce thickness up to 2 inches, and there is no limit to the thickness for sand casting. At the same time, dimensional precision and surface finish become progressively worse as we move from die casting, to permanent mold, to sand. Extrusion and rolling can be used to produce long parts with constant cross section. Powder metallurgy parts must be able to be ejected from a compacting die.

PHYSICAL PROPERTIES (ELECTRICAL, MAGNETIC, THERMAL, AND OPTICAL)

1. Are there any electrical requirements? Conductivity? Resistivity?
2. Are any magnetic properties desired?
3. Are thermal properties significant? Thermal conductivity? Changes in dimension with change in temperature?
4. Are there any optical requirements?
5. Is weight a significant factor?
6. How important is appearance? Is there a preferred color, texture, or feel?

ENVIRONMENTAL CONSIDERATIONS

1. What are the lowest, highest, and normal temperatures the product will see? Will temperature changes be cyclic? How fast will temperature changes occur?
2. What is the most severe environment that is anticipated as far as corrosion or deterioration of material properties is concerned?
3. What is the desired service lifetime for the product?
4. What is the anticipated level of inspection and maintenance during use?
5. Should the product be manufactured with disassembly, repairability, or *recyclability* in mind?

MANUFACTURING CONCERNS

A final area of consideration is the variety of factors that will directly influence the method of manufacture. Some of these *manufacturing concerns* are:

1. How many of the components are to be produced? At what rate? (*Note:* One-of-a-kind parts and small quantities are rarely made by processes that require dedicated patterns, molds, or dies, since the expense of the tooling is hard to justify. High-volume, high-rate products may require automatable processes.)
2. What is the desired level of quality compared to similar products on the market?
3. What are the quality control and inspection requirements?
4. Are there any assembly (or disassembly) concerns? Any key relationships or restrictions with respect to mating parts?
5. What are the largest and smallest section thicknesses?
6. Have standard sizes and shapes been specified wherever possible (both as finished shapes and as starting raw material)? What would be the preferred form of starting material (plate, sheet, foil, bar, rod, wire, powder, ingot)?
7. Has the design addressed the requirements that will facilitate ease of manufacture (machinability, castability, formability, weldability, hardenability)?
8. What is the potential liability if the product should fail?
9. Are there any end-of-use disposal concerns?

The considerations just mentioned are only a sample of the many questions that must be addressed when precisely defining what it is that we want to produce. While there is a natural tendency to want to jump to an answer, in this case a material and method of manufacture, time spent determining the various requirements will be well rewarded. Collectively, the requirements direct and restrict material and process selections. It is possible that several families of materials, and numerous members within those families, all appear to be adequate. In this case, selections may become a matter of preference. It is also possible, however, that one or more of the requirements will emerge as a dominant restrictor (such as the need for ultra-high strength, superior wear resistance, the ability to function at extreme operating temperatures, or the ability to withstand highly corrosive environments), and selection then becomes focused on those materials offering that specific characteristic.

It is important that *all* factors be listed and *all* service conditions and uses be considered. Many failures and product liability claims have resulted from engineering oversights or failure to consider the entire spectrum of conditions that a product might experience in its lifetime. Consider the failure of several large electric power transformers where fatigue cracks formed at the base of horizontal cooling fins that had been welded to the exterior of the casing. The subsequent loss of cooling oil through the cracks led to overheating and failure of the transformer coils. Since transformers operate under static conditions, fatigue was not considered in the original design and material selection. However, when the horizontal fins were left unsupported during shipping, the resulting vibrations were sufficient to induce the fatal cracks. It is also not uncommon for the most severe corrosion environment to be experienced during shipping or storage as opposed to normal operation. Products can also encounter unusual service conditions. Consider the numerous parts that failed on earthmoving and construction equipment when it was used in the construction of the trans-Alaskan oil pipeline. When this equipment was originally designed and the materials were selected, extreme subzero temperatures were not included as possible operating conditions.

Once we complete a thorough evaluation of the required properties, it may be helpful to assign a relative importance to the various needs. Some requirements may be absolutes, while others may be *relative*. Absolute requirements are those for which there can be no compromise. The consequence of not meeting them will be certain failure of the product. Materials that fall short of absolute requirements should be automatically eliminated. For example, if a component must possess good electrical conductivity, most plastics and ceramics would not be appropriate. Relative or compromisable properties are those that frequently differentiate "good," "better," and "best," where all would be considered as acceptable.

■ 9.5 ADDITIONAL FACTORS TO CONSIDER

When evaluating candidate materials, an individual is often directed to handbook-type data that has been obtained through standardized materials characterization tests. It is important to note the conditions of these tests in comparison with those of the proposed application. Significant variations in factors such as temperature, rates of loading, or surface finish can lead to major changes in a material's behavior. In addition, one should keep in mind that the handbook values often represent an average or mean and that actual material properties may vary to either side of that value. Where vital information is missing or the data may not be applicable to the proposed use, one is advised to consult with the various materials producers or qualified materials engineers.

At this point it is probably appropriate to introduce *cost* as an additional factor. Because of competition and marketing pressures, economic considerations are often as important as technical ones. However, we have chosen to adopt the philosophy that cost should not be considered until a material has been shown to meet the necessary requirements. If acceptable candidates can be identified, cost will certainly become an important part of the selection process, and both material cost and the cost of fabrication should be considered.[4] Often, the final decision involves some form of compromise among material cost, ease of fabrication, and performance or quality. Numerous questions might be asked, such as:

1. Is the material too expensive to meet the marketing objectives?

2. Is a more expensive material justifiable if it offers improved performance?

3. How much additional expense might be justified to gain ease of fabrication?

In addition, it is important that the appropriate cost figures be considered. Material costs are most often reported in the form of dollars per pound or some other form of cost per unit weight. If the product has a fixed size, however, material comparisons should probably be based on cost per unit volume. For example, aluminum has a density about one-third that of steel. For products where the size is fixed, 1 pound of aluminum

---

[4]A more appropriate cost consideration might be total lifetime cost, which begins with the starting material, the energy to produce it, and the environmental impact of its production. To this are added the cost of converting it into the desired product, the cost of operating or using the product through its full lifetime, and finally the cost of disposal or recycling.

can be used to produce three times as many parts as 1 pound of steel. If the per-pound cost of aluminum were less than three times that of steel, aluminum would actually be the cheaper material. Whenever the densities of materials are quite different, as with magnesium and stainless steel, the relative rankings based on cost per pound and cost per cubic inch can be radically different.

Material availability is another important consideration. The material selected may not be available in the size, quantity, or shape desired, or it may not be available in any form at all. The diversity and reliability of supply may be additional factors that will facilitate competitive pricing and avoid production bottlenecks. If availability or supply may be a problem, one should be prepared to recommend alternative materials, provided that they, too, are feasible candidates for the specific use.

Still other factors to be considered when making material selections include:

1. Are there possible misuses of the product that should be considered? If the product is to be used by the general public, one should definitely anticipate the worst. Screwdrivers are routinely used as chisels and pry bars (different forms of loading from the intended torsional twist). Scissors may be used as wire cutters. Other products are similarly misused.

2. Have there been any failures of this or similar products? If so, what were the identified causes and have they been addressed in the current product? Failure analysis results should definitely be made available to the designers, who can directly benefit from them.

3. Has the material (or class of materials) being considered established a favorable or unfavorable performance record? Under what conditions was unfavorable performance noted?

4. Has an attempt been made to benefit from material standardization, whereby multiple components are manufactured from the same material or by the same manufacturing process? Although function, reliability, and appearance should not be sacrificed, one should not overlook the potential for savings and simplification that standardization has to offer.

■ 9.6 CONSIDERATION OF THE MANUFACTURING PROCESS

The overall attractiveness of an engineering material depends not only on its physical and mechanical properties but also on our ability to shape it into useful objects in an economical and timely manner. Without the necessary shape, parts cannot perform, and without economical production, the material will be limited to a few high-value applications. For this reason, our material selection should be further refined by considering the possible fabrication processes and the suitability of each "prescreened" material to each of those processes. Familiarity with the various manufacturing alternatives is a necessity, together with a knowledge of the associated limitations, economics, product quality, surface finish, precision, and so on. All processes are not compatible with all materials. Steel, for example, cannot be fabricated by die casting. Titanium can be forged successfully by isothermal techniques but generally not by conventional drop hammers. Wrought alloys cannot be cast, and casting alloys are not attractive for forming.

Certain fabrication processes have distinct ranges of product size, shape, and thickness, and these should be compared with the requirements of the product. Each process has its characteristic precision and surface finish. Since secondary operations, such as machining, grinding, and polishing, all require the handling, positioning, and processing of individual parts, as well as additional tooling, they can add significantly to manufacturing cost. Usually it is best to hit the target with as few operations as possible. Some processes require prior heating or subsequent heat treatment. Still other considerations include production rate, production volume, desired level of automation, and the amount of labor required, especially if it is skilled labor. All of these concerns will be reflected in the cost of fabrication. There may also be additional constraints, such as the need to design a product so that it can be produced with existing equipment or facilities, or with a minimum of lead time, or with a minimal expenditure for dedicated tooling.

It is not uncommon for a certain process to be implied by the geometric details of a component design, such as the presence of cored features in a casting, the magnitude of draft allowances, or the recommended surface finish. The designer often specifies these features prior to consultation with manufacturing experts. It is best, therefore, to consider all possible methods of manufacture and, where appropriate, work with the designer to incorporate changes that would enable a more attractive means of production.

■ 9.7 ULTIMATE OBJECTIVE

The real objective of this activity is to develop a manufacturing system—a combination of material and process (or sequence of processes) that is the best solution for a given product. Figure 9-5 depicts a series of activities that move from a well-defined set of needs and objectives through material and process selection to the manufacture and evaluation of a product. Numerous decisions are required, most of which are judgmental in nature. For example, we may have to select among "good," "better," and "best," where "better" and "best" carry increments of added cost, or make compromises when all of the requirements cannot be simultaneously met.

While Figure 9-5 depicts the various activities as having a definite, sequential pattern, one should be aware that they are often rearranged and are definitely interrelated. Figure 9-6 shows a modified form, where material selection and process selection have been moved to be parallel instead of sequential. It is not uncommon for one of the two selections to be dominant and the other to become dependent or secondary. For example, the production of a large quantity of small, intricate parts with thin walls, precise dimensions, and smooth surfaces is an ideal candidate for die casting. Material selection, therefore, may be limited to die-castable materials—assuming feasible alternatives are available. In a converse example, highly restrictive material properties, such as the ability to endure extremely elevated temperatures or severe corrosive environments, may significantly limit the material options. Fabrication options will tend to be limited to those processes that are compatible with the candidate materials.

In both models, decisions in one area generally impose restrictions or limitations in another. As shown in Figure 9-7 selection of a material may limit processes, and selection

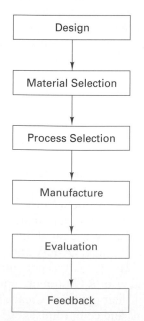

FIGURE 9-5 Sequential flow chart showing activities leading to the production of a part or product.

FIGURE 9-6 Alternative flow chart showing parallel selection of material and process.

Process \ Material	Irons	Steel	Aluminum	Copper	Magnesium	Nickel	Refractory Metals	Titanium	Zinc
Sand Casting	X	X	X	X	X	X			0
Permanent Mold Cast	X	0	X	0	X	0			0
Die Casting			X	0	X				X
Investment Casting		X	X	X	0	0			
Closed-Die Forging		X	0	0	0	0	0	0	
Extrusion		0	X	X	X	0	0	0	
Cold Heading		X	X	X		0			
Stamping, Deep Draw		X	X	X	0	X		0	0
Screw Machine	0	X	X	X	0	X	0	0	0
Powder Metallurgy	X	X	0	X		0	X	0	

Key: X = Routinely performed
       0 = Performed with difficulty, caution, or some sacrifice (such as die life)
       Blank = Not recommended

FIGURE 9-7 Compatibility chart of materials and processes. Selection of a material may restrict possible processes. Selection of a process may restrict possible materials.

of a process may limit material. Each material has its own set of performance characteristics, both strengths and limitations. The various fabrication methods impart characteristic properties to the material, and all of these may not be beneficial (consider anisotropy, porosity, or residual stresses). Processes designed to improve certain properties (such as heat treatment) may adversely affect others. Economics, environment, energy, efficiency, recycling, inspection, and serviceability all tend to influence decisions.

On rare occasions, a single solution will emerge as the obvious choice. More likely, several combinations of materials and processes will all meet the specific requirements, each with its own strengths and limitations. Compromise, opinion, and judgment all enter into the final decision making, where our desire is to achieve the best solution while not overlooking a major requirement. Listing and ranking the required properties will help ensure that all of the necessary factors were considered and weighed in making the ultimate decision. If no material–process combination meets the requirements, or if the compromises appear to be too severe, it may be necessary to redesign the product, adjust the requirements, or develop new materials or processes.

The individuals making materials and manufacturing decisions must understand the product, the materials, the manufacturing processes, and all of the various interrelations. This often requires multiple perspectives and diverse expertise, and it is not uncommon to find the involvement of an entire team. Design engineers ensure that each of the requirements is met and that any compromise or adjustment in those requirements is acceptable. Materials specialists bring expertise in candidate materials and the effects of various processing. Manufacturing personnel know the capabilities of processes, the equipment available, and the cost of associated tooling. Quality and environmental specialists add their perspective and expertise. Failure analysis personnel can share valuable experience gained from past unsuccessful efforts. Customer representatives or marketing specialists may also be consulted for their opinions. Clear and open communication is vital to the making of sound decisions and compromises.

The design and manufacture of a successful product is an iterative, evolving, and continual process. The failure of a component or product may have revealed deficiencies in design, poor material selection, material defects, manufacturing defects, improper assembly, or improper or unexpected product use. The costs of both material and processing continually change, and these changes may prompt a reevaluation. The availability of new materials, technological advances in processing methods, increased restrictions in environment or energy, or the demand for enhanced performance of an existing product all provide a continuing challenge. Materials availability may also have become an issue. A change in material may well require companion changes in the manufacturing process. Improvements in processing may warrant a reevaluation of the material.

■ 9.8 MATERIALS SUBSTITUTION

As new technology is developed or market pressures arise, it is not uncommon for new materials to be substituted into an existing design or manufacturing system. Quite often, the substitution brings about improved quality, reduced cost, ease of manufacture, simplified assembly, or enhanced performance. When making a *material substitution*, however, it is also possible to overlook certain requirements and cause more harm than good.

Consider the efforts related to the production of lighter-weight, more fuel-efficient, less emission-producing automobiles. The development of high-strength low-alloy steel sheets (HSLA) provided the opportunity to match the strength of traditional body panels with thinner-gage material. Once some of the early forming and fabrication problems were overcome, the substitution appeared to be a natural one. However, it is important to consider the total picture and become aware of any possible compromises. While strength was indeed increased, corrosion resistance and elastic stiffness (rigidity) remained essentially unaltered. The thinner sheets would corrode in a shorter time, and previously unnoticed vibrations could become a significant problem. Measures to retard corrosion and design modifications to reduce vibration would probably be necessary before the new material could be effectively substituted. Aluminum sheet has replaced steel panels, enabling a 50% reduction in weight, but the vibration problems associated with the lower elastic modulus required special design consideration.

Aluminum castings might be considered as an alternative to cast iron for engine blocks and transmission housings. Corrosion resistance would be enhanced and weight savings would be substantial. However, the mechanical properties must be ensured to be adequate, and consideration would also have to be given to the area of noise and vibration. Gray cast iron has excellent damping characteristics and effectively eliminates these undesirable features. Aluminum transmits noise and vibration, and its use in transmission housings would probably require the addition of some form of sound isolation material. When aluminum was first used for engine blocks, the transmitted vibrations required a companion redesign of the engine support system.

Polymeric materials have been used successfully for body panels, bumpers, fuel tanks, pumps, and housings. Composite-material drive shafts have been used in place of metal. Cast metal, powder metallurgy products, and composite materials have all been used for connecting rods. Ceramic and reinforced plastic components have been used for engine components. Magnesium is being used for instrument panels and steering wheels. Fiber-reinforced polymer composite has been used to produce the cargo beds for pickup trucks. When making a material substitution in a successful product, however, it is important to first consider all of the design requirements. Approaching a design or material modification as thoroughly as one approaches a new problem may well avoid costly errors.

Table 9-2 summarizes some of the weight-saving material substitutions that have been used on automobiles and calls attention to the fact that many of these substitutions are accompanied by an increase in cost, where total cost incorporates both cost of the material itself and the cost of converting that material into the desired product.

■ 9.9 EFFECT OF PRODUCT LIABILITY ON MATERIALS SELECTION

Product liability actions, court awards, and rising insurance costs have made it imperative that designers and manufacturers employ the very best procedures in selecting and processing materials. Although many individuals feel that the situation has grown to absurd proportions, there have also been many instances where sound procedures were not

TABLE 9-2	Material Substitutions to Reduce Weight in an Automobile[a]		
New Material	Previous Material	Weight Reduction	New Relative Cost
High-strength steel	Mild steel	10%	100% (no change)
Aluminum	Steel or cast iron	40–60%	130–200%
Magnesium	Steel or cast iron	60–75%	150–250%
Magnesium	Aluminum	25–35%	100–150%
Glass fiber reinforced plastic	Steel	25–35%	100–150%

[a] Data taken from "Automotive Materials in the 21st Century," by William F. Powers, published in *Advanced Materials and Processes*, May 2000.

used in selecting materials and methods of manufacture. In today's business and legal climate, such negligence cannot be tolerated.

An examination of recent product liability claims has revealed that the five most common causes have been:

1. Failure to know and use the latest and best information about the materials being specified
2. Failure to foresee, and account for, all reasonable uses of the product
3. Use of materials for which there were insufficient or uncertain data, particularly with regard to long-term properties
4. Inadequate and unverified quality control procedures
5. Material selection made by people who were completely unqualified

An examination of these faults reveals that there is no good reason for them to exist. Consideration of each, however, is good practice when seeking to ensure the production of a quality product and can greatly reduce the number and magnitude of product liability claims.

■ 9.10 AIDS TO MATERIAL SELECTION

From the discussion in this chapter, it is apparent that those who select materials should have a broad, basic understanding of the nature and properties of materials and their processing characteristics. Providing this background is a primary purpose of this text. The number of engineering materials is so great, however, and the mass of information that is both available and useful is so large, that a single book of this type and size cannot be expected to furnish all that is required. Anyone who does much work in material selection needs to have ready access to many sources of data.

It is almost imperative that one have access to the information contained in the various volumes of *Metals Handbook*, published by ASM International. This multivolume series contains a wealth of information about both engineering metals and associated manufacturing processes. The one-volume *Metals Handbook Desk Edition* provides the highlights of this information in a less voluminous, more concise format. A parallel *ASM Engineered Materials Handbook* series and one-volume *Desk Edition* provides similar information for composites, plastics, adhesives, and ceramics. These resources are also available on computer CD-ROMs and directly via the Internet through paid subscription.

ASM also offers a one-volume *ASM Metals Reference Book* that provides extensive data about metals and metalworking in tabular or graphic form. *Smithells Metal Reference Book* provides nearly 2000 pages of useful information and data. Additional handbooks are available for specific classes of materials, such as titanium alloys, stainless steels, tool steels, plastics, and composites. Various technical magazines often provide annual issues that serve as information databooks. Some of these include *Modern Plastics*, *Industrial Ceramics*, and ASM's *Advanced Materials and Processes*.

Persons selecting materials and processes should also have available several of the handbooks published by various materials organizations, technical societies, and trade associations. These may be material related (such as the Aluminum Association's *Aluminum Standards and Data* and the Copper Development Association's *Standards Handbook: Copper, Brass, and Bronze*), process related (such as the *Steel Castings Handbook* by the Steel Founder's Society of America and the *Heat Treater's Guide* by ASM International), or profession related (such as the *SAE Handbook* by the Society of Automotive Engineers, the *ASME Handbook* by the American Society for Mechanical Engineers, and the *Tool and Manufacturing Engineers Handbook* by the Society for Manufacturing Engineers). These may be supplemented further by a variety of supplier-provided information. While the latter is excellent and readily available, the user should recognize that supplier information might not provide a truly objective viewpoint.

It is also important to have accurate information on the cost of various materials. Since these tend to fluctuate, it may be necessary to consult a daily or weekly publication such as the *American Metal Market* newspaper or online service. Costs associated with various processing operations are more difficult to obtain and can vary greatly from one company to another. These costs may be available from within the firm or may have to be estimated from outside sources. A variety of texts and software packages are available.

Each of the above references provides focused information about a class of materials or a specific type of process. A number of texts have attempted to achieve integration with a focus on design and material selection. Possibly the most well known is the work of M. F. Ashby, with his *Materials Selection in Mechanical Design* text and tools, and the Cambridge Materials Selector database that was developed to use them.

With the evolution of high-speed computers with large volumes of searchable memory, materials selection can now be computerized. Most of the textbook and handbook references are now available on CDs or directly on the Internet, and all of the information in an entire handbook series can be accessed almost instantaneously. Programs have been written to utilize information databases and actually perform materials selection. The various property requirements can be specified and the entire spectrum of engineering materials can be searched to identify possible candidates. Search parameters can then be tightened or relaxed so as to produce a desired number of candidate materials. In a short period of time, a wide range of materials can be considered, far greater than could be considered in a manual selection. Process simulation packages can then be used to verify the likelihood of producing a successful product.

While the capabilities of computers and computer software are indeed phenomenal, the knowledge and experience of trained individuals should not be overlooked. Experienced personnel should reevaluate the final materials and manufacturing sequence to ensure full compliance with the needs of the product.

The appendix titled "Selected References For Additional Study" provides on extensive list of additional resources.

■ Key Words

absolute requirement	environmental considerations	material selection	production design
case history	functional design	material substitution	prototype
conceptual design	geometric requirements	mechanical properties	recyclability
cost	manufacturing concerns	physical properties	relative requirement
design	material availability	product liability	service environment

■ Review Questions

1. What is the objective of a manufacturing operation, and what are some of the details in meeting this objective?
2. What are some possible undesirable features of significantly exceeding the requirements of a product?
3. In a manufacturing environment, why should the selection and use of engineering materials be a matter of constant reevaluation?
4. How have different materials enabled advances and specializations in bicycle manufacture?
5. Discuss the interrelation between engineering material and the fabrication processes used to produce the desired shape and properties.
6. What is design?

7. What are the three primary stages of product design, and how does the consideration of materials differ in each?
8. What is the benefit of requiring prototype products to be manufactured from the same materials that will be used in production and by the same manufacturing techniques?
9. What sequence of activities is common to nearly every engineered component or product?
10. What are some of the possible pitfalls in the case-history approach to materials selection?
11. What is the most frequent pitfall when seeking to improve an existing product?
12. What should be the first step in any materials selection problem?
13. In what ways do the concept of shape or geometry go beyond a dimensioned sketch?
14. How might temperature enter into the specification of mechanical properties?
15. What are "physical properties" of materials?
16. What are some of the important aspects of the service environment to be considered when selecting an engineering material?
17. What are some of the possible manufacturing concerns that should be considered?
18. Why is it important to resist jumping to the answer and first perform a thorough evaluation of product needs and requirements, considering all factors and all service conditions?
19. What is the difference between an absolute and relative requirement?
20. What are some possible pitfalls when using handbook data to assist in materials selection?

21. Why might it be appropriate to defer cost considerations until after evaluating the performance capabilities of various engineering materials?
22. Give an example of a product or component where material cost should be compared on a cost-per-pound basis. Give a contrasting example where cost per unit volume would be more appropriate.
23. In what way might failure analysis data be useful in a material selection decision?
24. Why should consideration of the various fabrication process possibilities be included in material selection? What aspects of a manufacturing process should be considered?
25. Why might it be better to perform material selection and process selection in a parallel, as opposed to sequential, fashion?
26. Give an example of where selection of a material may limit processes and where selection of a process may limit materials.
27. Why is it likely that multiple individuals will be involved in the material and process selection activity?
28. Why should the design and manufacture of a successful product be an iterative, evolving, and continual process?
29. Give an example of an unexpected problem that occurred in a materials substitution.
30. What are some of the most common causes of product liability losses?
31. How have high-speed, high-capacity computers changed materials selection? Have they replaced trained individuals?

■ Problems

1. One simple tool that has been developed to assist in materials selection is a rating chart, such as the one shown in Figure 9-A. Absolute properties are identified and must be present for a material to be considered. The various relative properties are weighted as to their significance, and candidate materials are rated on a scale such as 1 to 5 or 1 to 10 with regard to their ability to provide that property. A rating number is then computed by multiplying the property rating by its weighted significance and summing the results. Potential materials can then be compared in a uniform, unbiased manner, and the best candidates can often be identified. In addition, by placing all the requirements on a single sheet of paper, the designer is less likely to overlook a major requirement. Finalist materials should then be reevaluated to assure that no key requirement has been overlooked or excessively compromised.

   Three materials,—X, Y, and Z,—are available for a certain use. Any material selected must have good weldability. Tensile strength, stiffness, stability, and fatigue strength have also been identified as key requirements. Fatigue strength is considered the most important of these requirements, and stiffness is least important. The three materials can be rated as follows:

	X	Y	Z
Weldability	Excellent	Poor	Good
Tensile strength	Good	Excellent	Fair
Stiffness	Good	Good	Good
Stability	Good	Excellent	Good
Fatigue strength	Fair	Good	Excellent

   Develop a rating chart such as that in Figure 9-A to determine which material you would recommend.

2. The chalk tray on a classroom chalkboard has very few performance requirements. As a result, it can be made from a wide spectrum of materials. Wood, aluminum, and even plastic have been used in this application. Discuss the performance and durability requirements and the pros and cons of the three listed materials. Chalk trays have a continuous cross section, but the processes used to produce such a configuration may vary with material. Discuss how a chalk tray might be mass-produced from each of the three materials classifications. Might this be a candidate for some form of wood by-product similar to particle board? Since the product demands are low, might some form of recycled material be considered?

3. Examine the properties of wood, aluminum, and extruded vinyl as they relate to household window frames. Discuss the pros and cons of each, considering cost, ease of manufacture, and aspects of performance, including strength, energy efficiency, thermal expansion and contraction, response to moisture and humidity, durability, rigidity, appearance (the ability to be finished in a variety of colors), ease of maintenance, property changes with low and high extremes of temperature, and any other factor that you feel is important. Which would be your preference for your particular location? Might your preference change if you were located in the dry Southwest (e.g., Arizona), New England, Alaska, or Hawaii? Can you imagine some means of combining materials to produce windows that might be superior to any single material? Which of the features above would apply to residential home siding?

4. Automobile body panels have been made from carbon steel, high-strength steel, aluminum, and various polymer-based molding compounds (both thermoplastic and thermoset). Discuss the key material properties and the relative performance characteristics of each, considering both use and manufacture.

Rating chart for selecting materials

Material	Go-No-Go** screening			Relative rating number (†rating number x *weighting factor)								Material rating number
	Corrosion	Weldability	Brazability	Strength (5)*	Toughness (5)	Stiffness (5)	Stability (5)	Fatigue (4)	As-welded strength (4)	Thermal stresses (3)	Cost (1)	$\frac{\Sigma \text{ rel rating no.}}{\Sigma \text{ rating factors}}$

*Weighting factor = 1 lowest to 5 most important
† Range = 1 poorest to 5 best
**Code = S = satisfactory
       U = unsatisfactory

FIGURE 9-A   Rating chart for comparing materials for a specific application.

For what type of vehicle might you prefer the various materials? Consider low-volume versus high-volume production, family versus commercial versus performance, low-cost versus luxury, and so on. How might preferences change if recyclability were required?

5. Consider the two-wheel bicycle frame and the variety of materials that have been used in its construction—low-carbon plain-carbon steel, somewhat higher-carbon chrome-moly alloy steel, cold-drawn aluminum tubing (strengthened by cold work), age-hardened aluminum tubing (strengthened by the age-hardening heat treatment), titanium alloy, fiber-reinforced composite, and still others. Some can be assembled by conventional welding or brazing. Others require low-temperature joining methods, since exposure to high temperature will compromise material strength. For still others, a one-piece structure (no joints) may be feasible. Select a material other than steel, and discuss the possible methods of manufacture and concerns you might have. Would your solution be appropriate for high- or low-production bicycles? Would it be good for pleasure bikes? Rugged mountain bikes? Racing bikes? What would be its unique selling features?

6. Go to the local hardware or building supply store and examine a specific class of fastener (nail, screw, bolt, rivet, etc.). Is it available in different grades or classes based on strength or intended use? What are they and how do they differ? How

might the materials and methods of manufacture be different for these identified groups? Summarize your findings.

7. Decorative fence posts for a residential home have been made from wood, extruded PVC, recycled polyethylene, decorative concrete, and various metals. Discuss the key material requirements and the pros and cons of the various potential materials.

8. The individual turbine blades used in the exhaust region of jet engines must withstand high temperatures, high stresses, and highly corrosive operating conditions. These demanding conditions severely limit the material possibilities, and most jet engine turbine blades have been manufactured from one of the high-temperature superalloys. The fabrication processes are limited to those that are compatible with both the material and the desired geometry. Through the 1960s and early 1970s the standard method of production was investment casting, and the resultant product was a polycrystalline solid with thousands of polyhedral crystals. In the 1970s production shifted to unidirectional solidification, where elongated crystals ran the entire length of the blade. More recently, advances have enabled the production of single-crystal turbine blades. Investigate this product to determine how the various material and processing conditions produce products with differing performance characteristics.

www.wiley.com/college/degarmo

Chapter 9 CASE STUDY

Material Selection

This study is designed to get you to question why parts are made from a particular material and how they could be fabricated to their final shape. For one or more of the products listed below, write a brief evaluation that addresses the following questions.

QUESTIONS:

1. What are the normal use or uses of this product or component? What are the normal operating conditions in terms of temperatures, loadings, impacts, corrosive media, and so on? Are there any unusual extremes?
2. What are the major properties or characteristics that the material must possess in order for the product to function?
3. What material (or materials) would you suggest and why?
4. How might you propose to fabricate this product?
5. Would the product require heat treatment? For what purpose? What kind of treatment?
6. Would this product require any surface treatment or coating? For what purpose? What would you recommend?
7. Would there be any concerns relating to environment? Recycling? Product liability?

PRODUCTS:

A. The head of a carpenter's claw hammer
B. The exterior of an office filing cabinet
C. A residential interior doorknob
D. A paper clip
E. Staples for an office stapler
F. A pair of scissors
G. A moderate to high-quality household cook pot or frying pan
H. A case for a jeweler-quality wristwatch
I. A jet engine turbine blade to operate in the exhaust region of the engine
J. A standard open-end wrench
K. A socket-wrench socket to install and remove spark plugs
L. The frame of a 10-speed bicycle
M. Interior panels of a microwave oven
N. Handle segments of a retractable blade utility knife with internal storage for additional blades
O. The outer skin of an automobile muffler
P. The exterior case for a classroom projector
Q. The basket section of a grocery store shopping cart
R. The body of a child's toy wagon
S. A decorative handle for a kitchen cabinet
T. An automobile radiator
U. The motor housing for a chain saw
V. The blade of a household screwdriver
W. Household dinnerware (knife, fork, and spoon)
X. The blades on a high-quality cutlery set
Y. A shut-off valve for a ½-in. household water line
Z. The base plate (with heating element) for an electric steam iron
AA. The front sprocket of a 10-speed bicycle
BB. The load-bearing structure of a child's outdoor swing set
CC. The perforated spin tub of a washing machine
DD. A commemorative coin for a corporation's 100th anniversary
EE. The keys for a commercial-quality door lock
FF. The exterior canister for an automobile oil filter

MEASUREMENT AND INSPECTION AND TESTING

■ 10.1 INTRODUCTION

Measurement, the act of measuring or being measured, is the fundamental activity of testing and inspection. The intent of *inspection* is to ensure that what is being manufactured will conform to the specifications of the product. *Testing* evaluates product quality or performance; trying to ensure there are no defects to impair performance as is often a form of final inspection

Most products are manufactured to standard sizes and shapes. For example, the base of a 60-W light bulb has been standardized so that when one bulb burns out, the next will also fit the socket in the lamp. The socket in the lamp has also been designed and made to accept the standard bulb size. Christmas tree light bulbs are made to a different standard size. Standardization is a necessity for interchangeable parts and is also important for economic reasons. A 69-W light bulb cannot be purchased because that is not a standard wattage. Light bulbs are manufactured only in standard wattages so that they can be mass-produced in large volumes by high-speed automated equipment. This results in a low unit cost.

Large-scale manufacturing based on the principles of standardization of sizes and interchangeable parts became common practice early in the twentieth century. Size control must be built into machine tools and workholding devices through the precision manufacture of these machines and their tooling. The output of the machines must then be checked carefully (1) to determine the capability of specific machines and (2) for the control and maintenance of the quality of the product. A designer who specifies the dimensions and tolerances of a part often does so to enhance the function of the product, but the designer is also determining the machines and processes needed to make the part. Frequently, the design engineer has to alter the design or the specifications to make the product easier or less costly to manufacture, assemble, or inspect (or all of these). Designers should always be prepared to do this provided that they are not sacrificing functionality, product reliability, or performance.

ATTRIBUTES VERSUS VARIABLES

The examination of the product during or after manufacture, manually or automatically, falls in the province of *inspection*. Basically, inspection of items or products can be done in two ways:

1. By *attributes*, using gages to determine if the product is good or bad, resulting in a yes/no, go/no-go decision.

2. By *variables*, using calibrated instruments to determine the actual dimensions of the product for comparison with the size desired.

In an automobile, a speedometer and oil pressure gage are variable types of measuring instruments, and an oil pressure light is an attributes-type of gage. As is typical of an attributes gage, the driver does not know *what* the pressure actually is if the light goes on, only that it is not good. On the factory floor, *measurement* is the generally accepted industrial term for inspection by variables. *Gaging* (or *gauging*) is the term for determining whether the dimension or characteristic is larger or smaller than the established standard or is within some range of acceptability. Variable types of inspection generally take more time and are more expensive than attribute inspection, but they yield more information because the magnitude of the characteristic is known in some standard unit of measurement.

■ 10.2 STANDARDS OF MEASUREMENT

The four fundamental measures on which all others depend are *length, time, mass*, and *temperature*. Three of these basic measures are defined in terms of material constants, as shown in Table 10-1, along with the original definitions. These four measures, along with the *ampere* and the *candela*, provide the basis for all other units of measurement, as shown in Figure 10-1, along with the original definitions. Most mechanical measurements involve combinations of units of mass, length, and time. Thus, the newton, a unit of force, is derived from Newton's second law of motion ($f = ma$) and is defined as the force that gives an acceleration of 1m/sec/sec to a mass of 1 kilogram. Figure 10-2 and Table 10-2 provide basic metric-to-English conversions.

LINEAR STANDARDS

When people first sought a unit of length, they adopted parts of the human body, mainly the hands, arms, or feet. Such tools were not very satisfactory because they were not universally standard in size. Satisfactory measurement and gaging must be based on

---

TABLE 10-1 International System of Units, Founded on Seven Base Quantities on Which All Others Depend

Quantity	Name of Base	Symbol	Definition or Comment
Length	Meter (or metre)	m	Original: 1/10,000,000 of quadrant of earth's meridian passing through Barcelona and Dunkirk. Present: 1,650,763.73 wavelengths in vacuum of transition between energy levels $2p_{10}$ and $5d_5$ of krypton-86 atoms, excited at triple point of nitrogen ($-210°C$).
Mass	Kilogram	kg	Original: Mass of 1 cubic decimeter (1000 cubic centimeters) of water at its maximum density ($4°C$).
			Present: Mass of Prototype Kilogram No. 1 kept at International Bureau of Weights and Measures at Sèvres, France.
Time	Second	s	Original: 1/86,400t of mean solar day. Present: 9,192,631,770 cycles of frequency associated with transition between two hyperfine levels of isotope cesium-133.
Electric current	Ampere		Present: The rate of motion of charge in a circuit is called the *current*. The unit of current is the *ampere*. One ampere exists when the charge flows at a rate of 1 coulomb per second.
Thermodynamic	Degree Celsius	°C (K)	Present: 1/273.16 of the thermodynamic temperature of the triple point of temperature (Kelvin) water ($0.01°C$).
Amount of substance	Mole	mol	Present: A mole is an artificially chosen number ($N_0 = 6.02 \times 10^{23}$) that measures the number of molecules.
Luminous intensity	Candle		Present: One lumen per square foot is a footcandle.

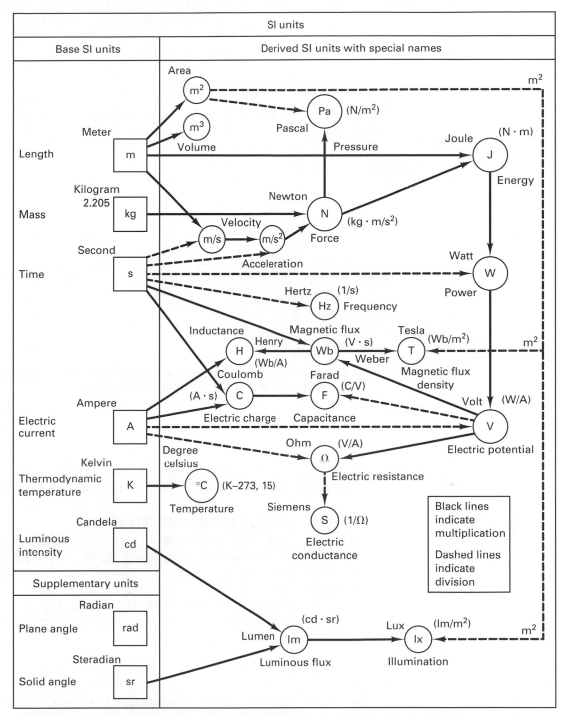

FIGURE 10-1 Relationship of secondary physical quantities to basic SI units. Solid lines signify multiplication; dashed lines signify division.

reliable, preferably universal, standards. These have not always existed. For example, although the musket parts made in Eli Whitney's shop were interchangeable, they were not interchangeable with parts made by another contemporary gunmaker *from the same drawings* because the two gunsmiths *had different foot rulers*. Today, the entire industrialized world has adopted the *international meter* as the standard of linear measurement. The inch, used by both the United States and Great Britain, has been defined officially as 2.54 centimeters. The U.S. standard inch is 41,929.399 wavelengths of the orange-red light from krypton-86.

Although *officially* the United States is committed to conversion to the metric (SI) system of measurement, which uses millimeters for virtually all linear measurements in

CF = conversion factor; examples: 10 kg = (2.203) (10) lb; 10 in. = 2.54 (10) mm; 1 newton ≈ 0.1 kg ≈ 0.22 lb; 1 micron = 40 microinches

FIGURE 10-2 Metric-to-English conversion of some widely used measures in engineering.

manufacturing, the English system of feet and inches is still being used by many manufacturing plants, and its use will probably continue for some time.

LENGTH STANDARDS IN INDUSTRY

Gage blocks provide industry with linear standards of high accuracy that are necessary for everyday use in manufacturing plants. The blocks are small, rectangular, square, or round in cross section and are made from steel or carbide with two very flat and parallel surfaces that are certain specified distances apart (see Figure 10-3). These gage blocks were first conceived by Carl E. Johansson in Sweden just before 1900. By 1911 he was able to produce sets of such blocks on a very limited scale, and they came into limited but significant use during World War I. Shortly after the war, Henry Ford recognized the importance of having such gage blocks generally available. He arranged for Johansson to come to the United States, and through facilities provided by the Ford Motor Company, methods were devised for the large-scale production of gage block sets. Today, gage block sets of excellent quality are produced by a number of companies in this country and abroad.

Steel gage blocks are made of alloy steel, hardened to R_c65 and carefully heat treated (*seasoned*) to relieve internal stresses and to minimize subsequent dimensional change. Carbide gage blocks provide extra wear resistance. The measuring surfaces of each block are surface-ground to approximately the required dimension and are then lapped and mirror polished to bring the block to the final dimension and to produce a very flat and smooth surface finish, 0.4 millionths of an inch (0.01 μm). (Surface finish is discussed in Chapter 36.)

Gage blocks are commonly made in grades with various tolerances as given in Table 10-3, which shows grade block grades according to NBS Standard No. 731/222 131 (ANSI/ASME B89.1OM-1984). Blocks up to 1 in. in length have absolute accuracies as stated, whereas the tolerances are per inch of length for blocks larger than 1 in. Some companies supply blocks in AA quality (which corresponds to Grade 1) and in A+ quality (which corresponds to Grade 2).

TABLE 10-2   Metric–English Conversions

Measurement	Metric Symbol	Metric Unit	English Conversion
Linear dimensions	m	meter	1 in. = 0.0254 m[a]
			1 ft = 0.348 m[a]
	cm	centimeter	1 in. = 2.54 cm[a]
	mm	millimeter	1 in. = 25.4 mm[a]
	μm	micrometer	1 μin. = 0.0254 μm[a]
Area	m^2	square meter	1 ft^2 = 0.093 m^{2b}
	cm^2	square centimeter	1 $in.^2$ = 6.45 cm^{2b}
	mm^2	square millimeter	1 $in.^2$ = 645.16 mm^{2a}
Volume and capacity	m^3	cubic meter	1 ft^3 = 0.028 m^{3c}
	cm^3	cubic centimeter	1 $in.^3$ = 16.39 cm^3
	mm^3	cubic millimeter	1 $in.^3$ = 16.387.06 mm^{3b}
	L	liter	1 ft^3 = 28.32 L[c]
			1 U.S. gal = 3.79 L[c]
Velocity, acceleration, and flow	m/s	meters per second	1 ft/s = 0.3048 m/s[a]
	m/min	meters per minute	1 ft/min = 0.3048 m/mm[a]
	m/s^2	meters per second squared	1 $in./s^2$ = 0.0254 m/s^{2c}
	L/mm	liters per minute	1 ft^3/min = 28.3 l/min[c] and
			1 gallon (U.S. liquid)/min = 3.785 l/min[c]
Mass	g	gram	1 oz = 28.36 g[c]
	kg	kilogram	1 lb = 0.45 kg[c]
	t	metric ton	1 short ton (2000 lb) = 0.7072t[d]
Force	N	newton	1 lb-force = 4.448 N[c]
	kN	kilonewton	1 short ton-force (2000 lb) = 8.896 kN[c]
Bending moment or torque	N · m	newton-meter	1 oz-force/in. = 0.007 N. m[c]
			1 lb-force/in. = 0.113 N. m[c]
			1 lb-force/ft = 1.3558 N. m[c]
Pressure	Pa	pascal	1 lb/ft^2 = 47.88 Pa[c]
	kPa	kilopascal	1 $lb/in.^2$ = 6.895 kPa[c]
Energy, work, or quantity of heat	J	joule	1 Btu^d = 1055.056[c]
	kJ	kilojoule	1 Btu^d = 1.055 kJ[c]
Power	W	watt 1	1 hp (550 ft-lb/s) = 745.7 W[c]
			1 hp (electric) = 746 W[a]
	kW	kilowatt	1 hp (550 ft-lb/s) = 0.7457 kW[b]
Temperature	C	Celsius	degrees C = $\dfrac{\text{degrees F} - 32}{1.8}$
Frequency	Hz	hertz	1 cycle per second = 1 Hz
	kHz	kilohertz	1000 cycles per second = 1 kHz
	MHz	megahertz	1,000,000 cycles per second = 1 MHz

[a] Comments on the metric system:

Metric symbols are usually presented the same way in singular and in plural (1 mm, 100 mm), and periods are not used after symbols, except at the end of sentences. Degree (instead of radian) continues to be used for plane angles, but angles are expressed with decimal subdivisions rather than minutes and seconds. Surface finishes are specified in micrometres.

Accuracy of conversion. Multiplying an English measurement by an exact metric conversion factor often provides an accuracy not intended by the original value. In general, use one less significant digit to the right of the decimal point than was given in the original value; 0.032 in., (0.81) in., (0.008 mm), etc. Fractions of an inch are converted to the nearest tenth of a millimeter: $\frac{1}{8}$ in. (3.2 mm).

Weight, mass, and force. Confusion exists in the use of the term *weight* as a quantity to mean either force or mass. In nontechnical circles, the term *weight* nearly always means mass, and this use will probably persist. Weight is a force generated by mass under the influence of gravity. Mass is expressed in gram (g), kilogram (kg), or metric ton (t) units, and force in newton (N) or kilonewton (kN) units.

Pressure. Kilopascal (kPa) is the recommended unit for fluid pressure. Absolute pressure is specified either by using the identifying phrase *absolute pressure* or by adding the word *absolute* after the unit symbol, separating the two by a comma or a space.

[b] Exact.   [c] Approximate.   [d] International Table.

TABLE 10-3   Grade Block Grades According to NBS Standard No. 731/222 131 (ANSI/ASME B89.1OM-1984)

Grade		Inches	Millimeters	Recalibration Period
0.5	(laboratory)	±0.000001	±0.000 03	Annually
1	(laboratory)	±0.000002	±0.000 05	Annually
2	(precision)	+0.000004	+0.000 10	Monthly to semiannually
		−0.000002	−0.000 05	
3	(working)	+0.000008	+0.000 20	Monthly to quarterly
		−0.000004	−0.000 10	

What's in the box

9 Blocks 0.1001 through 0.1009 in. in steps of 0.0001 in.
49 Blocks 0.101 through 0.149 in. in steps of 0.001 in.
19 Blocks 0.050 through 0.950 in. in steps of 0.050 in.
4 Blocks 1.000 through 4.000 in. in steps of 1.000 in.

◄— Lapped and mirror polished to
a very low micro-surface finish.

FIGURE 10-3 Standard set of rectangular gage blocks with 0.000050-in. accuracy; three individual blocks are shown.

FIGURE 10-4
Seven gage blocks wrung together to build up a desired dimension. *(Courtesy of DoALL Company.)*

Dial indicator gage

Gage blocks

FIGURE 10-5 Wrung-together gage blocks in a special holder, used with a dial gage to form an accurate comparator. *(Courtesy of DoALL Company.)*

Grade 0.5 (grand-master) blocks are used as a basic reference standard in calibration laboratories. Grade 1 (laboratory-grade) blocks are used for checking and calibrating other grades of gage blocks. Grade 2 (precision-grade) blocks are used for checking Grade 3 blocks and master gages. Grade 3 (B or working-grade) blocks are used to calibrate or check routine measuring devices, such as micrometers, or in actual gaging operations.

The dimensions of individual blocks are established by light-beam interferometry, with which it is possible to calibrate these blocks routinely with an uncertainty as low as one part per million.

Gage blocks usually come in sets containing various numbers of blocks of various sizes, such as those shown in Figure 10-3. By "wringing the blocks together" in various combinations, as shown in Figure 10-4, any desired dimension can be obtained. For example, if the last two blocks on the stack are 0.100 and 0.05 in., what is the total length of the wrung-together stack of gage blocks?

Gage blocks are wrung together by sliding one past another using hand pressure. They will adhere to one another with considerable force and must not be left in contact for extended periods of time. Gage blocks are available in different shapes (squares, angles, rounds, and pins), so standards of high accuracy can be obtained to fill almost any need. In addition, various auxiliary clamping, scribing, and base block attachments are available that make it possible to form very accurate gaging devices, such as the setup shown in Figure 10-5.

STANDARD MEASURING TEMPERATURE

Because all the commonly used metals are affected dimensionally by temperature, a standard measuring temperature of 68°F (20°C) has been adopted for precision measuring work. All gage blocks, gages, and other precision-measuring instruments are calibrated at this temperature. Consequently, when measurements are to be made to accuracies greater than 0.0001 in. (0.0025 mm), the work should be done in a room in which the temperature is controlled at standard. Although it is true that to some extent both the workpiece and the measuring or gaging device *may* be affected to about the same extent by temperature variations, one should not rely on this. Measurements to even 0.0001 in. (0.0025 mm) should not be relied on if the temperature is very far from 68°F (20°C).

ACCURACY VERSUS PRECISION IN PROCESSES

It is vitally important that the difference between accuracy and precision be understood. *Accuracy* refers to the ability to hit what is aimed at (the bull's-eye of the target). *Precision* refers to the repeatability of the process. Suppose that five sets of five shots are fired at a target from the same gun. Figure 10-6 shows some of the possible outcomes. In Figure 10-6a, inspection of the target shows that this is a good process—accurate and precise. Figure 10-6b shows precision (repeatability) but poor accuracy. The agreement with a standard is not good. In Figure 10-6c the process is on the average quite accurate, as the X (average) is right in the middle of the bull's-eye, but the process has too much scatter or variability; it does not repeat. Finally, in Figure 10-6d, a failure to repeat accuracy between samples with respect to time is observed; the process is not stable. These four outcomes are typical but not all-inclusive of what may be observed.

In Chapter 36, more discussion on accuracy and precision is presented as they relate to process capability. This term is used to describe how well a manufacturing process performs in its part making.

In measuring instruments used in the factory, precision (or repeatability) is critical because the devices must be very repeatable as well as accurate. For manually operated instruments, the skill of the operator must also be considered—this is called reproducibility.

So accuracy, repeatability, and reproducibility are characteristics of what is called *gage capability*, which is also discussed in Chapter 36.

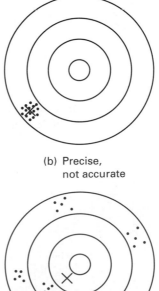

(a) Accurate and precise

(b) Precise, not accurate

FIGURE 10-6   Accuracy versus precision. Dots in targets represent location of shots. Cross (×) represents the location of the average position of all shots. (a) Accurate and precise; (b) precise, not accurate; (c) accurate, not precise; (d) precise within sample, not precise between samples, not accurate overall or within sample.

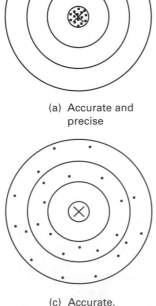

(c) Accurate, not precise

(d) Precise within sample
    Not precise between samples
    Not accurate overall or within sample

■ 10.3 ALLOWANCE AND TOLERANCE

If the desired fit between mating parts is to be obtained, the designer must specify two factors, allowance and tolerance. *Allowance* is the intentional, desired difference between the dimensions of two mating parts. It is the difference between the dimension of the largest interior-fitting part (shaft) and that of the smallest exterior-fitting part (hole). Figure 10-7 shows shaft A designed to fit into the hole in block B. This difference (0.5035–0.5025) thus determines the condition of *tightest* fit between mating parts. Allowance may be specified so that either *clearance* or *interference* exists between the mating parts. In the case of a shaft and mating hole, it is the difference in diameters of the largest shaft and the smallest hole. With clearance fits, the largest shaft is smaller than the smallest hole, whereas with interference fits, the hole is smaller than the shaft.

Tolerance is an undesirable but permissible deviation from a desired dimension. There is variation in all processes, and no part can be made *exactly* to a specified dimension, except by chance. Furthermore, such exactness is neither necessary nor economical. Consequently, it is necessary to permit the actual dimension to deviate from the desired theoretical dimension (called the *nominal*) and to control the degree of deviation so that satisfactory functioning of the mating parts will still be ensured.

Now we can see that the objective of *inspection*, by means of measurement techniques, is to provide feedback information on the actual size of the parts with reference to the size specified by the designer on the part drawing.

The manufacturing processes that make the shaft are different from those that make the hole, but both the hole and the shaft are subject to deviations in size because of variability in the processes and the materials. Thus, while the designer wishes ideally that all the shafts would be exactly 0.500 (Figure 10-8a) and all the holes 0.506, the reality of processing is that there will be deviations in size around these nominal or ideal sizes.

Most manufacturing processes result in products whose measurements of the geometrical features and sizes are distributed normally (Figure 10-8b). That is, most of the (0.0535–0.5025) measurements are clustered around the average dimension, $\overline{X}$ calculated as

$$\overline{X} = \frac{\sum\limits_{i1}^{n} X_i}{n} \quad \text{for } n \text{ items} \tag{10-1}$$

$\overline{X}$ will be equal to the nominal dimension only if the process is 100% accurate, that is, perfectly centered. More likely, parts will be distributed on either side of the average, and the process might be described (modeled) with a normal distribution. In normal distributions, as shown in Figure 10-8c, 99.73% of the measurements (X_i) will fall within plus or minus 3 standard deviations ($\pm 3\sigma$) of the mean, 95.46% will be within $\pm 2\sigma$, and 68.26% will be within $\pm 1\sigma$ where

$$\sigma = \sqrt{\frac{\sum\limits_{i=1}^{n} (X_i - \overline{X})^2}{n}} \tag{10-2}$$

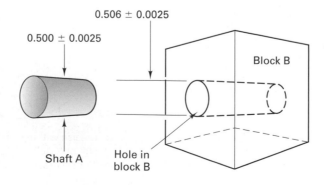

0.506 ± 0.0025

0.500 ± 0.0025

Block B

Shaft A

Hole in block B

FIGURE 10-7 When mating parts are designed, each shaft must be smaller than each hole for a clearance fit.

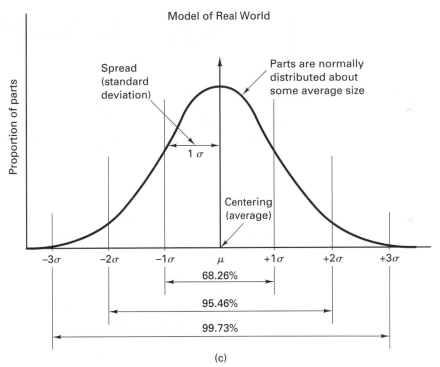

FIGURE 10-8 (a) In the ideal situation, the process would make all parts exactly the same size. (b) In the real world of manufacturing, parts have variability in size. (c) The distribution of sizes can often be modeled with a normal distribution.

    In summary, the designer applies nominal values to the mating parts according to the desired fit between the parts. Tolerances are added to those nominal values in recognition of the fact that all processes have some natural amount of variability.

    Assume that the data for both the hole and the shaft are normally distributed. The $\pm 3\sigma$ added to the mean (μ) gives the *upper and lower natural tolerance limits*, defined as

$$\mu + 3\sigma = UNTL$$
$$\mu - 3\sigma = LNTL$$

As shown in Figure 10-9a, the average fit of two mating parts is equal to the difference between the mean of the shaft distribution and the mean of the hole distribution. The *range of fit* would be the difference between the minimum diameter shaft and the maximum diameter hole. The minimum *clearance* would be the difference between the

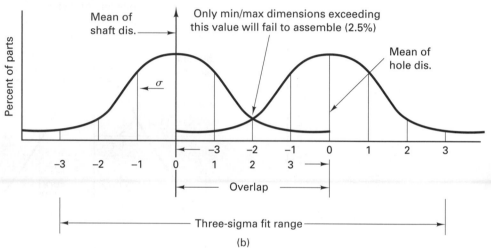

FIGURE 10-9   (a) The manner in which the distributions of the two mating parts interact determines the fit. UNTL, (upper natural tolerance limit) = $\mu + 3\sigma$; LNTL, lower natural tolerance limit = $\mu - 3\sigma$ (b) Shifting the means of the distributions toward each other results in some interface fits.

smallest hole and the largest shaft. The tighter the distributions (i.e., the more precise the process), the better the fit between the parts.

During machining processes, the cutting tools wear and change size. If tool wear is considered, the diameter of the shaft will tend to get larger as the tool wears. However, the diameter of the hole in block B of Figure 10-7 decreases in size with tool wear. If no corrective action is taken, the means of the distributions will drift toward each other, and the fit will become increasingly tight (the clearance will be decreased). If the hole and the shaft distributions overlap by 2 standard deviations, as shown in Figure 10-9b, 6 parts out of every 10,000 could not be assembled. As the shaft and the hole distributions move closer together, more interference between parts will occur, and the fit will become tighter, eventually becoming an interference fit.

The designer must specify the tolerances according to the function of the mating parts. Suppose that the mating parts are the cap and the body of an ink pen. The cap must fit snugly but must be able to be easily removed by hand. A snug fit would be too tight for a dead bolt in a door lock. A snug fit is also too tight for a high-speed bearing for rotational parts but is not tight enough for permanently mounting a wheel on an axle. In the next section, the manner in which tolerances and allowances are specified will be introduced, but design engineers are expected to have a deeper understanding of this topic.

SPECIFYING TOLERANCE AND ALLOWANCES
Tolerance can be specified in four ways: bilateral, unilateral, limits, and geometrically. *Bilateral* tolerance is specified as a plus or minus deviation from the nominal size, such

as 2.000 ± 0.002 in. More modern practice uses the *unilateral* system, where the deviation is in one direction from the basic size, such as

$$50.8 \text{ mm} \begin{array}{c} +0.1 \text{ mm} \\ -0.0 \text{ mm} \end{array} \quad \text{or} \quad 2.000 \text{ in.} \begin{array}{c} +0.004 \text{ in.} \\ -0.000 \text{ in.} \end{array} \quad \text{in metric}$$

In the first case, that of bilateral tolerance, the dimension of the part could vary between 1.998 and 2.002 in., a total tolerance of 0.004 in. For the example of unilateral tolerance, the dimension could vary between 2.000 and 2.004 in., again a tolerance of 0.004 in. Obviously, to obtain the same maximum and minimum dimensions with the two systems, different basic sizes must be used. The maximum and minimum dimensions that result from application of the designated tolerance are called *limit dimensions*, or *limits*. (*Geometric* tolerances are discussed in the next section.)

There can be no rigid rules for the amount of clearance that should be provided between mating parts; the decision must be made by the designer, who considers how the parts are to function. The American National Standards Institute, Inc. (ANSI) has established eight classes of fits that serve as a useful guide in specifying the allowance and tolerance for typical applications and that permit the amount of allowance and tolerance to be determined merely by specifying a particular class of fit. These classes are as follows:

Class 1: *Loose fit:* large allowance. Accuracy is not essential.

Class 2: *Free fit:* liberal allowance. For running fits where speeds are above 600 rpm and pressures are 4.1 MPa (600 psi) or above.

Class 3: *Medium fit:* medium allowance. For running fits under 600 RPM and pressures less than 4.1 MPa (600 psi) and for sliding fits.

Class 4: *Snug fit:* zero allowance. No movement under load is intended, and no shaking is wanted. This is the tightest fit that can be assembled by hand.

Class 5: *Wringing fit:* zero to negative allowance. Assemblies are selective and not interchangeable.

Class 6: *Tight fit:* slight negative allowance. An interference fit for parts that must not come apart in service and are not to be disassembled or are to be disassembled only seldomly. Light pressure is required for assembly. Not to be used to withstand other than very light loads.

Class 7: *Medium force fit:* an interference fit requiring considerable pressure to assemble; ordinarily assembled by heating the external member or cooling the internal member to provide expansion or shrinkage. Used for fastening wheels, crank disks, and the like to shafting. The tightest fit that should be used on cast iron external members.

Class 8: *Heavy force and shrink fits:* considerable negative allowance. Used for permanent shrink fits on steel members.

The allowances and tolerances that are associated with the ANSI classes of fits are determined according to the theoretical relationship shown in Table 10-4. The actual resulting dimensional values for a wide range of basic sizes can be found in tabulations in drafting and machine design books.

TABLE 10-4 ANSI Recommended Allowances and Tolerances

Class of Fit	Allowance	Average Interference	Hole Tolerance	Shaft Tolerance
1	$0.0025\sqrt[3]{d^2}$	—	$+0.0025\sqrt[3]{d}$	$-0.0025\sqrt[3]{d}$
2	$0.0014\sqrt[3]{d^2}$	—	$+0.0013\sqrt[3]{d}$	$-0.0013\sqrt[3]{d}$
3	$0.0009\sqrt[3]{d^2}$	—	$+0.0008\sqrt[3]{d}$	$-0.0008\sqrt[3]{d}$
4	0	—	$+0.0006\sqrt[3]{d}$	$-0.0004\sqrt[3]{d}$
5	—	0	$+0.0006\sqrt[3]{d}$	$+0.0004\sqrt[3]{d}$
6	—	$0.00025d$	$+0.0006\sqrt[3]{d}$	$+0.0006\sqrt[3]{d}$
7	—	$0.0005d$	$+0.0006\sqrt[3]{d}$	$+0.0006\sqrt[3]{d}$
8	—	$0.001d$	$+0.0006\sqrt[3]{d}$	$+0.0006\sqrt[3]{d}$

In the ANSI system, the hole size is always considered basic, because the majority of holes are produced through the use of standard-size drills and reamers. The internal member, the shaft, can be made to any one dimension as readily as to another. The allowance and tolerances are applied to the basic hole size to determine the limit dimensions of the mating parts. For example, for a basic hole size of 2 in. and a Class 3 fit, the dimensions would be:

Allowance	0.0014 in.
Tolerance	0.0010 in.
Hole	
Maximum	2.0010 in.
Minimum	2.0000 in.
Shaft	
Maximum	1.9986 in.
Minimum	1.9976 in.

It should be noted that for both clearance and interference fits, the permissible tolerances tend to result in a looser fit.

The *ISO System of Limits and Fits* (Figure 10-10a) is widely used in a number of leading metric countries. This system is considerably more complex than the ANSI system just discussed. In this system each part has a *basic size*. Each limit of size of a part, high and low, is defined by its *deviation* from the basic size, the magnitude and sign being obtained by subtracting the basic size from the limit in question. The difference between the two limits of size of a part is called *tolerance*, an absolute amount without sign.

There are three classes of fits: (1) *clearance fits*, (2) *transition fits* (the assembly may have either clearance or interference), and (3) *interference fits*. Either a *shaft-* or *hole-basis system* may be used (Figure 10-10b). For any given basic size, a range of tolerances and deviations may be specified with respect to the line of zero deviation, called the *zero line*. The tolerance is a function of the basic size and is designated by a number symbol, called the *grade* (e.g., the *tolerance grade*). The *position* of the tolerance with respect to the zero line, also a function of the basic size, is indicated by a letter symbol (or two letters)—a capital letter for holes and a lowercase letter for shafts—as illustrated in Figure 10-10c.[1] Thus the specification for a hole and a shaft having a basic size of 45 mm might be 45 H8/g7.

Eighteen standard grades of tolerances are provided, called IT 01, IT 0, and IT 1 through IT 16, providing numerical values for each nominal diameter, in arbitrary steps up to 500 mm (i.e., 0-3, 3-6, 6-10, ..., 400-500 mm). The valve of the tolerance unit, *i*, for grades 5–16 would be

$$i = 0.45\sqrt{D} + 0.001\,D$$

where *i* is in micrometers and *D* in millimeters.

Standard shaft and hole deviations are provided by similar sets of formulas. However, for practical application, both tolerances and deviations are provided in three sets of rather complex tables. Additional tables give the values for basic sizes above 500 mm and for "commonly used shafts and holes" in two categories: "general purpose" and "fine mechanisms and horology" (horology is the art of making timepieces).

GEOMETRIC TOLERANCES

Geometric tolerances state the maximum allowable deviation of a form or a position from the perfect geometry implied by a drawing. These tolerances specify the diameter or the width of a tolerance zone necessary for a part to meet its required accuracy.

---

[1]It will be recognized that the "position" in the ISO system essentially provides the "allowance" of the ANSI system.

(a) Basic size, deviation and tolerance in the ISO system.

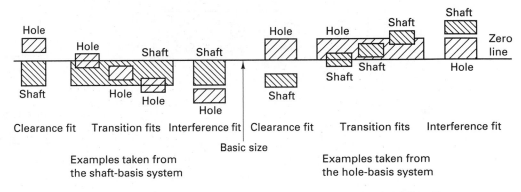

(b) Shaft-basis and hole-basis system for specifying fits in the ISO system.

(c) Position of the various tolerance zones for a given diameter in the ISO system.

FIGURE 10-10 The ISO System of Limits and Fits. *(By permission from Recommendations R286-1962, System of Limits and Fits, copyright 1962, American Standards Institute, NJ.)*

	Tolerance	Characteristic	Symbol
Individual features	Form	Straightness	—
		Flatness	▱
		Circularity	○
		Cylindricity	⌭
Individual or related features	Profile	Line	⌒
		Surface	⌓
Related features	Orientation	Angularity	∠
		Perpendicularity	⊥
		Parallelism	//
	Location	Position	⊕
		Concentricity	◎
	Runout	Circular runout	↗
		Total runout	↗↗

Notes	⌀	Ⓜ	Ⓛ	Ⓢ
	DIA	MMC	LMC	RFS

(a)

(b)

(c)

(d)

FIGURE 10-11   (a) Geometric tolerancing symbols; (b) feature control symbols for part drawings; (c) how a geometric tolerance for flatness is specified; (d) what the specification means.

Figure 10-11 shows the various symbols used to specify the required geometric characteristics of dimensioned drawings. A modifier is used to specify the limits of size of a part when applying geometric tolerances. The *maximum material condition* (MMC) indicates that a part is made with the largest amount of material allowable (e.g., a hole at its smallest permitted diameter or a shaft at its largest permitted diameter). The least material condition (LMC) is the converse of the maximum material condition. *Regardless of feature size* (RFS) indicates that tolerances apply to a geometric feature for any size it may be. Many geometric tolerances or *feature control symbols* are stated with respect to a particular datum or reference surface. Up to three datum surfaces can be given to specify a tolerance. Datum surfaces are generally designated by a letter symbol.

Figure 10-11b gives examples of the symbols used for datum planes and feature control symbols.

There are four tolerances that specify the permitted variability of forms: *flatnesss, straightness, roundness*, and *cylindricity*. Form tolerances describe how an actual feature may vary from a geometrically ideal feature.

The surface of a part is ideally flat if all its elements are coplanar. The *flatness* specification describes the tolerance zone formed by two parallel planes that bound all the elements on a surface. A 0.5-mm tolerance zone is described by the feature control symbol in Figure 10-11c. The distance between the highest point on the surface to the lowest point on the surface may not be greater than 0.5 mm, as shown in Figure 10-11d.

In Section 10.7, on coordinate measuring machines, additional examples of geometric tolerances can be found.

■ 10.4 INSPECTION METHODS FOR MEASUREMENT

The field of *metrology*, even limited to geometrical or dimensional measurements, is far too large to cover here. This chapter concentrates on basic linear measurements and the measurement and testing devices most commonly found in a company's metrology or quality control facility. At a minimum, such labs would typically contain optical flats: one or two granite measuring tables; an assortment of indicators, calipers, micrometers, and height gages; an optical comparator; a set or two of Grade 1 gage blocks; a coordinate measuring machine; a laser scanning device; a laser interferometer; a toolmaker's microscope; and pieces of equipment specially designed to inspect and test the company's products.

Table 10-5 provides a summary of inspection methods, listing five basic kinds of devices: air, light optical and electron optical, electronic, and mechanical. The variety seems to be endless, but digital electronic readouts connected to any of the measuring devices are becoming the preferred method.

The discrete digital readout on a clear liquid-crystal display (LCD) eliminates reading interpretations associated with analog scales and can be entered directly into dedicated microprocessors or computers for permanent recording and analysis. The added speed and ease of use for this type of equipment have allowed it to be routinely used on the plant floor instead of in the metrology lab. In summary, the trend toward tighter tolerances (greater precision) and accuracy associated with the need for superior quality and reliability has greatly enhanced the need for improved measurement methods.

TABLE 10-5   Five Basic Kinds of Inspection Method

Method	Typical Accuracy	Major Applications	Comments
Air	0.5–10 μin. or 2 to 3% of scale range	Gaging holes and shafts using a calibrated difference in air pressure or airflow, with magnifications of 20,000–40,000 to 1; also used for machine control, sorting, and classifying.	High precision and flexibility; can measure out-of-round, taper, concentricity, camber, squareness, parallelism, and clearance between mating parts; noncontact principle good for delicate parts.
Optical light energy	0.2–2 μin. or better with laser interferometry 0.5–1 second of arc in autocollimation optical comparators	Interferometry; checking flatness and size of gage blocks; finding surface flaws; measuring spherical shapes, flatness of surface plates, accuracy of rotary index tables; includes all light microscopes and devices common on plant floor	Largest variety of measuring equipment; autocollimators are used for making precision angular measurements; lasers are used to make precision in-process measurements; laser scanning.
Optical electron energy	100 Å	Precision measurement in scanning electron microscopes of microelectronic circuits and other small precision parts.	Part size restricted by vacuum chamber size; electron beam can be used for processing and part testing of electronic circuits.
Electronics	0.5–10 μin.	Widely used for machine control, on-line inspection, sorting, and classification; ODs, IDs, height, surface, and geometrical relationships, profile tracing for roundness, surface roughness, contours, etc.; most devices are comparators with movement of stylus or spindle producing an electronic signal that is amplified electronically; commonly connected to microprocessors and minicomputers for process adjustment.	Electronic gages come in many forms but usually have a sensory head or detector combined with an amplifier; capable of high magnification with resolution limited by size or geometry for sensory head; readouts commonly have multiple magnification steps; solid-state digital electronics make these devices small, portable, stable, and extremely flexible, with extremely fast response time.
Mechanical	1–10 μin.	Large variety of external and internal measurements using dial indicators, micrometers, calipers, and the like; commonly used for bench comparators for gage calibration work.	Moderate cost and ease of use make many of these devices and workhorses of the shop floor; highly dependent on workers' skills and often subject to problems of linkages.

FACTORS IN SELECTING INSPECTION EQUIPMENT

Many inspection devices use electronic output to communicate directly to microprocessors. Inspection devices are being built into the processes themselves and are often computer-aided. In-process inspection generates feedback sensory data from the process or its output to the computer control of the machine, which is the first step in making the processes responsive to changes (adaptive control). In addition to in-process inspection, many other quality checks and measurements of parts and assemblies are needed. In general, six factors should be considered when selecting equipment for an inspection job by measurement techniques.

1. *Gage capability.* The measurement device (or working gage) should be 10 times more precise than the tolerance to be measured. This is known on the factory floor as "the rule of 10." The rule actually applies to all stages in the inspection sequence, as shown in Figure 10-12. The master gage should be 10 times more precise than that of the inspection device. The reference standard used to check the master gage should be 10 times more precise than the master gage. The application of the rule greatly reduces the probability of rejecting good parts or accepting bad components and performing additional work on them. Additional discussion of gage capability is found in Chapter 36.

2. *Linearity.* This factor refers to the calibration accuracy of the device over its full working range. Is it linear? What is its degree of nonlinearity? Where does it become nonlinear, and what, therefore, is its real linear working region?

3. *Repeat accuracy.* How repeatable is the device in taking the same reading over and over on a given standard?

4. *Stability.* How well does this device retain its calibration over a period of time? Stability is also called *drift.* As devices become more accurate, they often lose stability and become more sensitive to small changes in temperature and humidity.

5. *Magnification.* This refers to the amplification of the output portion of the device over the actual input dimension. The more accurate the device, the greater must be its magnification factor, so that the required measurement can be read out (or observed) and compared with the desired standard. Magnification is often confused with resolution, but they are not the same thing.

6. *Resolution.* This is sometimes called *sensitivity* and refers to the smallest unit of scale or dimensional input that the device can detect or distinguish. The greater the resolution of the device, the smaller will be the things it can detect or identify (resolve) and the greater will be the magnification required to expand these measurements up to the point where they can be observed by the naked eye.

Some other factors of importance in selecting inspection devices include the type of measurement information desired; the range or the span of sizes the device can handle versus the size and geometry of the workpieces; the environment; the cost of the

FIGURE 10-12 The rule of 10 states that for reliable measurements each successive step in the inspection sequence should have 10 times the *precision* of the preceding step.

	gage		
Tolerance needed on part ± 0.001 on hole diameter	Precision needed on gage ± 0.0001 in.	To check and set the air gage, needs to be ± 0.00001 in.	In the manufacture of the master gage, a standard of precision of at least ± 0.000001 in. is needed
Workpiece	Air gage or working gage	Master gage	Reference end standard

device; and the cost of installing, training, and using the device. The last factor depends on the speed of measurement, the degree to which the system can be automated, and the functional life of the device in service.

■ 10.5 MEASURING INSTRUMENTS

Because of the great importance of measuring in manufacturing, a great variety of instruments are available that permit measurements to be made routinely, ranging in accuracy from $\frac{1}{64}$ to 0.00001 in. and from 0.5 to 0.0003 mm. Machine-mounted measuring devices (probes and lasers) for automatically inspecting the workpiece during manufacturing are beginning to compete with post-process gaging and inspection, in which the part is inspected, automatically or manually, after it has come off the machine. In-process inspection for automatic size control has been used for some years in grinding to compensate for the relatively rapid wear of the grinding wheel. Touch trigger probes, with built-in automatic measuring systems, are being used on CNC machine tools to determine cutting tool offsets and compensations for tool wear. These systems are discussed in Chapter 26.

For manually operated analog instruments, the ease of use, precision, and accuracy of measurements can be affected by (1) the least count of the subdivisions on the instrument, (2) line matching, and (3) the parallax in reading the instrument. Elastic deformation of the instrument and workpiece and temperature effects must be considered. Some instruments are more subject to these factors than others. In addition, the skill of the person making the measurements is very important. Digital readout devices in measuring instruments lessen or eliminate the effect of most of these factors, simplify many measuring problems, and lessen the chance of making a math error.

LINEAR MEASURING INSTRUMENTS

Linear measuring instruments are of two types: direct reading and indirect reading. *Direct-reading instruments* contain a line-graduated scale so that the size of the object being measured can be read directly on this scale. *Indirect-reading instruments* do not contain line graduations and are used to transfer the size of the dimension being measured to a direct-reading scale, thus obtaining the desired size information indirectly.

The simplest and most common direct-reading linear measuring instrument is the *machinist's rule*, shown in Figure 10-13. Metric rules usually have two sets of line graduations on each side, with divisions of $\frac{1}{2}$ and 1 mm; English rules have four sets, with divisions of $\frac{1}{16}, \frac{1}{32}, \frac{1}{64}$, and $\frac{1}{100}$ in. Other combinations can be obtained in each type.

The machinist's rule is an end- or line-matching device. For the desired reading to be obtained, an end and a line, or two lines, must be aligned with the extremities of the object or the distance being measured. Thus, the accuracy of the resulting reading is a

FIGURE 10-13 Machinist's rules: (a) metric and (b) inch graduations; 10ths and 100ths on one side, 32nds and 64ths on the opposite side. *(Courtesy of L.S. Starrett Company.)*

FIGURE 10-14 Combination set. *(Courtesy of MTI Corporation.)*

FIGURE 10-15 This vernier caliper can make measurements using both inside (for holes) and outside (shafts) anvils.

function of the alignment and the magnitude of the smallest scale division. Such scales are not ordinarily used for accuracies greater than $\frac{1}{64}$ in. (0.01 in.), or about $\frac{1}{2}$ mm.

Several attachments can be added to a machinist's rule to extend its usefulness. The *square head* (Figure 10-14) can be used as a miter or tri-square or to hold the rule in an upright position on a flat surface for making height measurements. It also contains a small bubble-type level so that it can be used by itself as a level. The *bevel protractor* permits the measurement or layout of angles. The *center head* permits the center of cylindrical work to be determined.

The *vernier caliper* (Figure 10-15) is an end-measuring instrument, available in various sizes, that can be used to make both outside and inside measurements to theoretical accuracies of 0.01 mm or 0.001 in. End-measuring instruments are more accurate and somewhat easier to use than line-matching types because their jaws are placed against either end of the object being measured, so any difficulty in aligning edges or lines is avoided. However, the difficulty remains in obtaining uniform contact pressure, or "feel," between the legs of the instrument and the object being measured.

A major feature of the vernier caliper is the auxiliary scale (Figure 10-16). The caliper shown has a graduated beam with a metric scale on the top, a metric vernier plate, and an English scale on the bottom with an English vernier. The manner in which readings are made is explained in the figure.

Figure 10-17 also shows a vernier depth gage for measuring the depth of holes or the length of shoulders on parts and a vernier height gage for rmaking height measurements. Figure 10-18 shows calipers that have a dial indicator or a digital readout that replace the vernier. The latter two calipers are capable of making inside and outside measurements as well as depth measurements.

The *micrometer caliper*, more commonly called a *micrometer*, is one of the most widely used measuring devices. Until recently, the type shown in Figure 10-19 was virtually standard. It consists of a fixed anvil and a movable spindle. When the thimble is rotated on the end of the caliper, the spindle is moved away from the anvil by means of an accurate screw thread. On English types, this thread has a lead of 0.025 in., and one revolution of the thimble moves the spindle this distance. The barrel, or sleeve, is calibrated in 0.025-in. divisions, with each $\frac{1}{10}$ of an inch being numbered. The circumference at the edge of the thimble is graduated into 25 divisions, each representing 0.001 in.

A major difficulty with this type of micrometer is making the reading of the dimension shown on the instrument. To read the instrument, the division on the thimble that coincides with the longitudinal line on the barrel is added to the largest reading exposed on the barrel.

FIGURE 10-16 Vernier caliper graduated for English and metric (direct) reading. The metric reading is 27 + 0.42 = 27.42 mm.

Refer to the upper bar graduations and metric vernier plate. Each bar graduation is 1.00 mm. Every tenth graduation is numbered in sequence—10 mm, 20 mm, 30 mm, 40 mm, etc.—over the full range of the bar. This provides for direct reading in millimeters.

In the picture, the vernier plate zero line is one inch (1.000") plus one-twentieth (0.50") beyond the zero line on the bar, or 1.050". The 29th graduation on the vernier plate coincides with a line on the bar (as indicated by stars). 29 × 0.001 (.029") is therefore added to the 1.050" bar reading, and the total is 1.079".

Vernier height gage

FIGURE 10-17 Variations in the vernier caliper design result in other basic gages.

Vernier caliper with inch or metric scales and 0.001-in. accuracy

Dial caliper with 0.001-in. accuracy

FIGURE 10-18 Three styles of calipers in common use today: (a) Vernier caliper with inch or metric scales and 0.001-in. accuracy; (b) dial caliper with 0.001-in. accuracy; (c) digital electronic caliper with 0.001-in. (0.03-mm) accuracy and 0.0001-in. resolution with inch/metric conversion.

Digital electronic caliper with 0.001-in. (0.03-mm) accuracy and 0.0001-in. resolution with inch/metric conversion.

FIGURE 10-19 Micrometer caliper graduated in ten-thousandths of an inch with insets A, B, and C showing two example readings. (*Courtesy Starrett Bulletin No. 1203.*)

Micrometers graduated in ten-thousandths of an inch are the same as those graduated in thousandths, except that an additional vernier scale is placed on the sleeve so that a reading of ten-thousandths is obtained and added to the thousandths reading. The vernier consists of 10 divisions on the sleeve, shown in B, which occupy the same space as 9 divisions on the thimble. Therefore, the difference between the width of one of the 10 spaces on the vernier and one of the 9 spaces on the thimble is one-tenth of a division on the thimble, or one-tenth of one-thousandth, which is one ten-thousandth. To read a ten-thousandths micrometer, first obtain the thousandths reading, then see which of the lines on the vernier coincides with a line on the thimble. If it is the line marked "1," add one ten-thousandth; if it is the line marked "2," add two ten-thousandths; and so on.

EXAMPLE: REFER TO INSETS A AND B IN FIGURE 10-19.

The "2" line on sleeve is visible, representing	0.200 in.
Two additional lines, each representing 0.025"	$2 \times .025" = -0.050$ in.
Line "O" on the thimble coincides with the reading line on the sleeve, representing	0.000 in.
The "O" lines on the vernier coincide with lines on the thimble, representing	0.000 in.
The micrometer reading is	0.2500 in.

Now you try to read inset C.

The "2" line on sleeve is visible, representing	0.200 in.
Two additional lines, each representing 0.025"	$2 \times 0.025' = 0.050$ in.
The reading line on the sleeve lies between the "O" and "1" on the thimble, so ten-thousandths of an inch is to be added as read from the vernier.	
The "7" line on the vernier coincides with a line on the thimble, representing	$7 \times 0.0001" = 0.007$ in.
The micrometer reading is	0.2507 in.

However, owing to the lack of pressure control, micrometers can seldom be relied on for accuracy beyond 0.0005 in., and such vernier scales are not used extensively. On metric micrometers the graduations on the sleeve and thimble are usually 0.5 mm and 0.01 mm, respectively (see the Problems at the end of the chapter).

Many errors have resulted from the ordinary micrometer being misread, the error being ±0.025 or ±0.5 mm. Consequently, direct-reading micrometers have been developed. Figure 10-20 shows a digital outside micrometer that reads to 0.001 in. on the digit counter and 0.0001 in. on the vernier on the sleeve. The range of a micrometer is limited to 1 in. Thus a number of micrometers of various sizes are required to cover a wide range of dimensions. To control the pressure between the anvil, the spindle, and the piece being measured, most micrometers are equipped with a ratchet or a friction device, as shown in Figures 10-19 and 10-20. Calipers that do not have this device may be overtightened and sprung by several thousandths by applying excess torque to the thimble. Micrometer

FIGURE 10-20 Digital micrometer for measurements from 0 to 1 in., in 0.0001-in. graduations.

calipers should not usually be relied on for measurements of greater accuracy than 0.01 mm or 0.001 in., unless they are of the new digital design.

Micrometer calipers are available with a variety of specially shaped anvils and/or spindles, such as point, balls, and disks, for measuring special shapes, including screw threads. Micrometers are also available for inside measurements, and the micrometer principle is also incorporated into a *micrometer depth gage*.

Bench micrometers with direct readout to data processors are becoming standard inspection devices on the plant floor (see Figure 10-21). The data processor provides a record of measurement as well as control charts and histograms. Direct-gaging height gages, calipers, indicators, and micrometers are also available with statistical analysis capability. See Chapter 38 for more discussion on control charts and process capability.

Larger versions of micrometers, called *supermicrometers*, are capable of measuring 0.0001 in. when equipped with an indicator that shows that a selected pressure between the anvils has been obtained. The addition of a digital readout permits the device to measure to ±0.00005 in. (0.001 mm) directly when it is used in a controlled-temperature environment.

The *toolmaker's microscope*, shown in Figure 10-22, is a versatile instrument that measures by optical means; no pressure is involved. Thus it is very useful for making ac-

Micrometer

Bench micrometers

Height gage

Data analysis

A record of measurement	$\bar{x}$-R control chart (data preparation)	Histogram
Differential calculation		Constant variable calculation

Printer

Plotter

Host computer

Data logger transmitter

FIGURE 10-21 Direct-gaging system for process control and statistical analysis of inspection data. *(Courtesy of MITUTOYO.)*

FIGURE 10-22   Toolmaker's microscope with digital readouts for X and Y table movements. *(Courtesy of Nikon.)*

FIGURE 10-23   Optical comparator, measuring the contour on a workpiece. Digital indicators with conversions add to the utility of optical comparators.

curate measurements on small or delicate parts. The base, on which the special microscope is mounted, has a table that can be moved in two mutually perpendicular, horizontal directions (X and Y) by means of accurate micrometer screws that can be read to 0.0001 in. or, if so equipped, by means of the digital readout. Parts to be measured are mounted on the table, the microscope is focused, and one end of the desired part feature is aligned with the cross-line in the microscope. The reading is then noted, and the table is moved until the other extremity of the part coincides with the cross-line. From the final reading, the desired measurement can be determined. In addition to a wide variety of linear measurements, accurate angular measurements can also be made by means of a special protractor eyepiece. These microscopes are available with digital readouts.

The *optical projector* or *comparator* (Figure 10-23) is a large optical device on which both linear and angular measurements can be made. As with the toolmaker's microscope, the part to be measured is mounted on a table that can be moved in X and Y directions by accurate micrometer screws. The optical system projects the image of the part on a screen, magnifying it from 5 to more than 100 times. Measurements can be made directly by means of the micrometer dials, the digital readouts, or the dial indicators, or on the magnified image on the screen by means of an accurate rule. A very common use for this type of instrument is the checking of parts, such as dies and screws. A template is drawn to an enlarged scale and is placed on the screen. The projected contour of the part is compared to the desired contour on the screen. Some projectors also function as low-power microscopes by providing surface illumination.

MEASURING WITH LASERS

One of the earliest and most common metrological uses of low-power lasers has been in interferometry. The interferometer uses light interference bands to determine distance and thickness of objects (Figure 10-24). First, a beam splitter divides a beam of light into a measurement beam and a reference beam. The measurement beam travels to a reflector (optical glass plate A), resting on the part whose distance is to be measured, while the reference beam is directed at fixed reflector B. Both beams are reflected back through the beam splitter, where they are recombined into a single beam before traveling

FIGURE 10-24 Interference bands can be used to measure the size of objects to great accuracy. *(Based on the Michelson interferometer, invented in 1882.)*

FIGURE 10-25 Method of calibrating gage block by light-wave interference.

to the observer. This recombined beam produces interference fringes, depending on whether the waves of the two returning beams are in phase (called *constructive interference*) or out of phase (termed *destructive interference*). In-phase waves produce a series of bright bands, and out-of-phase waves produce dark bands. The number of fringes can be related to the size of the object, measured in terms of light waves of a given frequency. The following example will explain the basics of the method.

To determine the size of object U in Figure 10-25, a calibrated reference standard S, an optical flat, and a toolmaker's flat are needed, along with a monochromatic light source. *Optical flats* are quartz or special glass disks, from 2 to 10 in. (50 to 250 mm) in diameter and about $\frac{1}{2}$ to 1 in. (12 to 25 mm) thick, whose surfaces are very nearly true planes and nearly parallel. Flats can be obtained with the surfaces within 0.000001 in. (0.00003 mm) of true flatness. It is not essential that both surfaces be accurate or that they be exactly parallel, but one must be certain that only the accurate surface is used in making measurements. A *toolmaker's flat* is similar to an optical flat but is made of steel and usually has only one surface that is accurate. A *monochromatic light source*, light of a single wavelength, must be used. Selenium, helium, or cadmium light sources are commonly used along with helium-neon lasers.

The block to be measured is U and the calibrated block is S. Distances a and b must be known but do not have to be measured with great accuracy. By counting the number of *interference bands* shown on the surface of block U, the distance c–d can be determined. Because the difference in the distances between the optical flat and the surface of U is one-half wavelength, each dark band indicates a change of one-half wavelength in the elevation. If a monochromatic light source having a wavelength of 23.2 μin. (0.589 mm) is used, each interference band represents 11.6 μin. (0.295 μm). Then, by simple geometry, the difference in the heights of the two blocks can be computed. The same method is applicable for making precise measurements of other objects by comparing them with a known gage block.

Accurate measurement of distances greater than a few inches was very difficult until the development of laser interferometry, which permits accuracies of ±0.5 part per million over a distance of 6.1 m with 0.01 μm resolution. Such equipment is particularly useful in checking the movement of machine tool tables, aligning and checking large assembly jigs, and making measurements of intricate machined parts such as tire-tread molds.

The Hewlett-Packard laser interferometer (Figure 10-26) uses a helium–neon laser beam split into two beams, each of different frequency and polarized. When the beams are recombined, any relative motion between the optics creates a Doppler shift in the frequency. This shift is then converted into a distance measurement. The laser light has less tendency to diverge (spread out) and is also monochromatic (of the same wavelength). A process that has been largely confined to the optical industry and the metrology lab is now suitable for the factory, where its extremely precise distance-measuring capabilities have been applied to the alignment and calibration of machine tools.

The company's first two-frequency interferometer calibration system was introduced in 1970 to overcome workplace contamination by thermal gradients, air turbulence, oil mist, and so on, which affect the intensity of light. Doppler laser interferometers are relatively insensitive to such problems. The system can be used to measure linear distances, velocities, angles, flatness, straightness, squareness, and parallelism in machine tools.

Lasers provide for accurate machine tool alignment. Large, modern machine tools can move out of alignment in a matter of months, causing production problems often attributed to the cutting tools, the workholders, the machining conditions, or the numerical control part program.

(a)

(b)   Linear interferometer combination

Optical setup for *x*-axis linear positioning measurement

Linear retroreflector shown with height adjuster, post and base

(c)

FIGURE 10-26   (Top) Calibrating the *x*-axis linear table displacement of a vertical spindle milling machine; (middle) schematic of optical setup; (bottom) schematic of components of a two-frequency laser interferometer. *(Courtesy of Hewlett-Packard.)*

FIGURE 10-27 Scanning laser measuring system. *(Courtesy of ZYGO Corporation.)*

Light interference also makes it possible to determine easily whether a surface is exactly flat. The achievement of interference fringes is largely dependent on the coherence of the light used. The availability of highly coherent *laser* light (in-phase light of a single frequency) has made interferometry practical in far less restrictive environments than in the past. The sometimes arduous task of extracting usable data from a close-packed series of interference fringes has been taken over by microprocessors.

The most widely used laser technique for inspection and in-process gaging is known as *laser scanning*. At its most basic level, the process consists of placing an object between the source of the laser beam and a receiver containing a photodiode. A microprocessor then computes the object's dimensions based on the shadow that the object casts (Figure 10-27).

The noncontact nature of laser scanning makes it well suited to in-process measurement, including such difficult tasks as the inspection of hot-rolled or extruded material, and its comparative simplicity has led to the development of highly portable systems. The bench gage versions can measure to resolutions of 0.0001 mm.

■ 10.6 VISION SYSTEMS FOR MEASUREMENT

If a picture is worth a thousand words, then vision systems are the tome of inspection methods (see Figure 10-28). Machine vision is used for visual inspection, for guidance and control, or for both. Normal TV image formation on photosensitive surfaces or arrays is used, and the video signals are analyzed to obtain information about the object. Each picture frame represents the object at some brief interval of time. Each frame must be dissected into picture elements (called *pixels*). Each pixel is *digitized* (has binary numbers assigned to it) by fixing the brightness or gray level of each pixel to produce a *bit-map* of the object (Figure 10-29). That is, each pixel is assigned a numerical value based on its shade of gray. Image preprocessing improves the quality of the image data by removing unwanted detail. The bit-map is stored in a buffer memory. By analyzing and processing the digitized and stored bit-map, the patterns are extracted, edges located, and dimensions determined.

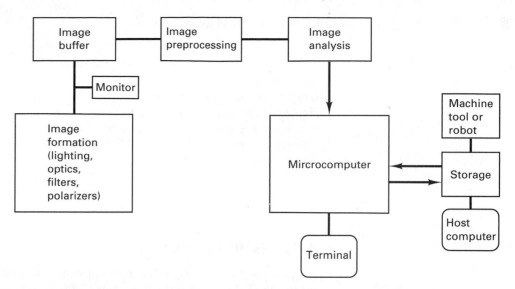

FIGURE 10-28 Schematic of elements of a machine vision system.

Sophisticated computer algorithms using artificial intelligence have greatly reduced the computer operations needed to achieve a result, but even the most powerful video-based systems currently require one to two seconds to achieve a measurement. This may be too long a time for many on-line production applications. Table 10.6 provides a comparison of vision systems to laser scanning.

With the recent emphasis on quality and 100% inspection, applications for inspection by machine vision have increased markedly. Vision systems can check hundreds of parts per hour for multiple dimensions. Resolutions of ±0.01 in. have been demonstrated, but 0.02 in. is more typical for part location. Machine vision is useful for robot guidance in material handling, welding, and assembly, but nonrobotic inspection and part location applications are still more typical. The use of vision systems in inspection, quality control, sorting, and machining tool monitoring will continue to expand. Systems can cost $100,000 or more to install and must be justified on the basis of improved quality rather than labor replacement.

FIGURE 10-29 Vision systems use a gray scale to identify objects. (a) Object with three different gray values. (b) One frame of object (pixels). (c) Each pixel assigned a gray-scale number.

(a) Object   (b) Video picture frame   (c) Digitized frame

TABLE 10-6	Laser Scanning versus Vision Systems	
Variable	Laser-Scanning Systems	Video-Based Systems
Ambient lighting	Independent	Dependent
Object motion	Object usually stationary	Multiple cameras or strobe lighting may be required
Adaptability to robot systems	Readily adapted; some limitations on robot motion speed or overall system operation	Readily adapted; image-processing delays may delay system operation
Signal processing	Simple; computers often not required	Requires relatively powerful computers with sophisticated software
Cycle time	Very fast	Seconds of computer time may be needed
Applicability to simple tasks	Readily handled; edges and features produce sharp transitions in signal	Requires extensive use of sophisticated software algorithms to identify edges
Sizing capability	Can size an object in a single scan per axis	Can size on horizontal axis in one scan; other dimensions require full-frame processing
Three-dimensional capability	Limited three dimensionality; needs ranging capability	Uses two views of two cameras with sophisticated software or structured light
Accuracy and precision	Submicrometer 0.001 to 0.0001 in. or better accuracy; highly repeatable	Depends on resolution of cameras and distance between camera and object; systems with 0.004-in. precision and 0.006-in. accuracy are typical

■ 10.7 COORDINATE MEASURING MACHINES

Precision measurements in three-dimensional Cartesian coordinate space can be made with *coordinate measuring machines* (CMM) of the design shown in Figure 10-30. The parts are placed on a large granite flat or the table. The vertical arm carries a probe that can be precisely moved in *x–y–z* directions to produce 3D measurements. In this design, the vertical column rides on a bridge beam and carries a touch-trigger probe. Such machines use digital readouts, air bearings, computer controls, and granite tables to achieve accuracies of the order of 0.0002 to 0.0004 in. over spans of 10 to 30 in. or more. These systems may have computer routines that give the best fit to feature measurements and that provide the means of establishing geometric tolerances, discussed earlier in this chapter. Figure 10-31 gives a partial listing of the results one can achieve with these machines.

■ 10.8 ANGLE-MEASURING INSTRUMENTS

Accurate angle measurements are usually more difficult to make than linear measurements. Angles are measured in degrees (a degree is $\frac{1}{360}$ part of a circle) and decimal subdivisions of a degree (or in minutes and seconds of arc). The SI system calls for measurements of plane angles in radians, but degrees are permissible. The use of degrees will continue in manufacturing, but with minutes and seconds of arc possibly being replaced by decimal portions of a degree.

The bevel protractor (Figure 10-32) is the most general angle-measuring instrument. The two movable blades are brought into contact with the sides of the angular part, and the angle can be read on the vernier scale to 5 minutes of arc. A clamping device is provided to lock the blades in any desired position so that the instrument can be used for both direct measurement and layout work. As indicated previously, an angle attachment on the combination set can also be used to measure angles, similar to the way a bevel protractor is used but usually with somewhat less accuracy.

FIGURE 10-30 Coordinate measuring machine with inset showing probe and a part being measured.

Granite table

Digital readouts computer

Part being inspected by probe

Straightness

Straightness
Measured or previously calculated points may be used to determine a "best fit" line. The form routine establishes two reference lines that are parallel to the "best fit" line and that just contain all of the measured or calculated points.
Straightness is defined as the distance D between these two reference lines.

Flatness

Flatness
Measured or previously calculated points may be used to determine the "best fit" plane. The form routine establishes two reference planes that are parallel to the "best fit" plane and that just contain all of the measured or calculated points.
Flatness is defined as the distance D between these two reference planes.

Roundness

Roundness
Measured or previously calculated points may be used to determine the "best fit" circle. The form routine establishes two reference circles that are concentric with the "best fit" circle and that just contain all of the measured or calculated points.
Roundness is defined as the difference D in radius of these two reference circles.

Cylindricity

Cylindricity
Measured or previously calculated points may be used to determine the "best fit" cylinder. The form routine establishes two reference cylinders that are co-axial to the "best fit" cylinder and that just contain all of the measured or calculated points.
Cylindricity is the difference D in radius of these two reference cylinders. Also applicable to stepped cylinders.

Conicity

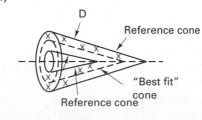

Conicity
Measured or previously calculated points may be used to determine the "best fit" cone. The form routine establishes two reference cones that are co-axial with and similar to the "best fit" cone and that just contain all of the measured or calculated points.
Conicity is defined as the distance D between the side of these two reference cones.

FIGURE 10-31 Examples of geometric form tolerances developed by probing surface with a CMM.

FIGURE 10-32 Measuring an angle on a part with a bevel protractor. *(Courtesy of Brown & Sharpe Mfg. Co.)*

The toolmaker's microscope is very satisfactory for making angle measurements, but its use is restricted to small parts. The accuracy obtainable is 5 minutes of arc. Similarly, angles can be measured on the optical contour projector. Angular measurements can also be made by means of an angular interferometer with the laser system.

A *sine bar* may be used to obtain accurate angle measurements if the physical conditions will permit. This device (Figure 10-33) consists of an accurately ground bar on which two accurately ground pins of the same diameter are mounted an exact distance apart. The distances used are usually either 5 or 10 in., and the resulting instrument is called a 5- or 10-in. sine bar. Sine bars are also available with millimeter dimensions. Measurements are made by using the principle that the sine of a given angle is the ratio of the opposite side to the hypotenuse of the right triangle.

The part being measured is attached to the sine bar, and the inclination of the assembly is raised until the top surface is exactly parallel with the surface plate. A stack of gage blocks is used to elevate one end of the sine bar, as shown in Figure 10-33. The height of the stack directly determines the difference in height of the two pins. The difference in height of the pins can also be determined by a dial indicator gage or any other type of gage. The difference in elevation is then equal to either 5 or 10 times the sine of the angle being measured, depending on whether a 5- or 10-in. bar is being used. Tabulated values of the angles corresponding to any measured elevation difference for 5- or 10-in. sine bars are available in various handbooks. Several types of sine bars are available to suit various requirements.

Accurate measurements of angles to 1 second of arc can be made by means of *angle gage blocks*. These come in sets of 16 blocks that can be assembled in desired combinations. Angle measurements can also be made to ±0.001° on rotary indexing tables having suitable numerical control.

■ 10.9 GAGES FOR ATTRIBUTES MEASURING

In manufacturing, particularly in mass production, it may not be necessary to know the exact dimensions of a part, only that it is within previously established limits. Limits can often be determined more easily than specific dimensions by the use of attribute-type instruments called *gages*. They may be of either fixed type or deviation type, may be used for both linear and angular dimensions, and may be used manually or mechanically (automatically).

FIXED-TYPE GAGES

Fixed-type gages are designed to gage only one dimension and to indicate whether it is larger or smaller than the previously established standard.

They do not determine how much larger or smaller the measured dimension is than the standard. Because such gages fulfill a simple and limited function, they are relatively inexpensive and usually quick and easy to use.

Gages of this type are ordinarily made of hardened steel of proper composition and are heat treated to produce dimensional stability. Hardness is essential to minimize wear and maintain accuracy. Because steels of high hardness can become dimensionally un-

FIGURE 10-34   Plain plug gage having the go member on the left end (1.1250-in. diameter) and no-go member on the right end. *(Courtesy of Sheffield.)*

FIGURE 10-35   Step-type plug gage with go and no-go elements on the same end. *(Courtesy of Sheffield.)*

stable, some fixed gages are made of softer steel, then given a hard chrome plating to provide surface hardness. Chrome plating can also be used for reclaiming some worn gages. Where gages are to be subjected to extensive use, they may be made of tungsten carbide at the wear points.

One of the most common fixed gages is the *plug gage.* As shown in Figure 10-34, plug gages are accurately ground cylinders used to gage internal dimensions, such as holes. The gaging element of a *plain plug gage* has a single diameter. To control the minimum and maximum limits of a given hole, two plug gages are required. The smaller, or *go gage,* controls the minimum because it must go (slide) into any hole that is larger than the required minimum. The larger, or *no-go gage,* controls the maximum dimension because it will not go into any hole unless that hole is over the maximum permissible size. The go and no-go plugs are often designed with two gages on a single handle for convenience in use. The no-go plug is usually much shorter than the go plug; it is subjected to little wear because it seldom slides into any holes. Figure 10-35 shows a *step-type go/no-go gage* that has the go and no-go diameters on the same end of a single plug, the go portion being the outer end. The user knows that the part is good if the *go* gage goes into the hole but the *no-go* gage does not go. Such gages require careful use and should never be forced into (or onto) the part. Obviously these plug gages were specially designated and made for checking a specific hole on a part.

In designing plug and snap ring gages, the key principle is: *it is better to reject a good part than declare a bad part to be within specifications.* All gage design decisions are made with this principle in mind. Gages must have tolerances like any manufactured components. All gages are made with gage and wear tolerances. Gage tolerance allows for the permissible variation in the manufacture of the gage. It is typically 5 to 20% (depending on the industry) of the tolerance on the dimension being gaged. Wear tolerances compensate for the wear of the gage surface as a result of repeated use. Wear tolerance is applied only to the go side of the gage because the no-go side should seldom see contact with a part surface. It is typically 5 to 20% of the dimensional tolerance.

Plug-type gages are also made for gaging shapes other than cylindrical holes. Three common types are *taper plug gages, thread plug gages,* and *spline gages.* Taper plug gages gage both the angle of the taper and its size. Any deviation from the correct angle is indicated by looseness between the plug and the tapered hole. The size is indicated by the depth to which the plug fits into the hole, the correct depth being denoted by a mark on the plug. Thread plug gages come in go and no-go types. The go gage must screw into the threaded holes, and the no-go gage must not enter.

Ring gages are used to check shafts or other external round members. These are also made in go and no-go types, as shown in Figure 10-36. Go ring gages have plain knurled exteriors, whereas no-go ring gages have a circumferential groove in the knurling, so that they can easily be distinguished. *Ring thread gages* are made to be slightly adjustable because it is almost impossible to make them exactly to the desired size. Thus they are adjusted to exact, final size after the final grinding and polishing have been completed.

Snap gages are the most common type of fixed gage for measuring external dimensions. As shown in Figure 10-37, they have a rigid, U-shaped frame on which are two or three gaging surfaces, usually made of hardened steel or tungsten carbide. In the adjustable type shown, one gaging surface is fixed, and the other(s) may be adjusted over a small range and locked at the desired position(s). Because in most cases one wishes to control both the maximum and the minimum dimensions, the *progressive* or *step-type snap gage* is used most frequently. These gages have one fixed anvil and two

FIGURE 10-36   Go and no-go (on right) ring gages for checking a shaft. *(Courtesy of Automation and Measurement Division, Bendix Corporation.)*

FIGURE 10-38 Flush-pin gage being used to check height of step.

adjustable surfaces to form the outer go and the inner no-go openings, thus eliminating the use of separate go and no-go gages.

Snap gages are available in several types and a wide range of sizes. The gaging surfaces may be round or rectangular. They are set to the desired dimensions with the aid of gage blocks.

Many types of special gages are available or can be constructed for special applications. The *flush-pin gage* (Figure 10-38) is an example for gaging the depth of a shoulder. The main section is placed on the higher of the two surfaces, with the movable step pin resting on the lower surface. If the depth between the two surfaces is sufficient but not too great, the top of the pin, but not the lower step, will be slightly above the top surface of the gage body. If the depth is too great, the top of the pin will be below the surface. Similarly, if the depth is not great enough, the lower step on the top of the pin will be above the surface of the gage body. When a finger or fingernail is run across the top of the pin, the pin's position with respect to the surface of the gage body can readily be determined.

Several types of *form gages* are available for use in checking the *profile* of various objects. Two of the most common types are *radius gages* (Figure 10-39) and *screw-thread pitch gages* (Figure 10-40).

DEVIATION-TYPE GAGES

A large amount of gaging, and some measurement, is done through the use of *deviation-type gages*, which determine the amount by which a measured part deviates, plus or minus, from a standard dimension to which the instrument has been set. In most cases, the deviation is indicated directly in units of measurement, but in some cases, the gage shows only whether the deviation is within a permissible range. A good example of a deviation-type gage is a flashlight battery checker, which shows whether the battery is good (green), bad (red), or borderline (yellow) but not how much voltage or current is generated. Such gages use mechanical, electrical, or fluidic amplification techniques so that very small linear deviations can be detected. Most are quite rugged, and they are available in a variety of designs, amplifications, and sizes.

Dial indicators, as shown in Figure 10-41, are a widely used form of deviation-type gage. Movement of the gaging spindle is amplified mechanically through a rack and pinion and a gear train and is indicated by a pointer on a graduated dial. Most dial indicators have a spindle travel equal to about $2\frac{1}{2}$ revolutions of the indicating pointer and are read in either 0.001 or 0.0001 in. (or 0.02 or 0.002 mm).

The dial can be rotated by means of the knurled bezel ring to align the zero point with any position of the pointer. The indicator is often mounted on an adjustable arm to permit its being brought into proper relationship with the work. It is important that the axis of the spindle be aligned exactly with the dimension being gaged if accuracy is to be achieved. Digital dial indicators are also readily available.

FIGURE 10-39 Set of radius gages, showing how they are used. *(Courtesy of MTI Corporation.)*

FIGURE 10-40   Thread pitch gages. *(Courtesy of L.S. Starrett Company.)*

FIGURE 10-41   Digital dial indicator with 1-in. range and 0.0001-in. accuracy. *(Courtesy of CDI.)*

Dial indicators should be checked occasionally to determine if their gage capability has been lost through wear in the gear train. Also, it should be remembered that the pressure of the spindle on the work varies because of spring pressure as the spindle moves into the gage. This spring pressure normally causes no difficulty unless the spindles are used on soft or flexible parts.

Linear variable-differential transformers (LVDT) are used as sensory elements in many electronic gages, usually with a solid-state diode display or in automatic inspection setups. These devices can frequently be combined into multiple units for the simultaneous gaging of several dimensions. Ranges and resolutions down to 0.0005 and 0.00001 in. (0.013 and 0.00025 mm, respectively) are available.

Air gages have special characteristics that make them especially suitable for gaging holes or the internal dimensions of various shapes. A typical gage of this type, shown earlier in Figure 10-12, indicates the clearance between the gaging head and the hole by measuring either the volume of air that escapes or the pressure drop resulting from the airflow. The gage is calibrated directly in 0.0001-in. or 0.02-mm divisions. Air gages have an advantage over mechanical or electronic gages for this purpose in that they detect not only linear size deviations but also out-of-round conditions. Also, they are subject to very little wear because the gaging member is always slightly smaller than the hole and the airflow minimizes rubbing. Special types of air gages can be used for external gaging.

■ 10.10   TESTING

A variety of tests have been developed to evaluate product quality and ensure the absence of any performance-impairing flaws. *Destructive testing* provides one such means of product assessment. Components or assemblies are selected and then subjected to conditions that induce failure. Determining the specific conditions where failure occurs can provide insight into the performance characteristics and quality of the remaining products. Statistical methods are used to determine the probability that the remaining products would exhibit similar behavior. For example, assume that 100 parts are produced and then one is selected (randomly) and tested to failure. Is it safe to assume that the remaining 99 will perform the same way? A satisfactory test of another randomly selected part (or, more typically, the first and last of the 100 parts) would further increase our confidence in the remaining 98. Additional tests would enhance this confidence, but the cost of destroying each of the tested (i.e., destroyed) products must be borne by the remaining quantity. Regardless of the amount of testing, there will still be some degree of uncertainty since none of the remaining products have actually been subjected to any form of property assessment.

Proof testing is another means of ensuring product quality. Here a product is subjected to a load or pressure of some determined magnitude (generally equal to or greater than the designed capacity or the condition expected during operation). If the part remains intact, there is reason to believe that it will subsequently perform in an adequate fashion, provided it is not subjected to abuse or service conditions that exceed its rated level. Proof tests can be conducted under laboratory conditions or at the site of installation or assembly, as with large manufactured assemblies such as pressure vessels.

In some situations, *hardness tests* can be used to provide insight into the quality of a product. With the correct material and proper heat treatment, the resulting hardness values should fall within a well-defined range of values. Abnormal results usually indicate some form of manufacturing error, such as improper material, missed operations, or poorly controlled processes. Hardness tests can be performed quickly, and the surface indentations are often small enough that they can be concealed or easily removed from a product. The results, however, relate only to the surface strength of the product and bear no correlation to defects such as cracks or voids.

Table 10-7 provides a summary of the advantages and limitations of destructive testing and compares that approach with *nondestructive testing*. In nondestructive testing, the product is examined in a manner that retains its usefulness for future service. Tests can be performed on parts during or after manufacture, or even on parts that are already in service. An entire production lot can be inspected, or representative samples can be taken. Different tests can be applied to the same item, either simultaneously or sequentially, and the same test can be repeated on the same specimen for additional verification. Little or no specimen preparation is required, and the equipment is often portable, permitting on-site testing in most locations.

Nondestructive tests can detect internal or surface flaws, measure a product's dimensions, determine a material's structure or chemistry, or evaluate a material's physical or mechanical properties. In general, nondestructive tests incorporate the following aspects: (1) some means of probing a material or product; (2) a means by which a flaw, defect, material property, or specimen feature interacts with or modifies whatever is probing; (3) a sensor to detect the response; (4) a device to indicate or record the response; and (5) a way to interpret and evaluate quality.

TABLE 10-7   Advantages and Limitations of Destructive and Nondestructive Testing

Destructive Testing

Advantages
1. Provides a direct and reliable measurement of how a material or component will respond to service conditions.
2. Provides quantitative results, useful for design.
3. Does not require interpretation of results by skilled operators.
4. Usually finds agreement as to meaning and significance of test results.

Disadvantages
1. Applied only to a sample; must show that the sample is representative of the group.
2. Tested parts are destroyed during testing.
3. Usually cannot repeat a test on the same item or use the same specimen for multiple tests.
4. May be restricted for costly or few-in-number parts.
5. Hard to predict cumulative effect of service usage.
6. Difficult to apply to parts in use; if done, testing terminates their useful life.
7. Extensive machining or preparation of test specimens is often required.
8. Capital equipment and labor costs are often high.

Nondestructive Testing

Advantages
1. Can be performed directly on production items without regard to cost or quantity available.
2. Can be performed on 100% of production lot (when high variability is observed) or a representative sample (if sufficient similarity is noted).
3. Different tests can be applied to the same item, and a test can be repeated on the same specimen.
4. Can be performed on parts that are in service; the cumulative effects of service life can be monitored on a single part.
5. Little or no specimen preparation is required.
6. The test equipment is often portable.
7. Labor costs are usually low.

Disadvantages
1. Results often require interpretation by skilled operators.
2. Different observers may interpret the test results differently.
3. Properties are measured indirectly, and results are often qualitative or comparative.
4. Some test equipment requires a large capital investment.

Regardless of the specific test, nondestructive testing can be a vital element in good manufacturing practice. Its potential value has been widely recognized as productivity and production rates increase, consumers demand higher-quality products, and product liability continues to be a concern. Rather than being an added manufacturing cost, nondestructive testing can actually expand profit by ensuring product reliability and customer satisfaction. In addition to its role in quality control, nondestructive testing can also be used as an assessment aid in product design. Periodic testing can provide a means of controlling a manufacturing process and reducing overall manufacturing costs by preventing the continued manufacture of out-of-specification, defective, or poor-quality parts.

■ 10.11 VISUAL INSPECTION

Probably the simplest and most widely used nondestructive testing method is *visual inspection*, summarized in Table 10-8. The human eye is a very discerning instrument and, with training, the brain can readily interpret the signals. Optical aids such as mirrors, magnifying glasses, and microscopes can expand the capabilities of this system. Video cameras and computer systems, such as digital image analyzers, can be used to automate the inspection and perform quantitative geometrical evaluations. Bore scopes and similar tools can provide accessibility to otherwise inaccessible locations. Only the surfaces of a product can be examined, but that is often sufficient to reveal corrosion, contamination, surface-finish flaws, and a wide variety of surface discontinuities.

TABLE 10-8 Visual Inspection	
Principle	Illuminate the test specimen and observe the surface. Can reveal a wide spectrum of surface flaws and geometric discontinuities. Use of optical aids or assists (such as magnifying glass, microscopes, illuminators, and mirrors) is permitted. While most inspection is by human eye, video cameras and computer-vision systems can be employed.
Advantages	Simple, easy to use, relatively inexpensive.
Limitations	Depend on skill and knowledge of inspector. Limited to detection of surface flaws.
Material limitations	None.
Geometrical limitations	Any size or shape providing viewing accessibility of surfaces to be inspected.
Permanent record	Photographs or videotapes are possible. Inspectors' reports also provide valuable records.
Remarks	Should always be the initial and primary means of inspection and is the responsibility of everyone associated with parts manufacture.

■ 10.12 LIQUID PENETRANT INSPECTION

Liquid penetrant testing, also called dye penetrant inspection, is an effective method of detecting surface defects in metals and other nonporous materials; it is illustrated schematically in Figure 10-42. The piece to be tested is first subjected to a thorough cleaning and is dried prior to the test. Then a *penetrant*, a liquid material capable of wetting the entire surface and being drawn into fine openings, is applied to the surface of the workpiece by dipping, spraying, or brushing. Sufficient time is given for capillary action to draw the penetrant into any surface discontinuities, and the excess penetrant liquid is then removed by wiping, water wash, or solvent. The surface is then coated with a thin film of *developer*, an absorbent material capable of drawing traces of penetrant from the defects back onto the surface. Brightly colored dyes or fluorescent materials that glow under ultraviolet light are generally added to the penetrant to make these

(a)   (b)   (c)   (d)

FIGURE 10-42 Liquid penetrant testing: (a) initial surface with open crack; (b) penetrant is applied and is pulled into the crack by capillary action; (c) excess penetrant is removed; (d) developer is applied, some penetrant is extracted, and the product is inspected.

TABLE 10-9	Liquid Penetrant Inspection
Principle	A liquid penetrant containing fluorescent material or dye is drawn into surface flaws by capillary action and subsequently revealed by developer material in conjunction with visual inspection.
Advantages	Simple, inexpensive, versatile, portable, easily interpreted, and applicable to complex shapes.
Limitations	Can only detect flaws that are open to the surface; surfaces must be cleaned before and after inspection; deformed surfaces and surface coatings may prevent detection; and the penetrant may be wiped or washed out of large defects. Cannot be used on hot products.
Material limitations	Applicable to all materials with a nonporous surface.
Geometrical limitations	Any size or shape permitting accessibility of surfaces to be inspected.
Permanent record	Photographs, videotapes, and inspectors' reports provide the most common records.

traces more visible, and the developer is often selected to provide a contrasting background. Radioactive tracers can also be added and used in conjunction with photographic paper to produce a permanent image of the defects. Cracks, laps, seams, lack of bonding, pinholes, gouges, and tool marks can all be detected. After inspection, the developer and residual penetrant are removed by a second cleaning operation.

To be successful, the inspection for surface defects must be correlated with the manufacturing operations. If previous processes such as shot peening, honing, burnishing, machining, or various forms of cold working produced plastic deformation of the surface material, a chemical etching may be required to remove material that might be covering critical flaws. An alternative procedure is to perform a penetrant test before any surface-finishing operations, when significant defects will still be open and available for detection. Penetrant inspection systems can range from aerosol spray cans of cleaner, penetrant, and developer (for portable applications), to automated, mass-production equipment using sophisticated computer vision systems. Table 10-9 is a summary of the process and its advantages and limitations.

■ 10.13 MAGNETIC PARTICLE INSPECTION

Magnetic particle inspection, summarized in Table 10-10, is based on the principle that ferromagnetic materials (such as the alloys of iron, nickel, and cobalt), when magnetized, will have distorted magnetic fields in the vicinity of material defects. As shown in Figure 10-43, surface and subsurface flaws, such as cracks and inclusions, will produce magnetic anomalies that can be mapped with the aid of magnetic particles on the specimen surface. As with the previous method, the specimen must be cleaned prior to inspection. A suitable magnetic field is then established in the part. As shown in Figure 10-44, orientation can be quite important. For a flaw to be detected, it must produce a significant disturbance of the magnetic field at or near the surface. If a bar of

TABLE 10-10	Magnetic Particle Inspection
Principle	When magnetized, ferromagnetic materials will have a distorted magnetic field in the vicinity of flaws and defects. Magnetic particles will be strongly attracted to regions where the magnetic flux breaks the surface.
Advantages	Relatively simple, fast, easy-to-interpret; portable units exist; can reveal both surface and subsurface flaws and inclusions (as much as 6-mm deep) and small, tight cracks.
Limitations	Parts must be relatively clean; alignment of the flaw and the field affects the sensitivity so that multiple inspections with different magnetizations may be required; can only detect defects at or near surfaces; must demagnetize part after test; high current source is required; some surface processes can mask defects; postcleaning may be required.
Material limitations	Must be ferromagnetic; nonferrous metals such as aluminum, magnesium, copper, lead, tin, and titanium and the ferrous (but not ferromagnetic) austenitic stainless steels cannot be inspected.
Geometrical limitations	Size and shape are almost unlimited; most restrictions relate to the ability to induce uniform magnetic fields within the piece; hard to use on rough surfaces.
Permanent record	Photographs, videotapes, and inspectors' reports are most common. In addition, the defect pattern can be preserved on the specimen by an application of transparent lacquer or transferred to a piece of transparent tape that has been applied to the specimen and peeled off.

FIGURE 10-43 (a) Magnetic field showing disruption by a surface crack; (b) magnetic particles are applied and are preferentially attracted to field leakage; (c) subsurface defects can also produce surface-detectable disruptions if they are sufficiently close to the surface.

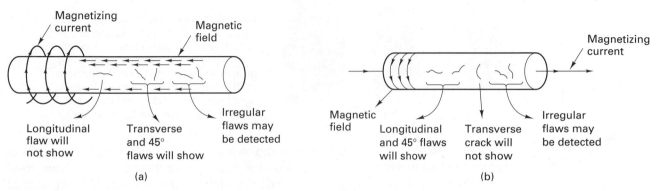

FIGURE 10-44 (a) A bar placed within a magnetizing coil will have an axial magnetic field. Defects parallel to this field may go unnoticed, while those that disrupt the field and are sufficiently close to a surface are likely to be detected. (b) When magnetized by a current passing through it, the bar has a circumferential magnetic field and the geometries of detectable flaws are reversed.

steel is placed within an energized coil, a magnetic field will be produced whose lines of flux travel along the axis of the bar. Any defect perpendicular to this axis will significantly alter the field. If the perturbation is sufficiently large and close enough to the surface, the flaw can be detected. However, if the flaw is in the form of a crack aligned with the specimen axis, there will be little perturbation of the lines of flux and the flaw is likely to go undetected. If the cylindrical specimen is then magnetized by passing a current through it, a circumferential magnetic field will be produced. Any axial defect now becomes a significant perturbation, and a defect perpendicular to the axis will likely go unnoticed. To fully inspect a product, therefore, a series of inspections may be required using various forms of magnetization. Passing a current between various points of contact is a popular means of inducing the desired fields. Electromagnetic coils of various shapes and sizes are also used. Alternating-current methods are most sensitive to surface flaws, while direct-current inspections are better for detecting subsurface defects, such as nonmetallic inclusions.

If the cylindrical specimen is then magnetized by passing a current through it, a circumferential magnetic field will be produced. Any axial defect now becomes a significant perturbation, and a defect perpendicular to the axis will likely go unnoticed. To fully inspect a product, therefore, a series of inspections may be required using various forms of magnetization. Passing a current between various points of contact is a popular means of inducing the desired fields. Electromagnetic coils of various shapes and sizes are also used. Alternating-current methods are most sensitive to surface flaws, while direct-current inspections are better for detecting subsurface defects, such as nonmetallic inclusions.

Once the specimen has been subjected to a magnetic field, magnetic particles are applied to the surface in the form of either a dry powder or a suspension in a liquid carrier. These particles are attracted to places where the lines of magnetic flux break the surface, revealing anomalies that can then be interpreted. To better reveal the orientation of the lines of flux, the particles are often made in an elongated form. They can also be

FIGURE 10-45 Front-axle king pin for a truck. (a) As manufactured and apparently sound; (b) inspected under conventional magnetic particle inspection to reveal numerous grinding-induced cracks; (c) fluorescent particles and ultraviolet light make the cracks even more visible. *(Courtesy of Magnaflux Corporation.)*

(a)   (b)   (c)

treated with a fluorescent material to enhance observation under ultraviolet light or coated with a lubricant to prevent oxidation and enhance their mobility. Figure 10-45 shows a component of a truck front-axle assembly: as manufactured, under straight magnetic particle inspection, and under ultraviolet light with fluorescent particles.

■ 10.14 ULTRASONIC INSPECTION

Sound has long been used to provide an indication of product quality. A cracked bell will not ring true, but a fine crystal goblet will have a clear ring when lightly tapped. Striking an object and listening to the characteristic ring is an ancient art but is limited to the detection of large defects because the wavelength of audible sound is rather large compared to the size of most defects. By reducing the wavelength of the signal to the ultrasonic range, typically between 100,000 and 25 million hertz, ultrasonic inspection can be used to detect rather small defects and flaws.

As shown in Table 10-11, *ultrasonic inspection* involves sending high-frequency waves through a material and observing the response. Within the specimen, sound waves can be affected by voids, impurities, changes in density, delaminations, interfaces with materials having a different speed of sound, and other imperfections. At any interface, part of the ultrasonic wave will be reflected and part will be transmitted. If the incident beam

TABLE 10-11	Ultrasonic Inspection
Principle	High-frequency sound waves are propagated through a test specimen, and the transmitted or reflected signal is monitored and interpreted.
Advantage	Can reveal internal defects; high sensitivity to most cracks and flaws; high-speed test with immediate results; can be automated and recorded; portable; high penetration in most important materials (up to 60 ft in steel); indicates flaw size and location; access to only one side is required; can also be used to measure thickness, Poisson's ratio, or elastic modulus; presents no radiation or safety hazard.
Limitations	Difficult to use with complex shapes; external surfaces and defect orientation can affect the test (may need dual transducer or multiple inspections); a couplant is required; the area of coverage is small (inspection of large areas requires scanning); trained, experienced, and motivated technicians may be required.
Material limitations	Few can be used on metals, plastics, ceramics, glass, rubber, graphite, and concrete, as well as joints and interfaces between materials.
Geometric limitations	Small, thin, or complex-shaped parts or parts with rough surfaces and nonhomogeneous structure pose the greatest difficulty.
Permanent record	Ultrasonic signals can be recorded for subsequent playback and analysis. Strip charts can also be used.

is at an angle to an interface where materials change, the transmitted portion of the beam will be bent to a new angle by the phenomenon of refraction. By receiving and interpreting either transmitted or reflected signals, ultrasonic inspection can be used to detect flaws within the material, measure thickness from only one side, or characterize metallurgical structure.

An ultrasonic inspection system begins with a pulsed oscillator and *transducer*, a device that transforms electrical energy into mechanical vibrations. The pulsed oscillator generates a burst of alternating voltage, with a characteristic principal frequency, duration, profile, and repetition rate. This burst is then applied to a sending transducer, which uses a piezoelectric crystal to convert the electrical oscillations into mechanical vibrations. Because air is a poor transmitter of ultrasonic waves, an acoustic *coupling medium*—generally a liquid such as oil or water—is required to link the transducer to the piece to be inspected and transmit the vibrations into the part. The pulsed vibrations then propagate through the part with a velocity that depends on the density and elasticity of the test material. A receiving transducer is then used to convert the transmitted or reflected vibrations back into electrical signals. The receiving transducer is often identical to the sending unit, and the same transducer can actually perform both functions. A receiving unit then amplifies, filters, and processes the signal for display, possible recording, and final interpretation. An electronic clock is generally integrated into the system to time the responses and provide reference signals for comparison purposes.

Depending on the test objectives and part geometry, several different inspection methods can be employed:

1. In the *pulse-echo technique*, an ultrasonic pulse is introduced into the piece to be inspected, and the echoes from opposing surfaces and any intervening flaws are detected by the receiver. The time interval between the initial emitted pulse and the various echoes can be displayed on the horizontal axis of a display screen. Defects are identified by the position and amplitude of the various echoes. Figure 10-46 shows a schematic of a single-transducer pulse-echo inspection and the companion signal as it would appear on a display. Figure 10-47a depicts a dual-transducer pulse-echo examination. Both cases require access to only one side of the specimen.

FIGURE 10-46   (a) Ultrasonic inspection of a flat plate with a single transducer; (b) plot of sound intensity or transducer voltage versus time showing the initial pulse and echoes from the bottom surface and intervening defect.

FIGURE 10-47   (a) Dual-transducer ultrasonic inspection in the pulse-echo mode; (b) dual transducers in through-transmission configuration.

2. The *through-transmission technique* requires separate sending and receiving transducers. As shown in Figure 10-47b, a pulse is emitted by the sending transducer and detected by a receiver on the opposite surface. Flaws in the material decrease the amplitude of the transmitted signal because of back-reflection and scattering.

3. *Resonance testing* can be used to determine the thickness of a plate or sheet from one side of the material. Input pulses of varying frequency are fed into the material. When resonance is detected by an increase in energy at the transducer, the thickness can be calculated from the speed of sound in the material and the time of traverse. Ultrasonic thickness gages can be calibrated to provide direct digital readout of the thickness of a material.

Reference standards—specimens of known thickness or containing various types and sizes of machined "flaws"—are often used to ensure consistent results and aid in interpreting any indications of internal discontinuities.

■ 10.15 RADIOGRAPHY

Radiographic inspection, summarized in Table 10-12, employs the same principles and techniques as those of medical X-rays. A shadow pattern is created when certain types of radiation (X-rays, gamma rays, or neutron beams) penetrate an object and are differentially absorbed due to variations in thickness, density, or chemistry, or the presence of defects in the specimen. The transmitted radiation is registered on a photographic film that provides a permanent record and a means of analyzing the component. Fluorescent screens can provide direct conversion of radiation into visible light and enable fast and inexpensive viewing without the need for film processing. The fluorescent image, however, usually does not offer the sensitivity of the photographic methods.

Various types of radiation can be used for inspection. X-rays are an extremely short wavelength form of electromagnetic radiation that are capable of penetrating many materials that reflect or absorb visible light. They are generated by a high-voltage electrical apparatus—the higher the voltage, the shorter the X-ray wavelength and the greater the energy and penetrating power of the beam. Gamma rays, another useful form of electromagnetic radiation, are emitted during the disintegration of radioactive nuclei. Various radioactive isotopes can be selected as the radiation source. Neutron beams for radiography can be obtained from nuclear reactors, nuclear accelerators, or radioisotopes. For most applications it is necessary to moderate the energy and collimate the beam before use.

The absorption of X-rays and gamma rays depends on the thickness, density, and atomic structure of the material being inspected. The higher the atomic number, the greater the attenuation of the beam. Figure 10-48 shows a radiograph of the historic Liberty Bell. The famous crack is clearly visible, along with the internal spider (installed to support the clapper in 1915) and the steel beam and bolts installed in the wooden yoke in 1929. Other radiographs disclosed previously unknown shrinkage separations and additional cracks in the bell, as well as a crack in the bell's clapper.

TABLE 10-12	Radiography
Principle	Some form of radiation (X-ray, gamma ray, or neutron beam) is passed through the sample and is differentially absorbed depending on the thickness, type of material, and the presence of internal flaws or defects.
Advantages	Probes the internal regions of a material; provides a permanent record of the inspection; can be used to determine the thickness of a material; very sensitive to density changes.
Limitations	Most costly of the NDT methods (involves expensive equipment); radiation precautions are necessary (potentially dangerous to human health); the defect must be at least 2% of the total section thickness to be detected (thin cracks can be missed if oriented perpendicular to the beam); film processing requires time, facilities, and care; the image is a two-dimensional projection of a three-dimensional object, so the location of an internal defect requires a second inspection at a different angle; complex shapes can present problems; a high degree of operator training is required.
Material limitations	Applicable to most engineering materials.
Geometric limitations	Complex shapes can present problems in setting exposure conditions and obtaining proper orientation of source, specimen, and film. Two-side accessibility is required.
Permanent record	A photographic image is part of the standard test procedure.

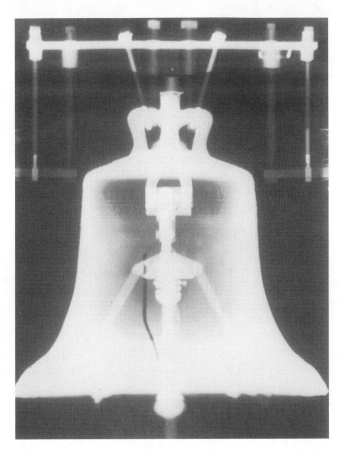

FIGURE 10-48   Radiograph of the Liberty Bell. The photo reveals the famous crack, as well as the iron spider installed in 1915 to support the clapper and the steel beam and supports, which were set into the yoke in 1929. *(Courtesy of Eastman Kodak Company.)*

In contrast to X-ray absorption, neutron absorption varies widely from atom to atom, with no pattern in terms of atomic number. Unusual contrasts can be obtained that would be impossible with other inspection methods. For example, hydrogen has a high neutron absorption. The presence of water in a product can be easily detected by neutron radiography. X-rays, on the other hand, are readily transmitted through water, and its presence could be missed.

When a radiation beam is passed through an object, part of the radiation is scattered in all directions. This scatter produces an overall "fogging" of the radiograph, reducing the contrast and sharpness of the image. The thicker the material, the more troublesome the scattered radiation becomes. Photographic considerations relating to the exposure time and development also affect the quality of the radiographic image. Image-enhancing computer software can help reveal the subtle but important variations in photographic density.

A standard test piece, or *penetrameter*, is often included in a radiographic exposure. Penetrameters are made of the same or similar material as the specimen and contain features with known dimensions. The image of the penetrameter is compared to the image of the product being inspected. Regions of similar intensity are considered to be of similar thickness.

Radiography is expensive, however. Many users, therefore, recommend extensive use only during the development of a new product or process, followed by spotchecks and statistical methods during subsequent production.

■ 10.16   EDDY-CURRENT TESTING

When an electrically conductive material is exposed to an alternating magnetic field such as that generated by a coil of wire carrying an alternating current, small electric currents are induced on or near the surface of the material (Figure 10-49). These induced *eddy currents*, in turn, generate their own opposing magnetic field, which then reduces the strength of the field from the coil. This change in magnetic field causes a change in the *impedance* of the coil, which in turn changes the magnitude

FIGURE 10-49 Relation of the magnetizing coil, magnetizing current, and induced eddy currents. The magnetizing current is actually an alternating current, producing a magnetic field that forms, collapses, and re-forms in the opposite direction. This dynamic magnetic field induces the eddy currents, and the changes in the eddy currents produce a secondary magnetic field that interacts with the sensor coil or probe.

FIGURE 10-50 Eddy currents are constrained to travel within the conductive material, but the magnitude and path of the currents will be affected by defects and changes in material properties. By focusing on the magnitude of the eddy currents, features such as differences in heat treatment can be detected.

of the current flowing through it. By monitoring the impedance of the exciting coil, or a separate indicating coil, eddy-current testing can be used to detect any condition that would affect the current-carrying ability (or conductivity) of the test specimen. Figure 10-50 shows how the eddy-current paths would be forced to alter around a crack, thereby changing the characteristics of the induced magnetic field in that vicinity.

Eddy-current testing, summarized in Table 10-13, can be used to detect surface and near-surface flaws, such as cracks, voids, inclusions, and seams. Stress concentrations, differences in metal chemistry, or variations in heat treatment (i.e., microstructure and hardness) will all affect the magnetic permeability and conductivity of a metal and therefore alter the eddy-current characteristics. Material mix-ups and processing errors can therefore be detected. Specimens can be sorted by hardness, case depth, residual stresses, or any other structure-related property.

TABLE 10-13	Eddy-Current Testing
Principle	When an electrically conductive material is brought near an alternating-current coil that produces an alternating magnetic field, surface currents (eddy currents) are generated in the material. These surface currents generate their own magnetic field, which interacts with the original, modifying the impedance of the originating coil. Various material properties and/or defects can affect the magnitude and direction of the induced eddy currents and can be detected by the electronics.
Advantage	Can detect both surface and near-surface irregularities; applicable to both ferrous and nonferrous metals; versatile—can detect flaws; variations in alloy or heat treatment; variations in plating or coating thickness, wall thickness, and crack depth; intimate contact with the specimen is not required; can be automated; electrical circuitry can be adjusted to select sensitivity and function; pass–fail inspection is easily conducted; high speed; low cost; no final cleanup required.
Limitations	Response is sensitive to a number of variables, so interpretation may be difficult; sensitivity varies with depth, and depth of inspection depends on the test frequency; reference standards are needed for comparison; trained operators are generally required.
Material limitations	Only applicable to conductive materials, such as metals; some difficulties may be encountered with ferromagnetic materials.
Geometric limitations	Depth of penetration is limited; must have accessibility of coil or probe; constant separation distance between coils and specimen is required for good results.
Permanent record	Electronic signals can be recorded using devices such as strip-chart recorders.

Thickness (or variation in thickness) of platings, coatings, or even corrosion can be detected and measured.

Eddy-current test equipment can range from simple, portable units with hand-held probes to fully automated systems with computer control and analysis. Each system includes a source of changing magnetic field capable of inducing eddy currents in the part being tested, a means of sensing the field, and a means of measuring and interpreting the resulting impedance changes. When comparing alternative techniques, eddy current is usually not as sensitive as penetrant testing in detecting small, open flaws, but it requires none of the cleanup operations and is noticeably faster. In a similar manner, it is not as sensitive as magnetic particle inspection to small subsurface flaws, but it can be applied to all metals (ferromagnetic and nonferromagnetic alike). In addition, eddy-current testing offers capabilities that cannot be duplicated by the other methods, such as the ability to differentiate between various chemistries and heat treatments.

■ 10.17 ACOUSTIC EMISSION MONITORING

Materials experiencing the dynamic events of deformation or fracture emit stress waves in frequencies as high as 1 MHz. While these sounds are inaudible to the human ear, they are detectable through the use of sophisticated electronics. Transducers, amplifiers, filters, counters, and computers can be used to isolate and analyze the sonic emissions of a cracking or deforming material. Much like the warning sound of ice cracking underneath boots or skates, the acoustic emissions of materials can be used to provide a warning of impending danger. They can detect deformations as small as 10^{-12} in./in. (which occur in short intervals of time), initiation or propagation of cracks (including stress–corrosion cracking), delamination of layered materials, and fiber failure in composites. By using multiple sensors, it is possible to accurately pinpoint the source of these sounds by a triangulation method similar to that used to locate seismic sources (earthquakes) in the earth.

Acoustic emission monitoring, summarized in Table 10-14, involves listening for indications of failure. Temporary monitoring can be used to detect the formation of cracks in materials during production operations, such as welding and subsequent cooling of the weld region. Monitoring can also be employed to ensure the absence of plastic deformation during preservice proof testing. Continuous surveillance may be used when the product or component is particularly critical, as with bridges and nuclear reactor pressure vessels. The sensing electronics can be coupled to an alarm and safety system to protect and maintain the integrity of the structure.

In contrast to the previous inspection methods, acoustic emission cannot detect an existing defect in a static product. Instead, it is a monitoring technique designed to detect a dynamic change in the material, such as the formation or growth of a crack or defect, or the onset of plastic deformation.

TABLE 10-14	Acoustic Emission Monitoring
Principle	Almost all materials will emit high-frequency sound (acoustic emissions) when stressed, deformed, or undergoing structural changes, such as the formation or growth of a crack or defect. These emissions can now be detected and provide an indication of dynamic change within the material.
Advantages	The entire structure can be monitored with near-instantaneous detection and response; continuous surveillance is possible; defects inaccessible to other methods can be detected; inspection can be in harsh environments; and the location of the emission source can be determined.
Limitations	Only growing or "active" flaws can be detected (the mere presence of defects is not detectable); background signals may cause difficulty; there is no indication of the size or shape of the flaw; expensive equipment is required; and experience is required to interpret the signals.
Material limitations	Virtually unlimited, provided that they are capable of transmitting sound.
Geometric limitations	Requires continuous sound-transmitting path between the source and the detector. Size and shape of the component affect the strength of the emission signals that reach the detector.

■ 10.18 OTHER METHODS OF NONDESTRUCTIVE TESTING AND INSPECTION

LEAK TESTING

Leak testing is a form of nondestructive testing designed to determine the existence or absence of leak sites and the rate of material loss through the leaks. Various testing methods have been developed, ranging from the rather crude bubble-emission test (pressurize, immerse, and look for bubbles), through simple pressure drop tests with either air or liquid as the pressurized media, to advanced techniques involving tracers, detectors, and sophisticated apparatus. Each has its characteristic advantages, limitations, and sensitivity. Selection should be on the basis of cost, sensitivity, reliability, and compatibility with the specific product to be tested.

THERMAL METHODS

Temperature-sensing devices (including thermometers, thermocouples, pyrometers, temperature-sensitive paints and coatings, liquid crystals, infrared scanners, infrared film, and others) can also be used to evaluate the soundness of engineering materials and components. Parts can be heated and then inspected during cool-down to reveal abnormal temperature distributions that are the result of faults or flaws. The identification of "hot spots" on an operating component is often an indication of a flaw or defect and may provide advanced warning of impending failure. For example, faulty electrical components tend to be hotter than defect-free devices. Composite materials (difficult to inspect by many standard techniques) can be subjected to brief pulses of intense heat and then inspected to reveal the temperature pattern produced by the subsequent thermal conductivity. Thermal anomalies tend to appear in areas where the bonding between the components is poor or incomplete. In another technique, ultrasonic waves are used to produce heat at internal defects, which are then detected by infrared examination.

STRAIN SENSING

Although used primarily during product development, strain-sensing techniques can also be used to provide valuable insight into the stresses and stress distribution within a part. Brittle coatings, photoelastic coatings, or electrical resistance strain gages can be applied to the external surfaces of a part, which are then subjected to an applied stress. The extent and nature of cracking, the photoelastic pattern produced, or the electrical resistance changes then provide insight into the strain at various locations. X-ray diffraction methods and extensometers have also been used.

ADVANCED OPTICAL METHODS

Although visual inspection is often the simplest and least expensive of the nondestructive inspection methods, there are also several advanced optical methods. Monochromatic laser light can be used to detect differences in the backscattered pattern from a part and a master. The presence or absence of geometrical features such as holes or gear teeth is readily detected. Holograms can provide three-dimensional images of an object, and holographic interferometry can detect minute changes in the shape of an object under stress.

RESISTIVITY METHODS

The *electrical resistivity* of a conductive material is a function of its chemistry, processing history, and structural soundness. Measurement of resistivity can therefore be used for alloy identification, flaw detection, or the ensurance of proper processing. Tests can be developed to evaluate the effects of heat treatment, the amount of cold work, the integrity of welds, or the depth of case hardening. The development of sensitive microohmmeters has greatly expanded the possibilities in this area.

COMPUTED TOMOGRAPHY

While X-ray radiography provides a single image of the X-ray intensity being transmitted through an object, X-ray *computed tomography* (CT) is an inspection technique that provides a cross-sectional view of the interior of an object along a plane parallel to the X-ray beam. This is the same technology that has revolutionized medical diagnostic imaging (CAT scans), with the process parameters (such as the energy of the X-ray source) being adapted to permit the nondestructive probing of industrial products. Basic systems include an X-ray source, an array of detectors, a mechanical system to move and rotate the test object, and a dedicated computer system. The intensity of the received signal is recorded at each of the numerous detectors with the part in a variety of orientations. Complex numerical algorithms are then used to construct an image of the interior of the component. Internal boundaries and surfaces can be determined clearly, enabling inspection and dimensional analysis of a product's interior. The presence of cracks, voids, or inclusions can also be detected, and their precise location can be determined.

CT inspections are slow and costly, so they are currently used only when the component is critical and the more standard inspection methods prove to be inadequate due to features such as shape complexity, thick walls, or poor resolution of detail. The video images of the CT technique also permit easy visualization and interpretation.

Acoustic holography is another computer reconstruction technique, this time based on ultrasound reflections from within the part.

CHEMICAL ANALYSIS AND SURFACE TOPOGRAPHY

While nondestructive inspection is usually associated with the detection of flaws and defects, various nondestructive techniques can also be employed to determine the chemical and elemental analysis of surface and near-surface material. These techniques include Auger electron spectroscopy (AES), energy-dispersive X-ray analysis (EDX), electron spectroscopy for chemical analysis (ESCA), and various forms of secondary-ion mass spectroscopy (SIMS). Because of its large depth of focus, the scanning electron microscope has become an extremely useful tool for observing the surfaces of materials. More recently, the atomic-force microscope and scanning tunneling microscope have extended this capability and can now provide information about surface topography with resolution to the atomic scale.

■ 10.19 DORMANT VERSUS CRITICAL FLAWS

There was a time when the detection of a flaw was considered to be sufficient cause for rejecting a material or component, and material specifications often contained the term *flaw-free*. Such a criterion, however, is no longer practical, because the sensitivity of detection methods has increased dramatically. If materials were rejected upon detection of a flaw, we would find ourselves rejecting nearly all commercial engineering materials. If a defect is sufficiently small, it is possible for it to remain dormant throughout the useful lifetime of a product, never changing in size or shape. Such a defect is clearly allowable. Larger defects, or defects of a more undesirable geometry, may grow or propagate under the same (cyclic) conditions of loading, often causing sudden or catastrophic failure. These flaws would be clearly unacceptable. The objective (or challenge), therefore, is to identify the conditions below which a flaw remains *dormant* and above which it becomes *critical* and a cause for rejection. This issue is addressed in the section on "Fracture Toughness and the Fracture Mechanics Approach" in Chapter 2.

■ Key Words

accuracy	coupling medium	impedance	magnification
acoustic holography	critical flaw	interference bands	mass
acoustic emission	destructive testing	interference fit	metrology
allowance	dormant flaw	laser interferometer	micrometer caliper
ampere	drift	lay	nondestructive testing (also
attributes	eddy-current testing	leak testing	nondestructive inspection)
candela	electrical resistivity	length	optical comparator
clearance fit	flaw-free	linearity	penetrameter
computed tomography	gage blocks	liquid penetrant testing	penetrant
coordinate measuring	geometric tolerances	machinist's rule	plug gage
machine	hardness testing	magnetic particle inspection	precision

proof test
pulse-echo method
radiographic inspection
resolution
resonance testing
ring gage
rule of 10

sine bar
snap gage
stability
super micrometer
temperature
through transmission
   technique

time
tolerance
tomography
toolmaker's flat
toolmaker's microscope
transducer
ultrasonic inspection

variables
vernier caliper
vision system
visual inspection

■ Review Questions

1. What are some of the advantages to the consumer of standardization and of interchangeable parts?
2. *DFM* stands for "design for manufacturing." Why is it important for designers to interface with manufacturing as early as possible with the design phase?
3. Explain the difference between attributes and variables inspection.
4. Why have so many variable-type devices in autos been replaced with attribute-type devices?
5. What are the four basic measures upon which all others depend?
6. What is a pascal, and how is it made up of the basic measures?
7. What are the different grades of gage blocks, and why do they come in sets?
8. When gage blocks are "wrung together," what keeps them together?
9. What is the difference between tolerance and allowance?
10. Here is a table that provides a description of fits from clearance to interference. Try to think of an example of each of these fits.

	ISO Symbol			Example
	Hole Basis	Shaft Basis		
Clearance Fits	H11/c11	C11/h11	*Loose-running fit:* for wide commercial tolerances or allowances on external members	
	H9/d9	D9/h9	*Free-running fit:* not for use where accuracy is essential, but good for large temperature variations	
	H8/f7	F8/h7	*Close-running fit:* for running on accurate machines and for accurate location at moderate speeds and journal pressures	
	H7/g6	G7/h6	*Sliding fit:* not intended to run freely, but to move and turn freely and locate accurately	
	H7/h6	H7/h6	*Locational-clearance fit:* provides snug fit for locating stationary parts, but can be freely assembled and disassembled	
Transition Fits	H7/k6	K7/h6	*Locational-transition fit:* for accurate location; a compromise between clearance and interference	
	H7/n6	N7/h6	*Locational-transition fit:* for more accurate location where greater interference is permissible	
	H7/p6	P7/h6	*Locational-interference fit:* for parts requiring rigidity and alignment with prime accuracy of location, but without special bore pressure requirements	
Interference Fits	H7/s6	S7/h6	*Medium-drive fit:* for ordinary steel parts or shrink fits on light sections; the tightest fit usable with cast iron	
	H7/u6	U7/h6	*Force fit:* for highly stressed parts or for shrink fits where the heavy pressing forces required are impractical	

11. What type of fit would describe the following situations.
   a. The cap of a ball-point pen
   b. The lead in a mechanical lead pencil, at the tip
   c. The bullet in a barrel of a gun
12. What does the word *shrink* imply in a shrink fit?
13. Why might you use a shrink fit to join the wheels of trains to the axle rather than welding them?
14. Explain the difference between accuracy and precision.
15. When measuring time, is it more important to be accurate or precise? Why?
16. Into which of the five basic kinds of inspection does interferometry fall?
17. What factors should be considered in selecting measurement equipment?
18. Explain what is meant by the statement that usable magnification is limited by the resolution of the device.
19. What is parallax? (Why do linesmen in tennis sit looking down the line?)
20. What is the rule of 10?
21. How does the vernier caliper work to make measurements?
22. What are the two most likely sources of error in using a micrometer caliper?
23. What is the major disadvantage of a micrometer caliper as compared with a vernier caliper?
24. What is the main advantage of a micrometer over the vernier caliper?
25. What would be the major difficulty in obtaining an accurate measurement with a micrometer depth gage if it were not equipped with a ratchet or friction device for turning the thimble?
26. Why is the toolmaker's microscope particularly useful for making measurements on delicate parts?

27. In what two ways can linear measurements be made using an optical projector?
28. What type of instrument would you select for checking the accuracy of the linear movement of a machine tool table through a distance of 50 inches?
29. What are the chief disadvantages of using a vision system for measurement compared to laser scanning?
30. What is a CMM (coordinate measuring machine)?
31. What is the principle of a sine bar?
32. How can the no-go member of a plug gage be easily distinguished from the go member?
33. What is the primary precaution that should be observed in using a dial gage?
34. What tolerances are added to gages when they are being designed?
35. Explain how a go/no-go ring gage works for check a shaft.
36. Why are air gages particularly well suited for gaging the diameter of a hole?
37. Explain the principle of measurement by light-wave interference.
38. How does a toolmaker's flat differ from an optical flat?
39. Why must quality decisions derived from destructive testing be made on a statistical basis?
40. What is a proof test, and what assurance does it provide?
41. What quality-related features can a hardness test reasonably ensure?
42. What exactly is nondestructive testing, and what are some attractive features of the approach?
43. What are some possible objectives of nondestructive testing?
44. What are some factors that should be considered when selecting a nondestructive testing method?
45. How might the costs of nondestructive testing actually be considered as an asset rather than a liability?
46. Why should visual inspection be considered as the initial and primary means of inspection?
47. What is the primary limitation of a visual inspection?
48. What types of defects can be detected in a liquid penetrant test?

49. What is the primary materials-related limitation of magnetic particle inspection?
50. Describe how the orientation of a flaw with respect to a magnetic field can affect its detectability during magnetic particle inspection.
51. What is the major limitation of sonic testing, where one listens to the characteristic ring of a product in an attempt to detect defects?
52. What is the role of a coupling medium in ultrasonic inspection?
53. What are three types of ultrasonic inspection methods?
54. What types of radiation can be used in radiographic inspection of manufactured products?
55. What are penetrameters, and how are they used in radiographic inspection?
56. While radiographs offer a graphic image that looks like the part being examined, the technique has some significant limitations. What are some of these limitations?
57. Why would we not use eddy-current inspection with ceramics or polymeric materials?
58. What types of detection capabilities are offered by eddy-current inspection that cannot be duplicated by the other methods?
59. Why can't acoustic emission methods be used to detect the presence of an existing but static defect?
60. How can acoustic emission be used to determine the location of a flaw or defect?
61. How can temperature be used to reveal defects?
62. What kinds of product features can be evaluated by electrical resistivity methods?
63. What type of information can be obtained through computed tomography?
64. What are some of the techniques that can be used to determine the chemical composition of surface and near-surface material?
65. Why is it important to determine the distinction between allowable and critical flaws, as opposed to rejecting all materials that contain detectable flaws?

■ Problems

1. Read the 25-division vernier graduated in English (Figure 10-A).

2. Read the 25-division vernier graduated in metric (direct reading) (Figure 10-B).

English Reading

FIGURE 10-A

Metric Reading

FIGURE 10-B

FIGURE 10-C

FIGURE 10-D

3. Convert the larger of the two readings to units of the smaller and subtract.
4. Suppose that in Figure 10-33 the height of the gage blocks are 3.2500 in. What is the angle θ assuming that the dial indicator is reading zero?
5. What is the estimated error in this measurement, given that Grade 3 working gage blocks are being used?
6. In Figure 10-C, the sleeve-thimble region of three micrometers graduated in thousandths of an inch are shown. What are the readings for these three micrometers? (*Hint:* Think of the various units as if you were making change from a $10 bill. Count the figures on the sleeve as dollars, the vertical lines on the sleeve as quarters, and the divisions on the thimble as cents. Add up your change, and put a decimal point instead of a dollar sign in front of the figures.)
7. Figure 10-D shows the sleeve-thimble region of two micrometers graduated in thousandths of an inch with a vernier for an additional ten-thousandths. What are the readings?
8. In Figure 10-E, two examples of a metric vernier micrometer are shown. The micrometer is graduated in hundredths of a millimeter (0.01 mm), and an additional reading in two-thousandths of a millimeter (0.002 mm) is obtained from vernier on the sleeve. What are the readings?

9. Progress in machine tool technology over the last 100 years has led to the continual redefinition of precision as shown in Figure 10-F, developed by Taniguchi. The trend here is very clear—that precision in machining continues to improve over time and approaches some limit. Discuss this figure, addressing such issues as
   9.1 What is the limit in machining precision?
   9.2 The vertical axis of the plot uses the term *accuracy*, while the curves use the term *precision*. Are these the same? Explain.
   9.3 What is nanoprocessing? Give some examples.
   9.4 What is the current industrial level of precision for machine tools?
   9.5 What is the correct title for the vertical axis on this figure?

FIGURE 10-E

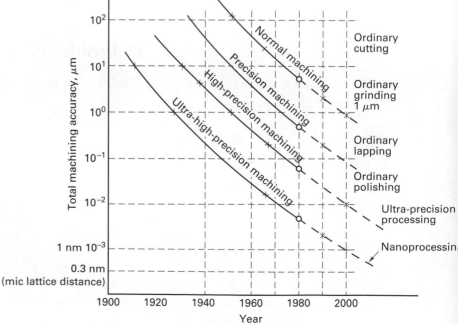

FIGURE 10-F

10. What processes might be grouped into the nanotechnology field? For example, what level of precision is needed in a CD player or an artificial joint?

11. Suppose you had a 2-ft steel bar in your supermicrometer. Could you detect a length change if the temperature of the bar changed by 20°F?

12. Figure 10-G shows a section of a vernier caliper. What is the reading for the outside caliper?

Upper vernier reads

0.02 mm

0.001 In

Lower vernier reads

FIGURE 10-G

13. For each of the inspection methods listed below, cite one majo limitation to its use.
    a. Visual inspection
    b. Liquid penetrant inspection
    c. Magnetic particle inspection
    d. Ultrasonic inspection
    e. Radiography
    f. Eddy-current testing
    g. Acoustic emission monitoring

14. Which of the major nondestructive inspection methods might you want to consider if you want to detect (1) surface flaws and (2) internal flaws in products made from each of the following materials?
    a. Ceramics
    b. Polymers
    c. Fiber-reinforced composites with (i) polymer matrix and (ii) metal matrix (Consider various fiber materials.)

15. Discuss the application of nondestructive inspection methods to powder metallurgy (metallic) products with low, average, and high density.

www.wiley.com/college/degarmo

C*hapter 10* **CASE STUDY**

Measuring an Angle

Figure CS-10a, shows a part drawing. After the part is made, θ needs to be inspected. The quality engineer, Kavit, suggested the setup shown in Figure CS-10b. (No sine plate was available.)

(a) Determine the angle from the part drawing and the value of *X* for the stack of gage blocks.

(b) What blocks would you use in the stack to get the total to "*X*"? (You will have to find a box of gage blocks.)

(c) Suggest another way to check this angle (no sine plate, or gage blocks).

(d) Show the setup you would use if you had a sine plate.

FIGURE CS-10A Part drawing

FIGURE CS-10B Setup for checking the angle θ

CHAPTER 11

FUNDAMENTALS OF CASTING

■ 11.1 INTRODUCTION TO MATERIALS PROCESSING

Almost every manufactured product (or component of a product) goes through a series of activities that include (1) design, defining what we want to produce, (2) material selection, (3) process selection, (4) manufacture, (5) inspection and evaluation, and (6) feedback. Previous chapters have presented the fundamentals of *materials engineering,* the study of the structure, properties, processing, and performance of engineering materials and the systems interactions among these aspects. Other chapters address the use of heat treatment to achieve desired properties and the use of surface treatments to alter features such as wear or corrosion resistance. In this chapter, we begin a focus on *materials processing,* the science and technology through which a material is converted into a useful shape with structure and properties that are optimized for the proposed service environment. A less technical definition of materials processing might be "whatever must be done to convert stuff into things."

A primary objective of materials processing is the production of a desired shape in the desired quantity. Shape-producing processes are often grouped into four basic "families," as indicated in Figure 11-1. *Casting processes* exploit the properties of a liquid as it flows into and assumes the shape of a prepared container, and then solidifies upon cooling. The *material removal processes* remove selected segments from an initially oversized piece. Traditionally, these processes have often been referred to as *machining*, a term used to describe the mechanical cutting of materials. The more general term, *material removal,* includes a wide variety of techniques, including those based on chemical, thermal, and physical processes. *Deformation processes* exploit the ductility or plasticity of certain materials, mostly metals, and produce the desired shape by mechanically moving or rearranging the solid. *Consolidation processes* build a desired shape by putting smaller pieces together. Included here are welding, brazing, soldering, adhesive bonding, and mechanical fasteners. *Powder metallurgy* is the manufacture of a desired shape from particulate material, a definite form of consolidation, but can also involve aspects of casting and forming.

Each of the four basic families has distinct advantages and limitations, and the various processes within the families have their own unique characteristics. For example, cast products can have extremely complex shapes, but also possess structures that are produced by solidification and are therefore subject to such defects as shrinkage and porosity. Material removal processes are capable of outstanding dimensional precision but produce scrap when material is cut away to produce the desired shape. Deformation processes can have high rates of production but generally require powerful equipment and dedicated tools or dies. Complex products can often be assembled from simple shapes, but the joint areas are often affected by the joining process and may possess characteristics different from the original base material.

Family	Subgroup	Typical processes

Casting
- Expendable mold
  - Sand casting
  - Shell casting
  - Investment casting
  - Lost-foam casting
- Multiple-use mold
  - Die casting
  - Permanent-mold casting

Material removal
- Mechanical machining
  - Turning
  - Milling
  - Drilling
  - Boring
  - Sawing
- Nontraditional machining
  - Etching
  - Electropolishing
  - Electrodischarge machining
  - Water jet
  - Laser beam

Deformation processes
- Hot bulk forming
  - Forging
  - Rolling
  - Extrusion
- Cold forming
  - Wire drawing
  - Swaging
  - Roll forming
  - Deep drawing

Consolidation processes
- Welding
  - Oxyfuel
  - Arc
  - Plasma
  - Resistance
  - Laser
- Brazing
- Soldering
- Adhesive bonding
- Mechanical joining
  - Discrete fasteners
  - Integral fasteners
  - Press fit
  - Shrink fit

FIGURE 11-1   The four materials processing families, with subgroups and typical processes.

When selecting the process or processes to be used in obtaining a desired shape and achieving the desired properties, decisions should be made with the knowledge of all available alternatives and their associated assets and limitations. A large portion of this book is dedicated to presenting the various processes that can be applied to engineering materials. They are grouped according to the four basic categories, with powder metallurgy being included at the end of the section on deformation process. The emphasis is on process fundamentals, descriptions of the various alternatives, and an assessment of associated assets and limitations. We will begin with a survey of the casting processes.

■ 11.2 INTRODUCTION TO CASTING

In the *casting* processes, a material is first melted, heated to proper temperature, and sometimes treated to modify its chemical composition. The molten material is then poured into a cavity or mold that holds it in the desired shape during cool-down and solidification. In a single step, simple or complex shapes can be made from any material that can be melted. By proper design and process control, the resistance to working stresses can be optimized and a pleasing appearance can be produced.

Cast parts range in size from a fraction of a centimeter and a fraction of a gram (such as the individual teeth on a zipper) to over 10 meters and many tons (as in the huge propellers and stern frames of ocean liners). Moreover, the casting processes have distinct advantages when the production involves complex shapes, parts having hollow sections or internal cavities, parts that contain irregular curved surfaces (except those that can be made from thin sheet metal), very large parts, or parts made from metals that are difficult to machine.

It is almost impossible to design a part that cannot be cast by one or more of the commercial casting processes. However, as with all manufacturing techniques, the best results and lowest cost are only achieved if the designer understands the various options and tailors the design to use the most appropriate process in the most efficient manner. The variety of casting processes use different mold materials (sand, metal, or various ceramics) and pouring methods (gravity, vacuum, low pressure, or high pressure). All share the requirement that the material should solidify in a manner that will maximize the properties and avoid the formation of defects, such as shrinkage voids, gas porosity, and trapped inclusions.

BASIC REQUIREMENTS OF CASTING PROCESSES

Six basic steps are present in most casting processes:

1. A container must be produced with a *mold cavity*, having the desired shape and size, with due allowance for shrinkage of the solidifying material. Any geometrical feature desired in the finished casting must be present in the cavity. The mold material must provide the desired detail and also withstand the high temperatures and not contaminate the molten material that it will contain. In some processes, a new mold is prepared for each casting (single-use molds) while in other processes, the mold is made from a material that can withstand repeated use, such as metal or graphite. The *multiple-use molds* tend to be quite costly and are generally employed with products where large quantities are desired. The more economical *single-use molds* are usually preferred for the production of smaller quantities but may be required when casting the higher-melting-temperature materials.

2. A *melting process* must be capable of providing molten material at the proper temperature, in the desired quantity, with acceptable quality, and at a reasonable cost.

3. A *pouring technique* must be devised to introduce the molten metal into the mold. Provision should be made for the escape of all air or gases present in the cavity prior to pouring, as well as those generated by the introduction of the hot metal. The molten material must be free to fill the cavity, producing a high-quality casting that is fully dense and free of defects.

4. The *solidification process* should be properly designed and controlled. Castings should be designed so that solidification and solidification shrinkage can occur without producing internal porosity or voids. In addition, the molds should not provide excessive restraint to the shrinkage that accompanies cooling, a feature that may cause the casting to crack when it is still hot and its strength is low.

5. It must be possible to remove the casting from the mold (i.e., *mold removal*). With single-use molds that are broken apart and destroyed after each casting, mold removal presents no serious difficulty. With multiple-use molds, however, the removal of a complex-shaped casting may be a major design problem.

6. Various *cleaning, finishing, and inspection* operations may be required after the casting is removed from the mold. Extraneous material is usually attached where the metal entered the cavity, excess material may be present along mold parting lines, and mold material may adhere to the casting surface. All of these must be removed from the finished casting.

Each of the six steps will be considered in more detail as we move through the chapter. The fundamentals of solidification, pattern design, gating, and risering will all be developed. Various defects will also be considered, together with their causes and cures.

■ 11.3 CASTING TERMINOLOGY

Before we proceed to the process fundamentals, it is helpful to first become familiar with a bit of casting vocabulary. Figure 11-2 shows a two-part mold, its cross section, and a variety of features or components that are present in a typical casting process. To produce a casting, we begin by constructing a *pattern*, an approximate duplicate of the final casting. *Molding material* will then be packed around the pattern and the pattern is removed to create all or part of the mold cavity. The rigid metal or wood frame that holds the molding aggregate is called a *flask*. In a horizontally parted two-part mold, the top half of the pattern, flask, mold, or core is called the *cope*. The bottom half of any of these features is called the *drag*. A *core* is a sand (or metal) shape that is inserted into a mold to produce the internal features of a casting, such as holes or passages for water cooling. Cores are produced in wood, metal, or plastic tooling, known as *core boxes*. A *core print* is a feature that is added to a pattern, core, or mold and is used to locate and support a core within the mold. The mold material and the cores then combine to produce a completed *mold cavity*, a shaped hole into which the molten metal is poured and solidified to produce the desired casting. A *riser* is an additional void in the mold that also fills with molten metal. Its purpose is to provide a reservoir of additional liquid that can flow into the mold cavity to compensate for any shrinkage that occurs during solidification. By designing so the riser contains the last material to solidify, shrinkage voids should be located in the riser, not the final casting.

FIGURE 11-2   Cross section of a typical two-part sand mold, indicating various mold components and terminology.

The network of connected channels used to deliver the molten metal to the mold cavity is known as the *gating system*. The *pouring cup* (or pouring basin) is the portion of the gating system that receives the molten metal from the pouring vessel and controls its delivery to the rest of the mold. From the pouring cup, the metal travels down a *sprue* (the vertical portion of the gating system), then along horizontal channels, called *runners*, and finally through controlled entrances, or *gates*, into the mold cavity. Additional channels, known as *vents*, may be included in a mold or core to provide an escape for the gases that are originally present in the mold or are generated during the pour. (These and other features of a gating system will be discussed later in the chapter and are illustrated in Figure 11-9.)

The *parting line* or *parting surface* is the interface that separates the cope and drag halves of a mold, flask, or pattern, and also the halves of a core in some core-making processes. *Draft* is the term used to describe the taper on a pattern or casting that permits it to be withdrawn from the mold. The draft usually expands toward the parting line. Finally, the term *casting* is used to describe both the process and the product when molten metal is poured and solidified in a mold.

◼ 11.4 THE SOLIDIFICATION PROCESS

Casting is a *solidification process* where the molten material is poured into a mold and then allowed to freeze into the desired final shape. Many of the structural features that ultimately control product properties are set during solidification. Furthermore, many casting defects, such as *gas porosity* and *solidification shrinkage*, are also solidification phenomena, and they can be reduced or eliminated by controlling the solidification process.

Solidification is a two-stage, nucleation and growth, process, and it is important to control both of these stages. *Nucleation* occurs when stable particles of solid form from within the molten liquid. When a material is at a temperature below its melting point, the solid state has a lower energy than the liquid. As solidification occurs, internal energy is released. At the same time, however, interface surfaces must be created between the new solid and the parent liquid. Formation of these surfaces requires energy. In order for nucleation to occur, there must be a net reduction or release of energy. As a result, nucleation generally begins at a temperature somewhat below the equilibrium melting point (the temperature where the internal energies of the liquid and solid are equal). The difference between the melting point and the actual temperature of nucleation is known as the amount of *undercooling*.

If nucleation can occur on some form of existing surface, it no longer requires the creation of a full, surrounding interface, and the required energy is reduced. Such surfaces are usually present in the form of mold or container walls, or solid impurity particles contained within the molten liquid. When ice cubes are formed in a tray, the initial solid forms on the walls of the container. The same phenomena can be expected with metals and other engineering materials.

Each nucleation event produces a crystal or grain in the final casting. Since fine-grained materials (many small grains) possess enhanced mechanical properties, efforts may be made to promote nucleation. Particles of existing solid may be introduced into the liquid before it is poured into the mold. These particles provide the surfaces required for nucleation and promote the formation of a uniform, fine-grained product. This practice of introducing solid particles is known as *inoculation* or *grain refinement*.

The second stage in the solidification process is *growth*, which occurs as the heat of fusion is extracted from the liquid material. The direction, rate, and type of growth can be controlled by the way in which this heat is removed. *Directional solidification*, in which the solidification interface sweeps continuously through the material, can be used to assure the production of a sound casting. The molten material on the liquid side of the interface can flow into the mold to continuously compensate for the shrinkage that occurs as the material changes from liquid to solid. The relative rates of nucleation and growth control the size and shape of the resulting crystals. Faster rates of cooling generally produce products with finer grain size and superior mechanical properties.

FIGURE 11-3 Cooling curve for a pure metal or eutectic-composition alloy (metals with a distinct freezing point), indicating major features related to solidification.

COOLING CURVES

Cooling curves, such as those introduced in Chapter 4, can be one of the most useful tools for studying the solidification process. By inserting thermocouples into a casting and recording the temperature versus time, one can obtain valuable insight into what is happening in the various regions.

Figure 11-3 shows a typical cooling curve for a pure or eutectic-composition material (one with a distinct melting point) and is useful for depicting many of the features and terms related to solidification. The *pouring temperature* is the temperature of the liquid metal when it first enters the mold. *Superheat* is the difference between the pouring temperature and the freezing temperature of the material. Most metals are poured at temperatures of 100–200°C (200–400°F) above the temperature where solid begins to form. The higher the superheat, the more time is given for the material to flow into the intricate details of the mold cavity before it begins to freeze. The *cooling rate* is the rate at which the liquid or solid is cooling and can be viewed as the slope of the cooling curve at any given point. The *thermal arrest* is the plateau in the cooling curve that occurs during the solidification of a material with fixed melting point. At this temperature, the energy or heat being removed from the mold comes from the latent heat of fusion that is being released during the solidification process. The time from the start of pouring to the end of solidification is known as the *total solidification time*. The time from the start of solidification to the end of solidification is the *local solidification time*.

If the metal or alloy being cast does not have a distinct melting point, such as the one shown in Figure 11-4, solidification will occur over a range of temperatures. The *liquidus* temperature is the lowest temperature where the material is all liquid, and the *solidus* temperature is the highest temperature where it is all solid. The region between the *liquidus* and *solidus* temperatures is known as the *freezing range*. The onset and termination of solidification appear as slope changes in the cooling curve.

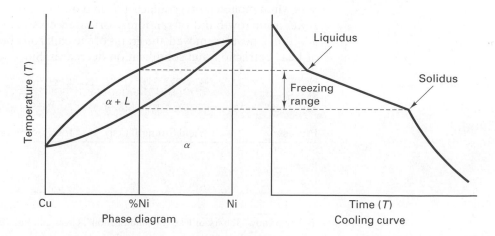

FIGURE 11-4 Phase diagram and companion cooling curve for an alloy with a freezing range. The slope changes indicate the onset and termination of solidification.

The actual form of a cooling curve will depend on the type of material being poured, the nature of the nucleation process, and the rate and means of heat removal from the mold. By analyzing experimental cooling curves, we can gain valuable insight into both the casting process and the cast product. Fast cooling rates and short solidification times generally lead to finer structures and improved mechanical properties.

PREDICTION OF SOLIDIFICATION TIME: CHVORINOV'S RULE

The amount of heat that must be removed from a casting to cause it to solidify depends upon both the amount of superheating and the volume of metal in the casting. Conversely, the ability to remove heat from a casting is directly related to the amount of exposed surface area through which the heat can be extracted and the environment surrounding the molten material (i.e., the mold and mold surroundings). These observations are reflected in *Chvorinov's rule*,[1] which states that the total solidification time, t_s, can be computed by:

$$t_s = B\ (V/A)^n \text{ where } n = 1.5 \text{ to } 2.0$$

The total solidification time, t_s, is the time from pouring to the completion of solidification; V is the volume of the casting; A is the surface area through which heat is extracted; and B is the *mold constant*. The mold constant, B, incorporates the characteristics of the metal being cast (heat capacity and heat of fusion), the mold material (heat capacity and thermal conductivity), the mold thickness, initial mold temperature, and the amount of superheat.

Test specimens can be cast to determine the value of B for a given mold material, casting material, and condition of casting. This value can then be used to compute the solidification times for other castings made under the same conditions. Since a riser and casting both lie within the same mold and fill with the same metal under the same conditions, Chvorinov's rule can be used to compare the solidification times of each and thereby ensure that the riser will solidify after the casting. This condition is absolutely essential if the liquid within the riser is to effectively feed the casting and compensate for solidification shrinkage. Aspects of riser design, including the use of Chvorinov's rule, will be developed later in this chapter.

Different cooling rates and solidification times can produce substantial variation in the structure and properties of the resulting casting. Die casting, for example, uses water-cooled metal molds, and the faster cooling produces higher-strength products than sand casting, where the mold material is more thermally insulating. Even variations in the type and condition of sand can produce different cooling rates. Sands with high moisture contents extract heat faster than ones with low moisture. Table 11-1 presents a comparison of the properties of aluminum alloy 443 cast by the three different processes of sand casting (slow cool), permanent mold casting (intermediate cooling rate), and die casting (fast cool).

THE CAST STRUCTURE

The products that result when molten metal is poured into a mold and permitted to solidify may have as many as three distinct regions or zones. The rapid nucleation that occurs when molten metal contacts the cold mold walls results in the production of a *chill zone,* a narrow band of randomly oriented crystals on the surface of a casting. As additional heat is removed, the grains of the chill zone begin to grow inward, and the rate of heat extraction and solidification decreases. Since most crystals have directions of

TABLE 11-1	Comparison of As-Cast Properties of 443 Aluminum Cast by Three Different Processes		
Process	Yield Strength (ksi)	Tensile Strength (ksi)	Elongation (%)
Sand cast	8	19	8
Permanent mold	9	23	10
Die cast	16	33	9

[1]N. Chvorinov, "Theory of Casting Solidification", Giesserei, Vol. 27, 1940, pp. 177–180, 201–208, 222–225.

FIGURE 11-5 Internal structure of a cast metal bar showing the chill zone at the periphery, columnar grains growing toward the center, and a central shrinkage cavity.

rapid growth, a selection process begins. Crystals with rapid-growth direction perpendicular to the casting surface grow fast and shut off adjacent grains whose rapid-growth direction is at some intersecting angle. The favorably oriented crystals continue to grow, producing the long, thin columnar grains of a *columnar zone*. The properties of this region are highly directional, since the selection process has converted the purely random structure of the surface into one of parallel crystals of similar orientation. Figure 11-5 shows a cast structure containing both chill and columnar zones.

In many materials, new crystals then nucleate in the interior of the casting and grow to produce another region of spherical, randomly oriented crystals, known as the *equiaxed zone*. Low pouring temperatures, alloy additions, and the addition of inoculants can be used to promote the formation of this region, whose isotropic properties (uniform in all directions) are far more desirable than those of columnar grains.

MOLTEN METAL PROBLEMS

Castings begin with molten metal, and there are a number of chemical reactions that can occur between molten metal and its surroundings. These reactions and their products can often lead to defects in the final casting. For example, oxygen and molten metal can react to produce metal oxides (a nonmetallic or ceramic material), which can then be carried with the molten metal during the pouring and filling of the mold. Known as *dross* or *slag*, this material can become trapped in the casting and impair surface finish, machinability, and mechanical properties. Material eroded from the linings of furnaces and pouring ladles and loose sand particles from the mold surfaces can also contribute nonmetallic components to the casting.

Dross and slag can be controlled by using special precautions during melting and pouring, as well as by good mold design. Lower pouring temperatures or superheat slows the rate of dross-forming reactions. Fluxes can be used to cover and protect molten metal during melting, or the melting and pouring can be performed under a vacuum or protective atmosphere. Measures can be taken to agglomerate the dross and cause it to float to the surface of the metal, where it can be skimmed off prior to pouring. Special ladles can be used that extract metal from beneath the surface, such as those depicted in Figure 11-6. Gating systems can be designed to trap any dross, sand, or eroded mold material and keep it from flowing into the mold cavity. In addition, ceramic *filters* can be inserted into the feeder channels of the mold. These filters are available in a variety of shapes, sizes, and materials.

Liquid metals can also contain significant amounts of dissolved gas. When these materials solidify, the solid structure cannot accommodate the gas, and the rejected atoms form bubbles, or *gas porosity*, within the casting. Figure 11-7 shows the maximum

FIGURE 11-6 Two types of ladles used to pour castings. Note how each extracts molten material from the bottom, avoiding transfer of the impure material from the top of the molten pool.

FIGURE 11-7 The maximum solubility of hydrogen in aluminum as a function of temperature.

solubility of hydrogen in aluminum as a function of temperature. Note the substantial decrease that occurs as the material goes from liquid to solid. Figure 11-8 shows a small demonstration casting that has been made from aluminum that has been saturated with dissolved hydrogen

Several techniques can be used to prevent or minimize the formation of gas porosity. One approach is to prevent the gas from initially dissolving in the molten metal. Melting can be performed under vacuum, in an environment of low-solubility gases, or under a protective flux that excludes contact with the air. Superheat temperatures can be kept low to minimize gas solubility. In addition, careful handling and pouring can do much to control the flow of molten metal and minimize the turbulence that brings air and molten metal into contact.

Another approach is to remove the gas from the molten metal before it is poured into castings. *Vacuum degassing* sprays the molten metal through a low-pressure environment. Spraying creates a large amount of surface area, and the amount of dissolved gas is reduced as the material seeks to establish equilibrium with its new surroundings. (See a discussion of Sievert's law in any basic chemistry text.) Passing small bubbles of inert or reactive gas through the melt, known as *gas flushing*, can also be effective. In seeking equilibrium, the dissolved gases enter the flushing gas and are carried away. Bubbles of nitrogen or chlorine, for example, are particularly effective in removing hydrogen from molten aluminum.

The dissolved gas can also be reacted with something to produce a low-density compound, which then floats to the surface and can be removed with the dross or slag. Oxygen can be removed from copper by the addition of phosphorus. Steels can be deoxidized with addition of aluminum or silicon. The resulting phosphorus, aluminum, or silicon oxides are then removed by skimming or are left on the top of the container as the remaining high-quality metal is extracted from beneath the surface.

FLUIDITY AND POURING TEMPERATURE

When molten metal is poured to produce a casting, it should first *flow* into all regions of the mold cavity and then *freeze* into this new shape. It is vitally important that these two functions occur in the proper sequence. If the metal begins to freeze before it has completely filled the mold, defects known as *misruns* and *cold shuts* are produced.

The ability of a metal to flow and fill a mold, its "runniness," is known as *fluidity*, and casting alloys are often selected for this property. Fluidity affects the minimum section thickness that can be cast, the maximum length of a thin section, the fineness of detail, and the ability to fill mold extremities. While no single method has been accepted to measure fluidity, various "standard molds" have been developed where the results are sensitive to metal flow. One popular approach produces castings in the form of a long, thin spiral that progresses outward from a central sprue. The length of the final casting will increase with increased fluidity.

Fluidity is dependent on the composition, freezing temperature, and freezing range of the metal or alloy as well as the surface tension of oxide films. The most important controlling factor, however, is usually the *pouring temperature* or the amount of *superheat*. The higher the pouring temperature, the higher the fluidity. Excessive temperatures should be avoided, however. At high pouring temperatures, chemical reactions between the metal and the mold, and the metal and its pouring atmosphere, are all accelerated, and larger amounts of gas can be dissolved.

If the metal is too runny, it may not only fill the mold cavity, but also flow into the small voids between the particles that compose a sand mold. The surface of the resulting casting then contains small particles of embedded sand, a defect known as *penetration*.

THE ROLE OF THE GATING SYSTEM

When molten metal is poured into a mold, the gating system conveys the material and delivers it to all sections of the mold cavity. The speed or rate of metal movement is important as well as the amount of cooling that occurs while it is flowing. Slow filling and high loss of heat can result in misruns and cold shuts. Rapid rates of filling, on the other hand, can produce erosion of the gating system and mold cavity, and might result in the entrapment of mold material in the final casting. It is imperative that the cross-sectional

FIGURE 11-8 Demonstration casting made from aluminum that has been saturated in dissolved hydrogen. Note the extensive gas porosity.

FIGURE 11-9 Typical gating system for a horizontal parting plane mold, showing key components involved in controlling the flow of metal into the mold cavity.

areas of the various channels be selected to regulate flow. The shape and length of the channels affect the amount of temperature loss. When heat loss is to be minimized, short channels with round or square cross sections (minimum surface area) are the most desirable. The gates are usually attached to the thickest or heaviest sections of a casting to control shrinkage and to the bottom of the casting to minimize turbulence and splashing. For large castings, multiple gates and runners may be used to introduce metal to more than one point of the mold cavity.

Gating systems should be designed to minimize *turbulent flow*, which tends to promote absorption of gases, oxidation of the metal, and erosion of the mold. Figure 11-9 shows a typical gating system for a mold with a horizontal parting line and can be used to identify some of the key components that can be optimized to promote the smooth flow of molten metal. Short sprues are desirable, since they minimize the distance that the metal must fall when entering the mold and the kinetic energy that the metal acquires during that fall. Rectangular pouring cups prevent the formation of a vortex or spiraling funnel, which tends to suck gas and oxides into the sprue. Tapered sprues also prevent vortex formation. A large *sprue well* can be used to dissipate the kinetic energy of the falling stream and prevent splashing and turbulence as the metal makes the turn into the runner.

The *choke*, or smallest cross-sectional area in the gating system, serves to control the rate of metal flow. If the choke is located near the base of the sprue, flow through the runners and gates is slowed and flow is rather smooth. If the choke is moved to the gates, the metal might enter the mold cavity with a fountain effect, an extremely turbulent mode of flow, but the small connecting area would enable easier separation of the casting and gating system.

Gating systems can also be designed to trap dross and sand particles and keep them from entering the mold cavity. Given sufficient time, the lower-density contaminants will rise to the top of the molten metal. Long, flat runners can be beneficial (but these promote cooling of the metal), as well as gates that exit from the lower portion of the runners. Since the first metal to enter the mold is most likely to contain the foreign matter (dross from the top of the pouring ladle and loose particles washed from the walls of the gating system), *runner extensions and wells* (see Figure 11-9) can be used to catch and trap this first metal and keep it from entering the mold cavity. These features are particularly effective with aluminum castings since aluminum oxide has approximately the same density as molten aluminum.

Screens or ceramic *filters* of various shapes, sizes, and materials can also be inserted into the gating system to trap foreign material. Wire mesh can often be used with the nonferrous metals, but ceramic materials are generally required for irons and steel. Figure 11-10 shows several ceramic filters and depicts the two basic types—extruded and foam. The pores on the extruded ceramics are uniform in size and shape and provide

FIGURE 11-10 Various types of ceramic filters that may be inserted into the gating systems of metal castings.

parallel channels. The foams contain interconnected pores of various size and orientation, forcing the material to change direction as it negotiates its passage through the filter. Since these devices can also restrict the fluid velocity, streamline the fluid flow, or reduce turbulence, proper placement is an important consideration. To ensure removal of both dross and eroded sand, the filter should be as close to the mold cavity as possible, but since a filter can also act as the choke, it may be positioned at other locations, such as at the base of the pouring cup, at the base of the sprue, or in one or more of the runners.

The specific details of a gating system often vary with the metal being cast. Turbulent-sensitive metals (such as aluminum and magnesium) and alloys with low melting points generally employ gating systems that concentrate on eliminating turbulence and trapping dross. Turbulent-insensitive alloys (such as steel, cast iron, and most copper alloys) and alloys with a high melting point generally use short, open gating systems that provide for quick filling of the mold cavity.

SOLIDIFICATION SHRINKAGE

Once they enter the mold cavity and begin to cool, most metals and alloys undergo a noticeable volumetric contraction. Figure 11-11 shows the typical changes experienced by a metal column as the material goes from superheated liquid to room-temperature solid. There are three principal stages of *shrinkage*: (1) *shrinkage of the liquid* as it cools to the temperature where solidification begins, (2) *solidification shrinkage* as the liquid turns into solid, and (3) *solid metal contraction* as the solidified material cools to room temperature.

FIGURE 11-11 Dimensional changes experienced by a metal column as the material cools from a superheated liquid to a room-temperature solid. Note the significant shrinkage that occurs upon solidification.

TABLE 11-2	Solidification Shrinkage of Some Common Engineering Metals (Expressed in Percent)
Aluminum	6.6
Copper	4.9
Magnesium	4.0
Zinc	3.7
Low-carbon steel	2.5–3.0
High-carbon steel	4.0
White cast iron	4.0–5.5
Gray cast iron	−1.9

The amount of liquid metal contraction depends on the coefficient of thermal contraction (a property of the metal being cast) and the amount of superheat. Liquid contraction is rarely a problem, however, because the metal in the gating system continues to flow into the mold cavity as the liquid already in the cavity cools and contracts.

As the metal changes state from liquid to crystalline solid, the new atomic arrangement is usually more efficient, and significant amounts of shrinkage can occur. The actual amount of shrinkage varies from alloy to alloy, as shown in Table 11-2. As indicated in that table, not all metals contract upon solidification. Some actually expand, such as gray cast iron, where low-density graphite flakes form as part of the solid structure.

When solidification shrinkage does occur, however, it is important to control the form and location of the resulting void. Metals and alloys with short freezing ranges, such as pure metals and eutectic alloys, tend to form large cavities or pipes. These can be avoided by designing the casting to have directional solidification where freezing begins farthest away from the feed gate or riser and moves progressively toward it. As the metal solidifies and shrinks, the shrinkage void is continually filled with additional liquid metal. When the flow of additional liquid is exhausted and solidification is complete, we hope that the final shrinkage void is located external to the desired casting in either the riser or the gating system.

Alloys with large freezing ranges have a period of time when the material is in a slushy (liquid plus solid) condition. As the material cools between the liquidus and solidus, the relative amount of solid increases and tends to trap small, isolated pockets of liquid. It is almost impossible for additional liquid to feed into these locations, and the resultant casting tends to contain small but numerous shrinkage pores dispersed throughout. This type of shrinkage is far more difficult to prevent by means of gating and risering, and a porous product may be inevitable. If a gas- or liquid-tight product is required, these castings may need to be impregnated (the pores filled with a resinous material or lower-melting-temperature metal) in a subsequent operation. Castings with dispersed porosity tend to have poor ductility, toughness, and fatigue life.

After solidification is complete, the casting will contract further as it cools to room temperature. This solid metal contraction is often called patternmaker's contraction, since compensation for these dimensional changes should be made when the mold cavity or pattern is designed. Examples of these compensations will be provided later in this chapter. Concern arises, however, when the casting is produced in a rigid mold, such as the metal molds used in die casting. If the mold provides constraint during the time of contraction, tensile forces can be generated within the hot, weak casting, and cracking can occur (*hot tears*). It is often desirable, therefore, to eject the hot castings as soon as solidification is complete.

RISERS AND RISER DESIGN

Risers are added reservoirs designed to fill with liquid metal, which is then fed to the casting as a means of compensating for solidification shrinkage. To effectively perform this function, the risers must solidify after the casting. If the reverse were true, liquid metal would flow from the casting toward the solidifying riser and the casting shrinkage would be even greater. Hence, castings should be designed to produce directional solidification that sweeps from the extremities of the mold cavity toward the riser. In this way, the riser can continuously feed molten metal and will compensate for the solidification shrinkage of the entire mold cavity. Figure 11-12 shows a three-level step block cast in aluminum with and without a riser. Note that the riser is positioned so directional solidification moves from thin to thick and the shrinkage void is moved from the casting to the riser. If a single directional solidification is not possible, multiple risers may be required, with various sections of the casting each solidifying toward their respective riser.

The risers should also be designed to conserve metal. If we define the *yield* of a casting as the casting weight divided by the total weight of metal poured (complete gating

FIGURE 11-12 A three-tier step-block aluminum casting made with (top) and without (bottom) a riser. Note how the riser has moved the shrinkage void external to the desired casting.

system, risers, and casting), it is clear that there is a motivation to make the risers as small as possible, yet still able to perform their task. This is usually done through proper consideration of riser size, shape, and location, as well as the type of connection between the riser and casting.

A good shape for a riser would be one that has a long freezing time. According to Chvorinov's rule, this would favor a shape with small surface area per unit volume. While a sphere would make the most efficient riser, this shape presents considerable difficulty to both patternmaker and moldmaker. The most popular shape for a riser, therefore, is a cylinder, where the height-to-diameter ratio is varied depending upon the nature of the alloy being cast, the location of the riser, the size of the flask, and other variables. A one-to-one height-to-diameter ratio is generally considered to be ideal.

Risers should be located so that directional solidification occurs from the extremities of the mold cavity back toward the riser. Since the thickest regions of a casting will be the last to freeze, risers should feed directly into these locations. Various types of risers are possible. A *top riser* is one that sits on top of a casting. Because of their location, top risers have shorter feeding distances and occupy less space within the flask. They give the designer more freedom for the layout of the pattern and gating system. *Side risers* are located adjacent to the mold cavity, displaced horizontally along the parting line. Figure 11-13 depicts both a top and a side riser. If the riser is contained entirely within the mold, it is known as a *blind riser*. If it is open to the atmosphere, it is called an *open riser*. Blind risers are usually larger than open risers because of the additional heat loss that occurs where the top of the riser is in contact with mold material.

Live risers (also known as hot risers) receive the last hot metal that enters the mold and generally do so at a time when the metal in the mold cavity has already begun to cool and solidify. Thus, they can be smaller than *dead (or cold) risers*, which fill with metal that has already flowed through the mold cavity. As shown in Figure 11-13, top risers are almost always dead risers. Risers that are part of the gating system generally live risers.

FIGURE 11-13 Schematic of a sand casting mold, showing (a) an open-type top riser and (b) a blind-type side riser. The side riser is a live riser, receiving the last hot metal to enter the mold. The top riser is a dead riser, receiving metal that has flowed through the mold cavity.

Top riser (open-type)   Side riser (blind-type)

Mold cavity   Mold cavity

(a)   (b)

The minimum size of a riser can be calculated from Chvorinov's rule by setting the total solidification time for the riser to be greater than the total solidification time for the casting. Since both cavities receive the same metal and are in the same mold, the mold constant, B, will be the same for both regions. Assuming that $n = 2$ and that a safe difference in solidification time is 25% (the riser takes 25% longer to solidify than the casting), we can write this condition as

$$t_{\text{riser}} = 1.25 t_{\text{casting}} \tag{11-1}$$

or

$$(V/A)^2_{\text{riser}} = 1.25 \, (V/A)^2_{\text{casting}} \tag{11-2}$$

Calculation of the riser size then requires selection of a riser geometry, which is generally cylindrical. For a cylinder of diameter D and height H, the volume and surface area can be written as:

$$V = \pi D^2 H / 4$$
$$A = \pi D H + 2(\pi D^2/4)$$

Selecting a specific height-to-diameter ratio for the riser then enables equation 11-2 to be written as a simple expression with one unknown, D. The volume-to-area ratio for the casting is computed for its particular geometry, and the equation can then be solved to provide the size of the required riser. One should note that if the riser and casting share a surface, as with a blind top riser, the area of the common surface should be subtracted from both components since it will not be a surface of heat loss to either. It should be noted that there are actually a number of methods to calculate riser size. The Chvorinov's rule method will be the only one presented here.

A final aspect of riser design is the connection between the riser and the casting. Since the riser must ultimately be separated from the casting, it is desirable that the connection area be as small as possible. On the other hand, the connection area must be sufficiently large so that the link does not freeze before solidification of the casting is complete. If the risers are placed close to the casting with relatively short connections, the mold material surrounding the link receives heat from both the casting and the riser. It should heat rapidly and remain hot throughout the cast, thereby preventing solidification of the metal in the channel.

RISERING AIDS

Various methods have been developed to assist the risers in performing their job. Some are intended to promote directional solidification, while others seek to reduce the number and size of the risers, thereby increasing the yield of a casting. These techniques generally work by either speeding the solidification of the casting (*chills*) or retarding the solidification of the riser (*sleeves* or *toppings*).

External chills are masses of high-heat-capacity, high-thermal-conductivity material (such as steel, iron, graphite, or copper) that are placed in the mold, adjacent to the casting, to absorb heat and accelerate the cooling of various regions. Chills can effectively promote directional solidification or increase the effective feeding distance of a riser. They can also be used to reduce the number of risers required for a casting. External chills are frequently covered with a protective wash, silica flour, or other refractory material to prevent bonding with the casting.

Internal chills are pieces of metal that are placed within the mold cavity to absorb heat and promote more rapid solidification. When the molten metal of the pour surrounds the chill, it absorbs heat as it seeks to come to equilibrium with its surroundings. Internal chills ultimately become part of the final casting, so they must be made from an alloy that is the same as or compatible with the alloy being cast.

The cooling of risers can be slowed by methods that include (1) switching from a blind riser to an open riser, (2) placing *insulating sleeves* around the riser, and (3) surrounding the sides or top of the riser with *exothermic material* that supplies added heat to just the riser segment of the mold. The objective of these techniques is generally to reduce the riser size rather than promote directional solidification.

It is important to note that risers are not always necessary or functional. For alloys with large freezing ranges, risers would not be particularly effective, and one generally accepts the fine, dispersed porosity that results. For processes such as die casting, low-pressure permanent molding, and centrifugal casting, the positive pressures associated with the process provide the feeding action that is required to compensate for solidification shrinkage.

■ 11.5 PATTERNS

Casting processes can be divided into two basic categories: (1) those for which a new mold must be created for each casting (the *expendable-mold processes*) and (2) those that employ a permanent, *reusable mold*. Most of the expendable-mold processes begin with some form of reusable *pattern*—a duplicate of the part to be cast, modified dimensionally to reflect both the casting process and the material being cast. Patterns can be made from wood, metal, foam, or plastic, with urethane now being the material of choice for nearly half of all casting patterns.

The dimensional modifications that are incorporated into a pattern are called *allowances*, and the most important of these is the *shrinkage allowance*. Following solidification, a casting continues to contract as it cools to room temperature, the amount of this contraction being as much as 2% or 1/4 in./ft. To produce the desired final dimensions, the pattern (which sets the dimensions upon solidification) must be slightly larger than the room-temperature casting. The exact amount of this shrinkage compensation, which depends on the metal that is being cast, can be estimated by the equation Δ length = length $\alpha\,\Delta T$, where α is the coefficient of thermal expansion and ΔT is the difference between the freezing temperature and room temperature. Typical allowances for some common engineering metals are:

Cast iron	0.8–1.0%
Steel	1.5–2.0%
Aluminum	1.0–1.3%
Magnesium	1.0–1.3%
Brass	1.5%

Shrinkage allowances are often incorporated into a pattern through use of special *shrink rules*—measuring devices that are larger than a standard rule by an appropriate shrink allowance. For example, a shrink rule for brass would designate 1 foot at a length that is actually 1 foot 3/16 inch, since the anticipated 1.5% shrinkage will reduce the length by 3/16 inch. A complete pattern made to shrink rule dimensions will produce a proper size casting after cooling.

Caution should be exercised when using shrink rule compensations, however, for thermal contraction may not be the only factor affecting the final dimensions. The various phase transformations discussed in Chapter 4 are often accompanied by significant dimensional expansions or contractions. Examples include eutectoid reactions, martensitic reactions, and graphitization.

In many casting processes, mold material is formed around the pattern and the pattern is then extracted to create the mold cavity. To facilitate pattern removal, molds are often made in two or more sections that separate along mating surfaces called the *parting line* or *parting plane*. A flat parting line is usually preferred, but the casting design or molding practice may dictate the use of irregular or multiple parting surfaces.

If the pattern contains surfaces that are perpendicular to the parting line (parallel to the direction of pattern withdrawal), friction between the pattern and the mold material as well as any horizontal movement of the pattern during extraction could induce damage to the mold. This damage could be particularly severe at the corners, where the mold cavity intersects the parting surface. Such extraction damage can be minimized by incorporating a slight taper, or *draft*, on all pattern surfaces that are parallel to the direction of withdrawal. A slight withdrawal of the pattern will free it from the mold material on all surfaces, and it can then be further removed without damage to the mold. Figure 11-14 illustrates the use of draft to facilitate pattern removal.

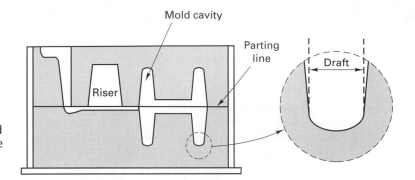

FIGURE 11-14   Two-part mold showing the parting line and the incorporation of a draft allowance on vertical surfaces.

The size and shape of the pattern, the depth of the mold cavity, the method used to withdraw the pattern, the pattern material, the mold material, and the molding procedure all influence the actual amount of draft required. Draft is seldom less than 1° or 1/8 in./ft., with a minimum taper of about 1/16 inch over the length of any surface. Since draft allowances increase the size of a pattern (and thus the size and weight of a casting), it is generally desirable to keep them to the minimum that will permit satisfactory pattern removal. Molding procedures that produce higher-strength molds and the use of mechanical pattern withdrawal can often enable reductions in draft allowances. By reducing the taper, casting weight and the amount of subsequent machining can both be reduced.

When smooth machined surfaces are required, it may be necessary to add an additional *machining allowance*, or *finish allowance*, to the pattern. The amount of this allowance depends to a great extent on the casting process and the mold material. Ordinary sand castings have rougher surfaces than those of shell-mold castings. Die castings have smooth surfaces that may require little or no metal removal, and the surfaces of investment castings are even smoother. It is also important to consider the location of the desired machining and the presence of other allowances, since the draft allowance may provide part or all of the extra metal needed for machining.

Some casting shapes require an additional allowance for *distortion*. Consider a U-shaped section where the arms are restrained by the mold at a time that the base of the U is free to shrink. The result will be a final casting with outwardly sloping arms. If the arms are designed to originally slope inward, however, the subsequent distortion will produce the desired straight-shape casting. Distortion depends greatly on the particular configuration of the casting, and casting designers must use experience and judgment to provide an appropriate distortion allowance.

Figure 11-15 illustrates the manner in which the various allowances are incorporated into a casting pattern. Similar allowances are applied to the cores that create the holes or interior passages of a casting.

If a casting is to be made in a multiuse metal mold, all of the "pattern allowances" discussed above should be incorporated into the machined cavity. The dimensions of this cavity will further change, however, as sequential casts raise the mold temperature to an equilibrium level. An additional correction should be added to compensate for this effect.

Original outline with shrink rule

3 mm ($\frac{1}{8}$") all around for machining

V slot to be machined

$1\frac{1}{2}$° draft allowance

FIGURE 11-15   Various allowances incorporated into a casting pattern.

■ 11.6 DESIGN CONSIDERATIONS IN CASTINGS

To produce the best-quality product at the lowest possible cost, it is important that the designers of castings give careful attention to several process requirements. It is not uncommon for minor and readily permissible changes in design to greatly facilitate and simplify the casting of a component and also reduce the number and severity of defects.

One of the first features that must be considered by a designer is the *location and orientation of the parting plane*, an important part of all processes that use segmented or separable molds. The location of the parting plane can affect (1) the number of cores, (2) the method of supporting the cores, (3) the use of effective and economical gating, (4) the weight of the final casting, (5) the final dimensional accuracy, and (6) the ease of molding.

In general, it is desirable to minimize the use of cores. A change in the location or orientation of the parting plane can often assist in this objective. The change illustrated in Figure 11-16 not only eliminates the need for a core but can also reduce the weight of the casting by eliminating the need for draft. Figure 11-17 shows another example of how a core can be eliminated by a simple design change. Figure 11-18 shows how the specification of draft can act to fix the parting plane. This figure also shows that simply noting the desired shape and the need to provide sufficient draft can provide considerable design freedom. Since mold closure may not always be consistent, consideration should also be given to the fact that dimensions across the parting plane are subject to greater variation than those that lie entirely within a given segment of the mold.

Controlling the solidification process is of prime importance in obtaining quality castings, and this control is also related to design. Those portions of a casting that have a high ratio of surface area to volume will experience more rapid cooling, and will be stronger and harder than the other regions. Thicker or heavier sections will cool more slowly, and

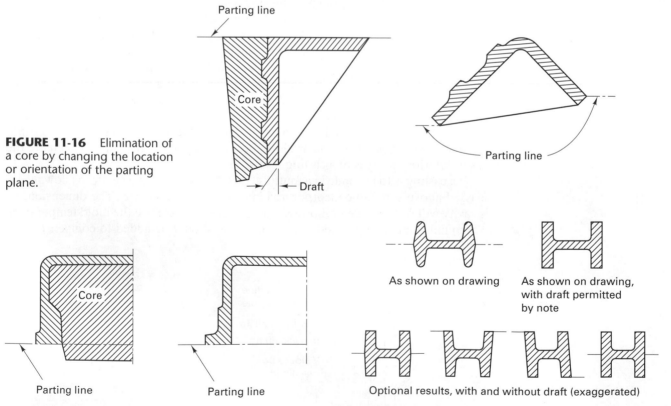

FIGURE 11-16 Elimination of a core by changing the location or orientation of the parting plane.

FIGURE 11-17 Elimination of a dry-sand core by a change in part design.

FIGURE 11-18 (Top left) Design where the location of the parting plane is specified by the draft. (Top right) Part with draft unspecified. (Bottom) Various options to produce the top-right part, including a no-draft design.

FIGURE 11-19   Typical guidelines for section change transitions in castings.

If $D > 1.5''$ and $d < 2D/3$, then $r = D/3$ with a 15° slope between the two parts

If $D > 1.5''$ and $d > 2/3\ D$, then $r = D/3$

If $D < 1.5''$ and $d > 2/3\ D$, then $r = 1/2''$

may contain shrinkage cavities and porosity, or have weaker, large grain-size structures. Ideally, a casting should have uniform thickness at all locations. Instead of thicker sections, ribs or other geometric features can often be used to impart additional strength while maintaining uniform wall thickness. When the section thickness must change, it is best if these changes are gradual, as indicated in the recommendations of Figure 11-19.

When sections of castings intersect, as in Figure 11-20a, two problems can arise. The first of these is *stress concentration*. Generous *fillets* (inside radii) at all interior corners can better distribute stresses and help to minimize potential problems, including shrinkage cracks. If the fillets are excessive, however, the additional material can aug-

FIGURE 11-20   (a) The "hot spot" at section r_2 is caused by intersecting sections. (b) An interior fillet and exterior radius lead to more uniform thickness and more uniform cooling.

(a)   (b)

ment the second problem, known as *hot spots*. Thick sections, like those at the intersection in Figure 11-20a and those illustrated in Figure 11-21, cool more slowly than other locations and tend to be sites of localized shrinkage. Shrinkage voids can be sites of subsequent failure and should be prevented if at all possible. Where thick sections must exist, an adjacent riser is often used to feed the section during solidification and shrinkage. If the riser is designed properly, the shrinkage cavity will lie totally within the riser, as illustrated in Figure 11-22, and will be removed when the riser is cut off. Sharp exterior corners tend to cool faster than the other sections of a casting. If an exterior radius is provided, the surface area can be reduced and cooling slowed to be more consistent with the surrounding material. Figure 11-20b shows a recommended modification to Figure 11-20a.

FIGURE 11-21   Hot spots often result from intersecting sections of various thickness.

FIGURE 11-22   Attached risers can move the shrinkage cavity external to the actual casting.

FIGURE 11-23 Using staggered ribs to prevent cracking during cooling.

When sections intersect to form continuous ribs, like those in Figure 11-23, contraction occurs in opposite directions as each of the arms cools and shrinks. As a consequence, cracking frequently occurs at the intersections. By staggering the ribs, as shown in the second portion of Figure 11-23, there is opportunity for distortion to occur that would provide relaxation to the high residual stresses that might otherwise induce cracking.

The location of the parting line may also be an appearance consideration. A small amount of fin, or flash, is often present at the parting line, and when the flash is removed (or left in place if it is small enough), a line of surface imperfection results. If the location is in the middle of a flat surface, it will be clearly visible in the product. If the parting line can be moved to coincide with a corner, however, the associated "defect" will go largely unnoticed.

Thin-walled castings are often desired because of their reduced weight, but thin walls can often present manufacturing problems related to mold filling (premature freezing before complete fill). Minimum section thickness should always be considered when designing castings. Specific values are rarely given, however, because they tend to vary with the shape and size of the casting, the type of metal being cast, the method of casting, and the practice of an individual foundry. Table 11-3 presents typical minimum-thickness values for several cast materials and casting processes. Zinc die casting can now produce walls as thin as 0.5 mm.

Casting design can often be aided by *computer simulation*. The mathematics of fluid flow can be applied to *mold filling*, and the principles of heat transfer can be used for *solidification modeling*. The mathematical tools of finite element or finite difference calculations can be coupled with the use of high-speed computers to permit beneficial design changes before the manufacture of patterns or molds.

TABLE 11-3	Typical Minimum Section Thickness for Various Engineering Metals and Casting Processes		
Casting Method	Minimum Section Thickness (mm)		
	Aluminum	Magnesium	Steel
Sand casting	3.18	3.96	4.75
Permanent mold	2.36	3.18	—
Die cast	1.57	2.36	—
Investment cast	1.57	1.57	2.36
Plaster mold	2.03	—	—

■ 11.7 THE CASTING INDUSTRY

The U.S. metal-casting industry ships over 14 million pounds of castings every year, with gray iron, ductile iron, aluminum alloys, and copper-base metals comprising the major portion. Thirty-five percent of the market is directed toward automotive and light truck manufacture. The average 2005 passenger car and light truck contained 75 kg (160 pounds) of iron castings and 115 kg (250 pounds) of cast aluminum. Magnesium castings are also beginning to achieve a presence.

Metal castings form primary components in: agricultural implements; construction equipment; mining equipment; valves and fittings; metalworking machinery; power tools; pumps and compressors; railroad equipment; power transmission equipment; and heating, refrigeration, and air-conditioning equipment. Ductile iron pipe is a mainstay for conveying pressurized fluids, and household appliances and electronics all utilize metal castings.

■ Key Words

allowance	Chvorinov's rule	cooling rate	deformation processes
blind riser	cold shut	cope	directional solidification
casting	columnar zone	core	distortion
chill	computer simulation	core box	draft
chill zone	consolidation processes	core print	drag
choke	cooling curve	dead riser	dross

equiaxed zone
expendable-mold process
external chill
fillet
filters
flask
fluidity
freezing range
gas flushing
gas porosity
gate
gating system
grain refinement
growth
hot spot
hot tears
inoculation

insulating sleeve
internal chill
liquidus
live riser
local solidification time
machining
machining allowance
material removal
materials processing
misruns
mold cavity
mold constant
mold material
multiple-use mold
nucleation
open riser
parting line (parting surface)

pattern
penetration
pouring cup
pouring temperature
powder metallurgy
reuseable mold
riser
runner
runner extension
shrink rule
shrinkage allowance
side riser
single-use mold
slag
sleeves
solidification
solidification modeling

solidification shrinkage
solidus
sprue
sprue well
stress concentrators
superheat
thermal arrest
top riser
total solidification time
turbulent flow
undercooling
vacuum degassing
vent
yield

■ Review Questions

1. What are the six activities that are conducted on almost every manufactured product?
2. What is "materials processing"?
3. What are the four basic families of shape-production processes? Cite one advantage and one limitation of each family.
4. Describe the capabilities of the casting process in terms of size and shape of the product.
5. How might the desired production quantity influence the selection of a single-use or multiple-use molding process?
6. Why is it important to provide a means of venting gases from the mold cavity?
7. What types of problem or defect can occur if the mold material provides too much restraint to the solidifying and cooling metal?
8. What is a casting pattern? Flask? Core? Mold cavity? Riser?
9. What are some of the components that combine to make up the gating system of a mold?
10. What is a parting line or parting surface?
11. What is draft and why is it used?
12. What are the two stages of solidification, and what occurs during each?
13. Why is it that most solidification does not begin until the temperature falls somewhat below the equilibrium melting temperature (i.e., undercooling is required)?
14. Why might it be desirable to promote nucleation in a casting through inoculation or grain refinement processes?
15. Heterogeneous nucleation begins at preferred sites within a mold. What are some probable sites for heterogeneous nucleation?
16. Why might directional solidification be desirable in the production of a cast product?
17. Describe some of the key features observed in the cooling curve of a pure metal.
18. What is superheat?
19. What is the freezing range for a metal or alloy?
20. Discuss the roles of casting volume and surface area as they relate to the total solidification time and Chvorinov's rule.
21. What characteristics of a specific casting process are incorporated into the mold constant, B, of Chvorinov's rule?
22. What is the correlation between cooling rate and final properties of a casting?
23. What is the chill zone of a casting, and why does it form?
24. Which of the three regions of a cast structure is least desirable? Why are its properties highly directional?
25. What is dross or slag, and how can it be prevented from becoming part of a finished casting?
26. What are some of the possible approaches that can be taken to prevent the formation of gas porosity in a metal casting?
27. What is fluidity, and how can it be measured?
28. What is a misrun or cold shut, and what causes them to form?
29. What defect can form in sand castings if the pouring temperature is too high and fluidity is too great?
30. Why is it important to design the geometry of the gating system to control the rate of metal flow as it travels from the pouring cup into the mold cavity?
31. What are some of the undesirable consequences that could result from turbulence of the metal in the gating system and mold cavity?
32. What is a choke, and how does its placement affect metal flow?
33. What features can be incorporated into the gating system to aid in trapping dross and loose mold material that is flowing with the molten metal?
34. What features of the metal being cast tend to influence whether the gating system is designed to minimize turbulence and reduce dross, or promote rapid filling to minimize temperature loss?
35. What are the three stages of contraction or shrinkage as a liquid is converted into a finished casting?
36. Why is it more difficult to prevent shrinkage voids from forming in metals or alloys with large freezing ranges?
37. What type of flaws or defects form during the cooling of an already-solidified casting?
38. Why is it desirable to design a casting to have directional solidification sweeping from the extremities of a mold toward a riser?
39. Based on Chvorinov's rule, what would be an ideal shape for a casting riser? A desirable shape from a practical perspective?
40. What is "yield," and how does it relate to the number and size of the specified risers?
41. Define the following riser-related terms: *top riser, side riser, open riser, blind riser, live riser*, and *dead riser*.
42. What assumptions were made when using Chvorinov's rule to calculate the size of a riser in the manner presented in the text? Why is the mold constant, B, not involved in the calculations?

43. Discuss aspects relating to the connection between a riser and the casting.
44. What is the purpose of a chill? Of an insulating sleeve? Of exothermic material?
45. What types of modifications or allowances are generally incorporated into a casting pattern?
46. What is a shrink rule, and how does it work?
47. What is the purpose of a draft or taper on pattern surfaces?

48. Why is it desirable to make the pattern allowances as small as possible?
49. What are some of the features of the casting process that are directly related to the location of the parting plane?
50. What are "hot spots" and what sort of design features cause them to form?
51. Are metal castings used in passenger cars and light trucks? To what extent?

■ Problems

1. Using Chvorinov's rule as presented in the text, with $n = 2$, calculate the dimensions of an effective riser for a casting that is a rectangular plate 2 in. by 4 in. by 6 in. with the dimensions. Assume that the casting and riser are not connected, except through a gate and runner, and that the riser is a cylinder of height/diameter ratio $H/D = 1.5$. The finished casting is what fraction of the combined weight of the riser and casting?
2. Reposition the riser in Problem 1 so that it sits directly on top of the flat rectangle, with its bottom circular surface being part

of the surface of the casting, and recompute the size and yield fraction. Which approach is more efficient?
3. A rectangular casting having the dimensions 3 in. by 5 in. by 10 in. solidifies completely in 11.5 minutes. Using $n = 2$ in Chvorinov's rule, calculate the mold constant, B. Then compute the solidification time of a casting with the dimensions 0.5 in. by 8 in. by 8 in. poured under the same conditions.

www.wiley.com/college/degarmo

*C*hapter 11   CASE STUDY

The Cast Oil-Field Fitting

A cast iron, T-type fitting is being produced for the oil drilling industry, using an air-set or no-bake sand for both the mold and the core. A silica sand has been used in combination with a catalyzed alkyd-oil/urethane binder. The figure shows a cross section of the mold with the core in place (a) and a cross section of the finished casting (b). The final casting contains several significant defects. Gas bubbles are observed in the bottom section of the horizontal tee. A penetration defect is observed near the bottom of the inside diameter, and there is an enlargement of the casting at location C.

1. What is the most likely source of the gas bubbles? Why are they present only at the location noted? What might you recommend as a solution?

2. What factors may have caused the penetration defect? Why is the defect present on the inside of the casting but not on the outside? Why is the defect near the bottom of the casting but not near the top?

3. What factors led to the enlargement of the casting at point C? What would you recommend to correct this problem?

4. Another producer has noted penetration defects on all surfaces of his castings, both interior and exterior. What would be some possible causes? What could you recommend as possible cures?

5. Could these molds and cores be reclaimed (i.e., recycled) after breakout? Discuss.

(a)

(b)

EXPENDABLE-MOLD CASTING PROCESSES

■ 12.1 INTRODUCTION

The versatility of metal casting is made possible by a number of distinctly different processes, each with its own set of characteristic advantages and benefits. Selection of the best process requires a familiarization with the various options and capabilities as well as an understanding of the needs of the specific product. Some factors to be considered include the desired dimensional precision and surface quality, the number of castings to be produced, the type of pattern and core box that will be needed, the cost of making the required mold or die, and restrictions imposed by the selected material.

As we begin to survey the various casting processes, it is helpful to have some form of process classification. One approach focuses on the molds and patterns and utilizes the following three categories:

1. Single-use molds with multiple-use patterns
2. Single-use molds with single-use patterns
3. Multiple-use molds

Categories 1 and 2 are often combined under the more general heading *expendable-mold casting processes*, and these processes will be presented in this chapter. Sand, plaster, ceramics, or other refractory materials are combined with binders to form the mold. Those processes where a mold can be used multiple times will be presented in Chapter 13. The multiple-use molds are usually made from metal.

Since the casting processes are primarily used to produce metal products, the emphasis of the casting chapters will be on metal casting. The metals most frequently cast are iron, steel, stainless steel, aluminum alloys, brass, bronze and other copper alloys, magnesium alloys, certain zinc alloys, and nickel-based superalloys. Among these, cast iron and aluminum are the most common, primarily because of their low cost, good fluidity, adaptability to a variety of processes, and the wide range of product properties that are available. The processes used to fabricate products from polymers, ceramics (including glass), and composites, including casting processes, will be discussed in Chapter 15.

■ 12.2 SAND CASTING

Sand casting is by far the most common and possibly the most versatile of the casting processes, accounting for over 90% of all metal castings. Granular refractory material (such as silica, zircon, olivine, or chromite sand) is mixed with small amounts of other materials, such as clay and water, and is then packed around a pattern that has the shape of the desired casting. Because the grains can pack into thin sections and can be economically used in large quantities, products spanning a wide range of sizes and detail can be made by this method. If the pattern is to be removed before pouring, the mold is usually made in two or more segments. An opening called a *sprue hole* is cut from the top of the mold through the sand and connected to a system of channels called *runners*. The molten metal is poured down the sprue hole, flows through the runners, and enters the mold cavity through one or more openings, called *gates*. Gravity flow is the most common means of introducing the metal into the mold. The metal is allowed to solidify, and the mold is then broken to permit removal of the finished casting. Because the mold is destroyed in product removal, a new mold must be made for each casting. Figure 12-1 shows the essential

FIGURE 12-1 Sequential steps in making a sand casting. (a) A pattern board is placed between the bottom (drag) and top (cope) halves of a flask, with the bottom side up. (b) Sand is then packed into the bottom or drag half of the mold. (c) A bottom board is positioned on top of the packed sand, and the mold is turned over, showing the top (cope) half of pattern with sprue and riser pins in place. (d) The upper or cope half of the mold is then packed with sand. (e) The mold is opened, the pattern board is drawn (removed), and the runner and gate are cut into the bottom parting surface of the sand. (e') The parting surface of the upper or cope half of the mold is also shown with the pattern and pins removed. (f) The mold is reassembled with the pattern board removed, and molten metal is poured through the sprue. (g) The contents are shaken from the flask and the metal segment is separated from the sand, ready for further processing.

steps and basic components of a sand casting process. A two-part cope-and-drag mold is illustrated, and the casting incorporates both a core and a riser (discussed in Chapter 11).

PATTERNS AND PATTERN MATERIALS

The first step in making a sand casting is the design and construction of a *pattern*. This is a duplicate of the part to be cast, modified in accordance with the requirements of the casting process, the metal being cast, and the particular molding technique that is being used. Selection of the pattern material is determined by the number of castings to be made, the size and shape of the casting, the desired dimensional precision, and the molding process. Wood patterns are relatively easy to make and are frequently used when small quantities of castings are required. Wood, however, is not very dimensionally stable. It may warp or swell with changes in humidity, and it tends to wear with repeated use. Metal patterns are more expensive but are more stable and durable. Hard plastics, such as urethanes, offer another alternative and are often preferred with processes that use strong, organically bonded sands that tend to stick to other pattern materials. In the full-mold and lost-foam processes, expanded polystyrene (EPS) is used, and investment casting uses patterns made from wax. In the latter processes, both the pattern and the mold are single-use, each being destroyed when a casting is produced.

TYPES OF PATTERNS

Many types of patterns are used in the foundry industry, with selection being based on the number of duplicate castings required and the complexity of the part.

One-piece or *solid patterns*, such as the one shown in Figure 12-2, are the simplest and often the least expensive type. They are essentially a duplicate of the part to be cast, modified only by the various allowances discussed in Chapter 11 and by the possible addition of core prints. One-piece patterns are relatively cheap to construct, but the subsequent molding process is usually slow. As a result, they are generally used when the shape is relatively simple and the number of duplicate castings is rather small.

If the one-piece pattern is simple in shape and contains a flat surface, it can be placed directly on a *follow board*. The entire mold cavity will be created in one segment of the mold, with the follow board forming the parting surface. If the parting plane is to be more centrally located, special follow boards are produced with inset cavities that position the one-piece pattern at the correct depth for the parting line. Figure 12-3 illustrates this technique, where the follow board again forms the parting surface.

Split patterns are used when moderate quantities of a casting are desired. The pattern is divided into two segments along what will become the parting plane of the mold. The bottom segment of the pattern is positioned in the drag portion of a flask, and the bottom segment of the mold is produced. This portion of the flask is then inverted, and the upper segment of the pattern and flask are attached. Tapered pins in the cope half of the pattern align with holes in the drag segment to assure proper positioning. Mold material is then packed around the full pattern to form the upper segment (*cope*) of the mold. The two segments of the flask are separated, and the pattern pieces are removed to produce the mold cavity. Sprues and runners are cut, and the mold is then reassembled, ready for

FIGURE 12-2   Single-piece pattern for a pinion gear.

Pattern   Parting line   Pattern

Follow board

FIGURE 12-3   Method of using a follow board to position a single-piece pattern and locate a parting surface. The final figure shows the flask of the previous operation (the drag segment) inverted in preparation for construction of the upper portion of the mold (cope segment).

FIGURE 12-4   Split pattern, showing the two sections together and separated. The light-colored portions are core prints.

FIGURE 12-5   Match-plate pattern used to produce two identical parts in a single flask. (Left) Cope side; (right) drag side. (*Note:* The views are opposite sides of a single-pattern board.)

pour. Figure 12-4 shows a split pattern that also contains several core prints (lighter color).

Match-plate patterns, like the one shown in Figure 12-5, further simplify the process and can be coupled with modern molding machines to produce large quantities of duplicate molds. The cope and drag segments of a split pattern are permanently fastened to opposite sides of a wood or metal *match plate*. The match plate is positioned between the upper and lower flask segments. Mold material is then packed on both sides of the match plate to form the cope and drag segments of a two-part mold. The mold sections are then separated and the match-plate pattern is removed. The pins and guide holes ensure that the cavities in the cope and drag will be in proper alignment upon reassembly. The necessary gates, runners, and risers are usually incorporated on the match plate as well. This guarantees that these features will be uniform and of the proper size in each mold, thereby reducing the possibility of defects. Figure 12-5 further illustrates the common practice of including more than one pattern on a single match plate.

When large quantities of identical parts are to be produced, or when the casting is quite large, it may be desirable to have the cope and drag halves of split patterns attached to separate pattern boards. These *cope-and-drag patterns* enable independent molding of the cope and drag segments of a mold. Large molds can be handled more easily in separate segments, and small molds can be made at a faster rate if a machine is only producing one segment. Figure 12-6 shows the mating pieces of a typical cope-and-drag pattern.

When the geometry of the product is such that a one-piece or two-piece pattern could not be removed from the molding sand, a *loose-piece pattern* can sometimes be de-

FIGURE 12-6   Cope-and-drag pattern for producing two heavy parts. (Left) Cope section; (right) drag section. (*Note:* These are two separate pattern boards.)

FIGURE 12-7 Loose-piece pattern for molding a large worm gear. After sufficient sand has been packed around the pattern to hold the pieces in position, the wooden pins are withdrawn. The mold is then completed, after which the pieces of the pattern can be removed in a designated sequence.

veloped. Separate pieces are joined to a primary pattern segment by beveled grooves or pins (Figure 12-7). After molding, the primary segment of the pattern is withdrawn. The hole that is created then permits the remaining segments to be sequentially extracted. Loose-piece patterns are expensive. They require careful maintenance, slow the molding process, and increase molding costs. They do, however, enable the sand casting of complex shapes that would otherwise require the full-mold, lost-foam, or investment processes.

SANDS AND SAND CONDITIONING

The sand used to make molds must be carefully prepared if it is to provide satisfactory and uniform results. Ordinary silica (SiO_2), zircon, olivine, or chromite sands are compounded with additives to meet four requirements:

1. *Refractoriness*: the ability to withstand high temperatures without melting, fracture, or deterioration
2. *Cohesiveness* (also referred to as *bond*): the ability to retain a given shape when packed into a mold
3. *Permeability*: the ability of mold cavity, mold, and core gases to escape through the sand
4. *Collapsibility*: the ability to accommodate metal shrinkage after solidification and provide for easy removal of the casting through mold disintegration (*shakeout*)

Refractoriness is provided by the basic nature of the sand. Cohesiveness, bond, or strength is obtained by coating the sand grains with clays, such as bentonite, kaolinite, or illite, that become cohesive when moistened. Collapsibility is sometimes enhanced by adding cereals or other organic materials, such as cellulose, that burn out when they come in contact with the hot metal. The combustion of these materials reduces both the volume and strength of the restraining sand. Permeability is a function of the size of the sand particles, the amount and type of clay or bonding agent, the moisture content, and the compacting pressure.

Good molding sand always represents a compromise between competing factors. The size of the sand particles, the amount of bonding agent (such as clay), the moisture content, and the organic additives are all selected to obtain an acceptable compromise among the four basic requirements. The overall composition must be carefully controlled to ensure satisfactory and consistent results. Since molding material is often reclaimed and recycled, the temperature of the mold during pouring and solidification is also important. If organic materials have been incorporated into the mix to provide collapsibility, a portion will burn during the pour. Adjustments will be necessary, and ultimately some or all of the mold material may have to be discarded and replaced with new.

A typical green-sand mixture contains about 88% silica sand, 9% clay, and 3% water. To achieve good molding, it is important for each grain of sand to be coated uniformly with the proper amount of additive agents. This is achieved by putting the ingredients through a *muller*, a device that kneads, rolls, and stirs the sand. Figure 12-8 shows both a continuous and batch-type muller, with each producing the desired mixing

FIGURE 12-8 Schematic diagram of a continuous (left) and batch-type (right) sand muller. Plow blades move and loosen the sand, and the muller wheels compress and mix the components. *(Courtesy of ASM International. Metals Park, OH.)*

through the use of rotating blades that lift, fluff, and redistribute the material and wheels that compress and squeeze. After mixing, the sand is often discharged through an aerator, which fluffs it for further handling.

SAND TESTING

If a foundry is to produce high-quality products, it is important that it maintain a consistent quality in its molding sand. The sand itself can be characterized by grain size, grain shape, surface smoothness, density, and contaminants. Blended molding sand can be characterized by moisture content, clay content, and *compactability*. Key properties of compacted sand or finished molds include *mold hardness*, *permeability*, and *strength*. Standard tests and procedures have been developed to evaluate many of these properties.

Grain size can be determined by shaking a known amount of clean, dry sand downward through a set of 11 standard screens or sieves of decreasing mesh size. After being shaken for 15 minutes, the amount of material remaining on each sieve is weighed, and these weights are used to compute an AFS (American Foundry Society) grain fineness number.

Moisture content can be determined by a special device that measures the electrical conductivity of a small sample of compressed sand. A more direct method is to measure the weight lost by a 50-gram sample after it has been subjected to a temperature of about 110°C (230°F) for sufficient time to drive off all the water.

Clay content is determined by washing the clay from a 50-gram sample of molding sand, using water that contains sufficient sodium hydroxide to make it alkaline. Several cycles of agitation and washing may be required to fully remove the clay. The remaining sand is then dried and weighed to determine the amount of clay removed from the original sample.

Permeability and strength tests are conducted on compacted sands, using a *standard rammed specimen*. An amount of sand is first placed into a 2-inch-diameter steel tube. A 14-pound weight is then dropped on it three times from a height of 2 inches, and the height of the resulting specimen must be within $\frac{1}{32}$ inch of a targeted 2-inch height.

Permeability is a measure of how easily gases can pass through the narrow voids between the sand grains. Air in the mold before pouring, plus the steam that is produced when the hot metal contacts the moisture in the sand along with various combustion gases, must all be allowed to escape, rather than prevent mold filling or be trapped in the casting as porosity or blowholes. During the permeability test, shown schematically in Figure 12-9, a sample tube containing the standard rammed specimen is subjected to an air pressure of 10 g/cm². By means of either a flow rate determination or a measurement of the steady-state pressure between the orifice and the sand specimen, an *AFS permeability number*[1] can be computed. Most test devices are now calibrated to provide a direct readout of this number.

FIGURE 12-9 Schematic of a permeability tester in operation. A standard sample in a metal sleeve is sealed by an O-ring onto the top of the unit while air is passed through the sand. *(Courtesy of Dietert Foundry Testing Equipment Inc, Detroit, MI)*

Sand
Ramming cylinder
Pressure measured between orifice and sand
Calibrated orifice
O-ring
Air blown or forced in under a constant pressure

---

[1]The AFS permeability number is defined as:

$$\text{AFS number} = (V \times H)/(P \times A \times T)$$

where V is the volume of air (2000 cm³), H is the height of the specimen (5.08 cm), P is the pressure (10 g/cm²), A is the cross-section area of the specimen (20.268 cm²), and T is the time in seconds to pass a flow of 2000 cm³ of air through the specimen. Substituting each of the above constants, the permeability number becomes equal to $3000.2/T$.

FIGURE 12-10   Sand mold hardness tester. *(Courtesy of Dietert Foundry Testing Equipment Inc., Detroit, MI)*

All molding material must have sufficient strength to retain the integrity of the mold cavity while the mold is being handled between molding and pouring. The mold material must also withstand the erosion of the liquid metal as it flows into the mold and the pressures induced by a column of molten metal. The *compressive strength* of the sand (also referred to as *green compressive strength*) is a measure of the mold strength at this stage of processing. It is determined by removing the rammed specimen from the compacting tube and placing it in a mechanical testing device. A compressive load is then applied until the specimen breaks, which usually occurs in the range of 10 to 30 psi (0.07 to 0.2 MPa). If there is too little moisture in the sand, the grains will be poorly bonded and strength will be poor. If there is excess moisture, the extra water acts as a lubricant and strength is again poor. In between, there is a condition of maximum strength with an optimum water content that will vary with the content of other materials in the mix. A similar optimum also applies to permeability, since unwetted clay blocks vent passages, as does excess water. Sand coated with a uniform thin film of moist clay provides the best molding properties. A ratio of one part water to three parts clay (by weight) is often a good starting point.

The *hardness* of compacted sand can give additional insight into the strength and permeability characteristics of a mold. Hardness can be determined by the resistance of the sand to the penetration of a 0.2-in. (5.08-mm)-diameter spring-loaded steel ball. A typical test instrument is shown in Figure 12-10.

Compactibility is determined by sifting loose sand into a steel cylinder, leveling off the column, striking it three times with the standard weight (as in making a standard rammed specimen), and then measuring the final height. The *percent compactibility* is the change in height divided by the original height, times 100%. This value can often be correlated with the moisture content of the sand, where a compactibility of around 45% indicates a proper level of moisture. A low compactibility is usually associated with too little moisture.

SAND PROPERTIES AND SAND-RELATED DEFECTS

The characteristics of the sand granules themselves can be very influential in determining the properties of foundry molding material. Round grains give good permeability and minimize the amount of clay required because of their low surface area. Angular sands give better green strength because of the mechanical interlocking of the grains. Large grains provide good permeability and better resistance to high-temperature melting and expansion, while fine-grained sands produce a better surface finish on the final casting. Uniform-size sands give good permeability, while a distribution of sizes enhances surface finish.

Silica sand is cheap and lightweight, but when hot metal is poured into a silica sand mold, the sand becomes hot, and at or about 585°C (1085°F) it undergoes a phase transformation that is accompanied by a substantial expansion in volume. Because sand is a poor thermal conductor, only the sand that is adjacent to the mold cavity becomes hot and expands. The remaining material stays fairly cool, does not expand, and often provides a high degree of mechanical restraint. Because of this uneven heating, the sand at the surface of the mold cavity may buckle or fold. Castings with large, flat surfaces are more prone to *sand expansion defects* since a considerable amount of expansion must occur in a single direction.

Sand expansion defects can be minimized in a number of ways. Certain particle geometries permit the sand grains to slide over one another, thereby relieving the expansion stresses. Excess clay can be added to absorb the expansion, or volatile additives, such as cellulose, can be added to the mix. When the casting is poured, the cellulose burns, creating voids that can accommodate the sand expansion. Another alternative is the use of olivine or zircon sand in place of silica. Since these sands do not undergo phase transformations upon heating, their expansion is only about one-half that of silica sand. Unfortunately, these sands are much more expensive and heavier in weight than the more commonly used silica.

Trapped or evolved gas can create gas-related *voids* or *blows* in finished castings. The most common causes are low sand permeability (often associated with angular, fine, or wide-size distribution sands, fine sand additives, and overcompaction) and large

TABLE 12-1	Desirable Properties of a Sand-Based Molding Material

1. Is inexpensive in bulk quantities
2. Retains properties through transportation and storage
3. Uniformly fills a flask or container
4. Can be compacted or set by simple methods.
5. Has sufficient elasticity to remain undamaged during pattern withdrawal
6. Can withstand high temperatures and maintains its dimensions until the metal has solidified
7. Is sufficiently permeable to allow the escape of gases
8. Is sufficiently dense to prevent metal penetration
9. Is sufficiently cohesive to prevent wash-out of mold material into the pour stream
10. Is chemically inert to the metal being cast
11. Can yield to solidification and thermal shrinkage, thereby preventing hot tears and cracks
12. Has good collapsibility to permit easy removal and separation of the casting
13. Can be recycled

amounts of evolved gas due to high mold-material moisture or excessive amounts of volatiles. If adjustments to the mold composition are not sufficient to eliminate the voids, vent passages may have to be cut into the mold, a procedure that may add significantly to the mold-making cost.

Molten metal can also penetrate between the sand grains, causing the mold material to become embedded in the surface of the casting. This defect, known as *penetration*, can be the result of high pouring temperatures (excess fluidity), high metal pressure (possibly due to excessive cope height or pouring from too high an elevation above the mold), or the use of high-permeability sands with coarse, uniform particles. Fine-grained materials, such as silica flour, can be blended in to fill the voids, but this reduces permeability and increases the likelihood of both gas and expansion defects.

Hot tears or *cracks* can form in castings made from metals or alloys with large amounts of solidification shrinkage. As the metal contracts during solidification and cooling to room temperature, it may find itself restrained by a strong mold or core. Tensile stresses can develop while the metal is still partially liquid or fully solidified but still hot and weak. If these stresses become great enough, the casting will crack. Hot tears are often attributed to poor mold collapsibility. Additives, such as cellulose, can be used to improve the collapsibility of sand molds.

Table 12-1 summarizes the many desirable properties of a sand-based molding material.

THE MAKING OF SAND MOLDS

When only a few castings are to be made, *hand ramming* is often the preferred method of packing sand to make a sand mold. Hand ramming, however, is slow, labor intensive, and usually results in nonuniform compaction. For normal production, sand molds are generally made using specially designed molding machines. The various methods differ in the type of flask, the way the sand is packed within the flask, whether mechanical assistance is provided to turn or handle the mold, and whether a flask is even required. In all cases, however, the molding machines greatly reduce the labor and required skill, and also lead to castings with good dimensional accuracy and consistency.

Molding usually begins with a pattern, like the match-plate pattern discussed earlier, and a *flask*. The flasks may be straight-walled containers with guide pins or removable jackets, and they are generally constructed of lightweight aluminum or magnesium. Figure 12-11 shows a snap flask, so named because it is designed to snap open for easy removal after the mold material has been packed in place.

The mixed sand (mold material) can be packed in the flask by one or more basic techniques. A *sand slinger* uses a rotating impeller to fling or throw sand against the

FIGURE 12-11 Bottom and top halves of a snap flask. (Left) drag segment in closed position; (right) cope segment with latches opened for easy removal.

FIGURE 12-12 Jolting a mold section. (*Note:* The pattern is on the bottom, where the greatest packing is expected.)

FIGURE 12-13 Squeezing a sand-filled mold section. While the pattern is on the bottom, the highest packing will be directly under the squeeze head.

pattern. The slinger is manipulated to progressively deposit compacted sand into the mold. Sand slinging is a common method of achieving uniform sand compaction when making large molds and large castings.

In a method known as *jolting*, a flask is positioned over a pattern, filled with sand, and the pattern, flask, and sand are then lifted and dropped several times, as shown in Figure 12-12. The weight and kinetic energy of the sand produces optimum packing at the bottom of the mass, that is, around the pattern. Jolting machines can be used on the first half of a match-plate pattern or on both halves of a cope-and-drag operation.

Squeezing machines use an air-operated squeeze head, a flexible diaphragm, or small, individually activated squeeze heads to compact the sand. The squeezing motion provides firm packing adjacent to the squeeze head, with density diminishing as you move farther into the mold. Figure 12-13 illustrates the squeezing process, and Figure 12-14 compares the density achieved by squeezing with a flat plate and squeezing with a flexible diaphragm.

In match-plate molding, a combination of jolting and squeezing is often used to produce a more uniform density throughout the mold. The match-plate pattern is positioned between the cope and drag sections of a flask, and the assembly is placed drag side up on the molding machine. A parting compound is sprinkled on the pattern, and the drag section of the flask is filled with mixed sand. The entire assembly is then jolted a specified number of times to pack the sand around the drag side of the pattern. A squeeze head is then swung into place, and pressure is applied to complete the drag portion of the mold. The entire flask is then inverted, and a squeezing operation is performed to compact loose sand in the cope segment. (*Note:* Jolting here might cause the already-compacted sand to break free of the inverted drag section of the pattern!) Since the drag segment sees both jolting and squeezing, while the cope is only squeezed, the pattern side with the greatest detail is generally placed in the drag. If the cope and drag segments of a mold are made on separate machines (using separate cope-and-drag patterns), the combination of jolting and squeezing can be performed on each segment of the mold.

The sprue hole is most often cut by hand, with this operation being performed before removal of the pattern to prevent loose sand from falling into the mold cavity. The

FIGURE 12-14 Schematic diagram showing relative sand densities obtained by flat-plate squeezing, where all areas get vertically compressed by the same amount of movement (left) and by flexible-diaphragm squeezing, where all areas flow to the same resisting pressure (right).

FIGURE 12-15 Activity sequence for automatic match-plate molding. Green sand is blown from the side and compressed vertically. The final mold is ejected from the flask and poured in a flaskless condition. (From "Five Considerations for Automatic Matchplate Molding," *Engineered Castings Solutions,* Winter, 2001, American Foundry Society, Schaumberg, IL.)

pouring basin may also be hand cut, or it may be shaped by a protruding segment on the squeeze board. The gates and runners are usually included on the pattern.

The pattern board is removed and the segments of the mold are reassembled ready for pour. Heavy metal weights are often placed on top of the molds to prevent the cope section from rising and "floating" when the hydrostatic pressure of the molten metal presses upward. The weights are left in place until solidification is complete, and they are then moved to other molds.

For mass-production molding, a number of automatic mold-making methods have been developed. These include *automatic match-plate molding,* automatic cope-and-drag molding, and methods that produce some form of stacked segments. Figure 12-15 shows the production sequence for one of the variations of automatic match-plate molding, where the sand is introduced into the cope-and-drag mold segments from the side and then vertically compressed. The two-part cope-and-drag mold is produced in one station, with a single pattern and one machine squeeze cycle. The compressed blocks are extracted from the molding machine and are poured in a flaskless condition.

Figure 12-16 depicts the *vertically parted flaskless molding* process, where the pattern has been rotated into a vertical position and the cope-and-drag impressions are now incorporated into opposing sides of a compaction machine. Molding sand is deposited between the patterns and squeezed with a horizontal motion. The patterns are withdrawn, cores are set, and the mold block is then joined to those that were previously molded. Since each block contains both a right-hand cavity and a left-hand cavity, an entire mold is made with each cycle of the machine. (*Note*: Previous techniques required two separate molding operations to produce the individual cope and drag segments of a two-part mold.) A vertical gating system is usually included on one side of the pattern, and the

FIGURE 12-16 Vertically parted flaskless molding with inset cores. Note how one mold block now contains both the cope-and-drag impressions.

assembled molds are usually poured individually. If a common horizontal runner is used to connect multiple mold segments, the method is known as the *H-process*. Since metal cools as it travels through long runners, the individual cavities of the H-process often fill with different-temperature metal. To assure product uniformity, most producers reject the H-process, preferring to pour their vertically parted molds individually.

In *stack molding*, sections containing a cope impression on the bottom and a drag impression on the top are piled vertically on top of one another. Metal is poured down a common vertical sprue, which is connected to horizontal gating systems at each of the parting planes.

For molds that are too large to be made either by hand ramming or by one of the previously discussed molding processes, large flasks can be placed directly on the foundry floor. Various types of mechanical aids, such as a sand slinger, can then be used to add and pack the sand. Pneumatic rammers can provide additional tamping. Even larger molds can be constructed in sunken pits. Because of the size, complexity, and need for strength, pit molds are often constructed by assembling smaller sections of baked or dried sand. Added binders may be required to provide the strength required for these large molds.

GREEN-SAND, DRY-SAND, AND SKIN-DRIED MOLDS

Green-sand casting (where the term *green* implies that the mold material has not been fired or cured) is the most widely used process for casting both ferrous and nonferrous metals. The mold material is composed of sand blended with clay, water, and additives, and the molds fill by gravity feed. Tooling costs are low, and the entire process is one of the least expensive of the casting methods. Almost any metal can be cast, and there are few limits on the size, shape, weight, and complexity of the products. Over the years, green-sand casting has evolved from a manually intensive operation to a mechanized and automated system capable of producing over 300 molds per hour. As a result, it can be economically applied to both small and large production runs.

Design limitations are usually related to the rough surface finish and poor dimensional accuracy—and the resulting need for finish machining. Still other problems can be attributed to the low strength of the mold material and the moisture that is present in the clay-and-water binder. Table 12-2 provides a process summary for green-sand casting, and Figure 12-17 shows a variety of parts that have been produced in aluminum.

Some of the problems associated with the green-sand process can be reduced if we heat the mold to a temperature between 150° and 300°C (300° to 575°F) and bake it until most of the moisture is driven off. This drying strengthens the mold and reduces the volume of gas generated when the hot metal enters the cavity. *Dry-sand molds* are very durable and may be stored for a relatively long period of time. They are not very popular, however, because of the long time required for drying, the added cost of that operation, and the availability of alternative processes. An attractive compromise may be the production of a *skin-dried mold*, drying only the sand that is adjacent to the mold cavity. Torches are often used to perform the drying, and the water is usually removed to a depth of about 13 mm ($\frac{1}{2}$ inch).

TABLE 12-2 Green-Sand Casting

Process: Sand, bonded with clay and water, is packed around a wood or metal pattern. The pattern is removed, and molten metal is poured into the cavity. When the metal has solidified, the mold is broken and the casting is removed.

Advantages: Almost no limit on size, shape, weight, or complexity; low cost; almost any metal can be cast.

Limitations: Tolerances and surface finish are poorer than in other casting processes; some machining is often required; relatively slow production rate; a parting line and draft are needed to facilitate pattern removal; due to sprues, gates, and risers, typical yields range from 50% to 85%.

Common metals: Cast iron, steel, stainless steel, and casting alloys of aluminum, copper, magnesium, and nickel.

Size limits: 30 g to 3000 kg (1 oz to 6000 lb).

Thickness limits: As thin as 0.25 cm ($\frac{3}{32}$ in.), with no maximum.

Typical tolerances: 0.8 mm for first 15 cm ($\frac{1}{32}$ in. for first 6 in.), 0.003 cm for each additional cm; additional increment for dimensions across the parting line

Draft allowances: 1–3°.

Surface finish: 2.5–25 microns (100–1000 μin.) rms.

FIGURE 12-17 A variety of sand cast aluminum parts. *(Courtesy of Bodine Aluminum Inc., St. Louis, MO)*

The molds used for the casting of large steel parts are almost always skin dried, because the pouring temperatures for steel are significantly higher than those for cast iron. These molds may also be given a high-silica wash prior to drying to increase the refractoriness of the surface, or the more thermally stable zircon sand may be used as a facing. Additional binders, such as molasses, linseed oil, or corn flour, can be added to the facing sand to enhance the strength of the skin-dried segment.

SODIUM SILICATE–CO$_2$ MOLDING

Molds (and cores) can also be made from sand that receives its strength from the addition of 3% to 6% *sodium silicate*, an inorganic liquid binder, commonly known as *water glass*. The sand can be mixed with the liquid sodium silicate in a standard muller and can be packed into flasks by any of the methods discussed previously in this chapter. It remains soft and moldable until it is exposed to a flow of CO$_2$ gas. It then hardens in a matter of seconds by the reaction:

$$Na_2SiO_3 + CO_2 \rightarrow Na_2CO_3 + SiO_2 \quad \text{(colloidal)}$$

The CO$_2$ gas is nontoxic, nonflammable, and odorless, and no heating is required to initiate or drive the reaction. The sands achieve a tensile strength of about 40 psi (0.3 MPa) after five seconds of CO$_2$ gassing, with strength increasing to 100–200 psi (0.7–1.4 MPa) after 24 hours of aging. The hardened sands, however, have extremely poor collapsibility, making shakeout and core removal quite difficult. Unlike most other sands, the heating that occurs as a result of the pour actually serves to make the mold stronger (a phenomenon similar to the firing of a ceramic material). Additives that will burn out during the pour are frequently used to enhance the collapsibility of sodium silicate molds. Care must also be taken to prevent the carbon dioxide in the air from hardening the premixed sand before the mold-making process is complete.

A modification of this process can be used when certain portions of a mold require better accuracy, thinner sections, or deeper draws than can be achieved with ordinary molding sand. Sand mixed with sodium silicate is packed around a special metal pattern to a thickness of about 1 in., followed by regular molding sand as a backing material. After the sand is fully compacted, CO_2 is introduced through vents in the metal pattern. The adjacent sand is further hardened, and the pattern can be withdrawn with less possibility of damage to the mold.

NO-BAKE, AIR-SET, OR CHEMICALLY BONDED SANDS

An alternative to the sodium silicate–CO_2 process involves room-temperature chemical reactions that can occur between organic or inorganic resin binders and liquid curing agents or catalysts. The two or more components are mixed with sand just prior to the molding operation, and the curing reactions begin immediately. The molds (or cores) are then made in a reasonably rapid fashion, since the mix remains workable for only a short period of time. After a few minutes to a few hours at room temperature (depending on the specific binder and curing agent), the sands harden sufficiently to permit removal from the pattern without concern for distortion. After time for additional curing and the possible application of a refractory coating, the molds are then ready for pour.

No-bake molding can be used with virtually all engineering metals over a wide range of product sizes and weights. Since the time for mold curing slows production, no-bake molding is generally limited to low to medium-production quantities. The cost of no-bake molding is about 20–30% greater than green-sand molding, so no-bake is generally used where offsetting savings can be achieved. Products can also be designed with thinner sections, deeper draws, and smaller draft, and the rigid molds enable high dimensional precision, along with good surface finish. Since no-bake sand can be compacted by only light vibrations, patterns can often be made from wood, plastic, fiberglass, or even Styrofoam, thereby reducing pattern cost.

A wide variety of *no-bake sand systems* are available, with selection being based on the metal being poured, the cure time desired, the complexity and thickness of the casting, and possible desire for sand reclamation. Like the molds produced by the sodium silicate process, no-bake offers good hot strength and high resistance to mold-related casting defects. In contrast to the sodium silicate material, however, the no-bake molds decompose readily after the metal has been poured, providing excellent shakeout characteristics. Permeability must be good, since the heat causes the resins to decompose to hydrogen, water vapor, carbon oxides, and various hydrocarbons—all gases that must be vented.

Air-set molding and *chemically bonded sands* are other terms that have been used to describe the no-bake process.

SHELL MOLDING

Another popular sand casting process is *shell molding*, the basic steps of which are described below and illustrated in Figure 12-18.

1. The individual grains of fine silica sand are first precoated with a thin layer of thermosetting phenolic resin and heat-sensitive liquid catalyst. This material is then dumped, blown, or shot onto a metal pattern (usually some form of cast iron) that has been preheated to a temperature between 230° and 315°C (450° and 600°F). During a period of sustained contact, heat from the pattern partially cures (polymerizes and crosslinks) a layer of material. This forms a strong, solid-bonded region adjacent to the pattern. The actual thickness of cured material depends on the pattern temperature and the time of contact but typically ranges between 10 and 20 mm (0.4 to 0.8 in.).

2. The pattern and sand mixture are then inverted, allowing the excess (uncured) sand to drop free. Only the layer of partially cured material remains adhered to the pattern.

3. The pattern with adhering shell is then placed in an oven, where additional heating completes the curing process.

FIGURE 12-18 Schematic of the dump-box version of shell molding. (a) A heated pattern is placed over a dump box containing granules of resin-coated sand. (b) The box is inverted, and the heat forms a partially cured shell around the pattern. (c) The box is righted, the top is removed, and the pattern and partially cured sand is placed in an oven to further cure the shell. (d) The shell is stripped from the pattern. (e) Matched shells are then joined and supported in a flask ready for pouring.

4. The hardened shell, with tensile strength between 350 and 450 psi (2.4–3.1 MPa), is then stripped from the pattern.

5. Two or more shells are then clamped or glued together with a thermoset adhesive to produce a mold, which may be poured immediately or stored almost indefinitely.

6. To provide extra support during the pour, shell molds are often placed in a pouring jacket and surrounded with metal shot, sand, or gravel.

Because the shell is formed and partially cured around a metal pattern, the process offers excellent dimensional accuracy. Tolerances of 0.08 to 0.13 mm (0.003 to 0.005 in.) are quite common. Shell-mold sand is typically finer than ordinary foundry sand and, in combination with the plastic resin, produces a very smooth casting surface. Cleaning, machining, and other finishing costs can be significantly reduced, and the mold process offers an excellent level of product consistency.

Figure 12-19 shows a set of metal patterns, the two shells before clamping, and the resulting shell-mold casting. Machines for making shell molds vary from simple ones for small operations to large, completely automated devices for mass production. The cost of a metal pattern is often rather high, and its design must include the gate and runner system, since these cannot be cut after molding. Large amounts of expensive binder are required, but the amount of material actually used to form a thin shell is not that great. High productivity, low labor costs, smooth surfaces, and a level of precision that reduces the amount of subsequent machining all combine to make the process economical for even moderate quantities. The thin shell provides for the easy escape of gases that evolve during the pour, and the volume of evolved gas is rather low because of the absence of moisture in the mold material. When the shell becomes hot, some of the resin binder burns out, providing excellent collapsibility and shakeout characteristics. In addition, both the molding sand and completed shells can be stored for indefinite periods of time. Table 12-3 summarizes the features of shell molding.

OTHER SAND-BASED MOLDING METHODS

Over the years, a variety of processes have been proposed to overcome some of the limitations or difficulties of the more traditional methods. While few have become commercially significant, several are included here to illustrate the nature of these efforts.

FIGURE 12-19 (Top) Two halves of a shell-mold pattern. (Bottom) The two shells before clamping, and the final shell-mold casting with attached pouring basin, runner, and riser. *(Courtesy of Shalco Systems, Lansing, MI.)*

In the *V-process* or *vacuum molding*, a vacuum performs the role of the sand binder. Figure 12-20 depicts the production sequence, which begins by draping a thin sheet of heat-softened plastic over a special vented pattern. A vacuum is applied within the pattern, drawing the sheet tight to its surface. A special vacuum flask is then placed over the pattern; the flask is filled with vibrated dry, unbonded sand; a sprue and pouring cup are formed; and a second sheet of plastic is placed over the mold. A vacuum is then drawn on the flask itself, compacting the sand to provide the necessary strength and hardness. The pattern vacuum is released, and the pattern is then withdrawn. The other segment of the two-part cope-and-drag mold is made in a similar fashion, and the mold halves are assembled to produce a plastic-lined cavity. The mold is then poured with a

TABLE 12-3   Shell-Mold Casting

Process: Sand coated with a thermosetting plastic resin is dropped onto a heated metal pattern, which cures the resin. The shell segments are stripped from the pattern and assembled. When the poured metal solidifies, the shell is broken away from the finished casting.

Advantages: Faster production rate than sand molding, high dimensional accuracy with smooth surfaces.

Limitations: Requires expensive metal patterns. Plastic resin adds to cost; part size is limited.

Common metals: Cast irons and casting alloys of aluminum and copper.

Size limits: 30 g (1 oz) minimum; usually less than 10 kg (25 lb); mold area usually less than 0.3 m^2 (500 in^2).

Thickness limits: Minimums range from 0.15 to 0.6 cm ($\frac{1}{16}$ to $\frac{1}{4}$ in.), depending on material.

Typical tolerances: Approximately 0.005 cm/cm or in/in.

Draft allowance: $\frac{1}{4}$ or $\frac{1}{2}$ degree.

Surface finish: $\frac{1}{3}$–4.0 microns (50–150 μin.) rms.

FIGURE 12-20 Schematic of the V-process or vacuum molding. (a) A vacuum is pulled on a pattern, drawing a heated shrink-wrap plastic sheet tightly against it. (b) A vacuum flask is placed over the pattern and filled with dry unbonded sand, a pouring basin and sprue are formed; the remaining sand is leveled; a second heated plastic sheet is placed on top; and a mold vacuum is drawn to compact the sand and hold the shape. (c) With the mold vacuum being maintained, the pattern vacuum is then broken and the pattern is withdrawn. The cope and drag segments are assembled, and the molten metal is poured.

vacuum of 300–600 torr being maintained in both the cope and drag segments of the flask. During the pour, the thin plastic film melts and vaporizes and is replaced immediately by metal, allowing the vacuum to continue holding the sand in shape until the casting has cooled and solidified. When the vacuum is released, the sand reverts to its loose, unbonded state and falls away from the casting.

With the vacuum serving as the binder, there is a total absence of moisture-related defects; binder cost is eliminated; and the loose, dry sand is completely and directly reusable. With no clay, water, or other binder to impair permeability, finer sands can be used, resulting in better surface finish in the resulting castings. With no burning binders, there are no fumes generated during the pouring operation. Shakeout characteristics are exceptional, since the mold collapses when the vacuum is released. Unfortunately, the process is relatively slow because of the additional steps and the time required to pull a sufficient vacuum. The V-process is used primarily for the production of prototype, frequently modified, or low- to medium-volume parts (more than 10 but less than 15,000).

In the *Eff-set process*, wet sand with just enough clay to prevent mold collapse is packed around a pattern. The pattern is removed, and the surface of the mold is sprayed with liquid nitrogen. The ice that forms serves as the binder, and the molten metal is poured into the mold while the surface is in its frozen condition. This process offers low binder cost and excellent shakeout but is not being used in a commercial operation.

■ 12.3 CORES AND CORE MAKING

Casting processes are unique in their ability to easily incorporate complex internal cavities or reentrant sections. To produce these features, however, it is often necessary to use *cores* as part of the mold. Figure 12-21 shows an example of a product that makes extensive use of cores to produce the various cylinders, cooling passages, and other internal features. While cores constitute an added cost, they significantly expand the capabilities of the process.

Cores can often be used to improve casting design and optimize processes. Consider the simple belt pulley shown schematically in Figure 12-22. Various methods of fabrication are suggested in the four sketches, beginning with the casting of a solid form and the subsequent machining of the through-hole for the drive shaft. A large volume of metal would have to be removed by a secondary machining process. A more economical approach would be to make the pulley with a cast-in hole. In Figure 12-22b each half of the pattern includes a tapered hole, which fills with the same green sand being used for the remainder of the mold. These protruding sections are an integral part

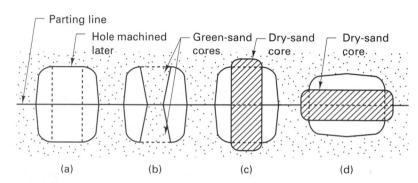

FIGURE 12-21   V-8 engine block (bottom center) and the five dry-sand cores that are used in the construction of its mold. *(Courtesy of General Motors Corporation, Detroit, MI.)*

FIGURE 12-22   Four methods of making a hole in a cast pulley. Three involve the use of a core.

of the mold, but they are also known as *green-sand cores*. Green-sand cores have a relatively low strength. If the protrusions are long or narrow, it might be difficult to withdraw the pattern without breaking them, or they might not have enough strength to even support their own weight. For long cores, a considerable amount of machining may still be required to remove the draft that must be provided on the pattern. In addition, green-sand cores are not an option for more complex shapes, where it might be impossible to withdraw the pattern.

Dry-sand cores can overcome some of the cited difficulties. These cores are produced separate from the remainder of the mold and are then inserted into core prints that hold them in position. The sketches in Figures 12-22c and 12-22d show dry-sand cores in the vertical and horizontal positions. Dry-sand cores can be made in a number of ways. In each, the sand, mixed with some form of binder, is packed into a wood or metal core box that contains a cavity of the desired shape. A *dump-core box* such as the one shown in Figure 12-23 offers the simplest approach. Sand is packed into

FIGURE 12-23   (Upper right) A dump-type core box; (bottom) two core halves ready for baking; and (upper left) a completed core made by gluing two opposing halves together.

the cavity and scraped level with the top surface (which acts like the parting line in a traditional mold). A wood or metal plate is then placed over the top of the box, and the box is inverted and lifted, leaving the molded sand resting on the plate. After baking or hardening, the core segments are assembled with hot-melt glue or some other bonding agent. Rough spots along the parting line are removed with files or sanding belts, and the final core may be given a thin coating to provide a smoother surface or greater resistance to heat. Graphite, silica, or mica can be sprayed or brushed onto the surface.

Single-piece cores can be made in a *split-core box*. Two halves of a core box are clamped together, with an opening in one or both ends through which sand is introduced and rammed. After the sand is compacted, the halves of the box are separated to permit removal of the core. Cores with a uniform cross section can be formed by a core-extruding machine and cut to the desired length as the product emerges. The individual cores are then placed in core supports for subsequent hardening. More complex cores can be made in core-blowing machines that use separating dies and receive the sand in a manner similar to injection molding or die casting.

Cores are frequently the most fragile part of a mold assembly. To provide the necessary strength, the various core-making processes utilize a number of special binders. In the *core-oil process*, sand is blended with about 1% vegetable or synthetic oil, along with 2–4% water and about 1% cereal or clay to help develop green strength (i.e., to help retain the shape prior to curing). The wet sand is blown or rammed into a relatively simple core box at room temperature. The fragile uncured cores are then gently transferred to flat plates or special supports and placed in convection ovens at 200° to 260°C (400° to 500°F) for curing. The heat causes the binder to cross-link or polymerize, producing a strong organic bond between the grains of sand. While the process is simple and the materials are inexpensive, the dimensional accuracy of the resultant cores is often difficult to maintain.

In the *hot-box method*, sand blended with a liquid thermosetting binder and catalyst is packed into a core box that has been heated to around 230°C (450°F). When the sand is heated, the initial stages of curing begin within 10 to 30 seconds. After this brief period, the core can be removed from the pattern and will hold its shape during subsequent handling. For some materials, the cure completes through an exothermic curing reaction. For others, further baking is required to complete the process.

In the above methods, cores must be handled in an uncured or partially cured state, and breakage or distortion is not uncommon. Processes that produce finished cores while still in the core box and do not require heating operations would appear to offer distinct advantages.

In the *cold-box process*, binder-coated sand is first blown into a room-temperature core box, which can now be made from wood, metal, or even plastic. The box is sealed, and a gas or vaporized catalyst is then passed through the permeable sand to polymerize the resin. In a variation of the process, hollow cores are produced by introducing small amounts of curing gas through holes in the core-box pattern, with the uncured sand in the center being dumped and reused. Unfortunately, the required gases tend to be either toxic (an amine gas) or odorous (SO_2), making special handling of both incoming and exhaust gas a process requirement.

Room-temperature cores can also be made with the *air-set* or *no-bake sands*. These systems eliminate the gassing operation of the cold-box process through the use of a reactive organic resin and a curing catalyst. As discussed previously, there is only a brief period of time to form the core once the components have been mixed. *Shell molding* is another core-making alternative, producing hollow cores with excellent strength and permeability.

Selecting the actual method of core production is usually based on a number of considerations, including production quantity, production rate, required precision, required surface finish, and the metal being poured. Certain metals may be sensitive to gases that are emitted from the cores when they come into contact with the hot metal. Other materials with low pouring temperatures may not break down the binder sufficiently to provide collapsibility and easy removal from the final casting.

To function properly, casting cores must have the following characteristics:

1. Sufficient strength before hardening if they will be handled in the "green" condition.

2. Sufficient hardness and strength after hardening to withstand handling and the forces of the casting process. As metal fills the mold, most cores want to "float." The cores must be strong enough to resist the induced stresses, and the supports must be sufficient to hold them in place. Flowing metal can also cause surface erosion. Compressive strength should be between 100 and 500 psi (0.7 to 3.5 MPa).

3. A smooth surface.

4. Minimum generation of gases when heated by the pour.

5. Adequate permeability to permit the escape of gas. Since cores are largely surrounded by molten metal, the gases must escape through the core.

6. Adequate refractoriness. Being surrounded by hot metal, cores can become quite a bit hotter than the adjacent mold material. They should not melt or adhere to the casting.

7. Collapsibility. After pouring, the cores must be weak enough to permit the casting to shrink as it cools, thereby preventing cracking. In addition, the cores must be easily removed from the interior of the finished product via shakeout.

Various techniques have been developed to enhance the natural properties of cores and core materials. Additional strength can be imparted by the addition of internal wires or rods. Collapsibility can be enhanced by producing hollow cores or by placing a material such as straw in the center. Hollow cores may be used to provide for the escape of trapped or evolved gases. Vent holes can be formed by pushing small wires into the core, and coke or cinders are sometimes placed in the center of large cores to enhance venting.

Since the gases must be expelled from the casting, and the core material itself must be removed to produce the desired hole or cavity, the cores must be connected to the outer surfaces of the mold cavity. Recesses at these connection points, known as *core prints*, are used to support the cores and hold them in proper position during mold filling. The dry-sand cores in Figures 12-22c and 12-22d are supported by core prints.

If the cores do not pass completely through the casting, where they can be supported on both ends, a single core print may not be able to provide sufficient support. Additional measures may also be necessary to support the weight of large cores or keep lighter ones from becoming buoyant as the molten metal fills the cavity. Small metal supports, called *chaplets*, can be placed between cores and the surfaces of a mold cavity, as illustrated in Figure 12-24. Because the chaplets are positioned within the mold cavity, they become an integral part of the finished casting. Chaplets should therefore be of the same, or at least comparable, composition as the material being poured. They should be large enough that they do not completely melt and permit the core to move, but small enough that their surface melts and fuses with the

Cope

Drag

FIGURE 12-24   (Left) Typical chaplets. (Right) Method of supporting a core by use of chaplets (relative size of the chaplets is exaggerated).

FIGURE 12-25 Method of making a reentrant angle or inset section by using a three-piece flask.

metal being cast. Since chaplets are one more source of possible defects and may become a location of weakness in the finished casting, efforts are generally made to minimize their use.

Additional sections of mold material can also be used to produce castings with reentrant angles. Figure 12-25 depicts a round pulley with a recessed groove around its perimeter. By using a third segment of flask, called a *cheek,* and adding a second parting plane, the entire mold can be made by conventional green-sand molding around withdrawable patterns. While additional molding operations are required, this may be an attractive approach for small production runs.

If we want to produce a large number of identical pulleys, rapid machine molding of a simple green-sand mold might be preferred. As shown in Figure 12-26, the pattern would be modified to include a seat for an inserted ring-shaped core. Molding time is reduced at the expense of a core box and a separate core-making operation.

FIGURE 12-26 Molding an inset section using a dry-sand core.

■ 12.4 OTHER EXPENDABLE-MOLD PROCESSES WITH MULTIPLE-USE PATTERNS

PLASTER MOLD CASTING

In *plaster molding* the mold material is plaster of paris (also known as calcium sulfate or gypsum), combined with various additives to improve green strength, dry strength, permeability, and castability. Talc or magnesium oxide can be added to prevent cracking and reduce the setting time. Lime or cement helps to reduce expansion during baking. Glass fibers can be added to improve strength, and sand can be used as a filler.

The mold material is first mixed with water, and the creamy slurry is then poured over a metal pattern (wood patterns tend to warp or swell) and allowed to set. Hydration of the plaster produces a hard mold that can be easily stripped from the pattern. (*Note:* Flexible rubber patterns can be used when complex angular surfaces or reentrant angles are required. The plaster is strong enough to retain its shape during pattern removal.) The plaster mold is then baked to remove excess water, assembled, and poured.

With metal patterns and plaster mold material, surface finish and dimensional accuracy are both excellent. Cooling is slow, since the plaster has low heat capacity and low thermal conductivity. The poured metal stays hot and can flow into thin sections and replicate fine detail, which can often reduce machining cost. Unfortunately, plaster casting is limited to the lower-melting-temperature nonferrous alloys (such as aluminum, copper, magnesium, and zinc). At the high temperatures of ferrous metal casting, the plaster would first undergo a phase transformation and then melt, and the water of hydration can cause the mold to explode. Table 12-4 summarizes the features of plaster mold casting.

TABLE 12-4   Plaster Casting

Process: A slurry of plaster, water, and various additives is poured over a pattern and allowed to set. The pattern is removed, and the mold is baked to remove excess water. After pouring and solidification, the mold is broken and the casting is removed.

Advantages: High dimensional accuracy and smooth surface finish; can reproduce thin sections and intricate detail to make net- or near-net-shaped parts.

Limitations: Lower-temperature nonferrous metals only; long molding time restricts production volume or requires multiple patterns; mold material is not reusable; maximum size is limited.

Common metals: Primarily aluminum and copper.

Size limits: As small as 30 g (1 oz) but usually less than 7 kg (15 lb).

Thickness limits: Section thickness as small as 0.06 cm (0.025 in.).

Typical tolerances: 0.01 cm on first 5 cm (0.005 in. on first 2 in.), 0.002 cm per additional cm (0.002 in. per additional in.)

Draft allowance: $\frac{1}{2}$–1 degree.

Surface finish: 1.3–4 microns (50–125 μin.) rms.

The Antioch process is a variation of plaster mold casting where the mold material is comprised of 50% plaster and 50% sand, mixed with water. An autoclave process is used to prepare the molds, which offer improved permeability and reduced solidification time. The addition of a foaming agent to a plaster–water mix can add fine air bubbles that increase the material volume by 50–100%. The resulting molds have much improved permeability compared to the conventional process.

CERAMIC MOLD CASTING

Ceramic mold casting (summarized in Table 12-5) is similar to plaster mold casting, except that the mold is now made from a ceramic material that can withstand the higher-melting-temperature metals. Much like the plaster process, ceramic molding can produce thin sections, fine detail, and smooth surfaces, thereby eliminating a considerable amount of finish machining. These advantages, however, must be weighed against the greater cost of the mold material. For large molds, the ceramic can be used to produce a facing around the pattern, which is then backed up by a less expensive material such as reuseable fireclay.

One of the most popular of the ceramic molding techniques is the *Shaw process*. A reusable pattern is positioned inside a slightly tapered flask, and a slurry-like mixture of refractory aggregate, hydrolyzed ethyl silicate, alcohol, and a gelling agent is poured on top. This mixture sets to a rubbery state that permits removal of both the pattern and the flask. The mold surface is then ignited with a torch. Most of the volatiles are consumed during the "burn-off," and a three-dimensional network of microscopic cracks (microcrazing) forms in the ceramic. The gaps are small enough to prevent metal penetration but large enough to provide venting of air and gas (permeability) and to accommodate both the thermal expansion of the ceramic particles during the pour and the subsequent shrinkage of the solidified metal. A baking operation then removes all of the remaining volatiles, making the mold hard and rigid. Ceramic molds are often preheated prior to pouring to ensure proper filling and to control the solidification characteristics of the metal.

TABLE 12-5   Ceramic Mold Casting

Process: Stable ceramic powders are combined with binders and gelling agents to produce the mold material.

Advantages: Intricate detail, close tolerances, and smooth finish.

Limitations: Mold material is costly and not reusable.

Common metals: Ferrous and high-temperature nonferrous metals are most common; can also be used with alloys of aluminum, copper, magnesium, titanium, and zinc.

Size limits: 100 grams to several thousand kilograms (several ounces to several tons).

Thickness limits: As thin as 0.13 cm (0.050 in.); no maximum.

Typical tolerances: 0.01 cm on the first 2.5 cm (0.005 in. on the first in.), 0.003 cm per each additional cm (0.003 in. per each additional in.).

Draft allowances: 1° preferred.

Surface finish: 2–4 microns (75–150 μin.) rms.

FIGURE 12-27 Group of intricate cutters produced by ceramic mold casting. *(Courtesy of Avnet Shaw Division of Avnet, Inc., Phoenix, AZ)*

Like plaster molding, the ceramic molding process can effectively produce small-size castings in small to medium quantities. Figure 12-27 shows a set of intricate cutters that were produced by this process.

EXPENDABLE GRAPHITE MOLDS

For metals such as titanium, which tend to react with many of the more common mold materials, powdered *graphite* can be combined with additives, such as cement, starch, and water, and compacted around a pattern. After "setting," the pattern is removed and the mold is fired at 1000°C (1800°F) to consolidate the graphite. The casting is poured, and the mold is broken to remove the product.

RUBBER-MOLD CASTING

Artificial elastomers can also be compounded in liquid form and poured over a pattern to produce a semirigid mold. These molds are sufficiently flexible to permit stripping from an intricate shape or patterns with reverse-taper surfaces. Unfortunately, rubber molds are generally limited to small castings and low-melting-point materials. The wax patterns used in investment casting are often made by rubber-mold casting, as are small quantities of finished parts made from plastics or metals that can be poured at temperatures below 250°C (500°F).

■ 12.5 EXPENDABLE-MOLD PROCESSES USING SINGLE-USE PATTERNS

INVESTMENT CASTING

Investment casting is actually a very old process—used in ancient China and Egypt and more recently performed by dentists and jewelers for a number of years. It was not until the end of World War II, however, that it attained a significant degree of industrial importance. Products such as rocket components and jet engine turbine blades required the fabrication of high-precision complex shapes from high-melting-point metals that are not easily machined. Investment casting offers almost unlimited freedom in both the complexity of shapes and the types of materials that can be cast, and millions of investment castings are now produced each year.

Investment casting uses the same type of molding aggregate as the ceramic molding process and typically involves the following sequential steps:

1. *Produce a master pattern*—a modified replica of the desired product made from metal, wood, plastic, or some other easily worked material.

2. *From the master pattern, produce a master die*. This can be made from low-melting-point metal, steel, or possibly even wood. If a low-melting-point metal is used, the die may be cast directly from the master pattern. Rubber molds can also be made directly from the master pattern. Steel dies are often machined directly, eliminating the need for step 1.

3. *Produce wax patterns*. Patterns are made by pouring molten wax into the master die, or injecting it under pressure (injection molding), and allowing it to harden. Release agents, such as silicone sprays, are used to assist in pattern removal. Plastic and frozen mercury are alternate pattern materials. The polystyrene plastic may be preferred for producing thin and complex surfaces, where its higher strength and greater durability are desired. Frozen mercury is seldom used because of its cost, handling prob-

lems, and toxicity. If cores are required, they can generally be made from soluble wax or ceramic. The soluble wax cores are dissolved out of the patterns prior to further processing, while the ceramic cores remain and are not removed until after solidification of the metal casting.

4. *Assemble the wax patterns onto a common wax sprue.* Using heated tools and melted wax, a number of wax patterns can be attached to a central sprue and runner system to create a pattern cluster, or a *tree*. If the product is sufficiently complex that its pattern could not be withdrawn from a single master die, the pattern may be made in pieces and assembled prior to attachment.

5. *Coat the cluster or tree with a thin layer of investment material.* This step is usually accomplished by dipping into a watery slurry of finely ground refractory material. A thin but very smooth layer of investment material is deposited onto the wax pattern, ensuring a smooth surface and good detail in the final product.

6. *Form additional investment around the coated cluster.* After the initial layer has dried, the cluster can be redipped, but this time the wet ceramic is coated with a layer of sand or coarse refractory, a process called stuccoing. After drying, the process is repeated until the investment coating has the desired thickness (typically 5 to 15 mm or $\frac{3}{16}$ to $\frac{5}{8}$ inch with up to eight layers). As an alternative, the single-dipped cluster can be placed upside down in a flask and liquid investment material poured around it. The flask is then vibrated to remove entrapped air and ensure that the investment material now surrounds all surfaces of the cluster.

7. *Allow the investment to fully harden.*

8. *Remove the wax pattern from the mold by melting or dissolving.* Molds or trees are generally placed upside down in an oven where the wax can melt and run out, and any residue subsequently vaporizes. This step is the most distinctive feature of the process because it enables a complex pattern to be removed from a single-piece mold. Extremely complex shapes can be readily cast. (*Note:* In the early years of the process, only small parts were cast, and when the molds were placed in the oven, the molten wax was absorbed into the porous investment. Because the wax "disappeared," the process was called the *lost-wax process*, and the name is still used.)

9. *Heat the mold in preparation for pouring.* Heating to 550° to 1100°C (1000° to 2000°F) ensures complete removal of the mold wax, cures the mold to give added strength, and allows the molten metal to retain its heat and flow more readily into all of the thin sections and details. Mold heating also gives better dimensional control because the mold and the metal can shrink together during cooling.

10. *Pour the molten metal.* While gravity pouring is the simplest, other methods may be used to ensure complete filling of the mold. When complex, thin sections are involved, mold filling may be assisted by positive air pressure, evacuation of the air from the mold, or some form of centrifugal process.

11. *Remove the solidified casting from the mold.* After solidification, techniques such as mechanical chipping or vibration, high-pressure water jet, or sand blasting are used to break the mold and remove the mold material from the metal casting.

Figure 12-28 depicts the investment procedure, where the investment material fills the entire flask, and Figure 12-29 shows the shell-investment method. Table 12-6 summarizes the features of investment casting.

Compared to other methods of casting, investment casting is a complex process and tends to be rather expensive. However, its unique advantages can often justify its use, and many of the steps can be easily automated. Extremely complex shapes can be cast as a single piece. Thin sections, down to 0.40 mm (0.015 in.), can be produced. Excellent dimensional precision can be achieved in combination with very smooth as-cast surfaces. Machining can often be completely eliminated or greatly reduced. When machining is required, allowances of as little as 0.4 to 1 mm (0.015 to 0.040 in.) are usually ample. These capabilities are especially attractive when making products from the high-melting-temperature, difficult-to-machine metals that cannot be cast with plaster- or metal-mold processes.

3
A metal flask is placed around the pattern cluster.

4
Flask is filled with investment-mold slurry.

5
After mold material has set and dried, patterns are melted out of mold.

6
Hot molds are filled with metal by gravity, pressure, vacuum, or centrifugal force.

7
Mold material is broken away from castings.

1
Wax or plastic is injected into die to make a pattern.

2
Patterns are gated to a central sprue.

8
Castings are removed from sprue, and gate stubs are ground off.

FIGURE 12-28 Investment-casting steps for the flask-cast method. *(Courtesy of Investment Casting Institute, Dallas, TX.)*

Wax or plastic is injected into die to make a pattern.

Patterns are gated to a central sprue.

Castings are removed from sprue, and gate stubs are ground off.

1

2

8

To shipping

3
Pattern clusters are dipped in ceramic slurry.

4
Refractory grain is sifted onto coated patterns. Steps 3 and 4 are repeated several times to obtain desired shell.

5
After mold material has set and dried, patterns are melted out of mold.

6
Hot molds are filled with metal by gravity, pressure, vacuum, or centrifugal force.

7
Mold material is broken away from castings.

FIGURE 12-29 Investment-casting steps for the shell-casting procedure. *(Courtesy of Investment Casting Institute, Dallas, TX.)*

TABLE 12-6	Investment Casting

Process: A refractory slurry is formed around a wax or plastic pattern and allowed to harden. The pattern is then melted out and the mold is baked. Molten metal is poured into the mold and solidifies. The mold is then broken away from the casting.

Advantages: Excellent surface finish; high dimensional accuracy; almost unlimited intricacy; almost any metal can be cast; no flash or parting line concerns.

Limitations: Costly patterns and molds; labor costs can be high; limited size.

Common metals: Just about any castable metal. Aluminum, copper, and steel dominate; also performed with stainless steel, nickel, magnesium, and the precious metals.

Size limits: As small as 3 g ($\frac{1}{10}$ oz) but usually less than 5 kg (10 lb).

Thickness limits: As thin as 0.06 cm (0.025 in.), but less than 7.5 cm (3.0 in.).

Typical tolerances: 0.01 cm for the first 2.5 cm (0.005 in. for the first inch) and 0.002 cm for each additional cm (0.002 in. for each additional in.).

Draft allowances: None required.

Surface finish: 1.3–4 microns (50 to 125 μin.) rms.

While most investment castings are less than 10 cm (4 in.) in size and weigh less than $\frac{1}{2}$ kg (1 lb), castings up to 1 m (36 in.) and 35 kg (80 lb) have been produced. Products ranging from stainless steel or titanium golf club heads to superalloy turbine blades have become quite routine. Figure 12-30 shows some typical investment castings. One should note that a high degree of shape complexity is a common characteristic of investment cast products.

The high cost of dies to make the wax patterns has traditionally limited investment casting to large production quantities. Recent advances in rapid prototyping, however, now enable the production of wax-like patterns directly from CAD data. The absence of part-specific tooling now enables the economical casting of one-of-a-kind or small-quantity products using the investment methods. The majority of investment castings now fall within the range of 100 to 10,000 pieces per year.

COUNTER-GRAVITY INVESTMENT CASTING

Counter-gravity investment casting turns the pouring process upside down. In one variation of the process, a ceramic shell mold is placed in an open-bottom chamber with the sprue end down. The open end of the sprue is lowered into a pool of molten metal, and the bottom of the chamber is set against a seal. A vacuum is then induced within the chamber. As the air is withdrawn, the vacuum draws metal up through the central sprue and into the mold. The castings are allowed to solidify, the vacuum is released, and any unsolidified metal flows back into the melt. In another variation, a low-pressure inert gas is used to push the molten metal upward into the mold. This approach is discussed in more detail and is also illustrated in the section on low-pressure permanent-mold casting in Chapter 13.

FIGURE 12-30 Typical parts produced by investment casting. *(Courtesy of Haynes International, Kokomo, IN.)*

The counter-gravity processes have a number of distinct advantages. Because the molten metal is withdrawn from below the surface of its ladle, it is generally free of slag and dross and has a very low level of inclusions. The vacuum or low-pressure filling allows the metal to flow with little turbulence, further enhancing metal quality. The reduction in metallic inclusions improves machinability and enables mechanical properties to approach or equal those of wrought material.

Since the gating system does not need to control turbulence, simpler gating systems can be used, reducing the amount of metal that does not become product. In the counter-gravity process, between 60% and 95% of the withdrawn metal becomes cast product, compared to a 15% to 50% level for gravity-poured castings. The pressure differential enables metal to flow into thinner sections, and lower "pouring" temperatures can be used, resulting in improved grain structure and better surface finish.

EVAPORATIVE PATTERN (FULL-MOLD AND LOST-FOAM) CASTING

Several limitations are common to most of the casting processes that have been presented. Some form of pattern is usually required, and this pattern may be costly to design and fabricate. Pattern costs may be hard to justify, especially when the number of identical castings is rather small or the part is extremely complex. In addition, reuseable patterns must be withdrawn from the mold, and this withdrawal often requires some form of design modification or compromise, division into multiple pieces, or special molding procedures. Investment casting overcomes the withdrawal limitations through the use of patterns that can be removed by melting and vaporization. Unfortunately, investment casting has its own set of limitations, including a large number of individual operations and the need to remove the investment material from the finished casting.

In the *evaporative pattern* processes, the pattern is made of *expanded polystyrene (EPS)*, or expanded polymethylmethacrylate (EPMMA), and remains in the mold. During the pour, the heat of the molten metal melts and burns the polystyrene, and the metal fills the space that was previously occupied by the pattern.

When small quantities are required, patterns can be cut by hand or machined from pieces of foamed polystyrene (a material similar to that used in Styrofoam drinking cups). This material is extremely light in weight and can be cut by a number of methods, including ones as simple as an electrically heated wire. Preformed material in the form of a pouring basin, sprue, runner segments, and risers can be attached with hot-melt glue to form a complete gating and pattern assembly. Small products can be assembled into clusters or trees, similar to investment casting.

When producing larger quantities of identical parts, a metal mold or die is generally used to mass-produce the evaporative patterns. Hard beads of polystyrene are first preexpanded and stabilized. The preexpanded beads are then injected into a heated metal die or mold, usually made from aluminum. A steam cycle causes them to further expand, fill the die, and fuse, after which they are cooled in the mold. The resulting pattern, a replica of the product to be cast, consists of about 2.5% polymer and 97.5% air. Pattern dies can be quite complex, and large quantities of patterns can be accurately and rapidly produced. When size or complexity is great, or geometry prevents easy removal, the pattern can be divided into multiple segments, or slices, which are then assembled by hot-melt gluing. The ideal glue should be strong, fast setting, and produce a minimum amount of gas when it decomposes or combusts.

FIGURE 12-31 Schematic of the full-mold process. (Left) An uncoated expanded polystyrene pattern is surrounded by bonded sand to produce a mold. (Right) Hot metal progressively vaporizes the expanded polystyrene pattern and fills the resulting cavity.

After a polystyrene gating system is attached to the polystyrene pattern, there are several options for the completion of the mold. In the *full-mold process*, shown schematically in Figure 12-31, green sand or some type of chemically bonded (no-bake) sand is compacted around the pattern and gating system, taking care not to crush or distort it. The mold is then poured like a conventional sand-mold casting.

In the *lost-foam* process, depicted schematically in Figure 12-32, the polystyrene assembly is first dipped into a water-based ceramic that wets both external and internal surfaces and forms a thin refractory coating. The coating must be thin enough and sufficiently

FIGURE 12-32 Schematic of the lost-foam casting process. In this process, the polystyrene pattern is dipped in a ceramic slurry, and the coated pattern is then surrounded with loose, unbonded sand.

permeable to permit the escape of the molten and gaseous pattern material, but rigid enough to prevent mold collapse during pouring. After the coating dries, the pattern assembly is suspended in a one-piece flask and surrounded by fine unbonded sand. Vibration ensures that the sand compacts around the pattern and fills all cavities and passages. During the pour, molten metal melts, vaporizes, and replaces the expanded polystyrene, while the coating isolates the metal from the loose, unbonded sand. After the casting has cooled and solidified, the loose sand is then dumped from the flask, freeing the casting and attached gating system. The backup sand can then be reused, provided the coating residue is removed and the organic condensates are periodically burned off. Figure 12-33 shows the series of operations used in producing a rather complex lost-foam casting.

The full-mold and lost-foam processes can produce castings of any size in both ferrous and nonferrous metals. Since the pattern need not be withdrawn, no draft is required in the design. Complex patterns can be produced to make shapes that would ordinarily require multiple cores, loose-piece patterns, or extensive finish machining. Multicomponent assemblies can often be replaced by a single casting. Because of the high precision and smooth surface finish, machining and finishing operations can often be reduced or totally eliminated. Fragile or complex-geometry cores are no longer required, and the absence of parting lines eliminates the need to remove associated lines or fins on the metal casting.

As the molten metal progresses through the pattern, it loses heat due to the melting and volatilizing of the foam. As a result, the material farthest from the gate is the coolest,

FIGURE 12-33 The stages of lost-foam casting, proceeding counterclockwise from the lower left: polystyrene beads → expanded polystyrene pellets → three foam pattern segments → an assembled and dipped polystyrene pattern → a finished metal casting that is a metal duplicate of the polystyrene pattern. *(Courtesy of Saturn Corporation, Spring Hill, TN.)*

TABLE 12-7 Lost-Foam Casting

Process: A pattern containing a sprue, runners, and risers is made from single or multiple pieces of foamed plastic, such as polystyrene. It is dipped in a ceramic material, dried, and positioned in a flask, where it is surrounded by loose sand. Molten metal is poured directly onto the pattern, which vaporizes and is vented through the sand.

Advantages: Almost no limits on shape and size; most metals can be cast; no draft is required and no flash is present (no parting lines).

Limitations: Pattern cost can be high for small quantities; patterns are easily damaged or distorted because of their low strength.

Common metals: Aluminum, iron, steel, and nickel alloys; also performed with copper and stainless steel.

Size limits: 0.5 kg to several thousand kg (1 lb to several tons).

Thickness limits: As small as 2.5 mm (0.1 in.) with no upper limit.

Typical tolerances: 0.003 cm/cm (0.003 in./in.) or less.

Draft allowance: None required.

Surface finish: 2.5–25 microns (100–1000 μin.) rms.

and solidification tends to proceed in a directional manner back toward the gate. For many castings, risers are not required. Metal yield (product weight versus the weight of poured metal) tends to be rather high. For these and other reasons, evaporative-pattern casting has grown rapidly in popularity and use. Table 12-7 summarizes the process and its capabilities.

■ 12.6 SHAKEOUT, CLEANING, AND FINISHING

In each of the casting processes presented in this chapter, the final step involves separating the castings from the molds and mold material. *Shakeout* operations are designed to separate the molds and sand from the flasks (i.e., containers), separate the castings from the molding sand, and separate or remove the cores from the castings. Punchout machines can be used to force the entire contents of a flask (both molding sand and casting) from the container. Vibratory machines, which can operate on either the entire flasks or the extracted contents, are available in a range of styles, sizes, and vibratory frequencies. Rotary separators remove the sand from castings by placing the mold contents inside a slow-turning, large-diameter, rotating drum. The tumbling action breaks the gates and runners from the castings, crushes lumps of sand, and extracts the cores. Because of possible damage to lightweight or thin-sectioned castings, rotary tumbling is usually restricted to cast iron, steel, and brass castings of reasonable thickness.

Processes such as blast cleaning can be used to remove adhering sand, oxide scale, and parting-line burrs. Compressed air or centrifugal force is used to propel abrasive particles against the surfaces of the casting. The propelled media can be metal shot (usually iron or steel), fine aluminum oxide, glass beads, or naturally occurring quartz or silica. The blasting action may be combined with some form of tumbling or robotic manipulation to expose the various surfaces. Additional finishing operations may include grinding, trimming, or various forms of machining.

■ 12.7 SUMMARY

Liquids have the characteristic property that they assume the shape of their container. A number of processes have been developed to create shaped containers and then utilize liquid fluidity and subsequent solidification to produce desired shapes. Each process has its unique set of capabilities, advantages, and limitations, and the selection of the best method for a given application requires an understanding of all possible options. This chapter has presented processes that produce castings with a single-use (expendable) mold. The following chapter will supplement this knowledge with a survey of multiple-use mold processes.

■ Key Words

air-set sand	ceramic mold	cold-box process	cope-and-drag-pattern
automatic match-plate molding	chaplets	collapsibility	core
	cheek	compactibility	core-oil process
blows	cohesiveness	compressive strength	core prints

counter-gravity investment
  casting
dump-core box
dry-sand mold
Eff-set process
evaporative-pattern casting
expanded polystyrene
expendable mold
flask
follow board
full-mold casting
gates
graphite mold
green compressive strength
green sand

green-sand cores
H-process
hand ramming
hardness
hot-box method
hot tears
investment casting
jolting
loose-piece pattern
lost-foam casting
lost-wax process
match plate
mold hardness
muller
no-bake sand

one-piece pattern
pattern
penetration
permeability
plaster mold
refractoriness
rubber-mold casting
runner
sand expansion defects
sand slinger
shakeout
Shaw process
shell molding
silica sand
single-piece cores

skin-dried mold
sodium silicate–CO_2 molding
split pattern
sprue
squeezing
stack molding
standard rammed specimen
V-process
vacuum molding
vertically parted flaskless
  molding
water glass

■ Review Questions

1. What are some of the factors that influence the selection of a specific casting process as a means of making a product?
2. What are the three basic categories of casting processes when classified by molds and patterns?
3. What metals are frequently cast into products?
4. Which type of casting is the most common and most versatile?
5. What is a casting pattern?
6. What are some of the materials used in making casting patterns? What features should be considered when selecting a pattern material?
7. What is the simplest and least expensive type of casting pattern?
8. What is a match plate, and how does it aid molding?
9. How is a cope-and-drag pattern different from a match-plate pattern? When might this be attractive?
10. For what types of products might a loose-piece pattern be required?
11. What are the four primary requirements of a molding sand?
12. In what ways might a molding sand be a compromise material?
13. What is a muller, and what function does it perform?
14. What are some of the properties or characteristics of foundry sands that can be evaluated by standard tests?
15. What is a standard rammed specimen for evaluating foundry sands, and how is it produced?
16. What is permeability, and why is it important in molding sands?
17. How does the ratio of water to clay affect the compressive strength of green sand?
18. How does the size and shape of the sand grains relate to molding sand properties?
19. What is a sand expansion defect, and what is its cause?
20. How can sand expansion defects be minimized?
21. What causes a "blow" to form in a casting, and what can be done to minimize their occurrence?
22. What features can cause the penetration of molten metal between the grains of the molding sand?
23. What are hot tears, and what can cause them to form?
24. Describe the distribution of sand density after compaction by jolting, squeezing, and a jolt–squeeze combination.
25. How can the use of vertically parted flaskless molding reduce the number of mold sections required to produce a series of castings?
26. What is stack molding?
27. How might extremely large molds be made?
28. What is green sand?
29. What are some of the limitations or problems associated with green sand as a mold material?
30. What restricts the use of dry sand molding?
31. What are some of the advantages and limitations of the sodium silicate–CO_2 process?
32. What is the primary feature of no-bake sands?
33. What material serves as the binder in the shell-molding process, and how is it cured?
34. Why do shell molds have excellent permeability and collapsibility?
35. What is the sand binder in the V-process? The Eff-set process?
36. What types of geometric features might require the use of cores?
37. What is the primary limitation of green-sand cores?
38. What is the sand binder in the core-oil process, and how is it cured?
39. What is the binder in the hot-box core-making process?
40. What is the primary attraction of the cold-box core-making process? The primary negative feature?
41. What is an attractive feature of shell-molded cores?
42. Why is it common for greater permeability, collapsibility, and refractoriness to be required of cores than for the base molding sand?
43. What is the role of chaplets, and why is it important that they not completely melt during the pouring and solidification of a casting?
44. Why are plaster molds suitable only for the lower-melting-temperature nonferrous metals and alloys?
45. What is the primary performance difference between plaster and ceramic molds?
46. For what materials might a graphite mold be required?
47. What materials are used to produce the expendable patterns for investment casting?
48. Describe the progressive construction of an investment-casting mold.
49. Why are investment-casting molds generally preheated prior to pouring?
50. Why are investment castings sometimes called "lost-wax" castings?
51. What are some of the attractive features of investment casting?
52. What are some of the advantages of counter-gravity investment casting over the conventional gravity pour approach?

53. What are some of the benefits of not having to remove the pattern from the mold (as in investment casting, full-mold casting, and lost-foam casting)?
54. Since both use expanded polystyrene as a pattern, what is the primary difference between full-mold and lost-foam casting?

55. What are some of the attractive features of the evaporative-pattern processes?
56. What are some of the objectives of a shakeout operation?
57. How might castings be cleaned after shakeout?

■ Problems

1. While cores increase the cost of castings, they also provide a number of distinct advantages. The most significant is the ability to produce complex internal passages. They can also enable the production of difficult external features, such as undercuts, or allow the production of zero-draft walls. Cores can reduce or eliminate additional machining, reduce the weight of a casting, and reduce or eliminate the need for multipiece assembly. Answer the following questions about cores.
   a. The cores themselves must be produced, and generally they have to be removed from core boxes or molds. What geometric limitations might this impose? How might these limitations be overcome?
   b. Cores must be positioned and supported within a mold. Discuss some of the limitations associated with core positioning and orientation. Consider the weight of a core, prevention of core fracture, minimization of core deflection, and possible buoyancy.
   c. Since cores are internal to the casting, adequate venting is necessary to eliminate or minimize porosity problems. Discuss possible features to aid in venting.
   d. How might core behavior vary with different materials being cast—steel versus aluminum, for example?
   e. Core removal is another design concern. Discuss how several different core-making processes might perform in the area of removal. What are some ways to assist or facilitate core removal?

www.wiley.com/college/degarmo

*C*hapter 12 CASE STUDY

Movable and Fixed Jaw Pieces for a Heavy-Duty Bench Vise

The figure presents a cutaway sketch of the movable and fixed jaw pieces of a heavy-duty vise that might see use in vocational schools, factories, and machine shops. The vise is intended to have a rated maximum clamping force of 15 tons. The slide of the moving jaw has been designed to be a 2-in. box channel. The jaw width is 5 in., the maximum jaw opening is 6 in., and the depth of the throat is 4 in. The designer has elected to use replaceable, serrated jaws and suggests that the material used for the receiving jaw pieces have a yield strength in excess of 35 ksi, with at least 15% elongation in a uniaxial tensile test (to ensure that an overload or hammer impact would not produce brittle fracture).

1. Determine some possible combinations of material and process that could fabricate the desired shapes with the required properties. Of the alternatives presented, which would you prefer and why?

2. Would the components require some form of subsequent heat treatment? Consider the possibilities of stress relief, homogenization, or the establishment of desired final properties. What would you recommend?

3. One of your colleagues has suggested that the slides be finished with a coat of paint. Do you think a surface treatment is necessary or desirable for your selected material and process? If so, what would you recommend? If not, defend your recommendation.

Replaceable jaws

Movable jaw piece — Fixed jaw piece

Slide

Note: shaded surfaces have been produced by cross-sectional cuts

MULTIPLE-USE-MOLD CASTING PROCESSES

■ 13.1 INTRODUCTION

In each of the expendable-mold casting processes discussed in Chapter 12, a separate mold had to be created for each pour. Variations in mold consistency, mold strength, moisture content, pattern removal, and other factors contribute to dimensional and property variation from casting to casting. In addition, the need to create and then destroy a separate mold for each pour results in rather low production rates.

The multiple-use-mold casting processes overcome many of these limitations, but they, in turn, have their own assets and liabilities. Since the molds are generally made from metal, many of the processes are restricted to casting the lower-melting-point nonferrous metals and alloys. Part size is often limited, and the dies or molds can be rather costly.

■ 13.2 PERMANENT-MOLD CASTING

In the *permanent-mold casting* process, also called gravity die casting, a reusable mold is machined from gray cast iron, alloy cast iron, steel, bronze, graphite, or other material. The molds are usually made in segments, which are often hinged to permit rapid and accurate opening and closing. After preheating, a refractory or mold coating is applied to the preheated mold, and the mold is clamped shut. Molten metal is then poured into the pouring basin, and it flows through the feeding system into the mold cavity by simple gravity flow. After solidification, the mold is opened and the product is removed. Since the heat from the previous cast is usually sufficient to maintain mold temperature, the process can be immediately repeated, with a single refractory coating serving for several pouring cycles. Aluminum-, magnesium-, zinc-, lead-, and copper-based alloys are the metals most frequently cast, along with gray cast iron. If graphite is used as the mold material, iron and steel castings can also be produced.

Numerous advantages can be cited for the permanent-mold process. Near-net shapes can be produced that require little finish machining. The mold is reusable, and a good surface finish is obtained if the mold is in good condition. Dimensions are consistent from part to part, and dimensional accuracy can often be held to within 0.25 mm (0.010 in.). Directional solidification can be achieved through good design or can be promoted by selectively heating or chilling various portions of the mold or by varying the thickness of the mold wall. The result is usually a sound, defect-free casting with good mechanical properties. The faster cooling rates of the metal mold produce a finer grain structure, reduced porosity, and higher-strength products than would result from a sand casting process. Cores, both expendable sand or plaster or retractable metal, can be used to increase the complexity of the casting, and multiple cavities can often be included in a single mold. When sand cores are used, the process is often called *semipermanent mold casting*.

On the negative side, the process is generally limited to the lower-melting-point alloys, and high mold costs can make low production runs prohibitively expensive. The useful life of a mold is generally set by molten metal erosion or thermal fatigue. When making products of steel or cast iron, mold life can be extremely short. For the lower-temperature metals, one can usually expect somewhere between 10,000 and 120,000 cycles. The actual mold life will depend upon the following:

1. *Alloy being cast.* The higher the melting point, the shorter the mold life.
2. *Mold material.* Gray cast iron has about the best resistance to thermal fatigue and machines easily. Thus it is used most frequently for permanent molds.
3. *Pouring temperature.* Higher pouring temperatures reduce mold life, increase shrinkage problems, and induce longer cycle times.
4. *Mold temperature.* If the temperature is too low, one can expect misruns and large temperature differences in the mold. If the temperature is too high, excessive cycle times result and mold erosion is aggravated.
5. *Mold configuration.* Differences in section sizes of either the mold or the casting can produce temperature differences within the mold and reduce its life.

The permanent molds contain the mold cavity, pouring basin, sprue, runners, risers, gates, possible core supports, alignment pins, and some form of ejection system. The molds are usually heated at the beginning of a run, and continuous operation then maintains the mold at a fairly uniform elevated temperature. This minimizes the degree of thermal fatigue, facilitates metal flow, and controls the cooling rate of the metal being cast. Since the mold temperature rises when a casting is produced, it may be necessary to provide a mold-cooling delay before the cycle is repeated. Refractory washes or graphite coatings can be applied to the mold walls to control or direct the cooling, prevent the casting from sticking, and prolong the mold life by minimizing thermal shock and fatigue. When pouring cast iron, an acetylene torch is often used to apply a coating of carbon black to the mold.

Since the molds are not permeable, special provision must be made for *venting*. This is usually accomplished through the slight cracks between mold halves or by very small vent holes that permit the escape of trapped air but not the passage of molten metal. Since gravity is the only means of inducing metal flow, risers must still be employed to compensate for solidification shrinkage, and with the necessary sprues and runners, yields are generally less than 60%.

Mold complexity is often restricted because the rigid cavity offers no collapsibility to compensate for the solid-state shrinkage of the casting. As a best alternative, it is common practice to open the mold and remove the casting immediately after solidification. This prevents the formation of hot tears that may form if the product is restrained during the shrinkage that occurs during cooldown to room temperature.

For permanent-mold casting, high-volume production is usually required to justify the high cost of the metal molds. Automated machines can be used to coat the mold, pour the metal, and remove the casting. Figure 13-1 shows a variety of automobile and truck pistons that were manufactured by the permanent-mold process, which is summarized in Table 13-1.

SLUSH CASTING

Hollow castings can be produced by a variant of permanent-mold casting known as *slush casting*. Hot metal is poured into the metal mold and is allowed to cool until a shell of desired thickness has formed. The mold is then inverted and the remaining liquid is poured out. The resulting casting is a hollow shape with good surface detail but variable wall thickness. Common applications include the casting of ornamental objects such as candlesticks, lamp bases, and statuary from the low-melting-temperature metals.

FIGURE 13-1   Truck and car pistons are mass-produced by the millions using permanent-mold casting. *(Courtesy of General Motors Corporation, Detroit, MI.)*

TABLE 13-1   Permanent-Mold Casting

Process: Mold cavities are machined into mating metal die blocks, which are then preheated and clamped together. Molten metal is then poured into the mold and enters the cavity by gravity flow. After solidification, the mold is opened and the casting is removed.

Advantages: Good surface finish and dimensional accuracy; metal mold gives rapid cooling and fine-grain structure; multiple-use molds (up to 120,000 uses); metal cores or collapsible sand cores can be used.

Limitations: High initial mold cost; shape, size, and complexity are limited; yield rate rarely exceeds 60%, but runners and risers can be directly recycled; mold life is very limited with high-melting-point metals such as steel.

Common metals: Alloys of aluminum, magnesium, and copper are most frequently cast; irons and steels can be cast into graphite molds; alloys of lead, tin, and zinc are also cast.

Size limits: 100 grams to 75 kilograms (several ounces to 150 pounds).

Thickness limits: Minimum depends on material but generally greater than 3 mm ($\frac{1}{8}$ in.); maximum thickness about 50 mm (2.0 in.).

Geometric limits: The need to extract the part from a rigid mold may limit certain geometric features. Uniform section thickness is desirable.

Typical tolerances: 0.4. mm for the first 2.5. cm (0.015 in. for the first inch) and 0.02 mm for each additional centimeter (0.002 in. for each additional inch); 0.25mm (0.01 in.) added if the dimension crosses a parting line.

Draft allowance: 2°–3°.

Surface finish: 2.5 to 7.5 μm (100–250 μin.) rms.

LOW-PRESSURE AND VACUUM PERMANENT-MOLD CASTING

Gravity pouring is the oldest, simplest, and most traditional form of permanent-mold casting. In a variation known as *tilt-pour permanent-mold casting*, the molten metal is placed in the pouring basin and the mold then rotates to induce flow into the mold cavity. In this way, turbulence is minimized as the metal flows through the gating system and into the mold.

In low-pressure and vacuum permanent-mold casting, the mold is turned upside down and positioned above a sealed, airtight chamber that contains a crucible of molten metal. A small pressure difference then causes the molten metal to flow upward into the die cavity. In the *low-pressure permanent-mold* (LPPM) process, illustrated in Figure 13-2, a low-pressure gas (3 to 15 psi) is introduced into a sealed chamber, driving molten metal up through a refractory fill tube and into the gating system or cavity of a metal mold. This metal is exceptionally clean, since it flows from the center of the melt and is fed directly into the mold (a distance of about 10 cm, or 3 to 4 in.), never passing through the atmosphere. Product quality is further enhanced by the nonturbulent mold filling, which helps to minimize gas porosity and dross formation.

Through design and cooling, the products directionally solidify from the top down. The molten metal in the pressurized fill tube acts as a riser to continually feed the casting

FIGURE 13-2 Schematic of the low-pressure permanent-mold process. *(Courtesy of Amsted Industries, Chicago, IL.)*

during solidification. When solidification is complete, the pressure is released and the unused metal in the feed tube simply drops back into the crucible. The reuse of this metal, coupled with the absence of additional risers, leads to yields that are often greater than 85%.

Nearly all low-pressure permanent-mold castings are made from aluminum or magnesium, but some copper-based alloys can also be used. Mechanical properties are typically about 5% better than those of conventional permanent-mold castings. Cycle times are somewhat longer, however, than those of conventional permanent molding.

Figure 13-3 depicts a similar variation of permanent-mold casting, where a vacuum is drawn on the die assembly and atmospheric pressure in the chamber forces the metal upward. All of the benefits and features of the low-pressure process are retained, including the subsurface extraction of molten metal from the melt, the bottom feed to the mold, the minimal metal disturbance during pouring, the self-risering action, and the downward directional solidification. Thin-walled castings can be produced with high metal yield and excellent surface quality. Because of the vacuum, the cleanliness of the metal and the dissolved gas content are superior to that of the low-pressure process. Final castings typically range from 0.2 to 5 kg (0.4 to 10 lb) and have mechanical properties that are even better than those of the low-pressure permanent-mold products.

■ 13.3 DIE CASTING

In the *die-casting* process, or more specifically pressure die casting, molten metal is forced into metal molds under pressures of several thousand pounds per square inch (tens of MPa) and held under high pressure during solidification. Because of the combination of metal molds or dies and high pressure, fine sections and excellent detail can be achieved, together with long mold life. Most die castings are made from nonferrous metals and alloys, with special zinc-, copper-, magnesium-, and aluminum-based alloys having been designed to produce excellent properties when die cast. Ferrous-metal die castings are possible but are generally considered to be uncommon. Production rates are high, the products exhibit good strength, shapes can be quite intricate, and dimensional precision and surface qualities are excellent. There is almost a complete elimination of subsequent machining. Most die castings can be classified as small- to medium-sized parts, but the size and weight of die castings are continually increasing. Parts can now be made with weights up to 10 kg (20 lb) and dimensions as large as 600 mm (24 in.).

FIGURE 13-3 Schematic illustration of vacuum permanent-mold casting. Note the similarities to the low-pressure process.

Die temperatures are usually maintained at about 150° to 250°C (300° to 500°F) below the solidus temperature of the metal being cast in order to promote rapid freezing. Since cast iron cannot withstand the high casting pressures, die-casting dies are usually made from hardened hot-work tool steels and are typically quite expensive. As shown in Figure 13-4, the dies may be relatively simple, containing only one or two mold cavities, or they may be complex, containing multiple cavities of the same or different products, or even be an assembly of multiple subcomponents. The rigid dies must separate into at least two pieces to permit removal of the casting. It is not uncommon, however, for complex die castings to require multiple-segment dies that open and close in several different directions. Die complexity is further increased as the various sections incorporate water-cooling passages, *retractable cores*, and moving pins to knock out or eject the finished casting.

Die life is usually limited by wear (or erosion), which is strongly dependent on the temperature of the molten metal. Surface cracking can also occur in response to the large number of heating and cooling cycles that are experienced by the die surfaces. If the rate of temperature change is the dominant feature, the problem is called *heat checking*. If the number of cycles is the primary cause, the problem is called *thermal fatigue*.

In the basic die-casting process, water-cooled dies are first lubricated and clamped tightly together. Molten metal is then injected under high pressure. Since high injection pressures cause turbulence and air entrapment, the specified values of pressure

Single-cavity die   Multiple-cavity die   Combination die   Unit die

FIGURE 13-4 Various types of die-casting dies. *(Courtesy of American Die Casting Institute, Inc., Des Plaines, IL.)*

FIGURE 13-5 Principal components of a hot-chamber die-casting machine. *(Adapted from Metals Handbook, 9th ed., Vol 15, p. 287, ASM International, Metals Park, OH.)*

and the time and duration of application vary considerably. The pressure need not be constant, and there has been a trend toward the use of larger gates and lower injection pressures, followed by the application of higher pressure after the mold has completely filled and the metal has started to solidify. By reducing turbulence and solidifying under high pressure, this cycle reduces both the porosity and inclusion content of the finished casting. After solidification is complete, the pressure is released, the dies separate, and ejector pins extract the finished casting along with its attached runners and sprues.

There are two basic types of die-casting machines. Figure 13-5 schematically illustrates the *hot-chamber*, or *gooseneck*, variety. A gooseneck chamber is partially submerged in a reservoir of molten metal. With the plunger raised, molten metal flows through an open port and fills the chamber. A mechanical plunger then forces the metal up through the gooseneck, through the runners and gates, and into the die, where it rapidly solidifies. Retraction of the plunger then allows the gooseneck to refill as the casting is being ejected, and the cycle repeats at speeds up to 100 shots per minute.

Hot-chamber die-casting machines offer fast cycling times (set by the ability of the water-cooled dies to cool and solidify the metal) and the added advantage that the molten metal is injected from the same chamber in which it is melted (i.e., there is no handling or transfer of molten metal). Unfortunately, the hot-chamber design cannot be used for the higher-melting-point metals, and it is unattractive for aluminum since the molten aluminum tends to pick up some iron during the extended time of contact with the casting equipment. Hot-chamber machines, therefore, see primary use with zinc-, tin-, and lead-based alloys.

Zinc die castings can also be made by a process known as *heated-manifold direct-injection die casting* (also known as direct-injection die casting or runnerless die casting). The molten zinc is forced through a heated manifold and then through heated mini-nozzles directly into the die cavity. This approach totally eliminates the need for sprues, gates, and runners. Scrap is reduced, energy is conserved (less molten metal per shot and no need to provide excess heat to compensate for cooling in the gating system), and product quality is increased. Existing die-casting machines can be converted through the addition of a heated manifold and modification of the various dies.

Cold-chamber machines are usually employed for the die casting of materials that are not suitable for the hot-chamber design. These include alloys of aluminum, magnesium, and copper as well as high-aluminum zinc. As illustrated in Figure 13-6, metal that has been melted in a separate furnace is transported to the die-casting machine, where a measured quantity is fed into an unheated shot chamber (or injection cylinder) and subsequently driven into the die by a hydraulic or mechanical plunger. The pressure is then maintained or increased until solidification is complete. Since molten metal must be transferred to the chamber for each shot, the cold-chamber process has a longer operating cycle compared to hot-chamber machines. Nevertheless, productivity is still high.

FIGURE 13-6 Principal components of a cold-chamber die-casting machine. *(Adapted from* Metals Handbook, *9th ed., Vol 15, p. 287, ASM International, Metals Park, OH.)*

In all variations of the process, die-casting dies fill with metal so fast that there is little time for the air in the runner system and mold cavity to escape, and the metal molds offer no permeability. The air can become trapped and cause a variety of defects, including blowholes, porosity, and misruns. To minimize these defects, it is crucial that the dies be properly vented, usually by wide, thin (0.13-mm or 0.005-in.) vents positioned along the parting line. Proper positioning is a must, since all of the air must escape before the molten metal contacts the vents. The long thin slots allow the escape of gas but promote rapid freezing of the metal and a plugging of the hole. The metal that solidifies in the vents must be trimmed off after the casting has been ejected. This can be done with special trimming dies that also serve to remove the sprues and runners.

Risers are not used in the die-casting process since the high injection pressures ensure the continuous feed of molten metal from the gating system into the casting. The porosity that is often found in die castings is not shrinkage porosity; it is more likely to be the result of either entrapped air or the turbulent mode of die filling. This porosity tends to be confined to the interior of castings, and its formation can often be minimized by smooth metal flow, good venting, and proper application of pressure. The rapidly solidified surface is usually harder and stronger than the slower-cooled interior and is usually sound and suitable for plating or decorative applications.

Sand cores cannot be used in die casting because the high pressures and flow rates cause the cores to either disintegrate or have excessive metal penetration. As a result, metal cores are required, and provisions must be made for their retraction, usually before the die is opened for removal of the casting. As with all mating segments and moving components, a close fit must be maintained to prevent the pressurized metal from flowing into the gap. Loose core pieces (also metal) can also be positioned into the die at the beginning of each cycle and then removed from the casting after its ejection. This procedure permits more complex shapes to be cast, such as holes with internal threads, but production rate is slowed and costs increase.

Cast-in *inserts* can also be incorporated in the die-casting process. Examples include prethreaded bosses, electrical heating elements, threaded studs, and high-strength bearing surfaces. These high-temperature components are positioned in the die before the lower-melting-temperature metal is injected. Suitable recesses must be provided in the die for positioning and support, and the casting cycle tends to be slowed by the additional operations.

Table 13-2 summarizes the key features of the die-casting process. Attractive aspects include smooth surfaces and excellent dimensional accuracy. For aluminum-, magnesium-, zinc-, and copper-based alloys, linear tolerances of 3 mm/m (0.003 in./in.) are not uncommon. Thinner sections can be cast than with either sand or permanent-mold casting. The minimum section thickness and draft vary with the type of metal, with typical values as follows:

Metal	Minimum Section	Minimum Draft
Aluminum alloys	0.89 mm (0.035 in.)	1:100 (0.010 in./in.)
Brass and bronze	1.27 mm (0.050 in.)	1:80 (0.015 in./in.)
Magnesium alloys	1.27 mm (0.050 in.)	1:100 (0.010 in./in.)
Zinc alloys	0.63 mm (0.025 in.)	1:200 (0.005 in./in.)

TABLE 13-2	Die Casting

Process: Molten metal is injected into closed metal dies under pressures ranging from 10 to 175 MPa (1500–25,000 psi). Pressure is maintained during solidification, after which the dies separate and the casting is ejected along with its attached sprues and runners. Cores must be simple and retractable and take the form of moving metal segments.

Advantages: Extremely smooth surfaces and excellent dimensional accuracy; rapid production rate; product tensile strengths as high as 415 Mpa (60 ksi).

Limitations: High initial die cost; limited to high-fluidity nonferrous metals; part size is limited; porosity may be a problem; some scrap in sprues, runners, and flash, but this can be directly recycled.

Common metals: Alloys of aluminum, zinc, magnesium, and lead; also possible with alloys of copper and tin.

Size limits: Less than 30 grams (1 oz) up through about 7 kg (15 lb) most common.

Thickness limits: As thin as 0.75 mm (0.03 in.), but generally less than 13 mm ($\frac{1}{2}$ in.).

Typical tolerances: Varies with metal being cast; typically 0.1mm for the first 2.5 cm (0.005 in. for the first inch) and 0.02 mm for each additional centimeter (0.002 in. for each additional inch).

Draft allowances: 1°–3°.

Surface finish: 1–2.5 μm (40–100 μin.) rms.

Because of the precision and finish, most die castings require no finish machining except for the removal of excess metal fin, or flash, around the parting line and the possible drilling or tapping of holes. Production rates are high, and a set of dies can produce many thousands of castings without significant change in dimensions. While die casting is most economical for large production volumes, quantities as low as 2000 can be justified if extensive secondary machining or surface finishing can be eliminated.

Thin-wall zinc die casting is now considered to be a significant competitor to plastic injection molding. The die castings are stronger, stiffer, more dimensionally stable, and more heat resistant. In addition, the metal parts are more resistant to ultraviolet radiation, weathering, and stress cracking when exposed to various reagents.

Figure 13-7 presents a variety of aluminum and zinc die castings. Table 13-3 compares the key features of the four dominant families of die-casting alloys, and Table 13-4 compares the mechanical properties of various die-cast alloys with the properties of other engineering materials.

■ 13.4 SQUEEZE CASTING AND SEMISOLID CASTING

Squeeze casting and semisolid casting are methods that enable the production of high-quality, near-net-shape, thin-walled parts with good surface finish and dimensional precision as well as properties that approach those of forgings. Both processes can be viewed as derivatives of conventional high-pressure die casting, since they employ tool steel dies and apply high pressure during solidification. While the majority of applications involve alloys of aluminum, each of the processes has been successfully applied to magnesium, zinc, copper, and a limited number of ferrous alloys.

TABLE 13-3	Key Properties of the Four Major Families of Die-Cast Metal
Metal	Key Properties
Aluminum	Lowest cost per unit volume; second lightest to magnesium; highest rigidity; good machinability, electrical conductivity, and heat-transfer characteristics.
Magnesium	Lowest density, faster production than aluminum since hot-chamber cast, highest strength-to-weight ratio, good vibration damping, best machinability, can provide electromagnetic shielding.
Zinc	Attractive for small parts; tooling lasts 3–5 times longer than for aluminum; heaviest of the die-castable metals but can be cast with thin walls for possible weight savings; good impact strength, machinability, electrical conductivity, and thermal conductivity.
Zinc–Aluminum	Highest yield and tensile strength, lighter than conventional zinc alloys, good machinability.

FIGURE 13-7   Variety of aluminum (left) and zinc (right) die castings. *(Courtesy of Yoder Die Casting Corporation, Dayton, OH.)*

TABLE 13-4   Comparison of Properties (Die-Cast Metals vs. Other Engineering Materials)

Material	Yield Strength		Tensile Strength		Elastic Modulus	
	MPa	ksi	MPa	ksi	GPa	10^6 psi
Die-cast alloys						
360 aluminum	170	25	300	44	71	10.3
380 aluminum	160	23	320	46	71	10.3
AZ91D magnesium	160	23	230	34	45	6.5
Zamak 3 zinc (AG4OA)	221	32	283	41	—	—
Zamak 5 zinc (AC41A)	269	39	328	48	—	—
ZA-8 (zinc–aluminum)	283–296	41–43	365–386	53–56	85	12.4
ZA-27 (zinc–aluminum)	359–379	52–55	407–441	59–64	78	11.3
Other metals						
Steel sheet	172–241	25–35	276	40	203	29.5
HSLA steel sheet	414	60	414	60	203	29.5
Powdered iron	483	70	—	—	120–134	17.5–19.5
Plastics						
ABS	—	—	55	8	7	1.0
Polycarbonate	—	—	62	9	7	1.0
Nylon 6[a]	—	—	152	22	10	1.5
PET[a]	—	—	145	21	14	2.0

[a] 30% glass reinforced.

In the *squeeze casting* process, molten metal is introduced into the die cavity of a metal mold, using large gate areas and slow metal velocities to avoid turbulence. When the cavity has filled, high pressure (20 to 175 MPa, or 3000 to 25,000 psi) is then applied and maintained during the subsequent solidification. Parts must be designed to directionally solidify toward the gates, and the gates must be sufficiently large that they freeze after solidification in the cavity, thereby allowing the pressurized runner to feed additional metal to compensate for shrinkage.

Intricate shapes can be produced at lower pressures than would normally be required for hot or cold forging. Both retractable and disposable cores can be used to create holes and internal passages. Gas and shrinkage porosity are substantially reduced, and mechanical properties are enhanced. While the squeeze casting process is most commonly applied to aluminum and magnesium castings, it has also been adapted to the production of metal-matrix composites where the pressurized metal is forced around or through foamed or fiber reinforcements that have been positioned in the mold.

For most alloy compositions, there is a range of temperatures where liquid and solid coexist, and several techniques have been developed to produce shapes from this *semisolid* material. In the *rheocasting* process, molten metal is cooled to the semisolid state with constant stirring. The stirring or shearing action breaks up the dendrites, producing a slurry of rounded particles of solid in a liquid melt. This slurry, with about a 30% solid content, can be readily shaped by high-pressure injection into metal dies. Because the slurry contains no superheat and is already partially solidified, it freezes quickly.

In the *thixocasting* variation, there is no handling of molten metal. The material is first subjected to special processing (stirring during solidification as in rheocasting) to produce solid blocks or bars with a nondendritic structure. When reheated to the semisolid condition, the *thixotropic material* can be handled like a solid but flows like a liquid when agitated or squeezed. The solid material is then cut to prescribed length, reheated to a semisolid state where the material is about 40% liquid and 60% solid, mechanically transferred to the shot chamber of a cold-chamber die-casting machine, and injected under pressure. In a variation of the process, solid metal granules or pellets are fed into a barrel chamber, where a rotating screw shears and advances the material through heating zones that raise the temperature to the semisolid region. When a sufficient volume of thixotropic material has accumulated at the end of the barrel, a shot system drives it into the die or mold at velocities of 1 to 2.5 m/sec (40–100 in./sec). The injection system of this process is a combination of the screw feed used in plastic injection molding and the plunger used in conventional die casting.

In all of the semisolid casting processes, the absence of turbulent flow during the casting operation minimizes gas pickup and entrapment. Because the material is already partially solid, the lower injection temperatures and reduced solidification time act to extend tool life. The prior solidification coupled with further solidification under pressure results in a significant reduction in solidification shrinkage and related porosity. The minimization of porosity enables the use of high-temperature heat treatments, such as the T6 solution treatment and artificial aging of aluminum, to further enhance strength. Since the thixocasting process does not use molten metal, both wrought and cast alloys have been successfully shaped. Walls have been produced with thickness as low as 0.2 mm (0.01 in.).

■ 13.5 CENTRIFUGAL CASTING

The inertial forces of rotation or spinning are used to distribute the molten metal into the mold cavity or cavities in the *centrifugal casting* processes, a category that includes true centrifugal casting, semicentrifugal casting, and centrifuging. In *true centrifugal casting*, a dry-sand, graphite, or metal mold is rotated about either a horizontal or vertical axis at speeds of 300 to 3000 rpm. As the molten metal is introduced, it is flung to the surface of the mold, where it solidifies into some form of hollow product. The exterior profile is usually round (as with gun barrels, pipes, and tubes), but hexagons and other symmetrical shapes are also possible.

No core or mold surface is needed to shape the interior, which will always have a round profile because the molten metal is uniformly distributed by the centrifugal forces. When rotation is about the horizontal axis, as illustrated in Figure 13-8, the inner surface is always cylindrical. If the mold is oriented vertically, as in Figure 13-9, gravitational forces cause the inner surface to become parabolic, with the exact shape being a function of the speed of rotation. Wall thickness can be controlled by varying the amount of metal that is introduced into the mold.

During the rotation, the metal is forced against the outer walls of the mold with considerable force, and solidification begins at the outer surface. Centrifugal force continues to feed molten metal as solidification progresses inward. Since the process compensates for shrinkage, no risers are required. The final product has a strong, dense exterior with all of the lighter impurities (including dross and pieces of the refractory mold coating) collecting on the inner surface of the casting. This surface is often left in the final casting, but for some products, it may be removed by a light boring operation.

Products can have outside diameters ranging from 7.5 cm to 1.4 m (3 to 55 in.) and wall thickness up to 25 cm (10 in.). Pipe (up to 12 m, or 40 ft, in length),

FIGURE 13-8 Schematic representation of a horizontal centrifugal casting machine. *(Courtesy of American Cast Iron Pipe Company, Birmingham, AL.)*

FIGURE 13-9 Vertical centrifugal casting, showing the effect of rotational speed on the shape of the inner surface. Paraboloid A results from fast spinning, whereas slower spinning will produce paraboloid B.

pressure vessels, cylinder liners, brake drums, the starting material for bearing rings, and all of the parts illustrated in Figure 13-10 can be manufactured by centrifugal casting. The equipment is rather specialized and can be quite expensive for large castings. The permanent molds can also be expensive, but they offer a long service life, especially when coated with some form of refractory dust or wash. Since no sprues, gates, or risers are required, yields can be greater than 90%. Composite products can also be made by centrifugal casting of a second material on the inside surface of an already-cast product. Table 13-5 summarizes the features of the centrifugal casting process.

In *semicentrifugal casting* (Figure 13-11) the centrifugal force assists the flow of metal from a central reservoir to the extremities of a rotating symmetrical mold. The rotational speeds are usually lower than for true centrifugal casting, and the molds may be either expendable or multiple-use. Several molds may also be stacked on top of one another, so they can be fed by a common pouring basin and sprue. In general, the mold shape is more complex than for true centrifugal casting, and cores can be placed in the mold to further increase the complexity of the product.

The central reservoir acts as a riser and must be large enough to ensure that it will be the last material to freeze. Since the lighter impurities concentrate in the center, however, the process is best used for castings where the central region will ultimately be hollow. Common products include gear blanks, pulley sheaves, wheels, impellers, and electric motor rotors.

Centrifuging, or *centrifuge centrifugal casting* (Figure 13-12), uses centrifugal action to force metal from a central pouring reservoir or sprue, through spoke-type runners, into separate mold cavities that are offset from the axis of rotation. Relatively low rotational speeds are required to produce sound castings with thin walls and intricate shapes. Centrifuging is often used to assist in the pouring of multiple-product investment casting trees.

FIGURE 13-10 Electrical products (collector rings, slip rings, and rotor end rings) that have been centrifugally cast from aluminum and copper. *(Courtesy of The Electric Materials Company, North East, PA.)*

TABLE 13-5   Centrifugal Casting

Process: Molten metal is introduced into a rotating sand, metal, or graphite mold and held against the mold wall by centrifugal force until it is solidified.

Advantages: Can produce a wide range of cylindrical parts, including ones of large size; good dimensional accuracy, soundness, and cleanliness.

Limitations: Shape is limited; spinning equipment can be expensive.

Common metals: Iron; steel; stainless steel; and alloys of aluminum, copper, and nickel.

Size limits: Up to 3 m (10 ft) in diameter and 15 m (50 ft) in length.

Thickness limits: Wall thickness 2.5 to 125 mm (0.1–5 in.).

Typical tolerances: O.D. to within 2.5 mm (0.1 in.); I.D. to about 4 mm (0.15 in.).

Draft allowance: 10 mm/m ($\frac{1}{8}$ in./ft).

Surface finish: 2.5–12.5 μm (100–500 μin.) rms.

FIGURE 13-11   Schematic of a semicentrifugal casting process.

FIGURE 13-12   Schematic of a centrifuging process. Metal is poured into the central pouring sprue and spun into the various mold cavities. *(Courtesy of American Cast Iron Pipe Company, Birmingham, AL.)*

Centrifuging can also be used to drive pewter, zinc, or wax into spinning rubber molds to produce products with close tolerances, smooth surfaces, and excellent detail. These can be finished products or the low-melting-point patterns that are subsequently assembled to form the "trees" for investment casting.

■ 13.6 CONTINUOUS CASTING

As discussed in Chapter 6 and depicted in Figure 6-5, *continuous casting* is usually employed in the solidification of basic shapes that become the feedstock for deformation processes such as rolling and forging. By producing a special mold, continuous casting can also be used to produce long lengths of complex cross-section product, such as the one depicted in Figure 13-13. Since each product is simply a cutoff section of the continuous strand, a single mold is all that is required to produce a large number of pieces. Quality is high as well, since the metal can be protected from contamination during melting and pouring, and only a minimum of handling is required.

FIGURE 13-13 Gear produced by continuous casting. (Left) As-cast material; (right) after machining. *(Courtesy of ASARCO, Tucson, AZ.)*

■ 13.7 MELTING

All casting processes begin with molten metal. Ideally, the molten metal should be available in an adequate amount, at the desired temperature, with the desired chemistry and minimum contamination. The melting furnace should be capable of holding material for an extended period of time without deterioration of quality, be economical to operate, and be capable of being operated without contributing to the pollution of the environment. Except for experimental or very small operations, virtually all foundries use cupolas, air furnaces (also known as direct fuel-fired furnaces), electric-arc furnaces, electric resistance furnaces, or electric induction furnaces. In locations such as fully integrated steel mills, molten metal may be taken directly from a steelmaking furnace and poured into casting molds. This practice is usually reserved for exceptionally large castings. For small operations, gas-fired crucible furnaces are common, but these have rather limited capacities.

Selection of the most appropriate melting method depends on such factors as (1) the temperature needed to melt and superheat the metal, (2) the alloy being melted and the form of available charge material, (3) the desired melting rate or the desired quantity of molten metal, (4) the desired quality of the metal, (5) the availability and cost of various fuels, (6) the variety of metals or alloys to be melted, (7) whether melting is to be batch or continuous, (8) the required level of emission control, and (9) the various capital and operating costs.

The feedstock entering the melting furnace may take several forms. While prealloyed ingot may be purchased for remelt, it is not uncommon for the starting material to be a mix of commercially pure primary metal and commercial scrap, along with recycled gates, runners, sprues, and risers, as well as defective castings. The chemistry can be adjusted through alloy additions in the form of either pure materials or master alloys that are high in a particular element but are designed to have a lower melting point than the pure material and a density that allows for good mixing. Preheating the metal being charged is another common practice, and it can increase the melting rate of a furnace by as much as 30%.

CUPOLAS

A significant amount of gray, nodular, and white cast iron is still melted in *cupolas*, although many foundries have converted to electric induction furnaces. A cupola is a refractory-lined, vertical steel shell into which alternating layers of coke (carbon), iron (pig iron and/or scrap), limestone or other flux, and possible alloy additions are charged and melted under forced air draft. The operation is similar to that of a blast furnace, with the molten metal collecting at the bottom of the cupola to be tapped off either continuously or at periodic intervals.

Cupolas are simple and economical, can be obtained in a wide range of capacities, and can produce cast iron of excellent quality if the proper raw materials are used and good control is practiced. Control of temperature and chemistry can be somewhat difficult, however. The nature of the charged materials and the reactions that occur within the cupola can all affect the product chemistry. Moreover, by the time the final chemistry is determined through analysis of the tapped product, a substantial charge of material is already working its way through the furnace. Final chemistry adjustments, therefore, are often performed in the ladle, using the various techniques of ladle metallurgy discussed in Chapter 6.

Various methods can be used to increase the melting rate and improve the economy of a cupola operation. In a hot-blast cupola, the stack gases are put through a heat exchanger to preheat the incoming air. Oxygen-enriched blasts can also be used to increase the temperature and accelerate the rate of melting. Plasma torches can be employed to melt the iron scrap. With typical enhancements, the melting rate of a continuously operating cupola can be quite high, such that production of 120 tons of hot metal per hour is not uncommon.

INDIRECT FUEL-FIRED FURNACES (OR CRUCIBLE FURNACES)

Small batches of nonferrous metal are often melted in *indirect fuel-fired furnaces* that are essentially crucibles or holding pots whose outer surface is heated by an external flame. The containment crucibles are generally made from clay and graphite, silicon carbide, cast iron, or steel. Stirring action, temperature control, and chemistry control are often poor, and furnace size and melting rate are limited. Nevertheless, these furnaces do offer low capital and operating cost.

Better control of temperature and chemistry can be obtained, however, if the crucible furnaces are heated by electrical resistance heating.

DIRECT FUEL-FIRED FURNACES OR REVERBERATORY FURNACES

Direct fuel-fired furnaces, also known as *reverberatory furnaces*, are similar to small open-hearth furnaces but are less sophisticated. As illustrated in Figure 13-14, a fuel-fired flame passes directly over the pool of molten metal, with heat being transferred to the metal through both radiant heating from the refractory roof and walls and convective heating from the hot gases. Capacity is significantly greater than that of the crucible furnace, but the operation is still limited to the batch melting of nonferrous metals and the holding of cast iron that has been previously melted in a cupola. The rate of heating and melting and the temperature and composition of the molten metal are all easily controlled.

ARC FURNACES

Arc furnaces are the preferred method of melting in many foundries because of the (1) rapid melting rates, (2) ability to hold the molten metal for any desired period of time, and (3) greater ease of incorporating pollution control equipment. The basic features and operating cycle of a *direct-arc furnace* can be described with the aid of Figure 13-15. The top of the wide, shallow unit is first lifted or swung aside to permit the introduction of charge material. The top is then repositioned, and the electrodes are lowered to create an arc between the electrodes and the metal charge. The path of the heating current is usually through one electrode, across an arc to the metal charge, through the metal charge, and back through another arc to another electrode.

FIGURE 13-14 Cross section of a direct fuel-fired furnace. Hot combustion gases pass across the surface of a molten metal pool.

FIGURE 13-15   Schematic diagram of a three-phase electric-arc furnace.

FIGURE 13-16   Electric-arc furnace, tilted for pouring. *(Courtesy of Lectromelt Corporation, Pittsburgh, PA.)*

Fluxing materials are usually added to create a protective slag over the pool of molten metal. Reactions between the slag and the metal serve to further remove impurities and are efficient because of the large interface area and the fact that the slag is as hot as the metal. Because the metal is covered and can be maintained at a given temperature for long periods of time, arc furnaces can be used to produce high-quality metal of almost any desired composition. They are available in sizes up to about 200 tons (but capacities of 25 tons or less are most common), and up to 50 tons per hour can be melted conveniently in batch operations. Arc furnaces are generally used with ferrous alloys, especially steel, and provide good mixing and homogeneity to the molten bath. Unfortunately, the noise and level of particle emissions can be rather high, and the consumption of electrodes, refractories, and power results in high operating costs. Figure 13-16 shows the pouring of an electric-arc furnace. Note the still-glowing electrodes at the top of the furnace.

INDUCTION FURNACES

Because of their very rapid melting rates and the relative ease of controlling pollution, electric *induction furnaces* have become another popular means of melting metal. There are two basic types of induction furnaces. The *high-frequency*, or *coreless* units, shown schematically in Figure 13-17, consist of a crucible surrounded by a water-cooled coil of copper tubing. A high-frequency electrical current passes through the coil, creating an alternating magnetic field. The varying magnetic field induces secondary electrical currents in the metal being melted, which bring about a rapid rate of heating.

Coreless induction furnaces are used for virtually all common alloys, with the maximum temperature being limited only by the refractory and the ability to insulate against heat loss. They provide good control of temperature and composition and are available in a range of capacities up to about 65 tons. Because there is no contamination from the heat source, they produce very pure metal. Operation is generally on a batch basis.

Low-frequency or *channel-type* induction furnaces are also seeing increased use. As shown in Figure 13-18, only a small channel is surrounded by the primary (current-carrying or heating) coil. A secondary coil is formed by a loop, or channel, of molten metal, and all the liquid metal is free to circulate through the loop and gain heat. To start, enough molten metal must be placed into the furnace to fill the secondary coil, with the remainder of the charge taking a variety of forms. The heating rate is high, and the temperature can be accurately controlled. As a result, channel-type furnaces are often preferred as holding furnaces, where the molten metal is maintained at a constant temperature for an extended period of time. Capacities can be quite large, up to about 250 tons.

FIGURE 13-17   Schematic showing the basic principle of a coreless induction furnace.

FIGURE 13-18 Cross section showing the principle of the low-frequency or channel-type induction furnace.

■ 13.8 POURING PRACTICE

Some type of pouring device, or ladle, is usually required to transfer the metal from the melting furnace to the molds. The primary considerations for this operation are (1) to maintain the metal at the proper temperature for pouring and (2) to ensure that only high-quality metal is introduced into the molds. The specific type of *pouring ladle* is determined largely by the size and number of castings to be poured. In small foundries, a handheld, shank-type ladle is used for manual pouring. In larger foundries, either bottom-pour or teapot-type ladles are used, like the ones illustrated in Figure 11-6. These are often used in conjunction with a conveyor line that moves the molds past the pouring station. Because metal is extracted from beneath the surface, slag and other impurities that float on top of the melt are not permitted to enter the mold.

High-volume, mass-production foundries often use automatic pouring systems, like the one shown in Figure 13-19. Molten metal is transferred from a main melting furnace to a holding furnace. A programmed amount of molten metal is further transferred into individual pouring ladles and is then poured into the corresponding molds

FIGURE 13-19 Automatic pouring of molds on a conveyor line. *(Courtesy of Roberts Sinto Corporation, Lansing, MI.)*

as they traverse by the pouring station. Laser-based control units position the pouring ladle over the sprue and control the flow rate into the pouring cup.

■ 13.9 CLEANING, FINISHING, AND HEAT TREATING OF CASTINGS

CLEANING AND FINISHING
After solidification and removal from the mold, most castings require some additional cleaning and finishing. Specific operations may include all or several of the following:

1. Removing cores
2. Removing gates and risers
3. Removing fins, flash, and rough spots from the surface
4. Cleaning the surface
5. Repairing any defects

Cleaning and finishing operations can be quite expensive, so consideration should be given to their minimization when designing the product and selecting the specific method of casting. In addition, consideration should also be directed toward the possibility of automating the cleaning and finishing.

Sand cores can usually be removed by mechanical shaking. At times, however, they must be removed by chemically dissolving the core binder. On small castings, sprues, gates, and risers can sometimes be knocked off. For larger castings, a cutting operation is usually required. Most nonferrous metals and cast irons can be cut with an abrasive cutoff wheel, power hacksaw, or band saw. Steel castings frequently require an oxy-acetylene torch. Plasma arc cutting can also be used.

The specific method of cleaning often depends on the size and complexity of the casting. After the gates and risers have been removed, small castings are often tumbled in barrels to remove fins, flash, and sand that may have adhered to the surface. Tumbling may also be used to remove cores and, in some cases, gates and risers. Metal shot or abrasive material is often added to the barrel to aid in the cleaning. Conveyors can be used to pass larger castings through special cleaning chambers, where they are subjected to blasts of abrasive or cleaning material. Extremely large castings usually require manual finishing, using pneumatic chisels, portable grinders, and manually directed blast hoses.

While defect-free castings are always desired, flaws such as cracks, voids, and laps are not uncommon. In some cases, especially when the part is large and the production quantity is small, it may be more attractive to repair the part rather than change the pattern, die, or process. If the material is weldable, repairs are often made by removing the defective region (usually by chipping or grinding) and filling the created void with deposited weld metal. Porosity that is at or connected to free surfaces can be filled with resinous material, such as polyester, by a process known as *impregnation*. If the pores are filled with a lower-melting-point metal, the process becomes *infiltration*. (See Chapter 16 for a further discussion of these processes.)

HEAT TREATMENT AND INSPECTION OF CASTINGS
Heat treatment is an attractive means of altering properties while retaining the shape of the product. Steel castings are frequently given a full anneal to reduce the hardness and brittleness of rapidly cooled, thin sections and to reduce the internal stresses that result from uneven cooling. Nonferrous castings are often heat treated to provide chemical homogenization or stress relief as well as to prepare them for subsequent machining. For final properties, virtually all of the treatments discussed in Chapter 5 can be applied. Ferrous-metal castings often undergo a quench-and-temper treatment, and many nonferrous castings are age hardened to impart additional strength. The variety of heat treatments is largely responsible for the wide range of properties and characteristics available in cast metal products.

Virtually all of the nondestructive *inspection techniques* can be applied to cast metal products. X-ray radiography, liquid penetrant inspection, and magnetic particle inspection are extremely common.

■ 13.10 AUTOMATION IN FOUNDRY OPERATIONS

Many of the operations that are performed in a foundry are ideally suited for robotic automation since they tend to be dirty, dangerous, or dull. Robots can dry molds, coat cores, vent molds, and clean or lubricate dies. They can tend stationary, cyclic equipment, such as die-casting machines, and if the machines are properly grouped, one robot can often service two or three machines. In the finishing room, robots can be equipped with plasma cutters or torches to remove sprues, gates, and runners. They can perform grinding and blasting operations, as well as various functions involved in the heat treatment of castings.

In the investment-casting process, robots can be used to dip the wax patterns into refractory slurry and produce the desired molds. In a similar manner, robots have been used to dip the Styrofoam patterns of the full-mold and lost-foam processes in their refractory coating and hang them on conveyors to dry. In a fully automated lost-foam operation, robots could be used to position the pattern, fill the flask with sand, pour the metal, and use a torch to remove the sprue.

■ 13.11 PROCESS SELECTION

As shown in the individual process summaries that have been included throughout Chapters 12 and 13, each of the casting processes has a characteristic set of capabilities, assets, and limitations. The requirements of a particular product (such as size, complexity, required dimensional precision, desired surface finish, total quantity to be made, and desired rate of production) often limit the number of processes that should be considered as production candidates. Further selection is usually based on cost.

Some aspects of product cost, such as the cost of the material and the energy required to melt it, are somewhat independent of the specific process. The cost of other features, such as patterns, molds, dies, melting and pouring equipment, scrap material, cleaning, inspection, and all related labor, can vary markedly and be quite dependent on the process. For example, pattern and mold costs for sand casting are quite a bit less than the cost of die-casting dies. Die casting, on the other hand, offers high production rates and a high degree of automation. When a small quantity of parts is desired, the cost of the die or tooling must be distributed over the total number of parts, and unit cost (or cost per casting) is high. When the total quantity is large, the tooling cost is distributed over many parts, and the cost per piece decreases.

Figure 13-20 shows the relationship between unit cost and production quantity for a product that can be made by both sand and die casting. Sand casting is an expendable mold process. Since an individual mold is required for each pour, increasing quantity does not lead to a significant drop in unit cost. Die casting involves a multiple-use mold, and the cost of the die can be distributed over the total number of parts. As shown in the figure, sand casting is often less expensive for small production runs, and processes such as die casting are preferred for large quantities. One should note that while the die-casting curve in Figure 13-20 is a smooth line, it is not uncommon for an actual curve to contain abrupt discontinuities. If the lifetime of a set of tooling is 50,000 casts, the

FIGURE 13-20 Typical unit cost of castings comparing sand casting and die casting. Note how the large cost of a die-casting die diminishes as it is spread over a larger quantity of parts.

cost *per part* for 45,000 pieces, using one set of tooling, would actually be less than for 60,000 pieces, since the latter would require a second set of dies.

In most cases, multiple processes are reasonable candidates for production, and the curves for all of the options should be included. The final selection is often based on a combination of economic, technical, and management considerations.

Table 13-6 presents a comparison of casting processes, including green-sand casting, chemically bonded sand molds (shell, sodium silicate, and air-set), ceramic mold and investment casting, permanent-mold casting, and die casting. The processes are compared on the basis of cost for both small and large quantities, thinnest section, dimensional precision, surface finish, ease of casting a complex shape, ease of changing the design while in production, and range of castable materials.

TABLE 13-6 Comparison of Casting Processes					
Property or Characteristic	Green-Sand Casting	Chemically Bonded Sand (Shell, Sodium Silicate, Air-Set)	Ceramic Mold and Investment Casting	Permanent-Mold Casting	Die Casting
Relative cost for small quantity	Lowest	Medium high	Medium	High	Highest
Relative cost for large quantity	Low	Medium high	Highest	Low	Lowest
Thinnest section (inches)	$\frac{1}{10}$	$\frac{1}{10}$	$\frac{1}{16}$	$\frac{1}{8}$	$\frac{1}{32}$
Dimensional precision (+/– in inches)	0.01–0.03	0.005–0.015	0.01–0.02	0.01–0.05	0.001–0.015
Relative surface finish	Fair to good	Good	Very good	Good	Best
Ease of casting complex shape	Fair to good	Good	Best	Fair	Good
Ease of changing design while in production	Best	Fair	Fair	Poor	Poorest
Castable metals	Unlimited	Unlimited	Unlimited	Low-melting-point metals	Low-melting-point metals

■ Key Words

arc furnace	heat checking	inspection	slush casting
centrifugal casting	heat treatment	low-pressure permanent-mold	squeeze casting
centrifuging	heated-manifold direct-	casting	thermal fatigue
cold-chamber die-casting	injection die casting	permanent-mold casting	thixocasting
machine	hot-chamber die-casting	pouring ladle	thixotropic material
continuous casting	machine	retractable cores	true centrifugal casting
cupola	impregnation	reverberatory furnaces	vacuum permanent-mold
die casting	indirect fuel-fired furnace	rheocasting	casting
direct fuel-fired furnace	induction furnace	semicentrifugal casting	venting
gooseneck die-casting	infiltration	semipermanent mold casting	
machine	inserts	semisolid casting	

■ Review Questions

1. What are some of the major disadvantages of the expendable-mold casting processes?
2. What are some possible limitations of multiple-use molds?
3. What are some common mold materials for permanent-mold casting? What are some of the metals more commonly cast?
4. Describe some of the process advantages of permanent-mold casting.
5. Why might low production runs be unattractive for permanent-mold casting?
6. What features affect the life of a permanent mold?

7. How is venting provided in the permanent-mold process?
8. Why are permanent-mold castings generally removed from the mold immediately after solidification has been completed?
9. What types of products would be possible candidates for manufacture by slush casting?
10. How does low-pressure permanent-mold casting differ from the traditional gravity-pour process?
11. What are some of the attractive features of the low-pressure permanent-mold process?
12. What are some additional advantages of vacuum permanent-mold casting over the low-pressure process?
13. Contrast the feeding pressures on the molten metal in low-pressure permanent molding and die casting.
14. Contrast the materials used to make dies for gravity-pour permanent-mold casting and die casting. Why is there a notable difference?
15. By what mechanisms do die-casting dies typically fail?
16. Why might it be advantageous to vary the pressure on the molten metal during the die-casting cycle?
17. For what types of materials would a hot-chamber die-casting machine be appropriate?
18. What metals are routinely cast with cold-chamber die-casting machines?
19. How does the air in the mold cavity escape in the die-casting process?
20. Are risers employed in die casting? Can sand cores be used?
21. What are some of the attractive features of die casting compared to alternative casting methods?
22. When might low quantities be justified for the die-casting process?
23. Describe the squeeze casting process.
24. What is a thixotropic material? How does it provide an attractive alternative to squeeze casting or rheocasting?
25. What are some of the attractive features of semisolid casting?
26. Contrast the structure and properties of the outer and inner

surfaces of a centrifugal casting.
27. What are the key differences between true centrifugal casting, semicentrifugal casting, and centrifuging?
28. How can continuous casting be used in the direct production of products?
29. What are some of the factors that influence the selection of a furnace type or melt procedure in a casting operation?
30. What are some of the possible feedstock materials that may be put in foundry melt furnaces?
31. What types of metals are commonly melted in cupolas?
32. What are some of the ways that the melting rate of a cupola can be increased?
33. What are some of the pros and cons of indirect fuel-fired furnaces?
34. What are some of the attractive features of arc furnaces in foundry applications?
35. Why are channel induction furnaces attractive for metal-holding applications where molten metal must be held at a specified temperature for long periods of time?
36. What are the primary functions of a pouring operation?
37. What are some of the typical cleaning and finishing operations that are performed on castings?
38. What are some common ways to remove cores from castings? To remove runners, gates, and risers?
39. What are some of the alternative methods of cleaning and finishing castings?
40. How might defective castings be repaired to permit successful use in their intended applications?
41. What are some of the ways that industrial robots can be employed in metal-casting operations?
42. Describe some of the features that affect the cost of a cast product. Why might the cost vary significantly with the quantity to be produced?
43. What are some of the key factors that should be considered when selecting a casting process?

www.wiley.com/college/degarmo

Chapter 13 CASE STUDY

Baseplate for a Household Steam Iron

The item depicted in the figure is the baseplate of a high-quality household steam iron. It is rated for operation at up to 1200 watts and is designed to provide both steady steam and burst of steam features. Incorporated into the design is an integral electrical resistance heating "horseshoe" that must be thermally coupled to the baseplate but remain electrically insulated. (This component often takes the form of a resistance heating wire, surrounded by ceramic insulation, all encased in a metal tube.) The steam emerges through a number of small vent holes in the base, each about 1/16 inch in diameter. There are about a dozen larger threaded recesses, about 1/8 inch in diameter, that are used in assembling the various components.

1. Discuss the various features that this component must possess in order to function in an adequate fashion. Consider strength, impact resistance, thermal conductivity, corrosion resistance, weight, and other factors.

2. What material or materials would appear to be strong candidates?

3. What are some possible means of producing the desired shape? Which would you prefer? Could the heating element assembly be incorporated during manufacture, or does it have to be added as a secondary operation? What are the major advantages of the method you propose?

4. Could all of the design features (holes, webs, and recesses) be incorporated in the initial manufacturing operation, or would secondary processing be required? If secondary processing is required, for what features, and how would you recommend that they be produced?

5. Some commercial irons have baseplates for which the bottom surfaces have been finished by a simple buff and polish, while others have a Teflon coating or have been anodized. If your desire is to produce a high-quality product, what form of surface finishing would you recommend?

CHAPTER 14

FABRICATION OF PLASTICS, CERAMICS, AND COMPOSITES

■ 14.1 INTRODUCTION

In Chapters 6, 7, and 8, *plastics, ceramics*, and *composites* were shown to be substantially different from metals in both structure and properties. It is reasonable to expect, therefore, that the principles of material selection and product design, as well as the fabrication processes, will also be somewhat different. In addition, there will also be some similarities. The specific material will still be selected for its ability to provide the required properties and the fabrication processes for their ability to produce the desired shape in an economical and practical manner.

In terms of differences, plastics, ceramics, and composites tend to be used closer to their design limits, and many of the fabrication processes convert the raw material into a finished product in a single operation. Large, complex shapes can often be formed as a single unit, eliminating the need for multipart assembly operations. Materials in these classes can often provide integral and variable color, and the processes used to manufacture the shape can frequently produce the desired finish and precision. As a result, finishing operations are often unnecessary—an attractive feature since, for many of these materials, altering the final dimensions or surface would be both difficult and costly. The joining and fastening operations used with these materials also tend to be different from those used with metals.

As with metals, the properties of these materials are affected by the processes used to produce the shape. The fabrication of an acceptable product, therefore, involves the selection of both (1) an appropriate material and (2) a companion method of processing, such that the resulting combination provides the desired shape, properties, precision, and finish.

■ 14.2 FABRICATION OF PLASTICS

The manufacture of a successful plastic product requires satisfying the various mechanical and physical property requirements through the use of the most economical resin or compound that will perform satisfactorily, and coupling it with a manufacturing process that is compatible with both the part design and the selected material.

Chapter 8 presented material about the wide variety of plastics or polymers that are currently used as engineering materials. As we move our attention to the fabrication of parts and shapes, we find that there are also a variety of processes from which to choose. Determination of the preferred method depends on the desired size, shape, and quantity, as well as whether the polymer is a *thermoplastic, thermoset*, or *elastomer*. Thermoplastic polymers can be heated to produce either a soft, formable solid or a liquid. The material can then be cast, injected into a mold, or forced into or through dies to produce a desired shape. Thermosetting polymers have far fewer options, because once the polymerization has occurred, the framework structure is established and no further deformation can occur. Thus the polymerization reaction must take place during the shape-forming operation. Elastomers are sufficiently unique that they will be treated in a separate section of this chapter.

Casting, blow molding, compression molding, transfer molding, cold molding, injection molding, reaction injection molding, extrusion, thermoforming, rotational molding, and *foam molding* are all processes that are used to shape polymers. Each has its distinct set of advantages and limitations that relate to part design, compatible materials, and production cost. To make optimum selections, we must be familiar with the shape capabilities of a process as well as how the process affects the properties of the material.

CASTING

Casting is the simplest of the shape-forming processes because no fillers are used and no pressure is required. While not all plastics can be cast, there are a number of castable thermoplastics, including acrylics, nylons, urethanes, and PVC plastisols. The thermoplastic polymer is simply melted, and the liquid is poured into a container having the shape of the desired part. Several variations of the process have been developed. Small products can be cast directly into shaped molds. Plate glass can be used as a mold to cast individual pieces of thick plastic sheet. Continuous sheets and films can be produced by injecting the liquid polymer between two moving belts of highly polished stainless steel, the width and thickness being set by resilient gasket strips on either end of the gap. Thin sheets can be made by ejecting molten liquid from a gap-slot die onto a temperature-controlled chill roll. The molten plastic can also be spun against a rotating mold wall (centrifugal casting) to produce hollow or tubular shapes.

Some thermosets (such as phenolics, polyesters, epoxies, silicones, and urethanes) can also be cast, as well as any resin that will polymerize at low temperatures and atmospheric pressure. Because of the need for curing, the casting of thermoset resins usually involves additional processing, often some form of heating while in the mold. Figure 14-1 depicts a process where a steel pattern is dipped into molten lead, withdrawn, and allowed to cool. A thin lead sheath is produced when the pattern is removed, and this becomes the mold for the plastic resin. Curing occurs, either at room temperature or by heating for long times at temperatures in the range of 65° to 95°C (150° to 200°F). After curing, the product is removed, and the lead sheaths can be reused.

Since cast plastics contain no fillers, they have a distinctly lustrous appearance, and a wide range of transparent and translucent colors are available. Since the product is shaped as a liquid, fiber or particulate reinforcement can be easily incorporated. The process is relatively inexpensive because of the comparative lack of costly dies, equipment, and controls. Typical products include sheets, plates, films, rods, and tubes, as well

Steel
mandrel

Plastic

To oven

Molten lead   Lead shell

FIGURE 14-1   Steps in the casting of plastic parts using a lead shell mold.

FIGURE 14-2 Steps in blow molding plastic parts: (1) a tube of heated plastic is placed in the open mold; (2) the mold closes over the tube, simultaneously sealing the bottom; (3) air expands the tube against the sides of the mold; and (4) after sufficient cooling, the mold opens to release the product.

as small objects, such as jewelry, ornamental shapes, gears, and lenses. While dimensional precision can be quite high, quality problems can occur because of inadequate mixing, air entrapment, gas evolution, and shrinkage.

BLOW MOLDING

A variety of *blow molding* processes have been developed, the most common being used to convert thermoplastic polyethylene, polyvinyl chloride (PVC), polypropylene, and PEEK resins into bottles and other hollow-shape containers. A solid-bottom, hollow-tube preform, known as a *parison*, is made from heated plastic by either extrusion or injection molding. The heated preform is then positioned between the halves of a split mold, the mold closes, and the preform is expanded against the mold by air or gas pressure. The mold is then cooled, the halves separated, and the product is removed. Any flash is then trimmed for direct recycling. Figure 14-2 depicts a form of this process where the starting material is a simple tube and the solid bottom is created by the pinching action of die closure. Blow molding has recently expanded to include the engineering thermoplastics and has been used to produce products as diverse as automotive fuel tanks, seat backs, ductwork, and bumper beams.

Variations of blow molding have been designed to provide both axial and radial expansion of the plastic (for enhanced strength) as well as to produce multilayered products. In one process, a sheet of heated plastic is placed between upper and lower cavities, the lower one having the shape of the product. Both cavities are then pressurized to 2 to 4 MPa (300 to 600 psi) with a nonreactive gas such as argon. When the pressure in the lower segment is then vented, the gas in the upper segment "blows" the material into the lower die cavity.

Because the thermoplastics must be cooled before removal from the mold, the molds for blow molding must contain the desired cavity as well as a cooling system, venting system, and other design features. The mold material must provide thermal conductivity and durability while being inexpensive and compatible with the resins being processed. Beryllium copper, aluminum, tool steels, and stainless steels are all popular mold materials.

COMPRESSION MOLDING OR HOT-COMPRESSION MOLDING

In *compression molding*, illustrated schematically in Figure 14-3, solid granules or preformed tablets of *unpolymerized* plastic are introduced into an open, heated cavity. A heated plunger then descends to close the cavity and apply pressure. As the material melts and becomes fluid, it is driven into all portions of the cavity. The heat and pressure are maintained until the material has "set" (i.e., cured or polymerized). The mold is then opened and the part is removed. A wide variety of heating systems and mold materials are used, and multiple cavities can be placed within a mold to produce more than one part in a single pressing. The process is simple and used primarily with the thermosetting polymers, although recent developments permit the shaping of thermoplastics and composites. Cycle times are set by the rate of heat transfer and the reaction or curing rate of the polymer. They typically range from under 1 minute to as much as 20 minutes or more.

FIGURE 14-3 The hot-compression molding process: (1) solid granules or a preform pellet is placed in a heated die; (2) a heated punch descends and applies pressure; and (3) after curing (thermosets) or cooling (thermoplastics), the mold is opened and the part is removed.

The tool and machinery costs for compression molding are often lower than for competing processes, and the dimensional precision and surface finish are high, thereby reducing or eliminating secondary operations. Compression molding is most economical when it is applied to small production runs of parts requiring close tolerances, high impact strength, and low mold shrinkage. It is a poor choice when the part contains thick sections (the cure times become quite long) or when large quantities are desired. Most products have relatively simple shapes because the flow of material is rather limited. Typical compression-molded parts include gaskets, seals, exterior automotive panels, aircraft fairings, and a wide variety of interior panels.

More recently, compression molding has been used to form fiber-reinforced plastics, both thermoplastics and thermosets, into parts with properties that rival the engineering metals. In the thermoset family, polyesters, epoxies, and phenolics can be used as the base of fiber-containing sheet-molding compound, bulk-molding compound, or sprayed-up reinforcement mats. These are introduced into the mold and shaped and cured in the normal manner. Cycle times range from about 1 to 5 minutes per part, and typical products include wash basins, bathtubs, equipment housings, and various electrical components.

If the starting material is a fiber-containing thermoplastic, precut blanks are first heated in an infrared oven to produce a soft, pliable material. The blanks are then transferred to the press, where they are shaped and cooled in specially designed dies. Compared to the thermosets, cycle times are reduced and the scrap is often recyclable. In addition, the products can be joined or assembled using the thermal "welding" processes applied to plastics.

Compression molding equipment is usually rather simple, typically consisting of a hydraulic or pneumatic press with parallel platens that apply the heat and pressure. Pressing areas range from 15 cm^2 (6 in.^{2}) to as much as 2.5 m^2 (8 ft^2), and the force capacities range from 6 to 9000 metric tons. The molds are usually made of tool steel and are polished or chrome plated to improve material flow and product quality. Mold temperatures typically run between 150° and 200°C (300° and 400°F) but can go as high as 650°C (1200°F). They are heated by a variety of means, including electric heaters, steam, oil, and gas.

TRANSFER MOLDING

Transfer molding is sometimes used to reduce the turbulence and uneven flow that can result from the high pressures of hot-compression molding. As shown in Figure 14-4, the *unpolymerized* raw material is now placed in a plunger cavity, where it is heated until molten. The plunger then descends, forcing the molten plastic through channels or runners into adjoining die cavities. Temperature and pressure are maintained until the thermosetting resin has completely cured. To shorten the cycle and extend the lifetime of the cavity, plunger, runner, and gates, the charge material may be preheated before being placed in the plunger cavity.

Because the material enters the die cavities as a liquid, there is little pressure until the cavity is completely filled. Thin sections, excellent detail, and good tolerances and finish are all characteristics of the process. In addition, inserts can be incorporated into the products of transfer molding. They are simply positioned within the cavity and maintained in place as the liquid resin is introduced around them.

FIGURE 14-4   Diagram of the transfer molding process. Molten or softened material is first formed in the upper heated cavity. A plunger then drives the material into an adjacent die.

Transfer molding is attractive for producing small to medium-sized parts with relatively complex shapes. It combines elements of both compression molding and injection molding (to be discussed), and enables some of the advantages of injection molding to be utilized with thermosetting polymers. The thermosetting resins can be reinforced with fillers, such as cellulose, glass, silica, alumina, or mica, to improve the mechanical or electrical properties and reduce shrinkage or warping. The main limitation of the process is the loss of material. The resin left in the pot or well, sprue, and runners also cures, and must now be discarded. Common products include electrical switchgear and wiring devices, parts of household appliances that require heat resistance, structural parts that require hardness and rigidity under load, under-hood automotive parts, and parts that require good resistance to chemical attack.

COLD MOLDING

In *cold molding*, the uncured thermosetting material is pressed to shape while cold and is then removed from the mold and cured in a separate oven. While the process is faster and more economical, the resulting products generally lack good surface finish and dimensional precision.

INJECTION MOLDING

Injection molding is the most widely used process for the high-volume production of relatively complex thermoplastic parts. Figure 14-5 illustrates one approach to the process, where granules of raw material are fed by gravity from a hopper into a cavity that lies ahead of a moving plunger. As the plunger advances, the material is forced through a preheating chamber and on through a torpedo section, where it is mixed, melted, and superheated. The superheated material is then driven through a nozzle that seats against a mold. Other types of injection units control the flow of material and generate the injection pressure with screws that have both rotation and axial movements, or combinations of screws and plungers. Alternative methods of heating the material include heated barrels and the shearing action as material moves through the screws.

Sprues and runners then channel the molten material into one or more closed-die cavities. Since the dies remain cool, the plastic solidifies almost as soon as the mold is filled. Premature solidification would cause defective parts, so the material must be rapidly forced into the mold cavities by pressures in the range 35 to 140 MPa (5 to 20 ksi), which are maintained during solidification. The mold halves must clamp tightly together during molding and then be easily separated for part ejection. Impact forces should be minimized during die closure, since they can adversely affect die life. Various types of clamping designs have been developed, including toggle, hydraulic, and hydro-mechanical.

Control systems coordinate all of the functions of the process, including the time required for cooling within the mold. By heating the material for the next part as the mold is separating for part ejection, a molding cycle can be completed in 1 to 30 seconds. The process is quite similar to the die casting of molten metal, and the result is usually a finished product needing no further work before assembly or use.

Some injection molding machines incorporate a hot-runner distribution system to transfer the material from the injection nozzle to the mold cavities. If the runners are

FIGURE 14-5 Schematic diagram of the injection molding process. A moving plunger advances material through a heating region (in this case, through a heated manifold and over a heated torpedo) and further through runners into a mold where the molten thermoplastic cools and solidifies.

cold, the material in the runner solidifies with each cycle and needs to be ejected and reprocessed or disposed of. With hot runners, the thermoplastic material is maintained in a liquid state until it reaches the gate. The material in the runners can be used in the subsequent shot, thereby reducing shot size and cycle time, since less material must be heated. Quality is improved, since all material enters the mold at the same temperature, recycled sprues and runners are not incorporated into the charge, and there is less turbulence since pressurized material is not injected into empty runners. Hot runners do add an additional degree of complexity to the design, operation, and control of the system, so the additional cost must be weighed against the cited benefits.

Injection molding can also be applied to the thermosetting materials, but the process must be modified to provide the temperature, pressure, and time required for curing. The injection chamber is now cold, and the mold is heated. The time in the heated mold must be sufficient to complete the curing process, and the relatively long cycle times are the major deterrent to the injection molding of the thermosets.

REACTION INJECTION MOLDING

Figure 14-6 depicts the *reaction injection molding* process, in which two or more liquid reactants are metered into a unit where they are intimately mixed by the impingement of liquid streams that have been pressurized to a value between 13 and 20 MPa (2000 and 3000 psi). The combined material flows through a pressure-reducing chamber and exits the mixhead directly into a mold. An exothermic chemical reaction takes place between the two components, resulting in thermoset polymerization. Since no heating is required, the production rates are set primarily by the curing time of the polymer, which is often less than 1 minute. Molds are made from steel, aluminum, or nickel shell, with selection being made on the basis of number of parts to be made and the desired quality. The molds are generally clamped in low-tonnage presses.

At present, the dominant materials for reaction injection molding are polyurethanes, polyamides, and composites containing short fibers or flakes. Properties can span a wide range, depending on the combination and percentage of base chemicals and the additives that are used. Different formulations can result in elastomeric or flexible, structural foam (foam core with a hard, solid outer skin), solid (no foam core), or composite products. Part size can range from $\frac{1}{2}$ to 50 kg (1 to 100 lb), shapes can be quite complex (with variable wall thickness), and surface finish is excellent. Automotive applications include steering wheels, airbag covers, instrument panels, door panels, armrests, headliners, and center consoles, as well as body panels, bumpers, and wheel covers. Rigid polyurethanes are also used in such products as computer housings, household refrigerators, water skis, hot-water heaters, and picnic coolers.

From a manufacturing perspective, reaction injection molding has a number of attractive features. The low processing temperatures and low injection pressures make the process attractive for molding large parts, and the large size can often enable parts consolidation. Thermoset parts can generally be fabricated with less energy than the

FIGURE 14-6 The reaction injection molding process. (Left) Measured amounts of reactants are combined in the mixing head and injected into the split mold. (Right) After sufficient curing, the mold is opened and the component is ejected.

injection molding of thermoplastics, with similar cycle times and a similar degree of automation. The metering and mixing equipment and related controls tend to be quite sophisticated and costly, but the lower temperatures and pressures enable the use of cheaper molds, which can be quite large.

EXTRUSION

Long plastic products with uniform cross sections can be readily produced by the *extrusion* process depicted in Figure 14-7. Thermoplastic pellets or powders are fed through a hopper into the barrel chamber of a screw extruder. A rotating screw propels the material through a preheating section, where it is heated, homogenized, and compressed, and then forces it through a heated die and onto a conveyor belt. To preserve its newly imparted shape, the material is cooled and hardened by jets of air or sprays of water. It continues to cool as it passes along the belt and is then either cut into lengths or coiled, depending on whether the material is rigid or flexible and on the desires of the customer. The process is continuous and provides a cheap and rapid method of molding. Common production shapes include a wide variety of constant cross-section profiles, such as window and trim molding, as well as tubes, pipes, and even coated wires and cables. Thermoplastic foam shapes can also be produced. If an emerging tube is expanded by air pressure, allowed to cool, and then rolled, the product can be a double layer of sheet or film.

THERMOFORMING

In the *thermoforming* process, thermoplastic sheet material is first heated to a working temperature. The starting material can be either discrete sheets or a continuous roll of material. If continuous material is used, it is usually heated by passing through an oven or other heating device. The material emerges over a male or female mold and is formed by the application of vacuum, pressure, or another mechanical tool. Cooling occurs upon contact with the mold, and the product hardens in its new shape. After sufficient cooling, the part can be removed from the mold and trimmed, and the unused strip material is diverted for recycling.

Figure 14-8 shows the process using a female mold cavity and discrete sheets of material. Here the material is placed directly over the die or pattern and is heated in place. Pressure and/or vacuum is then applied, causing the material to draw into the cavity. The female die imparts both the dimensions and finish or texture to the exterior surface. The sheet material can also be stretched over male form blocks, and here the tooling controls dimensions and finish on the interior surface. Mating male and female dies can also be used. An entire cycle requires only a few minutes.

While the starting material is a uniform-thickness sheet, the thickness of the products will vary as various regions undergo stretching. Typical products tend to be simple-shaped, thin-walled parts, such as plastic luggage, plastic trays, panels for light fixtures, or even pages of Braille text for the blind.

ROTATIONAL MOLDING

Rotational molding can be used to produce hollow, seamless products of a wide variety of sizes and shapes, including storage tanks, bins and refuse containers, doll parts, footballs, helmets, and even boat hulls. The process begins with a closed mold or cavity that

FIGURE 14-7 A screw extruder producing thermoplastic product. Some units may have a changeable die at the exit to permit production of different-shaped parts.

Heat lamps on   Heat lamps off

Vacuum pump off   Vacuum pump on

FIGURE 14-8   A type of thermoforming where thermoplastic sheets are shaped using a combination of heat and vacuum.

has been filled with a premeasured amount of thermoplastic powder or liquid. The molds are either preheated or placed in a heated oven and are then rotated simultaneously about two perpendicular axes. Other designs rotate the mold about one axis while tilting or rocking about another. In either case, the resin melts and is distributed in the form of a uniform-thickness coating over all of the surfaces of the mold. The mold is then transferred to a cooling chamber, where the motion is continued and air or water is used to slowly drop the temperature. After the material has solidified, the mold is opened and the uniform-thickness, hollow product is removed. All of the starting material is used in the product; no scrap is generated. The lightweight rotational molds are frequently made from cast aluminum, but sheet metal is often used for larger parts; electroformed or vaporformed nickel is used when fine detail is to be reproduced.

FOAM MOLDING

Foamed plastic products have become an important and widely used form of polymer. In foam molding, a foaming agent is mixed with the plastic resin and releases gas or volatilizes when the material is heated during molding. The materials expand to 2 to 50 times their original size, resulting in products with densities ranging from 32 to 640 g/L (2 to 40 lb/ft^3). *Open-cell foams* have interconnected pores that permit the permeability of gas or liquid. *Closed-cell foams* have the property of being gas- or liquid-tight.

Both rigid and flexible foams have been produced using both thermoplastic and thermosetting materials. The rigid type is useful for structural applications (including housings for computers and business machines), packaging, and shipping containers; as patterns for the full-mold and lost-foam casting processes (see Chapter 12); and for injection into the interiors of thin-skinned metal components, such as aircraft fins and stabilizers. Flexible foams are used primarily for cushioning.

Variations of the conventional molding processes can be used to produce a variety of unusual products. By introducing foaming material into the interior of a mold that has partly filled, parts can be produced with a solid outer skin and a rigid foam core. Figure 14-9 shows the cross section of a plastic gear with this type of dual structure.

FIGURE 14-9   A plastic gear with a solid outer skin and a rigid foam core. (*From* American Machinist.*)*

OTHER PLASTIC-FORMING PROCESSES

In the *calendering* process, a mass of dough-like thermoplastic is forced between and over two or more counter-rotating rolls to produce thin sheets or films of polymer, which are then cooled to induce hardening. Product thicknesses generally range between 0.3 and 1.0 mm (0.01 to $\frac{1}{16}$ in.) but can be reduced further to as low as 0.05 mm (0.002 in.) by subsequent stretching. Embossed designs can be incorporated into the rolls to produce products with textures or patterns.

Conventional *drawing* can be used to produce fibers, and *rolling* can be performed to change the shape of thermoplastic extrusions. In addition to changing the product dimensions, these processes can also serve to induce crystallization, or produce a preferred orientation to the thermoplastic polymer chains.

Filaments, fibers, and yarns can be produced by *spinning*, a modified form of extrusion. Molten thermoplastic polymer is forced through a die containing many small holes. Where multistrand yarns or cables are desired, the dies can rotate or spin to produce the twists and wraps.

The various plastic-forming processes are often combined in either sequential or integrated forms to produce specific products. For example, the closed-bottom parison for blow molding can be formed as a separate injection molding or as an integrated extrusion operation that is then followed by blow molding. Another example is the manufacture of thin plastic bags, such as those that are used as kitchen or bathroom trash can liners. Polymer granules flow through a hopper and enter the barrel of a screw extruder. As the screw drives the material forward, it is melted and driven through an open-ended metal die that forms a thin-walled plastic tube. Air flows through the center of the die, causing the diameter of the tube to expand substantially as it emerges from the die constraint. Air jets around the circumference of the expanded tube then cool the thin plastic material, after which it is passed through flattening rolls. The flattened tube is then periodically seam welded and perforated, and wound on a roll for easy dispensing.

MACHINING OF PLASTICS

Plastics can be milled, sawed, drilled, and threaded much like metals, but their properties are so variable that it is impossible to give descriptions that would be correct for all. It may be more important to consider some of the general characteristics of plastics that affect their machinability. Since plastics tend to be poor thermal conductors, most of the heat generated during chip formation remains near the cut interface and is not conducted into the material or carried away in the chips. Thermoplastics tend to soften and swell, and they occasionally bind or clog the cutting tool. Considerable elastic flexing can also occur, and this couples with material softening to reduce the precision of final dimensions. Because the thermosetting polymers have higher rigidity and reduced softening, they generally machine to more precise dimensions.

The high temperatures that develop at the point of cutting also cause the tools that are machining plastic to run very hot, and they may fail more rapidly than when cutting metal. Carbide tools may be preferred over high-speed tool steels if the cuts are of moderate duration or if high-speed cutting is performed. Coolants can often be used advantageously if they do not discolor the plastic or induce gumming. Water, soluble oil and water, and weak solutions of sodium silicate have been used effectively.

The tools that are used to machine plastics should be kept sharp at all times. Drilling is best done by means of straight-flute drills or by "dubbing" the cutting edge of a regular twist drill to produce a zero rake angle. These configurations are shown in Figure 14-10.

FIGURE 14-10 Straight-flute drill (left) and "dubbed" drill (right) used for drilling plastics.

Rotary files, saws, and milling cutters should be run at high speeds to improve cooling but with the feed carefully adjusted to avoid clogging the cutter.

Laser machining may be an attractive alternative to mechanical cutting. Because it vaporizes the material instead of forming chips, precise cuts can be achieved. Minute holes can be drilled, such as those in the nozzles of aerosol cans. Abrasive materials, such as filled and laminated plastics, can be machined in a manner that also eliminates the fine machining dust that is often considered to be a health hazard.

FINISHING AND ASSEMBLY OPERATIONS

Polymeric materials frequently offer the possibility of integral color, and the as-formed surface is often adequate for final use. Some of the finishing processes that can be applied to plastics include printing, hot stamping, vacuum metallizing, electroplating, and painting. Chapter 36 presents the processes of surface finishing and surface engineering.

Thermoplastic polymers can often be joined by heating the relevant surfaces or regions. The joining heat can be provided by a stream of hot gases, applied through a tool like a soldering iron, or generated by ultrasonic vibrations. The welding techniques that are applied to plastics are presented in Chapter 34. Adhesive bonding, another popular means of joining plastic, is presented in Chapter 35. Because of the low modulus of elasticity, plastics can also be easily flexed, and *snap-fits* are another popular means of assembling plastic components. Because of the softness of some polymeric materials, self-tapping screws can also be used.

DESIGNING FOR FABRICATION

The primary objective of any manufacturing activity is the production of satisfactory components or products, and this involves the selection of an appropriate material or materials. When polymers are selected as the material of construction, it is usually as a result of one or more of their somewhat unique properties, which include light weight, corrosion resistance, good thermal and electrical insulation, ease of fabrication, and the possibility of integral color. While these properties are indeed attractive, one should also be aware of the more common limitations, such as softening or burning at elevated temperatures, poor dimensional stability, and the deterioration of properties with age.

The basic properties and characteristics of polymeric materials have been described in Chapter 8. One should note, however, that property evaluation tests are conducted under specific test conditions. While a standard tensile test may show a polymer to have a moderately high strength value, a reduction in loading rate by two or three orders of magnitude may reduce this strength by as much as 80%. Conversely, an increase in loading rate can double or triple tensile strength. Polymers are often speed-sensitive materials. The can also be extremely sensitive to changes in temperature. Strength values can vary by a factor of 10 over a temperature range of as little as 200°F (100°C). Materials should be selected with full consideration given to the specific conditions of temperature, loads and load rates, and operating environments that will be encountered.

A second area of manufacturing concern is selecting the process or processes to be used in producing the shape and establishing the desired properties. Each of the wide variety of fabrication processes has distinct advantages and limitations, and efforts should be made to utilize their unique features. Once a process has been selected, the production of quality products further requires an awareness of all of the various aspects of that process. For example, consider a molding process in which a liquid or semifluid polymer is introduced into a mold cavity and allowed to harden. The proper amount of material must be introduced and caused to flow in such a way as to completely fill the cavity. Air that originally occupied the cavity needs to be vented and removed. Shrinkage will occur during solidification and/or cooling and may not occur in a uniform manner. Heat transfer must be provided to control the cooling and/or solidification. Finally, a means must be provided for part removal or ejection from the mold. Surface finish and appearance, the resultant engineering properties, and the ultimate cost of production are all dependent on good design and proper execution of the molding process. Product properties can be significantly affected by such factors as melt temperature, direction of flow, pressure during molding, thermal degradation, and cooling rate.

In all molded products, it is important to provide adequate fillets between adjacent sections to ensure smooth flow of the plastic into all sections of the mold and to eliminate stress concentrations at sharp interior corners. These fillets also make the mold less expensive to produce and reduce the danger of mold fracture during use. Even the exterior edges should be rounded where possible. A radius of 0.25 to 0.40 mm (0.010 to 0.015 in.) is scarcely noticeable but will do much to prevent an edge from chipping. Sharp corners should also be avoided in products that will be used for electrical applications, since they tend to increase voltage gradients, which can lead to product failure.

Wall or section thickness is also very important, since the hardening or curing time of a polymer is determined by the thickest section. If possible, sections should be kept nearly uniform in thickness, since nonuniformity can lead to serious warpage and dimensional control problems. As a general rule, one should use the minimum thickness that will provide satisfactory end-use performance. The specific value will be determined primarily by the size of the part and, to some extent, the process and the type of plastic being used. Recommended minimum thicknesses for molded plastics are as follows:

Small parts	1.25 mm (0.050 in.)
Average-sized parts	2.15 mm (0.085 in.)
Large parts	3.20 mm (0.125 in.)

Thick corners should also be avoided because they can lead to gas pockets, undercuring, or cracking. When extra strength is needed in a corner, it can usually be provided by incorporating ribs into the design.

Economical production is also facilitated by appropriate dimensional tolerances. A minimum tolerance of 0.08 mm (0.003 in.) should be allowed in directions that are parallel to the parting line of a mold or contained within a mold segment. In directions that cross a parting surface, a minimum tolerance of 0.25 mm (0.010 in.) is desirable. In both cases, increasing these values by about 50% can simultaneously reduce manufacturing difficulty as well as cost.

Since most molds are reusable, careful attention should also be given to the removal of the part. Rigid metal molds should be designed so that they can be easily opened and closed. A small amount of unidirectional taper should be provided to facilitate part withdrawal. Undercuts should be avoided whenever possible, since they will prevent part removal unless additional mold sections are used. These must move independently of the major segments of the mold, adding to the costs of mold production and maintenance, and slowing the rate of production.

INSERTS

Metal *inserts*, usually of brass or steel, are often incorporated into plastic products to provide enhanced performance or unique features. Since molded threads are difficult to produce, machined threads require additional processing, and both types tend to chip or deform, threaded metal inserts are frequently used when assemblies require considerable strength or when frequent disassembly and reassembly are anticipated. Figure 14-11 depicts one form of threaded insert, along with other types that provide pins or holes for alignment or mounting.

The successful use of inserts requires careful attention to design since they are generally held in place by only a mechanical bond that must resist both rotation and pullout. Knurling or grooving is often required to provide suitable sites for gripping. A medium or coarse knurl is usually adequate to resist torsional loads and moderate axial forces. Circumferential grooves are excellent for axial loads but offer little resistance to torsional rotation. Axial grooves resist rotation but do little to prevent pullout. Other

FIGURE 14-11 Typical metal inserts used to provide threaded cavities, holes, and alignment pins in plastic parts.

FIGURE 14-12   Various ways of anchoring metal inserts in plastic parts (*left to right*): bending, splitting, notching, swaging, noncircular head, and grooves and shoulders. Knurling is depicted in Figure 14-11.

means of anchoring include bending, splitting, notching, and swaging. Headed parts with noncircular heads may be used as formed. Combinations of notches, grooves, and shoulders are also common. Figure 14-12 depicts some common means of insert attachment.

If an insert is to act as a boss for mounting or serve as an electrical terminal, it should protrude slightly above the surface of the plastic. This permits a firm connection to be made without creating an axial load that would tend to pull the insert from its surroundings. If the insert serves to hold two mating parts together, it should be flush with the surface. In this way, the parts can be held together snugly without danger of loosening the insert. In all cases, the wall thickness of the surrounding plastic must be sufficient to support any load that may be transmitted through the insert. For small inserts, the wall thickness should be at least half the diameter of the insert. For inserts larger than 13 mm ($\frac{1}{2}$ in.) in diameter, the wall thickness should be at least 6.5 mm ($\frac{1}{4}$ in.).

DESIGN FACTORS RELATED TO FINISHING

Because plastics are frequently used where consumer acceptance is of great importance, special attention should be given to finish and appearance. In many cases, plastic parts can be designed to require very little finishing or decorative treatment. For small parts, fins and rough spots can often be removed by a barrel tumbling with suitable abrasives or polishing agents. Smoothing and polishing occur in the same operation.

By etching the surfaces of a mold, decorations or letters can be produced that protrude approximately 0.01 mm (0.004 in.) above the surface of the plastic. When higher relief is required, the mold can be engraved, but this adds significantly to mold cost. Whenever possible, depressed letters or designs should be avoided. These features, when transferred to the mold, become raised above the surrounding surface. Mold making then requires a considerable amount of intricate machining as the surrounding material must be cut away from the design or letters. When recessed features are absolutely required, manufacturing cost can be reduced if they can be incorporated into a small area that is raised above the primary surface.

When designing plastic parts, a prime objective is often the elimination of secondary machining, especially on surfaces that would be exposed to the customer. Even when fillers are used (as they are in most plastics), the surfaces of molded parts have a thin film of pure resin. This film provides the high luster that is characteristic of polymeric products. Machining cuts through the surface, exposing the underlying filler. The result is a poor appearance, as well as a site for the absorption of moisture.

One location that frequently requires machining is the parting line that is produced where the mold segments come together. Since perfect mating is difficult to achieve, a small fin, or "flash," is usually produced around the part perimeter, as illustrated in Figure 14-13. When the flash is trimmed off, the resulting line of exposed filler may be objectionable. By locating the parting line along a sharp corner, it is easier to maintain satisfactory mating of the mold sections, and the exposed filler that is created by flash removal will be confined to a corner, where it is less noticeable.

FIGURE 14-13   Trimming the flash from a plastic part ruptures the thin layer of pure resin along the parting line and creates a line of exposed filler.

FIGURE 14-14 *Stiffness can be imparted to large surfaces of plastic parts through the use of ribbing or doming.*

Since plastics have a low modulus of elasticity, large flat areas are not rigid and should be avoided whenever possible. Ribbing or doming, like that illustrated in Figure 14-14, can be used to provide the required stiffness. In addition, flat surfaces tend to reveal flow marks from the molding operation, as well as scratches that occur during handling or service. External ribbing then serves the dual function of increasing strength and rigidity while masking any surface flaws. Dimpled or textured surfaces can also be used to provide a pleasing appearance and conceal scratches.

Holes that are formed by pins protruding from the mold often require special consideration. During the mold closure and filling stages of compression molding, these pins can be subjected to considerable bending. When they are supported only at one end, the length should not exceed twice the diameter. In processes with reduced filling pressures, the length can be as much as five times the diameter without excessive problems.

Holes that are to be threaded or used to receive self-tapping screws should be countersunk. This not only assists in starting the tap or screw but also reduces chipping at the outer edge of the hole. If the threaded hole is less than 6.5 mm ($\frac{1}{4}$ in.) in diameter, it is best to cut the threads after molding, using some form of thread tap. For diameters greater than 6.5 mm, the threads can be molded or an insert should be used. If the threads are molded, however, special provisions must be made to remove the part from the mold. Since the additional operations extend the molding time and reduce productivity, they are generally considered to be uneconomical.

■ 14.3 PROCESSING OF RUBBER AND ELASTOMERS

Rubber and elastomeric products can be produced by a variety of fabrication processes. Relatively thin parts with uniform wall thickness, such as boots, gloves, and fairings, are often made by some form of *dipping*. A master form is first produced, usually from some type of metal. This form is then immersed into a liquid preparation or compound (usually based on natural rubber, neoprene, or silicone), then removed and allowed to dry. With each dip, a certain amount of the liquid adheres to the surface, with repeated dips being used to produce a final desired thickness. After vulcanization, usually in steam, the products are stripped from the molds.

The dipping process can be accelerated by using electrostatic charges. A negative charge is introduced to the latex particles, and the form or mold receives a positive charge, either through an applied voltage or by a coagulant coating that releases positive ions when dipped into the solution. The attraction and neutralization of the opposite charges causes the elastomeric particles to be deposited on the form at a faster rate and in thicker layers than the basic process. With electrostatic deposition, many products can be made in a single immersion.

When the parts are thicker or complex-shaped solids, the first step is the compounding of elastomeric resin, vulcanizers, fillers, antioxidants, accelerators, and pigments. This is usually done in some form of mixer, which blends the components to form a homogeneous mass. Adaptations of the processes previously discussed for plastics are frequently used to produce the desired shapes. Injection, compression, and transfer molding are used, along with special techniques for foaming. Urethanes and silicones can also be directly cast to shape.

Rubber compounds can be made into sheets using *calenders*, like that shown in Figure 14-15. The sheet coming from the calender is often rolled with a fabric liner to prevent the material from sticking. Three- or four-roll calenders can also be used to place a rubber or elastomer covering over cord or woven fabric. In the three-roll geometry, only one side of the fabric is coated in a single pass. The four-roll arrangement, shown in Figure 14-16, enables both sides to be coated simultaneously.

Products such as inner tubes, garden hoses, tubing, and strip moldings can be produced by the *extrusion* process. The compounded elastomer is forced through a die by a screw device similar to that described for plastics.

FIGURE 14-15 (a) Three-roll calender used for producing rubber or plastic sheet. (b) Schematic diagram showing the method of making sheets of rubber with a three-roll calender. [*(a) (Courtesy of Farrel-Birmingham Company, Inc. Ansonia, CT.)*]

Rubber

Rubber sheet

Conveyer

3-roll calender

(a)   (b)

FIGURE 14-16 Arrangement of the rolls, fabric, and coating material for coating both sides of a fabric in a four-roll calender.

Rubber

Uncoated fabric

Coated fabric

4-roll calender

Rubber or artificial elastomers can be bonded to metal, such as brass or steel, using a variety of polymeric *adhesives*. Only moderate pressures and temperatures are required to obtain excellent adhesion.

■ 14.4 Processing of Ceramics

The fabrication processes applied to ceramic materials generally fall into two distinct classes, based on the properties of the material. *Glasses* can be manufactured into useful articles by first heating the material to produce a molten or viscous state, shaping the material by means of *viscous flow*, and then cooling the material to produce a solid product. *Crystalline ceramics* have a characteristically brittle behavior and are normally manufactured into useful components by pressing moist aggregates or powder into a shape, followed by drying, and then bonding by one of a variety of mechanisms, which include chemical reaction, *vitrification* (cementing with a liquefied material), and *sintering* (solid-state diffusion).

FABRICATION TECHNIQUES FOR GLASSES

Glass is generally shaped at elevated temperatures, where the viscosity can be controlled. A number of the processes begin with material in the liquid or molten condition. Sheet and plate glass is formed by processes such as extruding through a narrow slit, rolling through water-cooled rolls, or floating on a bath of molten tin. Glass shapes can be produced by pouring the molten material directly into a mold. The cooling rate is then controlled, usually as slow as possible, to minimize residual stresses and the tendency for cracking. Constant-cross-section shapes can be made by extrusion, and glass fibers can be produced by forcing liquid glass through multiple openings in an extrusion die.

Hot gob

FIGURE 14-17 Viscous glass can be easily shaped by mating male and female die members.

Other glass-forming processes begin with viscous masses and use mating male and female die members to press the material into the desired shape, as illustrated in Figure 14-17. Bottles, containers, and shapes like incandescent light bulbs are made by a process similar to the *blow molding* of plastics. Cup-shaped pieces of viscous material are expanded against the outside of heated dies, as illustrated in Figure 14-18.

Special heat treatments can also be applied to glass material. By applying forced cooling to the exposed surfaces, a residual stress pattern of surface compression can be induced. The surface layers cool and contract, and the softer interior flows to conform. This is followed by the cooling of the interior, which tries to contract but is restrained by the already cold surface, creating the surface compression. The resulting product, called *tempered glass*, is stronger and more fracture resistant, since cracks tend to initiate on free surfaces. When unfavorable residual stresses are present that might lead to cracking, *annealing* treatments can be used to reduce their magnitude.

Glass-ceramics form a unique class of materials that are part crystalline and part glass. They are fabricated into shape as a glass and are then subjected to a special heat treatment (*devitrification*) that controls the nucleation and growth of the crystalline component. Because of the dual structure, the final properties include good strength and toughness, along with low thermal expansion. Typical products include cookware (such as the white CorningWare products), ceramic stove tops, and materials used in electrical and computer components.

FABRICATION OF CRYSTALLINE CERAMICS

Crystalline ceramics are hard, brittle materials with high melting points. As a result, they cannot be formed by techniques requiring either plasticity (i.e., forming methods) or melting (i.e., casting methods). Instead, these materials are generally processed in the solid state by techniques that utilize particles or aggregates and resemble those used in powder metallurgy. The particles can also be blended with additives that impart plasticity or flow and enable the forming or casting processes to be used.

Dry powders can be compacted and converted into useful shapes by pressing at either environmental or elevated temperatures. *Dry pressing* with rigid tooling, *isostatic pressing*, and *hot-isostatic pressing (HIP)* with flexible molds are common techniques and exhibit features and limitations similar to those discussed in Chapter 18.

Clay products are based on special types of ceramics blended with water and various additives to produce a material that can be shaped by most of the traditional forming methods. *Plastic forming* can also be applied to other ceramics if the ceramic particles are combined with additives that impart plasticity when subjected to pressure and heat. *Wet pressing* can be used to produce shapes at lower pressures than dry pressing. *Extrusion* can be used to produce products with constant cross sections.

Injection molding was discussed earlier in this chapter as a means of forming plastics, and metal injection molding (MIM) is presented in Chapter 19 as a way of producing small, complex-shaped metal parts. A form of injection molding can also be used to form complex, three-dimensional shapes from ceramic materials. Ceramic powder is mixed with polymer material, and heated material (125°–150°C or 250°–300°F) is then injected into an aluminum die under pressures on the order of 30–100 MPa (5–15 ksi). The mixture cools, and after about 30 seconds, it is hard enough to permit ejection from the die. Additive materials are then removed by

Parison

Mandrel

Air

FIGURE 14-18 Thin-walled glass shapes can be produced by a combination of pressing and blow molding.

Hot gob

thermal, solvent, catalytic, or wicking methods, and the remaining ceramic is fused together by a firing operation. As with metal injection molding, the die forms a part that is considerably oversized, and controlled shrinkage during firing produces the final dimensions. The major dimensions of most parts are less than 10 cm (4 in.); the wall thickness is less than 6 mm ($\frac{1}{4}$ in.); and tolerances are on the order of 1% or 0.1 mm (0.005 in.), whichever is greater. Most parts are made from the oxide ceramics, such as alumina or zirconia, but the process has also been used with silicon carbide and silicon nitride.

Several casting processes can be used to produce ceramic shapes, beginning with a pourable slurry that strengthens by partial removal of the liquid or the gellation, polymerization, or crystallization of a matrix phase. In the *slip casting* process, ceramic powder is mixed with a liquid to form a slurry, which is then cast into a mold containing very fine pores. Capillary action pulls the liquid from the slurry, allowing the ceramic particles to arrange into a "green" body with sufficient strength for subsequent handling. Pressure applied to the slurry, vacuum applied to the mold, or centrifugal pressure can all aid in liquid removal. Hollow shapes can be produced by pouring out the remaining slurry once a desired thickness of solid has formed on the mold walls. Slip casting has been used to produce a variety of porcelain products, including bathroom fixtures, fine china and dinnerware, and ceramic products for the chemical industry.

In the *tape casting* process, a controlled film of slurry is formed on a substrate. Evaporation of the liquid during controlled drying produces a thin, flexible, rubbery tape or sheet that has smooth surfaces and uniform thickness. These products are widely used in the multilayer construction of electronic circuits and capacitors.

In other casting-type processes, slurries containing bonding agents can be used to produce cast-in-place products, such as furnace linings or dental fillings. When mixed with a sticky binder, the material can be blown through a pipe to apply ceramic coatings or build up refractory linings.

The numerous variations of *sol-gel processing* can be used to produce ceramic films and coatings, fibers, and bulk shapes. These processes begin with a solution or colloidal dispersion (sol), which undergoes a molecular polymerization to produce a gel, which is then dried. This approach offers higher purity and homogeneity, lower firing temperatures, and finer grain size at the expense of higher raw-material cost, large volume shrinkages during processing, and longer processing times.

Table 14-1 summarizes some of the primary processes used to fabricate shapes from crystalline ceramics.

PRODUCING STRENGTH IN PARTICULATE CERAMICS

Each of the processes just described can be used to produce useful shapes from ceramic materials, but useful strength generally requires a subsequent heating operation, known as *firing* or *sintering*. Slurry-type materials must first be dried in a manner that is designed to control dimensional changes and minimize stresses, distortion, and cracking. The material is then heated to temperatures between 0.5 and 0.8 times the absolute melting point, where diffusion processes act to fuse the

TABLE 14-1   Processes Used to Form Products from Crystalline Ceramics

Process	Starting material	Advantages	Limitations
Dry axial pressing	Dry powder	Low cost; can be automated	Limited cross sections; density gradients
Isostatic processing	Dry powder	Uniform density; variable cross sections; can be automated	Long cycle times; small number of products per cycle
Slip casting	Slurry	Large sizes; complex shapes; low tooling cost	Long cycle times; labor-intensive
Injection molding	Ceramic–plastic blend	Complex cross sections; fast; can be automated; high volume	Binder must be removed; high tool cost
Forming processes (e.g., extrusion)	Ceramic–binder blend	Low cost; variable shapes (such as long lengths)	Binder must be removed; particles oriented by flow
Clay products	Clay, water, and additives	Easily shaped by forming methods; wide range of size and shape	Requires controlled drying

particles together and impart the desired mechanical and physical properties. The temperature and time are selected to control the resulting grain size, pore size, and pore shape. In some firing operations, surface melting (*liquid-phase sintering*) or component reactions (*reaction sintering*) can produce a substantial amount of liquid material (*vitrification*). The liquid then flows to produce a glassy bond between the ceramic particles and either solidifies as a glass or crystallizes.

Cementation is an alternative method of producing strength that does not require elevated temperature. A liquid binder material is used to coat the ceramic particles, and a subsequent chemical reaction converts the liquid to a solid, forming strong, rigid bonds.

Prototypes or small production quantities of ceramic products have been made by the *laser sintering* of ceramic powders. Successive layers of material are fused together by the laser sintering (or laser melting) of thin layers of heat-fusible powder. For ceramic parts, the powder particles are actually coated with a very thin thermoplastic polymer binder. The laser then acts on the polymer coating to produce the bond. After the laser bonding, the parts then undergo conventional debinding and sintering to about 55 to 65% of theoretical density. Isostatic pressing prior to sintering can raise the final density to 90 to 99% of ideal.

MACHINING OF CERAMICS

Most ceramic materials are brittle, and the techniques used to cut metals will generally produce uncontrolled or catastrophic cracks. In addition, ceramics are typically hard materials. Since ceramics are often used as abrasives or coatings on cutting tools, the tools needed to cut them have to be even harder.

Direct production to the desired final shape is clearly the most attractive alternative, but there are times when a material removal operation is necessary. Such machining can be performed before or after the final firing. Before firing, the material tends to be rather weak and fragile. While fracture is always a concern, a more significant consideration might be the dimensional changes that will occur upon subsequent firing. Shrinkage may be as much as 30%, so it may be difficult to achieve or maintain close final tolerances. For this reason, machining before firing, known as *green machining*, is usually rough machining designed to reduce the amount of finishing that will be required after firing.

When machining is performed after firing, the processes are generally ones we might consider to be nonconventional. Grinding, lapping, and polishing with diamond abrasives, drilling with diamond-tipped tooling, cutting with diamond saws, ultrasonic machining, laser and electron-beam machining, water-jet machining, and chemical etching have all been used. When mechanical forces are applied, material support is quite critical (since ceramic materials are almost always brittle). Because of the hardness of the ceramic, the tools must be quite rigid. Selection and use of coolants are also important issues.

Materials producers have developed "machinable" ceramics that lend themselves to precision shaping by more traditional machining operations. It should be noted, however, that these are indeed special materials and not characteristic of ceramics as a whole.

JOINING OF CERAMICS

When we consider joining operations, the unique properties of ceramics once again introduce fabrication limitations. Brittle ceramics cannot be joined by fusion welding or deformation bonding, and threaded assemblies should be avoided whenever possible. Therefore, most joining utilizes some form of adhesive bonding, brazing, diffusion bonding, or special cements. Even with these methods, the stresses that develop on the surfaces can lead to premature failure. As a result, most ceramic products are designed to be monolithic (single-piece) structures rather than multipart assemblies.

DESIGN OF CERAMIC COMPONENTS

Since ceramics are brittle materials, special care should be taken to minimize bending and tensile loading as well as design stress raisers. Sharp corners and edges should be avoided where possible. Outside corners should be chamfered to reduce the possibility of edge chips. Inside corners should have fillets of sufficient radius to minimize crack initiation. Undercuts are difficult to produce and should be avoided. Specifications should generally use the largest possible tolerances, since these can often be met with products in the as-fired condition. Extremely precise dimensions usually require hand grinding, and costs can escalate significantly. In addition, consideration should be given to surface-finish requirements, since grinding, polishing, and lapping operations can increase production cost substantially.

■ 14.5 FABRICATION OF COMPOSITE MATERIALS

As shown in Chapter 8, composite materials can be designed to offer a number of attractive properties. In some market areas, such as aerospace and sporting goods, their acceptance and growth have been phenomenal. Use can only occur, however, if the material can be produced in useful shapes at an acceptable cost and rate of production. Many of the manufacturing processes designed for composites are slow, and some require considerable amounts of hand labor. There is often a large degree of variability between nominally identical products, and inspection and quality control methods are not as well developed as for other materials. While these limitations may be acceptable for certain applications, they often restrict the use of composites for high-volume, mass-produced items. Faster production speeds, increased use of automation, reduced variability, and integrated quality control continue to be important issues in the expanded use of composite materials.

In Chapter 8, composite materials were classified by their basic geometry as particulate, laminar, and fiber-reinforced. Since the fabrication processes are often unique to a specific type of composite, they will also be grouped in the same manner.

FABRICATION OF PARTICULATE COMPOSITES

Particulate composites usually consist of discrete particles dispersed in a ductile, fracture-resistant polymer or metal matrix. Their fabrication, however, rarely requires processes unique to composite materials. Instead, the particles are simply dispersed in the matrix by introduction into a liquid melt or slurry, or by blending the various components as solids, using powder metallurgy methods. Subsequent processing generally follows the conventional methods of casting or forming, or utilizes the various techniques of powder metallurgy. These processes have been presented elsewhere in the text and will not be repeated here.

Reinforcement particles have been successfully blended into the highly viscous slurries of rheocast material, the semisolid mixtures that are viscous when agitated but retain their shape when static. Particle reinforcements have also been produced by spray forming multicomponent feeds.

FABRICATION OF LAMINAR COMPOSITES

Laminar composites include coatings and protective surfaces, claddings, bimetallics, laminates, and a host of other materials. Their production generally involves processes designed to form a high-quality bond between distinct layers of different materials. When the layers are metallic, as in claddings and bimetallics, the composites can be produced by hot or cold *roll bonding*. Sheets of the various materials are passed simultaneously through the rolls of a conventional rolling mill. If the amount of deformation is great enough, surface oxides and contaminants are broken up and dispersed, metal-to-metal contact is established, and the two surfaces become joined by a solid-state bond. U.S. coinage is a common example of a roll-bonded material.

Explosive bonding is another means of joining layers of metal. A sheet of explosive material progressively detonates above the layers to be joined, causing a pressure wave to sweep across the interface. A small open angle is maintained between the two surfaces. As the pressure wave propagates, any surface films are liquefied or scarfed off and are jetted out the open interface. Clean metal surfaces are then forced together at high pressures, forming a solid-state bond with a characteristically wavy configuration at the interface. Large areas, wide plates (too wide to roll bond conveniently), and dissimilar materials with large differences in mechanical properties are attractive candidates for explosive bonding.

Both metallic and nonmetallic materials can be joined by *adhesive bonding*. By gluing the layers of plywood at various orientations, the directional effects of wood grain can be minimized within the plane of the sheet. Later in this chapter, we will discuss the lamination of polymer-matrix composites where each ply is a fiber-reinforced or woven layer. Films of unpolymerized resin are created between the layers. Pressing at elevated temperature cures the resin and completes the bond. In a manner similar to adhesive bonding, layers of metal can be joined by brazing to form laminar composites that can withstand moderately elevated temperatures.

In *sandwich structures*, such as corrugated cardboard or the honeycomb shown in Figure 14-19, thin layers of facing material are bonded, usually by adhesive, to a lightweight filler material. Special fabrication methods may be employed to produce the foam, corrugated, or honeycomb filler.

FABRICATION OF FIBER-REINFORCED COMPOSITES

In the fiber-reinforced composites, the matrix and fiber reinforcement provide a system that offers properties not attainable by the individual components acting alone. The fiber reinforcement produces a significant increase in strength and stiffness, while the matrix functions as a binder, transfers the stresses, and provides protection against abrasion and environmental effects.

A number of processes have been developed to produce and shape the fiber-reinforced composites, with key differences relating to the orientation of the fibers, the length of continuous filaments, and the geometry of the final product. Each process seeks to embed the *fibers* in a selected *matrix* with the proper alignment and spacing necessary to produce the desired properties. Discontinuous fibers can be combined with a matrix to provide either a random or a preferred orientation. Continuous fibers are normally aligned in a unidirectional fashion in rods or tapes, woven into fabric layers, wound around a mandrel, or woven into a three-dimensional shape.

FIGURE 14-19 Fabrication of a honeycomb sandwich structure using adhesive bonding to join the facing sheets to the lightweight honeycomb filler. *(Courtesy of ASM International. Metals Park, OH.)*

Some of the fiber-reinforced processes are identical to those previously described for unreinforced plastics: compression, transfer and injection molding, extrusion, rotational molding, and thermoforming. Others are standard processes with simple modifications, such as reinforced reaction injection molding and resin transfer molding. Still others are specific to fiber-reinforced composites, such as hand lay-up, spray-up, vacuum-bag, pressure-bag, and autoclave molding; filament winding; and pultrusion.

Production of Reinforcing Fibers.   A number of processes have been developed to produce the various types of reinforcement fibers used in composites. Metallic fibers, glass fibers, and many polymeric fibers (including the popular Kevlar) are produced by variations of conventional wire drawing and extrusion. Boron, carbon, and ceramic fibers such as silicon carbide are too brittle to be produced by the deformation methods. Boron fibers are produced by chemical vapor deposition around a tungsten filament. Carbon (graphite) fibers can be made by carbonizing (decomposing) an organic material that is more easily formed to the fiber shape.

The individual fine filaments are often bundled into *yarns* (twisted assemblies of filaments), *tows* (untwisted assemblies of fibers), and *rovings* (untwisted assemblies of yarns or tows). Fibers can also be chopped into short lengths, usually 12 mm ($\frac{1}{2}$ in.) or less, for incorporation into the various sheet-or bulk-molding compounds. In these materials the fibers usually assume a random orientation.

Processes Designed to Combine Fibers and a Matrix.   A variety of processes have been developed to combine the fiber and the matrix into a unified material suitable for further processing. If the matrix material can be liquefied and the temperature is not harmful to the fibers, casting-type processes can be an attractive means of coating the reinforcement. The pouring of concrete around a steel reinforcing rod is a crude example of this method. In the case of the high-tech, fiber-reinforced plastics and metals, the liquid can be introduced between the fibers by means of *capillary action, vacuum infiltration*, or *pressure casting*. In a modification of *centrifugal casting*, resin is introduced into the center of a rotating mold and is then uniformly forced against and into the reinforcing material. Yet another alternative is to draw the fibers through a bath of molten material and combine them into aligned bundles before the liquid solidifies.

Prepregs, or pre-impregnated reinforcements, are sheets of unidirectional fibers or woven fabric that have been infiltrated with a matrix material. *Mats* are sheets of nonwoven, randomly oriented fibers in a matrix. When the matrix is a polymeric material, the resin in the prepreg or mat is usually only partially cured. Later fabrication then involves the stacking of layers and the application of heat and pressure to further cure the resin and bond the layers into a continuous solid matrix. Prepreg layers can be stacked in various orientations to provide various directional properties.

Individual filaments can be coated with a matrix material by drawing through a molten bath, plasma spraying, vapor deposition, electrodeposition, or other techniques. The coated fibers can then be used, either individually or in various assemblies. They can also be wound around a mandrel with a specified spacing and then cut to produce *tapes* that contain continuous, unidirectionally aligned filaments. These tapes are generally 1 fiber diameter in thickness and can be up to 1.2 m (48 in.) wide.

When the temperatures of the molten matrix become objectionable or potentially damaging to the fiber, matrix-fiber bonding can often be achieved through diffusion or deformation bonding (hot pressing or rolling). A common arrangement is to position aligned or woven fibers between sheets of foil material. Loosely woven fibers can also be infiltrated with a particulate matrix, which is then compacted at high pressures and sintered to form a continuous solid.

Sheet-molding compounds (SMC) are composed of chopped fibers (usually glass in lengths of 25 to 50 mm or 1 to 2 in.) and partially cured thermoset resin, along with fillers, pigments, catalysts, thickeners, and other additives, in sheets approximately 2.5 to 5 mm (0.1 to 0.2 in.) thick. With strengths in the range of 35 to 70 MPa (5 to 10 ksi) and the ability to be press formed in heated dies, these materials offer a feasible alternative to sheet metal in applications where light weight, corrosion resistance, and integral color are attractive features.

After initial compounding and a few days of curing, sheet-molding compounds generally take on the consistency of leather, making them easy to handle and mold. When they are placed in a heated mold, the viscosity is quickly reduced and the material flows easily under pressures of about 7 MPa (1000 psi). The elevated temperatures accelerate the chemical reactions, and final curing can often be completed in less than 60 seconds. As an added benefit, sheet-molding compounds can be easily recycled. One possible disadvantage, however, is that polymer flow may orient the reinforcing fibers, making the final orientation nonrandom and difficult to predict and control.

Bulk-molding compounds (BMC) are fiber-reinforced, thermoset, molding materials, where short fibers (6 to 12 mm or $\frac{1}{4}$ to $\frac{1}{2}$ in.) are distributed in random orientation. The starting material is usually a bulk material with the consistency of putty or modeling clay, although pellets and granules are also possible. The final shape is usually produced by compression molding in heated dies, but transfer molding and injection molding are other possibilities.

Fabrication of Final Shapes from Fiber-Reinforced Composites. A number of processes have been developed for the production of finished products from fiber-reinforced material. Many are simply extensions or adaptations of processes that are used to shape the matrix material (usually metals or polymers). Others are unique to the family of fiber-reinforced composites. The dominant techniques will be discussed individually in the sections that follow.

PULTRUSION *Pultrusion* is a continuous process that is used to produce long lengths of relatively simple shapes with uniform cross section, such as round, rectangular, tubular, plate, sheet, and structural products. As shown in Figure 14-20, bundles of continuous reinforcing fibers are drawn through a bath of thermoset polymer resin, and the impregnated material is then gathered to produce a desired cross-sectional shape. This material is then pulled through one or more heated dies, which further shapes the product and cures the resin. When it emerges from the heated dies, the product is cooled by air or water and then cut to length. Some products, such as structural shapes, are complete at this stage, while others are further fabricated into products such as fishing poles, golf club shafts, and ski poles. Extremely high strengths and stiffnesses are possible since the reinforcement can be as much as 75% of the final structure. Tensile strengths of 210 MPa (30 ksi) and elastic modulus of 17 Gpa (2.5 × 10^6 psi) are coupled with densities about 20% that of steel or 60% that of aluminum. Cross sections can be as much as 1.5 m (60 in.) wide and 0.3 m (12 in.) thick.

FILAMENT WINDING Resin-coated or resin-impregnated, high-strength, continuous filaments, bundles, or tapes made from fibers of glass, graphite, boron, Kevlar (aramid), or similar materials can be used to produce cylinders, spheres, cones, and other container-type shapes with exceptional strength-to-weight ratios. The filaments are wound over a rotating form or mandrel, using longitudinal, circumferential, or helical patterns, or a combination of these, designed to take advantage of their highly directional strength properties. By adjusting the density of the filaments in various locations and selecting the orientation of the wraps, products can be designed to have strength where needed and lighter weight in less critical regions. After winding, the part and mandrel are placed in an oven for curing, after which the product is stripped

FIGURE 14-20 Schematic diagram of the pultrusion process. The heated dies cure the thermoset resin.

FIGURE 14-21 A large tank being made by filament winding. *(Courtesy of Rohr Inc., Chula Vista, CA.)*

from the form. The matrix, often an epoxy-type polymer, binds the structure together and transmits the stresses to the fibers.

Figure 14-21 shows a large tank being produced by filament winding. Products such as pressure tanks and rocket motor casings can be made in virtually any size, some as large as 4.5 m (15 ft) in diameter and 20 m (65 ft) long. Smaller parts include helicopter rotor blades, baseball bats, and light poles. Moderate production quantities are feasible, and because the process can be highly mechanized, uniform quality can be maintained. A new form block is all that is required to produce a new size or design. Because the tooling is so inexpensive, the process offers tremendous potential for cost savings and flexibility. With advancements in computer software, equipment, and control, parts no longer need to be axisymmetric. Filament-wound products can now be made with changing surfaces, nonsymmetric cross sections, and compound curvatures.

LAMINATION AND LAMINATION-TYPE PROCESSES In the lamination process, prepregs, mats, or tapes are stacked to produce a desired thickness and cured under pressure and heat. The resulting products possess unusually high strength as a result of the integral fiber reinforcement. Because the surface is a thin layer of pure resin, laminates usually possess a smooth, attractive appearance. If the resin is transparent, the fiber material is visible and can impart a variety of decorative effects. Other decorative laminates use a separate patterned face sheet that is bonded to the laminate structure.

Laminated materials can be produced as sheets, tubes, and rods. Flat sheets can be made using the method illustrated in Figure 14-22. Prepreg sheets or reinforcement sheets saturated in resin are stacked and then compressed under pressures on the order of 7 MPa (1000 psi). Figure 14-23 depicts the technique used to produce rods or tubes. For tubing, the impregnated stock is wound around a mandrel of the desired internal diameter. Solid rods are made by using a small-diameter mandrel, which is removed prior to curing, or by wrapping the material tightly about itself. Sheet laminating can also be a continuous process in which multiple reinforcement sheets are passed through a resin bath and then through squeeze rolls.

FIGURE 14-22 Method of producing multiple sheets of laminated plastic material.

FIGURE 14-23 Method of producing laminated plastic tubing. In the final operation, the rolled tubes are cured by being held in heated tooling.

In all of the above cases, the final operation is curing, usually involving elevated temperature and possibly applied pressure. Because of their excellent strength properties, plastic laminates find a wide variety of uses. Some sheets can be easily blanked and punched. Gears machined from thick laminated sheets have unusually quiet operating characteristics when matched with metal gears.

Many laminated products are not flat but contain relatively simple curves and contours. Manufacturing processes that require zero to moderate pressures and relatively low curing temperatures can be used to produce boat bodies, automobile body panels, aerospace panels, safety helmets, and similar products. The only required tooling is often a female mold or male form block that can be made from metal, hardwood, or even particle board. The layers of prepreg or resin-dipped fabric are stacked in various orientations until the desired thickness is obtained. Care must be taken to avoid the entrapment of air bubbles and ensure that no impurities (such as oil, dirt, or other contaminants) are introduced between the layers. In the *vacuum-bag molding* process, the entire assembly (mold and material) is placed in a nonadhering, flexible bag, and the contained air is evacuated. Air pressure then eliminates entrapped air, expels excess resin, and holds the laminate against the mold while the resin is cured. While curing may occur at room temperature, moderately elevated temperatures may also be used. In *pressure-bag molding*, a flexible membrane is positioned over the female mold cavity and is pressurized to force the individual plies together and drive out entrapped air and excess resin. Pressures usually range from 0.2 to 0.4 MPa (30 to 50 psi) but can be as high as 2 MPa (250 psi). This pressing is coupled with room- or low-temperature curing. Pressure-bag molding has been used to produce extremely large components, such as the skins of military aircraft, large air deflectors for tractor-trailers, and body panels for trucks.

Higher heats and pressures can be used when parts are cured in an *autoclave*. The supporting molds and vacuum-bagged lay-ups are placed inside a heated pressure vessel, where curing occurs under elevated temperatures and pressures in the range 0.4 to 0.7 MPa (50 to 100 psi). Denser, void-free moldings are produced, and the properties can be further enhanced through the use of matrix resins that require higher-temperature cures. The size of the autoclave limits the size of the product.

When production quantities are large and quality needs to be high, matched metal dies can be substituted for the mold and bag. The process then becomes a modification of polymer *compression molding*. Sheet-molding compound, bulk-molding compound, or preformed mat is placed in the press, and heat and pressure are applied. Temperatures typically range from 110° to 160°C (225° to 325°F), coupled with pressures from 1 to 7 MPa (250 to 1000 psi). With heated dies, the thermoset resin cures during the compression operation, with cycles repeating every 1 to 5 minutes.

Resin-transfer molding is a low-pressure process that is intermediate to the slow, labor-intensive lay-up processes and the faster compression molding or injection molding processes, which generally require more expensive tooling. Continuous fiber mat or woven material (usually employing glass fiber) is positioned dry in the bottom half of a matching mold, which is then closed and clamped. A low-viscosity catalyzed resin is then injected into the mold, where it displaces the air, permeates the

FIGURE 14-24 Aerodynamic styling and smooth surfaces characterize the hood and fender of Ford Motor Company's AeroMax truck. This one-piece panel was produced as a resin-transfer molding by Rockwell International. *(Courtesy of ASM International, Metals Park, OH.)*

reinforcement, and subsequently cures at low temperatures. Because of the low pressures employed in the process, the mold tooling does not need to be steel but can be electroformed nickel shells, epoxy composite, or aluminum. In addition, low-capacity presses can be used to clamp the mold segments, and inflatable bags can be used to produce simple holes or hollows, in much the same way that cores are used in conventional casting. The resulting products can have excellent surfaces on both sides, since both mold surfaces can be precoated with a pigmented gel. Large parts can often be made as a single unit with a relatively low capital investment. Cycle times range from a few minutes to a few hours, depending on the part size and the resin system being used. The aerodynamic hood and fender assembly for Ford Motor Company's AeroMax heavy-duty truck (Figure 14-24) is an example of a large resin-transfer molding.

When the quality demands are not as great, the reinforcement-to-resin ratio is not exceptionally high, and only one surface needs to be finished to high quality, the pressing operations can often be eliminated. In a process known as *hand lay-up* or *open-mold processing*, depicted in Figure 14-25, successive layers of pliable resin-coated cloth are simply placed in an open mold or draped over a form. Squeegees or rollers are used to manually ensure good contact and remove any entrapped air, and the assembly is then allowed to cure, generally at room temperature. If prepreg layers are not used, a layer of mat, cloth, or woven roving can be put in place and a layer of resin brushed, sprayed, or poured on. This process can then be repeated to build the desired thickness.

While the hand lay-up process is slow and labor intensive and has part-to-part and operator-to-operator variability, the tooling costs are sufficiently low that single items or small quantities become economically feasible. Molds or forms can be made from wood, plaster, plastics, aluminum, or steel, so design changes and the associated tool modifications are rather inexpensive, and manufacturing lead time can be quite short. In addition, large parts can be produced as single units, significantly reducing

FIGURE 14-25 Schematic of the hand lay-up lamination process.

the amount of assembly, and various types of reinforcement can be incorporated into a single product, expanding design options. High-quality surfaces can be produced by applying a pigmented gel coat to the mold before the lay-up.

SPRAY MOLDING   When continuous or woven fibers are not required to produce the desired properties, sheet-type parts can be produced by mixing chopped fibers, fillers, and catalyzed resin and spraying the combination into or onto a mold form, as shown in Figure 14-26. Rollers or squeegees can be used to remove entrapped air and work the resin into the reinforcement. Room-temperature curing is usually preferred, but elevated temperatures are sometimes used to accelerate the cure. As with the hand lay-up process, an initial gel coat can be used to produce a smooth, pigmented surface.

SHEET STAMPING   Thermoplastic sheets that have been reinforced with nonwoven fiber can often be heated and press formed in a manner similar to conventional sheet metal forming. Precut blanks are heated and placed between the halves of a matched metal mold that is mounted in a vertical press. Ribs, bosses, and contours can be formed in parts with essentially uniform thickness. Cycle times range from 25 to 50 seconds for most parts.

INJECTION MOLDING   The injection molding of fiber-reinforced plastics is a process that competes with metal die castings and offers comparable properties at considerably reduced weight. In its simplest form, chopped or continuous fibers are placed in a mold cavity that is then closed and injected with resin. An improved method utilizes chopped fibers, up to 6 mm ($\frac{1}{4}$ in.) in length, which are premixed with the heated thermoplastic (often nylon) prior to injection. Another variation uses a feedstock of discrete pellets that have been manufactured by slicing continuous-fiber pultruded rods. The benefits of adding fiber reinforcement (compared to conventional plastic molding) include increased rigidity and impact strength, reduced possibility of brittle failure during impact, better dimensional stability at elevated temperatures and in humid environments, improved abrasion resistance, and better surface finish due to the reduced dimensional contraction and absence of related sink marks. The molding process is quite rapid, and the final parts can be both precise and complex.

BRAIDING, THREE-DIMENSIONAL KNITTING, AND THREE-DIMENSIONAL WEAVING   The primary causes of failure in lamination-type composites are interlaminar cracking and delamination (layer separation) upon impact. To overcome these problems, the high-strength reinforcing fibers can also be interwoven into three-dimensional preforms by processes that include weaving, braiding, and stitching through the thickness of stacked two-dimensional preforms. Resin is then injected into the assembly, and the resultant product is cured for use. Complex shapes can be produced, with the fiber orientations selected for optimum properties. Computers can be used to design and control the weaving, making the process less expensive than many of the more labor-intensive techniques.

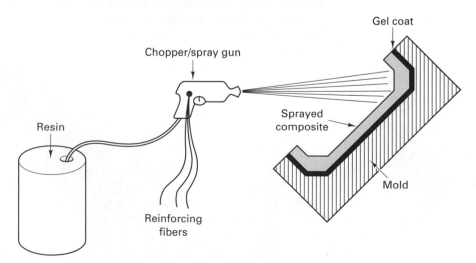

FIGURE 14-26   Schematic diagram of the spray forming of chopped-fiber-reinforced polymeric composite.

Fabrication of Fiber-Reinforced Metal-Matrix Composites. Continuous-fiber *metal-matrix composites* can be produced by variations of filament winding, extrusion, and pultrusion. Fiber-reinforced sheets can be produced by electroplating, plasma spray deposition coating, or vapor deposition of metal onto a fabric or mesh. These sheets are then shaped and bonded, often by some form of hot pressing. Diffusion bonding of foil–fabric sandwiches, roll bonding, and coextrusion are other means of producing fiber-reinforced metal products. Various casting processes have been adapted to place liquid metal around the fibers by means of capillary action, gravity, pressure (die casting and squeeze casting), or vacuum (countergravity casting). Products that incorporate discontinuous fibers can also be produced by powder metallurgy or spray-forming techniques and further fabricated by hot pressing, superplastic forming, forging, or some types of casting. In general, efforts are made to reduce or eliminate the need for finish machining, which would often require the use of diamond or carbide tools, or methods such as electrodischarge machining (EDM).

A critical concern with metal-matrix composites is the possibility of reactions between the reinforcement and the matrix during processing at the high temperatures required to melt and form metals as well as the temperatures of subsequent service. These concerns often limit the kinds of materials that can be combined. Barrier coatings have been employed to isolate reactive fibers.

In terms of properties, graphite-reinforced aluminum has been shown to be twice as stiff as steel and one-third to one-fourth the weight, with practically zero thermal expansion. Aluminum reinforced with silicon carbide exhibits increased strength (tension, compression, and shear at both room and elevated temperature) as well as increased hardness, fatigue strength, and elastic modulus. Thermal creep and thermal expansion are both reduced, but ductility, thermal conductivity, and electrical conductivity are also decreased. Magnesium, copper, and titanium alloys, as well as the superalloys, have also been used as the matrix in fiber-reinforced metal-matrix composites.

Fabrication of Fiber-Reinforced Ceramic-Matrix Composites. Unlike polymeric- or metal-matrix composites, where failures originate in or along the reinforcement fibers, *ceramic-matrix composites* often fail due to flaws in the matrix. If the reinforcement is bonded strongly to the matrix, a matrix crack might propagate right through the fibers. To impart toughness to the assembly, it is often desirable to promote a weak bond between the fiber and matrix. Cracks are redirected along the fiber–matrix interface rather than through the fiber and the remaining matrix.

The matrix materials and reinforcement fibers for ceramic-ceramic composites have been discussed in Chapter 8, along with some of the unique property combinations that can be achieved. Fabrication techniques are often quite different from the other composite families. One approach is to pass the fibers or mats through a slurry mixture that contains the matrix material. The impregnated material is then dried, assembled, and fired. Other techniques include the chemical vapor deposition or chemical vapor infiltration of a coated fiber base, where the coating serves to weaken an otherwise strong bond. Silicon nitride matrices can be formed by reaction bonding. The reinforcing fibers are dispersed in silicon powder, which is then reacted with nitrogen. Hot-pressing techniques can also be used with the various ceramic matrices. When the matrix is a glass, the heated material behaves much like a polymer, and the processing methods are often similar to those used for polymer-matrix composites.

Secondary Processing and Finishing of Fiber-Reinforced Composites. The various fiber-reinforced composites can often be processed further with conventional equipment (sawed, drilled, routed, tapped, threaded, turned, milled, sanded, and sheared), but special considerations should be exercised because composites are not uniform materials. Cutting some materials may be like cutting multilayer cloth, and precautions should be used to prevent the formation of splinters and cracks as well as frayed or delaminated edges. Sharp tools, high speeds, and low feeds are generally required. Cutting debris should be removed quickly to prevent the cutters from becoming clogged.

In addition, many of the reinforcing fibers are extremely abrasive and quickly dull most conventional cutting tools. Diamond or polycrystalline diamond tooling may be required to achieve realistic tool life. Abrasive slurrys can be used in conjunction with rigid tooling to assure the production of smooth surfaces. Lasers and water jets are alternative cutting tools. Lasers, however, can burn or carbonize the material or produce undesirable heat-affected zones. Water jets can create moisture problems with some plastic resins, and pressurized water can cause delaminations, but the low heat and light cutting force are attractive characteristics. Elastic deflections are minimized during the cut. Parts can often be held in place by simple vacuum cups, and water jets also minimize the generation of dust, which may be toxic.

When fiber-reinforced materials must be joined, the major concern is the lack of continuity of the fibers in the joint area. Thermoplastics can be softened and welded by applying pressure with heated tools, combining pressure and ultrasonic vibration, or using pressure and induction heating. Thermoset materials generally require the use of mechanical joints or adhesives, with each method having its characteristic advantages and limitations. Metal-matrix composites are often brazed.

■ Key Words

adhesive bonding	dipping	matrix	snap-fit
annealing	dry pressing	mats	sol-gel processing
autoclave	elastomer	metal-matrix composites	spinning
blow molding	explosive bonding	open-mold processing	spray molding
braiding	extrusion	parison	tape casting
bulk molding compound	fibers	plastics	tapes
compression molding	filament winding	prepregs	tempered glass
calendering	firing	pressure-bag molding	thermoforming
casting	foam molding	pultrusion	thermoplastic polymer
cementation	glass	reaction injection molding	thermosetting polymer
ceramics	glass ceramic	resin-transfer molding	tows
ceramic-matrix composites	hand lay-up	roll bonding	transfer molding
clay products	hot-isostatic pressing	rotational molding	vacuum-bag molding
cold molding	injection molding	rovings	viscous flow
composites	inserts	sandwich structures	vitrification
compression molding	isostatic pressing	sheet-molding compound	wet processing
crystalline ceramics	lamination	sintering	yarns
devitrification	laser sintering	slip casting	

■ Review Questions

1. Why are the fabrication processes applied to plastics, ceramics, and composites often different from those applied to metals? What are some of the key differences?
2. How does the fabrication of a thermoplastic polymer differ from the processing of a thermosetting polymer?
3. What are some of the ways that plastic sheet, plate, and tubing can be cast?
4. Why do cast plastic resins typically have a lustrous appearance?
5. What types of polymers are most commonly blow molded?
6. Why do blow molding molds typically contain a cooling system?
7. For what types of parts and production volumes would compression molding be an appropriate process?
8. What are typical mold temperatures for compression molding? What is the most common mold material?
9. What are some of the attractive features of the transfer molding process?
10. Cold molding is faster and more economical than other types of molding. What limits its use?
11. What is the most widely used process for the fabrication of thermoplastic materials (in terms of number of parts produced)?
12. In what ways is injection molding of plastic similar to the die casting of metal?
13. What is the benefit of a hot-runner distribution system in plastic injection molding?
14. Why is the cycle time for the injection molding of thermosetting polymers significantly longer than that for the thermoplastics?
15. How are the individual components mixed in the reaction injection molding process?
16. What are some of the attractive consequences of the low temperatures and low pressures of the reaction injection molding process?

17. What are some of the typical production shapes that are produced by the extrusion of plastics?
18. For what types of materials and products might thermoforming be considered attractive?
19. What types of products are produced by rotational molding?
20. What is the difference between open-cell and closed-cell foamed plastics?
21. What are some typical applications for rigid-type foamed plastics?
22. What type of products are produced by the spinning process?
23. What are some of the general properties of plastics that affect their machinability?
24. What are some of the attractive features of laser machining plastic?
25. What property of plastics is responsible for making snap-fit assembly a popular alternative for plastic products?
26. What are some of the attractive properties of plastics that favor their selection? What are some of the common limitations?
27. What are some of the design concerns when specifying and setting up a plastic molding process?
28. Why should adequate fillets be included between adjacent sections of a mold? What is a major benefit of rounding exterior corners?
29. Why is it most desirable to have uniform wall thickness in plastic products?
30. Why are product dimensions less precise when they cross a mold parting line?
31. Why might threaded inserts be preferred over other means of producing threaded holes in a plastic component?
32. What are some of the ways in which metal inserts are held in place in a plastic part?
33. When designing a decorative surface (design or lettering) on a plastic product, why is it desirable that the details be raised on the product rather than depressed?
34. Why does locating a parting line on a sharp corner make that feature less noticeable?
35. What is the benefit of countersinking holes that are to be threaded or used for self-tapping screws?
36. What types of products can be produced from elastomeric materials using the dipping process?
37. What process or equipment is used to form rubber compounds into sheets?
38. What method is generally used to bond elastomers to other materials?
39. What are the two basic classes of ceramic materials, and how does their processing differ?
40. What are some of the most common processes used to shape glass?
41. What are some of the special heat-treatment operations performed on glass products?

42. What are glass-ceramics? How are they produced?
43. What are some of the techniques that can be used to impart some degree of plasticity to crystalline ceramic materials?
44. Describe the differences between the injection molding of plastics and the injection molding of ceramics.
45. What is the difference between slip casting and tape casting?
46. What is the purpose of the firing or sintering operations in the processing of crystalline ceramic products?
47. How does cementation differ from sintering?
48. What are the benefits and limitations of machining ceramic materials before firing versus after firing?
49. What are some of the nonconventional methods used to machine ceramics?
50. Why are joining operations usually avoided when fabricating products from ceramic materials?
51. Discuss some of the design guidelines that relate to the production of parts from ceramic material.
52. Why are the processes used to fabricate particulate composites essentially the same as those used for conventional material?
53. What are some of the processes that can be used to produce a high-quality bond between the layers of a laminar composite?
54. List several fabrication processes for fiber-reinforced products that are essentially the same as for unreinforced plastics. List several that are unique to reinforced materials.
55. What types of materials are used as reinforcing fibers in fiber-reinforced composites?
56. What are some of the forms in which reinforcement fibers appear in composite materials?
57. What is a prepreg?
58. What are sheet-molding compounds (SMCs)? Bulk-molding compounds (BMCs)?
59. In what way is pultrusion similar to wire drawing?
60. What are some typical products that are made by filament winding?
61. What are some of the various molding processes that can be used to shape products from laminated sheets of woven fibers?
62. What are the benefits of using an autoclave instead of room-temperature and low-pressure curing?
63. What form of reinforcing fibers can be incorporated in the spray-molding process? Injection molding?
64. What is the major benefit of three-dimensional fiber reinforcement?
65. Describe some of the ways in which a metal matrix can be introduced into a fiber-reinforced composite.
66. Why might it be desirable to have a weak bond between a reinforcing fiber and a ceramic-matrix material?
67. Discuss some of the techniques used to cut fiber-reinforced composites.
68. What is the major concern when considering the joining of fiber-reinforced composites?

■ Problems

1. Consider some of the more prominent sporting goods that are fabricated from composite materials, such as skis, snowboards, tennis rackets, golf club shafts, bicycle frames, and body panels for racing cars. For two specific products, identify composite materials that are currently being used and the companion shape-producing fabrication methods.

2. Figure 14-A depicts the handles of two large wrenches, a ratchet wrench and a pipe wrench. These components are traditionally forged from ferrous alloy or made from a cast steel or cast iron. For various reasons, alternative materials may be desired. The ratchet wrench is quite long, and reduced weight may be a reasonable desire. Both of these tools could

be used in areas, such as a gas leak, where a nonsparking safety tool would be required. Current specifications for the ratchet handle call for a yield strength in excess of 50 ksi and a minimum of 2% elongation in all directions to ensure prevention of brittle fracture. The pipe wrench most likely has similar requirements.

a. Could a plastic or composite material be used to make a quality product with these additional properties? (*NOTE*: Metal jaw inserts can be used in the pipe wrench, enabling the other components to be considered as separate pieces.)

b. If so, how would you propose to manufacture the new handles?

FIGURE 14-A Handle segment of a large ratchet wrench and components of a pipe wrench.

www.wiley.com/college/degarmo

Chapter 14 CASE STUDY

Fabrication of Lavatory Wash Basins

Lavatory wash basins (bathroom sinks) have been successfully made from a variety of engineering materials, including cast iron, steel, stainless steel, ceramics, and polymers (such as melamine). Your company, Diversified Household Products, Inc., is considering a possible entrance into this market and has assigned you the tasks of (1) assessing the competition and (2) recommending the "best" approach toward producing this product.

1. For each of the materials (or families of materials), describe the material properties that are attractive for a wash basin application. What are the primary limitations or disadvantages?

2. For each of the materials (or families of materials), describe possible means of fabricating lavatory wash basins. Consider sheet metal forming, casting, molding, joining, and other types of fabrication processes. If multiple options exist, which one do you consider to be most attractive? Comment on the attractive features of the proposed system (materials and process) as well as the relative quality and cost.

3. Wash basins generally require a surface that is nonporous and stain resistant, scratch resistant, corrosion resistant, and attractive (and possibly available in a variety of colors). One approach to providing these properties on a steel or cast iron substrate is a coating of porcelain enamel. For each of the systems discussed in Question 2, discuss the need for additional surface treatment. What type of treatment would you recommend?

4. Most sinks contain an overflow feature that diverts excess water to the drain at a location beneath the stoppered basin. Discuss how this feature can be incorporated into each of your material–process manufacturing systems.

5. If your company were to consider producing lavatory wash basins on a competitive basis, which of the alternative manufacturing systems (material and manufacturing process) would you recommend? What features make it the most attractive?

FUNDAMENTALS OF METAL FORMING

■ 15.1 INTRODUCTION

Chapters 11 through 14 have already presented a variety of methods for producing a desired shape from an engineering material. Each of those methods had its characteristic set of capabilities, advantages, and limitations. If we are to select the best method to make a given product, however, we must have a reasonable understanding of the entire spectrum of available techniques for shape production and their related features.

The next several chapters will further our study of shape production methods by considering the family of *deformation processes*. These processes have been designed to exploit a remarkable property of some engineering materials (most notably metals) known as *plasticity*, the ability to flow as solids without deterioration of their properties. Since all processing is done in the solid state, there is no need to handle molten material or deal with the complexities of solidification. Since the material is simply moved (or rearranged) to produce the shape, as opposed to cutting away unwanted regions, the amount of waste can be substantially reduced. Unfortunately, the forces required are often high. Machinery and tooling can be quite expensive, and large production quantities may be necessary to justify the approach.

The overall usefulness of metals is largely due to the ease of fabrication into useful shapes. Nearly all metal products undergo metal deformation at some stage of their manufacture. By rolling, cast ingots, strands, and slabs are reduced in size and converted into basic forms such as sheets, rods, and plates. These forms can then undergo further deformation to produce wire or the myriad of finished products formed by processes such as forging, extrusion, sheet metal forming, and others. The deformation may be *bulk flow* in three dimensions, simple *shearing*, simple or compound *bending*, or complex combinations of these. The stresses producing these deformations can be tension, compression, shear, or any of the other varieties included in Table 15-1. Table 15-2 depicts a wide variety of specific processes and identifies the primary state of stress responsible for the deformation. For most of these processes, a wide range of speeds, temperatures, tolerances, surface finishes, and amounts of deformation are possible.

TABLE 15-1   Classification of States of Stress

(1) Simple uniaxial tension

(2) Biaxial tension

(3) Triaxial tension

(4) Biaxial tension, compression

(5) Biaxial tension and compression

(6) Uniaxial compression

(7) Biaxial compression

(8) Biaxial compression, tension

(9) Triaxial compression

(10) Pure shear

(11) Simple shear with triaxial compression

(12) Biaxial shear with triaxial compression

■ 15.2 FORMING PROCESSES: INDEPENDENT VARIABLES

Forming processes tend to be *complex systems* consisting of independent variables, dependent variables, and independent–dependent interrelations. *Independent variables* are those aspects of the process over which the engineer or operator has direct control, and they are generally selected or specified when setting up a process. Consider some of the independent variables in a typical forming process:

1. *Starting material.* When specifying the starting material, we may define not only the chemistry of that material but also its condition. In so doing, we define the initial properties and characteristics. These may be chosen entirely for ease of fabrication, or they may be restricted by the desire to achieve the required final properties upon completion of the deformation process.

2. *Starting geometry of the workpiece.* The starting geometry may be dictated by previous processing, or it may be selected from a variety of available shapes. Economic considerations often influence this decision.

3. *Tool or die geometry.* This is an area of major significance and has many aspects, such as the diameter and profile of a rolling mill roll, the bend radius in a sheet-forming operation, the die angle in wire drawing or extrusion, and the cavity details when forging. Since the tooling will induce and control the metal flow as the material goes from starting shape to finished product, success or failure of a process often depends on tool geometry.

4. *Lubrication.* It is not uncommon for friction between the tool and the workpiece to account for more than 50% of the power supplied to a deformation process. Lubricants can also act as coolants, thermal barriers, corrosion inhibitors, and parting compounds. Hence, their selection is an important aspect in the success of a forming operation. Specification includes type of lubricant, amount to be applied, and method of application.

5. *Starting temperature.* Since material properties can vary greatly with temperature, temperature selection and control are often key to the success or failure of a metal-forming operation. Specification of starting temperatures may include the temperatures of both the workpiece and the tooling.

6. *Speed of operation.* Most deformation processing equipment can be operated over a range of speeds. Since speed can directly influence the forces required for deformation (see Figure 2-32), the lubricant effectiveness, and the time available for heat transfer, its selection affects far more than the production rate.

7. *Amount of deformation.* While some processes control this variable through the design of tooling, others, such as rolling, may permit its adjustment at the discretion of the operator.

TABLE 15-2 Classification of Some Forming Operations

Process	Schematic Diagram	State of Stress in Main Part During Forming[a]
Rolling		7
Forging		9
Extrusion		9
Shear spinning		12
Tube spinning		9
Swaging or kneading		7
Deep drawing		In flange of blank, 5 In wall of cup, 1
Wire and tube drawing	(a) (b)	8
Stretching		2
Straight bending		At bend, 2 and 7
Contoured flanging	(a) Convex	At outer flange, 6 At bend, 2 and 7
	(a) Concave	At outer flange, 1 At bend, 2 and 7

[a] Numbers correspond to those in parentheses in Table 15-1.

■ 15.3 DEPENDENT VARIABLES

After specification of the independent variables, the process in turn determines the nature and values of a second set of features. Known as *dependent variables*, these, in essence, are the consequences of the independent variable selection. Examples of dependent variables include:

1. *Force or power requirements.* A certain amount of force or power is required to convert a selected material from a starting shape to a final shape, with a specified lubricant, tooling geometry, speed, and starting temperature. A change in any of the independent variables will result in a change in the force or power required, but the effect is indirect. We cannot directly specify the force or power; we can only specify the independent variables and then experience the consequences of that selection.

    It is extremely important, however, that we be able to predict the forces or powers that will be required for any forming operation. Without a reasonable estimate of forces or power, we would be unable to specify the equipment for the process, select appropriate tool or die materials, compare various die designs or deformation methods, and ultimately optimize the process.

2. *Material properties of the product.* While we can easily specify the properties of the starting material, the combined effects of deformation and the temperatures experienced during forming will certainly change them. The starting properties of the material may be of interest to the manufacturer, but the customer is far more concerned with receiving the desired final shape with the desired final properties. It is important to know, therefore, how the initial properties will be altered by the shape-producing process.

3. *Exit (or final) temperature.* Deformation generates heat within the material. Hot workpieces cool when in contact with colder tooling. Lubricants can break down or decompose when overheated or may react with the workpiece. The properties of an engineering material can be altered by both the mechanical and thermal aspects of a deformation process. Therefore, if we are to control a process and produce quality products, it is important to know and control the temperature of the material throughout the deformation. (*Note:* The fact that temperature may vary from location to location within the product further adds to the complexity of this variable.)

4. *Surface finish and precision.* The surface finish and dimensional precision of the resultant product depend on the specific details of the forming process.

5. *Nature of the material flow.* In deformation processes, dies and tooling generally exert forces or pressures and control the movement of the external surfaces of the workpiece. While the objective of an operation is the production of a desired shape, the internal flow of material may actually be of equal importance. As will be shown later in this chapter, product properties can be significantly affected by the details of material flow, and that flow depends on all the details of a process. Customer satisfaction requires not only the production of a desired geometric shape but also that the shape possess the right set of companion properties, without any surface or internal defects.

■ 15.4 INDEPENDENT–DEPENDENT RELATIONSHIPS

Figure 15-1 serves to illustrate the major problem facing metalforming personnel. On the left side are the *independent variables—those aspects of the process for which control is direct and immediate.* On the right side are the *dependent variables—those aspects for which control is entirely indirect.* Unfortunately, it is the dependent variables that we want to control, but their values are determined by the process, as complex consequences of the independent variable selection. If we want to change a dependent variable, we must determine which independent variable (or combination of independent variables) is to be changed, in what manner, and by how much. To make appropriate decisions, therefore, it is important for us to develop an understanding of the *independent variable–dependent variable interrelations.*

Understanding the links between independent and dependent variables is truly the most important area of knowledge for a person in metalforming. Unfortunately, this

Independent variables	Links	Dependent variables
Starting material		Force or power requirements
Starting geometry	-Experience-	
Tool geometry		Product properties
Lubrication	-Experiment-	Exit temperature
Starting temperature		Surface finish
Speed of deformation	-Modeling-	Dimensional precision
Amount of deformation		Material flow details

FIGURE 15-1 Schematic representation of a metalforming system showing independent variables, dependent variables, and the various means of linking the two.

knowledge is often difficult to obtain. Metalforming processes are complex systems composed of the material being deformed, the tooling performing the deformation, lubrication at surfaces and interfaces, and various other process parameters such as temperature and speed. The number of different forming processes (and variations thereof) is quite large. In addition, different materials often behave differently in the same process, and there are multitudes of available lubricants. Some processes are sufficiently complex that they may have 15 or more interacting independent variables.

We can gain information on the interdependencies of independent and dependent variables in three distinct ways:

1. *Experience.* Unfortunately, this generally requires long-time exposure to a process and is often limited to the specific materials, equipment, and products encountered during past contact. Younger employees may not have the experience necessary to solve production problems. Moreover, a single change in an area such as material, temperature, speed, or lubricant may make the bulk of past experience irrelevant.

2. *Experiment.* While possibly the least likely to be in error, direct experiment can be both time consuming and costly. Size and speed of deformation are often reduced when conducting laboratory studies. Unfortunately, lubricant performance and heat transfer behave differently at different speeds and sizes, and their effects are generally altered. The most valid experiment, therefore, is one conducted under full-size and full-speed production conditions—generally too costly to consider to any great degree. While laboratory experiments can provide valuable insight, caution should be exercised when extrapolating lab-scale results to more realistic production conditions.

3. *Process modeling.* Here one approaches the process through high-speed computing and one or more mathematical models. Numerical values are selected for the various independent variables, and the models are used to compute predictions for the dependent outcomes. Most techniques rely on the applied theory of plasticity with various simplifying assumptions. Alternatives vary from crude, first-order approximations to sophisticated, computer-based methods, such as finite element analysis. Various models may incorporate strain hardening, thermal softening, heat transfer, and other phenomena. Solutions may be algebraic relations that describe the process and reveal trends and relations between the variables or simply numerical values based on the specific input features.

■ 15.5 PROCESS MODELING

Metalforming simulations using the finite element modeling method became common in the 1980s but generally required high-power minicomputers or engineering workstations. By the mid-1990s, the rapid increase in computing power made it possible to model complex processes on desktop personal computers. With the continued expansion of computing power and speed, process simulations are now quick, inexpensive, and quite accurate. As a result, modeling is being used in all areas of manufacturing, including part design, manufacturing process design, heat-treatment and surface-treatment optimization, and others. Models can predict how a material will respond to a rolling process, fill a forging die, flow through an extrusion die, or solidify in a casting.

Entire heat treatments can be simulated, including cooling rates in various quenchants. Models can even predict the strain distribution, residual stresses, microstructure, and final properties at all locations within a product.

Advanced simulation techniques can provide a clear and thorough understanding of a process, eliminating costly trial-and-error development cycles. Product design and manufacturing methods can be optimized for quality and reliability, while reducing production costs and minimizing lead times. When coupled with appropriate sensors, the same models can be used to determine the type of adjustments needed to provide on-line process control. Process models can also serve as laboratory tools to explore new ideas or new products. New employees can become familiar with what works and what doesn't in a quick and inexpensive manner.

It is important to note, however, that the accuracy of any model can be no better than that of the input variables. For example, when modeling a metalforming operation, the mechanical properties of the deforming material (i.e., yield strength, ductility, etc.) must be known for the specific conditions of temperature, strain (amount of prior deformation), and strain rate (speed of deformation) being considered. The mathematical descriptions of material behavior as a function of the process conditions are known as constitutive relations. The development of such relationships is not an easy task, however, because the same material may respond differently to the same conditions if its microstructure is different. A 1040 steel that has been annealed (ferrite and pearlite) will not have the same properties as a quenched-and-tempered (tempered martensite) steel of the same chemistry. Microstructure and its effects on properties are difficult to describe in quantitative terms that can be input to a model.

Another rather elusive variable is the friction between the tool and the workpiece. Studies have shown that friction depends on contact pressure, contact area, surface finish, lubricant, speed, and the mechanical properties of the two contacting materials. We know that these parameters often vary from location to location and also change with time during a process, but many models tend to describe friction with a single variable of constant magnitude. Any variations with time and location are simply ignored in favor of mathematical simplicity or because of a lack of any better information.

At first glance, problems such as those just discussed appear to be a significant barrier to the use of mathematical models. It should be noted, however, that the same difficulties apply to the person trying to document, characterize, and extrapolate the results of experience or experiments. Process modeling often reveals features that might otherwise go unnoticed and can be quite useful when attempting to prevent or eliminate defects, optimize performance, or extend a process into a previously unknown area.

■ 15.6 GENERAL PARAMETERS

While much metalforming knowledge is specific to a given process, there are certain features that are common to all processes, and these will be presented here.

It is extremely important to characterize the *material being deformed*. What is its strength or resistance to deformation at the relevant conditions of temperature, speed of deformation, and amount of prior straining? What are the formability limits and conditions of anticipated fracture? What is the effect of temperature or variations in temperature? To what extent does the material strain-harden? What are the recrystallization kinetics? Will the material react with various environments or lubricants? These and many other questions must be answered to assess the suitability of a material to a given deformation process. Since the properties of engineering materials vary widely, the details will not be presented at this time. The reader is referred to the various chapters on engineering materials as well as the more in-depth references cited in Chapter 9 and the reference appendix.

Another general parameter is the *speed of deformation* and the various related effects. Some rate-sensitive materials may shatter or crack if impacted but will deform plastically when subjected to slow-speed loadings. Other materials appear to be stronger when deformed at higher speeds. For these *speed-sensitive materials*, more energy is needed to produce the same result if we wish to do it faster, and stronger tools may be required. Mechanical data obtained from slow strain rates in tensile tests may be totally useless if the deformation process operates at a significantly greater rate of deformation. Speed sensitivity is also greatest when the material is at elevated temperature, a condition that is frequently encountered in metalforming operations. The selection of hammer or press for the hot forging of a small product may well depend on the speed sensitivity of the material being forged.

In addition to the changes in mechanical properties, faster deformation speeds tend to promote improved lubricant efficiency. Faster speeds also reduce the time for heat transfer and cooling. During hot working, workpieces stay hotter and less heat is transferred to the tools.

Other general parameters include *friction and lubrication* and *temperature*. Both of these are of sufficient importance that they will be discussed in some detail.

■ 15.7 FRICTION AND LUBRICATION UNDER METALWORKING CONDITIONS

High forces or high pressures are applied through tools to induce the deformation of a material. Because of the relative motion between the workpiece and the tool, an important consideration in metal deformation processes is the friction that exists at this interface. For some processes, more than 50% of the input energy is spent in overcoming friction. Changes in lubrication can alter the mode of material flow during forming, create or eliminate defects, alter the surface finish and dimensional precision of the product, and modify product properties. Production rates, tool design, tool wear, and process optimization all depend on the ability to determine and control friction between the tool and workpiece.

In most cases, we want to economically reduce the effects of friction. However, some deformation processes, such as rolling, can only operate when sufficient friction is present. Regardless of the process, friction effects are hard to measure. As previously noted, the specific friction conditions depend on a number of variables, including contact area, contact pressure, surface finish, speed, lubricant, and temperature. Because of the many variables, the effects of friction are extremely difficult to scale down for laboratory testing, or extrapolate from laboratory tests to production conditions.

It should be noted that friction under metalworking conditions is significantly different from the friction encountered in most mechanical devices. The friction conditions of gears, bearings, journals, and similar components generally involve (1) two surfaces of similar material and similar strength, (2) experiencing elastic loads such that neither body undergoes permanent change in shape, (3) with wear-in cycles that produce surface compatibility, and (4) low to moderate operating temperatures. Metalforming operations, on the other hand, involve a hard, nondeforming tool interacting with a soft workpiece at pressures sufficient to cause plastic flow in the weaker material. Only a single pass is involved as the tool and workpiece interact, the workpiece is often at elevated temperature, and the contact area is frequently changing as the workpiece deforms.

Figure 15-2 shows a typical relationship between frictional resistance and contact pressure. For light, elastic loads, friction is directly proportional to the applied pressure, with

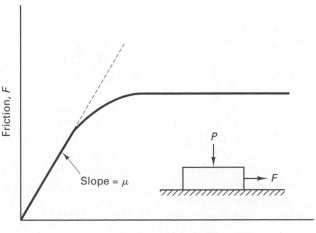

FIGURE 15-2   The effect of contact pressure on the frictional resistance between two surfaces.

the proportionality constant, μ, being known as the *coefficient of friction* or, more specifically, the Coulomb coefficient of friction. At high pressures, friction becomes independent of contact pressure and is more closely related to the strength of the weaker material.

An understanding of these results can be obtained from modern friction theory, whose primary premise is that "flat surfaces are not flat" but have some degree of roughness. When two irregular surfaces interact, sufficient contact is established to support the applied load. At the lightest of loads, only three points of contact may be necessary to support a plane. As the load is increased, the contacting points deform and the contact area increases, initially in a linear fashion. As the load continues to increase, more area comes into contact. Finally, at some high value of load, there is full contact between the surfaces. Additional loads can no longer bring additional areas into contact, and friction can now be described by a constant, independent of pressure.

Friction is the resistance to sliding along an interface. From a mechanistic viewpoint, this resistance can be attributed to (1) abrasion, the force necessary to plow the peaks of a harder material through a softer one, and/or (2) *adhesion*, the force necessary to rip apart microscopic weldments that form between the two materials. Since the weldment tears generally occur in the weaker of the two materials, it is reasonable to assume that the resistance attributed to both features would be proportional to the strength of the weaker material and also to the actual area of metal-to-metal contact. Thus, the curve depicted in Figure 15-2 could also be viewed as a plot of actual contact area at the interface versus contact pressure. Unfortunately, Figure 15-2, and the associated theory, applies only to unlubricated metal-to-metal contact. The addition of a lubricant, as well as any variation in its type or amount, can significantly alter the frictional response.

Surface deterioration or wear is another phenomenon that is directly related to friction. Since the workpiece only interacts with the tooling during a single forming operation, any wear experienced by the workpiece is usually not objectionable. In fact, a shiny, fresh-metal surface produced by wear is often viewed as desirable. Manufacturers whose processes retain most or all of the original dull finish may be accused of selling old or substandard products. Wear on the tooling, however, is quite the reverse. Tooling is expensive and it is expected to shape many workpieces. Tooling wear will generally result in change of workpiece dimensions. Tolerance control will be lost, and at some point the tools will have to be replaced. Other consequences of tool wear include increased frictional resistance (increased required power and decreased process efficiency), poor surface finish on the product, and loss of production during tool changes.

Lubrication is a key to success in many metalforming operations. While lubricants are generally selected for their ability to reduce friction and suppress tool wear, secondary considerations may include the ability to act as a thermal barrier, keeping heat in the workpiece and away from the tooling; the ability to act as a coolant, removing heat from the tools; and the ability to retard corrosion if left on the formed product. Other influencing factors include ease of application and removal; lack of toxicity, odor, and flammability; reactivity or lack of reactivity with material surfaces; adaptability over a useful range of pressure, temperature, and velocity; surface wetting characteristics; cost; availability; and the ability to flow or thin and still function as a lubricant. Lubricant selection is further complicated by the fact that lubricant performance may change with any change in the interface conditions. The exact response is often dependent on such factors as the finish of both surfaces, the area of contact, the applied load, the speed, the temperature, and the amount of lubricant.

The ability to select an appropriate lubricant can be a critical factor in determining whether a process is successful or unsuccessful, efficient or inefficient. For example, if a lubricant layer can prevent mechanical contact between the tool and the workpiece (full-fluid or solid layer separation), the forces and power required may decrease by as much as 30 to 40%, and tool wear becomes almost nonexistent. Considerable effort, therefore, has been directed to the study of friction and lubrication, a subject known as *tribology*, as it applies to both general metalworking conditions and specific metalforming processes. A substantial information base has been developed that can aid in optimizing the use of lubricants in metalworking.

■ 15.8 TEMPERATURE CONCERNS

In metalworking operations, workpiece temperature can be one of the most important process variables. The role of temperature in altering the properties of a material has been discussed in Chapter 2. In general, an increase in temperature brings about a decrease in strength, an increase in ductility, and a decrease in the rate of strain hardening—all effects that would tend to promote ease of deformation.

Forming processes tend to be classified as hot working, cold working, or warm working based on both the temperature and the material being formed. In hot working, the deformation is performed under conditions of temperature and strain rate where recrystallization occurs simultaneously with the deformation. To achieve this, the temperature of deformation is usually in excess of 0.6 times the melting point of the material on an absolute temperature scale (Kelvin or Rankine). Cold working is deformation under conditions where the recovery processes are not active. Here the working temperatures are usually less than 0.3 times the workpiece melting temperature. Warm working is deformation under the conditions of transition (i.e., a working temperature between 0.3 and 0.6 times the melting point).

HOT WORKING

Hot working is defined as the plastic deformation of metals at a temperature above the recrystallization temperature. It is important to note, however, that the recrystallization temperature varies greatly with different materials. Tin is near hot-working conditions at room temperature, steels require temperatures near 2000°F, and tungsten does not enter the hot-working regime until about 4000°F. Thus the term *hot working* does not necessarily correlate with high or elevated temperature, although such is usually the case.

As shown in Figures 2-30 and 2-31, elevated temperatures bring about a decrease in the yield strength of a metal and an increase in ductility. At the temperatures of hot working, recrystallization eliminates the effects of strain hardening, so there is no significant increase in yield strength or hardness, or corresponding decrease in ductility. The true stress–true strain curve is essentially flat once we exceed the yield point, and deformation can be used to drastically alter the shape of a metal without fear of fracture and without the requirement of excessively high forces. In addition, the elevated temperatures promote diffusion that can remove or reduce chemical inhomogeneities, pores can be welded shut or reduced in size during the deformation, and the metallurgical structure can often be altered through recrystallization to improve the final properties. An added benefit is observed for steels, where hot working involves the deformation of the weak, ductile, face-centered-cubic austenite structure, which then cools and transforms to the stronger body-centered-cubic ferrite or much stronger nonequilibrium structures, such as martensite.

From a negative perspective, the high temperatures of hot working may promote undesirable reactions between the metal and its surroundings. Tolerances are poorer due to thermal contractions, and warping or distortion can occur due to nonuniform cooling. The metallurgical structure may also be nonuniform, since the final grain size depends on the amount of deformation, the temperature of the last deformation/recrystallization, the cooling history after the deformation, and other factors, all of which may vary throughout a workpiece.

While recrystallization sets the minimum temperature for hot working, the upper limit for hot working is usually determined by factors such as excess oxidation, grain growth, or undesirable phase transformations. To keep the forming forces as low as possible and enable hot deformation to be performed for a reasonable amount of time, the starting temperature of the workpiece is usually set at or near the highest temperature for hot working.

Structure and Property Modification by Hot Working. When metals solidify into the large sections that are typical of ingots or continuously cast slabs or strands, coarse structures tend to form with a certain amount of chemical segregation. The size of the

FIGURE 15-3 Cross section of a 4-in.-diameter cast copper bar polished and etched to show the as-cast grain structure.

grains is usually not uniform, and undesirable grain shapes can be quite common, such as the columnar grains that have been revealed in Figure 15-3. Small gas cavities or shrinkage porosity can also form during solidification.

If a cast metal is reheated without prior deformation, it will simply experience grain growth and the accompanying deterioration in engineering properties. However, if the metal experiences a sufficient amount of deformation, the distorted structure will be rapidly replaced by new strain-free grains. This *recrystallization* is then followed by either (1) grain growth, (2) additional deformation and recrystallization, or (3) a drop in temperature that will terminate diffusion and "freeze in" the recrystallized structure. The structure in the final product is that formed by the last recrystallization and the thermal history that follows. By replacing the initial structure with a new one consisting of fine, spherical-shaped grains, it is possible to produce an increase not only in strength but also in ductility and toughness—a somewhat universal enhancement of properties.

Engineering properties can also be improved through the reorientation of inclusions or impurity particles that are present in the metal. With normal melting and cooling, many impurities tend to locate along grain boundary interfaces. If these are unfavorably oriented or intersect surfaces, they can initiate a crack or assist its propagation through a metal. When a metal is plastically deformed, the impurities tend to flow along with the base metal or fracture into rows of fragments (*stringers*) that are aligned in the direction of working. These nonmetallic impurities do not recrystallize with the base metal but retain their distorted shape and orientation. The product exhibits a *flow structure*, like the one shown in Figure 15-4, and final properties tend to exhibit directional variation. Through proper design of the deformation, impurities can often be reoriented into a "crack-arrestor" configuration where they are perpendicular to the direction of crack propagation. The outer lobe of the forging in Figure 15-4, for example, has excellent fracture resistance since all flow lines are parallel to the external surfaces. The impurities appear as crack initiators or crack propagators only at the top and bottom of the inner lobe, which hopefully are low-stress or noncritical locations.

Figure 15-5 schematically compares a machined thread and a rolled thread in a threaded fastener. If the axial defects in the starting wire or rod are reoriented to be parallel to the thread profile, the rolled thread offers improved strength and fracture resistance.

FIGURE 15-4 Flow structure of a hot-forged gear blank. Note how flow is parallel to all critical surfaces. *(Courtesy of Bethlehem Steel Corporation, Bethlehem, PA.)*

FIGURE 15-5 Schematic comparison of the grain flow in a machined thread (a) and a rolled thread (b). The rolling operation further deforms the axial structure produced by the previous wire- or rod-forming operations, while machining simply cuts through it.

Temperature Variations. The success or failure of a hot deformation process often depends on the ability to control the temperatures within the workpiece. Over 90% of the energy imparted to a deforming workpiece will be converted into heat. If the deformation process is sufficiently rapid, the temperature of the workpiece may actually increase. More common, however, is the cooling of the workpiece in its lower-temperature environment. Heat is lost through the workpiece surfaces, with the majority of the loss occurring where the workpiece is in direct contact with lower-temperature tooling. Nonuniform temperatures are produced, and flow of the hotter, weaker, interior may well result in cracking of the colder, less ductile, surfaces. Thin sections cool faster than thick sections, and this may further complicate the flow behavior.

To minimize problems, it is desirable to keep the workpiece temperatures as uniform as possible. Heated dies can reduce the rate of heat transfer, but die life tends to be compromised. For example, dies are frequently heated to 325° to 450°C (600° to 850°F) when used in the hot forming of steel. Tolerances could be improved and contact times could be increased if the tool temperatures could be raised to 550° to 650°C (1000° to 1200°F), but tool life drops so rapidly that these conditions become quite unattractive.

A final concern is the cool-down from the temperatures of hot working. Nonuniform cooling can introduce significant amounts of *residual stress* in hot-worked products. Associated with these stresses may be warping or distortion, and possible cracking.

COLD WORKING

The plastic deformation of metals below the recrystallization temperature is known as *cold working*. Here, the deformation is usually performed at room temperature, but mildly elevated temperatures may be used to provide increased ductility and reduced strength. From a manufacturing viewpoint, cold working has a number of distinct advantages, and the various cold-working processes have become quite prominent. Recent advances have expanded their capabilities, and a trend toward increased cold working appears likely to continue.

When compared to hot working, the advantages of cold working include the following:

1. No heating is required.
2. Better surface finish is obtained.
3. Superior dimensional control is achieved since the tooling sets dimensions at room temperature. As a result, little, if any, secondary machining is required.
4. Products possess better reproducibility and interchangeability.
5. Strength, fatigue, and wear properties are all improved through strain hardening.
6. Directional properties can be imparted.
7. Contamination problems are minimized.

Some disadvantages associated with cold-working processes include the following:

1. Higher forces are required to initiate and complete the deformation.
2. Heavier and more powerful equipment and stronger tooling are required.
3. Less ductility is available.
4. Metal surfaces must be clean and scale-free.
5. Intermediate anneals may be required to compensate for the loss of ductility that accompanies strain hardening.
6. The imparted directional properties may be detrimental.
7. Undesirable residual stresses may be produced.

The strength levels induced by strain hardening are often comparable to those produced by the strengthening heat treatments. Even when the precision and surface finish of cold working are not required, it may be cheaper to produce a product by cold working a less expensive alloy (achieving the strength by strain hardening) than by heat treating parts that have been hot formed from a heat-treatable alloy. In addition, better and more ductile metals and an improved understanding of plastic flow have done much to reduce the difficulties often experienced during cold forming. As an added benefit, most cold-working processes eliminate or minimize the production of waste material and the need for subsequent machining—a significant feature with today's emphasis on conservation and materials recycling.

Because the cold-forming processes require powerful equipment and product-specific tools or dies, they are best suited for large-volume production of precision parts where the quantity of products can justify the cost of the equipment and tooling. Considerable effort has been devoted to developing and improving cold-forming machinery along with methods to enable these processes to be economically attractive for modest production quantities. By grouping products made from the same starting material and using quick-change tooling, cold-forming processes can often be adapted to small-quantity or just-in-time manufacture.

Metal Properties and Cold Working.   The suitability of a metal for cold working is determined primarily by its tensile properties, and these are a direct consequence of its metallurgical structure. Cold working then alters that structure, thereby altering the tensile properties of the resulting product. It is important for both the incoming and outgoing properties to be considered when selecting metals that are to be processed by cold working.

Figure 15-6 presents the true stress–true strain curves for both a low- and a high-carbon steel. Focusing on the low-carbon material, we note that plastic deformation cannot occur until the strain exceeds the strain associated with the elastic limit, point *a* on the stress–strain curve. Plastic deformation then continues until the strain reaches the value x_4, where the metal ruptures. From the viewpoint of cold working, two features are significant: (1) the magnitude of the yield-point stress, which determines the force required to initiate permanent deformation, and (2) the extent of the strain region from x_1 to x_4, which indicates the amount of plastic deformation (or ductility) that can be achieved without fracture. If a considerable amount of deformation is desired, a material like the low-carbon steel is more desirable than the high-carbon variety. Greater ductility would be available and less force would be required to initiate and continue the deformation. The curve on the right, however, has a higher strain-hardening coefficient (see Chapter 2 for discussion). If strain hardening is being used to impart strength, this material would have a greater increase in strength for the same amount of cold work. In addition, the material on the right would be more attractive for shearing operations and might be easier to machine (see Chapter 20).

Springback is another cold-working phenomenon that can be explained with the aid of a stress–strain diagram. When a metal is deformed by the application of a load, part of the resulting deformation is elastic. For example, if a metal is stretched to point x_1 in Figure 15-6 and the load is removed, it will return to its original size and shape

FIGURE 15-6 Use of true stress–true strain diagrams to assess the suitability of two metals for cold working.

True stress

Low-carbon steel

True strain

Increase in tensile strength due to work hardening. The high-carbon steel will also have more springback.

True stress

True strain

High-carbon steel

because all of the deformation is elastic. If, on the other hand, the metal is stretched by an amount x_3, corresponding to point b on the stress–strain curve, the total strain is made up of two parts, a portion that is elastic and another that is plastic. When the deforming load is removed, the stress relaxation will follow line bx_2, and the final strain will only be x_2. The decrease in strain, $x_3 - x_2$, is known as *elastic springback*.

In cold-working processes, springback can be extremely important. If a desired size is to be achieved, the deformation must be extended beyond that point by an amount equal to the springback. Since different materials have different elastic moduli, the amount of springback from a given load will change from one material to another. A substitution in material, therefore, may well require adjustments in the forming process. Fortunately, springback is a predictable phenomenon, and most difficulties can be prevented by proper design procedures.

Initial and Final Properties in a Cold-Working Process.   The quality of the starting material is often key to the success or failure of a cold-working operation. To obtain a good surface finish and maintain dimensional precision, the starting material must be clean and free of oxide or scale that might cause abrasion and damage to the dies or rolls. Scale can be removed by pickling, a process in which the metal is dipped in acid and then washed. In addition, sheet metal and plate are sometimes given a light cold rolling prior to the major deformation. The rolling operation not only assures uniform starting thickness but also produces a smooth starting surface.

The light cold-rolling pass can also serve to remove the *yield-point phenomenon* and the associated problems of nonuniform deformation and surface irregularities in the product. Figure 15-7 presents an expansion of the left-hand region of Figure 2-6 or Figure 15-6, a stress–strain curve that is typical of many low-carbon steels. After loading to the upper yield point, the material exhibits a *yield-point runout* wherein the material can strain up to several percent with no additional force being required. Consider a piece of sheet metal that is to be formed into an automotive body panel. If a segment of that panel were to receive a total stretch less than the magnitude of the yield-point runout, it would be induced by a stress equal to the yield-point stress. Since the stress is constant in the runout region, the material is free to not deform at all, to deform the entire amount of the yield-point runout, or to select some point in between. It is not uncommon for some regions to deform the entire amount and thin correspondingly, while adjacent regions resist deformation and retain the original thickness. The resulting ridges and valleys, shown in Figure 15-7, are referred to as Luders bands or stretcher strains and are very difficult to remove or conceal. By first cold rolling the material to a strain near or past the yield-point runout, all subsequent forming occurs in a region where a well-defined strain corresponds to each value of stress. If the body panel were shaped from pre-rolled material, the deformation and thinning would be uniform throughout the piece.

FIGURE 15-7   (Left) Stress–strain curve for a low-carbon steel showing the commonly observed yield-point runout; (Right) Luders bands or stretcher strains that form when this material is stretched to an amount less than the yield-point runout.

FIGURE 15-8 Mechanical properties of pure copper as a function of the amount of cold work (expressed in percent).

Figure 15-8 shows how the mechanical properties of pure copper are affected by cold working. Individual tensile tests were conducted on specimens that had experienced progressively greater amounts of cold work. As the graph shows, yield strength and tensile strength increase with increased deformation. Hardness is not presented on the graph but generally follows tensile strength. Since the ductility decreases, the amount of cold working is generally limited by the onset of fracture. Reduction in area would show a decline similar to elongation, as would electrical conductivity and corrosion resistance.

In order to maximize the amount of starting ductility, an *annealing* heat treatment is often applied to a metal prior to cold working. If the required amount of deformation exceeds the fracture limit, however, *intermediate anneals* may be performed to restore ductility (set the amount of cold work back to zero), thereby enabling further working without the risk of fracture. If the desired final properties coincide with a given amount of cold work, the last anneal can be judiciously positioned in the deformation cycle. In this way, the desired shape can be produced along with the mechanical properties that accompany the amount of cold work imparted following that anneal. In all annealing operations, care should be exercised to control the grain size of the resulting material. Grain sizes that are too large or too small can both be detrimental.

Cold working, like hot working, also produces an anisotropic structure—one whose properties vary with direction. Here, the *anisotropy* is related to the distorted crystal structure and is not simply a function of the nonmetallic inclusions. Also associated with cold working is the generation of residual stresses. While anisotropy and residual stresses can be beneficial, they can also be quite harmful. Since they occur as a consequence of cold working, their effect on performance should always be considered.

WARM FORMING

Deformation produced at temperatures intermediate to hot and cold forming is known as *warm forming*. Compared to cold forming, warm forming offers the advantages of reduced loads on the tooling and equipment, increased material ductility, and a possible reduction in the number of anneals due to a reduction in the amount of strain hardening. The use of higher forming temperatures can often expand the range of materials and geometries that can be formed by a given process or piece of equipment. High-carbon steels may be formed without a spheroidization treatment.

Compared to hot forming, the lower temperatures of warm working produce less scaling and decarburization, and enable production of products with better dimensional precision and smoother surfaces. Finish machining is reduced and less material is converted into scrap. Because of the finer structures and the presence of some strain hardening, the as-formed properties may be adequate for many applications, enabling the elimination of final heat-treatment operations. The warm regime generally requires less energy than hot working due to the decreased energy in heating the workpiece (lower temperature), energy saved through higher precision (less material being heated), and the possible elimination of postforming heat treatments. Although the tools must exert 25 to 60% higher forces, they last longer since there is less thermal shock and thermal fatigue.

When energy was cheap, metalforming was usually conducted in either the hot- or cold-working regimes, and warm working was largely ignored. Even today, material behavior is less well characterized for the warm-working temperatures (the warm-working temperatures for steel are between 550° and 800°C or 1000° and 1500°F). Lubricants have not been as fully developed for the warm-working temperatures and pressures, and die design technology is not as well established. Nevertheless, the pressures of energy and material conservation, coupled with the other cited benefits, strongly favor the continued development of warm working. Cold forming is still the preferred method for fabricating small components, but warm forming is considered to be attractive for larger parts (up to about 10 lb) and steels with more than 0.35% carbon and/or high alloy content.

Hot working and warm forming are usually applied to bulk forming processes, like forging and extrusion. For sheet material, the surface-to-volume ratio is sufficiently large that the workpiece can rapidly lose its heat. As the major auto manufacturers seek to increase fuel efficiency, there has been significant interest in aluminum sheet as a replacement for steel. Unfortunately, the formability of high-strength aluminum is much lower than that of low-carbon steels of similar strength. If the steel is simply converted to aluminum, and the design and tooling remain unchanged, fracture often occurs in the more heavily worked regions. If the material, die, and blank holder are all heated to 200° to 300°C (400° to 575°F), however, aluminum sheet shows a significant increase in formability, and satisfactory parts can generally be produced.

ISOTHERMAL FORMING

Figure 15-9 shows the relationship between yield strength (or forging pressure) and temperature for several engineering metals. The 1020 and 4340 steels show a moderate increase in strength with decreasing temperature. In contrast, the strength of the titanium alloy (open circles) and the A-286 nickel-based superalloy (solid circles) shows a much stronger variation. Within the range of typical hot-working temperatures, cooling of as little as 100°C (200°F) could result in a doubling in strength. During hot forming, cooling surfaces surround a hotter interior. Any variation in strength can result in nonuniform deformation and cracking of the less ductile surface.

To successfully deform temperature-sensitive materials, deformation may have to be performed under *isothermal* (constant-temperature) conditions. The dies or tooling

FIGURE 15-9 Yield strength of various materials (as indicated by pressure required to forge a standard specimen) as a function of temperature. Materials with steep curves may require isothermal forming. *(From "A Study of Forging Variables," ML-TDR-64-95, March 1964; courtesy of Battelle Columbus Laboratories, Columbus, OH.)*

must be heated to the same temperature as the workpiece, sacrificing die life for product quality. Deformation speeds must be slowed so that any heat generated by deformation can be removed in a manner that would maintain a uniform and constant temperature. Inert atmospheres may be required because of the long times at elevated temperature. Although such methods are indeed costly, they are often the only means of producing satisfactory products from certain materials. Because of the uniform temperatures and slow deformation speeds, isothermally formed components generally exhibit close tolerances, low residual stresses, and fairly uniform metal flow.

■ Key Words

abrasion	dependent variable	lubrication	strain hardening
adhesion	dimensional precision	Luders bands	stretcher strains
anisotropy	elastic springback	oriented structure	stringers
annealing	flow structure	plasticity	surface finish
bending	friction	recrystallization	tooling
bulk flow	hot working	residual stresses	tribology
coefficient of friction	independent variable	shearing	warm working
cold working	intermediate anneal	speed sensitivity	wear
constitutive relation	isothermal forming	springback	yield-point runout
deformation processes			

■ Review Questions

1. What is plasticity?
2. What are some of the general assets of the metal deformation processes? Some general liabilities?
3. Why might large production quantities be necessary to justify metal deformation as a means of manufacture?
4. What is an independent variable in a metalforming process?
5. What is the significance of tool and die geometry in designing a successful metalforming process?
6. Why is lubrication often a major concern in metalforming?
7. What are some of the possible roles of a lubricant in addition to reducing friction?
8. What are some of the secondary effects that may occur when the speed of a metalforming process is varied?
9. What is a dependent variable in a metalforming process?
10. Why is it important to be able to predict the forces or powers required to perform specific forming processes?
11. Why is it important to know and control the thermal history of a metal as it undergoes deformation?
12. Why is it often difficult to determine the specific relationships between independent and dependent variables?
13. What are the three distinct ways of determining the interrelation of independent and dependent variables?
14. What features limit the value of laboratory experiments in modeling metalforming processes?
15. What features have contributed to the expanded use of process modeling?
16. What are some of the uses or applications of process models?
17. What is a constitutive relation for an engineering material?
18. What features may limit the accuracy of a mathematical model?
19. What simplifying assumptions are often made regarding friction between the tool and workpiece?

20. What type of information about the material being deformed may be particularly significant to a metalforming engineer?
21. How might a material's performance vary with changes in the speed of deformation?
22. Why is friction such an important parameter in metalworking operations?
23. Why are friction effects in metalworking difficult to scale down for laboratory testing or scale up from laboratory conditions to production conditions?
24. What are several ways in which the friction conditions during metalworking differ from the friction conditions found in most mechanical equipment?
25. According to modern friction theory, frictional resistance can be attributed to what two physical phenomena?
26. Discuss the significance of wear in metalforming: (a) wear on the workpiece and (b) wear on the tooling.
27. Lubricants are often selected for properties in addition to their ability to reduce friction. What are some of these additional properties?
28. What are some of the benefits that can be obtained by fully separating a tool and workpiece by an intervening layer of lubricant?
29. If the temperature of a material is increased, what changes in properties might occur that would promote the ease of deformation?
30. Define the various regimes of cold working, warm working, and hot working in terms of the melting point of the material being formed.
31. What is an acceptable definition of hot working? Is a specific temperature involved?
32. What are some of the attractive manufacturing and metallurgical features of hot-working processes?
33. What are some of the negative aspects of hot working?

34. How can hot working be used to improve the grain structure of a metal?
35. If the deformed grains recrystallize during hot working, how can the process impart an oriented or flow structure (and directionally dependent properties)?
36. Why are heated dies or tools often employed in hot-working processes?
37. What generally restricts the upper temperature to which dies or tooling is heated?
38. What is the primary cause of residual stresses in hot-worked products?
39. Compared to hot working, what are some of the advantages of cold-working processes?
40. What are some of the disadvantages of cold-forming processes?
41. How could cold working be used to reduce the cost of a moderate to high strength product?
42. How can the tensile test properties of a metal be used to assess its suitability for cold forming?

43. Why is elastic springback an important consideration in cold-forming processes?
44. What are Luders bands or stretcher strains, and what causes them to form? How can they be eliminated?
45. What engineering properties are likely to decline during the cold working of a metal?
46. How can the selective placement of the final intermediate anneal be used to establish desired final properties in a cold-formed product?
47. Is the anisotropy induced by cold working an asset or a liability? What about the residual stresses?
48. What are some of the advantages of warm forming compared to cold forming? Compared to hot forming?
49. What material feature is considered to be the driving force for isothermal forming?
50. Why is isothermal forming considerably more expensive than conventional hot forming?

■ Problems

1. Copper is being reduced from a hot-rolled 3/8-in.-diameter rod to a final diameter of 0.100 in. by wire drawing through a series of dies. The final wire should have a yield strength in excess of 50,000 psi and an elongation greater than 10%. Use Figure 15-8 to determine a desirable amount of final cold work. Compute the placement of the last intermediate anneal so that the final product has both the desired size and the desired properties.

2. a. List and discuss the various economic factors that should be considered when evaluating a possible switch from cold forming to warm forming.
   b. Repeat part a for a possible conversion from hot forming to warm forming.

3. An advertisement for automobile spark plugs has cited the superiority of rolled threads over machined threads. Figure 15-5 shows such a comparison for hot forming, where the deformation process reorients flaws and defects without significantly changing the structure and properties of the metal. The spark plug threads, however, were formed by cold rolling. Do the same benefits apply? Discuss the assets and liabilities of the cold rolling of threads compared to thread formation by conventional machining.

4. Computer modeling of metal deformation processes is a powerful and extremely useful tool. At the same time, there are several areas of limitation that can significantly compromise or even invalidate the final results. Consider each of the following areas of limitation, investigating what is currently being used or what current options are available:

   a. A mathematical description of material behavior (a constitutive equation). In almost all cases, some simplification of actual flow behavior is assumed. For accurate modeling, flow behavior should be known and mathematically characterized as a function of strain, strain rate, and temperature.

   b. Interfacial friction between the tooling and the workpiece. How is this being modeled? Does it consider the effects of surface finish, sliding velocity, interface temperature, and numerous other factors. As the process commences, lubricants may thin or be wiped from surfaces, forces and pressures change, temperatures change, and surface roughness or texture is modified. Does the model reflect any of these changes? Some models assign a single value to friction over the entire contact surface. This value may also remain constant throughout the entire operation.

   c. Assignment of boundary conditions. Often the mathematical solutions must conform to assigned features, such as defined motions or stresses at specified surfaces. The boundary conditions have a profound effect on the results that are calculated. Poor choices or choices made so as to facilitate easy analysis can often produce misleading or erroneous results.

Chapter 15 CASE STUDY

Repairs to a Damaged Propeller

The propeller of a moderately large pleasure boat has been cast from a nickel-aluminum-bronze alloy that contains 82% Cu, 9% Al, 4% Ni, 4% Fe, and 1% Mn. It is approximately 13 inches in diameter with three 10-pitch blades and has been designed for both fresh-water and saltwater usage.

1. One of the blades has struck a rock and is badly bent. A replacement propeller is quite expensive and cannot be obtained for several weeks. An attractive alternative, therefore, may be to repair the existing piece. Would you recommend such a repair, and how would you proceed? Can it simply be hammered back into shape? Would you recommend any additional processing, either before or after the repair? What is the rationale for your recommendations?

2. A second propeller, identical to the one above, has also been damaged by an impact. This time, however, the damage is in the form of a crack at the base of one of the blades, as shown in the figure. Since the crack does not penetrate into the hub, it is proposed that a repair be made using some form of welding or brazing process. Would you recommend such a repair? If so, how would you suggest the repair be made? Explain the rationale for your recommendations and outline the procedure that should be followed. Would there be any sacrifice in quality or performance with the repaired propeller?

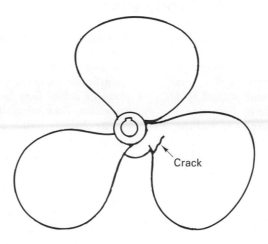

Crack

CHAPTER 16

BULK FORMING PROCESSES

■ 16.1 INTRODUCTION

The shaping of metal by deformation is as old as recorded history. The Bible, in the fourth chapter of Genesis, introduces Tubal-cain and cites his ability as a worker of metal. While we have no description of his equipment, it is well established that metal forging was practiced long before written records. Processes such as rolling and wire drawing were common in the Middle Ages and probably date back much further. In North America, by 1680 the Saugus Iron Works near Boston had an operating drop forge, rolling mill, and slitting mill.

Although the basic concepts of many forming processes have remained largely unchanged throughout history, the details and equipment have evolved considerably. Manual processes were converted to machine processes during the Industrial Revolution. The machinery then became bigger, faster, and more powerful. Water wheel power was replaced by steam and then by electricity. More recently, computer-controlled, automated operations have become the norm.

■ 16.2 CLASSIFICATION OF DEFORMATION PROCESSES

A wide variety of processes have been developed to mechanically shape material, and a number of classification methods have been proposed. One approach divides the processes into *primary* and *secondary*. Primary processes reduce a cast material into intermediate shapes, such as slabs, plates, or billets. Secondary processes further convert these shapes into finished or semifinished products. Unfortunately, some processes clearly fit both categories, depending on the particular product being made.

In Chapter 15, we discussed the temperature of deformation and presented the various regimes based on the temperature of the workpiece. These included cold working, warm working, hot forming, and isothermal deformation. This classification has also become somewhat blurred, especially with the increased emphasis on energy conservation. Processes that were traditionally performed hot are now being performed cold, and cold-forming processes can often be enhanced by some degree of heating. Warm working has experienced considerable growth.

Chapters 16 and 17 utilize a division that focuses on the size and shape of the workpiece and how that size and shape is changed. *Bulk deformation processes* are those where the thicknesses or cross sections are reduced or shapes are significantly changed. Since the volume of the material remains constant, changes in one dimension require proportionate changes in others. Thus the enveloping surface area changes significantly, usually increasing as the product lengthens or the shape becomes more complex. The bulk forming operations can be performed in all of the temperature regimes. Common processes include: rolling; forging; extrusion; cold forming; and wire, rod, and tube drawing.

In contrast, *sheet-forming operations* involve the deformation of a material where the thickness and surface area remain relatively constant. Common processes include shearing or blanking, bending, and deep drawing. Because of the large surface-to-volume ratio, sheet material tends to lose heat rapidly, and most sheet-forming operations are performed cold.

Even this division is not without confusion, however. Coining, for example, begins with sheet material but alters the thickness in a complex manner that is essentially bulk deformation. The bulk deformation processes will be presented in Chapter 17. Sheet-forming processes can be found in Chapter 18.

■ 16.3 BULK DEFORMATION PROCESSES

The bulk deformation processes that will be presented in this chapter include:

1. Rolling

2. Forging

3. Extrusion

4. Wire, rod, and tube drawing

5. Cold forming, cold forging, and impact extrusion

6. Piercing

7. Other squeezing processes

These processes can be further divided in several ways. One grouping separates the processes by focusing on the size and shape of the deforming region. In continuous flow processes, such as forging, the size and shape are continually changing, and process analysis must reflect this change. In processes such as rolling or wire drawing, material moves through the deforming region, but the size and shape of that region remain unchanged. Some form of steady-state analysis can often be applied.

In all of the bulk forming processes, the primary deformation stress is compression. This may be applied directly by tools or dies that squeeze the workpiece or indirectly as in wire drawing, where the workpiece is pulled in tension but the resisting die generates compression in the region undergoing deformation.

■ 16.4 ROLLING

Rolling operations reduce the thickness or change the cross section of a material through compressive forces exerted by rolls. As shown in Figure 16-1, rolling is often the first process that is used to convert material into a finished wrought product. Thick starting stock can be rolled into blooms, billets, or slabs, or these shapes can be obtained directly from continuous casting. A *bloom* has a square or rectangular cross section, with a thickness greater than 15 cm (6 in.) and a width no greater than twice the thickness. A *billet* is usually smaller than a bloom and has a square or circular cross section. Billets are usually produced by some form of deformation process, such as rolling or extrusion. A *slab* is a rectangular solid where the width is greater than twice the thickness. Slabs can be further rolled to produce *plate*, *sheet*, and *strip*. Plates have thickness greater than 6 mm ($\frac{1}{4}$ inch), while sheet and strip range from 6 mm to 0.1 mm ($\frac{1}{4}$ inch to 0.004 inch).

These hot-rolled products often form the starting material for subsequent processes, such as cold forming or machining. Sheet and strip can be fabricated into products or

Slabs   Blooms   Billets

Plates   Structural shapes   Rails   Bars

Skelp   Strip   Tube rounds   Wire rods

Pipe and tubing   Wire

Large diameter pipe   Tin plate   Seamless pipe   Cold drawn bars

FIGURE 16-1 Flow chart for the production of various finished and semifinished steel shapes. Note the abundance of rolling operations. *(Courtesy of American Iron and Steel Institute, Washington, D.C.)*

further cold rolled into thinner, stronger material or even into foil (thicknesses less than 0.1 mm). Blooms and billets can be further rolled into finished products, such as *structural shapes* or railroad rail, or they can be processed into semifinished shapes, such as *bar*, *rod*, *tube*, or *pipe*.

From a tonnage viewpoint, rolling is clearly predominant among all manufacturing processes, with approximately 90% of all metal products experiencing at least one rolling operation. Rolling equipment and rolling practices are sufficiently advanced that standardized, uniform-quality products can be produced at relatively low cost. Because shaped rolls are both massive and costly, shaped products are only available in standard forms and sizes where there is sufficient demand to permit economical production.

BASIC ROLLING PROCESS

In the basic rolling process, shown in Figure 16-2, metal is passed between two rolls that rotate in opposite directions, the gap between the rolls being somewhat less than the thickness of the entering metal. Because the rolls rotate with a surface velocity that exceeds the speed of the incoming metal, friction along the contact interface acts to propel the metal forward. The metal is then squeezed and elongates to compensate for the decrease in thickness or cross-sectional area. The amount of deformation that can be achieved in a single pass between a given pair of rolls depends on the friction conditions along the interface. If too much is demanded, the rolls cannot advance the material and simply skid over its surface. If too little deformation is taken, the operation will be successful, but the additional passes required to produce a given part will increase the cost of production.

HOT ROLLING AND COLD ROLLING

In hot rolling, as with all hot-working processes, temperature control is required for success. The starting material should be heated to a uniform elevated temperature. If the temperature is not uniform, the subsequent deformation will not be uniform. Consider a piece being reheated for rolling. If the soaking time is insufficient, the hotter exterior will flow in preference to the cooler, stronger interior. Conversely, if a uniform-temperature material is allowed to cool prior to working or has cooled during previous working operations,

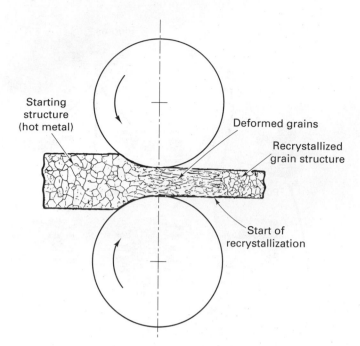

Starting structure (hot metal)

Deformed grains

Recrystallized grain structure

Start of recrystallization

FIGURE 16-2 Schematic representation of the hot-rolling process, showing the deformation and recrystallization of the metal being rolled.

the cooler surfaces will tend to resist deformation. Cracking and tearing of the surface may result as the hotter, weaker interior tries to deform.

It is not uncommon for high-volume producers to begin with continuous-cast feedstock. The cooling from solidification is controlled so as to enable direct insertion into a hot-rolling operation without additional handling or reheating. For smaller operations or secondary processing, the starting material is often a room-temperature solid, such as an ingot, slab, or bloom. This material must first be brought to the desired rolling temperature, usually in gas- or oil-fired soaking pits or furnaces. For plain-carbon and low-alloy steels, the soaking temperature is usually about 1200°C (2200° F). For smaller cross sections, induction coils may be used to heat the material prior to rolling.

Hot-rolling operations are usually terminated when the temperature falls to about 50° to 100°C (100° to 200°F) above the recrystallization temperature of the material being rolled. Such a *finishing temperature* ensures the production of a uniform fine grain size and prevents the possibility of unwanted strain hardening. If additional deformation is required, a period of reheating will be necessary to reestablish desirable hot-working conditions.

Cold rolling can be used to produce sheet, strip, bar, and rod products with extremely smooth surfaces and accurate dimensions. Cold-rolled *sheet* and *strip* can be obtained in various conditions, including *skin-rolled*, *quarter-hard*, *half-hard*, and *full-hard*. Skin-rolled metal is subjected to only a 0.5 to 1% reduction to produce a smooth surface and uniform thickness, and to remove or reduce the yield-point phenomenon (i.e., prevent formation of Luders bands upon further forming). This material is well suited for subsequent cold-working operations where good ductility is required. Quarter-hard, half-hard, and full-hard sheet and strip experience greater amounts of cold reduction, up to 50%. Their yield points are higher, properties have become directional, and ductility has decreased. Quarter-hard steel can be bent back on itself across the grain without breaking. Half-hard and full-hard can be bent back 90° and 45°, respectively, about a radius equal to the material thickness.

For products with a uniform cross section and cross-sectional dimensions less than about 5 cm or 2 inches, cold rolling of rod or bar may be an attractive alternative to extrusion or machining. Strain hardening can provide up to 20% additional strength to the material, and the process offers the smooth surfaces and high dimensional precision of cold working. Like the rolling of structural shapes, however, the process generally requires a series of shaping operations. Separate roll passes (and roll grooves) may be required for sizing, breakdown, roughing, semiroughing, semifinishing, and finishing. While the various grooves may be in a single set of rolls, a minimum order of several tons of product may be required to justify the cost of tooling.

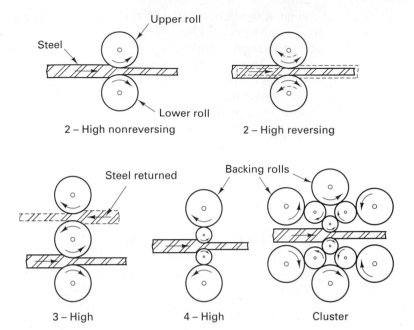

FIGURE 16-3 Various roll configurations used in rolling operations.

ROLLING MILL CONFIGURATIONS

As illustrated in Figure 16-3, rolling mill stands are available in a variety of roll configurations. Early reductions, often called primary, roughing, or breakdown passes, usually employ a two- or three-high configuration with rolls 60 to 140 cm (24 to 55 in.) in diameter. The *two-high nonreversing mill* is the simplest design, but the material can only pass through the mill in one direction. A *two-high reversing mill* permits back-and-forth rolling, but the rolls must be stopped, reversed, and brought back to rolling speed between each pass. A *three-high mill* eliminates the need for roll reversal but requires some form of elevator on each side of the mill to raise or lower the material and mechanical manipulators to turn or shift the product between passes.

As shown in Figure 16-4, smaller-diameter rolls produce less length of contact for a given reduction and therefore require lower force and less energy to produce a given change in shape. The smaller cross section, however, provides reduced stiffness, and the rolls are prone to flex elastically since they are supported on the ends and pressed apart by the metal passing through the middle (a condition known as three-point bending). *Four-high* and *cluster* arrangements use backup rolls to support the smaller work rolls. These configurations are used in the hot rolling of wide plate and sheets, and in cold

FIGURE 16-4 The effect of roll diameter on length of contact for a given reduction.

rolling, where even small deflections in the roll would result in an unacceptable variation in product thickness. Foil is almost always rolled on *cluster mills* since the small thickness requires small-diameter rolls. In a cluster mill, the roll in contact with the work can be as small as 6 mm ($\frac{1}{4}$ inch) in diameter. To counter the need for even smaller rolls, some foils are produced by *pack rolling*, a process where two or more layers of metal are rolled simultaneously as a means of providing a thicker input material. Household aluminum foil is usually rolled as a double sheet, as evidenced by the one shiny side (in contact with the roll) and one dull side (in contact with the other piece of foil).

In the rolling of nonflat or shaped products, such as structural shapes and railroad rail, the sets of rolls contain contoured grooves that sequentially form the desired shape, reduce the cross-sectional area, and control the metal flow. Figure 16-5 shows some typical roll-pass sequences used in the production of structural shapes. Length increases as the cross section is reduced.

CONTINUOUS (OR TANDEM) ROLLING MILLS

When the volume of a product justifies the investment, rolling may be performed on a *continuous* or *tandem rolling mill*. Billets, blooms, or slabs are heated and fed through an integrated series of nonreversing rolling mill stands. Continuous mills for the hot rolling of steel strip, for example, often consist of a roughing train of approximately four four-high mill stands and a finishing train of six or seven additional four-high stands. In a continuous structural mill, the rolls in each stand contain only one set of shaped grooves, in contrast to the multigrooved rolls used when the product is produced by back-and-forth passes through a single stand.

If a single piece of material is in multiple rolling stations at the same time, it is imperative that the same volume pass through each stand in the same amount of time. If the cross section is reduced, speed must be increased proportionately. Therefore, as a material is reduced in size, the rolls of each successive stand must turn faster than those of the preceding one. If a subsequent stand is running too slow, material will accumulate between stands. If the demand for incoming material exceeds the output of the previous stand, the material is placed in tension and may tear or rupture.

The synchronization of six or seven mill stands is not an easy task, especially when key variables such as temperature and lubrication may vary during a single run and the product may be exiting the final stand at speeds in excess of 110 kilometers per hour (70 miles per hour). Computer control is basic to successful rolling, and modern mills are equipped with numerous sensors to provide the needed information. When continuous casting units feed directly into continuous rolling mills, the time lapse from final solidification to finished rolled product is often a matter of a few minutes.

FIGURE 16-5 Typical roll-pass sequences used in producing structural shapes.

FIGURE 16-6 Schematic of a horizontal ring rolling operation. As the thickness of the ring is reduced, its diameter will increase.

RING ROLLING

Ring rolling is a special rolling process where one roll is placed through the hole of a thick-walled ring and a second roll presses in from the outside (Figure 16-6). As the rolls squeeze and rotate, the wall thickness is reduced and the diameter of the ring increases. Shaped rolls can be used to produce a wide variety of cross-section profiles. The resulting seamless rings have a circumferential grain orientation and find application in products such as rockets, turbines, airplanes, pipelines, and pressure vessels. Diameters can be as large as 8 m (25 ft) with face heights as great as 2 m (80 in.).

THREAD ROLLING

Thread rolling is a deformation alternative to the cutting of threads; it is illustrated in Figure 16-5 and discussed in Chapter 29.

CHARACTERISTICS, QUALITY, AND PRECISION OF ROLLED PRODUCTS

Because hot-rolled products are formed and finished above their recrystallization temperature, they have little directionality in their properties and are relatively free of deformation-induced residual stresses. These characteristics may vary, however, depending on the thickness of the product and the presence of complex sections. Nonmetallic inclusions do not recrystallize, so they may impart some degree of directionality. In addition, residual stresses can be induced by nonuniform cooling from the temperatures of hot working. Thin sheets often show directional characteristics, whereas thicker plate (above 20 mm or 0.8 in.) will usually have very little. Because of high residual stresses in the rapidly cooled edges, a complex shape, such as an I- or H-beam, may warp in a noticeable fashion if a portion of one flange is cut away.

As a result of the hot deformation and the good control that is maintained during processing, hot-rolled products are normally of uniform and dependable quality. It is quite unusual to find any voids, seams, or laminations when produced by reliable manufacturers. The surfaces of hot-rolled products are usually a bit rough, however, and are originally covered with a tenacious high-temperature oxide, known as *mill scale*. This can be removed by an acid pickling operation, resulting in a surprisingly smooth surface finish. The dimensional tolerances of hot-rolled products vary with the kind of metal and the size of the product. For most products produced in reasonably large tonnages, the tolerances are within 2 to 5% of the specified dimension (either height or width).

Cold-rolled products exhibit superior surface finish and dimensional precision, and they can offer the enhanced strength obtained through strain hardening.

FLATNESS CONTROL AND ROLLING DEFECTS

If we are rolling a flat product with uniform thickness, the gap between the rolls must be a uniform one. Attaining such an objective, however, may be difficult. Consider the upper roll in a set that is rolling sheet or plate. As shown in Figure 16-7, the material presses upward in the middle of the roll, while the roll is held in place by bearings that are mounted on either end and supported in the mill frame. The roll, therefore, is loaded in three-point bending and tends to flex in a manner that produces a thicker center and

FIGURE 16-7 (a) Loading on a rolling mill roll. The top roll is pressed upward in the center while being supported on the ends. (b) The elastic response to the three-point bending.

FIGURE 16-8 Use of a "crowned" roll to compensate for roll flexure. When the roll flexes in three-point bending, the crowned roll flexes into flatness.

thinner edge. Since the thicker center will not lengthen as much as the thinner edge, the result will be a product with either a wavy edge or a fractured center.

If the rolls are always used to reduce the same material at the same temperature by the same amount, the forces and deflections can be predicted and the roll can be designed to have a specified profile. If a "crowned," or barrel-shaped, roll is subjected to the designed load, it will deflect into flatness, as illustrated in Figure 16-8. If the applied load is not of the designed magnitude, however, the resulting profile will not be flat and defects may result. If the correction is insufficient, for example, wavy edges or center fractures will still occur. If the correction is excessive, the center becomes thinner and longer, resulting in a wavy center or cracking of the edges.

Since roll deflections are proportional to the forces applied to the rolls, product flatness can also be improved by measures that reduce these forces. If possible, friction could be reduced, smaller-diameter rolls could be used, and smaller reductions could be employed. Heating the workpiece generally makes it weaker, so increased workpiece temperature will also reduce the force on the rolls. Horizontal tensions can be applied to the piece as it is being rolled (strip tension in sheet metal rolling). Since these tensions combine with the vertical compression to deform the piece (stretching while squeezing), the roll forces and associated deflections are less. Other techniques to improve flatness include increasing the elastic modulus of the rolls themselves through material selection or providing some form of backup support to oppose deflection, as with the four-high and cluster mill configurations.

Successful rolling requires the balancing of many factors relating to the material being rolled, the variables of the rolling process, and lubrication between the workpiece and the rolls. Common defects include the nonuniform thickness previously discussed, dimensional variations caused by changes in workpiece temperature, surface flaws (such as rolled-in scale and roll marks), laps, seams, and various types of distortions.

THERMOMECHANICAL PROCESSING AND CONTROLLED ROLLING

As with most deformation processes, rolling is generally considered to be a way of changing the shape of a material. While heat may be used to reduce forces and promote plasticity, the thermal processes that produce or control product properties (heat treatments) are usually performed as subsequent operations. *Thermomechanical processing*, of which *controlled rolling* is an example, consists of integrating deformation and thermal processing into a single process that will produce not only the desired shape but also the desired properties, such as strength and toughness. The heat for the property modification is the same heat used in the rolling operation, and subsequent heat treatment becomes unnecessary.

A successful thermomechanical operation begins with process design. The starting material must be specified and the composition closely maintained. Then a time–temperature–deformation system must be developed to achieve the desired objective. Possible goals include producing a uniform fine grain size; controlling the nature, size, and distribution of the various transformation products (such as ferrite, pearlite, bainite, and martensite in steels); controlling the reactions that produce solid-solution strengthening or precipitation hardening; and producing a desired level of toughness. Starting structure (controlled by composition and prior thermal treatments), deformation details, temperature during the various stages of deformation, and the conditions of cool-down from the working temperature must all be specified and controlled. Moreover, the attainment of uniform properties requires uniform temperatures and deformations throughout the product. Computer-controlled facilities are an absolute necessity if thermomechanical processing is to be successfully performed.

Possible benefits of thermomechanical processing include improved product properties; substantial energy savings (by eliminating subsequent heat treatment); and the possible substitution of a cheaper, less alloyed metal for a highly alloyed one that responds to heat treatment.

■ 16.5 FORGING

Forging is a term applied to a family of processes that induce plastic deformation through localized compressive forces applied through dies. The equipment can take the form of hammers, presses, or special forging machines. While the deformation can be performed in all temperature regimes (hot, cold, warm, or isothermal), most forging is done with workpieces above the recrystallization temperature.

Forging is clearly the oldest known metalworking process. From the days when prehistoric peoples discovered that they could heat sponge iron and beat it into a useful implement by hammering with a stone, forging has been an effective method of producing many useful shapes. Modern forging is simply an extension of the ancient art practiced by the armor makers and immortalized by the village blacksmith. High-powered hammers and mechanical presses have replaced the strong arm and the hammer, and tool steel dies have replaced the anvil. Metallurgical knowledge has supplemented the art and skill of the craftsman, as we seek to control the heating and handling of the metal. Parts can range in size from ones whose largest dimension is less than 2 cm (1 in.) to others weighing more than 170 metric tons (450,000 lb).

The variety of forging processes currently offers a wide range of capabilities. A single piece can be economically fashioned by some methods, while others can mass-produce thousands of identical parts. The metal may be (1) *drawn out* to increase its length and decrease its cross section, (2) *upset* to decrease the length and increase the cross section, or (3) *squeezed in closed impression dies* to produce multidirectional flow. As indicated in Table 15-2, the state of stress in the work is primarily uniaxial or multiaxial compression.

Common forging processes include:

1. Open-die drop-hammer forging
2. Impression-die drop-hammer forging
3. Press forging
4. Upset forging
5. Automatic hot forging
6. Roll forging
7. Swaging
8. Net-shape and near-net-shape forging

OPEN-DIE HAMMER FORGING

In concept, *open-die hammer forging* is the same type of forging done by the blacksmith of old, but massive mechanical equipment is now used to impart the repeated blows. The metal is first heated to the proper temperature using a furnace or electrical induction heating. An impact is then delivered by some type of mechanical hammer. The simplest industrial hammer is a *gravity drop* machine, where a free-falling ram strikes the workpiece, and the energy of the blow is varied by adjusting the height of the drop. Most forging hammers now employ some form of energy augmentation, however, where pressurized air, steam, or hydraulic fluids are used to raise and propel the hammer. Higher striking velocities are achieved, with more control of striking force, easier automation, and the ability to shape pieces up to several tons. *Computer-controlled hammers* can provide blows of differing impact speed (energy) for different products or each of the various stages of a given operation. Their use can greatly increase the efficiency of the process and also minimize the amount of noise and vibration, which are the most common outlets for the excess energy not absorbed in the deformation of the workpiece. Figure 16-9 shows a large double-frame hammer along with a labeled schematic.

Open-die forging does not fully control the flow of metal. To obtain the desired shape, the operator must orient and position the workpiece between blows. The hammer may contact the workpiece directly, or specially shaped tools can be inserted to assist in making concave or convex surfaces, forming holes, or performing a cutoff operation. Manipulators may be used to position larger workpieces, which may weigh several tons. While some finished parts can be made by this technique, open-die forging is usually employed to preshape metal in preparation for further operations. For example, consider parts like turbine rotors and generator shafts with dimensions up to 20 m (70 ft) in length and up to 1 m (3 ft) in diameter. Open-die forging induces oriented plastic flow and

FIGURE 16-9 (Left) Double-frame drop hammer. *(Courtesy of Erie Press Systems, Erie, PA.)* (Right) Schematic diagram of a forging hammer.

1 Preform mounted on saddle/mandrel.

2 Metal displacement– reduce preform wall thickness to increase diameter.

3 Progressive reduction of wall thickness to produce ring dimensions.

4 Machining to near net shape.

FIGURE 16-10   (Top) Illustration of the unrestrained flow of material in open-die forging. Note the barrel shape that forms due to friction between the die and material. (Middle) Open-die forging of a multidiameter shaft. (Bottom) Forging of a seamless ring by the open-die method. *(Courtesy of Forging Industry Association, Cleveland, OH.)*

minimizes the amount of subsequent machining. Figure 16-10 illustrates the unrestricted flow of material along with the open-die forging of a multidiameter cylindrical shaft and a seamless metal ring.

IMPRESSION-DIE HAMMER FORGING

Open-die hammer forging (or smith forging, as it has been called) is a simple and flexible process, but it is not practical for large-scale production. It is a slow operation, and the shape and dimensional precision of the resulting workpiece is dependent on the skill of the operator. As shown in Figure 16-11, *impression-die* or *closed-die forging* overcomes these difficulties by using shaped dies to control the flow of metal. Figure 16-12 shows a typical set of multicavity dies. The upper piece attaches to the hammer and the lower piece to the anvil. Heated metal is positioned in the lower cavity and struck one

FIGURE 16-11   Schematic of the impression-die forging process, showing partial die filling and the beginning of flash formation in the center sketch and the final shape with flash in the right-hand sketch.

FIGURE 16-12 Impression drop-forging dies and the product resulting from each impression. The flash is trimmed from the finished connecting rod in a separate trimming die. The sectional view shows the grain flow resulting from the forging process. *(Courtesy of Forging Industry Association, Cleveland, OH.)*

or more blows by the upper die. The hammering causes the metal to flow and completely fill the die cavity. Excess metal is squeezed out along the parting line to form a *flash* around the periphery of the cavity. This material cools rapidly, increases in strength, and, by resisting deformation, effectively blocks the formation of additional flash. By trapping material within the die, the flash then ensures the filling of all of the cavity details. The flash is ultimately trimmed from the part in a final forging operation.

In *flashless forging*, also known as true closed-die forging, the metal is deformed in a cavity that provides total confinement. Accurate workpiece sizing is required since complete filling of the cavity must be ensured with no excess material. Accurate workpiece positioning is also necessary, along with good die design and control of lubrication. The major advantage of this approach is the elimination of the scrap generated during flash formation, an amount that is often between 20 and 45% of the starting material.

Most conventional forgings are impression-die with flash and are produced in dies with a series of cavities, where one or more blows of the hammer are used for each step in the sequence. The first impression is often an *edging, fullering,* or *bending* impression to distribute the metal roughly in accordance with the requirements of the later cavities. Edging gathers material into a region, while fullering moves material away. Intermediate impressions are for *blocking* the metal to approximately its final shape, with generous corner and fillet radii. For small production lots, the cost of further cavities may not be justified, and the blocker-type forgings are simply finished by machining. More often, the final shape and size are imparted by an additional forging operation in a *final* or *finisher impression*, after which the flash is trimmed from the part. Figure 16-12 shows an example of these steps and the shape of the part at the conclusion of each. Since every part is shaped in the same die cavities, each mass-produced part is a close duplicate of all the others.

Conventional closed-die forging begins with a simple hot-rolled shape and utilizes reheating and working to progressively convert it into a more complex geometry. The shape of the various cavities controls the flow of material, and the flow, in turn, imparts

High pressure striking air | Tong | Stock | High pressure striking air | Low pressure return air | Forging | Low pressure return air

Impeller | Impeller

Piston head | Piston head | Open to exhaust | Guide | Die | Die | Guide | Open to exhaust

STRIKING | RETURNING

FIGURE 16-13   Schematic diagram of an impactor in the striking and returning modes. *(Courtesy of Chambersburg Engineering Company, Chambersburg, PA)*

the oriented structure discussed in Chapter 16. (Grain flow that follows the external contour of the component is in the crack-arrestor orientation, improving strength, ductility, and resistance to impact and fatigue.) Through forging, we can also control the size and shape of various cross sections, so the metal can be distributed as needed to resist the applied loads. Couple these factors with a fine recrystallized grain structure (hot working) and the absence of voids (compressive forming stresses), and we see why forgings often have about 20% higher strength-to-weight ratios compared with cast or machined parts of the same material.

Board hammers, steam hammers, and air hammers have all been used in impression-die forging. An alternative to the hammer and anvil arrangement is the *counterblow machine*, or *impactor*, illustrated in Figure 16-13. These machines have two horizontal hammers that simultaneously impact a workpiece that is positioned between them. Excess energy simply becomes recoil, in contrast to the hammer and anvil arrangement, where energy is lost to the machine foundation, and a heavy machine base is required. Impactors also operate with less noise and less vibration and produce distinctly different flows of material, as illustrated in Figure 16-14.

Heat-treatment costs can be reduced by the direct quenching or controlled cooling of the hot parts as they emerge from the forging operation. Energy conservation can also be achieved through several processes that have been designed to produce a product that is somewhere between a conventional forging and a conventional casting. In one approach, a forging preform is cast from liquid metal, removed from the mold while still hot, and then finish forged in a single-cavity die. The flash is then trimmed and the part is quenched to room temperature. Forging preforms can also be produced by the spray deposition of metal droplets into shaped containers, as described in Section 19-10. These preforms are then removed from the mold, and the final shape and properties are imparted by a final forging operation. Still another approach is semisolid forging, discussed in Chapter 13.

PRESS FORGING

In hammer or impact forging, the metal flows to dissipate the energy imparted in the hammer–workpiece collision. Speeds are high, so the forming time is short. Contact times under load are on the order of milliseconds. There is little time for heat transfer and cooling of the workpiece, and the adiabatic heating that occurs during deformation helps to minimize chilling. It is possible, however, that all of the energy can be dissipated by deformation of just the surface of the metal (coupled with additional absorption by the anvil and foundation), and the interior of the workpiece remains essentially undeformed. Consider the deformation of a metal wood-splitting wedge after it has been struck repeatedly by a sledge hammer. The top is usually "mushroomed," while the remainder retains the original geometry and taper.

If large pieces or thick products are to be formed, *press forging* may be required. The deformation is now analyzed in terms of forces or pressures (rather than energy), and the slower squeezing action penetrates completely through the metal, producing a more uniform deformation and flow. New problems can arise, however, because of the longer

(a)

Conventional forged disk with paths of flow

(b)

Disk formed by impacter with paths of flow

FIGURE 16-14   A comparison of metal flow in conventional forging and impacting.

time of contact between the dies and the workpiece. As the surface of the workpiece cools, it becomes stronger and less ductile, and may crack if deformation is continued. Heated dies are generally used to reduce heat loss, promote surface flow, and enable the production of finer details and closer tolerances. Periodic reheating of the workpiece may also be required. If the dies are heated to the same temperature as the workpiece and pressing proceeds at a slow rate, *isothermal forging* can be used to produce near-net-shape components with uniform microstructure and mechanical properties.

Forging presses are of two basic types, mechanical and hydraulic, and are usually quite massive. *Mechanical presses* use cams, cranks, or toggles to produce a preset and reproducible stroke. Because of their mechanical drives, different forces are available at the various stroke positions. Production presses are quite fast, capable of up to 50 strokes per minute, and are available in capacities ranging from 300 to 18,000 tons (3 to 160 MN). *Hydraulic presses* move in response to fluid pressure in a piston and are generally slower, more massive, and more costly to operate. On the positive side, hydraulic presses are much more flexible and can have greater capacity. Since motion is in response to the flow of pressurized drive fluids, hydraulic presses can be programmed to have different strokes for different operations and even different speeds within a stroke. Presses can be used to perform all types of forging, including open die and impression die. Impression-die press forgings usually require less draft than drop forgings and have higher dimensional accuracy. In addition, press forgings can often be completed in a single closing of the dies as opposed to the multiple blows of a hammer. Machines with capacities up to 50,000 tons (445 MN) are currently in operation in the United States.

A third type of press is the *screw press*, which in many ways acts like a hammer. A large flywheel stores a predetermined amount of energy. This energy is then transmitted to a vertical screw, which drives a descending ram. Downward motion stops when all of the energy from the flywheel has been dissipated.

Additional information about the various types of presses and drive mechanisms can be found in the closing section of Chapter 17.

DESIGN OF IMPRESSION-DIE FORGINGS AND ASSOCIATED TOOLING

The geometrical possibilities for impression-die forging are quite numerous, with complex shapes like connecting rods, crankshafts, wrenches, and gears being commonly produced. Figure 16-15 shows a forged-and-machined steel automotive crankshaft that significantly outperformed similar components made of austempered ductile cast iron. Parts typically range from under 3 lb. up to about 750 lb., with the major dimension being between 7 and 20 in. (20 and 50 cm). Steels, stainless steels, and alloys of aluminum, copper, and nickel can all be forged with fair to excellent results.

The forging dies are usually made of high-alloy or tool steel and can be expensive to design and construct. Impact resistance, wear resistance, strength at elevated temperature, and the ability to withstand cycles of rapid heating and cooling must all be outstanding. In addition, considerable care is required to produce and maintain a smooth and accurate cavity and parting plane. Better and more economical results will be obtained if the following rules are observed:

1. The dies should part along a single, flat plane if at all possible. If not, the parting plane should follow the contour of the part.

2. The parting surface should be a plane through the center of the forging, not near an upper or lower edge.

3. Adequate draft should be provided—at least 3° for aluminum and 5° to 7° for steel.

4. Generous fillets and radii should be provided.

5. Ribs should be low and wide.

6. The various sections should be balanced to avoid extreme differences in metal flow.

7. Full advantage should be taken of fiber flow lines.

8. Dimensional tolerances should not be closer than necessary.

FIGURE 16-15   A forged-and-machined automobile engine crankshaft that has been formed from microalloyed steel. Performance is superior to cranks of cast ductile iron.

The various design details, such as the number of intermediate steps, the shape of each, the amount of excess metal required to ensure die filling, and the dimensions of the flash at each step, are often a matter of experience. Each component is a new design entity and brings its own unique challenges. Computer-aided design has made notable advances, however, and the development and accessibility of high-speed, immense-memory computers have enabled the accurate modeling of many complex shapes.

Good dimensional accuracy is a characteristic of impression-die forging. With reasonable care, the dimensions for steel products can be maintained within the tolerances of 0.02 to 0.03 in. (0.50 to 0.75 mm). It should be noted, however, that the dimensions across the parting plane are affected by closure of the dies and are therefore dependent on die wear and the thickness of the final flash. Dimensions contained entirely within a single die segment can be maintained at a significantly greater level of accuracy. Surface-finish values range from 80 to 300 μin. Draft angles can sometimes be reduced, occasionally approaching zero, but this is not recommended for general practice.

Selection of a lubricant is also critical to successful forging. The lubricant not only affects the friction and wear and associated metal flow, but it may also be expected to act as a thermal barrier (restricting heat flow from the workpiece to the dies) and a parting compound (preventing the part from sticking in the cavities).

UPSET FORGING

Upset forging involves increasing the diameter of a material by compressing its length. Because of its use with a myriad of fasteners, it is the most widely used of all forging processes when evaluated in terms of the number of pieces produced. Parts can be upset forged both hot and cold, with the operation generally being performed on special high-speed machines. The forging motion is usually horizontal, and the workpiece is rapidly moved from station to station. While most operations start with wire or rod, some machines can upset bars up to 25 cm (10 in.) in diameter.

Upset forging generally employs split dies that contain multiple positions or cavities, as seen in the typical die set of Figure 16-16. The dies separate enough for the bar to advance between them and move into position. They are then clamped together

FIGURE 16-16   Set of upset-forging dies and punches. The product resulting from each of the four positions is shown along the bottom. *(Courtesy of Ajax Manufacturing Company Euclid, OH.)*

Applications of rule 1   Applications of rule 2   Applications of rule 3

FIGURE 16-17 Schematics illustrating the rules governing upset forging. *(Courtesy of National Machinery Company, Tiffin, OH.)*

Violation of rule 1   Violation of rule 2   Violation of rule 3

and a heading tool or ram moves longitudinally against the bar, upsetting it into the cavity. Separation of the dies then permits transfer to the next position or removal of the product. If a new piece is started with each die separation and an operation is performed in each cavity simultaneously, a finished product can be made with each cycle of the machine. By including a shearing operation as the initial piece moves into position, the process can operate with continuous coil or long-length rod as its incoming feedstock.

Upset-forging machines are often used to form heads on bolts and other fasteners as well as to shape valves, couplings, and many other small components. The upset region can be on the end or central portion of the workpiece, and the final diameter may be up to three times the original. The following three rules, illustrated in Figure 16-17, should be followed when designing parts that are to be upset forged:

1. The length of unsupported metal that can be gathered or upset in one blow without injurious buckling should be limited to three times the diameter of the bar.

2. Lengths of stock greater than three times the diameter may be upset successfully provided that the diameter of the upset is not more than 1 times the diameter of the bar.

3. In an upset requiring stock length greater than three times the diameter of the bar, and where the diameter of the cavity is not more than 1 times the diameter of the bar (the conditions of rule 2), the length of unsupported metal beyond the face of the die must not exceed the diameter of the bar.

AUTOMATIC HOT FORGING

Several equipment manufacturers now offer highly automated upset equipment in which mill-length steel bars (typically 7 m or 24 ft long) are fed into one end at room temperature and hot-forged products emerge from the other end at rates of up to 180 parts per minute (86,400 parts per eight-hour shift). These parts can be solid or hollow, round or symmetrical, up to 6 kg (12 lb) in weight, and up to 18 cm (7 in.) in diameter.

The process begins with the lowest-cost steel bar stock: hot-rolled and air-cooled carbon or alloy steel. The bar is first heated to 1200° to 1300°C (2200° to 2350°F) in under 60 seconds as it passes through high-power induction coils. It is then descaled by rolls, sheared into individual blanks, and transferred through several successive forming stages, during which it is upset, preformed, final forged, and pierced (if necessary). Small parts can be produced at up to 180 parts per minute, and larger parts at rates on the order of 90 parts per minute. Figure 16-18 shows a typical deformation sequence and a variety of hot-forged ferrous products.

Sheared billet   Upset pancake   Blocker forging   Finished gear blank

(a)

(b)

FIGURE 16-18 (a) Typical four-step sequence to produce a spur-gear forging by automatic hot forging. The sheared billet is progressively shaped into an upset pancake, blocker forging, and finished gear blank. (b) Samples of ferrous parts produced by automatic hot forging at rates between 90 and 180 parts per minute. *(Courtesy of National Machinery Company, Tiffin, OH.)*

The *automatic hot-forging* process has a number of attractive features. Low-cost input material and high production speeds have already been cited. Minimum labor is required, and since no flash is produced, material usage can be as much as 20 to 30% greater than with conventional forging. With a consistent finishing temperature near 1050°C (1900°F), an air cool can often produce a structure suitable for machining, eliminating the need for an additional anneal or normalizing treatment. Tolerances are generally within 0.3 mm (0.012 in.), surfaces are clean, and draft angles need only be 0.5 to 1° (as opposed to the conventional 3° to 5°). Tool life is nearly double that of conventional forging because the contact times are only on the order of $\frac{6}{100}$ of a second.

Automatic hot formers can also be coupled with high-rate, cold-forming operations. Preform shapes can be hot formed at rates that approach 180 parts per minute. These products can then be cold formed to final shape on machines that operate at speeds near 90 parts per minute. The benefits of the combined operations include high-volume production at low cost, coupled with the precision, surface finish, and strain hardening that are characteristic of a cold-finished material.

To justify an automatic hot-forging operation, however, large quantities of a given product must be required. A single production line may well require an initial investment in excess of $10 million.

FIGURE 16-19 (Left) Roll-forging machine in operation. (Right) Rolls from a roll-forging machine and the various stages in roll forging a part. *(Courtesy of Ajax Manufacturing Company, Euclid, OH.)*

ROLL FORGING

In *roll forging*, round or flat bar stock is reduced in thickness and increased in length to produce such products as axles, tapered levers, and leaf springs. As illustrated in Figure 16-19, roll forging is performed on machines that have two cylindrical or semi-cylindrical rolls, each containing one or more shaped grooves. A heated bar is inserted between the rolls. When the bar encounters a stop, the rolls rotate, and the bar is progressively shaped as it is rolled out toward the operator. The piece is then transferred to the next set of grooves (or rotated and reinserted in the same groove), and the process repeats until the desired size and shape are produced. Figure 16-19 also shows a set of rolls and the product formed by each set of grooves. Figure 16-20 shows the cross section of one set of grooves and a piece being formed. In most cases there is no flash, and the oriented structure imparts favorable forging-type properties.

SWAGING

Swaging (also known as rotary swaging or radial forging) uses external hammering to reduce the diameter or produce tapers or points on round bars or tubes. Figure 16-21 shows a typical swaging machine, and Figure 16-22 shows a schematic of its internal components. The dies, located in the center of the apparatus, consist of two blocks of

Stock

Roll

Guide flange

FIGURE 16-20 Schematic of the roll-forging process showing the two shaped rolls and the stock being formed. *(Courtesy of Forging Industry Association, Cleveland, OH.)*

FIGURE 16-21 Tube being reduced in a rotary swaging machine. *(Courtesy of the Timkin Company, Canton, OH.)*

FIGURE 16-22 Basic components and motions of a rotary swaging machine. (*Note:* The cover plate has been removed to reveal the interior workings.) *(Courtesy of the Timkin Company, Canton, OH.)*

hardened tool steel. They combine to form a central hole that generally has a conical input transitioning to a cylinder. An external motor drives a large, massive flywheel, which is connected to the central spindle of the machine. High-speed rotation of the central unit generates centrifugal force, which causes the matching die segments and backing blocks to separate. As the spindle rotates, the backing blocks are driven into opposing rollers that have been mounted in a massive machine housing. To pass beneath the rollers, the backer blocks must squeeze the dies tightly together. Once the assembly clears the rollers, the dies once again separate and the cycle repeats, generating between 1000 and 3000 blows per minute.

With the machine in motion, the operator simply inserts a rod or tube between the dies and advances it during the periods of die separation. Because the dies rotate, the repeated blows are delivered from various angles, reducing the diameter and increasing the length. Since the rotating spindle is usually hollow, the workpiece can be passed completely through the machine or withdrawn after a preset length has been reduced.

Swaging operations can also be used to form tubular products with internal cavities of constant cross section. A shaped mandrel is inserted into a thick-walled tube (or hollow-end workpiece), and the metal is collapsed around it to simultaneously shape and size both the interior and exterior of the product. Swaging over a mandrel can be

FIGURE 16-23 A variety of swaged parts, some with internal details. *(Courtesy of Cincinnati Milacron, Inc., Cincinnati, OH.)*

used to form parts with internal gears, splines, recesses, or sockets. Figure 16-23 shows a variety of swaged products, many of which contain shaped holes.

The term *swaging* has also been applied to a process where material is forced into a confining die to reduce its diameter. This process is usually performed on heated material. Figure 16-24 shows a hot-swaging sequence being used to form the end of a pressurized gas cylinder.

NET-SHAPE AND NEAR-NET-SHAPE FORGING

As much as 80% of the cost of a forged gear can be incurred during the machining operations that follow forging, and a finished aerospace wing spar may contain as little as 4% of the original billet (the remaining 96% being lost as scrap in the forging and subsequent machining operations). To minimize both the expense and waste, considerable effort has been made to develop processes that can form parts close enough to final dimensions that little or no final machining is required. These are known as *net-shape*, or *near-net-shape*, operations and may also be referred to as *precision forging*. Cost savings often result from the reduction or elimination of secondary machining (and the associated handling, positioning, and fixturing), the companion reduction in scrap, and an overall decrease in the amount of energy required to produce the product.

Precision or near-net-shape forgings can now be produced with draft angles of less than 1° (or even zero draft). Complex shapes can be forged with such close tolerances that little or no finish machining is required. Since the design and implementation of net-shape processing can be rather expensive, application is usually reserved for parts where a significant cost reduction can be achieved.

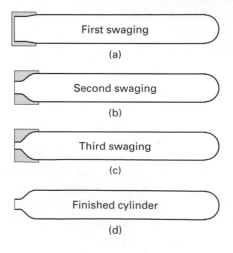

First swaging
(a)

Second swaging
(b)

Third swaging
(c)

Finished cylinder
(d)

FIGURE 16-24 Steps in swaging a tube to form the neck of a gas cylinder. *(Courtesy of United States Steel Corp., Pittsburgh, PA.)*

FIGURE 16-25 Direct extrusion schematic showing the various equipment components. *(Courtesy of Danieli Wean United, Cranberry Township, PA.)*

Direct extrusion

1 Extrusion
2 Die backer
3 Die
4 Billet

5 Dummy block
6 Pressing ram
7 Container liner
8 Container body

■ 16.6 EXTRUSION

In the *extrusion* process, metal is compressed and forced to flow through a suitably shaped die to form a product with reduced but constant cross section. Although extrusion may be performed either hot or cold, hot extrusion is commonly employed for many metals to reduce the forces required, eliminate cold-working effects, and reduce directional properties. Basically, the extrusion process is like squeezing toothpaste out of a tube. In the case of metals, a common arrangement is to have a heated billet placed inside a confining chamber. A ram advances from one end, causing the billet to first upset and conform to the confining chamber. As the ram continues to advance, the pressure builds until the material flows plastically through the die and *extrudes*, as depicted in Figure 16-25. The stress state within the material is one of triaxial compression.

Aluminum, magnesium, copper, lead, and alloys of these metals are commonly extruded, taking advantage of the relatively low yield strengths and low hot-working temperatures. Steels, stainless steels, nickel-based alloys, and titanium are far more difficult to extrude. Their yield strengths are high, and the metals tend to weld to the walls of the die and confining chamber under the required conditions of temperature and pressure. With the development and use of phosphate-based and molten glass lubricants, however, hot extrusions can be routinely produced from these high-strength, high-temperature metals. These lubricants are able to withstand the required temperatures and adhere to the billet, flowing and thinning in a way that prevents metal-to-metal contact throughout the process.

As shown in the left-hand segment of Figure 16-26, almost any cross-sectional shape can be extruded from the nonferrous metals. Size limitations are few because

FIGURE 16-26 Typical shapes produced by extrusion. (Left) Aluminum products. *(Courtesy of Aluminum Company of America, Pittsburgh, PA.)* (Right) Steel products. *(Courtesy of Allegheny Ludlum Steel Corporation, Pittsburgh, PA.)*

FIGURE 16-27 Direct and indirect extrusion. In direct extrusion, the ram and billet both move and friction between the billet and the chamber opposes forward motion. For indirect extrusion, the billet is stationary. There is no billet–chamber friction, since there is no relative motion.

presses are now available that can extrude any shape that can be enclosed within a circle of 75-cm (30-in.) diameter. In the case of steels and the other high-strength metals, the shapes and sizes are a bit more limited, but, as the right-hand segment of Figure 16-26 shows, considerable freedom still exists.

Extrusion has a number of attractive features. Many shapes can be produced as extrusions that are not possible by rolling, such as ones containing reentrant angles or longitudinal holes. No draft is required, so extrusions can offer savings in both metal and weight. Since the deformation is compressive, the amount of reduction in a single step is limited only by the capacity of the equipment. Billet-to-product cross-sectional area ratios can be in excess of 100-to-1 for the weaker metals. In addition, extrusion dies can be relatively inexpensive, and one die may be all that is required to produce a given product. Conversion from one product to another requires only a single die change, so small quantities of a desired shape can be produced economically. The major limitation of the process is the requirement that the cross section be uniform for the entire length of the product.

Extruded products have good surface finish and dimensional precision. For most shapes, tolerances of 0.003 cm/cm or in./in. with a minimum of 0.075 mm (0.003 in.) are easily attainable. Grain structure is typical of other hot-worked metals, but strong directional properties (longitudinal versus transverse) are usually observed. Standard product lengths are about 6 to 7 m (20 to 24 ft), but lengths in excess of 12 m (40 ft) have been produced. Since little scrap is generated, billet-to-product yields are rather high.

EXTRUSION METHODS

Extrusions can be produced by various techniques and equipment configurations. Hot extrusion is usually done by either the direct or indirect method, both of which are illustrated in Figure 16-27. In *direct extrusion*, a solid ram drives the entire billet to and through a stationary die and must provide additional power to overcome the frictional resistance between the surface of the moving billet and the confining chamber. With *indirect extrusion*, also called reverse, backward, or inverted extrusion, a hollow ram pushes the die back through a stationary, confined billet. Since there is no relative motion, friction between the billet and the chamber is eliminated. The required force is lower, and longer billets can be used with no penalty in power or efficiency.

Figure 16-28 shows the ram force versus ram position curves for both direct and indirect extrusion. The areas below the lines have units of newton-meters or foot-pounds, and are therefore proportional to the work required to produce the part. The area between the two curves is the work required to overcome the billet–chamber friction during direct extrusion, an amount that can be saved by converting to indirect extrusion. Unfortunately, the added complexity of the indirect process (applying force through a hollow ram, extracting the product through the hollow, and removing residual billet material at the end of the stroke) serves to increase the purchase price and maintenance cost of the required equipment.

With either process, the speeds of hot extrusion are usually rather fast, so as to minimize the cooling of the billet within the chamber. Extruded products can emerge at rates up to 300 m/min (1000 ft/min). The extrusion speed may be restricted, however, by

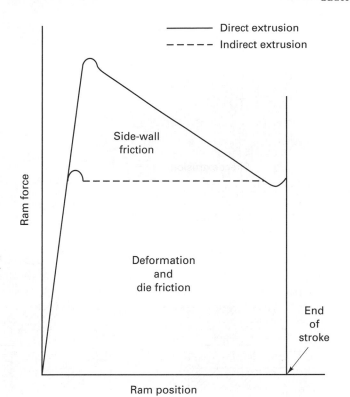

- Direct extrusion
- - - - - Indirect extrusion

Ram force

Side-wall friction

Deformation and die friction

End of stroke

Ram position

FIGURE 16-28 Diagram of the ram force versus ram position for both direct and indirect extrusion of the same product. The area under the curve corresponds to the amount of work (force × distance) performed. The difference between the two curves is attributed to billet–chamber friction.

the large amounts of heat that are generated by the massive deformation and the associated rise in temperature. Sensors are often used to monitor the temperature of the emerging product and feed this information back to a control system. For materials whose properties are not sensitive to strain rate, ram speed may be maintained at the highest level that will keep the product temperature below some predetermined value.

Lubrication is another important area of concern. If the reduction ratio (cross section of billet to cross section of product) is 100, the product will be 100 times longer than the starting billet. If the product has a complex cross section, its perimeter can be significantly greater than a circle of equivalent area. Since the surface area of the product is the length times the perimeter, this value can be more than an order of magnitude greater than the surface area of the original billet. A lubricant that is applied to the starting piece must thin considerably as the material passes through the die and is converted to product. An acceptable lubricant is expected to reduce friction and act as a barrier to heat transfer at all stages of the process.

METAL FLOW IN EXTRUSION

The flow of metal during extrusion is often complex, and some care must be exercised to prevent surface cracks, interior cracks, and other flow-related defects. Metal near the center of the chamber can often pass through the die with little distortion, while metal near the surface undergoes considerable shearing. In direct extrusion, friction between the forward-moving billet and both the stationary chamber and die serves to further impede surface flow. The result is often a deformation pattern similar to the one shown in Figure 16-29. If the surface regions of the billet undergo excessive cooling, surface deformation is further impeded, often leading to the formation of surface cracks. If quality is to be maintained, process control must be exercised in the areas of design, lubrication, extrusion speed, and temperature.

EXTRUSION OF HOLLOW SHAPES

Hollow shapes, and shapes with multiple longitudinal cavities, can be extruded by several methods. For tubular products, the stationary or moving *mandrel* processes of Figure 16-30 are quite common. The die forms the outer profile, while the mandrel shapes and sizes the interior.

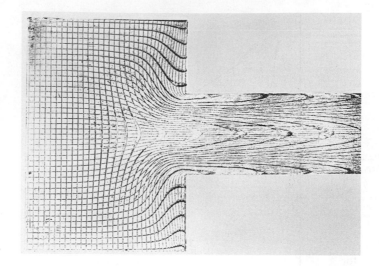

FIGURE 16-29 Grid pattern showing the metal flow in a direct extrusion. The billet was sectioned and the grid pattern was engraved prior to extrusion.

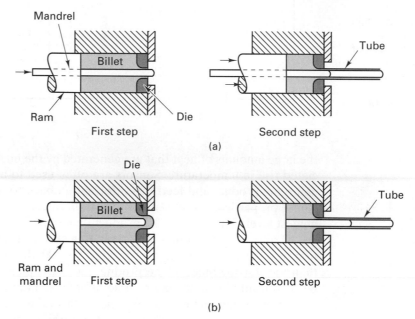

FIGURE 16-30 Two methods of extruding hollow shapes using internal mandrels. In part (a) the mandrel and ram have independent motions; in part (b) they move as a single unit.

For products with multiple or more complex cavities, a *spider-mandrel* die (also known as a porthole, bridge, or torpedo die) may be required. As illustrated in Figure 16-31, metal flows around the arms of a "spider," and a further reduction then forces the material back together. Since the metal is never exposed to contamination, perfect welds result. Unfortunately, lubricants cannot be used since they will contaminate the surfaces to be welded. The process is therefore limited to materials that can be extruded without lubrication and can also be easily pressure welded.

Since additional tooling is required, hollow extrusions will obviously cost more than solid ones, but a wide variety of continuous cross-section shapes can be produced that cannot be made economically by any other process.

HYDROSTATIC EXTRUSION

Another type of extrusion, known as *hydrostatic extrusion*, is illustrated schematically in Figure 16-32. Here high-pressure fluid surrounds the workpiece and applies the force necessary to extrude it through the die. The product emerges into either atmospheric pressure or a lower-pressure fluid-filled chamber. The process resembles direct extrusion, but the pressurized fluid surrounding the billet prevents any upsetting. Since the billet does not come into contact with the surrounding chamber, billet–chamber friction is eliminated. In addition, the pressurized fluid can also emerge between the billet and the die, acting in the form of a lubricant.

FIGURE 16-31 Hot extrusion of a hollow shape using a spider-mandrel die. Note the four arms connecting the external die and the central mandrel.

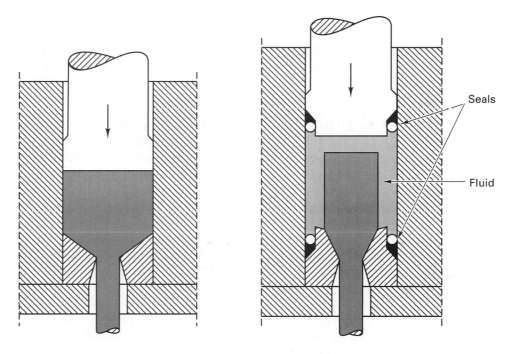

FIGURE 16-32 Comparison of conventional (left) and hydrostatic (right) extrusion. Note the addition of the pressurizing fluid and the O-ring and miter-ring seals on both the die and ram.

While the efficiency can be significantly greater than most other extrusion processes, there are problems related to the fluid and the associated high pressures (which typically range between 900 and 1700 MPa or 125 to 250 ksi). Temperatures are limited since the fluid acts as a heat sink, and many of the pressurizing fluids (typically light hydrocarbons and oils) burn or decompose at moderately low temperatures. Seals must be designed to contain the pressurized fluid without leaking, and measures must be taken to prevent the complete ejection of the product, often referred to as *blowout*. Because of these features, hydrostatic extrusion is usually employed only where the process offers unique advantages that cannot be duplicated by the more conventional methods.

Pressure-to-pressure extrusion is one of the unique capabilities. In this variant, the product emerges from one pressurized chamber into a second high-pressure chamber. In effect, the metal deformation is performed in a highly compressed environment. Crack formation begins with void formation, void growth, and void coalescence. Since voids are suppressed in a compressed environment, the result is a phenomenon known as *pressure-induced ductility*. Relatively brittle materials such as molybdenum, beryllium, tungsten, and various intermetallic compounds can be plastically deformed without fracture, and materials with limited ductility become highly formable. Products can be made that could not be otherwise produced, and materials can be considered that would otherwise have been rejected because of their limited ductility at room temperature and atmospheric pressure.

FIGURE 16-33   Cross-sectional schematic of the Conform continuous extrusion process. The material upsets at the abutment and extrudes. Section *x—x* shows the material in the shoe.

CONTINUOUS EXTRUSION

Conventional extrusion is a discontinuous process, converting finite-length billets into finite-length products. If the pushing force could be applied to the periphery of the feed-stock, rather than the back, continuous feedstock could be converted into continuous product, and the process could become one of *continuous extrusion*. The first continuous extrusion of solid metal feedstock was performed in 1970. Since then, a number of techniques have been proposed with varying degrees of success. In terms of commercial application, the most significant is probably the *Conform process*, illustrated schematically in Figure 16-33. Continuous feedstock is inserted into a grooved wheel and is driven by surface friction into a chamber created by a mating die segment. Upon impacting a protruding abutment, the material upsets to conform to the chamber, and the increased wall contact further increases the driving friction. Upsetting continues until the pressure reaches a value sufficient to extrude the material through a die opening that has been provided in either the shoe or abutment. At this point, the rate of material entering the machine equals the rate of product emerging, and a steady-state continuous process is established.

Since surface friction is the propulsion force, the feedstock can take a variety of forms, including solid rod, metal powder, punchouts from other forming operations, or chips from machining. Metallic and nonmetallic powders can be intimately mixed and co-extruded. Rapidly solidified material can be extruded without exposure to the elevated temperatures that would harm the properties. Polymeric materials and even fiber-reinforced plastics have been successfully extruded. The most common feed, however, is coiled aluminum or copper rod.

Continuous extrusion complements and competes with wire drawing and shape rolling as a means of producing nonferrous products with small, but uniform, cross sections. It is particularly attractive for complex profiles and cross sections that contain one or more holes. Since extrusion operations can perform massive reductions through a single die, one Conform operation can produce an amount of deformation equivalent to 10 conventional drawing or cold-rolling passes. In addition, sufficient heat can be generated by the deformation that the product will emerge in an annealed condition, ready for further processing without intermediate heat treatment.

■ 16.7 WIRE, ROD, AND TUBE DRAWING

Wire-, rod-, and tube-drawing operations reduce the cross section of a material by pulling it through a die. In many ways, the processes are similar to extrusion, but the applied stresses are now tensile, pulling on the product rather than pushing on the workpiece. *Rod* or *bar*

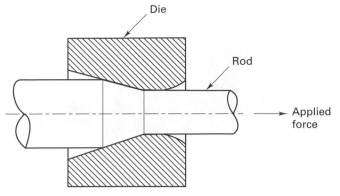

FIGURE 16-34 Schematic diagram of the rod- or bar-drawing process.

FIGURE 16-35 Diagram of a chain-driven multiple-die-draw bench used to produce finite lengths of straight rod or tube. *(Courtesy of Danieli Wean United, Cranberry Township, PA.)*

drawing, illustrated schematically in Figure 16-34, is probably the simplest of these operations. One end of a rod is reduced or pointed, so that it can pass through a die of somewhat smaller cross section. The protruding material is then placed in grips and pulled in tension, drawing the remainder of the rod through the die. The rods reduce in section, elongate, and become stronger (strain harden). Since the product cannot be readily bent or coiled, straight-pull *draw benches* are generally employed with finite-length feedstock. Hydraulic cylinders can be used to provide the pull for short-length products, while chain drives, as depicted in Figure 16-35, can be used to draw products up to 30 m (100 ft) in length.

The reduction in area is usually restricted to between 20 and 50%, since higher values require higher pulling forces that may exceed the tensile strength of the reduced product. To produce a desired size or shape, multiple draws may be required through a series of progressively smaller dies. Intermediate anneals may also be required to restore ductility and enable further deformation.

Tube drawing can be used to produce high-quality tubing where the product requires the smooth surfaces, thin walls, accurate dimensions, and added strength (from the strain hardening) that are characteristic of cold forming. Internal mandrels are often used to control the inside diameter of tubes, which range from about 12 to 250 mm (0.5 to 10 in.) in diameter. As shown in Figure 16-36, these mandrels are inserted through the incoming stock and are held in place during the drawing operation.

FIGURE 16-36 Cold-drawing smaller tubing from larger tubing. The die sets the outer dimension while the stationary mandrel sizes the inner diameter.

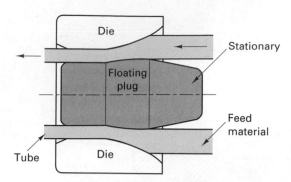

FIGURE 16-37 Tube drawing with a floating plug.

Thick-walled tubes and those less than 12 mm (0.5 in.) in diameter are often drawn without a mandrel in a process known as *tube sinking*. Precise control of the inner diameter is sacrificed in exchange for process simplicity and the ability to draw long lengths of product. If a controlled internal diameter must be produced in a long-length product, it is possible to utilize a *floating plug*, like the one shown in Figure 16-37. This plug must be designed for the specific conditions of material, reduction, and friction. If the friction on the plug surface is too great, the flowing tube will pull it too far forward, pinching off or fracturing the tube wall. If the amount of friction is insufficient, the plug will chatter or vibrate within the tube and will not assume a stable position. If properly designed, the floating plug will assume a stable position within the die, and size the internal diameter, while the external die shapes and sizes the outside of the tube.

The drawing of bar stock can also be used to make products with shaped cross sections. By using cold drawing instead of hot extrusion, the material emerges with precise dimensions and excellent surface finish. Inexpensive materials strengthened by strain hardening can often replace stronger alloys or ones that would require additional heat treatment. Small parts with complex but constant cross sections can be economically made by sectioning long lengths of cold-drawn shaped bars to produce the individual products. Steels, copper alloys, and aluminum alloys have all been cold drawn into shaped bars.

Wire drawing is essentially the same process as bar drawing except that it involves smaller-diameter material. Because the material can now be coiled, the process can be conducted in a somewhat continuous manner on rotating draw blocks, like the one illustrated schematically in Figure 16-38. Wire drawing usually begins with large coils of hot-rolled rod stock approximately 9 mm ($\frac{3}{8}$ in.) in diameter. After descaling or other forms of surface preparation, one end of the coil is pointed, fed through a die, gripped, and the drawing process begins.

Wire dies generally have a configuration similar to the one shown in Figure 16-39. The contact regions are usually made of wear-resistant tungsten carbide or polycrystalline, manufactured diamond. Single-crystal diamonds can be used for the drawing of very fine wire, and wear-resistant and low-friction coatings can be applied to the various

FIGURE 16-38 Schematic of wire drawing with a rotating draw block. The rotating motor on the draw block provides a continuous pull on the incoming wire.

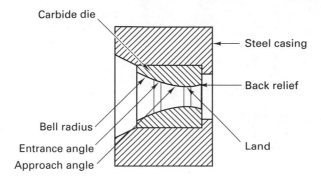

FIGURE 16-39 Cross section through a typical carbide wire-drawing die showing the characteristic regions of the contour.

FIGURE 16-40 Schematic of a multistation synchronized wire-drawing machine. To prevent accumulation or breakage, it is necessary to ensure that the same volume of material passes through each station in a given time. The loops around the sheaves between the stations use wire tensions and feedback electronics to provide the necessary speed control.

die material substrates. Lubrication boxes often precede the individual dies to help reduce friction drag and prevent wear of the dies.

Because the tensile load is applied to the already reduced product, the amount of reduction is severely limited. Multiple draws are usually required to affect any significant change in size. To convert hot-rolled rod stock to the fine wire that is used in household telephone lines requires passes through as many as 20 or 30 individual dies. To minimize handling and labor, these operations are usually performed on tandem machines, like the one shown schematically in Figure 16-40. Between 3 and 12 dies are mounted in a single machine, and the material moves continuously from one station to another in a synchronized manner that prevents any localized accumulation or tension that might induce fracture.

After passing through all the dies in a tandem machine, the material usually requires an intermediate anneal before it can be subjected to further deformation. By controlling the placement of the last anneal so the final product has a selected amount of cold work, wires can be made with a wide range of strengths (or tempers). When maximum ductility and conductivity are desired, the wire should be annealed in controlled-atmosphere furnaces after the final draw.

■ 16.8 COLD FORMING, COLD FORGING, AND IMPACT EXTRUSION

Large quantities of products are now being made by *cold forming*, a family of processes in which slugs of material are squeezed into or extruded from shaped die cavities to produce finished parts of precise shape and size. Workpiece temperature varies from room temperature to several hundred degrees Fahrenheit.

Cold heading is a form of the previously discussed upset forging. As illustrated in Figure 16-41, it is used for making enlarged sections on the ends of rod or wire, such as the heads of nails, bolts, rivets, or other fasteners. Two variations of the process are

FIGURE 16-41 Typical steps in a shearing and cold-heading operation.

common. In the first, a piece of rod is first sheared to a preset length and then transferred to a holder–ejector assembly. Heading punches then strike one or more blows on the exposed end to perform the upsetting. If intermediate shapes are required, the piece is transferred from station to station, or the various heading punches sequentially rotate into position. When the heading is completed, the ejector stop advances and expels the product. In the second variation, a continuous rod (or wire) is fed forward to produce a preset extension, clamped, and the head is formed. The rod is then advanced to a second preset length and sheared, and the cycle repeats. This procedure is particularly attractive for producing nails, since the point can be formed in the shearing or cutoff operation. Enlarged sections can also be produced at locations other than the ends of a rod or wire, in the manner illustrated in Figure 16-42.

While cold heading generally produces symmetrical parts, the expanded regions can also be square, hexagonal, or even offset. Production speeds tend to vary with the diameter of the incoming material. When the blanks are less than 6 mm (0.25 in.) in diameter, speeds of 400 to 600 pieces per minute are typical. For larger diameters, the speeds may reduce to 40 to 100 pieces per minute. Alloys of aluminum and copper have excellent formability, while mild steel and stainless steel are rated fair to good. Alloy steels are a bit more difficult because of their higher strength and lower ductility.

A variety of extrusion operations, commonly called *impact extrusion*, can also be incorporated into cold forming. In these processes, a metal slug of predetermined size is positioned in a die cavity, where it is struck a single blow by a rapidly moving punch. The metal may flow forward through the die, backward around the punch, or in a combination mode. Figure 16-43 illustrates the *forward* and *backward* variations, using both open and closed dies. In forward extrusion, the diameter is decreased while the length increases. Backward extrusion shapes hollow parts with a solid bottom. The punch controls the inside shape, while the die shapes the exterior. The wall thickness is determined by the clearance between the punch and die, and the bottom thickness is set by the stop position of the punch. Figure 16-44 provides additional schematics of forward, backward, and combination impacting. Typical production speeds range from 20 to 60 strokes per minute.

FIGURE 16-42 Method of upsetting the center portion of a rod. The stock is supported in both dies during upsetting.

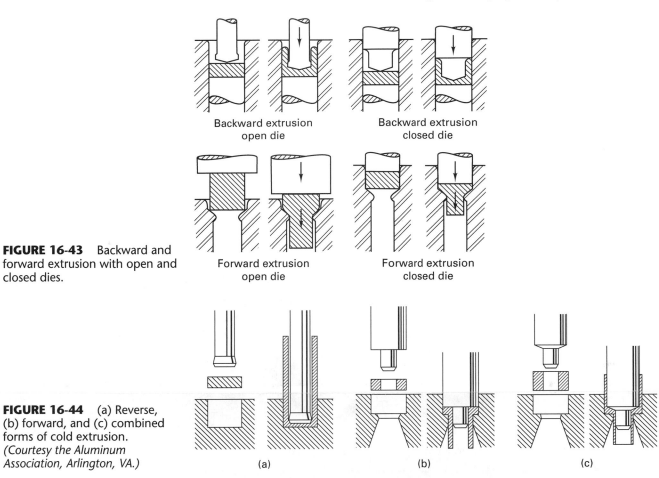

FIGURE 16-43 Backward and forward extrusion with open and closed dies.

FIGURE 16-44 (a) Reverse, (b) forward, and (c) combined forms of cold extrusion. *(Courtesy the Aluminum Association, Arlington, VA.)*

FIGURE 16-45 Steps in the forming of a bolt by cold extrusion, cold heading, and thread rolling. *(Courtesy of National Machinery Co., Tiffin, OH.)*

The impact extrusion processes were first used to shape low-strength metals such as lead, tin, zinc, and aluminum into products such as collapsible tubes for toothpaste, medications, and other creams; small "cans" for shielding electronic components; zinc cases for flashlight batteries; and larger cans for food and beverages. In recent years, impact extrusion has expanded to the forming of mild steel parts, where it is often used in combination with cold heading, as in the example of Figure 16-45. When heading alone is used, there is a definite limit to the ratio of the head and stock diameters (as presented in Figure 16-17 and related discussion). The combination of forward extrusion and cold heading overcomes this limitation by using an intermediate starting diameter. The shank portion is reduced by forward extrusion while upsetting is used to increase the diameter of the head.

By using various types of dies and combining high-speed operations such as heading, upsetting, extrusion, piercing, bending, coining, thread rolling, and knurling, a wide

FIGURE 16-46 Cold-forming sequence involving cutoff, squaring, two extrusions, an upset, and a trimming operation. Also shown are the finished part and the trimmed scrap. *(Courtesy of National Machinery Co., Tiffin, OH.)*

variety of relatively complex parts can be cold formed to close tolerances. Figure 16-46 illustrates an operation that incorporates two extrusions, a central upset, and a final operation to shape and trim that upset. Figure 16-47 presents an array of upset and extruded products. The larger parts are generally hot formed and machined, while the smaller ones are cold formed.

Since cold forming is a chipless manufacturing process, producing parts by deformation that would otherwise be machined from bar stock or hot forgings, the material

FIGURE 16-47 Typical parts made by upsetting and related operations. *(Courtesy of National Machinery Co., Tiffin, OH.)*

Cutting (74% waste)   Cold forming (6% waste)

is used more efficiently and waste is reduced. Figure 16-48 compares the manufacture
of a spark plug body by machining from hexagonal bar stock with manufacture by cold
forming. Material is saved, machining time and cost are reduced, and the product is
stronger, due to cold work, and tougher, as illustrated by the flow lines revealed in
Figure 16-49. By converting from screw machining to cold forming, a manufacturer of
cruise-control housings was able to reduce material usage by 65%, while simultaneously increasing production rate by a factor of 5.

   While cold forming is generally associated with the manufacture of small, symmetrical parts from the weaker nonferrous metals, the process is now used extensively on steel and stainless steel, with parts up to 45 kg (100 lb) in weight and 18 cm (7 in.)
in diameter. At the small end of the scale, microformers are now cold forming
extremely small electronic components with dimensional accuracies within 0.005 mm
(0.0002 in.).

   Cold-formed shapes are usually axisymmetric or those with relatively small
departures from symmetry. Production rates are high; dimensional tolerances and surface
finish are excellent; and there are no draft angles, parting lines, or flash to trim off.
There is almost no material waste, and a considerable amount of machining can often
be eliminated when used in place of alternate processes. Strain hardening can provide
additional strength (up to 70% stronger than machined parts), and favorable grain flow
can enhance toughness and fatigue life. As a result, parts can often be made smaller or
thinner, or from lower-cost materials. Unfortunately, the cost of the required tooling,
coupled with the high production speed, generally requires large-volume production,
typically in excess of 50,000 parts per year.

■ 16.9   PIERCING

Thick-walled *seamless tubing* can be made by *rotary piercing*, a process illustrated
in Figure 16-50. A heated billet is fed longitudinally into the gap between two large,
convex-tapered rolls. These rolls are rotated in the same direction, but the axes of
the rolls are offset from the axis of the billet by about 6°, one to the right and the
other to the left. The clearance between the rolls is preset at a value less than the
diameter of the incoming billet. As the billet is caught by the rolls, it is simultaneously
rotated and driven forward. The reduced clearance between the rolls forces the
billet to deform into a rotating ellipse. As shown in the right-hand segment of
Figure 16-50, rotation of the elliptical section causes the metal to shear about the
major axis. A crack tends to form down the center axis of the billet, and the cracked
material is then forced over a pointed mandrel that enlarges and shapes the opening
to create a seamless tube. The result is a short length of thick-walled seamless tubing,
which can then be passed through sizing rolls to reduce the diameter and/or wall
thickness. Seamless tubes can also be expanded in diameter by passing them
over an enlarging mandrel. As the diameter and circumference increase, the walls
correspondingly thin.

FIGURE 16-50 (Left) Principle of the Mannesmann process of producing seamless tubing. *(Courtesy of American Brass Company, Cleveland, OH.)* (Right) Mechanism of crack formation in the Mannesmann process.

The *Mannesmann mills* commonly used in hot piercing can be used to produce tubing up to 300 mm (12 in.) in diameter. Larger-diameter tubes can be produced on *Stiefel mills*, which use the same principle but replace the convex rolls of the Mannesmann mill with larger-diameter conical disks.

■ 16.10 OTHER SQUEEZING PROCESSES

ROLL EXTRUSION

Thin-walled cylinders can be produced from thicker-wall material by the *roll-extrusion* process. In the variant depicted in Figure 16-51a, internal rollers expand the internal diameter as they squeeze the rotating material against an external confining ring. The tube elongates as the wall thickness is reduced. In Figure 16-51b, the internal diameter is maintained as external rollers squeeze the material against a rotating mandrel. Although the process has been used to produce cylinders from 2 cm to 4 m (0.75 to 156 in.) in diameter, most products have diameters between 7.5 and 50 cm (3 and 20 in.).

SIZING

Sizing involves squeezing all or selected regions of forgings, ductile castings, or powder metallurgy products to achieve a prescribed thickness or enhanced dimensional precision. By incorporating sizing, designers can make the initial tolerances of a part more liberal, enabling the use of less costly production methods. Those dimensions that must be precise are then set by one or more sizing operations that are usually performed on simple, mechanically driven presses.

(a)

(b)

FIGURE 16-51 The roll-extrusion process: (a) with internal rollers expanding the inner diameter; (b) with external rollers reducing the outer diameter.

FIGURE 16-52   Joining components by riveting.

RIVETING

In *riveting*, an expanded head is formed on the shank end of a fastener to permanently join sheets or plates of material. Although riveting is usually done hot in structural applications, it is almost always done cold in manufacturing. Where there is access to both sides of the work, the method illustrated in Figure 16-52 is commonly used. The shaped punch may be driven by a press or contained in a special, hand-held riveting hammer. When a press is used, the rivet is usually headed in a single squeezing action, although the heading punch may also rotate so as to shape the head in a progressive manner, an approach known as *orbital forming*. Special riveting machines, like those used in aircraft assembly, can punch the hole, place the rivet in position, and perform the heading operation—all in about 1 second.

It is often desirable to use riveting in situations where there is access to only one side of the assembly. Figure 16-53 shows two types of special rivets that can be used for one-side-access applications. The shank on the "blind" side of an *explosive rivet* expands upon detonation to form a retaining head when a heated tool is touched against the exposed segment. In the pull type, or *pop-rivet*, a pull-up pin is used to expand a tubular shank. After performing its function, the pull pin breaks or is cut off flush with the head.

FIGURE 16-53   Rivets for use in "blind" riveting: (left) explosive type; (right) shank-type pull-up. *(Courtesy of Alcoa Fastening Systems, Pittsburgh, PA.)*

FIGURE 16-54   Permanently attaching a shaft to a plate by staking.

STAKING

Staking is a method of permanently joining parts together when a segment of one part protrudes through a hole in the other. As shown in Figure 16-54, a shaped punch is driven into the exposed end of the protruding piece. The deformation causes radial expansion, mechanically locking the two pieces together. Because the tooling is simple and the operation can be completed with a single stroke of a press, staking is a convenient and economical method of fastening when permanence is desired and the appearance of the punch mark is not objectionable. Figure 16-54 includes some of the decorative punch designs that are commonly used.

FIGURE 16-55   The coining process.

COINING

The term *coining* refers to the cold squeezing of metal while all of the surfaces are confined within a set of dies. The process, illustrated schematically in Figure 16-55, is used to produce coins, medals, and other products where exact size and fine detail are required and where thickness varies about a well-defined average. Because of the total confinement (there is no possibility for excess metal to escape from the die), the input material must be accurately sized to avoid breakage of the dies or press. Coining pressures may be as high as 1400 MPa or 200,000 psi.

FIGURE 16-56 Hubbing a die block in a hydraulic press. Inset shows close-up of the hardened hub and the impression in the die block. The die block is contained in a reinforcing ring. The upper surface of the die block is then machined flat to remove the bulged metal.

HUBBING

Hubbing[1] is a cold-working process that is used to plastically form recessed cavities in a workpiece. As shown in Figure 16-56, a male hub (or master) is made with the reverse profile of the desired cavity. After hardening, the hub is pressed into an annealed block (usually by a hydraulic press) until the desired impression is produced. (*Note*: Production of the cavity can often be aided by machining away some of the metal in regions where large amounts of material would be displaced.) The hub is withdrawn, and the displaced metal is removed by a facing-type machining cut. The workpiece, which now contains the desired cavity, is then hardened by heat treatment.

Hubbing is often more economical than die sinking (machining the cavity), especially when multiple impressions are to be produced. One hub can be used to form a number of identical cavities, and it is generally easier to machine a male profile (with exposed surfaces) than a female cavity (where you are cutting in a hole).

■ 16.11 SURFACE IMPROVEMENT BY DEFORMATION PROCESSING

Deformation processes can also be used to improve or alter the surfaces of metal products. *Peening* is the mechanical working of surfaces by repeated blows of impelled shot or a round-nose tool. The highly localized impacts flatten and broaden the metal surface, but the underlying material restricts spread, resulting in a surface with residual compression. Since the net loading on a material surface is the applied load minus the residual compression, peening tends to enhance the fracture resistance and fatigue life of tensile-loaded components. For this reason, shot impellers are frequently used to peen shafting, crankshafts, connecting rods, gear teeth, and other cyclic-loaded components.

Manual or pneumatic hammers are frequently used to peen the surfaces of metal weldments. Solidification shrinkage and thermal contraction produce surfaces with residual tension. Peening can reduce or cancel this effect, thereby reducing associated distortion and preventing cracking.

Burnishing involves rubbing a smooth, hard object (under considerable pressure) over the minute surface irregularities that are produced during machining or shearing. The edges of sheet metal stampings can be burnished by pushing the stamped parts through a slightly tapered die having its entrance end a little larger than the workpiece and its exit slightly smaller. As the part rubs along the sides of the die, the pressure is sufficient to smooth the slightly rough edges that are characteristic of a blanking operation (see Figure 18-2).

---

[1] This process should not be confused with "hobbing," a machining process used for cutting gears.

FIGURE 16-57 Tool for roller burnishing. The burnishing rollers move outward by means of a taper. *(Courtesy of Madison Industries, Inc., Sumter, SC.)*

Roller burnishing, illustrated in Figure 16-57, can be used to improve the size and finish of internal and external cylindrical and conical surfaces. The hardened rolls of a burnishing tool press against the surface and deform the protrusions to a more-nearly-flat geometry. The resulting surfaces possess improved wear and fatigue resistance, since they have been cold worked, and are now in residual compression.

■ Key Words

automatic hot forging	drop-hammer forging	mill scale	sheet
bar	extrusion	near-net-shape forging	sheet forming
billet	finisher impression	net-shape forging	sizing
blocking	finishing temperature	open-die forging	slab
bloom	flash	pack rolling	spider-mandrel die
bulk deformation processes	flashless forging	piercing	staking
closed-die forging	floating plug	pipe	Stiefel mill
cluster mill	forging	plate	strip
coining	hammer	press forging	structural shape
cold forming	hubbing	pressure-induced ductility	swaging
cold heading	hydraulic press	pressure-to-pressure extrusion	thermomechanical processing
Conform process	hydrostatic extrusion	ring rolling	thread rolling
continuous extrusion	impactor	riveting	tube
controlled rolling	impression-die forging	rod	tube drawing
counterblow machine	indirect extrusion	roll extrusion	tube sinking
crowned roll	isothermal forging	roll forging	upset
direct extrusion	mandrel	rolling	upset forging
draw bench	Mannesmann mill	rotary piercing	wire drawing
drawing	mechanical press	seamless tubing	

■ Review Questions

1. Briefly describe the evolution of forming equipment from ancient to modern.
2. What are some of the possible means of classifying metal deformation processes?
3. Why might the method of analysis be different for a process like forging and a process like rolling?
4. What are some of the common terms applied to the various shapes of rolled products?
5. Why are hot-rolled products generally limited to standard shapes and sizes?
6. Why is it undesirable to minimize friction between the workpiece and tooling in a rolling operation?
7. Why is it desirable to have uniform temperature when hot rolling a material?
8. Why is it important to control the finishing temperature of a hot-rolling operation?
9. What are some of the attractive attributes of cold rolling?
10. Discuss the relative advantages and typical uses of two-high rolling mills with large-diameter rolls, three-high mills, and four-high mills.
11. Why is foil almost always rolled on a cluster mill?
12. Why is speed synchronization of the various rolls so vitally important in a continuous or multistand rolling mill?
13. What types of products are produced by ring rolling?

14. Explain how hot-rolled products can have directional properties and residual stresses.
15. Discuss the problems in maintaining uniform thickness in a rolled product and some of the associated defects.
16. Why is a "crowned" roll always designed for a specific operation on a specific material?
17. What are some of the methods of improving the thickness uniformity of rolled products?
18. What is thermomechanical processing, and what are some of its possible advantages?
19. What are some of the types of flow that can occur in forging operations?
20. Why are steam or air hammers more attractive than gravity drop hammers for hammer forging?
21. What are some of the attractive features of computer-controlled forging hammers?
22. What is the difference between open-die and impression-die forging?
23. Why is open-die forging not a practical technique for large-scale production of identical products?
24. What additional controls must be exercised to perform flash-less forging?
25. What is a blocker impression in a forging sequence?
26. What attractive features are offered by counterblow forging equipment, or impactors?
27. For what types of forging products or conditions might a press be preferred over a hammer?
28. Why are heated dies generally employed in hot-press forging operations?
29. Describe some of the primary differences among hammers, mechanical presses, and hydraulic presses.
30. Why are different tolerances usually applied to dimensions contained within a single die cavity and dimensions across the parting plane?
31. What are some of the roles played by lubricants in forging operations?
32. What is upset forging?
33. What are some of the typical products produced by upset-forging operations?
34. What are some of the attractive features of automatic hot forging? What is a major limitation?
35. How does roll forging differ from a conventional rolling operation?
36. What is *swaging*? What kind of products are produced?

37. How can the swaging process impart different sizes and shapes to an interior cavity and the exterior of a product?
38. What are some possible objectives of near-net-shape forging?
39. What metals can be shaped by extrusion?
40. What are some of the attractive features of the extrusion process?
41. What is the primary shape limitation of the extrusion process?
42. What is the primary benefit of indirect extrusion?
43. What property of a lubricant is critical in extrusion that might not be required for processes such as forging?
44. What types of products are made using a spider-mandrel die? Why can lubricants not be used in spider-mandrel extrusion?
45. What are some of the unique capabilities and special limitations of hydrostatic extrusion?
46. What is the unique capability provided by pressure-to-pressure hydrostatic extrusion?
47. How is the feedstock pushed through the die in continuous extrusion processes?
48. Why are rods generally drawn on draw benches, while wire is drawn on draw block machines?
49. What is the difference between tube drawing and tube sinking?
50. For what types of products might a floating plug be employed?
51. Why are multiple passes usually required in wire-drawing operations?
52. What types of products are produced by cold heading?
53. What is impact extrusion and what variations exist?
54. If a product contains a large-diameter head and a small-diameter shank, how can the processes of cold extrusion and cold heading be combined to save metal?
55. What are some of the attractive properties or characteristics of cold-forming operations?
56. How might cold forging be used to substantially reduce material waste?
57. What processes can be used to produce seamless pipe or tubing?
58. What type of products can be made by the roll-extrusion process?
59. What types of rivets can be used when there is access to only one side of a joint?
60. Why might hubbing be an attractive way to produce a number of identical die cavities?
61. How might a peening operation increase the fracture resistance of a product?
62. What is burnishing?

■ Problems

1. Some snack foods, such as rectangular corn chips, are often formed by a rolling-type operation and are subject to the same types of defects common to rolled sheet and strip. Obtain a bag of such a snack and examine the chips to identify examples of rolling-related defects such as those discussed in Section 16.4 and shown in Figure 16-A.

FIGURE 16-A Some typical defects that occur during rolling: wavy edges, edge cracking, and center cracking.

2. Consider the extrusion of a cylindrical billet, and compute the following.
   a. Assume the starting billet to have a length of 0.3 m and a diameter of 15 cm. This is extruded into a cylindrical product that is 3 cm in diameter and 7.5 m long (a reduction ratio of 25). Neglecting the areas on the two ends, compute the ratio between the product surface area (wrap around cylinder) and the surface area of the starting billet.
   b. How would this ratio change if the product were a square with the same cross-sectional area as that of the 3-cm-diameter circle?
   c. Consider a cylinder-to-cylinder extrusion with a reduction ratio of R. Derive a general expression of the relative sur-

face areas of product to billet as a function of R. (*Hint:* Start with a cylinder with length and diameter both equal to 1 unit. Since the final area will be $1/R$ times the original, the final length will be R units and the final diameter will be proportional to $1/\sqrt{R}$).

   d. If the final product had a more complex cross-sectional shape than a cylinder, would the final area be greater than or less than that computed in part c?

   e. Relate your answers above to a consideration of lubrication during large-reduction extrusion operations.

3. The force required to compress a cylindrical solid between flat parallel dies (see Figure 16-10) has been estimated (by a theory of plasticity analysis) to be:

$$\text{force} = \pi R^2 \sigma_o \frac{1 + 2\,mR}{3\,\sqrt{3}\,T}$$

where:

   R = radius of the cylinder
   T = thickness of the cylinder
   σ_o = yield strength of the material
   m = friction factor (between 0 and 1 where 0 is frictionless and 1 is complete sticking)

An engineering student is attempting to impress his date by demonstrating some of the neat aspects of metalforming. He places a shiny penny between the platens of a 60,000-lb-capacity press and proceeds to apply pressure. Assume that the coin has a $\frac{3}{4}$-in. diameter and is $\frac{1}{16}$ in. thick. The yield strength is estimated as 50,000 psi, and since no lubricant is applied, friction is that of complete sticking, or $m = 1.0$.

   a. Compute the force required to induce plastic deformation.

   b. If this force is greater than the capacity of the press (60,000 lb), compute the pressure when the full-capacity force of 60,000 lb is applied.

   c. If the press surfaces are made from thick plates of steel with a yield strength of 120,000 psi, describe the results of the demonstration.

   d. A simple model of forging force uses the equation:

$$\text{force} = K\sigma_o\,A$$

where:

   K = a dimensionless multiplying factor
   σ_o = yield strength of the material
   A = projected area of the forging

K is assigned a value of 3–5 for simple shapes without flash, 5–8 for simple shapes with flash, and 8–12 for complex shapes with flash. Consider the two equations for forging force and discuss their similarities and differences.

4. Mathematical analysis of the rolling of flat strip reveals that the roll-separation force (the squeezing force required to deform the strip) is directly proportional to the term:

$$1 + \frac{K_1 mL}{t_{av}}$$

where:

   K_1 = geometric constant
   m = friction factor
   L = length of contact
   t_{av} = average thickness of the strip in the roll bite

Since L is proportional to the roll radius, R, K_1L can be replaced by K_2R, so the force becomes proportional to the term:

$$1 + \frac{K_2 mR}{t_{av}}$$

If K_2 and m are both positive numbers, how will the roll-separation force change as the strip becomes thinner? How can this effect be minimized? Relate your observations to the types of rolling mills used for various thicknesses of product.

5. In Figure 16-28, the vertical axis is force and the horizontal axis is position. If force is measured in pounds and position in feet, the area under the curves has units of foot-pounds and is a measure of the work performed. If the area under the indirect extrusion curve is proportional to the work required to extrude a product without billet–chamber frictional resistance, how could the relative regions of the direct extrusion curve be used to determine a crude measure of the mechanical "efficiency" of direct extrusion?

6. Compare the forming processes of wire drawing, conventional extrusion, and continuous extrusion with respect to continuity, reduction in area possible in a single operation, possible materials, speeds, typical temperatures, and other important processing variables.

7. Figure 16-B shows the rolling of a wide, thin strip where the width remains constant as thickness is reduced. Material enters the mill at a rate equal to $t_o w_o v_o$ and exits at a rate of $t_f w_o v_f$. Since material cannot be created or destroyed, these rates must be equal, and the w_o terms will cancel. As a result, v_f is equal to $(t_o/t_f)\,v_o$. The material enters at velocity v_o and accelerates to velocity v_f as the material passes through the mill. For stable rolling, the velocity of the roll surface, v_r, which is a constant, must be a value between v_o and v_f. For these conditions, describe the relative sliding between the strip and the rolls as the strip moves through the region of contact.

FIGURE 16-B   Strip rolling where the width of the strip remains unchanged. The lines across the workpiece block the area of contact with the rolls. The top roll has been removed for ease of visualization.

Chapter 16 CASE STUDY

Handle and Body of a Large Ratchet Wrench

Figure 15-A has already presented the handle and body segment of a relatively large ratchet wrench, such as those used with conventional socket sets. The design specifications require a material with a minimum yield strength of 50,000 psi and an elongation of at least 2% in all directions. Additional consideration should be given to weight minimization (because of the relatively large size of the wrench), corrosion resistance (due to storage and use environments), machinability (if finish machining is required), and appearance.

1. Based on the size and shape of the product, describe several methods that could be used to produce the component. For each method, briefly discuss the relative pros and cons.
2. What types of engineering materials might be able to meet the requirements? What would be the pros and cons of each general family?

3. For each of the shape generation methods in question 1, select an appropriate material from the alternatives discussed in question 2, making sure that the process and material are compatible.
4. Which of the combinations do you feel would be the "best" solution to the problem? Why?
5. For this system, outline the specific steps that would be required to produce the part from reasonable starting material.
6. For your proposed solution, would any additional heat treatment or surface treatment be required? If so, what would you recommend?
7. If a variation of this tool were to be marketed as a "safety tool" that could be used in areas of gas leaks where a spark might be fatal, how would you modify your previous recommendations? Discuss briefly.

CHAPTER 17

SHEET-FORMING PROCESSES

■ 17.1 INTRODUCTION

The various classification schemes for metal deformation processes have been presented at the beginning of Chapter 16, with the indication that our text will be grouping by bulk (Chapter 16) and sheet (Chapter 17). Bulk forming uses heavy machinery to apply three-dimensional stresses, and most of the processes are considered to be primary operations. Sheet metal processes, on the other hand, generally involve plane stress loadings and lower forces than bulk forming. Almost all sheet metal forming is considered to be secondary processing.

The classification into bulk and sheet is far from distinct, however. Some processes can be considered as either, depending on the size, shape, or thickness of the workpiece. The bending of rod or bar is often considered to be bulk forming, while the bending of sheet metal is sheet forming. Tube bending can be either, depending on the wall thickness and diameter of the tube. Similar areas of confusion can be found in deep drawing, roll forming, and other processes. The squeezing processes were described in Chapter 16. Presented here will be the processes that involve *shearing, bending*, and *drawing*. Table 17-1 lists some of the processes that fit these categories.

■ 17.2 SHEARING OPERATIONS

Shearing is the mechanical cutting of materials without the formation of chips or the use of burning or melting. It is often used to prepare material for subsequent operations, and its success helps to ensure the accuracy and precision of the finished product. When the two cutting blades are straight, the process is called *shearing*. When the blades are curved, the processes have special names, such as *blanking, piercing, notching*, and *trimming*. In terms of tool design and material behavior, however, all are shearing-type operations.

TABLE 17-1 Classification of the Nonsqueezing Metalforming Operations

Shearing	Bending	Drawing and Stretching
1. Simple shearing	1. Angle bending	1. Spinning
2. Slitting	2. Roll bending	2. Shear forming or flow turning
3. Piercing	3. Draw bending	3. Stretch forming
4. Blanking	4. Compression bending	4. Deep drawing and shallow drawing
5. Fineblanking	5. Press bending	5. Rubber-tool forming
6. Lancing	6. Tube bending	6. Sheet hydroforming
7. Notching	7. Roll forming	7. Tube hydroforming
8. Nibbling	8. Seaming	8. Hot drawing
9. Shaving	9. Flanging	9. High-energy-rate forming
10. Trimming	10. Straightening	10. Ironing
11. Cutoff		11. Embossing
12. Dinking		12. Superplastic sheet forming

A simple type of shearing operation is illustrated in Figure 17-1. As the punch (or upper blade) pushes on the workpiece, the metal responds by flowing plastically into the die (or over the lower blade). Because the clearance between the two tools is small, usually between 5 and 10% of the thickness of the metal being cut, the deformation occurs as highly localized shear. As the punch pushes downward on the metal, the material flows into the die, with the opposite surface bulging slightly. An instability arises when the penetration is between 15 and 60% of the metal thickness, the actual amount depending on the strength and ductility of the material. The applied stress exceeds the shear strength of the remaining material, and the metal tears or ruptures through the rest of its thickness, creating an inwardly inclined fracture and a ragged edge or burr. As shown in Figure 17-2, the two distinct stages of the shearing process, deformation and fracture, are often visible on the edges of sheared parts.

Because of the normal inhomogeneities in a metal and the possibility of nonuniform clearance between the shear blades, the final shearing does not occur in a uniform manner. Fracture and tearing begin at the weakest point and proceed progressively or intermittently to the next-weakest location. This usually results in a rough and ragged edge, which combined with possible microcracks and work hardening of the sheared edge can adversely affect subsequent forming processes.

Changing the clearance between the punch and the die can greatly change the condition of the cut edge. If the punch and die (or upper and lower shearing blades) have proper alignment and clearance, and are maintained in good condition, sheared edges can be produced that have sufficient smoothness to permit use without further finishing. The quality of the sheared edge can often be improved by clamping the starting stock firmly against the die (from above) and restraining the movement of the sheared piece by a plunger or rubber die cushion that applies opposing pressure from below. Each of these measures causes the shearing to take place more uniformly around the perimeter of the cut.

If the entire shearing operation is performed in a compressive environment, fracture is suppressed and the relative fraction of smooth edge (produced by deformation) is increased. Above a certain pressure, no fracture occurs and the entire edge is smooth, deformed metal. Figure 17-3 shows one method of producing a compressive environment. In the *fineblanking* process, a V-shaped protrusion is incorporated into the hold-down or pressure plate at a location slightly external to the contour of the cut. As pressure is applied to the hold-down or pressure plate, the protrusion is driven into the material, compressing the region to be cut. Matching upper and lower punches then squeeze the material from above and below, and descend in unison, extracting the desired segment. With punch-die clearances of about $\frac{1}{10}$ those of conventional blanking, the sheared edges are now both smooth and square, as shown in Figure 17-4.

FIGURE 17-1   Simple blanking with a punch and die.

FIGURE 17-2   (Top) Conventionally sheared surface showing the distinct regions of deformation and fracture, and (bottom) magnified view of the sheared edge. *(Courtesy of Feintool Equipment Corp., Cincinnati, OH.)*

FIGURE 17-3   Method of obtaining a smooth edge in shearing by using a shaped pressure plate to put the metal into localized compression and a punch and opposing punch descending in unison.

FIGURE 17-4   Fineblanked surface of the same component as shown in Figure 17-2. *(Courtesy of Feintool Equipment Corp., Cincinnati, OH.)*

FIGURE 17-5 Method of smooth shearing a rod by putting it into compression during shearing.

Fineblanked parts are usually less than 6 mm ($\frac{1}{4}$ inch) in thickness and typically have complex-shaped perimeters. Dimensional accuracy is often within 0.05 mm (0.002 in.), and holes, slots, bends, and semipierced projections can be incorporated as part of the fineblanking operation. Secondary edge finishing can often be eliminated, and the work hardening that occurs during the shearing process enhances wear resistance. In fineblanking, however, a triple-action press is generally required. The fineblanking force is about 40% greater than conventional blanking of the same contour, and the extra material required for the impinging protrusion often forces a greater separation between nested parts.

Figure 17-5 illustrates another means of shearing under compression. Bar stock is pressed against the closed end of a feed hole, placing the stock in a state of compression. A transverse punch then shears the material into smooth-surface, burr-free slugs, ready for further processing.

SIMPLE SHEARING

When sheets of metal are to be sheared along a straight line, *squaring shears*, like the one shown in Figure 17-6, are frequently used. As the upper ram descends, a clamping bar or set of clamping fingers presses the sheet of metal against the machine table to hold it firmly in position. A moving blade then comes down across a fixed blade and shears the metal. On larger shears, the moving blade is often set at an angle or "rocks" as it descends, so the cut is made in a progressive fashion from one side of the material to the other, much like a pair of household scissors. This action significantly reduces the amount of cutting force required, replacing a high force–short stroke operation with one of lower force and longer stroke.

The upper blade may also be inclined about 0.5° to 2.5° with respect to the lower blade and descend along this line of inclination. While squareness and edge quality may be compromised, this action helps to ensure that the sheared material does not become wedged between the blades.

SLITTING

Slitting is the lengthwise shearing process used to cut coils of sheet metal into several rolls of narrower width. Here the shearing blades take the form of cylindrical rolls with circumferential mating grooves. The raised ribs of one roll match the recessed grooves on

FIGURE 17-6 A 3-m (10-ft) power shear for 6.5 mm ($\frac{1}{4}$-in.) steel. *(Courtesy of Cincinnati Incorporated, Cincinnati, OH.)*

Workpiece

Scrap

Blanking   **Piercing**

FIGURE 17-7 Schematic showing the difference between piercing and blanking.

FIGURE 17-8 (Left to right) Piercing, lancing, and blanking precede the forming of the final ashtray. The small round holes assist positioning and alignment.

the other. The process is now continuous and can be performed rapidly and economically. Moreover, since the distance between adjacent shearing edges is fixed, the resultant strips have accurate and constant width, more consistent than that obtained from alternative cutting processes.

PIERCING AND BLANKING

Piercing and *blanking* are shearing operations where a part is removed from sheet material by forcing a shaped punch through the sheet and into a shaped die. Any two-dimensional shape can be produced, with one surface having a slightly rounded edge and the other surface containing a slight burr. Since both processes involve the same basic cutting action, the primary difference is one of definition. Figure 17-7 shows that in blanking, the piece being punched out becomes the workpiece. In piercing, the punchout is the scrap and the remaining strip is the workpiece. Piercing and blanking are usually done on some form of mechanical press.

Several variations of piercing and blanking are known by specific names. *Lancing* is a piercing operation that forms either a line cut (slit) or hole, like those shown in the left-hand portion of Figure 17-8. The primary purpose of lancing is to permit the adjacent metal to flow more readily in subsequent forming operations. In the case illustrated in Figure 17-8, the lancing makes it easier to shape the recessed grooves, which were formed before the ashtray was blanked from the strip stock and shallow drawn. *Perforating* consists of piercing a large number of closely spaced holes. *Notching* is used to remove segments from along the edge of an existing product.

In *nibbling*, a contour is progressively cut by producing a series of overlapping slits or notches, as shown in Figure 17-9. In this manner, simple tools can be used to cut a complex shape from sheets of metal up to 6 mm ($\frac{1}{4}$ in.) thick. The process is widely used when the quantities are insufficient to justify the expense of a dedicated blanking die. Edge smoothness is determined by the shape of the tooling and the degree of overlap in successive cuts.

Shaving is a finishing operation in which a small amount of metal is sheared away from the edge of an already blanked part. Its primary use is to obtain greater dimensional accuracy, but it may also be employed to produce a squared or smoother edge. Because only a small amount of metal is removed, the punches and dies must be made with very little clearance. Blanked parts, such as small gears, can be shaved to produce dimensional accuracies within 0.025 mm (0.001 in.).

In a *cutoff* operation, a punch and die are used to separate a stamping or other product from a strip of stock. The contour of the cutoff frequently completes the periphery of the workpiece. Cutoff operations are quite common in progressive die sequences, like several to be presented shortly.

FIGURE 17-9 Shearing operation being performed on a nibbling machine. *(Courtesy of Pacific Press Technologies, Mt. Carmel, IL.)*

FIGURE 17-10 The dinking process.

Dinking is a modified shearing operation that is used to blank shapes from low-strength materials, such as rubber, fiber, or cloth. As illustrated in Figure 17-10, the shank of a die is either struck with a hammer or mallet or the entire die is driven downward by some form of mechanical press.

TOOLS AND DIES FOR PIERCING AND BLANKING

As shown in Figure 17-11, the basic components of a piercing and blanking die set are a *punch*, a *die*, and a *stripper plate*, which is attached above the die to keep the strip material from ascending with the retracting punch. The position of the stripper plate and the size of its hole should be such that it does not interfere with either the horizontal motion of the strip as it feeds into position or the vertical motion of the punch.

Theoretically, the punch should fit within the die with a uniform clearance that approaches zero. On its downward stroke, it should not enter the die but should stop just as its base aligns with the top surface of the die. In general practice, the clearance is from 5 to 7% of the stock thickness and the punch enters slightly into the die cavity.

If the face of the punch is normal to the axis of motion, the entire perimeter is cut simultaneously. By tilting the punch face on an angle, a feature known as *shear* or *rake angle*, the cutting force can be reduced substantially. As shown in Figure 17-12, the periphery is now cut in a progressive fashion, similar to the action of a pair of scissors or the opening of a "pop-top" beverage can. Variation in the shear angle controls the amount of cut that is made at any given time and the total stroke that is necessary to complete the operation. Adding shear reduces the force but increases the stroke. It is an attractive way to cut thicker or stronger material on an existing piece of equipment.

Punches and dies should also be in proper alignment so that a uniform clearance is maintained around the entire periphery. The die is usually attached to the bolster plate of the press, which, in turn, is attached to the main press frame. The punch is attached to the movable ram, enabling motion in and out of the die with each stroke of the press. Punches and dies can also be mounted on a separate *punch holder* and *die shoe*, like the one shown in Figure 17-13, to create an *independent die set*. The holder and shoe are permanently aligned and guided by two or more guide pins. By aligning a punch and die, and fastening them to the die set, an entire unit can be inserted into a press without having to set or check the tool alignment. This can significantly reduce the amount of production time lost during tool change. Moreover, when a given punch and die are no longer needed, they can be removed and new tools attached to the shoe and holder assembly.

In most cases the punch holder attaches directly to the ram of the press, and ram motion acts to both raise and lower the punch. On smaller die sets, springs can be incorporated to provide the upward motion. The ram simply pushes on the top of the punch holder, forcing it downward. When the ram retracts, the springs cause the punch to return to its starting position. This form of construction makes the die set fully self-contained. It is simply positioned in the press and can be easily removed, thereby reducing setup time.

FIGURE 17-11 The basic components of piercing and blanking dies.

FIGURE 17-12   Blanking with a square-faced punch (left) and one containing angular shear (right). Note the difference in maximum force and contact stroke. The total work (the area under the curve) is the same for both processes.

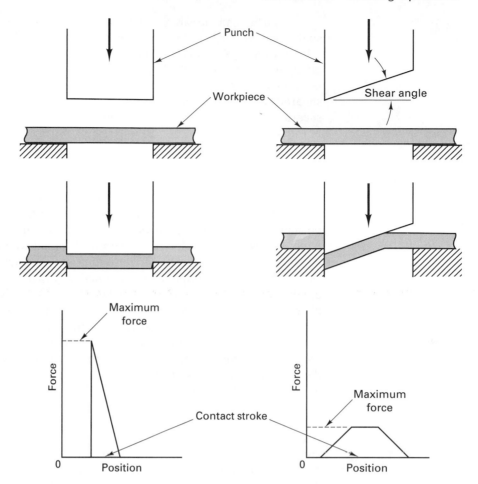

A wide variety of standardized, self-contained die sets have been developed. Known as *subpress dies* or *modular tooling*, these can often be assembled and combined on the bed of a press to pierce or blank large parts that would otherwise require large and costly complex die sets. Figure 17-14 shows an assembly of subpress dies where a piece of sheet is inserted between the tooling and the downward motion of the press produces a variety of holes and slots, all in proper relation to one another.

FIGURE 17-13   Typical die set having two alignment guideposts. *(Courtesy of Danly IEM, Cleveland, OH.)*

FIGURE 17-14   A piercing and blanking setup using self-contained subpress tool units. *(Courtesy of Strippit Division, Houdaille Industries, Inc., Akron, NY.)*

Punches and dies are usually made from low-distortion or air-hardenable tool steel so they can be hardened after machining with minimal warpage. The die profile is maintained for a depth of about 3 mm ($\frac{1}{8}$ inch) from the upper face, beyond which an angular clearance or back relief is generally provided (see Figure 17-11) to reduce friction between the part and the die and to permit the part to fall freely from the die after being sheared. The 3-mm depth provides adequate strength and sufficient metal so that the shearing edge can be resharpened by grinding a few thousandths of an inch from the face of the die.

Dies can be made as a single piece, or they can be made in component sections that are assembled on the punch holder and die shoe. The component approach simplifies production and enables the replacement of single sections in the event of wear or fracture. Complex dies like the one shown in Figure 17-15 can often be assembled from the many standardized punch and die components that are available. Substantial savings can often be achieved by modifying the design of parts to enable the use of standard die components. A further advantage of this approach is that when the die set is no longer needed, the components can be removed and used to construct tooling for another product.

When the cut periphery is composed of simple lines, and the material being cut is either soft metal or other soft material (such as plastics, wood, cork, felt, fabrics, and cardboard), "steel-rule" or "cookie-cutter" dies can often be used. The cutting die is fashioned from hardened steel strips, known as steel rule, that are mounted on edge in grooves that have been machined in the upper die block. The mating piece of tooling can be either a flat piece of hardwood or steel, a male shape that conforms to the part profile (such that the protruding strips descend around it), or a set of matching grooves into which the upper die can descend. Rubber pads are usually inserted between the strips to replace the stripper plate. During the compression stroke, the rubber compresses and allows the cutting action to proceed. As the ram ascends, the rubber then expands to push the blank free of the steel-rule cavity. Steel-rule dies are usually less expensive to construct than solid dies and are quite attractive for producing small quantities of parts.

Many parts require multiple cutting-type operations, and it is often desirable to produce a completed part with each cycle of a press. Several types of dies have been designed to accomplish this task. For simplicity, their operations are discussed in terms of manufacturing simple, flat washers from a continuous strip of metal.

The *progressive die* set, depicted in Figure 17-16, is the simpler of the two types. Basically, it consists of two or more sets of punches and dies mounted in tandem. Strip stock is fed into the first die, where a hole is pierced as the ram descends. When the ram raises, the stock advances and the pierced hole is positioned under the blanking punch.

FIGURE 17-15 A progressive piercing, forming, and cutoff die set built up mostly from standard components. The part produced is shown at the bottom. *(Courtesy of Oak Manufacturing Company, Los Angeles, CA.)*

FIGURE 17-16 Progressive piercing and blanking die for making a square washer. Note that the punches are of different length.

Upon the second descent, a pilot on the bottom of the blanking punch enters the hole that was pierced on the previous stroke to ensure accurate alignment. Further descent of the punch blanks the completed washer from the strip and, at the same time, the first punch pierces the hole for the next washer. As the process continues, a finished part is completed with each stroke of the press.

Progressive dies can be used for many combinations of piercing, blanking, forming, lancing, and drawing, as shown by the examples in Figures 17-8, 17-15, and 17-17. They are relatively simple to construct and are economical to maintain and repair, since a defective punch or die does not require replacement of the entire die set. The material moves through the operations in the form of a continuous strip. As the products are shaped, they remain attached to the strip or carrier until a final cutoff operation. While the attachment may restrict some of the forming operations and prevents part reorientation between steps, it also enables the quick and accurate positioning of material in each of the die segments.

If individual parts are mechanically moved from operation to operation within a single press, the dies are known as *transfer dies*. Part handling must operate in harmony with the press motions to move, orient, and position the pieces as they travel through the die.

In *compound dies*, like the one shown schematically in Figure 17-18, piercing and blanking, or other combinations of operations, occur sequentially during a single stroke of the ram. Dies of this type are usually more expensive to construct and are more susceptible to breakage, but they generally offer more precise alignment of the sequential operations.

If many holes of varying sizes and shapes are to be placed in sheet components, numerically controlled *turret-type punch presses* may be specified. In these machines, as many as 60 separate punches and dies are contained within a turret that can quickly be rotated to provide the specific tooling required for an operation. Between operations, the workpiece is repositioned through numerically controlled movements of the worktable. This type of machine is particularly attractive when a variety of materials and thicknesses (0.4 to 8.0 mm) are being processed. Still greater flexibility can be achieved by single machines that combine punch pressing with laser cutting or water-jet cutting.

FIGURE 17-17 The various stages of an 11-station progressive die. *(Courtesy of the Minster Machine Company, Minster, OH.)*

FIGURE 17-18 Method for making a simple washer in a compound piercing and blanking die. Part is blanked (a) and subsequently pierced (b) in the same stroke. The blanking punch contains the die for piercing.

DESIGN FOR PIERCING AND BLANKING

The construction, operation, and maintenance of piercing and blanking dies can be greatly facilitated if designers of the parts to be fabricated keep a few simple rules in mind:

1. Diameters of pierced holes should not be less than the thickness of the metal, with a minimum of 0.3 mm (0.025 in.). Smaller holes can be made, but with difficulty.
2. The minimum distance between holes, or between a hole and the edge of the stock, should be at least equal to the metal thickness.
3. The width of any projection or slot should be at least 1 times the metal thickness and never less than 2.5 mm ($\frac{3}{32}$ in.).
4. Keep tolerances as large as possible. Tolerances below about 0.075 mm (0.003 in.) will require shaving.
5. Arrange the pattern of parts on the strip to minimize scrap.

■ 17.3 BENDING

Bending is the plastic deformation of metals about a linear axis with little or no change in the surface area. Multiple bends can be made simultaneously, but to be classified as true bending, and treatable by simple bending theory, each axis must be linear and independent of the others. If multiple bends are made with a single die, the process is often called *forming*. When the axes of deformation are not linear or are not independent, the processes are known as *drawing* and/or *stretching*, and these operations will be treated later in the chapter.

As shown in Figure 17-19, simple bending causes the metal on the outside to be stretched while that on the inside is compressed. The location that is neither stretched nor compressed is known as the *neutral axis* of the bend. Since the yield strength of metals in compression is somewhat higher than the yield strength in tension, the metal on the outer side yields first, and the neutral axis is displaced from the midpoint of the material. The neutral axis is generally located between one-third and one-half of the way from the inner surface, depending on the bend radius and the material being bent. Because of this lack of symmetry and the dominance of tensile deformation, the metal is generally thinned at the bend. In a linear bend, thinning is greatest in the center of the sheet and less near the free edges, where inward movement can provide some compensation.

On the inner side of a bend, the compressive stresses can induce upsetting and a companion thickening of material. While this thickening somewhat offsets the thinning of the outer section, the upsetting can also produce an outward movement of the free edges. This contraction of the tensile segment and expansion of the compression segment can produce significant distortion of the edge surfaces that terminate a linear bend. This distortion is particularly pronounced when bends are produced across the width of thick but narrow plates.

Still another consequence of the combined tension and compression is the elastic recovery that occurs when the bending load is removed. The stretched region contracts

FIGURE 17-19 (Top) Nature of a bend in sheet metal showing tension on the outside and compression on the inside. (Bottom) The upper portion of the bend region, viewed from the side, shows how the center portion will thin more than the edges.

labels: Direction of travel — Workpiece — Cam roller — Wing — Cam — Folding leaf — Jaw — Shoe — Folding bar — Wedge adjustment for raising and lowering wing — Adjusting screw for thickness of material

FIGURE 17-20 Phantom section of a bar folder, showing position and operation of internal components. *(Courtesy of Niagara Machine and Tool Works, Buffalo, NY)*

and the compressed region expands, resulting in a small amount of "unbending," known as *springback*. To produce a product with a specified angle, the metal must be overbent by an amount equal to the subsequent springback. The actual amount of springback will vary with a number of factors, including the type of material and material thickness.

ANGLE BENDING (BAR FOLDER AND PRESS BRAKE)

Machines like the *bar folder*, shown in Figure 17-20, can be used to make angle bends up to 150° in sheet metal under 1.5 mm ($\frac{1}{16}$ in.) thick. The workpiece is inserted under the folding leaf and aligned in the proper position. Raising the handle then actuates a cam, causing the leaf to clamp the sheet. Further movement of the handle bends the metal to the desired angle. These manually operated machines can be used to produce linear bends up to about 3.5 m (12 ft) in length.

Bends in heavier sheet or more complex bends in thin material are generally made on *press brakes*, like the one shown in Figure 17-21. These are mechanical or hydraulic presses with a long, narrow bed and short strokes. The metal is bent between interchangeable dies that are attached to both the bed and the ram. As illustrated in Figures 17-21 and 17-22, different dies can be used to produce many types of bends. The metal can be repositioned between strokes to produce complex contours or repeated

FIGURE 17-21 (Left) Press brake with CNC gauging system. *(Courtesy of DiAcro Division, Acrotech Inc., Lake City, MN.)* (Right) Close-up view of press brake dies forming corrugations. *(Courtesy of Cincinnati Incorporated, Cincinnati, OH.)*

FIGURE 17-22 Press brake dies can form a variety of angles and contours. *(Courtesy of Cincinnati Incorporated, Cincinnati, OH.)*

FIGURE 17-23 Dies and operations used in the press brake forming of a roll bead. *(Courtesy of Cincinnati Incorporated, Cincinnati, OH.)*

bends, such as corrugations. Figure 17-23 shows how a roll bead can be formed with repeated strokes, repositioning, and multiple sets of tooling. Seaming, embossing, punching, and other operations can also be performed with press brakes, but these operations can usually be done more efficiently on other types of equipment.

The tools and support structures on a press brake are often loaded in the same three-point bending discussed in Chapter 16 for rolling mill rolls. Elastic deflections can cause a variety of bend deviations and defects, and a number of means have been developed to overcome the problems.

DESIGN FOR BENDING

Several factors must be considered when designing parts that are to be shaped by bending. One of the primary concerns is determining the smallest bend radius that can be formed without metal cracking (i.e., the *minimum bend radius*). This value is dependent on both the ductility of the metal (as measured by the percent reduction in area observed in a standard tensile test) and the thickness of the material being bent. Figure 17-24 shows how the ratio of the minimum bend radius R to the thickness of the material t varies with material ductility. As this plot reveals, an extremely ductile material is required if we wish to produce a bend with radius less than the thickness of the metal. If possible, bends should be designed with large bend radii. This permits easier forming and allows the designer to select from a wider variety of engineering materials.

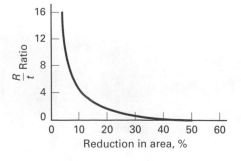

FIGURE 17-24 Relationship between the minimum bend radius (relative to thickness) and the ductility of the metal being bent (as measured by the reduction in area in a uniaxial tensile test).

Transverse bends
(poor)

Rolling
direction

Longitudinal bends
(good)

Poor

Rolling
direction

Good

If the punch radius is large and the bend angle is shallow, large amounts of spring-back are often encountered. The sharper the bend, the more likely the surfaces will be stressed beyond the yield point. Less severe bends have large amounts of elastically stressed material and large amounts of springback. In general, when the bend radius is greater than 4 times the material thickness, the tooling or process must provide springback compensation.

If the metal has experienced previous cold work or has marked directional properties, these features should be considered when designing the bending operation. Whenever possible, it is best to make the bend axis perpendicular to the direction of previous working, as shown in the upper portion of Figure 17-25. The explanation for this recommendation has little to do with the grain structure of the metal, but is more closely related to the mechanical loading applied to the weak, oriented inclusions. Cracks can easily start along tensile-loaded inclusions and propagate to full cracking of the bend. If intersecting or perpendicular bends are required, it is often best to place each at an angle to the rolling direction, as shown in the lower portion of Figure 17-25, rather than have one longitudinal and one transverse.

Another design concern is determining the dimensions of a flat blank that will produce a bent part of the desired precision. As discussed earlier in the chapter, metal tends to thin and lengthen when it is bent. The amount of lengthening is a function of both the stock thickness and the bend radius. Figure 17-26 illustrates one method that has been found to give satisfactory results for determining the blank length for bent products. In addition, the minimum length of any protruding leg should be at least equal to the bend radius plus 1.5 times the thickness of the metal.

$L = \ell_1 + \ell_2 - D$

$L = \ell_1 + \ell_2 + \ell_3 - 2D$

For R equal	Let D equal
t	$1.7\,t$
$2\,t$	$2.0\,t$
$3\,t$	$2.5\,t$

FIGURE 17-26 One method of determining the starting blank size (L) for several bending operations. Due to thinning, the product will lengthen during forming. l_1, l_2, and l_3 are the desired product dimensions. See table to determine D based on size of radius R where t is the stock thickness.

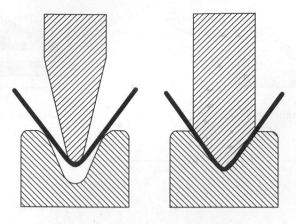

FIGURE 17-27   Comparison of air-bend (left) and bottoming (right) press brake dies. With the air-bend die, the amount of bend is controlled by the bottoming position of the upper die.

Whenever possible, the tolerance on bent parts should not be less than 0.8 mm ($\frac{1}{32}$ in.). Bends of 90° or greater should not be specified without first determining whether the material and bending method will permit them. Parts with multiple bends should be designed with most (or preferably all) of them of the same bend radius. This will reduce setup time and tooling costs. Consideration should also be given to providing regions for adequate clamping or support during manufacture. Bending near the edge of a material will distort the edge. If an undistorted edge is required, additional material must be included and a trimming operation performed after bending.

AIR-BEND, BOTTOMING, AND COINING DIES

Yet another design decision is the use of air-bend, bottoming, or coining dies. As shown in Figure 17-27, *bottoming dies* contact and compress the full area within the tooling. The angle of the resulting bend is set by the geometry of the tooling, adjusted for subsequent springback, and the inside bend radius is that machined on the nose of the punch. Bottoming dies are designed for a specific material and material thickness, and they form bends of a single configuration. If the results are outside specifications, or the material is changed and produces a different amount of springback, the geometry of the tooling will have to be modified. Once the geometry of the tool is successfully set, however, reproducibility of the bend geometry is excellent, provided there is consistency within the size and properties of the material being bent.

In contrast, *air-bend dies* produce the desired geometry by simple three-point bending. Since the resulting angle is controlled by the bottoming position of the upper die, a single set of tooling can produce a range of bend geometries from 180° through the included angle of the die. Air bending can also accommodate a variety of materials in a range of material gages, and it requires the least force of the three options. Product reproducibility depends on the ability to control the stroke of the press. Adaptive control and on-the-fly corrections are frequently used with air-bend tooling.

If bottoming dies continue to move beyond the full-contact position, the thickness of the bent material is reduced. Because of the extensive plastic deformation, the operation becomes one of *coining*. Springback can be significantly reduced, and more consistent results can be achieved with materials having variation in structure and thickness. Unfortunately, the loading is greatly increased on both the press and the tools.

Reproducible-stroke mechanical presses are generally used for bottom bending and coining, while adjustable-stroke hydraulic presses are preferred for air bending.

ROLL BENDING

Roll bending is a continuous form of three-point bending where plates, sheets, beams, pipe, and even rolled shapes and extrusions are bent to a desired curvature using forming rolls. As shown in Figure 17-28, roll-bending machines usually have three rolls in the form of a triangle. The two lower rolls are driven, and the position of the upper roll is adjusted to control the degree of curvature in the product. The rolls on the machine pictured in Figure 17-28 are supported on only one end. When wider material is being formed, the longer rolls often require support on both ends. The support frame on one

FIGURE 17-28 (Left) Schematic of the roll-bending process; (right) the roll bending of an I-beam section. Note how the material is continuously subjected to three-point bending. *(Courtesy of Buffalo Forge Company, Buffalo, NY.)*

end may be swung clear, however, to permit the removal of closed circular shapes or partially rolled product. Because of the variety of applications, roll-bending machines are available in a wide range of sizes, some being capable of bending plate up to 25 cm (10 in.) thick.

DRAW BENDING, COMPRESSION BENDING, AND PRESS BENDING

Bending machines can also utilize clamps and pressure tools to bend material against a form block. In *draw bending*, illustrated in Figure 17-29, the workpiece is clamped against a bending form and the entire assembly is rotated to draw the workpiece along a stationary pressure tool. In *compression bending*, also illustrated in Figure 17-29, the bending form remains stationary and the pressure tool moves along the surface of the workpiece.

Press bending, also shown in Figure 17-29, utilizes a downward-descending bend die, which pushes into the center of material that is supported on either side by wing dies. As the ram descends, the wing dies pivot up, bending the material around the form on the ram. The flexibility of each of the above processes is somewhat limited because a certain length of the product must be used for clamping.

TUBE BENDING

Quite often the material being bent is a tube or pipe, and this geometry presents additional problems. Key parameters are the outer diameter of the tube, the wall thickness, and the radius of the bend. Small-diameter, thick-walled tubes usually present little difficulty. As the outer diameter increases, the wall thickness decreases or the bend radius becomes smaller, the outside of the tube tends to pull to the center, flattening the tube, and the inside surface may wrinkle. For many years, a common method of overcoming these problems was to pack the tube with wet sand, produce the bend, and then remove the sand from the interior. Flexible mandrels have now replaced the sand and are currently available in a wide variety of styles and sizes.

(a) Draw bending   (b) Compression bending   (c) Press bending

FIGURE 17-29 (a) Draw bending, in which the form block rotates; (b) compression bending, in which a moving tool compresses the workpiece against a stationary form; (c) press bending, where the press ram moves the bending form.

(a)

(b)

FIGURE 17-30 (a) Schematic representation of the cold roll-forming process being used to convert sheet or plate into tube. (b) Some typical shapes produced by roll forming.

ROLL FORMING

The continuous *roll forming* of flat strip into complex sections has become a highly developed forming technique that competes directly with press brake forming, extrusion, and stamping. As shown in Figures 17-30 and 17-31, the process involves the progressive bending of metal strip as it passes through a series of forming rolls at speeds up to 80 m/min (270 ft/min). Only bending takes place, and all bends are parallel to one another. The thickness of the starting material is preserved, except for thinning at the bend radii.

Any material that can be bent can be roll formed—including cold-rolled, hot-rolled, polished, prepainted, coated, and plated metals—in thicknesses ranging from 0.1 through 20 mm (0.005 through $\frac{3}{4}$ in.). A variety of moldings, channeling, gutters and downspouts, automobile beams and bumpers, and other shapes of uniform wall thickness and uniform cross section are now being formed.

By changing the rolls, a single roll-forming machine can produce a wide variety of different shapes. However, changeover, setup, and adjustment may take several hours, so a production run of at least 3000 m (10,000 ft) is usually required for any given product. To produce pipe or tubular products, a resistance welding unit or seaming operation is often integrated with the roll forming.

Section of stock   1st pass   2nd pass   3rd pass   4th pass   5th pass   6th pass   7th pass   8th pass   Roll-formed shape

FIGURE 17-31 Eight-roll sequence for the roll forming of a box channel. *(Courtesy of the Aluminum Association, Washington, DC.)*

FIGURE 17-32 Various types of seams used on sheet metal.

FIGURE 17-33 Method of straightening rod or sheet by passing it through a set of straightening rolls. For rods, another set of rolls is used to provide straightening in the transverse direction.

SEAMING AND FLANGING

Seaming is a bending operation that can be used to join the ends of sheet metal in some form of mechanical interlock. Figure 17-32 shows several of the more common seam designs that can be formed by a series of small rollers. Seaming machines range from small hand-operated types to large automatic units capable of producing hundreds of seams per minute. Common products include cans, pails, drums, and other similar containers.

   Flanges can be rolled on sheet metal in essentially the same manner as seams. In many cases, however, the forming of both flanges and seams is a drawing operation, since the bending can occur along a curved axis.

STRAIGHTENING

The objective of *straightening* or *flattening* is the opposite of bending, and these operations are often performed before subsequent forming to ensure the use of flat or straight material that is reasonably free of residual stresses. *Roll straightening* or *roller leveling*, illustrated in Figure 17-33, subjects the material to a series of reverse bends. The rod, sheet, or wire is passed through a series of rolls with progressively decreased offsets from a straight line. As the material is bent, first up and then down, the surfaces are stressed beyond their elastic limit, replacing any permanent set with a flat or straight profile. Tension applied along the length of the product can help induce the required deformation.

   Sheet material can also be straightened by a process called *stretcher leveling*. Finite-length sheets are gripped mechanically and stretched beyond the elastic limit to produce the desired flatness.

■ 17.4 DRAWING AND STRETCHING PROCESSES

The term *drawing* can actually refer to two quite different operations. The drawing of wire, rod, and tube, presented in Chapter 16, refers to processes that reduce the cross section of material by pulling it through a die. When the starting material is sheet, drawing refers to a family of operations where plastic flow occurs over a curved axis and the flat sheet is formed into a recessed, three-dimensional part with a depth more than several times the thickness of the metal. These operations can be used to produce a wide range of shapes, from small cups to large automobile and aerospace panels.

SPINNING

Spinning is a cold-forming operation where a rotating disk of sheet metal is progressively shaped over a male form, or mandrel, to produce rotationally symmetrical shapes, such as cones, hemispheres, cylinders, bells, and parabolas. A form block possessing the shape of the desired part is attached to a rotating spindle, such as the drive section of a simple lathe. A disk of metal is centered on the small end of the form and held in place by a pressure pad. As the disk and form rotate, localized pressure is applied through a round-ended wooden or metal tool or small roller that traverses the entire surface of the part, causing it to

Steps in spinning

Final shape

Original blank of sheet metal

Follower held in tailstock

FIGURE 17-34 Progressive stages in the spinning of a sheet metal product.

Form attached to headstock spindle

Roller

FIGURE 17-35 Two stages in the spinning of a metal reflector. *(Courtesy of Spincraft, Inc., New Berlin, WI.)*

flow progressively against the form. Figure 17-34 depicts the progressive operation, and Figure 17-35 shows a part at two stages of forming. Because the final diameter of the final part is less than that of the starting disk, the circumferential length decreases. This decrease must be compensated by either an increase in thickness, a radial elongation, or circumferential buckling. Control of the process is often dependent on the skill of the operator.

During spinning, the form block sees only localized compression, and the metal does not move across it under pressure. As a result, form blocks can often be made of hardwood or even plastic. The primary requirement is simply to replicate the shape with a smooth surface. As a result, tooling cost can be extremely low, making spinning an attractive process for producing small quantities of a single part. With automation, spinning can also be used to mass-produce such high-volume items as lamp reflectors, cooking utensils, bowls, and the bells of some musical instruments. When large quantities are produced, a metal form block is generally preferred.

Spinning is usually considered for simple shapes that can be directly withdrawn from a one-piece form. More complex shapes, such as those with reentrant angles, can be spun over multipiece or offset forms. Complex form blocks can also be made from frozen water, which is melted out of the product after spinning.

SHEAR FORMING OR FLOW TURNING

Cones, hemispheres, and similar shapes are often formed by *shear forming* or *flow turning*, a modification of the spinning process in which each element of the blank maintains its distance from the axis of rotation. Since there is no circumferential shrinkage, the metal flow is entirely by shear and no compensating stretch has to occur. As shown in Figure 17-36, the wall thickness of the product, t_c, will vary with the angle of the particular region according to the relationship: $t_c = t_b \sin \alpha$, where t_b is the thickness of the

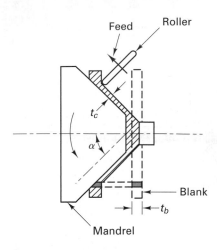

FIGURE 17-36 Schematic representation of the basic shear-forming process.

FIGURE 17-37 Forming a conical part by reverse shear forming.

FIGURE 17-38 Shear forming a cylinder by the direct process.

starting blank. If α is less than 30°, it may be necessary to complete the forming in two stages with an intermediate anneal in between. Reductions in wall thickness as high as 8:1 are possible, but the limit is usually set at about 5:1, or 80%.

Conical shapes are usually shear formed by the direct process depicted in Figure 17-36. The bottom of the product is held against the face of the form block or mandrel, while the material being formed moves in the same direction as the roller. Products can also be formed by a reverse process, like that illustrated in Figure 17-37. By controlling the position and feed of the forming rollers, the reverse process can be used to shape concave, convex, or conical parts without a matching form block or mandrel. Cylinders can be shear formed by both the direct and reverse processes. As shown in Figure 17-38, the direct process restricts the length of the product to the length of the mandrel. No schematic is provided for the reverse process because it is essentially the same as the roll-extrusion process, depicted in Figure 16-51.

STRETCH FORMING

Stretch forming, illustrated in principle in Figure 17-39, is an attractive means of producing large sheet metal parts in low or limited quantities. A sheet of metal is gripped by two or more sets of jaws that stretch it and wrap it around a single form block. Various combinations of stretching, wrapping, and block movements can be employed, depending on the shape of the part.

FIGURE 17-39 Schematic of a stretch-forming operation.

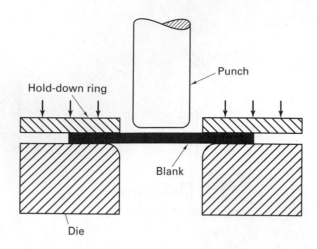

FIGURE 17-40 Schematic of the deep-drawing process.

Because most of the deformation is induced by the tensile stretching, the forces on the form block are far less than those normally encountered in bending or forming. Consequently, there is very little springback, and the workpiece conforms very closely to the shape of the tool. Since stretching accompanies bending or wrapping, wrinkles are pulled out before they occur. Because the forces are so low, the form blocks can often be made of wood, low-melting-point metal, or even plastic.

Stretch forming, or *stretch-wrap forming* as it is often called, is quite popular in the aircraft industry and is frequently used to form aluminum and stainless steel into cowlings, wing tips, scoops, and other large panels. Low-carbon steel can be stretch formed to produce large panels for the automotive and truck industry. If mating male and female dies are used to shape the metal while it is being stretched, the process is known as *stretch-draw forming*.

DEEP DRAWING AND SHALLOW DRAWING

The forming of solid-bottom cylindrical or rectangular containers from metal sheet is one of the most widely used manufacturing processes. When the depth of the product is less than its diameter (or the smallest dimension of its opening), the process is considered to be *shallow drawing*. If the depth is greater than the diameter, it is known as *deep drawing*.

Consider the simple operation of converting a circular disk of sheet metal into a flat-bottom cylindrical cup. Figure 17-40 shows the blank positioned over a die opening and a circular punch descending to pull or draw the material into the die cavity. The material beneath the punch remains largely unaffected and simply becomes the bottom of the cup. The cup wall is formed by pulling the remainder of the disk inward and over the radius of the die, as shown in Figure 17-41. As the material is pulled inward, its circumference decreases. Since the volume of material must remain constant, the decrease in circumferential dimension must be compensated by an increase in another dimension, such as thickness or radial length. Since the material is thin, an alternative is to relieve the circumferential compression by buckling or wrinkling. Wrinkle formation can be suppressed, however, by compressing the sheet between the die and a blankholder surface during the forming.

FIGURE 17-41 Flow of material during deep drawing. Note the circumferential compression as the radius is pulled inward.

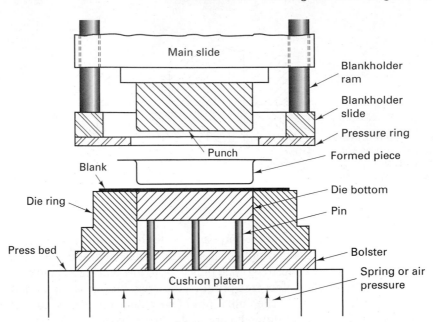

FIGURE 17-42 Drawing on a double-action press, where the blankholder uses the second press action.

In single-action presses, where there is only one movement that is available, springs or air pressure are often used to clamp the metal between the die and pressure ring. When multiple actions are available (two or more independent motions), as shown in Figure 17-42, the hold-down force can now be applied in a manner that is independent of the punch position. This restraining force can also be varied during the drawing operation. For this reason, multiple-action presses are usually specified for the drawing of more complex parts, while single-action presses can be used for the simpler operations.

Key variables in the deep-drawing process include the blank diameter and the punch diameter (which combine to determine the *draw ratio* and the height of the side walls), the die radius, the punch radius, the clearance between the punch and the die, the thickness of the blank, lubrication, and the hold-down pressure. Once a process has been designed and the tooling manufactured, the primary variable for process adjustment is the *hold-down pressure* or *blankholder force*. If the force or pressure is too low, wrinkling may occur at the start of the stroke. If it is too high, there is too much restraint, and the descending punch will simply tear the disk or some portion of the already-formed cup wall.

When drawing a shallow cup, there is little change in circumference, and a small area is being confined by the blankholder. As a result, the tendency to wrinkle or tear is low. As cup depth increases, there is an increased tendency for forming both of the defects. In a similar manner, thin material is more likely to wrinkle or tear than thick material. Figure 17-43 summarizes the effects of cup depth, blank thickness, and

FIGURE 17-43 Defect formation in deep drawing as a function of blankholder force, blank thickness, and cup depth.

FIGURE 17-44 Cup redrawing to further reduce diameter and increase wall height. (Left) forward redraw; (right) reverse redraw.

blankholder pressure. Note that for thin materials and deep draws, a defect-free product may not be possible in a single operation, as the defects simply transition from wrinkling, to wrinkle plus tear, to just tearing as blankholder force is increased. To produce a defect-free product, the draw ratio is often limited to values less than about 2.2.

The limitations of wrinkling and tearing can often be overcome by using multiple operations. Figure 17-44 shows two alternatives for converting drawn parts into deeper cups. In the *forward redraw* option, the material undergoes reverse bending as it flows into the die. In *reverse redrawing*, the starting cup is placed over a tubular die, and the punch acts to turn it inside out. Since all bending is in one direction, reverse redraws can produce greater changes in diameter.

When the part geometry becomes more complex, as with rectangular or asymmetric parts, it is best if the surface area and thickness of the material can remain relatively constant. Different regions may need to be differentially constrained. One technique that can produce variable constraint is the use of *draw beads*, vertical projections and matching grooves in the die and blankholder. The added force of bending and unbending as the material flows over the draw bead restricts the flow of material. The degree of constraint can be varied by adjusting the height, shape, and size of the bead and bead cavity.

Because of prior rolling and other metallurgical and process features, the flow of sheet metal is generally not uniform, even in the simplest drawing operation. Excess material may be required to assure final dimensions, and *trimming* may be required to establish both the size and uniformity of the final part. Figure 17-45 shows a shallow-drawn part before and after trimming. Trimming obviously adds to the production cost because it not only converts some of the starting material to scrap but also adds another operation to the manufacturing process, one that must be performed on an already-produced shape.

FORMING WITH RUBBER TOOLING OR FLUID PRESSURE

Blanking and drawing operations usually require mating male and female, or upper and lower, die sets, and process setup requires that the various components be properly positioned and aligned. When the amount of deformation is great, multiple operations may

FIGURE 17-45 Pierced, blanked, and drawn part before and after trimming.

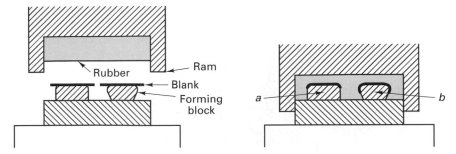

FIGURE 17-46   The Guerin process for forming sheet metal products.

be required, each with its own set of dedicated tooling, and intermediate anneals may also be necessary. Numerous processes have been developed that seek to (1) reduce tooling cost, (2) decrease setup time and expense, or (3) extend the amount of deformation that can be performed with a single set of tools. Although most of these methods have distinct limitations, such as complexity of shape or types of metal that can be formed, they also have definite areas of application.

Several forming methods replace either the male or female member of the die set with rubber or fluid pressure. The *Guerin process* (also known as rubber-die forming) is depicted in Figure 17-46. It is based on the phenomenon that rubber of the proper consistency, *when totally confined*, acts as a fluid and transmits pressure uniformly in all directions. Blanks of sheet metal are placed on top of form blocks, which can be made of wood, Bakelite, polyurethane, epoxy, or low-melting-point metal. The upper ram contains a pad of rubber 20 to 25 cm (8 to 10 in.) thick mounted within a steel container. As the ram descends, the rubber pad becomes confined and transmits force to the metal, causing it to bend to the desired shape. Since no female die is used and inexpensive form blocks replace the male die, the total tooling cost is quite low. There are no mating tools to align, process flexibility is quite high (different shapes can even be formed at the same time), wear on the material and tooling is low, and the surface quality of the workpiece is easily maintained, a feature that makes the process attractive for forming prepainted or specially coated sheet. When reentrant sections are produced (as with product *b* in Figure 17-46), it must be possible to slide the parts lengthwise from the form blocks or to disassemble a multipiece form from within the product.

The Guerin process was developed by the aircraft industry, where the production of small numbers of duplicate parts clearly favors the low cost of tooling. It can be used on aluminum sheet up to 3 mm ($\frac{1}{8}$ in.) thick and on stainless steel up to 1.5 mm ($\frac{1}{16}$ in.). Magnesium sheet can also be formed if it is heated and shaped over heated form blocks.

Most of the forming done with the Guerin process is multiple-axis bending, but some shallow drawing can also be performed. The process can also be used to pierce or blank thin gages of aluminum, as illustrated in Figure 17-47. For this application the blanking blocks are shaped the same as the desired workpiece, with a sharp face, or edge, of hardened steel. Round-edge supporting blocks are positioned a short distance from the blanking blocks to support the scrap skeleton and permit the metal to bend away from the sheared edges.

In *bulging*, fluid or rubber transmits the pressure required to expand a metal blank or tube outward against a split female mold or die. For simple shapes, rubber tooling can

FIGURE 17-47   Method of blanking sheet metal using the Guerin process.

FIGURE 17-48 Method of bulging tubes with rubber tooling.

be inserted, compressed, and then easily removed, as shown in Figure 17-48. For complicated shapes, fluid pressure may be required to form the bulge. More complex equipment is required since pressurized seals must be formed and maintained, while still enabling the easy insertion and removal of material that is required for mass production.

SHEET HYDROFORMING

Sheet hydroforming is really a family of processes in which a rubber bladder backed by fluid pressure, or pockets of pressurized liquid, replaces either the solid punch or female die of the traditional tool set. In a variant known as *high-pressure flexible-die forming*, or *flexforming*, the rubber pad of the Guerin process is replaced by a flexible rubber diaphragm backed by controlled hydraulic pressure at values between 140 and 200 MPa (20,000 and 30,000 psi). As illustrated in Figure 17-49, the solid punch drives the sheet into the resisting bladder, whose pressure is adjusted throughout the stroke.

In a variation of this process that shares similarity to stretch forming, the sheet is first clamped against the opening and the punch is retracted downward. The fluid is then pressurized, causing the workpiece to balloon downward toward the punch. Since the pressure is uniformly distributed over the workpiece, the sheet is uniformly stretched and uniformly thinned. The punch then moves upward, causing the prestretched metal to conform to its profile.

FIGURE 17-49 High-pressure flexible-die forming, showing (1) the blank in place with no pressure in the cavity; (2) press closed and cavity pressurized; (3) ram advanced with cavity maintaining fluid pressure; and (4) pressure released and ram retracted. *(Courtesy of Aluminum Association, Washington, DC)*

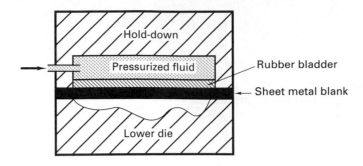

FIGURE 17-50   One form of sheet hydroforming.

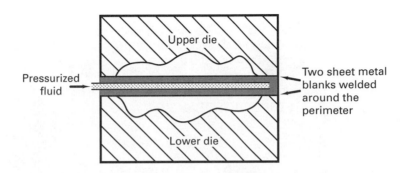

FIGURE 17-51   Two-sheet hydroforming, or pillow forming.

The flexible membrane can also be used to replace the hardened male punch, as shown in Figure 17-50. The ballooning action now causes the material to descend and conform fully to the female die, which may be made of epoxy or other low-cost material. *Parallel-plate hydroforming*, or *pillow forming*, extends the process to the simultaneous production of upper and lower contours. As shown in Figure 17-51, two sheet metal blanks are laser welded around their periphery or are firmly clamped between upper and lower dies. Pressurized fluid is then injected between the sheets, simultaneously forming both upper and lower profiles. This may be a more attractive means of producing complex sheet metal containers, since the manufacturer no longer has to cope with the problems of aligning and welding separately formed upper and lower pieces.

While the most attractive feature of sheet hydroforming is probably the reduced cost of tooling, there are other positive attributes. Because a more uniform distribution of strain is produced, materials exhibit greater formability. Drawing limits are generally about 1.5 times those of conventional deep drawing. Deeper parts can be formed without fracture, and complex shapes that require multiple operations can often be formed in a single pressing. Surface finish is excellent, and part dimensions are more accurate and more consistent.

Because the cycle times for sheet hydroforming are slow compared to mechanical presses, conventional deep drawing is preferred when the draw depth is shallow and the part is not complex. The reduced tool costs of hydroforming make the process attractive for prototype manufacturing and low-volume production (up to about 10,000 identical parts), such as that encountered in the aerospace industry. Sheet hydroforming has also attracted attention within the automotive community because of its ability not only to produce low-volume parts in an economical manner but also to successfully shape lower-formability materials, such as alloyed aluminum sheet and high-strength steels.

TUBE HYDROFORMING

Tube hydroforming, illustrated in Figure 17-52, has emerged as a significant process for manufacturing strong, lightweight, tubular components, which frequently replace an assembly of welded stampings. As shown in Figure 17-53, current automotive parts include engine cradles, frame rails, roof headers, radiator supports, and exhaust components.

In elementary terms, a tubular blank, either straight or preshaped, is placed in an encapsulating die, and the ends are sealed. A fluid is then introduced through one of the end plugs, achieving sufficient pressure to expand the material to the shape of the die. At the same time, the end closures may move inward to help compensate or

(a)

(b)

FIGURE 17-52   Tube hydroforming. (a) Process schematic; (b) actual copper product. Note the inward movement of the tube ends and the nonuniform wall thickness of the nonsymmetric product.

overcome the thinning that would otherwise accompany radial expansion. In actual operation, the process may use combinations or even sequences of internal pressure, axial motions, and even external counterpressure applied to bulging regions to control the flow and final thickness of the material. Product length is currently limited by the ability of end movements to create axial displacements within the die.

FIGURE 17-53   Use of hydroformed tubes in automotive applications. *(Courtesy of MetalForming, a publication of PMA Services, Inc., for the Precision Metalforming Association, Independence, OH.)*

In *low-pressure tube hydroforming* (pressures up to 35 MPa or 5000 psi), the tube is first filled with fluid and then the dies are closed around the tube. The primary purpose of the fluid is to act as a liquid mandrel that prevents collapsing as the tube is bent to the contour of the die. While the cross-section shape can be changed, the shapes must be simple, corner radii must be large, and there must be minimal expansion of the tube diameter. In *high-pressure tube hydroforming*, an internal pressure between 100 and 700 MPa (15,000 and 100,000 psi) is used to expand the diameter of the tube, forming tight corner radii and significantly altered cross sections. *Pressure-sequence hydroforming* begins by applying low internal pressure as the die is closing. This supports the inside wall of the tube and allows it to conform to the cavity. When the die is fully closed, high pressure is then applied to complete the forming of the tube walls.

Attractive features of tube hydroforming include the ability to use lightweight, high-strength materials; the increase in strength that results from strain hardening; and the ability to utilize designs with varying thickness or varying cross section. Welded assemblies can often be replaced by one-piece components, and secondary operations can often be reduced. Disadvantages include the long cycle time (low production rate) and relatively high cost of tooling and process setup.

An emerging alternative to tube hydroforming is *hot-metal-gas forming*. In this process, a straight or preformed tube is heated to forming temperature by induction heating, and shaped using the application of pressurized inert gas rather than fluid. Parts can then be quenched directly from the high-temperature forming operation. Expectations for the process include faster speed, lower cost, and greater flexibility compared to tube hydroforming.

HOT-DRAWING OPERATIONS

Because sheet material has a large surface area and small thickness, it cools rapidly in a lower-temperature environment. For this reason, most sheet forming is performed at room or mildly elevated temperature. Cold drawing uses relatively thin metal, changes the thickness very little or not at all, and produces parts in a wide variety of shapes. In contrast, hot drawing is used for forming relatively thick-walled parts of simple geometries, and the material thickness may change significantly during the operation.

As shown in Figure 17-54, hot-drawing operations are extremely similar to previously discussed processes. The upper-left schematic shows a simple disk-to-cup drawing operation without a hold-down. While the increased thickness of the material acts to resist wrinkling, the height of the cup wall is still restricted by defect formation. When smaller diameter and higher wall height are desired, redraws, like that depicted in the upper-right schematic, can be used. The lower schematic of Figure 17-54 illustrates an alternative where the cup is pushed through a series of dies with a single punch.

FIGURE 17-54 Methods of hot drawing a cup-shaped part. (Upper left) First draw. (Upper right) Redraw operation. (Lower) Multiple-die drawing. *(Courtesy of United States Steel Corp., Pittsburgh, PA.)*

If the drawn products are designed to utilize part of the original disk as a flange around the top of the cup, the punch does not push the material completely through the die but descends to a predetermined depth and then retracts. The partially drawn cup is then ejected upward, and the perimeter of the remaining flange is trimmed to the desired size and shape.

HIGH-ENERGY-RATE FORMING

A number of methods have been developed to form metals through the application of large amounts of energy in a very short time (high strain rate). These are known as *high-energy-rate forming* processes and often go by the abbreviation *HERF*. Many metals tend to deform more readily under the ultra-rapid load application rates used in these processes. As a consequence, HERF makes it possible to form large workpieces and difficult-to-form metals with less expensive equipment and cheaper tooling than would otherwise be required. HERF processes also produce less springback. This is probably associated with two factors: (1) the high compressive stresses that are created and (2) the elastic deformation of the die produced by the ultra-high pressure.

High-energy-release rates can be obtained by five distinct methods: (1) underwater explosions, (2) underwater spark discharge (electro-hydraulic techniques), (3) pneumatic-mechanical means, (4) internal combustion of gaseous mixtures, and (5) the use of rapidly formed magnetic fields (electromagnetic techniques). Specific processes were developed around each of these approaches and attracted considerable attention during the 1960s and 1970s when they were used to produce one-of-a-kind and small quantities of parts for the space program. They are currently playing a relatively insignificant role in manufacturing technology.

IRONING

Ironing is the name given to the process that thins the walls of a drawn cylinder by passing it between a punch and die where the gap is less than the incoming wall thickness. As shown in Figure 17-55, the walls reduce to a uniform thickness and lengthen, while the thickness of the base remains unchanged. The most common example of an ironed product is the thin-walled beverage can.

EMBOSSING

Embossing, shown in Figure 17-56, is a pressworking process in which raised lettering or other designs are impressed in sheet material. Basically, it is a very shallow drawing operation where the depth of the draw is limited to 1 to 3 times the thickness of the metal and the material thickness remains largely unchanged. A common example of an embossed product is the patterned or textured industrial stair tread.

SUPERPLASTIC SHEET FORMING

Conventional metals and alloys typically exhibit tensile elongations in the range of 10 to 30%. By producing sheet materials with ultra-fine grain size and performing the defor-

FIGURE 17-55 The ironing process.

FIGURE 17-56 Embossing.

mation at low strain rates and elevated temperatures, elongations can exceed 100% and may be as high as 2000 to 3000%. This *superplastic* behavior can be used to form material into large, complex-shaped products with compound curves. Deep or complex shapes can be made as single-piece, single-operation pressings rather than multistep conventional pressings or multipiece assemblies.

At the elevated temperatures required to promote superplasticity (about 900°C for titanium, between 450° and 520°C for aluminum, or generally above half of the melting point of the material on an absolute scale), the strength of material is sufficiently low that many of the superplastic forming techniques are adaptations of processes used to form thermoplastics (discussed in Chapter 14). The tooling doesn't have to be exceptionally strong, so form blocks can often be used in place of die sets. In thermoforming, a vacuum or pneumatic pressure causes the sheet to conform to a heated male or female die. Blow forming, vacuum forming, deep drawing, and combined superplastic forming and diffusion bonding are other possibilities. Precision is excellent, and fine details or surface textures can be reproduced accurately. Springback and residual stresses are almost nonexistent, and the products have a fine, uniform grain size.

The major limitation to superplastic forming is the low forming rate that is required to maintain superplastic behavior. Cycle times may range from 2 minutes to as much as 2 hours per part, compared to the several seconds that is typical of conventional presswork. As a result, applications tend to be limited to low-volume products such as those common to the aerospace industry. By making the products larger and eliminating assembly operations, the weight of products can often be reduced, there are fewer fastener holes to initiate fatigue cracks, tooling and fabrication costs are reduced, and there is a shorter production lead time.

PROPERTIES OF SHEET MATERIAL

A wide variety of materials have been used in sheet-forming operations, including hot- and cold-rolled steel, stainless steel, copper alloys, magnesium alloys, aluminum alloys, and even some types of plastics. Sheet material can also be coated or painted, or even a clad or laminated composite.

The success or failure of many sheet-forming operations is strongly dependent on the properties of the starting material. Sheet metal has already undergone a number of processes, such as casting, hot rolling, and cold rolling, and has acquired distinct properties and characteristics. A simple uniaxial tensile test of the sheet material can be quite useful by providing values for the yield strength, tensile strength, elongation, and strain-hardening exponent. The amount of elongation prior to necking, or the uniform elongation, is probably a more useful elongation number, since localized thinning is actually a form of sheet metal failure. A low-yield, high-tensile, and high-uniform elongation all combine to indicate a large amount of useful plasticity. A high value of the strain-hardening exponent, n (obtained by fitting the true stress and true strain to the equation: stress $= K$ (strain)^n), indicates that the material will have greater allowable stretch and a more uniform stretch.

In the uniaxial tensile test, the sheet is stretched in one direction and is permitted to contract and compensate in both width and thickness. Many sheet-forming operations subject the material to stretching in more than one direction, however. When the material is in biaxial tension, as in deep drawing or the various stretch-forming operations, all of the elongation must be compensated by a decrease in thickness. The strain to fracture is typically about one-third of the value observed in a uniaxial test. As a result, some form of biaxial stretching test may be preferred to assess formability, such as a dome-height or hydraulic bulge test.

Sheet metal is often quite anisotropic—properties varying with direction or orientation. A useful assessment of this variation can be obtained through the plastic strain ratio, R. During a uniaxial tensile test, as the length increases, the width and thickness both reduce. The R-value is simply the ratio of the width strain to the thickness strain. Materials with values greater than 1 tend to compensate for stretching by flow within the sheet and resist the thinning that leads to fracture. Hence materials with high R-values have good formability. Sheet materials can also have directional variations within the plane of the sheet. These are best evaluated by performing four uniaxial tensile tests, one cut longitudinally (generally along the direction of prior rolling), one transverse (perpendicular to the first), and one along each of the 45° axes. Values are then computed for the average normal anisotropy (the sum of the four values of R divided by 4) and the planar anisotropy, ΔR (computed as $R_{\text{longitudinal}} + R_{\text{transverse}} - R_{45\,\text{right}} - R_{45\,\text{left}}$, all divided by 2). Since a ΔR of zero means that the material is uniform in all directions within the plane of the sheet, an ideal drawing material would have a high value of R and a low value of ΔR. Unfortunately, the two values tend to be coupled, such that high formability is often accompanied by directional variations within the plane of the sheet. Due to these variations, disks being drawn into cylindrical cups will have variations in the final wall height.

Sheet-forming properties have also been shown to change significantly with variations in both temperature and speed of deformation, or strain rate. Low-carbon steels generally become stronger when deformed at higher speeds, while many aluminum alloys weaken and become more prone to failure during the operation.

DESIGN AIDS FOR SHEET METAL FORMING

The majority of sheet metal failures occur due to thinning or fracture, and both are the result of excessive deformation in a given region. A quick and economical means of evaluating the severity of deformation in a formed part is to use *strain analysis* and a *forming limit diagram*. A pattern or grid, such as the one in Figure 17-57, is placed on the surface of a sheet by scribing, printing, or etching. The circles generally have diameters between 2.5 and 5 mm (0.1 to 0.2 in.) to enable detection of point-to-point variations in strain distribution. During deformation, the circles convert into ellipses, and the distorted pattern can be measured and evaluated. Regions where the enclosed area has expanded are locations of sheet thinning and possible failure. Regions where the area has contracted have undergone sheet thickening and may be sites of possible buckling or wrinkles.

Using the ellipses on the deformed grid, the major strains (strain in the direction of the largest radius or diameter) and the associated minor strains (strain 90° from the major) can be determined for a variety of locations. These values can then be plotted on a forming limit diagram such as the one shown in Figure 17-57. If both major and minor strains are positive (right-hand side of the diagram), the deformation is known as *stretching*, and the sheet metal will definitely decrease in thickness. If the minor strain is negative, this contraction may partially or wholly compensate any positive stretching in the major direction. The combination of tension and compression is known as *drawing*, and the thickness may decrease, increase, or stay the same, depending on the relative magnitudes of the two strains. Regions where both strains are negative do not appear on the diagram, since its purpose is to reveal locations of possible fractures, and fractures only occur in a tensile environment.

FIGURE 17-57 (Left) Typical pattern for sheet metal deformation analysis; (right) forming limit diagram used to determine whether a metal can be shaped without risk of fracture. Fracture is expected when strains fall above the lines.

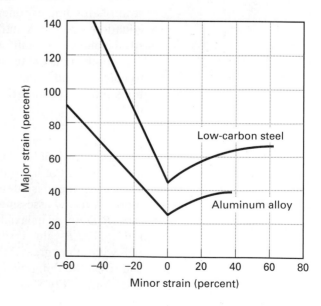

Those strains that fall above the forming limit line indicate regions of probable fracture. Possible corrective actions include modification of the lubricant, change in the die design, or variation in the clamping or hold-down pressure. Strain analysis can also be used to determine the best orientation of blanks relative to the rolling direction, assist in the design of dies for complex-shaped products, or compare the effectiveness of various lubricants. Because of the large surface-to-volume ratio in sheet material, lubrication can play a key role in process success or failure.

■ 17.5 ALTERNATIVE METHODS OF PRODUCING SHEET-TYPE PRODUCTS

ELECTROFORMING

Several manufacturing processes have been developed to produce sheet-type products by directly depositing metal onto preshaped forms or mandrels. In a process known as *electroforming*, the metal is deposited by plating. Nickel, iron, copper, or silver can be deposited in thicknesses up to 16 mm ($\frac{5}{8}$ in.). When the desired thickness has been attained, plating is stopped and the product is stripped from the mandrel.

A wide variety of sizes and shapes can be made by electroforming, and the fabrication of a product requires only a single pattern or mandrel. Low production quantities can be made in an economical fashion, with the principal limitation being the need to strip the product from the mandrel. Replication of the contact surface and profile are extremely good, but the uniformity of thickness and external profile may present problems. For applications like the production of multiple molds from a single master pattern, the interior surface is the critical one, and the wall thickness serves only to provide the necessary strength. Exterior irregularities are not critical, and various types of backup material can be employed to provide additional support. For applications where the exterior dimensions are also important, uniform deposition is required.

SPRAY FORMING

Similar parts can also be formed by *spray deposition*. One approach is to inject powdered material into a plasma torch (a stream of hot ionized gas with temperatures up to 11,000°C or 20,000°F). The particles melt and are propelled onto a shaped form or mandrel. Upon impact, the droplets flatten and undergo rapid solidification to produce a dense, fine-grained product. Multiple layers can be deposited to build up a desired size, shape, and thickness. Because of the high adhesion, the mandrel or form is often removed by machining or chemical etching. Most applications of plasma spray forming involve the fabrication of specialized products from difficult-to-form or ultra-high-melting-point materials.

The *Osprey process*, described in Chapter 18 (Section 18-10), can also be adapted to produce thin, spray-formed products. Here, molten metal flows through a nozzle, where it is atomized and carried by high-velocity nitrogen jets. The semisolid particles are propelled toward a target, where they impact and complete their solidification. Tubes, plates, and simple forms can be produced from a variety of materials. Layered structures can also be produced by sequenced deposition.

■ 17.6 PIPE WELDING

Large quantities of steel pipe are made by two processes that use the hot forming of steel strip coupled with deformation welding of the free edges. Both of these processes, *butt welding* of pipe and *lap welding* of pipe, utilize steel in the form of *skelp*—long strips with specified width, thickness, and edge configuration. Because the skelp was produced by hot rolling and the welding process produces further deformation and recrystallization, pipe welded by these processes tends to be very uniform in structure and properties.

BUTT-WELDED PIPE

In the manufacture of butt-welded pipe, steel skelp is heated to a specified hot-working temperature by passing it through a tunnel-type furnace. Upon exiting the furnace, the skelp is pulled through forming rolls that shape it into a cylinder and bring the free ends into contact. The pressure exerted between the opposite edges of the skelp is sufficient to upset the metal and produce a welded seam. Additional sets of rollers then size and shape the pipe, and it is cut to standard, preset lengths. Product diameters range from 3 mm ($\frac{1}{8}$ in.) to 75 mm (3 in.), and speeds can approach 150 m/min (500 ft/min).

LAP-WELDED PIPE

The lap-welding process for making pipe differs from the butt-welding technique in that the skelp now has beveled edges and the rolls form the weld by forcing the lapped edges down against a supporting mandrel. This process is used primarily for larger sizes of pipe, with diameters from about 50 mm (2 in.) to 400 mm (14 in.). Because the product is driven over an internal mandrel, product length is limited to about 6 to 7 m (20 to 25 ft).

■ 17.7 PRESSES

CLASSIFICATION OF PRESSES

The primary tool for performing most of the sheet-forming operations discussed in this chapter, and many of the bulk forming operations presented in Chapter 16, is some form of press, and successful manufacture often depends on using the right kind of equipment. When selecting a press for a given application, consideration should be given to the capacity required, the type of power (manual, mechanical, or hydraulic), the number of slides or drives, the type of drive, the stroke length for each drive, the type of frame or construction, and the speed of operation. Table 17-2 lists some of the major types of presses and groups them by the type of drive.

Manually operated presses such as foot-operated or *kick presses* are generally used for very light work such as shearing small sheets and thin material.

Moving toward larger equipment, we find that *mechanical drives* tend to provide fast motion and positive control of displacement. Once built, however, the flexibility of a mechanical press is limited, since the length of the stroke is set by the design of the drive. The available force usually varies with position, so mechanical presses are preferred for operations that require the maximum pressure near the bottom of the stroke, such as cutting, shallow forming, drawing (up to about 10 cm), and progressive and transfer die operations. Typical capacities range up to about 9000 metric tons.

Figure 17-58 depicts some of the basic types of mechanical press drive mechanisms. *Crank-driven* presses are the most common type because of their simplicity. They are used for most piercing and blanking operations and for simple drawing. Double-crank presses offer a means of actuating blankholders or operating multiple-action dies. *Eccentric or cam drives* are used where the ram stroke is rather short. Cam action can also provide a dwell at the bottom of the stroke and is often the preferred method of actuating the blankholder in deep-drawing processes. *Knuckle-joint drives* provide a very high mechanical advantage along with fast action. They are often preferred for coining, sizing, and Guerin forming. *Toggle mechanisms* are used principally in drawing presses to actuate the blankholder, and *screw-type drives* offer great mechanical advantage coupled with an action that resembles a drop hammer (but slower and with less impact). For this reason, screw presses have become quite popular in the forging industry.

In contrast to the mechanical presses, *hydraulic presses* produce motion as the result of piston movement, and longer or variable-length strokes can be programmed within the limitations of the cylinder (which may be as long as 250 cm or 100 in.). Forces and

TABLE 17-2	Classification of the Drive Mechanisms of Commercial Presses	
Manual	Mechanical	Hydraulic
Kick presses	Crank	Single slide
	Single	Multiple slide
	Double	
	Eccentric	
	Cam	
	Knuckle joint	
	Toggle	
	Screw	
	Rack and pinion	

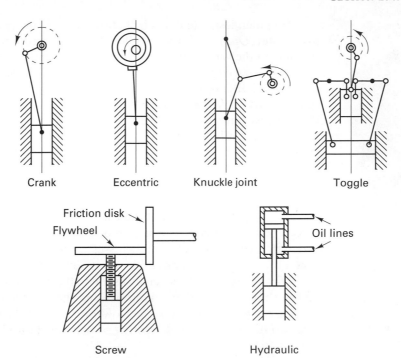

Crank   Eccentric   Knuckle joint   Toggle

Friction disk
Flywheel
Oil lines

FIGURE 17-58 Schematic representation of the various types of press drive mechanisms.

Screw   Hydraulic

pressures are more accurately controlled, and full pressure is available throughout the entire stroke. Speeds can be programmed to vary or remain constant during an operation. Since position is varied through fluid displacement, the reproducibility of position will have greater variation than a mechanical press. Hydraulic presses are available in capacities exceeding 50,000 metric tons and are preferred for operations requiring a steady pressure throughout a substantial stroke (such as deep drawing), operations requiring wide variation in stroke length, and operations requiring high or widely variable forces. In general, hydraulic presses tend to be slower than the mechanical variety, but some are available that can provide up to 600 strokes per minute in a high-speed blanking operation. By using multiple hydraulic cylinders, programmed loads can be applied to the main ram, while a separate force and timing are used on the blank holder.

TYPES OF PRESS FRAME

As shown in Table 17-3, presses should also be selected with consideration for the type of frame. Frame design often imposes limitations on the size and type of work that can be accommodated, how that work is fed and unloaded, the overall stiffness of the machine, and the time required to change dies.

Presses that have their frames in the shape of an arch (arch-frame presses) are seldom used today, except with screw drives for coining operations. *Gap-frame presses*, where the frames have the shape of the letter *C*, are among the most versatile and commonly preferred presses. They provide unobstructed access to the dies from three directions and permit large workpieces to be fed into the press. Gap-frame presses are available in a wide range of sizes, from small bench types of about 1 metric ton up to 300 metric tons or more.

TABLE 17-3	Classification of Presses According to Type of Frame	
Arch	Gap	Straight Sided
Crank or eccentric	Foot	Many variations, but all
Percussion	Bench	with straight-sided frames
	Vertical	
	Inclinable	
	Inclinable	
	Open back	
	Horn	
	Turret	

FIGURE 17-59 Inclinable gap-frame press with sliding bolster to accommodate two die sets for rapid change of tooling. *(Courtesy of Niagara Machine & Tool Works, Buffalo, NY.)*

FIGURE 17-60 Making a seam on a horn press. Note the protruding "horn" that replaces the lower press bed. *(Courtesy of Niagara Machine & Tool Works, Buffalo, NY.)*

FIGURE 17-61 A 200-ton (1800-kN) straight-sided press. *(Courtesy of Rousselle Corporation, West Chicago, IL.)*

Popular design features include open back, inclinability, adjustable bed, and sliding bolster. *Open-back presses* allow for the ejection of products or scrap through an opening in the back of the press frame. *Inclinable presses* can be tilted, so that ejection can be assisted by gravity or compressed air jets. As a result of these features, open-back inclinable (OBI) presses are the most common form of gap-frame press. The addition of an *adjustable bed* allows the base of the machine to raise or lower to accommodate different workpieces. A *sliding bolster* permits a second die to be set up on the press while another is in operation. Die changeover then requires only a few minutes to unclamp the punch segment of the active die, move the second die set into position, clamp the new punch to the press ram, and resume operation. Figure 17-59 shows an open-back inclinable gap-frame press with a sliding bolster.

A *horn press* is a special type of gap-frame press where a heavy cylindrical shaft, or "horn," appears in place of the usual bed. Curved or cylindrical workpieces can be placed over the horn for such operations as seaming, punching, and riveting, as illustrated in Figure 17-60. On some presses, both a horn and a bed are provided, with provision for swinging the horn aside when not needed.

Turret presses are especially useful in the production of sheet metal parts with numerous holes or slots that vary in size and shape. They usually employ a modified gap-frame structure and add upper and lower turrets that carry a number of punches and dies. The two turrets are geared together so that any desired tool set can be quickly rotated into position.

Straight-sided presses have frames that consist of a crown, two uprights, a base or bed, and one or more moving slides. Accessibility is generally from the front and rear, but openings are often provided in the side uprights to permit feeding and unloading of workpieces. Straight-sided presses are available in a wide variety of sizes and designs and are the preferred design for most hydraulic, large-capacity, or specialized mechanical-drive presses. As an added benefit, elastic deflections tend to be uniform across the working surface, as opposed to the angular deflections that are typical of the gap-frame design. Figure 17-61 shows a typical straight-sided press.

SPECIAL TYPES OF PRESSES

Presses have also been designed to perform specific types of operations. *Transfer presses* have a long moving slide that enables multiple operations to be performed simultaneously in a single machine. Multiple die sets are mounted side by side along the slide. After the completion of each stroke, a continuous strip is advanced or individual workpieces are transferred to the next station by a mechanism like the one shown in Figure 17-62. Transfer presses can be used to perform blanking, piercing, forming, trimming, drawing, flanging, embossing, and coining. Figure 17-63 illustrates the production of a part that incorporates a variety of these operations.

By using a single machine to perform multiple operations, transfer presses offer high production rates, high flexibility, and reduced costs (attributed to the reduced labor, floor space, energy, and maintenance). Since production is usually between 500 and 1500 parts per hour, these machines are usually restricted to operations where 4000 or more identical parts are required daily, each involving three or more separate operations. A total production run of 30,000 or more identical parts is generally desired between major changes in tooling. As a result, transfer presses are used primarily in industries such as automotive and appliances, where large numbers of identical products are being produced.

Four-slide or *multislide* machines like the one shown in Figure 17-64 are extremely versatile presses that are designed to produce small, intricately shaped parts from continuously fed wire or coil strip. The basic machine has four power-driven slides (or motions) set 90° apart. The attached tooling is controlled by cams and designed to operate in a progressive cycle. In the sheet metal variation, strip stock is fed into the machine, where it is straightened and progressively pierced, notched, bent, and cut off at the various slide stations. Figure 17-65 presents the operating mechanism of one such machine. As the material moves from right to left, it undergoes a straightening, two

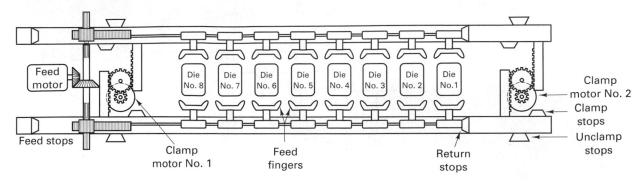

FIGURE 17-62   Schematic showing the arrangement of dies and the transfer mechanism used in transfer presses. *(Courtesy of Verson Allsteel Press Company, Chicago, IL.)*

FIGURE 17-63   Various operations can be performed during the production of stamped and drawn parts on a transfer press. *(Courtesy of U.S. Baird Corporation, Stratford, CT.)*

FIGURE 17-64   Multislide machine with guards and covers removed. *(Courtesy of U.S. Baird Corporation, Stratford, CT.)*

FIGURE 17-65 Schematic of the operating mechanism of a multislide machine. The material enters on the right and progresses toward the left as operations are performed. *(Courtesy of U.S. Baird Corporation, Stratford, CT.)*

FIGURE 17-66 Example of the piercing, blanking, and forming operations performed on a multislide machine. *(Courtesy of U.S. Baird Corporation, Stratford, CT.)*

successive pressing operations, various operations from all four directions, and a final cut-off. Figure 17-66 shows the carrier strip and the successive operations as flat strip is pierced, blanked, and formed into a folded sheet metal product. The strip stock may be up to 7.5 mm (3 in.) wide and 2.5 mm ($\frac{1}{32}$ in.) thick. Wires up to about 3 mm ($\frac{1}{8}$ in.) in diameter are also commonly processed. Products such as hinges, links, clips, and razor blades can be formed at very high rates on these machines. Setup times are long, so large production runs are preferred.

PRESS-FEEDING DEVICES

Although hand feeding may still be used in some press operations, operator safety and the desire to increase productivity have motivated a strong shift to feeding by some form of mechanical device. When continuous strip is used, it can be fed automatically by double-roll feeds mounted on the side of the press. Discrete products can be moved and positioned in a wide variety of ways. Dial-feed mechanisms enable an operator to insert workpieces into the front holes of a rotating dial, which then indexes with each stroke of the press to move the parts progressively into proper position between the punch and die. Lightweight parts can be fed by suction-cup mechanisms, vibratory-bed feeders, and similar devices. Robots are frequently used to place parts into presses and remove them after forming. See Chapter 36 for an expanded discussion of manufacturing automation.

■ Key Words

air-bend die	flanging	piercing	skelp
backward extrusion	flow turning	pillow forming	slitting
bar folder	forming	pipe welding	spinning
bending	forming limit diagram	planar anisotropy, ΔR	spray forming
blankholder force	forward extrusion	plastic strain ratio, R	springback
blanking	Guerin process	pop-rivet	squaring shears
bottoming die	high-energy-rate forming	press bending	steel-rule die
bulging	hold-down force	press brake	strain analysis
burnishing	hydraulic press	progressive die	strain-hardening exponent, n
cold forging	independent die set	punch	stretch forming
cold forming	ironing	rake angle	stretcher leveling
compound die	kick press	redrawing	stretching
compression bending	lancing	roll bending	strip
cutoff	mechanical press	roll forming	stripper plate
deep drawing	minimum bend radius	roll straightening	subpress die
dinking	modular tooling	roller burnishing	superplastic forming
draw bead	multislide press	rubber tooling	swaging
draw bending	neutral axis	seaming	thread rolling
draw ratio	nibbling	shallow drawing	transfer die
drawing	normal anisotropy	shaving	transfer press
electroforming	notching	shear	trimming
embossing	Osprey process	shear forming	tube bending
explosive forming	parallel-plate hydroforming	shearing	tube hydroforming
explosive rivet	peening	sheet	turret-type punch press
fineblanking	perforating	sheet hydroforming	

■ Review Questions

1. What distinguishes sheet forming from bulk forming?
2. What is a definition of shearing?
3. Why are sheared or blanked edges generally not smooth?
4. What measures can be employed to improve the quality of a sheared edge?
5. Why are fineblanking presses more complex than those used in conventional blanking?
6. Why might a long shearing cut be made in a progressive fashion?
7. What is a slitting operation?
8. What are the differences between piercing and blanking?
9. What are some types of blanking or piercing operations that have come to acquire specific names?
10. What is the purpose of having a shear angle on a punch?
11. Why is it important that a blanking punch and die be in proper alignment?
12. What is the major benefit of mounting punches and dies on independent die sets?
13. What is the major benefit of assembling a complex die set from standard subpress dies?
14. What is the benefit of making dies as a multipiece assembly?
15. What is a steel-rule die, and what types of materials can it cut?
16. What is a progressive die set?
17. What is the difference between progressive dies and transfer dies?
18. How do compound dies differ from progressive dies?
19. What is the attractive feature of a turret-type punch press?
20. When making bends in sheet metal, what is the distinction between bending, forming, and drawing?
21. What are the stress states on the exterior surface and interior surface of a bend?
22. Why does a metal usually become thinner in the region of a bend?
23. What is springback, and why is it a concern during bending?
24. What types of operations can be performed on a press brake?
25. What factors determine the minimum bend radius for a material?
26. If a right-angle bend is to be made in a cold-rolled sheet, should it be made with the bend lying along or perpendicular to the direction of previous rolling?
27. From a manufacturing viewpoint, why is it desirable for all bends in a product (or component) to have the same radius?
28. What is the difference between air-bend and bottoming dies? Which is more flexible? Which produces more reproducible bends?
29. What is the primary benefit of incorporating a coining action in bottom bending? The primary negative feature?
30. What type of products are produced by roll bending?
31. What is the role of the form block in draw bending and compression bending?
32. How can we prevent flattening or wrinkling when bending a tube?
33. What type of product geometry can be produced by cold-roll forming? Is the process appropriate for making short lengths of specialized products?
34. What are some methods for straightening or flattening rod or sheet?
35. What two distinctly different metalforming processes use the term *drawing*?
36. Why is the tooling cost for a spinning operation relatively low?

37. How is shear forming different from spinning?
38. For what types of products would stretch forming be an appropriate manufacturing technique?
39. What is the distinction between shallow drawing and deep drawing?
40. What is the function of the pressure ring or hold-down in a deep-drawing operation?
41. Explain why thin material may be difficult to draw into a defect-free cup.
42. How can redraw operations be used to produce a taller, smaller diameter cup than can be produced in a single deep-drawing operation?
43. What are draw beads, and what function do they perform?
44. Why is a trimming operation often included in a deep-drawing manufacturing sequence?
45. How does the Guerin process reduce the cost of tooling in a drawing operation?
46. How can fluid pressure or rubber tooling be used to perform bulging?
47. What is sheet hydroforming?
48. What explanation can be given for the greater formability observed during sheet hydroforming? How is this feature being used in the automotive industry?
49. What is the purpose of the inward movement of the end plugs during tube hydroforming?
50. What is the difference between low-pressure tube hydroforming and high-pressure tube hydroforming in terms of both pressures and the nature of the deformations produced?
51. What are some of the basic methods that have been used to achieve the high-energy-release rates needed in the HERF processes?
52. Why is springback rather minimal in high-energy-rate forming?

53. What are some well-known products that have been produced by processes including ironing? By embossing?
54. What material and process conditions are associated with superplastic forming?
55. What is the major limitation of the superplastic forming of sheet metal? What are some of the more attractive features?
56. What properties from a uniaxial tensile test can be used to assess sheet metal formability?
57. How is the formability in biaxial tension different from that in uniaxial tension?
58. What techniques can be used to assess "normal anisotropy" and "planar anisotropy" in sheet material?
59. How can strain analysis be used to determine locations of possible defects or failure in sheet metal components?
60. What is a forming limit diagram?
61. Explain the key difference between the right- and left-hand sections of a forming limit diagram. These sections correspond to stretching and drawing.
62. Describe two alternative methods of producing complex-shaped thin products without requiring sheet metal deformation techniques.
63. What two hot-forming operations can be used to produce pipe from steel strip?
64. What are the primary assets and limitations of mechanical press drives? Of hydraulic drives?
65. What are some of the common types of press frames?
66. What is the purpose of inclining or tilting a press?
67. Describe how multiple operations are performed simultaneously in a transfer press.
68. What types of products are produced on a four-slide or multislide machine?

■ Problems

1. The maximum punch force in blanking can be estimated by the equation:

$$\text{force} = S\,t\,L$$

where:

   S = the material shear strength
   t = the sheet thickness
   L = the total length of sheared edge (circumference or perimeter)

   a. How would this number change if the rake angle is equal to a $1t$ change across the width or diameter of the part being sheared?
   b. How would this number change if the rake were increased to a $3t$ change across the width or diameter?

2. Consider the various means of producing tubular products, such as extrusion, seam welding, butt welding during forming, piercing, and the various drawing operations. Describe the advantages, limitations, and typical applications of each.
3. Tube and sheet hydroforming have been undergoing rapid growth. Investigate current uses for these processes in automotive and other fields.
4. What are some of the techniques for minimizing the amount of springback in sheet-forming operations?
5. Select a forming process from either Chapter 16 or 17, and investigate the residual stresses that typically accompany or result from that process. If they are considered to be detrimental, how could these residual stresses be reduced or removed without damaging or deteriorating the product?

www.wiley.com/college/degarmo

Chapter 17 CASE STUDY

Fabrication of a One-Piece Brass Flashlight Case

The figure presents the sheet-metal-forming sequence to produce a one-piece flashlight case from CDA alloy 268, also known as yellow brass (66% copper, 34% zinc). As shown, the process involves a blanking operation, followed by three deep draws (the third of which is only a partial), and then several operations to shape the upper segment of the flashlight. The interior surface will be left as-fabricated, and the brass side walls of the case become part of the completed electrical circuit during operation. A decorative chrome plating is subsequently applied to the exterior.

The Copper Development Association's *Standards Handbook* cites a number of electrical applications for this alloy and particularly cites flashlight shells, so it would appear that a good material choice has been made. Among the cited "common fabrication processes" are blanking, drawing, forming and bending, spinning, and stamping. It would appear that the material is indeed compatible with the proposed processing activities. The "capacity for being cold worked" is rated as excellent, as is suitability for soldering and brazing.

In the annealed condition, the yield strength varies from 14 to 22 ksi, depending on grain size, while the companion tensile strength is 46 to 53 ksi and percent elongation ranges from 54 to 65%. Cold working brings about the following property changes:

The sheet metal fabrication sequence for a one-piece brass flashlight case.

Condition	Yield Strength(ksi)	Tensile Strength (ksi)	% Elongation
Quarter hard	40.0	54.0	43
Half hard	50.0	61.0	23
Full hard	60.0	74.0	8
Extra hard	62.0	85.0	5

1. What properties make this an appropriate material for this application?
2. What features would you want to specify when purchasing your starting material? Consider temper, grain size, surface condition, edge condition, and so on.
3. Discuss any significant features or concerns with regard to the blanking, first deep draw, redraw, and second partial redraw operations.
4. It is likely that some form of intermediate (and possibly final) anneal will be required. Prescribe a suitable procedure for this material, including the necessary details of time, temperature, cooling conditions, and so on. Would some form of protective atmosphere be required? If so, what would you recommend?
5. Since this material is capable of significant cold-work strengthening, would it be advantageous to market your flashlight in its cold-worked condition? What would be the advantages and disadvantages of this option? If you feel it is advantageous, what would be your desired final condition?
6. Yellow brass is quite susceptible to stress–corrosion cracking. What could you do to minimize this mode of failure?
7. Several alternative materials have been proposed for this application. Provide a brief evaluation of each, considering both fabrication and use.

   a. 5000 series aluminum sheet (the material often used in car bodies)—does not respond to age hardening
   b. 6061 aluminum sheet (the material used in canoe skins)—can be age hardened
   c. 1008 steel sheet that has been copper plated on both sides (inside for electrical conductivity and outside as a base for chrome plate)

CHAPTER 18

POWDER METALLURGY

■ 18.1 INTRODUCTION

Powder metallurgy is the name given to the process by which fine powdered materials are blended, pressed into a desired shape (compacted), and then heated (sintered) in a controlled atmosphere to bond the contacting surfaces of the particles and establish desired properties. The process, commonly designated as P/M, readily lends itself to the mass production of small, intricate parts of high precision, often eliminating the need for additional machining or finishing. There is little material waste, unusual materials or mixtures can be utilized, and controlled degrees of porosity or permeability can be produced. Major areas of application tend to be those for which the P/M process has strong economical advantage or where the desired properties and characteristics would be difficult to obtain by any other method. Because of its level of manufacturing maturity, powder metallurgy should actually be considered as a possible means of manufacture for any part where the geometry and production quantity are appropriate.

While a crude form of iron powder metallurgy existed in Egypt as early as 3000 B.C., and the ancient Incas made jewelry and other artifacts from precious metal powders, mass manufacturing of P/M products did not begin until the mid- or late-nineteenth century. At this time, powder metallurgy was used to produce copper coins and medallions, platinum ingots, lead printing type, and tungsten wires, the primary material for light bulb filaments. By the 1920s the tips of tungsten carbide cutting tools and nonferrous bushings were being produced. Self-lubricating bearings and metallic filters were other early products.

A period of rapid technological development occurred after World War II, based primarily on automotive applications, and iron and steel replaced copper as the dominant P/M material. Aerospace and nuclear developments created accelerated demand for refractory and reactive metals, materials for which powder processing is quite attractive. Full-density products emerged in the 1960s, and high-performance superalloy components, such as aircraft turbine engine parts, were a highlight of the 1970s. Developments in the 1980s and 1990s included the commercialization of rapidly solidified and amorphous powders as well as P/M injection molding technology.

Recent years have been ones of rapid growth for the P/M industry. From 1960 to 1980, the consumption of iron powder increased 10-fold. A similar increase occurred between 1980 and 1990, and the exponential growth continued through the 1990s. While most products are still under 50 mm (2 in.) in size, some have been produced with weights up to 45 kg (100 lb) with linear dimensions up to 500 mm (20 in.).

Automotive applications now account for nearly 70% of the powder metallurgy market. In 1990, the average U.S. automobile contained about 10 kg (21 lb) of P/M parts. By 1995, the amount had increased to over 13.6 kg (30 lb)—then to 16.3 kg (36 lb) by 2000 and to 19.5 kg (nearly 45 lb) by 2005. Some Ford, Chrysler, and General Motors V-8 engines contain between 27 and 33 pounds of P/M parts. A Chrysler V-6 engine contains 88 P/M parts with a total weight of 27 pounds. A high percentage of connecting rods are now produced by powder metallurgy. Automatic transmissions often contain between 15 and 25 pounds of P/M components.

Other areas where powder metallurgy products are used extensively include household appliances, recreational equipment, hand and power tools, hardware items, office equipment, industrial motors, and hydraulics. Areas of rapid growth include aerospace applications, advanced composites, electronic components, magnetic materials, metalworking tools, and a variety of biomedical and dental applications. Iron and low-alloy steels now account for 85% of all P/M usage, with copper and copper-based powders comprising about 7%. Stainless steel, high-strength and high-alloy steels, and aluminum and aluminum alloys are other high-volume materials. Titanium, magnesium, refractory metals, particulate composites, and intermetallics are seeing increased use.

■ 18.2 THE BASIC PROCESS

The powder metallurgy process generally consists of four basic steps: (1) powder manufacture, (2) mixing or blending, (3) compacting, and (4) sintering. Compaction is generally performed at room temperature, and the elevated-temperature process of sintering is usually conducted at atmospheric pressure. Optional secondary processing often follows to obtain special properties or enhanced precision. Figure 18-1 presents a simplified flow chart of the conventional die-compaction P/M process.

■ 18.3 POWDER MANUFACTURE

The properties of powder metallurgy products are highly dependent on the characteristics of the starting powders. Some important properties and characteristics include *chemistry and purity, particle size, size distribution, particle shape*, and the *surface texture* of the particles. Several processes can be used to produce powdered material, with each imparting distinct properties and characteristics to the powder and hence to the final product.

Over 80% of all commercial powder is produced by some form of melt *atomization*, where liquid material is fragmented into small droplets that cool and solidify into particles. Various methods have been used to form the droplets, several of which are

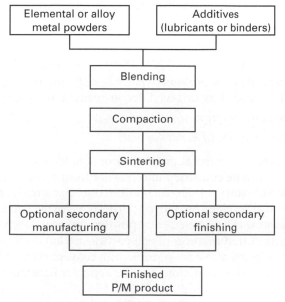

FIGURE 18-1 Simplified flow chart of the basic powder metallurgy process.

FIGURE 18-2   Two methods for producing metal powders: (a) melt atomization and (b) atomization from a rotating consumable electrode.

illustrated in Figure 18-2. Part (a) illustrates *gas atomization*, where jets of high-pressure gas (usually nitrogen, argon, or helium) strike a stream of liquid metal as it emerges from an orifice. Pressurized liquid (usually water) can replace the pressurized gas, converting the process to *liquid atomization* or *water atomization*. In part (b), an electric arc impinges on a rapidly rotating electrode. Centrifugal force causes the molten droplets to fly from the surface of the electrode and freeze in flight. Particle size is very uniform and can be varied by changing the speed of rotation.

Regardless of the specific process, atomization is an extremely useful means of producing *prealloyed powders*. By starting with an alloyed melt or prealloyed electrode, each powder particle has the desired alloy composition. Powders of aluminum alloys, copper alloys, stainless steel, nickel-based alloys (such as Monel), titanium alloys, cobalt-based alloys, and various low-alloy steels have all been commercially produced. The size, shape, and surface texture of the powder particles vary, depending on such process features as the velocity and media of the atomizing jets or the speed of electrode rotation, the starting temperature of the liquid (which affects the time that surface tension can act on the individual droplets prior to solidification), and the environment provided for cooling. When cooling is slow (such as in gas atomization) and surface tension is high, smooth-surface spheres can form before solidification. With the more rapid cooling of water atomization, irregular shapes tend to be produced.

Other methods of powder manufacture include:

1. *Chemical reduction of particulate compounds* (generally crushed oxides or ores). A large amount of iron powder is produced by reducing iron ore or rolling mill scale. The resulting powders are usually irregular in shape and spongy in texture.
2. *Electrolytic deposition* from solutions or fused salts with process conditions favoring the production of a powdery deposit that does not adhere to the cathode.
3. *Pulverization* or *grinding* of brittle materials (comminution).
4. *Thermal decomposition of particulate hydrides or carbonyls.* Iron and nickel powders are produced by carbonyl decomposition, resulting in small, spherical particles.
5. *Precipitation from solution.*
6. *Condensation of metal vapors.*

Almost any metal, metal alloy, or nonmetal (ceramic, polymer, or wax or graphite lubricant) can be converted into powder form by one or more of the powder production methods. Some methods can produce only elemental powder (often of high purity), while others can produce prealloyed particles. Alloying can also be achieved mechanically by processes that cause elemental powders to successively adhere and break apart. Material is transferred as traces of one particle are left on the other. Unusual compositions can be produced that are not possible with conventional melting. All of the powders may also undergo further operations, such as drying or heat treatment, prior to further processing.

■ 18.4 RAPIDLY SOLIDIFIED POWDER (MICROCRYSTALLINE AND AMORPHOUS)

Increasing the cooling rate of an atomized liquid can result in the formation of an ultra-fine or microcrystalline grain size. In these materials, a large percentage of the atoms are located in grain boundary regions, giving unusual properties (such as high diffusivity), expanded alloy possibilities, and good formability. If the cooling rate approaches or exceeds 10^6 °C/sec, metals can solidify without becoming crystalline. These *amorphous* or glassy metals can also exhibit unusual or unique properties, which include high strength, improved corrosion resistance, and reduced energy to induce and reverse a magnetization. Amorphous metal transformer cores lose 60 to 70% less energy in magnetization than conventional silicon steels. As a result, it is estimated that over half of all new power distribution transformers purchased in the United States will utilize amorphous metal cores.

Production of amorphous material, however, requires immensely high cooling rates and hence ultra-small dimensions. Atomization with rapid cooling and the "splat quenching" of a metal stream onto a cool surface to produce a continuous ribbon are two prominent methods. Since much of the ribbon material is further fragmented into powder, powder metallurgy is the primary means of fabricating useful products from amorphous material.

■ 18.5 POWDER TESTING AND EVALUATION

Key properties of powdered material include bulk chemistry, surface chemistry, particle size and size distribution, particle shape, surface texture, and internal structure. In addition, powders should also be evaluated for their suitability for further processing. *Flow rate* measures the ease by which powder can be fed and distributed into a die. Poor flow characteristics can result in nonuniform die filling as well as nonuniform density and properties in a final product.

Associated with the flow characteristics is the *apparent density*, a measure of a powder's ability to fill available space without the application of external pressure. A low apparent density means that there is a large fraction of unfilled space in the loose-fill powder. *Compressibility* tests evaluate the effectiveness of applied pressure in raising the density of the powder, and *green strength* is used to describe the strength of the pressed powder immediately after compacting. It is well established that higher product density correlates with superior mechanical properties, such as strength and fracture resistance. Good green strength is required to maintain smooth surfaces, sharp corners, and intricate details during ejection from the compacting die or tooling and the subsequent transfer to the sintering operation.

The overall objective is often to achieve a useful balance of the key properties. The smooth-surface spheres produced by gas atomization, for example, tend to pour and flow well, but the compacts have extremely low green strength, disintegrating easily during handling. The irregular particles of water-atomized powder have better compressibility and green strength but poorer flow characteristics. The sponge iron powders produced by chemical reduction of iron oxide are extremely porous and have highly irregular, extremely rough surfaces. They have poor flow characteristics and low compacted density, but green strength is quite high. Thus, the same material can have widely different performance characteristics, depending upon the specifics of powder manufacture.

■ 18.6 POWDER MIXING AND BLENDING

It is rare that a single powder will possess all of the characteristics desired in a given process and product. Most likely, the starting material will be a mixture of various grades or sizes of powder, or powders of different compositions, along with additions of *lubricants* or *binders*.

In powder products, the final chemistry is often obtained by combining pure metal or nonmetal powders, rather than starting with prealloyed material. To produce a uniform chemistry and structure in the final product, therefore, sufficient diffusion must occur during the sintering operation. Unique *composites* can also be produced, such as the distribution of an immiscible reinforcement material in a matrix, or the combination of metals and nonmetals in a single product such as a tungsten carbide–cobalt matrix cutting tool for high-temperature service.

Some powders, such as graphite, can even play a dual role, serving as a lubricant during compaction and a source of carbon as it alloys with iron during sintering to produce steel. Lubricants such as graphite or stearic acid improve the flow characteristics and compressibility at the expense of reduced green strength. Binders produce the reverse effect. Since most lubricants or binders are not wanted in the final product, they are removed (volatilized or burned off) in the early stages of sintering, leaving holes that are reduced in size or closed during subsequent heating.

Blending or *mixing* operations can be done either dry or wet, where water or other solvent is used to enhance particle mobility, reduce dust formation, and lessen explosion hazards. Large lots of powder can be homogenized with respect to both chemistry and distribution of components, sizes, and shapes. Quantities up to 16,000 kg (35,000 lb) have been blended in single lots to ensure uniform behavior during processing and the production of a consistent product.

■ 18.7 COMPACTING

One of the most critical steps in the P/M process is *compaction*. Loose powder is compressed and densified into a shape known as a green compact, usually at room temperature. High product density and the uniformity of that density throughout the compact are generally desired characteristics. In addition, the mechanical interlocking and cold welding of the particles should provide sufficient green strength for in-process handling and transport to the sintering furnace.

Most compacting is done with mechanical presses and rigid tools, but hydraulic and hybrid (combinations of mechanical, hydraulic, and pneumatic) presses can also be used. Figure 18-3 shows a typical mechanical press for compacting powders and a removable set of compaction tooling. The removable die sets allow the time-consuming alignment and synchronization of tool movements to be set up while the press is producing parts with another die set. Compacting pressures generally range between 3 and 120 tons/in². depending on material and application (see Table 18-1), with the range of 10 to 50 tons/in². being the most common. While most P/M presses have total capacities of less than 100 tons, increasing numbers are being purchased with higher capacity. Because of pressures and press capacity, powder metallurgy products are often limited to pressing areas of less than 10 square inches, but larger parts have become more common. Some P/M presses now have capacities up to 3000 tons and are capable of pressing areas up to 100 square inches. When even larger products are desired, compaction can be performed by dynamic methods, such as use of an explosively induced shock wave. Metalforming processes, such as rolling, forging, extrusion, and swaging, have also been adapted for powder compaction.

TABLE 18-1	Typical Compacting Pressures for Various Applications	
	Compaction Pressures	
Application	tons/in.²	Mpa
Porous metals and filters	3–5	40–70
Refractory metals and carbides	5–15	70–200
Porous bearings	10–25	146–350
Machine parts (medium-density iron & steel)	20–50	275–690
High-density copper and aluminum parts	18–20	250–275
High-density iron and steel parts	50–120	690–1650

FIGURE 18-3 (Left) Typical press for the compacting of metal powders. A removable die set (right) allows the machine to be producing parts with one die set while another is being fitted to produce a second product. (*Courtesy of Alfa Laval, Inc., Warminster, PA.*)

Figure 18-4 shows the typical compaction sequence for a mechanical press. With the bottom punch in its fully raised position, a feed shoe moves into position over the die. The feed shoe is an inverted container filled with powder, connected to a large powder container by a flexible feed tube. With the feed shoe in position, the bottom punch descends to a preset fill depth, and the shoe retracts, with its edges leveling the powder. The upper punch then descends and compacts the powder as it penetrates the die. The upper punch retracts and the bottom punch then rises to eject the green compact. As the die shoe advances for the next cycle, its forward edge clears the compacted product from the press, and the cycle repeats.

During uniaxial or one-direction compaction, the powder particles move primarily in the direction of the applied force. Since the loose-fill dimensions are 2 to 2.5 times the pressed dimensions, the amount of particle travel in the pressing direction can be substantial. The amount of lateral flow, however, is quite limited. In fact, it is rare to find a particle in the compacted product that has moved more than three particle diameters off of its original axis of pressing. Thus, the powder does not flow like a liquid; it simply

FIGURE 18-4 Typical compaction sequence for a single-level part, showing the functions of the feed shoe, die, core rod, and upper and lower punches. Loose powder is shaded; compacted powder is solid black.

FIGURE 18-5   Compaction with a single moving punch, showing the resultant nonuniform density (shaded), highest where particle movement is the greatest.

FIGURE 18-6   Density distribution obtained with a double-acting press and two moving punches. Note the increased uniformity compared to Figure 18-5. Thicker parts can be effectively compacted.

compresses until an equal and opposing force is created. This opposing force is probably a combination of (1) resistance by the bottom punch and (2) friction between the particles and the die surfaces. Densification occurs by particle movement as well as plastic deformation of the individual particles.

As illustrated in Figure 18-5, when the pressure is applied by only one punch, maximum density occurs below the punch and decreases as one moves down the column. It is very difficult to transmit uniform pressures and produce uniform density throughout a compact, especially when the thickness is large. By use of a double-action press, where pressing movements occur from both top and bottom (Figure 18-6), thicker products can be compacted to a more uniform density. Since side-wall friction is a key factor in compaction, the resulting density shows a strong dependence on both the thickness and width of the part being pressed. For uniform compaction, the ratio of thickness/width should be kept below 2.0 whenever possible. When the ratio exceeds 2.0, the products tend to exhibit considerable variation in density.

As shown in Figure 18-7, the average density of the compact depends on the amount of pressure that is applied, with the specific response being strongly dependent on the characteristics of the powder being compressed (its size, shape, surface texture, mechanical properties, etc.). The final density may be reported as either an absolute density in units such as grams per cubic centimeter or as a percentage of the pore-free or theoretical density. The difference between this percentage and 100% corresponds to the amount of void space remaining within the compact.

Figure 18-8 shows that a single displacement will produce different degrees of compaction in different thicknesses of powder. It is impossible, therefore, for a single punch to produce uniform density in a multithickness part. When more than one thickness is required, more complicated presses or compaction methods must be employed. Figure 18-9 illustrates two methods of compacting a dual-thickness part. By providing different amounts of motion to the various punches and synchronizing these movements to provide simultaneous compaction, a uniform-density product can be produced.

Since the complexity of the part dictates the complexity of equipment, powder metallurgy components have been grouped into classes. Class 1 components are the simplest and easiest to compact. They are thin, single-level parts that can be pressed with a force from one direction. The thickness is generally less than $\frac{1}{4}$ inch (6.35 mm). Class 2 parts are single-level parts of any thickness that require pressing from two directions.

FIGURE 18-7   Effect of compacting pressure on green density (the density after compaction but before sintering). Separate curves are for several commercial powders.

FIGURE 18-8 Compaction of a two-thickness part with only one moving punch. (a) Initial conditions; (b) after compaction by the upper punch. Note the drastic difference in compacted density.

3/4 original volume or 1-1/3 × original density

1/2 original volume or 2 × original density

Initial conditions

(a)

After compaction

(b)

Single lower punch

Double lower punch

FIGURE 18-9 Two methods of compacting a double-thickness part to near-uniform density. Both involve the controlled movement of two or more punches.

These are usually thicker parts. Class 3 parts are double-level parts that require pressing from two directions. Class 4 parts are the most complex of those produced by rigid die compaction. They are multilevel parts that require two or more pressing motions. These four classes are summarized in both Table 18-2 and Figure 18-10.

If an extremely complex shape is desired, the powder is generally encapsulated in a flexible mold, which is then immersed in a pressurized gas or liquid. This process is known as *isostatic* (uniform-pressure) *compaction*. Because the pressure is applied in all directions, lower compaction pressures produce densities higher than conventional punch-and-die compaction. Production rates are extremely low, but parts with weights up to several hundred pounds have been effectively compacted.

Warm compaction emerged as a common practice in the 1990s. By preheating the powder prior to pressing, the metal is softened and responds better to the applied pressures. The better compaction results in improved properties, both in the as-compacted state and after final processing.

TABLE 18-2	Features that Define the Various Classes of Press-and-Sinter P/M Parts	
Class	Levels	Press Actions
1	1	Single
2	1	Double
3	2	Double
4	More than 2	Double or multiple

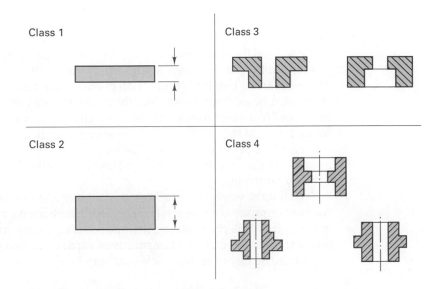

Class 1

Class 3

Class 2

Class 4

FIGURE 18-10 Sample geometries of the four basic classes of press-and-sinter powder metallurgy parts. Note the increased pressing complexity that would be required as class increases.

Compaction can also be enhanced by increasing the amount of lubricant in the powder. This reduces the friction between the powder and the die wall, as well as improving the transmission of pressure through the powder. If too much lubricant is used, however, the green strength may be reduced to the point where it is insufficient for part ejection and handling, or the final properties may become unacceptable.

While pressing rates vary widely, small mechanical presses can typically compact up to 100 pieces per minute. By means of bulk movement of particles, deformation of individual particles, and particle fracture or fragmentation, mechanical compaction can raise the density of loose powder to about 80% of an equivalent cast or forged metal. Sufficient strength can be imparted to retain the shape and permit a reasonable amount of careful handling. In addition, the compaction process sets both the nature and distribution of the porosity remaining in the product.

Because powder particles tend to be somewhat abrasive and high pressures are involved during compaction, wear of the tool components is a major concern. Consequently, compaction tools are usually made of hardened tool steel. For particularly abrasive powders, or for high-volume production, cemented carbides may be employed. Die surfaces should be highly polished and the dies should be heavy enough to withstand the high pressing pressures. Lubricants are also used to reduce die wear.

■ 18.8 SINTERING

In the *sintering* operation, the pressed-powder compacts are heated in a controlled atmosphere to a temperature below the melting point but high enough to permit solid-state diffusion, and held for sufficient time to permit bonding of the particles. Most metals are sintered at temperatures of 70 to 80% of their melting point, while certain refractory materials may require temperatures near 90%. Table 18-3 presents a summary of some common sintering temperatures. When the product is composed of more than one material, the sintering temperature may be above the melting temperature of one or more components. The lower-melting-point materials then melt and flow into the voids between the remaining particles, and the process becomes *liquid-phase sintering*.

Most sintering operations involve *three stages*, and many sintering furnaces employ three corresponding zones. The *first operation*, the *burn-off* or *purge*, is designed to combust any air, volatilize and remove lubricants or binders that would interfere with good bonding, and slowly raise the temperature of the compacts in a controlled manner. Rapid heating would produce high internal pressure from air entrapped in closed pores or volatilizing lubricants, and would result in swelling or fracture of the compacts. When the compacts contain appreciable quantities of volatile materials, their removal creates additional *porosity* and *permeability* within the pressed shape. The manufacture of products such as metal filters is designed to take advantage of this feature. When the products are load-bearing components, however, high amounts of porosity are undesirable, and the amount of volatilizing lubricant is kept to an optimized minimum. The *second*, or *high-temperature, stage* is where the desired solid-state diffusion and bonding between the powder particles take place. As the material seeks to lower its surface energy, atoms move toward the points of contact between the particles. The areas of contact become larger, and the part becomes a solid mass with small pores of various sizes and shapes. The mechanical bonds of compaction become true metallurgical bonds. The time in this stage must be sufficient to produce the desired density and final properties, usually varying from 10 minutes to several hours. Finally, a *cooling period* is required to lower the temperature of the products while retaining them in a controlled atmosphere. This feature serves to prevent oxidation that would occur upon direct discharge into air as well as possible thermal shock from rapid cooling. Both batch and continuous furnaces are used for sintering.

All three stages of sintering must be conducted in the oxygen-free conditions of a *vacuum* or *protective atmosphere*. This is critical because the compacted shapes typically have 10 to 25% residual porosity, and some of the internal voids are connected to exposed surfaces. At elevated temperatures, rapid oxidation would occur and significantly impair the quality of interparticle bonding. *Reducing atmospheres*, commonly based on

TABLE 18-3	Typical Sintering Temperatures for Some Common Metals and Materials	
	Sintering Temperature	
Metal	°C	°F
Aluminum alloys	590–620	1095–1150
Brass	850–950	1550–1750
Copper	750–1000	1400–1850
Iron/steel	1100–1200	2000–2200
Stainless steel	1200–1280	2200–2350
Cemented carbides	1350–1450	2450–2650
Molybdenum	1600–1700	2900–3100
Tungsten	2200–2300	4000–4200
Various ceramics	1400–2100	2550–3800

hydrogen, dissociated ammonia, or cracked hydrocarbons, are preferred since they can reduce any oxide already present on the particle surfaces and combust harmful gases that are liberated during the sintering. *Inert gases* cannot reduce existing oxides but will prevent the formation of any additional contaminants. *Vacuum* sintering is frequently employed with stainless steel, titanium, and the refractory metals. *Nitrogen atmospheres* are also common.

During the sintering operation, a number of changes occur in the compact. Metallurgical bonds form between the powder particles as a result of solid-state atomic diffusion, and strength, ductility, toughness, and electrical and thermal conductivities all increase. If different chemistry powders were blended, interdiffusion promotes the formation of alloys or intermetallic phases. As the pores reduce in size, there will be a concurrent increase in density and contraction in product dimensions. To meet final tolerances, the dimensional shrinkage will have to be compensated through the design of oversized compaction dies. During sintering, not all of the porosity is removed, however. Conventional pressed-and-sintered P/M products generally contain between 5 and 25% residual porosity.

Sinter brazing is a process in which two or more separate pieces are joined by brazing while they are also being sintered. The individual pieces are compacted separately, and are assembled with the braze metal positioned so it will flow into the joint. When the assembly is heated for sintering, the braze metal melts and flows between the joint surfaces to create the bond. As sintering continues, much of the braze metal diffuses into the surrounding metal, producing a final joint that is often stronger than the materials being joined.

■ 18.9 HOT-ISOSTATIC PRESSING

In conventional press-and-sinter powder metallurgy, the pressing or compaction is usually performed at room temperature and the sintering, at atmospheric pressure. *Hot-isostatic pressing* (HIP) combines powder compaction and sintering into a single operation that involves gas-pressure squeezing at elevated temperature. While this may seem to be an improvement over the two-step approach, it should be noted that heated powders may need to be "protected" or isolated from harmful environments, and the pressurizing media must be prevented from entering the voids between the particles. One approach to hot-isostatic pressing begins by sealing the powder in a flexible, airtight, evacuated container, which is then subjected to a high-temperature, high-pressure environment. Conditions for processing irons and steels involve pressures around 10,000 to 15,000 psi (70 to 100 MPa) coupled with temperatures in the neighborhood of 1250°C (2300°F). For the nickel-based superalloys, refractory metals, and ceramic powders, the equipment must be capable of 45,000 psi (310 MPa) and 1500°C (2750°F). Multiple pieces, totaling up to several tons, can now be processed in a single cycle that typically lasts several hours.

After processing, the products emerge at full density with uniform, isotropic properties that are often superior to those of other processes. Near-net shapes are possible, thereby reducing material waste and costly machining operations. Since the powder is totally isolated and compaction and sintering occur simultaneously, the process is attractive for reactive or brittle materials, such as beryllium, uranium, zirconium, and titanium. Since die compaction is not required, large parts are now possible, and shapes can be produced that would be impossible to eject from rigid compaction dies. Hot-isostatic pressing has also been employed to densify existing parts (such as those that have been conventionally pressed and sintered), heal internal porosity in castings, and seal internal cracks in a variety of products. The elimination or reduction of defects yields startling improvements in strength, toughness, fatigue resistance, and creep life.

Several aspects of the HIP process make it expensive and unattractive for high-volume production. The first is the high cost of *canning* the powder in a flexible isolating medium that can resist the subsequent temperatures and pressures, and then later removing this material from the product (*decanning*). Sheet metal containers are most common, but glass and even ceramic molds have been used. The second problem involves the relatively long time for the HIP cycle. While process advances have reduced cycle times from 24 hours to 6 to 8 hours, production is still limited to several loads a day, and the number of parts per load is limited by the ability to produce and maintain uniform temperature throughout the pressure chamber.

The *sinter-HIP* process and *pressure-assisted sintering* are techniques that have been developed to produce full-density powder products without the expense of canning and decanning. Conventionally compacted P/M parts are placed in a pressurizable chamber and sintered (heated) under vacuum for a time that is sufficient to seal the surface and isolate all internal porosity. (*Note:* This generally requires achieving a density greater than 92 to 95%.) While maintaining the elevated temperature, the vacuum is broken and high pressure is then applied for the remainder of the process. The sealed surface produced during the vacuum sintering acts as an isolating can during the high-pressure stage. Since these processes start with as-compacted powder parts, they eliminate the additional heating and cooling cycle that would be required if parts were first sintered in the conventional manner and then subjected to the HIP process for further densification.

■ 18.10 OTHER TECHNIQUES TO PRODUCE HIGH-DENSITY P/M PRODUCTS

The high-temperature metal deformation processes can also be used to produce high-density P/M parts. Sheets of sintered powder (produced by roll compaction and sintering) can be reduced in thickness and further densified by hot rolling in the process depicted in Figure 18-11. Rods, wires, and small billets can be produced by the hot extrusion of encapsulated powder or pressed-and-sintered slugs. Forging can be applied to form complex shapes from canned powder or simple-shaped sintered preforms. By using powdered material, these processes offer the combined benefits of powder metallurgy and the respective forming process, such as the production of fabricated shapes with uniform fine grain size, uniform chemistry, or unusual alloy composition.

FIGURE 18-11 One method of producing continuous sheet products from powdered feedstock.

The *Ceracon process* is another method of raising the density of conventional pressed-and-sintered P/M products without requiring encapsulation or canning. A heated preform is surrounded by hot granular material, usually smooth-surface ceramic particles. When the assembly is then compacted in a conventional hydraulic press, the granular material transmits a somewhat uniform pressure. Encapsulation is not required since the pressurizing medium is not capable of entering pores in the material. When the pressure cycle is complete, the part and the pressurizing medium separate freely, and the pressure-transmitting granules are reheated and reused.

Yet another means of producing a high-density shape from fine particles is *in situ compaction* or *spray forming* (also known as the *Osprey process*). Consider an atomizer similar to that of Figure 18-2a, in which jets of inert or harmless gas (nitrogen or carbon dioxide) propel molten droplets down into a collecting container. If the droplets solidify before impact, the container fills with loose powder. If the droplets remain liquid during their flight, the container fills with molten metal, which then solidifies into a conventional casting. However, if the cooling of the droplets is controlled so that they are semisolid (and computers can provide the necessary process control), they act as "slush balls" and flatten upon impact. The remaining freezing occurs quickly, and the resultant product is a uniform chemistry, fine grain size, high-density (in excess of 98%) solid. Depending on the shape of the collecting container, the spray-formed product can be a finished part, a strip or plate, a deposited coating, or a preform for subsequent operations, such as forging. Both ferrous and nonferrous products can be produced with deposition rates as high as 200 kilograms (400 pounds) per minute. Unique composites can be produced by the simultaneous deposition of two or more materials, injecting secondary particles into the stream and promoting in-stream reactions.

■ 18.11 METAL INJECTION MOLDING (MIM) OR POWDER INJECTION MOLDING (PIM)

For many years, injection molding has been used to produce small, complex-shaped components from plastic. A thermoplastic resin is heated to impart the necessary degree of fluidity and is then pressure injected into a die, where it cools and hardens. Die casting is a similar process for metals but is restricted to alloys with relatively low melting temperature, such as lead-, zinc-, aluminum-, and copper-based materials. Small, complex-shaped products of the higher-melting-point metals are generally made by more costly processes, which include investment casting, machining directly from metal stock, or conventional powder metallurgy. *Metal injection molding (MIM),* also called *powder injection molding (PIM),* is a rather recent extension of conventional powder metallurgy that combines the shape-forming capability of plastics, the precision of die casting, and the materials flexibility of powder metallurgy.

Since powdered material does not flow like a fluid, complex shapes are produced by first combining ultra-fine (usually in the range of 3 to 20 μm) spherical-shaped metal, ceramic, or carbide powder with a low-molecular-weight thermoplastic or wax material in a mix that is typically 60% powder by volume. This mixture is frequently produced in the form of pellets or granules, which become the feedstock for the injection process. After heating to a pastelike consistency (about 260°C or 500°F), the material is injected into a heated mold cavity under sufficient pressure (about 10,000 psi) to ensure die filling. After cooling and ejection, the binder material is removed by one of a variety of processes that include solvent extraction, controlled heating to above the volatilization temperature, or heating in the presence of a catalyst that breaks the binder down into removable products. Removing the binder is currently the most expensive and time-consuming part of the process. Heating rates, temperatures, and debinding times must be carefully controlled and adjusted for part thickness. The parts then undergo conventional sintering, where any remaining binder is first removed, and the diffusion processes then set the final properties of the product. During sintering, MIM parts typically shrink 15 to 25%, and the density increases from about 60% up to as much as 99% of ideal. (*Note:* Since MIM parts are molded without density variations, the subsequent shrinkage tends to be both uniform and repeatable.) Secondary processes may take the

FIGURE 18-12 Flow chart of the metal injection molding process (MIM) used to produce small, intricate-shaped parts from metal powder.

form of surface cleaning or finishing, plating, machining, or heat treating. The high final density enables the secondary processes to be conducted in the same manner as for wrought products. Figure 18-12 summarizes the full sequence of activities, and Table 18-4 provides a summary comparison of conventional powder metallurgy and MIM.

While the size of conventional P/M products is generally limited by press capacity, the size of P/M injection moldings is more limited by economics (cost of the fine powders) and binder removal. The best candidates for P/M injection molding are complex-shaped parts with thicknesses of less than $\frac{1}{4}$ inch, weights under 2 ounces (20 grams), and made from a metal that cannot be economically die cast. MIM parts compete with and frequently replace machined components or investment castings. The shapes are generally too complex to compact by conventional powder metallurgy, and the injection molding can often reduce or eliminate costly machining. Section thicknesses as small as 0.010 inch are possible because of the fineness of the powder. As a general rule, the smaller the part and the greater the complexity, the more likely MIM will be an attractive alternative to machining, casting, stamping, cold forming, or traditional powder metallurgy. Figure 18-13 shows a variety of MIM products.

Medium to large production volumes (more than 2000 to 5000 identical parts) are generally required to justify the cost of die design and manufacture. The relatively high final density (95 to 99% compared to 75 to 90% for conventional P/M parts), the uniformity of that density, the close tolerances (0.3 to 0.5%), and excellent surface finish (about 125 μ in.) all combine to make the process attractive for many applications. Parts can be made from a wide selection of metal alloys, including steels, stainless steel, tool steel, brass, copper, titanium, tungsten, nickel-based superalloys, ceramics, and many specialty materials. The final properties are superior to those of conventional powder metallurgy and are generally close to those of wrought or cast equivalents.

TABLE 18-4	Comparison of Conventional Powder Metallurgy and Metal Injection Molding	
Feature	P/M	MIM
Particle size	20–250 μm	<20 μm
Particle response	Deforms plastically	Undeformed
Porosity (% nonmetal)	10–20%	30–40%
Amount of binder/lubricant	0.5–2%	30–40%
Homogeneity of green part	Nonhomogeneous	Homogeneous
Final sintered density	<92%	>96%

FIGURE 18-13 Metal injection molding (MIM) is ideal for producing small, complex parts. *(Courtesy of Megamet Solid Metals, Inc., St. Louis, MO.)*

■ 18.12 SECONDARY OPERATIONS

Powder metallurgy products are often ready to use when they emerge from the sintering furnace. Many P/M products, however, utilize one or more secondary operations to provide enhanced precision, improved properties, or special characteristics.

During sintering, product dimensions shrink due to densification. In addition, warping or distortion may occur during nonuniform cool-down from elevated temperatures. As a result, a second pressing operation, known as *repressing, coining,* or *sizing,* may be required to restore or improve dimensional precision. The part is placed in a die and subjected to pressures equal to or greater than the initial pressing pressure. A small amount of plastic flow takes place, resulting in high dimensional accuracy, sharp detail, and improved surface finish. The associated cold working and increase in part density may combine to increase part strength by 25 to 50%. (*Note:* Because of the shrinkage that occurs during sintering, repressing cannot be performed with the same set of tooling that was used for the original powder compaction.)

If massive metal deformation takes place in the second pressing, the operation is known as *P/M forging.* Conventional powder metallurgy is used to produce a preform, which is one forging operation removed from the finished shape. The normal forging sequence of billet or bloom production, shearing, reheating, and sequential deformation is replaced by the manufacture of a comparatively simple-shaped powder metallurgy preform followed by a single hot-forging operation. The forging stage produces a more complex shape, adds precision, provides the benefits of metal flow, and increases the density (often up to 99%). The increase in density is accompanied by a significant improvement in mechanical properties. While protective atmospheres or coatings are required to prevent oxidation of the powder preform during heating and hot forging, the P/M process can often provide a significant reduction in scrap or waste. (By controlling preform weight to within 0.5%, flash-free forging can often be performed.) Forged products can benefit from the improved properties of powder metallurgy, such as the absence of segregation, the uniform fine grain size, and the use of novel alloys or unique composites. The conventional powder metallurgy process can be expanded to larger-size products with increased complexity. The tolerance requirements of cams, splines, and gears can often be met without subsequent machining. Figure 18-14 illustrates the reduction in scrap by comparing the same part made by conventional forging and the P/M forge approach. P/M forged connecting rods, like those shown in Figure 18-15, currently account for more than 60% of all connecting rods used in North America and are typical of the high-volume steel parts currently being produced.

Impregnation and infiltration are secondary processes that utilize the interconnected porosity or permeability of low-density P/M products. *Impregnation* refers to the forcing of oil or other liquid, such as a polymeric resin, into the porous network. This can be done by immersing the part in a bath and applying pressure or by a combination vacuum-pressure process. The most common application is that of oil-impregnated bearings. After impregnation, the bearing material will contain from 10 to 40% oil by volume, which will provide lubrication over an extended lifetime of operation. In a similar manner, P/M parts can be impregnated with fluorocarbon resin (such as Teflon) to produce products offering a combination of high strength and low friction.

When the presence of pores is undesirable, P/M products may be subjected to metal *infiltration*. In this process a molten metal or alloy with a melting point lower than the P/M constituent flows into the interconnected pores of the product under pressure or by capillary action. Steel parts are often infiltrated with copper, for example. After infiltration, the engineering properties such as strength and toughness are improved to a level where they are generally comparable to those of solid metal products. Infiltration can also be used to seal pores prior to plating, improve machinability or corrosion resistance, or make the components gas- or liquid-tight. Additional heating after infiltration can cause interdiffusion between the infiltrant and base metal, further enhancing mechanical properties.

FIGURE 18-14 Comparison of conventional forging and the forging of a powder metallurgy preform to produce a gear blank (or gear). Moving left to right, the top sequence shows the sheared stock, upset section, forged blank, and exterior and interior scrap associated with conventional forging. The finished gear is generally machined from the blank with additional generation of scrap. The bottom pieces are the powder metallurgy preform and forged gear produced entirely without scrap by P/M forging. *(Courtesy of GKN Sinter Metals, Auburn Hills, MI.)*

FIGURE 18-15 P/M forged connecting rods have been produced by the millions. *(Courtesy of Metal Powder Industries Federation, Princeton, NJ.)*

Powder metallurgy products can also be subjected to the conventional finishing operations of *heat treatment, machining*, and *surface treatment*. If the part is of high density (<10% porosity) or has been metal impregnated, conventional processing can often be employed. Special precautions must be taken, however, when processing low-density P/M products. During heat treatment, protective atmospheres must again be used and certain liquid quenchants should be avoided. Speeds and feeds must be adjusted when machining, and care should be taken to avoid pickup of lubricant or coolant. In general, P/M products should be machined using sharp tools, light cuts, and high feed rates. When a large amount of machining is required, special machinability-enhancing additions may be incorporated into the initial powder blend. Nearly all common methods of surface finishing can be applied to P/M products, including platings and coatings, diffusion treatments, surface hardening, and steam treatment (which is used to produce a hard, corrosion-resistant oxide on ferrous parts). As with the other secondary processes, some process modifications may be required if the part has a reasonable amount of porosity or permeability. Since most parts are small and are produced in large quantity, barrel tumbling is another common means of cleaning, deburring, and surface modification.

■ 18.13 PROPERTIES OF P/M PRODUCTS

Because the properties of powder metallurgy products depend on so many variables—type and size of powder, amount and type of lubricant, pressing pressure, sintering temperature and time, finishing treatments, and so on—it is difficult to provide generalized information. Products can range all the way from low-density, highly porous parts with tensile strengths as low as 10 ksi (70 MPa) up to high-density pieces with tensile strengths of 180 ksi (1250 MPa) or greater.

As shown in Figure 18-16, most mechanical properties exhibit a strong dependence on product density, with the fracture-limited properties of toughness, ductility, and fatigue life being more sensitive than strength and hardness. The voids in the P/M part act as stress concentrators and assist in starting and propagating fractures. The yield strength of P/M products made from the weaker metals is often equivalent to the same material in wrought form. If higher-strength materials are used or the fracture-related tensile strength is specified, the properties of the P/M product tend to fall below those of wrought equivalents by varying but usually substantial amounts. Table 18-5 shows the properties of a few powder metallurgy materials compared with those of wrought material of similar composition. When larger presses or processes such as P/M forging or hot-isostatic pressing are used to produce higher density, the strength of the P/M products approaches that of the wrought material. If the processing results in full density

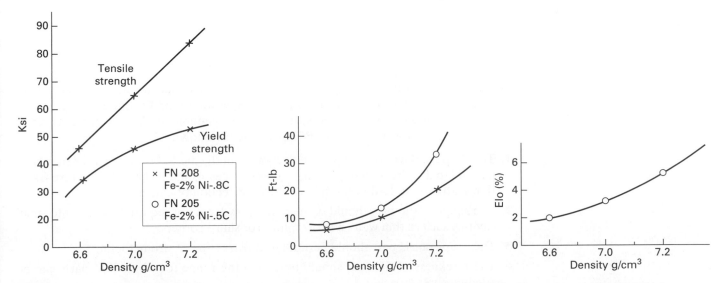

FIGURE 18-16 Mechanical properties versus as-sintered density for two iron-based powders. Properties depicted include yield strength, tensile strength, Charpy impact energy (shown in ft-lbs), and percent elongation in a 1-in. gage length.

TABLE 18-5 Comparison of Properties of Powder Metallurgy Materials and Equivalent Wrought Metals (Note how porosity diminishes mechanical performance)

Material[a]	Form and Composition	Condition[b]	Percent of Theoretical Density	Tensile Strength 10³ psi	Mpa	Elongation in 2 in. (%)
Iron	Wrought	HR	—	48	331	30
	P/M—49% Fe min	As sintered	89	30	207	9
	P/M—99% Fe min	As sintered	94	40	276	15
Steel	Wrought AISI 1025	HR	—	85	586	25
	P/M—0.25% C, 99.75% Fe	As sintered	84	34	234	2
Stainless steel	Wrought type 303	Annealed	—	90	621	50
	P/M type 303	As sintered	82	52	358	2
Aluminum	Wrought 2014	T6	—	70	483	20
	P/M 201 AB	T6	94	48	331	2
	Wrought 6061	T6	—	45	310	15
	P/M 601 AB	T6	94	36.5	252	2
Copper	Wrought OFHC	Annealed	—	34	234	50
	P/M copper	As sintered	89	23	159	8
		Repressed	96	35	241	18
Brass	Wrought 260	Annealed	—	44	303	65
	P/M 70% Cu-30% Zn	As sintered	89	37	255	26

[a]Equivalent wrought metal shown for comparison.   [b]HR, hot rolled; T6, age hardened.

with fine grain size, P/M parts can actually have properties that exceed the wrought or cast equivalents. Since the mechanical properties of powder metallurgy products are so dependent upon density, *it is important that products be designed and materials selected so that the final properties will be achieved with the anticipated amount of final porosity*.

Physical properties can also be affected by porosity. Corrosion resistance tends to be reduced due to the presence of entrapment pockets and fissures. Electrical, thermal, and magnetic properties all vary with density, usually decreasing with the presence of pores. Porosity actually increases the ability to damp both sound and vibration, however, and many P/M parts have been designed to take advantage of this feature.

■ 18.14 DESIGN OF POWDER METALLURGY PARTS

Powder metallurgy is a manufacturing system whose ultimate objective is to economically produce products for specific engineering applications. Success begins with good design and follows with good material and proper processing. In designing parts that are to be made by powder metallurgy, it must be remembered that P/M is a special manufacturing process and provision should be made for its unique factors. Products that are converted from other manufacturing processes without modification in design rarely perform as well as parts designed specifically for manufacture by powder metallurgy. Some basic rules for the design of P/M parts are as follows:

1. The shape of the part must permit ejection from the die. Side-wall surfaces should be parallel to the direction of pressing. Holes or recesses should have uniform cross section with axes and side walls parallel to the direction of punch travel.
2. The shape of the part should be such that powder is not required to flow into small cavities such as thin walls, narrow splines, or sharp corners.
3. The shape of the part should permit the construction of strong tooling.
4. The thickness of the part should be within the range for which P/M parts can be adequately compacted.
5. The part should be designed with as few changes in section thickness as possible.

6. Parts can be designed to take advantage of the fact that certain forms and properties can be produced by powder metallurgy that are impossible, impractical, or uneconomical to obtain by any other method.

7. The design should be consistent with available equipment. Pressing areas should match press capability, and the number of thicknesses should be consistent with the number of available press actions.

8. Consideration should also be made for product tolerances. Higher precision and repeatability are observed for dimensions in the radial direction (set by the die) than for those in the axial or pressing direction (set by punch movement).

9. Finally, design should consider and compensate for the dimensional changes that will occur after pressing, such as the shrinkage that occurs during sintering.

The ideal powder metallurgy part, therefore, has a uniform cross section and a single thickness that is small compared to the cross-sectional width or diameter. More complex shapes are indeed possible, but it should be remembered that uniform strength and properties require uniform density. Holes that are parallel to the direction of pressing are easily accommodated. Holes at angles to this direction, however, must be made by secondary processing. Multiple-stepped diameters, reentrant holes, grooves, and undercuts should be eliminated whenever possible. Abrupt changes in section, narrow deep flutes, and internal angles without generous fillets should also be avoided. Straight serrations can be readily molded, but diamond knurls cannot. Punches should be designed to eliminate sharp points or thin sections that could easily wear or fracture. Figure 18-17 illustrates some of these design recommendations and restrictions.

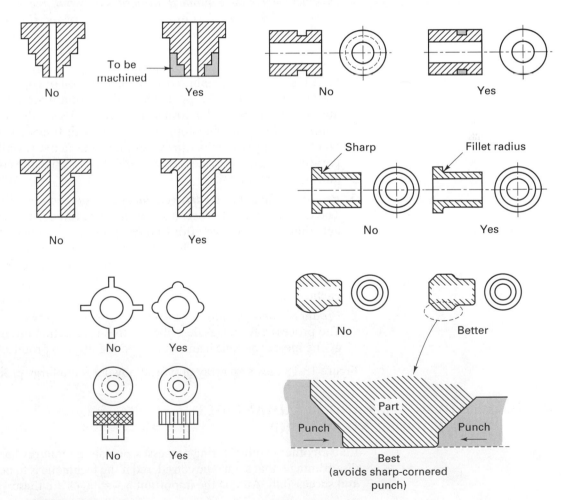

FIGURE 18-17   Examples of poor and good design features for powder metallurgy products. Recommendations are based on ease of pressing, design of tooling, uniformity of properties, and ultimate performance.

■ 18.15 POWDER METALLURGY PRODUCTS

The products that are commonly produced by powder metallurgy can generally be classified into six groups.

1. *Porous or permeable products, such as bearings, filters, and pressure or flow regulators.* Oil-impregnated bearings, made from either iron or copper alloys, constitute a large volume of P/M products. They are widely used in home appliance and automotive applications since they require no lubrication or maintenance during their service life. P/M filters can be made with pores of almost any size, some as small as 0.0025 mm (0.0001 in.). Unlike many alternative filters, powder metallurgy filters can withstand the conditions of elevated temperature, high applied stresses, and corrosive environments.

2. *Products of complex shapes that would require considerable machining when made by other processes.* Because of the dimensional accuracy and fine surface finish that are characteristic of the P/M process, many parts require no further processing, and others require only a small amount of finish machining. Tolerances can generally be held to within 0.1 mm (0.005 in.). Large numbers of small gears are currently being made by the powder metallurgy process. Other complex shapes, such as pawls, cams, and small activating levers, can be made quite economically.

3. *Products made from materials that are difficult to machine or materials with high melting points.* Some of the first modern uses of powder metallurgy were the production of tungsten lamp filaments and tungsten carbide cutting tools.

4. *Products where the combined properties of two or more metals (or metals and non-metals) are desired.* This unique capability of the powder metallurgy process is applied to a number of products. In the electrical industry, copper and graphite are frequently combined in applications like motor or generator brushes where copper provides the current-carrying capacity and graphite provides lubrication. Bearings have been made of graphite combined with iron or copper or from mixtures of two metals, such as tin and copper, where the harder material provides wear resistance and the softer material deforms in a way that better distributes the load. Electrical contacts often combine copper or silver with tungsten, nickel, or molybdenum. Here, the copper or silver provides high conductivity, while the high melting temperature material provides resistance to fusion when the contacts experience arcing and subsequent closure.

5. *Products where the powder metallurgy process produces clearly superior properties.* The development of processes that produce full density has resulted in P/M products that are clearly superior to those produced by competing techniques. In areas of critical importance such as aerospace applications, the additional cost of the processing may be justified by the enhancement of properties. As another example, consider the production of P/M magnets. A magnetic field can be used to align particles prior to sintering, resulting in a product with extremely high flux density.

6. *Products where the powder metallurgy process offers definite economic advantage.* The process advantages described in the next section often make powder metallurgy the most economical among competing ways to produce a part.

Figure 18-18 shows an array of typical powder metallurgy products.

■ 18.16 ADVANTAGES AND DISADVANTAGES OF POWDER METALLURGY

Like all other manufacturing processes, powder metallurgy has distinct advantages and disadvantages that should be considered if the technique is to be employed economically and successfully. Among the important advantages are these:

1. *Elimination or reduction of machining.* The *dimensional accuracy* and *surface finish* of P/M products are such that subsequent machining operations can be totally eliminated

FIGURE 18-18 Typical parts produced by the powder metallurgy process. *(Courtesy of PTX-Pentronix, Inc.)*

for many applications. If unusual dimensional accuracy is required, simple coining or sizing operations can often give accuracies equivalent to those of most production machining. Reduced machining is especially attractive for difficult-to-machine materials.

2. *High production rates.* All steps in the P/M process are simple and readily automated. Labor requirements are low, and product uniformity and reproducibility are among the highest in manufacturing.

3. *Complex shapes can be produced.* Subject to the limitations discussed previously, complex shapes can be produced, such as combination gears, cams, and internal keys. It is often possible to produce parts by powder metallurgy that cannot be economically machined or cast.

4. *Wide variations in compositions are possible.* Parts of very high purity can be produced. Metals and ceramics can be intimately mixed. Immiscible materials can be combined, and solubility limits can be exceeded. Compositions are available that are virtually impossible with any other process. In most cases the chemical homogeneity of the product exceeds that of all competing techniques.

5. *Wide variations in properties are available.* Products can range from low-density parts with controlled permeability to high-density parts with properties that equal or exceed those of equivalent wrought counterparts. Damping of noise and vibration can be tailored into a P/M product. Magnetic properties, wear properties, friction characteristics, and others can all be designed to match the needs of a specific application.

6. *Scrap is eliminated or reduced.* Powder metallurgy is the only common manufacturing process in which no material is wasted. In casting, machining, and press forming, the scrap can often exceed 50% of the starting material. This is particularly important where expensive materials are involved, and powder metallurgy may make it possible to use more costly materials without increasing the overall cost of the product. An example of such a product would be the rare earth magnets.

The major disadvantages of the powder metallurgy process are these:

1. *Inferior strength properties.* Because of the residual porosity, powder metallurgy parts generally have mechanical properties that are inferior to wrought or cast products of the same material. Their use may be limited when high stresses are involved. The required strength and fracture resistance, however, can often be obtained by using different materials or by employing alternate or secondary processing techniques that are unique to powder metallurgy.

2. *Relatively high tooling cost.* Because of the high pressures and severe abrasion involved in the process, the P/M dies must be made of expensive materials and be

relatively massive. Because of the need for part-specific tooling, production quantities of less than 10,000 identical parts are normally not practical.

3. *High material cost.* On a unit weight basis, powdered metals are considerably more expensive than wrought or cast stock. However, the absence of scrap and the elimination of machining can often offset the higher cost of the starting material. In addition, powder metallurgy is usually employed for rather small parts where the material cost per part is not very great.

4. *Size and shape limitations.* The powder metallurgy process is simply not feasible for many shapes. Parts must be able to be ejected from the die. The thickness/diameter (or thickness/width) ratio is limited. Thin vertical sections are difficult, and the overall size must be within the capacity of available presses. Few parts exceed 150 cm^2 (25 in^2.) in pressing area.

5. *Dimensions change during sintering.* While the actual amount depends on a variety of factors, including as-pressed density, sintering temperature, and sintering time, dimensional change can often be predicted and controlled.

6. *Density variations produce property variations.* Any nonuniform product density that is produced during compacting generally results in property variations throughout the part. For some products, these variations may be unacceptable.

7. *Health and safety hazards.* Many metals, such as aluminum, titanium, magnesium, and iron, are pyrophoric—they can ignite or explode when in particle form with large surface/volume ratios. Fine particles can also remain airborne for long times and can be inhaled by workers. To minimize the health and safety hazards, the handling of metal powders frequently requires the use of inert atmospheres, dry boxes, and hoods, as well as special cleanliness of the working environment.

■ 18.17 PROCESS SUMMARY

For many years, powder metallurgy products carried the stigma of "low strength" or "inferior mechanical properties." This label was largely the result of comparisons where "identical" parts were made of the same material, but by various methods of manufacture. In essence, the size, shape, *and material* were all specified. In such a comparison, any product with 10 to 25% residual porosity would naturally be inferior to a fully dense product made by casting, forming, or machining processes. Unfortunately, it is this type of comparison that is frequently made when one considers converting an existing design or existing part to P/M manufacture.

A far more valid comparison can be obtained by specifying size, shape, *and desired mechanical properties.* Each process can then be optimized by the selection of *both* material and process conditions. Powder metallurgy can use its unique materials, such as iron–copper blends for which there are no cast or wrought equivalents. The P/M products can be designed to provide the targeted properties while containing the typical amounts of residual porosity. Since all products will then possess the targeted mechanical properties, process comparison can then be based on economic factors, such as total production cost. On this basis, powder metallurgy has emerged as a significant manufacturing process, and its products no longer carry the stigma of "inferior mechanical properties."

Table 18-6 summarizes some of the important manufacturing features of four powder processing methods. Note the variations in product size, production rate, production quantity, mechanical properties, and cost.

	TABLE 18-6 Comparison of Four Powder Processing Methods			
Characteristic	Conventional Press and Sinter	Metal Injection Molding (MIM)	Hot-Isostatic Pressing (HIP)	P/M Forging
Size of workpiece	Intermediate <5 pounds	Smallest <1/4 pounds	Largest 1–1000 pounds	Intermediate <5 pounds
Shape complexity	Good	Excellent	Very good	Good
Production rate	Excellent	Good	Poor	Excellent
Production quantity	>5000	>5000	1–1000	>10,000
Dimensional precision	Excellent ±0.001 in./in.	Good ±0.003 in./in.	Poor ±0.020 in./in.	Very good ±0.0015 in./in.
Density	Fair	Very good	Excellent	Excellent
Mechanical properties	80–90% of wrought	90–95% of wrought	Greater than wrought	Equal to wrought
Cost	Low $0.50–5.00/lb	Intermediate $1.00–10.00/lb	High >$100.00/lb	Somewhat low $1.00–5.00/lb

■ Key Words

amorphous	gas atomization	permeability	sinter-HIP
apparent density	green strength	P/M forging	sintering
atomization	hot-isostatic pressing (HIP)	porosity	size distribution
binder	impregnation	powder injection molding	sizing
blending	infiltration	(PIM)	spray forming (Osprey
burn-off	isostatic compaction	powder metallurgy	process)
canning	liquid-phase sintering	prealloyed powder	surface texture
coining	lubricant	pressure-assisted sintering	vacuum sintering
compaction	metal injection molding	protective atmosphere	warm compaction
composites	(MIM)	rapidly solidified powder	water atomization
compressibility	mixing	repressing	
decanning	particle shape	secondary operations	
flow rate	particle size	sinter brazing	

■ Review Questions

1. What type of product would be considered to be a prospect for powder metallurgy manufacture?
2. What were some of the earliest powder metallurgy products?
3. What are some of the primary market areas for P/M products?
4. Which metal family currently dominates the powder metallurgy market?
5. What are the four basic steps that are usually involved in making products by powder metallurgy?
6. What are some of the important properties and characteristics of metal powders to be used in powder metallurgy?
7. What is the most common method of producing metal powders?
8. What are some of the other techniques that can be employed to produce particulate material?
9. Which of the powder manufacturing processes are likely to be restricted to the production of elemental (unalloyed) metal particles?
10. What are some of the unique properties of amorphous metals?
11. Why is powder metallurgy a key process in producing products from amorphous or rapidly solidified material?
12. Why is flow rate an important powder characterization property?

13. What is apparent density, and how is it related to the final density of a P/M product?
14. What is green strength, and why is it important to the manufacture of high-quality P/M products?
15. How do the various powder properties relate to the method of powder manufacture?
16. What are some of the objectives of powder mixing or blending?
17. How does the addition of a lubricant affect compressibility? Green strength?
18. How might the use of a graphite lubricant be fundamentally different from the use of wax or stearates?
19. What types of composite materials can be produced through powder metallurgy?
20. What are some of the objectives of the compaction operation?
21. What limits the cross-sectional area of most P/M parts to several square inches or less?
22. Describe the movement of powder particles during compaction. What feature is responsible for the fact that powder does not flow and transmit pressure like a liquid?
23. For what conditions might a double-action pressing be more attractive than compaction with a single moving punch?

24. How is the density of a P/M product typically reported?
25. Why is it more difficult to compact a multiple-thickness part?
26. Describe the four classes of conventional powder metallurgy products.
27. What is isostatic compaction? For what product shapes might it be preferred?
28. What is the benefit of warm compaction?
29. What is a reasonable compacted density? How much residual porosity is still present?
30. What types of materials are used in compaction tooling?
31. How do the common sintering temperatures compare to material melting points?
32. What are the three stages associated with most P/M sintering operations?
33. Why is it necessary to raise the temperature of P/M compacts slowly to the temperature of sintering?
34. Why is a protective atmosphere required during sintering? During the cool-down period?
35. What types of atmospheres are used during sintering?
36. What are some of the changes that occur to the compact during sintering?
37. What is the purpose of the sinter brazing process?
38. The combined heating and pressing of powder would seem to be an improvement over separate operations. What features act as deterrents to this approach?
39. What are some of the attractive properties of hot-isostatic pressed products?
40. What is the attractive feature of the sinter-HIP and pressure-assisted sintering processes?
41. What are some of the other methods that can produce high-density P/M products?
42. Describe the spray-forming process and the unique feature that enables production of high-density, fine grain size products.
43. How is the injection molding of powdered material similar to the injection molding of plastic or polymeric products?
44. In the MIM process, what is done to enable metal powder to flow like a fluid under pressure?
45. How is the metal powder used in metal injection molding (MIM) different from the metal powder used in a conventional press-and-sinter production?
46. What are some of the ways that the binder can be removed from metal injection molded parts?
47. Why are MIM products injection molded to sizes that are considerably larger than the desired product?
48. For what types of parts is P/M injection molding an attractive manufacturing process?
49. How does the final density of a MIM product compare to a press-and-sinter P/M part?
50. What is the purpose of repressing, coining, or sizing operations?
51. Why can't we use the original compaction tooling to perform repressing?
52. What is the major difference between repressing and P/M forging?
53. What is the difference between impregnation and infiltration? How are they similar?
54. Why might different conditions be required for the heat treatment, machining, or surface treatment of a powder metallurgy product?
55. The properties of P/M products are strongly tied to density. Which properties show the strongest dependence?
56. How do the physical properties of P/M products vary with density?
57. What advice would you want to give to a person who is planning to convert the manufacture of a component from die casting to powder metallurgy?
58. What is the shape of an "ideal" powder metallurgy product?
59. What are some P/M products that have been intentionally designed to use the porosity or permeability features of the process?
60. Give an example of a product where two or more materials are mixed to produce a composite P/M product with a unique set of properties.
61. What are the most cited assets of the powder metallurgy method of parts manufacture?
62. Why is finish machining such an expensive component in parts manufacture?
63. Describe some of the materials that can be made into P/M parts that could not be used for processes such as casting and forming.
64. Why is P/M not attractive for parts with low production quantities?
65. What features of the P/M process often compensate for the higher cost of the starting material?
66. How might you respond to the criticism that P/M parts have inferior properties?

■ Problems

1. When specifying the starting material for casting processes, the primary variables are chemistry and purity. Any structural features of the starting material will be erased by the melting. For forming processes, the material remains in the solid state, so the principal concerns relating to the starting material are chemistry and purity, ductility, yield strength, strain-hardening characteristics, grain size, and so on. What are some of the characteristics that should be specified for the starting powder to assure the success of a powder metallurgy process? In what ways are these similar to or different from those mentioned for casting and forming processes?

2. In conventional powder metallurgy manufacture, the material is compacted with applied pressure at room temperature and then sintered by elevated temperature at atmospheric pressure. With P/M hot pressing, the loose powder is subjected to pressure while it is also at elevated temperature. It would appear, therefore, that hot pressing could produce a finished part in a single operation and would be a more economical and attractive manufacturing process. What features have been overlooked in this argument that would tend to favor the press-and-sinter sequence for conventional manufacture?

3. Investigate the method(s) used to produce tungsten incandescent lamp filaments. How does the method used today compare to the method developed by Coolidge in the late 1800s?

*C*hapter 18   CASE STUDY

Impeller for an Automobile Water Pump

The component pictured in the figure is the impeller of a water pump used by a major automotive manufacturer. The outer diameter of the component is 2.75 in., and the total height of the six curved vanes is 0.75 in. (with a tolerance of 0.005 in.). The diameter of the center hole is 0.625 in., and the flat base is 0.187 in. thick. Vibration and balance considerations require accurate positioning and uniform thickness of the six curved vanes. A relatively smooth surface finish is desirable for good fluid flow.

The maximum operating temperature has been estimated at 300°F, and the contact fluid should be a water/antifreeze mixture with corrosion-resistant additives. The designer has provided a target tensile strength of 30,000 psi and notes that a minimum amount of fracture resistance is also desirable. Since there should be no direct metal-to-metal rubbing, enhanced wear resistance does not appear to be necessary. This is a high-volume component, however, so low total cost (material plus manufacturing) would appear to be a prime objective.

Similar components have been sand cast from cast iron, with a grinding operation being required to maintain controlled height. The manufacturer is interested in improving quality and lowering cost.

1. Is a ferrous material needed to provide the desired properties, or might a nonferrous metal be acceptable?
2. What processes would you want to consider to mass-produce such a shape? Are there more attractive casting processes? Is this a candidate for metal forming, and if so, which process or processes? Is powder metallurgy a possibility for this product? If so, can the desired properties be achieved with the densities that are common for a traditional press-and-sinter operation?
3. Investigate the various material–process combinations that would be candidates for production. Select and defend your "best choice."
4. Might this part be a candidate for manufacture from a nonmetal, such as molded nylon, some other polymer, or even some form of reinforced composite material? How would you suggest producing the desired shape if one or more of these materials were considered? Would you have to compromise on any of the performance requirements?

An automobile water pump impeller.

CHAPTER 19

ELECTRONIC ELECTROCHEMICAL CHEMICAL AND THERMAL MACHINING PROCESSES

■ 19.1 INTRODUCTION

Many material removal processes have been developed since World War II to address problems that can't be handled with conventional "chip-forming" machining processes. The advantages of these processes [often called nontraditional machining (NTM)] are as follows:

• Complex geometries beyond simple planar or cylindrical features can be machined.

• Parts with extreme surface-finish and tight tolerance requirements can be obtained.

• Delicate components that cannot withstand large cutting forces can be machined.

• Parts can be machined without producing burrs or inducing residual stresses.

• Brittle materials or materials with very high hardness can be easily machined.

• Microelectronic or integrated circuits are possible to mass-produce.

NTM processes can often be divided into four groups based upon the material removal mechanism: See Table 19-1.

1. *Chemical.* Chemical reaction between a liquid reagent and the workpiece results in etching.
2. *Electrochemical.* An electrolytic reaction at the workpiece surface is responsible for material removal.
3. *Thermal.* High temperatures in very localized regions evaporate materials.
4. *Mechanical.* High-velocity abrasives or liquids remove material (see Chapter 27).

Machining processes that involve chip formation have a number of inherent limitations. Large amounts of energy are expended to produce unwanted chips that must be removed and discarded. Much of the machining energy ends up as undesirable heat that often produces problems of distortion and surface cracking. Cutting forces require that the workpiece be held, which can also lead to distortion. Unwanted distortion, residual stress, and burrs caused by the machining process often require further processing. Finally, some geometries are too delicate to machine, while others are too complex.

When examining these processes, be aware that conventional end milling (see Chapter 25) has these typical machining parameters:

- Feed rate—25 to 5000 mm/min (5 to 200 in./min)
- Surface finish—1.5 to 3.75 μm (60 to 150 μin) AA
- Dimensional accuracy—0.025 to 0.05 mm (0.001 to 0.002 in.)
- Workpiece/feature size—61 cm $\times$ 61 cm (25 in. $\times$ 24 in.); 2.5 cm (1 in.) deep

In comparison, NTM processes typically have lower feed rates and require more power consumption when compared to machining. However, some processes permit batch processing, which increases the overall throughput of these processes and enables them to compete with machining. A major advantage of some NTM processes is that feed rate is independent of the material being processed. As a result, these processes are often used for difficult-to-machine materials. NTM processes typically have better accuracy and surface finish, with the ability of some processes to machine larger feature sizes at lower capital costs. In most applications, NTM requires part-specific tooling, while general-purpose cutting and workholding tools make machining very flexible. There are numerous hybrid forms of all these processes, developed for special applications, but only the main NTM processes are described here due to space limitations.

■ 19.2 CHEMICAL MACHINING PROCESSES

CHEMICAL MACHINING

Chemical machining (CHM) is the simplest and oldest of the chipless machining processes. The use of CHM dates back 4500 years to the Egyptians, who used it to etch jewelry. In modern practice, it is applied to parts ranging from very small microelectronic circuits to very large engravings up to 15 m (50 ft) long. Typically metals are chemically machined, although methods do exist for etching ceramics and even glass.

In CHM, material is removed from a workpiece by selectively exposing it to a chemical reagent or etchant. The mechanism for metal removal is the chemical reaction between the etchant and the workpiece, resulting in dissolution of the workpiece. One means for accomplishing CHM is called *gel milling*, where the etchant is applied to the workpiece in gel form. However, the most common method of CHM involves covering selected areas of the workpiece with a *maskant* (or etch resist) and imparting the remaining exposed surfaces of the workpiece to the etchant. The general material removal steps for CHM are:

1. *Cleaning.* Contaminants on the surface of the workpiece are removed to prepare for application of the maskant and permit uniform etching. This may include degreasing, rinsing, and/or pickling.

2. *Masking.* If selective etching is desired, an etch-resistant maskant is applied and selected areas of the workpiece are exposed through the maskant in preparation for etching.

3. *Etching.* The part is either immersed in an etchant or an etchant is continuously sprayed onto the surface of the workpiece. The chemical reaction is halted by rinsing.

4. *Stripping.* The maskant is removed from the workpiece and the surface is cleaned and desmutted as necessary.

Lateral dimensions in CHM are controlled in large part by the patterned maskant. Masking can be performed in one of several ways depending on the level of precision required in CHM. The simplest method of applying a maskant is the *cut-and-peel* method. In this procedure the maskant material, typically neoprene, polyvinyl chloride, or polyethylene, is applied to the entire surface of the workpiece by dipping or spraying. Once the coating dries, it is then selectively removed in those areas where etching is desired by scribing the maskant with a knife and peeling away the unwanted portions. When volume permits, scribing templates may be used to improve accuracy. Cut-and-peel

TABLE 19-1 Summary of NTM Processes

Process	Typical Penetration or Feed Rate, Mm/m (ipm)	Typical Surface Finish AA, μm (μin.)	Typical Accuracy, Mm (in.)	Typical Workpiece or Feature Size, cm (in.)	Comments
Chemical					
Chemical milling (cut-and-peel)	0.013 to 0.076 (0.0005 to 0.003)	1.6 to 6.35; as low as 0.2 (63 to 250; 8)	greater than 0.127 (0.005)	as large as 365 × 1524 (144 × 600); up to 1.27 (0.5) thick	No burrs; no surface stresses; tooling cost low
Photochemical machining	as above	as above	0.025 to 0.05 (0.001 to 0.002)	30 × 30 (12 × 12); up to 0.15 (0.06) thick	Limited to thin material; burr-free blanking of brittle material; tooling cost low; used in microelectronics
Electrochemical					
Electrochemical machining (ECM)	2.5 to 12.7 (0.1–0.5)	0.4 to 1.6 (16–63)	0.013 to 0.13 (0.0005–0.005); 0.05 (0.002) in cavities	30 × 30 (12 × 12); 5 (2) deep	Stress-free, burr-free metal removal in hard-to-machine metals; tool design expensive; disposal of wastes a problem; MRR independent of hardness; deep cuts will have tapered walls
Electrostream drilling	1.5 to 3 (0.06–0.12)	0.25 to 1.6 (10–63)	0.025 (0.001) or 5% of hole dia.	up to 0.5 (0.2) thick	Charged high-velocity stream of electrolyte; hole diameters down to 0.127 mm (0.005 in.); 40:1-hole aspect ratios possible
Shaped-tube electrolytic machining (STEM)	as above	0.8 to 3.1 (32–125)	0.025 to 0.125 (0.001–.005)	routinely up to 127 (5) thick	Special form of ECM using conductive tube with insulated surface and acidic electrolyte; 300:1-hole aspect ratios; hole diameters down to 0.5 mm (0.02 in.)
Thermal					
Electrical discharge machining (EDM)	up to 0.5 (0.02)	0.8 to 2.7 (32–105)	0.013 to 0.05 (0.0005–.002)	up to 200 × 200 (79 × 79); 5 (2) deep	Widely used and disseminated; dies expensive; cuts any conductive material regardless of hardness; forms recast layer
Electron-beam machining (EBM)	30 to 1500 (1.2–60)	0.8 to 6.35 (32–250)	0.005 to 0.025 (0.0002–0.001)	0.025 to 0.63 (0.01–0.25) thick	Capable of micromachining thin materials; hole sizes down to 0.05 mm (0.002 in.); 100:1 hole aspect ratios; requires high vacuum
Laser-beam machining (LBM)	100 to 2500 (4–100)	0.8 to 6.35 (32–250)	0.013 to 0.13 (0.0005–0.005)	up to 2.5 (1) thick	Capable of drilling holes down to 0.127 mm (0.005 in.) at 20:1 aspect ratio in seconds; has heat-affected zone and recast layers which may require removal
Plasma arc cutting (PAC)	250 to 5000 (10–200)	0.6 to 12.7 (25–500)	0.5 to 3.2 (0.02–0.125)	up to 15 (6) thick	Clean rapid cuts and profiles in almost all plates; 5° to 10° taper; cheaper capital equipment
Precision PAC	as above	as above	0.25 (0.01)	up to 1.5 (0.625) thick	Special form of PAC limited to thin sheets of material; straighter, smaller kerf
Wire EDM	100 to 250 (4–10)	0.8 to 1.6; as low as 0.38 (32–64; 15)	0.0025 to 0.1 (0.0001–0.004)	as large as 100 × 160 (40 × 64); up to 45 (18) thick	Special form of EDM using traveling wire; cuts straight narrow kerfs; wire diameters as small as 0.05 mm (0.002 in.); CNC machines permit complex geometries
Mechanical—see Chapter 29 on Abrasive Machining					
Abrasive jet machining	76 (3)	0.25 to 1.27 (10–50)	0.12 (0.005)	up to 0.15 (0.06) thick	Used for cutting brittle materials; produces tapers; inexpensive to implement; can cut up to 6.3 mm (0.25 in.) thick glass
Abrasive waterjet machining	15 to 450 (0.6–18)	2.0 to 6.35 (80–250)	0.13 to 0.38 (0.005–0.015)	up to 20 (8) thick	Use in glass, titanium, composites, nonmetals, and heat-sensitive or brittle materials; produces tapered walls in deep cuts; no burrs

TABLE 19-1	Continued				
Ultrasonic machining (impact grinding)	0.5 to 3.8 (0.02–.15)	0.4 to 1.6; as low as 0.15 (16–63; 6)	0.013 to 0.025 (0.0005–.001)	up to 100 cm^2 (16 in.^{2})	Most effective in hard materials, $R_C > 40$; tool wear and taper limit hole aspect ratio at 2.5:1
Water-jet machining	250 to 200,000 (10–7900) soft materials	1.27 to 1.9 (50–100)	0.13 to 0.38 (0.005–0.015)	up to 2.5 (1) thick	Used on leather, plastics, and other nonmetals; pressures of 60,000 psi and jet velocity of up to 3000 ft/sec

coatings are thick, ranging from 0.025 to 0.13 mm (0.001 to 0.005 in.). Because of this thickness, the maskant can withstand exposure to the etchant for the extended periods of time necessary to remove large volumes of material. This technique is generally preferred when the workpiece is not flat or is very large or for low-volume work where the development of screens or phototools necessary for other masking methods is not justified. The scribe-and-peel method for *stepped machining* is shown in Figure 19-1.

Another method used to apply maskants is *screen printing*, involving the use of traditional silk-screening technology. The method applies the maskant through a mask made from a fine silk mesh or stainless steel screen. Masks are typically formed by application, exposure, and development of a light-sensitive emulsion on top of the screen. The screen is pressed against the surface of the workpiece and the maskant is rolled on. Screen printing is good for high-volume, low-precision applications with tolerances typically in the 0.05 to 0.18 mm (0.002 to 0.007 in.) range. Etch depth is limited to about 1.5 mm (0.06 in.) by the thickness of the maskant, typically on the order of 0.05 mm (0.002 in.).

FIGURE 19-1 Steps required to produce a stepped contour by chemical machining.

FIGURE 19-2 Typical chemical milling defects: (a) overhang: deep cuts with improper agitation; (b) islands: isolated high spots from dirt, residual maskant, or work material inhomogeneity; (c) dishing: thinning in center due to improper agitation or stacking of parts in tank.

(a)   (b)   (c)

Etch rates in CHM are very slow compared with other nontraditional machining processes. However, etching proceeds on all exposed surfaces simultaneously, which significantly increases the overall material removal rate on large parts. The etch rate in CHM is directly proportional to the etchant concentration directly adjacent to the area being machined. For parts machined by immersion, the uniformity of the etchant concentration within the bath can be improved by agitation. If the bath is not agitated properly, several defect conditions can result, as shown in Figure 19-2. *Islands*, or isolated high spots, can be the result of improper agitation on large parts. Islands can also be formed due to inadequate cleaning or inhomogeneity with the work material.

Single-sided, blind etching of the part is called *chemical milling* or, when the photoresist method of applying maskants is used, *photochemical milling*. Chemical milling is so named because its earliest use was for replacing mechanical milling on large components. Chemical milling is often used to remove weight on aircraft components, as shown in Figure 19-3. Through-etching of the workpiece is called *chemical (or photochemical) blanking*. The process competes with blanking, laser cutting, and electrical discharge machining (EDM) for through-cutting of thin material sheets. Chemical blanking is typically performed using double-sided etching to increase production rates and minimize taper on the etched walls of the feature. A key requirement for chemical blanking is registration of top-side and bottom-side screens or phototools during masking. Because of the precision required, chemical blanking is not performed with the cut-and-peel method of masking.

FIGURE 19-3 Iconel 718 aircraft engine parts. These sheet metal parts for a jet fighter engine are chemically milled to remove weight. (Left) As-formed workpiece; (middle left) workpiece coated with liquid rubber, fiberglass scribing template in place; (middle right) scribed workpiece; (right) finished part. About 0.035 in. of stock is removed from the 0.070-in.-thick workpieces. Tolerances are held to ±0.004 in.

Advantages and Disadvantages of Chemical Machining.   Chemical machining has a number of distinct advantages when compared with other machining and forming processes. Except for the preparation of the artwork and phototool, screen or scribing template, the process is relatively simple, does not require highly skilled labor, induces no stress or cold working in the metal, and can be applied to almost any metal— aluminum, magnesium, titanium, and steel being the most common. Large areas can be machined; tanks for parts up to 12-by-50 ft and spray lines up to 10 ft wide are available. Machining can be done on parts of virtually any shape. Thin sections, such as honeycomb, can be machined because there are no mechanical forces involved.

The tolerances expected with CHM range from ±0.0005 in. on small etch depths up to ±0.004 in. in routine production involving substantial depths. Tolerances in chemical milling increase with the depth of the cut and with faster etch rates and vary for different materials. The surface finish is generally good to excellent for chemical polishing.

In using CHM, some disadvantages and limitations should be kept in mind. CHM requires the handling of dangerous chemicals and the disposal of potentially harmful by-products, although some recycling of chemicals may be possible. The metal removal rate is slow in terms of the unit area exposed, being about 0.2 to 0.04 lb/min per square foot exposed in the case of steel. However, because large areas can be exposed all at once, the overall removal rate may compare favorably with other metal removal processes, particularly when the work material is not machinable or the workpiece is thin and fragile, unable to sustain large cutting forces.

Photochemical Machining.   Figure 19-4 shows the specific steps that are involved when photochemical machining (PCM) is performed with the use of photoresists. These are as follows:

1. Clean the workpiece.

2. Coat the workpiece with a photoresist, usually by hot-roller lamination of dry-film photoresists, on both sides, although liquid photoresists may also be applied by dipping, flowing, rolling, or electrophoresis (i.e., migration of charged molecules in the presence of an electric field). For liquid photoresists, the coating is heated in an oven to remove solvents.

3. Prepare the artwork. A drawing of the workpiece is made on a computer-aided design (CAD) system.

FIGURE 19-4   Basic steps in photochemical machining (PCM).

4. Develop the phototool. The CAD file is used to derive a photographic negative of the workpiece. Several methods may be used. Typically, the CAD drawing is downloaded to a laser-imaging system that exposes the desired image directly onto photographic (e.g., silver halide) film. In the past, oversized artwork was used to increase the accuracy of the phototool through photographic reduction of the artwork.

5. Expose the photoresist. Bring the phototool in contact with the workpiece, using a vacuum frame to ensure good contact, and expose the workpiece to intense ultraviolet (UV) light.

6. Develop the photoresist. Exposure of the photoresist to intense UV light alters the chemistry of the photoresist, making it more resistant to dissolution in certain solvents. By placing the exposed maskant in the proper solvent, the unexposed areas of the resist are removed, exposing the underlying material for etching. All residue is rinsed away.

7. Spray the workpiece with (or immerse it in) the reagent.

8. Remove the remaining maskant.

PCM has been widely used for the production of small, complex parts, such as printed circuit boards, and very thin parts that are too small or too thin to be blanked or milled by ordinary sheet metal forming or machining operations, respectively. Refinements to the PCM process are used in the microeletronics fabrication.

Design Factors in Chemical Machining. When designing parts that are to be made by chemical machining, several unique factors related to the process must be kept in mind. First, if artwork is used, dimensional variations can occur through size changes in the artwork or phototool film due to temperature and humidity changes. These can be controlled or eliminated by putting the artwork on thicker polyester films or glass, by controlling the temperature or humidity in the artwork and phototool production areas, or by using a direct-write laser-imaging system.

The second item that must be considered is the *etch factor*, sometimes referred to as *etch radius*, which describes the undercutting of the maskant. The etchant acts isotropically on whatever surface is exposed. Areas that are exposed longer will have more metal removed from them. Consequently, as the depth of the etch increases, there is a tendency to undercut or etch under the maskant (Figure 19-5). The etch factor, *E*, in chemical machining is defined as

$$E = \frac{U}{d} \qquad (19\text{-}1)$$

where *d* is the depth of cut and *U* is the undercut as defined in Figure 19-5. In photochemical machining, the term *anisotropy* (sometimes referred to as etch factor) is used to describe the directionality of the cut. Anisotropy, *A*, of a material–etchant interaction in photochemical machining is defined as

$$A = \frac{d}{U} \qquad (19\text{-}2)$$

which is the inverse of equation 19-1. In many electrical and electronic products, anisotropies much greater than 1 are desirable in order to permit greater densities of electrical and electronic components and wires.

An allowance for the etch factor must be taken into account in designing the part and the artwork or scribing template. In the case of chemical milling, the width of the opening in the maskant must be reduced by an amount sufficient to compensate for the undercut under both sides of the maskant:

$$W_m = W_f - (E \cdot d) \qquad (19\text{-}3)$$

where W_f is the final desired width of the cut. This allowance is considered a minimum allowance; it has been found that results will vary based on etching conditions, and actual etch allowances will have to be somewhat greater and adapted to specific conditions.

In double-sided chemical blanking, a sharp edge remains along the line at which breakthrough occurs. Because such an edge is usually objectionable, etching ordinarily

(a)

(b)

Etching from one side

(c)

(d)

(e)

(f)

(g)

Etching from both sides

FIGURE 19-5 Undercutting of the mask or resist is defined by the etch factor, which must be accounted for in designing the part using the artwork on the scribing template.

is continued to produce nearly straight side walls, as shown in Figure 19-5g. In order to achieve nearly straight side walls, it is typical to allow the process to continue for the amount of time necessary to do a through-cut from one side.

The anisotropy of the cut defines the maximum limit for aspect ratio (i.e., ratio of depth to width) of the cut, which is a measure of how deep and narrow a cut can be made. In chemical blanking, by using a double-sided etch, the maximum aspect ratio of the cut may be effectively doubled if the process is stopped at breakthrough (Figure 19-5f). This is the effective maximum aspect ratio that may be achieved in CHM. Consequently, it is difficult to produce deep, narrow cuts in materials using CHM. Table 19-2 shows some etch rates and etch factors for common metal and etchant combinations in CHM.

TABLE 19-2	Etch Rates and Etch Factors for Some Common Metal–Etchant Combinations in CHM		
Metal	Preferred Etchant	Penetration Rate (mm/mm)	Etch Factor E
Aluminum	$FeCl_3$	0.025	1.7 : 1
Copper	$FeCl_3$	0.05	2.7 : 1
Nickel alloys	$FeCl_3$	0.018	2.0 : 1
Phosphor-bronze	Chromic acid	0.013	2.0 : 1
Silver	$FeNO_3$	0.02	1.5 : 1
Titanium	HF	0.25	2.0 : 1
Tool steel	HNO_3	0.018	1.5 : 1

Source: G. F. Benedict, *Nontraditional Manufacturing Processes*, Marcel Dekker, New York, 1987, p. 200.

The soundness and homogeneity of the metal are very important. Wrought materials should be uniformly heat treated and stress relieved prior to processing. Although chemical machining induces no stresses, it may release existing residual stresses in the metal and thus cause warpage. Castings can be chemically machined provided that they are not porous and have uniform grain size. Lack of the latter can cause nonuniform etching rates, producing islands. Because of the different grain structures that exist near welds, weldments usually are not suitable for chemical machining. Preferential etching due to *intergranular attack* can also be an issue in CHM.

CHEMICAL-MECHANICAL POLISHING

Chemical-mechanical polishing (CMP), or chemo-mechanical polishing, uses the synergy of chemistry and mechanical grinding to obtain flatness on the order of 50 nm. CMP is used in the fabrication of integrated circuits (ICs) to obtain planar surfaces after dielectric and metal depositions during interconnection of circuit components. As shown in Figure 19-6, the process involves rotation of the wafer as it is pressed against a slurry-filled abrasive pad that rotates counter to the wafer. The slurry contains both a particle abrasive as well as an etchant and is deposited directly onto the abrasive pad by a mechanical arm. The wafer is held in a carrier by a backing pad designed to distribute the mechanical force evenly over the surface of the wafer. Fused silica in a weak KOH solution is a common slurry for oxide polishing. Ferricyanide-phosphate with silica or alumina is used to polish tungsten. Raised features are etched much faster than flat regions, since mechanical pressure concentrated on the raised features enhances etching. For dielectric polishing, typical etch rates are on the order of 300 nm/minute.

FIGURE 19-6 Schematic of chemical-mechanical polishing (CMP).

PHOTOCHEMICAL MACHINING FOR ELECTRONICS

The most common and most precise method for creating maskants involves the use of UV light-sensitive emulsions, called *photoresists*. In this method, photoresists are applied to the surface of the workpiece and selectively exposed to an intense ray of UV light through a photographic negative of the image to be patterned. The use of photoresists, called *photochemical machining*, is widely used in the manufacture of integrated circuits in electronics. The entire sequence of operations for the manufacture of electronics is given in Figure 1-13 in Chapter 1.

Integrated circuits use semiconductor materials—such as silicon, gallium arsenide, and germanium—that can be made to be either electrically conducting or electrically insulating by changing the type and concentration of impurity atoms found within the material. Like metals, all semiconductors have crystalline microstructures exhibiting long-range order in the form of a lattice. However, unlike metals, semiconductor atoms are characterized as having half-filled valence shells, and so, when placed into a lattice, the semiconductor atoms form covalent bonds

At room temperature (25°C), silicon permits a small amount of electrical conductivity that is too small for most electronic applications. The electrical conductivity of semiconductors can be altered by inserting impurity atoms into the semiconductor lattice. The process of modifying the electrical properties of semiconductors by introducing impurity atoms is commonly referred to as *doping*.

The first level of electronic manufacturing involves the manufacture of the ICs or chips. This is a complex process involving many steps, the sequence of which depends on the particular electrical device. The initial steps of doping by diffusion or ion implantation and oxidation are performed in large machines that manipulate the wafers in and out of various vacuum chambers in the correct sequence and duration.

Doping can be accomplished in bulk by alloying at the time of crystal formation. However, selective doping is required for IC production. Selective doping in most early IC devices involved thermal diffusion; more recently, as device dimensions have continued to shrink, ion implantation has become more suitable to better control the depth and concentration of the dopant atoms in the silicon wafer.

Silicon is the most widely used semiconductor material. It is plentiful and can be readily produced in single-crystal form; see Figure 19-7 for details. Also, the native oxide, silicon dioxide, can be used as both a dieletric layer and a diffusion mask during processing.

(a) Seed being lowered down to melt

Seed
Melt
Noncontaminating liner
Graphite crucible

(b) Seed dipped in melt; freezing on seed just beginning

Solidification front
Melt

(c) Partially grown crystal slowly being withdrawn from melt

Growing crystal or boule

FIGURE 19-7   How the silicon wafer is made.
One of the key reasons that single crystal silicon is the most widely used semiconductor material is that it can be refined and grown economically in single-crystal form. Here's how it is done. Under equilibrium conditions, molten silicon, when cooled, produces a polycrystalline structure. However, under controlled conditions, silicon can be grown from a single seed in a large single-crystal ingot called a *boule*. The technique used most often for growing single-crystal silicon is called the *Czochralski method*. In the Czochralski method, a small *seed crystal* is lowered into molten silicon and raised slowly, allowing the crystal to grow from the seed. The size of the seed crystal is about 0.5 cm diameter by 10 cm long. Its crystallographic orientation is critical because it defines the crystallographic orientation of the boule which controls the electrical properties within the boule. The melt consists of electronic grade (99.999999999% pure) polycrystalline silicon (polysilicon). If desirable, dopant may be added to the melt, although alloying complicates the crystal growth process. The silicon is melted in a fused silica crucible within a furnace chamber backfilled with an inert gas such as argon. The crucible is heated to approximately 1500°C and maintained at slightly above the melting point with a graphite resistance heater.
Once grown, the boule is characterized for resistivity and crystallographic defects. The unusable end portions of the boule are cut off, and the outside of the body is ground into a cylindrical ingot. For diameters below 300 mm, *flats* are ground along the length of the ingot to identify the crystallographic orientation and the boule is sliced into wafers using wire or diamond saws. The wafers are lapped, chemical etched, and polished, ready for IC production.

1. Material preparation (P⁺ wafer)

2. Epitaxial growth (P⁻)

3. Mask oxide and photolithography

4. Etch and diffusion and oxide removal

5. Mask removal and fresh oxidation (gate oxide)

6. Deposited polysilicon

7. Photolithography

8. Etch

9. Photolithography

10. Ion implantation

11. Oxide deposition

12. Photolithography

13. Etch

14. Metallization

15. Photolithography

16. Etch

17. Final overcoat

FIGURE 19-8   The sequence of processes or steps required to manufacture a simple metal-oxide-semiconductor (MOS) field effect transistor device is shown here. *(Source: Texas Engineering Extension Service, Semiconductor Processing Overview, College Station, Texas: The Texas A&M University System, 1996.)*

HOW ICs ARE MADE

The ability to selectively modify the electrical properties of semiconductors is the backbone of microelectronic manufacturing.

As shown in Figure 19-8, the manufacturing fabrication sequence for making a simple bipolar diode has many steps, beginning with the production of a silicon wafer from a predoped, single-crystal ingot (*boule*) that is cut into wafers, lapped, and polished to produce silicon wafers. The wafers are placed in vacuum chambers, where an oxide layer is grown on the surface of the wafer to act as a mask during subsequent doping of the substrate. The oxide layer is patterned using photolithography in combination with etching (see Figure 19-9). Photolithography is used to produce a polymeric mask over the oxide layer, which will allow only select areas of the oxide layer to be etched. After etching (chemical machining), the polymeric mask is removed from the silicon dioxide layer, and the silicon is doped (by diffusion) with boron. After doping, the silicon dioxide mask is removed, and a second silicon dioxide layer is grown and patterned to establish openings in the silicon dioxide layer above the doped regions. Next, a thin metal film is deposited on top of the silicon dioxide to provide an electrical pathway allowing the positive and

FIGURE 19-9   (a) The photolithography process itself has multiple steps, and each step can have variation, as shown in (b), the exposure step (4) in part (a).

negative electron regions of the diode to be connected to an external power supply. Photolithography and etching are used once again to pattern the thin film into leads and contact pads large enough for biasing the device. To protect the final integrated device from mechanical damage and moisture, a final passivation coating is added.

This example shows the production of a single IC component. Typically, multiple components and, further, multiple circuits are produced in parallel during IC fabrication. Over the years IC fabrication has evolved from the original small-scale integration (SSI) architecture of the 1960s, with 2 to 50 electronic components per circuit, to the ultra-large-scale integration (ULSI) architectures of today, with tens of millions of components per circuit. The classification of ICs by scale of integration represents the successive advancement of semiconductor processing technologies to provide lower-cost, higher-performance ICs. Each increase in the number of components represented a breakthrough in miniaturization technology (e.g., photolithography and clean rooms) that permitted the fabrication of smaller IC components with improved performance. To achieve lower cost, manufacturing processing technology breakthroughs were needed to make miniaturization technologies possible and economical. Today this trend of seeking higher performance at lower cost continues.

The photolithography process involves the sequence of steps shown in Figure 19-9. First, a liquid photoresist is applied to the surface of the silicon oxide layer over the silicon wafer. Typically, this is done with a process known as *spin coating*. In spin coating, centrifugal forces are used to produce a photoresist layer of uniform thickness. Next, the coated wafer is *soft baked* on a hot plate or in an oven. In this step, solvents used to reduce the viscosity of the photoresist during spin coating are evaporated, and adhesion between the wafer and the photoresist is improved. After soft bake, the photoresist is *exposed*, using a photomask to transmit a pattern of electromagnetic radiation onto the surface of the photoresist. This step is performed using a machine called a *stepper*, because the lithographic pattern of the device is indexed or stepped across the wafer, subjecting it to repeated exposures—one for each chip being made. Once the resist has been exposed, the wafer is *developed* in a chemical solvent. Development chemically machines unwanted resist materials, exposing the underlying material to be etched. Next, the resist is *hard baked* to remove any remaining solvents after development and to further toughen the remaining resist against downstream etching or implantation processes. Hard bakes generally take longer and are done at slightly higher temperatures than soft bakes. Once the downstream etching or implantation has made use of the resist, a photoresist *stripping* step is necessary for removal of the resist.

Obviously, the most important requirement of the photoresist is that it resist the downstream etching or implantation process. Other requirements important to the function of resists are their *resolution* and *sensitivity*. Resolution refers to the smallest linewidth that can be reproduced repeatably by the resist. The resolution of the resist is strongly a function of the source of ionizing radiation or the exposure machine tool used. Sensitivity refers to the amount of ionizing energy required to modify the solubility of the resist.

IC MANUFACTURING AND ECONOMICS

By the start of the year 2000, ICs with over 10 million transistors had been produced. While it is true that small circuits are inexpensive, the cost of packaging, testing, and assembling the completed circuits into an electronic system must be taken into account. Once the ICs are separated into individual chips, each chip must be handled individually. Thus packaging and testing costs often dominate the other production costs in the fabrication of ICs.

One way to improve the economics of microelectronic manufacturing is to increase wafer sizes. The key benefit from processing larger wafers is an increase in the percentage of usable area. Larger wafers have a smaller proportion of the area being affected by edge losses and wafer dicing. Since the mid-1980s, wafer diameters have increased threefold from 100 mm to 300 mm, which required the development of new equipment throughout the semiconductor manufacturing process. A second strategy for improving semiconductor economics involved increasing the number of chips per wafer by decreasing IC dimensions. IC dimensions have decreased more than 50-fold in the past 30 years. The smallest feature size in 1971 was 10 microns. By 2001, transistors with gate features as small as 0.18 micron were made. Again, the catalyst for this improvement was an investment in the process technology—in particular, photolithography.

Die yield improvement is another more desirable way to improve economics without making large capital investments. The die yield depends on the wafer yield (the fraction of silicon wafers that started versus those that finished the process), which involves the processing yield (the fraction of good dies per wafer), the assembly yield (the fraction of dies that are packaged), and the burn-in yield (the fraction of packaged dies that survive wafer testing). A single, submicron dust particle trapped between the photoresist and reticle in a photolithographic step can cause a point defect that will result in the malfunction of an entire IC. As a result, all microelectronic manufacturing is conducted in *clean rooms*, where special clothing must be worn (to prevent contamination of wafers by dust particles). The air is continuously filtered and recirculated using high-efficiency particulate-arresting filters to keep the dust level at a minimum. Wafers are commonly processed in Class 100 clean rooms.

FIGURE 19-10 Tape-automated bonding (TAB) uses a polymer tape to carry the leads to the chip for bonding. *(Source: Jaeger, R. C., Introduction to Microelectronic Fabrication (Modular Series on Solid State Device, Volume 5), New York: Addison-Wesley, 1990.)*

Sprocket drive holes

Inner leads bonded to IC

Outer leads

IC

Window in tape to facilitate excising and expose outer leads beyond polymer support ring for outer lead bonding

Area of polymer support ring after excising by custom die

IC PACKAGING

Several levels of packaging and assembly are necessary to integrate the IC chip with other electronic devices to make it part of a fully functional commercial or military product. IC packaging serves to distribute electronic signals and power as well as provide mechanical interfacing to test equipment and printed circuit boards (PCBs). In addition to this interconnection role, IC packages protect the delicate circuitry from mechanical stresses and electrostatic discharge during handling and in corrosive environments during its operational life. Finally, because of the high density of the integrated circuits, dissipation of heat generated in the circuits has become more critical.

The first step in IC packaging is to attach the chip to the die using various techniques, including *wire bonding*, *tape-automated bonding* (TAB), and *flip-chip* technology. In wire bonding, also known as *chip-and-wire attachment*, the chip is attached to the tape with an adhesive, and wires are attached to bonding pads on the chip. See Figure 19-10. Gold wire as thin as 25 microns and aluminum wire as thin as 50 microns can be attached to the chip, and the other ends of the wires then become the leads on a die, so the die can not get packaged, as shown in Figure 19-11.

IC chips are mounted on a variety of packages made from a variety of materials. Figure 19-11 shows a cutaway view of the most well-known IC chip package; the dual in-line package (DIP) refers to the two sets of in-line pins that go into holes in the PCB.

Plastic DIP

No cavity between body and chip

Molded plastic body

Leadframe

Leads imbedded in plastic body

Wire bond from lead to chip

Pins

Lead pitch

FIGURE 19-11 The dual in-line package (DIP) has a leadframe and package body. The leads on the chips are connected to the pins on the DIP. *(Source: Seraphim, D. P., Lasky, R. C., and Li, C.-Y., Principles of Electronic Packaging, New York: McGraw-Hill, 1989.)*

The DIP, like all other IC packages, is made up of a leadframe and a package body. Typically composed of a copper alloy (sometimes with an aluminum coating), the leadframe provides electrical interface between the IC and the PCB. The DIP body is made from a low-cost epoxy, which facilitates mass production. In high-reliability applications (e.g., military), where hermetic (airtight) sealing of the package is important, ceramic package bodies are used.

Generally, IC packages are grouped mainly based on the arrangement, shape, and quantity of leads. *Lead pitch* refers to the center-to-center distance between leads on an IC package. In conformance to standard-setting bodies, such as the Electronics Industries Association (EIA) in the United States and EIA Japan, lead pitches above 20 mils (0.02 in.) are measured in inches. Below 20 mils, lead pitches are measured in millimeters.

There are two methods by which components are connected to the circuit on the PCB. The DIP is the leading example of *through-hole* (TH) technology, also known as *pin-in-hole* (PIH) technology, where IC packages and discrete components are inserted into metal-plated holes in the PCB and soldered from the underside of the PCB. In *surface mount* (SM) technology, electronic components are placed onto solder paste pads that have been dispensed onto the surface of the PCB. Figure 19-12 shows the cross section of solder joints for typical SM- and TH-packaged components on a PCB.

SM packages are more cost-effective in electronic assembly, and this SM technology has replaced a lot of the TH technology, but not entirely, since not all electronic components can be purchased in an SM package. SM packages are designed for automated production and allow for higher circuit board density than TH components. The manufacturing challenges associated with SM technology include weaker joint strength and solderabilty issues relating to lower in process lead temperatures. Also, TH components have only one lead geometry, whereas SM components have many different designs. The key packaging families for TH technology are dual in-line packages (DIPs) and *pin grid arrays* (PGAs).

In SM technology, IC packages cannot be discussed separately from lead geometry. Lead geometries affect the electrical performance, size constraints on the PCB, and ease of assembly of the IC package. The most basic form of SM lead is the *butt lead*, or *J-lead*. (See Figure 19-12.) Butt leads are normally formed by clipping the leads on TH components. This technique is sometimes used to convert an existing TH component to an SM component. Consequently, butt-leaded components do not typically save any space on the PCB. However, they can reduce costs by eliminating the need to perform TH soldering of the PCB after SM soldering. Butt-lead components tend to result in the lowest solder joint strengths, and therefore reliability is an issue.

Gull wing leads bend down and out, whereas *J-leads* bend down and in. Gull wing leads allow for thinner package sizes and smaller leads, which is important for compact applications such as laptop computers. In addition, packages with gull wing leads are compatible with most reflow soldering processes and have the ability to self-align during reflow if they are slightly misoriented. Gull wing leads are compatible with fine-pitch packages, but inspection of solder joints is difficult in the final soldered configuration. Gull wing leads are also susceptibile to lead damage and deviation from lead coplanarity. J-leads are sturdier than gull wings and stand up better in handling. The solder joint of J-leads face out, making inspection easier. J-leads have a higher profile than gull wings, which can be a disadvantage for compact applications. At the same time, this higher standoff makes postsolder cleaning easier. J-leads can be used for packages with between 20 and 84 leads.

Solder balls are increasingly being used to provide SM interconnection through *ball grid arrays* (BGAs). Figure 19-12 shows a BGA package. BGAs provide high lead density because the solder balls are arrayed across the entire bottom surface of the package. Lead counts on BGAs can go as high as 2400, with most in the 200 to 500 lead range. Because of their arrayed nature, BGAs do not need as fine a pitch (40 to 50 mils) as *quad flat packages* (QFPs), which can help in electronic assembly yields. To further boost yield, the solder balls on BGAs have excellent self-aligning capability during reflow and require less

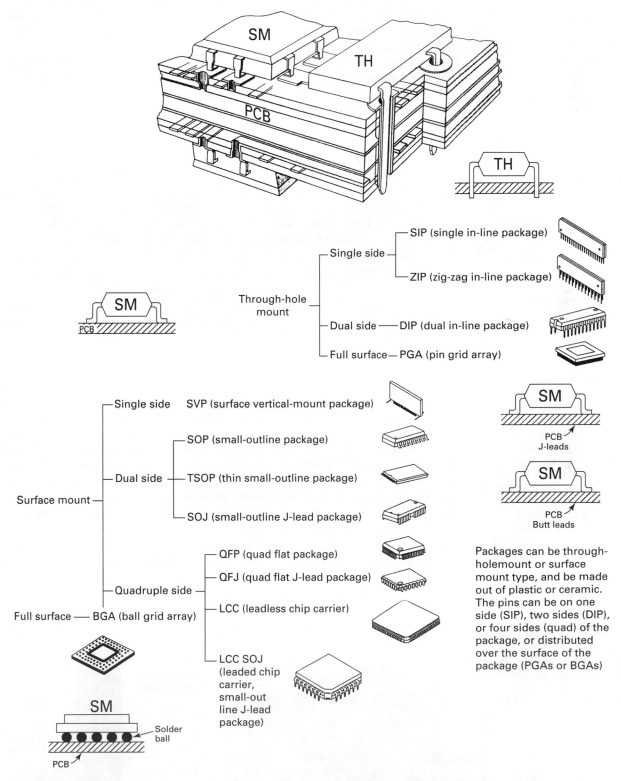

FIGURE 19-12   Here is a summary of the various types of packaging used for ICs. (*Source: Manzione, L. T., Plastic Packaging of Microelectronic Devices, New York: Van Nostrand Reinhold, 1990.*)

FIGURE 19-13 How printed circuit boards are made. The *printed circuit board* (PCB), or printed wiring board, connects the IC with other components to produce a functional circuit. Specifically, a PCB is a laminated set of dielectric layers or laminates of bulk sheet materials that have metallic circuits that are used to interconnect the various packaged components. As shown on the left, each PCB laminate has a base, tracks, and pads. The base material must be electrically insulating to provide support to all components making up the circuit. Pads on the laminate are connected by conductive tracks or traces (usually copper) that have been deposited onto the surface of the base. The metal for traces and pads is deposited by electroless plating and electroplating. Surface mount (SM) components are connected to the PCB at pads (lands) or, in the case of throughhole (TH) technology, at insertion holes.

Typical base materials used may be epoxy-impregnated fiberglass, polyimide, or ceramic. Epoxy-impregnated fiberglass is the cheapest substrate for interconnecting leaded packages. Fiberglass is used to increase the mechanical stiffness of the device for handling, while epoxy resin imparts better ductility. The fiberglass is impregnated on a continuous line where resin infiltrates the fiberglass mat in a dip basin, and the soaked fabric passes through a set of rollers to control thickness and an oven where the resin is partially cured. The resulting glass resin sheet is called *prepreg*. Multiple prepregs are then pressed together between electroformed copper foil under precise heat and pressure conditions to form a copper-clad laminate or PCB.

coplanarity (6 to 8 mils) than other leads. The downside of BGAs is the difficulty associated with cleaning, inspection, and rework of solder joints and the lack of compatibility with some reflow methods, since joints are out of sight beneath the package.

The next level of electronic manufacturing involves populating the PCBs with ICs and other devices using the surface mount or through-hole techniques. See Figure 19-13 for a brief description of the PCB fabrication process.

PCBs can be single sided, double sided, or multilayer. Single-sided PCBs simply have metallic circuits on one side of the laminate. Through-hole, single-sided PCBs have insertion holes that extend through the board to the other side, where TH components may be inserted into the board (Figure 19-14). Surface mount components are simply mounted onto the pads on the same side as the circuit and do not require through-holes. Double-sided PCBs are used in cases where circuits must "jump," or cross over, one another. In this case, *via holes* (or simply *vias*) are needed to route the circuits over one another. Vias are essentially metal-filled holes through the laminate material that connect a circuit on one side to the other. The metal inside of the via is electroplated. Vias that are also used as insertion holes are called plated through-holes (PTHs). As the number of packaged components on the board increases, the complexity of the circuits increases, giving rise to the need for multilayer PCBs in which multiple single- and double-sided boards are laminated together using prepreg. Vias that pass from an outermost track on one side of the board to the outermost track on the other side are called *through vias*. Vias within a laminate core on the inside of a multilayer PCB are called *buried vias*. Vias that come out on only one side of a multilayer PCB are called *blind or partially buried vias*. Multilayer PCBs can have as many as 20 layers, although 4 to 8 are more common.

ELECTRONIC ASSEMBLY

The term *electronic assembly* is generally reserved for the third level of electronics manufacturing involving the soldering of packaged ICs and other discrete components onto PCBs using either through-hole and/or surface mount. As explained in the section on IC packaging, TH technology refers to the insertion of packaged leads into plated through-holes in the PCB and soldering of the terminals from the backside. SM technology involves temporary attachment of components to the surface of the PCB via a flux-containing solder paste, which is reflowed within an oven. SM components are much smaller and have much different leads. Passive (non-IC) SM components have terminations rather than leads that permit better shock and vibration resistance as well as reduced inductance and capacitance losses.

The sequence of operations for SM and TH assembly is shown in Figure 19-15. Insertion can be performed either manually or with automatic insertion machines. After insertion, leads are generally clinched and trimmed if necessary to avoid *bridging* between joints during soldering, which can cause electrical shorting of the circuit. Generally, soldering of TH components is performed automatically through a process known as *wave soldering*. Wave soldering involves the conveyance of a preheated and prefluxed

PCB process to impregnate fiberglass with epoxy resin

FIGURE 19-14 PCBs can be single sided, double sided, or multilayer. *(Source: Judd, M., and Brindley, K., Soldering in Electronics Assembly, Boston: Reed International Books, 1992.)*

PCB over a standing wave of solder created by pumping action. The combination of capillary action and pumping action permits flow of the solder from the underside of the board into the joint. Cleanliness of the PCB is critical for wetting of the lead and PTH. A high-pressure air jet is used to blow off excess solder from the underside of the board to prevent solder bridging. Postsolder cleaning of the board includes degreasing and defluxing.

One key consideration for TH solder joints is joint strength. The trade-off is the clearance between the insertion lead and insertion hole. As the clearance decreases, joint strength increases. However, with smaller clearances it is more difficult to insert pins in holes. Clearances on the order of 0.25 mm are typical. Another factor affecting joint strength involves the *clinching* of leads. Clinched lead joints are much stronger than unclinched joints. Because the mechanical strength of TH joints is generally superior to SM joints, large, heavy components are generally attached with TH technology.

SM assembly involves application of solder paste to the lands on the surface of the PCB, placement of SM components on top of this paste, and reflow of the solder paste within an oven. Solder paste consists of small spherical particles of solder less than a tenth of a millimeter in diameter together with flux and solvents used to dissolve the flux (imparting tackiness) and thicken the paste. At the time of application, the paste has the consistency of peanut butter and is applied by screening, stenciling, or dispensing. In screening and stenciling, a solder paste printer is used to apply solder paste through a mask (screen or stencil) by running a squeegee over the surface of the mask. The mask

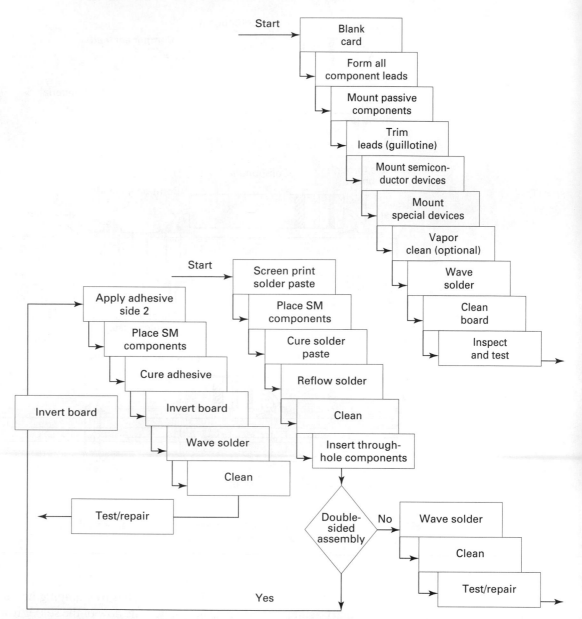

FIGURE 19-15 The assembly process steps for making a TH PCB (above). The steps for SM assembly are given below, along with the steps for mixed technologies—both TH and SM. *(Source: Haskard, M. R.,* Electronic Circuit Cards and Surface Mount Technology: A Guide to Their Design, Assembly, and Application, *New York: Prentice Hall, 1992.)*

is typically held off the surface by a distance on the order of 0.5 mm, known as the *snap-off distance* (see Figure 19-16). As the squeegee passes over the mask surface, the mask is pressed against the PCB, allowing contact between the paste and the lands on the board. After the squeegee has passed, the mask snaps back from the surface, leaving an island of solder paste on the PCB lands. Stencils are typically metal sheets or wire mesh that have been chemically etched using a lithographic process. Screens are typically formed by application, exposure, and development of a photosensitive emulsion on top of a wire mesh. Advantages of metal sheet stencils include longevity and multilevel (pads of varying thicknesses) printing, and the screens are cheaper to make. To decrease tooling costs during product development, pastes can also be dispensed without a mask through a syringe needle. Dispensing generally requires pastes with lower viscosity, which can lead to other problems, including solder paste *slump* (spreading out of the solder pastes after application).

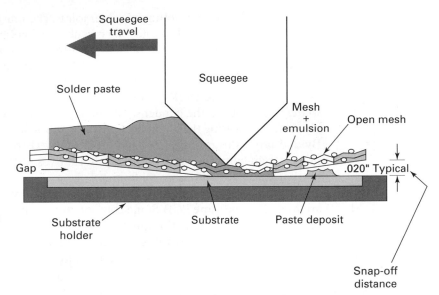

FIGURE 19-16 Schematic for applying solder paste on a substrate by squeegee in a screen printing process in SM technology. *(Source: Prasad, R., Surface Mount Technology: Principles and Practice, New York: Chapman & Hall, 1997, p. 493.)*

Once the solder paste is positioned on the board, a component placement machine, also known as a *pick-and-place* machine, is used to place the components onto the solder paste pads. The flux in the solder paste is tacky and holds positioned components in place until oven soldering. Components are fed to a robotic manipulator that has a vacuum chuck or a mechanical chuck, or both. Component feeders deliver components to the manipulator. Several types of feeders exist, including tape (or reel), bulk, tube (or stick), and waffle pack. The feeder system must be carefully selected based on the desired quantity per feeder, availability, part identification, component cost, inventory turns, and potential for damage during shipping and handling. Tape feeders are widely utilized and are most desirable for high-volume placement. Tube feeders are useful for smaller-volume assemblers, even though costs per component are higher. Waffle packs are flat-machined plates with inset pockets to hold various chips. In general, waffle packs increase the cost of assembly. However, some IC packages, like the bumperless, fine-pitch QFPs, require a high level of protection during handling to minimize lead damage, so this component requires the tape-feeding mechanism. Bulk feeding of IC components, through the use of a vibratory bowl, may be useful for prototyping environments.

The economics of SM technology are driven by component placement equipment, which determines the throughput of the SM line and is the source (at least partially) of most defects requiring rework. Further, placement equipment strongly influences start-up costs because it may involve as much as 50% of the capital equipment cost in setting up a line. Key criteria in the selection of placement equipment include placement accuracy, placement rate, maximum PCB size, types and sizes of components, and maximum number of feeders, among others. In general, placement equipment has been classified into four discrete types: (1) high throughput, (2) high flexibility, (3) high flexibility and high throughput, and (4) low cost and low throughput with high flexibility. High-throughput placement machines are called *chip shooters*. Chip shooters are typically dedicated to the placement of passive (resistors, capacitors, etc.) and small active (IC) components and can place components at rates up to 60,000 components per hour with linear repeatability around 0.05 to 0.1 mm and rotational accuracy of 0.2° to 0.5° over an area 350 by 450 mm. area.

After components are placed, the PCB is placed in a *reflow* oven where the solder paste melts, causing a fluxing action that permits the melted solder to wet the leads and the PCB lands. To achieve this, the PCB must be exposed to an appropriate *thermal profile*, or time–temperature curve, as it passes through the oven. At a minimum, the thermal profile must include at least four zones. The first zone, called preheating, is used to drive off any nonflux volatiles within the paste. The second zone, the soak zone, is used to bring the entire assembly up to just below the reflow temperature of the paste. The third (reflow) zone quickly raises the temperature of the solder paste above the reflow temperature, allowing for fluxing and wetting of solder joints. The fourth zone cools the assembly, permitting solidification. Reflow soldering is generally done in infrared (IR) reflow ovens, and heating involves both IR radiation and gas-forced convection. The minimum number of heating zones for a reflow oven must be 3 (the fourth is a cooling zone), but the oven can contain as many as 20 heating zones to provide better control over the thermal profile.

An alternative to IR reflow soldering is *vapor-phase soldering*, or *condensation soldering*, involving the condensation of a hot perfluorocarbon vapor onto the assembly surface, releasing the latent heat of vaporization into the solder joints and substrates. By and large, this process has been replaced by IR reflow soldering due to improved process reliability and control. Other alternatives to IR reflow soldering include laser, hot-bar, and hot-belt reflow soldering.

For a variety of reasons, SM technology and through-hole insertion technology are mixed on the same PCB. Some components are not available in SM packages. Some components are large and require the added strength provided by TH solder joints. Some components require more heat dissipations than SM can accommodate. Thus both methods will continue to be used in the future.

■ 19.3 ELECTROCHEMICAL MACHINING PROCESSES

ELECTROCHEMICAL MACHINING

Electrochemical machining, commonly designated ECM, removes material by anodic dissolution with a rapidly flowing electrolyte. It is basically a deplating process in which the tool is the cathode and the workpiece is the anode; both must be electrically conductive. The electrolyte, which can be pumped rapidly through or around the tool, sweeps away any heat and waste product (sludge) given off during the reaction. The sludge is captured and removed from the electrolyte through filtration. The shape of the cavity is defined by the tool, which is advanced by means of a servomechanism that controls the gap between the electrodes (i.e., the interelectrode gap) to a range from 0.003 to 0.03 in. (0.01 in. typical). The tool advances into the work at a constant feed rate, or penetration rate, that matches the deplating rate of the workpiece. The electrolyte is a highly conductive solution of inorganic salt—usually NaCl, KCl, and $NaNO_3$ (or other proprietary mixtures)—and is operated at about 75° to 150°F with flow rates ranging from 50 to 200 ft/sec. The temperature of the electrolyte is maintained through appropriate temperature controls. Tools are usually made of copper or brass and sometimes stainless steel. The process is shown schematically in Figure 19-17.

The behavior of the ECM process is governed by the laws of electrolysis (the use of electrical current to bring about chemical change). Faraday's first law of electrolysis states that the amount of chemical change (material removed) during electrolysis is proportional to the charge (number of electrons) passed. This can be expressed mathematically as

$$n = K \cdot I \cdot t \tag{19-4}$$

where n is the theoretical number of moles of material removed; I is the applied current, that is assumed to stay constant; and t is the time over which the current is applied. The term K is a proportionality constant, that is inversely proportional to the valence of the workpiece material.

To convert equation 19-4 into a useful expression for estimating the theoretical material removal rate, the number of moles, n, can be converted into a material volume by multiplying through the atomic weight, A_w, and dividing by the density, p, of the workpiece material. Further, by dividing through by the time, t, an expression for the volumetric material removal rate is obtained,

$$\text{MRR} = \left(\frac{K \cdot A_w}{p} \right) \cdot I = \text{MRR}_s \cdot I \tag{19-5}$$

where MRR_s is a proportionality constant called the specific material removal rate based on the valence, atomic weight, and density of the workpiece material. In suitable applications, material removal rates on the order of 0.1 in.^{3} min per 1000 A

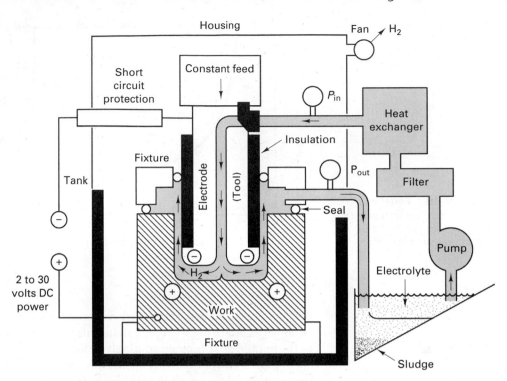

FIGURE 19-17   Schematic diagram of electrochemical machining process (ECM).

can be expected. Table 19-3 shows the specific material removal rate for some different materials.

An estimate for the theoretical feed rate, f_t, or penetration rate, in ECM can be made by dividing equation 19-5 by the area of workpiece material, A, exposed to the ECM tool at the interelectrode gap. This yields:

$$f_r = \text{MRR}_s \cdot \frac{I}{A} = \text{MRR}_s \cdot J \qquad (19\text{-}6)$$

Equations 19-4 through 19-6 assume that the efficiency, E, for metal removal by electrolysis is 100%, though normally it is in the range of 90% to 100% depending on the current, voltage, and other operating conditions. Therefore, equations 19-5 and 19-6 may be modified as follows:

$$\text{MRR} = E \cdot \text{MRR}_i \qquad (19\text{-}7)$$

$$f_r = E \cdot f_i \qquad (19\text{-}8)$$

where MRR and f_r are the actual material removal rate and actual feed rate, respectively.

As shown in equation 19-6, the penetration rate in ECM is primarily a function of the current density, J, and the physical properties of the material. Current densities from 50 to 1500 A/in.^{2} are used, and typical penetration rates in metals are from 0.02 to 0.75 in./min. Figure 19-18 shows that the penetration rate in ECM is also a function of the interelectrode gap. As the interelectrode gap is reduced, the resistance across the gap is also reduced, permitting a larger flow of electrons. One disadvantage of reducing the interelectrode gap is that the likelihood of plating anodic material onto the cathode is increased, thereby requiring higher electrolyte flow rates.

The importance of these implications is that the penetration rate, while a function of current density and interelectrode gap, is not directly affected by the

TABLE 19-3	Material Removal Rates for ECM of Alloys Assuming 100% Current Efficiency	

Alloy	Theoretical Removal Rates for 1000 Amperes per Square Inch	
	in.^{3}/min	cm^3/min
4340 steel	0.133	2.18
17-4 PH	0.123	2.02
A-286	0.117	1.92
M252	0.110	1.80
Rene 41	0.108	1.77
Udimet 500	0.110	1.80
Udimet 700	0.108	1.77
L605	0.107	1.75

Note: Rates listed were calculated using Faraday's law and valences as follows:

Aluminum	3	Copper	2	Silicon	0
Carbon	0	Iron	4	Titanium	4
Columbium	3	Manganese	3	Tungsten	6
Cobalt	2	Molybdenum	4	Vanadium	5
Chromium	3	Nickel	2		

FIGURE 19-18 Relationship of current density, penetration rate, and machining gap in electrochemical machining.

hardness or toughness of the work material. This makes ECM advantageous for the machining of certain high-strength materials with poor machinability. Penetration rates up to 0.1 in./min are obtained routinely in Waspalloy, a very hard metal alloy.

Pulsed-current ECM (PECM) has recently shown the potential to improve accuracies and surface finish in traditional ECM. In PECM, high-current densities (>100 A/cm^2) are pulsed on for durations on the order of 1 ms and pulsed off for intervals on the order of 10 ms. The relaxation (pulse off) interval permits reaction by-products to be removed from the interelectrode gap at low electrolyte flow rates without electrolytic deposition on the ECM tool. As a result, high-current densities can be used at small interelectrode distances, improving both removal rates and precision. Because PECM allows lower electrolyte flow rates, it has been proposed to remove recast layers from the surface of dies produced by electrical discharge machining (EDM). Some efforts are being made to integrate this technology into EDM platforms.

Pulsed currents have also been used in *electrochemical micromachining* (EMM). In one variation of the process, photolithographic masks are used to concentrate material removal in selective areas of thin films. In addition to providing smoother surfaces (on the order of electropolished surfaces) than those found in traditional ECM, the high-current density (~ 100 A/cm^2) results in less taper in etching profiles and more uniform material removal across the workpiece, which are important for shrinking feature sizes in micromachining.

Electrochemical polishing is a modification of the ECM process that operates essentially the same as ECM, but with a much slower penetration rate. Current density is lowered, which greatly reduces the material removal rate and produces a fine finish on the order of 10 min. Electrochemical polishing must be differentiated from *electropolishing*, which operates without the use of a part-specific hard tool.

Electrochemical Hole Machining. Several electrochemical hole-drilling processes have been developed for drilling small holes with high aspect (depth-to-diameter) ratios in difficult-to-machine materials. In *electrostream drilling*, large numbers of holes (over 50) can be simultaneously gang drilled in nickel and cobalt alloys with diameters down to 0.005 in. at aspect ratios of 50:1. Machining is performed by a high-velocity stream of charged, acidic electrolyte ejected from the capillary end of a drawn glass tube (the tool). An internal electrode (e.g., a small titanium wire) is fed into the large end of the tube and placed close to the capillary. The electrode is used to charge the electrolyte, which is pumped through the tube. An acidic electrolyte is used so that the sludge by-product goes into solution instead of clogging flow. Voltages used in the process are 10 to 20 times higher than that of typical ECM processes. This technique was originally developed to drill high-incident-angle, cooling holes in turbine blades for jet engines.

A second process, known as the *shaped-tube electrolytic machining* (STEM) process, was also created in response to unique challenges presented in the jet engine industry. Like the electrostream process, STEM is also capable of gang drilling small holes in difficult-to-machine materials. However, the STEM process is generally not capable of drilling holes smaller than about 0.02 in. STEM is capable of making shaped holes with aspect ratios as high as 300:1. Holes up to 24 in. in depth have been drilled. Like the electrostream process, it uses an acidic electrolyte to minimize clogging due to sludge buildup. The major differences between the STEM process and the electrostream process are the reduced voltage levels (5 to 10 V dc) and the special electrodes, which are long, straight, metallic tubes coated with an insulator (Figure 19-19). The insulator helps to eliminate taper by constraining the electrolytic action between the bottom of the tool and the workpiece. Titanium is often used for its ability to resist acids. The electrolyte is pressure-fed through the tube and returns through the gap (0.001 to 0.002 in.) between the insulated tube wall and the hole wall. Electrolyte concentrations may include up to 10% sulfuric acid. Lower concentrations may be used to increase tool life.

Electrochemical Grinding. *Electrochemical grinding*, commonly designated ECG, is a low-voltage, high-current variant of ECM in which the tool cathode is a rotating, metalbonded, diamond grit grinding wheel. The setup shown in Figure 19-20 for grinding a cutting tool is typical. As the electric current flows through the electrolyte between the workpiece and the wheel, some surface metal is removed electrochemically and some is changed to a metal oxide, which is ground away by the abrasives. As the oxide film is

FIGURE 19-19 The shaped-tube electrolytic machining (STEM) cell process is a specialized ECM technique for drilling small holes using a metal tube electrode or metal tube electrode with dielectric coating.

FIGURE 19-20 Equipment setup and electrical circuit for electrochemical grinding.

removed, new surface metal is exposed to electrolytic action. Most of the material removal is electrochemical, with only about 10% of the material being removed by grinding. The metal removal rate (MRR) is dependent on many variables. Table 19-4 gives some typical values for the MRR.

The wheels used in ECG must be electrically conductive abrasive wheels. For most metals, resin-bonded aluminum oxide wheels are recommended. The resin bond is loaded with copper to provide for negligible electrical resistance. The wheels are dressable, using a variety of wheel-dressing measures. Once dressed, the wheels can then be used for precision form-grinding operations.

The abrasive particles are hard, nonconducting materials such as aluminum oxide, diamond, or borazon (cubic boron nitride). In addition to increasing the efficiency of the process, the abrasives act as an insulating spacer, maintaining a separation of from 0.0005 to 0.003 in. (0.012 to 0.05 mm) between the electrodes. A dead short would result if the insulating particles were absent. The particles also serve to grind away any passivating oxide that should form during electrolysis. The process is used for shaping and sharpening carbide cutting tools, which cause high wear rates on expensive diamond wheels in normal grinding. ECG greatly reduces wheel wear. Fragile parts (e.g., honeycomb structures), surgical needles, and tips of assembled turbine blades have also been ECG-processed successfully. The lack of heat damage, burrs, and residual stresses is very beneficial, particularly when coupled with MRRs that are competitive with conventional grinding and far less wheel wear.

Electrochemical Deburring. *Electrochemical deburring* is a deburring process which works on the principle that electrolysis is accelerated in areas with small interelectrode gaps and prevented in areas with insulation between electrodes. The cathodic tool in electrochemical deburring is stationary and generally shaped as a negative of the workpiece to focus the electrolysis on the region of the workpiece where burrs are to be removed. Portions of the tool not used for deburring are coated with insulation to prevent the electrolytic reaction. The process does not require a feed mechanism. A fixture made of insulating material is used to position the workpiece with respect to the cathodic tool. Because of the small amount of material removed, electrochemical deburring generally requires short cycle times under one minute.

Design Factors in Electrochemical Machining. In general, current densities tend to concentrate at sharp edges or features and, therefore, produce rounded corners. Therefore, ECM processes should not be used to create sharp corners or pockets in the workpiece, although sharp corners are possible on through-features. Due to the need for an interelectrode gap, the actual feature size will be larger than the tool size. The overcut on the side of the tool is normally on the order of 0.005 in. Taper should be expected in all features due to electrolytic reaction between the side of the tool and the workpiece.

TABLE 19-4	Metal Removal Rates for ECG for Various Metals					
Metal	Valency	Density		Metal Removal Rate at 1000 A		
		lb/in^3	g/cm^3	lb/hr	in^3/min	cm^3/min
Aluminum	3	0.098	2.67	0.74	0.126	2.06
Beryllium	2	0.067	1.85	0.37	0.092	1.50
Chromium	2	0.260	7.19	2.14	0.137	2.25
	3			1.43	0.092	1.51
	6			0.71	0.046	0.75
Cobalt	2	0.322	8.85	2.42	0.125	2.05
	3			1.62	0.084	1.38
Niobium	3	0.310	8.57	2.55	0.132	2.16
(columbium)	4			1.92	0.103	1.69
	5			1.53	0.082	1.34
Copper	1	0.324	8.96	5.22	0.268	4.39
	2			2.61	0.134	2.20
Iron	2	0.284	7.86	2.30	0.135	2.21
	3			1.53	0.090	1.47
Magnesium	2	0.063	1.74	1.00	0.265	4.34
Manganese	2	0.270	7.43	2.26	0.139	2.28
	4			1.13	0.070	1.15
	7			0.65	0.040	0.66
Molybdenum	3	0.369	10.22	2.63	0.119	1.95
	4			1.97	0.090	1.47
	6			1.32	0.060	0.98
Nickel	2	0.322	8.90	2.41	0.129	2.11
	3			1.61	0.083	1.36
Silicon	4	0.084	2.33	0.58	0.114	1.87
Silver	1	0.379	10.49	8.87	0.390	6.39
Tin	2	0.264	7.30	4.88	0.308	5.05
	4			2.44	0.154	2.52
Titanium	3	0.163	4.51	1.31	0.134	1.65
	4			0.99	0.101	
Tungsten	6	0.697	19.3	2.52	0.060	0.98
	8			1.89	1.89	0.74
Uranium	4	0.689	19.1	4.90	0.117	1.92
	6			3.27	0.078	1.29
Vanadium	3	0.220	6.1	1.40	0.106	1.74
	5			0.84	0.064	1.05
Zinc	2	0.258	7.13	2.69	0.174	2.85

Source: 1985 SCTE Conference Proceedings, ASM, Metals Park, OH, 1986.

Depending on the tool design, taper can be held to 0.001 in./in. For micromachining applications, insulator material may be placed on the side of the tool to prevent the taper effects.

Control of the electrolyte flow can be difficult in parts with irregular shapes. Changes in electrolyte concentration due to varying flow patterns can change the local resistance across the interelectrode gap, resulting in local variations in removal rates and tolerances. In addition, high electrolytic flow rates can cause erosion of workpiece features. Therefore, complex geometries requiring tortuous interelectrode flow paths are discouraged.

Generally, no detrimental effects to the properties of the workpiece material are expected when using ECM. However, the lack of compressive residual stresses imparted to the surface of the workpiece during operation can have a negative impact on the fatigue resistance of the part when compared with mechanically machined parts. Further, if parameters are set improperly, the process may favor electrolysis at the grain boundaries, resulting in intergranular attack, which may also have a negative effect on the fatigue resistance of the part. If possible, shot peening or some similar process may be used to impart compressive stresses to the surface of the workpiece, thereby improving its fatigue resistance.

Advantages and Disadvantages of Electrochemical Machining. ECM is well suited for the machining of complex two-dimensional shapes into delicate or difficult-to-machine geometries made from poorly machinable but conductive materials. The principal tooling cost is for the preparation of the tool electrode, which can be time consuming and costly, requiring several cut-and-try efforts for complex shapes, since it is difficult to predict the precise final geometry due to variable current densities produced by certain electrode geometries (e.g., corners) or electrolyte flow variations. There is no tool wear during actual cutting, which suggests that the process becomes more economical with increasing volume. The process produces a stress-free surface, which can be advantageous, especially for small, thin parts. The ability to cut a large area simultaneously (as in chemical and electrical discharge machining) makes the production of small parts very productive. However, as in chemical machining, ECM requires the disposal of environmentally harmful by-products.

■ 19.4 ELECTRICAL DISCHARGE MACHINING

As early as the 1700s, Benjamin Franklin wrote of witnessing the removal of metal by electrical sparks. However, it was not until the 1940s that development of the *electrical discharge machining* (EDM) process, also known as electric or electro discharge machining, began in earnest. Today, it is one of the most widely used of the nontraditional machining processes.

EDM processes remove metal by discharging electric current from a pulsating DC power supply across a thin interelectrode gap between the tool and the workpiece. See Figure 19-21 for a schematic. The gap is filled by a dielectric fluid, which becomes locally ionized at the point where the interelectrode gap is the narrowest—generally, where a high point on the workpiece comes close to a high point on the tool. The ionization of the dielectric fluid creates a conduction path in which a spark is produced. The spark produces a tiny crater in the workpiece by melting and vaporization, and consequently tiny, spherical "chips" are produced by resolidification of the melted quantity of workpiece material. Bubbles from discharge gases are also produced. In addition to machining the workpiece, the high temperatures created by the spark also melt or vaporize the tool, creating tool wear. The dielectric fluid is pumped through the interelectrode gap and

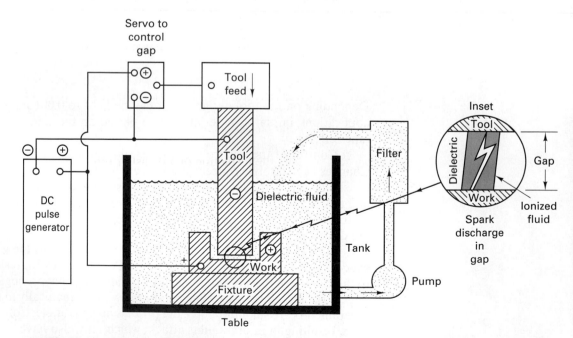

FIGURE 19-21 EDM or spark erosion machining of metal, using high-frequency spark discharges in a dielectric, between the shaped tool (cathode) and the work (anode). The table can make *X–Y* movements.

FIGURE 19-22 Schematic diagram of equipment for wire EDM using a moving wire electrode.

flushes out the chips and bubbles while confining the sparks. Once the highest point on the workpiece is removed, a subsequent spark is created between the tool and the next highest point, and so the process proceeds into the workpiece. Literally hundreds of thousands of sparks may be generated per second. This material removal mechanism is described as *spark erosion*.

Two different types of EDM exist based on the shape of the tool electrode used. In *ram EDM*, also known as *sinker EDM* or simply *EDM*, the electrode is a die in the shape of the negative of the cavity to be produced in a bulk material. By feeding the die into the workpiece, the shape of the die is machined into the workpiece.

In *wire EDM*, also known as *electrical discharge wire cutting*, the electrode is a wire used for cutting through-cut features, driving the workpiece with a computer numerical controlled (CNC) table (Figure 19-22).

Wire EDM uses a continuously moving conductive wire as the tool electrode. The tensioned wire of copper, brass, tungsten, or molybdenum is used only once, traveling from a take-off spool to a take-up spool while being "guided" to produce a straight, narrow kerf in plates up to 3 in. thick. The wire diameter ranges from 0.002 to 0.01 in., with positioning accuracy up to ±0.00002 in. in machines with numerical control (NC) or tracer control. The dielectric is usually deionized water because of its low viscosity. This process is widely used for the manufacture of punches, dies, and stripper plates, with modern machines capable of routinely cutting die relief, intricate openings, tight radius contours, and corners. See Figure 19-23 for some examples.

EDM processes are slow compared to more conventional methods of machining, and they produce a matte surface finish composed of many small craters. While feed rates in EDM are slow, EDM processes can still compete with conventional machining in producing complex geometries, particularly in hardened tool materials. As a result, one of the biggest applications of EDM processes is tool and die making. Another drawback of EDM is the formation of a recast or remelt layer on the surface and a heat-affected zone below the surface of the workpiece. Figure 19-24 shows a scanning electron micrograph of a recast layer on top of a ground surface. Note the small spheres in the lower-right corner attached to the surface, representing chips that did not escape the surface. Below the recast layer is a heat-affected zone on the order of 0.001 in. thick. The effect of the recast layer and heat-affected zone is poor surface finish as well as poor surface integrity and poor fatigue strength.

MRR and surface finish are both controlled by the spark energy. In modern EDM equipment, the spark energy is controlled by a DC power supply. The power supply works by pulsing the current on and off at certain frequencies (between 10 and 500 kHz). The on-time as a percentage of the total cycle time (inverse of the frequency) is called the *duty cycle*. EDM power supplies must be able to control the pulse voltage, current, duration, duty cycle, frequency, and electrode polarity. The power supply controls the spark energy mainly by two parameters: current on-time and discharge current.

EXAMPLES OF WORKPIECES

FIGURE 19-23 Examples of wire EDM workpieces made on NC machine (Hatachi).

FIGURE 19-24 SEM micrograph of EDM surface (right) on top of a ground surface in steel. The spheroidal nature of debris on the surface is in evidence around the craters (300×).

Figure 19-25 shows the effect of current on-time and discharge current on crater size. Larger craters are good for high MRRs. Conversely, small craters are good for finishing operations. Therefore, generally, higher duty cycles and lower frequencies are used to maximize MRR. Further, higher frequencies and lower discharge currents are used to improve surface finish while reducing the MRR. Higher frequencies generally cause increased tool wear.

FIGURE 19-25   The principles of metal removal for EDM.

An estimate for the MRR in ram EDM for a given workpiece material has been given in Weller:

$$\text{MRR} = \frac{C \cdot I}{T_m^{1,23}} \tag{19-9}$$

where MRR is the material removal rate in in.³/min (cm³/min); C is a proportionality constant equal to 5.08 in British standard units (39.86 in SI units); I is the discharge current in amps; and T_m is the melting temperature of the workpiece material, °F (°C). Melting points of selected metals are shown in Table 19-5. While this equation shows that the MRR is based mainly on the discharge current, it is recognized that MRR is more a function of current density. For graphite electrodes, a generally accepted rule of thumb for the maximum current density is 65 amps per square centimeter of surface area. Given these guidelines, an expression for the maximum MRR, MRR$_{max}$, can be derived:

$$\text{MRR}_{max} = \frac{C_{max} \cdot A_s}{T_m^{1,23}} \tag{19-10}$$

where C_{max} is a proportionality constant equal to 51.2 in British standard units (2591 in SI units) and is the bottom-facing surface area of the electrode in in.² (cm²). This formula suggests that the MRR in EDM is also a function of the tooling geometry. The larger and more highly contoured the electrode surface, the slower the MRR at the same discharge current.

TABLE 19-5	Melting Temperatures for Selected EDM Workpiece Materials

Metal	Melting Temperature, Metal °F (°C)
Aluminum	1200 (660)
Carbon steel	2500 (1371)
Cobalt	2696 (1480)
Copper	1980 (1082)
Manganese	2300 (1260)
Molybdenum	4757 (2625)
Nickel	2651 (1455)
Titanium	3308 (1820)
Tungsten	6098 (3370)

The size of the cavity cut by the EDM tool will be larger than the tool. That is, the distance between the surface of the electrode and the surface of the workpiece represents the *overcut* and is constrained by the minimum interelectrode distance necessary for a spark, which is essentially constant over all areas of the electrode, regardless of size or shape. Typical overcut values range from 0.0005 to 0.02 in. Overcut depends on the gap voltage plus the chip size, which varies with the amperage. EDM equipment manufacturers publish overcut charts for the different power supplies for their machines, and these values can be used by tool designers to determine the appropriate dimensions of the electrode. The dimensions of the tool are basically equal to the desired dimensions of the part less the overcut values.

While different materials are used for the tool electrodes, graphite is the most widely used, representing approximately 85% of electrodes. The choice of electrode material depends on its machinability and cost as well as the desired MRR, surface finish, and tool wear. Equations 19-9 and 19-10 show that the most important material characteristic for MRR is melting temperature. This relationship also applies to tool wear. The higher the melting temperature of the electrode, the less tool wear (i.e., material removed). In addition to its good machinability, graphite has a very high sublimation temperature (3500°C), which is good for minimizing tool wear. In addition to melting temperature, materials with high densities and high specific heats tend toward less tool wear. Cheaper metallic electrode materials with lower melting temperatures can be used in cases where low-temperature metals are to be machined. And in some cases requiring good surface finish, a case can be made for metallic electrodes, since spark energies are generally lower, resulting in reduced temperatures in the interelectrode gap. Copper, brass, copper–tungsten, aluminum, 70Zn-30Sn, and other alloys have all been employed as electrode (tool) materials for different reasons. In addition to tool material selection, tool wear may be minimized by using the proper polarity across the electrodes. To minimize tool wear in most electrode materials, the tool should be kept positive and the workpiece negative, although larger MRRs are possible with reversed polarity.

The dielectric fluid has four main functions: electrical insulation between the tool and workpiece; spark conductor; flushing medium; and coolant. The fluid must ionize to provide a channel for the spark and deionize quickly to become an insulator. Perhaps the most important factor in EDM is flushing of the interelectrode gap to remove residual materials and gas. Filters in the fluidic circuit are used to remove these wastes from the dielectric fluid. In addition, the fluid must carry away the heat produced in the process. A gross temperature change in the dielectric fluid significantly changes the properties of the fluid. Therefore, a heat exchanger is added to the fluidic circuit to remove heat from the dielectric fluid. Common dielectric fluids include paraffin, kerosene, and silicon-based dielectric oil. Polar compounds, such as glycerine water (90:10) with triethylene oil as an additive, have been shown to improve the MRR and decrease the tool wear when compared with traditional dielectric fluids, such as kerosene. The dielectric materials must be safe for inhalation and skin contact because operators are in constant in contact with the fluid.

Advantages and Disadvantages of EDM. EDM is applicable to all materials that are fairly good electrical conductors, including metals, alloys, and most carbides. The hardness, toughness, or brittleness of the material imposes no limitations. EDM provides a relatively simple method for making holes and pockets of any desired cross section in materials that are too hard or too brittle to be machined by most other methods. The process leaves no burrs on the edges. About 80 to 90% of the EDM work performed in the world is in the manufacture of tool and die sets for injection molding, forging, stamping, and extrusions. The absence of almost all mechanical forces makes it possible to EDM fragile or delicate parts without distortion. EDM has been used in micromachining to make feature sizes as small as 0.0004 in. (0.01 mm).

On most materials, the process produces a thin, hard recast surface, which may be an advantage or a disadvantage, depending on the use. When the workpiece material is one that tends to be brittle at room temperature, the surface may contain fine cracks caused by the thermally induced stresses. Consequently, some other finishing process is often used subsequent to EDM to remove the thin recast and heat-affected layers, particularly if the product will be fatigued. Fumes, resulting from the bubbles produced during spark erosion, are given off during the EDM process. Fumes can be toxic when electrical discharge machining boron carbide, titanium boride, and beryllium, posing a significant safety issue, although machining of these materials is hazardous in many other processes as well.

ELECTRON AND ION MACHINING

As a metals-processing tool, the electron beam is used mainly for welding, to some extent for surface hardening, and occasionally for cutting (mainly drilling). *Electron-beam machining* (EBM) is a thermal process that uses a beam of high-energy electrons focused on the workpiece to melt and vaporize metal. This process shown in Figure 19-26 is performed in a vacuum chamber. The electron beam is produced in the electron gun (also under vaccum) by thermionic emission. In its simplest form, a filament (tungsten or lanthanum-hexaboride) is heated to temperatures in excess of 2000°C, where a stream (beam) of electrons (more than 1 billion per second) is emitted from the tip of the filament. Electrostatic optics are used to focus and direct the beam. The desired beam path can be programmed with a computer to produce any desired pattern in the workpiece. The diameter of the beam is on the order of 0.0005 to 0.001 in., and holes or narrow slits with depth-to-width ratios of 100:1 can be "machined" with great precision in any material. The interaction of the beam with the surface produces dangerous X-rays; therefore, electromagnetic shielding of the process is necessary. The layer of recast material and the depth of the heat damage are very small. For micromachining applications, MRRs can exceed that of EDM or ECM. Typical tolerances are about 10% of the hole diameter or slot width. These machines require high voltages (50 to 200 kV) to accelerate the electrons to speeds of 0.5 to 0.8 the speed of light and should be operated by fully trained personnel.

Ion-beam machining (IBM) is a nano-scale (10^{-9}) machining technology used in the microelectronics industry to cleave defective wafers for characterization and failure analysis. IBM uses a focused ion beam created by thermionic emission similar to EBM to machine features as small as 50 nm. The ion beam may be focused down to a 50-nm diameter and is focused and positioned by an electrostatic optics column. Current densities up to 5 A cm^2 and voltages between 4 and 150 kV provide ion energies up to 300 A cm^2 keV. Target substrates as large as $7'' \times 7'' \times \frac{1}{4}''$ thick can be processed.

FIGURE 19-26 Electron-beam machining uses a high-energy electron beam (10^9 W/in.^{2}).

FIGURE 19-27 Schematic diagram of a laser-beam machine, a thermal NTM process that can micromachine any material.

LASER-BEAM MACHINING

Laser-beam machining (LBM) uses an intensely focused, coherent stream of light (a laser) to vaporize or chemically *ablate* materials. A schematic of the LBM process is shown in Figure 19-27. Lasers are also used for joining (welding, crazing, soldering), heat treating materials (see Chapters 5 and 36 for a discussion), and rapid prototyping (see Chapter 14). Power density and interaction time are the basic parameters in laser processing, as shown in Figure 19-28. Drilling requires higher power densities and shorter interaction times compared to most other applications.

The material removal mechanism in LBM is dependent on the wavelength of the laser used. At UV wavelengths (i.e., between about 200 and 400 nm), the material removal mechanism in polymers (for example) is generally thermal evaporation. Below 400 nm, polymeric material is typically removed by *chemical ablation*. In *ablation*, the chemical bonds between atoms are broken by the excess amount of laser energy absorbed by the valence electrons in the material. The advantage of chemical *ablation* is that since it is not a thermal process, it does not result in a heat-affected zone.

Laser light is produced within a laser cavity, which is a highly reflective cavity containing a laser rod and a high-intensity light source, or laser lamp. The light source is used to "pump up" the laser rod, which includes atoms of a lasing media that is capable of absorbing the particular wavelength of light produced by the light source. When an atom of lasing

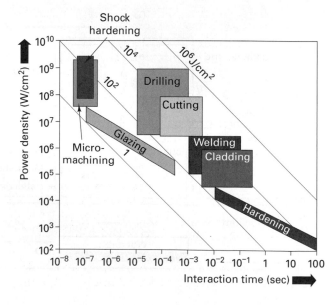

FIGURE 19-28 Power densities and interaction times in laser processing vary with the application.

TABLE 19-6 Commercial Lasers Available for Machining, Welding, and Trimming

Laser Type	Wavelength (μm)	Mode of Operation	Power (W)	Pulse Pulses per Second	Length of Time	Application	Comments
Argon	0.4880 0.5145	Pulse	20 peak; 0.005 average	60	50 μs	Scribing thin films	Power low
Ruby	0.6943	Pulse	2×10^5 peak	Low (5 to 10)	0.2–7 ms	Large material removal in one pulse, drilling diamond dies, spot welding	Often uneconomical
Nd-glass	1.06	Pulse	2×10^6 peak	Low (0.2)	0.5–10 ms	Large material removal in one pulse	Often uneconomical
Nd-YAG[a]	1.06	Continuous	1000	—	—	Welding	Compact, economical at low powers
Nd-YAG	1.06	Repetitively Q-switched	3×10^5 peak; 30 average	1–24,000	50–250 ns	Resistor trimming, electronic	Compact and economical
Nd-YAG	1.06	Pulse	400	300	0.5–7 ms	Spot weld, drill	
CO_2[b]	10.6	Continuous	15,000	—	—	Cutting organic materials, oxygen-assisted metal	Very bulky at high powers
CO_2	10.6	Repetitively Q-switched	75,000 peak 1.5 average	400	50–200 ns	Resistor trimming	Bulky but economical
CO_2	10.6	Pulse	100 average	100	100 μs and up	Welding, hole production, cutting	Bulky but economical
KrF (excimer)	0.248	Pulse	Up to several kW	100–1000	15–45 ns	Micromachining, industrial materials processing, and laser annealing	Short wavelength, high energy, and high average power
XeCl (excimer)	0.308	Pulse	200	300	40–50 ns		

Source: Modified from J. F. Ready, Selecting a laser for material working, *Laser Focus* (March 1970), p. 40.

[a] Neodymium-yttrium aluminum garnet.
[b] CO_2 plus He plus N_2 mixture.

media is struck by a photon of light, it becomes energized. When a second photon strikes the energized atom, the atom gives off two photons of identical wavelength, moving in the same direction and with the same phase. This process is called *stimulated emission*. As the two photons now stimulate further emission from other energized atoms, a cascading of stimulated emission ensues. To increase the number of stimulated emissions, the laser rod has mirrors on both ends that are precisely parallel to one another. Only photons moving perpendicular to these two mirrors stay within the laser rod, causing additional stimulated emission. One of the mirrors is partially transmissive and permits some percent of the laser energy to escape the cavity. The energy leaving the laser rod is the laser beam.

Table 19-6 lists some commercially available lasers for material processing. The most common industrial laser is the CO_2 laser. The CO_2 laser is a gas laser that uses a tube of helium and carbon dioxide as the laser rod. Output is in the far-infrared range (10.6 μm), and the power can be up to 10 kW. Nd:YAG lasers are called solid-state lasers. The laser rod in these lasers is a solid crystal of yttrium, aluminum, and garnet that has been doped with neodymium atoms (the lasing media). The output wavelength is in the near-infrared range (1064 nm), and power up to 500 W is common. For micromachining applications, modifications can be made to a 50-W Nd:YAG laser to output at one-half (532 nm), one-third (355 nm), and one-fourth (266 nm) of this wavelength, with roughly an order of magnitude reduction in power from 1064 to 532 to 266 nm. More recently, gas lasers, called *excimer* (from *exc*ited d*imer*) lasers, have been developed with laser rods consisting of excited complex molecules (usually noble gas halides) called dimers. Excimer lasers are pulsed lasers that output in the near and deep UV range at powers up to 100 W. Excimer lasers are significantly more expensive to purchase and operate than CO_2 or Nd:YAG lasers.

Lasers produce highly collimated, coherent (in-phase) light, which, when focused to a small diameter, produces high-power densities that are good for machining. It is generally accepted that in order to evaporate materials, infrared power densities in excess of 10^5 W/mm^2 are needed. For CO_2 lasers, these levels are directly achievable. However, in Nd:YAG lasers, these high power conditions would significantly decrease the life of the laser lamp. Therefore, Nd:YAG lasers make use of a Q-switch, which breaks up the continuous light stream into a series of higher power pulses. Pulsed Nd:YAG lasers may have peak powers in excess of 30 kW at 1064 nm and peak power densities in excess of 2 X 10^7 W/mm^2, which are easily enough for metal sublimation. With this magnitude of power density, the thermal effects of LBM are minimal.

Applications of LBM are widely varied. Most CO_2 industrial laser CNC machining centers can focus the beam down to a diameter of about 0.005 in. Applications for these systems range from cigarette paper cutting to drilling microholes in turbine engine blades to cutting steel plate for chain saw blades. For printed circuit board and chip scale packaging applications, Nd:YAG, excimer, and now pulsed lasers are being used to drill holes down to 0.001 in. in milliseconds in polyimide or polyester films. However, this drilling is limited to rather thin stock (0.01 in.), as the cutting speed drops off rapidly with penetration into the material. Hole depth-to-diameter ratios of 10:1 are common, and hole geometry is irregular. Recast and heat-affected zones exist adjacent to the cuts. Deep UV excimer and Nd:YAG lasers are ideal for micromachining applications. Deep UV lasers use primarily a chemical ablation mechanism for material removal. In addition to providing the potential for eliminating thermal effects in machining, deep UV lasers (having lower wavelengths) may be focused to tighter diameters than lasers with higher wavelengths. While less powerful than infrared lasers, deep UV lasers (because of chemical ablation mechanisms) experience much greater energy efficiencies in cutting. Holes as small as 60 μin. (1.5 μm) in diameter have been machined in thin-film materials with excimer lasers.

In LBM, the wavelength used to process the material is mainly determined by the optical characteristics (reflectivity, absorptivity, and transmissivity) of the workpiece. Not all materials can be machined by all lasers. Protective eyeware is necessary when working around laser equipment because of the potential damage to eyesight from either direct or scattered laser light.

PLASMA ARC CUTTING

Plasma arc cutting (PAC) uses a superheated stream of electrically ionized gas to melt and remove material (Figure 19-29). The 20,000° to 50,000°F plasma is created inside a water-cooled nozzle by electrically ionizing a suitable gas such as nitrogen, hydrogen, argon, or mixtures of these gases. The process can be used on almost any conductive metal. The plasma arc is a mixture of free electrons, positively charged ions, and neutral atoms. The arc is initiated in a confined gas-filled chamber by a high-frequency spark. The high-voltage, DC power sustains the arc, which exits from the nozzle at near-sonic velocity. The workpiece is electrically positive. The high-velocity gases melt and blow away the molten metal "chips." Dual-flow torches use a secondary gas or water shield to assist in blowing the molten metal out of the kerf, giving a cleaner cut. The process may be performed underwater, using a large tank to hold the plates being cut. The water assists in confining the arc and reducing smoke. The main advantage of PAC is speed. Mild steel $\frac{1}{4}$ in. thick can be cut at 125 in./min. Speed decreases with thickness. Greater nozzle life and faster cutting speeds accompany the use of water-injection-type torches. Control of nozzle standoff from the workpiece is important. One electrode size can be used to machine a wide variety of materials and thicknesses by suitable adjustments to the power level, gas type, gas flow rate, traverse speed, and flame angle. PAC is sometimes called plasma-beam machining.

PAC can machine exotic materials at high rates. Profile cutting of metals, particularly of stainless steel and aluminum, has been the most prominent commercial application. However, mild steel, alloy steel, titanium, bronze, and most metals can be cut cleanly and

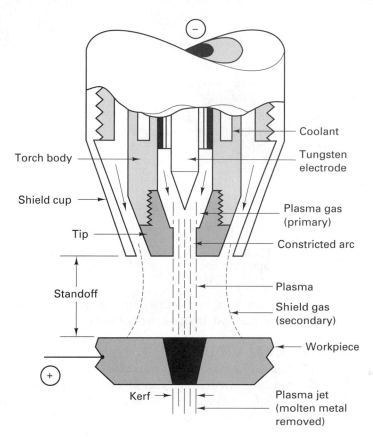

FIGURE 19-29 Plasma arc machining or cutting.

rapidly. Multiple-torch cuts are possible on programmed or tracer-controlled cutting tables on plates up to 6 in. thick in stainless steel. Smooth cuts free from contaminants are a PAC advantage. Well-attached dross on the underside of the cut can be a problem, and there will be a heat-affected zone (HAZ). The depth of the HAZ is a function of the metal, its thickness, and the cutting speed. Surface heat treatment and metal joining are beginning to use the plasma torch. See Chapter 31 for more discussion on plasma arc processes.

Some of the drawbacks of the PAC process include poor tolerances, tapered cuts, and double arcing, leading to premature wear on the nozzle. *Precision PAC*, also called *high-definition plasma* and *fine plasma cutting*, uses a special nozzle, where either a high-flow vortex or a magnetic field causes the plasma to spin rapidly and stabilizes the plasma pressure (Figure 19-30). The fast-spinning plasma results in a finely defined beam

Arc Current Density: 60,000 amps/sq. in.
Kerf Width: ~0.04″ (1 mm) on 1/8″ (3-mm) mild steel

FIGURE 19-30 Precision plasma cutting nozzle.

FIGURE 19-31 Thermochemical machining process for the removal of burrs and fins.

that cuts a narrow kerf with a perpendicular edge. Precision PAC also reduces the problems of HAZ and dross on the bottom of parts.

THERMAL DEBURRING

Thermal deburring, also known as the *thermal energy method*, has been developed for the removal of burrs and fins by exposing the workpiece to hot corrosive gases for a short period of time, typically on the order of a few milliseconds (Figure 19-31). The hot gases are formed by combusting (with a spark plug) explosive mixtures of oxygen and fuel (e.g., natural gas) in a chamber holding the workpieces. A thermal shock wave, moving at Mach 8 (2700 m/s; 6000 mph) with temperatures up to 3300°C (6000°F), vaporizes the burr in about 25 milliseconds. Because of the intense heat, the burrs and fins are unable to dissipate the heat fast enough to the surrounding workpiece and, consequently, sublimate. The workpiece remains unaffected and relatively cool because of its low surface-to-mass ratio and the short exposure time. Small amounts of metal are removed from all exposed surfaces, and this must be permissible if the process is to be used. Consequently, while large burrs and fins can be removed with this process, the procedure usually can be used only for removing small burrs. Care must be taken with parts with thin cross sections. Maximum burr thickness should be about the thinnest feature on the workpiece.

Thermal deburring will remove burrs or fins from a wide range of materials, but it is particularly effective with materials of low thermal conductivity, which easily oxidize. It will deburr thermosetting plastics, but not thermoplastic materials. Any workpiece of modest size requiring manual deburring or flash removal should be considered a candidate for thermal deburring. Die castings, gears, valves, rifle bolts, and similar small parts are deburred readily, including blind, internal, and intersecting holes in inaccessible locations. Carburetor parts are processed in automated equipment. The major advantage of thermal deburring is that fine burrs are removed much more quickly and cheaply than if they were removed by hand. Uniformity of results and greater quality assurance over hand deburring are also advantages of thermal deburring. One outcome of thermal deburring is that the part is coated with a fine oxide dust that can be removed easily by solvents. Capital costs can be several hundred thousand dollars, and the maximum workpiece dimensions are on the order of 250 mm (10 in.) by 690 mm (27 in.). See Chapter 36 for additional discussions on deburring.

■ Key Words

ablate	butt lead	chip	contact printing
ball grid arrays	chemical blanking/milling	chip shooter	Czochralski method
boule	chemical machining (CHM)	clinching	diamond sawing

die yield	etching	*p*-type semiconductors	seed crystal
doping	flip-chip	photochemical blanking	shaped-tube electrolytic
electrical discharge machining	gel milling	photochemical machining	machining (STEM)
(EDM)	gull wing	(PCM)	solder balls
electrochemical grinding	heat-affected zone (HAZ)	photomask	spark erosion
ECG)	high-definition plasma	photoresists	stepped machining
electrochemical machining	integrated circuit (IC)	pin grid arrays	surface mount
(ECM)	ion-beam machining (IBM)	pin-in-hole	thermal deburring
electrochemical polishing	J-leads	plasma arc cutting (PAC)	through-hole
electron-beam machining	laser-beam machining (LBM)	plasma etching	vapor-phase soldering
(EBM)	maskant	precision PAC	wafer testing
electronic assembly	nontraditional machining	printed circuit board	wafers
electropolishing	(NTM)	ram EDM	wire EDM (WEDM)
electrostream drilling	overcut	reactive ion etching	wire cutting
etch factor	overhang	screen printing	

■ Review Questions

1. How do the MRRs for most NTM processes compare to conventional metal cutting?
2. What are the steps in chemical machining using photosensitive resists?
3. Why is it preferable in chemical machining to apply the etchant by spraying instead of immersion?
4. What are the advantages of chemical blanking over regular blanking using punch and die methods?
5. How are multiple depths of cut (steps) produced by chemical machining?
6. Would it be feasible to produce a groove 2 mm wide and 3 mm deep by chemical machining?
7. A drawing calls for making a groove 23 mm wide and 3 mm deep by chemical machining. What should be the width of the opening in the maskant?
8. Could an ordinary steel weldment be chemically machined? Why or why not?
9. How would you produce a tapered section by chemical machining?
10. What is the principal application of thermochemical machining?
11. Is ECM related to chemical machining?
12. What effect does work material hardness have on the metal removal rate in ECM?
13. What is the principal cause of tool wear in ECM?
14. Would electrochemical grinding be a suitable process for sharpening ceramic tools? Why or why not?
15. Upon what factors does the metal removal rate depend in ECM?
16. Why is the tool insulated in the ECM schematic?
17. What is the nature of the surface obtained by electrodischarge machining?
18. What is the principal advantage of using a moving wire electrode in electrodischarge machining?
19. What effect would increasing the voltage have on the metal removal rate in electrodischarge machining? Why?
20. If the metal from which a part is to be made is quite brittle and the part will be subjected to repeated tensile loads, would you select ECM or electro discharge machining for making it? Why or why not?
21. If you had to make several holes in a large number of delicate parts, would you prefer ECM, EDM, EBM, or LBM? Why?
22. What process would you recommend to make many small holes in a very hard alloy where the holes will be used for cooling and venting?
23. Explain (using a little physics and metallurgy) why the "chips" in a thermal process like EDM are often hollow spheres?
24. What is a semiconductor?
25. In general, what technological breakthroughs were necessary to advance to each successive level of integration?
26. What is a silicon *boule*?
27. What is the most complicated, expensive, and critical step in microelectronics manufacturing?
28. List the photolithographic steps necessary to produce a resist mask on a silicon substrate.
29. List four requirements of a photoresist.
30. What is undercutting?
31. What are some possible defects that can result from underetching? From overetching?
32. What is meant by the term *chip* in electronics manufacturing versus EDM?
33. What drives the increase in component density and die area within microelectronic manufacturing?
34. Why are clean rooms so important to microelectronic processing?
35. What two subcomponents make up an IC package?
36. What are the advantages of surface mount technology versus through-hole (or pin-in-hole) technology for attachment of IC packages and discrete electrical components to boards?
37. Name the two key classes of TH packages.
38. Name the four different types of SM lead geometries, and discuss the advantages of each.
39. List the key steps involved in conventional IC packaging.
40. What is a printed circuit board (PCB)?

www.wiley.com/college/degarmo

Chapter 19 CASE STUDY

Fire Extinguisher Pressure Gage

As a materials engineer for the War Eagle Extinguisher Company, Carlos has recently been made aware of several potentially hazardous failures that have occurred in his company's products. Bourdon tube pressure gages (Figure CS-19) are used to monitor the internal pressure of the sodium bicarbonate (dry chemical) fire extinguishers. In several extinguishers, a longitudinal crack has formed along the axis of the bourdon tube, a curved tube of elliptical cross section that has been fabricated from phosphor bronze tubing. These cracks are particularly disturbing for several reasons. First, they allow the fire extinguisher to lose pressure and become inoperable. More significantly, however, the cracks allow the tube to deflect elastically in such a manner that the gage still indicates high internal pressure. Thus, while the extinguisher has actually lost all internal pressure and is useless, the reading on the gage

would cause the owner to believe that he still had an operable firefighting device.

Carlos is trying to determine the cause of these failures and suggest appropriate corrective measures. Can you help him?

1. What additional information would you like to have regarding the failed components, their fabrication history, and their service history? Why?
2. What might be some of the possible causes of these failures? What type of evidence would support each possibility? What types of additional tests or investigations might you propose to Carlos.
3. Could these failures have occurred in "normal" use, or is it likely that some form of negligence, abuse, or misuse was involved?
4. What possible corrective or preventative measures might you suggest to Carlos to prevent a recurrence?

(a) Schematic of a bourdon-tube pressure gage   (b) Cross section of bourdon tube

CHAPTER 20

FUNDAMENTALS OF MACHINING/ ORTHOGONAL MACHINING

■ 20.1 INTRODUCTION

Machining is the process of removing unwanted material from a workpiece in the form of chips. If the workpiece is metal, the process is often called *metal cutting* or *metal removal*. U.S. industries annually spend well over $100 billion to perform metal removal operations because the vast majority of manufactured products require machining at some stage in their production, ranging from relatively rough or nonprecision work, such as cleanup of castings or forgings, to high-precision work involving tolerances of 0.0001 in. or less and high-quality finishes. Thus machining undoubtedly is the most important of the basic manufacturing processes.

Beginning with the work of F. W. Taylor at Midvale steel in the 1880s, the process has been the object of considerable research and experimentation that have led to improved understanding of the nature of both the process itself and the surfaces produced by it. While this research effort led to marked improvements in machining productivity, the complexity of the process has resulted in slow progress in obtaining a complete theory of chip formation.

What makes this process so unique and difficult to analyze?

• Prior workhardening greatly affects the process.
• Different materials behave differently.
• The process is asymmetrical and unconstrained, bounded only by the cutting tool.
• The level of strain is very large.
• The strain rate is very high.
• The process is sensitive to variations in tool geometry, tool material, temperature, environment (cutting fluids), and process dynamics (chatter and vibration).

The objective of this chapter is to put all this in perspective for the practicing engineer.

FIGURE 20-1 The fundamental inputs and outputs to machining processes.

■ 20.2 FUNDAMENTALS

The process of metal cutting is complex because it has such a wide variety of inputs, which are listed in Figure 20-1. The variables are:

- The machine tool selected to perform the process
- The cutting tool selected (geometry and material)
- The properties and parameters of the workpiece
- The cutting parameters selected (speed, feed, depth of cut)
- The workpiece holding devices or fixtures or jigs

As we can see from Figure 20-1, the wide variety of inputs creates a host of outputs, most of which are critical to satisfactory performance of the component and product.

There are seven basic chip formation processes (see Figure 20-2): *turning, milling, drilling, sawing, broaching, shaping (planing), and grinding* (also called *abrasive machining*), discussed in Chapters 22–24 and 26–28. Chapter 25 describe workholding

FIGURE 20-2 The seven basic machining processes used in chip formation.

devices and Chapter 21 will provide additional insights on cutting tools. Usually the workpiece material is determined by the design engineer to meet the functional requirements of the part in service. The manufacturing engineer will often have to select the cutting-tool materials and workholder parameters and then cutting parameters based on that work material decision. Let us begin with the assumption that the workpiece material has been selected. To make the component, you decided to use a high-speed steel cutting tool for a turning operation (see Figure 20-3).

For all metal cutting processes, it is necessary to determine the parameters, speed, feed, and depth of cut. The turning process will be used to introduce these terms. In general, *speed (V)* is the primary cutting motion, which relates the velocity of the cutting tool relative to the *workpiece*. It is generally given in units of surface feet per minute (sfpm), inches per minute (in./min), meters per minute (m/m), or meters per second (m/s). *Speed (V)* is shown with the heavy dark arrow. *Feed (f_r)* is the amount of material removed per revolution or per pass of the tool over the workpiece. In turning, feed is in inches per revolution, and the tool feeds parallel to the rotational axis of the workpiece. Depending on the process, feed units are inches per revolution, inches per cycle, inches per minute, or inches per tooth. Feed is shown with dashed arrows. The *depth of cut* (DOC) represents the third dimension. In turning, it is the distance the tool is plunged into the surface. It is half the difference in the initial diameter, D_1, and the final diameter, D_2:

$$\text{DOC} = \frac{D_1 - D_2}{2} = d \qquad (20\text{-}1)$$

FIGURE 20-3 Turning a cylindrical workpiece on a lathe requires you to select the cutting speed, feed, and depth of cut.

NOTE
The rpm of the rotating workpiece is N_s. It establishes the cutting speed V, at the tool, according to $N_s = 12V/\pi D$.

The depth of cut, d, is equal to $(D_1 - D_2)/2$.
The length of cut is the distance the tool travels parallel to the axis, L.

The selection of the cutting speed V determines the surface speed of the rotating part that is related to the outer diameter of the workpiece.

$$V = \frac{\pi D_1 N_s}{12} \tag{20-2}$$

where D_1 is in inches, V is speed in surface feet per minute, and N_s is the revolutions per minute (rpm) of the workpiece. The input to the lathe will be in revolutions per minute of the spindle.

Figure 20-3 shows a typical *machine tool* for the turning process, a lathe. Workpieces are held in *workholding devices*. (See Chapter 25 for details on the design of workholders.) In this example, a three-jaw chuck is used to hold the workpiece and rotate it against the tool. The chuck is attached to the spindle, which is driven through gears by the motor.

Before the actual values for speed and feed are selected, the cutting-tool material and geometry must be selected. The *cutting tool* is used to machine (i.e., cut) the workpiece and is the most critical component. The geometry of a single point (single cutting edge) of a typical high-speed steel tool used in turning is found in Chapter 21. Various tool geometry is usually ground onto high speed steel blanks, depending on what material is being machined. Figure 20-4, taken from *Metcut's Machinability Data Handbook*, gives the MfE/IE starting values for cutting speed (sfpm or m/min) and feed (ipr or mm/r) for a given depth of cut, a given work material (hardness), and a given process (turning). Notice how speed decreases as DOC or feed increases, and cutting speeds increase with carbide and coated-carbide tool materials. To process different metals, the input parameters to the machine tools must be determined. For the lathe, the input parameters

Turning, Single Point and Box Tools

Material	Hardness Bhn	Condition	Depth of Cut* in / mm	High Speed Steel Tool Speed fpm / m/min	Feed ipr / mm/r	Tool Material AISI / ISO	Carbide Tool — Uncoated Speed Brazed fpm / m/min	Speed Indexable fpm / m/min	Feed ipr / mm/r	Tool Material Grade C / ISO	Carbide Tool — Coated Speed fpm / m/min	Feed ipr / mm/r	Tool Material Grade C / ISO
1. FREE MACHINING CARBON STEELS, WROUGHT (cont.) **Medium Carbon Leaded** (cont.) (materials listed on preceding page)	225 to 275	Hot Rolled, Normalized, Annealed, Cold Drawn or Quenched and Tempered	.040	160	.008	M2, M3	500	610	.007	C-7	925	.007	CC-7
			.150	125	.015	M2, M3	390	480	.020	C-6	600	.015	CC-6
			.300	100	.020	M2, M3	310	375	.030	C-6	500	.020	CC-6
			.625	80	.030	M2, M3	.240	290	.040	C-6	—	—	—
			1	49	.20	S4, S5	150	185	.18	P10	280	.18	CP10
			4	38	.40	S4, S5	120	145	.50	P20	185	.40	CP20
			8	30	.50	S4, S5	95	115	.75	P30	150	.50	CP30
			16	24	.75	S4, S5	73	88	1.0	P40	—	—	—
	275 to 325	Hot Rolled, Normalized, Annealed or Quenched and Tempered	.040	135	.007	T15, M42†	460	545	.007	C-7	825	.007	CC-7
			.150	105	.015	T15, M42†	350	425	.020	C-6	525	.015	CC-6
			.300	85	.020	T15, M42†	275	380	.030	C-6	425	020	CC-6
			.625	—	—		—	—			—	—	
			1	41	.18	S9, S11†	140	165	.18	P10	250	.18	CP10
			4	32	.40	S9, S11†	105	130	.50	P20	160	.40	CP20
			8	26	.50	S9, S11†	84	100	.75	P30	130	.50	CP30
			16	—	—		—	—			—	—	
	325 to 375	Quenched and Tempered	.040	100	.007	T15, M42†	390	480	.007	C-7	725	.007	CC-7
			.150	80	.015	T15, M42†	300	375	.020	C-6	475	.015	CC-6
			.300	65	.020	T15, M42†	230	290	.030	C-6	375	.020	CC-6
			.625	—	—		—	—			—	—	
			1	30	.18	S9, S11†	120	145	.18	P10	220	.18	CP10
			4	24	.40	S9, S11†	90	115	.50	P20	145	.40	CP20
			8	20	.50	S9, S11†	70	88	.75	P30	115	.50	CP30
			16										
	375 to 425	Quenched and Tempered	.040	70	.007	T15, M42†	325	400	.007	C-7	600	.007	CC-7
			.150	55	.015	T15, M42†	250	310	.020	C-6	400	.015	CC-6
			.300	45	.020	T15, M42†	200	240	.030	C-6	325	.020	CC-6
			.625	—	—		—	—			—	—	
			1	21	.18	S9, S11†	100	120	.18	P10	185	.18	CP10
			4	17	.40	S9, S11†	76	95	.50	P20	120	.40	CP20
			8	14	.50	S9, S11†	60	73	.75	P30	100	.50	CP30
			16	—	—		—	—			—	—	
2. CARBON STEELS, WROUGHT **Low Carbon** 1005 1010 1020 1006 1012 1023 1008 1015 1025 1009 1017	85 to 125	Hot Rolled, Normalized, Annealed or Cold Drawn	.040	185	.007	M2, M3	535	700	.007	C-7	1050	.007	CC-7
			.150	145	.015	M2, M3	435	540	.020	C-6	700	.015	CC-6
			.300	115	.020	M2, M3	340	420	.030	C-6	550	.020	CC-6
			.625	90	.030	M2, M3	265	330	.040	C-6	—	—	—
			1	56	.18	S4, S5	165	215	.18	P10	320	.18	CP10
			4	44	.40	S4, S5	135	165	.50	P20	215	.40	CP20
			8	35	.50	S4, S5	105	130	.75	P30	170	.50	CP30
			16	27	.75	S4, S5	81	100	1.0	P40	—	—	—
	125 to 175	Hot Rolled, Normalized, Annealed or Cold Drawn	.040	150	.007	M2, M3	485	640	.007	C-7	950	.007	CC-7
			.150	125	.015	M2, M3	410	500	.020	C-6	625	.015	CC-6
			.300	100	.020	M2, M3	320	390	.030	C-6	500	.020	CC-6
			.625	80	.030	M2, M3	245	305	.040	C-6	—	—	—
			1	46	.18	S4, S5	150	195	.18	P10	290	.18	CP10
			4	38	.40	S4, S5	125	150	.50	P20	190	.40	CP20
			8	30	.50	S4, S5	100	120	.75	P30	150	.50	CP30
			16	24	.75	S4, S5	75	95	1.0	P40	—	—	—
	175 to 225	Hot Rolled, Normalized, Annealed or Cold Drawn	.040	145	.007	M2, M3	460	570	.007	C-7	850	.007	CC-7
			.150	115	.015	M2, M3	385	450	.020	C-6	550	.015	CC-6
			.300	95	.020	M2, M3	300	350	.030	C-6	450	.020	CC-6
			.625	75	.030	M2, M3	235	265	.040	C-6	—	—	—
			1	44	.18	S4, S5	140	175	.18	P10	260	.18	CP10
			4	35	.40	S4, S5	115	135	.50	P20	170	.40	CP20
			8	29	.50	S4, S5	90	105	.75	P30	135	.50	CP30
			16	23	.75	S4, S5	72	81	1.0	P40	—	—	—
	225 to 275	Annealed or Cold Drawn	.040	125	.007	M2, M3	410	510	.007	C-7	750	.007	CC-7
			.150	95	.015	M2, M3	360	400	.020	C-6	500	.015	CC-6
			.300	75	.020	M2, M3	285	315	.030	C-6	400	.020	CC-6
			.625	60	.030	M2, M3	220	240	.040	C-6	—	—	—
			1	38	.18	S4, S5	125	155	.18	P10	230	.18	CP10
			4	29	.40	S4, S5	110	120	.50	P20	150	.40	CP20
			8	23	.50	S4, S5	87	95	.75	P30	120	.50	CP30
			16	18	.75	S4, S5	67	73	1.0	P40	—	—	—

See section 15.1 for Tool Geometry.
*Caution: Check Horsepower requirements on heavier depths of cut.
See section 16 for Cutting Fluid Recommendations.
†Any premium HSS (T15, M33, M41–M47) or (S9, S10, S11, S12).

FIGURE 20-4 Examples of a table for selection of speed and feed for turning. (*Source:* Metcut's Machinability Data Handbook.)

are DOC, the feed, and the rpm value of the spindle. The rpm value depends on the selection of the cutting speed V. Rewriting equation for N_s:

$$N_s = \frac{12V}{\pi D_1} \cong \frac{3.8V}{D_1} \qquad (20\text{-}3)$$

Cutting speed, feed, and DOC selection depend on many factors, and a great deal of experience and experimentation are required to find the best combinations. A good place to begin is by consulting tables of recommended values, as shown in Figure 20-4. Most tables are arranged according to the process being used, the material being machined, the hardness, and the cutting-tool material. The table given is a sample, to be used only for solving turning problems in the book. For industrial calculations, standard references listed at the end of the book or cutting-tool manufacturers should be consulted.

This table is for turning processes only. The amount of metal removed per pass determines the DOC. In practice, roughing cuts are heavier than finishing cuts in terms of DOC and feed and are run at a lower surface speed. Note that this table provides recommendations of V and f_r in both English and metric units based on the DOC needed to perform the job. Table values are usually conservative and should be considered starting points for determining the operational parameters for a process.

Once cutting speed V has been selected, equation 20-3 is used to determine the spindle rpm, N_s. The speed and feed can be used with the DOC to estimate the metal removal rate for the process, or MRR. For turning, the MRR is

$$MRR \cong 12V f_r d \qquad (20\text{-}4)$$

This is an approximate equation for MRR. For turning, MRR values can range from 0.1 to 600 in.^{3}/min. The MRR can be used to estimate the horsepower needed to perform a cut, as will be shown later. For most processes, the MRR equation can be viewed as the

Turning

Speed, stated in surface feet per minute (sfpm), is the peripheral speed at the cutting edge. Feed per revolution in turning is a linear motion of the tool parallel to the rotating axis of the workpiece. The depth of cut reflects the third dimension.

L = length of cut

$$T_m = \frac{L+A}{f_r N_s}$$

Boring

Enlarging hole of diameter D_1 to diameter D_2. Boring can be done with multiple cutting tools. Feed in inches per revolution, f_r.

Facing

Tool feeds to center of workpiece so $L = D/2$. The cutting speed is decreasing as the tool approaches the center of the workpiece.

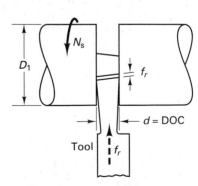

Grooving, parting, or cutoff

Tool feed perpendicular to the axis of rotation. The width of the tool produces the depth of cut (DOC).

FIGURE 20-5 Relationship of speed, feed, and depth of cut in turning, boring, facing, and cutoff operations typically done on a lathe.

volume of metal removed divided by the time needed to remove it

$$\text{MRR} = \frac{\text{volume of cut}}{T_m}$$

where T_m is the cutting time in minutes. For turning, the cutting time depends upon the length of cut L divided by the rate of traverse of the cutting tool past the rotating workpiece $f_r N_s$, as shown in Figure 20-5. Therefore,

$$T_m = \frac{L + \text{allowance}}{f_r N_s} \tag{20-5}$$

An allowance is usually added to the L term to allow for the tool to enter and exit the cut.

Turning is an example of a single-point tool process, as is shaping. Milling and drilling are examples of multiple-point tool processes. Figures 20-5 through 20-9 show the basic process schematically. Speed (V) is shown in these figures with a dark heavy arrow. Feed (f) is the amount of material removed per pass of the tool over the workpiece and is shown as a dashed arrow.

For many of the basic processes, the equations for T_m and MRR are given. These equations are commonly referred to as *shop equations* and are as fundamental as the processes themselves, so the student should be as familiar with them as with the basic processes. If one keeps track of the units and visualizes the process, the equations are, for the most part, straightforward. See Table 20-1 for summary.

In addition to turning, other operations can be performed on the lathe. For example, as shown in Figure 20-5, a flat surface on the rotating part can be produced by facing or a cutoff operation. Boring can produce an enlarged hole, and grooving puts a slot in the workpiece.

The process of milling requires two figures because it takes different forms

| TABLE 20-1 | Shop Formulas for Turning, Milling, Drilling, and Broaching (English Units) |

Parameter	Turning	Milling	Drilling	Broaching
Cutting speed, fpm	$V = 0.262 \times D_l \times$ rpm	$V = 0.262 \times D_m \times$ rpm	$V = 0.262 \times D_d \times$ rpm	V
Revolutions per minute, N_s	rpm $= 3.82 \times V_c/D_l$	rpm $= 3.82 \times V_c/D_m$	rpm $= 3.82 \times V_c/D_d$	—
Feed rate, in./min Feed per rev tooth pass, in./rev	$f_m = f_r \times$ rpm f_r	$f_m = f_t \times$ rpm f_t	$f_m = f_r \times$ rpm f_r	— —
Cutting time, min, T_m	$T_m = L/f_m$	$T_m = L/f_m$	$T_m = L/f_m$	$T_m = L/12V$
Rate of metal removal, in.³/min	MRR $= 12 \times d \times f_r \times V_c$	MRR $= w \times d \times f_m$	MRR $= \pi D^2 d/4 \times f_m$	MRR $= 12 \times w \times d \times V$
Horsepower required at spindle	hp $=$ MRR $\times$ HP$_s$	hp $=$ MRR $\times$ HP$_s$	hp $=$ MRR $\times$ HP$_s$	—
Horsepower required at motor	hp$_m$ $=$ MRR $\times$ HP$_s$/E	hp$_m$ $=$ MRR $\times$ HP$_s$/E	hp$_m$ $=$ MRR $\times$ HP$_s$/E	hp$_m$ $=$ MRR $\times$ HP$_s$/E
Torque at spindle	$t_s = 63,030$ hp/rpm	$t_s = 63,030$ hp/rpm	$t_s = 63,030$ hp/rpm	—

Symbols		
D_l = Diameter of workpiece in turning, inches D_m = Diameter of milling cutter, inches D_d = Diameter of drill, inches d = Depth of cut, inches E = Efficiency of spindle drive f_m = Feed rate, inches per minute f_r = Feed, inches per revolution f_t = Feed, inches per tooth hp$_m$ = Horsepower at motor MRR = Metal removal rate, in.³/min	hp = horsepower at spindle L = Length of cut, inches n = Number of teeth in cutter HP$_s$ = Unit power, horsepower per cubic inch per minute, specific horsepower N_s = Revolution per minute of work or cutter t_s = Torque at spindle, inch-pound T_m = Cutting time, minutes V = Cutting speed, feet per minute w = Width of cut, inches

Values for specific horsepower (unit power) are given in Table 20-4.

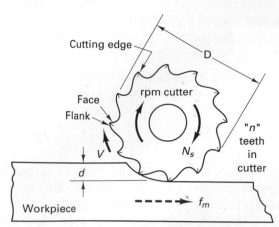

Slab milling – multiple tooth

Face milling
Multiple-tooth cutting

Slab milling is usually performed on a horizontal milling machine. Equations for T_m and MRR derived in Chapter 25.

The tool rotates at rpm N_s. The workpiece translates past the cutter at feed rate f_m, the table feed. The length of cut, L, is the length of workpiece plus allowance, L_A,

$$L_A = \sqrt{\frac{D^2}{4} - \left(\frac{D}{2} - d\right)^2} = \sqrt{d(D - d)} \text{ inches}$$

$$T_m = (L + L_A)/f_m$$

The MRR = Wdf_m where W = width of the cut and d = depth of cut.

Given a selected cutting speed V and a feed per tooth f_t, the rpm of the cutter is $N_s = 12V/\pi D$ for a cutting of diameter D. The table feed rate is $f_m = f_t n N_s$ for a cutter with n teeth.

The cutting time, $T_m = (L + L_A + L_o)/f_m$
where $L_o = L_A = \sqrt{W(D - W)}$ for $W < D/2$
or $L_o = L_A = D/2$ for $W \geqslant D/2$.
The MRR = Wdf_m where d = depth of cut.

FIGURE 20-6 Basics of milling processes (slab, face, and end milling) including equations for cutting time and metal removal rate (MRR).

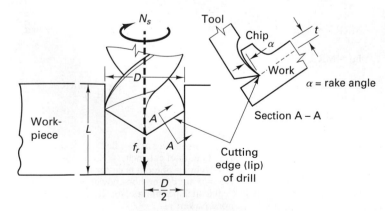

Drilling multiple-edge tool

Select cutting speed V, fpm and feed, f_r, in./rev. Select drill.
D = diameter of the drill which rotates 2 cutting edges at rpm N_s. V = velocity of outer edge of the lip of the drill.
$N_s = 12V/\pi D$. T_m = cutting time = $(L + A)/f_r N_s$ where f_r is the feed rate in in. per rev. The allowance $A = D/2$.
The MRR = $(\pi D^2/4)f_r N_s$ in.^{3}/min which is approximately $3DVf_r$.

FIGURE 20-7 Basics of the drilling (hole-making) processes, including equations for cutting time and metal removal rate (MRR).

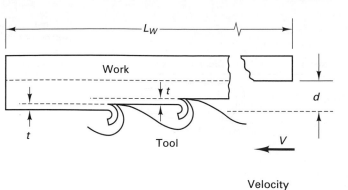

FIGURE 20-8 Process basics of broaching. Equations for cutting time and metal removal rate (MRR) are developed in Chapter 26.

The T_m for broaching is $T_m = L/12V$. The MRR (per tooth) is $12tWV$ in.^{3}/min where $V =$ cutting velocity in fpm, W is the width of cut, $t =$ rise per tooth.

(a)

The tool cuts at velocity V with a return velocity of V_R dictated by the rpm of the crank, N_s. The cutting speed $V = (l + A)N_s/12R_s$ where $R_s =$ stroke ratio $= 200°/360°$ and the length of stroke is $l = L + \text{ALLOW}$. The tool feed is f_c inches per stroke.
$T_m = W/N_s f_c$
$\text{MRR} = LdN_s f_c$ in.^{3}/min

Shaper (quick return) mechanism for driving tool past work.

(b)

FIGURE 20-9 (a) Basics of the shaping process, including equations for cutting time (T_m) and metal removal rate (MRR). (b) The relationship of the crank rpm N_s to the cutting velocity V.

depending on the selection of the machine tool and the cutting tool. Milling, a multiple-tooth process, has two feeds: the amount of metal an individual tooth removes, called the feed per tooth f_t, and the rate at which the table translates past the rotating tool, called the table feed rate f_m, in inches per minute. It is calculated from

$$f_m = f_t n N_s \tag{20-6}$$

where n is the number of teeth in a cutter and N_s is the rpm value of the cutter. Just as was shown for turning, standard tables of speeds and feeds for milling provide values for the recommended cutting speeds and feeds per tooth, f_t.

Operation	Block diagram	Most commonly used machines	Machines less frequently used	Machines seldom used
Turning		Lathe NC lathe machining center	Boring mill	Turret lathe
Grinding		Cylindrical grinder		Lathe (with special attachment)
Sawing (of plates and sheets)		Contour or band saw	Laser Flame cutting Plasma arc	
Drilling		Drill press Machining center (nc) Vert. milling machine	Lathe Horizontal boring machine	Horizontal milling machine Boring mill
Boring		Lathe Boring mill Horizontal boring machine Machining center		Milling machine Drill press
Reaming		Lathe Drill press Boring mill Horizontal boring machine Machining center	Milling machine	
Grinding		Cylindrical grinder		Lathe (with special attachment)
Sawing		Contour or band saw		
Broaching		Broaching machine	Arbor press (keyway broaching)	

FIGURE 20-10   Operations and machines used for machining cylindrical surfaces.

Figure 20-10 provides an overview of the basic machining processes in terms of typical machine tools that can generate cylindrical surfaces. Figure 20-11 provides an overview of the basic processes that can generate flat surfaces. Table 20-2 provides a summary on typical sizes (min–max), the production rates (part/hour), tolerances (precision or repeatability), and surface finish (roughness). Milling has pretty much replaced shaping and planing, although gear shaping is still a viable process. Milling combined with

Operation	Block diagram	Most commonly used machines	Machines less frequently used	Machines seldom used
Facing		Lathe	Boring mill	
Broaching		Broaching machine		Turret broach
Grinding		Surface grinder		Lathe (with special attachment)
Sawing		Cutoff saw	Contour saw	
Shaping		Horizontal shaper	Vertical shaper	
Planing		Planer		
Milling	slab milling	Milling machine	Lathe with special milling tools	
	face milling	Milling machine Machining center	Lathe with special milling tools	Drill press (light cuts)

FIGURE 20-11   Operations and machines used to generate flat surfaces.

other rotational multiple-edge tool processes (drilling or reaming) is often performed in machining centers rather than on milling machines. The turret lathe has been replaced by CNC turning centers[1] with multiple turrets in many factories.

■ 20.3 ENERGY AND POWER IN MACHINING

Most of the cutting operations process described to this point are examples of oblique, or three-force, cutting. The cutting force system in a conventional, oblique-chip formation process is shown schematically in Figure 20-12. Oblique cutting has three components:

1. F_c: Primary cutting force acting in the direction of the cutting velocity vector. This force is generally the largest force and accounts for 99% of the power required by the process.

2. F_f: Feed force acting in the direction of the tool feed. This force is usually about 50% of F_c but accounts for only a small percentage of the power required because feed rates are usually small compared to cutting speeds.

3. F_r: radial or thrust force acting perpendicular to the machined surface. This force is typically about 50% of F_r and contributes very little to power requirements because velocity in the radial direction is negligible. Figure 20-12 shows the general relationship between these forces and changes in speed, feed, and depth of cut. Note that these figures cannot be used to determine forces for a specific process.

---

[1]Machining centers and turning centers are NC or CNC machines, discussed in Chapter 27.

TABLE 20-2 Basic Machining Process

Applicable Process	Raw Material Form	Size — Maximum	Size — Minimum	Typical Production Rate	Material Choice	Typical Tolerance	Typical Surface Roughness
Turning (engine lathes)	Cylinders, preforms, castings, forgings	78 in. dia. × 73 in. long	$\frac{1}{64}$ in. typical	1–10 parts/hour	All ferrous and nonferrous material considered machinable	±0.002 in. on dia. common; ±0.001 in. obtainable	125–250
Turning (CNC)	Bar, rod, tube, preforms	36 in. dia. × 93 in. long	$\frac{1}{64}$ in. dia.	1–2 parts/minute to 1–4 parts/hour	Any material with good machinability rating	±0.001 in. on dia. where needed; ±0.0005 in. possible	63 or better
Turning (automatic screw machine)	Bar, rod	Generally 2 in. dia. × 6 in. long	$\frac{1}{16}$ in. dia. and less, weight less than 1 ounce	10–30 parts/minute	Any material with good machinability rating ±0.001 to ±0.003 in.	±0.0005 in. possible ±0.001 to ±0.003 in. common	63 average
Turning (Swiss automatic machining)	Rod	Collets adapt to $\frac{1}{2}$ in. dia.	Collets adapt to less than $\frac{1}{2}$ in.	12–30 parts/minute	Any material with good machinability rating	±0.0002 in. to ±0.001 in. common	63 and better
Boring (vertical)	Casting, preforms	98 in. × 72 in.	2 in. × 12 in.	2–20 hours/piece	All ferrous and nonferrous	±0.0005 in.	90–250
Milling	Bar, plate, rod, tube	4–6 ft long	Limited usually by ability to hold part	1–100 parts/hour	Any material with good machinability rating	±0.0005 in. possible; ±0.001 in. common	63–250
Hobbing (milling gears)	Blanks, preforms, rods	10-ft-dia. gears 14-in. face width	0.100 in. dia.	1 part/minute	Any material with good machinability rating	±0.001 in. or better	63
Drilling	Plate, bar, preforms	$3\frac{1}{2}$-in.-dia. drills (1-in.-dia. normal)	0.002-in. drill dia.	2–20 second/hole after setup	Any unhardened material; carbides needed for some case-hardened parts	±0.002–±0.010 in. common; ±0.001 in. possible	63–250
Sawing	Bar, plate, sheet	2-in. armor plate $\frac{1}{2}$ in. is preferred)	0.010 in. thick	3–30 parts/hour	Any nonhardened material;	±0.015 in. possible	250–1000
Broaching	Tube, rod, bar, plate	74 in. long	1 in.	300–400 parts/minute	Any material with good machinability rating	±0.0005–±0.001 in.	32–125
Grinding	Plate, rod, bars	36 in. wide × 7 in. dia.	0.020 in. dia.	1–1000 pieces/hour	Nearly all metallic materials plus many nonmetallic	0.0001 in. and less	16
Shaping	Bar, plate, casting	3 ft × 6 ft	Limited usually by ability to hold part	1–4 parts/hour	Low- to medium-carbon steels and nonferrous metals best; no hardened parts	±0.001–±0.002 in. (larger parts) ±0.0001–±0.0005 in. (small–medium parts)	63–250
Planing	Bar, plate, casting	42 ft wide × 18 ft high × 76 ft long	Parts too large for shaper work	1 part/hour	Low- to medium-carbon steels or nonferrous materials best	±0.001–±0.005 in.	63–125
Gear shaping	Blanks	120-in.-dia. gears 6-in. face width	1 in. dia.	1–60 parts/hour	Any material with good machinability rating	±0.001 in. or better at 200 D.P. to 0.0065 in. at 30 D.P.	63

FIGURE 20-12   Oblique machining has three measurable components of forces acting on the tool. The forces vary with speed, depth of cut, and feed.

3 Force

F_C = Cutting force (vertical)

F_R = Radial force (thrust)

F_F = Feed force

The power required for cutting is

$$P = F_c V \, (\text{ft-lb/min}) \qquad (20\text{-}7)$$

The horsepower at the spindle of the machine is therefore

$$\text{hp} = \frac{F_c V}{33,000} \qquad (20\text{-}8)$$

In metal cutting a very useful parameter is called the unit, or specific, horsepower HP_s, which is defined as

$$\text{HP}_s = \frac{\text{hp}}{\text{MRR}} (\text{hp/in.}^3/\text{min}) \qquad (20\text{-}9)$$

In turning, for example, where MRR $\cong 12Vf_r d$, then

$$\text{HP}_s = \frac{F_c}{396,000 f_r d} \qquad (20\text{-}10)$$

Thus this term represents the approximate power needed at the spindle to remove a cubic inch of metal per minute.

Values for specific horsepower HP_s, which is also called unit power, are given in Table 20-3. These values are obtained through orthogonal metalcutting experiments described later in this chapter.

TABLE 20-3 Values for Unit Power and Specific Energy (cutting stiffness)

Material		Unit Power (hp-min. in.^{3}) HP_s	Specific Energy (in.-lb/in.^{3}) K_s or U	Hardness Brinell HB
Nonalloy carbon steel	C 0.15%	.58	268,000	125
	C 0.35%	.58	302,400	150
	C.0.60%	.75	324,800	200
Alloy steel	Annealed	.50	302,400	180
	Hardened and tempered	0.83	358,400	275
	Hardened and tempered	0.87	392,000	300
	Hardened and tempered	1.0	425,000	350
High-alloy steel	Annealed	0.83	369,000	200
	Hardened	1.2	560,000	325
Stainless steel, annealed	Martensitic/ferritic	0.75	324,800	200
Steel castings	Nonalloy	0.62	257,000	180
	Low-alloy	0.67	302,000	200
	High-alloy	0.80	336,000	225
Stainless steel, annealed	Austenitic	0.73	369,600	180
Heat-resistant alloys	Annealed	0.78		200
	Aged—Iron based	—		280
	Annealed—Nickel or cobalt	1.10		250
	Aged	1.20		350
Hard steel	Hardened steel	1.4	638,400	55 HRC
	Manganese steel 12%	1.0	515,200	250
Malleable iron	Ferritic	0.42	156,800	130
	Pearlitic	—	257,600	230
Cast iron, low tensile		0.62	156,800	180
Cast iron, high tensile		0.80	212,800	260
Nodular SG iron	Ferritic	0.55	156,800	160
	Pearlitic	0.76	257,600	250
Chilled cast iron		—	492,800	400
Aluminum alloys	Non-heat-treatable	.25	67,200	60
	Heat-treatable	.33	100,800	100
Aluminum alloys (cast)	Non-heat-treatable	.25	112,000	75
	Heat-treatable	.33	123,200	90
Bronze-brass alloys	Lead alloys, Pb>1%	.25	100,800	110
	Brass, cartridge brass	1.8–2.0	112,000	90
	Bronze and lead-free copper	0.33–0.83		
	Includes Electrolytic copper	0.90	246,400	100
Zinc alloy	Diecast	0.25	—	—
Titanium		.034	250-275	

Values assume normal feed ranges and sharp tools. Multiply values by 1.25 for a dull tool.

Calculation of unit power (HP_s)

$\text{HP} = F_c V/33000$

$\text{HP}_s = \text{HP}/\text{MRR}$ Where

$\text{MRR} = 12Vtw$ for tube turning

$\text{HP}_s = F_c V/12Vtw \times 33000 = F_c/tw \times 396000$

Calculation of specific energy (U)

$U = F_c V/Vtw = F_c/tw$ for tube turning

Specific power can be used in a number of ways. First, it can be used to estimate the motor horsepower required to perform a machining operation for a given material. HP_s values from the table are multiplied by the approximate MRR for the process. The motor horsepower HP_m, is then

$$HP_m = \frac{HP_s \times MRR \times CF}{E} \qquad (20\text{-}11)$$

where E is the efficiency of the machine. The E factor accounts for the power needed to overcome friction and inertia in the machine and drive moving parts. Usually, 80% is used. Usually the maximum MRR is used in this calculation. Correction factors (CFs) may also be used to account for variations in cutting speed, feed, and rake angle. There is usually a tool wear correction factor of 1.25, used to account for the fact that dull tools use more power than sharp tools.

The primary cutting force F_c can be roughly estimated according to

$$F_c \cong \frac{HP_s \times MRR \times 33,000}{V} \qquad (20\text{-}12)$$

This type of estimate of the major force F_c is useful in analysis of deflection and vibration problems in machining and in the proper design of workholding devices, because these devices must be able to resist movement and deflection of the part during the process.

In general, increasing the speed, the feed, or the depth of cut will increase the power requirement. Doubling the speed doubles the horsepower directly. Doubling the feed or the depth of cut doubles the cutting force F_c. In general, increasing the speed does not increase the cutting force F_c, a surprising experimental result. However, speed has a strong effect on tool life because most of the input energy is converted into heat, which raises the temperature of the chip, the work, and the tool, to the latter's detriment. Tool life (or tool death) is discussed in Chapter 22.

Equation 20-12 can be used to estimate the maximum depth of cut, d, for a process as limited by the available power.

$$d_{max} = \frac{HP_m \times E}{12 HP_s\, V F_r\, (CF)} \qquad (20\text{-}13)$$

Another handbook value useful in chatter or vibration calculations is cutting stiffness K_s. In this text, the term *specific energy U* will be used interchangeably with cutting stiffness K_s.

It is interesting to compute the total specific energy in the process and determine how it is distributed between the primary shear and the secondary shear that occurs at the interface between the chip and the tool. It is safe to assume that the majority of the input energy is consumed by these two regions. Therefore,

$$U = U_s + U_f \qquad (20\text{-}14)$$

where specific energy (also called cutting stiffness) is

$$U = \frac{F_c V}{V f_r\, d} = \frac{F_c}{f_r d} = K_s\ (\text{turning}) \qquad (20\text{-}15)$$

The specific shear energy is

$$U_s = \frac{F_c V_s}{V f_r\, d} \qquad (20\text{-}16)$$

where V_s is the shear velocity and F_s is the shear force.
Specific friction energy is

$$U_f = \frac{F V_c}{V f_r\, d} = \frac{F r_c}{f_r\, d} \qquad (20\text{-}17)$$

where V_c is the chip velocity and r_c is the chip thickness ratio. See equation 20-18 for the calculation of r_c.

(a) OPM V (Front view) See Figure 21-14

(b) OTT (Top view) See Figure 21-15

(c) ODM (Top view)

FIGURE 20-13 Three ways to perform orthogonal machining. (a) Orthogonal plate machining on a horizontal milling machine, good for low-speed cutting. (b) Orthogonal tube turning on a lathe; high-speed cutting (see Figure 20-16). (c) Orthogonal disk machining on a lathe; very high-speed machining with tool feeding (ipr) in the facing direction.

Usually, 30 to 40% of the total energy goes into friction and 60 to 70% into the shear process.

Typical values for U are given in Table 20-3. This is experimental data developed by the orthogonal machining experiment described in the next section.

■ 20.4 ORTHOGONAL MACHINING (TWO FORCES)

Orthogonal machining (OM) is carried out mostly in research laboratories, in order to better understand this complex process. In OM, the tool geometry is simplified from the three-dimensional (oblique) geometry, as shown in Figure 20-1.

Using this simplified tool geometry, metals can be cut to test machining mechanics and theory. There are basically three orthogonal machining setups, as shown in Figure 20-13.

1. Orthogonal Plate Machining a plate in a milling machine—low-speed cutting

2. Orthogonal Tube Turning end-cutting a tube wall in a turning setup—medium-speed ranges

3. Orthogonal Disk Machining end-cutting a plate feeding in a facing direction—high-speed cutting

In *oblique* machining, as in shaping, drilling, and single-point turning, the cutting edge and the cutting motion are not perpendicular to each other. In the orthogonal case, the cutting velocity vector and the cutting edge are perpendicular. The OPM low-speed

(a) End view of OPM setup on horizontal mill. Table feed is used for cutting speed.

(b) Horizontal milling setup for OPM using QSD. Front view.

FIGURE 20-14 Schematics of the orthogonal plate machining setups. (a) End view of table, quick-stop device (QSD), and plate being machined for OPM. (b) Front view of horizontal milling machine. (c) Orthogonal plate machining with fixed tool, moving plate. The feed mechanism of the mill is used to produce low cutting speeds. The feed of the tool is t and the DOC is w, the width of the plate.

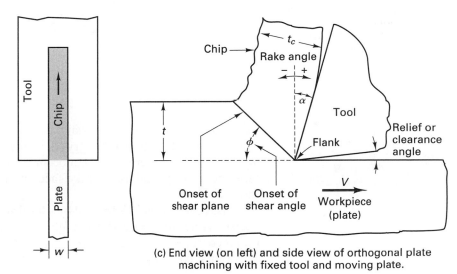

(c) End view (on left) and side view of orthogonal plate machining with fixed tool and moving plate.

plate machining is shown in more detail in Figure 20-14, using a modified horizontal milling machine where the table traverse provides the cutting speed and the tool is mounted in a tool holder in the overarm. As shown in Figure 20-15, OTT can be done on solid cylinders that have had a groove machined on the end to form a tube wall w, or a tubular workpiece can be used. The tubular workpieces can be mounted in a lathe and normal cutting speeds developed for the machining experiment. This setup has the advantage of being very easy to modify so that cutting-temperature experiments can be performed, using the tool/chip thermocouple method. The orthogonal case is more easily modeled for temperature experiments. Low-speed orthogonal plate machining uses a flat plate setup in a milling machine. The workpiece moves past the tool at velocity V. The feed of the work up into the tool is now called t, the uncut chip thickness. The DOC is the width of the plate w. The cutting edge of the tool is perpendicular to the direction of motion V. The angle that the tool makes with respect to a vertical from the workpiece is called the *back rake angle* α. A positive angle is shown in the schematic. The chip is formed by *shearing*. The *onset of shear* occurs at a low boundary deformed by angle ϕ with respect to the horizontal. This model is sufficient to allow us to consider

FIGURE 20-15 Orthogonal tube turning (OTT) produces a two-force cutting operation at speeds equivalent to those used in most oblique machining operations. The slight difference in cutting speed between the inside and outside edge of the chip can be neglected.

the behavior of the work material during chip formation, the influence of the most critical elements of the tool geometry (the edge radius of the cutting tool and the back rake angle α), and the interactions that occur between the tool and the freshly generated surfaces of the chip against the rake face and the new surface as rubbed by the flank of the tool.

Basically, the chip is formed by a localized shear process that takes place over a very narrow zone. This large-strain, high-strain-rate, plastic deformation evolves out of a radial compression zone that travels ahead of the tool as it passes over the workpiece. This radial compression zone has, like all plastic deformations, an elastic compression region that changes into a plastic compression region when the yield strength of the material is exceeded. The plastic compression generates dislocation tangles and networks in annealed metals. The applied stress level increases as the material approaches the tool, where the material has no recourse but to shear. The onset of the shear process takes place along the lower boundary of the shear zone defined by the shear angle ϕ. The shear lamella (microscopic shear planes) lie at the angle ψ to the shear plane.

This can be seen in the videograph in Figure 20-16 and the schematic made from the videograph (see Figure 20-17). The videograph was made by videotaping the orthogonal machining of an aluminum plate at over $100\times$ with a high-speed videotaping machine capable of 1000 frames per second. By machining at low speeds ($V = 8.125$ ipm), the behavior of the process was captured at high frame rates and then observed at playback at very slow frame rates. The uncut chip thickness was $t = 0.020$. The termination of the shear process as defined by the upper boundary cannot be observed in the still videograph but can easily be seen in the videos. Further videographic experiments revealed the increase in the shear angle with workpiece hardness. This directly agrees with the material behavior observed in tensile/compression testing—that yield (and ultimate) strength increases with hardness. In steels the correlation is so good, hardness tests are used to estimate ultimate tensile strength. So in tensile testing, we observe that the onset of plastic deformation (yielding) is delayed by increased hardness

FIGURE 20-16 Videograph made from the orthogonal plate machining process.

FIGURE 20-17 Schematic representation of the material flow, that is, the chip-forming shear process. ϕ defines the onset of shear or lower boundary. ψ defines the direction of slip due to dislocation movement.

(increased dislocation density). In metalcutting, we observe that the onset of shear (to form the chip) is delayed by increased hardness (so ϕ increases directly with hardness). As the material being machined gets harder, dislocation motion becomes more difficult and plastic deformation (with continuous chips) gives way to fracture (discontinuous chips) just as it does in tensile testing. See Figure 20-18 for examples of chips. If the work material has hard second-phase particles dispersed in it, they can act as barriers to the shear front dislocations, which cannot penetrate the particle. The dislocations create voids around the particles. If there are enough particles of the right size and shape, the chip will fracture through the shear zone, forming segmented chips. *Free-machining steels*, which have small percentages of hard second-phase particles added to them, use this metallurgical phenomenon to break up the chips for easier chip handling.

FIGURE 20-18 Three characteristic types of chips. (Left to right) Discontinuous, continuous, and continuous with built-up edge. Chip samples produced by quick-stop technique. *(Courtesy of Eugene Merchant (deceased) at Cincinnati Milacron, Inc., Ohio.)*

■ 20.5 MERCHANT'S MODEL

For the purpose of modeling chip formation, assume that the shear process takes place on a single narrow plane, shown in Figure 20-19 as A–B rather than on the set of shear fronts that actually comprise a narrow shear zone. Further, assume that the tool's cutting edge is perfectly sharp and no contact is being made between the flank of the tool and the new surface. The workpiece passes the tool with velocity V, the cutting speed. The uncut chip thickness is t. Ignoring the compression deformation, chips having thickness t_c are formed by the shear process. The chip has velocity V_c. The shear process then has velocity V_s and occurs at the onset of shear angle ϕ. The tool geometry is given by the back rake angle α and the clearance angle γ. The velocity triangle for V, V_c, and V_s is also shown (see Figure 20-19). The chip makes contact with the rake face of the tool over length l_c. The plate thickness is w.

From orthogonal machining experiments, the chip thickness is measured and used to compute the shear angle from the *chip thickness ratio*, r_c, defined as t/t_c;

$$r_c = \frac{t}{t_c} = \frac{AB \sin \phi}{AB \cos(\phi - \alpha)} \tag{20-18}$$

where AB is the length of the shear plane from the tool tip to the free surface.

Equation 20-18 may be solved for the *shear angle* ϕ as a function of the measurable chip thickness ratio by expanding the cosine term and simplifying:

$$\tan \phi = \frac{r_c \cos \alpha}{1 - r_c \sin \alpha} \tag{20-19}$$

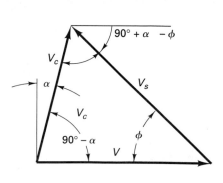

FIGURE 20-19 Velocity diagram associated with Merchant's orthogonal machining model.

There are numerous other ways to measure chip ratios and obtain shear angles both during (dynamically) and after (statically) the cutting process. For example, the ratio of the length of the chip L_c to the length of the cut L can be used to determine r_c. Many researchers use the chip compression ratio, which is the reciprocal of r_c, as a parameter. See Problem 2 at the end of the chapter for another method. The shear angle can be measured statically by instantaneously interrupting the cut through the use of *quick-stop devices*. These devices disengage the cutting tool from the workpiece while cutting is in progress, leaving the chip attached to the workpiece. Optical and scanning electron microscopy is then used to observe the direction of shear. Figure 20-14 shows a QSD on an OPM setup, and Figure 20-18 was made using a quick-stop device. High-speed motion pictures and high-speed videographic systems have also been used to observe the process at frame rates as high as 30,000 frames per second. Figure 20-16 is a high-speed videograph. Machining stages have been built that allow the process to be performed inside a scanning electron microscope and recorded on videotapes for high-resolution, high-magnification examination of the deformation process. Using sophisticated electronics and slow-motion playback, this technique can be used to measure the shear velocity. The vector sum of V_s and V_c equals V.

For consistency of volume, we observe that

$$r_c = \frac{t}{t_c} = \frac{\sin \phi}{\cos(\phi - \alpha)} = \frac{V_c}{V} \tag{20-20}$$

indicating that the chip ratio (and therefore the onset of shear angle) can be determined dynamically if a reliable means to measure V_c can be found.

The ratio of V_s to V is

$$\frac{V_s}{V} = \frac{\cos \alpha}{\cos(\phi - \alpha)} \tag{20-21}$$

These velocities are important in power calculations, heat and temperature calculations, and vibration analysis associated with chatter in chip formation.

■ 20.6 Mechanics of Machining (Statics)

Orthogonal machining has been defined as a two-force system. Consider Figure 20-20, which shows a free-body diagram of a chip that has been separated at a shear plane. It is assumed that the resultant force R acting on the back of the chip is equal and opposite to the resultant force R' acting on the shear plane. The resultant R is composed of the *friction force F* and the normal force N acting on the tool–chip interface contact area. The resultant force R' is composed of a *shear force F_s* and normal force F_n acting on the shear plane area A_s. Since neither of these two sets of forces can usually be measured, a third set is needed, which can be measured using a dynamometer (force transducer) mounted either in the workholder or the tool holder. Note that this set has resultant R, which is equal in magnitude to all the other resultant forces in the diagram. The resultant force R is composed of a *cutting force F_c* and a tangential (normal) force F_t. Now it is necessary to express the desired forces (F_s, F_n, F, N) in terms of the measured dynamometer components, F_c and F_t, and appropriate angles. To do this, a circular force diagram is developed in which all six forces are collected in the same force circle (Figure 20-21). The only symbol in this figure as yet undefined is β, which is the angle between the normal force N and the resultant R. It is called friction angle β and is used to describe the friction coefficient μ on the tool–chip interface area, which is defined as F/N so that

$$\beta = \tan^{-1}\mu = \tan^{-1}\frac{F}{N} \tag{20-22}$$

The friction force F and its normal N can be shown to be

$$F = F_c \sin \alpha + F_t \cos \alpha \tag{20-23}$$

$$F = F_c \cos \alpha - F_t \sin \alpha \tag{20-24}$$

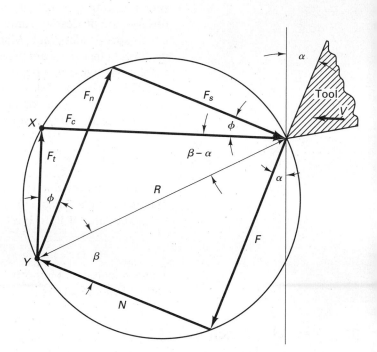

FIGURE 20-20 Free-body diagram of orthogonal chip formation process, showing equilibrium condition between resultant forces R and R'.

FIGURE 20-21 Merchant's circular force diagram used to derive equations for F_s, F_r, F_t, and N as functions of F_c, F_r, ϕ, α, and β.

and the resultant R is

$$R = \sqrt{F_c^2 + F_t^2} \tag{20-25}$$

Notice that in the special situation where the back rake angle is zero, $F = F_t$ and $N = F_c$, so that in this orientation, the friction force and its normal can be directly measured by the dynamometer.

The forces parallel and perpendicular to the shear plane can be shown from the circular force diagram (Figure 20-21) to be

$$F_s = F_c \cos\phi - F_t \sin\phi \tag{20-26}$$

$$F_n = F_c \sin\phi + F_t \cos\phi \tag{20-27}$$

F_s is of particular interest, because it is used to compute the shear stress on the shear plane. This shear stress is defined as

$$\tau_s = \frac{F_s}{A_s} \tag{20-28}$$

where A_s is the area of the shear plane, as

$$A_s = \frac{tw}{\sin\phi} \tag{20-29}$$

recalling that t was the uncut chip thickness and w was the width of the workpiece. The *shear stress (flow stress)* is, therefore,

$$\tau_s = \frac{F_c \sin\phi\cos\phi - F_t\sin^2\phi}{tw} \text{ psi} \tag{20-30}$$

For a given polycrystalline metal, this shear stress has been shown to be not sensitive to variations in cutting parameters, tool material, or the cutting environment. Figure 20-22 gives some typical values for the flow stress for a variety of metals, plotted against hardness.

Specific horsepower is related to and correlates well with shear stress for a given metal, which will be derived later. Unit power is sensitive to material properties (e.g., hardness), rake angle, depth of cut, and feed, whereas τ_s is sensitive to material properties only.

■ 20.7 SHEAR STRAIN γ AND SHEAR FRONT ANGLE φ

Using Merchant's chip formation bubble model, which emulates the videographic images, a new "stack-of-cards" model, as shown in Figure 20-23, can be developed. From this model, strain is expressed as

$$\gamma = \cos\alpha / [\sin(\phi + \varphi)\cos(\phi + \varphi - \alpha)] \tag{20-31}$$

where

$$\varphi = \text{the angle of the onset of the shear plane}$$

$$\psi = \text{the shear front angle}$$

Using the available machining data, ψ is observed to decrease, reach a minimum, and rise again for all rake angles for a given metal of some hardness.

The minimum energy principle has been reported to have use in various fields such as physics, metalforming processes, and machining processes. Applied to metal cutting, the specific shear energy (shear energy/volume) equals shear stress × shear strain

$$U_s = \tau \times \gamma \tag{20-32}$$

FIGURE 20-22 Shear stress τ_s variation with the Brinell hardness number for a group of steels and aerospace alloys. Data of some selected fcc metals are also included. *(Adapted with permission from S. Ramalingham and K. J. Trigger,* Advances in Machine Tool Design and Research, *1971, Pergamon Press.)*

Mechant's bubble
model of chip formation

Black-Huang stack-of-cards model

The shaded shear triangle
on the right is used to
develop the basic
equation for shear
strain, γ.

FIGURE 20-23   The Black–Huang "stack-of-cards" model for calculating shear strain in metal cutting is based on Merchant's bubble model for chip formation, shown on the left.

The minimum energy principle is used here, where ψ will take on values (shear directions) to reduce shear energy to a minimum.

That is

$$dU_s/d\psi = 0 \qquad (20\text{-}33)$$

The shear front angle is obtained by

$$\psi = 45° - \phi + \alpha/2 \qquad (20\text{-}34)$$

Substituting ψ in equation 20-34 into equation 20-31, the shear strain can be expressed as

$$\gamma = 2\cos\alpha/(1 + \sin\alpha) \qquad (20\text{-}35)$$

which shows that the *shear* strain is dependent only on the rake angle α. The agreement between the predicted shear strain from this model and measured shear strain obtained from metalcutting experiments is exceptionally good. Generally speaking, metalcutting strains are quite large compared to other plastic deformation processes, on the order of 1 to 2 in./in.

This large strain occurs, however, over very narrow regions, resulting in extremely high shear strain rates, ε, typically in the range of 10^4 to 10^8 in./in. per second. It is this combination of large strains and high strain rates operating within a process constrained only by the rake face of the tool that results in great difficulties in theoretical analysis of this process.

In order to verify equation 20-34, metalcutting experiments in copper with a hardness gradient ranging from dead soft to full hard were performed. The equation was experimentally verified to 99% confidence!

- The material begins to shear at the lower boundary of the shear zone, defined by the angle ϕ. As the hardness of a material increases, ϕ increases while ψ decreases, so $\psi + \phi = 45° + \alpha / 2$ is maintained for all levels of hardness.

- The material in the shear zone shears at an inclination angle ψ to the plane of the onset ϕ of shear plane for aluminum and steel.

- Shear strain and shear front angle can be determined by

$$\gamma = 2 \cos \alpha / (1 + \sin \alpha)$$

$$\psi + \phi = 45° + \alpha/2$$

where ϕ and ψ vary with hardness.

■ 20.8 MECHANICS OF MACHINING (DYNAMICS)

Machining is a dynamic process of large strain and high strain rates. All the process variables are dependent variables. The process is intrinsically a closed-loop interactive process as shown in Figure 20-24.

Starting at the top, inputs to the processes (speed, feed, depth of cut) determine the chip load on the tool. The chip load determines the cutting forces (magnitude and direction) (usually elastic), which alters the chip load on the tool. The altered chip load produces new forces. The cycle repeats, producing chatter and vibration.

Remember that plastic deformation is always preceded by elastic deformation. The elastic deflection behaves like a big spring. The mechanism by which a process

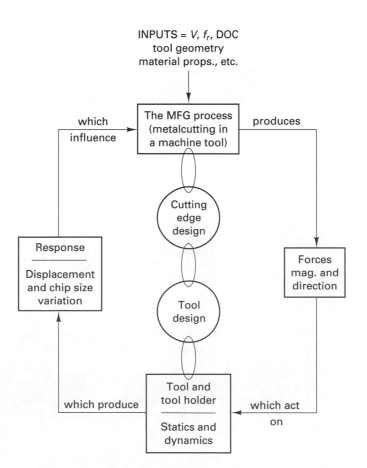

FIGURE 20-24 Machining dynamics is a closed-loop interactive process that creates a force-displacement response.

dissipates energy is called chatter or vibration. In machining, it has long been observed in practice that rotational speed may greatly influence process stability and chatter. Experienced operators commonly listen to machining noise and interactively modify the speed when optimizing a specific application. In addition, experience demonstrates that the performance of a particular tool may vary significantly based on the machine tool employed and other characteristics such as the workpiece, fixture holder, and the like. Today more than ever, the manufacturing industry is more competitive and responsive, characterized by both high-volume and small-batch production, seeking economies of scale. High productivity is achieved by increased machine and tooling capabilities along with the elimination of all non-value-added activities. Few companies can afford lengthy trial-and-error approaches to machining-process optimization or additional processes to treat the effect of chatter.

In metalcutting, chatter is a self-excited vibration that is caused by the closed-loop force-displacement response of the machining process. The process-induced variations in the cutting force may be caused by changes in the cutting velocity, chip cross section (area), tool–chip interface friction, built-up edge, workpiece variation, or, most commonly, process modulation resulting in regeneration of vibration. When more energy is input into the dynamic machining system than can be dissipated by mechanical work, damping, and friction, equilibrium (the state of minimum potential energy) is sought by the machining system through the generation of chatter vibration.

The proper classification of the type of vibration is the first step in identifying and solving the cause of unwanted vibration (see Figure 20-25).

- *Free vibration* is the response to any initial condition or sudden change. The amplitude of the vibration decreases with time and occurs at the natural frequency of the system. Interrupted machining is an example that often appears as lines or shadows following a surface discontinuity.

- *Forced vibration* is the response to a periodic (repeating with time) input. The response and input occur at the same frequency. The amplitude of the vibration remains constant for set input conditions and is linearly related to speed. Unbalance, misalignment, tooth impacts, and resonance of rotation systems are the most common examples.

- **Free Vibration** The response to an initial condition or sudden change. The amplitude of the vibration decreases with time and occurs at the natural frequency of the system often produced by interrupted machining. Often appears as lines or shadows following a surface discontinuity.

- **Forced Vibration** The response to a periodic (repeating with time) input. The response and input occur at the same frequency. The amplitude of the vibration remains constant for a set input condition and is nonlinearly related to speed. Unbalance, misalignment, tooth impacts, and resonance of rotating systems are the most common examples.

- **Self-Excited Vibration** The periodic response of the system to a constant input. The vibration may grow in amplitude (unstable) and occurs near the natural frequency of the system regardless of the input. Chatter due to the regeneration of surface waviness is the most common metal cutting example.

FIGURE 20-25   There are three types of vibration in machining.

- *Self-excited vibration* is the periodic response of the system to a constant input. The vibration may grow in amplitude (become unstable) and occurs near the natural frequency of the system regardless of the input. Chatter due to the regeneration of waviness in the machined surface is the most common metal cutting example.

How do we know chatter exists? Listen and look! Chatter is characterized by the following:

1. There is a sudden onset of vibration (a screech or buzz or whine) that rapidly increases in amplitude until a maximum threshold (saturation) is reached.
2. The frequency of chatter remains very close to a natural frequency (critical frequency) of the machining system and changes little with variation of process parameters. The largest force-displacement response occurs at resonance and therefore the greatest energy dissipation.
3. Chatter often results in unacceptable surface finish, exhibited by a helical or angular pattern (pearled or fish scaled) superimposed over normal feed marks.
4. Visible surface undulations are found in the feed direction and corresponding wavy or serrated chips with variable thickness.

Figure 20-26 shows some typical examples of chatter visible in the surface finish marks.

There are several important factors that influence the stability of a machining process:

- Cutting stiffness of the workpiece material (related to the machinability), K_s
- Cutting-process parameters (speed, feed, DOC, total width of chip)
- Cutter geometry (rake and clearance angles, edge prep, insert size and shape)
- Dynamic characteristics of the machining process (tooling, machine tool, fixture, and workpiece)

K_s, cutting stiffness, is closely aligned with flow stress but simpler to calculate in that ϕ is not used. Like flow stress, cutting stiffness can be viewed as a material property of the workpiece, dependent on hardness.

CHIP FORMATION AND REGENERATIVE CHATTER

In machining, the chip is formed due to the shearing of the workpiece material over the chip area (A = thickness $\times$ width = $t \times w$), which results in a cutting force F_c. The magnitude of the resulting cutting force is predominantly determined by the material cutting stiffness K_s and the chip area such that $F_c = K_s \times t \times w$. The direction of the cutting force F_c is influenced mainly by the geometries of the rake and clearance angles as well as the edge prep.

Machining operations require an overlap of cutting paths that generate the machined surface (see Figure 20-27). In single-point operations, the overlap of cutting paths does not occur until one complete revolution. In milling or drilling, overlap occurs in a fraction of a revolution, depending on the number of cutting edges on the tool.

The cutting force causes a relative displacement X between the tool and workpiece, which affects the uncut chip thickness t and, in turn, the cutting force. This coupled relationship between displacement in the Y-direction (modulation direction) and the resulting cutting force forms a closed-loop response system. The modulation direction is normal to the surface defining the chip thickness.

A phase shift ε between subsequent overlapping surfaces results in a variable chip thickness and modulation of the displacement, causing chatter vibration. The phase shift between overlapping cutting paths is responsible for producing chatter. However, there is a preferred speed that corresponds to a phase-locked condition ($\varepsilon = 0$) that results in a constant chip thickness t. A constant chip thickness results in a steady cutting force and the elimination of the feedback mechanism responsible for regenerative chatter. This what the operators are trying to achieve when they vary cutting speeds (see Figure 20-28).

FIGURE 20-26 Some examples of chatter that are visible on the surfaces of the workpiece.

HOW DO THE IMPORTANT FACTORS INFLUENCE CHATTER?

- *Cutting stiffness K_s.* This is a material property related to shear flow stress, hardness, and work hardening and is often described in a relative sense of the machinability of materials. Materials such as steel and titanium require much greater shear forces than aluminum or cast iron; therefore, the corresponding larger cutting forces lead to greater displacement in the Y-direction and less machining stability.

- *Speed.* The process parameters are the easiest factors to change chatter and its amplitude. The rotational *speed* of the tool affects the phase shift between overlapping sur-

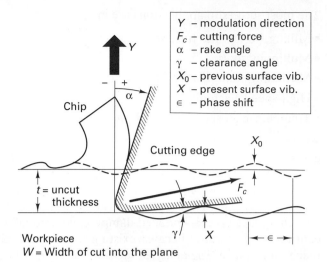

Y	– modulation direction
F_c	– cutting force
α	– rake angle
γ	– clearance angle
X_0	– previous surface vib.
X	– present surface vib.
$\in$	– phase shift

FIGURE 20-27 When the overlapping cuts get out of phase with each other, a variable chip thickness is produced, resulting in a change in F_c on the tool or workpiece.

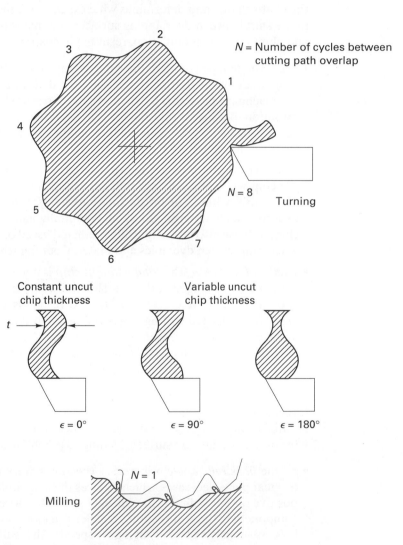

FIGURE 20-28 Regenerative chatter in turning and milling produced by variable uncut chip thickness.

faces and the regeneration of vibration. A handheld speed analyzer[2] that produces dynamically preferred speed recommendations is commercially available. When applied to processes exhibiting a relative rotational motion between the cutting tool and workpiece, it recommends a speed to eliminate chatter.

_____

[2]Best speed by Design Manufacturing Inc., Tampa, Florida.

The most successful applications are in

• Milling, boring, and turning
• Multipoint tools
• Machining aluminum and cast iron
• High-speed machining
• Thin-chip, high-speed die machining

At slow speeds (relative to the vibration frequency) process stability is mainly due to increased frictional losses occurring between the tool clearance angle (defined by γ) and the present surface vibration X. This interference and friction dissipates energy in the form of heat and is called *process damping*. As machining speeds are increased, the wavelength of the surface vibration also increases, which reduces the slope of the surface and eliminates process-induced damping. Additionally, chatter becomes more significant as speeds increase, because existing forces approach the natural frequencies of the machining system. The analyzer measures and identifies the vibrational frequencies of the chatter noise and determines which speeds will most closely result in $\varepsilon = 0$. A zero phase shift between overlapping surfaces eliminates the variation of the chip thickness and eliminates the modulation, resulting in chatter.

• *Feed.* The *feed* per tooth defines the average uncut chip thickness t and influences the magnitude of the cutting force. The feed does not greatly influence the stability of the machining process (i.e., whether chatter occurs) but does control the severity of the vibration. Because no cutting force exists if the vibration in the Y-direction results in the loss of contact between the tool and workpiece, the maximum amplitude of chatter vibration is limited by the feed.

• *DOC.* The *depth of cut* is the primary cause and control of chatter. The DOC defines the chip width and acts as the feedback gain in the closed-loop machining process. The stability limit (or borderline between stable machining and chatter) may be experimentally determined by incrementally increasing the DOC until the onset of chatter. It can also be analytically predicted based on a thorough understanding of the machining system dynamics and material cutting stiffness.

• *Total width of chip.* The *total width of chip* is equal to the DOC times the number of cutting edges engaged in the cut. The total width of cut directly influences the stability of the process. At a fixed DOC that corresponds to the stability limit, increasing the number of engaged cutting edges will result in chatter. The number of engaged teeth in the cut may be increased by adding inserts (using a fine-pitch cutter) or increasing the radial immersion of a milling cutter. Conversely, reducing the number of edges in the cut will have a stabilizing effect on the process.

The *cutting tool geometry* influences the magnitude and direction of the cutting force, especially the amount of the force component in the modulation direction Y. A greater projection of force in the Y-direction results in increased displacement and vibration normal to the surface, leading to potential chatter.

• As the *back rake angle* α increases (becomes more positive), the length of the onset of shear plane decreases, which reduces the magnitude of the cutting force, F_c. A more positive rake also directs the cutting force to be more tangential and reduces the force component in the Y-direction. In general, a more positive cutting geometry increases process stability, especially at higher speeds. An insufficient feed compared to the edge radius results in less efficient machining, greater tool deflection, and poorer machining stability.

• A *reduced clearance angle* γ, which increases the frictional contact between the tool and workpiece, may produce process damping. The stabilizing effect is due to energy dissipation in the form of heat, which potentially decreases tool life and may thermally distort the workpiece or increase the heat-affected zone in the workpiece. The initial wear of a new cutting edge may have a stabilizing effect on chatter.

FIGURE 20-29   Milling and boring operations can be made more stable by correct selection of insert geometry.

- The *size* (nose radius), *shape* (diamond, triangular, square, round), and *lead angle* of the insert all influence the chip area shape and the corresponding Y-direction (Figure 20-29). In milling, the feed direction is transverse to the tool axis (i.e., radial), and the DOC is defined by the axial immersion of the tool. In milling, as the lead angle of the cut increases or the shape of the insert becomes rounded (same effect as a large nose radius compared to a small DOC), the Y-orientation is directed away from the more flexible radial tool direction and toward the stiffer axial direction. The orientation of the modulation direction Y toward a dynamically more rigid direction results in decreased vibrational response and therefore greater process stability (less tendency for chatter). In boring, the feed direction is axial, and the DOC is determined by the radial direction. Therefore, in boring, a reduced lead angle or a less round (and smaller nose radius compared to the DOC) insert maintains a more axial (stiffer tool direction) orientation of Y, leading to greater stability.

Because the stability of the machining process is a direct result of the dynamic *force-displacement characteristics* between the tool and workpiece, all components of the machining system (tool, spindle, workpiece, fixture, machine tool) may, to varying degrees, influence chatter. Maximizing the dynamics (the product of the static stiffness and damping) of the machining system leads to increased process stability. For example, the static stiffness of an overhung (cantilever) tool with circular cross section varies nonlinearly with the diameter D and the unsupported length L. Machining stability is increased by having the tool with the largest possible diameter with the minimum overhand. The frequency of chatter occurs near the most flexible vibrational mode of the machining system.

STABILITY LOBE DIAGRAM

A stability lobe diagram (Figure 20-30) relates the total width of cut that can be machined to the rotational speed of the tool with a specified number of cutting edges. If the total width of cut is maintained below a minimum level (although this may be of limited practical value for some machining systems), then the process stability exhibits speed independence

FIGURE 20-30 Dynamic analysis of the cutting process produces a stability lobe diagram, which defines speeds that produce stable and unstable cutting conditions.

or "unconditional" stability. At slow speeds, increased stability may be achieved within the process damping region. The "conditional" stability lobe regions allow increased total width of cut (DOC × number of edges engaged in the cut) at dynamically preferred speeds at which the phase shift ε between overlapping or consecutive cutting paths approaches zero. The stability lobe number N indicates complete cycles of vibration that exist between overlapping surfaces. As can be seen by the diagram, the higher speeds correspond to lower lobe numbers and provide the greatest potential increase in the total width of cut and material removal rate (due to greater lobe height and width). If the total width of cut exceeds the borderline of stability, even if the process is operating at a preferred speed, chatter occurs. The greater the total width of cut above the stability limit, the more unstable and violent will be the chatter vibration.

When a chatter condition occurs, such as at point a on the stability lobe diagram, the rotational speed is adjusted to the first recommended speed ($N = 1$), which results in stable machining at point b on the diagram. The DOC may be incrementally increased until chatter again occurs as the stability border is crossed at point c. Using the analyzer again, chattering under the new operating conditions will result in a modified speed recommendation corresponding to point d. If desired, the DOC may again be incrementally increased (conservative steps promote safety) to point e. In general, do not attempt to maintain the DOC (and total width of cut) right up to the borderline of stability because workpiece variation affecting K_s, speed errors, or small changes in the dynamic characteristics of the machining system may result in crossing the stability limit into severe chatter. The amplitude of chatter vibration may be more safely limited by temporary reduction of the feed per tooth until a preferred speed and stable depth of cut have been established.

HEAT AND TEMPERATURE IN METALCUTTING

In metalcutting, the power put into the process (F_cV) is largely converted to heat, elevating the temperatures of the chip, the workpiece, and the tool. These three elements of the process, along with the environment (which includes the cutting fluid), act as the heat sinks. Figure 20-31 shows the distribution of the heat to these three sinks as a function of cutting speed. As speed increases, a greater percentage of the heat ends up in the chip to the point where the chips can be cherry red or even burn at high cutting speeds.

There are three main *sources* of heat. Listed in order of their heat-generating capacity, they are shown in Figure 20-32.

1. The shear front itself, where plastic deformation results in the major heat source. Most of this heat stays in the chip.
2. The tool–chip interface contact region, where additional plastic deformation takes place in the chip and considerable heat is generated due to sliding friction.
3. The flank of the tool, where the freshly produced workpiece surface rubs the tool.

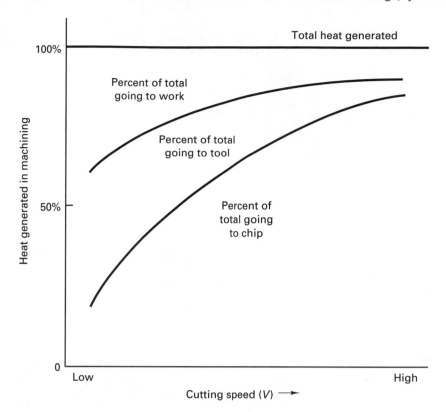

FIGURE 20-31   Distribution of heat generated in machining to the chip, tool, and workpiece. Heat going to the environment is not shown. Figure based on the work of A. O. Schmidt.

FIGURE 20-32   There are three main sources of heat in metal cutting. (1) Primary shear zone. (2) Secondary shear zone tool–chip (T–C) interface. (3) Tool flank. The peak temperature occurs at the center of the interface, in the shaded region.

There have been numerous experimental techniques developed to measure cutting temperatures and some excellent theoretical analyses of this "moving" multiple-heat-source problem. Space does not permit us to explore this problem in depth. Figure 20-33 shows the effect of cutting speed on the tool–chip interface temperature. The rate of wear of the tool at the interface can be shown to be directly related to temperature (see Figure 20-33b). Because cutting forces are concentrated in small areas near the cutting edge, these forces produce large pressures. The tool material must be hard (to resist wear) and tough (to resist cracking and chipping). Tools used in

FIGURE 20-33 The typical relationship of temperature at the tool–chip interface to cutting speed shows a rapid increase. Correspondingly, the tool wears at the interface rapidly with increased temperature, often created by increased speed.

interrupted cutting, such as milling, must be able to resist impact loading as well. Tool materials must sustain their hardness at elevated temperatures. The challenge to manufacturers of cutting tools has always been to find materials that satisfy these severe conditions. Cutting tool materials that do not lose hardness at the high temperatures associated with high speeds are said to have "hot hardness." Obtaining this property usually requires a trade-off in toughness, as hardness and toughness are generally opposing properties. In Chapter 21 cutting-tool materials will be addressed in more depth.

■ 20.9 SUMMARY

In this chapter, the basics of the machining processes have been presented. Chapters 22 through 29 provide additional information on the various operations and machine tools. Modern machine tools can often perform several basic operations. As described in Chapter 26, machining centers are now widely used. These chapters on basic processes must be carefully studied. The relationship between the basic processes and the machine tools that can be used to perform these processes has been presented. In Chapter 26, an anatomy for automation will be presented, which will show that most of the machines described in this chapter will be of the A(2) or A(3) level of automation. Machining centers, which are numerical control machines, are A(4). Machining centers have automatic tool-change capability and are usually capable of milling, drilling, boring, reaming, tapping (hole threading), and other minor machining processes. For particular machines, you will need to become familiar with new terminology, but in general all machining processes will need inputs concerning rpm (given that you selected the cutting speed), feeds, and depths of cut. Note also from these tables that the same process can be performed on two or more different machine tools. There are many ways to produce flat surfaces, internal and external cylindrical surfaces, and special geometries in parts. Generally, the quantity to be made is the driving factor in the selection of processes, as will be explained later.

This chapter also introduced orthogonal machining, a laboratory machining process used to experimentally determine values for specific horsepower, cutting stiffness, and flow stress for machining.

As noted previously, the properties of the work material are important in chip formation. High-strength materials produce larger cutting forces than materials of lower strength, causing greater tool and work deflection; increased friction, heat generation, and operating temperatures; the structure and composition also influence metal cutting. Hard or abrasive constituents, such as carbides in steel, accelerate tool wear.

Work hardness prior to machining is an important factor because it controls the onset of shear. Onset is delayed by increased hardness, so ϕ increases, as does τ_s. Highly ductile materials not only permit extensive plastic deformation of the chip during cutting, which increases heat generation and temperature, but also result in longer, "continuous" chips that remain in contact longer with the tool face, thus causing more frictional heat. Chips of this type are severely deformed and have a characteristic curl. On the other hand, materials that are already heavily work hardened or brittle, such as gray cast iron, lack the ductility necessary for appreciable plastic deformation. Consequently, the compressed material ahead of the tool fails in brittle fracture, sometimes along the shear front, producing small fragments. Such chips are termed *discontinuous* or *segmented*.

A variation of the continuous chip, often encountered in machining ductile materials, is associated with a *built-up edge* (BUE) formation on the cutting tool. The local high temperature and extreme pressure in the cutting zone cause the work material to adhere or pressure weld to the cutting edge of the tool forming the built-up edge, rather like a dead metal zone in the extrusion process. Although this material protects the cutting edge from wear, it modifies the geometry of the tool. BUEs are not stable and will break off periodically, adhering to the chip or passing under the tool and remaining on the machined surface. Built-up edge formation can be eliminated or minimized by reducing the depth of cut, altering the cutting speed, using positive rake tools, applying a coolant, or changing cutting-tool materials.

■ Key Words

back rake angle	depth of cut	oblique machining	shear velocity
boring	drilling	onset of shear	specific horsepower
broaching	dynamics	orthogonal machining	speed
built-up edge	feed	regenerative chatter	stability lobe diagram
chatter	flow stress	sawing	turning
chip ratio	friction force	self-excited vibration	vibration
chip velocity	grinding (abrasive machining)	shaping	workholding device
cutting force	machine tool	shear angles	workpiece
cutting stiffness	metalcutting	shear force	
cutting tool	milling	shear strain	

■ Review Questions

1. Why has the metalcutting process resisted theoretical solution for so many years?
2. What variables must be considered in understanding a machining process?
3. Which of the seven basic chip formation processes are single point, and which are multiple point? See Figure 20-2.
4. How is feed related to speed in the machining operations called turning?
5. Before you select speed and feed for a machining operation, what have you had to decide? (*Hint:* See Figure 20-4.)
6. Milling has two feeds. What are they, and which one is an input parameter to the machine tool?
7. What is the fundamental mechanism of chip formation?
8. What are the implications of Figure 20-17, given that this videograph was made at a very low cutting speed?
9. What is the difference between oblique machining and orthogonal machining?
10. Note that the units for the approximate equation for MRR for turning are not correct. When is the approximate equation not very good (yields a large error in MRR values)?
11. For orthogonal machining, the cutting edge radius is assumed to be small compared to the uncut chip thickness. Why?
12. How do the magnitude of the strain and strain rate values of metal cutting compare to those of tensile testing?
13. Why is titanium such a difficult metal to machine? (Note its high value of HP_s).
14. Explain why you get segmented or discontinuous chips when you machine cast iron.
15. Why is metal cutting shear stress such an important determination?
16. Which of the three cutting forces in oblique cutting consumes most of the power?
17. How is the energy in a machining process typically consumed?
18. Where does the energy consumed in metalcutting ultimately go?
19. State two ways of estimating the primary cutting force F_c.
20. How is cutting speed related to tool wear?
21. What is the relationship between hardness and temperature in metal cutting tool materials?
22. Why does the cutting force F_c increase with increased feed or DOC?

23. Why doesn't the cutting force F_c increase with increased speed V?
24. How do the selection of the machining parameters (speed, feed, DOC) influence chatter?

25. You had a machining operation (boring) running perfectly and you changed work materials. All of a sudden, you are getting lots of chatter. Why?
26. Explain Figure 20-31. Why is the percentage of total heat generated during machining changing as speed increases?

■ Problems

1. For a turning operation, you have selected an HSS tool and turning a hot rolled free machining steel, Bhn = 300. Your depth of cut will be 0.150 in. The diameter of the workpiece is 1.00 inches.
   a. What speed and feed would you select for this job?
   b. Using a speed of 105 sfpm and a feed of 0.015, calculate the spindle rpm for this operation.
   c. Calculate the metal removal rate.
   d. Calculate the cutting time for the operation with a length of cut of 4 in. and 0.10-in. allowance.

2. For a slab milling operation using a 5-in.-diameter, 11-tooth cutter (see Figure 20-6), the feed per tooth is 0.005 in./tooth with a cutting speed of 100 sfpm (HSS steel). Calculate the rpm of the cutter and the feed rate (f_m) of the table, then calculate the metal removal rate, MRR, where the width of the block being machined is 2 in. and the depth of cut is 0.25 in. Calculate the time to machine (T_m) a 6-in.-long block of metal with this setup. Suppose you switched to a coated-carbide tool, so you increase the cutting speed to 400 sfpm. Now recalculate the machining time (T_m) with all the other parameters the same.

3. The power required to machine metal is related to the cutting force (F_c) and the cutting speed. For Problem 1, estimate cutting force F_c for this turning operation. (*Hint:* You have to estimate a value of HPs for this material.)

4. In order to drill a hole in the material described in Problem 1 using an HSS drill, you have to select a cutting speed and a feed rate. Using a speed of 105 sfpm for the HSS drill, calculate the rpm for a $\frac{3}{4}$-in.-diameter drill and the MRR if the feed rate is 0.008 inches per revolution.

5. Explain how the constant 33,000 in equation 20-8 is obtained.

6. Explain how the constant 396,000 in equation 20-10 is obtained.

7. Suppose you have the following data obtained from a metal-cutting experiment (orthogonal machining). Compute the shear angle, the shear stress, the specific energy, the shear strain, and the coefficient of friction at the tool–chip interface.

How do your HP_s and τ_s values compare with the values found in Chapter 20?

8. For the data in Problem 7, determine the specific shear energy and the specific friction energy.

9. Derive equations for F and N using the circular force diagram. (*Hint:* Make a copy of the diagram. Extend a line from point X intersecting force F perpendicularly. Extend a line from point Y intersecting the previous line perpendicularly. Find the angle α made by these constructions.)

10. Derive equations for F_s and F_n using the circular force diagram. (*Hint:* Construct a line through X parallel to vector F_n. Extend vector F_s to intersect this line. Construct a line from X perpendicular to F_n. Construct a line through point Y perpendicular to the line through X.)

11. For the data in Problem 7, calculate the shear strain and compare it to $1/r_c$. Comment on the comparison $1/r_c = t_c/t = L/L_c$ assuming that $W = W_c$.

12. A manufacturing engineer needs an estimate of the cutting force F_c to estimate the loss of accuracy of a machining process due to deflection. The material being machined is Inconel 600 with a BHN value of 100. The cutting speed was 250 fpm, the feed was 0.020 in./rev, and the depth of cut was 0.250 in. The chip from the process measured 0.080 in. thick. Estimate the cutting force F_c assuming that $F_t = F_c/2$.

13. Using Figure 20-4 for input data, determine the maximum and minimum MRR values for rough machining (turning) a 1020 carbon steel with a BHN value of 200. Repeat for finish machining assuming a DOC value equal to 10% of the roughing DOC.

14. Estimate the horsepower needed to remove metal at 550 in.³/min with a feed of 0.005 in./rev at a DOC value of 0.675 in. The cutting force F_c was measured at 10,000 lb. Comment on these values.

15. For a turning process, the horsepower required was 24 hp. The metal removal rate was 550 in.³/min. Estimate the specific horsepower and compare to published values for 1020 steel at 200 BHN.

Machining data for 1020 steel, as-received, in air with a K3H carbide tool, orthogonally (tube cutting on lathe) with tube OD = 2.875. The cutting speed was 530 fpm. The tube wall thickness was 0.200 in. The back rake angle was zero for all cuts.

Run Number	F_c	F_t	Feed ipr ×1/1000	Chip Ratio r_c	ϕ	τ_s	U	HP_s	γ	μ
			Data							
1	330	295	4.89	0.331						
2	308	280	4.89	0.381						
3	410	330	7.35	0.426						
4	420	340	7.35	0.426						
5	510	350	9.81	0.458						
6	540	395	9.81	0.453						

Chapter 20 CASE STUDY

Orthogonal Plate Machining Experiment at Auburn University

Jeremy has just been to his ISNY3800 metalcutting lab where he learned about orthogonal machining.

This lab provided him with a hands-on-experience in material cutting analysis. The experiment varied his cutting speed and uncut chip thickness (UCT), two of the most important cutting parameters in orthogonal metalcutting.

An orthogonal plate machining setup was used in the experiment. A horizontal milling machine equipped with QSD and a dynamometer was used to perform the experiment. The material for the experiment was cartridge brass. Eighteen runs were performed using 2 levels of

speed, 3 levels of UCT, and 3 levels of positive rake angles. A toolmakers microscope were used to measure the thickness of the chip (t_c), and the shear angle. The cutting and thrust forces (Fc and Ft) were measured with a dynamometer. The data was recorded in Table CS-20a.

Your task is to make a table with headings for r_c, phi (ϕ), F, N, beta, Mu, F_s, Tau (τ_s), HPs, and specific energy (U). Complete the table and then discuss (using plots) the effect (on the forces and other calculated values) of the changing level in the input parameters.

TABLE CS20A	The Data for OPM					
RUN	ALPHA	to	V	Fc	Ft	tc
1	10	0.005	HI	92	18	0.019
2	10	0.005	LO	62	11	0.015
3	10	0.010	HI	140	35	0.020
4	10	0.010	LO	128	25	0.025
5	10	0.020	HI	270	45	0.044
6	10	0.020	LO	200	25	0.040
7	20	0.005	HI	48	4	0.011
8	20	0.005	LO	54	4	0.012
9	20	0.010	HI	110	15	0.027
10	20	0.010	LO	106	12	0.029
11	20	0.020	HI	251	30	0.041
12	20	0.020	LO	249	25	0.057
13	30	0.005	HI	55	4	0.011
14	30	0.005	LO	640	4	0.010
15	30	0.010	HI	98	14	0.028
16	30	0.010	LO	106	3	0.027
17	30	0.020	HI	170	22	0.051
18	30	0.020	LO	192	5	0.040

Hi = 16 in/min

Lo = 2 in/min

CHAPTER 21

CUTTING TOOLS FOR MACHINING

■ 21.1 INTRODUCTION

Success in metal cutting depends on the selection of the proper cutting tool (material and geometry) for a given work material. A wide range of cutting-tool materials are available with a variety of properties, performance capabilities, and costs. These include high-carbon steels and low-/medium-alloy steels, high-speed steels, cast cobalt alloys, cemented carbides, cast carbides, coated carbides, coated high-speed steels, ceramics, cermets, whisker-reinforced ceramics, sialons, sintered polycrystalline cubic boron nitride (CBN), sintered polycrystalline diamond, and single-crystal natural diamond. Figure 21-1 shows some of these common tool materials ranked by the cutting speeds used to machine a unit volume of steel materials, assuming equal tool lives. As the speed (feed rate and DOC) increases, so does the metal removal rate. The time required to remove a given unit volume of material therefore decreases. Notice the fivefold increase in speed that the AL_2O_3-coated carbide has over the WC/Co tool ($250 \rightarrow 1200$ sfpm). Today, approximately 85% of carbide tools are coated, almost exclusively by the *chemical vapor deposition* (CVD) process. The cutting tool (material and geometry) is the most critical aspect of the machining process. Clearly, the cutting-tool material, cutting parameters,

FIGURE 21-1 Improvements in cutting tool materials have led to significant increases in cutting speeds (and productivity) over the years.

and tool geometry selected directly influence the productivity of the machining operation. Figure 21-2 outlines the input variables that influence the tool material selection decision. The elements which influence the decision are:

- Work material characteristics, hardness, chemical and metallurgical state,
- Part characteristics (geometry, accuracy, finish, and surface-integrity requirements)
- Machine tool characteristics, including the workholders (adequate rigidity with high horsepower, and wide speed and feed ranges)
- Support systems (operator's ability, sensors, controls, method of lubrication, and chip removal)

Tool material technology is advancing rapidly, enabling many difficult-to-machine materials to be machined at higher removal rates and/or cutting speeds with greater performance reliability. Higher speed and/or removal rates usually improve productivity. Predictable tool performance is essential when machine tools are computer controlled with minimal operator interaction. Long tool life is desirable especially when machines become automatic or are placed in cellular manufacturing systems.

The cutting tool is subjected to severe operating conditions. Tool temperatures of 1000°C and high local stresses require that the tool have these characteristics.

1. High hardness (Figure 21-3)
2. High hardness temperature, *hot hardness* (refer to Figure 21-3)
3. Resistance to abrasion, wear due to severe sliding friction
4. Chipping of the cutting edges
5. High toughness (impact strength) (refer to Figure 21-4)
6. Strength to resist bulk deformation
7. Good chemical stability (inertness or negligible affinity with the work material)

FIGURE 21-2   The selection of the cutting-tool material and geometry followed by the selection of cutting conditions for a given application depends upon many variables.

Knoop hardness scale—1000 Kp/mm$_2$

(a)

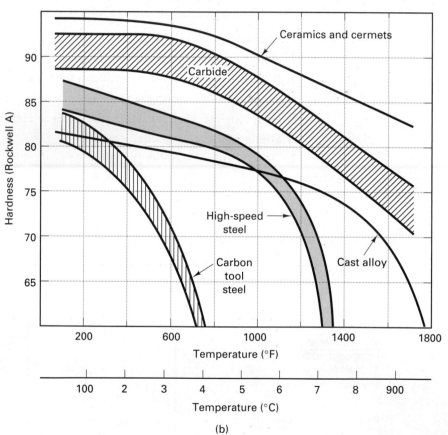

(b)

FIGURE 21-3 (a) Hardness of cutting materials and (b) decreasing hardness with increasing temperature, called hot hardness. Some materials display a more rapid drop in hardness above some temperatures. *(From* Metal Cutting Principles, *2nd ed. Courtesy of Ingersoll Cutting Tool Company.)*

8. Adequate thermal properties

9. High elastic modulus (stiffness)

10. Correct geometry and surface finish

Figure 21-5 compares these properties for various cutting-tool materials. Overlapping characteristics exist in many cases. Exceptions to the rule are very common. In many classes of tool materials, a wide range of compositions and properties are obtainable.

Methods of toughness testing

Toughness

Toughness (as considered for tooling materials) is the relative resistance of a material to breakage, chipping, or cracking under impact or stress. Toughness may be thought of as the opposite of brittleness. Toughness testing is not the same as standardized hardness testing. It may be difficult to correlate the results of different test methods. Common toughness tests include Charpy impact tests and bend fracture tests.

Conventional tool steel microstructure

Wear Resistance

Alloy elements (Cr, V, W, Mo) form hard carbide particles in tool steel microstructures. Amount & type present influence wear resistance.

Hardness of carbides:

• Hardened steel	• 60/65 HRC
• Chromium carbides	• 66/68 HRC
• Moly, tungsten carbides	• 72/77 HRC
• **Vanadium carbides**	• **82/84 HRC**

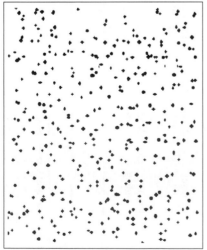

P/M tool steels microstructure

Microstructure of P/M tool steel versus conventional tool steels shows the fine carbide distribution, uniformly distributed.

FIGURE 21-4   The most important properties of tool steels are:
1.   Hardness—resistance to deforming and flattening
2.   Toughness—resistance to breakage and chipping
3.   Wear resistance—resistance to abrasion and erosion.

FIGURE 21-5 Salient Properties of Cutting Tool Materials[a]

	Carbon and Low-/Medium-Alloy Steels	High-Speed Steels	Sintered Cemented Carbides	Coated HSS	Coated Carbides	Ceramics	Polycrystalline CBN	Diamond
Toughness	◀——————————————— Decreasing —————————————————▶							
Hot hardness	◀——————————————— Increasing —————————————————▶							
Impact strength	◀——————————————— Decreasing —————————————————▶							
Wear resistance	◀——————————————— Increasing —————————————————▶							
Chipping resistance	◀——————————————— Decreasing —————————————————▶							
Cutting speed	◀——————————————— Increasing —————————————————▶							
Depth of cut	Light to medium	Light to heavy	Light to heavy	Light to heavy	Light to heavy	Light to heavy	Light to heavy	Very light for single-crystal diamond
Finish obtainable	Rough	Rough	Good	Good	Good	Very good	Very good	Excellent
Method of manufacture	Wrought	Wrought cast, HIP sintering	Cold pressing and sintering, PM	PVD[b] after forming	CVD[c]	Cold pressing and sintering or HIP sintering	High-pressure–high-temperature sintering	High-pressure–high-temperature sintering
Fabrication	Machining and grinding	Machining and grinding	Grinding	Machining and grinding, coating	Grinding before coating	Grinding	Grinding and polishing	Grinding and polishing
Thermal shock resistance	◀——————————————— Increasing —————————————————▶							
Tool material cost	◀——————————————— Increasing —————————————————▶							

[a]Overlapping characteristics exist in many cases. Exceptions to the rule are very common. In many classes of tool materials, a wide range of compositions and properties are obtainable.
[b]Physical vapor deposition.
[c]Chemical vapor deposition.

Figure 21-3 compares various tool materials on the basis of hardness, the most critical characteristic, and hot hardness (hardness decreases slowly with temperature). Figure 21-4 compares hot hardness with toughness, or the ability to take impacts during interrupted cutting. Naturally, it would be wonderful if these materials were also easy to fabricate, readily available, and inexpensive, since cutting tools are routinely replaced, but this is not usually the case. Obviously, many of the requirements conflict and therefore tool selection will always require trade-offs.

■ 21.2 Cutting-Tool Materials

In nearly all machining operations, cutting speed and feed are limited by the capability of the tool material. Speeds and feeds must be kept low enough to provide for an acceptable tool life. If not, the time lost changing tools may outweigh the productivity gains from increased cutting speed. Coated high-speed steel (HSS) and uncoated and coated carbides are currently the most extensively used tool materials.

Coated tools cost only about 15 to 20% more than uncoated tools, so a modest improvement in performance can justify the added cost. About 15 to 20% of all tool steels are coated, mostly by the *physical vapor deposition* (PVD) processes. Diamond and CBN are used for applications in which, despite higher cost, their use is justified. Cast cobalt alloys are being phased out because of the high raw-material cost and the increasing availability of alternate tool materials. New ceramic materials called *cermets* (ceramic material in a metal binder) are having a significant impact on future manufacturing productivity.

Tool requirements for other processes that use noncontacting tools, as in electrodischarge machining (EDM) and electrochemical machining (ECM), or *no tools at all* (as in laser machining), are discussed in Chapter 20. Grinding abrasives will be discussed in Chapter 29.

TOOL STEELS

Carbon steels and low-/medium-alloy steels, called *tool steels,* were once the most common cutting-tool materials. Plain-carbon steels of 0.90 to 1.30% carbon when hardened and tempered have good hardness and strength and adequate toughness and can be given a keen cutting edge. However, tool steels lose hardness at temperatures above 400°F because of tempering and have largely been replaced by other materials for metal cutting.

The most important properties for tool steels are hardness, hot hardness, and toughness. Low-/medium-alloy steels have alloying elements such as Mo and Cr, which improve hardenability, and W and Mo, which improve wear resistance. These tool materials also lose their hardness rapidly when heated to about their tempering temperature of 300° to 650°F, and they have limited abrasion resistance. Consequently, low-/medium-alloy steels are used in relatively inexpensive cutting tools (e.g., drills, taps, dies, reamers, broaches, and chasers) for certain low-speed cutting applications when the heat generated is not high enough to reduce their hardness significantly. High-speed steels, cemented carbides, and coated tools are also used extensively to make these kinds of cutting tools. Although more expensive, they have longer tool life and improved performance. These steels greatly benefit from P/M manufacturing due to uniformly distributed carbides.

HIGH-SPEED STEELS

First introduced in 1900 by F. W. Taylor and White, high-alloy steel is superior to tool steel in that it retains its cutting ability at temperatures up to 1100°F, exhibiting good "red hardness." Compared with tool steel, it can operate at about double or triple cutting speeds to about 100 sfpm with equal life, resulting in its name: *high-speed steel,* often abbreviated HSS.

Today's high-speed steels contain significant amounts of W, Mo, Co, V, and Cr besides Fe and C. W, Mo, Cr, and Co in the ferrite as a solid solution provide strengthening of the matrix beyond the tempering temperature, thus increasing the hot hardness. Vanadium (V), along with W, Mo, and Cr, improves hardness (R_c 65–70) and wear resistance. Extensive solid solutioning of the matrix also ensures good hardenability of these steels.

Although many formulations are used, a typical composition is that of the 18-4-1 type (tungsten 18%, chromium 4%, vanadium 1%), called T1. Comparable performance can also be obtained by the substitution of approximately 8% molybdenum for the tungsten, referred to as a tungsten equivalent (W_{eq}). High-speed steel is still widely used for drills and many types of general-purpose milling cutters and in single-point tools used in general machining. For high-production machining, it has been replaced almost completely by carbides, coated carbides, and coated HSS.

HSS main strengths are:

- Great toughness—superior transverse rupture strength
- Easily fabricated
- Best for sever applications where complex tool geometry is needed (gear cutters, taps, drills, reamers, dies)

High-speed steel tools are fabricated by three methods: cast, wrought, and sintered (using the powder metallurgy technique). Improper processing of cast and wrought products can result in carbide segregation, formation of large carbide particles and significant variation of carbide size, and nonuniform distribution of carbides in the matrix. The material will be difficult to grind to shape and will cause wide fluctuations of properties, inconsistent tool performance, distortion, and cracking.

To overcome some of these problems, a powder metallurgy technique has been developed that uses the hot-isostatic pressing (HIP) process on atomized, prealloyed tool steel mixtures. Because the various constituents of the P/M alloys are "locked" in place by the compacting procedure, the end product is a more homogeneous alloy, Figure 21-4. P/M high-speed steel cutting tools exhibit better grindability, greater toughness, better wear resistance, and higher red (or hot) hardness; they also perform more consistently. They are about double the cost of regular HSS.

TiN-COATED HIGH-SPEED STEELS

Coated high-speed steel provides significant improvements in cutting speeds, with increases of 10 to 20% being typical. First introduced in 1980 for gear cutters (hobs) and in 1981 for drills, TiN-coated HSS tools have demonstrated their ability to more than pay for the extra cost of the coating process.

In addition to hobs, gear-shaper cutters, and drills, HSS tooling coated by TiN now includes reamers, taps, chasers, spade-drill blades, broaches, bandsaw and circular saw blades, insert tooling, form tools, end mills, and an assortment of other milling cutters.

Physical vapor deposition has proved to be the most viable process for coating HSS, primarily because it is a relatively low-temperature process that does not exceed the tempering point of HSS. Therefore, no subsequent heat treatment of the cutting tool is required. Films 0.0001 to 0.0002 in. in thickness adhere well and withstand minor elastic, plastic, and thermal loads. Thicker coatings tend to fracture under the typical thermomechanical stresses of machining.

There are many variations to the PVD process, as outlined in Table 21-1. The process usually depends on gas pressure and is performed in a vacuum chamber. PVD processes are carried out with the workpieces heated to temperatures in the range of 400° to 900°F. Substrate heating enhances coating adhesion and film structure.

Because surface pretreatment is critical in PVD processing, tools to be coated are subject to a vigorous cleaning process. Precleaning methods typically involve degreasing, ultrasonic cleaning, and Freon drying. Deburring, honing, and more active cleaning methods are also used.

The advantages of TiN-coated HSS tooling include reduced tool wear. Less tool wear results in less stock removal during tool regrinding, thus allowing individual tools to be reground more times. For example, a TiN hob can cut 300 gears per sharpening; the uncoated tool would cut only 75 parts per sharpening. Therefore the cost per gear is reduced from 20 cents to 2 cents. Naturally, reduced tool wear means longer tool life.

Higher hardness, with typical values for the thin coatings, "equivalent" to R_c 80–85, as compared to R_c 65–70 for hardened HSS, means reduced abrasion wear. Relative inertness (i.e., TiN does not react significantly with most workpiece materials) results in

TABLE 21-1 Surface Treatments for Cutting Tools

Process	Method	Hardness[a] and Depth	Advantages	Limitations
Black oxide	HSS cutting tools are oxidized in a steam atmosphere at 1000°F	No change in prior steel hardness	Prevents built-up edge formations in machining of steel.	Strictly for HSS tools.
Nitriding case hardening	Steel surface is coated with nitride layer by use of cyanide salt at 900° to 1600°F, or ammonia, gas, or N$_2$ ions.	To 72 R$_c$; Case depth: 0.0001 to 0.100 in.	High production rates with bulk handling. High surface hardness. Diffuses into the steel surfaces. Simulates strain hardening.	Can only be applied to steel. Process has embrittling effect because of greater hardness. Post-heat treatment needed for some alloys.
Electrolytic electroplating	The part is the cathode in a chromic acid solution; anode is lead. Hard chrome plating is the most common process for wear resistance.	70-72 R$_c$; 0.0002 to 0.100 in.	Low friction coefficient, antigalling. Corrosion resistance. High hardness.	Moderate production; pieces must be fixtured. Part must be very clean. Coating does not diffuse into surface, which can affect impact properties.
Vapor deposition chemical vapor deposition (CVD)	Deposition of coating material by chemical reactions in the gaseous phase. Reactive gases replace a protective atmosphere in a vacuum chamber, At temperatures of 1800° to 1200°F, a thin diffusion zone is created between the base metal and the coating.	To 84 R$_c$; 0.0002 to 0.0004 in.	Large quantities per batch. Short reaction times reduce substrate stresses. Excellent adhesion, recommended for forming tools. Multiple coatings can be applied (TiN, TiC, AL$_2$O$_3$). Line-of-sight not a problem.	High temperatures can affect substrate metallurgy, requiring post-heat treatment, which can cause dimensional distortion (except when coating sintered carbides). Necessary to reduce effects of hydrogen chloride on material properties, such as impact strength. Usually not diffused. Tolerances of +0.001 required for HSS tools.
Physical vapor desposition (PVD sputtering)	Plasma is generated in a vacuum chamber by ion bombardment to dislodge particles from a target made of the coating material. Metal is evaporated and is condensed or attracted to substrate surfaces.	To 84 R$_c$; To 0.0002 in. thick	A useful experimental procedure for developing wear surfaces. Can coat substrates with metals, alloys, compounds, and refractories. Applicable for all tooling.	Not a high-production method. Requires care in cleaning. Usually not diffused.
PVD (electron beam)	A plasma is generated in vacuum by evaporation from a molten pool that is heated by an electron-beam gun.	To 84 R$_c$; To 0.0002 in. thick	Can coat reasonable quantities per batch cycle. Coating materials are metals, compounds, alloys, and refractories. Substrate metallurgy is preserved. Very good adhesion. Fine particle deposition. Applicable for all tooling.	Parts require fixturing and orientation in line-of-sight process. Ultra-cleanliness required.
PVD/ARC	Titanium is evaporated in a vacuum and reacted with nitrogen Gas. Resulting titanium nitride plasma is ionized and electrically attracted to the substrate surface. A high-energy process with multiple plasma guns.	To 85 R$_c$; To 0.0002 in. thick	Process at 900°F preserves substrate metallurgy. Excellent coating adhesion. Controllable deposition of grain size and growth. Dimensions, surface finish, and sharp edges are preserved. Can coat all high-speed steels without distortion.	Parts must be fixtured for line-of-sight process. Parts must be very clean. No by-products formed in reaction. Usually only minor diffusion.

[a]Rockwell hardness values above 68 are estimates.

greater tool life through a reduction in adhesion. TiN coatings have a low coefficient of friction. This can produce an increase in the shear angle, which in turn reduces the cutting forces, spindle power, and heat generated by the deformation processes. PVD coatings generally fail in high-stress applications such as cold extrusion, piercing, roughing, and high-speed machining.

CAST COBALT ALLOYS

Cast cobalt alloys, popularly known as *stellite tools*, are cobalt-rich, chromium–tungsten–carbon cast alloys having properties and applications in the intermediate range between high-speed steel and cemented carbides. Although comparable in room-temperature hardness to high-speed steel tools, cast cobalt alloy tools retain their hardness to a much higher temperature. Consequently, they can be used at higher cutting speeds (25% higher) than HSS tools. Cast cobalt alloys are hard as cast and cannot be softened or heat treated.

Cast cobalt alloys contain a primary phase of Co-rich solid solution strengthened by Cr and W and dispersion hardened by complex hard, refractory carbides of W and Cr. Other elements added include V, B, Ni, and Ta. The casting provides a tough core and elongated grains normal to the surface. The structure is not, however, homogeneous.

Tools of cast cobalt alloys are generally cast to shape and finished to size by grinding. They are available only in simple shapes, such as single-point tools and saw blades, because of limitations in the casting process and the expense involved in the final shaping (grinding). The high cost of fabrication is primarily due to the high hardness of the material in the as-cast condition. Materials machinable with this tool material include plain-carbon steels, alloy steels, nonferrous alloys, and cast iron.

Cast cobalt alloys are currently being phased out for cutting-tool applications because of increasing costs, shortages of strategic raw materials (Co, W, and Cr), and the development of other, superior tool materials at lower cost.

CARBIDE OR SINTERED CARBIDES

Carbide cutting-tool inserts are traditionally divided into two primary groups:

1. Straight tungsten grades, which are used for machining cast irons, austenitic stainless steel, and nonferrous and nonmetallic materials.

2. Grades containing major amounts of titanium, tantalum, and or columbium carbides, which are used for machining ferritic workpieces. There are also the titanium carbide grades, which are used for finishing and semifinishing ferrous alloys.

The classification of carbide insert grades employs a C-classification system in the United States and ISO P and M classification system in Europe and Japan. These classifications are based on application, rather than composition or properties. Each cutting-tool vendor can provide proprietary grades and recommended applications.

Carbides, which are nonferrous alloys, are also called *sintered* (or cemented) carbides because they are manufactured by powder metallurgy techniques. The P/M process is outlined in Figure 21-6. See Chapter 16 for details on powder metallurgy processes. These materials became popular during World War II, as they afforded a four- or five-fold increase in cutting speeds. The early versions had tungsten carbide as the major constituent, with a cobalt binder in amounts of 3 to 13%. Most carbide tools in use today

FIGURE 21-6 P/M process for making cemented carbide insert tools.

Tungsten is carburized in a high-temperature furnace, mixed with cobalt and blended in large ball mills. After ball milling, the powder is screened and dried. Paraffin is added to hold the mixture together for compacting. Carbide inserts are compacted using a pill press. The compacted powder is sintered in a high-temperature vacuum furnace. The solid cobalt dissolves some tungsten carbide, then melts and fills the space between adjacent tungsten carbide grains. As the mixture is cooled, most of the dissolved tungsten carbide precipitates onto the surface of existing grains. After cooling, inserts are finish ground and honed or used in the pressed condition.

are either straight WC or multicarbides of W-Ti or W-Ti-Ta, depending on the work material to be machined. Cobalt is the binder. These tool materials are much harder, are chemically more stable, have better hot hardness, have high stiffness, have lower friction, and operate at higher cutting speeds than HSS. They are more brittle and more expensive and use strategic metals (W, Ta, Co) more extensively.

Cemented carbide tool materials based on TiC have been developed primarily for auto industry applications using predominantly Ni and Mo as a binder. These are used for higher-speed (>1000 ft/min) finish machining of steels and some malleable cast irons.

Cemented carbide tools are available in insert form in many different shapes: squares, triangles, diamonds, and rounds. They can be either brazed or mechanically clamped onto the tool shank. Mechanical clamping (Figure 21-7) is more popular because when one edge or corner becomes dull, the insert is rotated or turned over to expose a new cutting edge. Mechanical inserts can be purchased in the as-pressed state or the insert can be ground to closer tolerances. Naturally, precision-ground inserts cost more. Any part tolerance less than ±0.003 normally cannot be manufactured without radial adjustment of the cutting tool, even with ground inserts. If no radial adjustment is performed, precision-ground inserts should be used only when the part tolerance is between ±0.006 and ±0.003. Pressed inserts have an application advantage because the cutting edge is unground and thus does not leave grinding marks on the part after machining. Ground inserts can break under heavy cutting loads because the grinding marks on the insert produce stress concentrations that result in brittle fracture. Diamond grinding is used to finish carbide tools. Abusive grinding can lead to thermal cracks and premature (early) failure of the tool. Brazed tools have the carbide insert brazed to the steel tool shank. These tools will have a more accurate geometry than the mechanical insert tools, but they are more expensive. Since cemented carbide tools are relatively brittle, a 90° corner angle at the cutting edge is desired. To strengthen the edge and prevent edge chipping, it is rounded off by honing, or an appropriate chamfer or a negative land (a T-land) on the rake face is provided. The preparation of the cutting edge can affect tool life. The sharper the edge (smaller edge radius), the more likely the edge is to chip or break. Increasing the edge radius will increase the cutting forces, so a trade-off is required. Typical edge radius values are 0.001 to 0.003 in.

FIGURE 21-7 Boring head with carbide insert cutting tools. These inserts have a chip groove that can cause the chips to curl tightly and break into small, easily disposed lengths.

A *chip groove* (see Figure 21-7) with a positive rake angle at the tool tip may also be used to reduce cutting forces without reducing the overall strength of the insert significantly. The groove also breaks up the chips for easier disposal by causing them to curl tightly.

For very low-speed cutting operations, the chips tend to weld to the tool face and cause subsequent microchipping of the cutting edge. Cutting speeds are generally in the range of 150 to 600 ft/min. Higher speeds (>1000 ft/min) are recommended for certain less-difficult-to-machine materials (such as aluminum alloys) and much lower speeds (100 ft/min) for more difficult-to-machine materials (such as titanium alloys). In interrupted cutting applications, it is important to prevent edge chipping by choosing the appropriate cutter geometry and cutter position with respect to the workpiece. For interrupted cutting, finer grain size and higher cobalt content improve toughness in straight WC-Co grades.

After use, carbide inserts (called disposable or throwaway inserts) are generally recycled in order to reclaim the Ta, WC, and Co. This recycling not only conserves strategic materials but also reduces costs. A new trend is to regrind these tools for future use where the actual size of the insert is not of critical concern.

COATED-CARBIDE TOOLS

Beginning in 1969 with TiC-coated WC, coated tools became the norm in the metalworking industry because coating can consistently improve tool life 200 or 300% or more. In cutting tools, material requirements at the surface of the tool need to be abrasion resistant, hard, and chemically inert to prevent the tool and the work material from interacting chemically with each other during cutting. A thin, chemically stable, hard refractory coating of TiC, TiN, or AL_2O_3 accomplishes this objective. The bulk of the tool is a tough, shock-resistant carbide that can withstand high-temperature plastic deformation and resist breakage. The result is a composite tool as shown in Figure 21-8.

To be effective, the coatings should be hard, refractory, chemically stable, and chemically inert to shield the constituents of the tool and the workpiece from interacting chemically under cutting conditions. The coatings must be fine grained, free of binders and porosity. Naturally, the coatings must be metallurgically bonded to the substrate. Interface coatings are graded to match the properties of the coating and the substrate. The coatings must be thick enough to prolong tool life but thin enough to prevent brittleness.

Coatings should have a low coefficient of friction so the chips do not adhere to the rake face. Coating materials include single coatings of TiC, TiN, Al_2O_3, HfN, or HfC. Multiple coatings are used, with each layer imparting its own characteristic to the tool. Successful coating combinations include TiN/TiC/TiCN/TiN and TiC/Al_2O_3/TiN. Chemical vapor deposition is used to obtain coated carbides. The coatings are formed by chemical reactions that take place only on or near the substrate. Like electroplating, CVD is a process in which the deposit is built up atom by atom. It is therefore capable of producing deposits of maximum density and of closely reproducing fine detail on the substrate surface.

Control of critical variables such as temperature, gas concentration, and flow pattern is required to assure adhesion of the coating to the substrate. The coating-to-substrate adhesion must be better for cutting-tool inserts than for most other coatings applications to survive the cutting pressure and temperature conditions without flaking off. Grain size and shape are controlled by varying temperature and/or pressure.

The purpose of multiple coatings is to tailor the coating thickness for prolonged tool life. Multiple coatings allow a stronger metallurgical bond between the coating and the substrate and provide a variety of protection processes for machining different work materials, thus offering a more general-purpose tool material grade. A very thin final coat of TiN coating (μm) can effectively reduce crater formation on the tool face by one to two orders of magnitude relative to uncoated tools.

Coated inserts of carbides are finding wide acceptance in many metalcutting applications. Coated tools have two or three times the wear resistance of the best uncoated tools with the same breakage resistance. This results in a 50 to 100% increase in speed for the same tool life. Because most coated inserts cover a broader application range, fewer

Titanium carbide remains as the basic material covering the substrate for strength and wear resistance. The second layer is aluminium oxide, which has proven chemical stability at high temperatures and resists abrasive wear. The third layer is a thin coating of titanium nitride to give the insert a lower coefficient of friction and to reduce edge build up.

Titanium nitride coating—low coefficient of friction

Aluminum oxide 2nd layer

Titanium carbide (TiCN) as first layer—strength and wear resistance

Tungsten carbide core

WC

Al_2O_3 Aluminum oxide 2nd layer—chemical stability at high temperature—resists abrasive wear

Relative thickness of coatings

Titanium nitride coating
Aluminum oxide–2nd layer
Titanium carbide–1st layer
Carbide substrate

TiN
Al_2O_3
TiCN
Special carbide substrate

FIGURE 21-8   Triple-coated carbide tools provide resistance to wear and plastic deformation in machining of steel, abrasive wear in cast iron, and built-up edge formation.

grades are needed, and therefore inventory costs are lower. Aluminum oxide coatings have demonstrated excellent crater wear resistance by providing a chemical diffusion reaction barrier at the tool–chip interface, permitting a 90% increase in cutting speeds in machining some steels.

Coated-carbide tools have progressed to the place where in the United States about 80 to 90% of the carbide tools used in metalworking are coated.

CERAMICS

Ceramics are made of pure *aluminum oxide*, Al_2O_3, or Al_2O_3 used as a metallic binder. Using P/M, very fine particles are formed into cutting tips under a pressure of 20 to 28 tons/in.^{2} (267 to 386 MPa) and sintered at about 1800°F (1000°C). Unlike the case with ordinary ceramics, sintering occurs without a vitreous phase.

Ceramics are usually in the form of disposable tips. They can be operated at two to three times the cutting speeds of tungsten carbide. They almost completely resist cratering, run with no coolant, and have about the same tool life at their higher speeds as tungsten carbide does at lower speeds. As shown in Table 21-2, ceramics are usually as hard as carbides but are more brittle (lower bend strength) and therefore require more rigid tool holders and machine tools in order to take advantage of their capabilities. Their hardness and chemical inertness make ceramics a good material for high-speed finishing and/or high-removal-rate machining applications of superalloys, hard-chill cast iron, and high-strength steels. Because ceramics have poor thermal and mechanical shock resistance, interrupted cuts and interrupted application of coolants can lead to premature tool failure. Edge chipping is usually the dominant mode of tool failure. Ceramics are not suitable for aluminum, titanium, and other materials that react chemically with alumina-based ceramics. Recently, whisker-reinforced ceramic materials that have greater transverse rupture strength have been developed. The whiskers are made from silicon carbide.

CERMETS

Cermets are a new class of tool materials best suited for finishing. Cermets are ceramic TiC, nickel, cobalt, and tantalum nitrides. TiN and other carbides are used for binders. Cermets have superior wear resistance, longer tool life, and can operate at higher cutting speeds with superior wear resistance. Cermets have higher hot hardness and oxidation resistance than cemented carbides. The better finish imparted by a cermet is due to its low level of chemical reaction with iron [less cratering and *built-up edge* (BUE)]. Compared to carbide, the cermet has less toughness, lower thermal conductivity, and greater thermal expansion, so thermal cracking can be a problem during interrupted cuts.

Cermets are usually cold pressed, and proper processing techniques are required to prevent insert cracking. New cermets are designed to resist thermal shocking during milling by using high nitrogen content in the titanium carbonitride phase (produces finer grain size) and adding WC and TaC to improve shock resistance. PVD-coated cermets have the wear resistance of cermets and the toughness range of a coated carbide, and they perform well with a coolant.

Figure 21-9 shows a comparison of speed feed coverage of typical cermets compared to ceramics, carbides, and coated carbides. The values illustrate that cermets can clearly cover a wide range of important metalcutting applications.

DIAMONDS

Diamond is the hardest material known. Industrial diamonds are now available in the form of polycrystalline compacts, which are finding industrial application in the ma-

TABLE 21-2	Properties of Cutting-Tool Materials Compared for Carbides, Ceramics, HSS, and Cast Cobalt[a]			
	Hardness Rockwell A or C	Transverse Rupture (bend) Strength ($\times 10^3$ psi)	Compressive Strength ($\times 10^3$ psi)	Modulus of Elasticity (e)($\times 10^6$ psi)
Carbide C1–C4	90–95 R_A	250–320	750–860	89–93
Carbide C5–C8	91–93 R_A	100–250	710–840	66–81
High-speed steel	86 R_A	600	600–650	30
Ceramic (oxide)	92–94 R_A	100–125	400–650	50–60
Cast cobalt	46–62 R_C	80–120	220–335	40

[a]Exact properties depend upon materials, grain size, bonder content, volume.

Tool Material Group	General Applications	Versus Cermet
PCD (polycrystal diamond)	High-speed machining of aluminum alloys, nonferrous metals, and nonmetals.	Cermets can machine same materials, but at lower speeds and significantly less cost per corner.
CBN (cubic boron nitride)	Hard workpieces and high-speed machining on cast irons.	Cermets cannot machine the harder workpieces that CBN can. Cermets cannot machine cast iron at the speeds CBN can. The cost per corner of cermets is significantly less.
Ceramics (cold press)	High-speed turning and grooving of steels and cast iron.	Cermets are more versatile and less expensive than cold press ceramics but cannot run at the higher speeds.
Ceramics (hot press)	Turning and grooving of hard workpieces; high-speed finish machining of steels and irons.	Cermets cannot machine the harder workpieces or run at the same speeds on steels and irons but are more versatile and less expensive.
Ceramics (silicon nitride)	Rough and semirough machining of cast irons in turning and milling applications at high speeds and under unfavorable conditions.	Cermets cannot machine cast iron at the high speeds of silicon nitride ceramics, but in moderate-speed applications cermets may be more cost effective.
Coated carbide	General-purpose machining of steels, stainless steels, cast iron, etc.	Cermets can run at higher cutting speeds and provide better tool life at less cost for semiroughing to finishing applications.
Carbides	Tough material for lower-speed applications on various materials.	Cermets can run at higher speeds, provide better surface finishes and longer tool life for semiroughing to finishing applications.

FIGURE 21-9 Comparison of cermets with various cutting-tool materials.

chining of aluminum, bronze, and plastics, greatly reducing the cutting forces as compared to carbides. Diamond machining is done at high speeds, with fine feeds for finishing, and produces excellent finishes. Recently, *single-crystal* diamonds, with a cutting-edge radius of 100 Å or less, have been used for precision machining of large mirrors. However, single-crystal diamonds have been used for years to machine brass watch faces, thus eliminating polishing. They have also been used to slice biological materials into thin films for viewing in transmission electron microscopes. (This process, known as ultramicrotomy, is one of the few industrial versions of orthogonal machining in common practice.)

The salient features of diamond tools include high hardness; good thermal conductivity; the ability to form a sharp edge of cleavage (single-crystal, natural diamond); very low friction; nonadherence to most materials; the ability to maintain a sharp edge for a long period of time, especially in machining soft materials such as copper and aluminum; and good wear resistance.

To be weighed against these advantages are some shortcomings, which include a tendency to interact chemically with elements of Group IVB to Group VIII of the periodic table. In addition, diamond wears rapidly when machining or grinding mild steel. It wears less rapidly with high-carbon alloy steels than with low-carbon steel and has occasionally machined gray cast iron (which has high carbon content) with long life. Diamond has a tendency to revert at high temperatures (700°C) to graphite and/or to oxidize in air. Diamond is very brittle and is difficult and costly to shape into cutting tools, the process for doing the latter being a tightly held industry practice.

The limited supply of, increasing demand for, and high cost of natural diamonds led to the ultra-high-pressure (50 Kbar), high-temperature (1500°C) synthesis of diamond from graphite at the General Electric Company in the mid-1950s and the subsequent development of *polycrystalline* sintered diamond tools in the late 1960s.

Polycrystalline diamond (PCD) tools consist of a thin layer (0.5 to 1.5 mm) of fine grain size diamond particles sintered together and metallurgically bonded to a cemented carbide substrate. A high-temperature/high-pressure process, using conditions close to those used for the initial synthesis of diamond, is needed. Fine diamond powder (1 to 30 μm) is first packed on a support base of cemented carbide in the press. At the appropriate sintering conditions of pressure and temperature in the diamond stable region, complete consolidation and extensive diamond-to-diamond bonding take place. Sintered diamond tools are then finished to shape, size, and accuracy by laser cutting and grinding. See Figure 21-10. The cemented carbide provides the necessary elastic support for the hard and brittle diamond layer above it. The main advantages of sintered polycrystalline tools over natural single-crystal tools are better quality, greater toughness, and improved wear resistance, resulting from the random orientation of the diamond grains and the lack of large cleavage planes.

Diamond tools offer dramatic performance improvements over carbides. Tool life is often greatly improved, as is control over part size, finish, and surface integrity.

Positive rake tooling is recommended for the vast majority of diamond tooling applications. If BUE is a problem, increasing cutting speed and using more positive rake angles may eliminate it. If edge breakage and chipping are problems, one can reduce the feed rate. Coolants are not generally used in diamond machining unless, as in the machining of plastics, it is necessary to reduce airborne dust particles. Diamond tools can be reground.

There is much commercial interest in being able to coat HSS and carbides directly with diamond, but getting the diamond coating to adhere reliably has been difficult. Diamond-coated inserts would deliver roughly the same performance as PCD tooling when cutting nonferrous materials but could be given more complex geometries and chip breakers while reducing the cost per cutting edge.

POLYCRYSTALLINE CUBIC BORON NITRIDES

Polycrystalline cubic boron nitride (PCBN) is a man-made tool material widely used in the automotive industry for machining hardened steels and superalloys. It is made in a compact form for tools by a process quite similar to that used for sintered polycrystalline diamonds. It retains its hardness at elevated temperatures (Knoop 4700 at 20°C, 4000 at 1000°C) and has low chemical reactivity at the tool–chip interface. This material can be used to machine hard aerospace materials like Inconel 718 and René 95 as well as chilled cast iron.

Although not as hard as diamond, PCBN is less reactive with such materials as hardened steels, hard-chill cast iron, and nickel- and cobalt-based superalloys. PCBN can be used efficiently and economically to machine these difficult-to-machine materi-

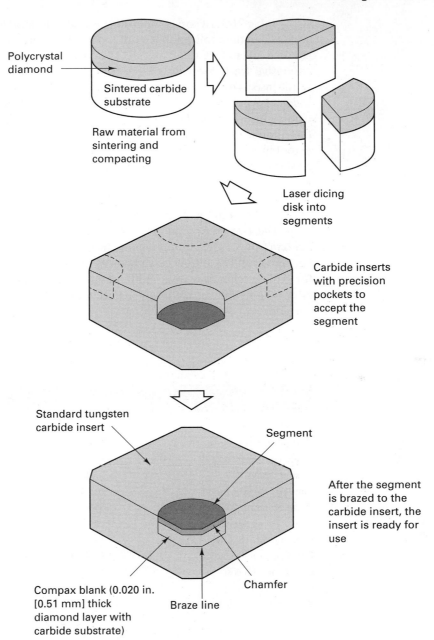

Polycrystal diamond

Sintered carbide substrate

Raw material from sintering and compacting

Laser dicing disk into segments

Carbide inserts with precision pockets to accept the segment

Standard tungsten carbide insert

Segment

After the segment is brazed to the carbide insert, the insert is ready for use

Chamfer

Braze line

Compax blank (0.020 in. [0.51 mm] thick diamond layer with carbide substrate)

FIGURE 21-10 Polycrystalline diamond tools are carbides with diamond inserts. They are restricted to simple geometries.

als at higher speeds (fivefold) and with a higher removal rate (fivefold) than cemented carbide, and with superior accuracy, finish, and surface integrity. PCBN tools are available in basically the same sizes and shapes as sintered diamond and are made by the same process. The cost of an insert is somewhat higher than either cemented carbide or ceramic tools, but the tool life may be five to seven times that of a ceramic tool. Therefore, to see the economy of using PCBN tools, it is necessary to consider all the factors.

Here is an industrial example of analysis of tooling economics, where a comparison is being made between two tool materials (insert tools). A manufacturer of diesel engines is producing an in-line six-cylinder engine block that is machined on a transfer line. Each cylinder hole must be bored to accept a sleeve liner. This operation has a depth of cut of 0.062 in. per side, for a total of 0.125 in. stock removal. The tolerance on this bore is ±0.001 in. and the spindle is operating at 2000 sfpm. Ceramic inserts are used on this operation, but with these inserts, wear was severe enough to require indexing after only 35 pieces. The ceramic insert was replaced with PCBN inserts made of a high-content BCN. Both inserts had a 0.001- to 0.002-in. radius hone for edge preparation.

Table 21-3 is a cost comparison between the ceramic and PCBN insert. The PCBN insert used in the application is a full-top PCBN insert, meaning that the entire top of the insert is a layer of PCBN material. At first glance the PCBN tool appears to be extremely expensive. Each insert costs $208.00 and provides only three usable edges, whereas the ceramic insert costs $14.90 and provides six usable edges. However, the ceramic tool must be indexed every 35 pieces. The PCBN tool is indexed every 500 pieces. The cost per bore, including insert cost and the cost of labor to perform indexing, comes to $0.125 per bore for the ceramic tool and $0.142 per bore for the PCBN tool. This appears to make the ceramic tool more cost-effective, but downtime for indexing has not been accounted for. In this application, the ceramic insert required 10.75 hours of downtime for indexing, whereas the PCBN tool required only 0.75 hour of downtime for indexing. Use of the PCBN cutting tool significantly reduces the total cost per piece by eliminating 10 hours of downtime of the machine. Later in this chapter the economics of machining will be addressed again.

The two predominant wear modes of PCBN tools are notching at the *depth-of-cut line* (DCL) and *microchipping*. In some cases, the tool will exhibit flank wear of the cutting edge. These tools have been used successfully for heavy interrupted cutting and for milling white cast iron and hardened steels using negative lands and honed cutting edges. See Table 21-4 for suggested applications of CBN and diamonds along with carbides and ceramics.

TABLE 21-3   Cost Comparison for Machining Liner Bores in 1500 Engine Blocks[a]

	Ceramic TNG-433	PCBN BTNG-433
Cost per insert	$14.90	$208.00
Edges per insert	6	3
Cost per edge	$2.48	$69.33
Time per index (6 tools)	0.25 hr	0.25 hr
Cost per index at $45 per hour	$11.25	$11.25
Indexes per 1500 blocks	43	3
Indexing cost (indexes × $11.25)	$483.75	$33.75
Insert cost for 6 spindles	$638.34	$1248.00
Labor and tool cost	$1122.09	$1281.00
Cost per bore	$.125	$.142
Total number of tool changes	43	3
Downtime for 1500 blocks	× 0.25 hr 10.75 hr	× 0.25 hr 0.75 hr

[a]To see the economy of using PCBN cutting tools, it is important to consider all factors of the operation, especially downtime for tool changing.

TABLE 21-4   Application of Cutting Tool Materials to Workpiece Materials

Workpiece Material	Carbide-Coated Carbide	Ceramic, Cermet	Cubic Boron Nitride	Diamond Compacts
Cast irons, carbon steels	X	uninterrupted finishing cuts X		
Alloy steels, alloy cast iron	X	X	X	
Aluminum, brass	X	X		X
High-silicon aluminum	X			X
Nickel-based	X	X	X	
Titanium	X			
Plastic composites	X		X	

Since diamond and PCBN are extremely hard but brittle materials, new demands are being placed on the machine tools and on machining practice in order to take full advantage of the potential of these tool materials. These demands include:

• Use of more rigid machine tools and machining practices involving gentle entry and exit of the cut in order to prevent microchipping

• Use of high-precision machine tools, because these tools are capable of producing high finish and accuracy

• Use of machine tools with higher power, because these tools are capable of higher metal removal rates and faster spindle speeds

■ 21.3 TOOL GEOMETRY

Figure 21-11 shows the cutting-tool geometry for a single-point tool (HSS) used in turning. The *back rake angle* affects the ability of the tool to shear the work material and form the chip. It can be positive or negative. Positive rake angles reduce the cutting forces, resulting in smaller deflections of the workpiece, tool holder, and machine. In machining hard work materials, the back rake angle must be small, even negative for carbide and diamond tools. Generally speaking, the higher the hardness of the workpiece, the smaller the back rake angle. For high-speed steels, back rake angle is normally chosen in the positive range, depending on the type of tool (turning, planing, end milling, face milling, drilling, etc.) and the work material.

FIGURE 21-11 Standard terminology to describe the geometry of single-point tools: (a) three dimensional views of tool, (b) oblique view of tool from cutting edge, (c) top view of turning with single-point tool, (d) oblique view from shank end of single-point turning tool.

For carbide tools, inserts for different work materials and tool holders can be supplied with several standard values of back rake angle: $-6°$ to $+6°$. The side rake angle and the back rake angle combine to form the effective rake angle. This is also called the true rake angle or resultant rake angle of the tool.

True rake inclination of a cutting tool has a major effect in determining the amount of chip compression and the shear angle. A small rake angle causes high compression, tool forces, and friction, resulting in a thick, highly deformed, hot chip. Increased rake angle reduces the compression, the forces, and the friction, yielding a thinner, less deformed, and cooler chip. Unfortunately, it is difficult to take much advantage of the desirable effects of larger positive rake angles, since they are offset by the reduced strength of the cutting tool, due to the reduced tool section, and by its greatly reduced capacity to conduct heat away from the cutting edge.

To provide greater strength at the cutting edge and better heat conductivity, zero or negative rake angles are commonly employed on carbide, ceramic, polydiamond, and PCBN cutting tools. These materials tend to be brittle, but their ability to hold their superior hardness at high temperatures results in their selection for high-speed and continuous machining operations. While the negative rake angle increases tool forces, it keeps the tool in compression and provides added support to the cutting edge. This is particularly important in making intermittent cuts and in absorbing the impact during the initial engagement of the tool and work.

In general, the power consumption is reduced by approximately 1% for each 1° in alpha (α). The end relief angle is called gamma (γ). The wedge angle Θ determines the strength of the tool and its capacity to conduct heat and depends on the values of α and γ. The relief angles mainly affect the tool life and the surface quality of the workpiece. To reduce the deflections of the tool and the workpiece and to provide good surface quality, larger relief values are required. For high-speed steel, relief angles in the range of 5° to 10° are normal, with smaller values being for the harder work materials. For carbides, the relief angles are lower to give added strength to the tool.

The side and end cutting-edge angles define the nose angle and characterize the tool design. The nose radius has a major influence on surface finish. Increasing the nose radius usually decreases tool wear and improves surface finish.

Tool nomenclature varies with different cutting tools, manufacturers, and users. Many terms are still not standard because of all this variety. The most common tool terms will be used in later chapters to describe specific cutting tools.

The introduction of coated tools has spurred the development of improved tool geometries. Specifically, *low-force groove* (LFG) geometries have been developed that reduce the total energy consumed and break up the chips into shorter segments. These grooves effectively increase the rake angle, which increases the shear angle and lowers the cutting force and power. This means that higher cutting speeds or lower cutting temperatures (and better tool lives) are possible.

As a chip breaker, the groove deflects the chip at a sharp angle and causes it to break into short pieces that are easier to remove and are not as likely to become tangled in the machine and possibly cause injury to personnel. This is particularly important on high-speed, mass-production machines.

The shapes of cutting tools used for various operations and materials are compromises, resulting from experience and research so as to provide good overall performance. For coated tools, edge strength is an important consideration. A thin coat enables the edge to retain high strength, but a thicker coat exhibits better wear resistance. Normally, tools for turning have a coating thickness of 6 to 12 μm. Edge strength is higher for multilayer coated tools. The radius of the edge should be 0.0005 to 0.005 in.

■ 21.4 TOOL COATING PROCESSES

The two most effective coating processes for improving the life and performance of tools are the chemical vapor deposition and physical vapor deposition of titanium nitride (TiN) and titanium carbide (TiC). The selection of the *cutting materials for cutting* tools depends on what property you are seeking. If you want

Oxidation and corrosion resistance; high-temperature stability	select	Al_2O_3, TiN, TiC
Crater resistance,	select	Al_2O_3, TiN, TiC
Hardness and edge retention,	select	TiC, TiN, Al_2O_3
Abrasion resistance and flank wear,	select	Al_2O_3, TiN, TiC
Low coefficient of friction and high lubricity,	select	TiN, Al_2O_3, TiC
Fine grain size,	select	TiN, TiC, Al_2O_3

The CVD process, used to deposit a protective coating onto carbide inserts, has been benefiting the metal removal industry for many years and is now being applied with equal success to steel. The PVD processes have quickly become the preferred TiN coating processes for high-speed steel and carbide-tipped cutting tools.

CVD

Chemical vapor deposition is an atmosphere-controlled process carried out at temperatures in the range of 950° to 1050°C (1740° to 1920°F). Figure 21-12 shows a schematic of the CVD process.

Cleaned tools ready to be coated are staged on precoated graphite work trays (shelves) and loaded onto a central gas distribution column (tree). The tree loaded with parts to be coated is placed inside the retort of the CVD reactor. The tools are heated under an inert atmosphere until the coating temperature is reached. The coating cycle is initiated by the introduction of titanium tetrachloride ($TiCL_4$), hydrogen, and methane (CH_4) into the reactor. $TiCL_4$ is a vapor and is transported into the reactor via a hydrogen carrier gas; CH_4 is introduced directly. The chemical reaction for the formation of TiC is:

$$TiCl_4 + CH_4 \rightarrow TiC + 4HCl \qquad (21\text{-}1)$$

To form titanium nitride, a nitrogen–hydrogen gas mixture is substituted for methane. The chemical reaction for TiN is:

$$2TiCl_4 + 2H_2 + N_2 \rightarrow 2TiN + 4HCl \qquad (21\text{-}2)$$

PVD

The simplest form of PVD is evaporation, where the substrate is coated by condensation of a metal vapor. The vapor is formed from a source material called the charge, which is heated to a temperature less than 1000°C. PVD methods currently being used include reactive sputtering, reactive ion plating, low-voltage electron-beam evaporation, triode high-voltage electron-beam evaporation, cathodic evaporation, and arc evaporation. In each of the methods, the TiN coating is formed by reacting free titanium ions with nitrogen away from the surface of the tool and relying on a physical means to transport the coating onto the tool surface.

FIGURE 21-12   Chemical vapor deposition is used to apply layers (TiC, TiN, etc.) to carbide cutting tools.

All of these PVD processes share the following common features:

1. The coating takes place inside a vacuum chamber under a hard vacuum with the workpiece heated to 200° to 405°C (400 to 900°F).

2. Before coating, all parts are given a final cleaning inside the chamber to remove oxides and improve coating adhesion.

3. The coating temperature is relatively low (for cutting and forming tools), typically about 842° F (450° C).

4. The metal source is vaporized in an inert gas atmosphere (usually argon), and the metal atoms react with gas to form the coating. Nitrogen is the reactive gas for nitrides, and methane or acetylene (along with nitrogen) is used for carbides.

5. All four are ion-assisted deposition processes. The ion bombardment compresses the atoms on the growing film, yielding a dense, well-adhered coating.

A typical cycle time for the coating of functional tools, including heat-up and cool down, is about six hours.

Of the three, PVD arc evaporation, shown in Figure 21-13, is the most recent development. The plasma sources are from several arc evaporators located on the sides and top of the vacuum chamber. Each evaporator generates plasma from multiple arc spots. In this way a highly localized electrical arc discharge causes minute evaporation of the material of the cathode, and a self-sustaining arc is produced that generates a high-energy and concentrated plasma.

The kinetic energy of deposition is much greater than that found in any other PVD method. During coating, this energy is of the order of 150 electron volts and more. Therefore, the plasma is highly reactive and the greater percentage of the vapor is atomic and ionized.

Coating temperatures can be selected and controlled so that metallurgy is preserved. This enables a coating of a wide variety of sintered carbide tools, for example, brazed tools, solid carbide tools such as drills, end mills, form tools, and inserts. The PVD arc evaporation process will preserve substrate metallurgy, surface finish, edge sharpness, geometrical straightness, and dimensions.

CVD AND PVD—COMPLEMENTARY PROCESSES

CVD and PVD are complementary coating processes. The differences between the two processes and resultant coatings dictate which coating process to use on different tools.

Since CVD is done at higher temperatures, the adhesion of these coatings tends to be superior to a PVD–CVD-deposited coating. CVD coatings are normally deposited thicker than PVD coatings (6 to 9 μm for CVD, 1 to 3 μm for PVD). See Figure 21-14.

FIGURE 21-13   Schematic of PVC arc evaporation process.

Comparison of PVD Process Characteristics

Process	Processing Temperature, °C	Throwing Power	Coating Materials	Coating Applications and Special Features
Vacuum evaporation	RT—700, usually <200	Line-of-sight	Chiefly metal, especially Al (a few simple alloys/ a few simple compounds)	Electronic, optical, decorative, simple masking.
Ion implantation	200–400, best <250 for N	Line-of-sight	Usually N (B, C)	Wear resistance for tools, dies, etc. Effect much deeper than original implantation depth. Precise area treatment, excellent process control.
Ion plating, ARE	RT—0.7 T_m of coating. Best at elevated temperatures.	Moderate to good	Ion plating: Al, other metals (few alloys) ARE: TiN and other compounds	Electronic, optical, decorative. Corrosion and wear resistance. Dry lubricants. Thicker engineering coatings.
Sputtering	RT—0.7 T_m of metal coatings. Best >200 for nonmetals.	Line-of-sight	Metals, alloys, glasses, oxides. TiN, and other compounds	Electronic, optical, wear resistance. Architectural (decorative). Generally thin coatings. Excellent process control.
CVD	300–2000, usually 600–1200	Very good	Metals, especially refractory TiN and other compounds; pyrolytic BN	Thin, wear-resistant films on metal and carbide dies, tools, etc. Free-standing bodies or refractory metals and pyrolytic C or BN.

RT= room temperature; ARE = activated reactive evaporation; T_m = absolute melting temperature. (a) Compounds: oxides, nitrides, carbides, silicides, and borides of Al, B, Cr, Hf, Mo, Nb, Ni, Re, Si, Ta, Ti, V, W, Zr.

Source: Advanced Materials and Processes, December 2001.

FIGURE 21-14 Comparison of PVD methods for depositing thin films on microelectronic devices as well as cutting tools.

With CVD multiple coatings, layers may be readily deposited but the tooling materials are restricted. CVD coated tools must be heat treated after coating. This limits the application to loosely toleranced tools. However, the CVD process, being a gaseous process, results in a tool that is coated uniformly all over; this includes blind slots and blind holes.

Since PVD is mainly a line-of-sight process, all surfaces of the part to be coated may be masked. PVD also requires fixturing of each part in order to effect the substrate bias.

APPLICATIONS
Applications for the two different processes are as follows:

CVD
- Loosely toleranced tooling
- Piercing and blanking punches, trim dies, phillips punches, upsetting punches
- AISI A, D, H, M, and air hardening and tool steel parts
- Solid carbide tooling

PVD
- All HSS, solid carbide, and carbide-tipped cutting tools
- Fine blanking punches, dies (0.001 in. tolerance or less)
- Non-composition-dependent process; virtually all tooling materials, including mold steels and bronze

■ 21.5 TOOL FAILURE AND TOOL LIFE

In metalcutting, the failure of the cutting tool can be classified into two broad categories, according to the failure mechanisms that caused the tool to die (or fail):

1. *Physical failures* mainly include gradual tool wear on the flank(s) of the tool below the cutting edge (called flank wear) or wear on the rake face of the tool (called *crater wear*) or both.

2. *Chemical failures*, which include wear on the rake face of the tool (called *crater wear*) are rapid, usually unpredictable, and often catastrophic failures resulting from abrupt, premature death of a tool.

Other modes of failure are outlined in Figure 21-15. The selection of failure criteria is also widely varied. Figure 21-15 also shows a sketch of a "worn" tool, showing *crater wear* and *flank wear*, along with wear of the tool nose radius and an outer-diameter groove (the DCL groove). Tools also fail by edge chipping and edge fracture.

As the tool wears, its geometry changes. This geometry change will influence the cutting forces, the power being consumed, the surface finish obtained, the dimensional accuracy, and even the dynamic stability of the process. Worn tools are duller, creating greater cutting forces and often resulting in chatter in processes that otherwise are usually relatively free of vibration. The actual wear mechanisms active in this high-temperature environment are abrasion, adhesion, diffusion, or chemical interactions. It appears that in metalcutting, any or all of these mechanisms may be operative at a given time in a given process.

Tool failure by plastic deformation, brittle fracture, fatigue fracture, or edge chipping can be unpredictable. Moreover, it is difficult to predict which mechanism will dominate and result in a tool failure in a particular situation. What can be said is that tools, like people, die (or fail) from a great variety of causes under widely varying conditions.

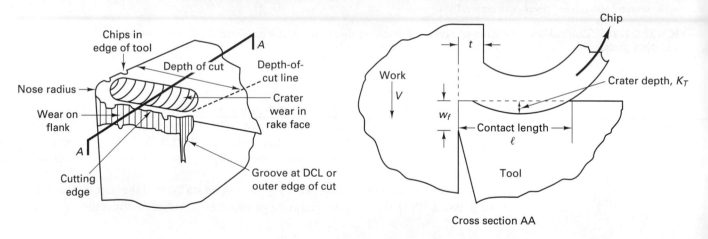

No.	Failure		Cause
1-3	Flank wear		Due to the abrasive effect of hard grains contained in the work material
4-5	Groove		Due to wear at the DCL or outer edge of the cut
6	Chipping	Physical	Fine chips caused by high-pressure cutting, chatter, vibration, etc.
7	Partial fracture		Due to the mechanical impact when an excessive force is applied to the cutting edge
8	Crater wear		Carbide particles are removed due to degradation of tool performances and chemical reactions at high temperature
9	Deformation	Chemical	The cutting edge is deformed due to its softening at high temperature
10	Thermal crack		Thermal fatigue in the heating and cooling cycle with interrupted cutting
1	Built-up edge		A portion of the workpiece material adheres to the insert cutting edge

FIGURE 21-15 Tools can fail in many ways. Tool wear during oblique cutting can occur on the flank or the rake face; t = uncut chip thickness; k_t = crater depth; w_f = flank wear land length; DCL = depth-of-cut line.

Therefore, tool life should be treated as a random variable, or probabilistically, not as a deterministic quantity.

■ 21.6 FLANK WEAR

During machining, the tool is performing in a hostile environment in which high contact stresses and high temperatures are commonplace, and therefore tool wear is always an unavoidable consequence. At lower speeds and temperatures, the tool most commonly wears on the flank. Suppose that the tool wear experiment were to be repeated 15 times without changing any of the input parameters. The result would look like Figure 21-16, which depicts the variable nature of tool wear and shows why tool wear must be treated as a random variable. In Figure 21-16 the average time is denoted as μ_T and the standard deviation as σ_T, where the wear limit criterion was 0.025 in. At a given time during the test, 35 minutes, the tool displayed flank wear ranging from 0.013 to 0.021 in, with an average of $\mu_w = 0.00175$ in. with standard deviation σ_w 0.001 in.

In Figure 21-17 four characteristic tool wear curves (average values) are shown for four different cutting speeds, V_1 through V_4; V_1 is the fastest cutting speed and therefore

w_f values for general life determination (for cemented carbides)	
Width of Wear in.	Applications
0.008	Finish cutting of nonferrous alloys, fine & light cut, etc.
0.016	Cutting of special steels
0.028	Normal cutting of cast irons, steels, etc.
0.040–0.050	Rough cutting of common cast irons

FIGURE 21-16   Tool wear on the flank displays a random nature, as does tool life. W_f = flank wear limit value.

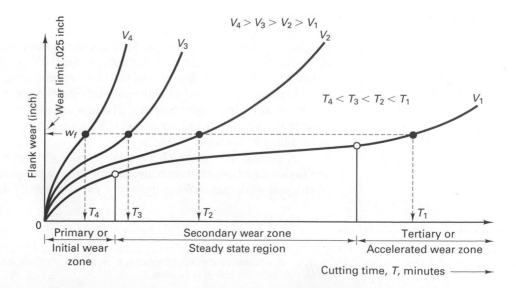

FIGURE 21-17 Typical tool wear curves for flank wear at different velocities. The initial wear is very fast, then it evens out to a more gradual pattern until the limit is reached; after that, the wear substantially increases.

FIGURE 21-18 Construction of the Taylor tool life curve using data from deterministic tool wear plots like those of Figure 21-17. Curves like this can be developed for both flank and crater wear.

generates the fastest wear rates. Such curves often have three general regions, as shown in the figure. The central region is a steady-state region (or the region of secondary wear). This is the normal operating region for the tool. Such curves are typical for both flank wear and crater wear. When the amount of wear reaches the value w_f, the permissible tool wear on the flank, the tool is said to be "worn out." w_f is typically set at 0.025 to 0.030 in. for flank wear for high-speed steels and 0.008 to 0.050 for carbides, depending on the application. For crater wear, the depth of the crater is used to determine tool failure.

Using the empirical tool wear data shown in Figure 21-17, which used the values of T (time in minutes) associated with V (cutting speed) for a given amount of tool wear, w_f (see the dashed-line construction), Figure 21-18 was developed. When V and T are plotted on log-log scales, a linear relationship appears, described by the equation

$$VT^n = \text{constant} = C \qquad (21\text{-}3)$$

This equation is called the Taylor tool life equation because in 1907, F. W. Taylor published his now-famous paper "On the Art of Cutting Metals" in ASME Transactions, wherein tool life (T) was related to cutting speed (V) and feed (f). This equation had the form

$$T = \frac{\text{constant}}{f_x V_y} \qquad (21\text{-}4)$$

which over the years took the more widely published form

$$VT^n = C^\ddagger$$

where n is an exponent that depends mostly on tool material but is affected by work material, cutting conditions, and environment and C is a constant that depends on all the input parameters, including feed. Table 21-5 provides some data on Taylor tool life constants.

Figure 21-19 shows typical tool life curves for one tool material and three work materials. Notice that all three plots have about the same slope, n. Typical values for n are 0.14 to 0.16 for HSS, 0.21 to 0.25 for uncoated carbides, 0.30 for TiC inserts, 0.33 for polydiamonds, 0.35 for TiN inserts, and 0.40 for ceramic-coated inserts.

It takes a great deal of experimental effort to obtain the constants for the Taylor equation, as each combination of tool and work material will have different constants. Note that for a tool life of 1 minute, $C = V$, or the cutting speed that yields about 1 minute of tool life for this tool.

A great deal of research has gone into developing more sophisticated versions of the Taylor equation, wherein constants for other input parameters (typically feed, depth of cut, and work material hardness) are experimentally determined, for example,

$$VT^n F^m d^p = K' \qquad (21\text{-}5)$$

---

‡Carl Barth, who was Taylor's mathematical genius, is generally thought to be the author of these formulations along with early versions of slide rules.

TABLE 21-5 Tool Life Information for Various Materials and Conditions

Source	Tool Material	Geometry	Workpiece Material	Size of Cut (in.) Depth	Feed	Cutting Fluid	$VT^n = C$ n	C
1	High-carbon steel	8.14, 6.6, 6.15, 3/64	Yellow brass (.60 Cu, 40 Zn, 85 NI. .006 Pb)	.050 .100	.0255 .0127	Dry Dry	.081 .096	242 299
1	High-carbon steel	8.14, 6.6, 6.15, 3/64	Bronze (.9 Cu, .1.5n)	.050 .100	.0255 .0127	Dry Dry	.086 .111	190 232
1	HSS-18-4-1	8.14, 6.6, 6.15, 3/64	Cast Iron 160 Bhn	.050	.0255	Dry	.101	172
			Cast iron, Nickel, 164 Bhn	.050	.0255	Dry	.111	186
			Cast iron, NI-Cr, 207 Bhn	.050	.0255	Dry	.088	102
1	HSS-18-4-1	8.14, 6.6, 6.15, 3/64	Stell, SAE B1113 C.D.	.050	.0127	Dry	.080	260
			Stell, SAE B1112 C.D.	.050	.0127	Dry	.105	225
			Stell, SAE B1120 C.D.	.050	.0127	Dry	.100	270
			Stell, SAE B1120 + Pb C.D.	.050	.0127	Dry	.060	290
			Stell, SAE B1035 C.D.	.050	.0127	Dry	.110	130
			Stell, SAE B1035 + Pb C.D.	.050	.0127	Dry	.110	147
1	HSS-18-4-1	8.14, 6.6, 6.15, 3/64	Stell, SAE 1045 C.D.	.100	.0127	Dry	.110	192
		8.14, 6.6, 6.13, 3/66	Stell, SAE 2340 185 Bhn	.100	.0125	Dry	.147	143
		8.14, 6.6, 6.15, 3/64	Stell, SAE 2345 198 Bhn	.050	.0255	Dry	.105	126
		8.14, 6.6, 6.15, 3/64	Stell, SAE 3140 190 Bhn	.100	.0125	Dry	.160	178
1	HSS-18-4-1	8.14, 6.6, 6.15, 3/64	Stell, SAE 4350 363 Bhn	.0125	.0127	Dry	.080	181
			Stell, SAE 4350 363 Bhn	.0125	.0255	Dry	.125	146
			Stell, SAE 4350 363 Bhn	.0250	.0255	Dry	.125	95
			Stell, SAE 4350 363 Bhn	.100	.0127	Dry	.110	78
			Stell, SAE 4350 363 Bhn	.100	.0255	Dry	.110	46
1	HSS-18-4-1	8.14, 6.6, 6.15, 3/64	Stell, SAE 4140 230 Bhn	.050	.0127	Dry	.180	190
			Stell, SAE 4140 271 Bhn	.050	.0127	Dry	.180	159
			Stell, SAE 6140 240 Bhn	.050	.0127	Dry	.150	197
1	HSS-18-4-1	8.22, 6.6, 6.15, 3/64	Monel metal 215 Bhn	.100	.0127	Dry	.080	170
				.150	.0255	Dry	.074	127
				.100	.0127	Em	.080	185
				.100	.0127	SMO	.105	189
1	Stellite 2400	0.0, 6.6, 6.0, 3/32	Steel, SAE 3240 annealed	.187	.031	Dry	.190	215
				.125	.031	Dry	.190	240
				.062	.031	Dry	.190	270
				.031	.031	Dry	.190	310
1	Stellite No. 3	0.0, 6.6, 6.0, 3/32	Cast iron 200 Bhn	.062	0.31	Dry	.150	205
1	Carbide (T 64)	6.12, 5.5, 10.45	Steel, SAE 1040 annealed	.062	.025	Dry	.156	800
			Steel, SAE 1060 annealed	.125	.025	Dry	.167	660
			Steel, SAE 1060 annealed	.187	.025	Dry	.167	615
			Steel, SAE 1060 annealed	.250	.025	Dry	.167	560
			Steel, SAE 1060 annealed	.062	.021	Dry	.167	880
			Steel, SAE 1060 annealed	.062	.042	Dry	.164	510
			Steel, SAE 1060 annealed	.062	.062	Dry	.162	400
			Steel, SAE 2340 annealed	.062	.025	Dry	.162	630
2	Ceramic	not available	AISI 4150	.160	.016	Dry	.400	2000
			AISI 4150	.160	.016	Dry	.200	620

Sources: 1- *Fundamentals of Tool Design.* ASTME. A.R. Koneeny, W. J. Potthoff   2 - *Theory of Metal Cutting*, P.N. Black

where n, m, and p are exponents and K' is a constant. Equations of this form are also deterministic and determined empirically.

The problem has been approached probabilistically in the following way. Since T depends on speed, feed, materials, and so on, one writes

$$T = \frac{K^{1/n}}{V^{1/n}} = \frac{K}{V^m} \qquad (21\text{-}6)$$

where K is now a random variable that represents the effects of all unmeasured factors and is an input variable.

The sources of tool life variability include factors such as:

1. Variation in work material hardness (from part to part and within a part)
2. Variability in cutting-tool materials, geometry, and preparation
3. Vibrations in machine tool, including rigidity of work and tool-holding devices
4. Changing surface characteristics of workpieces

FIGURE 21-19 Log-log tool life plots for three steel work materials cut with HSS tool material.

FIGURE 21-20 Tool life viewed as a random variable has a log normal distribution with a large coefficient of variation.

The examination of the data from a large number of tool life studies in which a variety of steels were machined shows that regardless of the tool material or process, tool life distributions are usually log normal and typically have a large standard deviation. As shown in Figure 21-20, tool life distributions have a large coefficient of variation, so tool life is not very predictable.

Other criteria that can be used to define tool death in addition to wear limits are:

• When surface finish deteriorates unacceptably
• When workpiece dimension is out of tolerance
• When power consumption or cutting forces increase to a limit
• Sparking or chip discoloration and disfigurement
• Cutting time or component quantity

In automated processes, it is very beneficial to be able to monitor the tool wear online so that the tool can be replaced prior to failure, wherein defective products may also result. The feed force has been shown to be a good, indirect measure of tool wear. That is, as the tool wears and dulls, the feed force increases more than the cutting force increases.

Once criteria for failure have been established, tool life is that time elapsed between start and finish of the cut, in minutes. Other ways to express tool life, other than time, include:

1. Volume of metal removed between regrinds or replacement of tool
2. Number of pieces machined per tool
3. Number of holes drilled with a given tool

Drilling tool failure is discussed more in chapter 23 and is very complex because of the varied and complex geometry of the tools and as shown here in Figure 21-21, the tool material.

RECONDITIONING CUTTING TOOLS
In the reconditioning of tools by sharpening and recoating, care must be taken in grinding the tool's surfaces. The following guidelines should be observed:

1. Resharpen to original tool geometry specifications. Restoring the original tool geometry will help the tool achieve consistent results on subsequent uses. Computer numerical control (CNC) grinding machines for tool resharpening have made it easier to restore a tool's original geometry.

2. Grind cutting edges and surfaces to a fine finish. Rough finishes left by poor and abusive regrinding hinder the performance of resharpened tools. For coated tools, tops of ridges left by rough grinding will break away in early tool use, leaving uncoated and unprotected surfaces that will cause premature tool failure.

3. Remove all burrs on resharpened cutting edges. If a tool with a burr is coated, premature failure can occur because the burr will break away in the first cut, leaving an uncoated surface exposed to wear.

4. Avoid resharpening practices that overheat and burn or melt (called *glazed over*) the tool surfaces, as this will cause problems in coating adhesion. Polishing or wire brushing of tools causes similar problems.

FIGURE 21-21 Tool life test data for various coated drills. TiN-coated HSS drills outperform uncoated drills. Life based on the number of holes drilled before drill failure.

Drill performance based on the number of holes drilled with 1/4-in.-diameter drills in T–1 structural steel.

The cost of each recoating is about one-fifth the cost of purchasing a new tool. By recoating, the tooling cost per workpiece can be cut by between 20 and 30%, depending on the number of parts being machined.

■ 21.7 ECONOMICS OF MACHINING

The cutting speed has such a great influence on the tool life compared to the feed or the depth of cut that it greatly influences the overall economics of the machining process. For a given combination of work material and tool material, a 50% increase in speed results in a 90% decrease in tool life, while a 50% increase in feed results in a 60% decrease in tool life. A 50% increase in depth of cut produces only a 15% decrease in tool life. Therefore, in limited-horsepower situations, depth of cut and then feed should be maximized while speed is held constant and horsepower consumed is maintained within limits. As cutting speed is increased, the machining time decreases but the tools wear out faster and must be changed more often. In terms of costs, the situation is as shown in Figure 21-22, which shows the effect of cutting speed on the cost per piece.

The total cost per operation is comprised of four individual costs: machining costs, tool costs, tool-changing costs, and handling costs. The machining cost is observed to decrease with increasing cutting speed because the cutting time decreases. Cutting time is proportional to the machining costs. Both the tool costs and the tool-changing costs increase with increases in cutting speeds. The handling costs are independent of cutting speed. Adding up each of the individual costs results in a total unit cost curve that is observed to go through a minimum point. For a turning operation, the total cost per piece C equals

$$C = C_1 + C_2 + C_3 + C_4 \tag{21-7}$$

= Machining cost + tooling cost + tool-changing cost + handling cost per piece

Expressing each of these cost terms as a function of cutting velocity will permit the summation of all the costs.

$C_1 = T_m \times C_o$ where C_o = operating cost ($/min)
$\qquad\qquad\qquad T_m$ = cutting time (min/piece)

$C_2 = \left(\dfrac{T_m}{T}\right)C_t$ where T = tool life (min/tool)

$\qquad\qquad\qquad C_t$ = initial cost of tool ($)

$C_3 = t_c \times C_o\left(\dfrac{T_m}{T}\right)$ where t_c = time to change tool (min)

$\qquad\qquad\qquad \dfrac{T_m}{T}$ = number of tool changes per piece

C_4 $\qquad\qquad\qquad$ labor, overhead, and machine tool costs consumed while part is being loaded or unloaded, tools are being advanced, machine has broken down, and so on.

Since $T_m = L/Nf_r$ for turning

$\qquad\qquad = \pi DL/12Vf_r$

and $T = (K/V)^{1/n}$, by rewriting equation 21-3,

and using "K" for the constant "C", the cost per unit, C, can be expressed in terms of V:

$$C = \frac{L\pi DC_o}{12Vf_r} + \frac{C_t V^{1/n}}{K^{1/n}} + \frac{t_c C_o V^{1/n}}{K^{1/n}} + C_4 \tag{21-8}$$

To find the minimum, take $dc/dV = 0$ and solve for V:

$$V_m = \left[\frac{1}{n} - 1\right]\left[\frac{C_t + (C_o \times t_c)}{C_o}\right] \tag{21-9}$$

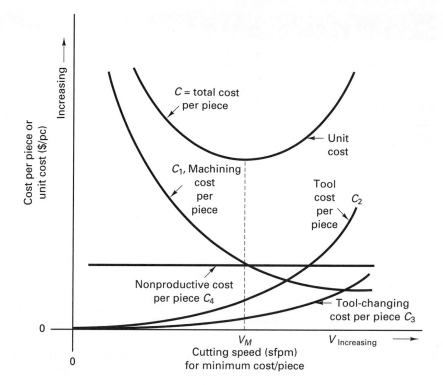

FIGURE 21-22 Cost per unit for a machining process versus cutting speed. Note that the "C" in this figure and related equations is not the same "C" used in the Taylor tool life (equation 21-3).

So V_m represents a cutting speed that will minimize the cost per unit, as depicted in Figure 21-22. However, a word of caution here is appropriate. Note that this derivation was totally dependent upon the Taylor tool life equation. Such data may not be available because they are expensive and time consuming to obtain. Even when the tool life data are available, this procedure assumes that the tool fails only by whichever wear mechanism (flank or crater) was described by this equation and by no other failure mechanism. Recall that tool life has a very large coefficient of variation and is probabilistic in nature. This derivation assumes that for a given V, there is one T—and this simply is not the case, as was shown in Figure 21-16. The model also assumes that the workpiece material is homogeneous, the tool geometry is preselected, the depth of cut and feed rate are known and remain unchanged during the entire process, sufficient horsepower is available for the cut at the economic cutting conditions, and the cost of operating time is the same whether the machine is cutting or not cutting.

Another example of tooling economics is summarized in Table 21-6, where a comparison is made between four different tools, all used for turning hot-rolled 8620 steel with triangular inserts. Operating costs for the machine tool are $60 per hour. The low-force groove insert has only three cutting edges available instead of six. It takes 3 minutes to change inserts and half a minute to unload a finished part and load in a new 6-in.-diameter bar stock. The length of cut is about 24 in. The student should study and analyze this table carefully so that each line is understood. Note that the cutting-tool cost per piece is three times higher for the low-force groove tool over the carbide but really of no consequence, since the major cost per piece comes from two sources: the machining cost per piece and the nonproductive cost per piece.

MACHINABILITY

Machinability is a much maligned term that has many different meanings but generally refers to the ease with which a metal can be machined to an acceptable surface finish. The principal definitions of the term are entirely different, the first based on material properties, the second based on tool life, and the third based on cutting speed.

1. Machinability is defined by the ease or difficulty with which the metal can be machined. In this light, specific energy, specific horsepower, and shear stress are used as

TABLE 21-6   Cost Comparison of Four Tool Materials, Based on Equal Tool Life of 40 Pieces per Cutting Edge

	Uncoated	TiC-Coated	Al_2O_3-Coated	Al_2O_3LFG
Cutting speed (surface ft/min)	400	640	1100	1320
Feed (in./rev)	0.020	0.022	0.024	0.028
Cutting edges available per insert	6	6	6	3
Cost of an insert ($/insert)	4.80	5.52	6.72	6.72
Tool life (pieces/cutting edge)	192	108	60	40
Tool-change time per piece (min)	0.075	0.075	0.075	0.075
Nonproductive cost per piece ($/pc)	0.50	0.50	0.50	0.50
Machining time per piece (min/pc)	4.8	2.7	1.50	1.00
Machining cost per piece ($/unit)	4.8	2.7	1.5	1.00
Tool-change cost per piece ($/pc)	0.08	0.08	0.08	0.08
Cutting-tool cost per piece ($/pc)	0.02	0.02	0.03	0.06
Total cost per piece ($/pc)	5.40	3.30	2.11	1.64
Production rate (pieces/hr)	11	18	29	38
Improvement in productivity based on pieces/hr (%)	0	64	164	245

Source: Data from T. E. Hale et al., "High Productivity Approaches to Metal Removal," *Materials Technology*, Spring 1980, p. 25.

measures, and, in general, the larger the shear stress or specific power values, the more difficult the material is to machine, requiring greater forces and lower speeds. In this definition, the material is the key.

2. Machinability is defined by the relative cutting speed for a given tool life while cutting some material, compared to a standard material cut with the same tool material. As shown in Figure 21-23, tool life curves are used to develop machinability ratings. In steels, the material chosen for the standard material was B1112 steel, which has a tool life of 60 min at a cutting speed of 100 sfpm. Material X has a 70% rating, which implies that steel X has a cutting speed of 70% of B1112 for equal tool life. Note that this definition assumes that the tool fails when machining X by whatever mechanism dominated the tool failure when machining the B1112. There is no guarantee that this will be the case. ISO standard 3685 has machinability index numbers based on 30 min of tool life with flank wear of 0.33 mm.

3. Cutting speed is measured by the maximum speed at which a tool can provide satisfactory performance for a specified time under specified conditions. See ASTM standard E 618-81: "Evaluating machining performance of ferrous metals using an automatic screw bar machine."

4. Other definitions of machinability are based on the ease of removal of the chips (chip disposal), the quality of the surface finish of the part itself, the dimensional stability of the process, or the cost to remove a given volume of metal.

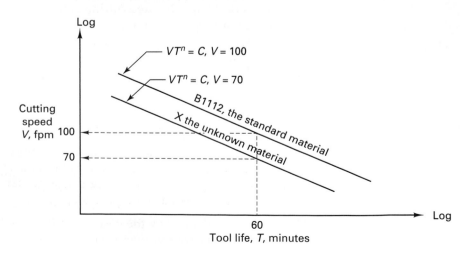

FIGURE 21-23 Machinability ratings defined by deterministic tool life curves.

Further definitions are being developed based on the probabilistic nature of the tool failure, in which machinability is defined by a tool reliability index. Using such indexes, various tool replacement strategies can be examined and optimum cutting rates obtained. These approaches account for the tool life variability by developing coefficients of variation for common combinations of cutting tools and work materials.

The results to date are very promising. One thing is clear, however, from this sort of research: although many manufacturers of tools have worked at developing materials that have greater tool life at higher speeds, few have worked to develop tools that have less variability in tool life at all speeds. The reduction in variability is fundamental to achieving smaller coefficients of variation, which typically are of the order of 0.3 to 0.4. This means that a tool with a 100-min average tool life has a standard deviation of 30 to 40 min, so there is a good probability that the tool will fail early. In automated equipment, where early, unpredicted tool failures are extremely costly, reduction of the tool life variability will pay great benefits in improved productivity and reduced costs.

21.8 CUTTING FLUIDS

From the day that Frederick W. Taylor demonstrated that a heavy stream of water flowing directly on the cutting process allowed the cutting speeds to be doubled or tripled, *cutting fluids* have flourished in use and variety and have been employed in virtually every machining process. The cutting fluid acts primarily as a coolant and secondly as a lubricant, reducing the friction effects at the tool–chip interface and the work flank regions. The cutting fluids also carry away the chips and provide friction (and force) reductions in regions where the bodies of the tools rub against the workpiece. Thus in processes such as drilling, sawing, tapping, and reaming, portions of the tool apart from the cutting edges come in contact with the work, and these (sliding friction) contacts greatly increase the power needed to perform the process, unless properly lubricated.

The reduction in temperature greatly aids in retaining the hardness of the tool, thereby extending the tool life or permitting increased cutting speed with equal tool life. In addition, the removal of heat from the cutting zone reduces thermal distortion of the work and permits better dimensional control. Coolant effectiveness is closely related to the thermal capacity and conductivity of the fluid used. Water is very effective in this respect but presents a rust hazard to both the work and tools and also is ineffective as a lubricant. Oils offer less effective coolant capacity but do not cause rust and have some lubricant value. In practice, straight cutting oils or emulsion combinations of oil and water or wax and water are frequently used. Various chemicals can also be added to serve as wetting agents or detergents, rust inhibitors, or polarizing agents to promote formation of a protective oil film on the work. The extent to which the flow of a cutting fluid washes the very hot chips away from the cutting area is an important factor in heat removal. Thus the application of a coolant should be copious and of some velocity.

The possibility of a cutting fluid providing lubrication between the chip and the tool face is an attractive one. An effective lubricant can modify the process, perhaps producing a cooler chip, discouraging the formation of a built-up edge on the tool, and promoting improved surface finish. However, the extreme pressure at the tool–chip interface and the rapid movement of the chip away from the cutting edge make it virtually impossible to maintain a conventional hydrodynamic lubricating film at the tool–chip interface. Consequently, any lubrication action is associated primarily with the formation of solid chemical compounds of low shear strength on the freshly cut chip face, thereby reducing chip–tool shear forces or friction. For example, carbon tetrachloride is very effective in reducing friction in machining several different metals and yet would hardly be classified as a good lubricant in the usual sense. Chemically active compounds, such as chlorinated or sulfurized oils, can be added to cutting fluids to achieve such a lubrication effect. Extreme-pressure lubricants are especially valuable in severe operations, such as internal threading (tapping), where the extensive tool–work contact results in high friction with limited access for a fluid. In addition to functional effectiveness as coolant and lubricant, cutting fluids should be stable in use and storage, noncorrosive to

TABLE 21-7	Cutting Fluid Contaminants	
Category	Contaminants	Effects
Solids	Metallic fines, chips Grease and sludge Debris and trash	Scratch product's surface Plug coolant lines Produce wear on tools and machines
Tramp fluids	Hydraulic oils (coolant) Water (oils)	Decrease cooling efficiency Cause smoking Clog paper filters Grow bacteria faster
Biologicals (coolants)	Bacteria Fungi Mold	Acidity coolant Break down emulsions Cause rancidity, dermatitis Require toxic biocides

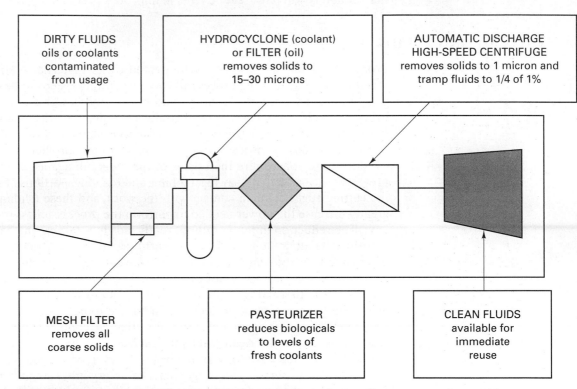

FIGURE 21-24   A well-designed recycling system for coolants will return more than 99% of the fluid for reuse.

work and machines, and nontoxic to operating personnel. The cutting fluid should also be restorable by using a closed recycling system that will purify the used coolant and cutting oils. Cutting fluids become contaminated in three ways (Table 21-7.) All these contaminants can be eliminated by filtering, hydrocycloning, pasteurizing, and centrifuging. Coolant restoration eliminates 99% of the cost of disposal and 80% or more of new fluid purchases. See Figure 21-24 for a schematic of a coolant recycling system.

■ Key Words

aluminum oxide	coated tools	flank wear	powder metallurgy
BUE (built-up edge)	crater wear	hardness	sintered carbides
carbides	cubic boron nitride (CBN)	hot hardness	stellite
cast cobalt alloy	cutting fluids	HSS (high-speed steel)	tool life
ceramics	cutting tool materials	machinability	titanium nitride
cermets	DCL (depth-of-cut line)	metal cutting	wear land
chemical vapor deposition	diamonds	physical vapor deposition	

■ Review Questions

1. For metalcutting tools, what is the most important material property (i.e., the most critical characteristic)? Why?
2. What is hot hardness compared to hardness?
3. What is impact strength and how is it measured?
4. Why is impact strength an important property in cutting tools?
5. What is HIP, and how is it used for tool fabrication?
6. What are the primary considerations in tool selection?
7. What is the general strategy behind coated tools?
8. What is a cermet?
9. How is a CBN tool manufactured?
10. F. W. Taylor was one of the discoverers of high-speed steel. What else is he well known for?
11. What casting process do you think was used to fabricate cast cobalt alloys?
12. Discuss the constraints in the selection of a cutting tool.
13. What does *cemented* mean in the manufacture of carbides?
14. What advantage do ground carbide inserts have over pressed carbide inserts?
15. What is a chip groove?
16. What is the DCL?
17. Suppose you made four beams out of carbide, HSS, ceramic, and cobalt. The beams are identical in size and shape, differing only in material. Which beam would do each of the following?
    a. Deflect the most, assuming the same load
    b. Resist penetration the most
    c. Bend the farthest without breaking
    d. Support the greatest compressive load
18. Multiple coats or layers are put on the carbide base for what different purposes?
19. What are the various ways a cutting tool can fail and what can be done to remedy this? See Figures 21-A and 21-15.

	Failure		Basic Remedy
Edge Failure	Excessive flank wear	Tool material Cutting conditions	• Use a more wear-resistant grade carbide → {coated / cermet} • Decrease speed
	Excessive crater wear	Tool material Tool design Cutting conditions	• Use a more wear-resistant grade carbide → {coated / cermet} • Enlarge the rake angle • Select the correct chip breaker • Decrease speed • Reduce the depth of cut and feed
	Cutting-edge chipping	Tool material Tool design Cutting conditions	• Use tougher grades If carbide, (AC2000 → AC3000) • If built-up edge occurs, change to a less susceptible grade (cermet) • Reinforce the cutting edge (honing) • Reduce the rake angle • Increase speed (if caused by edge build-up)
	Partial fracture of cutting edges	Tool material Tool design Cutting conditions	• Use tougher grades if carbide, (AC2000 → AC3000) • Use holder with a large approach angle • Use larger shank-size holder • Reduce the depth of cut and feed
	Built-up edge	Tool material Cutting conditions	• Change to a grade that is adhesion resistant • Increase the cutting speed and feed • Use cutting fluids
	Plastic deformation	Tool material Cutting conditions	• Change to highly thermal-resistant grades • Reduce the cutting speed and feed

FIGURE 21-A

20. What makes the process that makes TiC coatings for tools a problem? See equation 21-1.
21. Why does a TiN-coated tool consume less power than an uncoated HSS under exactly the same cutting conditions?
22. For what work material are CBN tools more commonly used and why?
23. Why is CBN better for machining steel than diamond?
24. What is the typical coefficient of variation for tool life data, and why is this a problem?
25. What is meant by the statement "Tool life is a random variable"?
26. The typical value of a coefficient of variation in metalcutting tool life distributions is 0.3. How could it be reduced?
27. Machinability is defined in many ways. Explain how a rating is obtained.

28. What are the chief functions of cutting fluids?
29. How are CVD tools manufactured?
30. Why is the PVD process used to coat HSS tools?
31. Why is there no universal cutting-tool material?
32. What is an 18-4-1 HSS composed of?
33. Over the years, tool materials have been developed that have allowed significant increases in MRR. Nevertheless, HSS is still widely used. Under what conditions might HSS be the material of choice?
34. Why is the rigidity of the machine tool an important consideration in the selection of the cutting-tool material?
35. Explain how it can be that the tool wears when it may be four times as hard as the work material.
36. What is a honed edge on a cutting tool and why is it done?

■ Problems

1. Figure 21-B gives data for cutting speed and tool life. Determine the constants for the Taylor tool life equation for these data. What do you think the tool material might have been?

FIGURE 21-B

2. Suppose you have a turning operation using a tool with a zero back rake and 5° end relief. The insert flank has a wear land on it of 0.020 in. How much has the diameter of the workpiece grown (increased) due to this flank wear, assuming the tool has not been reset to compensate for the flank wear?

3. In Figure 21-C, a single-point tool is shown. Identify points A through G using tool nomenclature.

A = _____
B = _____
C = _____
D = _____
E = _____
F = _____
G = _____

FIGURE 21-C

4. The following data have been obtained for machining an Si-Al alloy.

| Workpiece | Tool | Cutting Speed (m/min) for Tool Life (m/min) of | | |
Material	Material	20 min	30 min	60 min
Sand casting	Diamond polycrystal	731	642	514
Permanent-mold casting	Diamond polycrystal	591	517	411
PMC with flood cooling	Diamond polycrystal	608	554	472
Sand casting	WC-K-20	175	161	139

Compute the C and n values for the Taylor tool life equation.
How do these n values compare to the typical values?

5. In Figure 21-D, the insert at the top is set with a $0°$ side cutting-edge angle. The insert at the bottom is set so that the edge contact length is increased from a 0.250-in. depth of cut to 0.289 in. The feed was 0.010 ipr.
   a. Determine the side cutting-edge angle for the offset tool.
   b. What is the uncut chip thickness in the offset position?
   c. What effect will this have on the forces and the process?

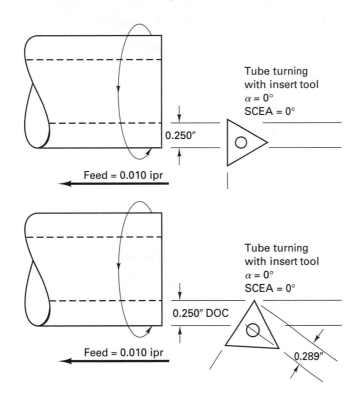

FIGURE 21-D

6. Tool cost is often used as the major criterion for justifying tool selection. Either silicon nitride or PCBN insert tips can be used to machine (bore) a cylinder block on a transfer line at a rate of 312,000 parts/yr (material: gray cast iron). The operation requires 12 inserts (2 per tool), as six bores are machined simultaneously. The machine was run at 2600 sfpm with a feed of 0.014 in. at 0.005-in. DOC for finishing. Here are some additional data.

	SiN	PCBN
Tips in use per part	12	12
Tool life (parts per tool)	200	4700
Cost per tip	$1.25	$28.50

   a. Which tool material would you recommend?
   b. On what basis?

7. A 2-in.-diameter bar of steel was turned at 284 rpm, and tool failure occurred in 10 min. The speed was changed to 132 rpm, and the tool failed in 30 min of cutting. Assume that a straight-line relationship exists. What cutting speed should be used to obtain a 60-min tool life of V_{60}?

8. Table 21-6 shows a cost comparison for four tool materials. Show how the data in this table were generated.

9. Refer to Problem 1. Show the relationship between cutting speed and tool temperature. What does this mean with regard to tool failure?

10. The outside diameter of a roll for a steel (AISI 1015) rolling mill is to be turned. In the final pass, the starting diameter = 26.25 in. and the length = 48.0 in. The cutting conditions will be feed = 0.0100 in./rev and depth of cut = 0.125 in. A cemented carbide cutting tool is to be used, and the parameters of the Taylor tool life equation for this setup are $n = 0.25$ and $C = 1300$. It is desirable to operate at a cutting speed such that

the tool will not need to be changed during the cut. Determine the cutting speed that will make the tool life equal to the time required to complete this turning operation. (Problem suggested by Groover, *Fundamentals of Modern Manufacturing: Materials, Processes, and Systems*, 2nd ed., John Wiley & Sons, 2002.)

11. Using data from Problems 8 and 10, determine the necessary horsepower for the machine tool to make this cut.

12. Figure 21-E shows a sketch of a single-point tool and its associated tool signature. Put the signature from the tool in Figure 21-C in the same order as shown in Figure 21-E. Which tool would produce the larger F_c given that both are cutting at the same $V, f_r,$ and DOC in the same material?

Tool Signature	10,	20,	8,	8,	20,	15,	125
Normal back rake angle							
Normal side rake angle							
Normal end relief angle							
Normal side relief angle							
End cutting-edge angle							
Side cutting-edge angle							
Nose radius							

FIGURE 21-E

*C*hapter 21 CASE STUDY

Comparing Tool Materials Based on Tool Life

The Linus Drilling Company is trying to decide on what kind of inserts to use in their indexable insert drills. These drills do not cut at dead center but rather form a thin center slug that is pushed out of the way by the drill body. These indexable drills often provide a marked productivity improvement compared to conventional HDD drills, particularly when they are combined with coated insert tools. The company is trying to determine which type of insert to use in the drill for machining some hot rolled 8620 steel shafts using triangular inserts. The operating cost of the machine tool is $60.00 per hour. It takes about three minutes to change the inserts and about 30 seconds to unload a finished part and load a new part in the machine. The company is currently using uncoated inserts at the following operating conditions—400 sfpm and 0.020 ipr. These speeds and feeds resulted in each cutting edge producing about 40 pieces before the tool's cutting edge became dull. The tool was then indexed. Since it was a triangular tool, each tool had 6 cutting edges available before it had to be replaced. At these speeds and feeds, the drilling time was 4.8 minutes and the production rate

was 11 parts per hour, while the machining cost per piece was $4.80 ($1.00/min × 4.8 min/pc.). The manufacturing engineer on the job, Brian Paul, has found three new tool materials being used in triangular insert tools. They are listed in Table CS-21 along with the data for the uncoated tool. The new materials are TiC-coated carbide, Al_2O_3-coated carbide and a ceramic-coated insert with a single side, low force groove geometry. The expected cutting conditions, speeds and feeds, are given in the table along with Brian's estimates of the production rates in pieces per hour for each of these new tool materials. The low force groove geometry tools can only be used three times—they cannot be flipped over—so only three cutting edges are available per insert before it has to be replaced. Brian has argued that even though the ceramic-coated inserts cost more, they result in a lower cost per piece, considering all the costs. Determine the machining cost per piece, the tool changing cost per piece, and the tool cost per piece which make up the total cost per piece, and verify Brian's belief that these coated tools will provide some cost savings.

TABLE CS.21   Cost Comparison of Four Tool Materials, Based on Equal Tool Life

	Uncoated	TiC-Coated	Al_2O_3-coated	Al_2O_3 LFG
Cutting speed (surface ft/min)	400	640	1100	1320
Feed (in/rev)	0.020	0.022	0.024	0.028
Cutting edges available per insert	6	6	6	3
Cost of an insert ($/insert)	4.80	5.52	6.72	6.72
Tool life (pieces/cutting edge)	192	108	60	40
Tool change time per piece (min)	0.075	0.075	0.075	0.075
Nonproductive cost per piece ($/pc)	0.50	0.50	0.50	0.50
Machining time per piece (min/pc)	4.8	2.7	1.50	1.00
Machining cost per piece ($/pc)	4.80			
Tool change cost per piece ($/pc)	0.08			
Cutting tool cost per piece ($/pc)	0.02			
Total cost per piece ($)	5.40			
Production rate (pieces/hr)	11	18	29	38
Improvement in productivity based on pieces/hr (%)	0	64	164	245

Source: Data from T. E. Hale et al., "High Productivity Approaches to Metal Removal," *Materials Technology*, Spring 1980, p. 25.

TURNING AND BORING PROCESSES

■ 22.1 INTRODUCTION

Turning is the process of machining external cylindrical and conical surfaces. It is usually performed on a machine tool called a lathe, as shown in Figure 22-1, using a cutting tool. The workpiece is held in a workholder. More details on lathes are shown later in this chapter. As indicated in Figure 22-2, relatively simple work and tool movements are involved in turning a cylindrical surface. The workpiece is rotated and the single-point cutting tool is fed longitudinally into the workpiece. If the tool is fed at an angle to the axis of rotation, an external conical surface results. This is called *taper turning*. If the tool is fed to the axis of rotation, using a tool that is wider than the depth of the cut, the operation is called *facing*, and a flat surface is produced on the end of the cylinder.

By using a tool having a specific form or shape and feeding it radially or inward against the work, external cylindrical, conical, and irregular surfaces of limited length can also be turned. The shape of the resulting surface is determined by the shape and size

FIGURE 22-1 Schematic of a standard engine lathe performing a turning operation, with the cutting tool shown in inset.

FIGURE 22-2 Basic turning machines can rotate the work and feed the tool longitudinally for turning and can perform other operations by feeding transversely. Depending on what direction the tool is fed and on what portion of the rotating workpiece is being machined, the operations have different names. The dashed arrows indicate the tool feed motion relative to the workpiece.

of the cutting tool. Such machining is called *form turning*. If the tool is fed all the way to the axis of the workpiece, it will be cut in two. This is called *parting* or *cutoff*, and a simple, thin tool is used. A similar tool is used for *necking* or *partial cutoff*.

Boring is a variation of turning. Essentially, boring is internal turning. Boring can use single-point cutting tools to produce internal cylindrical or conical surfaces. It does not create the hole but, rather, machines or opens the hole up to a specific size. Boring can be done on most machine tools that can do turning. However, boring also can be done using a rotating tool with the workpiece remaining stationary. Also, specialized machine tools have been developed that will do boring, drilling, and reaming but will not do turning. Other operations, like *threading* and *knurling*, can be done on machines used for turning. In addition, drilling, reaming, and tapping can be done on the rotation axis of the work.

In recent years, turning centers have been developed that use turrets to hold multiple-edge rotary tools in powered heads. Some new machine tools feature two opposing spindles with automatic transfer from one to the other and two turrets of tools.

■ 22.2 FUNDAMENTALS OF TURNING, BORING, AND FACING TURNING

Turning constitutes the majority of lathe work. The cutting forces, resulting from feeding the tool from right to left, should be directed toward the headstock to force the workpiece against the workholder and thus provide better work support.

If good finish and accurate size are desired, one or more roughing cuts are usually followed by one or more finishing cuts. Roughing cuts may be as heavy as proper chip thickness, cutting dynamics, tool life, lathe horsepower, and the workpiece permit. Large *depths of cut* and smaller feeds are preferred to the reverse procedure, because fewer cuts are required and less time is lost in reversing the carriage and resetting the tool for the following cut.

On workpieces that have a hard surface, such as castings or hot-rolled materials containing mill scale, the initial roughing cut should be deep enough to penetrate the hard materials. Otherwise, the entire cutting edge operates in hard, abrasive material throughout the cut, and the tool will dull rapidly. If the surface is unusually hard, the cutting speed on the first roughing cut should be reduced accordingly.

Finishing cuts are light, usually being less than 0.015 in. in depth, with the feed as fine as necessary to give the desired finish. Sometimes a special finishing tool is used, but often the same tool is used for both roughing and finishing cuts. In most cases, one finishing cut is all that is required. However, where exceptional accuracy is required, two finishing cuts may be made. If the diameter is controlled manually, a short finishing cut ($1/4$ in. long) is made and the diameter checked before the cut is completed. Because the previous micrometer measurements were made on a rougher surface, it may be necessary to reset the tool in order to have the final measurement, made on a smoother surface, check exactly.

In turning, the primary cutting motion is rotational, with the tool feeding parallel to the axis of rotation (Figure 22-3). The desired cutting speed established the necessary rpm (N_s) of the rotating workpiece. The feed f_r is given in inches per revolution (ipr).

The depth of cut is d, where

$$d = \text{DOC} = \frac{D_1 - D_2}{2} \text{ inches} \tag{22-1}$$

The length of cut is the distance traveled parallel to the axis L plus some allowance or overrun A to allow the tool to enter and/or exit the cut.

Here is how the inputs to the turning process are determined. First it is necessary that the engineer select the cutting speed, V, in feet per minute, the feed (f_r), and the depth of cut, based on what material is being machined, what tool material is being used,

FIGURE 22-3 Basics of the turning process normally done on a lathe. The dashed arrows indicate the feed motion of the tool relative to the work.

and other factors like what process is being performed. The rpm value for the machine tool can be determined by

$$N_s = \frac{12V}{\pi D_1} \tag{22-2}$$

(using the larger diameter), where the factor of 12 is used to convert feet to inches. The cutting time is

$$T_m = \frac{L + A}{f_r N_s} \tag{22-3}$$

where A is overrun allowance, and f_r is the selected feed in inches per revolution.

Example of Turning

The 1.78-inch diameter steel bar shown in Figure 1-10 is to be turned down to a 1.10-inch diameter on a standard engine lathe. The overall length of the bar is 18.750 inches, and the region to be turned is 16.50 inches. After turning, the bar looks like stage 2. The part is made from cold-drawn free-machining steel (this means the chips breakup nicely) with a Bhn of 250. Since you want to take the bar from 1.78 to 1.10, you have a total depth of cut, d, of 0.34 inch (0.68/2). You decide you want to make two cuts, a roughing pass and a finishing pass. Rough at $d = 0.300$ and finish with $d = 0.04$ inches. Looking at the table for selecting speed and feed (Figure 20-4), you select $V = 100$ fpm and feed = 0.020 ipr because you have selected high-speed steel cutting tools.

The bar is held in a chuck with a feed through the hole in the spindle and is supported on the right end with a live center. The ends of the bar have been center drilled. Allowance should be 0.50 inches for approach (no overtravel). Allow 1.0 minutes to reset the tool after the first cut. To determine the inputs to the lathe, we calculate the spindle rpm:

$$N_s = 12V/\pi D_1 = 12 \times 100/3.14 \times 1.78 = 214$$

But your lathe does not have this particular rpm, so you select the closest rpm, which is 200. You don't need any further calculations for lathe inputs as you input the feed in ipr directly.

The time to make the cut is

$$T_m = \frac{L + ALL}{f_r \times N_s} = \frac{16.50 + 0.50}{0.020 \times 200} = 4.25 \text{ min}$$

You could reduce this time by changing to a coated carbide tool that would allow you to increase the cutting speed to 925 sfpm. (See table). The time for the second cut will be different if you change the feed and/or the speed to improve the surface finish. Again from the table in Figure 21-4, speed = 925, feed = 0.007, so

$$N_s = 12 \times 925/3.14 \times 1.10 = 3213 \text{ with 3200 rpm the closest value:}$$

$$T_m = \frac{L + ALL}{0.007 \times 3200} = 0.75 \text{ min}$$

The metal removal rate is

$$\text{MRR} = \frac{\text{volume removed}}{\text{time}} = \frac{(\pi D_1^2 - \pi D_2^2)L}{4L/f_r N}$$

(omitting the allowance term). By rearranging and substituting N_s, an exact expression for MRR is obtained:

$$\text{MRR} = 12Vf_r \frac{(D_1^2 - D_2^2)}{4D_1} \tag{22-4}$$

Rewriting the last term

$$\frac{D_1^2 - D_2^2}{4D_1} = \frac{D_1 - D_2}{2} \times \frac{D_1 - D_2}{2D_1}$$

Therefore, since

$$d = \frac{D_1 - D_2}{2} \quad \text{and} \quad \frac{D_1 - D_2}{2D_1} \cong 1 \quad \text{for small } d$$

then,

$$\text{MRR} \cong 12\,Vf_r d \text{ in.}^3/\text{min} \tag{22-5}$$

Note that Equation (22-5) is an approximate equation that assumes that the depth of cut d is small compared to the uncut diameter D_1.

BORING

Boring always involves the enlarging of an existing hole, which may have been made by a drill or may be the result of a core in a casting. An equally important and concurrent purpose of boring may be to make the hole concentric with the axis of rotation of the workpiece and thus correct any eccentricity that may have resulted from the drill starting or drifting off the center line. Concentricity is an important attribute of bored holes.

When boring is done in a lathe, the work usually is held in a chuck or on a faceplate. Holes may be bored straight, tapered, with threads, or to irregular contours. Figure 22-4 shows the relationship of the tool and the workpiece for boring. Think of boring as internal turning while feeding the tool parallel to the rotation axis of the workpiece, with two important differences.

First, the relief and clearance angles on the tool should be larger and the tool overhand (length to diameter) must be considered with regard to stability and deflection problems.

Given V and f_r for a cut of length L, the cutting time is

$$T_m = \frac{L + A}{f_r N_s} \tag{22-6}$$

where $N_s = 12V/\pi D_1$ for D_1, the diameter of bore, and A, the overrun allowance. The metal removal rate is

$$\text{MRR} = \frac{L(\pi D_1^2 - \pi D_2^2)/4}{L/f_r N}$$

where D_2 is the original hole diameter.

$$\text{MRR} \cong 12\,Vf_r d \tag{22-7}$$

(omitting allowance term), where d is the depth of cut.

In most respects, the same principles are used for boring as for turning. Again, the tool should be set exactly at the same height as the axis of rotation. Larger end clearance angles help to prevent the heel of the tool from rubbing on the inner surface of the hole. Because the tool overhang will be greater, feeds and depths of cut may be reduced to reduce forces that cause tool vibration and chatter. In some cases, the boring bar may be made of tungsten carbide because of this material's greater stiffness.

FACING

Facing is the producing of a flat surface as the result of the tool being fed across the end of the rotating workpiece, as shown in Figure 22-4b. Unless the work is held on a mandrel, if both ends of the work are to be faced, it must be turned end for end after the first end is completed and the facing operation repeated.

FIGURE 22-4 Basic movement of boring, facing, and cutoff (or parting) process.

(a) Boring a drilled hole   (b) Facing from the cross side   (c) Cutoff or parting

The cutting speed should be determined from the largest diameter of the surface to be faced. Facing may be done either from the outside inward or from the center outward. In either case, the point of the tool must be set exactly at the height of the center of rotation. Because the cutting force tends to push the tool away from the work, it is usually desirable to clamp the carriage to the lathe bed during each facing cut to prevent it from moving slightly and thus producing a surface that is not flat.

In the facing of castings or other materials that have a hard surface, the depth of the first cut should be sufficient to penetrate the hard material to avoid excessive tool wear.

In facing, the tool feeds perpendicular to the axis of the rotating workpiece. Because the rpm is constant, the cutting speed is continually decreasing as the axis is approached. The length of cut L is $D_1/2$ or $(D_1 - D_2)/2$ for a tube.

$$T_m = \text{cutting time} = \frac{L + A}{f_r N} \text{ minutes}$$

$$= \frac{\dfrac{D_1}{2} + A}{f_r N}$$

$$\text{MRR} = \frac{\text{VOL}}{T_m} = \frac{\pi D_1^2 d f_r N}{4L} = 6V f_r d \quad \text{in.}^3/\text{min} \tag{22-8}$$

where d is the depth of cut and $L = D_1/2$ is the length of cut.

PARTING

Parting is the operation by which one section of a workpiece is severed from the remainder by means of a cutoff tool, as shown in Figure 22-4c. Because parting tools are quite thin and must have considerable overhand, this process is more difficult to perform accurately. The tool should be set exactly at the height of the axis of rotation, be kept sharp, have proper clearance angles, and be fed into the workpiece at a proper and uniform feed rate.

In parting or cutoff work, the tool is fed (plunged) perpendicular to the rotational axis, as it was in facing. The length of cut for solid bars is $D_1/2$. For tubes,

$$L = \frac{D_1 - D_2}{2}$$

In cutoff operations, the width of the tool is d in inches, the width of the cutoff operation. The equations for T_m and MRR are then basically the same as for facing.

DEFLECTION

In boring, facing, and cutoff operations, the speeds, feeds, and depth of cut selected are generally less than those recommended for straight turning because of the large overhang of the tool often needed to complete the cuts. Recall the basic equation for deflection of a cantilever beam, modifying for machining,

$$\delta = \frac{Pl^3}{3EI} = \frac{F_c l^3}{3EI} \tag{22-9}$$

In equation 22-9, l represents the overhang of the tool, which greatly affects the deflection, so it should be minimized whenever possible. In equation 22-9,

$$E = \text{modulus of elasticity}$$

$$I = \text{moment of inertia of cross section of tool}$$

$$P = F_c = \text{applied load or cutting force}$$

where $\quad I = \pi D_1^2/64 \text{ solid round bar}$

$$I = \pi(D_1^4 - (D_2^4))/64 \text{ bar with hole}$$

$$D_1 = \text{diameter of tube or bar}$$

$$D_2 = \text{inside diameter of the tube}$$

Deflection is proportional to the fourth power of the boring bar diameter and the third power of the bar overhand. Select the largest-diameter bar diameter and minimize the overhand. Use carbide shank boring bars ($E \cong 80,000,000$ psi), and select tool geometries that direct cutting forces into the feed direction to minimize chatter. The reduction of the feed or depth of cut reduces the forces operating on the tools. The cutting speed usually controls the occurrence of chatter and vibration. See the dynamics of machining discussion in Chapter 20.

Any imbalance in the cutting forces will deflect the tool to the side, resulting in loss of accuracy in cutoff lengths. At the outset, the forces will be balanced if there is no side rake on the tool. As the cutoff tool reaches the axis of the rotating part, the tool will be deflected away from the spindle, resulting in a change in the length of the part.

PRECISION BORING

Sometimes bored holes are slightly bell mouthed because the tool deflects out of the work as it progresses into the hole. This often occurs in castings and forgings where the holes have draft angles so that the depth of cut increases as the tool progresses down the bore. This problem can usually be corrected by repeating the cut with the same tool setting; however, the total cutting time for the part is increased. Alternately, a more robust setup can be used. Large holes may be precision bored using the setup shown in Figure 22-5, where a pilot bushing is placed in the spindle to mate with the hardened ground pilot of the boring bar. This setup eliminates the cantilever problems common to boring.

Because the rotational relationship between the work and the tool is a simple one and is employed on several types of machine tools, such as lathes, drilling machines, and milling machines, boring is very frequently done on such machines. However, several machine tools have been developed primarily for boring, especially in cases involving large workpieces or for large-volume boring of smaller parts. Such machines as these are also capable of performing other operations, such as milling and turning. Because boring frequently follows drilling, many boring machines also can do drilling, permitting both operations to be done with a single setup of the work.

DRILLING

Drilling, discussed in detail in Chapter 23, can be done on lathes with the drill mounted in the tailstock quill of engine lathes or the turret on turret lathes and fed against a

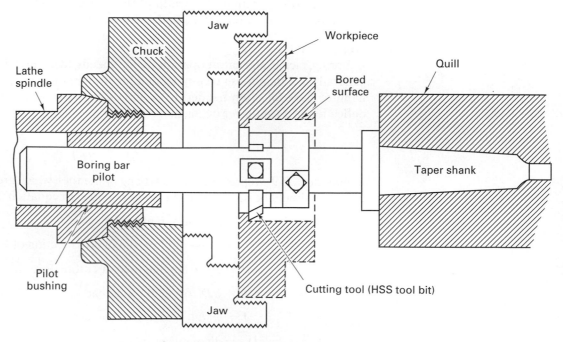

FIGURE 22-5   Pilot boring bar mounted in tailstock of lathe for precision boring large hole in casting. The size of the hole is controlled by the rotation diameter of the cutting tool.

rotating workpiece. Straight-shank drills can be held in Jacobs chucks, or drills with taper shanks mounted directly in the quill hole can drill holes online (center of rotation). Drills can also be mounted in the turrets of modern turret centers and fed automatically on the rotational axis of the workpiece or off axis with power heads. It also is possible to drill on a lathe with the drill bit mounted and rotated in the spindle while the work remains stationary, supported on the tailstock or the carriage of the lathe.

The usual speeds used for drilling should be selected for lathe work. Because the feed may be manually controlled, care must be exercised, particularly in drilling small holes. Coolants should be used where required. In drilling deep holes, the drill should be withdrawn occasionally to clear chips from the hole and to aid in getting coolant to the cutting edges. This is called peck drilling. See Chapter 23 for further discussion on drilling.

REAMING

Reaming on a lathe involves no special precautions. Reamers are held in the tailstock quill, taper-shank types being mounted directly and straight-shank types by means of a drill chuck. Rose-chucking reamers are usually used (see Chapter 23). Fluted-chucking reamers may also be used, but these should be held in some type of holder that will permit the reamer to float (i.e., have some compliance) in the hole and conform to the geometry created by the boring process.

KNURLING

Knurling produces a regularly shaped, roughened surface on a workpiece. Although knurling also can be done on other machine tools, even on flat surfaces, in most cases it is done on external cylindrical surfaces using lathes. Knurling is a chipless, cold-forming process. See Figures 22-6 and 22-18 for examples. The two hardened rolls are pressed against the rotating workpiece with sufficient force to cause a slight outward and lateral displacement of the metal to form the knurl in a raised, diamond pattern. Another type of knurling tool produces the knurled pattern by cutting chips. Because it involves less pressure and thus does not tend to bend the workpiece, this method is often preferred for workpieces of small diameter and for use on automatic or semiautomatic machines.

SPECIAL ATTACHMENTS

For engine lathes, taper turning and milling can be done on a lathe but require special attachments. The *milling attachment* is a special vise that attaches to the cross slide to hold work. The milling cutter is mounted and rotated by the spindle. The work is fed by means of the cross-slide screw. *Tool-post grinders* are often used to permit grinding to be done on a lathe. Taper turning will be discussed later. *Duplicating attachments* are available that, guided by a template, will automatically control the tool movements for turning irregularly shaped parts. In some cases, the first piece, produced in the normal manner, may serve as the template for duplicate parts. To a large extent, duplicating lathes using templates have been replaced by numerically controlled lathes and milling is done with power tools in NC turret lathes.

FIGURE 22-6   (a) Knurling in a lathe, using a forming-type tool, and showing the resulting pattern on the workpiece; (b) knurling tool with forming rolls. *(Courtesy of Armstrong Brothers Tool Company.)*

Two forming rolls

(a)   (b)

DIMENSIONAL ACCURACY

Dimensional accuracy in turning operations is controlled by many factors. Precision is influenced by deflection due to the cutting forces and surface roughness. Tool wear causes the workpiece dimension to change from the initial diameter when the tool is sharp to the diameter obtained after the tool has worn. The cutting forces increase as the tool wears, resulting in increased deflection between the workpiece and the cutting tool. A built-up edge (BUE) may form at the tip of the cutting tool. A BUE has the tendency to change the actual diameter of the workpiece. Thus, to hold close tolerances, the size of the wear land, the magnitude of the radial (thrust) force, and the elimination of the BUE should be taken into account. See Figure 22-7.

Dimensional accuracy will also be influenced by the workpiece shape, the material, the rigidity of all elements, the surface finish, and vibrations. For example, holding the dimensional accuracy of a boring operation on a deep hole is a problem, due to the deflection (rigidity) of the boring bar.

Turned surfaces display characteristic turning grooves that are produced by the feed and the tool tip corner radius, as shown in Figure 22-7. The roughness resulting from feed marks from a round-nosed tool can be approximated by the formula

$$y = CR - \frac{\sqrt{CR^2 - f_r^2}}{4} \cong \frac{f_r^2}{8CR} \qquad (22\text{-}10)$$

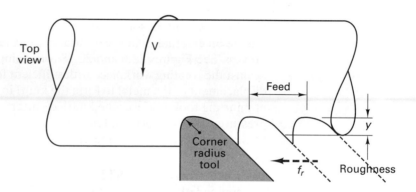

The feed and the corner radius (CR) of the cutting tool influence the surface roughness

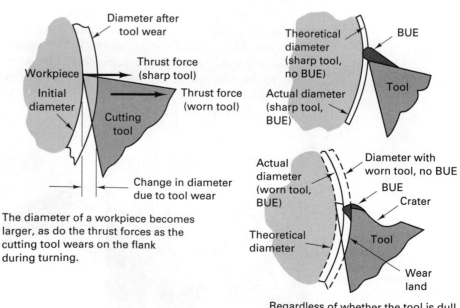

The diameter of a workpiece becomes larger, as do the thrust forces as the cutting tool wears on the flank during turning.

Regardless of whether the tool is dull or sharp, a built-up edge (BUE) causes the diameter of the workpiece to be smaller than desired.

FIGURE 22-7 Accuracy and precision in turning is a function of many factors, including tool wear and BUE.

where y is the roughness height, CR the corner radius of insert, and f_r the feed rate (in./rev). To improve the surface finish, reduce the feed and increase the corner radius.

Other factors like BUE formations, cutting-edge sharpness, and tool wear grooves in the flank wear area also affect the surface finish in turning. Flank wear and BUE can combine to affect both surface finish and accuracy, as shown in Figures 22-1 and 22-7. Wear on the corner radius may cause grooves and nicks, which produce additional surface roughness on the finish-turned surfaces. Thus, to hold the surface roughness within specified limits, minimize tool wear and use small feeds and large-corner-radius tools. To minimize BUE formation, employ cutting speeds higher than those used in rough turning operations.

■ 22.3 LATHE DESIGN AND TERMINOLOGY

Knowing the terminology of a machine tool is fundamental to understanding how it performs the basic processes, how the workholding devices are interchanged, and how the cutting tools are mounted and interfaced to the work. *Lathes* are machine tools designed primarily to do turning, facing, and boring. Very little turning is done on other types of machine tools, and none can do it with equal facility. Lathes also can do facing, drilling, and reaming, and recent designs permit milling and drilling operations using live (also called powered) spindles in multiple-tool turrets, so their versatility permits multiple operations to be done with a single setup of the workpiece. Consequently, the lathe is probably the most common machine tool, along with milling machines.

Lathes in various forms have existed for more than 2000 years, but modern lathes date from about 1797, when Henry Maudsley developed one with a leadscrew, providing controlled, mechanical feed of the tool. This ingenious Englishman also developed a change-gear system that could connect the motions of the spindle and leadscrew and thus enable threads to be cut.

LATHE DESIGN
The essential components of an *engine lathe* (Figure 22-8) are the bed, headstock assembly (which includes the spindle), tailstock assembly, carriage assembly, quick-change gearbox, and the leadscrew and feed rod. The *bed* is the base and backbone of a lathe.

FIGURE 22-8   Schematic diagram of an engine lathe, showing basic components.

The bed is usually made of well-normalized or aged gray or nodular cast iron and provides a heavy, rigid frame on which all the other basic components are mounted. Two sets of parallel, longitudinal *ways*, inner and outer, are contained on the bed. On modern lathes, the ways are surface hardened and precision machined, and care should be taken to assure that the ways are not damaged. Any inaccuracy in them usually means that the accuracy of the entire lathe is destroyed.

The *headstock*, mounted in a fixed position on the inner ways, provides powered means to rotate the work at various rpm values. Essentially, it consists of a hollow *spindle*, mounted in accurate bearings, and a set of transmission gears—similar to a truck transmission—through which the spindle can be rotated at a number of speeds. Most lathes provide from 8 to 18 choices of rpm. On modern lathes all the rpm rates can be obtained merely by moving from two to four levers or the lathe has a continuously variable spindle rpm using electrical or mechanical drives.

The accuracy of a lathe is greatly dependent on the *spindle*. It carries the workholders and is mounted in heavy bearings, usually preloaded tapered roller or ball types. The spindle has a hole extending through its length, through which long bar stock can be fed. The size of this hole is an important dimension of a lathe because it determines the maximum size of bar stock that can be machined when the materials must be fed through the spindle.

The spindle protrudes from the gearbox and contains means for mounting various types of workholding devices (chucks, face and dog plates, collets). Power is supplied to the spindle from an electric motor through a V-belt or silent-chain drive. Most modern lathes have motors of from 5 to 25 hp to provide adequate power for carbide and ceramic tools cutting hard materials at high cutting speeds.

For the classic engine lathe, the *tailstock* assembly consists, essentially, of three parts. A lower casting fits on the inner ways of the bed, can slide longitudinally, and can be clamped in any desired location. An upper casting fits on the lower one and can be moved transversely upon it, on some type of keyed ways, to permit aligning the tailstock and headstock spindles (for turning tapers). The third major component of the assembly is the *tailstock quill*. This is a hollow steel cylinder, usually about 2 to 3 in. in diameter, that can be moved longitudinally in and out of the upper casting by means of a handwheel and screw. The open end of the quill hole has a Morse taper. Cutting tools or a *lathe center* are held in the quill. A graduated scale is usually engraved on the outside of the quill to aid in controlling its motion in and out of the upper casting. A locking device permits clamping the quill in any desired position. In recent years, dual-spindle NC turning centers have emerged, where a subspindle replaces the tailstock assembly. Parts can be automatically transferred from the spindle to the subspindle for turning the back end of the part. See Chapters 31 and 36 for more on NC turning centers.

The *carriage assembly*, together with the apron, provides the means for mounting and moving cutting tools. The *carriage*, a relatively flat H-shaped casting, rides on the outer set of ways on the bed. The *cross slide* is mounted on the carriage and can be moved by means of a feed screw that is controlled by a small handwheel and a graduated dial. The cross slide thus provides a means for moving the lathe tool in the facing or cutoff direction.

On most lathes, the tool post is mounted on a *compound rest*. See Figure 22-9. The compound rest can rotate and translate with respect to the cross slide, permitting further positioning of the tool with respect to the work. The *apron*, attached to the front of the carriage, has the controls for providing manual and powered motion for the carriage and powered motion for the cross slide. The carriage is moved parallel to the ways by turning a handwheel on the front of the apron, which is geared to a pinion on the back side. This pinion engages a rack that is attached beneath the upper front edge of the bed in an inverted position.

Powered movement of the carriage and cross slide is provided by a rotating *feed rod*. The feed rod, which contains a keyway, passes through the two reversing bevel pinions (Figure 22-10) and is keyed to them. Either pinion can be activated by means of the feed reverse lever, thus providing "forward" or "reverse" power to the carriage. Suitable clutches connect either the rack pinion or the cross-slide screw to provide longitudinal motion of the carriage or transverse motion of the cross slide.

FIGURE 22-9 The cutting tools for lathe work are held in the tool post on the compound rest, which can translate and swivel.

(a)

(b)

FIGURE 22-10 The carriage, cross slide, and apron assembly for an engine lathe.

For cutting threads, a *leadscrew* is used. When a friction clutch is used to drive the carriage, motion through the leadscrew is by a direct, mechanical connection between the apron and the leadscrew. A *split nut* is closed around the leadscrew by means of a lever on the front of the apron directly driving the carriage without any slippage.

Modern lathes have *quick-change gearboxes*, driven by the spindle, that connect the feed rod and leadscrew. Thus when the spindle turns a revolution, the tool (mounted on the carriage) translates (longitudinally or transversely) a specific distance in inches—that is, inches per revolution (ipr). This revolutions per minute, *rpm* or N_s, times the feed, f_r, gives the feed rate, f, in inches per minute that the tool is moving. In this way, the calculations for turning rpm and feed in inches per revolution are "mechanically related." Typical lathes may provide as many as 48 feeds, ranging from 0.002 to 0.118 in. (0.05 to 3 mm) per revolution of the spindle, and, through the leadscrew, leads from to 92 threads per inch.

SIZE DESIGNATION OF LATHES

The size of a lathe is designated by two dimensions. The first is known as the *swing*. This is the maximum diameter of work that can be rotated on a lathe. Swing is approximately twice the distance between the line connecting the lathe centers and the nearest point on the ways. The maximum diameter of a workpiece that can be mounted between centers is somewhat less than the swing diameter because the workpiece must clear the carriage assembly as well as the ways. The second size dimension is the *maximum distance between centers*. The swing thus indicates the maximum workpiece diameter that can be turned in the lathe, while the distance between centers indicates the maximum length of workpiece that can be mounted between centers.

TYPES OF LATHES

Lathes used in manufacturing can be classified as speed, engine, toolroom, turret, automatics, tracer, and numerical control turning centers. Speed lathes usually have only a headstock, a tailstock, and a simple tool post mounted on a light bed. They ordinarily have only three or four speeds and are used primarily for wood turning, polishing, or metal spinning. Spindle speeds up to about 4000 rpm are common.

Engine lathes are the type most frequently used in manufacturing. Figure 22-1 and Figure 22-8 are examples of this type. They are heavy-duty machine tools with all the components described previously and have power drive for all tool movements except on the compound rest. They commonly range in size from 12- to 24-in. swing and from 24- to 48-in. center distances, but swings up to 50 in. and center distances up to 12 ft are not uncommon. Very large engine lathes (36- to 60-ft-long beds) are therefore capable of performing roughing cuts in iron and steel at depths of cut of $1/2$ to 2 in. and at cutting speeds at 50 to 200 sfpm with WC tools run at 0.010 to 0.100 in./rev. To perform such heavy cuts requires rigidity in the machine tool, the cutting tools, the workholder, and the workpiece (using steady rests and other supports) and large horsepower (50 to 100 hp).

Most engine lathes are equipped with chip pans and a built-in coolant circulating system. Smaller engine lathes, with swings usually not over 13 in., also are available in *bench type*, designed for the bed to be mounted on a bench or table.

Toolroom lathes have somewhat greater accuracy and, usually, a wider range of speeds and feeds than ordinary engine lathes. Designed to have greater versatility to meet the requirements of tool and die work, they often have a continuously variable spindle speed range and shorter beds than ordinary engine lathes of comparable swing, since they are generally used for machining relatively small parts. They may be either bench or pedestal type.

Several types of special-purpose lathes are made to accommodate specific types of work. On a *gap-bed lathe*, for example, a section of the bed, adjacent to the headstock, can be removed to permit work of unusually large diameter to be swung. Another example is the *wheel lathe*, which is designed to permit the turning of railroad-car wheel-and-axle assemblies.

Figure 22-11 shows a CNC vertical turning lathe. Vertical lathes are an excellent alternative to large horizontal CNC lathes. Gravity-aided seating of large/heavy work-

FIGURE 22-11 This CNC vertical turning center is used for turning large circular parts rotated under vertically mounted tools.

pieces allows a high degree of process repeatability. A smaller footprint, lower initial cost, and increased productivity are all advantages when compared to traditional horizontal lathes.

Although engine lathes are versatile and very useful, the time required for changing and setting tools and for making measurements on the workpiece is often a large percentage of the cycle time. Often, the actual chip-production time is less than 30% of the total cycle time. Methods to reduce setup and tool-change time are now well known, reducing setups to minutes and unload/load steps to seconds. The placement of single-cycle machinetools into interim or lean manufacturing cells will increase the productivity of the workers because they can run more than one machine. Turret lathes, screw machines, and other types of semiautomatic and automatic lathes have been highly developed and are widely used in manufacturing as another means to improve cutting productivity.

Turret Lathes   The basic components of a *turret lathe* are depicted in Figure 22-12. Basically, a longitudinally feedable, hexagon turret replaces the tailstock. The turret, on which six tools can be mounted, can be rotated about a vertical axis to bring each tool into operating position, and the entire unit can be translated parallel to the ways, either manually or by power, to provide feed for the tools. When the turret assembly is backed away from the spindle by means of a capstan wheel, the turret indexes automatically at the end of its movement, thus bringing each of the six tools into operating position in sequence.

The square turret on the cross slide can be rotated manually about a vertical axis to bring each of the four tools into operating position. On most machines, the turret can be moved transversely, either manually or by power, by means of the cross slide, and longitudinally through power or manual operation of the carriage. In most cases, a fixed tool holder also is added to the back end of the cross slide; this often carries a parting tool.

Through these basic features of a turret lathe, a number of tools can be set up on the machine and then quickly be brought successively into working position so that a complete part can be machined without the necessity for further adjusting, changing tools, or making measurements.

The two basic types of turret lathes are the ram-type turret lathe and the saddle-type turret lathe. In the *ram-type turret lathe*, the ram and turret are moved up to the cutting position by means of the capstan wheel, and the power feed is then engaged. As the ram is moved toward the headstock, the turret is automatically locked into position so that rigid tool support is obtained. Rotary stopscrews control the forward travel of

FIGURE 22-12 Block diagrams of ram- and saddle-turret lathe.

the ram, one stop being provided for each face on the turret. The proper stop is brought into operating position automatically when the turret is indexed. A similar set of stops is usually provided to limit movement of the cross slide. The *saddle-type turret lathe* provides a more rugged mounting for the hexagon turret than can be obtained by the ram-type mounting. In *saddle-type lathes*, the main turret is mounted directly on the saddle, and the entire saddle and turret assembly reciprocates. Larger turret lathes usually have this type of mounting. However, because the saddle-turret assembly is rather heavy, this type of mounting provides less rapid turret reciprocation. When such lathes are used with heavy tooling for making heavy or multiple cuts, a *pilot arm* attached to the headstock engages a pilot hole attached to one or more faces of the turret to give additional rigidity. Turret-lathe headstocks can shift rapidly between spindle speeds and brake rapidly to stop the spindle very quickly. They also have automatic stock-feeding for feeding bar stock through the spindle hole. If the work is to be held in a chuck, some type of air-operated chuck or a special clamping fixture is often employed to reduce the time required for part loading and unloading.

Single-Spindle Automatic Screw Machines. There are two common types of *single-spindle screw machines*. One, an American development and commonly called the turret type (Brown & Sharpe), is shown in Figure 22-13. The other is of Swiss origin and is referred to as the Swiss type. The *Brown & Sharpe screw machine* is essentially a small automatic turret lathe, designed for bar stock, with the main turret mounted in a vertical plane on a ram. Front and rear tool holders can be mounted on the cross slide. All motions of the turret, cross slide, spindle, chuck, and stock-feed mechanism are controlled

FIGURE 22-13   On the turret-type single-spindle automatic, the tools must take turns to make cuts.

by cams. The turret cam is essentially a program that defines the movement of the turret during a cycle. These machines are usually equipped with an automatic rod-feeding magazine that feeds a new length of bar stock into the collet (the workholding device) as soon as one rod is completely used.

Often, screw machines of the Brown & Sharpe type are equipped with a transfer or "picking" attachment. This device picks up the workpiece from the spindle as it is cut off and carries it to a position where secondary operation is performed by a small, auxiliary power head. In this manner screwdriver slots are put in screw heads, small flats are milled parallel with the axis of the workpiece, or holes are drilled normal to the axis.

On the *Swiss-type automatic screw machine*, the cutting tools are held and moved in radial slides (Figure 22-14). Disk cams move the tools into cutting position and provide feed into the work in a radial direction only; they provide any required longitudinal feed by reciprocating the headstock.

FIGURE 22-14   Close-up view of a Swiss-type screw machine, showing the tooling and radial tool sides, actuated by rocker arms controlled by a disk cam, shown in lower left. *(Courtesy of George Gorton Machine Corporation.)*

Most machining on Swiss-type screw machines is done with single-point cutting tools. Because they are located close to the spindle collet, the workpiece is not subjected to much deflection. Consequently, these machines are particularly well suited for machining very small parts.

Both types of single-spindle screw machines can produce work to close tolerances, the Swiss-type probably being somewhat superior for very small work. Tolerances of 0.0002 to 0.0005 in. are not uncommon. The time required for setting up the machine is usually an hour or two and can be much less. One person can tend many machines, once they are properly tooled. They have short cycle times, frequently less than 30 seconds per piece.

Multiple-Spindle Automatic Screw Machines. Single-spindle screw machines utilize only one or two tooling positions at any given time. Thus the total cycle time per workpiece is the sum of the individual machining and tool-positioning times. On *multiple-spindle screw machines*, sufficient spindles, usually four, six, or eight, are provided so that many tools can cut simultaneously. Thus, the cycle time per piece is equal to the maximum cutting time of a single tool position plus the time required to index the spindles from one position to the next.

The two distinctive features of multiple-spindle screw machines are shown in Figure 22-15. First, the six spindles are carried in a rotatable drum that indexes in order to bring each spindle into a different working position. Second, a nonrotating tool slide contains the same number of tool holders as there are spindles and thus provides and positions a cutting tool (or tools) for each spindle. Tools are fed by longitudinal reciprocating motion. Most machines have a cross slide at each spindle position so that an additional tool can be fed from the side for facing, grooving, knurling, beveling, and cutoff operations. These slides are also shown in Figure 22-15. All motions are controlled automatically.

Starting with the sixth position, follow the sequence of processing steps on the tooling sheet for making a part shown in Figure 22-15. With a tool position available on the end tool slide for each spindle (except for a stock-feed stop at position 6), when the slide moves forward, these tools cut essentially simultaneously. At the same time, the tools in the cross slides move inward and make their cuts. When the forward cutting motion of the end tool slide is completed, it moves away from the work, accompanied by the outward movement of the radial slides. The spindles are indexed one position, by rotation of the spindle carrier, to position each part for the next operation to be performed. At spindle position 5, finished pieces are cut off. Bar stock 1 in. in diameter is fed to correct length for the beginning of the next operation. Thus a piece is completed each time the tool slide moves forward and back. Multiple-spindle screw machines are made in a considerable range of sizes, determined by the diameter of the stock that can be accommodated in the spindles. There may be four, five, six, or eight spindles. The operating cycle of the end tool slide is determined by the operation that requires the longest time.

Once a multiple-spindle screw machine is set up, it requires only that the bar stock feed rack be supplied and the finished products checked periodically to make sure that they are within desired tolerances. One operator usually services many machines.

Most multiple-spindle screw machines use cams to control the motions. Setting up the cams and the tooling for a given job may require from 2 to 20 hours. However, once such a machine is set up, the processing time per part is very short. Often, a piece may be completed every 10 seconds. Typically, a minimum of 2000 to 5000 parts are required in a lot to justify setting up and tooling a multiple-spindle automatic screw machine. The precision of multiple-spindle screw machines is good, but seldom as good as that of single-spindle machines. However, tolerances from 0.0005 to 0.001 in. on the diameter are typical.

■ 22.4 CUTTING TOOLS FOR LATHES

LATHE CUTTING TOOLS

Most lathe operations are done using single-point *cutting tools*, such as the classic tool designs shown in Figure 22-16. On right-hand turning (and left-hand turning) and facing tools, the cutting usually takes place on the side of the tool; therefore, the side rake

All spindles on multiple-spindle automatic have the same tool path

The six spindle automatic

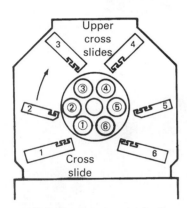

Spindle arrangement for 6-spindle automatic. The barstock is usually fed to a stop at position 6. The cutoff position is the one preceding the bar feed position.

Tooling sheet for making a part on a six-spindle.

FIGURE 22-15 The multiple-spindle, automatic screw machine makes all cuts simultaneously and then performs the noncutting functions (tool withdrawal, index, bar feed) at high speed.

FIGURE 22-16 Common types of forged tool holders: (a) right-hand turning, (b) facing, (c) grooving cutoff, (d) boring, (e) threading. *(Courtesy of Armstrong Brothers Tool Company.)*

angle is of primary importance, particularly when deep cuts are being made. On the round-nose turning tools, cutoff tools, finishing tools, and some threading tools, cutting takes place on or near the tip of the tool, and the back rake is therefore of importance. Such tools are used with relatively light depths of cut.

Because tool materials are expensive, it is desirable to use as little as possible. At the same time, it is essential that the cutting tool be supported in a strong, rigid manner to minimize deflection and possible vibration. Consequently, lathe tools are supported in various types of heavy, forged steel tool holders, as shown in Figure 22-16. The high-speed steel (HSS) tool bit should be clamped in the tool holder with minimum overhang; otherwise tool chatter and a poor surface finish may result.

In the use of carbide, ceramic, or coated carbides for higher speed cutting, throw-away inserts are used that can be purchased in a great variety of shapes, geometrics (nose radius, tool angles, and groove geometry), and sizes (see Figure 22-17 for some examples).

Insert shape	Available cutting edges	Typical insert holder
Round	4–10 on a side 8–20 total	15° Square insert
80°/100° diamond	4 on a side 8 total	
Square	4 on a side 8 total	0° Triangular insert
Triangle	3 on a side 6 total	
55° diamond	2 on a side 4 total	35° diamond
35° diamond	2 on a side 4 total	5°

FIGURE 22-17   Typical insert shapes, available cutting edges per insert, and insert holders for throwaway insert cutting tools. *(Adapted from* Turning Handbook of High Efficiency Metal Cutting, *courtesy of General Electric Company.)*

FIGURE 22-18   Quick-change tool post and accompanying toolholders. *(Courtesy of Armstrong Brothers Tool Company.)*

QUICK-CHANGE TOOL POST

TURNING, FACING, AND BORING TOOL HOLDER
V-Slot holds round boring bars as well as square tool bits

TURNING AND FACING TOOL HOLDER
Takes turning and facing tool bits

KNURLING TOOL HOLDER
Revolving head, self-centering.
3 pairs of knurls

When several different operations on a lathe are performed repeatedly in sequence, the time required for changing and setting tools may constitute as much as 50% of the total cycle time. Quick-change tool holders (Figure 22-18) are used to reduce manual tool-changing time. The individual tools, preset in their holders, can be interchanged in the special tool post in a few seconds. With some systems, a second tool may be set in the tool post while a cut is being made with the first tool and can then be brought into proper position by rotating the post.

In lathe work, the nose of the tool should be set exactly at the same height as the axis of rotation of the work. However, because any setting below the axis causes the work to tend to "climb" up on the tool, most machinists set their tools a few thousandths of an inch above the axis, except for cutoff, threading, and some facing operations.

Form turning

FIGURE 22-19 Circular and block types of form tools. *(Courtesy of Speedi Tool Company, Incorporated.)*

FORM TOOLS

In Figure 22-15, the use of form tools was shown in automatic lathe work. Form tools are made by grinding the inverse of the desired work contour into a block of HSS or tool steel. A threading tool is often a form tool. Although form tools are relatively expensive to manufacture, it is possible to machine a fairly complex surface with a single inward feeding of one tool. For mass-production work, adjustable form tools of either flat or rotary types, such as are shown in Figure 22-19, are used. These are expensive to make initially but can be resharpened by merely grinding a small amount off the face and then raising or rotating the cutting edge to the correct position.

The use of form tools is limited by the difficulty of grinding adequate rake angles for all points along the cutting edge. A rigid setup is needed to resist the large cutting forces that develop with these tools. Light feeds with sharp, coated HSS tools are used on multiple-spindle automatics, turret lathes, and transfer line machines.

TURRET-LATHE TOOLS

In turret lathes, the work is generally held in collets and the correct amount of bar stock is fed into the machine to make one part. The tools are arranged in sequence at the tool stations with depths of cut all preset. The following factors should be considered when setting up a turret lathe.

1. *Setup time*: time required to install and set the tooling and set the stops. Standard tool holders and tools should be used as much as possible to minimize setup time. Setup time can be greatly reduced by eliminating adjustment in the setup.

2. *Workholding time:* time to load and unload parts and/or stock.

3. *Machine-controlling time:* time required to manipulate the turrets. Can be reduced by combining operations where possible. Dependent on the sequence of operations established by the design of the setup.

4. *Cutting time:* time during which chips are being produced. Should be as short as is economically practical and represent the greatest percentage of the total cycle time possible.

5. *Cost:* cost of the tool, setup labor cost, lathe operator labor cost, and the number of pieces to be made.

There are essentially eleven tooling stations, as shown in Figure 22-20, with six in the turret, four in the indexable tool post, and one in the rear tool post. The tooling is more rugged in turret lathes because heavy, simultaneous cuts are often made. Tools mounted in the hex turret that are used for turning are often equipped with pressure rollers set on the opposite side of the rotating workpiece from the tool to counter the cutting forces.

Turret lathes are most economical in producing lots too large for engine lathes but too small for automatic screw machines or automatic lathes. In recent years much of this work has been assumed by numerical control lathes or turning centers. For example, the component (threaded shaft) shown in Figure 22-20 could have also been made on an NC turret lathe with some savings in cycle time.

FIGURE 22-20 Turret-lathe tooling setup for producing part shown. Numbers in circles indicate the sequence of the operations from 1 to 9. The letters A through F refer to the surfaces being machined. Operation 3 is a combined operation. The roll turner is turning surface F while tool 3 on the square post is turning surface B. The first operation stops the stock at the right length. The last operation cuts the finished bar off and puts a chamfer on the bar, which will next be advanced to the stock stop.

■ 22.5 WORKHOLDING IN LATHES

WORKHOLDING DEVICES FOR LATHES

Five methods are commonly used for supporting workpieces in lathes:

1. Held between centers

2. Held in a chuck

3. Held in a collet

4. Mounted on a faceplate

5. Mounted on the carriage

In the first four of these methods, the workpiece is rotated during machining. In the fifth method, which is not used extensively, the tool rotates while the workpiece is fed into the tool.

A general discussion of workholding devices is found in Chapter 25, and the student involved in designing working devices should study the reference materials under *Tool Design*. For lathes, workholding is a matter of selecting from standard tooling.

LATHE CENTERS

Workpieces that are relatively long with respect to their diameters are usually machined between centers. See Figure 22-21. Two *lathe centers* are used, one in the spindle hole and the other in the hole in the tailstock quill. Two types are used, called *dead* and *live*. Dead centers are *solid*, that is, made of hardened steel with a Morse taper on one end so that it will fit into the spindle hole. The other end is ground to a taper. Sometimes the tip of this taper is made of tungsten carbide to provide better wear resistance. Before a center

FIGURE 22-21 Work being turned between centers in a lathe, showing the use of a dog and dog plate. *(Courtesy of South Bend Lathe.)*

FIGURE 22-22 Live lathe center can rotate with the part.

Plain solid mandrel

Gang mandrel

Cone mandrel

FIGURE 22-23 Three types of mandrels, which are mounted between centers for lathe work.

is placed in position, the spindle hole should be carefully wiped clean. The presence of foreign material will prevent the center from seating properly, and it will not be aligned accurately.

Live centers are shown in Figure 22-22. A mechanical connection must be provided between the spindle and the workpiece to provide rotation. This is accomplished by a *lathe dog* and *dog plate*. The dog is clamped to the work. The *tail* of the dog enters a slot in the dog plate, which is attached to the lathe spindle in the same manner as a lathe chuck. For work that has a finished surface, a piece of soft metal, such as copper or aluminum, can be placed between the work and the dog setscrew clamp to avoid marring. Live centers are designed so that the end that fits into the workpiece is mounted on ball or roller bearings. It is free to rotate. No lubrication is required. Live centers may not be as accurate as the solid type and therefore are not often used for precision work.

Before a workpiece can be mounted between lathe centers, a center hole must be drilled in each end. This is typically done in a drill press or on the lathe with a tool held in the rear turret. A combination center drill and countersink ordinarily is used, with care taken that the center hole is deep enough so that it will not be machined away in any facing operation and yet is not drilled to the full depth of the tapered portion of the center drill (see Chapter 23).

Because the work and the center of the headstock end rotate together, no lubricant is needed in the center hole at this end. The center in the tailstock quill does not rotate; adequate lubrication must be provided. A mixture of white lead and oil is often used. Failure to provide proper lubrication at all times will result in scoring of the workpiece center hole and the center, and inaccuracy and serious damage may occur. Live centers are often used in the tailstock to overcome these problems.

The workpiece must rotate freely, yet no looseness should exist. Looseness will usually be manifested in chattering of the workpiece during cutting. The setting of the centers should be checked after cutting for a short time. Heating and thermal expansion of the workpiece will reduce the clearances in the setup.

MANDRELS

Workpieces that must be machined on both ends or are disk-shaped are often mounted on *mandrels* for turning between centers. Three common types of mandrels are shown in Figure 22-23. *Solid mandrels* usually vary from 4 to 12 in. in length and are accurately ground with a 1:2000 taper (0.006 in./ft). After the workpiece is drilled and/or bored, it is pressed on the mandrel. The mandrel should be mounted between centers so that the cutting force tends to tighten the work on the mandrel taper. Solid mandrels permit the work to be machined on both ends as well as on the cylindrical surface. They are available in stock sizes but can be made to any desired size.

Gang (or disk) mandrels are used for production work because the workpieces do not have to be pressed on and thus can be put in position and removed more rapidly. However, only the cylindrical surface of the workpiece can be machined when this type of mandrel is used. *Cone mandrels* have the advantage that they can be used to center workpieces having a range of hole sizes.

LATHE CHUCKS

Lathe chucks are used to support a wider variety of workpiece shapes and to permit more operations to be performed than can be accomplished when the work is held between centers. Two basic types of chucks are used (Figure 22-24).

Three-jaw self-centering chucks are used for work that has a round or hexagonal cross section. The three jaws are moved inward or outward simultaneously by the rotation of a spiral cam, which is operated by means of a special wrench through a bevel gear. If they are not abused, these chucks will provide automatic centering to within about 0.001 in. However, they can be damaged through use and will then be considerably less accurate.

Each jaw in a *four-jaw independent chuck* can be moved inward and outward independent of the others by means of a chuck wrench. Thus they can be used to support a wide variety of work shapes. A series of concentric circles engraved on the chuck face aid in adjusting the jaws to fit a given workpiece. Four-jaw chucks are heavier and more rugged than the three-jaw type, and because undue pressure on one jaw does not destroy the accuracy of the chuck, they should be used for all heavy work. The jaws on both three- and four-jaw chucks can be reversed to facilitate gripping either the inside or the outside of workpieces.

Combination four-jaw chucks are available in which each jaw can be moved independently or can be moved simultaneously by means of a spiral cam. Two-jaw chucks are also available. For mass-production work, special chucks are often used in which the jaws are actuated by air or hydraulic pressure, permitting very rapid clamping of

3-jaw
self-centering

Actuates
opening
and closing

Removable jaws

4-jaw
with each jaw
independent

FIGURE 22-24 The jaws on chucks for lathes (four-jaw independent or three-jaw self-centering) can be removed and reversed.

FIGURE 22-25 Hydraulically actuated through-hole three-jaw power chuck shown in section view to left and in the spindle of the lathe above connected to the actuator.

the work. See Figure 22-25 for a schematic. The rapid exchange of tooling is a key manufacturing strategy in manufacturing cells. Chuck jaw sets are dedicated and customized for specific parts. The first time a chuck jaw set is used, each jaw is marked with the number of the jaw slot where it was installed and an index mark that corresponds with the alignment of the jaw serrations and the first tooth on the chuck master jaw. The jaws can now be reinstalled on the chuck exactly where they were bored. The adjustability of the chuck body lets the operator dial in part concentrically without resetting the jaw.

COLLETS
Collets are used to hold smooth cold-rolled bar stock or machined workpieces more accurately than with regular chucks. As shown in Figure 22-26, collets are relatively thin tubular steel bushings that are split into three longitudinal segments over about two-thirds of their length. At the split end, the smooth internal surface is shaped to fit the piece of stock that is to be held. The external surface of the collet is a taper that mates with an internal taper of a collet sleeve that fits into the lathe spindle. When the collet is pulled inward into the spindle (by means of the draw bar), the action of the two mating tapers squeezes the collet segments together, causing them to grip the workpiece.

Collets are made to fit a variety of symmetrical shapes. If the stock surface is smooth and accurate, good collets will provide very accurate centering, with runout less than 0.0005 in. However, the work should be no more than 0.002 in. larger or 0.005 in. smaller than the nominal size of the collet. Consequently, collets are used only on drill-rod, cold-rolled, extruded, or previously machined stock.

Collets that can open automatically and feed bar stock forward to a stop mechanism are commonly used on automatic lathes and turret lathes. An example of a collet chuck is shown in Figure 22-27. Another type of collet similar to a Jacobs drill chuck has a greater size range than ordinary collets; therefore, fewer are required.

Round collet Square collet Hexagon collet Cutaway view of collet

FIGURE 22-26 Several types of lathe collets. *(Courtesy of South Bend Lathe.)*

Workpiece

Collet

Sleeve

Draw bar

Spindle

Workpiece

Force indicator pins

Mechanical grip

Full spindle bore

Standard collet pads

FIGURE 22-27 (a) Method of using a draw-in collet in lathe spindle. *(Courtesy of South Bend Lathe.)* (b) Schematic of a collet chuck in which the clamping force can be adjusted.

FIGURE 22-28 Cutting a thread on a long, slender workpiece, using a follow rest (left) and a steady rest (right) on an engine lathe. Note the use of a dog and face plate to drive the workpiece. *(Courtesy of South Bend Lathe.)*

FACEPLATES

Faceplates are used to support irregularly shaped work that cannot be gripped easily in chucks or collets. The work can be bolted or clamped directly on the faceplate or can be supported on an auxiliary fixture that is attached to the faceplate. The latter procedure is time saving when identical pieces are to be machined.

MOUNTING WORK ON THE CARRIAGE

When no other means is available, boring occasionally is done on a lathe by mounting the work on the carriage, with the boring bar mounted between centers and driven by means of a dog.

STEADY AND FOLLOW RESTS

If one attempts to turn a long, slender piece between centers, the radial force exerted by the cutting tool or the weight of the workpiece itself may cause it to be deflected out of line. Steady rests and follow rests (Figure 22-28) provide a means for supporting such work between the headstock and the tailstock. The steady rest is clamped to the lathe ways and has three movable fingers that are adjusted to contact the work and align it. A light cut should be taken before adjusting the fingers to provide a smooth contact-surface area.

A steady rest also can be used in place of the tailstock as a means of supporting the end of long pieces, pieces having too large an internal hole to permit using a regular dead center, or work where the end must be open for boring. In such cases the headstock end of the work must be held in a chuck to prevent longitudinal movement. Tool feed should be toward the headstock.

The follow rest is bolted to the lathe carriage. It has two contact fingers that are adjusted to bear against the workpiece, opposite the cutting tool, in order to prevent the work from being deflected away from the cutting tool by the cutting forces.

■ Key Words

apron	depth of cut	headstock	reaming
automatic lathe	dog	knurling	screw machine
bed	drilling	lathe centers	steady rest
boring	engine lathe	mandrels	tailstock
carriage	faceplates	metal removal rate	taper turning
chucks	facing	milling	turning
collets	feed	parting	turret lathe
cutoff	follow rest	quill	workholding
cutting tools			

■ Review Questions

1. How is the tool–work relationship in turning different from that in facing?
2. What different kinds of surfaces can be produced by turning versus facing?
3. How does form turning differ from ordinary turning?
4. What is the basic difference between facing and a cutoff operation?
5. Which machining operations shown in Figure 22-2 do not form a chip?
6. Why is it difficult to make heavy cuts if a form turning tool is complex in shape?
7. Show how equation 22-5 is an approximate equation.
8. Why is the spindle of the lathe hollow?
9. What function does a lathe carriage have?
10. Why is feed specified for a boring operation typically less than that specified for turning if the MRR equations are the same?
11. What function is provided by the leadscrew on a lathe that is not provided by the feed rod?
12. How can work be held and supported in a lathe?
13. How is a workpiece that is mounted between centers on a lathe driven (rotated)?
14. What will happen to the workpiece when turned, if held between centers, and the centers are not exactly in line?
15. Why is it not advisable to hold hot-rolled steel stock in a collet?
16. How does a steady rest differ from a follow rest?
17. What are the advantages and disadvantages of a four-jaw independent chuck versus a three-jaw chuck?
18. Why should the distance a lathe tool projects from the tool holder be minimized?
19. What is the difference between a ram- and a saddle-turret lathe?
20. How can a tapered part be turned on a lathe?
21. Why might it be desirable to use a heavy depth of cut and a light feed at a given speed in turning rather than the opposite?
22. If the rpm for a facing cut (assuming given work and tool materials) is being held constant, what is happening during the cut to the speed? To the feed?
23. Why is it usually necessary to take relatively light feeds and depths of cut when boring on a lathe?
24. How does the corner radius of the tool influence the surface roughness?
25. What effect does a BUE have on the diameter of the workpiece in turning?
26. How does the multiple-spindle screw machine differ from the single-spindle machine?
27. Why does boring ensure concentricity between the hole axis and the axis of rotation of the workpiece (for boring tool), whereas drilling does not?
28. Why are vertical spindle machines better suited for machining large workpieces than horizontal lathes?
29. What is the principal advantage of a horizontal boring machine over a vertical boring machine for large workpieces?
30. In which figures in this chapter is a workpiece being held in a three-jaw chuck?
31. How is the workpiece in Figure 22-14 being held?
32. In which figures in this chapter is a dead center shown?
33. In which figures in this chapter is a live center shown?
34. In which figures in this chapter showing setups do you find the following being used as a workholding device?
    a. Three-jaw chuck
    b. Collet
    c. Faceplate
    d. Four-jaw chuck
35. How many form tools are being utilized in the process shown in Figure 22-15 to machine the part?
36. From the information given in Figure 22-20, start with a piece of round bar stock and show how it progresses, operation by operation, into a finished part—a threaded shaft.

■ Problems

1. A cutting speed of 100 sfpm has been selected for a turning cut. At what rpm should a 3-in.-diameter bar be rotated?
2. Assume that the workpiece in Problem 1 is 8 in. (203.2 mm) long and a feed of 0.020 in. (0.51 mm) per revolution is used. What is the machining time for a cut across its entire length? Don't forget to add an allowance.
3. If the depth of cut in Problem 2 is 0.25 in., what is the metal removal rate (MRR) exactly? What is the MRR approximately?
4. Using the same recommended cutting speeds and feeds, calculate the machining time to cut off the bar in Problem 2.
5. The following data apply for machining a part on a turret lathe and on an engine lathe:

	Engine Lathe	Turret Lathe
Times, in minutes, to machine part	30 min	5 min
Cost of special tooling	0	$300
Time to set up the machine	30 min	3 hr
Labor rates	$8/hr	$8/hr
Machine rates (overhead)	$10/hr	$12/hr

a. How many pieces would have to be made for the cost of the engine lathe to just equal the cost of the turret lathe? This is the BEQ.
b. What is the cost per unit at the BEQ?

6. A finish cut for a length of 10 in. on a diameter of 3 in. is to be taken in 1020 steel with a speed of 100 fpm and a feed of 0.005 ipr. What is the machining time?
7. A workpiece 10 in. in diameter is to be faced down to a diameter of 2 in. on the right end. The CNC lathe (see Chapter 26) controls the spindle speed and maintains the cutting speed at 100 fpm throughout the cut by changing the rpm. What should be the time for the cut? Now suppose the spindle rpm for the workpiece is set to give a speed of 100 fpm for the 10-in. diameter and is not changed during the cut. What is the machining time for the cut now? The feed rate is 0.005 ipr.
8. A hole 89 mm in diameter is to be drilled and bored through a piece of 1340 steel that is 200 mm long, using a horizontal boring, drilling, and milling machine. High-speed tools will be used. The sequence of operations will be center drilling; drilling with an 18-mm drill followed by a 76-mm drill; then boring to size in one cut, using a feed of 0.50 mm/rev. Drilling feeds will be 0.25 mm/rev for the smaller drill and 0.64 mm/rev for the

larger drill. The center drilling operation requires 0.5 min. To set or change any given tool and set the proper machine speed and feed requires 1 min. Select the initial cutting speeds, and compute the total time required for doing the job. (Neglect setup time for the fixture.) This is often referred to as the run time or the cycle time. (*Hint:* Check in Chapter 21 for recommended speeds for turning.)

9. Figure 22-A shows the fixed and variable costs for a part being produced on an engine lathe. Figure 22-B has three plots of unit production cost ($/unit) versus production volume ($Q =$ build quantity). (Note that this plot is made on log-log paper.) Cost per unit for a particular process decreases with increased volume as fixed costs are spread out over more units. For a particular process there is no minimum cost but rather production volumes within which particular processes are most economical.

FIGURE 22-A

a. Each of these curves is a plot of the equation for total cost divided by quantity, which means each is the sum of the fixed cost per unit (mainly setup and tooling costs) and the variable costs per unit (direct labor, direct material). From the data on the plots, estimate the fixed costs for the engine lathe, the NC lathe, and the single-spindle automatic.
b. For what build quantities is the NC lathe most economical (approximately)?
c. What cost per unit does the NC lathe approach as the build quantity becomes very large?
d. Explain how to go from a cost vs. quantity plot to a cost/unit vs. quantity plot?
e. What happens to these plots if you plot them on regular Cartesian coordinates? Try it and comment on what you find.
f. Many Japanese manufacturers have found innovative ways to eliminate setup time in many of their processes. What is the impact of this on these kinds of plots, on cost per unit economics, and on job shop inventories?

10. The derivation of the approximate equation 22-5 for the MRR for turning process requires an assumption regarding the diameters of the parts being turned. Determine the error in the equation for Problem 3. What is the assumption?

FIGURE 22-B

Chapter 22 CASE STUDY

Estimating the Machining Time for Turning

As the plant manufacturing engineer at **BRC**, Inc., Jay Strom has been called into the production department to provide an expert opinion on a machining problem. Unfortunately, the only tool or instrument you have available at the time is a 1-inch micrometer.

Katrin Zachary, the production manager, would like to know the minimum time required to machine a large forging. The 8-foot-long forging is to be turned down from an original diameter of 10 in. to a final diameter of 6 in. The forging has a BHN of 300 to 400. The turning is to be performed on a heavy-duty lathe, which is equipped with a 50 HP motor and a continuously variable speed drive on the spindle. The work will be held between centers, and the overall efficiency of the lathe has been determined to be 75%. See **Chapter 21**.

The forging (or log) is made from medium-carbon, 4345 alloy steel. The steel manufacturer, some basic experimentation, and established knowledge of the product and its manufacture have provided the following information:

1. A tool-life equation developed for the most suitable type of tool material at a feed of 0.020 ipr and a rake angle of $\alpha = 10$ degrees. The equation $VT^n = C$ generally fits the data, with V = cutting speed and T = the time in minutes to tool failure. Two test cuts were run, one at $V = 60$ sfpm where $T = 100$ minutes and another at $V = 85$ sfpm, where $T = 10$ minutes.

2. According to the vendor, the dynamic shear strength of the material is on the order of 125,000 psi.

3. Jay decides to make two test cuts at the standard feed of 0.020 ipr. He assumes that the chip thickness ratio varies almost linearly between the speeds of 20 and 80 fpm, the values being 0.4 at the speed of 20 fpm and 0.6 at 80 fpm. The chip thickness values were determined by micrometer measurement in order to determine the value of r_C.

4. The machined forging (log) will be used as a roller in a newspaper press and must be precisely machined. If the log deflects during the cutting more than 0.005 in., the roll will end up barrel-shaped after final grinding and polishing.

How should Jay proceed to estimate the minimum time required to machine this forging, assuming that one finishing pass will be needed when the log has been reduced to 6 in. in diameter? The deflection due to cutting forces must be kept below 0.005 in. at the mid-log location.

You can assume that $F_C \times 0.5 = F_f$ and $F_f \times 0.5 = F_R$ and that F_R causes the deflection.

DRILLING AND RELATED HOLE-MAKING PROCESSES

■ 23.1 INTRODUCTION

In manufacturing it is probable that more holes are produced than any other shape, and a large proportion of these are made by *drilling*. Of all the machining processes performed, drilling makes up about 25%. Consequently, drilling is a very important process. Although drilling appears to be a relatively simple process, it is really a complex process. Most drilling is done with a tool having two cutting edges, or *lips*, as shown in Figure 23-1. This is a twist drill, the most common drill geometry. The cutting edges are at the end of a relatively flexible tool. Cutting action takes place inside the workpiece. The only exit for the chips is the hole that is mostly filled by the drill. Friction between the margin and the hole wall produces heat that is additional to that due to chip formation. The counterflow of the chips in the flutes makes lubrication and cooling difficult. There are four major actions taking place at the point of a drill.

1. A small hole is formed by the web—chips are not cut here in the normal sense.

2. Chips are formed by the rotating lips.

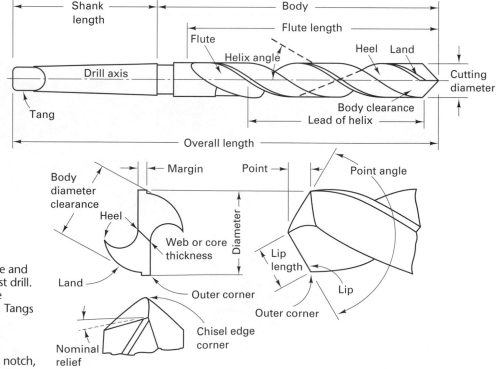

FIGURE 23-1 Nomenclature and geometry of conventional twist drill. Shank style depends upon the method used to hold the drill. Tangs or notches prevent slippage:
(a) straight shank with tang,
(b) tapered shank with tang,
(c) straight shank with whistle notch,
(d) straight shank with flat notch.

628

3. Chips are removed from the hole by the screw action of the helical flutes.

4. The drill is guided by lands or margins that rub against the walls of the hole.

In recent years, new drill-point geometries and TiN coatings have resulted in improved hole accuracy, longer life, self-centering action, and increased feed-rate capabilities. However, the great majority of drills manufactured are twist drills. One estimate has U.S. manufacturing companies consuming 250 million twist drills per year.

When high-speed steel (HSS) drills wear out, the drill can be reground to restore its original geometry. If regrinding is not done properly, the original drill geometry may be lost, and so will drill accuracy and precision. Drill performance also depends on the drilling machine tool, the workholding device, the drill holder, and the surface of the workpiece. Poor surface conditions (sand pockets and/or chilled hard spots on castings, or hard oxide scale on hot-rolled metal) can accelerate early tool failure and degrade the hole-drilling process.

■ 23.2 FUNDAMENTALS OF THE DRILLING PROCESS

The process of drilling creates two chips. A conventional two-flute drill, with drill of diameter D, has two principal cutting edges rotating at an rpm rate of N and feeding axially. The rpm of the drill is established by the selected cutting velocity or cutting speed

$$N_s = \frac{12V}{\pi D} \qquad (23\text{-}1)$$

with V in surface feet per minute and D in inches (mm). This equation assumes that V is the cutting speed at the outer corner of the cutting lip (point X in Figure 23-2).

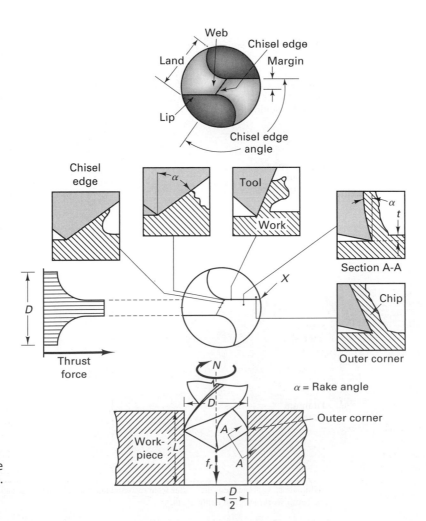

FIGURE 23-2   Conventional drill geometry viewed from the point showing how the rake angle varies from the chisel edge to the outer corner along the lip. The thrust force increases as the web is approached.

630   CHAPTER 23   Drilling and Related Hole-Making Processes

The feed, f_r, is given in inches per revolution. The depth of cut in drilling is equal to half the feed rate, or $t = f_r/2$ (see section A–A in Figure 23-2). The feed rate in inches per minute, f_m, is $f_r N_s$. The length of cut in drilling equals the depth of the hole, L, plus an allowance for approach and for the tip of drill, usually $A = D/2$. In drilling, the speed and feed depend upon the material being machined, the cutting tool material, and the size of drill. Table 23-1 gives some typical values for V and for carbide indexable insert drills, a type of drill shown later in the chapter.

After selecting the cutting speed and feed for drilling the hole, the rpm value of the spindle of the machine is determined from equation 23-1, the maximum velocity occurring at the extreme ends of the drill lips. The velocity is very small near the center of the chisel end of the drill. For drilling, cutting time is

$$T_m = \frac{(L + A)}{f_r N_s} = \frac{L + A}{f_m} \tag{23-2}$$

TABLE 23-1   Recommended Speeds and Feeds for Indexable-Insert Drills

Material Group	Size Range	Cutting Speed sfpm	Feed Rate (in./rev.)[a]
Cast iron: modular, ductile, or malleable	$\frac{13}{16}$–$1\frac{1}{8}$	165–300	0.004–0.008
	1–$1\frac{3}{8}$	165–300	0.005–0.010
	$1\frac{1}{4}$–$1\frac{5}{8}$	165–300	0.006–0.012
	$1\frac{1}{2}$–$2\frac{1}{2}$	165–300	0.008–0.014
	$2\frac{3}{8}$–$3\frac{1}{2}$	165–300	0.010–0.015
1000 Series steels like 1081, 1020, etc.	$\frac{13}{16}$–$1\frac{1}{8}$	300–400	0.003–0.005
	1–$1\frac{3}{8}$	350–450	0.003–0.006
	$1\frac{1}{4}$–$1\frac{5}{8}$	400–550	0.004–0.007
	$1\frac{1}{2}$–$2\frac{1}{2}$	450–600	0.004–0.007
	$2\frac{3}{8}$–$3\frac{1}{2}$	500–700	0.005–0.009
Low-carbon unalloyed case-hardening steels	$\frac{13}{16}$–$1\frac{1}{8}$	200–300	0.003–0.005
	1–$1\frac{3}{8}$	250–350	0.004–0.006
	$1\frac{1}{4}$–$1\frac{5}{8}$	300–425	0.005–0.007
	$1\frac{1}{2}$–$2\frac{1}{2}$	330–490	0.005–0.008
	$2\frac{3}{8}$–$3\frac{1}{2}$	350–550	0.006–0.010
High-carbon alloyed heat-treated steels	$\frac{13}{16}$–$1\frac{1}{8}$	200–300	0.003–0.005
	1–$1\frac{3}{8}$	250–325	0.004–0.006
	$1\frac{1}{4}$–$1\frac{5}{8}$	300–400	0.005–0.008
	$1\frac{1}{2}$–$2\frac{1}{2}$	335–450	0.005–0.008
	$2\frac{3}{8}$–$3\frac{1}{2}$	350–500	0.006–0.010
High-tensile steels	$\frac{13}{16}$–$1\frac{1}{8}$	165–250	0.004–0.005
	1–$1\frac{3}{8}$	195–300	0.004–0.006
	$1\frac{1}{4}$–$1\frac{5}{8}$	230–300	0.005–0.007
	$1\frac{1}{2}$–$2\frac{1}{2}$	265–390	0.006–0.008
	$2\frac{3}{8}$–$3\frac{1}{2}$	265–425	0.006–0.009
Stainless steels	$\frac{13}{16}$–$1\frac{1}{8}$	230–280	0.003–0.004
	1–$1\frac{3}{8}$	265–300	0.004–0.005
	$1\frac{1}{4}$–$1\frac{5}{8}$	280–345	0.004–0.005
	$1\frac{1}{2}$–$2\frac{1}{2}$	295–395	0.004–0.005
	$2\frac{3}{8}$–$3\frac{1}{2}$	300–400	0.004–0.006
Titanium steels	$\frac{13}{16}$–$1\frac{1}{8}$	100–135	0.003–0.004
	1–$1\frac{3}{8}$	100–150	0.004–0.007
	$1\frac{1}{4}$–$1\frac{5}{8}$	115–165	0.005–0.008
	$1\frac{1}{2}$–$2\frac{1}{2}$	130–175	0.006–0.009
	$2\frac{3}{8}$–$3\frac{1}{2}$	135–190	0.006–0.010

[a] Ultimate speeds and feeds may differ from recommended speeds and feeds depending on materials, rigidity of machine and setup, workpiece, and depth of cut.

Example of Drilling

A cast iron plate is 2 in. thick and needs 4-in.-diameter holes drilled in it. An indexable-insert drill has been selected. Looking at Table 23-1, we select a cutting speed of 200 fpm and a feed of 0.005 ipr.

The spindle rpm = $12V/\pi D$ = 12 × 200 / 3.14 × 1 = 764 rpm.

What if the machine does not have this specific rpm? Pick the closest value: Let's say it is 750 rpm.

The penetration rate or feed rate (in./min) = feed (ipr). The rpm = 0.005 × 750 = 3.75 in./min. The maximum chip load = feed (ipr)/2 = 0.005/2 = 0.0025in./rev.

What if the machine does not have the specific feed rate? Pick the next lowest value as a starting value. Let's say it is 3.5 in./min.

The material removal rate (in.³/min) = $(\pi/4)$ × $(D)^2$ × feed (ipr) × rpm = $(\pi/4)$ 1^2 × feed rate = 3.14/4 × 1^2 × 3.50 = 2.75 in.³/min.

The MRR can be used with the unit power for cast iron (see Chapter 20) to estimate the HP needed to drill the hole. Let HP_s = 0.33 for this Cl:

$$HP = HP_s \times MRR = 0.33 \times 2.75 = .90$$

This value would typically represent 80% of the total motor horsepower (HP) needed, so in this case, a horsepower motor greater than 1.5 or 2 would be sufficient.

In estimating the cost of a job, it is often necessary to determine the time to drill a hole.

$$\text{drill time/hole} = \frac{\text{length drilled + allowance}}{\text{feed rate (in./min)}}$$

$$+ \frac{\text{rapid traverse length of withdrawal}}{\text{rapid traverse rate}}$$

$$+ \text{prorated downtime to change drills per hole}$$

The last term prorated downtime is

$$\frac{\text{drill change downtime}}{\text{holes drilled per drill (tool life)}}$$

And the cost/hole is

$$\text{drilling time/holes} \times (\text{labor + machine rate}) + \text{prorated cost of drill/hole}$$

The metal removal rate is

$$MRR = \frac{\text{volume}}{T_m}$$

$$= \frac{\pi D^2 L/4}{L/f_r N_s} \text{(omitting allowances)} \tag{23-3}$$

which reduces to

$$MRR = (\pi D^2/4)f_r N_s \text{ in.}^3 \tag{23-4}$$

Substituting for N with equation 23-1, we obtain an approximate form

$$MRR \cong 3DVf_r \tag{23-5}$$

■ 23.3 TYPES OF DRILLS

The most common types of drills are *twist drills*. These have three basic parts: the *body*, the *point*, and the *shank*, shown in Figures 23-1 and 23-2. The body contains two or more spiral or helical grooves, called *flutes*, separated by *lands*. To reduce the friction between the drill and the hole, each land is reduced in diameter except at the leading edge, leaving a narrow *margin* of full diameter to aid in supporting and guiding the drill and thus aiding in obtaining an accurate hole. The lands terminate in the point, with the leading edge of each land forming a cutting edge. The flutes serve as channels through which the chips are withdrawn from the hole and coolant gets to the cutting edges. Although most drills have two flutes, some, as shown in Figure 23-3, have three, and some have only one.

FIGURE 23-3 Types of twist drills and shanks. Bottom to top: Straight-shank, three-flute core drill; straight-shank; taper-shank; bit-shank; straight-shank, high-helix angle; straight-shank, straight-flute; taper-shank, subland drill.

The principal rake angles behind the cutting edges are formed by the relation of the flute *helix angle* to the work. This means that the rake angle of a drill varies along the cutting edges (or lips), being negative close to the point and equal to the helix angle out at the lip. Because the helix angle is built into the twist drill, the primary rake angle cannot be changed by normal grinding. The helix angle of most drills is 24°, but drills with larger helix angles—often more than 30°—are used for materials that can be drilled very rapidly, resulting in a large volume of chips. Helix angles ranging from 0° to 20° are used for soft materials, such as plastics and copper. Straight-flute drills (zero helix and rake angles) are also used for drilling thin sheets of soft materials. It is possible to change the rake angle adjacent to the cutting edge by a special grinding procedure called *dubbing*.

The cone-shaped point on a drill contains the cutting edges and the various clearance angles. This cone angle affects the direction of flow of the chips across the tool face and into the flute. The 118° cone angle that is used most often has been found to provide good cutting conditions and reasonable tool life when drilling mild steel, thus making it suitable for much general-purpose drilling. Smaller cone angles—from 90° to 118°—are sometimes used for drilling more brittle materials, such as gray cast iron and magnesium alloys. Cone angles from 118° to 135° are often used for the more ductile materials, such as aluminum alloys. Cone angles less than 90° frequently are used for drilling plastics. Many methods of grinding drills have been developed that produce point angles other than 118°.

The drill produces a thrust force T and a torque M. Drill torque increases with feed (in./rev) and drill diameter, while the thrust force is influenced greatly by the web or chisel end design, as shown in Figures 23-2 and 23-4.

The relatively thin *web* between the flutes forms a metal column or backbone. If a plain conical point is ground on the drill, the intersection of the web and the cone produces a straight-line *chisel end*, which can be seen in the end view of Figure 23-2. The chisel point, which also must act as a cutting edge, forms a 56° negative rake angle with the conical surface. Such a large negative rake angle does not cut efficiently, causing excessive deformation of the metal. This results in high thrust forces and excessive heat being developed at the point. In addition, the cutting speed at the drill center is low, approaching zero. As a consequence, drill failure on a standard drill occurs both at the center, where the cutting speed is lowest, and at the outer tips of the cutting edges, where the speed is highest.

When the rotating, straight-line chisel point comes in contact with the workpiece, it has a tendency to slide or "walk" along the surface, thus moving the drill away from the desired location. The conventional point drill, when used on machining centers or high-speed automatics, will require additional supporting operations like center drilling, burr removal, and tool change, all of which increase total production time and reduce productivity.

Many special methods of grinding drill points have been developed to eliminate or minimize the difficulties caused by the chisel point and to obtain better cutting action and tool life (see Figure 23-4 for some examples).

The center core or slot-point drill shown in Figure 23-5 has twin carbide tips brazed on a steel shank and a hole (or slot) in the center. The work material in the slot is not machined but, rather, fractured away. The center core drill has a self-centering action and greatly relieves the thrust force produced by the chisel edge of conventional twist drills. This drill operates at about 30 to 50% less thrust than that of

Helical (S-shape chisel point)
Can eliminate center drilling on NC machining centers
Excellent hole geometry
Close relationship between drill size and hole size
Increased tool life
Lower thrust requirements
Leaves burr on breakthrough

Relieved helical
Reduces thrust force
Eliminates chisel end
Equal, rake angle

Four-facet
Good self-centering ability
Breaks up chips for deep-hole drilling
Can be generated in a single grinding operation: reduces thrust.
Eliminates center drilling in NC

Bickford
Combination of helical and Racon point features
Self-centering and reduced burrs
Excellent hole geometry
Increased tool life

Racon (radiused conventional point)
Increased feed rates
Increased tool life (8–10 times in C.I.)
Reduced burrs at breakthrough
Not self-centering

FIGURE 23-4 As the drill advances, it produces a thrust force. Variations in the drill-point geometry are aimed at reducing the thrust force.

FIGURE 23-5 Center core drills can greatly reduce the thrust force.

conventional drills. All rake angles of the cutting edge are positive, which further reduces the cutting force.

The conventional point also has a tendency to produce a burr on the exit side of a hole. Some type of chip breaker is often incorporated into drills. One procedure is to grind a small groove in the rake face, parallel with and a short distance back from the cutting edge. Drills with a special chip-breaker rib as an integral part of the flute are available. The rib interrupts the flow of the chip, causing it to break into short lengths.

The split-point drill is a form of *web thinning* to shorten the chisel edge. This design reduces thrust and allows for higher feed rates. Web thinning uses a narrow grinding wheel to remove a portion of the web near the point of the drill. Such methods have had varying degrees of success, and they require special drill-grinding equipment.

Also shown in Figure 23-4 is a four-facet self-centering point that works well in tougher materials. The facets refer to the number of edges on the clearance surfaces exposed to the cutting action. The self-centering drill lasts longer and saves machining time on numerical control (NC) centers as they can eliminate the need for center drills. A common aspect in drill-point terminology is total indicator runout (TIR). This is a measure of the cutting lips' relative side-to-side accuracy. The original drill point produced by the manufacturer lasts only until the first regrind; thereafter performance and life depend upon the quality of regrind. Proper regrinding (reconditioning) of a drill is a complex and important operation. If satisfactory cutting and hole size are to be achieved, it is essential that the point angle, lip clearance, lip length, and web thinning be correct. As illustrated in Figure 23-6, incorrect sharpening often results in unbalanced cutting forces at the tip, causing misalignment and oversized holes. Drills, even small drills, should always be machine ground, never hand ground. Drill grinders, often computer controlled, should be used to ensure exact reproduction of the geometry established by the manufacturer of the drill. This is extremely important when drills are used on mass-production or numerically controlled machines. Companies invest huge sums in NC machining centers but overlook the value of a top-quality drill-grinding machine.

Drill shanks are made in several types. The two most common types are the straight and the taper. *Straight-shank* drills are usually used for sizes up to $\frac{3}{8}$-in. diameter and must be held in some type of drill chuck. *Taper shanks* are available on larger drills and are common on drills above 1 in. Morse tapers are used on taper-shank drills, ranging from a number 1 taper to a number 6.

Taper-shank drills are held in a female taper in the end of the machine tool spindle. If the taper on the drill is different from the spindle taper, adapter sleeves are available. The taper assures the drill's being accurately centered in the spindle. The *tang* at the end

FIGURE 23-6 Typical causes of drilling problems.

Outer corners break down: Cutting speed too high; hard spots in material; no cutting compound at drill point; flutes clogged with chips

Cutting lips chip: Too much feed; lip relief too great

Checks or cracks in cutting lips: Overheated or too quickly cooled while sharpening or drilling

Chipped margin: Oversize jig bushing

Drill breaks: Point improperly ground; feed too heavy; spring or backlash in drill press, fixture, or work; drill is dull; flutes clogged with chips

Tang breaks: Imperfect fit between taper shank and socket caused by dirt or chips or by burred or badly worn sockets

Drill breaks when drilling brass or wood: Wrong type drill; flutes clogged with chips

Drill spilts up center: Lip relief too small; too much feed

Drill will not enter work: Drill is dull; web too heavy; lip relief too small

Hole rough: Point improperly ground or dull; no cutting compounds at drill point; improper cutting compound; feed too great; fixture not rigid

Hole oversize: Unequal angle of the cutting edges; unequal length of the cutting edges; see part (a)

Chip shape changes while drilling: Dull drill or cutting lips chipped

Large chip coming from one flute, small chip from the other: Point improperly ground, one lip doing all the cutting

(a) Angle unequal   (b) Length unequal

of the taper shank fits loosely in a slot at the end of the tapered hole in the spindle. The drill may be loosened for removal by driving a metal wedge, called a *drift*, through a hole in the side of the spindle and against the end of the tang. It also acts as a safety device to prevent the drill from rotating in the spindle hole under heavy loads. However, if the tapers on the drill and in the spindle are proper, no slipping should occur. The driving force to the drill is carried by the friction between the two tapered members. Standard drills are available in four size series, the size indicating the diameter of the drill body:

- *Millimeter series:* 0.01- to 0.50-mm increments, according to size, in diameters from 0.015 mm
- *Numerical series:* no. 80 to no. 1 (0.0135 to 0.228 in.)
- *Lettered series:* A to Z (0.234 to 0.413 in.)
- *Factional series:* to 4 in. (and over) by 64ths.

TiN coating of conventional drills greatly improves drilling performance. The increase in tool life of TiN-coated drills over uncoated drills in machining steel is more than 200 to 1000%.

DEPTH-TO-DIAMETER RATIO
The depth of the hole to be drilled divided by the diameter of the drill is the depth-to-diameter ratio. Most machinists consider a ratio of 3 to 1 to be deep-hole drilling, after which hole accuracy (location) drilling speed and tool life will be reduced. The bores of rifle barrels were once drilled using conventional drills. Today, *deep-hole drills*, or *gundrills*, are used when deep holes are to be drilled.

The oldest of these deep-hole techniques is *gundrilling*. The original *gundrills* were very likely half-round drills, drilled axially with a coolant hole to deliver cutting fluids to the cutting edge (see Figure 23-7). Modern gundrills typically consist of an alloy-steel-tubing shank with a solid carbide or carbide-edged tip brazed or mechanically fixed to it. Guide pads following the cutting edge by about 90° to 180° are also standard.

The gundrill is a single-lipped tool, and its major feature is the delivery of coolant through the tool at extremely high pressures—typically from 300 psi to 1800 psi, depending on diameter—to force chips back down the flute. Successful application of a gundrill depends almost entirely on the formation of small chips that can be effectively evacuated by the flow of cutting fluids.

Standard gundrills are made in diameters from 0.0078 in. (2 mm) to 2 in. or more. Depth-to-diameter ratios of 100 to 1 or more are possible.

FIGURE 23-7 The gundrill geometry is very different from that of conventional drills.

In gundrilling tolerances for diameters of drilled holes under 1 in. can be held to 0.0005-in. total tolerance, and, should not exceed 0.001 in. over all. According to one source, "roundness accuracies of 0.00008 inch can be attained." Because of the burnishing effect of the guide pads, excellent surface finishes can be produced.

Hole straightness is affected by a number of variables, such as diameter, depth, uniformity of workpiece material, condition of the machine, sharpness of the gundrill, feeds and speeds used, and the specific technique used (rotation of the tool, of the work, or both), but deviation should not exceed about 0.002 in. TIR in a 4-in. depth at any diameter, and it can be held to 0.002 in. per foot.

Basic setup for a gundrilling operation, which is generally horizontal, requires a drill bushing very close to the work entry surface and may involve rotating the work or the tool, or both. Best concentricity and straightness are achieved by the work and the tool rotating in opposite directions.

Other deep-hole drills are called *BTA* (Boring Trepanning Association) *drills* and *ejector drills*. A deep hole is one in which the length (or depth) of the hole is three or more times the diameter. Coolants can be fed internally through these drills to the cutting edges. See Figure 23-8 for schematic of an ejector drill and the machine tool used for gundrilling. The coolants flush the chips out the flutes. The special design of these drills reduces the tendency of the drill to drift, thus producing a more accurately aligned hole. The typical BTA deep-hole drilling tools are designed for single-lip end cutting of a hole in a single pass. Solid-deep-hole drills have alloy-steel shanks with a carbide-edged tip that is fixed to it mechanically. The cutting edge cuts through the center on one side of the hole, leaving no area of material to be extruded. The cutting is done by the outer and inner cutting angles, which meet at a point. Theoretically, the depth of the hole has no limit, but practically, it is restricted by the torsional rigidity of the shank.

Gundrills have a single-lip cutting action. Bearing areas and lifting forces generated by the coolant pressure counteract the radial and tangential loads. The single-lip construction forces the edge to cut in a true circular pattern. The tip thus follows the direction of its own axis. The *trepanning gundrill* leaves a solid core.

FIGURE 23-8 BTA drills for (a) boring, (b) trepanning, (c) counterboring, (d) deep-hole drilling with ejector drill, (e) horizontal deep-hole-drilling machine. *(S. Azad and S. Chandeashekar,* Mechanical Engineering, *Sept. 1985, pp. 62, 63.)*

TABLE 23-2 Drilling Processes Compared

	Twist Drill	Pivot (micro)	Spade (inserted blade)	Indexable-Insert Drill	Gundrill	BTA System	Ejector Drill	Trepanning
Diameter, in.	0.020–2	0.001–0.020	1–6	$5/8$–3	0.078–1	$7/16$–8	$3/4$–$2 1/2$	$1 3/4$–10
Typical range	0.0059	<0.0001	$5/8$ spec, 1	$5/8$	0.039	$3/16$	$3/4$	$1 3/4$
Min	$3 1/2$ std	$1/8$	std	3	$2 1/2$	12	7	>24
Max	6 spec		18					
Depth/Diameter Ratio								
Min. practical	No min	No min	No min	<1	1	1	1	10
Common max[a]	5–10	3–10	>40 (horiz)	2–3	100	100	50	100
Ultimate	>50	20	10 (vert)	—	200	>100	>50	>100
			>100 (horz)					

[a]Maximum depth/diameter ratios in this table are estimates of what can be achieved with special attention and under ideal conditions. Equality of tolerances should not be assumed for the different processes.

FIGURE 23-9 Hole cutter used for thin sheets. (*Courtesy of Armstrong-Blum Manufacturing Company.*)

Hole straightness is affected by variables such as diameter, depth, uniformity of the workpiece material, condition of the machine, sharpness of cutting edges, feeds and speeds used, and whether the tool or the workpiece is rotated or counterrotated.

Two-flute drills are available that have holes extending throughout the length of each land to permit coolant to be supplied, under pressure, to the point adjacent to each cutting edge. These are helpful in providing cooling and also in promoting chip removal from the hole in drilling to moderate depths. They require special fittings through which the coolant can be supplied to the rotating drill, and they are used primarily on automatic and semiautomatic machines. See Table 23-2 for comparison of drilling processes.

Larger holes in thin material may be made with a *hole cutter* (Figure 23-9), whereby the main hole is produced by the thin-walled, multiple-tooth cutter with saw teeth. Hole cutters are often called *hole saws*.

When starting to drill a hole, a drill can deflect rather easily because of the "walking" action of the chisel point. Hole location accuracy is lost. Consequently, to ensure that a hole is started accurately, a *center drill* (Figure 23-10) is used prior to a regular chisel-point twist drill. The center drill and countersink tool have a short, straight drill section extending beyond a 60° taper portion. The heavy, short body provides rigidity so that a hole can be started with little possibility of tool deflection. The hole should be drilled only partway up on the tapered section of countersink. The conical portion of the hole serves to guide the drill being used to make the main hole. Combination center drills are made in four sizes to provide a starting hole of proper size for any drill. If the drill is sufficiently

Step 1 Centering and countersinking with a combination center drill and countersink. (*Courtesy of Chicago-Latrobe*)

Step 2 Drilling with a standard twist drill.

Step 3 Truing hole by boring.

Step 4 Final sizing and finishing with a reamer.

FIGURE 23-10 To obtain a hole that is accurate as to size and aligned on center (located), this four-step sequence of operations is usual.

Subland drill

FIGURE 23-11 Special-purpose subland drill (above), and some of the operations possible with other combination drills (below).

| Drill multiple diameters | Multiple drill countersink and counterbore | Drill and countersink | Drill and counterbore | Drill and chamfer | Drill, countersink, and counterbore |

large in diameter, or if it is sufficiently short, satisfactory accuracy may often be obtained without center drilling. Special drill holders are available that permit drills to be held with only a very short length protruding.

Because of its flexibility and endpoint geometry, a drill may start or drift off centerline during drilling. The use of a center (start) drill will help to ensure that a drill will start drilling at the desired location. Nonhomogeneities in the workpiece and imperfect drill geometries may also cause the hole to be oversize or off-line. For accuracy, it is necessary to follow center drilling and drilling by boring and reaming. Boring corrects the hole alignment, and reaming brings the hole to accurate size and improves the surface finish.

Special *combination drills* can drill two or more diameters, or drill and countersink and/or counterbore, in a single operation (Figure 23-11). Countersinking and counterboring usually follow drilling. These operations are described in more detail later in this chapter. A *step drill* has a single set of flutes and is ground to two or more diameters. *Subland drills* have a separate set of flutes on a single body for each diameter or operation; they provide better chip flow, and the cutting edges can be ground to give proper cutting conditions for each operation. Combination drills are expensive and may be difficult to regrind but can be economical for production-type operations if they reduce work handling, setups, or separate machines and operations.

Spade drills (Figure 23-12) are widely used for making holes 1 in. or larger in diameter at low speeds or with high feeds (Table 23-3). The workpiece usually has an

Regular spade drill

Spade drill with oil holes

FIGURE 23-12 (Top) Regular spade drill; (middle) spade drill with oil holes; (bottom) spade drill geometry, nomenclature.

TABLE 23-3	Recommended Surface Speeds and Feeds for High-Speed Steel Spade Drills for Various Materials
Material	Surface Speed (ft per min)
Mild machinery steel 0.2 and 0.3 carbon	65–110
Steel, annealed 0.4 to 0.5 carbon	55–80
Tool steel, 1.2 carbon	45–60
Steel forging	35–50
Alloy steel	45–70
Stainless steel, free machining	50–70
Stainless steel, hard	25–40
Cast iron, soft	80–150
Cast iron, medium hard	55–100
Cast iron, hard, chilled	25–40
Malleable iron	79–90
Brass and bronze, ordinary	200–300
Bronze, high tensile	70–150
Monel metal	35–50
Aluminum and its alloys	200–300
Magnesium and its alloys	250–400

Feed Rates for Spade Drilling (inches per revolution)

Drill Size (inches)	Cast Iron Malleable Iron Brass Bronze	Medium Steel Stainless Steel Monel Metal Drop-Forged Alloys Tool Steel (annealed)	Tough Steel Drop Forging Aluminum
1 to $1\frac{1}{4}$	0.010–0.020	0.008–0.014	0.006–0.012
$1\frac{1}{4}$ to $\frac{3}{4}$	0.010–0.024	0.008–0.018	0.008–0.017
$1\frac{3}{4}$ to $2\frac{1}{2}$	0.010–0.030	0.010–0.024	0.010–0.017
$2\frac{1}{2}$ to 4	0.012–0.032	0.012–0.030	0.010–0.017
4 to 6	0.012–0.032	0.010–0.024	0.008–0.017

Source: Waukesha Cutting Tools, Inc.

existing hole, but a spade drill can drill deep holes in solids or stacked materials. Spade drills are less expensive because the long supporting bar can be made of ordinary steel. The drill point can be ground with a minimum chisel point. The main body can be made more rigid because no flutes are required, and it can have a central hole through which a fluid can be circulated to aid in cooling and in chip removal. The cutting blade is easier to sharpen; only the blades need to be TiN-coated.

Spade drills are often used to machine a shallow locating cone for a subsequent smaller drill and at the same time to provide a small bevel around the hole to facilitate later tapping or assembly operations. Such a bevel also frequently eliminates the need for deburring. This practice is particularly useful on mass-production and numerically controlled machines.

Carbide-tipped drills and drills with indexable inserts are also available (see Figure 23-13) with one- and two-piece inserts for drilling shallow holes in solid workpieces. *Indexable insert drills* can produce a hole four times faster than a spade drill because they run at high speeds/low feeds and are really more of a boring operation than a drilling process.

However, to use indexable drills, you must have an extremely rigid machine tool and setup, adequate horsepower, and lots of cutting fluid. Indexable drills are roughing tools generating hole tolerances of and surface finish of 250 rpms or greater. The tool is designed for the inboard insert to cut past the centerline of the tool so the inboard tool is positioned radially below the center. See Table 23-4 for an indexable drilling troubleshooting guide.

Insert

FIGURE 23-13 One- and two-lipped indexable insert drills are widely used for holes over 1 inch in diameter. *(Courtesy of Waukesha.)*

TABLE 23-4 Indexable Drilling Troubleshooting Guide[a]

Problem	Source	Solution
Insert chipping or breakage[b]	Off-center drill, caused by misalignment	Maintain proper alignment. Concentricity not to exceed ±0.005 TIR.
	Improper seating of tool in tool holder, spindle, or turret	Check tool shank and socket for nicks and dirt. Check parting line between tool shank and socket with feeler gage. Check to see if tool is locked tightly.
	Deflection because of too much overhand and lack of rigidity	With indicator, check if tool can be moved by hand. Check if tool can be held shorter.
	Improper seating of inserts in pockets	Clean pockets whenever indexing or changing inserts. Check pockets for nicks and burrs. Check if inserts rest completely on pocket bottoms.
	Damaged insert screws	Check head and thread for nicks and burns. Do not overtorque screws.
	Improper speeds and feeds	Check recommended guidelines for given materials.
	Insufficient coolant supply	Check coolant flow.
	Improper carbide grade in inboard station	Recommend straight grade for multiple-insert drills.
Grooving on back stroke; drill body rubbing hole wall; over- or undersize holes	Off-center drill	Maintain proper alignment and concentricity. Check bottom of hole or disk for center stub.
	Deflection	Check setup rigidity. Check speed and feed guidelines.
Poor hole surface finish	Vibrations	Check setup and part rigidity. Check seat in spindle or tool holder. Check speeds and feeds.
	Insufficient coolant pressure and volume	Increase coolant pressure and flow. Is coolant flow constant? Make sure coolant reaches inserts at all times.
	Recutting chips, causing drill to jump	Increase coolant flow. Add coolant grooves.
	Poor chip control; chips trapped in hole	Mostly speed or feed.
	Chatter	Mostly feed rate.
Very short, thick, flat chips	Feed rate too high in relation to cutting speed	Lower feed or increase speed.
Long and stringy chips	Feed rate too low in relation to cutting speed	Increase feed rate or decrease speed. Use dimple inserts.
Unable to loosen insert locking screws	Seized threads, caused by coolant or heat	Apply water and heat-resistant lubricant to threads.

[a]*Source:* "Fundamentals of Indexable Drilling," K. L. Anderson, *Machining Technology,* vol. 2, no. 3, 1991.
[b]If constant chipping occurs, especially on an inner insert, and conditions are optimum, try an uncoated-carbide insert or a grade with higher transverse rapture strength.

Sector	Typical drill types
A	Twist drill (HSS) Center core drill
B	Twist drill Gundrill BTA Ejector drill
C	Twist drill Indexable insert drill Spade drill Center core drill
D	BTA Ejector drill

FIGURE 23-14 Drill selection depends on hole diameter and hole depth.

A high-pressure, pulsating coolant system can generate pressures up to 300 psi and works well with indexable drilling. It can have disadvantages, however. High pressure with pulsating action can decrease chip control and cause drill deflection. A high-pressure coolant stream can flatten chips at the point of forming and forces them into the cut, causing recutting, insert chipping, and poor hole finishes. The pressure can force chips between the drill body and hole diameter, wrapping them around the drill. Friction then will weld the chips to the tool body or hole.

The diameter of the hole and the length/diameter ratio usually determine what kind of drill to use. Figure 23-14 explores how drill selection depends on the depth of the hole and the diameter of the drill: Section A shows the drilling areas of relatively shallow holes and small diameters. About half of all the drilling process falls within the category of this section. It is the section for which the majority of the work is done by twist drills and a very few cemented carbide drills. Section B is the drilling of deep holes for which cemented carbide gundrills are used. Section C is that of shallow holes having large diameters, for which spade drills are used. Section D is that of deep holes having large diameters, for which BTA tools are used.

MICRODRILLING

As the term suggests, microdrilling involves very-small-diameter cutting tools, including drills, end mills, routers, and other special tools. Drills from 0.002 in. (0.05 mm) and mills to 0.005 in. in diameter are used to produce geometries involving dimensions at which many workpiece materials no longer exhibit uniformity and homogeneity. Grain borders, inclusions, alloy or carbide segregates, and microscopic voids are problems in microdrilling, where holes of 0.02 to 0.0001 in. have been drilled using pivot drills, as shown in Figure 23-15.

Pivot drills are two-lipped (two-fluted), end-cutting tools of relatively simple geometry. Web thickness tapers toward the point, and a generous back-taper is incorporated. For softer workpiece materials, point angles are typically 118° and lip clearance is 15°. For steels and general use in harder metals, 135° points and 8° clearance are recommended. The chisel edge is similar to that of a twist drill. Pivot drills are made of tungsten-alloy tool steel in standard sizes from 0.0001 in. to 0.125 in. and of sintered tungsten carbide from 0.001 in. to 0.125 in.

Small drills easily deflect, and getting accurate and precise holes requires a machine with a high-quality spindle and very sensitive feeding pressure. In the medical components field, much of this machining work is performed on Computer Numerical Control (CNC) Swiss-type turning machines. Speeds and feeds are greatly reduced with frequent pecking to clear the chips. Use a light, lard-based, sulfurized cutting oil.

Microdrill

Dia.

Point angle,
118°–135°

Note back taper

Clearance angle
8°–15°

Shank

FIGURE 23-15   Pivot microdrill for drilling very-small-diameter holes.

■ 23.4 Tool Holders for Drills

Straight-shank drills must be held in some type of drill *chuck* (Figure 23-16). Chucks are adjustable over a considerable size range and have radial steel fingers. When the chuck is tightened by means of a chuck key, these fingers are forced inward against the drill. On smaller drill presses, the chuck often is permanently attached to the machine

3 jaw
Jacobs
chuck

Chuck key

(a)

Collet
chuck

Synthetic
rubber
support
for jaws

(b)

FIGURE 23-16   Two of the most commonly used types of drill chucks are the 3-jaw Jacobs chuck (above) and the collet chuck with synthetic rubber support for jaws. *(Courtesy of Jacobs Manufacturing Company.)*

spindle, whereas on larger drilling machines the chucks have a tapered shank that fits into the female Morse taper of the machine spindle. Special types of chucks in semiautomatic or fully automatic machines permit quite a wide range of sizes of drills to be held in a single chuck.

Chucks using chuck keys require that the machine spindle be stopped in order to change a drill. To reduce the downtime when drills must be changed frequently, *quick-change chucks* are used. Each drill is fastened in a simple round collet that can be inserted into the chuck hole while it is turning by merely raising and lowering a ring on the chuck body. With the use of this type of chuck, center drills, drills, counterbores, reamers, and so on can be manually changed in quick succession. For carbide drills, collet-type holders with thrust bearings are recommended (Figure 23-17). For drills using an internal coolant supply, a very rigid chuck with either an inducer or through-spindle coolant source is recommended.

Conventional holders such as keyless chucks cannot be used because the gripping strength is limited. Collet holders should be cleaned periodically with oil to remove small chips.

The entire flute length must protrude from the chuck. At maximum hole depth, the length of flute protruding from the hole must be at least 1 to 1.5 times the drill diameter. Radial runout at the drill tip must not exceed 0.001 in.

FIGURE 23-17 Here are some suggestions for correct chucking of carbide drills.

■ 23.5 WORKHOLDING FOR DRILLING

Work that is to be drilled is ordinarily held in a vise or in specially designed workholders called *jigs*. Workholding devices are the subject of Chapter 25, where the design of workholding devices is discussed. Many examples of drill jigs are shown.

With regard to safety, the work should not be held on the table by hand unless adequate leverage is available, even in light drilling operations. This is a dangerous practice and can lead to serious accidents, because the drill has a tendency to catch on the workpiece and cause it to rotate, especially when the drill exits the workpiece. Work that is too large to be held in a jig can be clamped directly to the machine table using suitable bolts and clamps and the slots or holes in the table. Jigs and workholding devices on indexing machines must be free from play and firmly seated.

■ 23.6 MACHINE TOOLS FOR DRILLING

The basic work and tool motions required for drilling—relative rotation between the workpiece and the tool, with relative longitudinal feeding—also occur in a number of other machining operations. Thus drilling can be done on a variety of machine tools such as lathes, horizontal and vertical milling machines, boring machines, and machining centers. This section will focus on those machines that are designed, constructed, and used primarily for drilling.

First of all, the machine tools must have sufficient power (torque) and thrust to perform the cut. It is the task of the engineer to select the correct machine or select the cutting parameters (speed and feed) based on the drill diameter, drill material, and work material (hardness). Because of the complex geometry of the drill, empirical equations are widely used. Figure 23-18 shows the type of information provided by cutting-tool manufacturers to calculate (estimate) thrust in drilling. Data with K_s (specific cutting force) and X and Y (empirical constants) are obtained from cutting-tool manufacturers. Much of this kind of data has been developed for high-speed-steel tools. When using solid carbide tools, rigid machines such as machining centers or NC turning machines are recommended, whereas a radial drilling machine is not recommended (not rigid enough).

Rigidity is especially important in avoiding chatter. A lack of rigidity in the cutting tool, the workpiece, or the machine tool permits the affected members to deflect due to the cutting forces developing the conditions for chatter (see Chapter 20), with the result that the cutting lips have a hammering action against the work. So, use the shortest tool possible.

In addition, backlash in the feed mechanism should be kept at a minimum to reduce strain on the drill when it breaks through the bottom of the hole.

The common name for the machine tool used for drilling is the *drill press*. Drill presses consist of a *base*, a *column* that supports a *powerhead*, a *spindle*, and a *worktable*. On small machines, the base rests on a workbench, whereas on larger machines it rests on the floor (Figure 23-19). The column may be either round or of box-type construction, the latter being used on larger, heavy-duty machines, except in radial types. The powerhead contains an electric motor and means for driving the spindle in rotation at several speeds. On small drilling machines this may be accomplished by shifting a belt on a step-cone pulley, but on larger machines a geared transmission is used.

The heart of any drilling machine is its spindle. In order to drill satisfactorily, the spindle must rotate accurately and also resist whatever side forces result from the drilling process. In virtually all machines the spindle rotates in preloaded ball or taper-roller bearings. In addition to powered rotation, provision is made so that the spindle can be moved axially to feed the drill into the work. On small machines the spindle is fed by hand, using the handles extending from the capstan wheel; on larger machines power feed is provided. Except for some small bench types, the spindle contains a hole with a Morse taper in its lower end into which taper-shank drills or drill chucks can be inserted.

The worktables on drilling machines may be moved up and down on the column to accommodate work of various sizes. On round-column machines the table can usually

MATERIAL	BRINELL HARDNESS		FEED (IPR)							
			0.004	0.005	0.006	0.008	0.010	0.012	0.016	0.020
		E	0.35	0.39	0.47	0.60	0.70	0.80	0.90	1.08
		C	3.3	3.5	3.8	4.4	4.4	4.4	4.0	3.6
PLAIN-CARBON STEEL	140–220		444230	435510	431150	426790	418070	409350	391910	374460
	220–300		493590	483900	479060	474210	464520	454830	435450	416070
FREE-MACHINING STEELS	120–180		296150	290340	287440	284530	278710	272900	261270	249640
	180–260		345510	338730	335340	331950	325160	318380	304820	291250
ALLOY STEELS	260–340		493590	483900	479060	474210	464520	454830	435450	416070
STAINLESS STEELS	150–200		370190	362930	359300	355660	348390	341120	326590	312050
	200–300		444230	435510	431150	426790	418070	409350	391910	374460
CAST IRON	180–250		345510	338730	335340	331950	325160	318380	304820	291250
ALUMINUM			148080	145170	143720	142260	139360	136450	130640	124820
TITANIUM			320830	314530	311390	308240	301940	295640	283040	270450
HIGH-TEMPERATURE ALLOYS			542950	532290	526970	521630	510970	500310	478990	457680

F_v = Axial thrust
$= D^{1.15} \times K_s \times f_r^{0.8}$

where

D = Drill diameter (inches)
K_s = Specific cutting energy from table (in-lb/in²)
f_r = Feed (in./rev)

Values in in.-lb/in².

FIGURE 23-18 Estimating the thrust force in drilling; example from Waukesha Cutting Tools.

be rotated out of the way so that workpieces can be mounted directly on the base. On some box-column machines the table is mounted on a subbase so that it can be moved in two directions in a horizontal plane by means of feed screws.

Figure 23-19 shows examples of common types of drilling machines used in production environments. Drilling machines usually are classified as bench, upright with single spindle, turret or NC turret, gang, multispindle, deep-hole, and transfer.

(a)

Power head

Speed control

Speed indicator

Feed control

Quill

Spindle

Crank to adjust table height

Worktable

Column (round)

Base

(b)

RAM

Column

10-tool turret

Z

CNC controller

X

Y

Table

Base

(c)

Gang-drilling machine

(d)

Arm-elevating screw

Column

Drive motor

Head moves in or out

Arm

Arm moves up or down

Arm swings about column

Spindle feeds down

Spindle and drill rotate

Column clamps

Work clamps to base

(e)

Multiple-spindle drilling machine

FIGURE 23-19 Examples of drilling machines: (a) upright column drilling machine; (b) CNC turret drilling machine; (c) gang-drilling machine; (d) radial drill press; (e) multiple-spindle drilling machine.

647

With bench drill presses, holes up to $^1/_2$ in. in diameter can be drilled. The same type of machine can be obtained with a long column so that it can stand on the floor. The size of bench and upright drilling machines is designated by *twice* the distance from the centerline of the spindle to the nearest point on the column, this being an indication of the maximum size of the work that can be drilled in the machines. For example, a 15-in. drill press will permit a hole to be drilled at the center of a workpiece 15 in. in diameter.

Sensitive drilling machines are essentially smaller, plain bench-type machines with more accurate spindles and bearings. They are capable of operating at higher speeds, up to 30,000 rpm. Very sensitive hand-operated feeding mechanisms are provided for use in drilling small holes. Such machines are used for tool and die work and for drilling very small holes, often less than a few thousandths of an inch in diameter, when high spindle speeds are necessary to obtain proper cutting speed and sensitive feel to provide delicate feeding to avoid breakage of the very small drills.

Upright drilling machines usually have spindle speed ranges from 60 to 3500 rpm and power feed rates, from 4 to 12 steps, from about 0.004 to 0.025 in. rev. Most modern machines use a single-speed motor and a geared transmission to provide the range of speeds and feeds. The feed clutch disengages automatically when the spindle reaches a preset depth.

Worktables on most upright drilling machines contain holes and slots for use in clamping work and nearly always have a channel around the edges to collect cutting fluid, when it is used. On box-column machines, the table is mounted on vertical ways on the front of the column and can be raised or lowered by means of a crank-operated elevating screw.

In mass production *gang-drilling machines* are often used when several related operations, such as drilling holes of different sizes, reaming, or counterboring, must be done on a single part. These consist essentially of several independent columns, heads, and spindles mounted on a common base and having a single table. The work can be slid into position for the operation at each spindle. They are available with or without power feed. One or several operators may be used. This machine would be an example of a simple small cell except that the machines are usually not single-cycle automatics.

Turret-type, upright drilling machines are used when a series of holes of different sizes, or a series of operations (such as center drilling, drilling, reaming, and spot facing), must be done repeatedly in succession. The selected tools are mounted in the turret. Each tool can quickly be brought into position merely by rotation of the turret. These machines automatically provide individual feed rates for each spindle and are often numerically controlled.

Radial drilling machine tools are used on large workpieces that cannot be easily handled manually. As shown in Figure 23-19, these machines have a large, heavy, round, vertical column supported on a large base. The column supports a radial arm that can be raised and lowered by power and rotated over the base. The spindle head, with its speed- and feed-changing mechanism, is mounted on the radial arm. It can be moved horizontally to any desired position on the arm. Thus the spindle can quickly be positioned properly for drilling holes at any point on a large workpiece mounted either on the base of the machine or even sitting on the floor.

Plain radial drilling machines provide only a vertical spindle motion. On *semiuniversal machines*, the spindle head can be pivoted at an angle to a vertical plane. On *universal machines*, the radial arm is rotated about a horizontal axis to permit drilling at any angle.

Radial drilling machines are designated by the radius of the largest disk in which a center hole can be drilled when the spindle head is at its outermost position. Sizes from 3 to 12 ft are available. Radial drilling machines have a wide range of speeds and feeds, can do boring, and include provisions for tapping (internal threading) (see Chapter 29).

Multiple-spindle drilling machines (Figure 23-19) are mass-production machines with as many as 50 spindles driven by a single powerhead and fed simultaneously into the work. Figure 23-20 shows an adjustable multiple-spindle head that can be mounted on a regular single spindle-drill press. Figure 23-20 shows the methods of driving and positioning the spindles, which permit them to be adjusted so that holes can be drilled

Adjustable drill head
Spindle: 6 Production: 50 pieces

Geared drill head
Spindle: 8 Production: 80,000 pieces

Gearless drill head
Spindle: 16 Production: 30,000 pieces

450 rpm
1050 rpm
575 rpm
1050 rpm
1050 rpm
375 rpm

An adjustable drill head should be considered for low-production jobs. However, many short-run jobs such as this would be required to justify a multiple-spindle head.

A geared drill head is most appropriate in this situation, where there is a large difference in sizes and a high daily production.

Only a gearless head can perform this operation in one pass, due to the close proximity of the spindle centers.

FIGURE 23-20 Three basic types of multiple-spindle drill heads: (left) adjustable; (middle) geared; (right) gearless. *(Courtesy of Zagar Incorporated.)*

at any location within the overall capacity of the head. Special drill jigs are often designed and built for each job to provide accurate guidance for each drill. Although such machines and workholders are quite costly, they can be cost-justified when the quantity to be produced will justify the setup cost and the cost of the jig. Reducing setup on these machines is difficult. Numerically controlled drill presses other than turret drill presses are not common because drilling and all its related processes can be done on vertical or horizontal NC machining centers equipped with automatic tool changers (see Chapter 31).

Special machines are used for drilling long (deep) holes, such as are found in rifle barrels, connecting rods, and long spindles. High cutting speeds, very light feeds, and a copious flow of cutting fluid ensure rapid chip removal. Adequate support for the long, slender drills is required. In most cases horizontal machines are used. The work is rotated in a chuck with steady rests providing support along its length, as required. The drill does not rotate and is fed into the work. Vertical machines are also available for shorter workpieces. Notice the similarity between this process and boring.

■ 23.7 CUTTING FLUIDS FOR DRILLING

For shallow holes, the general rules relating to cutting fluids, as given in Chapter 21, are applicable. When the depth of the hole exceeds one diameter, it is desirable to increase the lubricating quality of the fluid because of the rubbing between the drill margins and the wall of the hole. The effectiveness of a cutting fluid as a coolant is quite variable in drilling. While the rapid exit of the chips is a primary factor in heat removal, this action

TABLE 23-5	Cutting Fluids for Drilling
Work Material	**Cutting Fluid**
Aluminum and its alloys	Soluble oil, kerosene, and lard-oil compounds; light, nonviscous neutral oil; kerosene and soluble oil mixtures
Brass	Dry or a soluble oil; kerosene and lard-oil compounds; light, nonviscous neutral oil
Copper	Soluble oil, strained lard oil, oleic-acid compounds
Cast iron	Dry or with a jet of compressed air for cooling
Malleable iron	Soluble oil, nonviscous neutral oil
Monel metal	Soluble oil, sulfurized mineral oil
Stainless steel	Soluble oil, sulfurized mineral oil
Steel, ordinary	Soluble oil, sulfurized oil, high extreme-pressure-value mineral oil
Steel, very hard	Soluble oil, sulfurized oil, turpentine
Wrought iron	Soluble oil, sulfurized oil, mineral-animal oil compound

- Neat oil can be used effectively with the solid carbide drills for low-speed drilling (up to 130 sfpm).
- If the work surface becomes hard or blue in color, decrease the rpm and use neat oil.
- For heavy-duty cutting, emulsion-type oil containing some extreme pressure additive is recommended.
- A volume of 3.0 gal/min at a pressure of 37–62 lb/in.^{2} is recommended.
- A double stream supply of fluid is recommended.

also tends to restrict entry of the cutting fluid. This is of particular importance in drilling materials that have poor heat conductivity. Recommendations for cutting fluids for drilling are given in Table 23-5.

If the hole depth exceeds two or three diameters, it is usually advantageous to withdraw the drill each time it has drilled about one diameter of depth, to clear chips from the hole. Some machines are equipped to provide this "pecking" action automatically.

Where cooling is desired, the fluid should be applied copiously. For severe conditions, drills containing coolant holes have a considerable advantage. Not only is the fluid supplied near the cutting edges, but the coolant flow aids in flushing the chips from the hole. Where feasible, drilling horizontally has distinct advantages over drilling vertically downward.

■ 23.8 COUNTERBORING, COUNTERSINKING, AND SPOT FACING

Drilling is often followed by *counterboring, countersinking,* or *spot facing.* As shown in Figure 23-21, each provides a bearing surface at one end of a drilled hole. They are usually done with a special tool having from three to six cutting edges.

Counterboring provides an enlarged cylindrical hole with a flat bottom so that a bolt head, or a nut, will have a smooth bearing surface that is normal to the axis of the hole; the depth may be sufficient so that the entire bolt head or nut will be below the surface of the part. The pilot on the end of the tool fits into the drilled hole and helps to ensure concentricity with the original hole. Two or more diameters may be produced in a single counterboring operation. Counterboring also can be done with a single-point tool, although this method ordinarily is used only on large holes and essentially is a boring operation. Some counterboring tools are shown in Figure 23-21b.

Countersinking makes a beveled section at the end of a drilled hole to provide a proper seat for a flat-head screw or rivet. The most common angles are 60°, 82°, and 90°. Countersinking tools are similar to counterboring tools except that the cutting edges are elements of a cone, and they usually do not have a pilot because the bevel of the tool causes them to be self-centering.

Spot facing is done to provide a smooth bearing area on an otherwise rough surface at the opening of a hole and normal to its axis. Machining is limited to the minimum depth that will provide a smooth, uniform surface. Spot faces thus are somewhat easier and more economical to produce than counterbores. They are usually made with a multiedged end-cutting tool that does not have a pilot, although counterboring tools are frequently used.

Counterbore   Countersink   Spot face

FIGURE 23-21 (a) Surfaces produced by counterboring, countersinking, and spot facing. (b) Counterboring tools: (bottom to top) interchangeable counterbore; solid, taper-shank counterbore with integral pilot; replaceable counterbore and pilot; replaceable counterbore, disassembled. *(Courtesy of Ex-Cell-O Corporation and Chicago Latrobe Twist Drill Works.)*

■ 23.9 REAMING

Reaming removes a small amount of material from the surface of holes. It is done for two purposes: to bring holes to a more exact size and to improve the finish of an existing hole. Multiedge cutting tools are used, as shown in Figure 23-22. No special machines are built for reaming. The same machine that was employed for drilling the hole can be used for reaming by changing the cutting tool.

To obtain proper results, only a minimum amount of materials should be left for removal by reaming. As little as 0.005 in. is desirable, and in no case should the amount exceed 0.015 in. A properly reamed hole will be within 0.001 in. of correct size and have a fine finish.

The principal types of reamers are shown in Figures 23-22 and 23-23. *Hand reamers* are intended to be turned and fed by hand and to remove only a few thousandths of an inch of metal. They have a straight shank with a square tang for a wrench. They can have straight or spiral flutes and be solid or expandable. The teeth have relief along their edges and thus may cut along their entire length. However, the reamer is tapered from 0.005 to 0.010 in. in the first third of its length to assist in starting it in the hole, and most of the cutting therefore takes place in this portion.

Machine or *chucking reamers* are for use with various machine tools at slow speeds. The best feed is usually two to three times the drilling feed. Machine reamers have chamfers on the front end of the cutting edges. The chamfer causes the reamer to seat firmly and concentrically in the drilled hole, allowing the reamer to cut at full diameter. The longitudinal cutting edges do little or no cutting. Chamfer angles are usually 45°. Reamers have straight or tapered shanks and straight or spiral flutes. *Rose-chucking reamers* are ground cylindrical and have no relief behind the outer edges of the teeth. All cutting is done on the beveled ends of the teeth. *Fluted-chucking reamers*, on the other hand, have relief behind the edges of the teeth as well as beveled ends. They can therefore cut on all portions of the teeth. Their flutes are relatively short, and they are intended for light finishing cuts. For best results they should not be held rigidly but permitted to float and be aligned by the hole.

FIGURE 23-22 Standard nomenclature for hand and chucking reamers.

Shell reamers often are used for larger sizes in order to save cutting-tool material. The shell, made of tool steel for smaller sizes and with carbide edges for larger sizes or for mass-production work, is held on an arbor that is made of ordinary steel. One arbor may be used with any number of shells. Only the shell is subject to wear and needs to be replaced when worn. They may be ground as rose or fluted reamers.

Expansion reamers can be adjusted over a few thousandths of an inch to compensate for wear or to permit some variation in hole size to be obtained. They are available in both hand and machine types.

FIGURE 23-23 Types of reamers: (top to bottom) Straight-fluted rose reamer, straight-fluted chucking reamer, straight-fluted taper reamer, straight-fluted hand reamer, expansion reamer, shell reamer, adjustable insert-blade reamer.

Adjustable reamers have cutting edges in the form of blades that are locked in a body. The blades can be adjusted over a greater range than expansion reamers. This permits adjustment for size and to compensate for regrinding. When the blades become too small from regrinding, they can be replaced. Both tool steel and carbide blades are used.

Taper reamers are used for finishing holes to an exact taper. They may have up to eight straight or spiral flutes. Standard tapers, such as Morse, Jarno, or Brown & Sharpe, come in sets of two. The *roughing reamer* has nicks along the cutting edge to break up the heavy chips that result as a cylindrical hole is cut to a taper. The *finishing reamer* has smooth cutting edges.

REAMING PRACTICE

If the material to be removed is free-cutting, reamers of fairly light construction will give satisfactory results. However, if the material is hard, then tough, solid-type reamers are recommended, even for fairly large holes.

To meet quality requirements, including both finish and accuracy (tolerances on diameter, roundness, straightness, and absence of bell-mouth at ends of holes), reamers must have adequate support for the cutting edges, and reamer deflection must be minimal. Reaming speed is usually two-thirds the speed for drilling the same materials. However, for close tolerances and fine finish, speeds should be slower.

Feeds are usually much higher than those for drilling and depend upon material. A feed of between 0.0015 and 0.004 in. per flute is recommended as a starting point. Use the highest feed that will still produce the required finish and accuracy. Recommended cutting fluids are the same as those for drilling. Reamers, like drills, should not be allowed to become dull. The chamfer must be reground long before it exhibits excessive wear. Sharpening is usually restricted to the starting taper or chamfer. Each flute must be ground exactly even, or the tool will cut oversize.

Reamers tend to chatter when not held securely, when the work or workholder is loose, or when the reamer is not properly ground. Irregularly spaced teeth may help reduce chatter. Other cures for chatter in reaming are to reduce the speed, vary the feed rate, chamfer the hole opening, use a piloted reamer, reduce the relief angle on the chamfer, or change cutting fluid. Any misalignment between the workpiece and the reamer will cause chatter and improper reaming.

■ Key Words

center core drill	gang-drilling machine	radial drilling machine	thrust force
chisel end	gundrill	reaming, hand	trepanning
chuck	helix angle	reaming, machine	turret drilling machine
counterboring	indexable insert drill	shell reamer	twist drill
countersinking	jig	spade drill	web
deep-hole drilling	lip	spot facing	
drill press	multiple-spindle drilling	subland drill	
drilling	machine	tang	
flute			

■ Review Questions

1. What functions are performed by the flutes on a standard twist drill?
2. What determines the rake angle of a drill? See Figure 23-2.
3. Basically, what determines what helix angle a drill should have?
4. When a large-diameter hole is to be drilled, why is a smaller-diameter hole often drilled first?
5. Equation 23-4 for the MRR for drilling can be thought of as _____ times _____ where $f_r N_s$ is the feed rate of the drill bit.
6. Are the recommended surface speeds for spade drills given in Table 23-3 typically higher or lower than those recommended for twist drills? How about the feeds? Why?
7. What can happen when an improperly ground drill is used to drill a hole?
8. Why are most drilled holes oversize with respect to the nominally specified diameter?
9. What are the two primary functions of a combination center drill?
10. What is the function of the margins on a twist drill?

11. What factors tend to cause a drill to "drift" off the centerline of a hole?
12. The drills shown in Figure 23-13 have coolant passages in the flutes. What is the purpose of these holes?
13. In drilling, the deeper the hole, the greater the torque. Why?
14. Why do cutting fluids for drilling usually have more lubricating qualities than those for most other machining operations?
15. How does a gang-drilling machine differ from a multiple-spindle drilling machine?
16. How does a multiple-spindle drilling machine differ from a NC drilling machine with a tool changer that would hold all the drills found in the multiple-spindle machine?
17. How does the thrust force vary with feed? Why?
18. Holding the workpiece by hand when drilling is not a good idea. Why?
19. What is the rationale behind the operation sequence shown in Figure 23-10?
20. In terms of thrust, what is unusual about the slot-point drill compared to other drills?
21. What is the purpose of spot facing?
22. How does the purpose of counterboring differ from that of spot facing?
23. What are the primary purposes of reaming?
24. What are the advantages of shell reamers?
25. A drill that operated satisfactorily for drilling cast iron gave very short life when used for drilling a plastic. What might be the reason for this?
26. What precautionary procedures should be used when drilling a deep, vertical hole in mild steel when using an ordinary twist drill?
27. What is the advantage of a spade drill? Is it really a drill?
28. What is a "pecking" action in drilling?
29. Why does drill feed increase with drill size?
30. Suppose you specified a drilling feed rate that was too large. What kinds of problems do you think this might cause? See Figure 23-6 and Table 23-4 for help.

■ Problems

1. Suppose you wanted to drill a 1.5-in.-diameter hole through a piece of 1020 cold-rolled steel that is 2 in. thick, using an indexable insert drill. What values of feed and cutting speed will you specify, along with an appropriate allowance. Is this the correct tool? What other drill types could be used?
2. How much time will be required to drill the hole in Problem 1 using the insert drill?
3. What is the metal removal rate when a 1.5-in.-diameter hole, 2 in. deep, is drilled in 1020 steel at a cutting speed of 200 fpm with a feed of 0.010 ipr? What is the cutting time?
4. If the specific horsepower for the steel in Problem 3 is 0.9, what horsepower would be required, assuming 80% efficiency in the machine tool?
5. If the specific power of an AISI 1020 steel of 0.9, and 80% of the output of the 1.0-kW motor of a drilling machine is available at the tool, what is the maximum feed that can be used in drilling a 1-in.-diameter hole with a carbide drill? (Use the cutting speed suggested in Problem 3.)
6. Show how the approximate equation 23-5 for MRR in drilling was obtained. What assumption was needed?
7. A workpiece must have 10 holes finished in it. Manual layout time is $^1/_2$ hr/piece. To drill and ream all the holes requires 1 hour on the machine for each piece, not counting layout or setup. The labor rate is $10/hr and the machine rate is $20/hr. If a jig is used, the labor cost to lay out each piece can be saved. Both methods give the same-quality product, but this jig saves 40 min in processing time on the machine. How large a lot justifies the use of a jig that costs $150 to make (labor and materials)?
8. A part has two holes located for drilling by manual layout. If a drill jig is used, 0.5 min in processing time is saved for each piece. The labor rate is $9/hr. The overhead rate on the labor saved is 100%. Setup time is no more with than without the jig. The combined rate for interest, insurance, taxes, and maintenance is 35%. The cost of the jig is $500.
   a. How many pieces must be made in one lot to make the jig worthwhile?
   b. How many pieces must be made on the jig in one lot each month to earn the cost of the jig in two years?
9. Manufacturer's charts will help determine the best feed and speed to run the drills. For example, a 1.5-in hole is to be drilled in 4140 steel annealed to Bhn 275. For the spade drill, speed is 80 sfm; feed, 0.009 ipr; and spindle rotation, 204 rpm. For the indexable insert drill, speed is 358 sfm; feed, 0.007 ipr; and spindle rotation, 891 rpm. Typically, an indexable insert drill can produce a hole four times faster than a spade drill but may cost (with inserts) 50% to 75% more than the equivalent spade blade and holder. For making only a couple of holes, the extra cost is not usually justified. Determine the number of holes needed to justify the extra cost of the indexable insert drill. Some additional cost data are given below.

   Ignore tool life and assume that the blades and the indexable drills make about the same number of holes. (Why is this a reasonable assumption?) The holes are 3 in. deep, with no allowance needed. Cost of drills:

Spade drill	Indexable-insert Drill
$139.00 holder	$273.00 drill
+21.90 per blade	+12.80 per two inserts
$160.90	$285.80

   Assume for this example that a machine rate of $45/hr includes the cost of labor and machine burden.

10. Assume that you are drilling eight holes, equally spaced in a bolt-hole circle. That is, there would be holes at 12, 3, 6, and 9 o'clock and four more holes equally spaced between them. The diameter of the bolt hole circle is 6 in. The designer says that the holes must be 45° ± 1° from each other around the circle.
    a. Compute the tolerance between hole centers.
    b. Do you think a typical multiple-spindle drill setup could be used to make this bolt circle—using eight drills all at once? Why or why not?
    c. Do you think that the use of a jig may help improve the situation?
    d. Do you think a CNC drilling process could do the holes best?
11. A part with seven holes can be machined on a numerically controlled turret drill press in 3 min (estimated time based on

similar parts). The rate on the CNC machine for labor is $34/hr. Currently, the part is being machined on a gang drill press with a special jig in 10 minutes per piece. The jig for the gang drill costs $300; the combined rate for depreciation, interest, insurance, and taxes is 135%; and the hourly rate for the gang drill and operator is $16. Setup time is about the same for both machines. For how many pieces is it economical to switch to the CNC?

12. It is estimated that a jig for machining a part with three holes costs $400 and with it the operation takes 15 min per part. The operation can be done without a jig on a numerically controlled drill press in 5 min. Assume that any other conditions are the same as in Problem 11. How many pieces are needed to cost-justify the use of a jig?

*C*hapter 23   **CASE STUDY**

Bolt-down Leg on a Casting

Steve Hunter is a consulting engineer and has just received the drawing shown in Figure CS-23. This is one of four legs on a casting made by the CRS Company. These legs are used to attach the device to the floor. The section drawing to the right shows the typical loading to which the leg is subjected. The company is currently drilling the bolt hole and then counterboring the land, but manufacturing has experienced some difficulty in machining the four holes. They report a lot of drill breakage. Quality control reports that distances between the four holes are frequently too large. Sales has recently reported that a substantial number of in-service failures

have occurred with these legs. Steve has obtained a sketch from sales showing where the legs typically fail. This casting is manufactured from gray cast iron using the sand casting process.

1. What machining difficulties should Steve suspect this leg to have?
2. Why were the distances between the holes too large?
3. What should Steve recommend for solving these problems in the future in terms of materials, design, and manufacturing methods?
4. What should Steve recommend be done with the units in the field to stop the failures?

Figure CS-23 shows the design of one of four legs on a casting made by the BRC Company.

CHAPTER 24

MILLING

■ 24.1 INTRODUCTION

Milling is a basic machining process by which a surface is generated by progressive chip removal. The workpiece is fed into a rotating cutting tool. Sometimes the workpiece remains stationary, and the cutter is fed to the work. In nearly all cases, a multiple-tooth cutter is used so that the material removal rate is high. Often the desired surface is obtained in a single pass of the cutter or work and, because very good surface finish can be obtained, milling is particularly well suited and widely used for mass-production work. Many types of milling machines are used, ranging from relatively simple and versatile machines that are used for general-purpose machining in job shops and tool and die work (these are NC or CNC machines) to highly specialized machines for mass production. Unquestionably, more flat surfaces are produced by milling than by any other machining process.

The cutting tool used in milling is known as a *milling cutter*. Equally spaced peripheral teeth will intermittently engage and machine the workpiece. This is called *interrupted cutting*. The workpieces are typically held in fixtures, as described in Chapter 25.

■ 24.2 FUNDAMENTALS OF MILLING PROCESSES

Milling operations can be classified into two broad categories called peripheral milling and face milling. Each has many variations. In *peripheral milling* the surface is generated by teeth located on the periphery of the cutter body (Figure 24-1). The surface is parallel with the axis of rotation of the cutter. Both flat and formed surfaces can be produced by this method, the cross section of the resulting surface corresponding to the axial contour of the cutter. This process, often called *slab milling*, is usually performed on horizontal spindle milling machines. In slab milling, the tool rotates (mills) at some rpm (N_s) while the work feeds past the tool at a table feed rate f_m in inches per minute, which depends on the feed per tooth, f_t.

As in the other processes, the cutting speed V and feed per tooth are selected by the engineer or the machine tool operator. As before, these variables depend upon the work material, the tool material, and the specific process. The cutting velocity is that which occurs at the cutting edges of the teeth in the milling center. The rpm of the spindle is determined from the surface cutting speed, where D is the cutter diameter in inches according to

$$N_s = \frac{12V}{\pi D} \tag{24-1}$$

The depth of cut, called DOC or d in Figure 24-1, is simply the distance between the old and new machined surface.

(a) Horizontal-spindle milling machine

(b) Slab milling—multiple tooth

(c) Allowances for cutter approach

(d) Feed per tooth

FIGURE 24-1   Peripheral milling can be performed on a horizontal-spindle milling machine. The cutter rotates at rpm N_s, removing metal at cutting speed V. The allowance for starting and finishing the cut depends on the cutter diameter and depth of cut, d. The feed per tooth, f_t and cutting speed are selected by the operator or process planner.

The width of cut is the width of the cutter or the work, in inches, and is given the symbol W. The length of the cut L is the length of the work plus some allowance L_A for approach and overtravel. The feed of the table f_m, in inches per minute, is related to the amount of metal each tooth removes during a revolution, the feed per tooth f_t, according to

$$f_m = f_t N_s n \tag{24-2}$$

where n is the number of teeth in the cutter (teeth rev.).

The *cutting time is*

$$T_m = \frac{L + L_A}{f_m} \tag{24-3}$$

The length of approach is

$$L_A = \sqrt{\frac{D^2}{4} - \left(\frac{D}{2} - \text{DOC}\right)^2} = \sqrt{d(D-d)} \tag{24-4}$$

The metal removal rate is

$$\text{MRR} = \frac{\text{volume}}{T_m} = \frac{LWd}{T_m} = Wf_m d \text{ in.}^3/\text{min} \tag{24-5}$$

ignoring L_A. Values for f_t are given in Table 24-1, along with recommended cutting speeds in feet per minute.

TABLE 24-1 Suggested Starting Feeds and Speeds Using High-Speed Steel and Carbide Cutters[a]

Material	Feed (in./tooth) / Speed (fpm)	Carbide Cutters						High-Speed-Steel Cutters					
		Face Mills	Slab Mills	End Mills	Full and Half-Side Mills	Saws	Form Mills	Face Mills	Slab Mills	End Mills	Full and Half-Side Mills	Saws	Form Mills
Malleable iron	Feed per tooth	.005–.015	.005–.015	.005–.010	.005–.010	.003–.004	.005–.010	.005–.015	.005–.015	.003–.015	.006–.012	.003–.006	.005–.010
Soft/hard	Speed, fpm	200–300	200–300	200–350	200–300	200–350	175–275	60–100	60–90	60–100	60–100	60–100	60–80
Cast steel	Feed per tooth	.008–.015	.005–.015	.003–.010	.005–.010	.002–.004	.005–.010	.008–.015	.010–.015	.005–.010	.005–.010	.002–.005	.008–.012
Soft/hard	Speed, fpm	150–350	150–350	150–350	150–350	150–300	150–300	40–60	40–60	40–60	40–60	40–60	40–60
100–150	Feed per tooth	.010–.015	.008–.015	.005–.010	.008–.012	.003–.006	.004–.010	.010–.015	.008–.015	.003–.010	.010–.020	.003–.006	.008–.010
BHN steel	Speed, fpm	450–800	450–600	450–600	450–800	350–600	350–600	80–130	80–130	80–140	80–130	70–100	70–100
150–250	Feed per tooth	.010–.015	.008–.015	.005–.010	.007–.012	.003–.006	.004–.010	.010–.015	.008–.015	.003–.010	.010–.015	.003–.006	.006–.010
BHN steel	Speed, fpm	300–450	300–450	300–450	300–450	300–450	300–450	50–70	50–70	60–80	50–70	50–70	50–70
250–350	Feed per tooth	.008–.015	.007–.012	.005–.010	.005–.012	.002–.005	.003–.008	.005–.010	.005–.010	.003–.010	.005–.010	.002–.005	.005–.010
BHN steel	Speed, fpm	180–300	150–300	150–300	160–300	150–300	150–300	35–60	35–50	40–60	35–50	35–50	35–50
350–450	Feed per tooth	.008–.015	.007–.012	.004–.008	.005–.012	.001–.004	.003–.008	.003–.008	.005–.008	.003–.010	.003–.008	.001–.004	.003–.008
BHN steel	Speed fpm	125–180	100–150	100–150	125–180	100–150	100–150	20–35	20–35	20–40	20–35	20–35	20–35
Cast iron, hard	Feed per tooth	.005–.010	.005–.010	.003–.008	.003–.010	.002–.003	.005–.010	.005–.012	.005–.010	.003–.008	.005–.010	.002–.004	.005–.010
BHN 180–225	Speed, fpm	125–200	100–175	125–200	125–200	125–200	100–175	40–60	35–50	40–60	40–60	35–60	35–50
Cast iron, medium	Feed per tooth	.008–.015	.008–.015	.005–.010	.005–.012	.003–.004	.006–.012	.010–.020	.008–.015	.003–.010	.008–.015	.003–.005	.008–.012
BHN 180–225	Speed, fpm	200–275	175–250	200–275	200–275	200–250	175–250	60–80	50–70	60–90	60–80	60–70	50–60
Cast iron, soft	Feed per tooth	.015–.025	.010–.020	.005–.012	.008–.015	.003–.004	.008–.015	.015–.030	.010–.025	.004–.010	.010–.020	.002–.005	.010–.015
BHN 150–180	Speed, fpm	275–400	250–350	275–400	275–400	250–350	250–350	80–120	70–110	80–120	80–120	70–110	60–80
Bronze	Feed per tooth	.010–.020	.010–.020	.005–.010	.008–.012	.003–.004	.008–.015	.010–.025	.008–.020	.003–.010	.008–.015	.003–.005	.008–.015
Soft/hard	Speed, fpm	300–1000	300–800	300–1000	300–1000	300–1000	200–800	50–225	50–200	50–250	50–225	50–250	50–200
Brass	Feed per tooth	.010–.020	.010–.020	.005–.010	.008–.012	.003–.004	.008–.015	.010–.025	.008–.020	.005–.015	.008–.015	.003–.005	.008–.015
Soft/hard	Speed, fpm	500–1500	500–1500	500–1500	500–1500	500–1500	500–1500	150–300	100–300	150–350	150–350	150–300	100–300
Aluminum alloy	Feed per tooth	.010–.040	.010–.030	.003–.015	.008–.025	.003–.006	.008–.015	.010–.040	.015–.040	.015–.040	.010–.030	.004–.008	.010–.020
Soft/hard	Speed, fpm	2000 UP	2000 UP	2000 UP	2000 UP	2000 UP	2000 UP	300–1200	300–1200	300–1200	300–1200	300–1000	300–1200

[a]Generally, lower end of range used for inserted blade cutters, higher end of range for indexable insert cutters.

In *face milling* and *end milling*, the generated surface is at right angles to the cutter axis (Figure 24-2). Most of the cutting is done by the peripheral portions of the teeth, with the face portions providing some finishing action. Face milling is done on both horizontal- and vertical-spindle machines.

The tool rotates (face mills) at some rpm (N_s) while the work feeds past the tool. The rpm is related to the surface cutting speed V and the cutting tool diameter D, according to equation 24-1. The depth of cut is d, in inches, as shown in Figure 24-2b. The width of cut is W, in inches, and may be width of the workpiece or width of the cutter, depending on the setup. The length of cut is the length of the workpiece L plus an allowance L_A for approach and overtravel L_O, in inches. The feed rate of the table f_m, in inches per minute, is related to the amount of metal each tooth removes during a pass over the work, called the feed per tooth f_t, so $f_m = f_t N_s n$ where the number of teeth in the cutter is n. The cutting time is

$$T_m = \frac{L + L_A + L_O}{f_m} \text{ min} \qquad (24\text{-}6)$$

The *metal removal rate* is

$$\text{MRR} - \frac{\text{volume}}{T_m} = \frac{LWd}{T_m} = f_m W d \text{ in.}^3/\text{min}$$

When calculating the MRR, ignore L_O and L_A. The length of approach is usually equal to the length of overtravel, which usually equals $D/2$ in. For a setup where the tool does not completely pass over the workpiece,

$$L_O = L_A = \sqrt{W(D - W)} \text{ for } W < \frac{D}{2} \qquad (24\text{-}7)$$

$$L_O = L_A = \frac{D}{2} \text{ for } W \geq \frac{D}{2} \qquad (24\text{-}8)$$

FACE MILLING EXAMPLE

A 4-in.-diameter, six-tooth face mill is selected, using carbide inserts (Figure 24-3). The material being machined is low-alloy steel, annealed. Using cutting data recommendations, the cutting speed chosen is 400 sfpm with a feed of 0.008 in./tooth at a d of 0.12 inches. Determining rpm at the spindle,

$$N_s = \frac{12V}{\pi D} = \frac{12 \times 400}{3.14 \times 4} = 392 \text{ rpm}$$

Determining the feed rate of the table, $f_m = nN_s f_t$,

$$f_m = 0.008 \times 6 \times 392 = 19 \text{ in./min}$$

If slab or side milling were being performed, as shown in Figure 24-4, with the same parameters being selected as above, the setup would be different but the spindle rpm and table feed rate the same. The cutting time would be different because the allowances for face milling are greater than for slab milling. In milling, power consumption is usually the limiting factor. A thick chip is more power efficient than a thin chip.

END MILLING EXAMPLE

End milling is a very common operation performed on both vertical- and horizontal-spindle milling machines or machining centers. Figure 24-5 shows a vertical spindle end milling process, cutting a step in the workpiece. This cutter can cut on both the sides and ends of the tool. If you were performing this operation on a block of metal (for example, 430F stainless steel), you (the manufacturing engineer) would select a specific machine tool. You would have to determine how many passes (rough and finish cuts) were needed to produce the geometry specified in the design. Why? The number of passes determines the total cutting time for the job.

Using a vertical-spindle milling machine, an end mill can produce a step in the workpiece. In Figure 24-5, an end mill with six teeth on a 2-in. diameter is used to cut a step in

(a) Vertical-spindle milling machine
Top views

Column

Six-tooth face mill

(b) Face milling over part of surface

Table

Workpiece

(c) Allowance for partial coverage

Machined surface

(d) Allowance for full coverage

FIGURE 24-2 Face milling is often performed on a spindle milling machine using a multiple-tooth cutter ($n = 6$ teeth) rotating N_s at rpm to produce cutting speed V. The workpiece feeds at rate f_m in inches per minute past the tool. The allowance depends on the tool diameter and the width of cut.

Top view

Cutter

Spindle

Workpiece

Machined surface

f_m = feed rate
= 19 in./min

FIGURE 24-3 Face milling viewed from above with vertical spindle-machine.

D

d

f_m Workpiece

Machined surface

FIGURE 24-4 Slab or side milling being done as a down milling process with horizontal spindle-machine.

End View   (not to scale)   Side View

Rotating
Cutting
Tool – End Mill = 2 in. diam., 6 teeth

1.25 Inches DOI

0.375   ← d →   ← Work ←
                                        feed rate (in/min)

Fixture │ Workpiece │ Vise   Vise
or vise │ 430F   │   d = Depth of cut
        │ stainless │   DOI = Depth of immersion

FIGURE 24-5   End milling a step feature in a block using a flat-bottomed, end mill cutter in a vertical spindle-milling machine. On left, photo. In middle, end view, table moving the block into the cutter. On right, side view, workpiece feeding right to left into tool.

430F stainless. The d (depth of cut) is 0.375 in., and the depth of immersion (DOI) is 1.25 in. The tool deflects due to the cutting forces, so the cut needs to be made at full immersion; but there may not be enough power for a full DOC. Can the step be cut in one pass or will multiple cuts be necessary? The vertical milling machine tool available has a 5-hp motor with an 80% efficiency. The specific horsepower for 430F stainless is 1.3 hp/in.^{3}/min.

The maximum amount of material that can be removed per pass is usually limited by the available power. Using the hp equation from Chapter 21,

$$hp = HP_s \times MRR = HP_s \times f_m WD = HP_s f_m \times DOI \times d \qquad (24\text{-}9)$$

Select $f_t = 0.005$ ipt and $V = 250$ fpm from Table 24-1. Calculate the spindle rpm:

$$N_s = \frac{12 \times 250}{3.14 \times 2} = 477 \text{ rpm of cutter}$$

Next, assuming the machine tool has this rpm available, calculate the table feed rate:

$$f_m = f_t \times n \times N_s = 0.005 \times 6 \times 477 = 14.31 \text{ in./min}$$

But the actual table feed rates for the selected machine are 11 in./min or 16 in./min, so, being conservative, select

$$f_m = \text{table feed rate} = 11.00 \text{ in./min}$$

Next, assuming 80% of the available power is used for cutting, calculate the depth of cut from equation 24-9:

$$d = DOC \cong \frac{5 \times 0.8}{1.3 \times 11.00 \times 1.25} \cong 0.225 \text{ in. maximum}$$

Therefore, two passes are needed because (0.375/0.225 = 1.6):

$$0.375 - 0.225 = 0.150 \text{ in. second pass DOC}$$

2 passes: DOC = 0.225 rough cut
DOC = 0.150 finish cut
0.375 total DOC

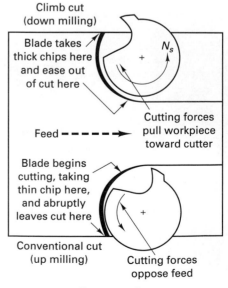

Note that for $d = 0.150$, the feed per tooth would be only slightly increased to 0.0051 ipt:

$$f_t = \frac{0.5 \times 0.8}{1.3 \times 6 \times 477 \times 0.150 \times 1.25} = 0.0051 \text{ in./tooth}$$

You may want to change f_t to improve the surface finish. With a smaller f_t, a better surface finish is usually obtained. However, there are other factors to consider, like machining time.

In general (for face, slab, or end milling), if machine power is lacking the following actions may help.

1. Use a cutter with a positive rake as this can be more efficient than one with a negative rake.
2. Use a cutter with a coarser pitch (fewer teeth).
3. Use a smaller cutter and take several passes (reduce d or DOI).

UP VERSUS DOWN MILLING

For either slab or end or face milling, surfaces can be generated by two distinctly different methods (Figure 24-6). *Up milling* is the traditional way to mill and is called *conventional* milling. The cutter rotates against the direction of feed of the workpiece. In *climb* or *down milling*, the cutter rotation is in the same direction as the feed rate. The method of chip formation is completely different in the two cases.

In up milling, the chip is very thin at the beginning, where the tooth first contacts the work; then it increases in thickness, becoming a maximum where the tooth leaves the work. The cutter tends to push the work along and lift it upward from the table. This action tends to eliminate any effect of looseness in the feed screw and nut of the milling machine table and results in a smooth cut. However, the action tends to loosen the work from the fixture. Therefore, greater clamping forces must be employed, with the danger of deflecting the part. In addition, the smoothness of the generated surface depends greatly on the sharpness of the cutting edges. In up milling, chips can be carried into the newly machined surface, causing the surface finish to be poorer (rougher) than in down milling and causing damage to the insert.

In down milling, maximum chip thickness occurs close to the point at which the tooth contacts the work. Because the relative motion tends to pull the workpiece into the cutter, any possibility of looseness in the table feed screw must be eliminated if down

FIGURE 24-6 Climb cut or down milling versus conventional cut or up milling for slab or face or end milling.

milling is to be used. It should never be attempted on machines that are not designed for this type of milling. Virtually all modern milling machines are capable of down milling, and it is a most favorable application for carbide cutting edges. Because the material yields in approximately a tangential direction at the end of the tooth engagement, there is less tendency (than when up milling is used) for the machined surface to show tooth-marks, and the cutting process is smoother, with less chatter. Another advantage of down milling is that the cutting force tends to hold the work against the machine table, permitting lower clamping forces. However, the fact that the cutter teeth strike against the surface of the work at the beginning of each chip can be a disadvantage if the workpiece has a hard surface, as castings sometimes do. This may cause the teeth to dull rapidly. Metals that readily workharden should be down milled, and many toolmakers recommend that down milling should always be the first choice.

MILLING SURFACE FINISH

The average surface finishes that can be expected on free-machining materials range from 60 to 150 μin. Conditions exist, however, that can produce wide variations on either side of these ranges. For example, some inserts are designed with wiper flats (short parallel surface behind the tool tip). If the feed per revolution [feed per tooth × number of teeth] of the cutter is smaller than the length of the wiper flat (the land on the tool), then the surface finish on the workpiece will be generated by the highest insert. In finishing cuts, keeping the depth of cut small will limit the axial cutting force, reducing vibrations and producing a superior finish. See Chapter 35 for discussions on measuring surface finish.

Milling is an interrupted cutting process wherein entering and leaving the cut subjects the tool to impact loading, cyclic heating, and cycle cutting forces. As shown in Figure 24-7, the cutting force, F_c, builds rapidly as the tool enters the work at A and progresses to B, peaks as the blade crosses the direction of feed at C, decreases to D, and then drops to zero abruptly upon exit. The diagram does not indicate the impulse loads caused by impacts. The interrupted-cut phenomenon explains in large part why milling cutter teeth are designed to have small positive or negative rakes, particularly when the tool material is carbide or ceramic. These brittle materials tend to be very strong in compression, and negative rake results in the cutting edges being placed in compression by the cutting forces rather than tension. Cutters made from high-speed steel (HSS) are made with positive rakes, in the main, but must be run at lower speeds. Positive rake tends to lift the workpiece, while negative rakes compress the workpiece and allow heavier cuts to be made. Table 24-2 summarizes some additional milling problems.

■ 24.3 MILLING TOOLS AND CUTTERS

Most milling work today is done with face mills and end mills. The face mills use indexable carbide insert tooling, while the end mills are either solid HSS or insert tooling (Figure 24-8). Basically, *mills* are shank-type cutters having teeth on the circumferential surface and one end. They thus can be used for facing, profiling, and end milling. The teeth

FIGURE 24-7 Conventional face milling (left) with cutting force diagram for F_c (right) showing the interrupted nature of the process. *(From* Metal Cutting Principles, *2nd ed., Ingersoll Cutting Tool Company.)*

TABLE 24-2 Probable Causes of Milling Problems

Problem	Probable Cause	Cures
Chatter (vibration)	1. Lack of rigidity in machine, fixtures, arbor, or workpiece 2. Cutting load too great 3. Dull cutter 4. Poor lubrication 5. Straight-tooth cutter 6. Radial relief too great 7. Rubbing, insufficient clearance	Use larger arbors. Change rpm (cutting speed). Decrease feed per tooth or number of teeth in contact with work. Sharpen or replace inserts. Flood coolant. Use helical cutter. Check tool angles.
Loss of accuracy (cannot hold size)	1. High cutting load causing deflection 2. Chip packing, between teeth 3. Chips not cleaned away before mounting new piece of work	Decrease number of teeth in contact with work or feed per tooth. Adjust cutting fluid to wash chips out of teeth.
Cutter rapidly dulls	1. Cutting load too great 2. Insufficient coolant	Decrease feed per tooth or number of teeth in contact. Add blending oil to coolant.
Poor surface finish	1. Feed too high 2. Tool dull 3. Speed too low 4. Not enough cutter teeth	Check to see if all teeth are set at same height.
Cutter digs in (hogs into work)	1. Radial relief too great 2. Rake angle too large 3. Improper speed	Check to see that workpiece is not deflecting and is securely clamped.
Work burnishing	1. Cut is too light 2. Tool edge worn 3. Insufficient radial relief 4. Land too wide	Enlarge feed per tooth. Sharpen cutter.
Cutter burns	1. Not enough lubricant 2. Speed too high	Add sulfur-based oil. Reduce cutting speed. Flood coolant.
Teeth breaking	1. Feed too high 2. Depth of cut too large	Decrease feed per tooth. Use cutter with more teeth. Reduce table feed rate.

Adapted from *Cutting Tool Engineering*, October 1990, p. 90, by Peter Liebhold, museum specialist, Division of Engineering and Industry, the Smithsonian Institute, Washington, DC.

FIGURE 24-8 Solid end mills are often coated. Insert tooling end mills come in a variety of sizes and are mounted on taper shanks.

may be either straight or helical, but the latter is more common. Small end mills have straight shanks, whereas taper shanks are used on larger sizes (Figure 24-8).

Plain end mills have multiple teeth that extend only about halfway toward the center on the end. They are used in milling slots, profiling, and facing narrow surfaces. *Two-lip mills* have two straight or helical teeth that extend to the center. Thus they may be sunk into material, like a drill, and then fed lengthwise to form a groove, a slot, or a pocket.

Shell end mills are solid multiple-tooth cutters, similar to plain end mills but without a shank. The center of the face is recessed to receive a screw head or nut for mounting the cutter on a separate shank or a stub arbor. One shank can hold any of several cutters and thus provides great economy for larger-sized end mills.

Hollow end mills are tubular in cross section, with teeth only on the end but having internal clearance. They are used primarily on automatic screw machines for sizing cylindrical stock, producing a short cylindrical surface of accurate diameter.

Face mills have a center hole so that they can be arbor mounted. Face milling cutters are widely used in both horizontal- and vertical-spindle machine tools and come in a wide variety of sizes (diameters and heights) and geometries (round, square, triangular, etc.), as shown in Figure 24-9.

The insert can usually be indexed four times and must be well supported. Either the power or the rigidity of the machine tool will be the limiting factor, although sometimes setup can be the limiting factor.

FIGURE 24-9 Face mills come in many different designs using many different insert geometries and different mounting arbors.

FIGURE 24-10 The side-milling cutter can cut on sides and ends of the teeth, so it makes slots or grooves. However, only a few teeth are engaged at any one point in time, causing heavy torsional vibrations. The average chip thickness, h_i, will be less than the feed per tooth, f_t. The actual feed per tooth f_a will be less than feed per tooth selected, F_t, according to $f_a = h_i \sqrt{\dfrac{D}{d}}$.

Another common type of arbor-mounted milling cutter is called a side mill because it cuts on the ends and sides of the cutters. Figure 24-10 shows the geometry of a staggered-tooth side milling cutter.

Staggered-tooth milling cutters are narrow cylindrical cutters having staggered teeth, and with alternate teeth having opposite helix angles. They are ground to cut only on the periphery, but each tooth also has chip clearance ground on the protruding side. These cutters have a free cutting action that makes them particularly effective in milling deep slots. *Staggered-tooth cutters* are really special *side-milling cutters*, which are similar to plain milling cutters except that the teeth extend radially part way across one or both ends of the cylinder toward the center. The teeth may be either straight or helical. Frequently, these cutters are relatively narrow, being disklike in shape. Two or more side milling cutters often are spaced on an arbor to straddle the workpiece (called *straddle milling*), and two or more parallel surfaces are machined at once.

In Figure 24-11 insert-tooth side mills are arranged in a gang-milling setup to cut three slots in the workpiece simultaneously. Thus the desired part geometry is repeatedly produced by the setup as the position of the cutters is fixed. However, in side- and face-milling operations only a few teeth are engaged at any point in time, resulting in heavy torsional vibrations detrimental to the resulting machined product. A flywheel can solve this problem and in many cases be the key to improved productivity.

FIGURE 24-11   Arbor (two views) used on a horizontal-spindle milling machine on left. On right, a gang-milling setup showing three side-milling cutters mounted on an arbor (A) with an outboard flywheel (B).

For the gang milling as shown in Figure 24-11 the diameter of the flywheel should be as large as possible. (The moment of inertia increases with the square of the radius.) The best position of the flywheel is inboard on the arbor at A, but depending on the setup, this may not be possible, so then position B should be chosen. It is important that the distance between the cutters and flywheel be as small as possible.

A flywheel can be built up from a number of carbon steel disks, each having a center hole and keyway to fit the arbor, so the weight can be easily varied.

Interlocking slotting cutters consist of two cutters similar to side mills but made to operate as a unit for milling slots. The two cutters are adjusted to the desired width by inserting shims between them.

Slitting saws are thin, plain milling cutters, usually from $1/32$ to $3/16$ in. thick, which have their sides slightly "dished" to provide clearance and prevent binding. They usually have more teeth per unit of diameter than ordinary plain milling cutters and are used for milling deep narrow slots and cutting-off operations.

In milling, the average chip thickness (h_i) is not the same as the feed per tooth. For example, a thickness (h_i) of 0.004 in. corresponds to 0.012 in. feed per tooth in most side and face milling operations.

If the radial depth of cut, d, is very small compared to the cutter diameter, D, use this formula:

$$\text{feed per tooth} = f_t = 0.004\sqrt{\frac{D}{d}}\ (\text{ipt})$$

(*Note:* For calculating the table feed, use half the number of inserts in a full side and face mill to arrive at the effective number of teeth.)

table feed rate (ipm) = rpm × number of effective teeth × feed per tooth

Another method of classification for face and end mill cutters relates to the direction of rotation. A *right-hand cutter* must rotate counterclockwise when viewed from the front end of the machine spindle. Similarly, a *left-hand cutter* must rotate clockwise. All other cutters can be reversed on the arbor to change them from one hand to the other. Positive rake angles are used on general-purpose HSS milling cutters. Negative rake angles are commonly used on carbide- and ceramic-tipped cutters employed in mass-production milling in order to obtain the greater strength and cooling capacity. TiN coating of these tools is quite common, resulting in significant increases in tool life.

Plain milling cutters used for plain or slab milling have straight or helical teeth on the periphery and are used for milling flat surfaces. *Helical mills* (Figure 24-12) engage the work gradually, and usually more than one tooth cuts at a given time. This reduces shock and chattering tendencies and promotes a smoother surface. Consequently, this type of cutter usually is preferred over one with straight teeth.

Angle milling cutters are made in two types: single angle and double angle. Angle cutters are used for milling slots of various angles or for milling the edges of workpieces to a desired angle. *Single-angle cutters* have teeth on the conical surface, usually at an

FIGURE 24-12 The chips are formed progressively by the teeth of a plain helical-tooth milling cutter during up milling.

angle of 45° to 60° to the plane face. *Double-angle cutters* have V-shaped teeth, with both conical surfaces at an angle to the end faces but not necessarily at the same angle. The V-angle usually is 45°, 60°, or 90°.

Form milling cutters have the teeth ground to a special shape—usually an irregular contour—to produce a surface having a desired transverse contour. They must be sharpened by grinding only the tooth face, thereby retaining the original contour as long as the plane of the face remains unchanged with respect to the axis of rotation. Convex, concave, corner-rounding, and gear-tooth cutters are common examples (Figure 24-13). Solid HSS cutters of simple shape and reasonably small size are usually more economical in initial cost than inserted-blade cutters. However, inserted-blade cutters may be lowest in overall cost on large production jobs.

Form-relieved cutters can be cost effective where intricately shaped cuts are needed. Solid or carbide insert tool cutters may need large volumes to be cost-justified by high-production requirements.

Most larger-sized milling cutters are of the *inserted-tooth type.* The cutter body is made of steel, with the teeth made of high-speed steel, carbides, or TiN carbides, fastened to the body by various methods. An insert tooth cutter uses indexable carbide or ceramic inserts, as shown in Figure 24-9. This type of construction reduces the amount of costly material that is required and can be used for any type of cutter, but it is most often used with face mills.

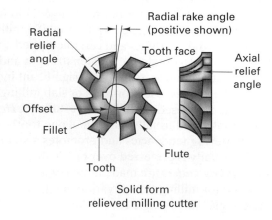

Solid form relieved milling cutter

FIGURE 24-13 Solid form relieved milling cutter, would be mounted on an arbor in a horizontal milling machine.

T-slot cutters are integral-shank cutters with teeth on the periphery and *both* sides. They are used for milling the wide groove of a T-slot. To use them, the vertical groove must first be made with a slotting mill or an end mill to provide clearance for the shank. Because the T-slot cutter cuts on five surfaces simultaneously, it must be fed with care.

Woodruff keyseat cutters are made for the single purpose of milling the semicylindrical seats required in shafts for Woodruff keys. They come in standard sizes corresponding to Woodruff key sizes. Those below 2 in. in diameter have integral shanks; the larger sizes may be arbor mounted.

Occasionally, *fly cutters* may be used for face milling or boring. Both operations may be done with a single tool at one setup. A single-point cutting tool is attached to a special shank, usually with provision for adjusting the effective radius of the cutting tool with respect to the axis of rotation. The cutting edge can be made in any desired shape and, because it is a single-point tool, is very easy to grind.

■ 24.4 MACHINES FOR MILLING

The four most common types of manually controlled milling machines are listed below in order of increasing power (and therefore metal removal capability):

1. Ram-type milling machines
2. Column-and-knee-type milling machines
   a. Horizontal spindle
   b. Vertical spindle
3. Fixed-bed-type milling machines
4. Planer-type milling machines

Milling machines whose motions are electronically controlled are listed in order of increasing production capacity and decreasing flexibility:

1. Manual data input milling machines
2. Programmable CNC milling machines
3. Machining centers (tool changer and pallet exchange capability)
4. Flexible Manufacturing Cell and Flexible Manufacturing System
5. Transfer lines

BASIC MILLING MACHINE CONSTRUCTION

Most basic milling machines are of *column-and-knee* construction, employing the components and motions shown in Figure 24-14. The column, mounted on the base, is the main supporting frame for all the other parts and contains the spindle with its driving mechanism. This construction provides controlled motion of the worktable in three mutually perpendicular directions: (1) through the *knee*, moving vertically on ways on the front of the column; (2) through the *saddle*, moving transversely on ways on the knee; and (3) through the *table*, moving longitudinally on ways on the saddle. All these motions can be imparted by either manual or powered means. In most cases, a powered rapid traverse is provided in addition to the regular feed rates for use in setting up work and in returning the table at the end of a cut.

The ram-type milling machine is one of the most versatile and popular milling machines, using the knee and column design. Ram-type machines have a head equipped with a motor-stopped pulley and belt drive as well as a spindle. The ram, mounted on horizontal ways at the top of the column, supports the head and permits positioning of the spindle with respect to the table. Ram-type milling machines are normally 10 hp or less and suitable for light-duty milling, drilling, reaming, and so on (Figure 24-15).

Milling machines having only the three mutually perpendicular table motions are called *plain column-and-knee type*. These are available with both horizontal and vertical spindles (Figure 24-14). On the older horizontal-spindle-type machines, an adjustable overarm provides an outboard bearing support for the end of the cutter arbor, which is shown in Figure 24-11. These machines are well suited for slab, side, or straddle milling.

FIGURE 24-14 Major components of a plain column-and-knee-type milling machine, which can have horizontal spindle shown on the left, or a turret type machine with a vertical spindle, shown on the right. The workpiece and workholder on the table can be translated in *X, Y,* and *Z* directions with respect to the tool.

In some vertical-spindle machines the spindle can be fed up and down, either by power or by hand. Vertical-spindle machines are especially well suited for face and end milling operations. They also are very useful for drilling and boring, particularly where holes must be spaced accurately in a horizontal plane, because of the controlled table motion.

Turret-type column-and-knee milling machines have dual heads that can be swiveled about a horizontal axis on the end of a horizontally adjustable ram. This permits milling to be done horizontally, vertically, or at any angle. This added flexibility is advantageous when a variety of work has to be done, as in tool and die or experimental shops. They are available with either plain or universal tables.

Universal column-and-knee milling machines differ from plain column-and-knee machines in that the table is mounted on a housing that can be swiveled in a horizontal plane, thereby increasing its flexibility. Helices, as found in twist drills, milling cutters, and helical gear teeth, can be milled on universal machines.

Ram-type
miller

Vertical spindle ram-type

FIGURE 24-15 The ram-type knee-and-column milling machine is one of the most versatile and popular milling machine tools ever designed.

BED-TYPE MILLING MACHINES

In production manufacturing operations, ruggedness and the capability of making heavy cuts are of more importance than versatility. *Bed-type milling machines* (Figure 24-16) are made for these conditions. The table is mounted directly on the bed and has only longitudinal motion. The spindle head can be moved vertically in order to set up the machine for a given operation. Normally, once the setup is completed, the spindle head is clamped in position and no further motion of it occurs during machining. However, on some machines vertical motion of the spindle occurs during each cycle.

After such milling machines are set up, little skill is required to operate them, permitting faster learning time for the operators. Some machines of this type are equipped with automatic controls so that all the operator has to do is load and unload workpieces into the fixture and set the machine into operation. For stand-alone machines, a fixture can be located at each end of the table so that one workpiece can be loaded while another is being machined.

Bed-type milling machines with single spindles are sometimes called *simplex milling machines*; they are made with both horizontal and vertical spindles. Bed-type machines also are made in *duplex* and *triplex* types, having two or three spindles respectively, permitting the simultaneous milling of two or three surfaces at a single pass.

PLANER-TYPE MILLING MACHINES

Planer-type milling machines (Figure 24-17) utilize several milling heads, which can remove large amounts of metal while permitting the table and workpiece to feed quite slowly. Often only a single pass of the workpiece past the cutters is required. Through the use of different types of milling heads and cutters, a wide variety of surfaces can be machined with a single setup of the workpiece. This is an advantage when heavy workpieces are involved.

ROTARY-TABLE MILLING MACHINES

Some types of face milling in mass-production manufacturing are often done on *rotary-table milling machines*. Roughing and finishing cuts can be made in succession as the workpieces are moved past the several milling cutters while held in fixtures on the rotating table. The operator can load and unload the work without stopping the machine.

Bed-type vertical spindle

FIGURE 24-16   Bed-type vertical-spindle heavy-duty production machine tools for milling usually have three axes of motion.

FIGURE 24-17 Large planer-type milling machine. Inset shows 90° head being used. *(Courtesy of Cosa Corporation.)*

PROFILERS AND DUPLICATORS

Milling machines that can duplicate external or internal geometries in two dimensions are called *profilers* or tracer-controlled machines. A tracing probe follows a two-dimensional pattern or template and, through electronic or hydraulic air-actuated mechanisms, controls the cutting spindles in two mutually perpendicular directions.

All hydraulic tracers work basically the same way, in that they utilize a stylus connected to a precision servomechanism for each axis of control. The servos are connected to hydraulic actuators on the machine slides. As the stylus traces a template, the servos control the motion of the slides so that the milling cutter duplicates the template shape onto the workpiece.

Duplicators produce forms in three dimensions and are widely used to machine molds and dies. Sometimes these machines are called *die-sinking machines*. They are used extensively in the aerospace industry to machine parts from wrought plate or bar stock as substitutes for forgings when the small number of parts required would make the cost of forging dies uneconomical. Many of these kinds of jobs are now done on NC- and CNC-type machines; their applications are discussed in Chapters 26.

MILLING MACHINE SELECTION

When purchasing or using a milling machine, consider the following issues:

1. Spindle orientation and rpm
2. Machine capability (accuracy and precision)
3. Machine capacity (size of workpieces)
4. Horsepower available at spindle (usually 70% of machine horsepower)
5. Automatic tool changing

The choice of spindle orientation, horizontal or vertical, depends on the parts to be machined. Relatively flat parts are usually done on vertical machines. Cubic parts are usually done on a horizontal machine, where chips tend to fall free of the part. Operations like slotting and side milling are best done on horizontal machines with outboard supports for the arbor. Use the largest-diameter arbor possible to reduce twist and deflection due to cutting forces.

Machine capability refers to the tolerances, while *capacity* refers to the size of parts and the power available.

FIGURE 24-18 End milling a helical groove on a horizontal-spindle milling machine using a universal dividing head and a universal milling attachment. *(Courtesy of Cincinnati Milacron, Inc.)*

Labels on figure: Universal milling attachment; Spindle; Index plate; Crank

As with all tooling applications, the tolerances that can be maintained in milling are dependent upon the rigidity of the workpiece, the accuracy and rigidity of the machine spindle, the precision and accuracy of the workholding device, and the quality of the cutting tool itself. Milling produces forces that contribute to chatter and vibration because of the intermittent cutting action. Soft materials tend to adhere to the cutter teeth and make it more difficult to hold tolerances. Materials such as cast iron and aluminum are easy to mill.

Within these criteria, properly maintained cutters used in rigid spindles on properly fixtured workpieces can expect to machine within tolerances with surface flatness tolerances of 0.001 in./ft. Such tolerances are also possible on "slotting" operations with milling cutters, but +0.001 in. to +0.002 in. is more probable. Flatness specifications are more difficult to maintain in steel and easier to maintain in some types of aluminum, cast iron, and other nonferrous material.

Part size is the primary factor in selecting the machine size, but the length of the tooling as mounted in the spindle must be considered. Horsepower required at the spindle depends on the MRR and the materials (unit horsepower, hp_s). Remember, coated inserts allow the MRR (the cutting speed) to be increased and available power may be exceeded.

Finally, the capacity of the tool changers on machining centers is limited by the number, size, and weight of the tools, especially if large-diameter tools are being employed. These often have to be stored in every other space in the storage mechanism.

ACCESSORIES FOR MILLING MACHINES

The usefulness of ordinary milling machines can be greatly extended by employing various accessories or attachments. Here are some examples.

A horizontal milling machine can be equipped with a vertical milling attachment to permit vertical milling to be done. Ordinarily, heavy cuts cannot be made with such an attachment.

The *universal milling attachment* (Figure 24-18) is similar to the vertical attachment but can be swiveled about both the axis of the milling machine spindle and a second, perpendicular axis to permit milling to be done at any angle.

The *universal dividing head* is by far the most widely used milling machine accessory, providing a means for holding and indexing work through any desired arc of rotation. The work may be mounted between centers (Figure 24-18) or held in a chuck that is mounted in the spindle hole of the dividing head. The spindle can be tiled from about 5° below horizontal to beyond the vertical position.

Basically, a dividing head is a rugged, accurate, 40:1 worm-gear reduction unit. The spindle of the dividing head is rotated one revolution by turning the input crank 40 turns. An index plate mounted beneath the crank contains a number of holes, arranged in concentric circles and equally spaced, with each circle having a different number of holes. A plunger pin on the crank handle can be adjusted to engage the holes of any circle. This permits the crank to be turned an accurate, fractional part of a complete circle as represented by the increment between any two holes of a given circle on the index plate. Utilizing the 40:1 gear ratio and the proper hole circle on the index plate, the spindle can be rotated a precise amount by the application of either of the following rules:

$$\text{number of turns of crank} = \frac{40}{\text{cuts per revolution of work}}$$

$$\text{holes to be indexed} = \frac{40 \times \text{holes index circle}}{\text{cuts per revolution of work}}$$

If the first rule is used, an index circle must be selected that has the proper number of holes to be divisible by the denominator of any resulting fractional portion of a turn of the crank. In using the second rule, the number of holes in the index circle must be such that the numerator of the fraction is an even multiple of the denominator. For example, if 24 cuts are to be taken about the circumference of a workpiece, the number of turns of the crank required would be $1\,^2/_3$. An index circle having 12 holes could be used with one full turn plus eight additional holes. The second rule would give the same result. Adjustable *sector arms* are provided on the index plate that can be set to a desired number of holes, less than a full turn, so that fractional turns can be made readily without the necessity for counting holes each time. Dividing heads are made having ratios other than 40:1. The ratio should be checked before using.

Because each full turn of the crank on a standard dividing head represents 360/40, or 9° of rotation of the spindle, indexing to a fraction of a degree can be obtained. Indexing can be done in three ways. *Plain indexing* is done solely by the use of the 40:1 ratio in the dividing head. In *compound indexing*, the index plate is moved forward or backward a number of hole spaces each time the crank handle is advanced. For *differential indexing* the spindle and the index plate are connected by suitable gearing so that as the spindle is turned by means of the crank, the index plate is rotated a proportional amount.

The dividing head can also be connected to the feed screw of the milling machine table by means of gearing. This procedure is used to provide a definite rotation of the workpiece with respect to the longitudinal movement of the table, as in cutting helical gears. This procedure is illustrated in Chapter 29.

■ Key Words

climb (down) milling	end milling	metal removal rate	slab milling
column-and-knee	face milling	milling	staggered-tooth milling cutter
milling machine	insert-tooth milling cutter	milling cutters	straddle milling
conventional (up) milling	interrupted cutting	milling machines	Woodruff keyseat
cutting time	machining center	peripheral milling	

■ Review Questions

1. Suppose you wanted to machine a cast iron with BHN of 275. The process to be used is face milling and an HSS cutter is going to be used. What feed and speed values would you select?

2. Explain how table feed (ipm) and spindle rpm are specified or computed for a milling machine after speed and feed per tooth are selected.

3. Why must the number of teeth on the cutter be known when calculating milling machine table feed, in in./min?

4. Why is the question of up or down milling more critical in horizontal slab milling than in vertical-spindle (end or face) milling?

5. For producing flat surfaces in mass production machining, how does face milling differ basically from peripheral milling?

6. Milling has a higher metal removal rate than planing. Why?

7. Which type of milling (up or down) is being done in Figures 24-1, Figure 24-2, and 24-7?

8. Why does down milling dull the cutter more rapidly than up milling when machining sand castings?

9. What parameters do you need to specify in order to calculate MRR in milling?

10. In Figure 24-2b the tool material is carbide. What would you change in the process?

11. What is the advantage of a helical-tooth cutter over a straight-tooth cutter for slab milling?
12. What would the cutting force diagram for F_c look like if the cutter were performing climb milling?
13. Could the stub arbor–mounted face mill shown in Figure 24-9 be used to machine a T-slot? Why or why not?
14. In a typical solid arbor milling cutter shown in Figure 24-10, why are the teeth staggered? (Check in Chapter 19 for discussion of dynamics.)
15. Make some sketches to show how you would you set up a plain column-and-knee milling machine to make it suitable for milling the top and sides of a large block.
16. Make some sketches to show how you would set up a horizontal milling machine to cut both sides of a block of metal simultaneously.
17. Explain how controlled movements of the work in three mutually perpendicular directions are obtained in column-and-knee-type milling machines.
18. What is the basic principle of a universal dividing head?
19. What is the purpose of the hole-circle plate on a universal dividing head?

■ Problems

1. You have selected a feed per tooth and a cutting speed for a face milling process. Reasonable values for feed and speed are 0.010 in. per tooth and 200 sfpm. The cutter is 8 in. in diameter, as shown in Figure 24-9. Compute the input values for the machine tool.
2. How much time will be required to face mill an AISI 1020 steel surface Bhn, 150, that is 12 in. long and 5 in. wide, using a 6-in-diameter, eight-tooth tungsten carbide inserted-tooth face mill cutter? Select values of feed per tooth and cutting speed from Table 24-1.
3. If the depth of cut is 0.35 in., what is the metal removal rate in Problem 2?
4. Estimate the power required for the operation of Problem 3. Do not forget to consider Figure 24-7.
5. Examine the part shown in Figure 1-6. The slot on the left end must be produced by machining. Provide a process plan (a description [sketch] of how the part would set up in the machine for machining the slot and the details regarding cutting tools, such as material, sizes, and so on). Specify (select) the type of milling machine, the cutting parameters, and any other information needed to make this component.
6. A gray cast iron surface 6 in. wide and 18 in. long may be machined on either a vertical milling machine, using an 8-in.-diameter face mill having eight inserted HSS teeth, or on a horizontal milling machine using an HSS slab mill with eight teeth on a 4-in. diameter. Which machine has the faster cutting time?
7. An operation is to be performed to machine three grooves on a number of parts shown in Figure 24-11. Setup time is 20 minutes on a shaper (not shown) and 30 minutes on the horizontal milling machine. The direct time to machine each piece on the shaper is 14 min and on the miller is 6 min. Labor costs $10/hr. The charge for the use of the shaper is $10/hr and for the milling machine $20/hr. What is the breakeven quantity, below which the shaper is more economical than the mill.

8. In Figure 24-12, the feed is 0.006 in. per tooth. The cutter is rotating at an rpm that will produce the desired surface cutting speed of 125 sfpm. The cutter diameter is 5 in. The depth of cut is 0.5 in. The block is 2 in. wide.
   a. What is the feed, in inches per minute, of the milling machine table?
   b. What is the MRR for this situation?
   c. What is horsepower (HP) consumed by this process, assuming an 80% efficiency and a HP_s value for this material of 1.8?
9. Suppose you want to do the job described in Problem 6 by slab milling. You have selected a 6-in.-diameter cutter with eight TiN-coated carbide teeth. The cutting speed will be 500 sfpm and the feed per tooth will be 0.010 in. per tooth. Determine the input parameters for the machine (rpm of arbor and table feed), then calculate the T_m and MRR. Compare these answers with what you got for face milling the block with HSS teeth.
10. The Bridgeport vertical-spindle milling machine is perhaps the single most popular machine tool. Virtually every factory (or shop) that does machining has one or more of these type machines. Go to your nearest machine shop and find a Bridgeport, make a sketch to show how it works, and explain what makes it so popular.

www.wiley.com/college/degarmo

Chapter 24 CASE STUDY

HSS versus Tungsten Carbide Milling

The K & C Machine works, which does job shop machining, has received an order to make 40 duplicate pieces, made of AISI 4140 steel, which will require 1 hour per piece of actual cutting time if a high-speed-steel (MI) milling cutter is used. Abigail Langley, a new machinist, says the cutting time could be reduced significantly if the company would purchase a suitable tungsten carbide milling cutter. Hugh Fellows, the foreman for the milling area, says he does not believe that Abigail's estimate is realistic, and he is not going to spend $450 (the current price from the vendor) of the company's money on a carbide cutter that probably would not be used again. The machine hour rate, including labor for your shop is $40 per hour. Abigail and Hugh have come to you, the supervisor of the shop, for a decision on whether or not to buy the cutter, which is readily available from a local supplier.

What factors should you consider in this situation? How much faster could the carbide cutter cut compared to the HSS cutter? See reference table. Based on your best guess as to the savings in actual cutting time per piece, who do you think is correct, Abigail or Hugh?

Representative Cutting Data							
Material						**Forces**	
Work	Tool	Back Rake (deg.)	Feed (ipt)	Width (in.)	Velocity (fpm)	Cutting (lb)	Thrust (lb)
AIS14140	HSS	0	0.0104	0.100	100	360	190
AIS14140	Carbide	0	0.011	0.15	540	540	156
AIS14145	Carbide	0	0.015	0.25	560	1190	560

WORKHOLDING DEVICES FOR MACHINE TOOLS

■ 25.1 INTRODUCTION

In the chapters on machining processes, the manner in which workpieces are mounted and held in the various machine tools was discussed. Workholding devices, often called jigs and fixtures, are critical components in the manufacturing of interchangeable parts. *Workholders* hold and locate the work in the machine tool with respect to the cutting tool. For example, Figure 25-1 shows an engine lathe and a CNC turning center with two spindles, each holding a chuck that holds the workpiece in the correct location with respect to the cutting tools. For many machine tools, fixtures hold the workpieces while providing location with respect to the cutting tools. With workholders, process accuracy and precision (repeatability) can be achieved that otherwise would be impossible with a given combination of cutting tools and machine tools. In this chapter, workholding devices (jigs and fixtures) will be considered as important production tools or adjuncts, with primary attention being directed toward their functional characteristics, their relationship to the machine tools, and the manufacturing processes.

In recent years, workholding devices have become more flexible; that is, (1) able to hold more than one part and (2) able to be quickly exchanged. Flexible workholders are critical elements in lean manufacturing cells, where components are made in families of parts (groups of parts of similar design). Further, being able to change from one device to another quickly to accommodate different parts means smaller lot sizes can be run, which reduces inventory levels in plants. These flexibility requirements add significantly to the complexity of conventional jig and fixture design. Let's begin with a discussion of the basics of jig and fixture design.

■ 25.2 CONVENTIONAL FIXTURE DESIGN

In the conventional method of fixture design, tool designers rely on their experience and intuition to design simple, single-purpose fixtures for specific machining operations, often using a trial-and-error method until the workholders perform satisfactorily. Of course, these designers should calculate the clamping forces or stress distributions in the fixturing elements to make sure that the loads will not deform the fixtures or the workpieces elastically or plastically. In the design of the workholding devices, two primary functions must be considered: *locating and clamping. Locating* refers to orienting and positioning the part in the machine tool with respect to the cutting tools to achieve

FIGURE 25-1 Two kinds of lathes. Above, a conventional engine lathe with principal parts named, including the workholder device, the chuck. Below, a CNC turning center with two chucks, turrets for cutting tools, and C-axis control for the main spindle. The C-axis control, on the spindle, can stop it in any orientation so the powered tools can operate on the workpiece.

the required specifications. *Clamping* refers to holding or maintaining the part in that location during the cutting operations (resisting the cutting forces).

Jigs and *fixtures* are specially designed and built workholding devices that hold the work during machining or assembly operations. In addition, a jig determines a location dimension that is produced by machining or fastening. For example, location dimensions determine the position of a hole on a plate (Figure 25-2). Consider the subject of dimensioning as used in drafting practice. Dimensions are of two types: *size* and

FIGURE 25-2 Drawing of a plate showing locating dimensions *(a, b, c, d)* versus sizing dimensions *(e, f, g, h)*.

location. Size dimensions denote the size of geometrical shapes—holes, cubes, parallelepipeds, and so on—of which objects are composed. *Location* dimensions, on the other hand, determine the position or location of these geometrical shapes *with respect to each other.* Thus *a* and *c* in Figure 25-2 are location dimensions, whereas *e* and *g* are size dimensions. With location dimensions in mind, one can precisely define a jig as follows: *A jig is a special workholding device that, through built-in features, determines location dimensions that are produced by machining or fastening operations.* The key requirement of a jig is that it determine a location dimension. Thus, jigs accomplish layout by means of their design.

In order to establish location dimensions, jigs may do a number of other things. They frequently guide tools, as in drill jigs, and thus determine the location of a component geometrical shape. However, they do not always guide tools. In the case of welding jigs, component parts are held (located) in a desired relationship with respect to each other while an unguided tool accomplishes the fastening. The guiding of a tool is not a necessary requirement of a jig.

Similarly, jigs usually hold the work that is to be machined, fastened, or assembled. However, in certain cases, the work actually supports the jig. Thus, although a jig *may* incidentally perform other functions, the basic requirement is that, through qualities that are built into it, certain critical dimensions of the workpiece are determined.

A *fixture* is *a special workholding device that holds work during machining or assembly operations and establishes size dimensions.* The key characteristic is that it is a *special* workholding device, designed and constructed for a particular part or shape. A general-purpose device, such as a chuck in a lathe or a clamp on a milling machine table, is usually not considered to be a fixture. Thus a fixture has as its specific objective the facilitating of *setup*, or making the part holding easier. Because many jigs hold the work while determining critical location dimensions, they usually meet all the requirements of a fixture. Alternatively, many fixtures are used in NC machines holding parts where holes are located and drilled according to a program. So the strict definition of jigs and fixtures has been blurred by the changes in technology.

In designing workholders, the designer must consider whether the part is a casting, forging, or bar stock. With castings and forgings, variations in shape and size must be accommodated in the design, and usually a machining operation is required to establish a reference surface (called the datum surface) to aid initial fixturing.

In parts cut from bar stock, allowances must be made for inaccuracies and irregularities produced by the cutoff operations. Table 25-1 provides a summary of design criteria for workholders for you to review. Obviously, it is impossible to meet all these design criteria for workholders. Compromise is inevitable. Still, it is useful to know the optimal design objectives to illustrate the positioning, holding, and supporting functions that fixtures must fulfill.

TABLE 25-1   Design Criteria for Workholders

Positive location A fixture must, above all else, hold the workpiece precisely in space to prevent each of 12 kinds of degrees of freedom—linear movement in either direction along the *X*-, *Y*-, and *Z*-axes and rotational movement in either direction about each axis.

Repeatability Identical workpieces should be located by the workholder in precisely the same space on repeated loading and unloading cycles. It should be impossible to load the workpiece incorrectly. This is called "foolproofing" the jig or fixture.

Adequate clamping forces The workholder must hold the workpiece immobile against the forces of gravity, centrifugal forces, inertial forces, and cutting forces but not distort the part. Milling and broaching operations, in particular, tend to pull the workpiece out of the fixture, and the designer must calculate these machining forces against the fixture's holding capacity. The device must be rigid.

Reliability The clamping forces must be maintained during machine operation every time the device is used. The mechanism must be easy to maintain and lubricate.

Ruggedness Workholders usually receive more punishment during the loading and unloading cycle than during the machining operation. The device must endure impact and abrasion for at least the life of the job. Elements of a device that are subject to damage and wear should be easily replaceable.

Design and construction ease Workholders should use standard elements as much as possible to allow the engineer to concentrate on function rather than on construction details. Modular fixtures epitomize this design rule as the entire workholder is made from standard elements, permitting a bolt-together approach for substantial time and cost savings over custom workholders.

Low profile Workholder elements must be clear of the cutting-tool path. Designing lugs on the part for clamping can simplify the fixture and allow proper tool clearance.

Workpiece accommodation Surface contours of castings or forgings vary from one part to the next. The device should tolerate these variations without sacrificing positive location or other design objectives.

Ergonomics and safety Clamps should be selected and positioned to eliminate pinch points and facilitate ease of operation. The workholder elements should not obstruct the loading or unloading of workpieces. In manual operations, the operator should not have to reach past the tool to load or unload parts. A rule sometimes used is that the operator can repeatedly exert a force of 30 to 40 lb to open or close a clamp but greater forces than this can cause ergonomic problems.

Freedom from part distortion Parts being machined can be distorted by gravity, the machining forces, or the clamping forces. Once clamped into the device, the part must be unstressed or, at least, undistorted. Otherwise, the newly machined surfaces take on any distortions caused by the clamping forces.

Flexibility The workholding device can locate and restrain more than one type (design) of part. Many different schemes are being proposed to provide workholder flexibility. Modular vise fixturing, programmable clamps using air-activated plungers, part encapsulation with a low-melting-point alloy, and NC-controlled clamping machines are some of the more recently developed systems. Despite their flexibilities, these clamping systems have some significant drawbacks. They are expensive, and the individual systems may not integrate well into individual machine tools. (See the discussions of intermediate jig concept and group jigs for additional thoughts on flexibility.)

■ 25.3 DESIGN STEPS

The classical design of a workholder (e.g., a drill jig) involves the following steps:

1. Analyze the drawing of the workpiece and determine (visualize) the machining operations required to machine it. Note the critical (size and location) dimensions and tolerances.

2. Determine the orientations of the workpiece in relation to the cutting tools and the movements of the tools and tables.

3. Perform an analysis to estimate the magnitude and direction of the cutting forces (see Chapter 21).

4. Study the standard devices available for workholders and for the clamping functions. Can an off-the-shelf device be modified? What standard elements can be used?

5. Form a mental picture of the workpiece in position in the workholder in the machine tool with the cutting tools performing the required operation(s). See the figures in chapters on machining for examples.

6. Make a three-dimensional sketch of the workpiece in the workholder in its required position to determine the location of all the elements: clamps, locator buttons, bushings, and so on.

7. Make a sketch of the workholder and workpiece in the machine tool to show the orientation of these elements with respect to the cutting tool in the machine tool.

3-2-1 LOCATION PRINCIPLE

After determining the orientation of the workpiece in the workholder, the next step is to locate it in that position. This location is also used for all similar workpieces. The designer must select or design locating devices (supports) *that ensure that every workpiece placed*

FIGURE 25-3 Workpiece location is based on the 3-2-1 principle. Three points will define a base surface, two points in a vertical plane will establish an end reference, and one point in a third plane will positively locate most parts.

in the device occupies the same position with respect to the cutting tools. Thus, when the machining operation is performed, the workpieces are processed identically. This is, of course, the key to making interchangeable parts. In locating the workpiece, the basic *3-2-1 principle* of location is used (Figure 25-3). For positive location, the fixture must position the workpiece in each of three perpendicular planes. Positioning processes can vary greatly, but workholder design always begins by defining the first plane of reference with three points. Once the object is defined in a single plane, supported at three points (like a three-legged stool on a floor), a second plane can be assigned that is perpendicular to the first. To do this, the object is brought up against any two points in the second plane. To continue the example, the stool is slid along the floor until two legs touch a wall.

A third plane, perpendicular to each of the other two, is then defined by designating one point on it. As long as an object is in contact with three points on the first plane, two points on the second, and a single point on the third, it is positively located in space. The location points within each plane should be selected as far apart as possible for maximum stability.

In practice, it is often necessary to support a workpiece on more points than this 3-2-1 formula dictates. The machining of a large rectangular plate, for example, typically requires support at four or more points. However, any extra points must be established carefully to support the workpiece in a plane defined by three—and only three—points.

Appropriate clamping devices are selected so that the clamping forces hold the workpiece in the proper location and resist the effects of the cutting forces, centrifugal forces, and vibrations. If possible, the machining forces should act into the location points, not into the clamps, so that smaller clamps can be used. In reality, the worker often determines clamping force when loading the part into the workholder. Fixtures are usually fastened to the table of the machine tool. Although used primarily on milling and

TABLE 25-2   Twenty Principles for Workholder Design

1. Determine the critical surfaces or points for the part.
2. Decide on locating points and clamping arrangements.
3. For mating parts, use corresponding locating points or surfaces to ensure proper alignment when assembled.
4. Try to use 3-2-1 location, with 3 assigned to largest surface. Additional points should be adjustable.
5. Locating points should be visible so that the operator can see if they are clean. Can they be replaced if worn?
6. Provide clamps that are as quick acting and easy to use as is economically justifiable for rapid loading and unloading.
7. Clamps should not require undue effort by the operator to close or to open, nor should they harm hands and fingers during use.
8. Clamps should be integral parts of device. Avoid loose parts that can get lost.
9. Avoid complicated clamping arrangements or combinations that can wear out or malfunction. Keep it simple.
10. Locate clamps opposite locaters (if possible) to avoid deflection/distortion during machining and spring-back afterward.
11. Take the thrust of the cutting forces on the locaters (if possible), not on the clamps.
12. Arrange the workholder so that the workpiece can easily be loaded and unloaded from the device and so that it can be loaded only in the correct manner (mistake-proof) and in such a way that the location can be found quickly (visually).
13. Consistent with strength and rigidity, make the workholder as light as possible.
14. Provide ample room for chip clearance and removal.
15. Provide accessibility for cleaning.
16. Provide for entrance and exit of cutting fluid (which may carry off chips) if one is to be used.
17. Provide four feet on all movable workholders.
18. Provide hold-down lugs on all fixed workholders.
19. Provide keys to align fixtures on machine tables so fixtures can be replaced in exactly the same position.
20. Do not sacrifice safety for production.

broaching machines, fixtures are also designed and used to hold workpieces for various operations on most of the standard machine tools and machining centers. Some additional design rules for fixture design are given in Table 25-2.

■ 25.4 CLAMPING CONSIDERATIONS

Clamping of the work is closely related to support of the work. Any clamping, of course, induces some stresses into the part that can cause some distortion of the workpiece, usually elastic. If this distortion is measurable, it will cause some inaccuracy in final dimensions, as illustrated in an exaggerated manner in Figure 25-4. The obvious solution is to spread the clamping forces over a sufficient area to reduce the stresses to a level that will not produce appreciable distortion. The clamping forces should direct the work against the points of location and work support. Clamped surfaces often have some irregularities that may produce force components in an undesired direction. Consequently, clamping forces should be applied in directions that will assure that the work will remain in the desired position.

Whenever possible, jigs and fixtures should be designed so that the forces induced by the cutting process act to hold the workpiece in position against the supports. These

FIGURE 25-4 Exaggerated illustration of the manner in which excessive clamping forces can affect the final dimensions of a workpiece.

Workpiece

Overclamped before machining; part distorted

After machining (still clamped)

Unclamped distortion

FIGURE 25-5 In (a) and (b), proper work support to resist the forces imposed by cutting tools is demonstrated. In (c), three buttons form a triangle for the work to rest on.

forces are predictable, and proper utilization of them can materially aid in reducing the magnitude of the clamping stresses required. In addition to locating the work properly, the stops or work-supporting areas must be arranged so as to provide adequate support against the cutting forces. As shown in Figure 25-5a, having the cutting force act against a fixed portion of the jig or fixture and not against a movable section permits lower clamping forces to be used. Figure 25-5b illustrates the principle of keeping the points of clamping as nearly as possible in line with the action forces of the cutting tool so as to reduce their tendency to pull the work from the clamping jaws. Compliance with this principle results both in lower clamping stresses and less massive clamping devices. Don't forget that down milling produces different forces than up milling. The location points should be as far apart as possible but positioned so as not to allow the cutting force to distort the work, as shown in Figure 25-5c. The cutting forces may distort the work, with resulting inaccuracy or broken tools. These design suggestions materially reduce vibration and chatter during the cutting process.

As many operations as are possible and practical should be performed with each clamping of the workpiece. This principle has both physical and economic aspects. Because some stresses result from each clamping, with the possibility of accompanying distortion, greater accuracy is achieved if multiple operations are performed with each clamping. From the economic viewpoint, if the number of jigs or fixtures is reduced, less capital will be required and less time will be spent handling the workpiece loading and unloading.

■ 25.5 CHIP DISPOSAL

When jigs or fixtures are used in connection with chip-making operations, adequate provision must be made for the easy removal of the chips. This is essential for several reasons. First, if chips become packed around the tool, heat will not be carried away and tool life can be decreased. Figure 25-6 illustrates how insufficient clearance between the end of a drill bushing and the workpiece can prevent the chips from escaping, whereas too much clearance may not provide accurate drill guidance and can result in broken drills.

FIGURE 25-6 Proper clearance between drill bushing and tool of workpiece is important.

Too much clearance permits tool drift

Correct

Drill fills with chips; too little clearance

A second reason chips must be removed is so that they do not interfere with the proper seating of the work in the jig or fixture (Figure 25-7). Even though chips and dirt always have to be cleaned from the locating and supporting surfaces by a worker or by automatic means, such as an air blast, the design details should be such that chips and other debris will not readily adhere to, or be caught in or on, the locating surfaces, corners, or overhanging elements and thereby prevent the work from seating properly. Such a condition results in distortion, high clamping stresses, and incorrect workpiece dimensions.

■ 25.6 UNLOADING AND LOADING TIME

The cost of the workholders must be justified by the quantities of production involved, and their primary purpose is to increase productivity and quality. While work is being put into or being taken out of jigs and fixtures, the machines with which they are used are not making chips. The loading and unloading time plus the machining time (also called the run time) plus any delay times equals the cycle time for a part. The loading and unloading time is greatly influenced by the choice of clamps.

There are many ways in which jigs and fixtures can be made easier to load and unload. Some clamping methods can be operated more readily than others. For example, in the drill jig shown in Figure 25-8, a *knurled clamping screw* is used to hold the block against the buttons at the end of the jig. To clamp or unclamp the block in this direction requires several motions. On the other hand, a *cam latch* is used to close the jig and hold the workpiece against the rear locating buttons. This type of latch can be operated with a single motion.

Certainly, the device should be designed so that the part cannot be loaded incorrectly. Defect prevention is often accomplished by the clamping device so that a part loaded improperly cannot be clamped. Ease of operation of workholders not only directly increases the productivity of such equipment but also results indirectly in better quality and fewer lost-time accidents.

The workholder is as critical as the machine tool and the cutting tool to the final quality of the part. The use of the workholder eliminates manual layout of the desired features of the part on the raw material. Manual layout requires a highly skilled worker and is very time consuming. The workholder permits a less skilled person to achieve quality and repeatable production with far greater efficiency.

■ 25.7 EXAMPLE OF JIG DESIGN

Several principles of work location and tool guidance are illustrated in Figure 25-8. The two mounting holes in the base of the bearing block are to be located and drilled. The dimensions A, B, and C are determined by the jig. While it is not specified on the drawing,

FIGURE 25-7 Methods of providing chip clearance to ensure proper seating of the work.

Chips
Not good

Chips
Good

Better

FIGURE 25-8   (Lower left) Part to be drilled; (lower right) box drill jig for drilling two holes; (upper left) jig in drill press; (upper right) drill being guided by drill bushing.

there is one other location dimension that must be controlled. The axes of the mounting holes must be at right angles to the bottom surface of the block, so the bottom must be machined (milled) prior to this drilling step.

The way in which the part dimensions are obtained in the finished workpiece is as follows. The surfaces marked with a large carats ∨ are reference (or location) surfaces and are finished (machined) prior to insertion of the part into the drill jig. The part rests in the jig on four buttons marked X in Figure 25-8. These buttons, made of hardened steel, are set into the bottom plate of the jig and are accurately ground so that their surfaces are in a single plane. The left-hand end of the part is held against another button Y. This locating button is built into the jig so that its surface is at right angles to the plane of the

X buttons. When the block is placed in the jig, its rear surface rests against three more buttons marked Z. These buttons are located and ground so that their surfaces lie in a plane that is at right angles to the planes of both the X and Y buttons. The part is held in its located position by the two clamps marked C.

The use of four buttons on the bottom of this jig (X buttons) appears not to adhere to the 3-2-1 principle stated previously. However, although only two X buttons would have been required for complete location, the use of only two buttons would not have provided adequate support during drilling. The thrust from the drills would have dislodged the part from the locators. Thus the 3-2-1 principle is a *minimum* concept and often must be exceeded.

To ensure that the mounting holes are drilled in their proper locations, the drill must be located and then guided during the drilling process. This is accomplished by the two drill bushings marked K. Such drill bushings are accurately made of hardened steel, with their inner and outer cylindrical surfaces concentric. The inner diameter is made slightly larger than the drill—usually 0.0005 to 0.002 in.—so that the drill can turn freely but not shift appreciably. The bushings are accurately mounted in the upper plate of the jig and positioned so that their axes are exactly perpendicular to the plane of the X buttons, at a distance A from the Z buttons and at distances B and C, respectively, from the plane of the Y button. Note that the bushings are sufficiently long that the drill is guided close to the surface where it will start drilling. Consequently, when the workpiece is properly placed and clamped in the jig, the drill will be located and guided by the bushings so that the critical dimensions on the workpiece will be correct. The right hole will be drilled in a vertical-spindle drill press (not running), and then the jig will be shifted (manually by the operator) to the right, and the left hole will be drilled. The box construction is rigid but open for chip removal.

■ 25.8 TYPES OF JIGS

Jigs are made in several basic forms and carry names that are descriptive of their general configurations or predominant features. Several of these are illustrated in Figure 25-9.

A *plate jig* is one of the simplest types, consisting only of a plate that contains the drill bushings and a simple means of clamping the work in the jig or the jig to the work. In the latter case, wherein the jig is clamped to the work, the device is sometimes called a *clamp-on jig*. Such jigs are frequently used on large parts, where it is necessary to drill one or more holes that must be spaced accurately with respect to each other, or to a corner of the part, but that need not have an exact relationship with other portions of the work.

Channel jigs also are simple and derive their name from the cross-sectional shape of the main member. They can be used only with parts having fairly simple shapes.

Ring jigs are used only for drilling round parts, such as pipe flanges. The clamping force must be sufficient to prevent the part from rotating in the jig. *Diameter jigs* provide a means of locating a drilled hole exactly on a diameter of a cylindrical or spherical piece.

Leaf jigs derive their name from the hinged leaf or cover that can be swung open to permit the workpiece to be inserted and then closed to clamp the work in position. Drill bushings may be located in the leaf as well as in the body of the jig to permit locating and drilling holes on more than one side of the workpiece. Such jigs are called *rollover jigs* or *tumble jigs* when they require turning to permit drilling from more than one side.

Box jigs are very common, deriving their name from their boxlike construction. They have five fixed sides and a hinged cover or leaf, which opens to permit loading the workpiece, and a cam that locks the workpiece in place. Usually, the drill bushings are located in the fixed sides to ensure retention of their accuracy. The fixed sides of the box are usually fastened by means of dowel pins and screws so that they can be taken apart and reassembled without loss of accuracy. Because of their more complex construction, box jigs are costly, but their inherent accuracy and strength can be justified when there is sufficient volume of production. They have two obvious disadvantages: (1) it is usually more difficult to put work into them than into simpler types, and (2) there is a greater tendency for chips to accumulate within them. Figure 25-8 shows a box-type jig.

FIGURE 25-9 Examples of some common types of workholders—jigs.

Because jigs must be constructed very accurately and be made sufficiently rugged so as to maintain their accuracy despite the use (and abuse) to which they inevitably are subjected, they are expensive. Consequently, several methods have been devised to aid in lowering the cost of manufacturing jigs. One way to reduce this cost is to use simple, standardized plate and clamping mechanisms called *universal jigs* (Figure 25-10).

FIGURE 25-10   Two types of universal jigs are manual (bottom) and power-actuated (center). A completed jig (on the top) made from unit right below. *(Courtesy of Cleveland Universal Jig Division, The Industrial Machine Company.)*

These can easily be equipped with suitable locating buttons and drill bushings to construct a jig for a particular job. Such universal jigs are available in a variety of configurations and sizes, and because they can be produced in quantities, their cost is relatively low. However, the variety of work that can be accommodated by such jigs obviously is limited. While the drill bushing should be spaced far enough from the work to allow chips to escape without entering the bushing, when drilling into an angled surface, the bushing should be very close. Once the drill has penetrated to at least one-half of the drill diameter, the bushing should be retracted to provide chip clearance. Design of the drill jig must not obstruct coolant flow to where it is needed. Bushing length should be 1.75 to 2.5 times the drill diameter.

■ 25.9 CONVENTIONAL FIXTURES

Many examples of conventional fixtures have appeared in the text. Production milling, broaching, and boring processes as performed on NC machines, conventional equipment, or machining centers routinely use fixtures to locate and hold the part properly with respect to the cutting tools on the machine tool. Like cutting tools, tooling for workholding is sold separately and is not usually supplied by the machine tool builder. Traditionally, beginning with Eli Whitney, manufacturers have designed and built custom-made, dedicated fixtures. Because of the pressure of shorter production runs and smaller lot sizes, many companies are turning to modified fixturing approaches. The greatest advantage of these systems is that the fixture can be constructed quickly.

Perhaps the most common fixture uses the *vise* as its base element. Figure 25-11 shows a schematic and photo of a typical commercially available vise that can be adapted for use as a fixture. As shown, the vise jaws are readily modified to conform to the 3-2-1 location principle and provide adequate clamping forces for almost every machining operation. Four vises (also shown in Figure 25-11) can be mounted on a subplate for rapid insertion and location in the machine, or four vices can be mounted on a tombstone for milling parts in a CNC machine.

		Smooth jaws for parts with sensitive clamping surfaces.
		Grooved jaws (standard jaws) with smaller surface for increasing the specific surface pressure.
		Soft basic jaws for manufacturing your own special jaws. Mat: 21 Mn Cr 5 G.
		Spring leaf pull-down jaws for parts with rough clamping surfaces. Spring leaves press the workpiece against the supporting surfaces.
		Roller pull-down jaws for parts with sensitive clamping surfaces. Jaw components adjacent to the workpiece "put" the workpiece to the supporting surface.
		Prismatic jaws for the horizontal and vertical clamping of round and flat workpieces.
		Hold-down jaws for horizontal clamping of round workpieces.
		Swivel jaws for nonparallel clamping surface, compensation of conicity up to approx. 7.5 in.
		Compensation jaws I for multiple clamping of workpieces. Compensates differences in dimensions up to +3 mm.
		Compensation jaws II for workpieces with greatly varying surfaces. Compensated differences in dimensions up to +3 mm.
		Pull-down jaws for horizontal clamping of round workpieces. Roller jaw "pulls" workpiece to the supporting surface.

FIGURE 25-11 The conventional or standard vise (top left and right) can be modified with removable jaw plates to adapt to different part geometries. These vices can be integrated into milling fixtures (right middle and bottom).

The chucks used in lathes are really general-purpose fixtures for rotational parts. Newer chuck designs have greatly improved their flexibility (the range of diameters the chuck can accommodate in a given setup and speed of setup). Figure 25-12 shows a complete change of top jaws for a three-jaw chuck being done in less than 5 minutes. The normal time for this part of the setup might exceed 15 minutes. New quick-change insert top jaws may even snap in by hand with no jaw nuts, keys, screws, or tools. Jaws that can be exchanged by robots can also be designed.

1. Pre-assembled Mini-System and top jaws.
2. Assembly being inserted into the Master Jaw.
3. Quickly retighten cap screw.
4. All 3 jaws changed in 5 minutes or less.

FIGURE 25-12 Quick-changing of the top jaws on a three-jaw chuck. *(Courtesy of Huron Machine Products.)*

Most producers of chucks use some variation of equation 25-1 to compute the maximum rpm rate at which the chuck can run:

$$S_m^2 = \frac{F_m}{3 \times (2.84 \times 10^{-5}) \times W \times D} \tag{25-1}$$

where

S_m = maximum rpm value at which gripping force equals $^1/_3$ F_m

F_m = maximum rate gripping force, at rest (lb)

W = combined weight of jaws (lb)

D = distance from spindle centerline to center of jaw mass (in.)

Thus, with this equation, a 10-in. power chuck with a published rating F_m of 13,200 lb would retain one-third of its initial gripping force at 2507 rpm. (Check this calculation using W = 8 lb, D = 3.1 in.) The higher the rpm value, the greater the centrifugal force factor. This is an important factor in high-speed machining operations in which the part is rotating.

■ 25.10 MODULAR FIXTURING

Modular fixtures have all the same design criteria as those of conventional fixtures, plus one more—*versatility*. Modular fixture elements must be useful for a variety of machining applications and easily adaptable to different workpiece geometries. Individual fixture designs can be photographed or entered into a computer-aided design (CAD) library for future reference. After the job is done, the fixture itself can be dismantled and the elements returned to the toolroom. The erector-set approach uses either T-slot or dowel-pin designs. Figures 25-13 and 25-14 show two examples of modular fixturing. The designs begin with base plates. Elements for locating and clamping are added to the subplate. Rectangular, square, and round are the typical patterns for the subplates. Also shown are the typical components for modular fixturing systems used for mounting points, locators, attachments, and so on. The standard elements needed to construct the fixture include riser blocks, vee blocks, angle plates, cubes, box parallels, and the like. Smaller elements such as locator pins, supports, pads, and clamps are added to the subplate on the larger structural elements. Mechanical clamping devices are shown, but power-assisted clamps are available. The base and fixturing elements are made to tolerances of ±0.0002 to 0.0004 in. in flatness, parallelism, and size. Figure 25-14 shows a part in a dedicated fixture compared to a modular fixture. The dedicated fixture represents a capital investment that must be absorbed by the job and must be maintained after the job is complete. The modular fixture is disassembled and the elements reused later in fixtures for other parts. Modular fixtures are commonly used for prototype tooling and small-batch production runs. They are being incorporated more frequently into regular production as users gain confidence in this approach.

FIGURE 25-13 Modular fixturing begins with a subplate (grid base) and adds locators and clamps.

FIGURE 25-14 Dedicated fixture on the left versus modular fixture on the right. *(From Manufacturing Engineering, January, 1984.)*

■ 25.11 SETUP AND CHANGEOVER

Every part coming out of the workholder should be the same, resulting in interchangeable parts. But what about the first part? Does it meet specifications? What about the initial setup of the workholder into the machine tool? In many cases, the setup operation takes hours and the machine is not producing anything during this time. Rapid exchanges of workholding devices is a key technique in modern manufacturing systems. (See Shingo, 1985 for more discussion of the elimination of setup and SMED.) Reducing setup times permits shorter production runs (smaller lot sizes). Do not confuse initial setup (of workholders) with part loading and unloading or tool changing. The trick with initial setup is to do it quickly and to get the first part out of the process as a good part, with no adjustment of the machine, the tooling, or setup workholder. Quick tool and die exchange is a critical component in the strategy for the factory with a future.

FIGURE 25-15 Master jig designed for a family of similar components. (a) Part family of rounds plates (six parts, A–F); (b) group jig for drilling, showing adapter and part A.

Another approach to rapid setup is shown in Figure 25-15, where instead of six different jigs, a master jig is made (also called a group jig) for a family of similar components and then a set of adapters is made that customizes the jig for each part in the part family. This concept of group jigs and fixtures originated from the group technology (GT) concept for master jigs as a method to form cellular manufacturing systems by determining a family of parts where an imaginary part, called the composite part, is designed that has all the key features of all the parts in the family. In other words, the composite part is an envelope, the shape of which encompasses the shapes of all the parts in the family. The theory is that if the tooling is designed for the composite part, any part that fits within the envelope could be machined without any tooling changes. This part is used for designing the workholder. The workholding devices should be able to accommodate all the parts within the parts family. For manufacturing cells, the workholders will also have to compensate for variation in cutting forces, centrifugal forces, and so on. Group workholders are designed to accept every part-family member, with or without adapters, that accommodate minor part variations.

INTERMEDIATE JIG CONCEPT

One way to achieve rapid fixture exchange is to employ the intermediate jig concept. This means that the workholding devices are designed so that they all appear the same to the machine tool but different to the parts. This usually requires one to construct intermediate jig or fixture plates to which the jig or fixture is attached. The jigs or fixtures are all different, but the plates are all identical.

The cassette tape for your VCR is an example of an intermediate workholder. To the VCR, every cassette appears to be the same and can be quickly loaded and unloaded with one handling—that is, one touch. From the outside, every tape appears to be the same, but on the inside, every tape is different. If you think about the workholding devices in terms of the intermediate jig concept, you can quickly achieve one-touch setups.

Figure 25-16 shows an example of the *intermediate jig concept*, applied to lathes and chucks. An adapter or intermediate fixture is bolted to the lathe's spindle and is a permanent part of the machine tool. The intermediate fixture will accept mating chucks that have been preset for the workpiece prior to insertion. Different chuck designs mount interchangeably on the common actuator. This method greatly reduces setup time and permits the operator to perform chuck maintenance and retooling (setup) while the machine is running. The chucks can be exchanged automatically.

Quick-change fixtures for CNC milling machines and machining centers (using the intermediate jig concept) are now available commercially.

| Lathe spindle | Intermediate fixture-actuator | Chuck Three types | Workpiece |

| Actuator | Diaphragm chuck | Collet chuck | Jaw chuck |

FIGURE 25-16 Example of the intermediate jig concept applied to lathe chucks. The actuator is mounted on the lathe and can quickly adapt to three different chuck types. *(Courtesy of Sheffer Collet Company.)*

FIGURE 25-17 Examples of basic types of clamps used for workholding. The clamp elements come in a wide variety of sizes.

■ 25.12 CLAMPS

When designing a jig or fixture, there are many choices to be made regarding the clamps. Manual clamps, which include screw, strap, swing, edge, cam, toggle, and C-clamps, each with certain strengths and weaknesses, are usually cheaper but slower. In Figure 25-17 typical types of clamps that are used in fixtures are shown. The *strap clamp* comes in many forms and sizes and is simple, low cost, and flexible. The force can be applied by a hand knob, a cam, or a wrench turning down a nut. A conventional *toggle clamp* accommodates only small thickness variation from part to part yet provides an excellent, consistent clamping force.

Power-actuated clamps (shown in Figure 25-18) provide more consistent clamping forces than do manual clamps, especially in applications that promote operator fatigue. The higher cost must be weighed against the capability for consistent and repeatable operation, automatic adjustment of holding forces, remote actuations, and automating sequencing of clamping actions. *Extending clamps* operate in a manner similar to that of a manual clamp-strap assembly. They extend forward horizontally, then clamp down. *Edge clamps* have a very low profile. They clamp down and forward simultaneously.

■ 25.13 OTHER WORKHOLDING DEVICES

ASSEMBLY JIGS

Because *assembly jigs* usually must provide for the introduction of several component parts and the use of some type of fastening equipment, such as welding or riveting, they commonly are of the open-frame type. Such jigs are widely used in automobile body welding and aircraft assembly. Large jigs of the type are shown in Figure 25-19 are used for the assembly and usually feature automatic clips. This jig is constructed mainly of reinforced concrete.

(a)

(b)

FIGURE 25-18 Examples of power-clamping devices: (a) extending clamp; (b) edge clamp.

FIGURE 25-19 Example of large assembly jig for an airplane wing. The body of the wing and flap are held in the correct location with each other and then the flap is mechanically attached.

MAGNETIC WORKHOLDERS

Because of the light cuts and low cutting forces, workpieces can be held in a different manner on surface grinders than on other machine tools. *Magnetic chucks* are used for ferromagnetic materials. To obtain high accuracy, it is desirable to reduce clamping forces and distribute them over the entire area of the workpiece. Also, grinding is frequently done on quite thin or relatively delicate workpieces, which would be difficult to clamp by normal methods. In addition, there is often the problem of grinding a number of small, duplicate workpieces. Magnetic chucks solve all these problems very satisfactorily. Magnetic chucks are available in disk or rectangular shapes. Dry-disk rectifiers are used to provide the necessary direct-current power. Some magnetic chucks utilize permanent magnets, as shown in Figure 25-20, and can be tilted so that angles can be ground. Magnetic chucks provide an excellent means of holding workpieces provided that the

FIGURE 25-20 Example of magnetic chuck. *(Courtesy of O. S. Walker.)*

cutting or inertial forces are not too great. The holding force is distributed over the entire contact surface of the work, the clamping stresses are low, and therefore there is little tendency for the work to be distorted. Consequently, pieces can be held and ground accurately. Also, a number of small pieces can be mounted on a chuck and ground at the same time. Magnetic chucks provide great part-to-part repeatability because the holding power from one part to the next is the same. Initial setup is usually fast, simple, and relatively inexpensive. Parts loading and unloading is also relatively easy.

It often is necessary to demagnetize work that has been held on a magnetic chuck. Some electrically powered chucks provide satisfactory demagnetization by reversing the direct current briefly when the power is shut off.

ELECTROSTATIC WORKHOLDERS

Magnetic chucks can be used only with ferromagnetic materials. Electrostatic chucks can be used with any electrically conductive material. This principle (Figure 25-21) directs that work be held by mutually attracting electrostatic fields in the chuck and the workpiece. These provide a holding force of up to 20 psi (21,000 Pa). Nonmetal parts can usually be held if they are flashed (i.e., coated) with a thin layer of metal. These chucks have the added advantage of not inducing residual magnetism in the work.

VACUUM CHUCKS

Vacuum chucks are also available. In one type, illustrated in Figure 25-22, the holes in the work plate are connected to a vacuum pump and can be opened or closed by means of valve screws. The valves are opened in the area on which the work is to rest. The other type has a porous plate on which the work rests. The workpiece and plate are covered

FIGURE 25-21 Principle of electrostatic chuck.

FIGURE 25-22   Cutaway view of a vacuum chuck. *(Courtesy of Dunham Tool Company, Inc.)*

with a polyethylene sheet. When the vacuum is turned on, the film forms around the workpiece, covering and sealing the holes not covered by the workpiece and thus producing a seal. The film covering the workpiece is removed or the first cut removes the film covering the workpiece. Vacuum chucks have the advantage that they can be used on both nonmetals and metals and can provide an easily variable force. Magnetic, electrostatic, and vacuum chucks are used for some light milling and turning operations.

As shown in Figure 25-23, T-slots are provided on milling machine tables so that workpieces can be clamped directly to the table. More often various workholding devices, called *vices* or *fixtures*, are utilized. Smaller workpieces are usually held in a vise mounted on the table. Fixtures designed to specifically hold a part in the correct location with respect to the tool are used for larger volumes. Fixtures reduce the time it takes to put the part in the machine and assure repeatable location with respect to the cutting tools. Fixtures provide clamping forces that counteract the cutting forces.

FIGURE 25-23   While the work can be clamped directly to the milling machines table (on the left), the workpiece on the right is located in a fixture mounted in a circular attachment located on the table, so a circular slot can be milled in the workpiece.

■ 25.14 ECONOMIC JUSTIFICATION OF JIGS AND FIXTURES

As discussed previously, workholders are expensive, even when designed and constructed by using standard components. Obviously, their cost is a part of the total cost of production, and one must determine whether they can be justified economically by the savings in labor and machine cost and improvements in quality that will result from their use. Often it is only through the use of such devices that the design specifications can be met and sustained from part to part. To determine the economic justification of any special tooling, the following factors must be considered:

1. The cost of the tooling
2. Interest or profit charges on the tooling cost
3. The savings resulting from the use of the tooling; can result from reduced cycle times or improved quality or lower-cost labor
4. The savings in machine cost due to increased productivity
5. The number of units that will be produced using the tooling

The economic relationship between these factors can be expressed in the following manner:

savings per piece (exclusive of tooling costs) ≥ additional cost per piece

$$\left\{ \begin{array}{c} \text{total cost per piece} \\ \text{without tooling} \end{array} \right\} - \left\{ \begin{array}{c} \text{total cost per piece} \\ \text{using tooling} \\ \text{(exclusive of tooling cost)} \end{array} \right\} \geq \text{tooling cost per piece}$$

$$\underbrace{\begin{array}{c}\text{labor cost} \\ \text{per piece} \\ \text{without} \\ \text{tooling} \end{array} + \begin{array}{c}\text{machine and} \\ \text{overhead cost} \\ \text{per piece} \\ \text{without tooling}\end{array}}_{} - \underbrace{\begin{array}{c}\text{labor cost} \\ \text{per piece} \\ \text{with} \\ \text{tooling}\end{array} + \begin{array}{c}\text{machine and} \\ \text{overhead cost} \\ \text{per piece} \\ \text{with tooling}\end{array}}_{} \geq \underbrace{\begin{array}{c}\text{cost} \\ \text{of} \\ \text{tooling}\end{array} + \begin{array}{c}\text{interest on} \\ \text{tooling cost}\end{array}}_{}$$

$$[(R)(t) + (R_m)(t)] - [(R_t)(t_t) + (R_m)(t_t)] \geq \frac{C_t + (C_t/2)(n)(i)}{N} \qquad (25\text{-}2)$$

where

R = labor rate per hour, without tooling

R_t = labor rate per hour, using tooling

t = hours per piece, without tooling

t_t = hours per piece, using tooling

R_m = machine cost per hour, including all overhead

C_t = cost of the special tooling

n = number of years tooling will be used

i = interest rate (or what invested capital is worth)

N = number of pieces that will be produced with the tooling

Equation 25-2 can be expressed in a simpler form:

$$(R + R_m)t - (R_t + R_m)t_t \geq \frac{C_t}{N}\left(1 + \frac{n \times i}{2}\right) \qquad (25\text{-}3)$$

This equation assumes straight-line depreciation and computes interest on the average amount of capital invested throughout the life of the tooling.[1] When the time over

[1]For the use of more sophisticated economic analysis, see C. S. Park, *Contempory Engineering Economics*, 2nd ed., New York, Wiley, 1999.

which the tooling is to be used is less than one year, companies often do not include an interest cost. If this factor is neglected, the right-hand term of equation 25-3 reduces to C_t/N.

The equations assume that the material cost will be the same regardless of whether or not special tooling is used. This is not always true. Although these equations are not completely accurate for all cases, they are satisfactory for determining tooling justification in most cases, because the life of tooling for machine tools seldom exceeds five years and often does not exceed two years. The equation does not include the cost of poor quality. This can be included by estimating the decrease in the number of defective parts when the workholder is used versus when it is not used.[2]

The following example illustrates the use of equation 25-3 to determine tooling justification for a dedicated jig. In drilling a series of holes on a radial drill, the use of a drill jig will reduce the time from $1/2$ hour per piece to 15 minutes per piece. If a jig is not used, a machinist, whose hourly rate is $18/hr, must be used. If the jig is used, the job can be done by a machinist whose rate is $12/hr. The hourly rate for the radial drill is $32/hr.

The cost of making the jig would include $350 for design, $150 for material, and 50 hours of toolmaker's labor, which is charged at the rate of $22/hr to include all machine and overhead costs in the toolmaking department. Investment capital is worth 16% to the company. It is estimated that the jig would last three years and that it would be used for the production of 300 parts over this period. Is the jig justified?

The cost of the jig, C_t, is estimated to be

$$C_t = \$350 + \$150 + \$150 + \$22 \times 50hr = \$1,600$$

Substituting the values given in equation 25-3, we find:

$$(18 + 32)0.5 - (12 + 32)0.25 \geq \$\frac{1600}{300}\left(1 + \frac{3 \times 0.16}{2}\right)$$

$$or\ 14.00 \cong 13.76$$

So this jig is not justified based on cost savings.

One could also ask how many parts would have to be produced with the jig to break even (i.e., increased costs just equal savings). By omitting the value 300 in the solution above and solving for N, it is found that at least 1627 pieces would have to be produced annually with the jig for it to break even.

This analysis assumes that the time (of the people and machines) saved by the use of the special tooling can be used for other productive work. If this is not the case, the cost analysis should be altered to take this important fact into account. Otherwise, the tooling justification may be substantially in error.

The application of group technology, NC machines, and lean manufacturing techniques may eliminate the need for designing and building a new jig or fixture every time a new part is designed. New measures of manufacturing productivity that include terms for quality and flexibility are being developed.

---

[2]For a discussion of new measures and methods on cost accounting, see C. S. Park, "Counting the Costs," January 1987, *Mechanical Engineering*, p. 66.

■ Key Words

3-2-1 principle	electrostatic workholder	location	toggle clamp
assembly jig	fixture	magnetic chuck	vacuum chuck
box jig	intermediate jig concept	plate jig	workholder
channel jigs	jig	ring jig	
clamping	leaf jig	strap clamp	

■ Review Questions

1. What are the two primary functions of a workholding device?
2. What distinguishes a jig from a fixture?
3. An early treatise defined a jig as "a device that holds the work and guides a tool." Why was this definition incorrect?
4. What modifications do you need to make to an ordinary vise so it can be a fixture?
5. What basic criteria should be considered in designing jigs and fixtures?
6. In any part drawing, what are the critical surfaces of a part (i.e., what makes a part surface critical)? (This question requires an understanding of basic part drawings.)
7. What difficulties can result from not keeping clamping stresses low in designing jigs and fixtures?
8. Explain the 3-2-1 concept for workpiece location in a workholder on a machine tool.
9. Which of the basic design principles relating to jigs and fixtures would most likely be in conflict with the 3-2-1 location concept?
10. What are two reasons for not having drill bushings actually touching the workpiece? How many of the designs shown in this chapter violate this rule? It is not uncommon to have conflicts and trade-offs in fixture design situations.
11. Why does the use of down milling often make it easier to design a milling fixture than if up milling were used?
12. Name another example of the intermediate workholder concept aside from the video cassette.
13. A large assembly jig for an airplane-wing component gave difficulty when it rested on four-point support. The assembled wing components were not consistent in shape. It was satisfactory when only three supporting points were used. Why?
14. Explain why the use of a given fixture may not be economical when used with one machine tool but may be economical when used in conjunction with another machine tool.
15. What are rollover jigs, and what advantages do they offer?
16. In the clamps shown in Figure 25-17, what is the purpose of the spherical washer?
17. What are other common types of clamps?
18. What is the purpose of dimensioning the strap-clamp assembly in Figure 25-17 with letters?
19. Figure 25-8 showed the part sitting on locator buttons. Why not have the part rest on the flat plate?
20. In Figure 25-8, why aren't there three points put on the x-plane, two points on the z-plane, and one point on the y-plane?
21. Which set of locators in Figure 25-8 establishes the A dimension on the part?
22. To prepare the workpiece shown in Figure 25-8, which surface would you have milled first—the bottom, the back, or the front?
23. For the part shown in Figure 25-8, why don't you drill the holes first, then mill? Why mill at all?
24. Notice that the holes in the part in Figure 28-8 need to be countersunk after they are drilled. How must the jig be designed to put the countersinks on the mounting holes while the part is in the jig, or would this operation be done afterward?

■ Problems

1. Using the following values, determine the number of pieces that would have to be made to justify the use of a jig costing $3000.

$$R = \$5.75$$
$$R_t = \$4.50$$
$$t_t = 1\,{}^1/_4$$
$$t = 2\,{}^1/_2$$
$$I = 10\%$$
$$R_m = \$4.50$$
$$N = 3$$

2. Suppose in the sample problem at the end of this chapter that modular fixturing is used, which reduces the toolmaker's labor to 4 hours and the design cost to $100 (4 hours at $25 per hour), and that the material cost (modular elements) for the subplate structural elements, clamps, and so on was $600. What is the break-even quantity for a modular fixture? (*Note:* The modular fixture is used for the job, then disassembled and returned to the tool room. The parts are reused in other workholders.)
3. Suppose the leaf jig in Figure 25-A could be improved by replacing the screw clamp with swing clamps. See Figure 25-A. Many things are needed to be able to cost justify the improvement in the jig. This problem requires that the engineer estimate or determine the following:

1. How much time is saved with a swing clamp (in the loading and unloading) cycle?
2. How much does a swing clamp cost?
3. How much will it cost to modify the existing jig?
Currently, for this job, the machining cycle time to drill the hole is 30 sec, the unload/load time is 30 sec, the operator is getting $12/hr, and the machine cost is $30/hr.
4. Examine Figure 25-23. The part has a circular slot in the top. How else could you produce this slot?

FIGURE 25-A
Forces on a 1500-lb work piece produced by face milling.

Chapter 25 CASE STUDY

Fixture versus No Fixture in Milling

Design engineering has sent down to manufacturing engineering a drawing that calls for a surface $4\frac{1}{2}$ inches wide by 10 in. long to be rough milled with a depth of cut of 0.30 in.

A 16-tooth cemented carbide face mill 150 mm (6 in.) in diameter has been selected for the job. The material is medium hard cast iron (220 to 260 Bhn).

This surface is to be milled on a large number of identical pieces. The estimated time to unload and load a piece in a fixture is 0.30 minutes. Here are three possible choices of machine setup you could use to do this job.

1. Face milling on a vertical spindle milling machine (NO fixture, 6 minutes to remove part from table and bolt up a new part).
2. String milling with three pieces 0.6 in. apart in a fixture.
3. Index base milling (0.15 minute required to index the base) on a vertical single spindle mill where the operator is unloading/loading the index table while the alternate piece is being machined (2 fixtures required).

Which arrangement should be selected based on your estimated operation time per piece and your estimated fixture cost per piece?

NUMERICAL CONTROL (NC) AND THE A(4) LEVEL OF AUTOMATION

■ 26.1 INTRODUCTION

The first numerically controlled (NC) machine tool was developed in 1952 at the Massachusetts Institute of Technology (MIT). It had three-axis positional feedback control and is generally recognized as the first NC machine tool. By 1958, the first NC *machining center* was being marketed by Kearney and Trecker. A machining center was a compilation of many machine tools capable of performing many processes (milling, drilling, tapping, and boring), as shown in Figure 26-1. This NC machine had a tool changer and could automatically change tools. Almost from the start, computers were needed to help program these machines. Within 10 years, NC machine tools had become computer numerical control (CNC) machine tools with onboard microprocessors and could be programmed directly.

With the advent of the NC type of machine (and, more recently, programmable robots), two types of automation were defined. *Hard* or *fixed automation* is exemplified by transfer machines or automatic screw machines controlled by a mechanical cam. *Flexible* or *programmable automation* is typified by CNC machines or robots that can be taught or programmed externally by means of computers. The control is in computer software rather than mechanical hardware.

You may not be familiar with the concept of feedback control, where some aspect (usually position) of the process is measured using a detection device (sensor). This information is fed back to an electronic comparator, housed in the machine control unit (MCU), which makes comparisons with the desired level of operation. If the output and input are not equal, an error signal is generated and the table is adjusted to reduce the error.

For a milling machine, Figure 26-2 shows the difference between an open-loop machine and a closed-loop machine, with feedback provided on the location of the table and the part with respect to the axis of the spindle of the cutting tool. Three position control schemes are shown.

In CNC turning machines, the feedback is on the tool tip with respect to the rotating part creating tool paths. Figure 26-3 shows how a part can be turned (machined) from a round bar in a CNC lathe. A program is written that directs the machine to execute the necessary roughing and finishing passes.

$$\text{number of rough passes} = \frac{\text{stock diameter} - \text{minimum diameter} + \text{finish}}{\text{depth of cut} \times 2}$$

In this case, eight roughing passes and one finishing cut were specified.

Vertical milling machine

Vertical boring machine

Upright drill press

Early NC machine

Vertical-spindle NC milling machine

Early CNC machine

CNC machine with integrated NC rotary table, pivotable through 90°

Early CNC machining center

Machining center with tool changer (20 tools), chip conveyor, pallet changer, and anti-splash booth

FIGURE 26-1   Early NC machine tools were controlled by paper tape. Soon onboard computers were added, followed by tool changers and pallet changers.

FIGURE 26-2 Open-loop NC versus three position control schemes for NC and CNC machine tools.

BRIEF HISTORY OF NUMERICAL CONTROL AND FLEXIBLE MANUFACTURING SYSTEMS (FMSs)

The advent and wide scale adoption of numerically (tape- and computer-) controlled machine tools has been the most significant development in machine tools in the past 60 years. These machines raised automation to a new level by providing positional feedback as well as programmable flexibility to machine tools. Numerical control of machine tools created entirely new concepts in manufacturing. Certain operations are now routine that previously were very difficult, if not impossible, to accomplish.

However, NC impacts on machine tools were greater than expected. Machine tools had to be redesigned to sustain more wear and tear on the drives, gears, and motors, and made better (more precise) because the operator no longer controlled the position of the work in respect to the cutting tool. The machines cost more (for controls and precision), with automatic tool changers and pallet changers quickly being added to make the NC machine a "machining center."

Point	X (in.) diameter	Z (in.)
2	4.5	0.1
A	1.8	0
B	1.8	–0.9
C	2.6	–2.6
D	2.6	–2.9
E	3.1	–2.9
F	3.7	–3.2
G	3.7	–3.8
H	4.5	–3.8

FIGURE 26-3   The tool paths necessary to rough and finish turn a part in a CNC lathe are computer generated using G codes.

In earlier years, highly trained NC programmers were required. The development of low-cost, solid-state microprocessing chips resulted in machines that can be quickly programmed by skilled machinists after only a few hours of training, using only simple machine shop language. As a consequence, today almost all manufacturing facilities, from the largest to the smallest job shops, have one or more numerically controlled machine tools in routine use.

NC came into being to fill a need. The U.S. Air Force (USAF) and the airframe industry were seeking a means to manufacture complex contoured aircraft components to close tolerances on a highly repeatable basis. John Parsons of the Parsons Corporation of Traverse, Michigan, had been working on a project to solve this problem; in 1947 he and his engineers developed a machine that would machine templates to be used for inspecting helicopter blades. He conceived of a machine (a jig borer) that was controlled by numerical data to make these templates and took his proposal to the USAF. Parsons convinced the USAF to fund the development of a machine. Subsequently, MIT was subcontracted to build the first NC machine in 1949. The prototype was a conventional two-axis tracer mill retrofitted with servomechanisms. As luck would have it, the servomechanism lab was located next to a lab where one of the very first digital computers (Whirlwind) was being developed. This computer generated the digital numerical data for the servomechanisms, and in 1952 a modified three-axis Cincinnati Hydrotel milling machine was demonstrated.

In 1962, 35 billion was spent on machine tools and NC machines accounted for about 10% of total. Early on, NC machines were continuous-path or contouring machines where the entire path of the tool was controlled with close accuracy in regard to position and velocity, and the large aerospace companies had many home-made NC machines. Today, milling machines, machining centers, laser beam and water-jet cutting machines, and lathes are popular applications of continuous-path control requiring feedback control. Next, point-to-point machines were produced in which the path taken between operations was relatively unimportant and therefore not monitored continuously. Point-to-point machines are used primarily for drilling, milling straight cuts, cutoff, and punching. Automatic tool changers, which require that the tools be precisely set to a given length prior to installation in the machines, permitted the merging of many into

FIGURE 26-4 Horizontal machining center with four-axis control (X, Y, Z, R table) receives inputs to the control panel from many sources.

one machine. Modern machines are often equipped with two pallets so that one can be set up while the other is working. The two-sided (or four-sided) "tombstone" fixture shown in Figure 26-4 has multiple mounting and locating holes for attaching part-dedicated fixture plates, which greatly extends the utility of a horizontal machining center.

During the early days of NC, the machine tools had to be programmed using a machine control language, a difficult, time-consuming task prone to error. Many companies developed computer languages to aid the NC programming task, each developing a different language to describe tool geometry, tool movements, and machining instructions. Confusion reigned.

The USAF sponsored the placement of many NC machines in the major aerospace companies. The companies soon concluded that a universally accepted NC programming language was needed. Under the auspices of the Aerospace Industries Association (AIA), these companies, with the USAF, sponsored additional research at MIT to develop a computer language that would use simple English-like statements to produce an output that would control the NC machines. This language, called *APT* (automatically programmed tools), was introduced by MIT in 1959, when computer technology was in its adolescence. Hundreds of thousands of parts have been programmed using APT, running initially on large mainframe computers. The output from the APT program had to be converted to the language of a particular machine. This was called *postprocessing*. Traditional postprocessing yields NC workpiece programs that are not exchangeable. To machine an identical workpiece on another machine, the program must be postprocessed again unless the machine and control are exactly the same.

With the arrival of high-resolution *computer-aided design* (CAD) graphics, many people thought that APT would be phased out. But when complex tool control is required for complex parts, APT or one of its many offspring is still used. The chief problems in NC programming were tool radius compensation and tool path interpolation (discussed later). Computer software with the capability to perform linear, circular, parabolic, and other kinds of interpolations were developed. The capabilities were included in APT, and such software programs are now routinely available on CNC machines.

In the 1960s, it was envisioned that a large computer could be used directly to control, in real time, a number of machines. A limited number of *direct numerical control* (DNC) systems were developed, with the idea that the programs were to be sent directly to the machines (eliminating paper tape handling). The mainframe computer would be shared on a real-time basis by many machine tools. The machine operator would have access to the main computer through a remote terminal at the machine, while management would have up-to-the-minute data on production status and machine utilization. This version of DNC had very few takers. Instead, NC machines became *computer numerical control* (CNC) machines through the development of small, inexpensive computer microprocessors with large memories. These onboard computers have every sort of input imaginable coupled with functions such as program storage, tool offset and tool compensation, program-editing capability, various degrees of computation, and the ability to send and receive data from a variety of sources, including remote locations. The computer can store multiple-part programs, recalling them as needed for different parts. As the software and controllers developed, it was immediately found that the machine tool operator could readily learn how to program these machines (manually) for many component parts, often eliminating the need for a part programmer.

In recent years the DNC concept has been revived with the small but powerful computers on the machine tools being networked to a larger computer to provide enhanced computer memory and computational capacity. Therefore, DNC now means *distributed numerical control,* with the distribution of NC programs by a central computer to individual CNC units. And so emerged a special type of manufacturing system called the flexible manufacturing system, or FMS. Historically speaking, the first examples of FMS systems appeared in the late 1960s, but few companies adopted them because of their high initial cost.

FLEXIBLE MANUFACTURING SYSTEMS

The development of FMSs began in England and the United States simultaneously in the 1960s. By combining the repeatability and productivity of the transfer line with the programmable flexibility of the NC machine, a variety of parts could be produced on the same set of machines. In the United States the first systems were called variable mission or flexible manufacturing systems. In the late 1960s Sundstrand installed a system for machining aircraft speed drive housings that was used for over 30 years. Overall, however, very few of these systems were sold until the late 1970s and early 1980s, when a worldwide FMS movement began. But even today international trade in FSM is not significant, and there are fewer than 2000 systems in the world (less than 0.1% of the machine tool population). There is also some evidence that the market for these large, expensive systems became saturated around the mid-1980s.

Essentially, the FMS permits (schedules) the products to take random paths through the machines. This system is fundamentally an automated, conveyorized, computerized job shop so the system is complex to schedule. Because the machining time for different parts varies greatly, the FMS is difficult to link to an integrated system and often remains an island of very expensive automation.

Application:

An aircraft parts manufacturer needed parts transfer mobility, in/out parts queue, cutting tool library, and quality control management for production of high-technology parts.

Wire-guided vehicles offer interdepartment transfer capability as well as in-cell transport. The qc center manages the machining accuracy for continuous flow of acceptable parts. Parts are scheduled in batch and/or random, controlled by a management computer.

The machines are equipped with telemetry probes, adaptive control, bulk tool storage, and complete tool management.

FIGURE 26-5 An example of a sophisticated FMS developed for machining aircraft parts. A wire-guided cart called an AGV (automated guided vehicle) is used to transport pallets from the unload/load station to the machines.

About 60 to 70% of FMS implementations are for components consisting of non-rotational (prismatic) parts such as crankcases and transmission housings. Figure 26-5 depicts an FMS with two machine tools serviced by a pallet system and an automated guided vehicle (AGV). The balance of FMS installations are for rotational parts or a mixture of both types of parts.

The number of machine tools in an FMS varies from 2 to 10, with 3 or 4 being typical. Annual production volumes for the systems are usually in the range of 3000 to 10,000 parts, the number of different kinds of parts ranging from 2 to 20, with 8 being typical. The lot sizes are typically 20 to 100 parts, and the typical part has a machining time of

TABLE 26-1	Common Features of Flexible Manufacturing Systems

Pallet changers

Multiple machine tools: NC or CNC

Automated material handling system (to deliver parts to machines)

Computer control for system: DNC

Multiple parts: Medium-sized lots (200–10,000) with families of parts

Random sequencing of parts to machines (optional)

Automatic tool changing

Inprocess inspection

Parts washing (optional)

Automated storage and retrieval (optional, to deliver parts to system)

about 30 minutes with a range of 6 to 90 minutes per part. Each part typically needs two or three chuckings or locating positions and 30 or 40 machining operations. Early on, NC machining centers were used, but in recent years, CNC machine tools have been favored, leading to a considerable number of systems being operated under direct numerical control. The machining centers always have tool changers. To overcome the limitation of a single spindle, some systems are being built with head changers. These are sometimes referred to as modular machining centers.

Common features of FMSs (see Table 26-1) are *pallet changers,* underfloor conveyor systems for the collection of chips (not shown), and a conveyor system that delivers parts to the machine. This is also an expensive part of the system, as the conveyor systems are either powered rollers, mechanical pallet transfer conveyors, or AGVs operating on underground towlines or buried guidance cables. The carts are more flexible than the conveyors. The AGVs also serve to connect the islands of automation, operating between FMSs and replacing human guided vehicles (forklift trucks).

Pallets are a significant cost item for the FMS because the part must be accurately located on the pallet and the pallet accurately located in the machine. Since many pallets are required for each different component, a lot of pallets are needed and they typically represent anywhere from 15 to 20% of the total system cost. FMSs cost about $1 million per machine tool. Thus the seven-machine FMS costs $6 million for hardware and software, with the transporter costing over $1 million.

Computer Control in FMS.   The CNC machines receive programs as needed from a host minicomputer, which acts as a supervisory computer for the system, tracking the status of any particular machine in the system. In recent years, in-process inspection, detection, and automatic tool position correction for tool wear and breakage have been added because a very common problem with these systems is the monitoring of the tool condition and performance. Most installations also incur problems in the performance and reliability of the software and the control systems. Because it may take just as long to debug software as it does to debug hardware, delays of two to six months in startup are not uncommon.

However, operators are typically needed to load workpieces, unload finished parts, change worn tools, and perform equipment maintenance and repairs. CNC and DNC functions are often incorporated into a single FMS. The system can usually monitor piece-part counts, tool changes, and machine utilization, with the computer also providing supervisory control of the production. The workpieces are launched randomly into the system, which identifies each part in the family and routes it to the proper machines. The systems generally display reduced manufacturing lead time, low in-process inventory, and high machine tool utilization, with reduced indirect and direct labor. The materials-handling system must be able to route any part to any machine in any order and provide each machine with a small queue of "banked parts" waiting to be processed so as to maximize machine utilization. Convenient access for loading and unloading parts, compatibility with the control system, and accessibility to the machine tools are other necessary design features for the material handling system.

The computer control system for an FMS typically has three levels. The master control monitors the entire system for tool failures or machine breakdowns, schedules the work, and routes the parts to the appropriate machine. The DNC computer distributes programs to the CNC machines and supervises their operations, selecting the required programs and transmitting them at the appropriate time. It also keeps track of the completion of the cutting programs and sends this information to the master computer. The bottom level of computer control is at the machines themselves.

It is difficult to design an FMS because it is, in fact, a very complex assembly of elements that must work together. Designing the FMS to be flexible is also difficult. Many companies have found that between the time they ordered their system and the time they had it installed and operational, design changes had eliminated a number of parts from the FMS. That is, the system was not as flexible as they thought. Figure 26-6 shows some typical FMS designs. Today a popular system has only one (or two) machining centers, with an automatic storage and retrieval system (ASRS) to provide the machine with a continuous supply of parts during all three shifts. This design is called a flexible manufacturing cell (FMC). FMSs are, in fact, classic examples of supermachines. Such large, expensive systems must be examined with careful and complete planning. It is important to remember that even though they are often marketed and sold as a *turnkey* installation (the buyer pays a lump sum and receives a system that can be turned on and run), this is only rarely possible with a system that has so many elements that must work together reliably. Taken in the context of integrated manufacturing systems, large FMSs may be difficult to synchronize to the rest of the system. The flexibility of the FMS requires variable speeds and cycles, numerical control, and a supervisory computer to coordinate cell operation. In the long run, smaller manned or unmanned cells may well be the better solution, in terms of system flexibility. Perhaps a better name for these systems would be *variable mission* or *random-path manufacturing systems*.

Many companies elect to identify the families of parts around which the FMS is designed, which greatly improves the utilization of the FMS. One might say that FMSs were developed before their time, since they are being more readily accepted since group technology has been used (at least conceptually) to identify families of parts for the system to produce.

As an FMS generally needs about three or four workers per shift to load and unload parts, change tools, and perform general maintenance, it cannot really be said to be self-operating. FMS systems are rarely left untended, as in third-shift operations. Other than the personnel doing the loading and unloading, the workers in the FMS are usually highly skilled and trained in NC and CNC. Most installations run fairly reliably (once they are debugged) over three shifts, with uptime ranging from 70 to 80%, and many are able to run on one shift untended.

■ 26.2 BASIC PRINCIPLES OF NUMERICAL CONTROL

NC uses a processing language to control the movement of the cutting tool or workpiece or both. The programs contain information about the machine tool and cutting tool geometry, the part dimensions (from rough to finish size), and the machining parameters (speeds and feeds and depth of cut). Thus NC machines can duplicate consecutive parts, and a part made at a later date will be the same as one made today. Repeatability and quality are improved over conventional (job shop) machines. Workholding devices can be made more universal, and setup time can be reduced, along with tool-change time, thus making programmable machines economical for producing small lots or even a single piece. When combined with the managerial and organizational strategies of group technology (GT) and cellular manufacturing, programmable machines lead to tremendous improvements in quality and productivity. GT basically leads to the creation of families of parts made in machining cells containing flexible, programmable machines. The compatibility of the components (similarly in process and sequences of processes) greatly enhances the productivity (utility) of the programmable equipment.

A side result has been the decrease in the non-chip-producing time of machine tools. The operator was relieved of the jobs of changing speeds and feeds and locating

FIGURE 26-6   Examples of machining centers—FMC and FMS designs.

the tool relative to the work. Even simple forms of NC and digital readout equipment have provided both greater productivity and increased accuracy. Most early NC machine tools were developed for special types of work where accuracies of as much as 0.00005 in. were required, and many NC machines are built to provide accuracies of at least 0.0001 in., regardless of whether or not it is needed. This forced many machine tool builders to redesign their machines and improve their quality because the operator was not available to compensate for the machine (positioning) error. While most NC machine tools today will provide greater accuracy than is required for most jobs, the trend is toward greater accuracy and precision (i.e., better quality) but at no increase in cost.

Therefore, NC machines will continue to be the very backbone of the machine tool business. Hopefully, all builders will one day arrive at a common language, so that if one can program one machine, one can program them all.

As the name implies, *numerical control* is a method of controlling the motion of machine components by means of numbers or coded instructions. Assume that three 1-in. holes in the part shown in Figure 26-7 are to be drilled and bored on a vertical-spindle machine. The centers of these holes must be located relative to each other and with respect to the left-hand edge (*X* direction) and the bottom edge (*Y* direction) of the workpiece. The depth of the hole will be controlled by the *Z*- (or *W*-) axis. For this part, this is the zero reference point for the part.

The holes will be produced by center drilling, hole drilling, boring, reaming, and counterboring (five tool changes). If this were done conventionally or in manned cells, three or four different machines might be required. On the NC machining center, it only requires changing the tool automatically. The movements of the table are controlled by coordinate systems for each direction. The machine tool has a zero point. The accurate positioning of the cutting tool with respect to the work is established by these zero points. The center of the table or a point along the edge of the traverse range is commonly used. The workpiece is positioned on the table with respect to this zero point on the table. For our example, the lower left-hand corner of the part is placed on the table 12 in. in the *X*+ direction, 4 in. in the *Y*+ direction, and 2 in. in the *Z*+ direction with respect to the machine zero point.

A software program is written that instructs the table to move with respect to the axis of the spindle to bring the holes to the correct location for machining. The machine shown in Figure 26-7 is called a five-axis machine because it has five movements (shown by the dark arrows) under numerical control. No fixture is shown in this example, but one would typically be used to obtain quick and repeatable location of the part on the table.

HOW CNC MACHINES WORK

NC and CNC machines can be subdivided into two types, shown in Figure 26-8. In *point-to-point machines,* the tool path is not controlled but tools can be moved in straight lines or parallel traverses at desired table feed rates, but only one axis drive is operated at a time. *Contouring* permits two or three axes to be controlled simultaneously, permitting two- or three-dimensional geometries to be generated. Another term for contouring is *continuous path*. Most milling and turning machining centers have contouring capability and are closed loop. The point-to-point machines can be open loop rather than closed loop.

The *X*-axis of the three-axis vertical spindle CNC machine tool shown in Figure 26-7 will be used to explain how the closed-loop positional control works.

Controlling a machine tool using variable input, such as a punched tape or a stored program, is known as numerical control and is defined by the Electronic Industries Association (EIA) as "a system in which actions are controlled by direct insertion of a numerical data at some point (the measured data is call the parts program). The system must automatically interpret at least some portion of the data."

Traditionally, NC machine tool has a *machine control unit* (MCU). The MCU is further divided into two elements: the data-processing unit (DPU) and the control-loops unit (CLU). The DPU processes the coded data that are read from the tape or some other input medium that it gives to the CPU, specifically the position of each axis, its direction of motion feed, and its auxiliary-function control signals. The CLU operates the drive mechanisms of the machine.

The CNC control system, shown schematically in Figure 26-9, uses a resolver or encoder to provide axis-position feedback to the MCU. A closed-loop control requires a transducer or sensing device to detect machine table position (and velocity for contouring) and transmit that information back to the MCU to compare the current status with the desired state. If they are different, the control unit produces a signal to the drive motors to move the table, reducing the error signal and ultimately moving the table to the desired position at the desired velocity. At this point, the command counter reaches zero, meaning that the correct number of pulses has been sent to move the table to the desired position. In a closed-loop system, a comparator is used to compare

FIGURE 26-7 The part (above) to be machined on the NC machine (below) has a zero reference point. The machine also has a zero reference point.

FIGURE 26-8 NC and CNC systems are subdivided into two basic categories: point-to-point controls or contouring controls.

Point-to-point control
(milling machines, welding)

Contouring control
(machining centers)

FIGURE 26-9 The table of the CNC machine (above) is translated with a ball screw mechanism, and its location is detected with a resolver. The schematic below shows how the table is located with respect to the spindle axis of the machine tool.

feedback pulses with the original value, generating an error signal. Thus, when the machine control unit receives a signal to execute this command, the table is moved to the specified location, with the actual position being monitored by the feedback transducer. Table motion ceases when the error signal has been reduced to zero and the function (drilling a hole) takes place. Closed-loop systems tend to have greater accuracy and respond faster to input signals but may exhibit stability problems (oscillating about a desired value instead of achieving it) not found in open-loop machines.

For most NC controls, the feedback signals are supplied by transducers actuated either by the feed screw or by the actual movement of the component. The transducers may provide either *digital* or *analog* information (signals). The resolver in Figure 26-9 measures (indirectly) the movement of the table by the rotation of the ball screw. A pulse disk on the end of the ball screw converts analog movement back into digital pulses, which are used to calculate table movement. The tachometer on the drive motor measures the table velocity.

Two basic types of digital transducers are used. One supplies *incremental* information and tells how much motion of the input shaft or table has occurred since the last time. The information supplied is similar to telling newspaper carriers that papers are to be delivered to the first, fourth, and eighth houses from a given corner on one side of the block. To follow the instructions, the carriers would need a means of counting the houses (pulses) as they passed them. They would deliver papers as they counted 1, 4, and 8. The second type of digital information is *absolute* in character, with each pulse corresponding to a specific location of the machine components. To continue the carrier analogy, this would correspond to telling the carriers to deliver papers to the houses having house numbers 2400, 2406, and 2414. In this case it would only be necessary for the carriers (machine component) to be able to read the house numbers (addresses) and stop to deliver a paper when arriving at a proper address. This *address* system is a common one in numerical control systems, because it provides absolute location information relative to a machine zero point.

When analog information is used, the signal is usually in the form of an electric voltage that varies as the input shaft is rotated or the machine component is moved, the variable output being a function of movement. The movement is evaluated by measuring, or matching, the voltage, or by measuring the ratios between the applied and feedback voltages; this eliminates the effect of supply-voltage variations.

The input information (i.e., the location of the holes) is given in binary form to the machine control unit in the form of a punched tape—a magnetic tape in a cassette on old machines or disks, as shown, on newer machines. The data are input directly via the control panel or a computer program. This command signal is converted into pulses by the machine control unit, which in turn drives the servomotor or stepper motor.

Alternating-current servomotors are rapidly replacing direct-current motors on new CNC machine tools. The reasons for change are better reliability, better performance-to-weight ratio, and lower power consumption.

If the system is point-to-point (or positioning), the control system disregards the paths between points. Some positioning systems provide for control of straight cuts along the machine axes and produce diagonal paths at 45° to the axes by maintaining one-to-one relationships between the motions of perpendicular axes. Contouring systems require directional changes at controlled velocity. Any lost motion in the system can distort the part. In contouring, it is usually necessary to control multiple paths between points by interpolating intermediate coordinate positions. As many of these systems as desired can be combined to provide control in several axes—two- and three-axis controls are most common, but some machines have as many as seven. In many, conversion to either English or metric measurement is available merely by throwing a switch.

The components required for such a numerical control system are now standardized items of hardware. In most cases the drive motor is electric, but hydraulic systems are also used. They are usually capable of moving the machine elements, such as tables, at high rates of speed, up to 200 in./min being common. Thus exact positioning can be achieved more rapidly than by manual means. The transducer can be placed on the drive motor or connected directly to the leadscrew, with special precautions being taken, such as the use of extra-large screws and ball nuts, to avoid backlash and to assure accuracy. In other systems, the encoder is attached to the machine table, providing direct measurement of the table position. Various degrees of accuracy are obtainable. Guaranteed positioning accuracies of 0.001 or 0.0001 in. are common, but greater accuracies can be obtained at higher cost. Most NC systems are built into the machines, but they can be retrofitted to some machine tools.

FIGURE 26-10 The ball lead screw shown in detail provides great accuracy and position to NC and CNC machine tools.

Initially, NC machines provided tool change; tool setting; and speed, feed, and depth-of-cut settings for positioning the work relative to the tool, with the remaining functions controlled by the operator. Gradually, the functions were incorporated into the control system so that the machines could change the tools automatically, change the speeds and feeds as needed for different operations, position the work relative to the tools, control the cutter path and velocity, reposition the tool rapidly between operations, and start and stop the sequence as needed.

Tables and tools are positioned using recirculating ball-screw drives (see Figure 26-10), or linear accelerators, which greatly reduce the backlash in the drive systems, helping to eliminate problems of servoloop oscillation and machine instability. Using such hardware, NC machines are manufactured with greater accuracy and repeatability and more rapid table movements than is possible for conventional machines.

Some of the functions in programmable machines require feedforward or preset loops. The machine must know in advance the rough dimensions of a casting or a forging so that it can determine how many roughing cuts are needed prior to the finishing cut. However, for most CNC work, the operator (or part programmer) still plans the sequence of operations; selects the cutting tools and workholding devices; and selects the speeds, feeds, and depths of cut. Common machining routines such as pocket milling or peck drilling have been programmed into many CNC machines. These are called *canned cycles.* As shown in Figure 26-11, the operator merely supplies the information requested by the control menu, and the machine fills in the necessary data for the previously written program to perform the desired machining routines.

Rectangular pocket milling
Control menu asks for:
• Setup clearance
• Milling depth
• Roughing depth
• Feed rate for roughing
• First side length
• Second side length
• Feed rate
• Direction or rotation

Peck drilling
Control menu asks for:
• Setup clearance
• Total hole depth
• Pecking depth
• Dwell time (seconds)
• Feed rate

Tapping
Control menu asks for:
• Setup clearance
• Total hole depth
• Dwell time (seconds)
• Feed rate

Source: Heidenhain Corp. (Elk Grove Village, IL.)

FIGURE 26-11 Canned or preprogrammed machining routines greatly simplify programming CNC machines. *(Courtesy of Heidenkain Corporation, Elk Grove Village, ILL.)*

To ensure accurate machining of a workpiece on a CNC machine, the control system has to *know* certain dimensions of the tools. These *tool dimensions* are referenced to a fixed *setting point* on the tool holder. For the milling cutter, the dimensions are length L and cutter radius. For the turning tool, the dimensions are length L and transverse overhang. These dimensions are part of the information given to the operator on the setting sheet (see Figure 26-12). The program assumes that the tools will have the specified dimensions.

FIGURE 26-12   The location of the corner of the end mill (left) or the tip of a single-point tool (right) must be known with respect to the tool setting points so that tool dimensions are accurately set.

(a) Workpiece drawing (*XY*-Plane)

(b) Workpiece drawing (*XZ*-Plane)

(c) Programming the outer contour

(d) Programming the slot

FIGURE 26-13 Example of programming a part in a vertical-spindle NC machine.

PART PROGRAMMING

Obviously, the preparation of the control program for use in NC machines is a critical step. Many standard languages and programs have been developed by the machine tool builders, and only the very basic aspect can be presented here.

The basic steps can be illustrated by reference to the part shown in Figure 26-13. The first step is to modify the drawing received from design engineering to establish the zero reference axes: the X and Y directions. The zero reference point is the lower left-hand corner of the part. The hole labeled 1 is located $+3.3437$ in the X direction and ±5.2882 in the Y direction, with respect to its zero point. The part should be dimensioned with respect to its zero reference point. This may require redrawing and redimensioning the part. Obviously, this step can be avoided if the original drawing is made in the desired form or a CAD design is used. The setup instructions given to the operator establish the position of the workpiece properly on the machine table with respect to the axis of the spindle or the machine zero reference point.

The second step is to develop a *part program*. The program (1) defines the sequence of operations required to fabricate the part; (2) gives the x, y, and z coordinate positions of the operations; and (3) specifies the spindle traverse that determines the depth of the cut, the spindle speed, and the feed rate, and also determines whether the same tool can continue the next operation or whether a tool change is required. The last four items are specified by code symbols, or NC words (see Table 26-2). The NC words are put together in a specified order to define a *block* of information needed to execute an operation. By convention, the data are usually arranged in blocks in the sequential order shown in the table.

Before starting with programming tool movements, the table movements and the workpiece drawing are studied. As a first step, the workpiece zero point is established. The workpiece drawing, two views, is redone into a coordinate system. In Figure 26-8 the workpiece zero point is located at the bottom left-hand corner of the workpiece drawing.

The three axes of the machine coordinate system is established in relation to the workpiece as follows:

- The x- and y-axes are used to show the workpiece geometry in the XY-plane.
- The z-axis shows the down feed or depth.

All the coordinates of the most important points on the part (in this case the points P1–P11) have been collated in the form of a table:

	P1	P2	P3	P4	P5	P6	P7	P8	P9	P10	P11
x	190	190	100	100	70	70	10	10	20	160	130
y	10	130	130	105	105	130	130	20	10	100	45
z	20	20	20	20	20	20	20	20	20	20	10

Preparation of the program includes the selection of the cutting tools (a 10-mm-diameter end mill) and the selection of desired cutting speeds and feed per tooth. As discussed in Chapter 25, these parameters, along with the depth of cut and depth of immersion, determine the spindle rpm and table feed rates.

For determining the outer contour, the tool radius must be considered because the path of the milling cutting along the outer contour is input into the control program, not the actual perimeter of the parts. To ensure that the control system can in fact control the milling cutter center axis correctly, the size of the milling cutter (radius or diameter) has to be known beforehand, as does the side of the finished contour (relative to the machining direction) on which the milling cutter is located. How these two items of information are input is not standardized for all machine tools, and it is therefore assumed here that the control system has already received this information.

Machining the outer contour begins with the milling cutter moving at rapid traverse rate—that is, function G00 X190 Y-5 to the starting point adjacent to the workpiece—whereupon down feed takes place with function G00 Z20 after switching on the spindle rotation.

Machining of the outer contour is then accomplished with the following G functions: The supplementary function F produces the necessary feed rate in mm/min. The down feed depth is taken up by the milling cutter at the starting point.

G01	Y130	F200			Straight line from starting point to P2
G01	X100				Straight line from P2 to P3
G01	X105	F150			Straight line from P3 to P4
G02	X70	Y105	R15		Radial arc, clockwise, with 15 radius
G01	Y130	F200			Straight line from P5 to P6
G01	X10				Straight line from P6 to P7
G01	Y20				Straight line from P7 to P8
G03	X20	Y10	R10	F150	Radial arc, counterclockwise with 10 radius
G01	X190	F200			Straight line from P9 to P1

TABLE 26-2   Definitions of Common NC Words

NC Word	Use
N	*Sequence number:* identifies the block of information
G	*Preparatory function:* requests different control functions, including preprogrammed machining routines
X, Y, Z, B	*Dimensional coordinate data:* linear and angular motion commands for the axis of the machine
F	*Feed function:* sets feed rate for this operation
S	*Speed function:* sets cutting speed for this operation
T	*Tool function:* tells the machine the location of the tool in the tool holder or tool turret
M	*Miscellaneous function:* turns coolant on or off, opens spindle, reverses spindle, tool change, etc.
EOB	*End of block:* indicates to the MCU that a full block of information has been transmitted and the block can be executed

For milling the slot, the milling cutter is first retracted by G01 Z35 from the workpiece to point P1 shown in Figure 26-13d.

Because the width of the slot is equal to the cutter diameter, the path of the cutter is easy to program. The tool radius compensation feature is switches off (by G40) so the following traverse instructions refer to the cutter center line axis.

Milling of the slot is accomplished by the following instructions:

G00	X160	X100		Rapid traverse to point P10
G01	Z20	F150		Down feed at point P10
G01	X130	Y45	Z10	Straight line from P10 to P11
G01	Z35	F200		Retraction from workpiece
G00	X300	Y300		Rapid traverse away from workpiece

After the program has been written, it is *verified*, which means the steps in making the part are graphically simulated. The verification step can use the computer monitor, which simulates the part being made by tracing all the toolwork paths as they would occur on the machine tool. Sometimes a sample part is machined in plastic or machinable wax for checking the part specifications from the drawing against the real part.

The program can be directly entered into the control panel of a CNC machine. Usually, longer programs are entered by tape or disk and short programs are entered manually. The inputs may be in binary code (in older machines) or in arabic numbers with verbal commands that the machine understands.

Historically, punched tape was used to control the movements. The idea came from the old player-piano roll, which was a form of tape control, and punched cards had been used for many years for controlling complicated weaving and business machines. Thus tape control of machine tools was an extension of an existing basic concept in which holes, representing information that has been punched into the tape, are read by sensing devices and used to actuate the devices that control various electrical or mechanical mechanisms.

Over the years, four basic types of tape format have been used for NC input to communicate dimensional and nondimensional information: *fixed-sequential format, block-address format, tab-sequential format,* and *word-address* format. Most new NC or CNC systems use the word-address format, which allows the words to be presented in any order and is the most flexible.

Today, CNC machines permit the user to program the interface between the machine tool and the control, greatly reducing the number of machining system components and interconnections. Current CNCs have extensive self-diagnostics and performance-monitoring systems. The greatest advances in CNC technology, however, are in part programming, where easy-to-use, menu-driven software makes programming almost as simple as setting up the machine manually. With older NC machines, the operator could override the program when necessary but could not reprogram the machine unless a new program was written. The CNC machine has the capability of reading a program into its computer memory, and the program can be modified at the machine like any other computer program.

On CNC machines, the machine tool operator may perform all the programming steps right at the console of the machine, programming the processing steps for the part directly into the computer memory. The program can be saved by having the machine print out a copy of the program, which can be used later for reorders of the same part. Features such as program edit, canned routines, program storage, diagnostics, constant surface speed, and tape punch are common on today's CNC machines.

As more and more design work is done on the computer (CAD) using databases and software that are compatible to the machine tools, there will be less dependence on tape for program storage and more utilization of floppy disks, hard disks, and other typical computer storage means. For example, a machining cell composed of NC machine tools designed for a family of 10 component parts is able to make the 10 different parts without needing retooling or refixturing, but it will still have 10 different programs for these parts for each machine. If the programs are stored in the computer, they can be readily accessed, but if they are stored on disk, delays will occur in dumping the different programs into or out of the control computer.

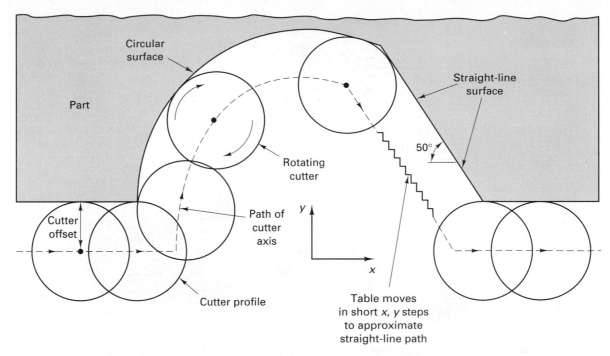

FIGURE 26-14 Two classic problems in NC programming are the determination of cutter offset and interpolation of cutter parts.

Cutter Offset and Interpolation in Numerical Control. Although the majority of NC and tape-controlled machine tools do not provide for machining contoured surfaces, most CNC and DNC machines do. The required curves and contours are generated approximately by a series of very short straight lines or segments of some type of regular curves, such as hyperbolas. This is called *interpolation*. The program fed to the machine is arranged to approximate the required curve within the desired accuracy. Figure 26-14 illustrates how a desired straight-line or curved surface can be approximated by means of short segments. Interpolation refers to the fact that curved surfaces as generated by machine tools must be approximated by a series of very short, straight-line movements in the x, y, and z directions. The length of the segment must be varied in accordance with the deviation permitted. Most NC machine tools with contouring capability will produce a surface that is within 0.001 in. of the one desired, and many will provide considerably better performance. Most contouring machines have either two- or three-axis capability, but five-axis capability as shown in Figures 26-7 and 26-15 are readily available.

In milling machines, the centerline of the cutter is offset from the desired surface by the radius of the cutter. The path that the cutter needs to take to generate the desired geometry is not simply the perimeter or profile of the part. Thus cutter offset programs must be included in the software. Obviously, contour machining (with cutter offset and interpolation) requires that complex information be entered into the computer because the number of straight-line or curved segments may be quite large. Manual programming of the tape can be quite laborious. Computer programming can translate simple commands into the complex information required by the machine.

■ 26.3 MACHINING CENTER FEATURES AND TRENDS

Computer numerical control is used on a wide variety of machine tools. These range from single-spindle drilling machines, which often have only two-axis control and can be obtained for about $10,000, to machining centers, such as shown in Figure 26-15. The machining center can do drilling, boring, milling, tapping, and so on, with four- or five-axis control. It can automatically select and change 40 to 180 preset tools. The table can move left/right and in/out, and the spindle can move up/down and in/out, with positioning accuracy in the range 0.00012 in. with repeatability to ±.00004 in. over 40 in. of travel.

Randomly selected tools switched in 3.5 seconds

Hi-Torque 33-hp, 7000-rpm spindle

Powerful 32-bit ACRAMATIC 950MC CNC

Dual pallets —

Two 500-mm-square, each with 2200-lb load capacity

Two 630-mm-square, each with 3000-lb load capacity

Compact 14.2-square-meter footprint (MAXIM 500)

Superior "T" design moves work to and from spindle like HBM

Integral load station/ workchanger switches pallets in about 10 seconds

Generous X-Y-Z range

29.5″ × 27.6″ × 29.5″

40.4″ × 31.5″ × 31.5″

FIGURE 26-15 Modern machining centers will typically have horizontal spindles with rpms up to 15,000, dual pallets, and cutting-tool magazines holding 40 to 100 tools.

Beyond accuracy and repeatability and the number of controlled axes, CNC machines are also categorized by spindle speed, horsepower, and the size of the workpiece. The accuracy and repeatability (precision) result from a combination of factors including the resolution of the control instrumentation and the accuracy of the hardware. The control resolution is the minimum length distinguishable by the control unit. It is called the base length unit (BLU) and is mainly a factor of the axis transducer and the quality of the leadscrew or linear translator. There are many sources of error in a machine, including wear in the machine sliding elements, machine tool assembly errors, spindle runout (wear), and leadscrew backlash.

Tool deflection due to the cutting forces produces dimensional error and chatter marks. Thermal error, caused by the thermal expansion of machine elements, is not uniform and is normally the greatest source of machine error. Methods used to remove heat from a machine include cutting fluids, locating drive motors away from the center of a machine, reducing friction from the ways and bearings, and spray cooling control element of the machine.

Most modern machining centers have automatic tool-change and automatic work-transfer capability, so that workpieces can be loaded and unloaded while machining is in process. Such a machine can cost over $200,000. Between these extremes are numerous machine tools that do less varied work than the highly sophisticated machining centers but that combine high output and minimum setup time with remarkable flexibility (large number of tool motions provided).

CNC TURNING CENTERS

The modern lathe has CNC control and tools mounted in turrets on slanted beds. The tailstock has been replaced by a live, powered spindle and chuck. On some lathes the concept of automatic tool changing has been implemented. The tools are held on a rotating tool magazine, and a gantry-type tool changer is used to change the tools. Each magazine

FIGURE 26-16   This CNC turning center has a multiple-axis capability with two spindles and a 12-tool turret with X, Y, and Z control as well as axis control of the spindles.

holds one type of cutting tool. This is an example of the trend of providing greater versatility along with high productivity in lathes. The versatility is being further increased by combining both rotary-work and rotary-tool operations—turning and milling in a single machine. Live, powered, or driven tools replace regular tools in the turrets and perform milling and drilling operations when the spindle is stopped. See Figure 26-16.

OTHER NC MACHINES

Numerical control has been applied to a wide variety of other production processes. NC turret punches with X-Y control on the table, CNC wire EDM machines, laser and water-jet abrasive machining, flame cutters, and many other machines are readily available. Some new trends are being observed in the development of machining centers, such as smaller, compact machining centers with higher spindle speeds. Machines with four- and five-axis capability are readily available. Modern machining centers have contributed significantly to improved productivity in many companies. They have eliminated the time lost in moving workpieces from machine to machine and the time needed for workpiece loading and unloading for separate operations. In addition, they have minimized the time lost in changing tools, carrying out gaging operations, and aligning workpieces on the machine.

The latest generation of machining centers is aimed at further improving utilization by reducing the time when machines are stopped, either during pauses in a shift or between shifts. Delays are caused by tool breakage, unforeseen tool wear, a limited number of tools, or an inadequate number of available workpieces. Machines are fitted with tool breakage monitors, tool wear compensating devices, and means for increasing the number of tools and workpieces available.

Probes on CNC machines can greatly improve the process capability of the machine tool. There is a big difference between the claimed program resolution for an NC machine and the accuracy and precision (the process capability) of the actual parts. As shown in Figure 26-17, true positioning accuracy and precision are affected by machine alignment, machine and fixture setup, variations in the workholding device, raw material variations, workpiece location in the fixture variations, and cutting-tool tolerances.

FIGURE 26-17 Process capability in NC machines is affected by many factors.

Thus the finished workpiece may be unacceptable even though the machine is more than capable of producing the part to the design specifications. The part program has no assurance that the part is properly located in the fixture or that the fixture is properly located on the table of the machine. However, a probe, carried in the tool storage magazine and mounted when needed in the spindle like a cutting tool, can establish the location of the surface features relative to each other and to the spindle axis within 0.0005 in. (Figure 26-18). The machine controller, using the probe data, will then shift the program reference data accordingly. The probe can be used to determine the amount of material on a rough casting, locate a corner of a part, define the center of a hole, or check for the presence or absence of a feature. All of the variability described in Figure 26-17 can be compensated for except for variations in the cutting-tool geometry or tool wear. A probe mounted on the machine tool can be used to automatically update tool-offset data in the control computer. Thus the machine tool can function as a coordinate measuring machine. By comparing the actual touched location with the programmed location, the measuring routine determines appropriate compensation.

(a)

FIGURE 26-18 (a) Probe carried in the tool changer can be mounted in the spindle (b) for checking the location of part features accurately.

(b)

Many of the new machining centers are equipped with novel automatic pallet changers and workpiece loading and unloading devices. Robots are increasingly being applied for workpiece handling in machine groups, including manufacturing cells. In some cases the robot is also used for tool-changing functions.

■ 26.4 Ultra-High-Speed Machining Centers (UHSMCs)

Anyone involved in the product development process knows that the longest lead-time path from a new product (like a car) design to a finished product coming off the final assembly line always includes the die-making process. For example, in the automotive industry, each new car design may have many die sets for forged and sheet metal parts. Machining of the dies is a key process in the conventional die-making process (see Figure 26-19), where the major steps are milling, electronic discharge machining (EDM), polishing, die assembly, tryout, and modification.

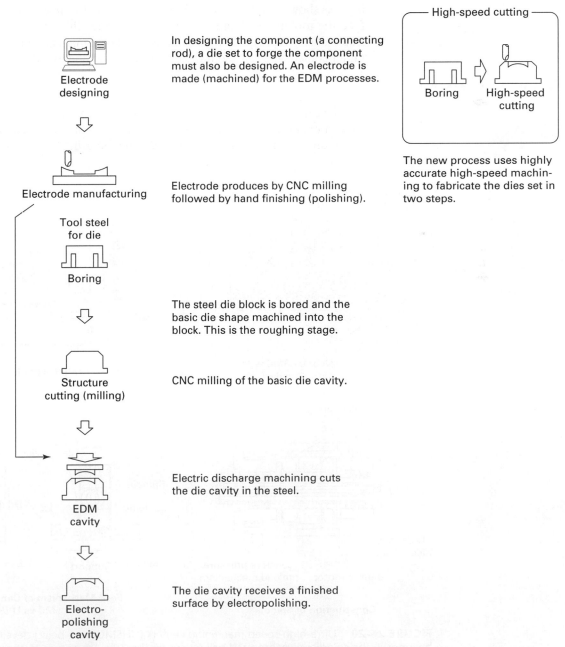

FIGURE 26-19   The process to manufacture dies for forging processes is shown on the left. Using ultra-high-speed machining centers reduces the sequence to two steps.

Ultra-high-speed machining centers (UHSMCs) are available to shorten the die-making process lead time. The new machining centers have highly accurate high-speed spindles capable of rpms of 30,000 to 50,000, cutting feed rates of 60 m/min, and cutting feed accelerations of 9.8 m/sec². The machine requires a spindle with high stiffness utilizing ceramic ball bearings with a constant-pressure preload mechanism and jet lubrication for superior performance.

Robust machining centers of this sort are capable of very high metal removal rates, particularly in materials like aluminum. However, before investing in a high-speed machining center (HSMC), many issues must be addressed. The new machine will usually require the purchase of new cutting tools and tool holders. At the higher rpms and feed rates, smaller-diameter tools are easier to balance. Tungsten carbide is the primary tool material. Make sure the insert retention mechanism is adequate in case of a failure. Other major problems will include chatter and removal of the large volume of chips.

UHSMCs require exceptionally accurate, stiff spindles using ceramic ball bearings, raising the possible rpm as compared to regular ball bearings, air bearings, or magnetic bearings. As shown schematically in Figure 26-20, synchronized sets of ball screws are used to feed the tool in the X and Y directions to reduce the errors caused by distortion in the frame during rapid acceleration of the tool. These machines have usually four or five axes under control and represent the pinnacle of machine tool development at this time.

■ 26.5 SUMMARY

Numerical control machines, robots, FMSs, and computers are critical elements of the advanced manufacturing technologies available for the next decade, which is being touted as the time when computer-integrated manufacturing (CIM) will become a widespread

FIGURE 26-20   Ultra-high-speed machining centers (UHSMCs) are being developed with ceramic ball bearings in the spindles, synchronized ball screws on the *X*-axis to reduce distortion (due to inertia) in the moving components. (*"Development of Ultra High Speed Machining Center"*, Toyota Technical Review, *vol. 49, No. 1, September 1999.*)

reality. Computer technology abounds: computer-aided design; computer-aided manufacturing with NC, CNC, or A/C and DNC; computer-aided process planning; computer-aided testing and inspection (CATI); artificial intelligence; smart robots; and much more. But a word of caution: Any company can buy computers, robots, and other pieces of automation hardware and software. The secret to manufacturing success lies in the design of a simple, unique manufacturing system that can achieve superior quality at low cost with on-time delivery and still be flexible. Flexibility means the system can readily adapt to changes in the customer demand (both volume and mix) while quickly implementing design changes. This requires a visionary management team and a change in culture at all levels, starting on factory floor with an empowered and involved workforce. Changing how people work in a manufacturing system means you have to redesign the system. No better example of this can be found than the Toyota Motor Company. Led by their vice president for manufacturing, Taiichi Ohno, who conceived, developed, and implemented Toyota's unique manufacturing system, this company has emerged as the world leader in car production. The implementation of this system has saved many companies (Harley Davidson, for example) and carried many others to the top position in their industry. The Toyota system is unique and as revolutionary today as were the American Armory System (job shops) and the Ford system (flow shops) in their day. It is significant that virtually every manufacturing system or technology cited in this chapter is practiced at Toyota. This new system is now being called *lean production* (to contrast it to mass production). Toyota does not use the word *CIM* because the computer is only a tool used in their system, a manufacturing system design that recognizes people as the most flexible element (Black and Hunter, 2003). Students of industrial, mechanical and manufacturing engineering are well advised to be knowledgeable of this new unique system.

■ Key Words

automated guided vehicle (AGV)
automation
canned cycles
closed-loop control
computer numerical control (CNC)
contouring
controller
direct numerical control (DNC)
distributed numerical control (DNC)
feedback device
feedforward
flexible manufacturing system (FMS)
machine control unit (MCU)
machining center
numerical control (NC)
open-loop control
pallet changer
part program
point-to-point machines
transfer machine
turning center
ultra-high-speed machining centers (UHSMCs)

■ Review Questions

1. What human attribute is replaced by an NC machine?
2. Give an everyday example of a household device or appliance that exhibits feedback in its control system.
3. Explain how a toaster could be made a closed-loop device.
4. The first NC machines were closed-loop control. Later some machines were open loop. What change did this require on the part of machine tool builders?
5. Can a continuous-path NC machine be open loop? Why or why not?
6. How is a machining center different from a milling machine?
7. What role did John Parsons play in the development of NC?
8. What was DNC as first practiced, and what does DNC usually mean today?
9. Many manufacturers have purchased large machining centers tied to an ASRS. Go on the Internet to find an example of an ASRS, and explain how such a system works.
10. How does feedforward differ from feedback in a process control system?
11. Why was it necessary for machine tool builders to improve the leadscrews on their machines when they made them into NC machines?
12. Explain what is meant by *interpolation* in NC programming.
13. Explain the problem of cutter offset in NC programming by making a sketch showing an end mill cutting a perimeter on a square plate.
14. Some of the functions performed by the operator in piece-part manufacturing are very difficult to automate completely. Name the functions and explain why.
15. Why do you think there are no NC shapers?
16. Why isn't manual programming used for continuous-path NC?
17. What are the three basic closed-loop feedback A(4) schemes used on CNC machines.
18. Which method for position feedback is the most accurate?
19. What is the difference between the zero reference point and the machine zero point?
20. What is an encoder?
21. What is a peck-drilling subroutine for a CNC machine?
22. What is pocket milling, and what kind of milling cutter is usually used to perform it?
23. How are probes used in CNC machines to improve process capability?
24. What are G words used for in NC?

25. Flexible manufacturing systems use CNC machines for processing and AGVs, robots, or conveyors to transport parts. So what differentiates the FMS from a transfer line?

26. What major changes are being introduced into UHSMCs?

■ Problems

1. What are the X and Y dimensions for the center position of holes 1, 2, and 3 in the part shown in Figure 26-7?

2. Configurations obtained from continuous-path machining are the result of a series of straight-line, parabolic-span or higher-order curves. The degree to which curved surfaces correspond to their design depends on how many lines or spans are used. Four equal chords in a circle describe a square (Figure 26-A). Six make a hexagon, and, as the number increases, the lines themselves come closer to a perfect circle. The number of lines needed is determined by a maximum tolerance allowed between the design of the curved section and the actual chord programmed. This is the dimension T. The program for a parabolic-span control unit requires enough spans for any deviation to stay within an acceptable tolerance. For a tolerance of T = 0.001 in., how long should the span be for a curve with a 5-in. radius.? Assume that the arc is part of a circle. What is the span angle here, in degrees?

FIGURE 26-A

3. In Problem 2, suppose that the acceptable tolerance was 0.0001 in. Determine the span angle.

4. Suppose that the plate shown in Figure 26-B was to be profile milled around the periphery with a 1-in.-diameter end milling cutter. The dashed line is the cutter path. The programmer must calculate an offset path to allow for cutter diameter. Since the programmed points are followed by the cutter centerline and the profile is made at the tool's periphery, the programmer called for a $1/2$-in. cutter offset. Working with computer assistance, the programmer would describe the part profile to be machined and specify the cutter. The computer would generate the cutter path. Complete the table below to specify the cutter path, starting with the

origin at the zero reference point. Move the tool around the plate counterclockwise.

Programmed Point Locations

PT	X	Y
1		
2		
3		
4		
5		
1		

5. Suppose that surface finish is very important for the profile milling job described in Problem 4. Thus down milling is going to be used. Rewrite the NC program points to accommodate this requirement. Show the new path on a sketch such as Figure 26-B. (Up versus down milling is discussed in Chapter 25.)

6. You have received the part drawing for a typical lathe part that will require turning, facing, grooving, boring, and threading as it is machined from a casting. See Figure 26-C. Unfortunately, you do not yet know how many of these parts will be ordered this year, so you do not know what the build quantity will be. To be prepared, you have developed some cost data for the manufacture of the part by four different lathe processes (see Table 26-A). Complete the table by determining the run cost per batch, the cost per unit at the various quantities, and the total cost per batch. Answer the following questions regarding this situation:

FIGURE 26-B

FIGURE 26-C

1. Of the four costs listed for each process, which costs are fixed and which are variable?
2. What is included in the cost per unit? Do you need to estimate the machining time per piece and the cycle time per piece, including the time to change parts and setups in order to compute run cost?
3. How would you go about estimating this time, and what time elements might be included in the cycle time in addition to the machining time?
4. How would you use this estimate of time in the cost table? Show the calculation.
5. Make a plot of cost in dollars versus quantity, with all four methods on one plot.
6. Make a plot of cost per unit versus quantity, again with all four methods on one plot. Find the break-even quantities. (*Hint:* Did you plot the data on log paper?)
7. Discuss the break-even quantities that you found in part 5 versus part 6. (They should be the same.)
8. When would you use the NC lathe? The turret lathe? When would you use the turret lathe if you have no NC lathe?

TABLE 26-A   Cost Data for Lathe Processes

	Make Quantity				
	10,000 Units	1,000 Units	100 Units	10 Units	1 Unit
Cost to produce on six-spindle automatic					
Total cost of batch	—	—	—	—	—
Engineering 2.5 hr at $40 /hr	50.00	50.00	50.00	50.00	50.00
Tooling (cutting tools and workholders)	600.00	600.00	600.00	600.00	600.00
Setup 8 hr at $15/hr	120.00	120.00	120.00	120.00	120.00
Run cost per batch: 50 per piece	5000.00	500.00	50.00	5.00	0.5
Cost each	—	—	—	—	—
Cost to produce on turret lathe					
Total cost of batch					
Engineering 2 hr at $20/ hr	—	—	—	—	—
Tooling	40.00	40.00	40.00	40.00	40.00
Setup 4 hr at $20/hr	150.00	150.00	150.00	150.00	150.00
Run cost per batch: $8 per piece	48.00	48.00	48.00	48.00	48.00
Cost each	—	—	—	—	—
	—	—	—	—	—
Cost to produce on engine lathe					
Total cost of batch	—	—	—	—	—
Engineering 1 hr at $20/hr	20.00	20.00	20.00	20.00	20.00
Tooling (no cost)	—	—	—	—	—
Setup 2 hr at $12.00/hr	24.00	24.00	24.00	24.00	24.00
Run cost per batch: $12	120,000.00	12000.00	120.00	120.00	12.00
Cost each	12.00				
Cost to produce on NC lathe					
Total cost of batch	—	—	—	—	—
Engineering and programming	150.00	150.00	150.00	150.00	150.00
Tooling	100.00	100.00	100.00	100.00	100.00
Setup 1 hr at $20/hr	20.00	20.00	20.00	20.00	20.00
Run cost per batch: $2 per piece	—	—	—	—	—
Cost each	—	—	—	—	—

www.wiley.com/college/degarmo

Chapter 26 CASE STUDY

Process Planning for the MfE

Figure CS 26a shows a part design for a small flange to be made out of 1020 steel. Prepare a sequence of operations or process plan to make this part. Note that you need to machine the top, bottom, and all the holes. Specify the machines and the tools you would use assuming you select bar stock as your raw material. The term Co bore in the drawing means counter bore. Check with your instructor for an example of a process plan. Assume the lot size here is 1200 parts/year made in lots of 120 every month.

Now assume that the designer provides you a design like CS 26b and calls for the flange to be made from cast iron. The casting will probably have a large hole in the center cored during the casting process. What casting process will you use? Prepare another process plan for the cast part. Will the sequence of operations be the same? What about the cost per unit? Which will be lower?

1. The part drawings failed to provide a critical specification in the design. What is it?
2. The drawing failed to show the final geometry (in the top view) correctly. Redraw the part to show these corrections.
3. As the process engineer, how would you advise the design engineer he screwed up twice?

Figure CS 26a

Figure CS 26b

Figure CS 26 Two different designs of a flange. The design on the right suggested by Doyle et al. in 3rd edition of *Manufacturing Process and Materials for Engineers*, Prentice-Hall.

OTHER MACHINING PROCESSES

■ 27.1 INTRODUCTION

While milling, drilling, and turning make up the bulk of the machining processes, there are many other chipmaking (metal removal) processes. This chapter will cover shaping, planing, broaching, sawing, and filing.

■ 27.2 INTRODUCTION TO SHAPING AND PLANING

The processes of *shaping* and *planing* are among the oldest single-point machining processes. Shaping has largely been replaced by milling and broaching as a production process, while planing still has applications in producing long flat cuts, like those in the ways of machine tools. From a consideration of the relative motions between the tool and the workpiece, shaping and planing both use a straight-line cutting motion with a single-point cutting tool to generate a flat surface.

In shaping, the workpiece is fed at right angles to the cutting motion between successive strokes of the tool, as shown in Figure 27-1, where f_c is the feed per stroke, V is the cutting speed, and d is the depth of cut (DOC). (In planing, discussed next, the workpiece is reciprocated and the tool is fed at right angles to the cutting motion.) For either shaping or planing, the tool is held in a clapper box, which prevents the cutting edge from being damaged on the return stroke of the tool. In addition to plain flat surfaces, the shapes most commonly produced on the shaper and planer are those illustrated in Figure 27-2. Relatively skilled workers are required to operate shapers and planers, and most of the shapes that can be produced on them can also be made by much more productive processes, such as milling, broaching, or grinding. Consequently, except for certain special types, planers that will do only planing have become obsolete. Today, shapers are used mainly in tool and die work, in very low volume production, or in the manufacture of gear teeth.

In shaping, the cutting tool is held in the tool post located in the ram, which reciprocates over the work with a forward stroke, cutting at velocity V and a quick return stroke at velocity V_R. The rpm of drive crank (N_s) drives the ram and determines the velocity of the operation (see Figure 27-1d). The stroke ratio is

$$R_s = \frac{\text{cutting stroke angle}}{360°} = \frac{200°}{360°} = \frac{5}{9} \tag{27-1}$$

(a) Basic geometry for shaping and planing

(c) Shaper tool holder, clapper box and workpiece

(b)

(d) Shaper (quick-return) mechanism for driving tool past work.

FIGURE 27-1   Basics of shaping and planing. (a) The cutting speed, V, and feed per stroke f_c. (b) Block diagram of the machine tool. (c) The cutting tool is held in a clapper box so the tool does not damage the workpiece on the return stroke. (d) The ram of the shaper carries the cutting tool at cutting velocity V and reciprocates at velocity V_R by the rotation of a bull wheel turning at rpm n_s.

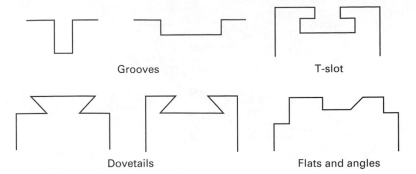

FIGURE 27-2 Types of surfaces commonly machined by shaping and planing.

Grooves

T-slot

Dovetails

Flats and angles

The tool is advancing 55% of the time. The number of strokes per minute is determined by the rpm of the drive crank. Feed is in inches per stroke and is at right angles to the cutting direction. As in other machining processes, speed and feed are selected by the operator.

The length of stroke l must be greater than the length of the workpiece (or length of cut) L, since velocity is position variant. Let l = twice the length of the block being cut, or $2L$. The cutting velocity V is assumed to be twice the average forward velocity V of the ram. The general relationship between cutting speed and rpm is

$$V = \frac{\pi D N_s}{12 R_s} \text{ft/min} \qquad (27\text{-}2)$$

where D is the diameter (of the rotating bull wheel) in inches. For shaping, the cutting speed is

$$V = \frac{2 l N_s}{12 R_s} \text{ft/min} \qquad (27\text{-}3)$$

Once a cutting speed is selected, the rpm of the machine can be calculated. Tables for suggested feed values, f_c, are in inches per stroke (or cycle), and recommended depths of cut are also available. The maximum depth of cut is based on the horsepower available to form the chips. This calculation requires that the metal removal rate (MRR) be known. The MRR is the volume of metal removed per unit time

$$\text{MRR} = \frac{L W d}{T_m} \text{in.}^3/\text{min} \qquad (27\text{-}4)$$

where W is the width of block being cut and L is the length of block being cut, so volume of cut = WLd, where d is the depth of cut and T_m is the time in minutes to cut that volume.

In general, T_m is the total length of the cut divided by the feed rate. For shaping, T_m is the width of the block divided by the feed rate f_c of the tool moving across the width. Thus, for shaping,

$$T_m = \frac{W}{N_s \times f_c} \qquad (27\text{-}5)$$

Also,

$$T_m = \frac{S}{N_s} \qquad (27\text{-}6)$$

where the number of strokes for the job is for a surface of width W.

MACHINE TOOLS FOR SHAPING

Shapers, as machine tools, are usually classified according to their general design features as follows:

1. Horizontal
   a. Push-cut
   b. Pull-cut or draw-cut shaper

FIGURE 27-3 The most widely used shaper is the horizontal push-cut machine tool, shown here with no tool in the tool post.

2. Vertical

    a. Regular or slotters

    b. Keyseaters

3. Special

They are also classified as to the type of drive employed: *mechanical drive* or *hydraulic drive*. Most shapers are of the *horizontal push-cut* type (Figure 27-3), where cutting occurs as the ram *pushes* the tool across the work.

On horizontal push-cut shapers, the work is usually held in a heavy vise mounted on the top side of the table. Shaper vises have a very heavy movable jaw, because the vise must often be turned so that the cutting forces are directed against this jaw.

In clamping the workpiece in a shaper vise, care must be exercised to make sure that it rests solidly against the bottom of the vise (on parallel bars) so that it will not be deflected by the cutting force and so that it is held securely yet not distorted by the clamping pressure.

Most shaping is done with simple high-speed-steel or carbide-tipped cutting tool bits held in a heavy, forged tool holder. Although shapers are versatile tools, the precision of the work done on them is greatly dependent on the operator. Feed dials on shapers are nearly always graduated in 0.001-in. divisions, and work is seldom done to greater precision than this. A tolerance of 0.002 to 0.003 in. is desirable on parts that are to be machined on a shaper, because this gives some provision for variations due to clamping, possible looseness or deflection of the table, and deflection of the tool and ram during cutting.

PLANING MACHINES

Planing can be used to produce horizontal, vertical, or inclined flat surfaces on large workpieces (too large for shapers). However, planing is much less efficient than other basic machining processes, such as milling, that will produce such surfaces. Consequently, planing and planers have largely been replaced by planer milling machines or machines that can do both milling and planing.

Figure 27-4 shows the basic components and motions of planers. In most planing, the action is opposite to that of shaping. The work is moved past one or more stationary single-point cutting tools. Because a large and heavy workpiece and table must be reciprocated at relatively low speeds, several tool heads are provided, often with multiple tools in each head. In addition, many planers are provided with tool heads arranged so that cuts occur on both directions of the table movement. However, because only single-point cutting tools are used and the cutting speeds are quite low, planers are low in productivity as compared with some other types of machine tools.

(a)

Block diagram
showing the basic components
of a double-housing planer.

(b)

Block diagram
of an open-side planer.

FIGURE 27-4 Schematic of planers. (a) Double-housing planer with multiple tool heads (4) and a large reciprocating table; (b) single-housing or open-sided planer; (c) interchangeable multiple tool holder for use in planers. *(Photograph courtesy Gebr Boehringer GmbH.)*

(c)

Four planer tools in tool holder

Figure 27-4 depicts the most common double-housing and single-housing types. The double-housing has a closed-housing structure, spanning the reciprocating worktable, with a cross rail supported at each end on a vertical column and carrying two tool heads. An additional tool head is usually mounted on each column, so that four tools (or four sets) can cut during each stroke of the table. Obviously, the closed-frame structure of this type of planer limits the size of the work that can be machined. Open-side planers have the cross rail supported on a single column. This design provides unrestricted access to one side of the table to permit wider workpieces to be accommodated. Some open-side planers are convertible, in that a second column can be attached to the bed when desired so as to provide added support for the cross rail.

WORKHOLDING AND SETUP ON PLANERS

Workpieces in planers are usually large and heavy. They must be securely clamped to resist large cutting forces and the high-inertia forces that result from the rapid velocity changes at the ends of the strokes. Special stops are provided at each end of the workpiece to prevent it from shifting.

Considerable time is usually required to set up the planer, thus reducing the time the machine is available for producing chips. Sometimes special setup plates are used for quick setup of the workpiece. Another procedure is to use two tables. Work is set up on one table while another workpiece is being machined on the other. The tables can be fastened together for machining long workpieces.

The large workpieces can usually support heavy cutting forces, so large depths of cut are recommended, which decrease the cutting time. Consequently, planer tools usually are quite massive and can sustain the large cutting forces. Usually, the main shank of the tools is made of plain-carbon steel, with tips made from high-speed steel or carbide. Chip breakers should be used to avoid long and dangerous chips in ductile materials.

Theoretically, planers have about the same precision as shapers. The feed and other dimension-controlling dials are usually graduated in 0.001-in. divisions. However, because larger and heavier workpieces are usually involved, and much longer beds and tables, the working tolerances for planer work are somewhat greater than for shaping.

■ 27.3 INTRODUCTION TO BROACHING

The process of *broaching* is one of the most productive of the basic machining processes. The machine tool is called a *broaching machine* or a *broach* and the cutting tool is also called the *broach*. Figure 27-5 shows the basic shape of a conventional pull broach. In this figure, P = pitch, n_s is the number of semiroughing teeth, and n_f is the number of finishing teeth where the rise per tooth gets smaller from rough to finish.

The feed per tooth in broaching is the change in height of successive teeth. This is called the rise per tooth (RPT or t_r; see Figure 27-6). Broaching looks similar to sawing except that the saw makes many passes through the cut, whereas the broach produces a finished part in one pass. The heart of this process lies in the broaching tool, in which roughing, semifinishing, and finishing teeth are combined into one tool, as shown in Figure 27-5. Broaching is unique in that it is the only one of the basic machining processes in which *feed*, which determines the chip thickness, is built into the cutting tool. The machined surface is always the inverse of the profile of the broach, and, in most cases,

P – pitch of teeth
D – depth of teeth (0.4P)
L – land behind cutting edge (0.25P)
R – radius of gullet (.25P)
α – hook angle or rake angle
γ – backoff angle or clearance angle
RPT – rise per tooth (chip load), t_r

FIGURE 27-5 (a) Photo of pull broach. (b) Basic shape and nomenclature for a conventional pull (hole) broach. Section A–A' shows the cross section of a tooth. P = pitch; n_r = number of roughing teeth; n_s = number of semifinishing teeth; n_f = number of finishing teeth.

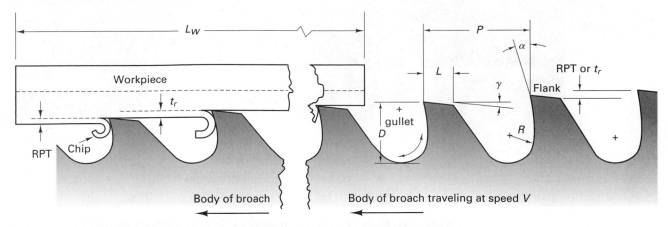

FIGURE 27-6 The feed in broaching depends on the rise per tooth t_r (RPT). The sum of the RPT gives the depth of cut, DOC. P = pitch of teeth; D = depth of teeth (0.4P); L = land behind cutting edge (0.25P); R = radius of gullet (0.25P); α = hook angle or rake angle; γ = backoff angle or clearance angle.

it is produced with a single linear stroke of the tool across the workpiece (or the workpiece across the broach).

Broaching competes economically with milling and boring and is capable of producing precision-machined surfaces. The broach finishes an entire surface in a single pass. Broaches are used in production to finish holes, splines, and flat surfaces. Typical workpieces include small to medium-sized castings, forgings, screw-machine parts, and stampings.

This rise per tooth (RPT), also known as *step* or the feed per tooth, determines the amount of material removed. No feeding of the broaching tool is required. The frontal contour of the teeth determines the shape of the resulting machined surface. As the result of these conditions being built into the tool, no complex motion of the tool relative to the workpiece is required and the need for highly skilled machine operators is minimized.

Figure 27-7 shows a *pull broach* in a vertical pull-down broaching machine. The pull end of the broach is passed through the part, and a key mates to the slot. The broach is pulled through the part. The broach is retracted (pulled up) out of the part. The part is transferred from the left fixture to the right fixture. One finished part is completed in every manufacturing cycle.

◾ 27.4 FUNDAMENTALS OF BROACHING

In broaching, the tool (or work) is translated past the work (or tool) with a single stroke of velocity V. The feed is provided by a gradual increase in height of successive teeth. The rise per tooth varies depending on whether the tooth is for roughing (t_r), semifinishing (t_s), or final sizing or finishing (t_f). In a typical broach there are three to five semifinishing and finishing teeth specified. The number of roughing teeth must be determined so that broach length, which is needed to estimate the cutting time, can be calculated. Other lengths needed for a typical pull broach are shown in Figure 27-6. The chip breakers in the first section of roughing teeth may be extended to more teeth if the cut is heavy or material difficult to machine. The distance between the teeth, called the pitch P, is important because it determines the tooth construction and strengths and the number of teeth actually cutting at a given instant. It is preferable that at least two teeth be in contact with the workpiece at any instant.

The pitch or distance between teeth is

$$P \cong 35\sqrt{L_w} \tag{27-7}$$

where length of cut usually equals L_w, as shown in Figure 27-5.

The number of roughing teeth is

$$n_r = \frac{\mathrm{DOC} - n_s t_s - n_f t_f}{t_r} \tag{27-8}$$

where DOC is the total amount of metal to be removed and is the rise per tooth.

(a) Rotor- or jump-tooth broach design.

(b) Round, push-type broach with chip-breaking notches on alternate teeth except at the finishing end.

(c) Notched tooth, flat broach

(d) Progressive surface broach. *(Courtesy of Detroit Broach & Machine Company)*

FIGURE 27-7 Methods to decrease force or break up chip rings in broaches. (a) Rotor or jump tooth; (b) notched tooth, round; (c) notched tooth, flat design (overlapping teeth permit large RPTs without increasing chip load); (d) progressive tooth design for flat broach.

The overall length of the broach for a pull broach is

$$L_B = (n_r + n_s + n_f) P + L_s + L_{RP} \tag{27-9}$$

The length of stroke is $L = L_B - L_w$, in inches, if the broach moves past work or $L_w + L_B$ if the work moves past the broach. The cutting time is

$$T_m = \frac{L}{12V} \tag{27-10}$$

where V is the cutting speed, in surface feet per minute.

The metal removal rate depends upon the number of teeth (roughing) contacting the work.

$$\text{MRR (per tooth)} = 12t_r WV \text{ in.}^3/\text{min per roughing tooth} \tag{27-11}$$

where W is the width of broach tooth.

The number of roughing teeth in contact with part $n \cong L_w/P$ for a broach longer than the part.

$$\text{MRR (for process)} = 12t_rWV \text{ in.}^3/\text{min} \qquad (27\text{-}12)$$

where n is usually rounded off to the next-largest whole number.

The pull broach must be strong enough so that it will not be pulled apart. The strength of a pull broach is determined by its minimum cross section, which occurs either at the root of the first tooth or at the pull end:

$$\text{allowable pull} = \frac{\text{area of minimum section} \times \text{Y.S. of broach material}}{\text{factor of safety}} \qquad (27\text{-}13)$$

where Y.S. is yield strength.

The *push broach* must be strong enough so that it will not buckle. If the length-to-diameter ratio, L/D_r, is greater than 25, the broach must be considered a long column that can buckle if overloaded. Let L = length from push end to first tooth, D_r = root diameter at $0.5L$, and S = factor of safety:

$$\text{allowable load} = \frac{13.5 \times 10^6 \times D_r^4}{SL^2} \qquad (27\text{-}14)$$

For L/D_r less than 25, the normal broach loads are not critical.

Calculation of the total push or pull load depends upon the number of teeth engaged, n estimated from L_w/P; the width of the cut, W; the RPT per tooth engaged, t_r; and the shear strength of the metal being machined.

The force necessary to operate a broach depends upon the material being broached, the conditions of the tool, and the nature of the process. An empirical constant is required in the force calculation to account for the large amount of rubbing (friction) between the tool, the chips captured in the tooth gullet, and the workpiece.

Let F_{CB} be the broach pull force in pounds:

$$F_{CB} \cong 5\tau_s nt_r W \qquad (27\text{-}15)$$

where τ_s is flow stress.

τ_s found in Chapter 20, depends upon the hardness for the metal. This force estimate can be used to estimate the horsepower needed for the broaching machine.

THE ADVANTAGES AND LIMITATIONS OF BROACHING

Because of the features built into a broach, it is a simple and rapid method of machining. There is a close relationship among the contour of the surface to be produced, the amount of material that must be removed, and the design of the broach. For example, the total depth of the material to be removed cannot exceed the total step provided in the broach, and the step of each tooth must be sufficient to provide proper chip thickness for the type of material to be machined. Consequently, either a special broach must be made for each job or the workpiece must be designed so that a standard broach can be used. Broaching is widely used and particularly well suited for mass production because the volume can easily justify the cost of the broaching tool, which can be easily $15,000 to $30,000 per tool. It is also used for certain simple and standardized shapes, such as keyways, where inexpensive standard broaches can be used.

Broaching was originally developed for machining internal keyways. However, its obvious advantages quickly led to its development for mass-production machining of various surfaces, such as flat, interior or exterior, cylindrical or semicylindrical, and many irregular surfaces. Because there are few limitations as to the contour form that broach teeth may have, there is almost no limitation in the shape of surfaces that can be produced by broaching. The only physical limitations are that there must be no obstruction to interfere with the passage of the entire tool over the surface to be machined and that the workpiece must be strong enough to withstand the forces involved. In internal broaching, a hole must exist in the workpiece into which the broach may enter. Such a hole can be made by drilling, boring, or coring.

Broaching usually produces better accuracy and finish than can be obtained by drilling, boring, or reaming. Although the relative motion between the broaching tool and the work usually is a single linear one, a rotational motion can be added to permit the broaching of spiral splines or gun-barrel rifling.

BROACH DESIGN (THE CUTTING TOOL)
Broaches commonly are classified by the following design features:

Purpose	Motion	Construction	Function
Single	Push	Solid	Roughing
Combination	Pull	Built-up	Sizing
	Stationary		Burnishing

Figure 27-5 shows the principal components of a pull broach and the shape and arrangement of the teeth. Each tooth is essentially a single-edge cutting tool, arranged much like the teeth on a saw except for the step, which determines the depth cut by each tooth, as shown in Figure 27-6. The rise per tooth, which determines the chip load, varies from about 0.006 in. for roughing teeth in machining free-cutting steel to a minimum 0.001 in. for finishing teeth. Typically the RPT is 0.003 to 0.006 in. in surface broaching and 0.0012 to 0.0025 in. on the diameter for internal broaching. The exact amount depends on several factors. Too-large cuts impose undue stresses on the teeth and the work; too-small cuts result in rubbing rather than cutting action. The strength and ductility of the metal being cut are the primary factors.

Where it is desirable for each tooth to take a deep cut, as in broaching castings or forgings that have a hard, abrasive surface layer, *rotor-cut* or *jump-cut* tooth design may be used (Figure 27-7). In this design, two or three teeth in succession have the same diameter, or height, but each tooth of the group is notched or cut away so that it cuts only a portion of the circumference or width. This permits deeper but narrower cuts by each tooth without increasing the total load per tooth. This tooth design also reduces the forces and the power requirements. Chip-breaker notches are also used on round broaches to break up the chips (Figure 27-7b).

A similar idea can be used for flat surfaces. Tooth loads and cutting forces also can be reduced by using the *double-cut* construction, shown in Figure 27-7c. Four consecutive teeth get progressively wider. The teeth remove metal over only a portion of their width until the fourth tooth completes the cut.

Another technique for reducing tooth loads utilizes the principle illustrated in Figure 27-7d. Employed primarily for broaching wide, flat surfaces, the first few teeth in *progressive* broaches completely machine the center, while succeeding teeth are offset in two groups to complete the remainder of the surface. Rotor, double-cut, and progressive designs require the broach to be made longer than if normal teeth were used, and they therefore can be used only on a machine having adequate stroke length.

The cutting edges of the teeth on surface broaches may be either normal to the direction of motion or at an angle of from 5° to 20°. The latter, *shear-cut* broaches, provide smoother cutting action with less tendency to vibrate. Other shapes that can be broached are shown in Figure 27-8 along with push- or pull-type broaches used for the job.

The pitch of the teeth and the gullet between them must be sufficient to provide ample room for the chips. All chips produced by a given tooth during its passage over the full length of the workpiece must be contained in the space between successive teeth.

At the same time, it is desirable to have the pitch sufficiently small so that at least two or three teeth are cutting at all times.

The *hook* determines the primary rake angle and is a function of the material being cut. It is 15° to 20° for steel and 6° to 8° for cast iron. *Back-off* or end clearance angles are from 1° to 3° to prevent rubbing.

Most of the metal removal is done by the *roughing teeth*. *Semifinishing teeth* provide surface smoothness, whereas *finishing teeth* produce exact size. On a new broach all the finishing teeth are usually the same size. As the first finishing teeth become worn, those behind continue the sizing function. On some round broaches, *burnishing teeth* are provided for finishing. These teeth have no cutting edges but are rounded disks of

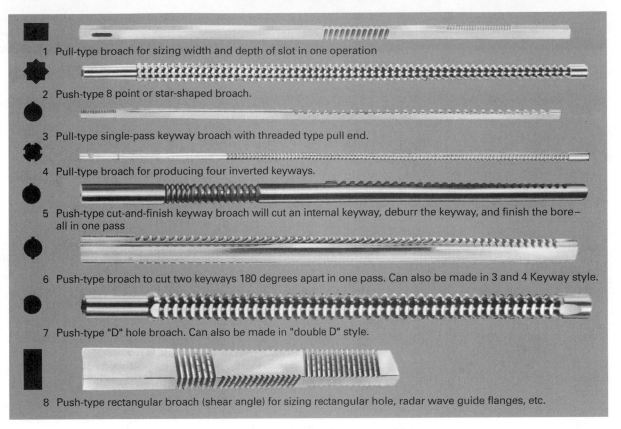

1   Pull-type broach for sizing width and depth of slot in one operation

2   Push-type 8 point or star-shaped broach.

3   Pull-type single-pass keyway broach with threaded type pull end.

4   Pull-type broach for producing four inverted keyways.

5   Push-type cut-and-finish keyway broach will cut an internal keyway, deburr the keyway, and finish the bore— all in one pass

6   Push-type broach to cut two keyways 180 degrees apart in one pass. Can also be made in 3 and 4 Keyway style.

7   Push-type "D" hole broach. Can also be made in "double D" style.

8   Push-type rectangular broach (shear angle) for sizing rectangular hole, radar wave guide flanges, etc.

FIGURE 27-8   Examples of push- or pull-type broaches. *(Courtesy of DuMont Corporation.)*

hard steel or carbide that are from 0.001 to 0.003 in. larger than the size of the hole. The resulting rubbing action smooths and sizes the hole. They are used primarily on cast iron and nonferrous metals.

The *pull end* of a broach provides a means of quickly attaching the broach to the pulling mechanism. The *front pilot* aligns the broach in the hole before it begins to cut, and the *rear pilot* keeps the tool square with the finished hole as it leaves the workpiece. *Shank length* must be sufficient to permit the broach to pass through the workpiece and be attached to the puller before the first roughing tooth engages the work. If a broach is to be used on a vertical machine that has a tool-handling mechanism, a *tail* is necessary.

A broach should not be used to remove a greater depth of metal than that for which it is designed—the sum of the steps of all the teeth. In designing workpieces, a minimum of 0.020 in. should be provided on surfaces that are to be broached, and about 0.025 in. is the practical maximum.

BROACHING SPEEDS, ACCURACY, FINISH

Depending on the metal being cut, cutting speeds for broaching range from low (25 to 20 sfpm) to high while completing the surface in a single stroke, so the productivity is high. A complete cycle usually requires only from 5 to 30 seconds, with most of that time being taken up by the return stroke, broach handling, and workpiece loading and unloading. Such cutting conditions facilitate cooling and lubrication and result in very low tool wear rates, which reduce the necessity for frequent resharpening and prolong the life of the expensive broaching tool.

For a given cutting speed and material, the force required to pull or push a broach is a function of the tooth width, the step, and the number of teeth cutting. Consequently, it is necessary to design or specify a broach within the stroke length and power limitations of the machine on which it is to be used. The average machining precision is typically ±0.001-in. (±0.02-mm) tolerance with surface finish 120 to 60 RMS or better. Burrs are minimal on the exit side of cuts.

FIGURE 27-9 Shell construction for a pull broach.

BROACHING MATERIALS AND CONSTRUCTION

Because of the low cutting speeds employed, most broaches are made of alloy or high-speed tool steel. Carbide-tipped broaches are seldom used for machining steel parts or forgings, as the cutting edges tend to chip on the first stroke, probably due to a lack of rigidity in the combination of machine tool and cutting tool. TiN coating of high-speed-steel (HSS) broaches is becoming more common, greatly prolonging the life of broaches. When they are used in continuous mass-production machines, particularly in surface broaching of cast iron, tungsten carbide teeth may be used, permitting the broach to be used for long periods of time without resharpening.

Internal broaches are usually solid but may be made of *shells* mounted on an arbor (Figure 27-9). When the broach (or a section of it) is subject to rapid wear, a single shell can be replaced. This will be much cheaper than replacing an entire solid broach. Shell construction, however, is initially more expensive than a solid broach of comparable size.

Small-surface broaches may be of solid construction, but larger ones usually use modular construction (Figure 27-10). Building in sections makes the broach easier and cheaper to construct and sharpen. It also often provides some degree of interchangeability of the sections for different parts, bringing down the tool cost significantly.

SHARPENING BROACHES

Most broaches are resharpened by grinding the hook faces of the teeth. The lands of internal broaches must not be reground because this would change the size of the broach. Lands of flat-surface broaches are sometimes ground, in which case all of them must be ground to maintain their proper relationship.

■ 27.5 BROACHING MACHINES

Because all the factors that determine the shape of the machined surface and that determine all cutting conditions except speed are built into the broaching tool, broaching machines are relatively simple. Their primary functions are to impart plain reciprocating motion to the broach and to provide a means for handling the broach automatically.

FIGURE 27-10 A modularly constructed broach is cheaper to build and can be sharpened in sections.

TABLE 27-1	Broaching Machines
Vertical	
Push-broaching	Arbor press with guided ram 5- to 50-ton capacity Internal broaching
Pull-down	Double-ram design most common Long changeover times
Pull-up	Ram above table pulling broach up Machines with multiple rams common
Surface	No handling of broach Multiple slides
Horizontal	Short Cycle Times
Pull	Longer strokes and broaches Basically vertical machines laid on side
Surface	Broaches stationary, work moves on conveyor Work held in fixtures
Continuous	Conveyor chain holds fixtures
Rotary	Rotary broach stationary, work translates beneath tool Work held in fixtures

Most broaching machines are driven hydraulically, although mechanical drive is used in a few special types. The major classification relates to whether the motion of the broach is vertical or horizontal, as given in Table 27-1.

The choice between vertical and horizontal machines is determined primarily by the length of the stroke required and the available floor space. Vertical machines seldom have strokes greater than 60 in. because of height limitations. Horizontal machines can have almost any length of stroke, but they require greater floor space. The most common machine is the vertical pull-down machine shown in Figure 27-11. The worktable, usually having a spherical-seated workholder, sits below the broach elevator, with a pulling mechanism below the table. When the elevator raises the broach above the table, the work can be placed into position. The elevator then lowers the pilot end of the broach through the hole in the workpiece, where it is engaged by the puller. The elevator then releases the upper end of the broach, and it is pulled through the workpiece. The workpieces are removed from the table, and the broach is raised upward to be engaged by the elevator mechanism. In some machines with two rams, one broach is being pulled down while the work is being unloaded and the broach raised at the other station. In Figure 27-11, the part is being broached in two passes, first on the left, then on the right.

■ 27.6 INTRODUCTION TO SAWING

Sawing is a basic machining process in which chips are produced by a succession of small cutting edges, or *teeth*, arranged in a narrow line on a saw "blade." As shown in Figure 27-12, each tooth forms a chip progressively as it passes through the workpiece. The chips are contained within the spaces between successive teeth until the teeth pass from the work. Because sections of considerable size can be severed from the workpiece with the removal of only a small amount of the material in the form of chips, sawing is probably the most economical of the basic machining processes with respect to the waste of material and power consumption, and in many cases with respect to labor.

In recent years vast improvements have been made in saw blades (design and materials) and sawing machines, resulting in improved accuracy and precision of the process. Most sawing is done to sever bar stock and shapes into desired lengths for use in other operations. There are many cases in which sawing is used to produce desired shapes. Frequently, and especially for producing only a few parts, contour sawing may be more economical than any other machining process.

SAW BLADES

Saw blades are made in three basic configurations. The first type, commonly called a *hacksaw* blade, is straight, relatively rigid, and of limited length, with teeth on one edge.

FIGURE 27-11 Vertical pull-down broaching machine shown with parts in position ready for the two broaches to be inserted. An extra part is shown lying at the front of the machine.

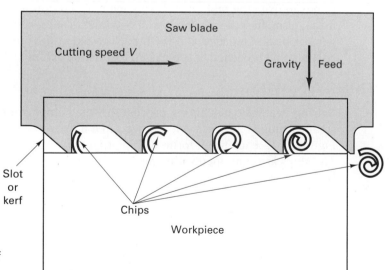

FIGURE 27-12 Formation of chips in sawing.

The second type, called a *bandsaw*, is sufficiently flexible that a long length can be formed into a continuous band with teeth on one edge. The third form is a rigid disk having teeth on the periphery; these are called *circular saws* or *cold saws*. Figure 27-13 gives the standard nomenclature for a saw blade.

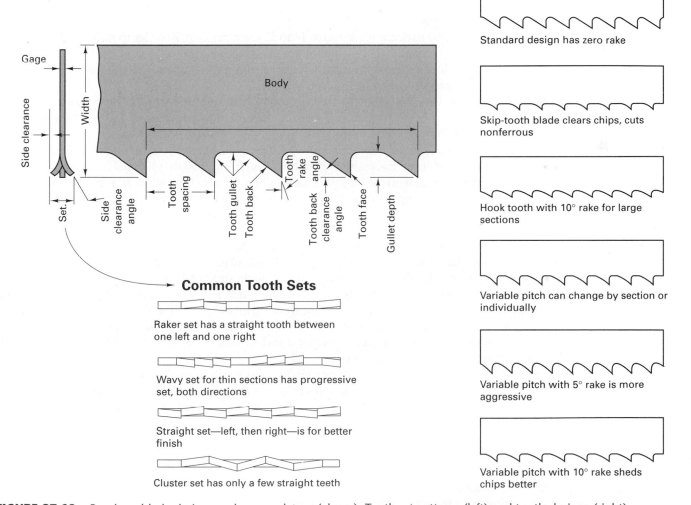

MATRIX MODIFIED MIX-TOOTH	M-42 COBALT WELDED-EDGE	M-2 HIGH-SPEED WELDED-EDGE	HARD BACK CARBON	FLEXIBLE BACK CARBON
The best all-purpose welded-edge blade for sawing varying sizes, shapes, and cross sections. Cobalt-tough for cutting wide range of materials. Welded to length and coil stock.	For high-production cutting of solids, superalloys, tool steels, high-temperature alloys. Welded to length and coil stock.	The original and widely used welded-edge band blade for general-purpose sawing. Welded to length and coil stock.	Hardened back provides greater beam strength for more accurate sawing. Welded to length and coil stock.	Recommended for contour saws running over 3000 SFPM. Welded to length and coil stock.

Standard design has zero rake

Skip-tooth blade clears chips, cuts nonferrous

Hook tooth with 10° rake for large sections

Variable pitch can change by section or individually

Variable pitch with 5° rake is more aggressive

Variable pitch with 10° rake sheds chips better

Common Tooth Sets

Raker set has a straight tooth between one left and one right

Wavy set for thin sections has progressive set, both directions

Straight set—left, then right—is for better finish

Cluster set has only a few straight teeth

FIGURE 27-13 Bandsaw blade designs and nomenclature (above). Tooth set patterns (left) and tooth designs (right).

All saw blades have certain common and basic features: (1) material, (2) tooth form, (3) tooth spacing, (4) tooth set, and (5) blade thickness or gage. Small hacksaw blades are usually made entirely of tungsten or molybdenum high-speed steel. Blades for power-operated hacksaws are often made with teeth cut from a strip of high-speed steel that has been electron-beam-welded to the heavy main portion of the blade, which is made from a tougher and cheaper alloy steel (see Figure 27-13). Bandsaw blades are frequently made with this same type of construction but with the main portion of the blade made of relatively thin, high-tensile-strength alloy steel to provide the required flexibility. Bandsaw blades are also available with tungsten carbide teeth and TiN coatings. The three most common *tooth forms* are regular, skip tooth, and hook. *Tooth spacing* is very important in all sawing because it determines three factors. First, it controls the size of the teeth. From the viewpoint of strength, large teeth are desirable. Second, tooth spacing determines the space (*gullet*) available to contain the chip that is formed. The chip cannot drop from this space until it emerges from the slot cut in the workpiece, called the *kerf*. *Tooth set* refers to the manner in which the teeth are offset from the centerline in order to make the kerf wider than the gage (the thickness of the back) of the blade. This allows the saw to move more freely in the kerf, reducing rubbing, friction, and heating. The kerf–gullet space must be such that there is no crowding of the chip. Chips should not become wedged between the teeth and not drop out of the gullet when the saw emerges from the cut.

Third, tooth spacing determines how many teeth will bear against the work. This is very important in cutting thin material, such as tubing. At least two teeth should be in contact with the work at all times. If the teeth are too coarse, only one tooth rests on the work at a given time, permitting the saw to rock, and the teeth may be stripped from the saw.

Hand hacksaw blades have 14 to 32 teeth per inch. In order to make it easier to start a cut, some hand hacksaw blades are made with a short section at the forward end having teeth of a special form with negative rake angles. Tooth spacing for power hacksaw blades ranges from 4 to 18 teeth per inch.

Raker-tooth saws are used in cutting most steel and iron. *Straight-set teeth* are used for sawing brass, copper, and plastics. Saws with *wave-set teeth* are used primarily for cutting thin sheets and thin-walled tubing.

The gage or *blade thickness* of nearly all hand hacksaw blades is 0.025 in. Saw blades for power hacksaws vary in thickness from 0.050 to 0.100 in. Hand hacksaw blades come in two standard lengths, 10 and 12 in. All are $1/2$ in. wide. Blades for power hacksaws vary in length from 12 to 24 in. and in width from 1 to 2 in. Wider and thicker blades are desirable for heavy-duty work. As a general rule, in hacksawing the blade should be at least twice as long as the maximum length of cut that is to be made.

Bandsaw blades are available in straight, raker, wave, or combination sets. In order to reduce the noise from high-speed bandsawing, it is becoming increasingly common to use blades that have more than one pitch, size of teeth, and type of set. Blade width is very important in bandsawing because it determines the minimum radius that can be cut. The most common widths are from $1/16$ in. to $1/2$ in., although wider blades can be obtained. Because wider blades are stronger, select the widest blade possible. However, cutting small radii requires a narrower and weaker blade. Bandsaw blades come in tooth spacings from 2 to 32 teeth per inch.

Circular saws for cutting metal are often called *cold saws* to distinguish them from friction-type disk saws. Friction saws do not make chips but rather heat the metal to the melting temperature at the point of metal removal. Cold saws cut rapidly and produce chips like a milling cutter while producing surfaces that are comparable in smoothness and accuracy with surfaces made by slitting saws in a milling machine or by a cutoff tool in a lathe.

Disk or *circular saws* necessarily differ somewhat from straight-blade forms. The sizes up to about 18 in. in diameter have an integral-tooth design with teeth cut directly into the disk (Figure 27-14). Larger saws use either *segmented* or *inserted* teeth. The teeth are made of high-speed steel or tungsten carbide. The remainder of the disk is made of ordinary, less expensive, and tougher steel. *Segmental* blades are composed of segments mounted around the periphery of the disk, usually fitted with a tongue and groove and fastened by means of screws or rivets. Each segment contains several teeth. If a single tooth is broken, only one segment needs be replaced to restore the saw to an operating condition.

As shown in Figure 27-14, circular saw teeth are usually *beveled*. A common tooth form has every other tooth beveled on both sides; that is, the first tooth is beveled on the left side, the second tooth on both sides, the third tooth on the right side, the fourth tooth on the left side, and so forth. Another method is to bevel the opposite sides of successive teeth. Beveling is done to produce a smoother cut. Precision circular saws made from carbide, which are becoming available, are very thin (0.03 in.) and have high cutting-off accuracy, around ± 0.00008 in., with negligible burrs.

FIGURE 27-14 Circular sawing a structural shape, using (*left* to *right*) an insert tooth, a segmental tooth, and an integral-tooth circular saw blade.

TYPES OF SAWING MACHINES

Metal-sawing machines may be classified as follows:

1. Reciprocating saw
   a. Manual hacksaw
   b. Power hacksaw (Figure 27-15)
   c. Abrasive disc
2. Bandsaw
   a. Vertical cutoff (Figure 27-16)
   b. Horizontal cutoff (Figure 27-17)
   c. Combination cutoff and contour (Figure 27-18)
   d. Friction
3. Circular saw (Figure 27-14)
   a. Cold saw
   b. Steel friction disk

FIGURE 27-15 Power hacksaw blade (above) and hacksaw with automatic bar feeding (right) cutting two pieces of round stock.

Blade changing

Easy blade loading from the top on all models and quick removal of guards.

FIGURE 27-16   Front view and rear view of a horizontal bandsawing machine sawing a cylinder of steel. Inset shows blade-changing operation.

FIGURE 27-17   Vertical bandsaw cutting a piece of pipe, showing head tilted 45°.

FIGURE 27-18 Contour bandsawing on vertical bandsawing machine, shown in inset.

POWER HACKSAWS

As the name implies, power hacksaws are machines that mechanically reciprocate a large hacksaw blade (Figure 27-15). These machines consist of a bed, a workholding frame, a power mechanism for *reciprocating* the saw frame, and some type of feeding mechanism. Because of the inherent inefficiency of cutting in only one stroke direction, they have often been replaced by more efficient, horizontal bandsawing machines.

BANDSAWING MACHINES

The earliest metalcutting bandsawing machines were direct adaptations from wood-cutting bandsaws. Modern machines of this type are much more sophisticated and versatile and have been developed specifically for metal cutting. To a large degree they were made possible by the development of vastly better and more flexible bandsaw blades and simple flash-welding equipment, which can weld the two ends of a strip of bandsaw blade together to form a band of any desired length. Three basic types of bandsawing machines are in common use.

Horizontal metalcutting bandsawing machines were developed to combine the flexibility of reciprocating power hacksaws and the continuous cutting action of vertical bandsaws. These heavy-duty automatic bandsaws feed the saw vertically by a hydraulic mechanism and have automatic stock feed that can be set to feed the stock laterally any desired distance after a cut is completed and automatically clamp it for the next cut. Such machines can be arranged to hold, clamp, and cut several bars of material simultaneously. Computer numerical control (CNC) bandsaws are available with automatic storage and retrieval systems for the bar stock. Smaller and less expensive types have swing-frame construction, with the bandsaw head mounted in a pivot on the rear of the machine. Feed is accomplished by gravity through rotation of the head about the pivot point. Because of their continuous cutting action, horizontal bandsawing machines are very efficient (Figure 27-16).

Upright, cutoff, bandsawing machines (Figure 27-17) are designed primarily for cutoff work on single stationary workpieces that can be held on a table. On many machines the blade mechanism can be tilted to about 45°, as shown, to permit cutting at an angle. They usually have automatic power feed of the blade into the work, automatic stops, and provision for supplying coolant.

Combination cutoff and contour bandsawing machines (Figure 27-18) can be used not only for cutoff work but also for contour sawing. They are widely used for cutting irregular shapes in connection with making dies and the production of small numbers of parts and are often equipped with rotary tables. Additional features on these machines include a table that pivots so that it can be tilted to any angle up to 45°. Usually these machines have a small flash butt welder on the vertical column, so that a straight length of bandsaw blade can be welded quickly into a continuous band. A small grinding wheel is located beneath the welder so that the flash can be ground from the weld to provide a smooth joint that will pass through the saw guides. This welding and grinding unit makes it possible to cut internal openings in a part by first drilling a hole, inserting one end of the saw blade through the hole, and then butt welding the two ends together. When the cut is finished, the band is again cut apart and removed from the opening. The cutting speed of the saw blade can be varied continuously over a wide range to provide correct operating conditions for any material. A method of power feeding the work is provided, sometimes gravity-actuated.

Contour-sawing machines are made in a wide range of sizes, the principal size dimension being the throat depth. Sizes from 12 to 72 in. are available. The speeds available on most machines range from about 50 to 2000 ft/min. Modern horizontal bandsaws are accurate to ±0.002 in. per vertical inch of cut but have feeding accuracy of only ±0.005 in., subject to the size of the stock and the feed rate. Repeatability from one feed to the next may be ±0.010 to ±0.020 in.

CNC-controlled sawing centers with microprocessor controls have opened up new automation aspects for sawing. Such control systems can improve accuracy to within ±0.005 in. over entire cuts by controlling saw speed, blade feed pressure, and feed rate.

Special bandsawing machines are available with very high speed ranges, up to 14,000 ft/min. These are known as *friction bandsawing machines*. Material is not cut by chip formation. Instead, the friction between the rapidly moving saw blade and the work is sufficient to raise the temperature of the material at the end of the kerf to or just below the melting point, where its strength is very low. The saw blade then pulls the molten, or weakened, material out of the kerf. Consequently, the blades do not need to be sharp; they frequently have no teeth—only occasional notches in the blade to aid in removing the metal.

Almost any material, including ceramics, can be cut by friction sawing. Because only a small portion of the blade is in contact with the work for an instant and then is cooled by its passage through the air, it remains cool. Usually, the major portion of the work, away from contact with the saw blade, also remains quite cool. The metal adjacent to the kerf is heat affected, recast, and sometimes harder than the bulk metal. It is also a very rapid method for trimming the flash from sheet metal parts, castings, and forgings.

CUTTING FLUIDS

Cutting fluids should be used for all bandsawing, with the exception that cast iron is always cut dry. Commercially available oils or light cutting oils will give good results in cutting ferrous materials. Beeswax or paraffin are common lubricants for cutting aluminum and aluminum alloys.

FEEDS AND SPEEDS

Because of the many different types of feed involved in bandsawing, it is not practical to provide tabular feed or pressure data. Under general conditions, however, an even pressure, without forcing the work, gives best results. A nicely curled chip usually indicates an ideal feed pressure. Burned or discolored chips indicate excessive pressure, which can cause tooth breakage and premature wear.

Most bandsaws provide recommended cutting speed information right on the machine, depending upon the material being sawed. In general, HSS blades are run at 200 to 300 ft/min when cutting 1-in.-thick, low- and medium-carbon steels. For high-carbon steels, alloy steels, and tool and die steels, the range is from 150 to 225 ft/min, and most stainless steels are cut at 100 to 125 ft/min.

CIRCULAR-BLADE SAWING MACHINES

Machines employing rotating circular or cold saw blades are used exclusively for cut-off work. These range from small, simple types, in which the saw is fed manually, to very large saws having power feed and built-in coolant systems, commonly used for cutting off hot-rolled shapes as they come from a rolling mill. In some cases friction saws are used for this purpose, having disks up to 6 ft in diameter and operating at surface speeds up to 25,000 ft/min. Steel sections up to 24 in. can be cut in less than 1 minute by this technique.

Although technically not a sawing operation, cutoff work up to about 6 in. is often done utilizing thin *abrasive* disks. The equipment used is the same as for sawing. It has the advantage that very hard materials that would be very difficult to saw can be cut readily. A thin rubber- or resinoid-bonded abrasive wheel is used. Usually a somewhat smoother surface is produced.

■ 27.7 INTRODUCTION TO FILING

Basically, the metal-removing action in filing is the same as in sawing, in that chips are removed by cutting teeth that are arranged in succession along the same plane on the surface of a tool, called a *file*. There are two differences: (1) the chips are very small, and therefore the cutting action is slow and easily controlled, and (2) the cutting teeth are much wider. Consequently, fine and accurate work can be done.

Files are classified according to the following:

1. The type, or *cut*, of the teeth
2. The degree of coarseness of the teeth
3. Construction
   a. Single solid units for hand use or in die-filing machines
   b. Band segments, for use in band-filing machines
   c. Disks, for use in disk-filing machines

Four types of *cuts* are available. *Single-cut files* have rows of parallel teeth that extend across the entire width of the file at the angle of from 65° to 85°. *Double-cut files* have two series of parallel teeth that extend across the width of the file. One series is cut at an angle of 40° to 45°. The other series is coarser and is cut at an opposite angle that varies from about 10° to 80°. A *vixen-cut file* has a series of parallel curved teeth, each extending across the file face. On a *rasp-cut file*, each tooth is short and is raised out of the surface by means of a punch. These four types of cuts are shown in Figure 27-19.

FIGURE 27-19 Four types of teeth (cuts) used in files. Left to right: Single, double, rasp, and curved (vixen). *(Courtesy of Nicholson File Company.)*

The coarseness of files is designated by the following terms, arranged in order of increasing coarseness: *dead smooth*, *smooth*, *second cut*, *bastard*, *coarse*, and *rough*. There is also a series of finer Swiss pattern files, designated by numbers from 00 to 8.

Files are available in a number of cross-sectional shapes: *flat*, *round*, *square*, *triangular*, and *half-round*. Flat files can be obtained with no teeth on one or both narrow edges, known as *safe edges*. Safe edges prevent material from being removed from a surface that is normal to the one being filed. Most files for hand filing are from 10 to 14 in. in length and have a pointed *tang* at one end on which a wood or metal handle can be fitted for easy grasping.

FILING MACHINES

An experienced operator can do very accurate work by hand filing, but it can be a difficult task. Therefore, three types of filing machines have been developed that permit quite accurate results to be obtained rapidly and with much less effort. *Die-filing machines* hold and reciprocate a file that extends upward through the worktable. The file rides against a roller guide at its upper end, and cutting occurs on the downward stroke; therefore, the cutting force tends to hold the work against the table. The table can be tilted to any desired angle. Such machines operate at from 300 to 500 strokes per minute, and the resulting surface tends to be at a uniform angle with respect to the table. Quite accurate work can be done. Because of the reciprocating action, approximately 50% of the operating time is nonproductive.

Band-filing machines provide continuous cutting action. Most band filing is done on contour bandsawing machines by means of a special band file that is substituted for the usual bandsaw blade. The principle of a band file is shown in Figure 27-20. Rigid, straight file segments, about 3 in. long, are riveted to a flexible steel band near their leading ends. One end of the steel band contains a slot that can be hooked over a pin in the other end to form a continuous band. As the band passes over the drive and idler wheels of the machine, it flexes so that the ends of adjacent file segments move apart. When the band becomes straight, the ends of adjacent segments move together and interlock to form a continuous straight file. Where the file passes through the worktable, it is guided and supported by a grooved guide, which provides the necessary support to

(a)

FIGURE 27-20 Band file segments (a) are joined together to form a continuous band (b) which runs on a band-filing machine (c). *(Courtesy of DoALL Co.)*

(b)

(c)

FIGURE 27-21 Disk-type filing machine and some of the available types of disk files. *(Courtesy of Jersey Manufacturing Company.)*

resist the pressure of the work against the file. Band files are available in most of the standard cuts and in several widths and shapes. Operating speeds range from about 50 to 250 ft/min.

Although band filing is considerably more rapid than can be done on a die-filing machine, it usually is not quite as accurate. Frequently, band filing may be followed by some finish filing on a die-filing machine.

Some *disk-filing machines* have files in the form of disks (Figure 27-21). These are even simpler than die-filing machines and provide continuous cutting action. However, it is difficult to obtain accurate results by their use.

■ Key Words

band-filing machine	circular saw	planning	sawing
bandsaw	cold saw	pull broach	shaping
broach	filing	push broach	surface broach
broaching	hacksaw	reciprocating saw	tooth set
burnishing teeth	kerf	rise per tooth	

■ Review Questions

1. What is unique about the broaching process compared to the other basic machining processes?
2. Can a thick saw blade be used as a broach? Why or why not?
3. Broaching machines are simpler in a basic design than most other machine tools. Why is this?
4. Why is broaching particularly well suited for mass production?
5. In designing a broach, what would be the first thing you have to calculate?
6. Why is it necessary to relate the design of a broach to the specific workpiece that is to be machined?
7. What two methods can be utilized to reduce the force and power requirements for a particular broaching cut?
8. For a given job, how would a broach having rotor-tooth design compare in length with one having regular, full-width teeth?
9. Why are the pitch and radius of the gullet between teeth on a broach of importance?
10. Why are broaching speeds usually relatively low, as compared with other machining operations?
11. What are the advantages of shell-type broach construction?
12. Why are most broaches made from alloy or high-speed steel rather than from tungsten carbide?
13. What are the advantages of TiN-coated broaching tools?
14. For mass-production operations, which process is preferred, pull-up broaching or pull-down broaching?
15. What is the difference between the roughing teeth and the finishing teeth in a typical pull broach?
16. The sides of a square, blind hole must be machined all the way to the bottom (who designed this part?). The hole is first drilled to full depth, then the bottom is milled flat. Is it possible to machine the hole square by broaching? Why or why not?

17. The interior, flat surfaces of socket wrenches, which have one "closed" end, often are finished to size by broaching. By examining one of these, determine what design modification was incorporated to make broaching possible.
18. Why is sawing one of the most efficient of the chip-forming processes?
19. Explain why tooth spacing (pitch) is important in sawing.
20. What is the tooth gullet used for on a saw blade?
21. Explain what is meant by the "set" of the teeth on a saw blade.
22. How is tooth set related to saw kerf?
23. Why can a bandsaw blade not be hardened throughout the entire width of the band?
24. What are the advantages of using circular saws?
25. Why have bandsawing machines largely replaced reciprocating saws?
26. Explain how the hole in Figure 27-18 is made on a contour bandsawing machine.

27. How would you calculate or estimate T_m for a horizontal bandsaw cutting a 3-in. round of 1040 steel?
28. What is the disadvantage of using gravity to feed a saw in cutting round bar stock?
29. To what extent is filing different from sawing?
30. What is a safe edge on a file?
31. Why is an end-filing machine more efficient than a die-filing machine?
32. How does a rasp-cut file differ from other types of files?
33. How does the process of shaping differ from planing?
34. How is feed per stroke in shaping related to feed per tooth in milling?
35. What are some ways to improve the efficiency of a planer? Do any of these apply to the shaper?

■ Problems

1. A surface 12 in. long is to be machined with a flat, solid broach that has a rise per tooth of 0.0047 in. What is the minimum cross-sectional area that must be provided in the chip gullet between adjacent teeth?
2. The pitch of the teeth on a simple surface broach can be determined by equation 27-1. If a broach is to remove 0.25 in. of material from a gray iron casting that is 3 in. wide and 17.75 in. long, and if each tooth has a rise per tooth of .004 in., what will be the length of the roughing section of the broach?
3. Estimate the (approximate maximum) horsepower needed to accomplish the operation described in Problem 2 at a cutting speed of 10 m/min. (*Hint:* First find the HP used per tooth and determine the maximum number of teeth engaged at any time. What are those units?)
4. Estimate the approximate force acting in the forward direction during cutting for the conditions stated in Problems 2 and 3.
5. In cutting a 6-in.-long slot in a piece of AISI 1020 cold-rolled steel that is 1 in. thick, the material is fed to a bandsaw blade with teeth having a pitch of 1.27 mm (20 pitch) at the rate of 0.0001 in. per tooth. Estimate the cutting time for the cut.
6. The strength of a pull broach is determined by its minimum cross section, which usually occurs either at the root of the first tooth or at the pull end. Suppose the minimum root diameter is D_r, the pull end diameter is D_p and the width of the pull slot is W. Write an equation for the allowable pull, in psi, using 200,000 as the yield strength for the broach material.
7. Suppose you want to shape a block of metal 7 in. wide and 4 in. long ($L = 4$ in.) using a shaper as set up in Figure 27-1. You have determined for this metal that the cutting speed should be 25 sfpm, the depth of cut needed here for roughing is 0.25 in., and the feed will be 0.1 in. per stroke. Determine the

approximate crank rpm and then estimate the cutting time and the MRR.
8. Could you have saved any time in Problem 7 by cutting the block in the 7-in. direction? Redo with $L = 7$ and $W = 4$ in.
9. Derive the equation for shaping cutting speed, equation 27-6.
10. How many strokes per minute would be required to obtain a cutting speed of 36.6 m (120 ft) per minute on a typical mechanical drive shaper if a 254-mm (10-in.) stroke is used?
11. How much time would be required to shape a flat surface 254 mm (1 in.) wide and 203 mm (8 in.) long on a hydraulic drive shaper, using a cutting speed of 45.7 m (150 ft) per minute, a feed of 0.51 mm (0.020 in.) per stroke, and an overrun of 12.7 mm ($^1/_2$ in.) at each end of the cut?
12. What is the metal removal rate in Problem 11 if the depth of cut is 6.35 mm ($^1/_4$ in.)?
13. Suppose you decide to mill the flat surface described in Problems 11 and 12. The work will be done on a vertical milling machine using a 1.25-in.-diameter end mill (four teeth) (HSS) cutting at 150 sfpm with a feed per tooth of 0.005 in. per tooth cutting at $d = 0.25$ in. Compare the milling time and MRR to that of shaping.
14. A planer has a 10-hp motor, and 75% of the motor output is available at the cutting tool. The specific power for cutting cast iron metal is 0.03 W/mm³, or 0.67 hp/in.³/ min. What is the maximum depth of cut that can be taken in shaping a surface in this material if the surface is 305 × 305 mm (12 × 12), the feed is 0.25 in. per stroke, and the cutting speed is 54.9 mm (180 ft) per minute?
15. Calculate the T_m for planing the block of cast iron in Problem 14 and then estimate T_m for milling the same surface. You will have to determine which milling process to use and select speeds and feeds for an HSS cutter.

Chapter 27 CASE STUDY

Cost Estimating—Planing vs. Milling

KC Kendoric works in the BRC factory as a manufacturing engineer. There are two machine tools available for a job the company is bidding on. One is a 48 × 48 X 10 double housing planer that originally costs $80,000, is depreciated over a 20-year period, and is operated about 6,000 hr/yr. The charge for the use of the machine is $2.50/hr. and labor and overhead in addition are charged at $20.00/hr. The other machine is a large vertical spindle CNC milling machine that costs $165,000 new, depreciated over a 20-year time period also. The charge for the use of the CNC milling machine is $6.00/hr., and labor and overhead in addition are charged at $32.00/hr. To machine the workpiece under consideration takes 10 hours on the planer and 5 hours on the CNC milling machine. The cutting tools consumed cost $5.00/piece for the planer and $31.00/piece for the mill.

Purchasing has not issued an order (quantity not yet decided) so KC needs to determine the BEQ so she knows which machine to use when the order is placed.

CHAPTER 28

ABRASIVE MACHINING PROCESSES

■ 28.1 INTRODUCTION

Abrasive machining is a material removal process that involves the interaction of abrasive grits with the workpiece at high cutting speeds and shallow penetration depths. The chips that are formed resemble those formed by other machining processes. Unquestionably, abrasive machining is the oldest of the basic machining processes. Museums abound with examples of utensils, tools, and weapons that ancient peoples produced by rubbing hard stones against softer materials to abrade away unwanted portions, leaving desired shapes. For centuries, only natural abrasives were available for grinding, while other more modern basic machining processes were developed using superior cutting materials. However, the development of manufactured abrasives and a better fundamental understanding of the abrasive machining process have resulted in placing abrasive machining and its variations among the most important of all the basic machining processes.

The results that can be obtained by abrasive machining range from the finest and smoothest surfaces produced by any machining process, in which very little material is removed, to rough, coarse surfaces that accompany high material removal rates. The abrasive particles may be (1) free; (2) mounted in resin on a belt (called *coated product*); or, most commonly (3) close packed into wheels or stones, with abrasive grits held together by bonding material (called *bonded product* or a grinding wheel). Figure 28-1 shows a surface grinding process using a grinding wheel. The depth of cut *d* is determined by the infeed and is usually very small, 0.002 to 0.005 in., so the arc of contact (and the chips) is small. The table reciprocates back and forth beneath the rotating wheel. The work feeds into the wheel in the cross-feed direction. After the work is clear of the wheel, the wheel is lowered and another pass is made, again removing a couple of thousandths of inches of metal. The metal removal process is basically the same in all abrasive machining processes but with important differences due to spacing of active grains (grains in contact with work) and the rigidity and degree of fixation of the grains. Table 28-1 summarizes the primary abrasive processes. The term *abrasive machining* applied to one particular form of the grinding process is unfortunate, because all these process are machining with abrasives.

Compared to machining, abrasive machining processes have three unique characteristics. First, each cutting edge is very small, and many of these edges can cut simultaneously. When suitable machine tools are employed, very fine cuts are possible, and fine surfaces and close dimensional control can be obtained. Second, because extremely

FIGURE 28-1 Schematic of surface grinding, showing infeed and cross feed motions along with cutting speeds V_S, and workpiece velocity V_W.

hard abrasive grits, including diamonds, are employed as cutting tool materials, very hard materials, such as hardened steel, glass, carbides, and ceramics, can readily be machined. As a result, the abrasive machining processes are not only important as manufacturing processes, they are indeed essential. Many of our modern products, such as modern machine tools, automobiles, space vehicles, and aircraft, could not be manufactured without these processes. Third, in grinding, you have no control over the actual tool geometry (rake angles, cutting edge radius) or all the cutting parameters (depth of cut). As a result of these parameters and variables, grinding is a complex process.

To get a handle on the complexity, Table 28-2 presents the primary grinding parameters, grouped by their independence or dependence. Independent variables are those that are controllable (by the machine operator) while the dependent variables are the resultant effects of those inputs. Not listed in the table is workpiece hardness, which has a significant effect on all the resulting effects. Workpiece hardness will be an input factor but it is not usually controllable.

■ 28.2 ABRASIVES

An *abrasive* is a hard material that can cut or abrade other substances. Natural abrasives have existed from the earliest times. For example, sandstone was used by ancient peoples to sharpen tools and weapons. Early grinding wheels were cut from slabs of sandstone, but because they were not uniform in structure throughout, they wore unevenly and did not produce consistent results. *Emery*, a mixture of alumina (Al_2O_3) and magnetite (Fe_3O_4), is another natural abrasive still in use today and is used on coated paper and cloth (emery paper). *Corundum* (natural Al_2O_3) and diamonds are other naturally occurring abrasive materials. Today, the only natural abrasives that have commercial

TABLE 28-1	Abrasive Machining Processes	
Process	Particle Mounting	Features
Grinding	Bonded	Uses wheels, accurate sizing, finishing, low MRR; can be done at high speeds (over 12,000 sfpm)
Creep feed grinding	Bonded open, soft	Uses wheels with long cutting arc, very slow feed rate, and large depth of cut
Abrasive machining	Bonded	High MRR, to obtain desired shapes and approximate sizes
Snagging	Bonded belted	High MRR, rough rapid technique to clean up and deburr castings, forgings
Honing	Bonded	"Stones" containing fine abrasives; primarily a hole-finishing process
Lapping	Free	Fine particles embedded in soft metal or cloth; primarily a surface-finishing process
Abrasive waterjet	Free in jet	Water jets with velocities up to 3000 sfpm carry abrasive particles (silica and garnet).
Ultrasonic	Free in liquid	Vibrating tool impacts abrasives at high velocity
Abrasive flow	Free in gel	Abrasives in gel flow over surface-edge finishing
Abrasive jet	Free in	A focused jet of abrasives in an inert gas at high velocity

TABLE 28-2 Grinding Parameters*

Independent Parameters/Controllable	Dependent Variables/Resulting Effects

Grinding wheel selection
  Abrasive type
  Grain size
  Hardness grade
  Openness of structure
  Bonding media
	Forces per unit width of wheel
	  Normal
	  Tangential
	Surface finish
	Material removal rate (MRR)
	Wheel wear (G, or grinding ratio)

Dressing of wheel
  Type of dressing tool
  Feed and depth of cut
  Sharpness of dressing tool
	Thermal effects
	Wheel surface changes
	Chemical effects
	Horsepower

Machine settings
  Wheel speed
  Infeed rate (depth of cut)
  Cross-feed rate
  Workpiece speed
  Rigidity of setup
  Type and quality of machine	

Grinding fluid
  Type
  Cleanliness
  Method of application	

importance are quartz, sand, garnets, and diamonds. For example, *quartz* is used primarily in coated abrasives and in air blasting, but artificial abrasives are also making inroads in these applications. The development of artificial abrasives having known uniform properties has permitted abrasive processes to become precision manufacturing processes.

Hardness, the ability to resist penetration, is the key property for an abrasive. Table 28-3 lists the primary abrasives and their approximate Knoop hardness (kg/mm²). The particles must be able to decompose at elevated temperatures. Two other properties are significant in abrasive grits—attrition and friability. *Attrition* refers to the abrasive wear action of the grits resulting in dulled edges, grit flattening, and wheel glazing. *Friability* refers to the fracture of the grits and is the opposite of toughness. In grinding, it is important that grits be able to fracture to expose new, sharp edges.

Artificial abrasives date from 1891, when E. G. Acheson, while attempting to produce precious gems, discovered how to make *silicon carbide* (SiC). Silicon carbide is made by charging an electric furnace with silica sand, petroleum coke, salt, and sawdust. By passing large amounts of current through the charge, a temperature of over 4000°F is maintained for several hours, and a solid mass of silicon carbide crystals results. After the furnace has cooled, the mass of crystals is removed, crushed, and graded (sorted) into

TABLE 28-3 Knoop Hardness Values for Common Abrasives

Abrasive Material	Year of Discovery	Hardness (Knoop)	Temperature of Decomposition in Oxygen (°C)	Comments and Uses
Quartz	?	320		Sand blasting
Aluminum oxide	1893	1600–2100	1700–2400	Softer and tougher than silicon carbide; used on steel, iron, brass, silicon
Carbide	1891	2200–2800	1500–2000	Used for brass, bronze, aluminum, and stainless and cast iron
Borazon [cubic boron nitride stainless (CBN)]	1957	4200–5400	1200–1400	For grinding hard, tough tool steels, stainless steel, cobalt and nickel based, superalloys, and hard coatings
Diamond (synthetic)	1955	6000–9000	700–800	Used to grind nonferrous materials, tungsten carbide, and ceramics

FIGURE 28-2 Loose abrasive grains at high magnification, showing their irregular, sharp cutting edges. *(Courtesy of Norton Company.)*

various desired sizes. As can be seen in Figure 28-2, the resulting grits, or grains, are irregular in shape, with cutting edges having every possible rake angle. Silicon carbide crystals are very hard (Knoop 2480), friable, and rather brittle. This limits their use. Silicon carbide is sold under the trade names Carborundum and Crystolon.

Aluminum oxide (Al_2O_3) is the most widely used artificial abrasive. Also produced in an arc furnace from bauxite, iron filings, and small amounts of coke, it contains aluminum hydroxide, ferric oxide, silica, and other impurities. The mass of aluminum oxide that is formed is crushed, and the particles are graded to size. Common trade names for aluminum oxide abrasives are Alundum and Aloxite. Although aluminum oxide is softer (Knoop 2100) than silicon carbide, it is considerably tougher. Consequently, it is a better general-purpose abrasive.

Diamonds are the hardest of all materials. Those that are used for abrasives are either natural, off-color stones (called *garnets*) that are not suitable for gems, or small, synthetic stones that are produced specifically for abrasive purposes. Manufactured stones appear to be somewhat more friable and thus tend to cut faster and cooler. They do not perform as satisfactorily in metal-bonded wheels. Diamond abrasive wheels are used extensively for sharpening carbide and ceramic cutting tools. Diamonds also are used for truing and dressing other types of abrasive wheels. Diamonds are usually used only when cheaper abrasives will not produce the desired results. Garnets are used primarily in the form of very finely crushed and graded powders for fine polishing.

Cubic boron nitride (CBN) is not found in nature. It is produced by a combination of intensive heat and pressure in the presence of a catalyst. CBN is extremely hard, registering at 4700 on the Knoop scale. It is the second-hardest substance created by nature or manufactured and is often referred to, along with diamonds, as a superabrasive. Hardness, however, is not everything.

CBN far surpasses diamond in the important characteristic of thermal resistance. At temperatures of 650°C, at which diamond may begin to revert to plain carbon dioxide, CBN continues to maintain its hardness and chemical integrity. When the temperature of 1400°C is reached, CBN changes from its cubic form to a hexagonal form and loses hardness. CBN can be used successfully in grinding iron, steel, alloys of iron, Ni-based alloys, and other materials. CBN works very effectively (long wheel life, high *G* ratio, good surface quality, no burn or chatter, low scrap rate, and overall increase in parts/shift) on hardened materials (R_c 50 or higher). It can also be used for soft steel in selected situations. CBN does well at conventional grinding speeds (6000 to 12,000 ft/min), resulting in lower total grinding in conventional equipment. CBN can also perform well at high grinding speeds (12,000 ft/min and higher) and will enhance the benefits from future machine tools. CBN can solve difficult-to-grind jobs, but it also generates cost benefits in many production grinding operations despite its higher cost. CBN is manufactured by the General Electric Company under the trade name of Borazon.

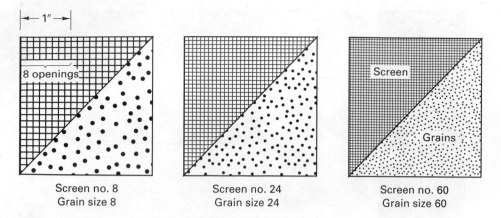

FIGURE 28-3 Typical screens for sifting abrasives into sizes. The larger the screen number (of opening per linear inch), the smaller the grain size. *(Courtesy of Carborundum Company.)*

Screen no. 8
Grain size 8

Screen no. 24
Grain size 24

Screen no. 60
Grain size 60

ABRASIVE GRAIN SIZE AND GEOMETRY

To enhance the process capability of grinding, abrasive grains are sorted into sizes by mechanical sieving machines. The number of openings per linear inch in a sieve (or screen) through which most of the particles of a particular size can pass determines the grain size (Figure 28-3).

A no. 24 grit would pass through a standard screen having 24 openings per inch but would not pass through one having 30 openings per inch. These numbers have since been specified in terms of millimeters and micrometers (see ANSI B74.12 for details). Commercial practice commonly designates grain sizes from 4 to 24, inclusive, as *coarse*; 30 to 60, inclusive, as *medium*; and 70 to 600, inclusive, as *fine*. Grains smaller than 220 are usually termed *powders*. Silicon carbide is obtainable in grit sizes ranging from 2 to 240 and aluminum oxide in sizes from 4 to 240. Superabrasive grit sizes normally range from 120 grit for CBN to 400 grit for diamond. Sizes from 240 to 600 are designated as *flour* sizes. These are used primarily for lapping, or in fine-honing stones for fine finishing tasks.

The grain size is closely related to the surface finish and metal removal rate. In grinding wheels and belts, coarse grains cut faster while fine grains provide better finish, as shown in Figure 28-4.

The grain diameter can be estimated from the screen number (S), which corresponds to the number of openings per inch. The mean diameter of the grain (g) is related to the screen number by $g \cong 0.7/S$.

Regardless of the size of the grain, only a small percentage (2 to 5%) of the surface of the grain is operative at any one time. That is, the depth of cut for an individual grain (the actual feed per grit) with respect to the grain diameter is very small. Thus the

FIGURE 28-4 MRR and surface finish versus grit size.

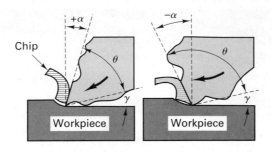

FIGURE 28-5   The rake angle of abrasive particles can be positive, zero, or negative.

chips are small. As the grain diameter decreases, the number of active grains per unit area increases and the cuts become finer because grain size is the controlling factor for surface finish (roughness). Of course, the MRR also decreases.

The grain shape is also important, because it determines the tool geometry—that is, the back rake angle and the clearance angle at the cutting edge of the grit (Figure 28-5). In the figure, γ is the clearance angle, θ is the wedge angle, and α is the rake angle. The cavities between the grits provide space for the chips, as shown in Figure 28-6. The volume of the cavities must be greater than the volume of the chips generated during the cut.

Obviously, there is no specific rake angle but rather a distribution of angles. Thus a grinding wheel can present to the surface rake angles in the range of $+45°$ to $−60°$ or greater. Grits with large negative rake angles or rounded cutting edges do not form chips but will rub or *plow* a groove in the surface (Figure 28-7). Thus abrasive machining is a

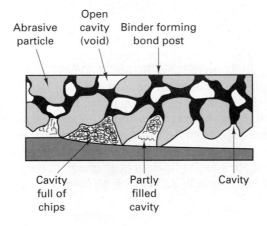

FIGURE 28-6   The cavities or voids between the grains must be large enough to hold all the chips during the cut.

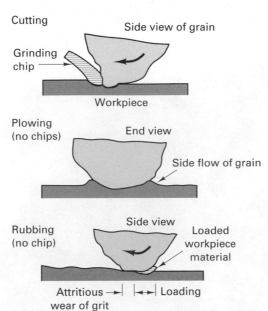

FIGURE 28-7   The grits interact with the surface in three ways: cutting, plowing, and rubbing.

FIGURE 28-8 SEM micrograph of a ground steel surface showing a plowed track (T) in the middle and a machined track (M) above. The grit fractured, leaving a portion of the grit in the surface (X), a prow formation (P), and a groove (G) where the fractured portion was pushed farther across the surface. The area marked (O) is an oil deposit.

mixture of *cutting, plowing, and rubbing*, with the percentage of each being highly dependent on the geometry of the grit. As the grits are continuously abraded, fractured, or dislodged from the bond, new grits are exposed and the mixture of cutting, plowing, and rubbing is changing continuously. A high percentage of the energy used for rubbing and plowing goes into the workpiece, but when chips are found, 95 to 98% of the energy (the heat) goes into the chip. Figure 28-8 shows a scanning electron microscope (SEM) micrograph of a ground surface with a plowing track.

In grinding, the chips are small but are formed by the same basic mechanism of compression and shear as discussed in Chapter 20 for regular metalcutting. Figure 28-9 shows steel chips from a grinding process at high magnification. They show the same structure as chips from other machining processes. Chips flying in the air from a grinding process often have sufficient heat energy to burn or melt in the atmosphere. Sparks observed during grinding steel with no cutting fluid are really burning chips, as shown in Figure 28-8.

The feeds and depths of cut in grinding are small while the cutting speeds are high, resulting in high specific horsepower numbers. Because cutting is obviously more efficient than plowing or rubbing, grain fracture and grain pullout are natural phenomena

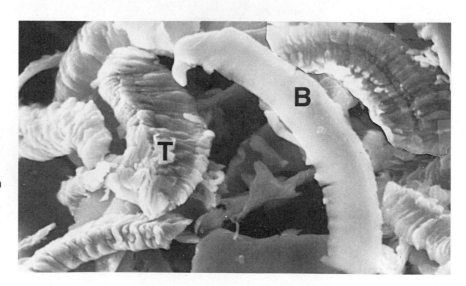

FIGURE 28-9 SEM micrograph of stainless steel chips from a grinding process. The tops (T) of the chips have the typical shear-front-lamella structure while the bottoms (B) are smooth where they slide over the grit 4800×.

used to keep the grains sharp. As the grains become dull, cutting forces increase, and there is an increased tendency for the grains to fracture or break free from the bonding material.

■ 28.3 GRINDING WHEEL STRUCTURE AND GRADE

Grinding, wherein the abrasives are bonded together into a wheel, is the most common abrasive machining process. The performance of grinding wheels is greatly affected by the bonding material and the spatial arrangement of the particles' grits.

The spacing of the abrasive particles with respect to each other is called *structure*. Close-packed grains have dense structure; open structure means widely spaced grains. Open-structure wheels have larger chip cavities but fewer cutting edges per unit area (Figure 28-10a).

The fracturing of the grits is controlled by the bond strength, which is known as the *grade*. Thus, grade is a measure of how strongly the grains are held in the wheel. It is really dependent on two factors: the strength of the bonding materials and the amount of the bonding agent connecting the grains. The latter factor is illustrated in Figure 28-10b. Abrasive wheels are really porous. The grains are held together with "posts" of bonding material. If these posts are large in cross section, the force required to break a grain free from the wheel is greater than when the posts are small. If a high dislodging force is required, the bond is said to be *hard*. If only a small force is required, the bond is said to be *soft*. Wheels are commonly referred to as hard or soft, referring to the net strength of the bond, resulting from both the strength of the bonding material and its disposition between the grains.

G RATIO

The loss of grains from the wheel means that the wheel is changing size. The grinding ratio, or *G* ratio, is defined as the cubic inches of stock removed divided by the cubic inches of wheel lost. In conventional grinding, the *G* ratio is in the range 20:1 to 80:1. The *G* ratio is a measure of grinding production and reflects the amount of work a wheel can do during its useful life. As the wheel loses material, it must be reset or repositioned to maintain workpiece size.

A typical vitrified grinding wheel will consist of 50 vol% abrasive particles, 10 vol% bond, and 40 vol% cavities; that is, the wheels have porosity. The manner in which the wheel performs is influenced by the following factors:

1. The mean force required to dislodge a grain from the surface (the grade of the wheel)
2. The cavity size and distribution of the porosity (the structure)

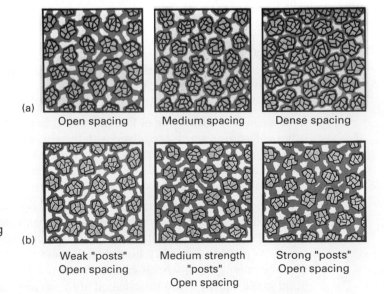

(a)   Open spacing   Medium spacing   Dense spacing

(b)   Weak "posts"   Medium strength   Strong "posts"
      Open spacing   "posts"   Open spacing
                      Open spacing

FIGURE 28-10   Meaning of terms *structure* and *grade* for grinding wheels. (a) The structure of a grinding wheel depends on the spacing of the grits. (b) The grade of a grinding wheel depends on the amount of bonding agent (posts) holding abrasive grains in the wheel.

3. The mean spacing of active grains in the wheel surface (grain size and structure)

4. The properties of the grain (hardness, attrition, and friability)

5. The geometry of the cutting edges of the grains (rake angles and cutting-edge radius compared to depth of cut)

6. The process parameters (speeds, feeds, cutting fluids) and type of grinding (surface, or cylindrical)

It is easy to see why grinding is a complex process, difficult to control.

BONDING MATERIALS FOR GRINDING WHEELS

Bonding material is a very important factor to be considered in selecting a grinding wheel. It determines the strength of the wheel, thus establishing the maximum operating speed. It determines the elastic behavior or deflection of the grits in the wheel during grinding. The wheel can be hard or rigid, or it can be flexible. Finally, the bond determines the force required to dislodge an abrasive particle from the wheel and thus plays a major role in the cutting action. Bond materials are formulated so that the ratio of bond wear matches the rate of wear of the abrasive grits. Bonding materials in common use are the following:

1. *Vitrified bonds* are composed of clays and other ceramic substances. The abrasive particles are mixed with the wet clays so that each grain is coated. Wheels are formed from the mix, usually by pressing, and then dried. They are then fired in a kiln, which results in the bonding material's becoming hard and strong, having properties similar to glass. Vitrified wheels are porous, strong, rigid, and unaffected by oils, water, or temperature over the ranges usually encountered in metal cutting. The operating speed range in most cases is 5500 to 6500 ft/min, but some wheels now operate at surface speeds up to 16,000 ft/min.

2. *Resinoid*, or phenolic resins, can be used. Because plastics can be compounded to have a wide range of properties, such wheels can be obtained to cover a variety of work conditions. They have, to a considerable extent, replaced shellac and rubber wheels. Composite materials are being used in rubber-bonded or resinoid-bonded wheels that are to have some degree of flexibility or are to receive considerable abuse and side loading. Various natural and synthetic fabrics and fibers, glass fibers, and nonferrous wire mesh are used for this purpose.

3. *Silicate* wheels use silicate of soda (waterglass) as the bond material. The wheels are formed and then baked at about 500°F for a day or more. Because they are more brittle and not so strong as vitrified wheels, the abrasive grains are released more readily. Consequently, they machine at lower surface temperatures than vitrified wheels and are useful in grinding tools when heat must be kept to a minimum.

4. *Shellac-bonded* wheels are made by mixing the abrasive grains with shellac in a heated mixture, pressing or rolling into the desired shapes, and baking for several hours at about 300°F. This type of bond is used primarily for strong, thin wheels having some elasticity. They tend to produce a high polish and thus have been used in grinding such parts as camshafts and mill rolls.

5. *Rubber* bonding is used to produce wheels that can operate at high speeds but must have a considerable degree of flexibility so as to resist side thrust. Rubber, sulfur, and other vulcanizing agents are mixed with the abrasive grains. The mixture is then rolled out into sheets of the desired thickness, and the wheels are cut from these sheets and vulcanized. Rubber-bonded wheels can be operated at speeds up to 16,000 ft/min. They are commonly used for snagging work in foundries and for thin cutoff wheels.

6. *Superabrasive* wheels are either electroplated (single layer of superabrasive plated to outside diameter of a steel blank) or a thin segmented drum of vitrified CBN surrounds a steel core. The steel core provides dimensional accuracy, and the replaceable segments provide durability, homogeneity, and repeatability while increasing wheel life. The latter type of wheels can use resin, metal, or vitrified bonding. Selection of bond grade and structure (also called abrasive concentration) is critical.

For the electroplated wheels, nickel is used to attach a single layer of CBN (or diamond) to the OD of an accurately ground or turned steel blank. For the vitrified wheel, superabrasives are mixed with bonding media and molded (or preformed and sintered) into segments or a ring. The ring is mounted on a split steel body. Porosity is varied (to alter structure) by varying preform pressure or by using "pore-forming" additives to the bond material that are vaporized during the sintering cycle. The steel-cored segmented design can rotate at 40,000 sfpm (200 m/s) whereas a plain vitrified wheel may burst at 20,000 fpm.

ABRASIVE MACHINING VERSUS CONVENTIONAL GRINDING VERSUS LOW-STRESS GRINDING

The condition wherein very rapid metal removal can be achieved by grinding is the one to which some have applied the term *abrasive machining*. The metal removal rates are compared with, or exceed, those obtainable by milling or turning or broaching, and the size tolerances are comparable. It is obviously just a special type of grinding, using abrasive grains as cutting tools, as do all other types of abrasive machining. Abrasive grinding done in an aggressive way can produce sufficient localized plastic deformation and heat in the surface so as to develop tensile residual stresses, layers of overtempered martensite (in steels), and even microcracks, because this process is quite abusive. See Figure 28-11 for a discussion of residual stresses produced by varius surface grinding processes.

Conventional grinding can be replaced by procedures that develop lower surface stresses when service failures due to fatigue or stress corrosion are possible. This is accomplished by employing softer grades of grinding wheels, reducing the grinding speeds and infeed rates, using chemically active cutting fluids (e.g., highly sulfurized oil or KNO_2 in water), as outlined in the table of grinding conditions in Figure 28-11. These procedures may require the addition of a variable-speed drive to the grinding machine. Generally, only about 0.005 to 0.010 in. of surface stock needs to be finish ground in this way, as the depth of the surface damage due to conventional grinding or abusive grinding is 0.005 to 0.007 in. High-strength steels, high-temperature nickel, and cobalt-based alloys and titanium alloys are particularly sensitive to surface deformation and cracking problems from grinding. Other postprocessing processes, such as polishing, honing, and chemical milling plus peening, can be used to remove the deformed layers in critically stressed parts. It is strongly recommended, however, that testing programs be used along with service experience on critical parts before these procedures are employed in production.

Grinding conditions			
	Abusive AG	Conventional CG	Low-stress LSG
Wheel	A46MV	A46KV	A46HV or A60IV
Wheel speed ft/min	6,000–18,000	4,500–6,500	2500–3000
Down feed in./pass	.002–.004	.001–.003	.0002–.005
Cross feed in./pass	.040–.060	.040–.060	.040–.060
Table speed ft/min	40–100	40–100	40–100
Fluid	Dry	Sol oil (1:20)	Sulfurized oil

FIGURE 28-11 Typical residual stress distributions produced by surface grinding with different grinding conditions for abusive, conventional, and low-stress grinding. Material is 4340 steel. *(From M. Field and W. P. Kosher, "Surface Integrity in Grinding," in New Developments in Grinding, Carnegie-Mellon University Press, Pittsburgh, 1972, p. 666.)*

FIGURE 28-12   Truing methods for restoring grinding geometry include nibs, rolls, disks, cups, and blocks.

In the casting and forging industries, the term often used for abrasive maching is *snagging*. *Snagging* is a type of rough manual grinding that is done to remove fins, gates, risers, and rough spots from castings or flash from forgings, preparatory to further machining. The primary objective is to remove substantial amounts of metal rapidly without much regard for accuracy, so this is a form of abrasive machining except that pedestal-type or *swing grinders* ordinarily are used. Portable electric or hand air grinders are also used for this purpose and for miscellaneous grinding in connection with welding.

Grinding wheels lose their geometry during use. *Truing* restores the original shape. A single-point diamond tool can be used to *true* the wheel while fracturing abrasive grains to expose new grains and new cutting edges on worn, glazed grains (Figure 28-12). Truing can also be accomplished by grinding the grinding wheel with a controlled-path or powered rotary device using conventional abrasive wheels. The precision in generating a trued wheel surface by these methods is poorer than by the method described earlier.

As the wheel is used, there is a tendency for the wheel to become *loaded* (metal chips become lodged in the cavities between the grains). Also, the grains dull or glaze (grits wear, flatten, and polish). Unless the wheel is cleaned and sharpened (or dressed), the wheel will not cut as well and will tend to plow and rub more. Figure 28-13 shows an arrangement for stick *dressing* a grinding wheel. The dulled grains cause the cutting forces on the grains to increase, ideally resulting in the grains' fracturing or being pulled out of the bond, thus providing a continuous exposure of sharp cutting edges. Such a continuous action ordinarily will not occur for light feeds and depths of cut. For heavier cuts, grinding wheels do become somewhat self-dressing, but the workpiece may become overheated and turn a bluish temper color (this is called *burn*) before the wheel reaches a

FIGURE 28-13   Schematic arrangement of stick dressing versus truing.

FIGURE 28-14 Continuous crush roll dressing and truing of a grinding wheel (form—truing and dressing throughout the process rather than between cycles) doing plunge cut grinding on a cylinder held between centers.

fully dressed condition. A burned surface, the consequence of an oxide layer formation, results in the scrapping of several workpieces before parts of good quality are ground.

Resin-bonded wheels can be trued by grinding with hard ceramics such as tungsten carbide. The procedure for truing and dressing a CBN wheel in a surface grinder might be as follows: Use 0.0002-in. downfeed per pass and cross feed slightly more than half the wheel thickness at moderate table speeds. The wheel speed is the same as the grinding speed. The grinding power will gradually increase, as the wheel is getting dull, while being trued. When the power exceeds normal power drawn during workpiece grinding, stop the truing operation. Dress the wheel face open using a J-grade stick, with abrasive one grit size smaller than CBN. Continue the truing. Repeat this cycle until the wheel is completely trued.

Modern grinding machines are equipped so that the wheel can be dressed and/or trued continuously or intermittently while grinding continues. A common way to do this is by *crush dressing* (Figure 28-14). Crush dressing consists of forcing a hard roll (tungsten carbide or high speed steel) having the same contour as the part to be ground against the grinding wheel while it is revolving—usually quite slowly. A water-based coolant is used to flood the dressing zone at 5 to 10 gal/min. The crushing action fractures and dislodges some of the abrasive grains, exposing fresh sharp edges, allowing free cutting for faster infeed rates. This procedure is usually employed to produce and maintain a special contour to the abrasive wheel. This is also called wheel profiling. Crush dressing is a very rapid method of dressing grinding wheels, and because it fractures abrasive grains, it results in free cutting and somewhat cooler grinding. The resulting surfaces may be slightly rougher than when diamond dressing is used.

■ 28.4 GRINDING WHEEL IDENTIFICATION

Most grinding wheels are identified by a standard marking system that has been established by the American National Standards Institute. This system is illustrated and explained in Figure 28-15. The first and last symbols in the marking are left to the discretion of the manufacturer.

GRINDING WHEEL GEOMETRY

The shape and size of the wheel are critical selection factors. Obviously, the shape must permit proper contact between the wheel and all of the surface that must be ground. Grinding wheel shapes have been standardized, and eight of the most commonly used types are shown in Figure 28-16. Types 1, 2, and 5 are used primarily for grinding external or internal cylindrical surfaces and for plain surface grinding. Type 2 can be mounted for grinding on either the periphery or the side of the wheel. Type 4 is used with tapered safety flanges so that if the wheel breaks during rough grinding, such as snagging, these

Standard bonded-abrasive wheel-marching system (ANSI *Standard* B74.13-1977)

Wheel-marching system for diamond and cubic boron nitride wheels (ANSI *Standard* B74.13-1977).

FIGURE 28-15 Standard marking systems for grinding wheels (ANSI standard B74. 13-1977).

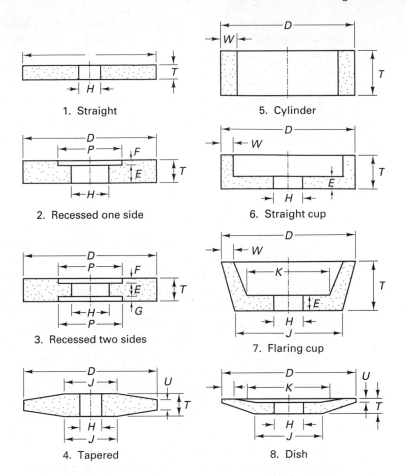

FIGURE 28-16 Standard grinding wheel shapes commonly used. *(Courtesy of Carborundum Company.)*

flanges will prevent the pieces of the wheel from flying and causing damage. Type 6, the straight cup, is used primarily for surface grinding but can also be used for certain types of offhand grinding. The flaring-cup type of wheel is used for tool grinding. Dish-type wheels are used for grinding tools and saws.

Type 1, the straight grinding wheels, can be obtained with a variety of standard faces. Some of these are shown in Figure 28-17.

The size of the wheel to be used is determined primarily by the spindle rpm values available on the grinding machine and the proper cutting speed for the wheel, as dictated by the type of bond. For most grinding operations the cutting speed is about 2500 to 6500 ft/min. Different types and grades of bond often justify considerable deviation from these speeds. For certain types of work using special wheels and machines, as in thread grinding and "abrasive machining," much higher speeds are used.

FIGURE 28-17 Standard face contours for straight grinding wheels. *(Courtesy of Carborundum Company.)*

The operation for which the abrasive wheel is intended will also influence the wheel shape and size. The major use categories are the following:

1. *Cutting off*: for slicing and slotting parts; use thin wheel, organic bond
2. *Cylindrical between centers*: grinding outside diameters of cylindrical workpieces
3. *Cylindrical, centerless*: grinding outside diameters with work rotated by regulating wheel
4. *Internal cylindrical*: grinding bores and large holes
5. *Snagging*: removing large amounts of metal without regard to surface finish or tolerances
6. *Surface grinding*: grinding flat workpieces
7. *Tool grinding*: for grinding cutting edges on tools such as drills, milling cutters, taps, reamers, and single-point high-speed-steel tools
8. *Offhand grinding*: work or the grinding tool is handheld

In many cases, the classification of processes coincides with the classification of machines that do the process. Other factors that will influence the choice of wheel to be selected include the workpiece material, the amount of stock to be removed, the shape of the workpiece, and the accuracy and surface finish desired. Workpiece material has a great impact on choice of the wheel. Hard, high-strength metals (tool steels, alloy steels) are generally ground with aluminum oxide wheels or cubic boron nitride wheels. Silicon carbide and CBN are employed in grinding brittle materials (cast iron and ceramics) as well as softer, low-strength metals such as aluminum, brass, copper, and bronze. Diamonds have taken over the cutting of tungsten carbides, and CBN is used for precision grinding of tool and die steel, alloy steels, stainless steel, and other very hard materials. There are so many factors that affect the cutting action that there are no hard-and-fast rules with regard to abrasive selection.

Selection of grain size is determined by whether coarse or fine cutting and finish are desired. Coarse grains take larger depths of cut and cut more rapidly. Hard wheels with fine grains leave smaller tracks and therefore are usually selected for finishing cuts. If there is a tendency for the work material to load the wheel, larger grains with a more open structure may be used for finishing.

BALANCING GRINDING WHEELS

Because of the high rotation speeds involved, grinding wheels must never be used unless they are in good balance. A slight imbalance will produce vibrations that will cause waviness in the work surface. It may cause a wheel to break, with the probability of serious damage and injury. The wheel should be mounted with proper bushings so that it fits snugly on the spindle of the machine. Rings of blotting paper should be placed between the wheel and the flanges to ensure that the clamping pressure is evenly distributed. Most grinding wheels will run in good balance if they are mounted properly and trued. Most machines have provision for compensating for a small amount of wheel imbalance by attaching weights to one mounting flange. Some have provision for semiautomatic balancing with weights that are permanently attached to the machine spindle.

SAFETY IN GRINDING

Because the rotational speeds are quite high, and the strength of grinding wheels is usually much less than that of the materials being ground, serious accidents occur much too frequently in connection with the use of grinding wheels. Virtually all such accidents could be avoided and are due to one or a combination of four causes. First, grinding wheels are occasionally operated at unsafe and improper speeds. All grinding wheels are clearly marked with the maximum rpm value at which they should be rotated. They are all tested to considerably above the designated rpm and are safe at the specified speed *unless abused. They should never, under any condition, be operated above the rated speed.* Second, a very common form of abuse, frequently accidental, is dropping the wheel or striking it against a hard object. This can cause a crack (which may not be readily visible), resulting in subsequent failure of the wheel while rotating at high speed under load. If a wheel is dropped or struck against a hard object, it should be discarded and never used unless tested at above the rated speed in a properly designed test stand. A third common cause of grinding wheel failure is

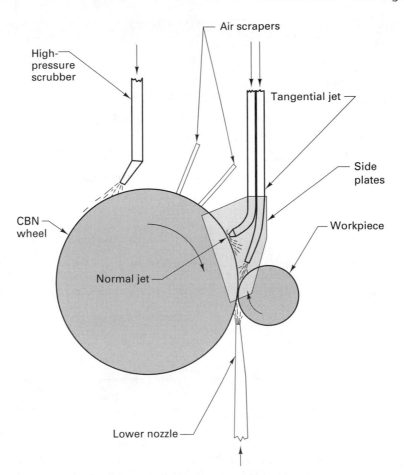

FIGURE 28-18   Coolant delivery system for optimum CBN grinding. (*Source: "Production Grinding with CBN," M. P. Hitchiner, CBN Grinding Systems Manager, Universal Beck, Romulus, MI,* Machining Technology, *Vol. 2, No. 2, 1991.)*

improper use, such as grinding against the side of a wheel that was designed for grinding only on its periphery. The fourth and most common cause of injury from grinding is the absence of a proper safety guard over the wheel and/or over the eyes or face of the operator. The frequency with which operators will remove safety guards from grinding equipment or fail to use safety goggles or face shields is amazing and inexcusable.

USE OF CUTTING FLUIDS IN GRINDING

Because grinding involves cutting, the selection and use of a cutting fluid is governed by the basic principles discussed in Chapter 21. If a fluid is used, it should be applied in sufficient quantities and in a manner that will ensure that the chips are washed away, not trapped between the wheel and the work. This is of particular importance in grinding horizontal surfaces. In hardened steel, the use of a fluid can help to prevent fine microcracks that result from highly localized heating. The air scraper shown in Figure 28-18 permits the cutting fluid (lubricant) to get onto the face of the wheel. Metal air scrapers disrupt the airflow. Upper and lower nozzles cool the grinding zone, while a high-pressure scrubber helps deter loading of the wheel.

Much snagging and off-hand grinding is done dry. On some types of material, dry grinding produces a better finish than can be obtained by wet grinding.

Grinding fluids strongly influence the performance of CBN wheels. Straight, sulfurized, or sulfochlorinated oils can enhance performance considerably when used with straight oils.

◼ 28.5 GRINDING MACHINES

Grinding machines commonly are classified according to the type of surface they produce. Table 28-4 presents such a classification, with further subdivision to indicate characteristic features of different types of machines within each classification. Grinding on all machines is done in three ways. In the first, the depth of cut (d_t) is obtained by *infeed*—moving the

TABLE 28-4	Grinding Machines	
Type of Machine	Type of Surface	Specific Types or Features

Cylindrical external	External surface on rotating, usually cylindrical parts	Work rotated between centers
		  Centerless
		  Centerless
		  Chucking
		  Tool post
		  Crankshaft, cam, etc.

Cylindrical internal	Internal diameters of holes	Chucking Planetary (work stationary) Centerless
Surface conventional	Flat surfaces	Reciprocating table or rotating table Horizontal or vertical spindle
Creep feed	Deep slots, profiles in hard steels, carbides, and ceramics using CBN and diamond	Rigid, chatter-free, creep feed rate Continuous dressing Heavy coolant flows NC or CNC control Variable speed wheel
Tool grinders	Tool angles and geometries	Universal Special
Other	Special or any of the above	Disk, contour, thread, flexible shaft, swing frame, snag, pedestal, bench

wheel down into the work or the work up into the wheel (Figure 28-19). The desired surface is then produced by traversing the wheel across (cross feed) the workpiece, or vice versa (Figure 28-19). In the second method, known as *plunge-cut* grinding, the basic movement is of the wheel being fed radially into the work while the latter revolves on centers. It is similar to form cutting on a lathe; usually a formed grinding wheel is used (Figure 28-14). In the third method, the work is fed very slowly past the wheel and the

FIGURE 28-19 Horizontal-spindle surface grinder, with insets showing movements of wheelhead.

FIGURE 28-20   Conventional grinding contrasted to creep feed grinding. Note that crush roll dressing is used here; see Figure 28-14.

total downfeed or depth (*d*) is accomplished in a single pass (Figure 28-20). This is called *creep feed grinding* (CFG). (Table 28-5 compares CFG to conventional and high-speed grinding for CBN applications.)

The CFG method, often done in the surface grinding mode, is markedly different from conventional surface grinding. The depth of cut is increased 1000 to 10,000 times, and the work feed ratio is decreased in the same proportion; hence the name *creep feed grinding*. The long arc of contact between the wheel and the work increases the cutting forces and the power required. Therefore, the machine tools to perform this type of grinding must be specially designed with high static and dynamic stability, stick-slip-free ways, adequate damping, increased horsepower, infinitely variable spindle speed, variable but extremely consistent table feed (especially in the low ranges), high-pressure cooling systems, integrated devices for dressing the grinding wheels, and specially designed (soft with open structure) grinding wheels. The process is mainly being applied to grinding deep slots with straight parallel sides or to grinding complex profiles in difficult-to-grind materials. The process is capable of producing extreme precision at relatively high metal removal rates. Because the process can operate at relatively low surface temperatures, the surface integrity of the metals being ground is good.

However, in CFG, the grinding wheels must maintain their initial profile much longer, so continuous dressing is used that is form-truing and dressing the grinding wheel throughout the process rather than between cycles. Continuous crush dressing results in higher MRRs, improved dimensional accuracy and form tolerance, reduced grinding

TABLE 28-5	Starting Conditions for CBN Grinding		
Grinding Variable	Conventional Grinding	Creep Feed Grinding	High-speed Grinding
Wheel speed (fpm)	5500–9500 versus 4500–6500 vitrified	5000–9000 versus 3000–5000	12000–25000
Table speed (fpm)	80–150	0.5–5	5–20
Feed (*f_t*) in./pass	0.0005–0.0015	0.100–0.250	250–.500
Grinding fluids	10% heavy-duty soluble oil or 3–5% light-duty soluble for light feeds	Sulfurized or sulfochlorinated straight grinding oil applied at 80 to 100 gal/min at 100 psi or more	

forces (and power), and reduced thermal effects while sacrificing wheel wear. Creep feed grinding eliminates preparatory operations such as milling or broaching, since profiles are ground into the solid workpiece. This can result in significant savings in unit part costs.

Grinding machines that are used for precision work have certain important characteristics that permit them to produce parts having close dimensional tolerances. They are constructed very accurately, with heavy, rigid frames to ensure permanency of alignment. Rotating parts are accurately balanced to avoid vibration. Spindles are mounted in very accurate bearings, usually of the preloaded ball-bearing type. Controls are provided so that all movements that determine dimensions of the workpiece can be made with accuracy—usually to 0.001 or 0.00001 in.

The abrasive dust that results from grinding must be prevented from entering between moving parts. All ways and bearings must be fully covered or protected by seals. If this is not done, the abrasive dust between moving parts becomes embedded in the softer of the two, causing it to act as lap and abrade the harder of the two surfaces, resulting in permanent loss of accuracy.

These special characteristics add considerably to the cost of these machines and require that they be operated by trained personnel. Production-type grinders are more fully automated and have higher metal removal rates and excellent dimensional accuracy. Fine surface finish can be obtained very economically.

CYLINDRICAL GRINDING

Center-type cylindrical grinding is commonly used for producing external cylindrical surfaces. Figures 28-14 and Figure 28-21 show the basic principles and motions of this process. The grinding wheel revolves at an ordinary cutting speed, and the workpiece rotates on centers at a much slower speed, usually from 75 to 125 ft/min. The grinding wheel and the workpiece move in opposite directions at their point of contact. The depth of cut is determined by infeed of the wheel or workpiece. Because this motion also determines the finished diameter of the workpiece, accurate control of this movement is required. Provision is made to traverse the workpiece with the wheel, or the work can be reciprocated past the wheel. In very large grinders, the wheel is reciprocated because of the massiveness of the work. For form or plunge grinding, the detail of the wheel is maintained by periodic crush roll dressing.

A *plain center-type cylindrical grinder* is shown in Figure 28-21. On this type the work is mounted between headstock and tailstock centers. Solid dead centers are always used in the tailstock, and provision is usually made so that the headstock center can be operated either dead or alive. High-precision work is usually ground with a dead headstock center, because this eliminates any possibility that the workpiece will run out of round due to any eccentricity in the headstock.

The table assembly can be reciprocated, in most cases, by using a hydraulic drive. The speed can be varied, and the length of the movement can be controlled by means of adjustable trip dogs.

Infeed is provided by movement of the wheelhead at right angles to the longitudinal axis of the table. The spindle is driven by an electric motor that is also mounted on the wheelhead. If the infeed movement is controlled manually by some type of vernier drive to provide control to 0.001 in. or less, the machine is usually equipped with digital readout equipment to show the exact size being produced. Most production-type grinders have automatic infeed with retraction when the desired size has been obtained. Such machines are usually equipped with an automatic diamond wheel-truing device that dresses the wheel and resets the measuring element before grinding is started on each piece.

The longitudinal traverse should be about one-fourth to three-fourths of the wheel width for each revolution of the work. For light machines and fine finishes, it should be held to the smaller end of this range. The depth of cut (infeed) varies with the purpose of the grinding operation and the finish desired. When grinding is done to obtain accurate size, infeeds of 0.002 to 0.004 in. are commonly used for roughing cuts. For finishing, the infeed is reduced to 0.00025 to 0.0005 in. The design allowance for grinding should be from 0.005 to 0.010 in. on short parts and on parts that are not to be hardened. On long or large parts and on work that is to be hardened, a grinding allowance of from

Movements
1. Wheel   2. Work (rotates)
3. Traverse   4. Infeed

FIGURE 28-21   Cylindrical grinding between centers.

0.015 to 0.030 in. is desirable. When grinding is used primarily for metal removal (called abrasive machining), infeeds are much higher, 0.020 to 0.040 in. being common. Continuous downfeed is often used, with rates up to 0.100 in./min being common.

Grinding machines are available in which the workpiece is held in a chuck for grinding both external and internal cylindrical surfaces. *Chucking-type external grinders* are production-type machines for use in rapid grinding of relatively short parts, such as ball-bearing races. Both chucks and collets are used for holding the work, the means dictated by the shape of the workpiece and rapid loading and removal.

In chucking-type internal grinding machines, the chuck-held workpiece revolves, and a relatively small, high-speed grinding wheel is rotated on a spindle arranged so that it can be reciprocated in and out of the workpiece. Infeed movement of the wheelhead is normal to the axis of rotation of the work (Figure 28-21).

CENTERLESS GRINDING

Centerless grinding makes it possible to grind both external and internal cylindrical surfaces without requiring the workpiece to be mounted between centers or in a chuck. This eliminates the requirement of center holes in some workpieces and the necessity for mounting the workpiece, thereby reducing the cycle time.

The principle of *centerless external grinding* is illustrated in Figure 28-22. Two wheels are used. The larger one operates at regular grinding speeds and does the actual grinding. The smaller wheel is the *regulating* wheel. It is mounted at an angle to the plane of the grinding wheel. Revolving at a much slower surface speed—usually 50 to 200 ft/min—the regulating wheel controls the rotation and longitudinal motion of the workpiece and is a usually a plastic- or rubber-bonded wheel with a fairly wide face.

A. Grinding wheel
B. Grinding face
C. Regulating wheel
D. Workpiece
E. Work rest blade

θ = Angle of tilt of regulating wheel

Movements
1. Grinding wheel   2. Work
3. Regulating wheel   4. Infeed
5. Traverse

FIGURE 28-22 Centerless grinding showing the relationship among the grinding wheel, the regulating wheel, and the workpiece in centerless method. *(Courtesy of Carborundum Company.)*

The workpiece is held against the work-rest blade by the cutting forces exerted by the grinding wheel and rotates at approximately the same surface speed as that of the regulating wheel. This axial feed is calculated approximately by the equation

$$F = ND \sin \phi \qquad (28\text{-}1)$$

where

F = feed (mm/min or in./min)

D = diameter of the regulating wheel (mm or in.)

N = revolutions per minute of the regulating wheel

ϕ = angle of inclination of the regulating wheel

Centerless grinding has several important advantages:

1. It is very rapid; infeed centerless grinding is almost continuous.
2. Very little skill is required of the operator.
3. It can often be made automatic (single-cycle automatic).
4. Where the cutting occurs, the work is fully supported by the work rest and the regulating wheel. This permits heavy cuts to be made.
5. Because there is no distortion of the workpiece, accurate size control is easily achieved.
6. Large grinding wheels can be used, thereby minimizing wheel wear.

Thus centerless grinding is ideally suited to certain types of mass-production operations. The major disadvantages are as follows:

1. Special machines are required that can do no other type of work.
2. The work must be round—no flats, such as keyways, can be present.
3. Its use on work having more than one diameter or on curved parts is limited.
4. In grinding tubes, there is no guarantee that the OD and Internal Diameter (ID) are concentric.

Special centerless grinding machines are available for grinding balls and tapered workpieces. The centerless grinding principle can also be applied to internal grinding, but the external surface of the cylinder must be finished accurately before the internal operation is started. However, it assures that the internal and external surfaces will be concentric. The operation is easily mechanized for many applications.

SURFACE GRINDING MACHINES

Surface grinding machines are used primarily to grind flat surfaces. However formed, irregular surfaces can be produced on some types of surface grinders by use of a formed wheel. There are four basic types of surface grinding machines, differing in the movement of their tables and the orientation of the grinding wheel spindles (Figure 28-23):

1. Horizontal spindle and reciprocating table
2. Vertical spindle and reciprocating table
3. Horizontal spindle and rotary table
4. Vertical spindle and rotary table

The most common type of surface grinding machine has a reciprocating table and horizontal spindle (Figures 28-19). The table can be reciprocated longitudinally either by handwheel or by hydraulic power. The wheelhead is given transverse (cross-feed) motion at the end of each table motion, again either by handwheel or by hydraulic power feed. Both the longitudinal and transverse motions can be controlled by limit switches. Infeed or downfeed on such grinders is controlled by handwheels or automatically. The size of such machines is determined by the size of the surface that can be ground.

In using such machines, the wheel should overtravel the work at both ends of the table reciprocation, so as to prevent the wheel from grinding in one spot while the table is being reversed. The transverse or cross-feed motion should be one-fourth to three-fourths of the wheel width between each stroke.

Movements
1. Wheel 2. Infeed
3. Work table traverse

A. Grinding wheel
B. Grinding face
C. Shaft
D. Workpiece
E. Magnetic chuck on table

Movements
1. Wheel 2. Work table rotation
3. Infeed 4. Cross feed

Movements
1. Wheel   2. Infeed
3. Work table rotation

FIGURE 28-23   Surface grinding: (a) horizontal surface grinding and reciprocating table; (b) vertical spindle with reciprocating table; (c) and (d) both horizontal- and vertical-spindle machines can have rotary tables. *(Courtesy of Carborundum Company.)*

Vertical-spindle reciprocating-table surface grinders differ basically from those with horizontal spindles only in that their spindles are vertical and that the wheel diameter must exceed the width of the surface to be ground. Usually, no transverse motion of either the table or the wheelhead is provided. Such machines can produce very flat surfaces.

Rotary-table surface grinders can have either vertical or horizontal spindles, but those with horizontal spindles are limited in the type of work they will accommodate and therefore are not used to a great extent. *Vertical-spindle rotary-table surface grinders* are primarily production-type machines. They frequently have two or more grinding heads, and therefore both rough grinding and finish grinding are accomplished in one rotation of the workpiece. The work can be held either on a magnetic chuck or in special fixtures attached to the table.

By using special rotary feeding mechanisms, machines of this type often are made automatic. Parts are dumped on the rotary feeding table and fed automatically onto work-holding devices and moved past the grinding wheels. After they pass the last grinding head, they are automatically unloaded.

DISK-GRINDING MACHINES

Disk grinders have relatively large side-mounted abrasive disks. The work is held against one side of the disk for grinding. Both single- and double-disk grinders are used; in the latter type the work is passed between the two disks and is ground on both sides simultaneously. On these machines, the work is always held and fed automatically. On small, single-disk grinders the work can be held and fed by hand while resting on a supporting table. Although manual disk grinding is not very precise, flat surfaces can be obtained quite rapidly with little or no tooling cost. On specialized, production-type machines, excellent accuracy can be obtained very economically.

TOOL AND CUTTER GRINDERS

Simple, single-point tools are often sharpened by hand on bench or pedestal grinders (*off-hand grinding*). More complex tools, such as milling cutters, reamers, hobs, and single-point tools for production-type operations require more sophisticated grinding machines, commonly called *universal tool and cutter grinders*. These machines are similar to small universal cylindrical center-type grinders, but they differ in four important respects:

1. The headstock is not motorized.
2. The headstock can be swiveled about a horizontal as well as a vertical axis.
3. The wheelhead can be raised and lowered and can be swiveled through at 360° rotation about a vertical axis.
4. All table motions are manual. No power feeds being provided.

Specific rake and clearance angles must be created, often repeatedly, on a given tool or on duplicate tools. Tool and cutter grinders have a high degree of flexibility built into them so that the required relationships between the tool and the grinding wheel can be established for almost any type of tool. Although setting up such a grinder is quite complicated and requires a highly skilled worker, after the setup is made for a particular job, the actual grinding is accomplished rather easily. Figure 28-24 shows several typical setups on a tool and cutter grinder.

Hand-ground cutting tools are not accurate enough for automated machining processes. Many numerically controlled (NC) machine tools have been sold on the premise that they can position work to very close tolerances—within ± 0.0001 to 0.0002 in.—only to have the initial workpieces produced by those machines out of tolerance by as much as 0.015 to 0.020 in. In most instances, the culprit was a poorly ground tool. For example, a twist drill with a point ground 0.005 in. off-center can "walk" as much as 0.015 in., thus causing poor hole location. Many companies are turning to computer numeric control (CNC) grinders to handle the regrinding of their cutting tools. A six-axis CNC grinder is capable of restoring the proper tool angles (rake and clearance), concentricity, cutting edges, and dimensional size.

FIGURE 28-24   Three typical setups for grinding single- and multiple-edge tools on a universal tool and cutter grinder. (a) Single-point tool is held in a device that permits all possible angles to be ground. (b) Edges of a large hand reamer are being ground. (c) Milling cutter is sharpened with a cupped grinding wheel.

MOUNTED WHEELS AND POINTS

Mounted wheels and points are small grinding wheels of various shapes that are permanently attached to metal shanks that can be inserted in the chucks of portable, high-speed electric or air motors. They are operated at speeds up to 100,000 rpm, depending on their diameters, and are used primarily for deburring and finishing in mold and die work. Several types are shown in Figure 28-25.

COATED ABRASIVES

Coated abrasives are being used increasingly in finishing both metal and nonmetal products. These are made by gluing abrasive grains onto a cloth or paper backing (Figure 28-26). Synthetic abrasives—aluminum oxide, silicon carbide, aluminum, zirconia, CBN, and diamond—are used most commonly, but some natural abrasives—sand, flint, garnet, and emery—also are employed. Various types of glues are utilized to attach the abrasive grains to the backing, usually compounded to allow the finished product to have some flexibility.

Coated abrasives are available in sheets, rolls, endless belts, and disks of various sizes. Some of the available forms are shown in Figure 28-26. Although the cutting action of coated abrasives basically is the same as with grinding wheels, there is one major difference: they have little tendency to be self-sharpened when dull grains are pulled from the backing. Consequently, when the abrasive particles become dull or the belt loaded, the belt must be replaced. Finer grades result in finer first cuts but slower material removal rates (MRR). This versatile process is now widely used for rapid stock removal as well as fine surface finishing.

FIGURE 28-25   Examples of mounted abrasive wheels and points. *(Courtesy of Norton Company.)*

Belt composition

Grit
Size coat
Glue or resin bond
Backing
Backing—Paper or Cloth (cotton, rayon, polyester)

Grit Size—grade

vs	Approx.	Finish (rms)
24	300	μin.
36	250	"
50	140	"
80	125	"
120	60-80	"
150	40-60	"

Bonds

Name	Make coat	Size coat	Backing
Glue bond	Glue	Glue	Non WP
Modified glue	Mod. glue	Mod. glue	"
Resin over glue	Glue	Resin	"
Resin over resin	Resin	Resin	"
Waterproof	Resin	Resin	WP

WP = waterproof

Platen grinder

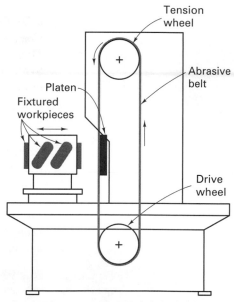

Tension wheel
Platen
Abrasive belt
Fixtured workpieces
Drive wheel

FIGURE 28-26 Belt composition for coated abrasives (top). Platen grinder (right) and examples of belts and disks for abrasive machining.

■ 28.6 HONING

Honing is a stock-removal process that uses fine abrasive stones to remove very small amounts of metal. Cutting speed is much lower than that of grinding. The process is used to size and finish bored holes, remove common errors left by boring (taper, waviness, and tool marks), or remove the tool marks left by grinding. The amount of metal removed is typically about 0.005 in. or less. Although honing is occasionally done by hand, as in finishing the face of a cutting tool, it usually is done with special equipment. Most honing is done on internal cylindrical surfaces, such as automobile cylinder walls. The honing stones are usually held in a honing head, with the stones being held against the work with controlled light pressure. The honing head is not guided externally but, instead, *floats* in the hole, being guided by the work surface (Figure 28-27).

FIGURE 28-27 Schematic of honing head showing the manner in which the stones are held. The rotary and oscillatory motions combine to produce a crosshatched lay pattern. Typical values for V_c and P_s are given below.

For:	Honing Parameters	Conventional Abrasives	Diamonds	CBN
High MRR	V_c (m/min)	20-30	40-70	35-90
	P_s (N/mm²)	1-2	2-8	2-4
Best-quality service	V_c (m/min)	5-30	40-70	20-60
	P_s (N/mm²)	0.5-1.5	1.0-3.0	1.0-2.0

The stones are given a complex motion so as to prevent a single grit from repeating its path over the work surface. Rotation is combined with an oscillatory axial motion. For external and flat surfaces, varying oscillatory motions are used. The length of the motions should be such that the stones extend beyond the work surface at the end. A cutting fluid is used in virtually all honing operations. The critical process parameters are rotational speed, V_r, oscillation speed, V_o, the length and position of stroke, and the honing stick pressure. V_c and the inclination angle are both products of V_o and V_r

Virtually all honing is done with stones made by bonding together various fine artificial abrasives. *Honing stones* differ from grinding wheels in that additional materials, such as sulfur, resin, or wax, are often added to the bonding agent to modify the cutting action. The abrasive grains range in size from 80 to 600 grit. The stones are equally spaced about the periphery of the tool. Reference values for V_c and honing stick pressure, P_s, for various abrasives are shown in Figure 28-27.

Single- and multiple-spindle honing machines are available in both horizontal and vertical types. Some are equipped with special sensitive measuring devices that collapse the honing head when the desired size has been reached.

For honing single, small, internal cylindrical surfaces, a procedure is often used wherein the workpiece is manually held and reciprocated over a rotating hone. If the volume of work is sufficient, honing is a fairly inexpensive process. A complete honing cycle, including loading and unloading the work, is often less than one minute. Size control within 0.0003 in. is achieved routinely.

■ 28.7 SUPERFINISHING

Superfinishing is a variation of honing that is typically used on flat surfaces. The process is:

1. Very light, controlled pressure, 10 to 40 psi
2. Rapid (over 400 cycles per minute), short strokes—less than $1/4$ in.
3. Stroke paths controlled so that a single grit never traverses the same path twice
4. Copious amounts of low-viscosity lubricant-coolant flooded over the work surface

FIGURE 28-28 In superfinishing and honing, a film of lubricant is established between the work and the abrasive stone as the work becomes smoother.

This procedure, illustrated in Figure 28-28, results in surfaces of very uniform, repeatable smoothness.

Superfinishing is based on the phenomenon that a lubricant of a given viscosity will establish and maintain a separating, lubricating film between two mating surfaces if their roughness does not exceed a certain value and if a certain critical pressure, holding them apart, is not exceeded. Consequently, as the minute peaks on a surface are cut away by the honing stone, applied with a controlled pressure, a certain degree of smoothness is achieved. The lubricant establishes a continuous film between the stone and the workpiece and separates them so that no further cutting action occurs. Thus, with a given pressure, lubricant, and honing stone, each workpiece is honed to the same degree of smoothness.

Superfinishing is applied to both cylindrical and plane surfaces. The amount of metal removed usually is less than 0.002 in., most of it being the peaks of the surface roughness. Copious amounts of lubricant-coolant maintain the work at a uniform temperature and wash away all abraded metal particles to prevent scratching.

LAPPING

Lapping is an abrasive surface finishing process wherein fine abrasive particles are *charged* (caused to become embedded) into a soft material, called a *lap*. The material of the lap may range from cloth to cast iron or copper, but it is always softer than the material to be finished, being only a holder for the hard abrasive particles. Lapping is applied to both metals and nonmetals.

As the charged lap is rubbed against a surface, the abrasive particles in the surface of the lap remove small amounts of material from the surface to be machined. Thus the abrasive does the cutting, and the soft lap is not worn away because the abrasive particles become embedded in its surface instead of moving across it. This action always occurs when two materials rub together in the presence of a fine abrasive: the softer one forms a lap, and the harder one is abraded away.

In lapping, the abrasive is usually carried between the lap and the work surface in some sort of a vehicle, such as grease, oil, or water. The abrasive particles are from 120 grit up to the finest powder sizes. As a result, only very small amounts of metal are removed, usually considerably less than 0.001 in. Because it is such a slow metal removing process, lapping is used only to remove scratch marks left by grinding or honing, or to obtain very flat or smooth surfaces, such as are required on gage blocks or for liquid tight seals where high pressures are involved.

Materials of almost any hardness can be lapped. However, it is difficult to lap soft materials because the abrasive tends to become embedded. The most common lap material is fine-grained cast iron. Copper is used quite often and is the common material for lapping diamonds. For lapping hardened metals for metallographic examination, cloth laps are used.

Lapping can be done either by hand or by special machines. In hand lapping, the lap is flat, similar to a surface plate. Grooves are usually cut across the surface of a lap to collect the excess abrasive and chips. The work is moved across the surface of the lap, using an irregular, rotary motion, and is turned frequently to obtain a uniform cutting action.

In lapping machines for obtaining flat surfaces, workpieces are placed loosely in holders and are held against the rotating lap by means of floating heads. The holders, rotating slowly, move the workpieces in an irregular path. When two parallel surfaces are to be produced, two laps may be employed, one rotating below and the other above the workpieces.

Various types of lapping machines are available for lapping round surfaces. A special type of centerless lapping machine is used for lapping small cylindrical parts, such as piston pins and ball-bearing races.

Because the demand for surfaces having only a few micrometers of roughness on hardened materials has become quite common, the use of lapping has increased greatly. However, it is a very slow method of removing metal, obviously costly compared with other methods, and should not be specified unless such a surface is absolutely necessary.

■ 28.8 FREE ABRASIVES

ULTRASONIC MACHINING

Ultrasonic machining (USM), sometimes called *ultrasonic impact grinding*, employs an ultrasonically vibrating tool to impel the abrasives in a slurry at high velocity against the workpiece. The tool is fed into the part as it vibrates along an axis parallel to the tool feed at an amplitude on the order of several thousandths of an inch and a frequency of 20 kHz. As the tool is fed into the workpiece, a negative of the tool is machined into the workpiece. The cutting action is performed by the abrasives in the slurry, which is continuously flooded under the tool. The slurry is loaded up to 60% by weight with abrasive particles. Lighter abrasive loadings are used to facilitate the flow of the slurry for deep drilling (up to 2 in. deep). Boron carbide, aluminum oxide, and silicon carbide are the most commonly used abrasives in grit sizes ranging from 400 to 2000. The amplitude of the vibration should be set approximately to the size of the grit. The process can use shaped tools to cut virtually any material but is most effective on materials with hardnesses greater than R_C 40, including brittle and nonconductive materials such as glass. Figure 28-29 shows a simple schematic of this process.

USM uses piezoelectric or magnetostrictive transducers to impart high-frequency vibrations to the tool holder and tool. Abrasive particles in the slurry are accelerated to great speed by the vibrating tool. The tool materials are usually brass, carbide, mild

FIGURE 28-29   Sinking a hole in a workpiece with an ultrasonically vibrating tool driving an abrasive slurry.

steel, or tool steel and will vary in tool wear depending on their hardness. Wear ratios (workpiece material removed versus tool material lost) from 1:1 (for tool steel) to 100:1 (for glass) are possible. Because of the high number of cyclic loads, the tool must be strong enough to resist fatigue failure.

The cut will be oversize by about twice the size of the abrasive particles being used, and holes will be tapered, usually limiting the hole depth-to-diameter ratio to about 3:1. Surface roughness is controlled by the size of the abrasive particles (finer finish with smaller particles). Holes, slots, or shaped cavities can be readily eroded in any hard material—conductive or nonconductive, metallic, ceramic, or composite. Advantages of the process include that it is one of the few machining methods capable of machining glass. Also, it is the safest machining method. Skin is impervious to the process because of its ductility. High-pitched noise can be a problem due to secondary vibrations. In addition to machining, ultrasonic energy has also been employed for coining, lapping, deburring, and broaching. Plastics can be welded using ultrasonic energy.

WATERJET CUTTING AND ABRASIVE WATERJET MACHINING

Waterjet cutting (WJC), also known as *waterjet machining* or *hydrodynamic machining*, uses a high-velocity fluid jet impinging on the workpiece to perform a slitting operation (Figure 28-30). Water is ejected from a nozzle orifice at high pressure (up to 60,000 psi). The jet is typically 0.003 to 0.020 in. in diameter and exits the orifice at velocities up to 3000 ft/sec. Key process parameters include water pressure, orifice diameter, water flow rate, and working distance (distance between the workpiece and the nozzle).

Nozzle materials include synthetic sapphire, due to its machinability and resistance to wear. Tool life on the order of several hundred hours is typical. Mechanisms for tool failure include chipping from contaminants or constriction due to mineral deposits. This emphasizes the need for high levels of filtration prior to pressure intensification. In the past, long-chain polymers were added to the water to make the jet more coherent (i.e., not come out of the jet dispersed). However, with proper nozzle design, a tight, coherent water jet may be produced without additives.

The advantages of WJC include the ability to cut materials without burning or crushing the material being cut. Figure 28-30 shows a comparison (end view) of cutting corrugated boxboard with a mechanical knife and with WJC. The mechanism for material removal is simply the impinging pressure of the water exceeding the compressive strength of the material. This limits the materials that can be cut by the process to leather, plastics, and other soft nonmetals, which is the major disadvantage of the process. Alternative fluids (alcohol, glycerine, cooking oils) have been used in processing meats, baked goods, and frozen foods. Other disadvantages include that the process is noisy and requires operators to have hearing protection.

The majority of the metalworking applications for waterjet cutting require the addition of abrasives. This process is known as *abrasive waterjet cutting* (AWC). A full range of materials, including metals, plastics, rubber, glass, ceramics, and composites, can be machined by AWC. Cutting feed rates vary from 20 in./min for acoustic tile to 50 in./min for epoxies and 500 in./min for paper products.

Abrasives are added to the waterjet in a mixing chamber on the downstream side of the waterjet orifice. A single, central waterjet with side feeding of abrasives into a mixing chamber is shown in Figure 28-30. In the mixing chamber, the momentum of the water is transferred to the abrasive particles, and the water and particles are forced out through the AWC nozzle orifice, also called the mixing tube. This design can be made quite compact; however, it also experiences rapid wear in the mixing tube. An alternate configuration is to feed the abrasives from the center of the nozzle with a converging set of angled water jets imparting momentum to the abrasives. This nozzle design produces better mixing of the water and abrasives as well as increased nozzle life. The inside diameter of the mixing tube is normally from 0.04 to 0.125 in. in diameter. These tubes are normally made of carbide.

Generally, the kerf of the cut is about 0.001 in. greater than the nozzle orifice. AWC requires control of additional process parameters over waterjet machining, including abrasive material (density, hardness, shape), abrasive size or grit, abrasive flow rate (pounds per minute), abrasive feed mechanism (pressurized or suction), and AWC

Abrasive cutting head

FIGURE 28-30   Schematic diagram of hydrodynamic jet machining. The intensifier elevates the fluid to the desired nozzle pressure while the accumulator smoothes out the pulses in the fluid jet. Schematic of an abrasive waterjet machining nozzle is shown on the right.

nozzle (design, orifice diameter, and material). Typical AWC systems operate under the following conditions: water pressures of 30,000 to 50,000 psi; water orifice diameters from 0.01 to 0.022 in.; and working distances of 0.02 to 0.06 in. Working distances are much smaller than in WJC to minimize the dispersion of the abrasive water jet prior to entering the material. Abrasive materials used include garnet, silica, silicon carbide, or aluminum oxide. Abrasive grit sizes range from 60 to 120 and abrasive flow rates from 0.5 to 3 lb/min. For many applications, the AWC tool is combined with a CNC controlled *X–Y* table, which permits contouring and surface engraving.

AWC can be used to cut any material through the appropriate choice of the abrasive, waterjet pressure, and feed rate. Table 28-6 gives cutting speeds for various metals. The ability of the abrasive waterjet to cut through thick materials (up to 8 in.) is attributed to the reentrainment of abrasive particles in the jet by the workpiece material. AWC is particularly suited for composites because the cutting rates are reasonable and they do not delaminate the layered material. In particular, AWC is used in the airplane industry to cut carbon-fiber composite sections of the airplane after autoclaving.

TABLE 28-6 Typical Values for Through-cutting Speeds for Simple Waterjet and Abrasive Waterjet of Machining Metals and Nonmetals.

Cutting speeds with abrasive waterjet

Material	Thickness (in.)	Nozzle speed (in./min)	Edge quality (comments)	Material	Thickness (in.)	Nozzle speed (in./min)	Edge quality (comments)
Aluminum	0.130	20-40	good	Titanium	2.0	0.5-1.0	125 RMS
Aluminum tube	0.220	50	burred	Tool steel	0.250	3-15	125 RMS
Aluminum casting	0.400	15		Tool steel	1.0	2-5	
Aluminum	0.500	6-10		**Nonmetals**			
Aluminum	3.0	0.5-5		Acrylic	0.375	15-50	good to fair
Aluminum	4.0	0.2-2		C-glass	0.125	100-200	shape dependent
Brass	0.125	18-20	good or small burr	Carbon/carbon comp.	0.125	50-75	good
Brass	0.500	4-5		Carbon/carbon comp.	0.500	10-20	good
Brass	0.75	0.75-3	striations at 1 +	Epoxy/glass composite	0.125	100-250	good
Bronze	1.100	1.0	good	Fiberglass	0.100	150-300	good
Copper	0.125	22	good	Fiberglass	0.250	100-150	good
Copper-nickel	0.125	12-14	fair edge	Glass (plate)	0.063	40-150	good
Copper-nickel	2.0	1.5-4.0	fair edge	Glass (plate)	0.75	10-20	125 RMS
Lead	0.25	10-50	good to striated	Graphite/epoxy	0.250	15-70	good to practical
Lead	2.0	3-8	slower = better	Graphite/epoxy	1.0	3-5	good
Magnesium	0.375	5-15	good	Kevlar (steel reinf.)	0.125	30-50	good
Armor plate	0.200	1.5-15	good	Kevlar	0.375/0.580	10-25	good
Carbon steel	0.250	10-12	good	Kevlar	1.0	3-5	good
Carbon steel	0.750	4-8	good to bad edge	Lexan	0.5	10	good
Carbon steel	3.0	0.4	good w. sm. nozzle	Phenolic	0.25-0.50	10-15	good
4130 carbon steel	0.5	3.0		Plexiglass	0.175/0.50	25	
Mild steel	7.5	0.017-0.05		Rubber belting	0.300	200	good
High-strength steel	3.0	0.38		**Ceramic matrix composites**			
Cast iron	1.5	1.0	good edge	Toughened zirconia	0.250	1.5	
Stainless steel	0.1	10-15	good to striated	SiC fiber in SiC	0.125	1.5	
Stainless steel	0.25	4-12	good to striated	Al$_2$O$_3$/CoCrAly (60%/40%)	0.125	2	
Stainless steel	1.0	1.0	65-150 RMS	SiC/TiB$_2$ (15%)	0.250	0.35	
15-5 PH stainless	4.0	0.3	striated	**Metal matrix composites**			
Inconel 718	1.25	0.5-1.0	good	Mg/B$_4$C (15%)	0.125	35	fair
Inconel	0.250	8-12	good to striated	Al/SiC (15%)	0.500	8-12	good to fair
Inconel	2-2.5	0.2	good to fair	Al/Al$_2$O$_3$ (15%)	0.250	15-20	good to fair
Titanium	0.025-0.050	5-50	good				
Titanium	0.500	1-6	65-150 RMS				

Table 2. Cutting speeds with simple waterjet

Material	Thickness (in.)	Nozzle speed (in./min)	Edge quality (comments)	Material	Thickness (in.)	Nozzle speed (in./min)	Edge quality (comments)
ABS plastic	0.087	20-50	100% separation	Lead	0.125	10	good, slight burr
Aluminum	0.050	2-5	burr	Plexiglass	0.118	30-35	fair
Cardboard	0.055	240-600	slits very well	Printed circuit bd.	0.050-0.125	50-5	good
Delrin	0.500	2-5	good to stringers	PVC	0.250	10-20	good to fair
Fiberglass	0.100	40-150	good to raggy	Rubber	0.050	2400-3600	good
Formica	0.040	1450		Vinyl	0.040	2000-2400	good
Graphite composite	0.060	25		Wood	0.125	40	fair
Kevlar	0.040-0.250	50-3	fair, some furring				

Comment on these tables: In trying to provide data on waterjet and abrasive waterjet cutting we have collected material from diverse sources. But we must note that most of the data presented is not from uniform tests. Also, note that in many cases data was largely absent on such parameters as pump horsepower, waterjet pressure, abrasive-particle rate of flow or type or size, and standoff distance. So these cutting rates vary widely in value—from laboratory control to shop floor ballpark estimates. Many of the top speeds cited either represent cuts made to illustrate speed alone, without regard to surface quality, or may reflect data from machines with very high power output. (*American Machinist, October 1989.*)

ABRASIVE JET MACHINING

One of the least expensive of the nontraditional processes is *abrasive jet machining* (AJM). AJM removes material by a focused jet of abrasives and is similar in many respects to AWC, with the exception that momentum is transferred to the abrasive particles by a jet of inert gas. Abrasive velocities on the order of 1000 ft/sec are possible with AJM. The small mass of the abrasive particles produces a microscale chipping action on the workpiece material. This makes AJM ideal for processing hard, brittle materials, including glass, silicon, tungsten, and ceramics. It is not compatible with soft, elastic materials.

Key process parameters include working distance, abrasive flow rate, gas pressure, and abrasive type. Working distance and feed rate are controlled by hand. If necessary, a hard mask can be placed on the workpiece to control dimensions. Abrasives are typically smaller than those used in AWC. Abrasives are typically not recycled, since the abrasives are cheap and are used only on the order of several hundred grams per hour. To minimize particulate contamination of the work environment, a dust-collection hood should be used in concert with the AJM system.

DESIGN CONSIDERATIONS IN GRINDING

Almost any shape and size of work can be finished on modern grinding equipment, including flat surfaces, straight or tapered cylinders, irregular external and internal surfaces, cams, antifriction bearing races, threads, and gears. For example, the most accurate threads are formed from solid cylindrical blanks on special thread grinding machines. Gears that must operate without play are hardened and then finish ground to close tolerances. Two important design recommendations are to reduce the area to be ground and to keep all surfaces that are to be ground in the same or parallel planes (Figure 28-31). This is an example of *design for manufacturing* (DFM).

Abrasive machining can remove scale as well as parent metal. Large allowances of material, needed to permit conventional metalcutting tools to cut below hard or abrasive inclusions, are not necessary for abrasive machining. An allowance of 0.015 in. is adequate, assuming, of course, that the part is not warped or out of round. This small allowance requirement results in savings in machining time, in material (often 60% less metal is removed), and in shipping of unfinished parts.

Original design of base plate

Original design of crankshaft bearing bracket

Redesigned to reduce weight and grinding time

Redesign eliminated shoulders and made part suitable for grinding in single setup

FIGURE 28-31 Reducing area to be ground and keeping all surface to be ground in the same or parallel planes are two important design recommendations. *(From* Machine Design, *June 1, 1972, p. 87.)*

■ Key Words

abrasive machining	cabic boron nitride (CBN)	grade	silicate bond
abrasive waterjet	cylindrical grinding	grinding	silicon carbide
aluminum oxide	diamond	honing	snagging
attrition	dressing	lapping	surface grinding
centerless grinding	emery	quartz	truing
coated abrasive	friability	resinoid bond	ultrasonic machining
corundum	*G* ratio	rubber bond	vitrified bond
creep feed grinding	garnet	shellac bond	waterjet machining
crush dressing			

■ Review Questions

1. What are machining processes that use abrasive particles for cutting tools called?
2. What is attrition in an abrasive grit?
3. Why is friability an important grit property?
4. Explain the relationship between grit size and surface finish.
5. Why is aluminum oxide used more frequently than silicon carbide as an abrasive?
6. Why is CBN superior to silicon carbide as an abrasive in some applications?
7. What materials commonly are used as bonding agents in grinding wheels?
8. Why is the grade of a bond in a grinding wheel important?
9. How does grade differ from structure in a grinding wheel?
10. What is crush dressing?
11. How does loading differ from glazing?
12. What is meant by the statement that grinding is a mixture of processes?
13. What is accomplished in dressing a grinding wheel?
14. How does abrasive machining differ from ordinary grinding?
15. What is a grinding ratio or *G* ratio?
16. How is the feed of the workpiece controlled in centerless grinding?
17. Why is grain spacing important in grinding wheels?
18. Why should a cutting fluid be used in copious quantities when doing wet grinding?
19. How does plunge-cut grinding compare to cylindrical grinding?
20. If grinding machines are placed among other machine tools, what precautions must be taken?
21. What is the purpose of low-stress grinding?
22. How is low-stress grinding done compared to conventional grinding?
23. The number of grains per square inch which actively contact and cut a surface decreases with increasing grain diameter. Why is this so?
24. Why are centerless grinders so popular in industry compared to center-type grinders?
25. Explain how a SEM micrograph is made. Check the Internet or the library to find the answer.
26. Why are vacuum chucks and magnetic chucks widely used in surface grinding but not in milling?
27. How does creep feed grinding differ from conventional surface grinding?
28. Why does a lap not wear, since it is softer than the material being lapped?
29. How do honing stones differ from grinding wheels?
30. What is meant by "charging" a lap?
31. Why is a honing head permitted to float in a hole that has been bored?
32. How does a coated abrasive differ from an abrasive wheel?
33. Figure out why the bottoms of chips shown in Figure 28-9 are so smooth. The magnification of the micrograph is 4800×. How thick are these chips?
34. What is the inclined angle in honing and what determines it?
35. What are the common causes of grinding accidents?
36. What other machine tool does a surface grinder resemble?
37. Figure 28-11 showed residual stress distributions produced by surface grinding. What is a residual stress?
38. In grinding, what is infeed versus cross feed?
39. One of the problems with waterjet cutting is that the process is very noisy. Why?
40. In AWC, what keeps the abrasive jet from machining the orifice?

■ Problems

1. Perhaps you have observed the following wear phenomena: A set of marble or wooden steps shows wear on the treads in the regions where people step when they climb (or descend) the steps. The higher up the steps, the less the wear on the tread. Given that soles of shoes (leather, rubber) are far softer than marble or granite, explain:
   a. Why and how the steps wear.
   b. Why the lower steps are more worn than the upper steps.

2. Explain why it is that a small particle of a material can be used to abrade a surface made of the same material (i.e., why does the small particle act harder or stronger than the bulk material)?

3. In grinding, both the wheel and workpiece are moving (or rotating). Using the data in Figure 28-11 and assuming that you are doing surface grinding (see Figure 28-1), what are some typical MRR values? How do these compare to MRR values for other machining processes, such as milling? What is the significance of this?

*C*hapter 28 CASE STUDY

Overhead Crane Installation

Nickolas has been given the task of installing an over-head crane (Figure CS 28) in one bay of the von-Turkovich Manufacturing Company and assembly plant. Brackets for the rails of the crane are to be mounted on eight columns, four on each side of the bay area facing each other. The rails for the crane will span four columns. Each bracket on each column will need six holes in a circular pattern. The holes must be accurately spaced within 5 minutes of arc of each other. The axis of the holes must be parallel and normal to the face of the columns. The center of the bolt hole circle must be at a height of at least 20 ft from the floor, but the centers of all eight bolt hole circles must be on the same parallel plane, so that the rails for the crane are level and parallel with each other. Four of the columns along the wall have their faces flush with the wall

surface so that mechanical clamping or attachments cannot be used. The building code will permit no welding of anything to these columns.

1. How would Nickolas proceed to get the bolt holes located in the right position on the beams?
2. How would he get the hole patterns located properly with respect to each other on all eight beams?
3. List the equipment he will need.
4. Make a sketch of any special tool you recommend. (Hint: Check Chapter 25 for drill jig designs and ask your favorite civil engineer for suggestions regarding surveying.)

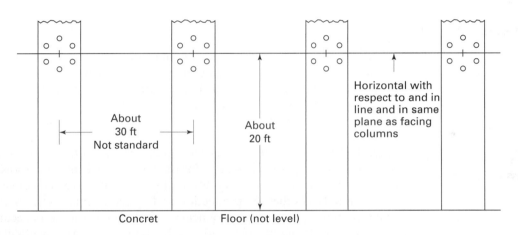

Figure CS 28 Overhead crane installation at the vonTurkovich Manufacturing Company.

CHAPTER 29

THREAD AND GEAR MANUFACTURING

■ 29.1 INTRODUCTION

Screw threads and gears are important machine elements. *Threading*, *thread cutting*, or *thread rolling* refers to the manufacture of threads on external diameters. *Tapping* refers to machining threads in (drilled) holes. Without these processes, our present technological society would come to a grinding halt. More screw threads are made each year than any other machined element. They range in size from those used in small watches to threaded shafts 10 in. in diameter. They are made in quantities ranging from one to several million duplicate threads. Their precision varies from that of inexpensive hardware screws to that of lead screws for the most precise machine tools. Consequently, it is not surprising that many very different procedures have been developed for making screw threads and that the production cost by the various methods varies greatly. Fortunately, some of the most economical methods can provide very accurate results. However, as in the design of most products, the designer can greatly affect the ease and cost of producing specified screw threads. Thus, understanding thread-making processes permits the designer to specify and incorporate screw threads into designs while avoiding needless and excessive cost.

Gears transmit power or motion mechanically between parallel, intersecting, or nonintersecting shafts. Although usually hidden from sight, gears are one of the most important mechanical elements in our civilization, possibly even surpassing the wheel, since most wheels would not be turning were power not being applied to them through gears. They operate at almost unlimited speeds under a wide variety of conditions. Millions are produced each year in sizes from a few millimeters up to more than 30 ft in diameter. Often the requirements that must be, and are routinely, met in their manufacture are amazingly precise. Consequently, many special machines and processes have been developed for producing gears. Let us begin by discussing threads.

SCREW-THREAD STANDARDIZATION AND NOMENCLATURE

A screw thread is a ridge of uniform section in the form of a helix on the external or internal surface of a cylinder, or in the form of a conical spiral on the external or internal surface of a frustrum of a cone. These are called *straight* or *tapered* threads, respectively. Tapered threads are used on pipe joints or other applications where liquid-tight joints are required. Straight threads, on the other hand, are used in a wide variety of applications, most commonly on fastening devices, such as bolts, screws, and nuts, and as integral

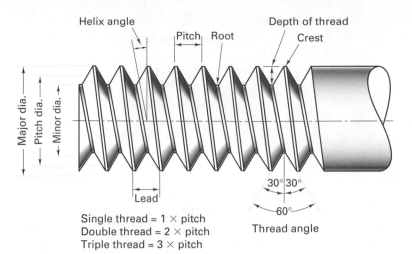

Single thread = 1 × pitch
Double thread = 2 × pitch
Triple thread = 3 × pitch

Thread angle

FIGURE 29-1   Standard screw-thread nomenclature.

elements on parts that are to be fastened together. But, as mentioned previously, they find very important applications in transmitting controlled motion, as in lead screws and precision measuring equipment.

The standard nomenclature for screw-thread components is illustrated in Figure 29-1. The symbol P, the pitch, refers to the distance from a point on one screw thread to the corresponding point on the next thread, measured parallel to the length axis of the part. In 1948, representatives of the United States, Canada, and Great Britain adopted the Unified and American Screw Thread Standards, based on the form shown in Figure 29-2. In 1968 the International Organization for Standards (ISO) recommended the adoption of a set of metric standards based on the basic thread profile. It appears likely that both types of threads will continue to be used for some time to come. In both the *Unified* and *ISO systems*, the crests of external threads may be flat or rounded. The root is usually made rounded to minimize stress concentrations at this critical area. The internal thread has a flat crest in order to mate with either a rounded or V-root of the external thread. A small round is used at the root to provide clearance for the flat crest of the external thread.

In the metric system, the *pitch* is always expressed in millimeters, whereas in the American (Unified) system, it is a fraction having as the numerator 1 and as the denominator the number of threads per inch. Thus, 16 threads per inch, $^1/_{16}$, is a 16 pitch. Consequently, in the Unified system, threads are more commonly described in terms of threads per inch rather than by the pitch.

While all elements of the thread form are based on the *pitch diameter*, screw-thread sizes are expressed in terms of the *outside*, or *major* diameter and the *pitch* or *number of threads per inch*. In threaded elements, *lead* refers to the axial advance of the element during one revolution; therefore, lead equals pitch on a single-thread screw.

TYPES OF SCREW THREADS

Eleven types, or series, of threads are of commercial importance, several having equivalent series in the metric system and Unified systems. See Figure 29-3. As has been indicated, the Unified threads are available in a coarse (UNC and NC), fine (UNF and NF), extra-fine (UNEF and NEF), and three-"pitch" (8, 12, and 16) series, the number of threads per inch in accordance with an arbitrary determination based on the major diameter.

Many nations have now adopted ISO threads into their national standards. Besides metric ISO threads, there are also inch-based ISO threads, namely the UN series with which people in the United States, Canada, and Great Britain are familiar. ISO offers a wide range of metric sizes. Individual countries have the choice of accepting all or a selection of the ISO offerings.

The size listings of metric threads start with "M" and continue with the outside diameter in millimeters. Most ISO metric thread sizes come in coarse, medium, and fine

(a) External thread

(b) Internal thread

(c) Metric ISO/R68-1969

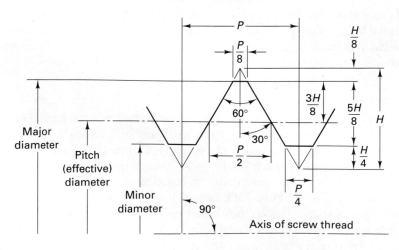

D = Major diameter of thread of nut ⎫ Normal
d = Major diameter of screw ⎭ diameter
D_1 = Minor diameter of thread of nut
d_1 = Minor diameter of screw
D_2 = Pitch diameter of thread of nut
d_2 = Pitch diameter of screw
H = Height of the complete theoretical thread profile
H_1 = Engagement
P = Pitch

$$H = \frac{\sqrt{3}}{2}P$$

FIGURE 29-2   Basic profiles of Unified and American screw threads: (a) external, (b) internal, and (c) metric.

Types of Screw Threads

1. *Coarse-thread series* (UNC and NC). For general use where not subjected to vibration.

2. *Fine-thread series* (UNF and NF). For most automotive and aircraft work.

3. *Extra-fine-thread series* (UNEF and NEF). For use with thin-walled material or where a maximum number of threads are required in a given length.

4. *Eight-thread series* (8UN and 8N). Eight threads per inch for all diameters from 1 to 6 in. Used primarily for bolts on pipe flanges and cylinder-head studs where an initial tension must be set up to resist steam or air pressures.

5. *Twelve-thread series* (12UN and 12N). Twelve threads per inch for diameters from $\frac{1}{2}$ through 6 in. Not used extensively.

6. *Sixteen-thread series* (16UN and 16N). Sixteen threads per inch for diameters from $\frac{3}{4}$ through 6 in. Used for a wide variety of applications that require a fine thread.

7. *American Acme thread.* This thread and the following three are used primarily in transmitting power and motion.

8. *Buttress thread.*

9. *Square thread.*

10. *29° Worm thread.*

11. *American, standard pipe thread.* This thread is the standard tapered thread used on pipe joints in this country. The taper on all pipe threads is $\frac{3}{4}$ in./ft.

FIGURE 29-3   Types of screw threads.

pitches. When a coarse thread is designated, it is not necessary to spell out the pitch. For example, a coarse 10-mm outside diameter (OD) thread is called out as "M 12." This thread has a pitch of 1.75 mm, but the pitch may be omitted from the call-out. A fine 12-mm OD thread is available. It has a 1.25-mm pitch and must be designated "M12 x 1.25" an extra-fine 12-mm OD thread having 0.75 mm-pitch would receive the designation "M 12 x 0.75."

The symbol "x" is not employed as a multiplication symbol in metric practice; rather, it is used to relate these two attributes of the threads. The full description of a thread fastener obviously includes information beyond the thread specification. Head type, length, length of thread, design of end, thread runout, heat treatment, applied finishes, and other data may be needed to fully specify a bolt besides the designation of the thread. The "x" should not be used to separate any of the other characteristics.

Here is another example of an ISO thread designation:

metric thread designation M6 x 0.75-5g6g

In the ISO system, tolerances are applied to "positions" and "grades." Tolerance positions denote the limits of pitch and crest diameters, using "e" (large), "g" (small), and "H" (no allowance) for internal threads. The grade is expressed by numerals 3 through 9. Grade 6 is roughly equivalent to U.S. grades 2A and B, medium-quality, general-purpose threads. Below 6 is fine quality and/or short engagement. Above 6 is coarse quality and/or long length of engagement.

In the Unified system, screw threads are designated by symbols as follows:

$$\frac{3}{8}'' - 16\,\text{UNC} - 2A$$

- class of thread
- thread series
- number of threads per inch
- nominal size

This type of designation applies to right-hand threads. For left-hand threads, the letters LH are added after the class of thread symbol.

In the Unified system, manufacturing tolerances are specified by three classes. Class 1 is for ordnance and other special applications. Class 2 threads are the normal production grade, and Class 3 threads have minimum tolerances where tight fits are required. The letters A and B are added after the class numerals to indicate external and internal threads, respectively.

The availability of fasteners, particularly nuts, containing plastic inserts to make them self-locking and thus able to resist loosening due to vibration, and the use of special coatings that serve the same purpose, have resulted in less use of finer-thread-series fasteners in mass production. Coarser-thread fasteners are easier to assemble and less subject to cross-threading (binding).

■ 29.2 THREAD MAKING

Three basic methods are used to produce threads: *cutting*, *rolling*, and *casting*. Although both external and internal threads can be cast, relatively few are made in this manner, primarily in connection with die casting, investment casting, or the molding of plastics. Today, by far the largest number of threads are made by rolling. Both external and internal threads can be made by rolling, but the material must be ductile. Because rolling is a less flexible process than thread cutting, it is restricted essentially to standardized and simple parts. Consequently, large numbers of external and internal threads still are made by cutting processes, including grinding and tapping.

External Thread Cutting Methods	Internal Thread Cutting Methods
Threading on an engine lathe	Threading (on an engine lathe or NC lathe)
Threading on an NC lathe	With a tap and holder (manual NC, machine,
With a die held in a stock (manual)	semiautomatic, or automatic)
With an automatic die (turret lathe	With a collapsible tap (turret lathe,
or screw machine) or NC lathe	screw machine, or special threading machine)
By milling	By milling
By grinding	

CUTTING THREADS ON A LATHE

Lathes provided the first method for cutting threads by machine. Although most threads are now produced by other methods, lathes still provide the most versatile and fundamentally simple method. Consequently, they often are used for cutting threads on special workpieces where the configuration or nonstandard size does not permit them to be made by less costly methods.

There are two basic requirements for thread cutting on a lathe. First, an accurately shaped and properly mounted tool is needed because thread cutting is a form-cutting operation. The resulting thread profile is determined by the shape of the tool and its position relative to the workpiece. Second, the tool must move longitudinally in a specific relationship to the rotation of the workpiece, because this determines the *lead* of the thread. This requirement is met through the use of the *lead screw* and the *split nut*, which provide positive motion of the carriage relative to the rotation of the spindle.

To cut a thread, it is also essential that a constant positional relationship be maintained among the workpiece, the cutting tool, and the lead screw. If this is not done, the tool will not be positioned correctly in the thread space on successive cuts. Correct relationship is obtained by means of a *threading dial* (Figure 29-4), which is driven directly by the lead screw through a worm gear. Because the workpiece and the lead screw are directly connected, the threading dial provides a means for establishing the desired positional relationship between the workpiece and the cutting tool. The threading dial is graduated into an even number of major and half divisions. If the feed mechanism is engaged in accordance with the following rules, correct positioning of the tool will result:

1. *For even-number threads:* at any line on the dial
2. *For odd-number threads:* at any numbered line on the dial
3. *For threads involving $\frac{1}{2}$ numbers:* at any odd-numbered line on the dial
4. *For $\frac{1}{4}$ or $\frac{1}{8}$ threads:* return to the original starting line on the dial

To start cutting a thread, the tool usually is fed inward until it just scratches the work, and the cross-slide dial reading is then noted or set at zero. The split nut is engaged and the tool permitted to run over the desired thread length. When the tool reaches the end of the thread, it is quickly withdrawn by means of the cross-slide control. The split nut is then disengaged and the carriage returned to the starting position, where the tool is clear of the workpiece. At this point the future thread will be indicated by a fine scratch line. This permits the operator to check the thread lead by means of a scale or thread gage to assure that all settings have been made correctly.

FIGURE 29-4 Cutting a screw thread on a lathe, showing the method of supporting the work and the relationship of the tool to the work with the compound swiveled. Inset shows face of threading dial.

Next, the tool is returned to its initial zero depth position by returning the cross slide to the zero setting. By using the compound rest, the tool can be moved inward the proper depth for the first cut. A depth of 0.010 to 0.025 in. is usually used for the first cut and smaller amounts on each successive cut, until the final cut is made with a depth of only 0.001 to 0.003 in. to produce a good finish. When the thread has been cut nearly to its full depth, it is checked for size by means of a mating nut or thread gage. Cutting is continued until a proper fit is obtained.

Figure 29-4 illustrates two methods of feeding the tool into the work. If the tool is fed radially, cutting takes place simultaneously on both sides of the tool. With this true form-cutting procedure, no rake should be ground on the tool, and the top of the tool must be horizontal and be set exactly in line with the axis of rotation of the work. Otherwise, the resulting thread profile will not be correct. An obvious disadvantage of this method is that the absence of side and back rake results in poor cutting (except on cast iron or brass). The surface finish on steel will usually be poor. Consequently, the second method commonly is used, with the compound swivelled 20°. The cutting then occurs primarily on the left-hand edge of the tool, and some side rake can be provided.

Proper speed ratio between the spindle and the lead screw is set by means of the gear-change box. Modern industrial lathes have ranges of ratios available so that nearly all standard threads can be cut merely by setting the proper levers on the quick-change gear box.

FIGURE 29-5 Canned subroutines called G codes are used on CNC lathes to produce threads. See Chapter 26 for CNC discussion.

X_a	Specifies the absolute X coordinate of the tool after axial infeed.
$G32$	Initiates the single-pass threading cycle.
Z_a	Specifies the absolute Z coordinate of the tool after the threading pass.
Fn	n specifies the feed rate
$X_s Z_s$	Specifies the absolute X and Z coordinates of the start point.

Cutting screw threads on a lathe is a slow, repetitious process that requires considerable operator skill. The cutting speeds usually employed are from one-third to one-half of regular speeds to enable the operator to have time to manipulate the controls and to ensure better cutting. The cost per part can be high, which explains why other methods are used whenever possible.

CUTTING THREADS ON A CNC LATHE

Computer numerical control (CNC) lathes and turning centers can be programmed to machine straight, tapered, or scroll threads. Threads machined using the same type of special tool have the thread shape shown in Figure 29-4. The tool is positioned at a specific starting distance from the end of the work (Figure 29-5). This distance will vary from machine to machine. Its value can be found in the machine's programming manual. The CNC software will have a set of preprogrammed machining routines (called G codes) specifically for threading. Beginning at the start point, the tool accelerates to the feed rate required to cut the threads. The tool creates the thread shape by repeatedly following the same path as axial infeed is applied. For standard V-threads, the infeed can be applied along a 0° or 29° angle. The depth of cut for the first pass is the largest. The cutting depth is then decreased for each successive pass until the required thread depth is achieved. A final finishing pass is then made with the tool set at the thread depth.

CUTTING THREADS WITH DIES

Straight and tapered external threads up to about $1\frac{1}{2}$ in. in diameter can be manually cut quickly by means of threading dies (Figure 29-6a). Basically, these dies are similar to hardened, threaded nuts with multiple cutting edges. The cutting edges at the starting end

FIGURE 29-6 (a) Solid threading die; (b) solid-adjustable threading die; (c) threading-die stock for round die (die removed). *(Courtesy of TRW-Greenfield Tap & Die.)*

are beveled to aid in starting the dies on the workpiece. As a consequence, a few threads at the inner end of the workpiece are not cut to full depth. Such threading dies are made of carbon or high-speed tool steel.

Solid-type dies are seldom used in manufacturing because they have no provision for compensating for wear. The solid-adjustable type (Figure 29-6b) is split and can be adjusted over a small range by means of a screw to compensate for wear or to provide a variation in the fit of the resulting screw thread. These types of threading dies are usually held in a *stock* for hand rotation. A suitable lubricant is desirable to produce a smoother thread and to prolong the life of the die, since there is extensive friction during the cutting process.

SELF-OPENING DIE HEADS

A major disadvantage of solid-type threading dies is that they must be unscrewed from the workpiece to remove them. They are therefore not suitable for use on high-speed, production-type machines, and *self-opening die heads* are used instead on turret lathes, screw machines, numerically controlled (NC) lathes, and special threading machines for cutting external threads.

There are three types of self-opening die heads, all having four sets of adjustable, multiple-point cutters that can be removed for sharpening or for interchanging for different thread sizes. This permits one head to be used for a range of thread sizes (see Figure 29-7). The cutters can be positioned radially or tangentially, resulting in less tool flank contact and friction rubbing. In some self-opening die heads, the cutters are circular, with an interruption in the circular form to provide an easily sharpened cutting face. The cutters are mounted on the holder at an angle equal to the helix angle of the thread.

As the name implies, the cutters in self-opening die heads are arranged to open automatically when the thread has been cut to the desired length, thereby permitting the die head to be quickly withdrawn from the workpiece. On die heads used on turret lathes, the operator must usually reset the cutters in the closed position before making the next thread. The die heads used on screw machines and automatic threading machines are provided with a mechanism that automatically closes the cutters after the heads are withdrawn.

Cutting threads by means of self-opening die heads is frequently called *thread chasing*. However, some people apply this term to other methods of thread cutting, even to cutting a thread in a lathe.

■ 29.3 INTERNAL THREAD CUTTING–TAPPING

The cutting of an internal thread by means of a multiple-point tool is called *thread tapping*, and the tool is called a *tap*. A hole of diameter slightly larger than the minor diameter of the thread must already exist, made by drilling/reaming, boring, or die casting. For small holes, solid *hand taps* (Figure 29-8) are usually used. The flutes create cutting edges on the thread profile and provide space for the chips and the passage of cutting fluid. Such taps are made of either carbon or high-speed steel and are now routinely coated with TiN. The flutes can be either straight, helical, or spiral.

Hand taps (Figure 29-8) have square shanks and are usually made in sets of three. The *taper tap* has a tapered end that will enter the hole a sufficient distance to help align the tap. In addition, the threads increase gradually to full depth, and therefore this type of tap requires less torque to use. However, only a through-hole can be threaded completely with a taper tap because it cuts to full depth only behind the tapered portion. A blind hole can be threaded to the bottom using three types of taps in succession. After the taper tap has the thread started in proper alignment, a *plug tap*, which has only a few tapered threads to provide gradual cutting of the threads to depth, is used to cut the threads as deep into the hole as its shape will permit. A *bottoming tap*, having no tapered threads, is used to finish the few remaining threads at the bottom of the hole to full depth. Obviously, producing threads to the full depth of a blind hole is time-consuming, and it also frequently results in broken taps and defective workpieces. Such configurations usually can be avoided if designers will give reasonable thought to the matter.

(a)

(b)

(c)

(d)

FIGURE 29-7   Self-opening die heads, with (a) radial cutter, (b) tangential cutters, (c) circular cutters, and (d) terminology of circular chasers and their relation to the work. *(Courtesy of Geometric Tool Company, Warner & Sawsey Company, National Acme Company, and TRW-Greenfield Tap & Die, respectively.)*

Taps operate under very severe conditions because of the heavy friction (high torque) involved and the difficulty of chip removal. Also, taps are relatively fragile. *Spiral-fluted taps* (Figure 29-9) provide better removal of chips from a hole, particularly in tapping materials that produce long, curling chips. They are also helpful in tapping holes where the cutting action is interrupted by slots or keyways. The *spiral point* cuts the thread with a shearing action that pushes the chips ahead of the tap so that they do not interfere with the cutting action and the flow of cutting fluid into the hole.

COLLAPSING TAPS

Collapsing taps are similar to self-opening die heads in that the cutting elements collapse inward automatically when the thread is completed. This permits withdrawing the tap from the workpiece without the necessity of unscrewing it from the thread. They can either be self-setting, for use on automatic machines, or require manual setting for each cycle. Figure 29-10 shows some of the types available.

HOLE PREPARATION

Drilling is the most common method of preparing holes for tapping, and when close control over hole size is required, reaming may also be necessary. The drill size determines the final thread contour and the drilling torque. Unless otherwise specified, the tap drill size for most materials should produce approximately 75% thread, that is, 75% of full thread depth.

FIGURE 29-8 Terminology for a plug tap with photographs of taper (t), plug (p), and bottoming (b) taps, which are used serially in threading holes. *(Courtesy of TRW-Greenfield Tap & Die.)*

FIGURE 29-9 (Left to right) Spiral-fluted tap; spiral-point tap; spiral-point tap cutting chips; fluteless bottoming tap and fluteless plug tap for cold-forming internal threads; cross section of fluteless forming tap. *(Courtesy of TRW-Greenfield Tap & Die.)*

FIGURE 29-10 Solid-adjustable and collapsible taps. *(Courtesy of Teledyne Landis Machine.)*

Solid adjustable   ALT collapsible   2LLS receding collapsible

LL receding collapsible   ELT collapsible   CBLM receding collapsible

TAPPING IN MACHINE TOOLS

Solid taps also are used in tapping operations on machine tools, such as lathes, drill presses, and special tapping machines. In tapping on a drill press, a tapping attachment is often used. These devices rotate the tap slowly when the drill press spindle is fed downward against the work. When the tapping is completed and the spindle raised, the tap is automatically driven in the reverse direction at a higher speed to reduce the time required to back the tap out of the hole. Some modern machine tools provide for extremely fast spindle reversal for backing taps out of holes.

When solid taps are used on a screw machine or turret, the tap is prevented from turning while it is being fed into the work. As the tap reaches the end of the hole, the tap is free to rotate with the work. The work is then reversed and the tap, again prevented from rotating, is backed out of the hole.

The machine should have adequate power, rigidity, speed and feed ranges, cutting fluid supply, and positive drive action. Chucks, tap holders, and collets should be checked regularly for signs of wear or damage. Accurate alignment of the tap holder, machine spindle, and workpiece is vital to avoid broken taps or bell-mouthed, tapered, or oversized holes.

TAPPING CUTTING TIME

The equation to calculate the cutting time for tapping is (approximately)

$$T_m = \frac{Ln}{N} = \frac{\pi DLn}{12V} + A_L + A_R \qquad (29\text{-}1)$$

where

N = spindle rpm

T_m = cutting time (min)

D = tap diameter (in.)

L = depth of tapped hole or length of cut (in.)

N = number of threads per inch (tpi)(feed rate)

V = cutting speed (sfpm)

A_L = allowance to start the tap (min)

A_R = allowance to withdraw the tap (min)

SPECIAL THREADING AND TAPPING MACHINES

Special machines are available for production threading and tapping. Threading machines usually have one or more spindles on which a self-opening die head is mounted, with suitable means for clamping and feeding the workpiece. Special tapping machines using self-collapsing taps substituted for the threading dies are also available. More commonly, tapping machines resemble drill presses, modified to provide spindle feeds both upward and downward, with the speed and feed more rapid on the upward motion.

COMMON TAPPING PROBLEMS

Tap overloading is often caused by poor lubrication, lands that are too wide, chips packed in the flutes, or tap wear. Surface roughness in the threads has many causes. A negative grind on the heel will prevent the tap from tearing the threads when backing out.

When a tap loses speed or needs more power, it generally indicates that the tap is dull (or improperly ground) or the chips are packed in flutes (loaded). The flutes may be too shallow or the lands too deep. When tapping soft ductile metals, loading can usually be overcome by polishing the tap before usage.

Improper hole size due to drill wear increases the percentage of threads being cut. Dull tools can also produce a rough finish or workharden the hole surface and cause the tap to dull more quickly. Check to see that the axis of the hole and tap are aligned. If the tap cuts when backing out, check to see if the hole is oversize.

TAPPING HIGH-STRENGTH MATERIALS

High-strength, thermal-resistant materials, sometimes called "exotics," cause special problems in tapping. A variety of materials are classified as exotics: stainless steel, precipitation-hardened stainless steel, high-alloy steels, iron-based superalloys, titanium, Inconcel, Hastelloy, Monel, and Waspalloy. Their most important attribute is their high strength-to-weight ratio.

Each material presents different problems to efficient tapping, but they all share certain similarities. Toughness and general abrasiveness top the list. It is also difficult to impart a good surface finish to exotics; heat tends to localize in the shear zone, and exotics tend to workharden and grab the tool.

A tap's chamfer and first full thread do virtually all the cutting. The remaining ground threads serve merely as chasers. Because of this, taps to thread exotics are increasingly manufactured with short threads and reduced necks. They diminish problems caused by material closure and provide more space for coolant and chip ejection.

When tapping exotics, the largest tap core diameter possible should be applied. Cutting 75 or 65% threads places less stress on the taps, lengthening tool life and reducing breakage. To cut threads in exotic alloys successfully, taps must combine geometries specifically tailored for those materials and be made of premium tool steels subjected to precisely controlled heat-treatment processes.

Stainless steel is known to workharden and to have slow heat-dissipation characteristics; stainless steel requires a tap geometry with a positive 6° to 9° rake, preventing workhardening and reducing torque. Grinding an appropriate eccentric thread and back-taper relief onto the tap will reduce friction. A surface treatment promotes lubricity. That is, the tool should be made from high-vanadium, high-cobalt tool steel and have a surface treatment so that coolant adheres to it. Stainless steel generates long chips and requires a tap with a 38° helix angle and adequate flute depth to promote chip evacuation. Proper hook and radial relief guarantee accurate thread-hole size and long tool life.

When tapping a titanium alloy, the material's tendency to concentrate heat in a small contact area must be considered. Concentrated heat often leads to excessive cutting-edge wear. Titanium generates average-to-short chips, is abrasive, and is prone to chip welding and high friction. These characteristics degrade tap performance and shorten tool life.

Taps for threading titanium are constructed of premium tool steel, nitrided for hardness. Titanium nitride (TiN) coatings cannot be used because they react chemically with the workpiece material, causing rapid tap failure. For tapping through-holes, a 3° to 5° rake and short thread design with high eccentric relief will promote efficient chip evacuation.

Nickel-based Inconel, Monel, Waspalloy, and Hastelloy present severe tapping problems. Among their machining characteristics are toughness, workhardening, heat retention, and built-up-edge (BUE).Taps designed to thread these materials need tremendous stability and a strong cross-sectional construction. The most popular tap materials for nickel alloys are high-vanadium, high-cobalt tool steel or powdered metal (PM) tool steel. Tapping blind holes in these alloys requires a 3° to 5° rake to shear and deflect the cutting forces downward toward the tap's root, its strongest area. A 26° helix angle promotes chip evacuation, and a nitride or TiN coating reduces friction and tool wear. Because of nickel alloys' toughness, taps to cut them should have the longest taper possible. This allows the cutting edges to progressively gain thread height before the first full thread begins its cut, distributing the load over a wider area. Spiral-pointed, straight-flute taps have a four- to five-thread taper. For tapping blind holes, the first two or three threads—more if possible—should be tapered.

CUTTING FLUID FOR TAPPING

Cutting fluids should be kept as clean as possible and should be supplied in copious quantities to reduce heat and friction and to aid in chip removal. Long tap life has been reported to result from routing high-pressure coolants through the tap to flush out the chips and cool the cutting edges. Recommended cutting fluids are listed in Table 29-1.

■ 29.4 THREAD MILLING

Highly accurate threads, particularly in larger sizes, are often form milled. Either a single- or a multiple-form cutter may be used. A single-form cutter having a single annular row of teeth is tilted at an angle equal to the helix angle of the thread and is fed inward radially to full depth while the work is stationary. The workpiece then is rotated slowly, and the cutter simultaneously is moved longitudinally, parallel with the axis of the work (or vice versa), by means of a lead screw, until the thread is completed. The thread can be completed in a single cut, or roughing and finish cuts can be used. This process is used primarily for large-lead or multiple-lead threads.

Some threads can be milled more quickly by using a multiple-form cutter having multiple rows of teeth set perpendicular to the cutter axis (the rows having no lead). The cutter must be slightly longer than the thread to be cut. It is set parallel with the axis of the workpiece and fed inward to full-thread depth while the work is stationary. The work then is rotated slowly for a little over one revolution, and the rotating cutter is simultaneously moved longitudinally with respect to the workpiece (or vice versa) according to

TABLE 29-1	Cutting Fluids for Tapping (HSS Tools)
Work Material	Cutting Fluid
Aluminum	Kerosene and lard oil; kerosene and light-base oil
Brass	Soluble oil or light-base oil
Naval brass	Mineral oil with lard or light-base oil
Manganese bronze	Mineral oil with lard or light-base oil
Phosphor bronze	Mineral oil with lard or light-base oil
Copper	Mineral oil with lard or light-base oil
Iron, cast malleable	Dry or soluble oil
	Soluble oil or sulfur-base oil
Magnesium	Light-base oil diluted with kerosene
Monel metal	Sulfur-base oil
Steels:	
Up to 0.25 carbon	Sulfur-base or soluble oil
Free machining	Sulfur-base or soluble oil
0.30–0.60 carbon, annealed	Chlorinated sulfur-base oil
0.30–0.60 carbon, heat treated	Sulfur-base oil
Tool, high-carbon, HSS	Chlorinated sulfur-base oil
Stainless	Chlorinated sulfur-base oil
Titanium	Chlorinated sulfur-base oil
Zinc die castings	Kerosene and lard oil

the thread lead. When the work has revolved one revolution, the thread is complete. This process cannot be used on threads having a helix angle greater than about 3°, because clearance between the sides of the threads and the cutter depends on the cutter diameter's being substantially less than that of the workpiece. Thus, although the process is rapid, its use is restricted to threads of substantial diameter and not more than about 2 in. long.

As shown in Figure 29-11, advances in CNC computer controls have led to thread milling on three-axis machines. Today's CNC can helically interpolate the axial feed controlling the thread pitch with circular feed controlling the circumference of the thread.

Thread milling on machining centers with multitooth indexable-carbide-insert cutters was introduced about twenty years ago. The cutter can produce a finished thread in one helical pass.

FIGURE 29-11 Thread milling on a three-axis NC machine can produce a complete thread in a single feed revolution. *(Fred Mason, American Machinist, November, 1988.)*

The cutter has teeth shaped like the desired thread form. The cutter rotates at high speeds while its axis slowly moves around the part in a planetary arc just over 360°. The cutter advances axially a distance equal to one pitch to generate the helical path. Thread milling has advantages in diameters over 1.5 in., including better surface finish and concentricity and the ability to produce right- or left-hand threads with the same tool. Thrilling (drilling plus threading) produces threaded holes by combining short hole drilling with thread milling using a combination tool with a drill point and a thread mill body. The details of the process are shown in Figure 29-12 and can be done on any CNC machining center. Compared to tapping, the process eliminates two tools, two tool holders, and two tool change cycles as the single tool combines the drill ream and tap functions into one tool. Threaded-hole depths are limited to about three hole diameters.

■ 29.5 THREAD GRINDING

Grinding can produce very accurate threads, and it also permits threads to be produced in hardened materials. Three basic methods are used. *Center-type grinding with axial feed* is the most common method, being similar to cutting a thread on a lathe. A shaped grinding wheel replaces the single-point tool. Usually, a single-ribbed grinding wheel is employed, but multiple-ribbed wheels are used occasionally. The grinding wheels are shaped by special diamond dressers or by crush dressing and must be inclined to the helix angle of the thread. Wheel speeds are in the high range. Several passes are usually required to complete the thread.

    Center-type infeed thread grinding is similar to multiple-form milling in that a multiple-ribbed wheel, as wide as the length of the desired thread, is used. The wheel is fed inward radially to full thread depth, and the thread blank is then turned through about $1\frac{1}{2}$ turns as the grinding wheel is fed axially a little more than the width of one thread. *Centerless thread grinding* is used for making headless setscrews. The blanks are hopper fed to the regulating wheel, which causes them to traverse the grinding wheel face, from which they emerge in completed form. Production rates of 60 to 70 screws of $\frac{1}{2}$–in. length per minute are possible.

■ 29.6 THREAD ROLLING

Thread rolling is used to produce threads in substantial quantities. This is a cold-forming process operation in which the threads are formed by rolling a thread blank between hardened dies that cause the metal to flow radially into the desired shape. Because no metal is removed in the form of chips, less material is required, resulting

FIGURE 29-12 The process of high-speed thrilling (drilling plus threading) a hole includes (1) approach, (2) drill plus chamfer, (3) retract one thread pitch, (4) radially ramp to the major thread diameter, (5) thread-mill with helical interpolation, (6) return the tool to the centerline of the hole, and (7) retract from the finished hole. At 20,000 rpm, a hole can be thrilled in aluminum in less than two seconds. *(Fred Mason, American Machinist, November, 1988.)*

in substantial savings. In addition, because of cold working, the threads have greater strength than cut threads, and a smoother, harder, and more wear-resistant surface is obtained. In addition, the process is fast, with production rates of one per second being common. The quality of cold-rolled products is consistently good. Chipless operations are cleaner and there is a savings in material (15% to 20% savings in blank stock weight is typical).

Thread rolling is done by four basic methods. The simplest of these employs one fixed and one movable flat rolling die (Figure 29-13). After the blank is placed in position on the stationary die, movement of the moving die causes the blank to be rolled between the two dies and the metal in the blank is displaced to form the threads. As the blank rolls, it moves across the die parallel with its longitudinal axis. Prior to the end of the stroke of the moving die, the blank rolls off the end of the stationary die, its thread being completed.

One obvious characteristic of a rolled thread is that its major diameter always is greater than the diameter of the blank. When an accurate class of fit is desired, the diameter of the blank is made about 0.002 in. larger than the thread-pitch diameter. If it is desired to have the body of a bolt larger than the outside diameter of the rolled thread, the blank for the thread is made smaller than the body.

Thread rolling can be done with cylindrical dies. Figure 29-13 illustrates the three-roll method commonly employed on turret lathes and screw machines. Two variations are used. In one, the rolls are retracted while the blank is placed in position. The rolls then move inward radially, while rotating, to form the thread. More commonly the three rolls are contained in a self-opening die head similar to the conventional type used for cutting external threads. The die head is fed onto the blank longitudinally and forms the thread progressively as the blank rotates. With this procedure, as in the case of cut threads, the innermost $1^1/_2$ to 2 threads are not formed to full depth because of the progressive action of the rollers.

The two-roll method is commonly employed for automatically producing large quantities of externally threaded parts up to 6 in. in diameter and 20 in. in length. The planetary-type machine is for mass production of rolled threads on diameters up to 1 in. Not only is thread rolling very economical, the threads are excellent as to form and strength. The cold working contributes to increased strength, particularly at the critical root areas. There is less likelihood of surface defects (produced by machining), which can act as stress raisers.

Large numbers of threads are rolled on thin, tubular products. In this case external and internal rolls are used. The threads on electric lamp bases and sockets are examples of this type of thread.

CHIPLESS TAPPING

Unfortunately, most internal threads cannot be made by rolling; there is insufficient space within the hole to permit the required rolls to be arranged and supported, and the required forces are too high. However, many internal threads, up to about $1^1/_2$ in. in diameter, are coldformed in holes in ductile metals by means of *fluteless taps*. Such a tap and its special cross section are shown in Figure 29-9. The forming action is essentially the same as in rolling external threads. Because of the forming involved and the high friction, the torque required is about double that for cutting taps. Also, the hole diameter must be controlled carefully to obtain full thread depth without excessive torque. However, fluteless taps produce somewhat better accuracy than cutting taps, and tap life is often greater than that of high-speed-steel (HSS) machine taps. A lubricating fluid should be used, water-soluble oils being quite effective. Fluteless taps are especially suitable for forming threads in dead-end holes because no chips are produced. They come in both plug and bottoming types.

MACHINING VERSUS ROLLING THREADS

Threads are machined or cut when full thread depth is needed (more than one pass necessary) for short production runs, when the blanks are not very accurate, when proximity to the shoulder in end threading is needed, for tapered threads, or when the workpiece material is not adaptable for rolling.

FIGURE 29-13 Roll forming threads using flat die thread rolling process shown in (a) and (b). The threads forming action is shown in (c) and the product in (d). Three variations of cylindrical rolling are shown in (e), (f), and (g).

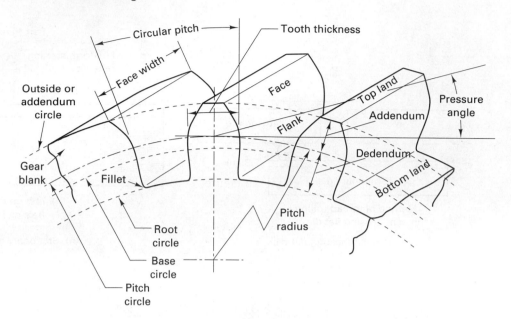

FIGURE 29-14 Gear-tooth nomenclature.

■ 29.7 GEAR MAKING

As with threads, we need to have an introductory understanding of the product before we can understand the process.

GEAR THEORY AND TERMINOLOGY

Basically, gears are modifications of wheels, with *gear teeth* added to prevent slipping and to ensure that their relative motions are constant. However, it should be noted that the relative surface velocities of the wheels (and shafts) are determined by the diameters of the wheels. Although wooden teeth or pegs were attached to disks to make gears in ancient times, the teeth of modern gears are produced by machining or forming teeth on the outer portion of the wheel. The *pitch circle* (Figures 29-14 and 29-15) corresponds to the diameter of the wheel. Thus the angular velocity of a gear is determined by the diameter of this imaginary pitch circle. All design calculations relating to gear performance are based on the pitch-circle diameter or, more simply, the *pitch diameter* (PD).

FIGURE 29-15 Tangent pitch circles between two gears produce a pitch point.

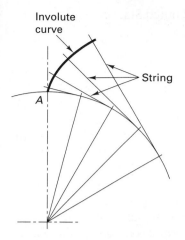

Involute curve

String

A

FIGURE 29-16 Method of generating an involute curve by unwinding a string from a cylinder.

For two gears to operate properly, their pitch circles must be tangential to each other. The point at which the two pitch circles are tangent, at which they intersect the centerline connecting their centers of rotation, is called the *pitch point*. The common normal at the point of contact of mating teeth must pass through the pitch point. This condition is illustrated in Figure 29-15.

To provide uniform pressure and motion and to minimize friction and wear, gears are designed to have rolling motion between mating teeth rather than sliding motion. To achieve this condition, most gears utilize a tooth form that is based on an *involute curve*. This is the curve that is generated by a *point* on a straight line when the line rolls around a *base circle*. A somewhat simpler method of developing an involute curve is that shown in Figure 29-16. By unwinding a tautly held string from a point on the base circle, point *A*, an involute curve is generated.

There are three other reasons for using the involute form for gear teeth. First, such a tooth form provides the desired pure rolling action. Second, even if a pair of involute gears is operated with the distance between the centers slightly too large or too small, the common normal at the point of contact between mating teeth will always pass through the pitch point. Obviously, the theoretical pitch circles in such cases will be increased or decreased slightly. Third, the *line of action* or *path of contact*—that is, the locus of the points of contact of mating teeth—is a straight line that passes through the pitch point and is tangent to the base circles of the two gears.

Cutting an involute shape in gear blanks can be done by simple form cutting (i.e., milling the shape into the workpiece) or by generating. Generating involves relative motion between the workpiece and the cutting tool. True involute tooth form can be produced by a cutting tool that has straight-sided teeth. This permits a very accurate involute tooth profile to be obtained through the use of a simple and easily made cutting tool. The straight-sided teeth are given a rolling motion relative to the workpiece to create the curved gear-tooth face, that is, the involute shape.

The basic size of gear teeth may be expressed in two ways. The common practice, especially in the United States and England, is to express the dimensions as a function of the *diametral pitch* (DP). DP *is the number of teeth (N) per unit of pitch diameter* (PD); thus (PD) = N/PD. Dimensionally, DP involves inches in the English system and millimeters in the SI system, and it is a measure of tooth size. Metric gears use the module system *(M)*, defined as *the pitch diameter divided by the number of teeth*, or M = PD/N. It thus is the reciprocal of diametral pitch and is expressed in millimeters. Any two gears having the same diametral pitch or module will mesh properly if they are mounted so as to have the correct distances and relationship. The important tooth elements can be specified in terms of the diametral pitch or the module and are as follows:

1. *Addendum:* the radial distance from the pitch circle to the outside diameter.

$$addendum = \frac{1}{DP}\ inches$$

2. *Dedendum:* the radial distance from the pitch circle to the root circle. It is equal to the addendum plus the *clearance*, which is provided to prevent the outer corner of a tooth from touching against the bottom of the tooth space.

3. *Circular pitch:* the distance between corresponding points of adjacent teeth, measured along the pitch circle:

$$\pi/diametral\ pitch$$

4. *Tooth thickness:* the thickness of a tooth, measured along the pitch circle. When tooth thickness and the corresponding *tooth space* are equal, no *backlash* exists in a pair of mating gears.

5. *Face width:* the length of the gear teeth in an axial plane.

6. *Tooth face:* the mating surface between the pitch circle and the addendum circle.

7. *Tooth flank:* the mating surface between the pitch circle and the root circle.

8. *Pressure angle:* the angle between a tangent to the tooth profile and a line perpendicular to the pitch surface.

Four shapes of involute gear teeth are used in the United States:

1. $14\frac{1}{2}^{\circ}$ pressure angle, full depth (used most frequently)
2. $14\frac{1}{2}^{\circ}$ pressure angle, composite (seldom used)
3. 20° pressure angle, full depth (seldom used)
4. 20° pressure angle, stub tooth (second most common)

In the $14\frac{1}{2}^{\circ}$ full-depth system, the tooth profile outside the base circle is an involute curve. Inward from the base circle, the profile is a straight radial line that is joined with the bottom land by a small fillet. With this system, the teeth of the basic rack have straight sides. The composite system and the 20° full-depth system provide somewhat stronger teeth. However, with the 20° full-depth system considerable undercutting occurs in the dedendum area; therefore, stub teeth often are used. The addendum is shortened by 20%, thus permitting the dedendum to be shortened a similar amount. This results in very strong teeth without undercutting. Table 29-2 gives the formulas for computing the dimensions of gear teeth in the $14^{1}/_{2}^{\circ}$ full-depth and 20° stub-tooth systems.

PHYSICAL REQUIREMENTS OF GEARS

A consideration of gear theory leads to five requirements that must be met in order for gears to operate satisfactorily:

1. The actual tooth profile must be the same as the theoretical profile.
2. Tooth spacing must be uniform and correct.
3. The *actual* and theoretical pitch circles must be coincident and be concentric with the axis of rotation of the gear.
4. The face and flank surfaces must be smooth and sufficiently hard to resist wear and prevent noisy operation.
5. Adequate shafts and bearings must be provided so that desired center-to-center distances are retained under operational loads.

The first four of these requirements are determined by the material selection and manufacturing process. The various methods of manufacture that are used represent attempts to meet these requirements to varying degrees with minimum cost, and their effectiveness must be measured in terms of the extent to which the resulting gears embody these requirements.

Before looking at the ways to manufacture gears, let's look at some examples of gears.

TABLE 29-2	Formula for Calculating the Standard Dimensions for Involute Gear Teeth	
	$14^{1}/_{2}^{\circ}$ Full Depth	20°, Stub Tooth
Pitch diameter (PD)	$\dfrac{N}{DP}$	$\dfrac{N}{DP}$
Addendum	$\dfrac{1}{DP}$	$\dfrac{0.8}{DP}$
Dedendum	$\dfrac{1.157}{DP}$	$\dfrac{1}{DP}$
Outside diameter	$\dfrac{N+2}{DP}$	$\dfrac{N+1.6}{DP}$
Clearance	$\dfrac{0.157}{DP}$	$\dfrac{0.2}{DP}$
Tooth thickness	$\dfrac{1.508}{DP}$	$\dfrac{1.508}{DP}$

DP = Number of teeth (N) per unit of pitch diameter (PD).

■ 29.8 GEAR TYPES

The more common types of gears are shown in Figure 29-17. *Spur gears* have straight teeth and are used to connect parallel shafts. They are the most easily made and the cheapest of all types.

The teeth on *helical gears* lie along a helix, the angle of the helix being the angle between the helix and a pitch cylinder element parallel with the gear shaft. Helical gears can connect either parallel or nonparallel nonintersecting shafts. Such gears are stronger and quieter than spur gears because the contact between mating teeth increases more gradually and more teeth are in contact at a given time. Although they usually are slightly more expensive to make than spur gears, they can be manufactured in several ways and are produced in large numbers.

Helical gears have one disadvantage. When they are in use, a side thrust is created that must be absorbed in the bearings. *Herringbone gears* neutralize this side thrust by having, in effect, two helical-gear halves, one having a right-hand and the other a left-hand helix. The *continuous* herringbone type is rather difficult to machine but is very strong. A modified herringbone type is made by machining a groove, or gap, around the gear blank where the two sets of teeth would come together. This provides a runout space for the cutting tool in making each set of teeth.

Different types of gears

Spur gear   Helical gears   Herringbone gears   Straight bevel gears

Skew bevels   Spiral bevels   Hypoid gears   Spiral gears

Worm gears   Internal gear

FIGURE 29-17   There are many different types of gears for transmitting motion, resulting in a wide variety of gear-making processes.

A *rack* is a gear with infinite radius, having teeth that lie on a straight line on a plane. The teeth may be normal to the axis of the rack or helical so as to mate with spur or helical gears, respectively.

A *worm* is similar to a screw. It may have one or more threads, the multiple-thread type being very common. Worms usually are used in conjunction with a *worm gear*. High gear ratios are easily obtainable with this combination. The axes of the worm and worm gear are nonintersecting and are usually at right angles. If the worm has a small helix angle, it cannot be driven by the mating worm gear. This principle is frequently employed to obtain nonreversible drives. Worm gears are usually made with the top land concave to permit greater area of contact between the worm and the gear. A similar effect can be achieved by using a *conical worm*, in which the helical teeth are cut on a double-conical blank, thus producing a worm that has an hourglass shape.

Bevel gears, teeth on a cone, are used to transmit motion between intersecting shafts. The teeth are cut on the surface of a truncated cone. Several types of bevel gears are made, the types varying as to whether the teeth are straight or curved and whether the axes of the mating gears intersect. On *straight-tooth* bevel gears the teeth are straight, and if extended all would pass through a common apex. *Spiral-tooth* bevel gears have teeth that are segments of spirals. Like helical gears, this design provides tooth overlap so that more teeth are engaged at a given time and the engagement is progressive. *Hypoid* bevel gears also have a curved-tooth shape but are designed to operate with nonintersecting axes. Rear-drive automobiles used hypoid gears in the rear axle so that the drive-shaft axis can be below the axis of the axle and thus permit a lower floor height. *Zerol* bevel gears have teeth that are circular arcs, providing somewhat stronger teeth than can be obtained in a comparable straight-tooth gear. They are not used extensively. When a pair of bevel gears are the same size and have their shafts at right angles, they are termed *miter gears*.

A *crown gear* is a special form of bevel gear having a 180° cone apex angle. In effect, it is a disk with the teeth on the side of the disk. It may also be thought of as a rack that has been bent into a circle so that its teeth lie in a plane. The teeth may be straight or curved. On straight-tooth crown gears the teeth are radial. Crown gears are seldom used, but they have the important quality that they will mesh properly with a bevel gear of any cone angle, provided that the bevel gear has the same tooth form and diametral pitch. This important principle is incorporated in the design and operation of two very important types of gear-generating machines that will be discussed later.

Most gears are of the external type, the teeth forming the outer periphery of the gear. Internal gears have the teeth on the inside of a solid ring, pointing toward the center of the gear.

■ 29.9 GEAR MANUFACTURING

Whether produced in large or small quantities, in cells, or in job shop batches, the sequence of processes for gear manufacturing requires four sets of operations. See Figure 29-18.

1. Blanking (turning)
2. Gear cutting (hobbing and shaving)
3. Heat treatment
4. Grinding

Blanking refers to the initial forming or machining operations that produce a semifinished part ready for gear cutting, starting from a piece of raw material. Turning on chuckers or lathes, facing and centering of shafts, milling, and sometimes grinding fall into this category of operations. Good-quality blanks are essential in precision gear manufacturing.

Hobbing, shaping, and shaving machines are the most frequently used machines for gear cutting, producing gears for automotive, truck, agricultural, and construction equipment. Other processes used in industrial gear production include broaching, rolling, grinding, milling, and shaving. The process selected depends on finding a cost-effective application based on quality specification, production volumes, and economic conditions.

FIGURE 29-18   The typical gear-making process (for very accurate gears) involves both hobbing and shaving followed by grinding after heat treating.

The gear-cutting or -machining operations can be divided into operations executed prior to heat treatment, when the material is still soft and easily machinable, and after heat treatment, performed on parts that have acquired high hardness and strength.

Heat treatment gives the material the strength and durability to withstand high loads and wear but results in a reduction in dimensional and geometrical accuracy. The metallurgical transformations that occur during hardening, quenching, and tempering cause a general quality deterioration in the gears. Therefore, precision grinding operations are used on external and internal bearing diameters, critical length dimensions, and fine surface finishes after heat treatment. Cylindrical grinders, angle-head grinders, internal grinders, and surface grinders are commonly used.

Gears are made in very large numbers by cold-roll forming; in addition, significant quantities are made by extrusion, by blanking, by casting, and some by powder metallurgy and by a forging process. However, it is only by machining that all types of gears can be made in all sizes, and although roll-formed gears can be made with accuracy sufficient for most applications, even for automobile transmissions, machining still is unsurpassed for gears that must have very high accuracy. Also, roll forming can be used only on ductile metals.

■ 29.10 MACHINING OF GEARS

FORM MILLING

Form cutting or *form milling* on a horizontal milling machine is illustrated in Figure 29-19. The multiple-tooth form cutter has the same form as the *space* between adjacent teeth. The tool is fed radially toward the center of the gear blank to the desired tooth depth, then across the tooth face to obtain the required tooth width. When one tooth space has been completed, the tool is withdrawn, the gear blank is indexed using a dividing head, and the next tooth space is cut. In machining gears by the form-cutting process, the form cutter is mounted on the machine arbor, and the gear blank is mounted on a mandrel held between the centers of some type of indexing device. Basically, form cutting is a simple and flexible method of machining gears. The equipment and cutters required are relatively

FIGURE 29-19 The basic method of machining a gear using a form cutter (left) to mill out the space between the teeth using the form cutter (middle) or the stocking cutter (right) to machine the gear. *(Courtesy of Brown and Sharpe Manufacturing Company.)*

simple, and standard machine tools (milling machines) are often used. However, in most cases the procedure is quite slow, and considerable care is required on the part of the operator. Therefore, this process is usually employed where only one or a few gears are to be made. When a helical gear is to be cut, the table must be set at an angle equal to the helix angle, and the dividing head is geared to the longitudinal feed screw of the table so that the gear blank will rotate as it moves longitudinally.

Standard cutters are usually employed in form-cutting gears. In the United States, these come in eight sizes for each diametral pitch and will cut gears having any number of teeth. A single cutter will not produce a theoretically perfect tooth profile for all sizes of gears in the range for which it is intended. However, the change in tooth profile over the range covered by each cutter is very slight, and most of the time satisfactory results can be achieved. When greater accuracy is required, half-number cutters can be obtained. Cutters are available for all common diametral pitches and for $14\frac{1}{2}°$ and $20°$ pressure angles. If the amount of metal that must be removed to form a tooth space is large, roughing cuts may be taken with a *stocking cutter*. The stepped sides of the stocking cutter remove most of the metal and leave only a small amount to be removed subsequently by the regular form cutter in a finish cut.

Straight-tooth bevel gears can be form cut on a milling machine, but this is seldom done. Because the tooth profile in bevel gears varies from one end of the tooth to the other, after one cut is taken to form the correct tooth profile at the smaller end, the relationship between the cutter and the blank must be altered. Shaving cuts are then taken on the side of each tooth to form the correct profile throughout the entire tooth length.

Although the form cutting of gears on a milling machine is a flexible process and is suitable for gears that are not to be operated at high speeds or that need not operate with extreme quietness, the process is slow and requires skilled labor. Semiautomatic machines are available for making gears by the form-cutting process. The procedure utilized is essentially the same as on a milling machine, except that, after setup, the various operations are completed automatically. Gears made on such machines are no more accurate than those produced on a milling machine, but the possibility of error is less, and they are much cheaper because of reduced labor requirements. For large quantities, however, form cutting is not used.

FIGURE 29-20 Blind gear spline broaching machine for producing internal gears. The machine has automated pick-and-place arms for load/unload. *(Courtesy of Apex Broach and Machine Company.)*

BROACHING

Broaching is another way to form cut teeth. All the tooth spaces are cut simultaneously, and the tooth is formed progressively. The circular table in Figure 29-20 holds 10 sets of progressive tooling. The table rotates, moving one set of tooling at a time under two workpieces. The arms load and unload a set of parts every 15 seconds, so the cycle time is very quick. Excellent gears can be made by *broaching*. However, a separate broach must be provided for each size of gear. The tooling tends to be expensive, restricting this method to large volumes.

GEAR GENERATING

Most high-quality gears that are made by machining are made by the *generating process*. This process is based on the principle that any two involute gears, or any gear and a rack, of the same diametral pitch will mesh together properly. Utilizing this principle, one of the gears (or the rack) is made into a cutter by proper sharpening. It can be used to cut into a mating gear blank and thus generate teeth on the blank. The two principal methods for gear generating are *shaping* and *hobbing*.

SHAPING

To carry out the shaping process, the cutter and the gear blank must be attached rigidly to their respective shafts, and the two shafts must be interconnected by suitable gearing so that the cutter and the blank rotate positively with respect to each other and have the same pitch-line velocities. To start cutting the gear, the cutter is reciprocated vertically and is fed radially into the blank between successive strokes. When the desired tooth depth has been obtained, the cutter and blank are then slightly indexed after each cutting stroke. The resulting generating action is indicated schematically in Figure 29-21a and shown in the cutting of an actual gear tooth in Figure 29-21b. Figure 29-22 shows a machine called a gear shaper. *Gear shapers* generate gears by a reciprocating tool motion. The gear blank is mounted on the rotating table (or vertical spindle) and the cutter on the end of a vertical, reciprocating spindle. The spindle and the table are connected by means of gears so that the cutter and gear blank revolve with the same pitch-line velocity. Cutting occurs on the downstroke (sometimes on the upstroke). At the end of each cutting stroke, the spindle carrying the blank retracts slightly to provide clearance between work and tool on the return stroke.

The conventional cutter for gear shaping is made from high-speed steel, which can be coated to improve wear life with a superhard layer of titanium nitride (TiN) using the physical vapor deposition (PVD) process. Recently, new throwaway disk-shaped insert tools have been developed for gear shaping, eliminating regrinding and recoating operations on the conventional tools. Regrinding the conventional cutter requires two adjustments (resetting operations) on the machine tool. Because the throwaway blades are all sized the same, machine resetting after cutting tool changeover is eliminated.

FIGURE 29-21 (Top) Generating action of a Fellows gear shaper cutter. (Bottom) Series of photographs showing various stages in generating one tooth in a gear by means of a gear shaper, action taking place from right to left, corresponding to the diagram above. One tooth of the cutter was painted white. *(Courtesy of Fellows Gear Shaper Company.)*

Either straight- or helical-tooth gears can be cut on gear shapers. To cut helical teeth, both the cutter and the blank are given an oscillating rotational motion during each stroke of the cutter, turning in one direction during the cutting stroke and in the opposite direction during the return stroke. Because the cutting stroke can be adjusted to end at any desired point, gear shapers are particularly useful for cutting cluster gears. Some machines can be equipped with two cutters simultaneously to cut two gears, often of different diameters. Gear shapers can also be adapted for cutting internal gears.

Special types of gear shapers have been developed for mass-production purposes. The *rotary gear shaper* is essentially 10 shaper units mounted on a rotating base and having a single drive mechanism. Nine gears are cut simultaneously while a finished gear is removed and a new blank is put in place on the tenth unit. *Planetary gear shapers* holding six gear blanks move in planetary motion about a large, central gear cutter. The cutter has no teeth in one portion to provide a space where the gear can be removed and a new blank placed on the empty spindle.

CNC gear shapers are now available with hydromechanical stroking systems that produce a uniform cutting velocity during the cutting portion of the downstroke. These machines can operate at 500 to 1700 rpm and use TiN-coated cutters to enhance tool life. *Vertical shapers* for gear generating have a vertical ram and a round table that can be rotated in a horizontal plane by either manual or power feed. These machine tools are sometimes called *slotters*. Usually, the ram is pivoted near the top so that it can be swung outward from the column through an arc of about 10°.

FIGURE 29-22   This machine tool is a gear shaper. The blank is rotating while the cutter is reciprocating vertically, as shown in the inset. The tool is very complex and is shown in detail below. (*Courtesy Fellows Gear Shaper Company.*)

Because one circular and two straight-line motions and feeds are available, vertical shapers are very versatile tools and thus find considerable use in one-of-a-kind manufacturing. Not only can vertical and inclined flat surfaces be machined, but external and internal cylindrical surfaces can be generated by circular feeding of the table between strokes. This may be cheaper than turning or boring for very small lot sizes. A vertical shaper can be used for generating gears or machining curved surfaces, interior surfaces, and arcs by using a stationary tool and rotating the workpiece. A *keyseater* is a special type of vertical shaper designed and used exclusively for machining keyways on the inside of wheel and gear hubs. For machining continuous herringbone gears, a *Sykes gear-generating machine* is used.

HOBBING

Involute gear teeth could be generated by a cutter that has the form of a rack. Such a cutter would be simple to make but has two major disadvantages. First, the cutter (or the blank) would have to reciprocate, with cutting occurring only during one stroke direction. Second, because the rack would have to move longitudinally as the blank rotated, the rack would need to be very long (or the gear very small) or the two would not be in mesh after a few teeth were cut. A *hob* overcomes the preceding two difficulties. As shown in Figure 29-23, a hob can be thought of, basically, as one long rack tooth that has been wrapped around a cylinder in the form of a helix and fluted at intervals to provide a number of cutting edges. Relief is provided behind each of the teeth. The cross section of each tooth, normal to the helix, is the same as that of a rack tooth. (A hob can also be thought of as a gashed worm gear.)

The action of a *hobbing* machine cutting a spur gear is illustrated in Figure 29-23. To cut a spur gear, the axis of the hob must be set off from the normal to the rotational axis of the blank by the helix angle of the hob. In cutting helical gears, the hob must be set over an additional amount equal to the helix angle of the gear. The cutting of a gear by means of a hob is a continuous action. The hob and the blank are connected by proper gearing so that they rotate in mesh. To start cutting a gear, the rotating hob is fed inward until the proper setting for tooth depth is obtained. The hob is then fed in a direction parallel with the axis of rotation of the blank. As the gear blank rotates, the teeth are generated and the feed of the hob across the face of the blank extends the teeth to the desired tooth-face width.

Hobbing is rapid and economical. More gears are cut by this process than by any other. The process produces excellent gears and can also be used for splines and sprockets. Single-, double-, and triple-thread hobs are used. Multiple-thread types increase the production rate but do not produce accuracy as high as single-thread hobs.

Gear-hobbing machines are made in a wide range of sizes. Machines for cutting accurate large gears are frequently housed in temperature-controlled rooms, and the temperature of the cutting fluid is controlled to avoid dimensional change due to variations in temperature.

COLD-ROLL FORMING

The manufacture of gears by *cold-roll forming* has been highly developed and widely adopted in recent years. Currently, millions of high-quality gears are produced annually by this process; many of the gears in automobile transmissions are made this way. As indicated in Figure 29-24, the process is basically the same as that by which screw threads are roll formed, except that in most cases the teeth cannot be formed in a single rotation of the forming rolls; the rolls are fed inward gradually during several revolutions.

Because of the metal flow that occurs, the top lands of roll-formed teeth are not smooth and perfect in shape; a depressed line between two slight protrusions can often be seen, as shown encircled in Figure 29-24. However, because the top land plays no part in gear-tooth action, if there is sufficient clearance in the mating gear, this causes no difficulty. Where desired, a light turning cut is used to provide a smooth top land and correct addendum diameter.

The hardened forming rolls are very accurately made, and the roll-formed gear teeth usually have excellent accuracy. In addition, because the severe cold working pro-

FIGURE 29-23 (a) Three views of hob and gear blanks (small spur gear). (b) Hob. (c) Schematic of mechanism of a hobbing machine, shown hobbing a large spur gear.

duces tooth faces that are much smoother and harder than those on the typical machined gear, they seldom require hardening or further finishing, and they have excellent wear characteristics.

The process is rapid (up to 50 times faster than gear machining) and easily mechanized. No chips are made and thus less material is needed. Less skilled labor is required. Small gears are often made by rolling a length of shaft and then slicing off the individual gear blanks. Usually, soft steel is required and 4 to 5 in. in diameter is about the limit, with fewer than six teeth, coarser than 12 diametral pitch, and no pressure angle less than 20°.

FIGURE 29-24　On the upper left, the method for cold forming gear teeth on a spline using three dies is shown. (Upper right) Worm gear being roll formed by means of two rotating rolling tools with typical worm made by rolling. (Lower left) Gear made by rolling. *(Courtesy of Landis Machine Company.)*

OTHER GEAR-MAKING PROCESSES

Gears can be made by the various casting processes. *Sand-cast gears* have rough surfaces and are not accurate dimensionally. They are used only for services where the gear moves slowly and where noise and inaccuracy of motion can be tolerated. Gears made by *die casting* are more accurate and have fair surface finish. They can be used to transmit light loads at moderate speeds. Gears made by *investment casting* may be accurate and have good surface characteristics. They can be made of strong materials to permit their use in transmitting heavy loads. In many instances, gears that are to be finished by machining are made from cast blanks, and in some larger gears the teeth can be cast to approximate shape to reduce the amount of machining.

Large quantities of gears are produced by *blanking* in a punch press. The thickness of such gears usually does not exceed about $1/16$ in. By shaving the gears after they are blanked, excellent accuracy can be achieved. Such gears are used in clocks, watches, meters, and calculating machines. *Fine blanking* is also used to produce thin, flat gears of good quality.

High-quality gears, both as to dimensional accuracy and surface quality, can be made by the *powder metallurgy process*. Usually, this process is employed only for small sizes, ordinarily less than 1 in. in diameter. However, larger and excellent gears are made by forging powder metallurgy preforms. This results in a product of much greater density and strength than usually can be obtained by ordinary powder metallurgy methods, and the resulting gears give excellent service at reduced cost. Gears made by this process often require little or no finishing.

Large quantities of plastic gears are made by *plastic molding*. The quality of such gears is only fair, and they are suitable only for light loads. Accurate gears suitable for heavy loads are frequently machined out of laminated plastic materials. When such gears are mated with metal gears, they have the quality of reducing noise.

Quite accurate small gears can be made by the *extrusion* process. Typically, long lengths of rod, having the cross section of the desired gear, are extruded. The individual gears are then sawed from this rod. Materials suitable for this process are brass, bronze, aluminum alloys, magnesium alloys, and, occasionally, steel.

Flame machining (oxyacetylene cutting) can be used to produce gears that are to be used for slow-moving applications wherein accuracy is not required.

A few gears are made by the hot-roll-forming process. In this process a cold master gear is pressed into a hot blank as the two are rolled together.

■ 29.11 GEAR FINISHING

To operate efficiently and have satisfactory life, gears must have accurate tooth profiles and the faces of the teeth must be smooth and hard. These qualities are particularly important when gears must operate quietly at high speeds. When they are produced rapidly and economically by most processes except cold-roll forming, the tooth profiles may not be as accurate as desired, and the surfaces are somewhat rough and subject to rapid wear. Also, it is difficult to cut gear teeth in a hardened gear blank, and therefore economy dictates that the gear be cut in a relatively soft blank and subsequently be heat treated to obtain greater hardness. Such heat treatment usually results in some slight distortion and surface roughness. Although most roll-formed gears have sufficiently accurate profiles, and the tooth faces are adequately smooth and frequently have sufficient hardness, this process is feasible only for relatively small gears. Consequently, a large proportion of high-quality gears are given some type of finishing operation after they have received primary machining or after heat treatment. Most of these finishing operations can be done quite economically because only minute amounts of metal are removed and they are fast and often automatic.

Gear shaving is the most commonly used method for finishing spur- and helical-gear teeth prior to hardening. The gear is run, at high speed, in contact with a shaving tool, usually of the type shown in Figure 29-25. Such a tool is a very accurate, hardened, and ground gear that contains a number of peripheral serrations, thus forming a series of sharp cutting edges on each tooth. The gear and shaving cutter are run in mesh with their

FIGURE 29-25 Rotary gear shaving cutter (see Figures 29-18 and 29-26).

FIGURE 29-26 Three methods for gear finishing are (a) shaving using a special cutter—see Figure 29-25, (b) grinding using form grinding wheels, and (c) using form rolls, that is, roll forming/finishing.

axes crossed at a small angle, usually about 10° (Figure 29-26). As they rotate, the gear is reciprocated longitudinally across the shaving tool (or vice versa). During this action, which usually requires less than one minute, very fine chips are *shaved* from the gear-tooth faces, thus eliminating any high spots and producing a very accurate tooth profile.

Rack shaving cutters are sometimes used for shaving small gears, the cutter reciprocating lengthwise, causing the gear to roll along it, as it is moved sideways across the cutter and fed inward.

Although shaving cutters are costly, they have a relatively long life because only a very small amount of metal is removed, usually 0.001 to 0.004 in. Some gear-shaving machines produce a slight crown on the gear teeth during shaving. Some gears are not hardened prior to shaving, although it is possible to remove very small amounts of metal from hardened gears if they are not too hard. However, modern heat-treating equipment makes it possible to harden gears after shaving without harmful effects, and therefore this practice is followed if possible.

Roll finishing is a cold-forming process that is used to finish helical gears. The unhardened gear is rolled with two hardened, accurately formed rolling dies. The center distance between the dies is reduced to cold work the surfaces and produce highly accurate tooth forms. High points on the unhardened gear are plastically deformed so that a smoother surface and more accurate tooth form are achieved. Because the operation is one of localized cold working, some undesirable effects may accrue, such as localized residual stresses and nonuniform surface characteristics. Surface finishes of 6 to 8 μm have been achieved. If roll finishing is to be used, attention must be paid to the prerolled geometry. Designers should consult the manufacturers of gear-rolling machines for specific recommendations.

Grinding is used to obtain very accurate teeth on hardened gears. Two methods are used. One employs a formed grinding wheel that is trued to the exact form of a tooth by means of diamonds mounted on a special holder and guided by a large template. The other method is involute-generation grinding, which uses straight-sided grinding wheels that simulate one side of a rack tooth. The surface of the gear tooth is ground as the gear rolls (and reciprocates) past the grinding wheels. Grinding produces very accurate gears, but because it is slow and expensive, it is used only on the highest-quality, hardened gears.

Lapping can also be used for finishing hardened gears. The gear to be finished is run in contact with one or more cast iron lapping gears under a flow of very fine abrasive in oil. Because lapping removes only a very small amount of metal, it is usually employed on gears that have previously been shaved and hardened. This combination of processes produces gears that are nearly equal to ground gears in quality, but at considerably lower cost.

■ 29.12 GEAR INSPECTION

As with all manufactured products, gears must be checked to determine whether the resulting product meets the design specifications and requirements. Because of their irregular shape and the number of factors that must be measured, inspection of gears is somewhat difficult. Among the factors to be checked are the linear tooth dimensions such as thickness, spacing, depth, and so on; tooth profile; surface roughness; and noise. Several special devices, most of them automatic or semiautomatic, are used for such inspection.

Gear-tooth vernier calipers can be used to measure the thickness of gear teeth on the pitch circle (Figure 29-27). CNC gear inspection machines (Figure 29-28) can quickly check several factors, including variations in circular pitch, involute profile, lead,

FIGURE 29-27 Using gear-tooth vernier calipers to check the tooth thickness at the pitch circle.

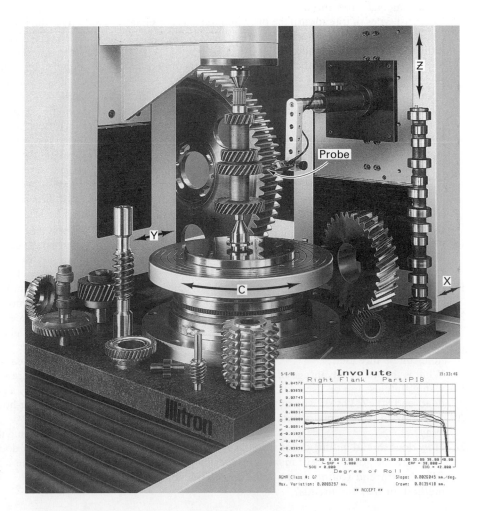

FIGURE 29-28 A CNC gear inspection machine has *X, Y, Z* motions plus a rotary table.

tooth spacing, and variations in pressure angle. The gear is usually mounted between centers. The probe is moved to the gear through X, Y, and Z translations. The gear is rotated between measurements. The inset in Figure 29-28 shows a typical display for an involute profile.

Because noise level is important in many applications, not only from the viewpoint of noise pollution but also as an indicator of probable gear life, special equipment for noise measurement is quite widely used, sometimes integrated into mass-production assembly lines.

■ Key Words

addendum	dedendum	gear teeth	plug tap
bevel gear	diametral pitch	herringbone gear	self-opening die head
blanking	fluteless tap	hob	spur gear
bottoming tap	form milling	hobbing	taper tap
broaching	gear finishing	involute curve	thread cutting
circular pitch	gear grinding	pitch	thread rolling
cold-roll forming	gear hobbing	pitch circle	thread tapping
collapsible tap	gear rolling	pitch diameter	worm gear
crown gear	gear shaping	planetary gear	

■ Review Questions

1. How does the pitch diameter differ from the major diameter for a standard screw thread?
2. For what types of threads are the pitch and the lead the same?
3. What is the helix angle of a screw thread?
4. Why are pipe threads tapered?
5. What are three basic methods by which external threads are produced?
6. Explain the meaning of $\frac{1}{4}''$ -20 UNC-3A.
7. What is meant by the designation M20 x 2.5 6g6g? (What does the "x" mean?)
8. What are two reasons fine-series threads are being used less now than in former years?
9. In cutting a thread on a lathe, how is the pitch controlled?
10. What is the function of a threading dial on a lathe?
11. In Chapter 22, what figure(s) showed a threading dial?
12. What controls the lead of a thread when it is cut by a threading die?
13. What is the basic purpose of a self-opening die head?
14. What is the reason for using a taper tap before a plug tap in tapping a hole?
15. What difficulties are encountered if full threads are specified to the bottom of the dead-end hole?
16. Can a fluteless tap be used for threading a hole in gray cast iron? Why or why not?
17. What provisions should a designer make so that a dead-end hole can be threaded?
18. What is the major advantage of a spiral-point tap?
19. Can a fluteless tap be used for threading to the bottom of a dead-end hole? Why or why not?
20. Is it desirable for a tapping fluid to have lubricating qualities? Why?
21. How does thread milling differ from thread turning?
22. What are the advantages of making threads by grinding?
23. Why has thread rolling become the most commonly used method for making threads?
24. How can you determine whether a thread has been produced by rolling rather than by cutting?
25. Why is the involute form used for gear teeth?
26. What is the diametral pitch of a gear?
27. What is the relationship between the diametral pitch and the module of a gear?
28. On a sketch of a gear, indicate the pitch circle, addendum circle, dedendum circle, and circular pitch.
29. What five requirements must be met for gears to operate satisfactorily? Which of these are determined by the manufacturing process?
30. What are the advantages of helical gears compared with spur gears?
31. What is the principal disadvantage of helical gears?
32. What difficulty would be encountered in hobbing a herringbone gear?
33. What is the only type of machine on which full-herringbone gears can be cut?
34. What modification in design is made to herringbone gears to permit them to be cut by hobbing?
35. Why aren't more gears made by broaching?
36. What is the most important property of a crown gear?
37. What are three basic processes for machining gears?
38. Which basic gear-machining process is utilized in a Fellows gear shaper?
39. When a helical gear is machined on a milling machine, the table lead screw and the universal dividing head have to be connected by a gear train. Why?
40. Why is a gear-hobbing machine much more productive than a gear shaper?
41. In gear shaping and gear hobbing, the tooth profiles are generated. What does that mean?
42. What are the advantages of cold-roll forming for making gears?

43. Why is cold-roll forming not suitable for making gray cast iron gears?
44. Under what conditions can shaving not be used for finishing gears?
45. What inherent property accrues from cold-roll forming of gears that may result in improved gear life?
46. Can lapping be used to finish cast iron gears?
47. What factors are usually checked in inspecting gears?
48. What are the basic methods for gear finishing?

■ Problems

1. Calculate the cutting time needed to tap a $^3/_4$-in.-diameter by 2-in.-deep hole using a cutting speed of 30 sfpm. The tap has 10 tpi.
2. The new manufacturing manager has recommended to you that chipless tapping be adopted for tapping holes in the deep, dead-end holes on the 2-cylinder engine blocks that the company makes. Chipless tapping can run at twice the speed of the conventional tapping process, works well on deep holes, and provides better quality and finish with longer tool life. The tapping process is the bottleneck process in machining all the cast iron engine blocks. In addition, 10% of the blocks have to be scrapped due to broken taps. What do you recommend?
3. A hob that has a pitch diameter of 76.2 mm is used to cut a gear having six teeth. If a cutting speed of 27.4 m/min is used, what will be the rpm value of the gear blank?
4. In Figure 29-19 a form-milling operation and cutters are shown. The gear is to be made from 4340 steel, R_c 50 prior to heat treat and final grind. Select the proper speeds and feeds for the job (the cutter is 4 in. in diameter) and compute the cutting time to mill this gear. Would you use up milling versus down milling?
5. A gear-broaching machine of the type shown in Figure 29-20 can do the gear in 15 seconds (about 240 parts per hour). How many additional gears per year are needed to cover the broaching tooling cost if each broach on the machine cost $250? Do you think the broach tool life is sufficient to handle that number of parts? What about TiN coating the broaches (cost $10.00 per broach)?
6. Assume that 10,000 spur gears, $1^1/_8$ in. in diameter and 1 in. thick, are to be made of 70-30 brass. What manufacturing method would you consider?
7. If only three gears described in problem 6 were to be made, what process would you select?
8. K.C. Stern, a design engineer for Boeing Commercial Airplane Group, is doing some design work and has a question regarding the hole size that is drilled before tapping internal threads. The book states, "A hole diameter slightly larger than the minor diameter of the thread must already exist, made by drilling/reaming, boring, or die casting."

    Conventional wisdom would lead one to believe that a hole slightly smaller should exist before tapping threads. The *Machinery's Handbook* seems to agree with this. Most tap drill sizes listed are smaller than the minimum minor diameter listed for a particular thread size in MIL-S-008879B. K.C. has found at least one exception to this, but the difference was only a couple of thousandths of an inch.

    K.C. consulted local sources on this matter, but no one would commit to the correctness or incorrectness of the book. A number of people are scratching their heads on this one. Your assistance in this matter is greatly appreciated. Should the hole be slightly smaller or slightly larger than the minor diameter of the thread and why?

www.wiley.com/college/degarmo

Chapter 29 CASE STUDY

Bevel Gear for a Riding Lawn Mower

Figure CS 29 is a bevel gear for the transaxle of a heavy duty riding lawn mower. The gear has an outer diameter of 3.50 inches and a maximum thickness of 1/2 inch. At the root of the teeth, the thickness is approximately 3/16 inch. The minimum material properties have been estimated to be a yield strength of 125 ksi and a surface hardness of Rockwell C 25. The maximum operating temperature should be less than 250°F.

Since the mower has a multi speed transmission, the transaxle gearing may be subject to sudden applications of load due to improper engagement of the clutch. The manufacturer has proposed a nonstandard test in which a 10 pound weight is dropped onto the gear from a height of 15 inches. While this does not translate to a Charpy test value, it is clear that impact resistance will be an important property.

The gear will operate in an oil-filled enclosure so corrosion resistance need not be outstanding. A smooth surface finish is desirable (especially on the teeth), and the total production run has been placed between 25,000 and 50,000 units.

1. Based on the size and shape of the product, describe at least three reasonable ways in which the component could be produced. For each method, briefly discuss its relative pros and cons.

2. What type of engineering materials might be able to meet the desired requirements? What would be the pros and cons of each general family?

3. For each of the shape generation methods in part 1, select an appropriate material from the alternatives discussed in part 2. (Note: Casting alloys should be matched for the casting processes, high machinability alloys would be favored for cutting applications, etc.).

4. Which of the above alternatives do you feel would be the "best" solution to the problem? Why? For this system, outline the specific steps that would be necessary to produce the part from reasonable starting material. Include any necessary heat treatments and/or surface treatments.

5. Your supervisor has asked you to evaluate the possibility of producing this part as a near net shape ferrous forging, with forged gear teeth. Investigate this process and determine if it would be a feasible approach. If so, would you recommend that the forging be performed cold, warm, or hot? Would subsequent heat treatment or surface treatment be required? What concerns might be associated with these processes?

Figure CS 29

CHAPTER 30

FUNDAMENTALS OF JOINING

■ 30.1 INTRODUCTION TO CONSOLIDATION PROCESSES

Large-size products, products with a high degree of shape complexity, or products with a wide variation in required properties are often manufactured as joined assemblies of two or more component pieces. These pieces may be smaller and therefore easier to handle, simpler shapes that are easier to manufacture, or segments that have been made from different materials. Assembly is an important part of the manufacturing process, and a wide variety of *consolidation processes* have been developed to meet the various needs.

Each of the methods has its own distinctive characteristics, strengths, and weaknesses. The metallurgical processes of welding, brazing, and soldering are usually used to join metals and often involve the solidification of molten material. The use of discrete fasteners (such as nuts, bolts, screws, and rivets) requires the creation of aligned holes and produces stress localization. While the holes may affect performance, disassembly and reassembly can often be performed with relative ease. Adhesive bonding has grown with new developments in polymeric materials and is being used extensively in automotive and aircraft production. Any material can be joined to any other material, and low-temperature joining is particularly attractive for composite materials. Production rates are often low, however, because of the time required for the adhesive to develop full strength. Lesser known joining techniques include shrink fits, slots and tabs, and a wide variety of other mechanical methods. From a technical viewpoint, powder metallurgy is another consolidation process, since the end product is built up by the joining of a multitude of individual particles.

Our survey of consolidation processes will begin with a spectrum of techniques known by the generic term of *welding*. Welding is the permanent joining of two materials, usually metals, by *coalescence*, which is induced by a combination of temperature, pressure, and metallurgical conditions. The particular combination of these variables can range from high temperature with no pressure to high pressure with no increase in temperature. Because welding can be accomplished under a wide variety of conditions, a number of different processes have been developed. Welding is the dominant method of joining in manufacturing, and a large fraction of metal products would have to be drastically modified if welding were not available.

Coalescence between two metals requires sufficient proximity and activity between the atoms of the pieces being joined to cause the formation of common crystals. The ideal metallurgical bond, for which there would be no noticeable or detectable interface, would require (1) perfectly smooth, flat, or matching surfaces; (2) surfaces that are clean and free from oxides, absorbed gases, grease, and other contaminants; (3) metals with no internal impurities; and (4) the joining of single crystals with identical crystallographic structure and orientation. These conditions would be difficult to obtain under laboratory conditions and are virtually impossible to achieve in normal production. Consequently, the various joining methods have been designed to overcome or compensate for the common deficiencies.

Surface roughness can be overcome either by force, causing plastic deformation and flattening of the high points, or by melting the two surfaces so that fusion occurs. The various processes also employ different approaches to cleaning the metal surfaces prior to welding and preventing further oxidation or contamination during the joining process. In solid-state welding, contaminated layers are generally removed by mechanical or chemical cleaning prior to welding or by causing sufficient metal flow along the interface so that new surface is created and existing impurities are displaced from the joint. In *fusion welding*, where molten material is produced and high temperatures accelerate the reactions between the metal and its surroundings, contaminants are often removed from the pool of molten metal through the use of fluxing agents. If welding is performed in a vacuum, the contaminants are removed much more easily, and coalescence is easier to achieve. In the vacuum of outer space, mating parts may weld under extremely light loads, even when welding is not intended.

When the process requires heat that is sufficient to induce melting, the structure of the metal may be significantly altered. Even when no melting occurs, the heating and cooling of the welding process can affect the metallurgical structure and quality of both the weld and the adjacent material. Since many of the changes are detrimental, the possible consequences of heating and cooling should be a major consideration when selecting a joining process.

To produce a high-quality weld, we will need (1) a source of satisfactory heat and/or pressure, (2) a means of protecting or cleaning the metals to be joined, and (3) caution to avoid, or compensate for, harmful metallurgical effects. These aspects will be developed in the sections that follow.

■ 30.2 CLASSIFICATION OF WELDING AND THERMAL CUTTING PROCESSES

Wherever possible, this text will utilize the nomenclature of the American Welding Society (AWS). The various welding processes have been classified in the manner presented in Figure 30-1, and letter symbols have been assigned to facilitate process designation. The variety of processes provide multiple ways of achieving coalescence and make it possible to produce effective and economical welds in nearly all metals and combinations of metals. Chapter 31 will present the gas and arc welding processes; Chapter 32 will cover resistance and solid-state welding; and Chapter 33 will present a variety of other processes, including brazing and soldering.

For many years, welding equipment (such as oxyfuel torches and electric arc units) has also been used to cut metal sheets and plates. Developed originally for salvage and repair work, then used for preparing plates for welding, this type of equipment is now widely used to cut sheets and plates into desired shapes for a variety of uses and operations. Laser- and electron-beam equipment can now cut both metals and nonmetals at speeds up to 25 m/min (1000 in./min), with accuracies of up to 0.25 mm (0.01 in.). Figure 30-2 summarizes the commonly used *thermal cutting* processes and provides their AWS designations. Since cutting is often an adaptation of welding, the welding and cutting capabilities will be presented together as the individual processes are discussed.

■ 30.3 SOME COMMON CONCERNS

Many of the problems that are inherent to welding and joining can be avoided by selecting the proper process and considering both general and process-specific characteristics and requirements. Proper design of the joint is extremely critical. Heating, melting, and resolidification can produce drastic changes in the properties of base and

FIGURE 30-1   Classification of common welding processes along with their AWS (American Welding Society) designations.

FIGURE 30-2   Classification of thermal cutting processes along with their AWS (American Welding Society) designations.

filler materials. Weld metal properties can also be changed by dilution of the filler by melted base metal, vaporization of various alloy elements, and gas–metal reactions.

Various types of weld defects can also be produced. These include cracks in various forms, cavities (both gas and shrinkage), inclusions (slag, flux, and oxides), *incomplete fusion* between the weld and base metals, *incomplete penetration* (insufficient weld depth), unacceptable weld shape or contour, arc strikes, spatter, undesirable metallurgical changes (aging, grain growth, or transformations), and excessive distortion. Figure 30-3 depicts several of these defects.

■ 30.4 Types of Fusion Welds and Types of Joints

Figure 30-4 illustrates four basic types of fusion welds. *Bead welds*, or surfacing welds, are made directly onto a flat surface and therefore require no edge preparation. Since the penetration depth is limited, bead welds are used primarily for joining thin sheets of metal, building up surfaces, and depositing hard-facing (wear-resistant) materials.

Incomplete fusion

Incomplete penetration

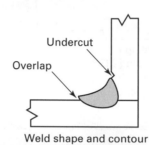

Undercut

Overlap

Weld shape and contour

FIGURE 30-3 Some common welding defects.

| Bead weld (or surfacing weld) | Groove weld | Fillet weld | Plug weld |

FIGURE 30-4 Four basic types of fusion welds.

Groove welds are used when full-thickness strength is desired on thicker material. Some sort of edge preparation is required to form a groove between the abutting edges. V, double-V (top and bottom), U, and J (one-sided V) configurations are most common and are often produced by oxyacetylene flame cutting. The specific type of groove usually depends upon the thickness of the joint, the welding process to be employed, and the position of the work. The objective is to obtain a sound weld throughout the full thickness with a minimum amount of additional weld metal. If possible, single-pass welding is preferred, but multiple passes may be required, depending upon the thickness of the material and the welding process being used. As shown in Figure 30-5, special consumable *inserts* can be used to ensure proper spacing between the mating edges and good quality in the root pass. These inserts are particularly useful in pipeline welding and other applications where welding must be performed from only one side of the work.

Fillet welds are used for tee, lap, and corner joints and require no special edge preparation. The size of the fillet is measured by the leg of the largest 45° right triangle that can be inscribed within the contour of the weld cross section. This is shown in

FIGURE 30-5 The use of a consumable backup insert in making a fusion weld. *(Courtesy Arcos Industries, Mount Carmel, PA).*

Insert in place   Insert tack-welded   Insert consumed   Completed weld

FIGURE 30-6   Preferred shape and the method of measuring the size of fillet welds.

Figure 30-6, which also depicts the proper shape for fillet welds to avoid excess metal deposition and reduce stress concentration.

Plug welds attach one part on top of another and are often used to replace rivets or bolts. A hole is made in the top plate and welding is started at the bottom of this hole.

Figure 30-7 shows five basic types of joints (*joint configurations*) that can be made with the use of bead, groove, and fillet welds, and Figure 30-8 shows some of the methods to construct these joints. In selecting the type of joint to be used, a primary consideration should be the type of loading that will be applied. A large portion of what are erroneously called "welding failures" can more accurately be attributed to inadequate consideration of loading. Cost and accessibility for welding are other important factors when specifying joint design but should be viewed as secondary to loading. Cost is affected by the amount of required edge preparation, the amount of weld metal that must be deposited, the type of equipment that must be used, and the speed and ease with which the welding can be accomplished.

FIGURE 30-7   Five basic joint designs for fusion welding.

FIGURE 30-8   Various weld procedures used to produce welded joints. *(Courtesy Republic Steel Corporation, Youngstown, OH).*

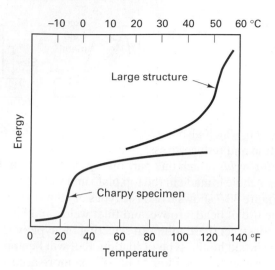

FIGURE 30-9 Effect of size on the transition temperature and energy-absorbing ability of a certain steel. While the larger structure absorbs more energy because of its size, it becomes brittle at a much higher temperature.

■ 30.5 DESIGN CONSIDERATIONS

Welding is a unique process that cannot be directly substituted for other methods of joining without proper consideration of its particular characteristics and requirements. Unfortunately, welding is also easy and convenient, and the considerations of proper design and implementation are often overlooked.

One very important fact is that welding produces *monolithic*, or one-piece, structures. When two pieces are welded together, they become one continuous piece. This can cause significant complications. For example, a crack in one piece of a multipiece structure may not be catastrophic, because it will seldom progress beyond the single piece in which it occurs. However, when a large structure, such as a ship hull, pipeline, storage tank, or pressure vessel, consists of many pieces welded together, a crack that starts in a single plate or weld can propagate for a great distance and cause complete failure. Obviously, this kind of failure is not the fault of the welding process itself but is simply a reflection of the monolithic nature of the product.

It is also important to note that a given material in small pieces may not behave as it does in a larger size. This feature is clearly illustrated in Figure 30-9, which shows the relationship between the energy required to fracture and temperature for the same steel tested as a small Charpy impact specimen (see Chapter 2 and Figure 2-20) and as a large, welded structure. In the form of a small Charpy bar, the material exhibits ductile behavior and good energy absorption at temperatures down to 25°F (4°C). When welded into a large structure, however, brittle behavior is observed at temperatures as high as 110°F (43°C). More than one welded structure has failed because the designer overlooked the effect of size on the notch-ductility of metal.

Another common error is to make welded structures too rigid, thereby restricting their ability to redistribute high stresses and avoid failure. Considerable thought may be required to design structures and joints that provide sufficient flexibility, but the multitude of successful welded structures attests to the fact that such designs are indeed possible.

Accessibility, welding position, component match-up, and the specific nature of a joint are other important considerations in welding design.

■ 30.6 HEAT EFFECTS

WELDING METALLURGY

Heating and cooling are essential and integral components of almost all welding processes and tend to produce metallurgical changes that are often undesirable. In *fusion welding*, the heat is sufficient to melt some of the *base metal*, and this is often followed by a rapid cooling. Thermal effects tend to be most pronounced for this type of welding, but they also exist to a lesser degree in processes where the heating–cooling cycle is less severe. If the thermal effects are properly considered, adverse results can usually be avoided or

FIGURE 30-10   Schematic of a butt weld between a plate of metal A and a plate of metal B, with a backing plate of metal C and filler of metal D. The resulting weld nugget becomes a complex alloy of all four metals.

minimized, and excellent service performance can be obtained. If they are overlooked, however, the results can be disastrous.

Because such a wide range of metals are welded and a variety of processes are used, welding metallurgy is an extensive subject, and the material presented here serves only as an introduction. In fusion welding, a pool of molten metal is created, with the molten metal coming from either the parent plate alone (*autogenous welding*) or a mixture of parent and filler material. Figure 30-10 shows a butt weld between plates of material A and material B. A backing strip of material C is used with filler metal of material D. In this situation, the molten pool is actually a complex alloy of all four materials. The molten material is held in place by a metal "mold" formed by the surrounding solids. Since the molten pool is usually small compared to the surrounding metal, fusion welding can often be viewed as *a small metal casting in a large metal mold*. The resultant structure and its properties can be best understood by first analyzing the casting and then considering the effects of the associated heat on the adjacent base material.

Figure 30-11 shows a typical microstructure produced by a fusion weld. In the center of the weld is a region composed of metal that has solidified from the molten state. The material in this *weld pool*, or *fusion zone*, is actually a mixture of parent metal and electrode or filler metal, with the ratio depending upon the particular process, the type of joint, and the edge preparation. Figure 30-12 compares two butt-weld designs, where the weld pool in the upper design would contain a large percentage of base metal and the weld pool in the lower design would be largely filler material. The metal in the fusion zone is cast material with a microstructure reflecting the cooling rate of the weld. This region cannot be expected to have the same properties and characteristics as the wrought material being welded, since their processing histories and resulting structures

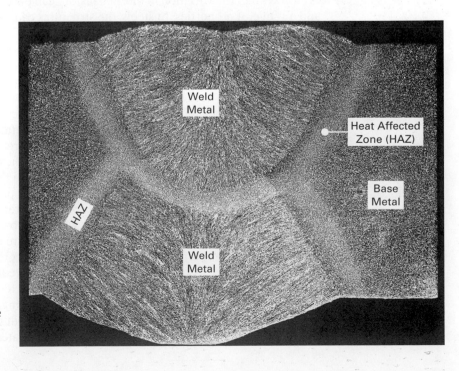

FIGURE 30-11   Grain structure and various zones in a fusion weld.

FIGURE 30-12 Comparison of two butt-weld designs. In the top weld, a large percentage of the weld pool is base metal. In the bottom weld, most of the weld pool is filler metal.

are usually different. Adequate mechanical properties, therefore, can only be achieved by selecting filler rods or electrodes, which have properties *in their as-deposited condition* that equal or exceed those of the wrought parent metal. It is not uncommon, therefore, for the filler metal to have a different chemistry from the metal being welded. The grain structure in the fusion zone may be fine or coarse, equiaxed or dendritic, depending upon the type and volume of weld metal and the rate of cooling, but most electrode and filler rod compositions tend to produce fine, equiaxed grains. The matching or exceeding of base metal strength, in the as-solidified condition, is the basis for several AWS specifications for electrodes and filler rods.

The pool of molten metal created by fusion welding is prone to all of the problems and defects associated with metal casting, such as gas porosity, inclusions, blowholes, cracks, and shrinkage. Since the amount of molten metal is usually small compared to the total mass of the workpiece, rapid solidification and rapid cooling of the solidified metal are quite common. Associated with these conditions may be the entrapment of dissolved gases, chemical segregation, grain size variation, grain shape problems, and orientation effects.

Adjacent to the fusion zone, and wholly within the base material, is the ever-present and generally undesirable *heat-affected zone (HAZ)*. In this region, the parent metal has not melted but has been subjected to elevated temperatures for a brief period of time. Since the temperature and its duration vary widely with location, fusion welding might be more appropriately described as "a metal casting in a metal mold, coupled with an abnormal and widely varying heat treatment." The adjacent metal may experience sufficient heat to bring about structure and property changes, such as phase transformations, recrystallization, grain growth, precipitation or precipitate coarsening, embrittlement, or even cracking. The variation in thermal history can produce a variety of microstructures and a range of properties. In steels, the structures can range from hard, brittle martensite all the way through coarse pearlite and ferrite.

Because of its altered structure, the heat-affected zone may no longer possess the desirable properties of the parent material, and since it was not melted, it cannot assume the properties of the solidified weld metal. Consequently, this is often the weakest area in the as-welded joint. Except where there are obvious defects in the weld deposit, most welding failures originate in the heat-affected zone. This region extends outward from the weld to the location where the base metal has experienced too little heat to be affected or altered by the welding process. Figure 30-13 presents a schematic of a fusion weld in steel, using standard terminology for the various regions and interfaces. Part of the heat-affected zone has been heated above the A_1 transformation temperature and could assume a totally new structure through phase transformation. The lower-temperature portion of the heat-affected zone (peak temperatures below the A_1 value) can experience diffusion-induced changes within the original structure.

Because of the melting, solidification, and exposure to a range of high temperatures, the structure and properties of welds can be extremely complex and varied. Through proper concern, however, associated problems can often be reduced or totally eliminated. Consideration should first be given to the thermal characteristics of the various processes. Table 30-1 classifies some of the more common welding processes with regard to their *rate of heat input*. Processes with low rates of heat input (slow heating) tend to produce large total heat content within the metal, slow cooling rates, large heat-affected zones, and

FIGURE 30-13 Schematic of a fusion weld in steel, presenting proper terminology for the various regions and interfaces. Part of the heat-affected zone has been heated above the transformation temperature and will form a new structure upon cooling. The remaining segment of the heat-affected zone experiences heat alteration of the initial structure. *(Courtesy Sandvik AB, Sandviken, Sweden)*

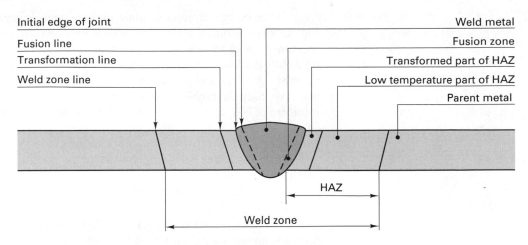

TABLE 30-1	Classification of Common Welding Processes by Rate of Heat Input

Low Rate of Heat Input	High Rate of Heat Input
Oxyfuel welding (OFW)	Plasma arc welding (PAW)
Electroslag welding (ESW)	Electron-beam welding (EBW)
Flash welding (FW)	Laser welding (LBW)
	Spot and seam resistance welding (RW)
	Percussion welding

Moderate Rate of Heat Input
Shielded metal arc welding (SMAW)
Flux cored arc welding (FCAW)
Gas metal arc welding (GMAW)
Submerged arc welding (SAW)
Gas tungsten arc welding (GTAW)

resultant structures with lower strength and hardness, but higher ductility. High-heat-input processes, on the other hand, have low total heats, fast cooling rates, and small heat-affected zones. The size of the heat-affected zone will also increase with increased starting temperature, decreased welding speed, increased thermal conductivity of the base metal, and a decrease in base metal thickness. Weld geometry is also important, with fillet welds producing smaller heat-affected zones than butt welds.

If the as-welded properties are unacceptable, the entire welded assembly might be heat treated after welding. Structure variations can be reduced or eliminated, but the results are restricted to those that can be produced through heat treatment. The structures and properties associated with cold working, for example, could not be achieved. In addition, problems may be encountered in trying to achieve controlled heating and cooling (heat treatment) within the large, complex-shaped structures commonly produced by welding. Moreover, furnaces, quench tanks, and related equipment may not be available to handle the full size of welded assemblies.

An alternative technique to reduce microstructural variation, or the sharpness of that variation, is to *preheat* either the entire base metal or material at least 10 centimeters (4 inches) on either side of the joint just prior to welding. This heating serves to reduce the cooling rate of both the weld deposit and the immediately adjacent metal in the heat-affected zone. The slower cooling produces a softer, more ductile structure and provides more time for the out-diffusion of harmful dissolved hydrogen. The welding stresses are distributed over a larger area, reducing the amount of weld distortion and the possibility of cracking. Preheating is more common with alloy steels and thicker sections, and it is particularly important with the high-thermal-conductivity metals, such as copper and aluminum, where the cooling rate would otherwise be extremely rapid.

If the carbon content of plain carbon steels is greater than about 0.3%, the cooling rates encountered in normal welding may be sufficient to produce hard, untempered martensite, with an accompanying loss of ductility. Since alloy steels possess higher hardenability, the likelihood of martensite formation will be even greater with these materials. Special pre- and postwelding heat cycles (*preheat* and *postheat*) may be required when welding the higher-carbon and alloy steels. For plain-carbon steels, a preheat temperature of 200° to 400°F (100° to 200°C) is usually adequate. Because they can be welded without the need for preheating or postheating, low-carbon, low-alloy steels are extremely attractive for welding applications.

In joining processes where little or no melting occurs, considerable pressure is often applied to the heated metal (as in forge or resistance welding). The weld region experiences deformation, and the resultant structure exhibits the characteristics of a wrought material.

Since steel is the primary metal that is welded, our discussion of metallurgical effects has largely focused on steel. It should be noted, however, that other metals also exhibit heat-related changes in their structure and properties. The exact effects of the heating and cooling associated with welding will depend upon the specific transformations and structural changes that can occur within the materials being joined.

THERMAL EFFECTS IN BRAZING AND SOLDERING

In brazing and soldering, there is no melting of the base metal, but the joint still contains a region of solidified liquid and heat-affected sections within the base material. For these processes, however, another thermal effect may become quite significant. The base and filler metals are usually of radically different chemistries, and the elevated temperatures of joining also promote interdiffusion. Intermetallic phases can form at the interface and alter the properties of the joint. If present in small amounts, they can enhance bonding and provide strength reinforcement. Most *intermetallic compounds*, however, are quite brittle. Too much intermetallic material can result in significant loss of both strength and ductility.

THERMAL-INDUCED RESIDUAL STRESSES

Another effect of heating and cooling is the introduction of *residual stresses*. In welding, these may be of two types and are most pronounced in fusion welding, where the greatest amount of heating occurs. Their effects can be observed in the form of dimensional changes, distortion, and cracking.

Residual welding stresses are the result of restraint to thermal expansion and contraction by the pieces being welded. Consider a rectangular bar of metal that is uniformly heated and cooled. When heated, the material expands and becomes larger in length, width, and thickness. Upon cooling, the material contracts, and each dimension returns to its original value. Now clamp the ends of the bar in a vise so that lengthwise expansion cannot take place and repeat the thermal cycle. Upon heating, all of the expansion is restricted to the width and thickness, but the contraction upon cooling will still occur uniformly. The resulting rectangle will be shorter, thicker, and wider than the original specimen.

Now apply these principles to a weld being made between two plates, like that illustrated in Figure 30-14. As the weld is produced, the liquid region conforms to the shape of the "mold," and the adjacent material becomes hot and expands. The molten pool can absorb expansion of the plate perpendicular to the weld line, but expansion along the length of the weld tends to be restrained by the adjacent plate material that is cooler and stronger. This resistance or restraint is often sufficient to induce deformation of the hot, weak, and thermally expanding heat-affected zone, which now becomes thicker instead of longer.

After the weld pool solidifies, both the weld metal and adjacent heat-affected region cool and contract. The surrounding metal now resists this contraction. The weld region wants to contract but is restrained and forced to remain in a "stretched" condition, known as residual tension (region *T*). The cooling weld, in turn, exerts forces that try to squeeze the adjacent material, producing regions of residual compression (regions *C*). While the

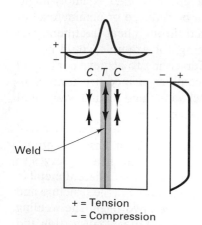

+ = Tension
− = Compression

FIGURE 30-14   Schematic of the longitudinal residual stresses in a fusion-welded butt joint.

FIGURE 30-15 Shrinkage of a typical butt weld in the transverse (a) and longitudinal (b) directions as the material responds to the induced stresses. Note that restricting transverse motion will place the entire weld in transverse tension.

net force must remain at zero (in keeping with the equilibrium laws of physics and mechanics), the localized tensions and compressions can be substantial. As Figure 30-14 depicts, a high residual tension is observed in the weld metal, which becomes compressive and then returns to zero as one moves away from the weld centerline. The magnitude of the residual tension will be relatively uniform along the weld line, except at the ends, where the stresses can be relieved by a pulling in of the edges.

During cooling, the thermal contractions occur both parallel (longitudinal) and perpendicular (transverse) to the weld line. Lateral movement of the material being welded can often compensate for the transverse contractions. The width of the welded assembly simply becomes less than that of the positioned components at the time of welding. Figure 30-15a depicts this reduction in width, while Figure 30-15b illustrates the longitudinal contractions that generate the complex stresses of Figure 30-14.

Components being joined during fabrication typically have considerable freedom of movement, but welds made on nearly completed structures or repair welds often join components that are somewhat restrained. If the welded plates in Figure 30-15a are restrained from horizontal movement, *additional stresses will be induced*. These residual stresses are known as *reaction stresses*, and they can cause cracking of the hot weld or heat-affected material, or contribute to failure during subsequent use. Their magnitude will be an inverse function of the length between the weld joint and the point of restraint, and can be as high as the yield strength of the parent metal (since yielding would occur to relieve any higher stresses).

EFFECTS OF THERMAL STRESSES

Distortion or warping of the assembly can easily result from the nonuniform temperatures and thermal stresses induced by welding. Figure 30-16 depicts some of the distortions that can occur during various welding configurations. Since the causing conditions

FIGURE 30-16 Distortions or warpage that may occur as a result of welding operations: (a) V-groove butt weld where the top of the joint contracts more than the bottom; (b) one-side fillet weld in a T-joint; (c) two-fillet weld in a T-joint with a high vertical web.

can vary widely, no fixed rules can be provided to assure the absence of warping. The following suggestions can help, however.

Total heat input to the weld should be minimized. Welds should be made with the least amount of weld metal necessary to form the joint. Overwelding is not an asset, since it actually increases residual stresses and distortion. Faster welding speeds reduce the welding time and also reduce the volume of metal that is heated. Welding sequences should be designed to use as few passes as possible, and the base material should be permitted to have a high freedom of movement. When constructing a multiweld assembly, it is beneficial to weld toward the point of greatest freedom, such as from the center to the edge.

The initial components can also be oriented out of position, so that the subsequent distortion will move them to the desired final shape. Another common procedure is to completely restrain the components during welding, thereby forcing some plastic flow in the joint and surrounding material. This procedure is used most effectively on small weldments where the reaction stresses will not be high enough to cause cracking.

Still another procedure is to balance the resulting thermal stresses by depositing the weld metal in a specified pattern, such as short lengths along a joint or on alternating sides of a plate. Warping can also be reduced by the use of *peening*. As the weld-bead surface is hammered with the peening tool or material, the metal is flattened and tries to spread. Being held back by the underlying material, the surface becomes compressed or squeezed. Surface rolling of the weld-bead area can have the same effect. In both processes, the compressive stresses induced by the surface deformation serve to offset the tensile stresses induced by welding.

Residual stresses should not have a harmful effect on the strength of weldments, except in the presence of notches or in very rigid structures where no plastic flow can occur. These two conditions should not exist if the welds have been properly designed and proper workmanship has been employed. Unfortunately, it is easy to inadvertently join heavy sections and produce rigid configurations that will not permit the small amounts of elastic or plastic movement required to reduce highly concentrated stresses. In addition, geometric notches, such as sharp interior corners, are often incorporated into welded structures. Other harmful "notches," such as gas pockets, rough beads, porosity, and arc "strikes," can serve as initiation sites for weld failures. These can generally be avoided by proper welding procedures, good workmanship, and adequate supervision and inspection.

The residual stresses of welding can also cause additional distortion when subsequent machining removes metal and upsets the stress equilibrium. For this reason, welded assemblies that are to undergo subsequent machining are frequently given a *stress-relief* heat treatment prior to that operation.

The reaction stresses that contribute to distortion are more often associated with cracking during or immediately following the welding operation (as the weld is cooling). This cracking is most likely to occur when there is great restraint to the shrinkage that occurs transverse to the direction of welding. When a multipass weld is being made, cracking tends to occur in the early beads where there is insufficient weld metal to withstand the shrinkage stresses. These cracks can be quite serious if they go undetected and are not chipped out and repaired, or melted and resolidified during subsequent passes. Figure 30-17 shows the various forms of cracking that can occur as a result of welding.

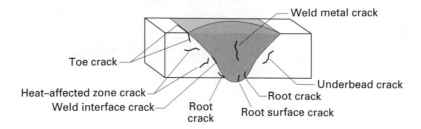

FIGURE 30-17 Various types and locations of cracking that can occur as a result of welding.

To minimize the possibility of fracture, welded joints should be designed to keep restraint to a minimum. The metals and alloys of the structure should be selected with welding in mind (more problems exist with higher-carbon steels, higher-alloy steels, and high-strength materials), and special consideration should be given when welding thicker materials. Crack-prevention efforts can also include maintaining the proper size and shape of the weld bead. While a concave fillet is desirable when machining, a concave weld profile has a greater tendency to crack upon cooling, since contraction actually increases the length of the surface. With a convex profile, the length of the surface will contract simultaneously with the volume, reducing the possibility of surface tension and cracking. Weld beads with high penetration (high depth–width ratio) are also more prone to cracking.

Still other methods to suppress cracking focus on reducing the stresses by making the cooling more uniform or relaxing them by promoting plasticity in the metals being welded. The metals to be welded may be preheated, and additional heat may be applied between the welding passes to retard cooling. Some welding codes also require the inclusion of a thermal stress relief after welding but prior to use. Hydrogen dissolved in the molten weld metal can also induce cracking. Slower welding and cooling will allow any hydrogen to escape, and the use of low-hydrogen electrodes and low-moisture fluxes will reduce the likelihood of hydrogen being present.

■ 30.7 WELDABILITY OR JOINABILITY

It is important to note that not all joining processes are compatible with all engineering materials. While the terms *weldability* or *joinability* imply a reliable measure of a material's ability to be welded or joined, they are actually quite nebulous. One process might produce excellent results when applied to a given material, whereas another might produce a dismal failure. Within a given process, the quality of results may vary greatly with variations in the process parameters, such as electrode material, shielding gases, welding speed, and cooling rate.

Table 30-2 shows the compatibility of the various joining processes with some of the major classes of engineering materials. In each case, the process is classified as recommended (R), commonly performed (C), performed with some difficulty (D), seldom used (S), and not used (N). It should be noted, however, that the classifications are generalizations, and exceptions often exist within both the family of materials (such as the

TABLE 30-2 Weldability or Joinability of Various Engineering Materials[a]

Material	Arc Welding	Oxyacetylene Welding	Electron-Beam Welding	Resistance Welding	Brazing	Soldering	Adhesive Bonding
Cast iron	C	R	N	S	D	N	C
Carbon and low-alloy steel	R	R	C	R	R	D	C
Stainless steel	R	C	C	R	R	C	C
Aluminum and magnesium	C	C	C	C	C	S	R
Copper and copper alloys	C	C	C	C	R	R	C
Nickel and nickel alloys	R	C	C	R	R	C	C
Titanium	C	N	C	C	D	S	C
Lead and zinc	C	C	N	D	N	R	R
Thermoplastics	Heated tool R	Hot gas R	N	Induction C	N	N	C
Thermosets	N	N	N	N	N	N	C
Elastomers	N	N	N	N	N	N	R
Ceramics	N	S	C	N	N	N	R
Dissimilar metals	D	D	C	D	D/C	R	R

[a]C, commonly performed; R, recommended (easily performed with excellent results); D, difficult; N, not used; S, seldom used.

various types of stainless steels) and the types of processes (arc welding here encompasses a large variety of specific processes). Nevertheless, the table can serve as a guideline to assist in process selection.

■ 30.8 SUMMARY

If the potential benefits of welding are to be obtained and harmful side effects are to be avoided, proper consideration should be given to (1) the selection of the process, the process parameters, and the filler material; (2) the design of the joint; and (3) the effects of heating and cooling on both the weld and parent material. Joint design should consider manufacturability, durability, fatigue resistance, corrosion resistance, and safety. Welding metallurgy helps determine the structure and properties across the joint, as well as the need for additional thermal treatments. Further attention may be required to control or minimize residual stresses and distortion.

Parallel considerations apply to brazing and soldering operations, with additional attention to the effects of interdiffusion between the filler and base metals. Flame and arc-cutting operations involve localized heating and cooling, and they also create altered structures in the heat-affected zone. Since many cut products undergo further welding or machining, however, the regions of undesirable structure may not be retained in the final product.

■ Key Words

autogenous weld	fusion zone	joint configuration	reaction stresses
base metal	groove weld	monolithic	residual stresses
bead weld (or surfacing weld)	heat-affected zone (HAZ)	peening	stress relief
coalescence	incomplete fusion	plug weld	thermal cutting
consolidation processes	incomplete penetration	postheat	weld metal (or weld pool)
distortion	insert	preheat	weldability
fillet weld	intermetallic compound	rate of heat input	welding
fusion weld			

■ Review Questions

1. What types of design features favor manufacture as a joined assembly?
2. What types of manufacturing processes fall under the classification of *consolidation processes*?
3. Define *welding*.
4. What four conditions are required to produce an ideal metallurgical bond?
5. What are some of the ways in which welding processes compensate for the inability to meet the conditions of an ideal bond?
6. What are some possible problems associated with the high temperatures that are commonly used in welding?
7. What are the three primary aspects required to produce a high-quality weld?
8. How are welding processes identified by the American Welding Society?
9. What is thermal cutting?
10. What are some of the common types of weld defects?
11. What are the four basic types of fusion welds?
12. What is the role of an insert in welding?
13. What types of weld joints commonly employ fillet welds?
14. What are the five basic joint types for fusion welding?
15. What are some of the factors that influence the cost of making a weldment?
16. Why is it important to consider welded products as monolithic structures?

17. How does the fracture resistance and temperature sensitivity of a steel vary with changes in material thickness?
18. How might excessive rigidity actually be a liability in a welded structure?
19. What is autogenous welding?
20. In what way is the weld-pool segment of a fusion weld like a small metal casting?
21. Why is it possible for the fusion zone to have a chemistry that is different from that of the filler metal?
22. Why is it not uncommon for the selected filler metal to have a chemical composition that is different from the material being welded?
23. What are some of the defects or problems that can occur in the molten metal region of a fusion weld?
24. Why can the resulting material properties vary widely within welding heat-affected zones?
25. What are some of the structure and property modifications that can occur in welding heat-affected zones?
26. Why do most welding failures occur in the heat-affected zone?
27. What are some of the characteristics and consequences of welding with processes that have low rates of heat input?
28. What are some of the difficulties or limitations encountered in heat treating large, complex welded structures?
29. What is the purpose of pre- and postheating in welding operations?

30. What heat-related metallurgical effects may produce adverse results when brazing or soldering?
31. What are some of the undesirable consequences of residual stresses?
32. What is the cause of reaction-type residual stresses?
33. How are reaction stresses affected by the distance between the weld and the point of fixed constraint?
34. What are some of the techniques that can reduce the amount of distortion in a welded structure?

35. Under what conditions might residual stresses have a harmful effect on load-bearing abilities?
36. Why might a welded structure warp if the structure is machined after welding?
37. What are some of the techniques that can be employed to reduce the likelihood of cracking in a welded structure?°
38. Why are the terms *weldability* and *joinability* somewhat nebulous?

■ Problems

1. Through the 1940s the hulls of oceangoing freighters were constructed by riveting plates of steel together. When the defense efforts of World War II demanded accelerated production of freighters to supply U.S. troops overseas, construction of the hulls was converted to welding. The resulting Liberty Ships proved quite successful but also drew considerable attention when minor or moderate impacts (usually under low-temperature conditions) produced cracks of lengths sufficient to scuttle the ship, often up to 50 feet or more. Since the material was essentially the same and the only significant process change had been the conversion from riveting to welding, the welding process was blamed for the failures.

    a. Is this a fair assessment?
    b. What do you think may have contributed to the problem?
    c. What evidence might you want to gather to support your beliefs?

2. Two pieces of AISI 1025 steel are being shielded-metal arc welded with E6012 electrodes. Some difficulty is being experienced with cracking in the weld beads and in the heat-affected zones. What possible corrective measures might you suggest?

3. Figure 30-A schematically depicts the design of a go-cart frame with cross bars and seat support. The assembly is to be constructed from hot-rolled, low-carbon, box-channel material with miter, butt, and fillet welds at the 12 numbered joints. Due to the solidification shrinkage and subsequent thermal contraction of the joint material, the welds are best made when

one or more of the sections are unrestrained. If the structure is too rigid at the time of welding, the associated dimensional changes are restricted, causing the generation of residual stresses that can lead to distortion, cracking, or tears.

    a. Consider the 12 welds in the proposed structure and recommend a welding sequence that would minimize the possibility of hot tears and cracks due to the welding of a restrained joint.
    b. Your company is developing a computer-assisted design program. Suggest one or more rules that may be programmed to aid in the selection of an acceptable weld sequence.

FIGURE 30-A

CHAPTER 31

GAS FLAME AND ARC PROCESSES

■ 31.1 OXYFUEL-GAS WELDING

OXYFUEL-GAS WELDING PROCESSES

Oxyfuel-gas welding (OFW) refers to a group of welding processes that use the flame produced by the combustion of a fuel gas and oxygen as the source of heat. It was the development of a practical *torch* to burn acetylene and oxygen, shortly after 1900, that brought welding out of the blacksmith's shop, demonstrated its potential, and started its development as a manufacturing process. Other processes have largely replaced gas-flame welding in large-scale manufacturing, but the process is still popular for small-scale and repair operations because of its portability, versatility (most ferrous and nonferrous metals can be welded), and the low capital investment required. Acetylene is still the principal fuel gas.

The combustion of oxygen and *acetylene* (C_2H_2) by means of a welding torch of the type shown in Figure 31-1 produces a temperature of about 3250°C (5850°F) in a

FIGURE 31-1 Typical oxyacetylene welding torch and cross-sectional schematic. *(Courtesy of Victor Equipment Company, Denton, TX)*

Mixer Oxygen Oxygen control unit

Fuel gas

Tip Fuel gas control valve

two-stage reaction. In the first stage, the supplied oxygen and acetylene react to produce carbon monoxide and hydrogen:

$$C_2H_2 + O_2 \rightarrow 2CO + H_2 + \text{heat}$$

This reaction occurs near the tip of the torch and generates intense heat. The second stage of the reaction involves the combustion of the CO and H_2 and occurs just beyond the first combustion zone. The specific reactions of the second stage are:

$$2CO + O_2 \rightarrow 2CO_2 + \text{heat}$$

$$H_2 + \frac{1}{2}O_2 \rightarrow H_2O + \text{heat}$$

The oxygen for these secondary reactions is generally obtained from the surrounding atmosphere.

The two-stage combustion process produces a flame having two distinct regions. As shown in Figure 31-2, the maximum temperature occurs near the end of the inner cone, where the first stage of combustion is complete. Most welding should be performed with the torch positioned so that this point of maximum temperature is just above the metal being welded. The outer envelope of the flame serves to preheat the metal and, at the same time, provides shielding from oxidation, since oxygen from the surrounding air is consumed in the secondary combustion.

Three different types of flames can be obtained by varying the oxygen–acetylene (or oxygen–fuel gas) ratio. If the ratio is between 1:1 and 1.15:1, all reactions are carried to completion and a *neutral flame* is produced. Most welding is done with a neutral flame, since it will have the least chemical effect on the heated metal.

A higher ratio, such as 1.5:1, produces an *oxidizing flame*, which is hotter than the neutral flame (about 3600°C or 6000°F) but similar in appearance. Such flames are used when welding copper and copper alloys but are generally considered harmful when

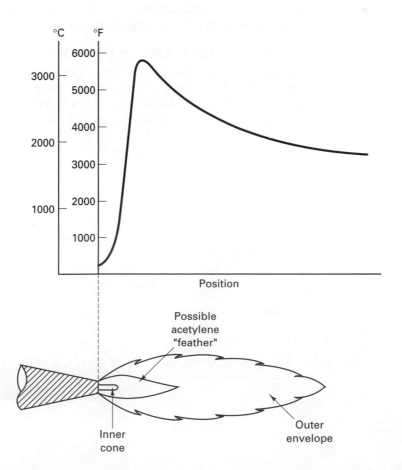

FIGURE 31-2 Typical oxyacetylene flame and the associated temperature distribution.

welding steel because the excess oxygen reacts with the carbon in the steel, lowering the carbon in the region around the weld.

Excess fuel, on the other hand, produces a *carburizing flame*. The excess fuel decomposes to carbon and hydrogen, and the flame temperature is not as great (about 3050°C or 5500°F). Flames with a slight excess of fuel are reducing flames. No carburization occurs, but the metal is well protected from oxidation. Flames of this type are used in welding Monel (a nickel–copper alloy), high-carbon steels, and some alloy steels, and for applying some types of hard-facing material.

For welding purposes, oxygen is usually supplied from pressurized tanks in a relatively pure form, but, in rare cases, air can also be used. The acetylene is usually obtained in portable storage tanks that hold up to 8.5 m³ (300 ft³) at 1.7 MPa (250 psi) pressure. Because acetylene is not safe when stored as a gas at pressures above 0.1 MPa (15 psi), it is usually dissolved in acetone. The storage cylinders are filled with a porous filler. Acetone is absorbed into the voids in the filler material and serves as a medium for dissolving the acetylene.

Alternative fuel gases include propane, propylene, and stabilized methyl acetylene propadiene, best known by the trade name of *MAPP* gas. While flame temperature is slightly lower, these gases can be safely stored in ordinary pressure tanks. Three to four times as much gas can be stored in a given volume, and cost per cubic foot can be less than acetylene. Butane, natural gas, and hydrogen have been used in combination with air or oxygen to weld the low-melting-temperature, nonferrous metals. They are generally not suited to the ferrous metals because the flame atmosphere is oxidizing and the heat output is too low.

The pressures used in gas-flame welding range from 0.006 to 0.1MPa (1 to 15 psi) and are controlled by pressure regulators on each tank. Because mixtures of acetylene and oxygen or air are highly explosive, precautions must be taken to avoid mixing the gases improperly or by accident. All acetylene fittings have left-hand threads, while those for oxygen are equipped with right-hand threads. This prevents improper connections.

The tip size, or orifice diameter of the torch, can be varied to control the shape of the inner cone and the flow rate of the gases. Larger tips permit greater flow of gases, resulting in greater heat input without the higher gas velocities that might blow the molten metal from the weld puddle. Larger torch tips are used for the welding of thicker metal.

USES, ADVANTAGES, AND LIMITATIONS

Almost all oxyfuel-gas welding is *fusion welding*. The metals to be joined are simply melted where a weld is desired and no pressure is required. Because a slight gap often exists between the pieces being joined, *filler metal* can be added in the form of a solid metal wire or rod. Welding rods come in standard sizes, with diameters from 1.5 to 9.5 mm ($^1/_{16}$ to $^3/_8$ in.) and lengths from 0.6 to 0.9 m (24 to 36 in.). They are available in standard grades that provide specified minimum tensile strengths or in compositions that match the base metal. Figure 31-3 shows a schematic of oxyfuel-gas welding using a consumable welding rod.

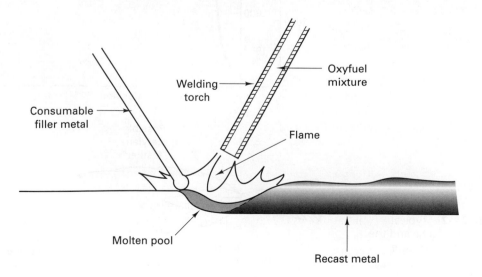

FIGURE 31-3   Oxyfuel-gas welding with a consumable welding rod.

TABLE 31-1	Process Summary: Oxyfuel-Gas Welding (OFW)
Heat source	Fuel gas—oxygen combustion
Protection	Gases produced by combustion
Electrode	None
Material joined	Best for steel and other ferrous metals
Rate of heat input	Low
Weld profile (Depth/Width)	$^1/_3$
Max. penetration	3 mm
Assets	Cheap, simple equipment, portable, versatile
Limitations	Large HAZ, slow

To promote the formation of a better bond, *fluxes* may be used to clean the surfaces and remove contaminating oxide. In addition, the gaseous shield produced by vaporizing flux can prevent further oxidation during the welding process, and the slag produced by solidifying flux can protect the weld pool as it cools. Flux can be added as a powder, the welding rod can be dipped in a flux paste, or the rods can be precoated.

The OFW processes can produce good-quality welds if proper caution is exercised. Welding can be performed in all positions, the temperature of the work can be easily controlled, and the puddle is visible to the welder. However, exposure of the heated and molten metal to the various gases in the flame and atmosphere makes it difficult to prevent contamination. Since the heat source is not concentrated, heating is rather slow. A large volume of metal is heated, and distortion is likely to occur. The thickness of the material being joined is usually less than 6.5 mm ($^1/_4$ in.). Thus, in production applications, the flame-welding processes have largely been replaced by arc welding. Nevertheless, flame welding is still quite common in field work, in maintenance and repairs, and in fabricating small quantities of specialized products.

Oxyfuel equipment is quite portable, relatively inexpensive, and extremely versatile.

A single set of equipment can be used for welding, brazing, and soldering, and as a heat source for bending, forming, straightening, and hardening. With the modifications to be discussed shortly, it can also perform flame cutting. Table 31-1 summarizes some of the key features of oxyfuel-gas welding. Table 31-2 shows its compatibility with some common engineering materials.

PRESSURE GAS WELDING

Pressure gas welding (PGW) is a process that uses equipment similar to the oxyfuel-gas process to produce butt joints between the ends of objects such as pipe and railroad rail. The ends are heated with a gas flame to a temperature below the melting point,

TABLE 31-2	Engineering Materials and Their Compatibility with Oxyfuel Welding
Material	Oxyfuel Welding Recommendation
Cast iron	Recommended with cast iron filler rods; braze welding recommended if there are no corrosion objections
Carbon and low-alloy steels	Recommended for low-carbon and low-alloy steels, using rods of the same material; more difficult for higher carbon
Stainless steel	Common for thinner material; more difficult for thicker
Aluminum and magnesium	Common for aluminum thinner than 1 in.; difficult for magnesium alloys
Copper and copper alloys	Common for most alloys; more difficult for some types of bronzes
Nickel and nickel alloys	Common for nickel, Monels, and Inconels
Titanium	Not recommended
Lead and zinc	Recommended
Thermoplastics, thermosets, and elastomers	Hot-gas welding used for thermoplastics, not used with thermosets and elastomers
Ceramics and glass	Seldom used with ceramics, but common with glass
Dissimilar metals	Difficult; best if melting points are within 50°F; concern for galvanic corrosion
Metals to nonmetals	Not recommended
Dissimilar nonmetals	Difficult

and the soft metal is then forced together under pressure. Pressure gas welding, therefore, is actually a form of solid-state welding where the gas flame simply softens the metal and coalescence is produced by pressure.

■ 31.2 OXYGEN TORCH CUTTING

PROCESSES

Oxyfuel-gas cutting (OFC), commonly called flame cutting, is the most common *thermal cutting* process. In some cases the metal is merely melted by the flame of the oxyfuel-gas torch and is blown away to form a gap, or *kerf*, as illustrated in Figure 31-4. When ferrous metal is cut, however, the process becomes one where the iron actually burns (or oxidizes) at high temperatures according to one or more of the following reactions:

$$Fe + O \rightarrow FeO + heat$$

$$3Fe + 2O_2 \rightarrow Fe_3O_4 + heat$$

$$4Fe + 3O_2 \rightarrow 2Fe_2O_3 + heat$$

Because these reactions do not occur until the metal is above 815°C (1500°F), the oxyfuel flame is first used to raise the metal to the temperature where burning can be initiated. Then a stream of pure oxygen is added to the torch (or the oxygen content of the oxyfuel mixture is increased) to oxidize the iron. The liquid iron oxide and any unoxidized molten iron are then expelled from the joint by the kinetic energy of the oxygen-gas stream. Because of the low rate of heat input and the need for preheating ahead of the cut, oxyfuel cutting produces a relatively large heat-affected zone and associated distortion compared to competing techniques. Therefore, the process is best used where the edge finish or tolerance is not critical and the edge material will either be subsequently welded or removed by machining. Cutting speeds are relatively slow, but the low cost of both the required equipment and its operation make the process attractive for many applications.

Theoretically, the heat supplied by the oxidation will be sufficient to keep the cut progressing, but additional heat is often necessary to compensate for losses to the atmosphere and the surrounding metal. If the workpiece is already hot from other processing, such as solidification or hot working, no supplemental heating is required and a supply of oxygen through a small pipe is all that is needed to initiate and continue a cut. This is known as *oxygen lance cutting* (OLC). A workpiece temperature of about 1200°C (2200°F) is required to sustain continuous cutting.

FIGURE 31-4 Flame cutting of a metal plate.

Oxyfuel-gas cutting works best on metals that oxidize readily but do not have high thermal conductivities. Carbon and low-alloy steels can be readily cut in thicknesses from 5 mm to in excess of 75 cm (30 in.). Stainless steels contain oxidation-resistant ingredients and are difficult to cut, as are aluminum and copper alloys.

FUEL GASES FOR OXYFUEL-GAS CUTTING

Acetylene is by far the most common fuel used in oxyfuel-gas cutting, and the process is often referred to as *oxyacetylene cutting* (OFC-A). Figure 31-5 shows a typical cutting torch. The tip contains a circular array of small holes through which the oxygen–acetylene mixture is supplied to form the heating flame. A larger hole in the center supplies a stream of oxygen and is controlled by a lever valve. The rapid flow of the cutting oxygen not only oxidizes the hot metal but also blows the formed oxides from the cut.

If the torch is adjusted and manipulated properly, it is possible to produce a relatively smooth cut. Cut quality, however, depends upon careful selection of the process variables, including preheat conditions, oxygen flow rate, and cutting speed. Oxygen purities over 99.5% are required for the most efficient cutting. If the purity drops to 98.5%, cutting speed will be reduced by 15%, oxygen consumption will increase by 25%, and the quality of the cut will diminish.

Cutting torches can be manipulated manually. However, when the process is applied to manufacturing, the desired path is usually controlled by mechanical or programmable means. Specialized equipment has been designed to produce straight cuts in flat stock and square-cut ends on pipe. The marriage of computer numerically controlled (CNC) machines and cutting torches has also proven to be quite popular. This approach, along with the use of robot-mounted torches, provides great flexibility along with good precision and control.

Fuel gases other than acetylene can also be used for oxyfuel-gas cutting, the most common being natural gas (OFC-N) and propane (OFC-P). While their flame temperatures are lower than that of acetylene, their use is generally a matter of economics and gas availability. For certain special work, hydrogen can also be used (OFC-H).

STACK CUTTING

When a modest number of duplicate parts are to be cut from thin sheet, but not enough to justify the cost of a blanking die, stack cutting may be the answer. The sheets should be flat, smooth, and free of scale, and they should be clamped together tightly so that there are no intervening gaps that could interrupt uniform oxidation or permit slag or

FIGURE 31-5 Oxyacetylene cutting torch and cross-sectional schematic. *(Courtesy of Victor Equipment Company, Denton, TX)*

FIGURE 31-6 Underwater cutting torch. Note the extra set of gas openings in the nozzle to permit the flow of compressed air and the extra control valve. *(Courtesy of Bastian-Blessing Company, Chicago, IL)*

molten metal to be entrapped. Obviously, stack cutting will produce a less accurate cut than could be achieved by using a blanking die.

METAL POWDER CUTTING, CHEMICAL FLUX CUTTING, AND OTHER THERMAL METHODS

When cutting hard-to-cut materials, modified torch techniques may be required. Metal powder cutting (POC) injects iron or aluminum powder into the flame to raise its cutting temperature. Chemical flux cutting (FOC) adds a fine stream of special flux to the cutting oxygen to increase the fluidity of the high-melting-point oxides. Both of these methods, however, have largely been replaced by plasma arc cutting (PAC), which is discussed as an extension of plasma arc welding, to be presented later in this chapter. Laser- and electron-beam cutting will be presented with their welding parallels in a future chapter.

UNDERWATER TORCH CUTTING

The thermal cutting of materials underwater presents a special challenge. A specially designed torch, like the one shown in Figure 31-6, is used to cut steel. An auxiliary skirt surrounds the main tip, and an additional set of gas passages conducts a flow of compressed air that provides secondary oxygen for the oxyacetylene flame and expels water from the zone where the burning of metal occurs. The torch is either ignited in the usual manner before descent or by an electric spark device after being submerged. Acetylene gas is used for depths up to about 7.5 m (25 ft). For greater depths, hydrogen is used, since the environmental pressure is too great for the safe use of acetylene.

■ 31.3 FLAME STRAIGHTENING

Flame straightening uses controlled, localized *upsetting* as a means of straightening warped or buckled material. Figure 31-7 illustrates the theory of the process. If a straight piece of metal is heated in a localized area, such as the shaded area of the upper diagram, the metal on side *b* will be upset (i.e., plastically deformed) as it softens and tries to expand against the cooler restraining metal. When the upset portion cools, it will con-

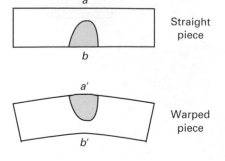

a | Straight piece
b

a' | Warped piece
b'

FIGURE 31-7 Schematic illustrating the theory of flame straightening.

tract, and the resulting piece will be shorter on side *b*, forcing it to bend to the shape in the lower diagram.

If the starting material is bent or warped, as in the lower segment of Figure 31-7, the upper surface can be heated. Upsetting and subsequent thermal contraction will shorten the upper surface at *a'*, bringing the plate back to a straight or flat configuration. This type of procedure can be used to restore structures that have been bent in an accident, such as automobile frames.

A similar process can be used to flatten metal plates that have become dished. Localized spots about 50 mm (2 in.) in diameter are quickly heated to the upsetting temperature while the surrounding metal remains cool. Cool water is then sprayed onto the plate, and the contraction of the upset spot brings the buckle into an improved degree of flatness. To remove large buckles, the process may have to be repeated at several spots within the buckled area.

Several cautions should be noted. When straightening steel, consideration should be given to the possible phase transformations that could occur during the heating and cooling. Since rapid cooling is used and martensite may form, a subsequent tempering operation may be required. In addition, one should also consider the residual stresses that are induced and their effect on subsequent cracking, stress–corrosion cracking, and other modes of failure. The effects of phase transformations and residual stresses have been discussed more fully in Chapter 31.

Also, flame straightening should not be attempted with thin material. For the process to work, the metal adjacent to the heated area must have sufficient rigidity to induce upsetting. If the material is too thin, localized heating and cooling will simply transfer the buckle from one area to another.

■ 31.4 ARC WELDING

With the development of commercial electricity in the late nineteenth century, it was soon recognized that an *arc* between two electrodes was a concentrated heat source that could produce temperatures approaching 4000°C (7000°F). As early as 1881, various attempts were made to use an arc as the heat source for fusion welding. A carbon rod was selected as one *electrode* and the metal workpiece became the other. Figure 31-8 depicts the basic electrical circuit. If needed, *filler metal* was provided by a metallic wire or rod that was independently fed into the arc. As the process developed, the filler metal replaced the carbon rod as the upper electrode. The metal wire not only carried the welding current but, as it melted in the arc, it also supplied the necessary filler.

The results of these early efforts were extremely uncertain. Because of the instability of the arc, a great amount of skill was required to maintain it, and contamination of the weld resulted from the exposure of hot metal to the atmosphere. There was little or no understanding of the metallurgical effects and requirements of arc welding. Consequently, while the great potential was recognized, very little use was made of the process until after World War I. Shielded metal electrodes were developed around 1920. These electrodes enhanced the stability of the arc by shielding it from the atmosphere and provided a fluxing action to the molten pool. The major problems of arc welding were overcome, and the process began to expand rapidly.

All *arc-welding* processes employ the basic circuit depicted in Figure 31-8. Welding currents vary from 1 to 4000 amps, with the range from 100 to 1000 being most typical.

FIGURE 31-8   The basic electrical circuit for arc welding.

Voltages are generally in the range of 20 to 50 volts. If direct current is used and the electrode is made negative, the condition is known as straight polarity (SPDC) or *DCEN*, for direct-current electrode-negative. Electrons are attracted to the positive workpiece, while ionized atoms in the arc column are accelerated toward the negative electrode. Since the ions are far more massive than the electrons, the heat of the arc is more concentrated at the electrode. DCEN processes are characterized by fast melting of the electrode (high metal deposition rates) and a shallow molten pool on the workpiece (weld penetration). If the work is made negative and the electrode positive, the condition is known as reverse polarity (RPDC) or *DCEP*, for direct-current electrode-positive. The positive ions impinge on the workpiece, breaking up any oxide films and giving deeper penetration. The metal deposition rate is lower, however. Sinusoidal *alternating current* provides a 50–50 average of the above two modes and is a popular alternative to the dc conditions. *Variable polarity* power supplies also alternate between DCEP and DCEN conditions, but they use rectangular waveforms to vary the fraction of time in each mode as well as the frequency of switching. Weld characteristics can now be varied over a continuous range between DCEN and DCEP conditions.

In one group of arc-welding processes, the electrode is consumed (*consumable-electrode processes*) and thus supplies the metal needed to fill the joint. Consumable electrodes have a melting temperature below the temperature of the arc. Small droplets are melted from the end of the electrode and pass to the workpiece. The size of these droplets varies greatly, and the transfer mechanism depends on the type of electrode, welding current, and other process parameters. Figure 31-9 depicts metal transfer by the globular, spray, and short-circuit transfer modes. As the electrode melts, the arc length and the electrical resistance of the arc path will vary. To maintain a stable arc and satisfactory welding conditions, the electrode must be moved toward the work at a controlled rate. Manual arc welding is almost always performed with shielded (covered) electrodes. Continuous bare-metal wire can be used as the electrode in automatic or semiautomatic arc welding, but this is always in conjunction with some form of shielding and arc-stabilizing medium and automatic feed control devices that maintain the proper arc length.

The second group of arc-welding processes employs a tungsten electrode, which is not consumed by the arc, except by relatively slow vaporization. In these *nonconsumable-electrode processes*, a separate metal wire is required to supply the filler metal.

Because of the wide variety of processes available, arc welding has become a widely used means of joining material. Each process and application, however, requires the selection or specification of the welding voltage, welding current, arc polarity (straight polarity, reversed polarity, or alternating), arc length, welding speed (how fast the electrode is moved across the workpiece), arc atmosphere, electrode or filler material, and flux. Filler materials must be selected to match the base metal with respect to properties and/or alloy content (chemistry). For many of the processes, the quality of the weld also depends on the skill of the operator. Automation and robotics are reducing this dependence, but the selection and training of welding personnel are still of great importance.

FIGURE 31-9 Three modes of metal transfer during arc welding. *(Courtesy of Republic Steel Corporation, Youngstown, OH)*

Globular   Spray   Short circuit

■ 31.5 CONSUMABLE-ELECTRODE ARC WELDING

Four processes make up the bulk of consumable-electrode arc welding:

1. Shielded metal arc welding (SMAW)
2. Flux-cored arc welding (FCAW)
3. Gas metal arc welding (GMAW)
4. Submerged arc welding (SAW)

These processes all have a medium rate of heat input and produce a fusion zone whose depth is approximately equal to its width. Because the fusion zone is composed of metal from both of the pieces being joined, plus melted filler (i.e., electrode), the electrode must be of the same material as that being welded; the processes cannot be used to join dissimilar metals or ceramics.

SHIELDED METAL ARC WELDING

Shielded metal arc welding (SMAW), also called *stick welding* or covered-electrode welding, is among the most widely used welding processes because of its versatility and because it requires only low-cost equipment. The key to the process is a finite-length electrode that consists of metal wire, usually from 1.5 to 6.5 mm in diameter and 20 to 45 cm in length. Surrounding the wire is a bonded coating containing chemical components that add a number of desirable characteristics, including all or many of the following:

1. Vaporize to provide a protective atmosphere (a gas shield around the arc and pool of molten metal).
2. Provide ionizing elements to help stabilize the arc, reduce weld metal spatter, and increase efficiency of deposition.
3. Act as a *flux* to deoxidize and remove impurities from the molten metal.
4. Provide a protective *slag* coating to accumulate impurities, prevent oxidation, and slow the cooling of the weld metal.
5. Add alloying elements.
6. Add additional filler metal.
7. Affect arc *penetration* (the depth of melting in the workpiece).
8. Influence the shape of the weld bead.

The coated electrodes are classified by the tensile strength of the deposited weld metal, the welding position in which they may be used, the preferred type of current and polarity (if direct current), and the type of coating. A four- or five-digit system of designation has been adopted by the American Welding Society (AWS) and is presented in Figure 31-10. As an example, type E7016 is a low-alloy steel electrode that will provide a deposit with a minimum tensile strength of 70,000 psi (485 MPa) in the non-stress-relieved condition; it can be used in all positions, with either alternating or reverse-polarity direct current; and it has a low-hydrogen plus potassium coating. To assist in identification, all electrodes are marked with colors in accordance with a standard established by the National Electrical Manufacturers Association. Electrode selection consists of determining the electrode coating, coating thickness, electrode composition, and electrode diameter. The current type and polarity are matched to the electrode.

A variety of electrode coatings have been developed. The cellulose and titania (rutile) coatings contain: SiO_2; TiO_2; small amounts of FeO, MgO, and Na_2O; and volatile matter. Upon decomposition, the volatile matter may release hydrogen, which can dissolve in the weld metal and lead to embrittlement or cracking in the joint. Low-hydrogen electrodes are available with compositions designed to provide shielding without the emission of hydrogen. Since many of the electrode coatings can absorb moisture, and this is another source of undesirable hydrogen, the coated electrodes are often baked just prior to use.

To initiate a weld, the operator briefly touches the tip of the electrode to the workpiece and quickly raises it to a distance that will maintain a stable arc. The intense heat quickly melts the tip of the electrode wire, the coating, and portions of the adjacent base

E (X) XX X X

		0–8		
	Current	Polarity	Type of coating
Minimum tensile strength, in 1.000 psi, as-deposited weld metal in non-stress-relieved condition.	AC or DC	Straight or reversed	0 Cellulosic
			1 Cellulosic + Ca and K
			2 Titania
Welding position			3 High titania–potassium
1 All			4 Titania + iron powder
2 Flat and horizontal			5 Low hydrogen (lime)
3 Flat only			6 Low hydrogen + potassium
45			7 Cellulosic + iron powder
60			8 Low hydrogen + iron powder
70			
80			
90			
100			
120			

FIGURE 31-10   Designation system for arc-welding electrodes.

metal. As part of the electrode coating melts and vaporizes, it forms a protective atmosphere of CO, CO_2, and other gases that stabilizes the arc and protects the molten and hot metal from contamination. Other coating components surround the metal droplets with a layer of liquid flux and slag. The fluxing constituents unite with any impurities in the molten metal and float them to the surface to be entrapped in the slag coating that forms over the weld. The slag coating then protects the cooling metal from oxidation and slows down the cooling rate to prevent the formation of hard, brittle structures. The glassy slag is easily chipped from the weld when it has cooled. Figure 31-11 illustrates the shielded metal arc welding process, and Figure 31-12 provides a schematic of metal deposition from a shielded electrode.

FIGURE 31-11   A shielded metal arc welding (SMAW) system.

FIGURE 31-12   Schematic diagram of shielded metal arc welding (SMAW). *(Courtesy of American Iron and Steel Institute, Washington, DC.)*

Iron powder can be added to the electrode coating to significantly increase the amount of weld metal that can be deposited with a given size electrode wire and current. Alloy elements can also be incorporated into the coating to adjust the chemistry of the weld.

Special contact or drag electrodes utilize coatings that are designed to melt more slowly than the filler wire. If these electrodes are tracked along the surface of the work, the faster-melting center wire will be recessed by the proper length to maintain a stable arc.

Since electrical contact must be maintained with the center wire, SMAW electrodes are finite-length "sticks." Length is limited since the current must be supplied near the arc, or the electrode will tend to overheat (by electrical resistance heating) and ruin the coating. Overheating also restricts the weld currents to values below 300 amps (generally about 40 amps per millimeter of electrode diameter). As a result, the arc temperatures are somewhat low, and penetration is generally less than 5 mm ($^3/_{16}$ in.). Welding of material thicker than 5 mm will require multiple passes, and the slag coating must be removed between each pass.

The shielded metal arc process is best used for welding ferrous metals; carbon steels, alloy steels, stainless steels, and cast irons can all be welded. Welds can be made in all positions. DCEP conditions are used to obtain the deepest possible penetration, with alternate modes being employed when welding a thin sheet. The mode of metal transfer is either globular or short circuit.

Shielded metal arc welding is a simple, inexpensive, and versatile process, requiring only a power supply, power cables, an electrode holder, and a small variety of electrodes. The equipment is portable and can even be powered by gasoline or diesel generators. Therefore, it is a popular process in job shops and is used extensively in repair operations. The electrode provides and regulates its own flux, and there is less sensitivity to wind and drafts than in the gas-shielded processes. Welds can be made in all positions. Unfortunately, the process is discontinuous, produces shallow welds, and requires slag removal after each welding pass. Table 31-3 presents a process summary for shielded metal arc welding.

FLUX-CORED ARC WELDING

Flux-cored arc welding (FCAW) overcomes some of the limitations of the shielded metal arc process by moving the powdered flux to the interior of a continuous tubular electrode (Figure 31-13). When the arc is established, the vaporizing flux again produces a protective atmosphere and also forms a slag layer over the weld pool that will require subsequent removal. Alloy additions (metal powders) can be blended into the flux to create a wide variety of filler metal chemistries. Compared to the stick electrodes of the shielded metal arc process, the flux-cored electrode is both continuous and less bulky, since binders are no longer required to hold the flux in place.

The continuous electrode is fed automatically through a welding gun, with electrical contact being maintained through the bare-metal exterior of the wire at a position near the exit of the gun. Overheating of the electrode is no longer a problem, and welding

TABLE 31-3	Process Summary: Shielded Metal Arc Welding (SMAW)
Heat source	Electric arc
Protection	Slag from flux and gas from vaporized coating material
Electrode	Discontinuous, consumable
Material joined	Best for steel
Rate of heat input	Medium
Weld profile (D/W)	1
Current	<300 amps
Max. penetration	3–6 mm
Assets	Cheap, simple equipment
Limitations	Discontinuous, shallow welds; requires slag removal

FIGURE 31-13   The flux-cored arc welding (FCAW) process. *(Courtesy of The American Welding Society, New York.)*

currents can be increased to about 500 A. The higher heat input increases penetration depth to about 1 cm ($^3/_8$ in.). The process is best used for welding steels, and welds can be made in all positions. Direct-current electrode-positive (DCEP) conditions are almost always used for the enhanced penetration. High deposition rates are possible, but the equipment cost is greater than that of SMAW because of the need for a controlled wire feeder and more costly power supply. Good ventilation is required to remove the fumes generated by the vaporizing flux.

In the basic flux-cored arc welding process, the shielding gas is provided by the vaporization of flux components. Better protection and cleaner welds can be produced by combining the flux with a flow of externally supplied shielding gas, such as CO_2.

Table 31-4 presents a process summary of flux-cored arc welding.

GAS METAL ARC WELDING

If the supplemental shielding gas flowing through the torch (described above) becomes the primary protection for the arc and molten metal, there is no longer a need for the volatilizing flux. The consumable electrode can now become a continuous, solid, uncoated metal wire or a continuous hollow tube with powdered alloy additions in the center, known as a metal-cored electrode. The resulting process, shown in Figure 31-14, was formerly called metal inert-gas, or MIG, welding and is now known as *gas metal arc welding (GMAW)*. The arc is still maintained between the workpiece and the automatically fed bare-wire electrode, which continues to provide the necessary filler metal.

TABLE 31-4	Process Summary: Flux-Cored Arc Welding (FCAW)
Heat source	Electric arc
Protection	Slag and gas from flux (optional secondary gas shield)
Electrode	Continuous, consumable
Material joined	Best for steel
Rate of heat input	Medium
Weld profile (D/W)	1
Current	<500 amps
Max. penetration	6–10 mm
Assets	Continuous electrode
Limitations	Requires slag removal

FIGURE 31-14 Schematic diagram of gas metal arc welding (GMAW). *(Courtesy of American Iron and Steel Institute, Washington, DC.)*

Electrode diameters range from 0.6 to 6.4 mm. The welding current, penetration depth, and process cost are all similar to the flux-cored process.

Because shielding is provided by the flow of gas, and fluxing and slag-forming agents are no longer required, the gas metal arc process can be applied to all metals. Argon, helium, and mixtures of the two are the primary shielding gases. When welding steel, some O_2 or CO_2 is usually added to improve the arc stability and reduce weld spatter. The cheaper CO_2 can also be used alone when welding steel, provided that a deoxidizing electrode wire is employed. Nitrogen and hydrogen may also be added to modify arc characteristics. Since these shielding gases only provide protection and do not remove existing contamination, starting cleanliness is critical to the production of a good weld.

The specific shielding gases can have considerable effect on the stability of the arc, the metal transfer from the electrode to the work, and also the heat transfer behavior, penetration, and tendency for undercutting (weld pool extending laterally beneath the surface of the base metal). Helium produces the hottest arc and deepest penetration. Argon is intermediate, and CO_2 yields the lowest arc temperatures and shallowest penetrations. Since argon is heavier than air, it tends to blanket the weld area, enabling the use of low gas flow rates. The lighter-than-air helium generally requires higher flow rates than either argon or carbon dioxide.

Electronic controls can be used to alter the welding current, enabling further control of the metal transfer mechanism, shown previously in Figure 31-9. *Short-circuit transfer* (GMAW-S) is promoted by the lowest currents and voltages (14 to 21 volts) and the use of CO_2 shield gas. The advancing electrode makes direct contact with the weld pool, and the short circuit causes a rapid rise in current. Big molten globs form on the tip of the electrode and then separate, forming a gap between the electrode and workpiece. This gap reinitiates a brief period of arcing, but the rate of electrode advancement exceeds the rate of melting in the arc, and another short circuit occurs. The power conditions oscillate between arcing and short circuiting at a rate of 20 to 200 cycles per second. Short-circuit transfer is preferred when joining thin materials and can be used in all welding positions, but it suffers from a high degree of spatter.

If the voltage and amperage are increased, the mode becomes one of *globular transfer*. The electrode melts from the heat of the arc, and metal drops form, with a diameter approximately equal to that of the electrode wire. Gravity and electromagnetic forces then transfer the drops to the workpiece at a rate of several per second. Since gravity plays a role in metal transfer, there is a definite limitation on the positions of welding. With even higher currents and voltages (25 to 32 volts and about 200 amps), argon gas shielding, and DCEP conditions, *spray transfer* (GMAW-ST) occurs. Small droplets emerge from a pointed electrode at a rate of hundreds per second. Because of their small size and the greater electromagnetic effects, the droplets are easily propelled across the arc in any direction, irrespective of the effects of gravity. Spray transfer is accompanied by deep penetration and low spatter. The biggest problem with out-of-position welding may be keeping the rather large molten weld pool in place until it solidifies.

Pulsed spray transfer (GMAW-P) was invented in the 1960s to overcome some of the limitations of conventional spray transfer. In this mode, a low welding current is first used to create a molten globule on the end of the filler wire. A burst of high current then "explodes" the globule and transfers the metal across the arc in the form of a spray. By alternating low and high currents at a rate of 60 to 600 times per second, the filler metal is transferred in a succession of rapid bursts, similar to the emissions of a rapidly squeezed aerosol atomizer. With the pulsed form of deposition, there is less heat input to the weld, and the weld temperatures are reduced. Thinner material can be welded, distortion is reduced, workpiece discoloration is minimized, heat-sensitive parts can be welded, high-conductivity metals can be joined, electrode life is extended, electrode cooling techniques may not be required, and fine microstructures are produced in the weld pool. Welds can be made in all positions, and the use of pulsed power lowers spattering and improves the safety of the process. The high speed of the process is attractive for productivity, and the energy or power required to produce a weld is lower than with other methods (reduced cost). Controls can be adjusted to alter the shape of the weld pool and vary the penetration.

In general, the gas metal arc process is fast and economical and currently accounts for over half of all weld metal deposition. There is no frequent change of electrodes as with the shielded metal arc process. No flux is required, and no slag forms over the weld. Thus, multiple-pass welds can be made without the need for intermediate cleaning. The process can be readily automated, and the lightweight, compact welding unit lends itself to robotic manipulation. A direct-current electrode-positive (DCEP) arc is generally used because of its deep penetration, spray transfer, and the ability to produce smooth welds with good profile. Process variables include type of current, current magnitude, shielding gas, electrode diameter, electrode composition, electrode stickout (extension beyond the gun), welding speed, welding voltage, and arc length. Table 31-5 provides a process summary for gas metal arc welding.

In a process modification known as advanced gas metal arc welding (AGMAW), a second power source is used to preheat the filler wire before it emerges from the welding torch. Less arc heating is needed to produce a weld, so less base metal is melted, producing less dilution of the filler metal and less penetration.

Another recent modification is the use of flat electrode wire, typically having a rectangular cross section of about 4 mm by 0.5 mm. By having a larger surface area participating in the arc, deposition rate is similar to a two-wire feed with only a single-wire delivery system. The arc is also asymmetric. Orienting the wire perpendicular to the weld seam produces a wide, shallow weld pool, suitable for bridging gaps and often eliminating the need to weave during deposition. A narrower, deeper weld pool results when the wire is parallel to the weld. Varying the angle between parallel and perpendicular generates a spectrum of weld-pool geometries.

SUBMERGED ARC WELDING

No shielding gas is used in the *submerged arc welding (SAW)* process, depicted in Figure 31-15. Instead, a thick layer of granular flux is deposited just ahead of a solid bare-wire consumable electrode, and the arc is maintained beneath the blanket of flux with only a few small flames being visible. A portion of the flux melts and acts to remove impurities from the rather large pool of molten metal, while the unmelted excess provides additional

TABLE 31-5	Process Summary: Gas Metal Arc Welding (GMAW)
Heat source	Electric arc
Protection	Externally supplied shielding gas
Electrode	Continuous, consumable
Material joined	All common metals
Rate of heat input	Medium
Weld profile (D/W)	1
Current	<500 amps
Max. penetration	6–10 mm
Assets	No slag to remove
Limitations	More costly equipment than SMAW or FCAW

FIGURE 31-15   (Top) Basic features of submerged arc welding (SAW). *(Courtesy of Linde Division, Union Carbide Corporation, Houston, TX)* (Bottom) Cutaway schematic of submerged arc welding. *(Courtesy of American Iron and Steel Institute, Washington, DC.)*

shielding. The molten flux solidifies into a glasslike covering over the weld. This layer, along with the flux that is not melted, provides good thermal insulation. The slow cooling of the weld metal helps to produce soft, ductile welds. Upon further cooling, the solidified flux cracks loose from the weld (due to the differential thermal contraction) and is easily removed. The unmelted granular flux is recovered by a vacuum system and reused.

Submerged arc welding is most suitable for making flat-butt or fillet welds in low-carbon steels (0.3% carbon). With some preheat and postheat precautions, medium-carbon and alloy steels and some cast irons, stainless steels, copper alloys, and nickel alloys can also be welded. The process is not recommended for high-carbon steels, tool steels, aluminum, magnesium, titanium, lead, or zinc. The reasons for this incompatibility are somewhat varied, including the unavailability of suitable fluxes, reactivity at high temperatures, and low sublimation temperatures.

Because the arc is totally submerged, high welding currents can be used (600 to 2000 A). High welding speeds, high deposition rates, deep penetration, and high cleanliness (due to the flux action) are all characteristic of submerged arc welding. A welding speed of 0.75 m/min in 2.5-cm-thick steel plate is typical. Single-pass welds can be made with penetrations up to 2.5 cm (1.0 in.), and greater thicknesses can be joined by multiple passes. Because the metal is deposited in fewer passes than with alternative processes, there is less possibility of entrapped slag or voids, and weld quality is further enhanced. For even higher deposition rates, multiple electrode wires can be employed.

Limitations to the process include the need for extensive flux handling, possible contamination of the flux by moisture (leading to porosity in the weld), the large volume of slag that must be removed, and shrinkage problems due to the large weld pool. The high heat inputs can produce large grain size structures, and the slow cooling rate

TABLE 31-6	Process Summary: Submerged Arc Welding (SAW)
Heat source	Electric arc
Protection	Granular flux provides slag and an isolation blanket
Electrode	Continuous, consumable
Material joined	Best for steel
Rate of heat input	Medium
Weld profile (D/W)	1
Current	<1000 amps
Max. penetration	25 mm
Assets	High-quality welds, high deposition rates
Limitations	Requires slag removal, difficult for overhead and out-of-position welding, joints often require backing plates

may enable segregation and possible hydrogen or hot cracking. Welding is restricted to the horizontal position, since the flux and slag are held in place by gravity. In addition, chemical control is quite important, since the electrode material often contributes over 70% of the molten weld region.

The electrodes are generally classified by composition and are available in diameters ranging from 1 to 10 mm (0.045 to $^3/_8$ in.). The larger electrodes can carry higher currents and enable more rapid deposition, but penetration is shallower. The welding of alloy steels can be performed in several ways: solid wire electrodes of the desired alloy, plain-carbon electrodes with the alloy additions being incorporated into the flux, or tubular metal electrodes with the alloy additions in the hollow core. Various fluxes are also available and are selected for compatibility with the weld metal. All are designed to have low melting temperatures, good fluidity, and brittleness after cooling.

In a modification of the submerged arc process known as *bulk welding*, iron powder is first deposited into the joint (ahead of the flux) as a means of increasing deposition rate. A single weld pass can then produce enough filler metal to be equivalent to seven or eight conventional submerged arc passes.

Table 31-6 provides a summary of the submerged arc welding process.

STUD WELDING

Stud welding (SW) is an arc-welding process used to attach studs, screws, pins, or other fasteners to a metal surface. A special gun is used, such as the one shown in Figure 31-16. The inserted stud acts as an electrode, and a DC arc is established between the end of

FIGURE 31-16   Diagram of a stud welding gun. (*Courtesy of American Machinist.*)

FIGURE 31-17   (Left) Types of studs used for stud welding. (Center) Stud and ceramic ferrule. (Right) Stud after welding and a section through a welded stud. *(Courtesy of Nelson Stud Welding Co, Elyria, OH)*

the stud and the workpiece. After a small amount of metal is melted, the two pieces are brought together under light pressure and allowed to solidify. Automatic equipment controls the arc, its duration, and the application of pressure to the stud.

Figure 31-17 shows some of the wide variety of studs that are specially made for this process. Many contain a recessed end that is filled with flux. A ceramic ferrule, such as the one shown in the center photo of Figure 31-17, may be placed over the end of the stud before it is positioned in the gun. During the arc, the ferrule serves to concentrate the heat and isolate the hot metal from the atmosphere. It also confines the molten or softened metal and shapes it around the base of the stud, as shown in the photo on the right of Figure 31-17. After the weld has cooled, the brittle ceramic is broken free and removed. Since burn-off or melting reduces the length of the stud, the original dimensions should be selected to compensate.

Stud welding requires almost no skill on the part of the operator. Once the stud and ferrule are placed in the gun and the gun positioned on the work, all the operator has to do is pull the trigger. The cycle is executed automatically and takes less than one second. Thus the process is well suited to manufacturing and can be used to eliminate the drilling and tapping of many special holes. Production-type stud welders can produce over 1000 welds per hour.

■ 31.6 NONCONSUMABLE-ELECTRODE ARC WELDING

GAS TUNGSTEN ARC WELDING

Gas tungsten arc welding (GTAW) was formerly known as tungsten inert-gas (TIG) welding, or Heliarc welding when helium was the shielding gas. A nonconsumable tungsten electrode provides the arc but not the filler metal. Inert gas (argon, helium, or a mixture of them) flows through the electrode holder to provide a protective shield around the electrode, the arc, the pool of molten metal, and the adjacent heated areas. (*Note:* CO_2 cannot be used in this process since it provides inadequate protection for the hot tungsten electrode.) While argon is the most widely used gas, and produces a smoother, more stable arc, helium may be added to increase the heat input (higher welding speeds and deeper penetration). Helium alone may be preferred for overhead welding since it is lighter than air and flows upward.

The composition, diameter, length, and tip geometry (balled, pointed, or truncated cone) of the tungsten or tungsten-alloy electrode are selected based on the material being welded, the thickness of the material, and the type of current being used. The tungsten is often alloyed with thorium oxide, zirconium oxide, cerium oxide, or lanthanum oxide to provide better current-carrying and electron-emission characteristics and longer electrode life. Since tungsten is not consumed at the temperatures of the arc, the arc length remains constant, and the arc is very stable and easy to maintain. Figure 31-18 shows a typical GTAW torch with cables and passages for gas flow, power, and cooling water.

In applications where there is a close fit between the pieces being joined, no filler metal may be needed. When filler metal is required, it is usually supplied as a separate rod, about a meter (3 ft) long and in various diameters, as illustrated in Figure 31-19. The filler metal is generally selected to match the chemistry and/or tensile strength of the metal being welded. When high deposition rates are desired, a separate resistance heating circuit can be provided to preheat the filler wire. As shown in Figure 31-20, the deposition rate of heated wire can be several times that of a cold wire. By oscillating the filler wire from side to side while making a weld pass, the deposition rate can be further increased. The hot-wire process is not practical when welding copper or aluminum, however, because it is difficult to preheat the low-resistivity filler wire.

With skilled operators, gas tungsten arc welding can produce high-quality welds that are very clean and scarcely visible. Since no flux is employed, no special cleaning or slag removal is required. However, the surfaces to be welded must be clean and free

FIGURE 31-20   Comparison of the metal deposition rates in GTAW with cold, hot, and oscillating-hot filler wire. (*Courtesy of* Welding Journal.)

of oil, grease, paint, and rust, because the inert gas does not provide any cleaning or fluxing action. It is also important to control the arc length throughout the process. Since the arc is somewhat bell-shaped, decreasing the standoff distance will decrease the melt and heat-affected widths on the workpiece. However, if the hot tungsten electrode comes into contact with the workpiece or molten pool, it will contaminate the electrode.

All metals and alloys can be welded by this process, and the use of inert gas makes it particularly attractive for the reactive metals, such as aluminum, magnesium, and titanium, as well as the high-temperature refractory metals. Maximum penetration is obtained with direct-current electrode-negative (DCEN) conditions, although alternating current may be specified to break up surface oxides (as when welding aluminum). DCEP or reverse-polarity conditions are used only when welding thin pieces where shallow penetrations are desired. Weld currents should be kept low, since this mode tends to melt the tungsten electrode. Weld voltage is typically 20 to 40 volts, and weld current varies from less than 125 amps for DCEP to 1000 amps for DCEN. A high-frequency, high-voltage, alternating current is often superimposed on the regular AC or DC welding current to make it easier to start and maintain the arc. The pulsed arc gas tungsten arc welding (GTAW-P) modification offers all of the advantages previously cited for pulsed gas metal arc welding, including the ability to weld thinner materials due to the lower heat input and lower temperatures.

GTAW costs more than SMAW and is slower than GMAW. However, it produces a high-quality weld in a very wide range of thicknesses, positions, and geometries. The process has a medium rate of heat input, and the welds have a depth that is approximately equal to the width. The materials being welded are generally thinner than 6.5 mm ($^1/_4$ in.). Table 31-7 provides a process summary.

TABLE 31-7 Process Summary: Gas Tungsten Arc Welding (GTAW)	
Heat source	Electric arc
Protection	Externally supplied shielding gas
Electrode	Nonconsumable
Material joined	All common metals
Rate of heat input	Medium
Weld profile (D/W)	1
Current	<500 amps
Max. penetration	3 mm
Assets	High-quality welds, no slag to be removed
Limitations	Slower than consumable electrode GMAW

FIGURE 31-21   Process schematic of spot welding by the inert-gas-shielded tungsten arc process.

GAS TUNGSTEN ARC SPOT WELDING

A variation of gas tungsten arc welding can be used to produce *spot welds* between two pieces of metal where access is limited to one side of the joint or where thin sheet is being attached to heavier material. The basic procedure is illustrated in Figures 31-21 and 31-22. A modified tungsten inert-gas gun is used with a vented nozzle on the end. The nozzle is pressed firmly against the material, holding the pieces in reasonably good contact. (The workpieces must be sufficiently rigid to sustain the contact pressure.) Inert gas, usually argon or helium, flows through the nozzle to provide a shielding atmosphere. Automatic controls then advance the electrode to initiate the arc and retract it to the correct distance for stabilized arcing. The duration of arcing is timed automatically to produce an acceptable spot weld. The depth and size of the weld nugget are controlled by the amperage, time, and type of shielding gas.

In arc spot welding, the weld nugget begins to form at the surface where the gun makes contact. This is in contrast to the more standard resistance spot-welding methods, where the weld nugget forms at the interface between the two members. Each technique has its characteristic advantages and disadvantages.

PLASMA ARC WELDING

In *plasma arc welding (PAW)* the arc is maintained between a nonconsumable electrode and either the welding gun (*nontransferred arc*) or the workpiece (*transferred arc*), as illustrated in Figure 31-23. The nonconsumable tungsten electrode is set back within the "torch" in such a way as to force the arc to pass through or be contained within a small-diameter nozzle. An inert gas (usually argon) is forced through this constricted arc, where it is heated to a high temperature and forms a hot, fast-moving *plasma*. The emerging gas then transfers its heat to the workpiece and melts the metal. This flow is called the *orifice gas*. A second flow of inert gas surrounds the plasma column and provides shielding to the weld pool. When filler metal is needed, it is provided by an external feed.

Figure 31-24 presents a comparison of the nonconstricted arc of the GTAW process and the constricted arc of plasma arc welding, and shows the differences in temperature distribution. Plasma arc welding is characterized by a high rate of heat input and temperatures on the order of 16,500°C (30,000°F). This in turn offers fast welding speeds, narrow welds with deep penetration (a depth-to-width ratio of about 3), a narrow heat-affected zone, reduced distortion, and a process that is insensitive to variations in arc length since the plasma column is cylindrical. Welds can be made in all positions, and nearly all metals and alloys can be welded.

FIGURE 31-22   Making a spot weld by the inert-gas-shielded tungsten arc process. *(Courtesy of Air Reduction Company Inc., New York, NY)*

FIGURE 31-23 Two types of plasma arc torches. (Left) Transferred arc; (right) nontransferred arc.

Tungsten electrode

Plasma gas

Shielding gas

Arc

Plasma jet

Workpiece

Transferred arc

Nontransferred arc

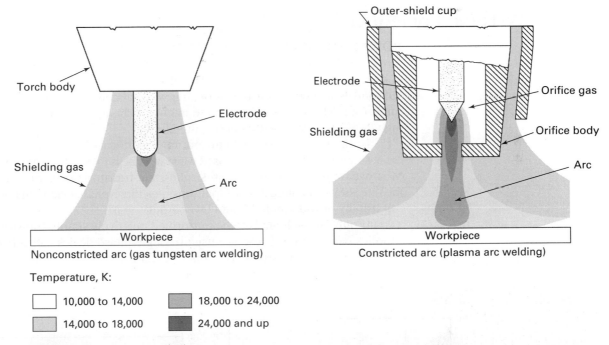

Torch body

Electrode

Shielding gas

Arc

Workpiece

Nonconstricted arc (gas tungsten arc welding)

Outer-shield cup

Electrode

Shielding gas

Orifice gas

Orifice body

Arc

Workpiece

Constricted arc (plasma arc welding)

Temperature, K:

☐	10,000 to 14,000	▨	18,000 to 24,000
▨	14,000 to 18,000	■	24,000 and up

FIGURE 31-24 Comparison of the nonconstricted arc of gas tungsten arc welding and the constricted arc of the plasma arc process. Note the level and distribution of temperature. *(Courtesy ASM International, Materials Park, OH.)*

With a low-pressure plasma and currents between 20 and 100 amps, the metal simply melts and flows into the joint. At higher pressures and currents in excess of 100 amps, a "keyhole" effect occurs in which the plasma gas creates a hole completely through the sheet (up to 20 mm thick) that is surrounded by molten metal. As the torch is moved, liquid metal flows to fill the keyhole. If the gas pressure is increased even further, the molten metal is expelled from the region and the process becomes one of plasma cutting, which is discussed later in this chapter.

Many plasma torches employ a small, nontransferred arc within the torch to heat the orifice gas and ionize it. The ionized gas then forms a good conductive path for the main transferred arc. This dual-arc technique permits instant ignition of a low-current

TABLE 31-8	Process Summary: Plasma Arc Welding (PAW)
Heat source	Plasma arc
Protection	Externally supplied shielding gas
Electrode	Nonconsumable
Material joined	All common metals
Rate of heat input	High
Weld profile (D/W)	3
Current	<500 amps
Max. penetration	12–18 mm
Assets	Can have long arc length
Limitations	High initial equipment cost, large torches may limit accessibility

arc, which is more stable than that of an ordinary plasma torch. Separate DC power supplies are used for the pilot and main arcs. Microplasma, or needle arc, torches can operate with very low currents (0.1 to 20 A) and still produce stable arcs. They are quite useful for welding very thin sheet.

Table 31-8 provides a process summary for plasma arc welding.

■ 31.7 WELDING EQUIPMENT

POWER SOURCES FOR ARC WELDING

Arc welding requires large electrical currents, often in the range of 100 to 1000 amps. The voltage is usually between 20 and 50 volts. Both DC and AC *power supplies* are available and generally employ the "drooping-voltage" characteristics shown in Figure 31-25. These characteristics are designed to minimize changes in welding current as the welding voltage fluctuates within anticipated limits.

In the past, most direct current units were gasoline- or diesel-powered motor-generator sets, and these are still used when welding is to be performed in remote locations. Most welding today, however, uses solid-state transformer-rectifier machines, such as the one shown in Figure 31-26. Operating on a three-phase electrical line, these machines can usually provide both AC and DC output.

If only AC welding is to be performed, relatively simple transformer-type power supplies can be used. These are usually single-phase devices with low power factors. When multiple machines are to be operated, as in a production shop, they are often connected to the various phases of a three-phase supply to help balance the load.

FIGURE 31-25 Drooping-voltage characteristics of typical arc-welding power supplies. (Left) Direct current; (right) alternating current.

FIGURE 31-26   Rectifier-type AC and DC welding power supply. *(Courtesy of Lincoln Electric Company, Cleveland, OH)*

Inverter-based power supplies, introduced in the 1980s, provide great flexibility. Through solid-state electronics, these AC machines can quickly modify the shape and frequency of the pulse waveform or momentarily change the power output. The square-wave technology currently being employed provides improved arc starts and more stable arcs. The percent of time in electrode negative or electrode positive can be adjusted, and output frequency can be varied from 20 to 250 Hz. Through feedback and logic control, the power supply can actually adjust to compensate for changes in a number of process variables.

JIGS, POSITIONERS, AND ROBOTS

Jigs or fixtures (also called *positioners*) are frequently used to hold the work in production welding. By positioning and manipulating the workpiece, the welding operations can often be performed in a more favorable orientation. Parts can also be mounted on numerically controlled (NC) tables that position them with respect to the welding tool.

Industrial robots have replaced humans for many welding applications. They can operate in hostile environments and are capable of producing high-quality welds in a repetitive mode.

■ 31.8 ARC CUTTING

While oxygen torch cutting has been discussed earlier in this chapter, and laser cutting will be covered in a future chapter, there are a number of arc-cutting methods. Virtually all metals can be cut by some form of electric arc. In these processes the material is melted by the intense heat of the arc and then permitted, or forced, to flow away from the region of the slit or notch (*kerf*). Most of the techniques are simply adaptations of the arc-welding procedures discussed in this chapter. Each has its inherent characteristics and capabilities, including tolerance, thickness capability, kerf width, edge squareness, size of the heat-affected zone, and cost. Selection depends upon factors such as tolerance requirements, the subsequent processes that will be performed on the cut part, and the end use of the product.

CARBON ARC AND SHIELDED METAL ARC CUTTING
The carbon arc cutting (CAC) and shielded metal arc cutting (SMAC) methods use the arc from a carbon or shielded metal arc electrode to melt the metal, which is then removed from the cut by gravity or the force of the arc itself. These processes are generally limited to small shops, garages, and homes, where there is limited investment in equipment.

AIR CARBON ARC CUTTING
In air carbon arc cutting, the arc is again maintained between a carbon electrode and the workpiece, but high-velocity jets of air are directed at the molten metal from holes in the electrode holder. While there is some oxidation, the primary function of the air is to blow the molten material from the cut. Air carbon arc cutting is particularly effective for cutting cast iron and preparing steel plates for welding. Speeds up to 0.6 m/min are possible, but the process is quite noisy, and hot metal particles tend to be blown over a substantial area.

OXYGEN ARC CUTTING
In oxygen arc cutting (AOC), an electric arc and a stream of oxygen are combined to make the cut. The electrode is a coated ferrous-metal tube. The coated metal serves to establish a stable arc, while oxygen flows through the bore and is directed on the area of incandescence. With easily oxidized metals, such as steel, the arc preheats the base metal, which then reacts with oxygen, becomes liquefied, and is expelled by the oxygen stream.

GAS METAL ARC CUTTING
If the wire feed rate and other variables of gas metal arc welding (GMAW) are adjusted so that the electrode penetrates completely through the workpiece, cutting rather than welding will occur, and the process becomes gas metal arc cutting (GMAC). The wire feed rate controls the quality of the cut, and the voltage determines the width of the slit or kerf.

GAS TUNGSTEN ARC CUTTING
Gas tungsten arc cutting (GTAC) employs the same basic circuit and shielding gas as used in gas tungsten arc welding, with a high-velocity jet of gas added to expel the molten metal.

PLASMA ARC CUTTING
The torches used in plasma arc cutting (PAC) produce the highest temperatures available from any practical source. With the nontransferred type of torch, the arc column is completely within the nozzle, and a temperature of about 16,500°C (30,000°F) is obtained. With the transfer-type torch, the arc is maintained between the electrode and the workpiece, and temperatures can be as high as 33,000°C (60,000°F). Ionized gases flowing at these temperatures and near supersonic speeds are capable of cutting virtually any electrically conductive material simply by melting it and blowing it away from the cut.

Early efforts to employ this technique showed that the speed, versatility, and operating cost were far superior to those of the oxyfuel cutting methods. However, the early systems could not constrict the arc sufficiently to produce the quality of cut needed to meet the demands of manufacturing. Therefore, plasma arc cutting was generally limited to those materials that could not be cut by the oxidation type of cutting

FIGURE 31-27 Cutting sheet metal with a plasma torch. *(Courtesy of GTE Sylvania, Danvers, MA)*

techniques. In the 1970s radial impingement of water on the arc was found to produce the desired constriction. It provides an intense, highly focused source of heat, and water-injected torches can now cut virtually any metal in any position. Magnetic fields have also been used to constrict the arc and can produce high-quality cuts without the need for water impingement.

Compared to oxyfuel cutting, plasma cutting is more economical (cost per cut is a fraction of oxyfuel), more versatile (can cut all metals as easily as mild steel), and much faster (typically, five to eight times faster than oxyfuel). Cutting speeds up to 7.5 m/min have been obtained in 6-mm-thick aluminum and up to 2.5 m/min in 12.5-mm-thick steel. The combination of the extremely high temperatures and jetlike action of the plasma produces narrow kerfs and remarkably smooth surfaces, nearly as smooth as can be obtained by sawing. Plasma-cut surfaces are often within 2° of vertical, and surface oxidation is nearly eliminated by the cooling effect of the water spray. In addition, the heat-affected zone in the metal is only one-third to one-fourth as large as that produced by oxyfuel cutting, and a preheat cycle is not required in the cutting of steel. Heat-related distortion is extremely small.

Transferred-arc torches are usually used for cutting metals, while the nontransferred type are employed with the low-conductivity nonmetals. Ordinary air or inexpensive nitrogen can be used as the plasma gas for the cutting of all types of metal. Oxygen plasma systems were introduced in the 1980s and are used on carbon and low-alloy steel products with thicknesses ranging from 2 to 32 mm (up to $1^{1}/_{4}$ in.). When cutting thick sections (greater than 12 mm), an argon–hydrogen mixture may be preferred to provide a deeper-penetrating arc. A secondary flow of shielding gas (nitrogen, air, or carbon dioxide) may be used to help cool the torch, blow the molten metal away, shield the arc, and prevent oxidation of the cut surface. The arc-constricting water flow can also serve as a shielding medium.

During the 1990s *high-density*, or *precision plasma*, systems began to appear. Various designs are used to restrict the orifice (i.e., superconstrict the plasma), producing vertical edges (less than 1° taper), close tolerances ($^{1}/_{3}$ that of conventional), and dross-free plasma cutting of thin materials. The lower-amperage torches (10 to 100 amps) are limited to cutting carbon and low-alloy steels less than 16 mm ($^{5}/_{8}$ in.) thick and higher-performance metals (such as stainless and high-strength steels, nickel alloys, titanium, and aluminum) less than 12 mm ($^{1}/_{2}$ in.) thick. In addition, the cutting speeds are slower than conventional plasma cutting, but there is no reduction in the size of the heat-affected zone. *Pulsed plasma arc cutting*, another recent development, can reduce heat input to the workpiece while producing kerfs that are 50% narrower and cleaner edges on the cuts.

Combining a plasma torch with CNC manipulation can provide fast, clean, and accurate cutting, like that shown in Figure 31-27. In this process, the cutting table may be placed underwater as a means of reducing noise, air pollution, dust, and arc glare (dyes are placed in the water). Plasma arc torches can also be incorporated into punch presses to provide a manufacturing machine with outstanding flexibility in producing cut and punched products from a variety of materials.

Plasma arc cutting is also suitable for robot application. A single robot system can be used for both cutting and welding of intricate shapes and contours. Water constriction of the manipulated arc is a problem, however, making it important to select the right process parameters and type of gas for the particular application.

Table 31-9 compares the features of oxyfuel cutting, plasma arc cutting, and laser cutting.

■ 31.9 METALLURGICAL AND HEAT EFFECTS IN THERMAL CUTTING

When used for cutting, the flame and arc processes expose materials to high localized temperatures and can produce harmful metallurgical effects. If the cut edges will be subsequently welded, or if they will be removed by machining, there is little cause for concern. When the edges are retained in the finished product, consideration should be given to the effects of cutting heat and their interaction with the applied loads. In some cases, additional steps may be required to avoid or overcome harmful consequences.

TABLE 31-9	Cutting Process Comparison: Oxyfuel, Plasma Arc, and Laser		
Feature	Oxyfuel Cutting	Plasma Arc Cutting	Laser Cutting
Preferred materials	Carbon steel and titanium	All electrically conductive metals	Metal, plastic, wood, textiles
Quality of cut	Average	Similar to oxyfuel Almost as good as laser on thin material	Good quality—best for plate material less than $1/2$-inch thick
Thickness range			
1. Steel	3/16 inch to unlimited	26 ga. to 3 inch	Foil to 1 inch
2. Stainless	not used	26 ga. to 5 inch	20 ga. to $3/4$ inch
3. Aluminum	not used	22 ga. to 6 inch	20 ga. to $3/4$ inch
Cutting speed or time	Long preheat is required	Fast cutting	Slower than plasma, but faster than oxyfuel

For carbon steels with less than 0.25% carbon, thermal cutting does not produce serious metallurgical effects. However, in steels of higher carbon content, the metallurgical changes can be quite significant, and preheating and/or postheating may be required. For alloy steels, additional consideration should be given to the effects of the various alloy elements.

Because of the low rate of heat input, oxyacetylene cutting will produce a rather large *heat-affected zone*. The arc-cutting methods produce intermediate effects that are quite similar to those of arc welding. Plasma arc cutting is so rapid, and the heat is so localized, that the original properties of a metal are only modified within 1.5 mm of the cut.

All of the thermal cutting processes produce some *residual stresses*, with the cut surface generally in tension. Except in the case of thin sheet, warping should not occur. However, if subsequent machining removes only a portion of the cut surface, or does not penetrate to a sufficient depth, the resulting imbalance in residual stresses can induce distortion. It may be necessary to remove all cut surfaces to a substantial depth to ensure good dimensional stability.

Thermal cutting can also introduce geometrical features into the edge. All flame- or arc-cut edges are rough to varying degrees and thus contain notches that can act as stress raisers and reduce the endurance or fracture strength. If cut edges are to be subjected to high or repeated tensile stresses, the cut surfaces and the heat-affected zone should be removed by machining or at least subjected to a stress-relief heat treatment.

■ Key Words

acetylene	flux-cored arc welding	oxidizing flame	short-circuit transfer
alternating current	fusion welding	oxyfuel-gas cutting	slag
arc	gas metal arc welding	oxyfuel-gas welding	spot weld
arc cutting	gas tungsten arc welding	oxygen lance cutting	spray transfer
arc welding	globular transfer	penetration	stack cutting
bulk welding	heat-affected zone	plasma	straight polarity
carburizing flame	jig	plasma arc welding	stud welding
consumable-electrode process	kerf	positioner	submerged arc welding
DCEN	MAPP	power supply	thermal cutting
DCEP	neutral flame	pulsed arc	torch
electrode	nonconsumable-electrode	pulsed spray transfer	transferred arc
filler metal	process	residual stresses	upsetting
flame straightening	nontransferred arc	reverse polarity	variable polarity
flux	orifice gas	shielded metal arc welding	

■ Review Questions

1. Why does an oxyfuel-gas welding torch usually have a flame with two distinct regions?
2. What is the location of the maximum temperature in an oxyacetylene flame?
3. What function or functions are performed by the outer zone of the welding flame?
4. What three types of flames can be produced by varying the oxygen–fuel ratio?
5. Which type of oxyfuel flame is most commonly used?
6. What are some of the attractive features of MAPP gas?
7. Why might a welder want to change the tip size (or orifice diameter) in an oxyacetylene torch?
8. What is filler metal, and why might it be needed to produce a joint?
9. What is the role of a welding flux?
10. Oxyfuel-gas welding has a low rate of heat input. What are some of the adverse features that result from the slow rate of heating?
11. What are some of the more attractive features of the oxyfuel-gas process?
12. How does pressure gas welding differ from the oxyfuel-gas process?
13. In what way does the torch cutting of ferrous metals differ from cutting nonoxidizing metals?
14. Why might it be possible to use only an oxygen lance to cut hot steel strands as they emerge from a continuous casting operation?
15. How does an oxyacetylene cutting torch differ from an oxyacetylene welding torch?
16. What are some of the ways in which cutting torches can be manipulated?
17. What modification must be incorporated into a cutting torch to permit it to cut metal underwater?
18. If a curved plate is to be straightened by flame straightening, should the heat be applied to the longer or shorter surface of the arc? Why?
19. Why does the flame-straightening process not work for thin sheets of metal?
20. What sorts of problems plagued early attempts to develop arc welding?
21. What are the three basic types of current and polarity that are used in arc welding?
22. What is the difference between a consumable and nonconsumable electrode? For which processes does a filler metal have to be added by a separate mechanism?
23. What are the three types of metal transfer that can occur during arc welding?
24. What are some of the process variables that must be specified when setting up an arc-welding process?
25. What are the four primary consumable electrode arc-welding processes?
26. What are some general properties of the consumable-electrode arc-welding processes?
27. What are some of the functions of the electrode coatings used in shielded metal arc welding?
28. How are welding electrodes commonly classified, and what information does the designation usually provide?
29. Why are shielded metal arc electrodes often baked just prior to welding?
30. What is the function of the slag coating that forms over a shielded metal arc weld?
31. What benefit can be obtair coating of shielded metal arc weld ferrous metals?
32. Why are shielded metal arc electro in length, forcing the process to be one operation?
33. Why is the shielded metal arc-welding process limite welding currents and shallow penetration?
34. What are some of the attractive features of the shielded metal arc-welding process?
35. What is the advantage of placing the flux in the center of an electrode (flux-cored arc welding) as opposed to a coating on the outside (shielded metal arc welding)?
36. What feature enables the welding current in FCAW to be higher than in SMAW?
37. What are some of the advantages of gas metal arc welding compared to the shielded metal arc process?
38. Describe the relative performance of argon, helium, and carbon dioxide gases in creating a high-temperature arc and promoting weld penetration.
39. Which of the metal transfer mechanisms is most used in arc welding?
40. Describe the metal transfer that occurs during pulsed arc gas metal arc welding.
41. What are some of the benefits that can be obtained by the reduced heating of the pulsed arc process?
42. What are some of the primary process variables in the gas metal arc-welding process?
43. What benefits can be gained by using a rectangular cross-section electrode wire as opposed to a round one?
44. What are some of the functions of the flux in submerged arc welding?
45. What are some of the attractive features of submerged arc welding? Major limitations?
46. What is the primary goal or objective in bulk welding?
47. What is the primary objective of stud welding?
48. What is the function of the ceramic ferrule placed over the end of the stud in stud welding?
49. What is the current (or proper) designation for MIG welding? TIG welding? Heliarc welding?
50. What types of shielding gases are used in the gas tungsten arc process?
51. What can be done to increase the rate of filler metal deposition during gas tungsten arc welding?
52. What are some of the attractive features of gas tungsten arc welding?
53. How are the spot welds produced by gas tungsten arc spot welding different from those made by conventional resistance spot welding?
54. How is the heating of the workpiece during plasma arc welding different from the heating in other arc-welding techniques?
55. What are the two different gas flows in plasma arc welding?
56. What are some of the attractive features of plasma arc welding?
57. What is the primary difference between plasma arc welding and plasma arc cutting?
58. What are the attractive features or benefits of an inverter-based power supply?
59. What is the kerf in thermal cutting operations?
60. What is the purpose of the oxygen in oxygen arc cutting?

is plasma arc cutting an attractive way of cutting high-
..ting-point materials?

.hat techniques can be used to constrict the arc in plasma arc
cutting, producing a narrower, more controlled cut?
Compared to oxyfuel cutting, what are some of the attractive
features of the plasma technique?

64. Describe the relative size of the heat-affected zone for the various cutting processes: oxyfuel, arc, and plasma.

65. Why might the residual stresses induced during cutting operations be objectionable?

66. Why might it be wise to machine away the thermally cut edge
and heat-affected zone of metal that will be used as a stressed
machine part?

www.wiley.com/college/degarmo

Chapter 31   CASE STUDY

Bicycle Frame Construction and Repair

As a new employee in a bicycle shop, you learn that
customers will frequently seek advice regarding various types of bicycle repair. Bent or broken frames appear
to be a common problem, and you learn that inexpensive
bicycles generally have frames made from welded low-carbon steel tubing. This material can be heated to straighten
and is easily repair welded using a wide variety of welding
techniques. As you move to the more costly lightweight or
high-performance models, you learn that repair is generally
not quite as simple.

a. In one case, the frame of a high-quality, lightweight bicycle had been fabricated from aluminum-alloy tubing, with
joints made by adhesive bonding of the tubes to connectors that incorporate either internal lugs or external
sleeves. When the frame fractured near one of the joints,
the owner contacted a local auto body shop and requested that a repair be made using conventional gas tungsten
arc welding. The welder was familiar with the welding of
aluminum, and the repair seemed to be of good quality.
Shortly thereafter, however, the frame broke again. This
time the fracture was adjacent to the repair weld and the
characteristics of the break were different. While the first
fracture was somewhat brittle in nature, the second appeared to be more ductile, with evidence of metal flow
prior to fracture. Since the second fracture occurred in the
tube material, not the weld itself, the welder felt that the
failure was not related to the attempted repair and that the
material in the tubing must be defective.

1. If the material had been cold-drawn aluminum tubing
(i.e., strain hardened), explain what may have occurred
during the repair. What is the probable cause of the

second fracture? Was the weld in any way defective?
Was the second failure related to the welding repair?

2. If the tubing had been strengthened by an age-hardening heat treatment, could the same results have
occurred? Explain.

3. Is there a better means of repairing the original fracture? What would have been your recommendation?

b. Titanium offers the strength of heat-treated steel at
approximately half of the weight. Magnesium, while not
as strong as steel or aluminum, is the lightest-weight
engineering metal. Would these materials be appropriate for bicycle frame construction? If so, how would
they be assembled?

c. If the bicycle frame deflects, motion of the cyclist and
related energy are wasted. Therefore, a rigid frame is
quite desirable. Beryllium is an extremely rigid, lightweight metal. Could it be used as a material for bicycle
frames? How would you fabricate the tubing and
assemble the frame?

d. Composite materials can be used to produce tailored
sets of properties. Fiber-reinforced composites can
have extremely high rigidity in the direction of fiber orientation, coupled with extremely light weight. If you
were to assemble a composite frame using fiber-
reinforced tubing, such as graphite fiber-reinforced
epoxy, how would you join the assembly?

e. Premium-quality racing bikes (such as Tour De France
models) have used one-piece carbon-fiber composite
frames (i.e. no joints at all!). What is the benefit of such
a design?

RESISTANCE AND SOLID-STATE WELDING PROCESSES

■ 32.1 INTRODUCTION

As indicated in the lists of Figures 30-1 and 30-2, there are a number of welding and cutting processes that utilize heat sources other than oxyfuel flames and electric arcs, and some use no heat source at all. We begin this chapter with a group of processes that use electrical resistance heating to form the joint. A second group of processes, known as solid-state welding processes, create joints without any melting of the workpiece or filler material.

■ 32.2 THEORY OF RESISTANCE WELDING

In *resistance welding*, heat and pressure are combined to induce coalescence. *Electrodes* are placed in contact with the material, and electrical resistance heating is used to raise the temperature of the workpieces and the interface between them. The same electrodes that supply the current also apply the pressure, which is usually varied throughout the weld cycle. A certain amount of pressure is applied initially to hold the workpieces in contact and thereby control the electrical resistance at the interface. When the proper temperature has been attained, the pressure is increased to induce coalescence. Because pressure is utilized, coalescence occurs at a lower temperature than that required for oxyfuel gas or arc welding. In fact, melting of the base metal does not occur in many resistance-welding operations. Resistance-welding processes might well be considered as a form of solid-state welding, although they are not officially classified as such by the American Welding Society.

In some resistance-welding processes, additional pressure is applied immediately after coalescence to provide a certain amount of forging action. Accompanying the deformation is a certain amount of grain refinement. Additional heating can also be employed after welding to provide tempering and/or stress relief.

The required temperature can often be attained, and coalescence can be achieved, in a few seconds or less. Resistance welding, therefore, is a very rapid and economical process, extremely well suited to automated manufacturing. No filler metal is required, and the tight contact maintained between the workpieces excludes air and eliminates the need for fluxes or shielding gases.

HEATING

The heat for resistance welding is obtained by passing a large electrical current through the workpieces for a short period of time. The amount of heat input can be determined by the basic relationship:

$$H = I^2 Rt$$

where:

> H = total heat input in joules
>
> I = current in amperes
>
> R = electrical resistance of the circuit in ohms
>
> t = length of time during which current is flowing in seconds

It is important to note that the workpieces actually form part of the electrical circuit, as illustrated in Figure 32-1, and that the total resistance between the electrodes consists of three distinct components:

1. The bulk resistance of the electrodes and workpieces—the upper electrode, upper workpiece, lower workpiece and lower electrode

2. The contact resistance between the electrodes and the workpieces—between the upper electrode and upper workpiece and the lower electrode and lower workpiece

3. The resistance between the surfaces to be joined, known as the *faying surfaces*

Since the maximum amount of heat is generated at the point of maximum resistance, it is desirable to have this be the location where the weld is to be made. Therefore, it is essential to keep components 1 and 2 as low as possible with respect to resistance 3. The bulk resistance of the electrodes is always quite low, and that of the workpieces is determined by the type and thickness of the metal being joined. Because of the large areas involved and the relatively high electrical conductivity of most metals, the workpiece resistances are usually much less than the contact or interface values. Resistance 2 (the resistance between the electrodes and workpieces) can be minimized by using electrode materials that are excellent electrical conductors and by controlling the size and shape of the electrodes and the applied pressure. Any change in the pressure between the electrodes and workpieces, however, also affects the contact between the faying surfaces. Therefore, only limited control of the electrode-to-work resistance can be obtained by pressure variation.

The final resistance, that between the faying surfaces, is a function of (1) the quality (surface finish or roughness) of the surfaces; (2) the presence of nonconductive scale, dirt, or other contaminants; (3) the pressure; and (4) the contact area. These factors must all be controlled if uniform resistance welds are to be produced.

As indicated in Figure 32-2, the objective of resistance welding is to bring both of the faying surfaces to the proper temperature, while simultaneously keeping the remaining material and the electrodes relatively cool. Water cooling is usually used to keep the electrode temperature low and thereby extend their useful life.

FIGURE 32-1 The basic resistance welding circuit.

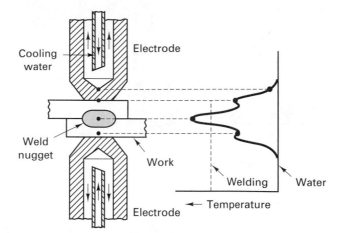

FIGURE 32-2   The desired temperature distribution across the electrodes and workpieces during resistance welding.

PRESSURE

Because the applied pressure promotes a forging action, resistance welds can be produced at lower temperatures than welds made by other processes. Controlling both the magnitude and timing of the pressure, however, is very important. If too little pressure is used, the contact resistance will be high and surface burning or pitting of the electrodes may result. If excessive pressure is applied, molten or softened metal may be expelled from between the faying surfaces or the electrodes may indent the softened workpiece. Ideally, moderate pressure should be applied to hold the workpieces in place and establish proper resistance at the interface prior to and during the passage of the welding current. The pressure should then be increased considerably just as the proper welding temperature is attained. This completes the coalescence and forges the weld to produce a fine-grained structure.

On small, foot-operated machines, only a single spring-controlled pressure is used. On larger, production-type welders, the pressure is generally applied through controllable air or hydraulic cylinders.

CURRENT AND CURRENT CONTROL

While surface conditions and pressure are important variables, the temperature achieved during resistance welding is primarily determined by the magnitude and duration of the welding current. The various resistances change as current flows and the material heats. The bulk resistances of metal increase as temperature rises, and the contact resistances decrease as the metal softens and pressure improves the contact. Since the best conditions are the initial ones, high currents and short time intervals are generally preferred. The weld location can attain the desired temperature while minimizing the amount of heat generated in or dissipated to the adjacent material.

In production-type welders, the magnitude, duration, and timing of both current and pressure can be programmed to follow specified cycles. Figure 32-3 shows a relatively simple cycle for a resistance weld that includes both forging and postheating operations.

FIGURE 32-3   A typical current and pressure cycle for resistance welding. This cycle includes forging and postheating operations.

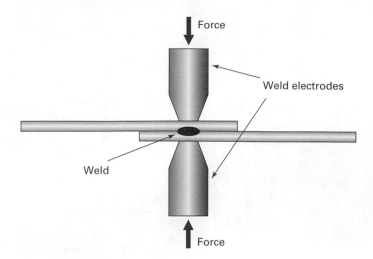

FIGURE 32-4 The arrangement of the electrodes and workpieces in resistance spot welding.

The quality of the final weld, therefore, often depends more on the development of a proper schedule and the subsequent setup, adjustment, and maintenance of equipment than it does on operator skill.

POWER SUPPLY

Since the overall resistance in the welding circuits can be quite low, high currents are generally required to produce a resistance weld. Power transformers convert the high-voltage, low-current line power to the high-current (up to 100,000 amps), low-voltage (0.5 to 10 volts) power required for welding. Smaller machines may utilize single-phase circuitry, but the larger units generally operate on three-phase power. Many resistance welders use DC welding current, obtained through solid-state rectification of the three-phase power. These machines reduce the current demand per phase, give a balanced load, and produce excellent welds.

■ 32.3 RESISTANCE WELDING PROCESSES

RESISTANCE SPOT WELDING

Resistance spot welding (RSW) is the simplest and most widely used form of resistance welding, providing a fast, economical means of joining overlapped materials that will not require subsequent disassembly. Even with all of the advances in technology, resistance spot welding is still the dominant method for joining sheet material, and the average steel-bodied automobile contains between 2000 and 5000 spot welds. Figure 32-4 presents a schematic of the process. Overlapped metal sheets are positioned between water-cooled electrodes, which have reduced areas at the tips to produce welds that are usually from 1.5 to 13 mm ($^1/_{16}$ to $^1/_2$ in.) in diameter. The electrodes close on the work, and the controlled cycle of pressure and current is applied to produce a weld at the metal interface. The electrodes are then opened, and the work is removed.

A satisfactory spot weld, like the one shown in Figure 32-5, consists of a *nugget* of coalesced metal formed between the faying surfaces. There should be little indentation

FIGURE 32-5 A spot-weld nugget between two sheets of 1.3-mm (0.05-in.) aluminum alloy. The nugget is not symmetrical because the radius of the upper electrode is greater than that of the lower electrode. *(Courtesy Lockheed Martin Corporation, Bethesda, MD.)*

FIGURE 32-6   Tear test of a satisfactory spot weld, showing how failure occurs outside of the weld.

of the metal under the electrodes. As shown in Figure 32-6, the strength of the weld should be such that in a tensile or tear test, the weld will remain intact while failure occurs in the heat-affected zone surrounding the nugget. Sound spot welds can be obtained with excellent consistency if proper current density and timing, electrode shape, electrode pressure, and surface conditions are maintained,

Spot-Welding Equipment.   A variety of spot-welding equipment is available to meet the needs of production operations. For light-production work where complex current–pressure cycles are not required, a simple *rocker-arm machine* is often used. The lower electrode arm is stationary, while the upper electrode, mounted on a pivot arm, is brought down into contact with the work by means of a spring-loaded foot pedal. Rocker-arm machines are available with throat depths up to about 1.2 m (48 in.) and transformer capacities up to 50 kVa. They are used primarily on steel.

Larger spot welders, and those used at high production rates, are generally of the press type, as shown in Figure 32-7. On these machines, the movable electrode is controlled by an air or hydraulic cylinder, and complex pressure cycles can be programmed. Capacities up to 500 kVa with a 1.5-m (60-in.) throat depth are quite common. Special-purpose press-type welders can employ multiple welding heads to make up to 200 simultaneous spot welds in less than 60 seconds.

Quite often, the desired products are too large to be manipulated and positioned on a welding machine. Portable spot-welding guns have been instrumental in extending the process to such applications. The guns are connected to a stationary power supply and control unit by flexible air hoses, electrical cables, and water-cooling lines. They can be used in a manual fashion or installed on industrial robots where programmed positioning enables quality spot welds to be produced in a highly automated fashion. Robotic spot welding is currently the most common means of joining sheet metal components in the automotive industry.

Electronic advances in the late 1980s enabled the welding transformer to be integrated into the welding gun. By transforming the power immediately adjacent to the area of use, the small integral transformer guns, or *transguns*, offer reduced power losses and enhanced process efficiency. However, if accurate positioning is required in an articulated system like an industrial robot, the added weight of the integral transformer may become a disadvantage. Servomotors have also been incorporated into a variety of spot-welding machines to control the electrode positioning, speed of closure, level of applied torque or pressure, and rate at which the load is applied.

Electrodes.   Resistance spot-welding electrodes must conduct the welding current to the work, set the current density at the weld location, apply force, and help dissipate heat during the noncurrent portions of the welding cycle. Electrical and thermal conductivity properties are important considerations for electrode selection. Hot compressive strength must be sufficient to resist electrode deformation during the application of pressure. In addition, the electrode should not melt under welding conditions and should be of a composition that does not alloy with the material being welded—a

FIGURE 32-7   Single-phase, air-operated, press-type resistance welder with microprocessor control. *(Courtesy Sciaky Inc., Chicago, IL.)*

TABLE 32-1	Metal Combinations That Can Be Spot Welded										
Metal	Aluminum	Brass	Copper	Galvanized Iron	Iron (Wrought)	Monel	Nichrome	Nickel	Nickel Silver	Steel	Tin Plate	Zinc
Aluminum	x										x	x
Brass		x	x	x	x	x	x	x	x	x	x	x
Copper		x	x	x	x	x	x	x	x	x	x	x
Galvanized Iron		x	x	x	x	x	x	x	x	x	x	
Iron (Wrought)		x	x	x	x	x	x	x	x	x	x	
Monel		x	x	x	x	x	x	x	x	x	x	
Nichrome		x	x	x	x	x	x	x	x	x	x	
Nickel		x	x	x	x	x	x	x	x	x	x	
Nickel Silver		x	x	x	x	x	x	x	x	x	x	
Steel		x	x	x	x	x	x	x	x	x	x	
Tin Plate	x	x	x	x	x	x	x	x	x	x		
Zinc	x	x	x									x

phenomenon that promotes sticking or galling and electrode wear. The Resistance Welder Manufacturers Association (RWMA) has standardized various electrode geometries and has approved a variety of electrode materials, including copper-base alloys, refractory metals, and refractory–metal composites.

Spot-Weldable Metals and Geometries. While steel is clearly the most common metal that is spot welded, one of the greatest advantages of the process is that virtually all of the commercial metals can be joined, and most of them can be joined to each other. In only a few cases do the welds tend to be brittle. Table 32-1 shows some of the many combinations of metals that can be successfully spot welded.

While spot welding is primarily used to join wrought sheet material, other forms of metal can also be welded. Sheets can be attached to rolled shapes and steel castings, as well as some types of nonferrous die castings. Most metals require no special preparation, except to be sure that the surface is free of corrosion and is not badly pitted. For best results, aluminum and magnesium should be cleaned immediately prior to welding by some form of mechanical or chemical technique. Metals that have high electrical conductivity require clean surfaces to ensure that the electrode-to-metal resistance is low enough for adequate temperature to be developed within the metal itself. Silver and copper are especially difficult to weld because of their high thermal conductivity. Higher welding currents coupled with water cooling of the surrounding material may be required if adequate welding temperatures are to be obtained.

When the two pieces being joined are of the same thickness, the practical limit for spot welding is about 3 mm ($^{1}/_{8}$ in.) for each sheet. Sheets of differing thickness can also be joined, and thin pieces can be attached to material that is considerably thicker than 3 mm. When metals of different thickness or different conductivity are to be welded, however, a larger electrode or one with higher conductivity is often used against the thicker or higher-resistance material to ensure that both workpieces will be brought to the desired temperature in a simultaneous fashion.

RESISTANCE SEAM WELDING

Resistance seam welds (RSEW) can be made by two distinctly different processes. In the first process, sheet metal segments are joined to produce gas- or liquid-tight vessels, such as gas tanks, mufflers, and simple heat exchangers. The weld is made between overlapping sheets of metal, and the seam is simply a series of overlapping spot welds, like

FIGURE 32-8 Seam welds made with overlapping spots of varied spacing. *(Courtesy Taylor-Winfield Corporation, Brookfield, OH.)*

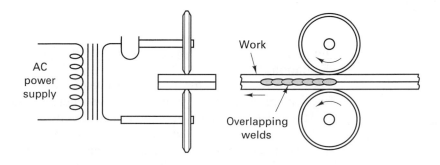

FIGURE 32-9 Schematic representation of the seam-welding process.

those shown in Figure 32-8. The basic equipment is the same as for spot welding, except that the electrodes now assume the form of rotating disks, like those shown schematically in Figure 32-9. As the metal passes between the electrodes, timed pulses of current form the overlapping welds. The timing of the welds and the movement of the work are controlled to ensure that the welds overlap and the workpieces do not get too hot. The welding current is usually a bit higher than in conventional spot welding (to compensate for the short circuit of the adjacent weld), and the workpiece is often cooled by a flow of air or water. In a variation of the process, a continuous current is passed through the rotating electrodes to produce a continuous seam. This form of seam welding is best suited for thin materials, but metals up to 6 mm ($^1/_4$ in.) can be joined. A typical welding speed is about 2 m/min for thin sheet.

The second type of resistance seam welding, known as *resistance butt welding*, is used to produce butt welds between thicker metal plates. The electrical resistance of the abutting metals is still used to generate heat, but high-frequency current (up to 450 kHz) is now employed to restrict the flow of current to the surfaces to be joined and their immediate surroundings. (*Note:* This is similar to the results obtained in the parallel process of high-frequency induction heating.) When the abutting surfaces attain the desired temperature, they are pressed together to form a weld.

Resistance butt welding is used extensively in the manufacture of pipes and tubes, as illustrated in Figure 32-10, but the process is also used to construct simple structural shapes from sections of plate. Material from 0.1 mm (0.004 in.) to more than 20 mm ($^3/_4$ in.) in thickness can be welded at speeds up to 80 m/min (250 fpm). The combination of high-frequency current and high welding speed produces a very narrow heat-affected zone. Almost any type or combination of metal can be welded, including difficult dissimilar metals and the high-conductivity metals, such as aluminum and copper.

PROJECTION WELDING

In a mass-production operation, conventional spot welding is plagued by two significant limitations. Because the small electrodes provide both the high currents and the required pressure, the electrodes generally require frequent attention to maintain their geometry. In addition, the process is designed to produce only one spot weld at a time. When increased strength is required, multiple welds are often needed, and this means

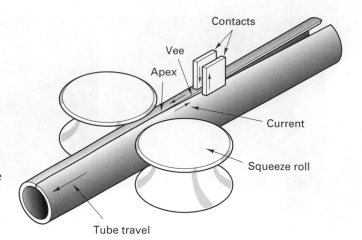

FIGURE 32-10 Using high-frequency AC current to produce a resistance seam weld in butt-welded tubing. Arrows from the contacts indicate the path of the high-frequency current.

multiple operations. *Projection welding* (RPW) provides a means of overcoming these limitations.

Figure 32-11 illustrates the principle of projection welding. A dimple is *embossed* into one of the workpieces at the location where a weld is desired. The two workpieces are then placed between large-area electrodes in a press machine, and pressure and current are applied as in spot welding. Since the current must flow through the points of contact (i.e., the dimples), the heating is concentrated where the weld is desired. As the metal heats and becomes plastic, the pressure causes the dimple to flatten and form a weld.

Because the projections are press formed into the sheet, they can often be produced during previous blanking and forming operations with virtually no additional cost. Moreover, the dimples or projections can be made in almost any shape—round, oval, or circular ring-shaped—to produce welds of shapes that optimize a given design. It is important, however, that the shape be such that the weld will form outward from the center of the projection. Since the heat tends to develop in the piece with the projections, it is best to incorporate them into the heavier of the two pieces to be joined or the metal with the higher conductivity.

Multiple dimples can be incorporated into a sheet, enabling multiple welds to be produced at one time. The number of projections is limited only by the ability of the machine to provide the required current and pressure and the need to uniformly distribute both. If more than three projections are to be made at one time, however, the height of all projections should be uniform to ensure uniform contact and heating.

An attractive feature of projection welding is the fact that it does not require special equipment. Conventional spot-welding machines can be converted to projection welding simply by changing the size and shape of the electrodes. In addition, projection welding leaves no indentation mark on the exterior surfaces, a definite advantage over spot welding when good surface appearance is required.

In a variation of the process, projection welding can also be used to attach bolts and nuts to other metal parts. Contact is made at a projection that has been machined or forged onto the bolt or nut. Current is applied, and the pieces are pressed together to form a weld.

FIGURE 32-11 Principle of projection welding (a) prior to application of current and pressure and (b) after formation of the welds.

■ 32.4 ADVANTAGES AND LIMITATIONS OF RESISTANCE WELDING

The resistance-welding processes have a number of distinct advantages that account for their wide use, particularly in mass-production operations:

1. They are very rapid.
2. The equipment can often be fully automated.
3. They conserve material, since no filler metal, shielding gases, or flux is required.
4. There is minimal distortion of the parts being joined.
5. Skilled operators are not required.
6. Dissimilar metals can be easily joined.
7. A high degree of reliability and reproducibility can be achieved.

The primary limitations of resistance welding include:

1. The equipment has a high initial cost.
2. There are limitations to the thickness of material that can be joined (generally less than 6 mm or $^{1}/_{4}$ inch) and the type of joints that can be made (mostly *lap joints*). Lap joints tend to add weight and material.
3. Access to both sides of the joint is usually required to apply the proper electrode force or pressure.
4. Skilled maintenance personnel are required to service the control equipment.
5. For some materials, the surfaces must receive special preparation prior to welding.

The resistance-welding processes are among the most common techniques for high-volume joining. The rapid heat inputs, short welding times, and rapid quenching by both the base metal and the electrodes can produce extremely high cooling rates in and around the weld. While these conditions can be quite attractive for most nonferrous metals, untempered martensite can form in steels containing more than 0.15% carbon. For these materials, some form of postweld heating is generally required to eliminate possible brittleness.

Table 32-2 provides a process summary for resistance welding.

■ 32.5 SOLID-STATE WELDING PROCESSES

FORGE WELDING

Being the most ancient of the welding processes, *forge welding* (FOW) has both historical and practical value, as it helps us to understand how and why the modern welding practices were developed. The armor makers of ancient times occupied positions of prominence in their society, largely because of their ability to join pieces of metal into single, strong products. The village blacksmith was a more recent master of forge welding. With his hammer and anvil, coupled with skill and training, he could create a wide variety of useful shapes from metal.

TABLE 32-2	Process Summary: Resistance Welding (RW)
Heat source	Electrical resistance heating with high current
Protection	None; isolation of weld site is adequate
Material joined	All common metals (steel, aluminum, and copper)
Rate of heat input	High
Weld profile (D/W)	Does not apply
Maximum penetration	Does not apply
Assets	High speed; Small HAZ; no flux, filler metal, or shielding gas required; adaptable to mass production
Limitations	Equipment is more expensive than arc welding; welds are weaker than arc welds; requires access to both sides of a joint

Using a charcoal forge, the blacksmith heated the pieces to be welded to a practical forging temperature and then prepared the ends by hammering so that they could be properly fitted together. The ends were then reheated and dipped into a borax flux. Heating was continued until the blacksmith judged (by color) that the workpieces were at the proper temperature for welding. They were then withdrawn from the heat and either struck on the anvil or hit by the hammer to remove any loose scale or impurities. The ends to be joined were then overlapped on the anvil and hammered to the degree necessary to produce an acceptable weld.

As the two pieces reduce in thickness, they spread in width, resulting in the creation of new, fresh, uncontaminated metal surface. As these surfaces are being created, the hammer blows also provide the necessary pressure to produce instant coalescence. Thus, by the correct combination of heat and deformation, a competent blacksmith could produce joints that might be every bit as strong as the original metal. However, because of the crudeness of the heat source, the uncertainty of temperature, and the difficulty in maintaining metal cleanliness, a great amount of skill was required and the results were highly variable.

FORGE-SEAM WELDING

Although forge welding has largely been replaced by other joining methods, a large amount of *forge-seam welding* is still used in the manufacture of pipe. As previously presented in Chapter 17, a heated strip of steel is first formed into a cylinder, and the edges are simply pressed together in either a lap or a butt configuration. Welding is the result of pressure and deformation when the metal is pulled through a conical welding bell or passed between welding rolls.

COLD WELDING

Cold welding is a variation of forge welding that uses no heating but produces metallurgical bonds by means of room-temperature plastic deformation. The surfaces to be joined are first cleaned and placed in contact. They are then subjected to high localized pressure, sufficient to cause about 30 to 50% cold work. While some heating will occur due to the severe deformation, the primary factor in producing coalescence is the high pressure acting on newly formed surface material. The cold-welding process is generally confined to the joining of small parts made from soft, ductile metal, such as the electrical connections shown in Figure 32-12.

ROLL WELDING OR ROLL BONDING

In the *roll-welding* or *roll-bonding* (ROW) process, two or more sheets or plates of metal are joined by passing them simultaneously through a rolling mill. As the materials are reduced in thickness, the length and/or width must increase to compensate. The newly

FIGURE 32-12 Small parts joined by cold welding. *(Courtesy of Koldweld Corporation, Willoughby, OH.)*

FIGURE 32-13 Examples of roll-bonded refrigerator freezer evaporators. Note the raised channels that have been formed between the roll-bonded sheets. *(Courtesy Olin Brass, East Alton, IL.)*

FIGURE 32-14 Sequence for making a friction weld. (a) Components with square surfaces are inserted into a machine where one part is rotated and the other is held stationary. (b) The components are pushed together with a low axial pressure to clean and prepare the surfaces. (c) The pressure is increased, causing an increase in temperature, softening, and possibly some melting. (d) Rotation is stopped and the pressure is increased rapidly, creating a forged joint with external flash.

created uncontaminated interfaces are pressed together by the rolls and coalescence is produced. Roll bonding can be performed either hot or cold and can be used to join either similar or dissimilar metals (such as the Alclad aluminums—a skin of high-corrosion-resistance aluminum over a core of high-strength aluminum—or conventional steel with a stainless steel cladding). The resulting bond can be quite strong, as evidenced by the roll-bonded "sandwich" material used in the production of various U.S. coins.

By precoating select portions of one interface surface with a material that prevents bonding, the roll-bonding process can be used to produce sheets that have both bonded and nonbonded areas. Subsequent heating in an oven or furnace can cause the no-bond coating to volatilize. The resulting pressure expands the no-bond regions, producing flow paths for gases or liquids. A common example of this technique is in the manufacture of refrigerator freezer panels, like those shown in Figure 32-13, where inexpensive sheet metal is used to produce structural panels that also serve to conduct the coolant.

FRICTION WELDING AND INERTIA WELDING

In *friction welding* (FRW) the heat required to produce the joint is generated by friction heating at the interface. The components to be joined are first prepared to have smooth, square-cut surfaces. As shown in Figure 32-14, one piece is then held stationary while the other is mounted in a motor-driven chuck or collet and rotated against the first piece at high speed. A low contact pressure may be applied initially to permit cleaning of the surfaces by a burnishing action. The pressure is then increased, and contact friction quickly generates enough heat to soften both components and raise the abutting surfaces to the welding temperature. As soon as this temperature is reached, rotation is stopped and the pressure is further increased to complete the weld. The softened metal is squeezed out to the edges of the joint, forming a *flash*, which can be removed by subsequent machining. Clean, uncontaminated material is left on the interface, and the force creates a "forged" structure in the joint. Friction welding has been used to join steel bars up to 20 cm (8 in.) in diameter and tubes of even larger diameter. The process is also ideal for welding dissimilar metals with very different melting temperatures and physical

FIGURE 32-15 Schematic diagram of the equipment used for friction welding. *(Courtesy of Materials Engineering.)*

properties, such as copper to aluminum, titanium to copper, and nickel alloys to steel. Figure 32-15 shows a schematic of the equipment required for friction welding.

Inertia welding is a modification of friction welding where the moving piece is attached to a rotating flywheel (Figure 32-16). The flywheel is brought to a specified rotational speed, storing a predetermined amount of kinetic energy, and is then separated from the driving motor. The rotating and stationary components are then pressed together, and the kinetic energy of the flywheel is converted into frictional heat at the interface between the two pieces. The weld is formed when the flywheel stops its motion and the pieces remain pressed together. Since the conditions of inertia welding are easily duplicated, welds of extremely consistent quality can be produced and the process can be readily automated.

With inertia welding, the time required to form a weld can be very short, often on the order of several seconds. Because of the high rate of heat input and the limited time for heat to flow away from the joint, both the weld and heat-affected zones are usually very narrow. Oxides and other surface impurities tend to be displaced radially into the upset *flash*, which is generally removed after welding. Because virtually all of the energy is converted to heat, the process is very efficient. No material is melted, so joints can be formed with a wide variety of metals or combinations of metals, including some not normally considered compatible, such as aluminum to steel. Graphite-bearing cast irons, free-machining metals, and some bearing materials must be excluded, since the graphite, lead, or free-machining additive smears across the surface, reducing the friction heating and preventing good solid-state bonding. One, or preferably both, of the components must be sufficiently ductile (when hot) to permit deformation during the forging stage. Grain size tends to be refined during the hot deformation, so the strength of the weld is about the same as that of the base metal. In addition, the friction processes are environmentally attractive since no smoke, fumes, or gases are generated, and no fluxes are required.

Because of the rotational motion, both of the above processes require that one of the components have rotational symmetry. They are used primarily to join round bars or tubes of the same size or connect bars or tubes to flat surfaces. By using linear, orbital, or angular reciprocating motions, friction welding can be extend to noncircular shapes, such as square and rectangular bars. Figure 32-17 shows some typical friction-welded parts.

FIGURE 32-16 Schematic representation of the various steps in inertia welding. The rotating part is now attached to a large flywheel.

FRICTION-STIR WELDING

A relatively new process, first performed by the Welding Institute of Great Britain in 1991, *friction-stir welding* (FSW) has matured rapidly and currently offers significant benefits compared to conventional methods. As illustrated in Figure 32-18, a nonconsumable welding tool (containing a shoulder and protruding cylindrical or tapered probe or pin) is rotated at several hundred revolutions per minute. It is then lowered into the interface between pieces of rigidly clamped material, and frictional heat is generated along the top surface (under the rotating shoulder) and along the surfaces of the rotating probe. After a period of time for heating and softening, the tool is driven along the material interface. As the probe traverses, a plasticized region is continually created. This softened material is swept along the periphery of the pin, flows to the back of the advancing probe, and coalesces to form a solid-state bond. The most common application is the formation of butt welds, usually between plates of the lower-melting-point metals (both wrought and cast alloys) or thermoplastic polymers.

FIGURE 32-17 Some typical friction-welded parts. (Top) Impeller made by joining a chrome–moly steel shaft to a nickel–steel casting. (Center) Stud plate with two mild steel studs joined to a square plate. (Bottom) Tube component where a turned segment is joined to medium-carbon steel tubing. *(Courtesy of Newcor Bay City, Division of Newcor, Inc., Royal Oak, MI.)*

Weld quality is excellent. The extensive plastic deformation creates a refined grain structure with no entrapped oxides or gas porosity. As a result, the strength, ductility, fatigue life, and toughness of the resulting weld are all quite good. Welds in aircraft aluminum are 30 to 50% stronger than those formed by arc welding. Since no material is melted, both wrought and cast alloys can be joined, and they can be joined to each other. No filler material or shielding gas is required, and the process is environmentally friendly (no fumes, weld spatter, or arc glare). Because of the high energy efficiency, total heat input and associated distortion and shrinkage are all low. Joint preparation is minimal, and surface oxides need not be removed. Welding can be performed in any position and requires access to only one side of the plate. Gaps up to 10% of the material thickness can be accommodated with no reduction in weld quality or performance. Weld speed, however, is slower than in most fusion processes.

As shown in Figure 32-18, the key process variables include probe geometry (diameter, depth, and profile), shoulder diameter, rotation speed, downward force, travel speed, and possible tilt to the tool.

Friction-stir welding has been used to weld nearly all of the wrought aluminum alloys, including some that are classified as "unweldable" by fusion processes. Aluminum plates up to 65 mm ($2\frac{1}{2}$ in.) thick have been successfully welded from a single side, and aluminum up to 75 mm (3 in.) thick has been welded by a two-sided process. Copper, lead, magnesium, titanium, and zinc have all been successfully welded, along with thin steel plate and some combinations of dissimilar materials.

As an indication of process capability, the Eclipse 500, a six-passenger, short-hop air taxi, built by Eclipse Aviation of Albuquerque, New Mexico, contains over 135 meters (5300 inches) of friction-stir welds on the aluminum fuselage and wing panel assemblies, eliminating over 60% of the usual rivets. Various thermoplastics, including polyethylene, polypropylene, nylon, polycarbonate, and ABS, have also been successfully welded, with several exhibiting as-welded strengths exceeding 95% of the tensile strength of the base material.

In an extension known as *friction-stir processing*, the thermomechanical features of friction-stir are used for purposes other than creating a joint. By tracking the stir tool through the material with overlapping passes, ultrafine grain size ($<10\ \mu$m) can be produced that enables superplastic forming at comparatively high strain rates. While superplasticity is usually limited to thin sheets, thicker material can now be made superplastic, and by stirring only selected locations, large parts can be made from less expensive conventional materials, with enhanced formability in only the needed locations. In other applications, key surfaces of large castings can be enhanced by the passage of the probe. The cast structure, with possible microporosity and segregation, is replaced by a fine, homogeneous, wrought microstructure. Strength, ductility, corrosion resistance, and fatigue resistance are all improved. Fusion welds can be stirred to replace the cast structure with a fine, worked structure, removing any weld defects and enhancing properties. Reinforcement particles can be stirred into a material to create a

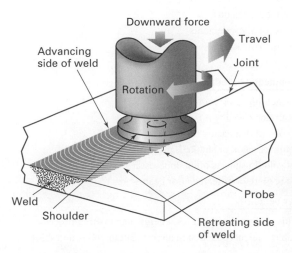

FIGURE 32-18 Schematic of the friction-stir welding process. The rotating probe generates frictional heat, while the shoulder provides additional friction heating and prevents expulsion of the softened material from the joint. (*Note*: To provide additional forging action and confine the softened material, the tool may be tilted so the trailing edge is lower than the leading segment.)

(a)

5mm

(b)

FIGURE 32-19 (a) Top surface of a friction-stir weld joining 1.5-mm- and 1.65-mm-thick aluminum sheets with 1500-rpm pin rotation. The welding tool has traversed left-to-right and has retracted at the right of the photo. (b) Metallurgical cross section through an alloy 356 aluminum casting that has been modified by friction-stir processing.

particle-reinforced composite surface on a standard alloy substrate. Powder products of unique composition can be brought to full density with attractive strength and ductility. In a modification known as friction-stir channeling, a slight upward helix is incorporated onto the surface of the rotating probe. As the probe rotates, material is displaced upward, creating a continuous subsurface channel, such as those used for conveying cooling water.

One of the photos in Figure 32-19 shows a friction-stir weld in aluminum plate, viewed from the top. The other shows the structural changes induced by a friction-stir pass through an aluminum-alloy casting. Note the significant refinement in structure. Table 32-3 summarizes some of the attractive features of the friction-stir process.

TABLE 32-3 Attractive Features of Friction-Stir Welding

Metallurgical benefits:

Excellent weld quality

Applicable to a wide range of materials, including some "nonweldable" by fusion methods

Solid-state process

Low distortion of the workpiece

High joint strength

No loss of alloy elements

Fine microstructure

No cracking

Environmental benefits:

No shielding gas is required

No surface cleaning is required

No solvent degreasing is used

No fumes, gases, or smoke is produced

Postweld finishing is often unnecessary

No arc glare or reflected laser beams

Energy benefits:

Welds produced with far less energy than in other processes

Enables weight reduction in aircraft, automobiles, and ships

FIGURE 32-20   Diagram of the equipment used in ultrasonic welding.

ULTRASONIC WELDING

In *ultrasonic welding* (USW), coalescence is produced by the localized application of high-frequency (10,000 to 200,000 Hz) shear vibrations to surfaces that are held together under rather light normal pressure. Although there is some heating at the faying surfaces, the interface temperature rarely exceeds one-half of the melting point of the material (on an absolute-temperature scale). Instead, it appears that the rapid reversals of stress along the contact interface break up and disperse the oxide films and surface contaminants, allowing the clean metal surfaces to coalesce into a high-strength bond.

Figure 32-20 depicts the basic components of the ultrasonic welding process. The ultrasonic transducer is essentially the same as that employed in ultrasonic machining. It is coupled to a force-application system that contains a welding tip on one end, either stationary for spot welds or rotating for seams. The pieces to be welded are placed between this tip and a reflecting anvil, thereby concentrating the vibratory energy.

Ultrasonic welding is restricted to the lap joint welding of thin materials—sheet, foil, and wire—or the attaching of thin sheets to heavier structural members. The maximum thickness is about 2.5 mm (0.1 in.) for aluminum and 1.0 mm (0.04 in.) for harder metals. As indicated in Table 32-4, the process is particularly attractive because of the number of metals and combinations of dissimilar metals that can be joined. It is even possible to bond metals to nonmetals, such as aluminum to ceramics or glass. Because the temperatures are low and no arcing or current flow is involved, the process is often preferred for heat-sensitive electronic components. Intermetallic compounds seldom form, and there is no contamination of the weld or surrounding area. The equipment is simple and reliable, and only moderate skill is required of the operator. Surface preparation is less than for most competing processes (such as resistance welding), and less energy is needed to produce a weld. Typical applications include joining the dissimilar metals in bimetallics, making microcircuit electrical contacts, welding refractory or reactive metals, bonding ultra-thin metal, and encapsulating explosives or chemicals.

Ultrasonic welding has also been used to produce spot and seam welds on thin plastics and to seal foil or plastic envelopes and pouches. Compared to joining methods that employ solvents or adhesives, the ultrasonic method is considerably faster and results in products with cleaner surfaces.

DIFFUSION WELDING

Diffusion welding (DFW) or *diffusion bonding* occurs when properly prepared surfaces are maintained in contact under sufficient pressure and time at elevated temperature.

TABLE 32-4   Metal Combinations Weldable by Ultrasonic Welding

Metal	Aluminum	Copper	Germanium	Gold	Molybdenum	Nickel	Platinum	Silicon	Steel	Zirconium
Aluminum	x	x	x	x	x	x	x	x	x	x
Copper		x		x		x	x		x	x
Germanium			x	x		x	x	x		
Gold				x		x	x	x		
Molybdenum					x	x			x	x
Nickel						x	x		x	x
Platinum							x		x	
Silicon										
Steel									x	x
Zirconium										x

In contrast to the deformation-welding methods, plastic flow is limited and the principal bonding mechanism is atomic diffusion. A well-prepared interface can be viewed as a planar grain boundary with intervening voids and impurities. Under low pressure and elevated temperature, atomic diffusion will provide the necessary void shrinkage and grain boundary migration to form a metallurgical bond.

The quality of a diffusion weld depends on the surface condition of the materials, temperature, time at temperature, pressure, and the possible use of intermediate material layers, which can either promote diffusion or prevent the formation of undesirable intermetallic compounds. Some intermediate layers are designed to melt and form a temporary liquid that significantly accelerates the rate of atom movement.

Diffusion bonding is frequently used to join dissimilar metals and composite materials. Furnaces with inert or protective atmospheres can be used to produce high-quality joints with the reactive metals, such as titanium, beryllium, and zirconium, and the high-temperature refractory metals. Since the bonding process is quite slow, multiple parts are generally loaded into a furnace or the application is restricted to low-volume production.

EXPLOSIVE WELDING

Explosive welding (EXW) is used primarily to bond sheets of corrosion-resistant metal to heavier plates of base metal (a cladding operation), particularly when large areas are involved. As shown in Figure 32-21, the bottom sheet or plate is positioned on a rigid base or anvil, and the top sheet is inclined to it with a small open angle between the surfaces to be joined. An explosive material, usually in the form of a sheet, is placed on top of the

FIGURE 32-21   (Left) Schematic of the explosive welding process. (Right) Explosive weld between mild steel and stainless steel, showing the characteristic wavy interface.

two layers of metal and detonated in a progressive fashion, beginning where the surfaces touch. A compressive stress wave, on the order of thousands of megapascals (hundreds of thousands of pounds per square inch), sweeps across the surface of the plates. Surface films are liquefied or scarfed off the metals and are jetted out of the interface. The clean metal surfaces are then thrust together under high contact pressure. The result is a low-temperature weld with an interface configuration consisting of a series of interlocking ripples. Since the bond strength is quite high, explosively clad plates can be subjected to a wide variety of subsequent processing, including further reduction in thickness by rolling. Because it is a solid-state welding process, numerous combinations of dissimilar metals can be joined.

■ Key Words

butt welding	flash	lap joint	roll bonding
cold welding	forge welding	nugget	solid-state welding
diffusion bonding	forge-seam welding	projection welding	spot-welding gun
electrodes	friction welding	resistance welding	transgun
embossed	friction-stir processing	resistance seam welding	ultrasonic welding
explosive welding	friction-stir welding	resistance spot welding	
faying surfaces	inertia welding	rocker-arm machine	

■ Review Questions

1. What are the two primary functions of the electrodes in resistance welding?
2. What are the two major roles of the applied pressure in resistance welding?
3. Why might resistance welding be considered as a form of solid-state welding?
4. Why is there no need for fluxes or shielding gases in resistance welding?
5. Based on the heat input equation, which term is most significant in providing heat—current, resistance, or time?
6. What are the three components that contribute to the total resistance between the electrodes?
7. What measures can be taken to reduce the resistance between the electrodes and the workpieces?
8. What are the possible consequences of too little pressure during the resistance-welding cycle? Too much pressure?
9. What is the ideal sequence for pressure application during resistance welding?
10. Why do the resistance-welding conditions become less favorable as the material heats and softens?
11. What magnitude of current may be used to produce resistance welds?
12. What is the simplest and most widely used form of resistance welding?
13. What is the typical size of a spot-weld nugget?
14. What are the two basic types of stationary spot-welding machines?
15. What is the major advantage of spot-welding guns?
16. What are some of the properties that must be possessed by resistance-welding electrodes?
17. What is the most common metal that is spot welded?
18. What is the practical limit of the thicknesses of material that can be readily spot welded?
19. What design features can be altered to permit the joining of different thicknesses or different conductivity metals?
20. What are the two methods used to produce resistance seam welds?
21. What two limitations of spot welding can be overcome by using the projection approach?
22. What limits the number of projection welds that can be formed in a single operation?
23. What are some of the attractive features of resistance welding when viewed from a manufacturing standpoint?
24. What are some of the primary limitations to the use of resistance welding?
25. What type of metallurgical problem might be encountered when spot welding medium- or high-carbon steels?
26. What were some of the limitations that made the forge welds of a blacksmith somewhat variable in terms of quality?
27. What features promote coalescence in cold welding?
28. Describe how the roll-bonding process can be used to fabricate products that contain pressure-tight, fluid-flow channels that once required the use of metal tubing.
29. How is inertia welding similar to friction welding? Different from friction welding?
30. How are surface impurities removed in the friction- and inertia-welding processes?
31. What are some of the geometric limitations of friction and inertia welding?
32. How does friction-stir welding differ from friction welding?
33. What are some of the attractive features of friction-stir welding?
34. What types of material or property modifications can be induced through friction-stir processing?
35. How do ultrasonic vibrations produce a weld?
36. What are some of the geometric limitations of ultrasonic welding?
37. What are some of the attractive features of ultrasonic welding?
38. What are the conditions necessary to produce high-quality diffusion welds?
39. How are surface contaminants removed during explosive welding?
40. If the interface of a weld is viewed in cross section, what is the distinctive geometric feature of an explosive weld?

■ Problems

1. Many advanced engineering products, as well as composite materials, require the joining of dissimilar materials. Select several of the processes discussed in this chapter and investigate the capability of the process to join dissimilar materials and the associated limitations.

2. Friction-stir processing is an interesting extension of friction-stir welding. Can you identify other examples of where a welding or joining process is currently being used for purposes other than those for which it was initially developed?

www.wiley.com/college/degarmo

C*hapter 32*   CASE STUDY

Field Repair to a Power Transformer

Electric power transformers do not operate at 100% efficiency and generally incorporate some means of cooling in their design. Large transformers are often submerged in oil-filled reservoirs, where the volume of oil provides a noncorrosive heat sink. In addition, it is not uncommon for additional features, such as horizontal cooling fins, to be added to the design to aid in dissipating heat from the reservoir.

The figure shows the exterior of a large transformer that has been installed in a rural, somewhat remote location. The reservoir housing has been constructed by welding low-carbon steel plates 9 and 12 mm ($\frac{3}{8}$ and $\frac{1}{2}$ in.) thick. While the transformer was in use, a service vehicle accidentally backed into the cooling-fin assembly, producing cracks in several of the fillet areas and a resulting loss of

oil. Overheating occurred, and a repair is now necessary. Because of the size of the transformer, some form of on-site repair is preferred. It is your job to determine the procedure and make the necessary arrangements.

1. Consider the full spectrum of welding processes and identify candidates that might be appropriate for this task.

2. For each of the candidate processes, identify its primary advantages and limitations.

3. Which of the candidate processes would you recommend? Why?

4. Describe the procedure that you would outline for such a repair. Are there any special concerns or precautions?

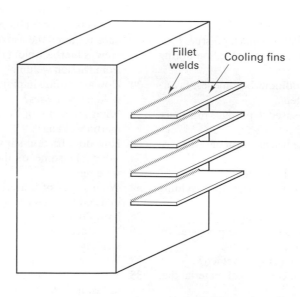

OTHER WELDING PROCESSES, BRAZING AND SOLDERING

◼ 33.1 INTRODUCTION

We have already surveyed gas-flame and arc welding (Chapter 31), as well as resistance and solid-state joining processes (Chapter 32). Other processes within the realm of welding include some that are quite old (thermit welding) and others that are among the newest in manufacturing (laser and electron beam). These and several others will be presented here, along with a brief section devoted to the application of welding and welding-related processes to surfacing and thermal spray coating.

There are also many joining or assembly operations where welding may not be the best choice. Perhaps the heat of welding is objectionable, the materials possess poor weldability, welding is too expensive, or the joint involves thin or dissimilar materials. In such cases low-temperature joining methods may be preferred. These include brazing, soldering, adhesive bonding, and the use of mechanical fasteners. Brazing and soldering will be explored in this chapter, while adhesive bonding and mechanical fasteners are deferred to Chapter 34.

◼ 33.2 OTHER WELDING AND CUTTING PROCESSES

THERMIT WELDING

Thermit welding (TW) is an extremely old process in which superheated molten metal and slag are produced from an exothermic chemical reaction between a metal oxide and a metallic reducing agent. The name *thermit* usually refers to a mechanical mixture of about one part (by weight) finely divided aluminum and three parts iron oxide, plus possible alloy additions. When this mixture is ignited by a magnesium fuse (the ignition temperature is about 1150°C or 2100°F), it reacts according to the following chemical equation:

$$8Al + 3Fe_3O_4 \rightarrow 9Fe + 4Al_2O_3 + heat$$

The temperature rises to over 2750°C (5000°F) in about 30 seconds, superheating the molten iron, which then flows by gravity into a prepared joint, providing both heat and filler metal. Runners and risers must be provided, as in a casting, to channel the molten metal and compensate for solidification shrinkage.

Steels and cast irons can be welded using the process described above. Copper, brass, and bronze can be joined using a starting mixture of copper oxide and aluminum. Nickel, chromium, and manganese oxides have also been used in the thermit welding of more exotic metals.

Thermit welding has been replaced by alternative methods to a large degree. Nevertheless, it is still effective and can be used to produce economical, high-quality welds in thick sections of material, particularly in remote locations or where more sophisticated welding equipment is not available. One such application is the field repair of large steel castings that have broken or cracked.

ELECTROSLAG WELDING

Electroslag welding (ESW), depicted in Figure 33-1, is a very effective process for welding thick sections of steel plate. There is no arc involved (except to start the weld), so the process is entirely different from submerged arc welding, and the electrical resistance of the metal being welded plays no part in producing the heat. Instead, heat is derived from the passage of electrical current through a pool of electrically conductive liquid slag. Resistance heating raises the temperature of the slag to around 1750°C (3200°F). The molten slag then melts the edges of the pieces that are being joined, as well as continuously fed solid or flux-cored electrodes. Multiple electrodes are often used to provide an adequate supply of filler metal and maintain the molten pool. Under normal operating conditions, there is a 65-mm (2.5-in.)-deep layer of molten slag, which serves to protect and cleanse the underlying 12- to 20-mm ($^1/_2$ to $^3/_4$ in.)-deep pool of molten metal. These liquids are confined to the gap by means of sliding, water-cooled *molding plates* that are usually made of copper. As the weld metal solidifies at the bottom of the pool, the molding plates move upward at a rate that is typically between 12 and 40 mm/min ($^1/_2$ to $1^1/_2$ in./min).

Since a vertical joint provides the easiest geometry for maintaining a deep slag bath, the process is used most frequently in this configuration. Circumferential joints can also be produced in large pipe by using special curved slag-holder plates and rotating the pipe to maintain the welding area in a vertical position.

Because large amounts of weld metal and heat can be supplied, electroslag welding is the best of all the welding processes for making welds in thick plates. The thickness

(a)   (b)

FIGURE 33-1 (a) Arrangement of equipment and workpieces for making a vertical weld by the electroslag process. (b) Cross section of an electroslag weld, looking through the water-cooled copper slide.

of the plates can vary from 12 to 900 mm ($^1/_2$ to 38 in.), and the length of the weld (amount of vertical travel) is almost unlimited. Edge preparation is minimal, requiring only squared edges separated by 25 to 35 mm. Applications have included building construction, shipbuilding, machine manufacture, heavy-pressure vessels, and the joining of large castings and forgings.

Solidification control is vitally important to obtaining a good electroslag weld, since slow cooling tends to produce a coarse grain structure. Cracking tendencies can be suppressed by adjusting the current, voltage, slag depth, number of electrodes, and electrode extension to produce a wide, shallow pool of molten metal. A large heat-affected zone and extensive grain growth are common features of the process. While these are undesirable metallurgical features, the long thermal cycle does serve to minimize residual stresses, distortion, and cracking in the heat-affected zone. Subsequent heat treatment of the welded structure may be necessary, however, if good fracture resistance is required.

ELECTRON-BEAM WELDING

In the *electron-beam welding* (EBW) process, the metal to be welded is heated by the impingement of a beam of high-velocity electrons. Originally developed for obtaining ultra-high-purity welds in reactive and refractory metals, the unique qualities of the process have led to a much wider range of applications.

Figure 33-2 presents the electron optical system. An electric current heats a tungsten filament to about 2200°C (4000°F), causing it to emit a stream of electrons by thermal emission. By means of a control grid, accelerating voltage, and focusing coils, these electrons are collected into a concentrated beam, accelerated, and directed to a focused

FIGURE 33-2   Schematic diagram of the electron-beam welding process.

spot between 0.8 and 3.2 mm. ($^1/_{32}$ to $^1/_8$ in.) in diameter. Since electrons accelerated at 150 kV achieve speeds nearly $^2/_3$ the speed of light, the electron beam is concentrated energy, capable of producing temperatures in excess of 1 million degrees Celsius when its kinetic energy is converted to heat. Since the beam is composed of charged particles, it can be positioned and moved by electromagnetic lenses. Unfortunately, the electrons cannot travel well through air. To be effective as a welding heat source, the beam must be generated and focused in a very high vacuum, typically at pressures of 0.01 Pa (1×10^{-4} mm Hg) or less.

In many operations, the workpiece must also be enclosed in the high-vacuum chamber, with provision for positioning and manipulation. The vacuum then ensures degasification and decontamination of the molten weld metal, and welds of very high quality are obtained. The size of the vacuum chamber, however, tends to impose serious limitations on the size of the workpiece that can be accommodated, and the need to break and reestablish the high vacuum as pieces are inserted and removed places a considerable restriction on productivity. As a consequence, electron-beam welding machines have been developed that operate at pressures considerably higher than those required for beam generation. Some permit the workpiece to remain outside the vacuum chamber, with the beam emerging through a small orifice in the vacuum chamber to strike an adjacent surface. High-capacity vacuum pumps are required to compensate for the leakage through the orifice. While these machines offer more production freedom, they do produce shallower, wider welds since the beam loses energy and diffuses as the pressure increases.

In general, two distinct ranges of accelerating voltage are employed in electron-beam welding. High-voltage equipment operates between 60 and 150 kV and produces a smaller spot size and greater penetration than does the lower-voltage type, which uses from 10 to 50 kV. Because of their high electron velocities, the high-voltage units emit considerable quantities of harmful X-rays and thus require expensive shielding and indirect viewing systems for observing the work. The X-rays produced by the low-voltage machines are sufficiently soft that the walls of the vacuum chamber absorb them, and the parts can be viewed directly through viewing ports.

Almost any metal can be welded by the electron-beam process, including those that are difficult to weld by other methods, such as zirconium, beryllium, and tungsten. Dissimilar metals, including those with extremely different melting points, can also be readily welded, since the intense beam will melt both metals simultaneously. Electron-beam welds typically exhibit a narrow profile and remarkable penetrations like those shown in Figure 33-3. The high power and heat concentrations can produce fusion zones with depth-to-width ratios up to 25:1. This is coupled with low total heat input, low distortion, and a very narrow heat-affected zone. Heat-sensitive materials can often be welded without damage to the base metal. Deep welds can be made in a single pass. High welding speeds are common; no shielding gas, flux, or filler metal is required; the process can be performed in all positions; and preheat or postheat is generally unnecessary.

FIGURE 33-3 (Left to right) Electron-beam welds in 19-mm-thick 7079 aluminum and 102-mm-thick stainless steel. *(Courtesy of Hamilton Standard Division of United Technologies Corporation, Hartford, CT.)*

TABLE 33-1	Process Summary: Electron-Beam Welding (EBW)
Heat source	High-energy electron beam
Protection	Vacuum
Electrode	None
Material joined	All common metals
Rate of heat input	High
Weld Profile (D/W)	20
Maximum penetration	175 mm (7 in.)
Assets	High precision; high quality; deep and narrow welds; small HAZ, low distortion; fast welding speed; beam is easily positioned and deflected; no filler metal, flux, or shielding gas required
Limitations	Beam and target must be within a vacuum; very expensive equipment; work piece size is limited by the vacuum chamber; significant edge preparation and alignment required; requires safety protection from X-ray and visible radiation

On the negative side, the equipment is quite expensive, and extensive joint preparation is required. Because of the deep and narrow weld profile, joints must be straight and precisely aligned over the entire length of the weld. Machining and fixturing tolerances are often quite demanding. The vacuum requirements tend to limit production rate, and the size of the vacuum chamber may restrict the size of the workpiece that can be welded.

The electron-beam process is best employed where welds of extremely high quality are required or where other processes will not produce the desired results. Electron-beam welds often exhibit joint strength 15 to 25% greater than arc welds in the same material. The unique capabilities have resulted in its routine use in a number of applications, particularly in the automotive and aerospace industries. Table 33-1 provides a process summary for electron-beam welding.

LASER-BEAM WELDING

Laser beams can be used as a heat source for welding, hole making, cutting, cladding, and heat treating a wide variety of engineering metals. When used for *laser-beam welding* (LBW), the beam of coherent light can be focused to a diameter of 0.1 to 1.0 mm, providing a power density in excess of 10^6 watts/mm^2. The high-intensity beam can be used to simply melt the material at the joint, but more often, it produces a very narrow column of vaporized metal (a "keyhole") with a surrounding liquid pool. As the beam traverses, the liquid flows into the joint to produce a weld with a depth-to-width ratio generally greater than 5:1. Because of the narrow weld-pool geometry, high travel speed of the beam (typically several meters per minute), and low total heat input, the molten metal solidifies quickly, producing a very thin heat-affected zone and little thermal distortion. Finishing costs are quite low. Since welds require only one-side access, many different joint configurations are possible.

Laser-beam welding is most effective for simple fusion welds without filler metal (*autogenous welds*), but careful joint preparation is required to produce the narrow gap and necessary level of cleanliness. Filler metal can be added if the gap is excessive, and a low-velocity flow of inert gas (generally helium or argon) may be used to protect the weld pool from oxidation. Figure 33-4 shows a typical laser-beam weld, and Table 33-2 provides a process summary for laser-beam welding.

As shown in Figure 33-5, laser-beam welding and electron-beam welding both offer some of the highest power densities of the welding processes. The well-collimated beam of intense laser energy can produce deep penetration welds that are similar to electron-beam welds, but the laser-beam technique offers several distinct advantages:

1. The beam can be transmitted through air (i.e., a vacuum environment is not required). There is no physical contact between the welding equipment and the workpiece. The originating laser can be a considerable distance removed.

FIGURE 33-4   Laser butt weld of 3-mm (0.125-in.) stainless steel, made at 1.5 m/min with a 1250-watt laser. *(Courtesy of Coherent, Inc., Santa Clara, CA.)*

TABLE 33-2	Process Summary: Laser-Beam Welding (LBW)
Heat source	Laser light
Protection	None, or externally supplied gas
Electrode	None
Material joined	All common metals
Rate of heat input	High
Weld profile (D/W)	5
Maximum penetration	25 mm (1 in.)
Assets	High heat-transfer efficiency; can weld any location that is light-accessible; small HAZ; low distortion; can accurately focus the beam with light optics; high welding speed
Limitations	Possible problems with reflectivity of some metals; good positioning and fit-up required

FIGURE 33-5 Comparison of the power densities of various welding processes. The high power densities of the electron-beam and laser-beam welding processes enable the production of deep, narrow welds with small heat-affected zones. Welds can be made quickly and at high travel speeds.

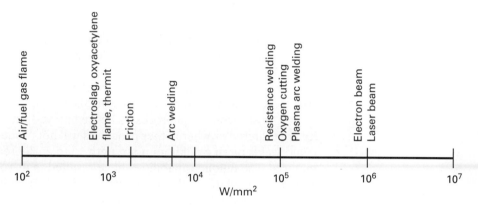

2. No harmful X-rays are generated.

3. The laser beam can be easily shaped, directed, and focused with both transmission and reflective optics (lenses and mirrors), and some beams can be transmitted through fiber-optic cables.

4. The only restriction on weld location is optical accessibility. Welds can be made in difficult-to-reach places, and materials can be joined within transparent containers, such as inside a vacuum tube.

A laser-welding system consists of an industrial laser, a means of guiding and focusing the beam, and a means of positioning and manipulating the parts to be welded. The traditional equipment for laser welding is a CO_2 laser with a power output range of 1.5 to 10 kW (or even higher). Because CO_2 lasers emit light with a far-infrared wavelength, they require mirror systems or special optical materials to focus and position the beam. They can be used to weld steel up to 25 mm (1 in.) thick at speeds ranging from 1 to 20 m/min. Nd:YAG (neodymium: yttrium–aluminum–garnet) lasers are more limited in power and capability but operate in a near-infrared wavelength that can utilize conventional glass lenses or delivery by flexible fiber-optic cable (as much as 3000 W of energy can be transmitted up to 150 m through a 0.6-mm-diameter fiber!).

The industrial lasers also lend themselves to automation and robotic manipulation. A single Nd:YAG laser with a fiber-optic system can distribute its beam to multiple workstations in either a simultaneous or time-sharing fashion, and the individual stations can be as much as 100 m (300 ft) from the laser. By using fiber-optic cables, laser energy can be piped directly to the end of a robot arm. This eliminates the need to mount and maneuver a heavy, bulky laser and, by reducing weight, enhances the speed and accuracy of both positioning and manipulation. Cutting, drilling, welding, and heat treating can all be performed with the same unit, and multiple axes of motion can provide a high degree of mobility and accessibility.

The equipment cost for a CO_2 or Nd:YAG laser-beam welding
but this cost can be somewhat offset by the faster welding spee
without filler metal, and low distortion, which enables a reduction
ening and machining. Caution should be used with such equipment, in
flected or scattered laser beams can be quite dangerous, even at great distan.
welding site. Eye protection is a must.

Because laser welds do not significantly reduce sheet metal formability, they ha
been used to produce tailored blanks for the production of sheet products. Different
types of steel or different thicknesses can be joined to produce single-piece products
with different properties at different locations. Laser welding has made great progress
in the welding of aluminum alloys and has replaced gas tungsten arc welding or riveting in a number of applications.

With a sharply focused beam and short exposure times, laser welds can be very
small and have a low total heat input, often on the order of 0.1 to 10 joules. These conditions are ideal for use in the electronics industry, and laser welding is frequently used
to connect lead wires to small electronic components. Lap, butt, tee, and cross-wire configurations can all be used. It is even possible to weld wires without removing the
polyurethane insulation. The laser simply evaporates the insulation and completes the
weld with the internal wire.

Lasers have also been used in *hybrid processes* that combine laser welding with arc
welding (GMAW, GTAW, or PAW), with both operating in one process zone and producing one weld pool. These hybrids combine the deep penetration, low distortion, and
high-welding-speed features of laser welding with the wider pool, gap-bridging capability of arc welding. The resulting weld pool is wide and shallow at the surface, transitioning to deep and narrow. In addition to the unique and flexible weld-pool geometry,
another benefit is the enhanced arc stabilization provided by the material that the laser
evaporates. Laser power can be reduced from that required for lasers operating alone,
and welds can be made faster than with just the arc-welding processes. The shielding
gas from the arc-welding process protects the entire weld pool.

LASER-BEAM CUTTING

Cutting small holes, narrow slots, and closely spaced patterns in a variety of materials,
or producing small quantities of complex-contoured sheet or plate, is another widely
used application of industrial lasers. *Laser-beam cutting* (LBC) begins by "drilling" a
hole through the material and then moving the beam along a programmed path. As
shown in Figure 33-6, the intense heat from the laser is used to melt and/or evaporate
the material being cut. A stream of *assist gas* blows the molten metal through the cut,
cools the workpiece, minimizes the heat-affected zone, and may participate in a combustion reaction with the material being cut.

Oxygen is the usual gas for cutting mild steel. The laser heats the metal to a
temperature where the iron and oxygen combine in an exothermic reaction. The
molten iron and iron oxide have a low viscosity and are easily blown away by the
flow of assist gas. In this exothermic cutting process, the assist gas actually contributes
additional heat. High cutting speeds are possible, the speed being limited by the rate
of material burning. Nitrogen is used with stainless steel and aluminum, and, because
of its high reactivity, titanium requires an inert gas, such as argon. Inert gases or air
are used when cutting nonmetallics. The latter processes are ones of endothermic
cutting, since the gas actually absorbs energy as it is heated. Cutting speed is set by
the rate at which the laser can melt and/or vaporize material. Exothermic cutting
produces an oxidized edge, while endothermic cutting (or clean cutting) results in
oxide-free surfaces.

Clean, accurate, square-edged cuts are characteristic of the laser cutting process,
and the *kerf* (typically as small as 0.25 mm) and heat-affected zone are narrower than
with any other thermal cutting process. No postcut finishing is required in many applications, even though the process does produce a thin recast surface. Figure 33-7
shows the edge of carbon steel that is 6 mm (0.25 in.) thick, laser cut at 1.8 m/min with
a 1250-W laser.

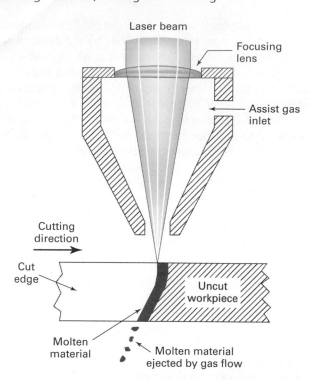

FIGURE 33-6   Schematic of laser-beam cutting. The laser provides the heat, and the flow of assist gas propels the molten droplets from the cut.

FIGURE 33-7   Surface of 6-mm-thick carbon steel cut with a 1250-watt laser at 1.8 m/min. *(Courtesy of Coherent, Inc., Santa Clara, CA.)*

CO_2 and YAG lasers have been used in both continuous and pulsed modes. While cutting speed depends on the material being cut and its thickness, it is greatest in the continuous mode that is preferred for straight and mildly contoured cuts. The pulsed mode is preferred for thin materials and enables tight corners and intricate details to be cut without excessive burning. Metal plates as well as a variety of nonmetals can be cut in thicknesses up to 30 mm ($1^{1}/_{4}$ in.). Cutting temperatures can be in excess of 11,000°C (20,000°F), and cutting speeds as high as 25 m/min have been observed with some nonmetals.

In addition to very common robotic applications, lasers have also been mounted on CNC-type machines or combined with traditional tools, such as punch presses, to produce extremely flexible hybrid equipment. Because no dedicated dies or tooling are required to produce a cut and there is no setup time, the laser is an economical alternative to blanking or nibbling for prototype or short-run products or for materials that are difficult to cut by conventional methods, such as plastics, wood, and composites.

Since lasers can cut a wide variety of metals and nonmetals, laser cutting has become a dominant process in the cutting of composite materials. The more uniform the thermal characteristics of the components, the better the cut and the less thermal damage to the material. Kevlar-reinforced epoxy cuts easiest and gives a narrow heat-affected zone. Glass-reinforced epoxy is more difficult because of the greater thermal differences, and graphite-reinforced epoxy is even worse because of the high dissociation temperature and thermal conductivity of the graphite. By the time the graphite has absorbed sufficient cutting heat, the epoxy matrix will have decomposed to a significant depth. The use of lasers for machining is discussed further in Chapter 27.

LASER SPOT WELDING

Lasers have also been used to produce spot welds in a manner that offers unique advantages when compared to the conventional resistance methods. A small clamping force is applied to ensure contact of the workpieces, and a fine-focused beam then scans the area of the weld. Welding is performed in the keyhole mode, where the laser produces a small hole through the molten puddle. As the beam is moved, molten metal flows into the hole and solidifies, forming a fusion-type nugget.

Laser spot welding can be performed with access to only one side of the joint. It is a noncontact process and produces no indentations. No electrodes are involved, so electrode wear is no longer a production problem. Weld quality is independent of material resistance, surface resistance, and electrode condition, and no water cooling is required. The total heat input is low, so the heat-affected zone is small. Speed of welding and strength of the resulting joint are comparable to resistance spot welds.

FLASH WELDING

Flash welding (FW) is a process used to produce butt welds between similar or dissimilar metals in solid or tubular form. The two pieces of metal are first secured in current-carrying grips and lightly touched together. An electric current may be passed through the joint to provide optional preheat, after which the pieces are withdrawn slightly. An intense flashing arc forms across the gap, which melts the material on both surfaces. The pieces are then forced together under high pressure (on the order of 70 MPa, or 10,000 psi), expelling the liquid and oxides, and upsetting the softened metal. The electric current is turned off, and the force is maintained until solidification is complete. If desired, the upset portion can then be removed by machining. Figure 33-8 shows a schematic of the flash-welding process, including both the equipment setup and the completed weld.

To produce a high-quality weld, it is important that the initial surfaces be flat and parallel so that the flashing is even across the area to be joined. The flashing action must be continued long enough to melt the interface and also soften the adjacent metal. Sufficient plastic deformation must occur during the upsetting to transfer the impurities and contaminants outward into the flash. The equipment required is generally large and expensive, but excellent welds can be made at high production rates.

Percussion welding (PEW) is a similar process, in which a rapid discharge of stored energy produces a brief period of arcing, which is followed by the rapid application of force to expel the molten metal and produce the joint. In percussion welding, the duration of the arc is on the order of 1 to 10 ms. The heat is intense but highly concentrated. Only a small amount of molten metal is produced, little or no upsetting occurs at the joint, and the heat-affected zone is quite small. Application is generally restricted to the butt welding of bar or tubing, where heat damage is a major concern.

Upset welding (UW) is also similar to flash welding, but there is no period of arcing. The equipment and geometries are similar, but the heating is achieved through electrical resistance. The parts are clamped in the machine, pressure is applied, and high current is passed through the joint. When the abutting surfaces have been heated to a suitable forging temperature, the current is stopped and an upsetting force is applied to produce coalescence. The initial conditions of interface flatness, finish, and alignment must create uniform contact if a good-quality weld is to be produced.

FIGURE 33-8   Schematic diagram of the flash-welding process. (a) Equipment and setup; (b) completed weld.

(a)

(b)

■ 33.3 SURFACE MODIFICATION BY WELDING-RELATED PROCESSES

SURFACING (INCLUDING HARD FACING)

Surfacing or overlaying is the process of depositing a layer of weld metal on the surface or edge of a different-composition base material. The usual objectives are to obtain improved resistance to wear, abrasion, heat, or chemical attack, without having to make the entire piece from an expensive material, one that is difficult to fabricate, or one that would not possess the desired bulk properties. Since the deposited surfaces are generally harder than the base metal, the process is often called *hard facing*. This is not always true, however, for in some cases a softer metal (such as bronze) is applied to a harder base material.

Surfacing Materials. The materials most commonly used for surfacing include (1) carbon and low-alloy steels; (2) high-alloy steels and irons; (3) cobalt-based alloys; (4) nickel-based alloys, such as Monel, Nichrome, and Hastelloy; (5) copper-based alloys; (6) stainless steels; and (7) ceramic and refractory carbides, oxides, borides, silicides, and similar compounds.

Surfacing Methods and Applications. Since some of the base metal melts during the deposition, surfacing is actually a variation of fusion welding and can be performed by nearly all of the gas-flame or arc-welding techniques, including oxyfuel gas, shielded metal arc, gas metal arc, gas tungsten arc, submerged arc, and plasma arc. Arc welding is frequently used for the deposition of high-melting-point alloys. Submerged arc welding is used when large areas are to be surfaced or a large amount of surfacing material is to be applied. The plasma arc process further extends the process capabilities because of its extreme temperatures. To obtain true fusion of the surfacing material, a transferred arc is used and the surfacing material is injected in the form of a powder. If a nontransferred arc is used, only a mechanical bond is produced, and the process becomes a form of *metallizing*. Lasers can also be used in surfacing operations.

THERMAL SPRAY COATING OR METALLIZING

The *thermal spray* processes offer a means of applying a coating of high-performance material (metals, alloys, ceramics, intermetallics, cermets, carbides, or even plastics) to more economical and more easily fabricated base metals. A wire or rod of the coating material is fed into a gas flame or arc, where it melts and becomes atomized by a stream of gas, such as argon, nitrogen, combustion gases, or compressed air. The gas stream propels the 0.01- to 0.05-mm (0.0004- to 0.002-in.)-diameter particles toward the target surface, where they impact ("splat"), cool, and bond. Very little heat is transferred to the substrate, whose peak temperatures generally range from 100 to 250°C (200 to 500°F). As a result, thermal spraying does not induce undesirable metallurgical changes or excessive distortion, and coatings can be applied to thin or delicate targets or to heat-sensitive materials such as plastics. The applied coating can range in thickness from 0.1 to 12 mm (0.004 to 0.5 in.).

Several of the thermal spray processes use adaptations of oxyfuel welding equipment. Figure 33-9 shows a schematic of an oxyacetylene metal spraying gun designed to utilize a solid wire feed. The flame melts the wire and a flow of compressed air disintegrates the molten material and propels it to the workpiece. An alternative type of oxyfuel gun uses material in the form of powder, which is gravity or pressure fed into the flame, where it is melted and carried by the flame gas onto the target. The powder feed permits the deposition of material that would be difficult to fabricate into wire, such as cermets, oxides, and carbides. In addition, the droplet size is controlled by the powder, not by the factors that control atomization.

The lower temperatures and lower particle velocities of the oxyfuel deposition methods result in coatings with high porosity and low cohesive strength. An adaptation of the process known as high-velocity oxyfuel (HVOF) spraying propels the droplets with a supersonic stream of hot gas. Because the particles impact with high kinetic energies, the resulting coating is dense and well bonded.

FIGURE 33-9 Schematic diagram of an oxyacetylene metal-spraying gun. *(Courtesy of Sulzer Metco, Winterthur, Switzerland.)*

The simplest of the electric arc methods is wire arc or electric arc spraying. Two oppositely charged electrode wires are fed through a gun, meeting at the tip, where they form an arc. A stream of atomizing gas flows through the gun, stripping off the molten metal to produce a high-velocity spray. Since all of the input energy is used to melt the metal, this process is extremely energy efficient.

Plasma spray metallizing, illustrated in Figure 33-10, is a more sophisticated technique. A plasma-forming gas serves as both the heat source and propelling agent for the coating material, which is usually fed in the form of powder. The molten particles attain high velocity and therefore produce a dense, strongly bonded coating. Since temperatures can reach 16,500°C (30,000°F), plasma spraying can be used to deposit materials with extremely high melting points. Metals, alloys, ceramics, carbides, cermets, intermetallics, and plastic-based powders have all been successfully deposited.

While thermal spraying or *metallizing* is similar to surfacing and is often applied for the same reasons, the coatings are usually thinner and the process is more suitable for irregular surfaces or heat-sensitive substrates. The deposition guns can be either handheld or machine-driven. A standoff distance of 0.15 to 0.25 m (6 to 10 in.) is usually maintained between the spray nozzle and the workpiece. Table 33-3 compares the features of five methods of thermal spray deposition.

Surface Preparation for Metallizing.   Unlike surfacing, metallizing does not melt the base metal. Adhesion is entirely mechanical, so it is essential that the base metal be prepared in a way that promotes good mechanical interlocking. The target material must first be clean and free of dirt, moisture, oil, and other contaminants. The surface is then roughened by one of a variety of methods to create minute crevices that can anchor the solidifying particles. Grit blasting with a sharp, abrasive grit is the most common technique, and a surface roughness of 2.5 to 7.5 microns is adequate for most applications.

FIGURE 33-10 Diagram of a plasma-arc spray gun. *(Courtesy of Sulzer Metco, Winterthur, Switzerland.)*

TABLE 33-3	Comparison of Five Thermal Spray Deposition Techniques					
Method	Heat		Deposited Materials	Particle Impact Velocity (m/sec)	Adhesion Strength	Maximum Spray Rate (kg/hr)
	Source	Temperature (°C)				
Flame spray						
Wire	Oxyfuel	3000	Metals	180	Medium	9
Powder	Oxyfuel	3000	Metals, ceramics, plastics	30	Low	7
High-velocity oxyfuel (HVOF)	Oxyfuel	3100	Metals, carbides	600–1000	Very high	14
Wire arc	DC arc	5500	Metals only	250	High	16
Plasma spray	DC arc	5500 to 16,500	All	250–1200	High to Very high	5-25

Characteristics and Applications of Sprayed Metals. During deposition, the atomized, molten, or semimolten particles mix with air and then cool rapidly upon impact with the base metal. The resultant coatings consist of bonded particles that span a range of size, shape, and degree of melting. Some particles become oxidized and interparticle voids can become entrapped. Compared to conventional wrought material, the coatings are harder, more porous (0.1 to 15% porosity), and more brittle. Thermal spray coatings add little, if any, additional strength to a part, since the strength of the porous coating is usually between one-third and one-half of its normal wrought strength. Applications, therefore, generally look to the coating to provide resistance to heat, wear, erosion, and/or corrosion, or to restore worn parts to original dimensions and specifications. Some typical applications include:

1. *Protective coatings.* Zinc and aluminum are sprayed on iron and steel to provide corrosion resistance—a process that may well extend the lifetime of bridges, buildings, and other infrastructure items. The interior surfaces of power boilers can be coated with high-chromium alloys to extend wall life by providing both heat-resistance and corrosion resistance.

2. *Building up worn surfaces.* Worn parts may be salvaged or their life extended by adding new metal to the depleted regions. The repair and restoration of aircraft engine components is probably the largest single use of thermal spraying.

3. *Hard surfacing.* Although metal spraying should not be compared to hard-facing deposits that are applied by welding techniques, it can be used when thin coatings are considered to be adequate. Typical applications might include automobile cylinder liners and piston rings; thread guides in textile plants; and critical parts within pumps, bearings, and seals.

4. *Applying coatings of expensive metals.* Metal spraying provides a simple method for applying thin coatings of noble metals to surfaces where conventional plating would not be economical.

5. *Electrical properties.* Because metal can be deposited on almost any surface, thermal spraying can be used to apply a conductive surface to an otherwise poor conductor or nonconductor. Copper, aluminum, or silver is frequently sprayed on glass or plastics for this purpose. Conversely, sprayed alumina (Al_2O_3) can be used to impart insulating or dielectric properties.

6. *Reflecting surfaces.* Aluminum, sprayed on the back of glass by a special fusion process, makes an excellent mirror.

7. *Decorative effects.* One of the earliest and still important uses of metal spraying was to obtain decorative effects. Because sprayed metal can be treated in a variety of ways, such as buffed, wire brushed, or left in the as-sprayed condition, it is frequently specified for finishing manufactured products and architectural materials.

8. *Tailored surface characteristics.* Porous coatings of cobalt or titanium alloys, or certain ceramic materials, have been applied to medical implants to help promote adhesion and in-growth of bone and tissue.

■ 33.4 BRAZING

In brazing and soldering, the surfaces to be joined are first cleaned, the components assembled or fixtured, and a low-melting-point nonferrous metal is then melted, drawn into the space between the two solids by capillary action, and allowed to solidify. *Brazing* is the permanent joining of similar or dissimilar metals or ceramics (or composites based on those two materials) through the use of heat and a filler metal whose melting temperature (actually, liquidus temperature) is above 450°C (840°F)[1] but below the melting point (or solidus temperature) of the materials being joined. The brazing process is different from welding in a number of ways:

1. The *composition* (or chemistry) of the brazing alloy is significantly different from that of the base metal.

2. The *strength* of the brazing alloy is usually lower than that of the base metal.

3. The *melting point* of the brazing alloy is lower than that of the base metal, so none of the base metal is melted.

4. Bonding requires *capillary action* to distribute the filler metal between the closely fitting surfaces of the joint. The specific flow is dependent upon the viscosity of the liquid, the geometry of the joint, and surface wetting characteristics.

Because of these differences, the brazing process has several distinct advantages:

1. A wide range of metallic and nonmetallic materials can be brazed. The process is ideally suited for joining dissimilar materials, such as ferrous metal to nonferrous metal, cast metal to wrought metal, metals with widely different melting points, or even metal to ceramic.

2. Since less heating is required than for welding, the process can be performed quickly and economically.

3. The lower temperatures reduce problems associated with heat-affected zones (or other material property alteration), warping, and distortion. Thinner and more complex assemblies can be joined successfully. Thin sections can be joined to thick. Metal as thin as 0.01 mm (0.0004 in.) and as thick as 150 mm (6 in.) can be brazed.

4. Assembly tolerances are closer than for most welding processes, and joint appearance is usually quite neat.

5. Brazing is highly adaptable to automation and performs well when mass-producing complex or delicate assemblies. Complex products can also be brazed in several steps using filler metals with progressively lower melting temperatures.

6. A strong permanent joint is formed.

Successful brazing or soldering requires that the parts have relatively good fit-up (i.e., small joint clearances) to promote capillary flow of the filler metal. The parts must be thoroughly cleaned prior to joining, and many parts will require flux removal after joining. It is also important to remember that any subsequent heating of the assembly can cause inadvertent melting of the braze metal, thereby weakening or destroying the joint.

Another concern with brazed joints is their enhanced susceptibility to corrosion. Since the *filler metal* is of different composition from the materials being joined, the brazed joint is actually a localized galvanic corrosion cell. Corrosion problems can often be minimized, however, by proper selection of the filler metal.

NATURE AND STRENGTH OF BRAZED JOINTS

Brazing, like welding. forms a strong metallurgical bond at the interfaces. Clean surfaces, proper clearance, good wetting, and good fluidity will all enhance the bonding.

---

[1]This temperature is an arbitrary one, selected to distinguish brazing from soldering.

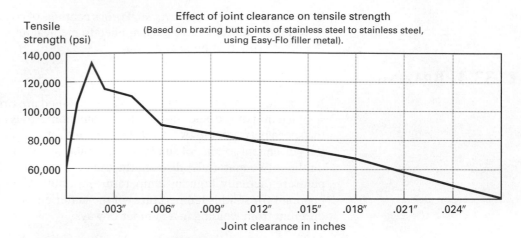

FIGURE 33-11 Typical variation of tensile strength with clearance in a butt-joint braze. *(Courtesy of Handy & Harman, Rye, NY.)*

The strength of the resulting joint can be quite high, certainly higher than the strength of the brazing alloy and often greater than the strength of the metal being brazed. Attainment of a high-strength joint, however, requires optimum processing and design.

Of all of the factors contributing to joint strength, *joint clearance* is the most important. If the joint is too tight, it may be difficult for the braze metal to flow into the gap (leaving unfilled voids), and flux may not be able to escape (remaining in locations that should be filled with braze material). There must be sufficient clearance for the braze metal to wet the joint and flow into it under the force of capillary action. As the gap is increased beyond an optimum value, however, the joint strength decreases rapidly, dropping off to the strength of the braze metal itself. If the gap becomes too great, the capillary forces may be unable to draw the material into the joint or hold it in place during solidification. Figure 33-11 shows the tensile strength of a butt-joint braze as a function of joint clearance.

Proper clearance can vary considerably, depending primarily on the type of braze metal being used. The ideal clearance is usually between 0.01 and 0.04 mm (0.0005 and 0.0015 in.), an "easy-slip" fit. A press fit can even be acceptable if fluxes are not used and surface roughness is sufficient to assure adequate flow of the filler metal into the joint. Clearances up to 0.075 mm (0.003 in.) can be accommodated with a more sluggish filler metal, such as nickel. When clearances range between 0.075 and 0.13 mm (0.003 and 0.005 in.), however, acceptable brazing becomes somewhat difficult, and joints with gaps in excess of 0.13 mm (0.005 in.) are almost impossible to braze. It should be noted that the specified gap should be maintained over the entire braze area—braze surfaces should be parallel.

It is also important to recognize that the dimensions cited above are the clearances that should exist *at the temperature of the brazing process*. Any effects of thermal expansion should be compensated when specifying the dimensions of the starting components. This is particularly significant when dissimilar materials are to be joined, for here the joint clearance will change as one material expands at a faster rate than the other. Consider a joint between brass and steel, like the one depicted in Figure 33-12. Brass expands more than steel when the temperature is increased. Therefore, if the insert tube is the brass component, the initial fit should be somewhat loose. The brass will expand more than the steel as the temperature is increased, and at the brazing temperature, the gap will assume the desired dimensions. Conversely, if a steel tube is to be inserted into a brass receiver, an initial force fit may be required since the interface will widen as the brass expands more than the steel. Problems can also occur when the reverse dimensional changes occur during cool-down. Significant residual stresses can form in the new joint, and tensile stresses can induce cracking.

Wettability is a strong function of the surface tensions between the braze metal and the base alloy. Generally, the wettability is good when the surfaces are clean and the two metals can form intermediate diffused alloys. Sometimes the wettability can be improved, as is done when steel is tin plated to accept a lead–tin solder, or plated with

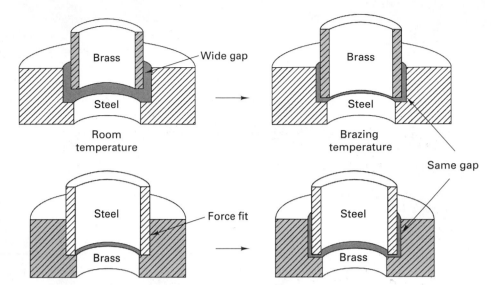

FIGURE 33-12 When brazing dissimilar metals, the initial joint clearance should be adjusted for the different thermal expansions (here brass expands more than steel). Proper brazing clearances should exist at the temperature where the filler metal flows.

nickel or copper to enhance brazing. *Fluidity* is a measure of the flow characteristics of the molten braze metal and is a function of the metal, its temperature, surface cleanliness, and clearance.

DESIGN OF BRAZED JOINTS

Because the strength of a braze filler metal is generally less than that of the metals being joined, a good joint design is required if one is to obtain adequate mechanical strength. The desired load-carrying ability is usually obtained by (1) ensuring proper joint clearance and (2) providing sufficient area for the bond. Figure 33-13 depicts the two most common types of brazed joints: *butt* and *lap*. Butt joints do not require additional thickness in the vicinity of the joint; they are most often used where the strength requirements are not that critical. The bonding area is limited to the cross-sectional area of the thinner or smaller member. In contrast, lap joints can provide bonding areas that are considerably larger than the butt configuration; they are often preferred when maximum strength is desired. If the joints are made very carefully, a lap of 1 to $1\frac{1}{4}$ times the material thickness can develop strength equal to that of the parent metal. For joints that are made by routine production, it is best to use a lap of 3 to 6 times the material thickness to ensure that failure will occur in the base metal, not in the brazed joint.

Variations of the two basic joint designs include the *butt-lap* and *scarf* configurations, shown in Figure 33-14. The butt-lap design is an attempt to combine the advantage of a uniform thickness with a large bonding area and companion high strength. Unfortunately, it also requires a higher degree of joint preparation. The scarf joint maintains uniform thickness and increases bonding area by tilting the butt joint interface. Careful joint preparation and component alignment is required to maintain the desired dimensions of joint clearance. Figure 33-14 shows relatively simple butt, lap, butt-lap, and scarf

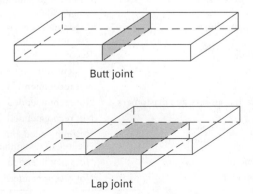

Butt joint

Lap joint

FIGURE 33-13 The two most common types of braze joints are butt and lap. Butt offers uniform thickness across the joint, whereas lap offers greater bonding area and higher strength.

Butt

Flat parts

Tubular parts

Lap

Flat parts

Tubular parts

Butt-lap

Flat parts

Tubular parts

Scarf

Flat parts

Tubular parts

FIGURE 33-14 Variations of the butt and lap configurations include the butt-lap and scarf. The four types are shown for both flat and tubular parts.

joints for both flat and tubular parts. Figure 33-15 shows good brazing designs for a variety of joint configurations.

The materials being brazed also need to be considered when designing a brazed joint. Table 33-4 summarizes the compatibility of various engineering materials with the brazing process.

FIGURE 33-15 Some common joint designs for assembling parts by brazing.

TABLE 33-4	Compatibility of Various Engineering Materials with Brazing
Material	Brazing Recommendation
Cast iron	Somewhat difficult
Carbon and low-alloy steels	Recommended for low- and medium-carbon materials; difficult for high-carbon materials; seldom used for heat-treated alloy steels
Stainless steel	Recommended; silver and nickel brazing alloys are preferred
Aluminum and magnesium	Common for aluminum alloys and some alloys of magnesium
Copper and copper alloys	Recommended for copper and high-copper brasses; somewhat variable with bronzes
Nickel and nickel alloys	Recommended
Titanium	Difficult, not recommended
Lead and zinc	Not recommended
Thermoplastics, thermosets, and elastomers	Not recommended
Ceramics and glass	Not recommended
Dissimilar metals	Recommended, but may be difficult, depending on degree of dissimilarity
Metals to nonmetals	Not recommended
Dissimilar nonmetals	Not recommended

FILLER METALS

The filler metal used in brazing can be any metal that melts between 450°C (840°F) and the melting point of the material being joined. Actual selection, however, considers a variety of factors, including compatibility with the base materials, brazing-temperature restrictions, restrictions due to service or subsequent processing temperatures, the brazing process to be used, the joint design, anticipated service environment, desired appearance, desired mechanical properties (such as strength, ductility, and toughness), desired physical properties (such as electrical, magnetic, or thermal), and cost. In addition, the material must be capable of flowing through small capillaries, "wetting" the joint surfaces, and partially alloying with the base metals. The most commonly used brazing metals are copper and copper alloys, silver and silver alloys, and aluminum alloys. Many of the brazing alloys are based on eutectic reactions (see Chapter 4), where the material melts at a single temperature that is lower than the melting points of the individual metals in the alloy. Table 33-5 presents some common braze metal families, the metals they are used to join, and the typical brazing temperatures.

Copper and copper alloys are the most commonly used braze metals. Unalloyed copper is used primarily for brazing steel and other high-melting-point materials, such as high-speed steel and tungsten carbide. Its melting point is rather high (about 1100°C), and tight-fitting joints are required (gaps less than 0.075 mm). Copper–zinc alloys offer lower melting points and are used extensively for brazing steel, cast irons, and copper. Copper–phosphorus alloys are used for the fluxless brazing of copper since the phosphorus can reduce the copper oxide film. These alloys should not be used with ferrous or nickel-based materials, however, since they form brittle compounds with phosphorus and the resulting joints may be brittle. Manganese bronzes can also be used as filler metal in brazing operations.

Pure silver can be used for brazing titanium. *Silver solders* (alloys based on silver and copper) have brazing temperatures significantly below that of pure copper and are used in joining steels, copper, brass, and nickel. While silver and silver alloys are expensive, only a small amount is required to make a joint, so the cost per joint is still low.

Aluminum–silicon alloys, containing between 6 and 12% silicon, are used for brazing aluminum and other aluminum alloys. By using a braze metal that is similar to the base metal, the possibility of galvanic corrosion is reduced. These brazing alloys, however, have melting points of about 610°C (1130°F), and the melting temperature of commonly brazed aluminum alloys, such as 3003, is around 670°C (1290°F). Therefore, control of the brazing temperature is critical. In brazing aluminum, proper fluxing action, surface cleaning, and/or the use of a controlled atmosphere or vacuum environment is required to assure adequate flow of the braze metal.

Nickel- and cobalt-based alloys are attractive for joining assemblies that will be subjected to elevated-temperature service conditions and/or extremely corrosive environments. The service temperature for brazed assemblies can be as high as 1200°C (2200°F). Gold and palladium alloys offer outstanding oxidation and corrosion resistance, as well as good electrical and thermal conductivity. Magnesium alloys can be used to braze other types of magnesium.

TABLE 33-5 Some Common Braze Metal Families, Metals They Are Used to Join, and Typical Brazing Temperatures

Braze Metal Family	Materials Commonly Joined	Typical Brazing Temperature (°C)
Aluminum-silicon	Aluminum alloys	565–620
Copper and copper alloys	Various ferrous metals as well as copper and nickel alloys and stainless steel	925–1150
Copper-phosphorus	Copper and copper alloys	700–925
Silver alloys	Ferrous and nonferrous metals, except aluminum and magnesium	620–980
Precious metals (gold-based)	Iron, nickel, and cobalt alloys	900–1100
Magnesium	Magnesium alloys	595–620
Nickel alloys	Stainless steel, nickel, and cobalt alloys	925–1200

A variety of brazing alloys are currently available in the form of amorphous foils, formed by cooling metal at extremely rapid rates. These foils are extremely thin (0.04 mm or 0.0015 in. being typical) and exhibit excellent ductility and flexibility, even when they are made from alloys whose crystalline form is quite brittle. Shaped inserts can be cut or stamped from the foil and inserted into the joint region. Since the braze material is fully dense, no shrinkage or movement occurs during the brazing operation.

One amorphous alloy, composed of nickel, chromium, iron, and boron, is used to produce assemblies that can withstand high temperatures. During the brazing operation, the boron diffuses into the base metal, raising the melting point of the remaining filler. The brazed assembly can then be heated to temperatures above the melting point of the original braze alloy, and the brazed joint will not melt.

FLUXES

In a normal atmosphere, the heat required to melt the brazing alloy would also cause the formation of surface oxides that oppose the wetting of the surface and subsequent bonding. *Brazing fluxes*, therefore, play an important part in the process by (1) dissolving oxides that may have formed on the surfaces prior to heating, (2) preventing the formation of new oxides during heating, and (3) lowering the surface tension between the molten brazing metal and the surfaces to be joined, thereby promoting the flow of the molten material into the joint. Ideally, the flux will melt and become active at a temperature below the solidus of the filler metal yet remain active throughout the entire range of temperatures encountered while making the braze.

Surface cleanliness is one of the most significant factors affecting the quality and uniformity of brazed joints. Although fluxes can dissolve modest amounts of oxides, *they are not cleaners*. Before a flux is applied, dirt, grease, oil, rust, and heat-treat scale should be removed from the surfaces that are to be brazed. Cleaning operations can involve water- or solvent-based techniques; high-temperature burn-off of oils, greases, and fuel residues; acid pickling; grit blasting with selected media; other mechanical methods; or exposure to high-temperature reducing atmospheres. The less cleaning the flux has to do, the more effective it will be during the brazing operation. Because the presence of surface graphite impairs wetting, cast iron materials often require special treatment. Graphite removal by chemical etching may be required before cast iron can be brazed.

Brazing fluxes usually take the form of chemical compounds in which the most common ingredients are borates, fused borax, fluoroborates, fluorides, chlorides, acids, alkalies, wetting agents, and water. The particular flux should be selected for compatibility with the base metal being brazed and the particular process being used. Paste fluxes are utilized for furnace, induction, and dip brazing, and they are usually applied by brushing. Either paste or powdered fluxes can be used with the torch-brazing process, where application is usually achieved by dipping the heated end of the filler wire into the flux material.

APPLYING THE BRAZE METAL

The brazing filler metal can be applied to joints in several ways. The oldest method (and still a common technique when torch brazing) uses brazing metal in the form of a rod or wire. The joint area is first heated to a temperature high enough to melt the braze alloy and ensure that it remains molten while flowing into the joint. The torch is then used to melt the braze metal, and capillary action draws it into the prepared gap.

The above method of braze metal application requires considerable labor, and care must be taken to ensure that the filler metal has flowed into the inner portions of the joint. To avoid these difficulties, the braze metal is often inserted into the joint prior to heating, usually in the form of wires, foils, shims, powders, or preformed rings, washers, disks, or slugs. Rings or shims can also be fitted into internal grooves in the joint before the parts are assembled.

When using preloaded joints, care must be exercised to ensure that the filler metal is not drawn away from the intended surface by the capillary action of another surface of contact. Capillary action will always pull the molten braze metal into the smallest clearance, regardless of whether that was the intended location. In addition, the flow of filler metal must not be cut off by inadequate clearances or the presence of entrapped or escaping air. Fillets and grooves within the joint can also act as reservoirs and trap the filler metal.

Yet another approach is to precoat one or both of the surfaces to be joined with the brazing alloy. Simply placing the materials in contact and heating forms the desired bond. By having the braze material already in place over the full area of contact, the joining operation does not have to rely on capillary action and metal flow. More complex assemblies can be produced than with conventional methods, and the thickness of the braze material is precisely controlled to provide maximum strength to the joint.

All of the components must maintain fixed positions during the brazing operation, and some form of restraint or fixturing is often required. Alignment and clearances can often be maintained by tack welding, riveting, staking, expanding or flaring, swaging, knurling, or dimpling. Shims, wires, ribbons, and screens can also be employed to assist in locating pieces or maintaining fit. For more complex components, special brazing *jigs and fixtures* are often used to hold the components during the heating. When these are used, however, it is necessary to provide springs that will compensate for thermal expansion, particularly when two or more dissimilar metals are being joined.

HEATING METHODS USED IN BRAZING

Since molten metal tends to flow toward the location of highest temperature, it is important that the heat sources used in brazing control both the temperature and the uniformity of that temperature throughout the joint. In specifying the heating method, a number of factors should be considered, including the size and shape of the parts being brazed, the type of material being joined, and the desired quantity and rate of production.

A common source of heat for brazing is a gas-flame torch. In the *torch-brazing* procedure, oxyacetylene, oxy-hydrogen, or other gas-flame combinations can be used. Most repair brazing is done in this manner because of its flexibility and simplicity, but the process is also widely used in production applications where specially shaped torches speed the heating and reduce the amount of skill required. Local heating permits the retention of most of the original material strength and enables large components to be joined with little or no distortion. The major drawbacks are the difficulty in controlling the temperature and maintaining uniformity of heating, as well as meeting the cost of skilled labor. A protective flux is usually required, and the flux residue must be removed after brazing.

If the flux and filler metal can be preloaded into the joints and the part can endure uniform heating, a number of assemblies can be brazed simultaneously in controlled-atmosphere or vacuum furnaces, a process known as *furnace brazing*. If the components are not likely to maintain their alignment, brazing jigs or fixtures must be used. Fortunately, for most assemblies that are to be furnace brazed, a light press fit is usually sufficient to maintain alignment. Figure 33-16 shows some typical furnace-brazed assemblies.

Because excellent control of the furnace temperature is possible and no skilled labor is required, furnace brazing is particularly well suited for mass-production operations, with either batch- or continuous-type furnaces being used. Furnace brazing heats the entire assembly in a uniform manner and therefore produces less warpage and distortion than processes that employ localized heating. Extremely complex assemblies can be produced, with multiple joints being formed in a single heating.

A variety of furnace atmospheres can be utilized to reduce oxide films and prevent both the base and filler metals from oxidizing during the brazing operation. A chemical flux may no longer be needed, and the parts emerge clean and free of contaminants. When reactive materials are to be joined or the joint must meet the highest of standards, a vacuum furnace may be preferred.

A third type of heating is *salt-bath brazing*, where the parts are preheated and then dipped in a bath of molten salt that is maintained at a temperature slightly above the melting point of the brazing metal. This process offers three distinct advantages:

1. The salt bath acts as the brazing flux, preventing oxidation and enhancing wettability.

2. The work heats very rapidly because it is in complete contact with the heating medium.

FIGURE 33-16 Typical furnace-brazed assemblies. *(Courtesy of Pacific Metals Company, a division of Reliance Steel & Aluminum, Los Angeles, CA.)*

3. Temperature can be accurately controlled, so thin pieces can be attached to thicker pieces without danger of overheating. This last feature makes the process well suited for brazing aluminum, where precise temperature control is often required.

In salt-bath brazing, the parts must be held in jigs or fixtures (or be prefastened in some manner), and the brazing metal must be preloaded into the joints. To assure that the bath remains at the desired temperature during the immersion process, its volume must be substantially larger than that of the assemblies to be brazed.

In *dip brazing* the assemblies are immersed in a bath of molten brazing metal. The bath thus provides both the heat and the metal for the joint. Since the braze metal will usually coat the entire workpiece, it is a somewhat wasteful process and is usually employed only for small products.

Induction brazing utilizes high-frequency induction currents as the source of heat and is therefore limited to the joining of electrically conductive materials. A variety of high-frequency AC power supplies is available in large and small capacities. These are coupled to a simple heating coil designed to fit around the joint. The heating coils are generally formed from copper tubing and typically carry a supply of cooling water. Although the filler metal can be added to the joint manually after it is heated, the usual practice is to use preloaded joints to speed the operation and produce more uniform bonds. Induction brazing offers the following advantages, which account for its extensive use:

1. The complete heating cycle is very rapid, usually only a few seconds in duration.

2. The operation can be made semiautomatic so that only semiskilled labor is required.

3. Heating can be confined to the specific area of the joint through use of specially designed coils, frequency control, and short heating times. This minimizes softening and distortion and reduces problems associated with scale and discoloration.

4. Uniform results are easily obtained.

5. By making new and relatively simple heating coils, a wide variety of work can be brazed with a single power supply.

Resistance brazing can be used to produce relatively simple joints in metals with high electrical conductivity. The parts to be joined are pressed between two electrodes and a current is passed through. Unlike resistance welding, the carbon or graphite electrodes provide most of the resistance in resistance brazing, and the heating of the joint is primarily by conduction from the hot electrodes.

Infrared heat lamps, lasers, and electron beams can also be used to provide the heat required for brazing. Recent studies have also shown microwave energy to be an efficient heat source. Silicon carbide plates are positioned around the joint and are heated by the microwaves. Heat is then transferred to the joint by radiation.

FLUX REMOVAL AND OTHER POSTBRAZE OPERATIONS

Since most brazing fluxes are corrosive, the flux residue should be removed from the work as soon as brazing is completed. Rapid and complete flux removal is particularly important in the case of aluminum, where chlorides can be particularly detrimental. Fortunately, many brazing fluxes are water soluble, and an immersion in a hot-water tank for a few minutes will often provide satisfactory results. Blasting with grit or sand is another effective method of flux removal, but this procedure may not be attractive if a good surface finish is to be maintained. Fortunately, such drastic treatment is seldom necessary.

Other postbraze operations may include heat treating, cleaning, and inspection. A visual examination is probably the simplest of the inspection techniques and is most effective when both sides of a brazed joint are accessible for examination. A proof test can be performed by subjecting the joint to loads in excess of those expected during service. Leak tests or pressure tests can assure gas- or liquid-tightness. Cracks and other flaws can be detected by dye penetrant, magnetic particle, ultrasonic, or radiographic examination. Destructive forms of evaluation include peel tests, tension or shear tests, and metallographic examination.

FLUXLESS BRAZING

Since the application and removal of brazing flux involves significant costs, particularly where complex joints and assemblies are involved, a large amount of work has been devoted to the development of procedures where a flux is not required. Controlled furnace atmospheres can make a flux unnecessary by reducing existing oxides and preventing the formation of new ones. Vacuum furnaces can also be used to create and preserve clean brazing surfaces. Special brazing metals have been developed with alloy additions, such as phosphorus, that can also fulfill the role of a flux.

BRAZE WELDING

Braze welding differs from straight brazing in that capillary action is not required to distribute the filler metal. Here the molten filler is simply deposited by gravity, as in oxyacetylene gas welding. Because relatively low temperatures are required and warping is minimized, braze welding is very effective for the repair of steel products and ferrous castings. It is also attractive for joining cast irons since the low heat does not alter the graphite shape, and the process does not require good wetting characteristics. Strength is determined by the braze metal being used and the amount applied. Considerable buildup may be required if full strength is to be restored to the repaired part.

Braze welding is almost always done with an oxyacetylene torch. The surfaces are first "tinned" with a thin coating of the brazing metal, and the remainder of the filler metal is then added. Figure 33-17 shows a schematic of braze welding.

■ 33.5 SOLDERING

Soldering is a brazing-type operation where the filler metal has a melting temperature (or liquidus temperature if the alloy has a freezing range) below 450°C (840°F). It is typically used for joining thin metals, connecting electronic components, joining metals while avoiding

FIGURE 33-17 Schematic of the braze-welding process.

exposure to high elevated temperatures, and filling surface flaws and defects. The process generally involves six important steps: (1) design of an acceptable joint; (2) selection of the correct solder for the job; (3) selection of the proper type of flux; (4) cleaning of the surfaces to be joined; (5) application of flux, solder, and sufficient heat to allow the molten solder to fill the joint by capillary action and solidify; and (6) removal of the flux residue.

DESIGN AND STRENGTH OF SOLDERED JOINTS

Soldering can be used to join a wide variety of sizes, shapes, and thicknesses; it is employed extensively to provide electrical coupling or gas- or liquid-tight seals. While the low joining temperatures are attractive for heat-sensitive materials, soldered joints seldom develop shear strengths in excess of 1.75 MPa (250 psi). Consequently, if appreciable strength is required, soldered joints should be avoided, the contact area should be large, or some form of mechanical joint, such as a rolled-seam lock, should be made prior to soldering. Butt joints should never be used, and designs where peeling action is possible should be avoided. Figure 33-18 shows some of the more common solder joint designs, including lap, flanged butt, and interlock.

As with brazing, there is an optimal clearance for best performance. For typical solder joints, a clearance of 0.025 to 0.13 mm (0.001 to 0.005 in.) provides for capillary flow of the solder, expulsion of the flux, and reasonable joint strength. In addition, the parts should be held firmly so that no movement can occur until the solder has cooled to well below the solidification temperature. Otherwise, the resulting joint may contain cracks and have very little strength.

METALS TO BE JOINED

Table 33-6 summarizes the compatibility of soldering with a variety of engineering materials. Copper, silver, gold, and tin-plated steels are all easily soldered. Aluminum has a strong, adherent oxide that makes soldering difficult. Special fluxes and modified techniques may be required, but adequate joints are indeed possible, as shown by the large number of soldered aluminum radiators currently in automotive use.

SOLDER METALS

Soldering alloys, the filler metals for soldering, are generally combinations of low-melting-temperature metals, such as lead, tin, bismuth, indium, cadmium, silver, gold, and germanium. Because of their low cost, acceptable mechanical and physical properties, and many years of use, the most common solders are alloys of lead and tin with the addition of small amounts of antimony, usually less than 0.5%. The three most common alloys contain 60, 50, and 40% tin, and all melt below 240°C (465°F). Because tin is expensive, those alloys having higher proportions of tin are used only where their higher fluidity, higher strength, and lower melting temperature are desired. For wiped joints and for filling dents and seams, where the primary desire is appearance and little strength is required, solders containing only 10 to 20% tin are preferred. Joints made with the 5%

Flanged butt   Flush lap

Flanged edge   Flanged bottom

Interlock   Pipe joint

FIGURE 33-18   Some common designs for soldered joints.

TABLE 33-6   Compatibility of Various Engineering Materials with Soldering

Material	Soldering Recommendation
Cast iron	Seldom used since graphite and silicon inhibit bonding
Carbon and low-alloy steels	Difficult for low-carbon materials; seldom used for high-carbon materials
Stainless steel	Common for 300 series; difficult for 400 series
Aluminum and magnesium	Seldom used; however, special solders are available
Copper and copper alloys	Recommended for copper, brass, and bronze
Nickel and nickel alloys	Commonly performed using high-tin solders
Titanium	Seldom used
Lead and zinc	Recommended, but must use low-melting-temperature solders
Thermoplastics, thermosets, and elastomers	Not recommended
Ceramics and glass	Not recommended
Dissimilar metals	Recommended, but with consideration for galvanic corrosion
Metals to nonmetals	Not recommended
Dissimilar nometals	Not recommended

tin alloy require higher temperatures to produce but will withstand service temperatures as high as 150°C (300°F).

Other soldering alloys may be specified for special purposes or where environmental or health concerns dictate the use of lead-free joints. Lead and lead compounds can be quite toxic. Since 1988, the use of lead-containing solders in drinking water lines has been prohibited in the United States, and concern has been expressed regarding other applications and industries. Japan and the European Union have banned the use of lead-containing solders in electronic equipment. If substitute solders are to be acceptable, however, they should not only be harmless to the environment, but should also exhibit desirable characteristics in the areas of melting temperature, wettability, electrical and thermal conductivity, thermal-expansion coefficient, mechanical strength, ductility, creep resistance, thermal fatigue resistance, corrosion resistance, manufacturability, and cost. At present, none of the *lead-free solders* meet all of these requirements, and most are deficient in more than one area. Compatible fluxes must also be identified, and assembly methods may need to be modified.

Most of the alternative solders have been proposed from other eutectic alloy systems. Tin–antimony and tin–copper alloys are useful in electrical applications and have good strength and creep resistance but high melting points. Bismuth alloys have very low melting points and good fluidity but suffer from poor wettability. Indium alloys offer low melting points, ductility that is retained even at cryogenic temperatures, and rapid creep that allows joints between dissimilar metals to adjust to changes in temperature without generating internal stresses. Tin–indium alloys have been used to join metal to glass and glass to glass. They have very low melting points and good wettability, but they are expensive and can be somewhat brittle. Aluminum is often soldered with tin–zinc, cadmium–zinc, or aluminum–zinc alloys. Tin–silver and tin–gold offer possibilities when a somewhat higher melting point is desired (typically above 205°C, or 400°F) coupled with good mechanical strength and creep resistance, but both systems are limited by the high cost of their components. Lead–silver and cadmium–silver alloys can also be used for higher-temperature service. The three-component tin–silver–copper system has received considerable attention for electronics applications.

Like the filler metal used in brazing and braze welding, solders are available as wire and paste, as well as in a variety of standard and special preshaped forms. Table 33-7 presents some of the more common solder alloys with their melting properties and typical applications.

SOLDERING FLUXES

As in brazing, soldering requires that the metal surfaces be clean and free of oxide so that the solder can wet the surfaces and be drawn into the joint to produce an effective bond. Soldering fluxes are used to remove surface oxides and prevent oxide formation

TABLE 33-7	Some Common Solders and Their Properties			
Composition (wt %)	Freezing Temperature (°C)			
	Liquidus	Solidus	Range	Applications
Lead-tin solders				
98 Pb–2 Sn	322	316	6	Side seams in three-piece can
90 Pb–10 Sn	302	268	34	Coating and joining metals
80 Pb–20 Sn	277	183	94	Filling and seaming auto bodies
70 Pb–30 Sn	255	183	72	Torch soldering
60 Pb–40 Sn	238	183	55	Wiping solder, radiator cores, heater units
50 Pb–50 Sn	216	183	33	General purpose
40 Pb–60 Sn	190	183	7	Electronic (low temperature)
Silver solders				
97.5 Pb–1 Sn–1.5 Ag	308	308	0	Higher–temperature service
36 Pb–62 Sn–2 Ag	189	179	10	Electrical
96 Sn–4 Ag	221	221	0	Electrical
Other alloys				
45 Pb–55 Bi	124	124	0	Low temperature
43 Sn–57 Bi	138	138	0	Low temperature
95 Sn–5 Sb	240	234	6	Electrical
50 Sn–50 In	125	117	8	Metal-to-glass
37.5 Pb–25 In–37.5 Sn	138	138	0	Low temperature
95.5Sn–3.9 Ag–0.6 CO	217	217	0	Electrical

during the soldering process, but it is essential that dirt, oil, and grease be removed before the flux is applied. This precleaning or surface preparation can be performed by a variety of chemical or mechanical means, including solvent or alkaline degreasers, acid immersion (pickling), grit blasting, sanding, wire brushing, and other mechanical abrasion techniques.

Soldering fluxes are generally classified as *corrosive* or *noncorrosive*. The most common noncorrosive flux is *rosin* (the residue after distilling turpentine) dissolved in alcohol. Rosin fluxes are suitable for making joints to copper and brass, and to tin-, cadmium-, and silver-plated surfaces, provided that the surfaces have been adequately cleaned prior to soldering. Aniline phosphate is a more active noncorrosive flux, but it has limited use because it emits toxic gases when heated. The wide variety of corrosive fluxes provide enhanced cleaning action but require complete removal after the soldering operation to prevent corrosion problems during service.

THE SOLDERING OPERATION

Soldering requires a source of sufficient heat and a means of transferring it to the metals being joined. Any method of heating that is suitable for brazing can be used for soldering, but furnace and salt-bath heating are seldom used. Most hand soldering is still done with soldering irons or small oxyfuel or air-fuel (acetylene, propane, butane, or MAPP) torches. Induction heating is used when large numbers of identical parts are to be soldered. For low-melting-point solders, infrared heat sources can also be employed. The joints can be preloaded with solder, or the filler metal can be supplied from a wire. The particular method of heating usually dictates which procedure is used.

Wave soldering, depicted in Figure 33-19, is a process used to solder wire ends, such as the multiple connectors that protrude through holes in electronic circuit boards. Molten solder is pumped upward through a submerged nozzle to create a wave or crest in a pool of molten metal. The circuit boards are then passed across this wave at a height where each of the pins sees contact with the molten metal. Wetting and capillary action pulls solder into each joint, and numerous connections are made as each board passes across the wave.

FIGURE 33-19 Schematic of wave soldering.

In the *vapor-phase soldering* process, a product with prepositioned solder is passed through a chamber containing hot, saturated vapor. The vapor condenses on the cooler product, transferring its heat of vaporization. This results in rapid and uniform heating, with excellent temperature control and the possibility of an oxygen-free environment. The soldering temperature is limited only by the boiling point of the fluid, with current materials operating in the range of 100° to 265°C (212°to 510°F). While the process can be used to cure epoxies and stress relieve metals, its primary application is the soldering of surface-mounted components to substrate materials. Because of the precise temperature control, multipass soldering is possible, using up to three different solder compositions with three different melting temperatures. Because the solder is prepositioned, this process is also known as *vapor-phase reflow soldering*.

Dip soldering, where the entire piece is immersed in molten metal, has been used to produce automobile radiators and "tinned" coatings.

FLUX REMOVAL

After soldering, the flux residues should be removed from the finished joints, either to prevent corrosion or for the sake of appearance. Flux removal is rarely difficult, provided that the type of solvent in the flux is known. Water-soluble fluxes can be removed with hot water and a brush. Alcohol will remove most rosin fluxes. However, when the flux contains some form of grease, as in most paste fluxes, a grease solvent must be used, followed by a hot-water rinse. In the past, solvents containing chlorofluorocarbons (CFCs) were the cleaners of choice, but since they have been implicated in the depletion of atmospheric ozone, an alternative means of flux removal should be employed or the process converted to fluxless soldering.

FLUXLESS SOLDERING

Several *fluxless-soldering* techniques have been developed using controlled atmospheres (such as hydrogen plasma), thermomechanical surface activation (such as plasma gas impingement), or protective coatings that prevent oxide formation and enhance wetting. Additional successes have been reported with both laser and ultrasonic soldering.

■ Key Words

assist gas	flash welding	lap joint	salt-bath brazing
autogenous weld	fluidity	laser spot welding	silver solder
braze welding	flux	laser-beam cutting	soldering
brazing	fluxless brazing	laser-beam welding	surfacing
butt joint	fluxless soldering	lead-free solder	thermal spray
capillary action	furnace brazing	metallizing	thermit
corrosive flux	hard facing	molding plates	thermit welding
dip brazing	hybrid processes	noncorrosive flux	torch brazing
dip soldering	induction brazing	percussion welding	upset welding
electron-beam welding	jigs and fixtures	reflow soldering	vapor-phase soldering
electroslag welding	joint clearance	resistance brazing	wave soldering
filler metal	kerf	rosin	wettability

■ Review Questions

1. In what ways is a thermit weld similar to the production of a casting?
2. What is the source of the welding heat in thermit welding?
3. For what types of applications might thermit welding be attractive?
4. What is the source of the welding heat in electroslag welding?
5. What are some of the various functions of the slag in electroslag welding?
6. Electroslag welding would be most attractive for the joining of what types of geometries and thicknesses?
7. Why is a high vacuum required in the electron-beam chamber of an electron-beam welding machine?
8. What types of production limitations are imposed by the high-vacuum requirements of electron-beam welding? What compromises are made when welding is performed on pieces outside the vacuum chamber?
9. What are the major assets and negative features of high-voltage electron-beam welding equipment?
10. What are some of the attractive features of electron-beam welding? Negative features?
11. What is unique about the fusion zone geometry of electron-beam welds?
12. What are some of the ways in which laser-beam welding is more attractive than electron-beam welding?
13. Which type of laser light can be transmitted through fiber-optic cable?
14. Why is laser-beam welding an attractive process for producing tailored blanks for sheet metal forming? For use on small electronic components?
15. What are the attractive properties of hybrid processes that combine laser and arc welding?
16. What is the function of the *assist gas* in laser-beam cutting?
17. How do the cut edges differ with endothermic laser cutting and exothermic laser cutting?
18. What features have made lasers a common means of cutting composite materials?
19. What are some of the attractive features of laser spot welding?
20. In the flash-welding process, why is it important to have a sufficient duration of arcing and sufficient amount of upsetting?
21. What are some common objectives of surfacing operations?
22. What types of materials are applied by surfacing methods?
23. What are some of the primary methods by which surfacing materials can be deposited onto a metal substrate?
24. What are some of the techniques that can be used to apply a thermal spray coating?
25. How is thermal spraying similar to surfacing? How is it different?
26. Why is surface preparation such a critical feature of metallizing?
27. What are some of the more common applications of sprayed coatings?
28. Provide a reasonable definition of brazing.
29. What are some key differences between brazing and fusion welding?
30. Why is brazing an attractive process for joining dissimilar materials?
31. What advantages can be gained by the lower temperatures of the brazing process?
32. Why do brazed joints have an enhanced susceptibility to corrosion?
33. What is the most important factor contributing to the strength of a brazed joint?
34. How does capillary action relate to joint clearance?
35. Why is it necessary to adjust the initial room-temperature clearance of a joint between two significantly different metals?
36. What is wettability? Fluidity? How does each relate to brazing?
37. What are the two most common types of brazed joints and the attractive features of each?
38. What are some important considerations when selecting a brazing alloy?
39. What are some of the most commonly used brazing metals?
40. What special measures should be taken when brazing aluminum?
41. What are the three primary functions of a brazing flux?
42. Why is it important to preclean brazing surfaces before applying the flux?
43. In what ways might braze metal be preloaded into joints?
44. What is the purpose of brazing jigs and fixtures?
45. What is the primary attraction of furnace-brazing operations?
46. Why might reducing atmospheres or a vacuum be employed during furnace-brazing operations?
47. Why is dip brazing usually restricted to use with small parts?
48. What are some of the attractive features of induction brazing?
49. Why is flux removal a necessary part of many brazing operations?
50. What benefits can be achieved through fluxless brazing?
51. How does braze welding differ from traditional brazing?
52. What is the primary difference between brazing and soldering?
53. Why is soldering unattractive if a high-strength joint is desired?
54. For many years, the most common solders were alloys of what two base metals?
55. What is driving the conversion to lead-free solders?
56. What are some of the difficulties encountered when attempting a conversion to lead-free solder?
57. What are the two basic families of soldering flux?
58. What are some of the more common heat sources for producing a soldered joint?

■ Problems

1. A common problem with brazed or soldered joints is galvanic corrosion, since the joint usually involves dissimilar metals in direct metal-to-metal electrical contact.
   a. For each of the various solder or braze joints described below, determine which material will act as the corroding anode.
      (1) Two pieces of low-carbon steel being brazed with a copper-base brazing alloy
      (2) A copper wire being soldered to a steel sheet using lead–tin solder
      (3) Pieces of tungsten carbide being brazed into recesses in a carbon-steel plate
   b. How do the various lead-free solders compare to the conventional lead–tin solders with regard to their potential for galvanic corrosion?
   c. If galvanic corrosion becomes a significant and chronic problem in a brazed assembly, what changes might you suggest that could possibly reduce or eliminate the problem?

ADHESIVE BONDING, MECHANICAL FASTENING, AND JOINING OF NONMETALS

■ 34.1 ADHESIVE BONDING

The *ideal adhesive* bonds to any material, needs no surface preparation, cures rapidly, and maintains a high bond strength under all operating conditions. It also doesn't exist. However, tremendous advances have been made in the development of adhesives that are stronger, easier to use, less costly, and more reliable than many of the alternative methods of joining. From early applications, such as plywood, the use of structural adhesives has grown rapidly. Adhesives are everywhere—in construction, packaging, furniture, appliances, electronics, bookbinding, product assembly, and even medical and dental applications. They are used to bond metals, ceramics, glass, plastics, rubbers, composite materials, woods, and even a variety of roofing materials. Even such quality- and durability-conscious fields as the automotive and aircraft industries now make extensive use of adhesive bonding. Adhesives in the automotive industry have advanced from the attaching of interior and exterior trim to the joining of major components, such as door, hood, and trunk assemblies, and the installation of the nonmoving front and rear windows. Adhesive bonding has become the preferred means of assembly for polymeric body panels made from sheet-molding compounds and reaction-injection-molded (RIM) materials. Moreover, since adhesive bonding has the ability to bond such a wide variety of materials, its use has grown significantly with the ever-expanding applications of plastics and composites.

ADHESIVE MATERIALS AND THEIR PROPERTIES

In *adhesive bonding*, a nonmetallic material (the *adhesive*) is used to fill the gap and create a joint between two surfaces. The actual adhesives span a wide range of material types and forms, including *thermoplastic* resins, *thermosetting* resins, artificial *elastomers*, and even some ceramics. They can be applied as drops, beads, pellets, tapes, or coatings (films) and are available in the form of liquids, pastes, gels, and solids. *Curing* can be induced by the use of heat, radiation or light (photoinitiation), moisture, activators, catalysts, multiple-component reactions, or combinations thereof. Applications can be full load bearing (structural adhesives), light-duty holding or fixturing, or simply sealing (the forming of liquid- or gas-tight joints). With such a wide range of possibilities, the selection of the best adhesive for the task at hand can often be quite challenging.

The *structural adhesives* are selected for their ability to effectively transmit load across the joint; they include epoxies, cyanoacrylates, anaerobics, acrylics, urethanes,

915

silicones, high-temperature adhesives, and hot melts. Both strength and rigidity may be important, and the bond must be able to be stressed to a high percentage of its maximum load for extended periods of time without failure.

1. *Epoxies*. The thermosetting epoxies are the oldest, most common, and most diverse of the adhesive systems; they can be used to join most engineering materials, including metal, glass, and ceramic. They are strong, versatile adhesives that can be designed to offer high adhesion, good tensile and shear strength, toughness, high rigidity, creep resistance, easy curing with little shrinkage, good chemical resistance, and tolerance to elevated temperatures. Various epoxies can be used over a temperature range from $-50°$ to $+250°C$ ($-60°$ to $+500°F$). After curing at room temperature, shear strengths can be as high as 35 to 70 MPa (5000 to 10,000 psi).

   Single-component epoxies use heat as the curing agent. Most epoxies, however, are two-component blends involving a resin and a curing agent, plus possible additives such as accelerators, plasticizers, and fillers that serve to enhance cure rate, flexibility, peel resistance, impact resistance, or other characteristics. Heat may again be required to drive or accelerate the cure.

   Low peel strength and poor flexibility limit epoxy adhesives, and the bond strength can be sensitive to moisture and surface contamination. Epoxies are often brittle at low temperatures, and the rate of curing is comparatively slow. Sufficient strength for structural applications is generally achieved in 8 to 12 hours, with full strength often requiring two to seven days.

2. *Cyanoacrylates*. These are liquid monomers that polymerize when spread into a thin film between two surfaces. Trace amounts of moisture on the surfaces promote curing at amazing speeds, often in as little as two seconds. Thus, the cyanoacrylates offer a one-component adhesive system that cures at room temperature with no external impetus. Commonly known as *superglues*, this family of adhesives is now available in the form of liquids and gels of varying viscosity, toughened versions designed to overcome brittleness, and even nonfrosting varieties.

   The cyanoacrylates provide excellent tensile strength, fast curing, and good shelf life, and they adhere well to most commercial plastics, metals, and rubbers. They are limited by their high cost, poor peel strength, and brittleness. Bond properties are poor at elevated temperatures, and effective curing requires good component fit (gaps must be smaller than 0.25 mm, or 0.010 in.).

3. *Anaerobics*. These one-component, thermosetting, polyester acrylics remain liquid when exposed to air. When confined to small spaces and shut off from oxygen, as in a joint to be bonded or along the threads of an inserted fastener, the polymer becomes unstable. In the presence of iron or copper, it polymerizes into a bonding-type resin, without the need for elevated temperature. Curing can occur across gaps as large as 1 mm, or 0.04 in. Additives can reduce odor, flammability, and toxicity and can speed the curing operation. Slow-curing anaerobics require 6 to 24 hours to attain useful strength. With selected additives and heat, however, curing can be reduced to as little as five minutes.

   The anaerobics are extremely versatile and can bond almost anything, including oily surfaces. The joints resist vibrations and offer good sealing to moisture and other environmental influences. Unfortunately, they are somewhat brittle and are limited to service temperatures below 150°C (300°F).

4. *Acrylics*. The acrylic-based adhesives offer good strength, toughness, and versatility, and they are able to bond a variety of materials, including plastics, metals, ceramics, and composites, even oily or dirty surfaces. Most involve application systems where a catalyst primer (curing agent) is applied to one of the surfaces to be joined and the adhesive is applied to the other. The pretreated parts can be stored separately for weeks without damage. Upon assembly, the components react to produce a strong bond at room temperature. Heat can often accelerate the curing, and at least one va-

riety cures with ultraviolet (UV) light. In comparison to other varieties of adhesives, the acrylics offer strengths comparable to the epoxies, flexible bonds, good resistance to water and humidity, and the added advantages of room-temperature curing and a no-mix application system. Major limitations include poor strength at high temperatures, flammability, and an unpleasant odor when still uncured.

5. *Urethanes.* Urethane adhesives are a large and diverse family of polymers that are generally targeted for applications that involve temperatures below 65°C (150°F) and components that require great flexibility. Both one-part thermoplastic and two-part thermosetting systems are available. Urethanes cure quickly to handling strength but are slow to reach the full-cure condition. Two minutes to handling with 24 hours to complete cure is common at room temperature.

    Compared to other structural adhesives, the urethanes offer good flexibility and toughness, even at low operating temperatures. They are somewhat sensitive to moisture, degrade in many chemical environments, and can involve toxic components or curing products.

6. *Silicones.* The silicone thermosets cure from the moisture in the air or adsorbed moisture from the surfaces being joined. They form low-strength structural joints and are usually selected when considerable expansion and contraction are expected in the joint; flexibility is required (as in sheet metal parts); or good gasket, gap-filling, or sealing properties are necessary. Metals, glass, paper, plastics, and rubbers can all be joined. The adhesives are relatively expensive, and curing is slow, but the bonds that are produced can resist moisture, hot water, oxidation, and weathering, and they retain their flexibility at low temperature.

7. *High-temperature adhesives.* When strength must be retained at temperatures in excess of 300°C (500°F), high-temperature structural adhesives should be specified. These include epoxy phenolics, modified silicones or phenolics, polyamides, and some ceramics. High cost and long cure times are the major limitations for these adhesives, which see primary application in the aerospace industry.

8. *Hot melts.* Hot-melt adhesives can be used to bond dissimilar substrates, such as plastics, rubber, metals, ceramics, glass, wood, and fibrous materials like paper, fabric, and leather. They can produce permanent or temporary bonds, seal gaps, and plug holes. While generally not considered to be true structural adhesives, the hot melts are being used increasingly to transmit loads, especially in composite-material assemblies. The joints can withstand exposure to vibration, shock, humidity, and numerous chemicals, and they offer the added features of sound deadening and vibration damping.

    Most hot-melt adhesives are thermoplastic resins that are solid at room temperature but melt abruptly when heated into the range of 100° to 150°C (200° to 300°F). They are usually applied as heated liquids (between 160° and 180°C) and form a bond as the molten adhesive cools and resolidifies. Another method of application is to position the adhesive in the joint prior to operations, such as the paint-bake process in automobile manufacture. During the baking, the adhesive melts, flows into seams and crevices, and seals against the entry of corrosive moisture. These adhesives contain no solvents and do not need time to cure or dry. Hot melts achieve over 80% of their bond strength within seconds of solidification, but they do soften and creep when subsequently exposed to elevated temperatures and can become brittle when cold.

Additives also play a large part in the success of industrial adhesives. They can impart or enhance properties like toughness, joint durability, moisture resistance, adhesion, and flame retardance. Rheological additives and plasticizers control viscosity and flow. Adhesives must penetrate the surfaces to be bonded but should not flow in an uncontrollable fashion. Fillers and extenders provide bulk and reduce cost.

U.S. Consumption of adhesives and sealants by product, 2003 (percentages by dollar value)

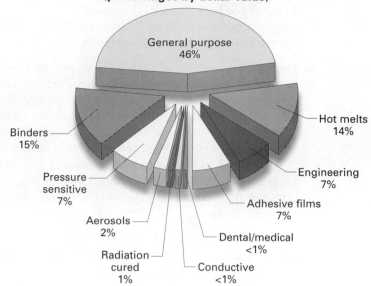

FIGURE 34-1 Distribution among the common types of adhesives and sealants. *(Reprinted with permission from* The Rauch Guide to the US Adhesives & Sealants Industry, *Fifth Edition, 2006, Grey House Publishing, Millerton, NY).*

U.S. Consumption of adhesives and sealants by end-use market, 2003 (percentages by dollar value)

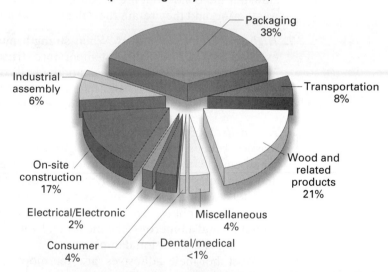

FIGURE 34-2 Distribution of adhesives and sealants by end-use areas. *(Reprinted with permission from* The Rauch Guide to the US Adhesives & Sealants Industry, *Fifth Edition, 2006, Grey House Publishing, Millerton, NY).*

Figure 34-1 shows the distribution of various types of adhesive and sealant products for a recent year, and Figure 34-2 classifies adhesives by end-use markets. Table 34-1 lists some popular structural adhesives, along with their service and curing temperatures and expected strengths. Table 34-2 presents the advantages and disadvantages of various curing processes.

NONSTRUCTURAL AND SPECIAL ADHESIVES

There are a number of other types of adhesives whose limited load-bearing capabilities place them in a nonstructural classification. Nevertheless, they still play roles in manufacturing through a variety of uses, such as labeling and packaging. The hot-melt adhesives are often placed in this category but can be used for applications in both classifications. *Evaporative adhesives* use an organic solvent or water base, coupled with vinyls, acrylics, phenolics, polyurethanes, or various types of rubbers. Some common evaporative adhesives are rubber cements and floor waxes. *Pressure-sensitive adhesives* are usually based on various rubbers, compounded with additives to bond at room

TABLE 34-1 Some Common Structural Adhesives, Their Cure Temperatures, Maximum Service Temperatures, and Strengths under Various Types of Loadings

Adhesive Type	Cure Temperature (°F)	Service Temperature (°F)	Lap Shear Strength (psi at °F)[a]	Peel Strength at Room Temperature (lb/in.)
Butyral-phenolic	275 to 350	−60 to 175	1000 at 175	10
			2500 at RT	
Epoxy				
Room-temperature cure	60 to 90	−60 to 180	1500 at 180	4
			2500 at RT	
Elevated-temperature cure	200 to 350	−60 to 350	1500 at 350	5
			2500 at RT	
Epoxy-nylon	250 to 350	−420 to 180	2000 at 180	70
			6000 at RT	
Epoxy-phenolic	250 to 350	−420 to 500	1000 at 175	10
			2500 at RT	
Neoprene-phenolic	275 to 350	−60 to 180	1000 at 180	15
			2000 at RT	
Nitrile-phenolic	275 to 350	−60 to 250	2000 at 250	60
			4000 at RT	
Polyimide	550 to 650	−420 to 1000	1000 at 1000	3
			2500 at RT	
Urethane	75 to 250	−420 to 175	1000 at 175	50
			2500 at RT	

[a]RT, room temperature.

TABLE 34-2 Advantages and Disadvantages of Various Structural Adhesive Curing Processes

Curing Process	Advantages	Disadvantages
Mixing reactive components	Good shelf life, unlimited depth of cure, accelerated with heat	High processing costs, mix ratio critical to performance
Anaerobic cure	Single-component adhesive, good shelf life	Poor depth of cure, require primer on many surfaces, sensitive to surface contaminants
Heat cure	Unlimited depth of cure, heat can aid adhesion	Expenses for oven energy cost, heat can adversely affect some substrates
Moisture cure	Room-temperature process, one component, no curing equipment required	Long cure cycles (12–72 hr), minimum % humidity required, limited depth of cure
Light cure	Rapid cure, cure on demand	Expenses for UV light source, limited depth of cure, most allow light to reach bond
Surface-initiated cure	Rapid cure	Poor depth of cure

temperature with a brief application of pressure. No cure is involved, and the tacky adhesive-coated surfaces require no activation by water, solvents, or heat. Peel-and-stick labels, cellophane tape, and Post-it notes are examples of this group of adhesives. *Delayed-tack adhesives* are similar to the pressure-sensitive systems but are nontacky until activated by exposure to heat. Once heated, they remain tacky for several minutes to a few days to permit use or assembly.

While most adhesives are electrical and thermal insulators, *conductive adhesives* can be produced by incorporating selected fillers, such as silver, copper, or aluminum, in the form of flakes or powder. Certain ceramic oxide fillers can be used to provide thermal conductivity coupled with electrical insulation.

Still another group of commercial adhesives are those designed to cure by exposure to radiation, such as visible, infrared, or ultraviolet light; microwaves; or electron beams. These *radiation-curing adhesives* offer rapid conversion from liquid to solid at room temperature and a curing mechanism that occurs throughout, rather than progressing from exposed surfaces (as with the competing low-temperature air or moisture cures). Current applications include a wide variety of dental amalgams that can fill cavities or seal surfaces while matching the color of the remaining tooth. In the manufacturing realm, heat-sensitive materials can be effectively bonded, and the rapid cure time significantly reduces the need for fixturing.

DESIGN CONSIDERATIONS

The structural adhesives have been used for a wide range of applications in fields as diverse as automotive, aerospace, appliances, biomedical, electronics, construction, machinery, and sporting goods. Proper selection and use, however, requires consideration of a number of factors, including the following:

1. What materials are being joined? What are their surface finishes, hardnesses, and porosities? Will the thermal expansions or contractions be different?

2. How will the joined assembly be used? What type of joint is proposed, what will be the bond area, and what will be the applied stresses? How much strength is required? Will there be mechanical vibration, acoustical vibration, or impacts?

3. What temperatures might be required to affect the cure, and what temperatures might be encountered during service? Consideration should be given to the highest temperature, lowest temperature, rates of temperature change, frequency of change, duration of exposure to extremes, properties required at the various conditions, and differential expansions or contractions.

4. Will there be subsequent exposure to solvents, water or humidity, fuels or oils, light, ultraviolet radiation, acid solutions, or general weathering?

5. What is the desired level of flexibility or stiffness? How much toughness is required?

6. Over what length of time is stability desired? What portion of this time will be under load?

7. Is appearance important?

8. How will the adhesive be applied? What equipment, labor, and skill are required?

9. Are there restrictions relating to storage or shelf life? Cure time? Disposal or recyclability?

10. What will it cost?

Because there is such a large difference in bonding area, adhesive-bonded joints are often classified as either continuous surface or core-to-face. In *continuous-surface bonds*, both of the adhering surfaces are relatively large and are of the same size and shape. *Core-to-face bonds* have one *adherend* area that is very small compared to the other, like when the edges of lightweight honeycomb core structures are bonded to the face sheets (see Figure 15-19).

A major design consideration for both types is the nature of the stresses that the joint will experience. As shown in Figure 34-3, applied stresses can subject the joint to *tension, compression, shear, cleavage*, and *peel*. Most of the structural adhesives are significantly weaker in peel and cleavage than they are in shear or tension. Therefore, adhesively bonded joints should be designed so as much of the stress as possible is in shear, tension, or compression, where all of the bonded area shares equally in bearing the load. The shear strengths of structural adhesives range from 14 to 40 MPa (2000 to 6000 psi) at room temperature, while the tensile strengths are only 4 to 8 MPa (600 to 1200 psi). The best adhesive-bonded joints, therefore, will be those that are designed to utilize the superior shear strengths. Creep, vibration and associated fatigue, thermal shock, and mechanical shocks can all induce additional stresses. When vibration or shock loading is expected, the elastomeric adhesives are quite attractive, since they can provide valuable damping.

Figure 34-4 shows some commonly used joint designs and indicates their relative effectiveness. The butt joint is unsatisfactory because it offers only a minimum of bond surface area and little resistance to cleavage. Useful strength is generally obtained by

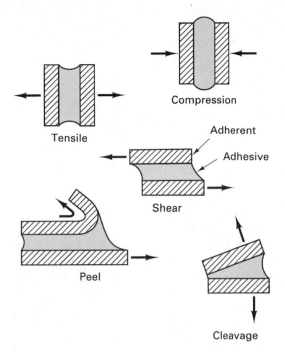

FIGURE 34-3   Types of stresses in adhesive-bonded joints.

increasing the bond area through the addition of straps or the conversion to some form of lap design. The scarf joint, shown previously in Figure 34-14, is also used when uniform thickness is required. Figure 34-5 shows some recommended designs for corner and angle joints. Adhesives can also be used in combination with welding, brazing, or mechanical fasteners. Spot welds or rivets can provide additional strength or simply prevent movement of the components when the adhesive is not fully cured or is softened by exposure to elevated temperature.

To obtain satisfactory and consistent quality in adhesive-bonded joints, it is essential that the surfaces be properly prepared. Procedures vary widely but frequently include cleaning of the surfaces to be joined. Contaminants, such as oil, grease, rust, scale, or even mold-release agents, must be removed to ensure adequate wetting of the surfaces by the adhesive. Solvent or vapor cleaning is usually adequate. Chemical alteration of the surface to form a new intermediate layer, chemical etching, steam cleaning, or abrasive techniques may also be employed to further enhance wetting and bonding. While thick or loose oxide films are detrimental to adhesive bonding, a thin porous oxide or surface primer can often provide surface roughness and enhance adhesion.

FIGURE 34-4   Possible designs of adhesive-bonded joints and a rating of their performance in service.

FIGURE 34-5 Adhesively bonded corner and angle joint designs.

The destructive testing of adhesive joints, or the examination of joint failures, can reveal much about the effectiveness of an adhesive system. If failure occurs by separation at the adhesive–substrate interface, as shown in Figure 34-6a, it is indicative of a bonding or adhesion problem. If the failure lies entirely within the adhesive, as in Figure 34-6b, then the bonding with the substrate is adequate, but the strength of the adhesive may need to be enhanced. Finally, if failure occurs within the substrate materials, as in Figure 34-6c, the joint is good, and failure is unrelated to the adhesive bonding operation.

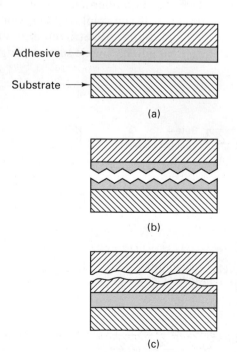

FIGURE 34-6 Failure modes of adhesive joints. (a) Adhesive failure, (b) cohesive failure within the adhesive, and (c) cohesive failure within the substrate.

ADVANTAGES AND LIMITATIONS

Adhesive bonding has many obvious advantages. Almost any material or combination of materials can be joined in a wide variety of sizes, shapes, and thicknesses. For most adhesives, the curing temperatures are low, seldom exceeding 180°C (350°F). A substantial number cure at room temperature or slightly above and can provide adequate strength for many applications. As a result, very thin or delicate materials, such as foils, can be joined to each other or to heavier sections. Heat-sensitive materials can be joined without damage, and heat-affected zones are not present in the product. When joining dissimilar materials that experience changes in temperature, the adhesive often provides a bond that can tolerate the stresses of differential expansion and contraction.

Because adhesives bond the entire joint area, good load distribution and fatigue resistance are obtained, and stress concentrations (such as those observed with screws, rivets, and spot welds) are avoided. Because of the high extension and recovery properties of flexible adhesives, the fatigue resistance can be up to 20 times that of riveted or spot-welded assemblies. The large contact areas that are usually employed provide a total joint strength that compares favorably with alternative methods of joining or attachment. Shear strengths of industrial adhesives can exceed 20 MPa, or 3000 psi, and additives can be incorporated to enhance strength, increase flexibility, or provide resistance to various environments.

Adhesives are generally inexpensive and frequently weigh less than the fasteners needed to produce a comparable-strength joint. In addition, an adhesive can also provide thermal and electrical insulation; act as a damper to noise, shock, and vibration; stop a propagating crack; and provide protection against galvanic corrosion when dissimilar metals are joined. By providing both a joint and a seal against moisture, gases, and fluids, adhesive-bonded assemblies often offer improved corrosion resistance throughout their useful lifetime. When used to bond polymers or polymer-matrix composites, the adhesive can be selected from the same family of materials to ensure good compatibility.

From a manufacturing viewpoint, the formation of a joint does not require the capillary-induced flow of material, as in brazing and soldering. The bonding adhesive is applied directly to the surfaces, and the joint is then formed by the application of heat and/or pressure. Most adhesives can be applied quickly, and useful strengths are achieved in a short period of time. Some curing mechanisms take as little as two to three seconds! Surface preparation may be reduced since bonding can occur with an oxide film in place, and rough surfaces are actually beneficial because of the increased contact area. Tolerances are less critical since the adhesives are more forgiving than alternative methods of bonding. The adhesives are often invisible; exposed surfaces are not defaced; smooth contours are not disturbed; and holes do not have to be made, as with rivets or bolts. These factors contribute to reduced manufacturing costs, which can be further reduced through the elimination of the mechanical fasteners and the absence of highly skilled labor. Bonding can often be achieved at locations that would prevent the access of many types of welding apparatus. Robotic dispensing systems can often be utilized.

The major disadvantages of adhesive bonding are the following:

1. There is no universal adhesive. Selection of the proper adhesive is often complicated by the wide variety of available options.
2. Most industrial adhesives are not stable above 180°C (350°F). Oxidation reactions are accelerated, thermoplastics can soften and melt, and thermosets decompose. While some adhesives can be used up to 260°C (500°F), elevated temperatures are usually a cause for concern.
3. Some adhesives shrink significantly during curing.
4. High-strength adhesives are often brittle (poor impact properties). Resilient ones often creep. Some become brittle when exposed to low temperatures.
5. Surface preparation and cleanliness, adhesive preparation, and curing can be critical if good and consistent results are to be obtained. Some adhesives are quite sensitive to the presence of grease, oil, or moisture on the surfaces to be joined. Surface roughness and wetting characteristics must be controlled.
6. Assembly times may be greater than for alternative methods, depending upon the curing mechanism. Elevated temperatures may be required, as well as specialized fixtures.

7. It is difficult to determine the quality of an adhesive-bonded joint by traditional nondestructive techniques, although some inspection methods have been developed that give good results for certain types of joints.

8. Some adhesives contain objectionable chemicals or solvents, or produce them upon curing.

9. Many structural adhesives deteriorate under certain operating conditions. Environments that may be particularly hostile include heat, ultraviolet light, ozone, acid rain (low pH), water and humidity, salt, and numerous solvents. Thus, long-term durability and reliability may be questioned, and life expectancy is hard to predict.

10. Adhesively bonded joints cannot be readily disassembled.

Nevertheless, the extensive and successful use of adhesive bonding provides ample evidence that these limitations can be overcome if adequate quality control procedures are adopted and followed.

■ 34.2 MECHANICAL FASTENING

INTRODUCTION AND METHODS

Mechanical fastening is a classification that includes a wide variety of techniques and fasteners designed to suit the individual requirements of a multitude of joints and assemblies. Included within this family are integral fasteners, threaded discrete fasteners (which includes screws, bolts, studs, and inserts), nonthreaded discrete fasteners (such as rivets, pins, retaining rings, nails, staples, and wire stitches), special-purpose fasteners (such as the quick-release and tamper-resistant types), shrink and expansion fits, press fits, seams, and others. Selection of the specific fastener or fastening method depends primarily on the materials to be joined, the function of the joint, strength and reliability requirements, weight limitations, dimensions of the components, and environmental conditions. Other considerations include cost, installation equipment and accessibility, appearance, and the need or desire for disassembly. When disassembly and reassembly are desired (as for parts replacement, maintenance, or repair), threaded fasteners, snap-fits, or other fasteners that can be removed quickly and easily should be specified. Such fasteners should not have a tendency to loosen after installation, however. If disassembly is not necessary, permanent fasteners are often preferred, or threaded fasteners can be coupled with anaerobic adhesives that cure to full strength at room temperature and "lock" the fastener in place.

A mechanical joint acquires its strength through either mechanical interlocking or interference of the surfaces as a result of a clamping force. No fusion or adhesion of the surfaces is required. The fasteners and fastening processes should be selected to provide the required strength and properties in view of the nature and magnitude of subsequent loading. Consider the possibility of vibrations and/or cyclic stresses that might promote loosening over a period of time. Added weight may be a significant factor in certain applications, such as aerospace and automotive. The need to withstand corrosive environments, operate at high or low temperatures, or face other severe conditions may be additional constraints to the selection of fasteners or fastening processes.

The effectiveness of a mechanical fastener often depends upon (1) the material of the fastener, (2) the fastener design (including the load-bearing area of the head), (3) hole preparation, and (4) the installation procedure. The general desire is to achieve a uniform load transfer, a minimum of stress concentration, and uniformity of installation torque or interference fit. Various means are available for achieving these goals, as described in the following paragraphs.

Integral fasteners are formed areas of a component that interfere or interlock with other components of the assembly; they are most commonly found in sheet metal products. Examples include lanced or shear-formed tabs, extruded hole flanges, embossed protrusions, edge seams, and crimps. Figure 34-7 shows some of these techniques, each of which involve some form of metal shearing and/or forming. The common beverage can includes several of these joints—an edge seam to join the top of the can to the body (as in Figure 34-7e) and an embossed protrusion that is subsequently flattened to attach the opener-tab (as in Figure 34-7d).

FIGURE 34-7 Several types of integral fasteners: (a) lanced tab to fasten wires or cables to sheet or plate; (b) and (c) assembly through folded tabs and slots for different types of loading; (d) use of a flattened embossed protrusion; (e) single-lock seam.

Cutaway section

Nail   Screw   Bolt and nut

Insert through holes in plates   After   Plates   Fracture pin

Before   After
Two-side rivet

Pull
Before
One-side rivet

Locked   Unlocked (turned 90°)   Snap-fit
Quick-release fastener

FIGURE 34-8 Various types of discrete fasteners, including a nail, screw, nut and bolt, two-side access supported rivet, one-side access blind rivet, quick-release fastener, and snap-fit.

Discrete fasteners, like those illustrated in Figure 34-8, are separate pieces whose function is to join the primary components. These include bolts and nuts (with accessory washers, etc.), screws, nails, rivets, quick-release fasteners, staples, and wire stitches. Over 150 billion discrete fasteners are consumed annually in the United States, with a variety so immense that the major challenge is usually selection of an appropriate, and hopefully optimum, fastener for the task at hand. Fastener selection is further complicated by inconsistent nomenclature and identification schemes. Some fasteners are identified by their specific product or application, while others are classified by the material from which they are made, their size, their shape, their strength, or primary operational features.[1] The commercial availability of such a wide range of standard and special types, sizes, materials, strengths, and finishes virtually ensures that an appropriate fastener can be found for most all joining needs. Discrete fasteners are easy to install, remove, and replace. In addition, most standard varieties are interchangeable. Steel is the most common material due to its high strength and low cost. Various finishes and coatings can be applied to withstand a multitude of service conditions.

Shrink and expansion fits form another major class of mechanical joining. Here, a dimensional change is introduced to one or both of the components by heating or cooling (heating one part only, heating one and cooling the other, or cooling one). Assembly is then performed, and a strong interference fit is established when the temperatures return to uniformity. Joint strength can be exceptionally high. A corrosion-resistant cladding or lining can also be applied to a less costly bulk material.

---

[1] Discussion of the primary terms used in identifying discrete fasteners can be found in ANSI Standard B18.12. Both the Society of Automotive Engineers (SAE) and the American Society of Testing and Materials (ASTM) have formalized various "grades" of bolts and identified them by markings that are stamped on the bolt head. These grades often relate to allowable stress and temperatures of operation.

Press fits are similar to shrink and expansion fits, but the results are obtained through mechanical force instead of differential temperatures.

REASONS FOR SELECTION
Mechanical fastening offers a number of attractive features, among them the following:

1. They are easy to disassemble and reassemble. The threaded fasteners are noteworthy for this feature, and semipermanent fasteners (such as rivets) can be drilled out for a major disassembly.
2. They have the ability to join similar or different materials in a wide variety of sizes, shapes, and joint designs. Some joint designs, such as hinges and slides, permit limited motion between the components.
3. Manufacturing cost is low. The fasteners are usually small formed components that cost little compared to the components being joined. They are readily available in a variety of mass-produced sizes.
4. Installation does not adversely affect the base materials, as is often the case with techniques involving the application of heat and/or pressure.
5. Little or no surface preparation or cleaning is required.

MANUFACTURING CONCERNS
Many mechanical fasteners require that the components contain aligned holes. Castings, forgings, extrusions, and powder metallurgy components can be designed to include integral holes. Holes can also be produced by such techniques as punching, drilling, and electrical, chemical, or laser-beam machining. Each of these techniques produces holes with characteristic surface finish, dimensional features, and properties. Secondary operations, such as shaving, deburring, reaming, and honing, can be used to improve precision and surface finish. Hole making and the proper positioning and alignment of the holes are major considerations in mechanical fastening.

Some fasteners, such as bolts coupled with nuts, require access to both sides of an assembly during joining. In contrast, screws offer one-side joining. If a bolt can be inserted into a threaded (tapped) hole, however, the nut can be eliminated, and only one-side access is required. If the bolt or screw is sufficiently hard, the fastener can often form its own threads, thereby eliminating the need for a threaded receptacle. This self-tapping feature is particularly attractive when assembling plastic products.

Stapling is a fast way of joining thin materials and does not require prior hole making. Rivets offer good strength but produce permanent or semipermanent joints. Snap-fits utilize the elasticity of one of the components, but the necessary elastic deformation must be possible without fracture.

DESIGN AND SELECTION
The design and selection of a fastening method requires numerous considerations, including the possible means of joint failure. When a product is assembled with fastened joints, the fasteners are extremely vulnerable sites. Mechanical joints generally fail because of oversight or lack of control in one of four areas: (1) the design of the fastener itself and the manufacturing techniques used to make it, (2) the material from which the fastener is made, (3) joint design, or (4) the means and details of installation. Fasteners may have insufficient strength or corrosion resistance or may be subject to stress corrosion cracking or hydrogen embrittlement. They may be unable to withstand the temperature extremes (both high and low) experienced by the final assembly. Metal fasteners provide electrical conductivity between the components, and an inappropriate choice of fastener or component material can cause severe galvanic corrosion. Nonmetallic fasteners (such as threaded nylon) can be used for low-strength applications where corrosion is a concern, but creep under load is a concern for these materials. Since mechanical fasteners only join at discrete points, gases or liquids can easily penetrate the joint area and further aggravate conditions.

Many failures are the result of poor joint preparation or improper fastener installation. A high percentage of the cracks in aircraft structures originate at fastener holes, and fatigue of fasteners is the largest single cause of fastener failure. Installation frequently imparts too much or too little preload (too tight or too loose). The joint surfaces may not be flat or parallel, and the area under the fastener head may be insufficient to bear the load. Vibrational loosening enhances fastener fatigue. The details of joint design should further consider stress distribution, since much of the load will be concentrated on the fasteners (in contrast to the previously discussed adhesive joints that distribute the load uniformly over the entire joint area).

Nearly all fastener failures can be avoided by proper design and fastener selection. Consideration should be given to the operating environment, required strength, and magnitude and frequency of vibration. Fastener design should incorporate a shank-to-head fillet whenever possible. Rolled threads can be specified for their superior strength and fracture resistance. Corrosion-resistant coatings can be employed for enhanced performance. Joint design should seek to avoid such features as offset or oversized holes. Proper installation and tightening are critical to good performance. Standard sizes, shapes, and grades should be used whenever possible, with as little variety as is absolutely necessary.

■ 34.3 JOINING OF PLASTICS

Mechanical fasteners, adhesives, and welding processes can all be employed to form joints between engineering plastics. Fasteners are quick and are suitable for most materials, but they may be expensive to use, they generally do not provide leak-tight joints, and the localized stresses may cause them to pull free of the polymeric material. Threaded metal inserts may have to be incorporated into the plastic components to receive the fasteners, further increasing the product cost. Adhesives can provide excellent properties and fully sound joints, but they are often difficult to handle and relatively slow to cure. In addition, considerable attention is required in the areas of joint preparation and surface cleanliness. In a modification of adhesive bonding, solvents may be used to soften surfaces, which are then pressed together to form a bond. Welding can be used to produce bonded joints with mechanical properties that approach those of the parent material. Unfortunately, only the *thermoplastic polymers* can be welded, since these materials can be melted or softened by heat without degradation and good bonds can be formed with the subsequent application of pressure. The *thermosetting polymers* do not soften with heat, tending only to char or burn, and must be joined by alternative methods, such as mechanical fasteners, adhesives, snap-fits, or possible co-curing (placing the components together and curing while in contact).

Because the thermoplastics soften and melt at such low temperatures, the heat required to weld these materials is significantly less than that required in the welding of metals. The processes used to weld plastics can be divided into two groups: (1) those that utilize mechanical movement and friction to generate heat, such as ultrasonic welding, spin welding, and vibration welding, and (2) those that involve external heat sources, such as hot-plate welding, hot-gas welding, and resistive and inductive implant welding. In both groups, it is important to control the rate of heating. Plastics have low thermal conductivity, and it is easy to induce burning, charring, or other material degradation before softening has occurred to the desired depth.

Ultrasonic welding of plastics uses high-frequency mechanical vibrations to create the bond. Parts are held together and are subjected to ultrasonic vibrations (20 to 40 kH frequency and 10 to 100 micron amplitude) perpendicular to the area of contact. The high-frequency stresses generate heat at the joint interface sufficient to produce a high-quality weld in a period of $1/2$ to $1^1/2$ seconds. The process can be readily automated, but the tools are expensive and large production runs are generally required. Ultrasonic welding is usually restricted to small components where relative movement is restricted and weld lengths do not to exceed a few centimeters.

In *vibration welding*, or linear friction welding, relative movement between the two parts is again used to generate the heat, but the direction of movement is now parallel to the interface and aligned with the longest dimension of the joint. The vibration

FIGURE 34-9 Using a hot-gas torch to make a weld in plastic pipe.

amplitudes are 10 times larger than in ultrasonic welding, and the frequencies are considerably less (on the order of 100 to 240 Hz). When molten material is produced, the vibration is stopped, parts are aligned, and the weld region is allowed to cool and solidify. The entire process takes about one to five seconds. Long-length, complex joints can be produced at rather high production rates. Nearly all thermoplastics can be joined, independent of whether their prior processing was by injection molding, extrusion, blow molding, thermoforming, foaming, or stamping.

The *friction welding* of plastics (also called *spin welding*) is similar to vibration welding, except the relative motion is now continuous and rotational. The process is essentially the same as the friction welding of metals, but melting now occurs at the joint interface. High-quality welds are produced with good reproducibility, and little end preparation is required. The major limitation is that at least one of the components must exhibit circular symmetry, and the axis of rotation must be perpendicular to the mating surface. Weld strengths vary from 50 to 95% of the parent material in bonds of the same plastic. Joints between dissimilar materials generally have poorer strengths. Butt welds can be made between plates of thermoplastic using the *friction-stir welding* process described in Chapter 32.

Hot-plate welding uses an external heat source and is probably the simplest of the mass-production techniques used to join plastics. The parts to be joined are held in fixtures and pressed against the opposite sides of an electrically heated tool. Contact is maintained until the surfaces have melted and the adjacent material has softened to a specified distance from the interface. The parts separate, the tool is removed, and the two prepared surfaces are pressed together and allowed to cool. Contaminated surface material is usually displaced into a flash region. Weld times are comparatively slow, ranging from 10 seconds to several minutes. The joint strength can be equal to that of the parent material, but the joint design is usually limited to a square-butt configuration, like that encountered when joining sections of plastic pipe. If the bond interface has a non-flat profile, shaped heating tools can be employed. Heated-tool welding can also be used to produce lap seams between flexible plastic sheets. Rollers apply pressure after the material has passed over a heater.

The *hot-gas welding* of plastics is similar to the oxyacetylene welding of metals, and V-groove or fillet welds are the most common joint configuration. A gas (compressed air, nitrogen, hydrogen, oxygen, or carbon dioxide) is heated by an electric coil as it passes through a welding gun, like the one shown in Figure 34-9. The hot-gas stream emerges from the gun at 200° to 300°C (400° to 570°F) and impinges on the joint area. Thin rods of plastic material are heated along with the workpiece and are then forced into the softened joint area, providing both the filler material and the pressure needed to produce coalescence. Because this process is usually slow and the results are generally dependent on operator skill, it is seldom used in production applications. It is, however, a popular process for the repair of thermoplastic materials.

Extrusion welding is similar to the hot-gas process, except that the external filler material rod is replaced by a stream of fully molten polymer that emerges from the weld tool as it moves along the joint. This is similar to the wire feed in the metal-welding processes.

In the *implant welding* of plastics, metal inserts are placed between the parts to be joined and are then heated by means of induction (radio frequency) or resistance heating. The resistance method requires that a current-carrying path exist to conduct the current to the implants, a feature that is not required for induction heating. The thermoplastic material melts around the heated implants and flows to form a joint. Since a weld forms only in the vicinity of the implants, the process resembles spot welding and produces joints that are considerably weaker than those formed by processes that bond the entire contact area. When bonding is desired over larger areas, tapes, rods, or gaskets of thermoplastic material laced with iron oxide or metal particles can be used to concentrate the heat at the interface and provide filler material. In all cases of implant welding, the metal implant material remains as an integral part of the final assembly.

Still other processes to weld thermoplastics are based on *infrared radiation* or *microwave heating*. Laser welding has also been performed on plastics.

The most common method of joining plastics to one another and joining metal to plastic is through *mechanical fasteners*. It is important that the plastic be able to withstand the strain of fastener insertion and the localized stresses around the fastener. Conventional machine screws are rarely used, except with extremely strong plastics. Instead, there are a number of fasteners designed specifically for use with plastics. Threaded fasteners work best with thick sections. Self-tapping, thread-cutting screws are used on hard plastics, and thread-forming screws are used with softer materials. If the joint is to undergo disassembly and reassembly, threaded metal inserts may be incorporated into the part to receive the fasteners. Because of the low elastic modulus of plastic materials, snap-fit assemblies are often an attractive alternative to the use of fasteners.

As described previously in this chapter, adhesives provide an attractive means of joining plastics. Since adhesives are polymeric materials, the joint material can be selected for compatibility with the material being joined. Probably the most common application is the adhesive joining of PVC (polyvinyl chloride) pipes and fittings in household plumbing.

■ 34.4 JOINING OF CERAMICS AND GLASS

The properties of ceramic materials are significantly different from the engineering metals, and these differences restrict or limit the processes that can be used for joining. High melting temperatures can be a significant deterrent to fusion welding. More significant, however, are the effects of low thermal conductivity and brittleness. Heating and cooling will likely result in nonuniform temperatures, and the thermally induced stresses are likely to result in cracking or fracture. The lack of useful ductility virtually eliminates any form of deformation bonding. Mechanical fasteners, and their associated threads and holes, create high concentrated stresses, and these stresses often lead to material fracture. As a result, most ceramic materials are joined by some form of adhesive bonding, brazing, diffusion or sinter bonding, or ceramic cements.

Adhesives and cements are probably the most common methods of joining ceramics to ceramics, ceramics to glasses, and ceramics to metals and other materials. The inserted material (polymer adhesive, glass or glass-ceramic frit, or ceramic cement or mortar) will bond to the surfaces and bridge what are often radically different compositions and structures.

Brazing and soldering use a low-melting-point metal or lower-melting ceramic as the intermediate material. Some materials, such as indium solders, directly wet ceramic surfaces. To promote adhesion and bonding with other joint materials, it may be necessary to first coat the ceramic with some form of metallized or deposited layer. These coatings bond to the ceramic, and the braze or solder material bonds to the coating.

Sinter bonding is a means of joining ceramic materials during their initial production. As the component pieces are held together and co-fired, diffusion bonds form across the interface while similar bonds form within the components. This process is best performed when the components are of identical material or materials with similar composition and structure. Various intermediate materials have been used to assist the joining of dissimilar ceramics.

The joining of glass is a much easier operation. Heating softens the two materials, which are then pressed together and cooled. A wide variety of heating methods are used, depending on the size, shape, and quantity of components to be joined.

■ 34.5 JOINING OF COMPOSITES

The joining of composite materials can be an extremely complex subject, especially when one considers the variety of composites and the fact that the joint interface is likely to be a distinct disruption to the continuity of structure and properties.

Particulate composites may have the least structural difference at an interface. Laminar composites will certainly behave differently if the joint surface is a core (or multilayer) surface or a single-material exposed face. Fiber-reinforced composites, regardless of the type of fiber and fiber configuration, will certainly lack fiber continuity across the joint.

The usual joining methods tend to be those used with the matrix of the composite. Metal-matrix composites can be welded, brazed, or soldered, or joined with screws or bolts or any of the other techniques applied to metals. The techniques for polymer-matrix and ceramic-matrix follow those for plastics and ceramics, as discussed previously in this chapter.

Theoretically, all composites can be adhesively bonded, but there may be limits set by the applied stresses, operating temperatures, or size of the workpiece (since many adhesives require a thermal cure). When working with polymer-matrix composites, the adhesive is often selected to match or be compatible with the matrix polymer. In many cases, however, the adhesive is being asked to bond to already-cured polymer surfaces.

■ Key Words

adherend	elastomer	integral fasteners	spin welding
adhesive	epoxies	mechanical fastening	structural adhesive
adhesive bonding	evaporative adhesives	peel	superglue
anaerobic	extrusion welding	press fit	thermoplastic
cleavage	friction welding	pressure-sensitive adhesives	thermosetting
continuous-surface bonds	hot-gas welding	radiation-curing adhesive	ultrasonic welding
core-to-face bonds	hot-melt adhesives	shear	vibration welding
curing	hot-plate welding	shrink fit	
discrete fasteners	implant welding	sinter bonding	

■ Review Questions

1. What would be some of the characteristics of an ideal adhesive?
2. What are some of the newer applications that have helped promote increased use of adhesive bonding?
3. What are some of the types of materials that have been used as industrial adhesives?
4. What are some of the ways in which adhesives can be cured?
5. What is a structural adhesive?
6. Characterize the temperature range over which epoxies might be used, typical values of shear strength, and commonly observed curing times.
7. What promotes the curing of cyanoacrylates? Of anaerobics?
8. What features or characteristics might favor the selection of a silicone adhesive?
9. What are some common applications of hot-melt adhesives?
10. What features or properties are provided or enhanced by adhesive additives?
11. What are some types of nonstructural or special adhesives?
12. How can polymeric adhesives be made electrically or thermally conductive?
13. What are some of the temperature considerations that apply when selecting an adhesive?
14. What are some of the environmental conditions that might reduce the performance or lifetime of a structural adhesive?
15. Why is it desirable for adhesive joints to be designed so the adhesive is loaded in shear, tension, or compression?
16. Why are butt joints unattractive for adhesive bonding?
17. What types of joints provide large bonding areas?

18. What are some common techniques by which surfaces are prepared for adhesive bonding?
19. Why are the structural adhesives an attractive means of joining dissimilar metals or materials? Different sizes or thicknesses?
20. What are some of the other attractive properties of structural adhesives?
21. In what ways might a structural adhesive offer manufacturing ease or reduced manufacturing cost?
22. In view of the relatively low strengths of the structural adhesives, how can adhesively bonded joints attain strengths comparable to other methods of joining?
23. Why are adhesive joints unattractive for applications that involve exposure to elevated temperature? Is low temperature a concern?
24. What factors would influence the selection of a specific type of mechanical fastener or fastening method?
25. What types of fasteners are attractive if the application requires the ability to disassemble and reassemble the product?
26. What factors determine the overall effectiveness of a mechanical fastener?
27. What is an integral fastener? Provide an example.
28. What are some of the primary types of discrete fasteners?
29. How are press fits similar to shrink or expansion fits? How do they differ?
30. What are some of the major assets of mechanical fasteners?
31. What are some of the ways that fastener holes can be made in manufactured products?

32. What are some of the common causes for failure of mechanically fastened joints?
33. From a manufacturing viewpoint, why is it desirable to use standard fasteners and minimize the variety of fasteners within a given product?
34. Why can the thermoplastic polymers be welded, but not the thermosetting varieties?
35. Describe several of the plastic joining processes that use mechanical movement or friction to generate the required heat.
36. What are some of the external heat sources that can be used in the welding of plastic materials?
37. Why are the crystalline ceramic materials particularly difficult to join?
38. What is sinter bonding, and how does it join ceramic materials?
39. When materials are joined, we create interfaces between the various components. Describe the structural features that might result at the interface when we join: (a) particulate composites, (b) laminar composites, and (c) fiber-reinforced composites.

■ Problems

1. Some automakers are using adhesives and sealants that cure under the same conditions used for the paint-bake operation. Determine the conditions used for paint-bake, and identify some adhesives and sealants that could be used. What are some of the pros and cons of such an integration?
2. A contractor has installed aluminum siding on a house with steel nails. Use the galvanic series to evaluate the corrosion properties of this assembly. (*Note*: The aluminum is exposed to air, so it should be considered to be in its passive condition.) What do you expect will be the outcome of this fastener selection? Can you recommend a better alternative?
3. Mechanical fasteners are an attractive means of joining composite materials because they avoid exposing the composite to heat and/or high pressure. Assume that the composite is a polymer-based fiber-reinforced material with either uniaxial or woven fibers. For this particular system, what are some possible fastener-related problems? Consider joint preparation, assembly, and possible service failure.
4. The heat-resisting tiles on the U.S. space shuttle are made from heat-resisting ceramics. Determine the method or methods used to attach them to the structure. What difficulties or problems have been encountered relating to this bonding?
5. The bicycle frames used by riders in recent Tour de France races have been single-piece fiber-reinforced composites. What difficulties or property compromises might be associated with a fabrication method that uses joints? What methods might be available to join the carbon fiber–epoxy composite materials commonly used in these bicycles?
6. The processes described for the joining of plastics focused almost exclusively on the thermoplastic polymers. What types of joining techniques could be applied to thermosetting polymers? To elastomeric polymers? To ceramic materials?

Chapter 34 CASE STUDY

Golf Club Heads with Insert

Y ou are employed by a small manufacturer of sporting goods equipment, and your design team has recently proposed a new line of high-performance golf clubs. While the club head of the irons is to be a "standard" AISI 431 martensitic stainless steel investment casting, the striking face will incorporate a metal insert, as shown in the figure. This insert will be produced by powder metallurgy and will consist of a copper-based alloy laced with presized particles of tungsten carbide. After a mild acid etching of the copper matrix, the carbide particles will protrude sufficiently to better grip the surface of the ball, imparting an enhanced amount of backspin to better control the "bite" of the ball upon landing. Since its purpose is to modify the striking face, the insert is rather thin, about 1.5 to 3 mm ($^1/_{16}$ to $^1/_8$ in.). It is important that the insert be incorporated into the club face in a manner that does not dampen the impact or compromise the "feel" of the club.

1. You must devise a means of incorporating the proposed insert into the face of the club. What are some possible means of joining or bonding the dissimilar materials? What are the advantages and limitations of each of your alternatives? What would you recommend?
2. An additional joint occurs where the club head is attached to the shaft. If a graphite fiber–reinforced epoxy is being considered for the shaft, how might it be attached to the stainless steel head? If the composite shaft were selected, which joining method would you recommend? If the shafts were metal, what other methods might be possible?
3. For production simplicity, it might be preferable to use the same joining procedure at all locations. In view of your answers to Questions 1 and 2, does this appear to be a possibility for this product?
4. If the bonding process and resulting interface prove to be problematic, there may be other ways to produce a raised-carbide surface on a stainless steel golf club face. Consider processes such as thermal spray, friction-stir to embed particles, and others. What do you see as the advantages and limitations for each of these alternatives, considering both manufacturing and performance? Would you expect them to be cheaper or more expensive than the proposed insert?
5. If quality and performance were the primary objective, which of the options in Question 4 would you recommend? Why?
6. If cost minimization were to become important as you seek an edge over competitors, what would be your recommendation and why?

CHAPTER 35

SURFACE ENGINEERING

■ 35.1 INTRODUCTION

Surface engineering is a multidisciplinary activity intended to tailor the properties of the surfaces of manufactured components so that their function and serviceability can bc improvcd. Processes include solidification treatments such as hot-dip coatings, weld-overlay coatings, and thermal spray surfaces; deposition surface treatments such as electrodeposition, chemical vapor deposition, and physical vapor deposition; and heat treatment coatings such as diffusion coatings and surface hardening. Electroplating means the electrodeposition of an adherent metallic coating onto an object that serves as the cathode in an electrochemical reaction. The resulting surface provides wear resistance, corrosion resistance, high-temperature resistance, or electrical properties different from those in the bulk material.

Many manufacturing processes influence surface properties, which in turn may significantly affect the way the component functions in service. The demands for greater strength and longer life in components often depend on changes in the surface properties rather than the bulk properties. These changes may be mechanical, thermal, chemical, and/or physical and therefore are difficult to describe in general terms. For example, two different surface finishes on Inconel 718 can have a marked effect on the fatigue life, changing the fatigue limit from 60 ksi after gentle grinding to as low as 22 ksi using electrical discharge machining (Figure 35-1).

Many metalcutting processes specified by the manufacturing engineer to produce a specific geometry can often have the effect of producing alterations in the surface material of the component, which, in turn, produces changes in performance.

The term *surface integrity* was coined by Field and Kahles in 1964 in reference to the nature of the surface condition that is produced by the manufacturing process. If we view the process as having five main components (workpiece, tool, machine tool, environment, and process variables), we see that surface properties can be altered by all of these parameters (see Table 35-1) by producing the following:

• High temperatures involved in the machining process

• Plastic deformation of the work material (residual stress)

• Surface geometry (roughness, cracks, distortion)

• Chemical reactions, particularly between the tool and the workpiece

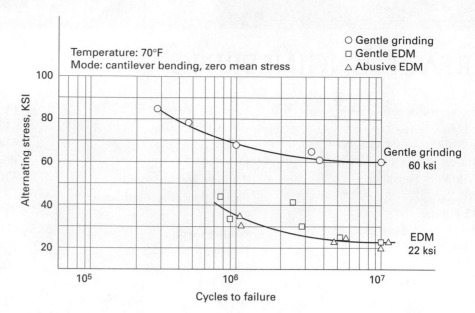

FIGURE 35-1 Fatigue strength of Inconel 718 components after surface finishing by grinding or EDM. *(Field and Kahles, 1971).*

More specifically, *surface integrity* refers to the impaired or enhanced surface condition of a component or specimen that influences its performance in service. Surface integrity has two aspects: topography characteristics and surface-layer characteristics. Topography is made up of surface roughness, waviness, errors of form, and flaws (Figure 35-2).

R_T = Maximum roughness depth (peak to valley) along l_m

R_A = Arithmetic roughness average

FIGURE 35-2 Machining processes produce surface flaws, waviness, and roughness that can influence the performance of the component.

TABLE 35-1   Characteristics of Manufacturing Processes That Affect Surface Integrity

Workpiece–Tool–Machine–Environment–Process Variables

Workpiece characteristics	Tool characteristics
Geometry	Tool body
Shape	Type of tool
Dimensions	Size
Material	Shape
Type	Number of cutting edges
Route of manufacture	Cutting edge
Mechanical properties	Shape (angles)
Elastic constants	Nose geometry/topography
Plastic constants	Microgeometry
Physical properties	Wear
Melting point	Material
Thermal diffusivity, conductivity, capacity	Type
Coefficient of thermal expansion	Coating
Phase transformations	Type
Chemical properties	Thickness
Chemical composition	Number and kind of layers
Chemical affinity to tool material and	Mechanical properties
environment	Elastic constants
Metallurgical properties	Plastic properties
Structure	Physical properties
Grain size	Thermal diffusivity, conductivity,
Hardness	capacity
	Coefficient of thermal expansion
	Chemical properties
Environment characteristics	Chemical composition
Type of medium (gas, fluid, mist)	Chemical affinity to tool material
Lubricity	Metallurgical properties
Cooling ability	Structure
Flow rate	Grain size
Temperature	
Chemical composition	**Process variables**
	Speed
Machine tool characteristics	Feed
Error motions	Depth of cut

Source: Advanced Manufacturing Engineering, Vol. 1, July 1989.

A typical roughness profile includes the peaks and valleys that are considered separately from waviness. Flaws also add to texture but should be measured independent of it. Changes in the surface layer, as a result of processing, include plastic deformation, residual stresses, cracks, and other metallurgical changes (hardness, overaging, phase changes, recrystallization, intergranular attack, and hydrogen embrittlement). See Figure 30-2. The surface layer will always contain local surface deformation due to any machining passes.

The material removal processes generate a wide variety of surfaces textures, generally referred to as *surface finish.* The cutting processes leave a wide variety of surface patterns on the materials. *Lay* is the term used to designate the direction of the predominant surface pattern produced by the machining process. In addition, certain other terms and symbols have been developed and standardized for specifying the surface quality. The most important terms are *surface roughness, waviness,* and *lay* (Figure 35-3). *Roughness* refers to the finely spaced surface irregularities. It results from machining operations in the case of machined surfaces. *Waviness* is surface irregularity of greater spacing than in roughness. It may be the result of warping, vibration, or the work being deflected during machining.

FIGURE 35-3 (a) Terminology used in specifying and measuring surface quality; (b) symbols used on drawing by part designers, with definitions of symbols; (c) lay symbols; (d) lay symbols applied on drawings.

A variety of instruments are available for measuring surface roughness and surface profiles. The majority of these devices use a diamond stylus that is moved at a constant rate across the surface, perpendicular to the lay pattern. The rise and fall of the stylus is detected electronically [often by a Linear Variable Differential Transformer Device (LVTD)], is amplified and recorded on a strip-chart, or is processed electronically to produce average or root-mean-square readings for a meter (Figure 35-4). The unit containing the stylus and the driving motor may be handheld or supported by skids that ride on the workpiece or some other supporting surface.

Roughness is measured by the height of the irregularities with respect to an average line. These measurements are usually expressed in micrometers or microinches.

FIGURE 35-4   (a) Schematic of stylus profile device for measuring surface roughness and surface profile with two readout devices shown: a meter for AA or rms values and a strip chart recorder for surface profile. (b) Profile enlarged. (c) Examples of surface profiles.

In most cases, the arithmetical average (AA) is used. In terms of the measurements, the AA would be as follows:

$$AA = \frac{\sum_{i=1}^{n} y_i}{n}$$

Cutoff refers to the sampling length used for the calculation of the roughness height. When it is not specified, a value of 0.030 in. (0.8 mm) is assumed. In the previous equation, y_i is a vertical distance from the centerline and n is the total number of vertical measurements taken within a specified cutoff distance. This average roughness value is also called R_a, occasionally used is the *root-mean-square* (rms) value, which is defined as

$$rms = \sqrt{\frac{\sum_{i=1}^{n} y_i^2}{n}}$$

The resolution of stylus profile devices is determined by the radius or the diameter of the tip of the stylus. When the magnitude of the geometric features begins to

FIGURE 35-5 Typical machined steel surface as created by face milling and examined in the SEM. A micrograph (same magnification) of a 0.00005-in. stylus tip has been superimposed at the top.

approach the magnitude of the tip of the stylus, great caution should be used in interpreting the output from these devices. As a case in point, Figure 35-5 shows a scanning electron micrograph of a face-milled surface on which has been superimposed (photographically) a scanning electron micrograph of the tip of a diamond stylus (tip radius of 0.0005 in.). Both micrographs have the same final magnification. Surface flaws of the same general size as the roughness created by the machining process are difficult to resolve with the stylus-type device, where both these features are about the same size as the stylus tip.

This example points out the difference between resolution and detection. Stylus tracing devices can often detect the presence of a surface crack, step, or ridge on the part but cannot resolve the geometry of the defect when the defect is of the same order of magnitude as the stylus tip or smaller.

Another problem with these devices is that they produce a reading (a line on the chart) where the stylus tip is not touching the surface, as is demonstrated in Figure 35-6a, which shows the S from the word *TRUST* on a U.S. dime. The scanning electron microscope (SEM) micrograph was made after the topographical map of Figure 35-6b had been made. Both figures are at about the same magnification. The tracks produced by the stylus tip are easily seen in the micrograph. Notice the difference between the features shown in the micrograph and the trace, indicating that the stylus tip was not in contact with the surface many times during its passage over the surface (left no track in the surface), yet the trace itself is continuous.

FIGURE 35-6 (a) SEM micrograph of a U.S. dime, showing the S in the word *TRUST* after the region has been traced by a stylus-type machine. (b) Topographical map of the S region of the word *TRUST* from a U.S. dime [compare to part (a)].

(a)   (b)

The range of surface roughnesses that are typically produced by various manufacturing processes is indicated in Figure 35-7, which is a very general picture of typical ranges associated with these processes. However, one can usually count on its being more expensive to generate a fine finish (low roughness). To aid designers, metal samples with various levels of surface roughness are available.

All of the processes used to manufacture components are important if their effects are present in the finished part. It is convenient to divide processes that are used to manufacture parts into three categories: traditional, nontraditional, and finishing treatments. In *traditional processes* the tool contacts the workpiece. Examples are grinding, milling, and turning. These material removal processes will inflict damage to the surface if improper parameters are used. Examples of improper parameters are dull tools, excessive infeed, inadequate coolant, and improper grinding wheel hardness. The *nontraditional processes* have intrinsic characteristics that, even if well controlled, will change the surface. In these processes the workpiece does not touch the tool. Electrochemical machining (ECM), electrical discharge machining (EDM), laser machining, and chemical milling are examples of

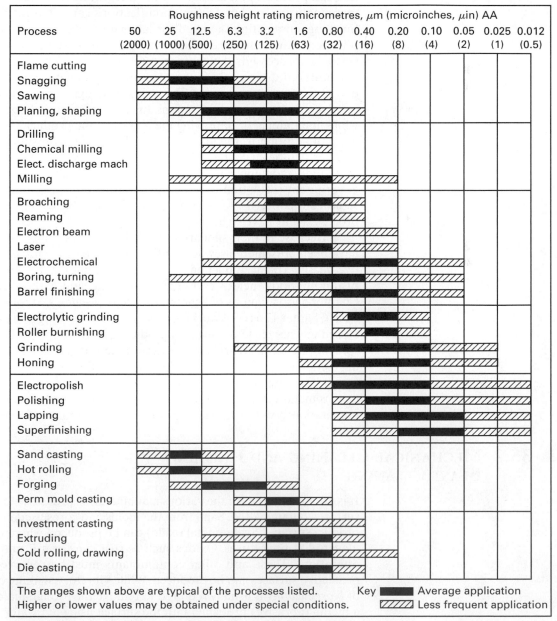

Extracted from General Motors Drafting Standards, June 1973 revision

FIGURE 35-7   Comparison of surface roughness produced by common production processes. *(Courtesy of* American Machinist.*)*

nontraditional methods. Such methods can leave stress-free surfaces, remelted layers, and excessive surface roughness. *Finishing treatments* can be used to negate or remove the impact of both traditional and nontraditional processes as well as providing good surface finish. For example, residual tensile stresses can be removed by shot peening or roller burnishing. Chemical milling can remove the recast layer left by EDM.

The objectives of the surface-modification processes can be quite varied. Some are designed to clean surfaces and remove the kinds of defects that occur during processing or handling (such as scratches, pores, burrs, fins, and blemishes). Others further improve or modify the products' appearance, providing features such as smoothness, texture, or color. Numerous techniques are available to improve resistance to wear or corrosion, or to reduce friction or adhesion to other materials. Scarce or costly materials can be conserved by making the interior of a product from a cheaper, more common material and then coating or plating the product surface.

As with all other processes, surface treatment requires time, labor, equipment, and material handling, and all of these have an associated cost. Efficiencies can be realized through process optimization and the integration of surface treatment into the entire manufacturing system. Design modifications can often facilitate automated or bulk finishing, eliminating the need for labor-intensive or single-part operations. Process selection should further consider the size of the part, the shape of the part, the quantity to be processed, the temperatures required for processing, the temperatures encountered during subsequent use, and any dimensional changes that might occur due to the surface treatment. Through knowledge of the available processes and their relative advantages and limitations, finishing costs can often be reduced or eliminated while maintaining or improving the quality of the product.

In addition to the above, the field of surface finishing has recently undergone another significant change. Many chemicals that were once "standard" to the field, such as cyanide, cadmium, chromium, and chlorinated solvents, have now come under strict government regulation. Wastewater treatment and waste disposal have also become significant concerns. As a result, processes may have to be modified or replacement processes may have to be used.

Because of their similarity to other processes, many surface finishing techniques have been presented elsewhere in the book. The *case hardening* techniques, both selective heating (flame, induction, and laser hardening) and altered surface chemistry (diffusion methods such as carburizing, nitriding, and carbonitriding), are presented in Chapter 5 as variations of heat treating. *Shot peening* and *roller burnishing* are presented in Chapter 17 as cold-working processes. *Roll bonding* and explosive bonding are discussed in Chapter 15 as means of producing laminar composites. *Hard facing* and metal spraying are included in Chapter 33 as adaptations of welding techniques. Chemical vapor deposition and physical vapor deposition are discussed in Chapters 19 and 21. Sputtering and ion implantation are also discussed in Chapter 19 as processes needed in electronics manufacturing. In this chapter we focus on techniques for cleaning and surface preparation as well as the remaining methods of surface finishing or surface modification.

■ 35.2 MECHANICAL CLEANING AND FINISHING BLAST CLEANING

It is not uncommon for the various manufacturing processes to produce certain types of surface contamination. Sand from the molds and cores used in casting often adheres to product surfaces. Scale (metal oxide) can be produced whenever metal is processed at elevated temperatures. Oxides such as rust can form if material is stored between operations. These and other contaminants must be removed before decorative or protective surfaces can be produced. While vibratory shaking can be useful, some form of *blast cleaning* is usually required to remove the foreign material. Blast cleaning uses a media (abrasive) propelled into the surface using air, water, or even a wheel (wheel blasting uses a high-rpm blocked wheel to deliver the media). The bulk of the work is done by kinetic energy of the impacting media; $KE = \frac{1}{2} MV^2$ where m = mass of the median and v = the velocity. Abrasives, steel grit, metal shot, fine glass shot, plastic beads, and even CO_2 are mechanically impelled against the surface to be cleaned. When sand is used, it should be clean, sharp-edged silica sand. Steel grit tends to clean more rapidly and generates much less dust, but it is more expensive and less flexible.

When the parts are large, it may be easier to bring the cleaner to the part rather than the part to the cleaner. A common technique for such applications is *sand blasting* or *shot blasting,* where the abrasive particles are carried by a high-velocity blast of air emerging from a nozzle with about a $^3/_8$-in . opening. Air pressures between 60 and 100 psi, producing particle speeds of 400 mph, are common when cleaning ferrous metals, and 10 to 60 psi is common for nonferrous metals. The abrasive may be sand or shot, or materials such as walnut shells, dry-ice pellets, or even baking soda. Pressurized water can also be used as a carrier medium.

When production quantities are large or the parts are small, the operation can be conducted in an enclosed hood, with the parts traveling past stationary nozzles. For large parts or small quantities, the blast may be delivered manually. Protective clothing and breathing apparatus must be provided and precautions taken to control the spread of the resulting dust. The process may even require a dedicated room or booth that is equipped with integrated air pollution control devices.

From a manufacturing perspective, these processes are limited to surfaces that can be reached by the moving abrasive (line-of-sight) and cannot be used when sharp edges or corners must be maintained (since the abrasive tends to round the edges).

BARREL FINISHING OR TUMBLING

Barrel finishing or *tumbling* is an effective means of finishing large numbers of small parts. In the Middle Ages, wooden casks were filled with abrasive stones and metal parts and were rolled about until the desired finish was obtained. Today, modifications of this technique can be used to deburr, radius, descale, remove rust, polish, brighten, surface-harden, or prepare parts for further finishing or assembly. The amount of stock removal can vary from as little as 0.0001 to as much as 0.005 in.

In the typical operation, the parts are loaded into a special barrel or drum until a predetermined level is reached. Occasionally, no other additions are made, and the parts are simply tumbled against one another. In most cases, however, additional media of metal slugs or abrasives (such as sand, granite chips, slag, or ceramic pellets) are added. Rotation of the barrel causes the material to rise until gravity causes the uppermost layer to cascade downward in a "landslide" movement, as depicted in Figure 35-8. The sliding produces abrasive cutting that can effectively remove fins, flash, scale, and adhered sand. Since only a small portion of the load is exposed to the abrasive action, long times may be required to process the entire contents.

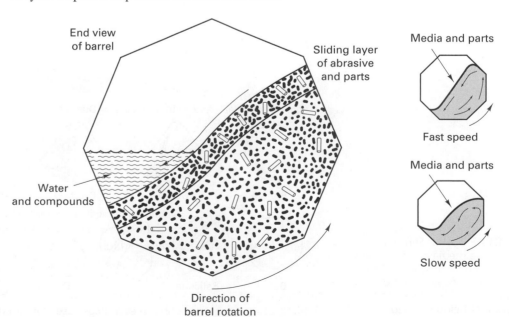

FIGURE 35-8 Schematic of the blow of material in tumbling or barrel finishing. The parts and media mass typically account for 50 to 60% of capacity.

Increasing the speed of rotation adds centrifugal forces that cause the material to rise higher in the barrel. The enhanced action can often accelerate the process, provided that the speed is not so great as to destroy the cascading action and that the additional action does not damage the workpiece. By a suitable selection of abrasives, filler, barrel size, ratio of workpieces to abrasive, fill level, and speed, a wide range of parts can be tumbled successfully. Delicate parts may have to be attached to racks within the barrel to reduce their movement while permitting the media to flow around them.

Natural and synthetic abrasives are available in a wide range of sizes and shapes, including those depicted in Figure 35-9, that enable the finishing of complex parts with irregular openings. The various media are often mixed in a given load, so that some will reach into all sections and corners to be cleaned.

Tumbling is usually done dry, but it can also be performed with an aqueous solution in the barrel. Chemical compounds can be added to the media to assist in cleaning, or descaling, or to provide features such as rust inhibition. Support equipment usually assists with loading and unloading the barrels as well as, with the separation of the workpieces from the abrasive media. The latter operation often uses mesh screens with selected size openings.

Barrel tumbling can be a very inexpensive way to finish large quantities of small parts and produce rounded edges and corners. Unfortunately, the abrasive action occurs on all surfaces and cannot be limited to selected areas. The cycle time is often long, and the process can be quite noisy.

In the *barrel burnishing* process, no cutting action is desired. Instead, the parts are tumbled against themselves or with media such as steel balls, shot, rounded-end pins, or ballcones. If the original material is free of visible scratches and pits, the combination of peening and rubbing will reduce minute irregularities and produce a smooth, uniform surface.

Barrel burnishing is normally done wet, using a solution of water and lubricating or cleaning agents, such as soap or cream of tartar. Because the rubbing action between the work and the media is very important, the barrel should not be loaded more than half full, and the volume ratio of media to work should be about 2:1 so the workpieces rub against the media, not each other. The speed of rotation should be set to maintain the cascading action and not fling the workpieces free of the tumbling mass.

Centrifugal barrel tumbling places the tumbling barrel at the end of a rotating arm. This adds centrifugal force to the weight of the parts in the barrel and can accelerate the process by as much as 25 to 50 times.

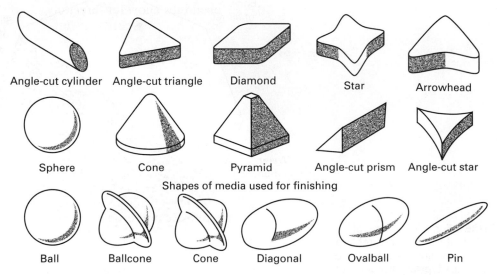

FIGURE 35-9 Synthetic abrasive media are available in a wide variety of sizes and shapes. Through proper selection, the media can be tailored to the product being cleaned.

Angle-cut cylinder Angle-cut triangle Diamond Star Arrowhead

Sphere Cone Pyramid Angle-cut prism Angle-cut star

Shapes of media used for finishing

Ball Ballcone Cone Diagonal Ovalball Pin

Steel media shapes used for burnishing

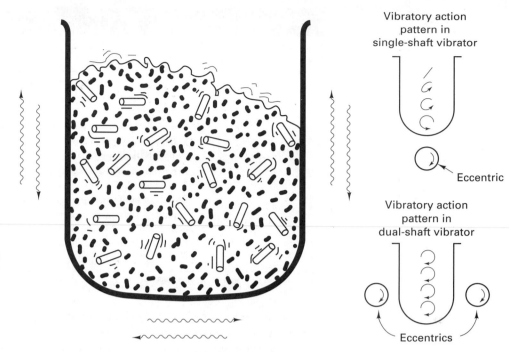

FIGURE 35-10 Schematic diagram of a vibratory-finishing tub loaded with parts and media. The single eccentric shaft drive provides maximum motion at the bottom, which decreases as one moves upward. The dual-shaft design produces more uniform motion of the tub and reduces processing time.

In *spindle finishing* the workpieces are attached to rotating shafts, and the assembly is immersed in media moving in a direction opposite to part rotation. This process is commonly applied to cylindrical parts and avoids the impingement of workpieces on one another. The abrasive action is accelerated, but time is required for fixturing and removal of the parts.

VIBRATORY FINISHING

Vibratory finishing is a versatile process widely used for deburring, radiusing, descaling, burnishing, cleaning, brightening, and fine finishing. In contrast to the barrel processing, *vibratory finishing* is performed in open containers. As illustrated in Figure 35-10, tubs or bowls are loaded with workpieces and media and are vibrated at frequencies between 900 and 3600 cycles per minute. The specific frequency and amplitude are determined by the size, shape, weight, and material of the pan, as well as the media and compound. Because the entire load is under constant agitation, cycle times are less than with barrel operations. The process is less noisy and is easily controlled and automated. In addition, the open tubs allow for direct observation during the process, which can also deburr or smooth internal recesses or holes.

MEDIA

The success of any of the mass-finishing processes depends greatly on *media* selection and the ratio of media to parts, as presented in Table 35-2. The media may prevent the parts from impinging upon one another as it simultaneously cleans and finishes. Fillers, such as scrap punchings, minerals, leather scraps, and sawdust, are often added to provide additional bulk and cushioning.

Natural abrasives include slag, cinders, sand, corundum, granite chips, limestone, and hardwood shapes, such as pegs, cylinders, and cubes. Synthetic media typically contain 50 to 70 wt% of abrasives, such as alumina (Al_2O_3), emery, flint, and silicon carbide. This material is embedded in a matrix of ceramic, polyester, or resin plastic, which is softer than the abrasive and erodes, allowing the exposed abrasive to perform the work. The synthetics are generally produced by some form of casting operation, so their sizes and shapes are consistent and reproducible (as opposed to the random sizes and shapes of the natural media). Steel media with no added abrasive are frequently specified for burnishing and light deburring.

TABLE 35-2   Typical Media-to-Part Ratios for Mass Finishing

Media/Part Ratio by Volume	Typical Application
0:1	Part-on-part processing or burr removal without media
1:1	Produces very rough surfaces and is suitable for parts in which part-on-part damage is not a problem
2:1	Somewhat less severe part-on-part damage, but more action from less media
3:1	May be acceptable for very small parts and very small media. Part-on-part contact is likely on larger and heavier parts
4:1	In general, a good average ratio for many parts; a good ratio for evaluating a new deburring process
5:1	Better for nonferrous parts subject to part-on-part damage
6:1	Suitable for nonferrous parts, especially preplate surfaces on zinc parts with resin-bonded media
8:1	For improved preplate surfaces with resin-bonded media
10:1	Produces very fine finishes

Source: American Machinist, August 1983.

Media selection should also be correlated with part geometry, since the abrasives should be able to contact all critical surfaces without becoming lodged in recesses or holes. This requirement has resulted in a wide variety of sizes and shapes, including those presented in Figure 35-9. The different abrasives, sizes, and shapes can be selected or combined to perform tasks ranging from light deburring with a very fine finish to heavy cutting with a rough surface.

COMPOUNDS

A variety of functions are performed by the *compounds* that are added in addition to the media and workpieces. These compounds can be liquid or dry, abrasive or nonabrasive, and acid, neutral, or alkaline. They are often designed to assist in deburring, burnishing, and abrasive cutting, as well as to provide cleaning, descaling, or corrosion inhibition.

In deburring and finishing, many small particles are abraded from both the media and the workpieces, and these must be suspended in the compound solution to prevent them from adhering to the parts. Deburring compounds also act to keep the parts and media clean and to inhibit corrosion. Burnishing compounds are often selected for their ability to develop desired colors and enhance brightness.

Cleaning compounds such as dilute acids and soaps are designed to remove excessive soils from both the parts and media and are often specified when the incoming materials contain heavy oil or grease. Corrosion inhibitors can be selected for both ferrous and nonferrous metals and are particularly important when steel media are being used.

Another function of the compounds may be to condition the water when aqueous solutions are being used. Consistent water quality, in terms of "hardness" and metal ion content, is important to ensure uniform and repeatable finishing results. Liquid compounds may also provide cooling to both the workpieces and the media.

SUMMARY OF MASS-FINISHING METHODS

The barrel and vibratory finishing processes are really quite simple and economical and can process large numbers of parts in a batch procedure. Soft, nonferrous parts can be finished in as little as 10 minutes, while the harder steels may require 2 hours or more. Sometimes the operations are sequenced, using progressively finer abrasives. Figure 35-11 shows a variety of parts before and after the mass-finishing operation, using the triangular abrasive shown with each component.

Despite the high volume and apparent success, these processes may still be as much art as science. The key factors of workpiece, equipment, media, and compound are all interrelated, and the effect of changes can be quite complex. Media, equipment, and compounds are often selected by trial and error, with various approaches being tested until the desired result is achieved. Even then, maintenance of consistent results may still be difficult.

FIGURE 35-11   A variety of parts before and after barrel finishing with triangular-shaped media. *(Courtesy of Norton Company.)*

BELT SANDING

In the *belt sanding* operation, the workpieces are held against a moving abrasive belt until the desired degree of finish is obtained. Because of the movement of the belt, the resulting surface contains a series of parallel scratches with a texture set by the grit of the belt. When smooth surfaces are desired, a series of belts may be employed, with progressively finer grits.

The ideal geometry for belt sanding is a flat surface, for the belt can be passed over a flat table where the workpiece can be held firmly against it. Belt sanding is frequently a hand operation and is therefore quite labor intensive. Furthermore, it is difficult to apply when the geometry includes recesses or interior corners. As a result, belt sanding is usually employed when the number of parts is small and the geometry is relatively simple. See Chapter 28.

WIRE BRUSHING

High-speed rotary *wire brushing* is sometimes used to clean surfaces and can also impart some small degree of material removal or smoothing. The resulting surface consists of a series of uniform curved scratches. For many applications, this *may* be an acceptable final finish. If not, the scratches can easily be removed by barrel finishing or buffing.

Wire brushing is often performed by hand application of a small workpiece to the brush or the brush to a larger workpiece. Automatic machines can also be used where the parts are moved past a series of rotating brushes. In another modification the brushes are replaced with plastic or fiber wheels that are loaded with abrasive.

BUFFING

Buffing is a polishing operation in which the workpiece is brought into contact with a revolving cloth wheel that has been charged with a fine abrasive, such as polishing rouge. The "wheels", which are made of disks of linen, cotton, broadcloth, or canvas, achieve the desired degree of firmness through the amount of stitching used to fasten the layers of cloth together. When the operation calls for very soft polishing or polishing into interior corners, the stitching may be totally omitted, the centrifugal force of the wheel rotation being sufficient to keep the layers in the proper position. Various types of polishing compounds are also available, with many consisting of ferric oxide particles in some form of hinder or carrier.

The buffing operation is very similar to the lapping process that was discussed in Chapter 27. In buffing, however, the abrasive removes only minute amounts of metal from the workpiece. Fine scratch marks can be eliminated and oxide tarnish can be removed. A smooth, reflective surface is produced. When soft metals are buffed, a small amount of metal flow may occur, which further helps to reduce high spots and produce a high polish.

In manual buffing, the workpiece is held against the rotating wheel and manipulated to provide contact with all critical surfaces. Once again, the labor costs can be quite extensive. If the workpieces are not too complex, semiautomatic machines can be used, where the workpieces are held in fixtures and move past a series of individual buffing wheels. By designing the part with buffing in mind, good results can be obtained quite economically.

ELECTROPOLISHING

Electropolishing is the reverse of electroplating (discussed later in this chapter) since material is removed from the surface rather than being deposited. A DC electrolytic circuit is constructed with the workpiece as the anode. As current is applied, material is stripped from the surface, with material removal occurring preferentially from any raised location. Unfortunately, it is not economical to remove more than about 0.001 in. of material from any surface. However, if the initial surface is sufficiently smooth (less than 8 in. rms), and the grain size is small, the result will be a smooth polish with irregularities of less than 2 μin.—a mirrorlike finish.

Electropolishing was originally used to prepare metallurgical specimens for examination under the microscope. It was later adopted as a means of polishing stainless steel sheets and other stainless products. It is particularly useful for polishing irregular shapes that would be difficult to buff.

■ 35.3 CHEMICAL CLEANING

Chemical cleaning operations are effective means of removing oil, dirt, scale, or other foreign material that may adhere to the surface of a product, as a preparation for subsequent painting or plating. Because of environmental, health, and safety concerns, however, many processes that were once the industrial standard have now been eliminated or substantially modified. While the major concern with the mechanical methods has usually been airborne particles, the chemical methods often require the disposal of spent or contaminated solutions, and they occasionally use hazardous, toxic, or environmentally unfriendly materials. Chlorofluorocarbons (CFCs) and carbon tetrachloride, for example, have been identified as ozone-depleting chemicals and have been phased out of commercial use. Process changes to comply with added regulations can significantly shift process economics. Manufacturers must now ask themselves if a part really has to be cleaned, what soils have to be removed, how clean the surfaces have to be, and how much they are willing to pay to accomplish that goal. Selection of the cleaning method will depend on cost of the equipment, power, cleaning materials, maintenance and labor, plus the cost of recycling and disposal of materials. Specific processes will depend on the quantity of parts to be processed (part per hour), part configuration, part material, desired surface finish, temperature of the process, and flexibility. Manufacturers want machines they can integrate with manufacturing cells so changes in products can be quickly handled.

ALKALINE CLEANING

Alkaline cleaning is basically the "soap and water" approach to parts cleaning and is a commonly used method for removing a wide variety of soils (including oils, grease, wax, fine particles of metal, and dirt) from the surfaces of metals. The cleaners are usually complex solutions of alkaline salts, additives to enhance cleaning or surface modification, and surfactants or soaps that are selected to reduce surface tension and displace, emulsify, and disperse the insoluble soils. The actual cleaning occurs as a result of one or more of the following mechanisms: (1) saponification, the chemical reaction of fats and other organic compounds with the alkaline salts; (2) displacement, where soil particles are lifted from the surface; (3) dispersion or emulsification of insoluble liquids; and (4) dissolution of metal oxides.

Alkaline cleaners can be applied by immersion or spraying, and they are usually heated to accelerate the cleaning action. The cleaning is then followed by a water rinse to remove all residue of the cleaning solution, as well as flush away some small amounts of remaining soil. A drying operation may also be required since the aqueous cleaners do not evaporate quickly, and some form of corrosion inhibitor (or rust preventer) may be required, depending on subsequent use.

Environmental issues relating to alkaline cleaning include (1) reducing or eliminating phosphate effluent, (2) reducing toxicity and increasing biodegradability, and (3) recycling the cleaners to extend their life and reduce the volume of discard.

SOLVENT CLEANING

In *solvent cleaning*, oils, grease, fats, and other surface contaminants are removed by dissolving them in organic solvents derived from coal or petroleum, usually at room temperature. The common solvents include petroleum distillates (such as kerosene, naphtha, and mineral spirits), chlorinated hydrocarbons (such as methylene chloride and trichloroethylene), and liquids such as acetone, benzene, toluene, and the various alcohols. Small parts are generally cleaned by immersion, with or without assisting agitation, or by spraying. Products that are too large to immerse can be cleaned by spraying or wiping. The process is quite simple, and capital equipment costs are rather low. Drying is usually accomplished by simple evaporation.

Solvent cleaning is an attractive means of cleaning large parts, heat-sensitive products, materials that might react with alkaline solutions (such as aluminum, lead, and zinc), and products with organic contaminants (such as soldering flux or marking crayon). Virtually all common industrial metals can be cleaned, and the size and shape of the workpiece are rarely a limitation. Insoluble contaminants, such as metal oxides, sand, scale, and the inorganic fluxes used in welding, brazing, and soldering, cannot be removed by solvents. In addition, resoiling can occur as the solvent becomes contaminated. As a result, solvent cleaning is often used for preliminary cleaning.

Many of the common solvents have been restricted because of health, safety, and environmental concerns. Fire and excessive exposure are common hazards. Adequate ventilation is critical. Workers should use respiratory devices to prevent inhalation of vapors and wear protective clothing to minimize direct contact with skin. In addition, solvent wastes are often considered to be hazardous materials and may be subject to high disposal cost.

VAPOR DEGREASING

In *vapor degreasing*, the vapors of a chlorinated or fluorinated solvent are used to remove oil, grease, and wax from metal products. A nonflammable solvent, such as trichloroethylene, is heated to its boiling point, and the parts to be cleaned are suspended in its vapors. The vapor condenses on the work and washes the soluble contaminants back into the liquid solvent. Although the bath becomes dirty, the contaminants rarely volatilize at the boiling temperature of the solvent. Therefore, vapor degreasing tends to be more effective than cold solvent cleaning, since the surfaces always come into contact with clean solvent. Since the surfaces become heated by the condensing solvent, they dry almost instantly when they are withdrawn from the vapor.

Vapor degreasing is a rapid, flexible process that has almost no visible effect on the surface being cleaned. It can be applied to all common industrial metals, but the solvents may attack rubber, plastics, and organic dyes that might be present in product assemblies. A major limitation is the inability to remove insoluble soils, forcing the process to be coupled with another technique, such as mechanical or alkaline cleaning. Since hot solvent is present in the system, the process is often accelerated by coupling the vapor cleaning with an immersion or spray using the hot liquid.

Unfortunately, environmental issues have forced the almost complete demise of the process. While the vapor degreasing solvents are chemically stable, have low toxicity, are nonflammable, evaporate quickly, and can be recovered for reuse, the CFC materials have been identified as ozone-depleting compounds and have essentially been banned from use. Solvents that can be used in the same process, or in a replacement process that offers the necessary cleaning qualities, include chlorinated solvents (methylene chloride, perchloroethylene, and trichloroethylene), most manufacturers have converted to some form of water-based process using alkaline, neutral, or acid cleaners or to a process using chlorine-free, hydrocarbon-based solvents. Sealed chamber machines use non-VOC, nonchlorinated solvents that are continuously recycled.

ULTRASONIC CLEANING

When high-quality cleaning is required for small parts, *ultrasonic cleaning* may be preferred. Here the parts are suspended or placed in wire-mesh baskets that are then immersed in a liquid cleaning bath, often a water-based detergent. The bath contains an ultrasonic transducer that operates at a frequency that causes *cavitation* in the liquid. The bubbles that form and implode provide the majority of the cleaning action, and if gross dirt, grease, and oil are removed prior to the immersion, excellent results can usually be obtained in 60 to 200 seconds. Most systems operate at between 10 and 40 kHZ. Because of the ability to use water-based solutions, ultrasonic cleaning has replaced many of the environmentally unfriendly solvent processes.

ACID PICKLING

In the *acid-pickling* process, metal parts are first cleaned to remove oils and other contaminants and then dipped into dilute acid solutions to remove oxides and dirt that are left on the surface by the previous processing operations. The most common solution is a 10% sulfuric acid bath at an elevated temperature between 150° and 185° F. Muriatic acid is also used, either cold or hot. As the temperature increases, the solutions can become more dilute.

After the parts are removed from the pickling bath, they should be rinsed to flush the acid residue from the surface and then dipped in an alkaline bath to prevent rusting. When it will not interfere with further processing, an immersion in a cold *milk of lime* solution is often used. Caution should be used to avoid overpickling, since the acid attack can result in a roughened surface.

■ 35.4 COATINGS

Each of the surface finishing methods previously presented has been a material removal process, designed to clean, smooth, and otherwise reduce the size of the part. Many other techniques have been developed to add material to the surface of a part. If the material is deposited as a liquid or organic gas (*or* from a liquid or gas medium), the process is called *coating*. If the added material is a solid during deposition, the process is known as *cladding*.

PAINTING, WET OR LIQUID

Paints and enamels are by far the most widely used finish on manufactured products, and a great variety are available to meet the wide range of product requirements. Most of today's commercial paints are synthetic organic compounds that contain pigments and dry by polymerization or by a combination of polymerization and adsorption of oxygen. Water is the most common carrying vehicle for the pigments. Heat can be used to accelerate the drying, but many of the synthetic paints and enamels will dry in less than an hour without the use of additional heat. The older oil-based materials have a long drying time and require excessive environmental protection measures. For these reasons they are seldom used in manufacturing applications.

Paints are used for a variety of reasons, usually to provide protection and decoration but also to fill or conceal surface irregularities, change the surface friction, or modify the light or heat absorption or radiation characteristics. Table 35-3 provides a list of some of the more commonly used organic finishes, along with their significant characteristics. *Nitrocellulose lacquers* consist of thermoplastic polymers dissolved in organic solvent. Although fast drying (by the evaporation of the solvent) and capable of producing very beautiful finishes, they are not sufficiently durable for most commercial applications. The *alkyds* are a general-purpose paint but are not adequate for hard-service applications. *Acrylic enamels* are widely used for automotive finishes and may require catalytic or oven curing. *Asphaltic paints*, solutions of asphalt in a solvent, are used extensively in the electrical industry, where resistance to corrosion is required and appearance is not of prime importance.

When considering a painted finish, the temptation is to focus on the outermost coat, to the exclusion of the underlayers. In reality, painting is a complex system that includes the substrate material, cleaning and other pretreatments (such as anodizing, phosphating, and various conversion coatings), priming, and possible intermediate layers. The method of application is another integral feature to be considered.

TABLE 35-3	Commonly Used Organic Finishes and Their Qualities		
Material	Durability (Scale of 1–10)	Relative Cost (Scale of 1–10)	Characteristics
Nitrocellulose lacquers	1	2	Fast drying; low durability
Epoxy esters	1	2	Good chemical resistance
Akyd-amine	2	1	Versatile; low adhesion
Acrylic lacquers	4	1.7	Good color retention; low adhesion
Acrylic enamels	4	1.3	Good color retention; though; high baking temperature
Vinyl solutions	4	2	Flexible; good chemical resistance; low solids
Silicones	4–7	5	Good gloss retention; low flexibility
Flouropolymers	10	10	Excellent durability; difficult to apply

PAINT APPLICATION METHODS

In manufacturing, almost all painting is done by one of four methods: *dipping, hand spraying, automatic spraying,* or *electrostatic spray finishing.* In most cases, at least two coats are required. The first (or *prime*) coat serves to (1) ensure adhesion, (2) provide a leveling effect by filling in minor porosity and other surface blemishes, and (3) improve corrosion resistance and thus prevent later coatings from being dislodged in service. These properties are less easily attainable in the more highly pigmented paints that are used in the final coats to promote color and appearance. When using multiple coats, however, it is important that the carrying vehicles for the final coats do not unduly soften the underlayers.

Dipping is a simple and economical means of paint application when all surfaces of the part are to be coated. The products can be manually immersed into a paint bath or passed through the bath while on or attached to a conveyor. Dipping is attractive for applying prime coats and for painting small parts where spray painting would result in a significant waste due to overspray. Conversely, the process is unattractive where only some of the surfaces require painting or where a very thin, uniform coating would be adequate, as on automobile bodies. Other difficulties are associated with the tendency of paint to run, producing both a wavy surfacc and a final drop of paint attached to the lowest drip point. Good-quality dipping requires that the paint be stirred at all times and be of uniform viscosity.

Spray painting is probably the most widely used paint application process because of its versatility and the economy in the use of paint. In the conventional technique, the paint is atomized and transported by the flow of compressed air. In a variation known as *airless spraying,* mechanical pressure forces the paint through an orifice at pressures between 500 and 4500 psi. This provides sufficient velocity to produce atomization and also propel the particles to the workpiece. Because no air pressure is used for atomization, there is less spray loss (paint efficiency may be as high as 99%) and less generation of gaseous fumes.

Hand spraying is probably the most versatile means of application but can be quite costly in terms of labor and production time. When air or mechanical means provide the atomization, workers must exercise considerable skill to obtain the proper coverage without allowing the paint to "run" or "drape." Only a very thin film can be deposited at one time, usually less than 0.001 in. As a result, several coats may be required with intervening time for drying.

One means of applying thicker layers in a single application is known as *hot spraying.* Special solvents are used that reduce the viscosity of the material when heated. Upon atomization, the faster-evaporating solvents are removed, and the drop in temperature produces a more viscous, run-resistant material that can be deposited in thicker layers.

When producing large quantities of similar or identical parts, some form of automatic system is usually employed. The simplest automatic equipment consists of some form of parts conveyor that transports the parts past a series of stationary spray heads.

While the concept is simple, the results may be unsatisfactory. A large amount of paint is wasted, and it is difficult to get uniform coverage.

Industrial robots can be used to move the spray heads in a manner that mimics the movements of a human painter, maintaining uniform separation distance and minimizing waste. This is an excellent application for the robot, since a monotonous and repetitious process can be performed with consistent results. In addition, use of a robot removes the human from an unpleasant, and possibly unhealthy, environment. Nowdays, cars are painted almost exclusively with robots.

Both manual and automatic spray painting can benefit from the use of *electrostatic deposition*. A DC electrostatic potential is applied between the atomizer and the workpiece. The atomized paint particles assume the same charge as the atomizer and are therefore repelled. The oppositely charged workpiece then attracts the particles, with the actual path of the particle being a combination of the kinetic trajectory and the electrostatic attraction. The higher the DC voltage, the greater the electrostatic attraction. Overspraying can be reduced by as much as 60 to 80%, as can the generation of airborne particles and other emissions. Unfortunately, part edges and holes receive a heavier coating than flat surfaces due to the concentration of electrostatic lines of force on any sharp edge. Recessed areas will receive a reduced amount of paint, and a manual touch-up may be required using conventional spray techniques. Despite these limitations, electrostatic spraying is an extremely attractive means of painting complex-shaped products where the geometry would tend to create large amounts of overspray.

In an electrostatic variation of airless spraying, the paint is fed onto the surface of a rapidly rotating cone or disk that is also one electrode of the electrostatic circuit. Centrifugal force causes the thin film of paint to flow toward the edge, where charged particles are spun off without the need for air assist. The particles are then attracted to the workpiece, which serves as the other electrode of the electrostatic circuit. Because of the effectiveness of the centrifugal force, paints can be used with high-solids content, reducing the amount of volatile emissions and enabling a thicker layer to be deposited in a single application.

Electrocoating or *electrodeposition* applies paint in a manner similar to the electroplating of metals. As shown schematically in Figure 35-12, the paint particles are suspended in an aqueous solution and are given an electrostatic charge by applying a DC voltage between the tank (cathode) and the workpiece (anode). As the electrically conductive workpiece enters and passes through the tank, the paint particles are attracted to it and deposit on the surface, creating a uniform, thin coating that is more than 90% resin and pigment. When the coating reaches a desired thickness, determined by the bath conditions, no more paint is deposited. The workpiece is then removed from the tank, rinsed in a water spray, and baked at a time and temperature that depends on the particular type of paint. Baking of 10 to 20 minutes at 375°F is somewhat typical.

Electrocoating combines the economy of ordinary dip painting with the ability to produce thinner, more uniform coatings. The process is particularly attractive for applying the prime coat to complex structures, such as automobile bodies, where good corrosion resistance is a requirement. Hard-to-reach areas and recesses can be effectively coated.

Direct-current power supply

Emulsion of water & paint

Water & detergent solution

Oven

Electrocoating:
Time cycle
1 minute

Rinsing

Baking:
Time cycle
10 minutes – 380°F

FIGURE 35-12 Basic steps in the electrocoating process.

Since the solvent is water, no fire hazard exists (as with the use of many solvents), and air and water pollution is reduced significantly. In addition, the process can be readily adapted to conveyor line production.

DRYING

Most paints and enamels used in manufacturing require from 2 to 24 hours to dry at normal room temperature. This time can be reduced to between 10 minutes and 1 hour if the temperature can be raised to between 275° and 450°F. As a result, elevated-temperature drying is often preferred. Parts can be batch processed in ovens or continuously passed through heated tunnels or under panels of infrared heat lamps.

Elevated-temperature drying is rarely a problem with metal parts, but other materials can be damaged by exposure to the moderate temperatures. For example, when wood is heated, the gases, moisture, and residual sap are expanded and driven to the surface beneath the hardening paint. Small bubbles tend to form that roughen the surface, or break, producing small holes in the paint.

POWDER COATING

Powder coating is yet another variation of electrostatic spraying, but here the particles are solid rather than liquid. Several coats, such as primer and finish, can be applied and then followed by a single baking, in contrast to the baking after each coat that is required in the conventional spray processes. In addition, the overspray powder can often be collected and reused. While volatilized solvents are no longer a concern, operators must now address the possibility of powder explosion, as well as the health hazards of airborne particles.

Modern powder technology can produce a high-quality finish with superior surface properties and usually at a lower cost than liquid painting. Powder painting is more efficient in the use of materials (the overspray can be captured and reused) and lower energy requirements. The economic advantages must be weighed against the limitations of powder coating. Dry systems have a longer color change time than wet systems. The process is not good for large objects (massive tanks) or heat-sensitive objects. It is not easy to produce film thickness less than 1 mil (0.03 mm).

Table 35-4 provides details on powders that are used in powder coatings. Thermoplastics can also be used, but thermosetting powders are most common. The elements

TABLE 35-4   Thermosetting Powder Coatings (Dry Painting) Have a Wide Variety of Properties and Applications

Properties	Epoxy	Epoxy/Polyester Hybrid	TGIC Polyester	Polyester Urethane	Acrylic Urethane
Application thickness	0.5–20 mils[a]	0.5–10 mils	0.5–10 mils	0.5–10 mils	0.5–10 mils
Cure cycle (metal temperatures)[b]	450°F—3 min 250°F—30 min	450°F—3 min; 325°F—25 min	400°F—7 min; 310°F—20 min	400°F — 7 min; 350°F—17 min	400°F—7 min; 360°F—25 min
Outdoor weatherability	Poor	Poor	Very good	Very good	Excellent
Pencil hardness	HB-5H	HB-2H	HB-2H	HB-3H	H-3H
Direct impact resistance, in lb[c]	80–160	80–160	80–160	80–160	20–60
Chemical resistance	Excellent	Very good Least expensive	Good	Good	Very good Most expensive
Cost (relative)	2	1	3	4	5
Applications	Furniture, cars, ovens, appliances	Water heaters, radiators, office furniture	Architectural aluminum, outdoor furniture, farm equipment	Car wheels/rims, playground equipment	Washing machines, refrigerators, ovens

[a]Thickness up to 150 mils can be applied via multiple coats in a fluidized bed.
[b]Time and temperature can be reduced, by utilizing accelerated curing mechanisms, while maintaining the same general properties.
[c]Tested at a coating thickness of 2.0 mils.

Powder application equipment

FIGURE 35-13 A schematic of a powder coating system. The wheels on the color modules permit it to be exchanged with a spare module to obtain the next color.

of a powder coating system are shown in Figure 35-13. The following aspects of the process must be considered:

- Types of guns—corona charged or tribo charged
- Number of guns—depends on many factors, such as parts per hour, size of parts, line speed, and powder types
- Color change time/frequency
- Safety
- Curing oven—coated parts put in ovens to melt, flow, and cure the powder

HOT-DIP COATINGS

Large quantities of metal products are given corrosion-resistant coatings by direct immersion into a bath of molten metal. The most common coating materials are zinc, tin, aluminum, and tene (an alloy of lead and tin).

Hot-dip galvanizing is the most widely used method of imparting corrosion resistance to steel. (The zinc acts as a sacrificial anode, protecting the underlying iron.) After the products, or sheets, have been cleaned to remove oil, grease, scale, and rust, they are fluxed by dipping into a solution of zinc ammonium chloride and dried. Next, the article is completely immersed in a bath of molten zinc. The zinc and iron react metallurgically to produce a coating that consists of a series of zinc–iron compounds and a surface layer of nearly pure zinc.

The coating thickness is usually specified in terms of weight per unit area. Values between 0.5 and 3.0 oz/ft^2 are typical, with the specific value depending on the time of immersion and speed of withdrawal. Thinner layers can be produced by incorporating some form of air jet or mechanical wiping as the product is withdrawn. Since the corrosion resistance is provided through the sacrificial action of the zinc, the thin layers do not provide long-lasting protection. Extremely heavy coatings, on the other hand, may tend to crack and peel. The appearance of the coating can be varied through both the process conditions and alloy additions of tin, antimony, lead, and aluminum. When the coatings are properly applied, bending or forming can often follow galvanizing without damage to the integrity of the coating. Zinc-galvanized sheet can be heat treated with a zinc–iron alloy coating. The 10% iron content adds strength and makes for good corrosion and pitting/chipping resistance. In auto applications, galvannealing beats out pure zinc on several counts: spot weldability, pretreatability, and ease of painting. Electrogalvanized zinc–nickel coatings that contain 10 to 15% Ni can be used in thinner layers (5–6 microns) and are easier to form and spot weld.

The primary limitations to hot-dip galvanizing are the size of the product (which is limited to the size of the tank holding the molten zinc) and the "damage" that might occur when a metal is exposed to the temperatures of the molten material (approximately).

Tin coatings can also be applied by immersing in a bath of molten tin with a covering of flux material. Because of the high cost of tin and the relatively thick coatings applied by hot dipping, most tin coatings are now applied by electroplating. *Terne coating* utilizes an alloy of 15 to 20% tin and the remainder lead. This material is cheaper than tin and can provide satisfactory corrosion resistance for many applications.

CHEMICAL CONVERSION COATINGS

In *chemical conversion coating,* the surface of the metal is chemically treated to produce a nonmetallic, nonconductive surface that can impart a range of desirable properties. The most popular types of conversion coatings are chromate and phosphate. Aluminum, magnesium, zinc, and copper (as well as cadmium and silver) can all be treated by a *chromate* conversion process that usually involves immersion in a chemical bath. The surface of the metal is convened into a layer of complex chromium compounds that can impart colors ranging from bright clear through blue, yellow, brown, olive drab, and black. Most of the films are soft and gelatinous when they are formed but harden upon drying. They can be used to (1) impart exceptionally good corrosion resistance; (2) act as an intermediate bonding layer for paint, lacquer, or other organic finishes; or (3) provide specific colors by adding dyes to the coating when it is in its soft condition.

Phosphate coatings are formed by immersing metals (usually steel or zinc) in baths where metal phosphates (iron, zinc, and manganese phosphates are all common) have been dissolved in solutions of phosphoric acid. The resultant coatings can be used to precondition surfaces to receive and retain paint or enhance the subsequent bonding with rubber or plastic. In addition, phosphate coatings are usually rough and can provide an excellent surface for holding oils and lubricants. This feature can be used in manufacturing, where the coating holds the lubricants that assist in forming, or in the finished product, as with black-color bolts and fasteners, whose corrosion resistance is provided by a phosphate layer impregnated with wax or oil.

BLACKENING OR COLORING METALS

Many steel parts are treated to produce a black, iron oxide coating—a lustrous surface that is resistant to rusting when handled. Since this type of oxide forms at elevated temperatures, the parts are usually heated in some form of special environment, such as spent carburizing compound or special blackening salts.

Chemical solutions can also be used to blacken, blue, and even "brown" steels. Brown, black, and blue colors can also be imparted to tin, zinc, cadmium, and aluminum through chemical bath immersions or wipes. The surfaces of copper and brass can be made to be black, blue, green, or brown, with a full range of tints in between.

ELECTROPLATING

Large quantities of metal *and plastic* parts are electroplated to produce a metal coating that imparts corrosion or wear resistance, improves appearance (through color or luster), or increases the overall dimensions. Virtually all commercial metals can be plated, including aluminum, copper, brass, steel, and zinc-based die castings. Plastics can be electroplated, provided that they are first coated with an electrically conductive material.

The most common platings are zinc, chromium, nickel, copper, tin, gold, platinum, and silver. The electrogalvanized zinc platings are thinner than the hot-dip coatings and can be produced without subjecting the base metal to the elevated temperatures of

FIGURE 35-14 Basic circuit for an electroplating operation, showing the anode, cathode (workpiece), and electrolyte (conductive solution).

molten zinc. Nickel plating provides good corrosion resistance but is rather expensive and does not retain its lustrous appearance. Consequently, when lustrous appearance is desired, a chromium plate is usually specified. Chromium is seldom used alone, however. An initial layer of copper produces a leveling effect and makes it possible to reduce the thickness of the nickel layer that typically follows to less than 0.0006 in. The final layer of chromium then provides the attractive appearance. Gold, silver, and platinum platings are used in both the jewelry and electronics industries, where the thin layers impart the desired properties while conserving the precious metals.

Hard chromium plate, with Rockwell hardnesses between 66 and 70, can be used to build up worn parts to larger dimensions and to coat tools and other products that need reduced surface friction and good resistance to both wear and corrosion. Hard chrome coatings are always applied directly to the base material and are usually much thicker than the decorative treatments, typically ranging from .003 to .010 in. thick. Even thicker layers are used in applications such as diesel cylinder liners. Since hard chrome plate does not have a leveling effect, defects or roughness in the base surface will be amplified. If smooth surfaces are desired, subsequent grinding and polishing may be necessary.

Figure 35-14 depicts the typical electroplating process. A DC voltage is applied between the parts to be plated (which is made the cathode) and an anode material that is either the metal to be plated or an inert electrode. Both of these components are immersed in a conductive electrolyte, which may also contain dissolved salts of the metal to be plated as well as additions to increase or control conductivity. In response to the applied voltage, metal ions migrate to the cathode, lose their charge, and deposit on the surface. While the process is simple in its basic concept, the production of a high-quality plating requires selection and control of a number of variables, including the electrolyte and the concentrations of the various dissolved components, the temperature of the bath, and the electrical voltage and current. The interrelation of these features adds to the complexity and makes process control an extremely challenging problem.

The surfaces to be plated must also be prepared properly if satisfactory results are to he obtained. Pinholes, scratches, and other surface defects must be removed if a smooth, lustrous finish is desired. Combinations of degreasing, cleaning, and pickling are used to ensure a chemically clean surface, one to which the plating material can adhere.

As shown in Figure 35-15, the plated metal tends to be preferentially attracted to corners and protrusions. This makes it particularly difficult to apply a uniform plating to irregular shapes, especially ones containing recesses, corners, and edges. Design features can be incorporated to promote plating uniformity, and improved results can often be obtained through the use of multiple spaced anodes or anodes whose shape resembles that of the workpiece.

Electroplating is frequently performed as a continuous process, where the individual parts to be plated are hung from conveyors. As they pass through the process, they

Convex surfaces: Plate uniformly especially if edges are rounded

Concave recesses: Platability depends on dimensions

Flat surfaces: Not desirable. Use slight crown to hide undulations.

Slots: Narrow slots and holes should have rounded corners.

Blind holes: Must be exempted from minimum thickness requirements. Require vent hole at blind end.

V-shaped grooves: Difficult to plate. Should be avoided.

Sharply angled edges: Plating is thinner in center areas. Round all areas.

Fins: Increase plating time and costs. Reduce durability of finish.

FIGURE 35-15 Design recommendations for electroplating operations.

are lowered into successive plating, washing, and fixing tanks. Ordinarily, only one type of workpiece is plated at a time, because the details of solutions, immersion times, and current densities are usually changed with changes in workpiece size and shape.

In the *electroforming* process, the coating becomes the final product. Metal is electroplated onto a mandrel (or mold) to a desired thickness and is then stripped free to produce small quantities of molds or other intricate-shaped sheet-metal type products.

ANODIZING

Anodizing is an electrochemical process, that is somewhat the reverse of electroplating, which produces a conversion-type coating on aluminum that can improve corrosion and wear resistance and impart a variety of decorative effects. If the workpiece is made the anode of an electrolytic cell, instead of a plating layer being deposited on the surface, a reaction progresses inward, increasing the thickness of the hard hexagonal aluminum oxide crystals on the surface. The hardness depends on thickness, density, and porosity of the coating, which are controlled by the cycle time and applied currents along with the chemistry, concentration, and temperature of the electrolyte. The surface texture very nearly duplicates the prefinishing texture, so a buffing prefinish produces a smooth, lustrous coating while sand blasting produces a grainy or satiny coating.

The flow diagram in Figure 35-16 shows the anodizing process. Coating thicknesses range from 0.1 mils to 0.25 mils. Note that the product dimensions will increase, however, because the aluminum oxide coating occupies about twice the volume of the metal from which it formed.

The nature of the developed coating is controlled by the electrolyte. If the oxide coating is not soluble in the anodizing solution, it will grow until the resistance of the oxide prevents current from flowing. The resultant coating, which is thin, nonporous, and nonconducting, is used in a variety of electrical applications.

If the oxide coating is slightly soluble in the anodizing solution, dissolution competes with oxide growth and a porous coating will be produced, where the pores provide for continued current flow to the metal surface. As the coating thickens, the growth rate decreases until it achieves steady state, where the growth rate is equal to the rate of dissolution. This condition is determined by the specific conditions of the process, including voltage, current density, electrolyte concentration, and electrolyte temperature. Sulfuric, chromic, oxalic, and phosphoric acids all produce electrolytes that dissolve oxide, with a sulfuric acid solution being the most common.

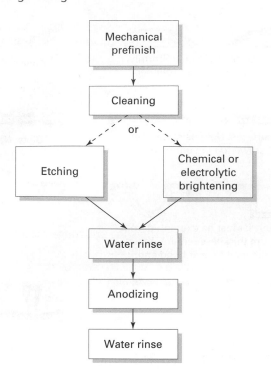

FIGURE 35-16   The anodizing process has many steps.

In a process variation known as *color anodizing,* a sulfuric acid bath is used to produce a layer of microscopically porous oxide that is transparent on pure aluminum and somewhat opaque on alloys. When this material is immersed in a dye solution, capillary action pulls the dye into the pores. The dye is then trapped in place by a sealing operation, usually performed simply by immersing the anodized metal in a bath of hot water. The aluminum oxide coating is converted to a monohydrate, with accompanying increase in volume. The pores close and become resistant to further staining or the leaching out of the dye.

While most people are familiar with the variety of colors in aluminum athletic goods, such as softball hats, the actual applications range from giftware, through automotive trim, to architectural use. Aluminum can be made to look like gold, copper, or brass, or it can take on a variety of colors with a combined metallic luster that cannot be duplicated by other methods.

If PTFE (Teflon) is introduced into the pores, coatings can be produced that couple high hardness and low friction. The porous oxide layer can also be used to enhance the adhesion of an additional layer of material, such as paint, or carry lubricant during a subsequent forming operation. Since the coating is integral to the part, subsequent operations can often be performed without destroying its integrity or reducing its protective qualities.

Anodizing can also be performed on other metals, such as magnesium, and the process is similar to the passivation of stainless steel.

ELECTROLESS PLATING

When using electroplating, it is almost impossible to obtain a uniform plating thickness on even moderately complex shapes, the platings cannot be applied to nonconductors, and a large amount of energy is required. For these reasons, a substantial effort has been directed toward the development of plating techniques that do not require an external source of electricity. These methods are known as *electroless,* or *autocatalytic, plating.* Considerable success has been achieved with nickel, but copper and cobalt, as well as some of the precious metals, can also be deposited.

In the electroless process, complex plating solutions (containing metal salts, reducing agents, complexing agents, pH adjusters, and stabilizers) are brought into contact with a substrate surface that acts as a catalyst or has been pretreated with catalytic material. The metallic ion in the plating solution is reduced to metal and deposits on the surface. Since the deposition is purely a chemical process, the coatings are uniform in thickness, independent of part geometry. Unfortunately, the rate of deposition is considerably slower than with electroplating.

Probably the most popular of the electroless coatings is electroless nickel, and various methods exist for its deposition using both acid and alkaline solutions. The coatings offer good corrosion resistance, as well as hardnesses between Rockwell C 49 and 55. In addition, the hardness can be increased further to as high as Rockwell C 80 by subsequent heat treatment.

ELECTROLESS COMPOSITE PLATING

A very useful adaptation of the electroless process has been developed wherein minute particles are co-deposited along with the electroless metal to produce composite-material coatings. Finely divided solid particles, with diameters between 1 and 10 in., are added to the plating bath and deposit up to 50 vol% with the matrix. While it may appear that a large variety of materials could be co-deposited, commercial applications have largely been limited to diamond, silicon carbide, aluminum oxide, and Teflon (PTFE).

Figure 35-17 shows a deposit of silicon carbide particles in a nickel-alloy matrix, where the particles constitute about 25% by volume. The coating offers the same corrosion resistance as nickel, but the high hardness of the silicon carbide particles (about 4500 on the Vickers scale, where tungsten carbide is 1300 and hardened steel is about 900) contributes outstanding resistance to wear and abrasion. Since the deposition is electroless, the thickness of the coating is not affected by the shape of the part. Applications include the coating of plastic-molding dies, for use where the polymer resin contains significant amounts of abrasive filler.

MECHANICAL PLATING

Mechanical plating, also known as peen plating or impact plating, is an adaptation of barrel finishing in which coatings are produced by cold-welding soft, malleable metal powder onto the substrate. Numerous small products are first cleaned and may be

FIGURE 35-17   (Left) Photomicrograph of nickel carbide plating produced by electroless deposition. Notice the uniform thickness coating on the irregularly shaped product. (Right) High-magnification cross section through the coating. *(Courtesy of Electro-Coatings Inc.)*

given a thin galvanic coating of either copper or tin. They are then placed in a tumbling barrel, along with a water slurry of the metal powder to be plated, glass or ceramic tumbling media, and chemical promoters or accelerators. The media particles peen the metal powder onto the surface, producing uniform-thickness deposits (possibly a bit thinner on edges and thicker in recesses—the opposite of electroplating!). Any metal that can be made into fine powder can be deposited, but the best results are obtained for soft materials, such as cadmium, tin, and zinc. Since the material is deposited mechanically, the coatings can be layered or involve mixtures with bulk chemistries that would be chemically impossible due to solubility limits. The fact that the coatings are deposited at room temperature, and in an environment that does not induce hydrogen embrittlement, makes mechanical plating an attractive means of coating hardened steels.

PORCELAIN ENAMELING

Metals can also be coated with a variety of glassy, inorganic materials that impart resistance to corrosion and abrasion, decorative color, electrical insulation, or the ability to function in high-temperature environments. Multiple coats may be used, with the first or ground coat being selected to provide adhesion to the substrate and the cover coat to provide the surface characteristics. The material is usually applied in the form of a multicomponent suspension or slurry (by dipping or spraying), which is then dried and fired. An alternative dry process uses electrostatic spraying of powder and subsequent firing. During the firing operation, which may require temperatures in the range of 800° to 8000°F, the coating materials melt, flow, and resolidify. Porcelain enamel is often found on the inner, perforated tubs of many washing machines and may be used to impart the decorative exterior on cookpots and frying pans.

■ 35.5 VAPORIZED METAL COATINGS

Vapor deposition processes can be classified into two main categories: *physical vapor deposition* (PVD) and *chemical vapor deposition* (CVD). While sometimes used as though it were a specific process, the term *PVD* applies to a group of processes in which the material to be deposited is carried physically to the surface of the workpiece. Vacuum metallizing and sputtering are key PVD processes, as are complex variations, such as ion plating. All are carried out in some form of vacuum, and most are line-of-sight processes in which the target surfaces must be positioned relative to the source. In contrast, the CVD processes deposit material through chemical reactions and generally require significantly higher temperatures. Tool steels treated by CVD may have to be heat treated again, while most PVD processes can be conducted below normal tempering temperatures. See Chapter 21 for additional discussions on PVD and CVD processes.

■ 35.6 CLAD MATERIALS

Clad materials are actually a form of composite in which the components are joined as solids, using techniques such as roll bonding, explosive welding, and extrusion. The most common form is a laminate, where the surface layer provides properties such as corrosion resistance, wear resistance, electrical conductivity, thermal conductivity, or improved appearance, while the substrate layer provides strength or reduces overall cost. Alclad aluminum is a typical example. Here surface layers of weaker but more corrosion-resistant single-phase aluminum alloys are applied to a base of high-strength but less corrosion-resistant, age-hardenable material. Aluminum-clad steel meets the same objective but with a heavier substrate, and stainless steel can be used to clad steels, reducing the need for nickel- and chromium-alloy additions throughout.

Wires and rods can also be made as *claddings*. Here the surface layer often imparts conductivity, while the core provides strength or rigidity. Copper-clad steel rods that can be driven into the ground to provide electrical grounding for lightning rod systems are one example.

■ 35.7 TEXTURED SURFACES

While technically not the result of a surface finishing process or operation, *textured surfaces* can be used to impart a number of desirable properties or characteristics. The types of textures that are often rolled onto the sheets used for refrigerator panels serve to conceal dirt, smudges, and fingerprints. Embossed or coined protrusions can enhance the grip of metal stair treads and walkways. Corrugations provide enhanced strength and rigidity. Still other textures can be used to modify the optical or acoustical characteristics of a material.

■ 35.8 COIL-COATED SHEETS

Traditionally, sheet metal components, such as panels for appliance cabinets, have been fabricated from bare-metal sheets. Pans are blanked and shaped by the traditional metalforming operations, and the shaped panels are then finished on an individual basis. This requires individual handling and the painting or plating of geometries that contain holes, bends, and contours. In addition, there is the time required to harden, dry, or cure the applied surface finish.

An alternative approach is to apply the finish to the sheet material after rolling but before coiling. *Coatings* can be applied continuously to one or both sides of the material while it is in the form of a flat sheet. Thus the coiled material is effectively prefinished, and efforts need to be taken to protect the surface during the blanking and forming operations used to produce the final shape. Various paints have been applied successfully, as well as a full spectrum of metal coatings and platings. The sheared edges will not be coated, but if this feature can be tolerated, the additional measures to protect the surface may be an attractive alternative to the finishing of individual components. A second sequence that has some advantages takes the coils of steel that have been cut to length and stamps the holes and notches into them to create blanks. The blanks are pretreated, dried, powder coated, cured, and restacked. Then they are postformed to shape them into the back, side, and front panels of appliance cabinets.

The manufacturers call this *blank coating*. The coating thickness is about 1.5 mils ± 0.2 mil versus 2 mils ± 0.5 mil (less powder, better quality), and rusting at the corners of the holes is eliminated.

■ 35.9 EDGE FINISHING AND BURRS

Burrs are the small, sometimes flexible projections of material that adhere to the edges of workpieces that are formed by cutting, punching, or grinding, like the exit-side burrs formed in the milled slot of Figure 35-18. Dimensionally, they are typically only 0.003 in. thick and 0.001 to 0.005 in. in height, but if not removed, they can lead to assembly failures, short circuits, injuries to workers, or even fatigue failures.

FIGURE 35-18 Schematic showing the formation of heavy burrs on the exit side of a milled slot. *(From L. X. Gillespie, American Machinist, November 1985.)*

The most basic way to detect a burr is to run your finger or fingernail over the edges of the part. Probes and visual inspection techniques (microscopes) are used to find burrs as well.

A number of different processes have been used for *burr removal,* including some discussed previously in this chapter and others presented as special types of machining. These include grinding, chamfering, barrel tumbling, vibratory finishing, centrifugal and spindle finishing, abrasive jet machining, water jet cutting, wire brushing, belt sanding, chemical machining, electropolishing, buffing, electrochemical machining, filing, ultrasonic machining, and abrasive flow machining (see Chapter 26).

Other burr removal methods may be quite specialized, such as thermal-energy deburring. Here the parts are loaded into a chamber, which is then filled with a combustible gas mixture. When the gas is ignited, the short-duration wavefront heats the small burrs to as much as 6000°F, while the remainder of the workpiece rarely exceeds 300°F. The burrs are vaporized in less than 20 ms, including those in inaccessible or difficult-to-

TABLE 35-5 Recommended Allowances for Deburring Processes[a]

Process	Edge Radius, mm (in.)	Stock Loss, mm (in.)	Surface Finish, μmAA (μin.AA[b])
Barrel tumbling	0.08–0.5 (0.003–0.020)	0.0025 (0–0.001)	1.5–0.5 (60–20)
Vibratory deburring	0.08–0.5 (0.003–0.020)	0–0.025 (0–0.001)	1.8–0.9 (70–35)
Centrifugal barrel tumbling	0.08–0.5 (0.003–0.020)	0–0.025 (0–0.001)	1.8–0.5 (70–20)
Spindle finishing	0.08–0.5 (0.003–0.020)	0–0.025 (0–0.001)	1.8–0.5 (70–20)
Abrasive-jet deburring	0.08–0.25 (0.003–0.010)	0–0.05 (0–0.002)[c]	0.8–1.3 (30–50)
Water-jet deburring	0–0.13 (0–0.005)(p)	0(p)	
Liquid hone deburring	0–0.13 (0–0.005)	0–0.013 (0–0.0005)	
Abrasive-flow deburring	0.025–0.5 (0.001–0.020)	0.025–0.13 (0.001–0.005)[d]	1.8–0.5 (70–20)
Chemical deburring	0–0.5 (0–0.002)	0–0.025 (0–0.001)	1.3–0.5 (50–20)
Ultrasonic deburring	0–0.05 (0–0.002)	0–0.025 (0–0.001)	0.5–0.4 (20–15)
Electrochemical deburring	0.05–0.25 (0.002–0.010)	0.025–0.08 (0.001–0.003)[e]	
Electropolish deburring	0–0.25 (0–0.010)	0.025–0.08 (0.001–0.003)[e]	0.8–0.4 (30–15)
Thermal-energy deburring	0.05–0.5 (0.002–0.020)	0	1.5–1.3(p) (60–50)
Power brushing	0.08–0.5 (0.003–0.020)	0–0.013 (0–0.0005)	
Power sanding	0.08–0.8 (0.003–0.030)[f]	0.013–0.08 (0.0005–0.003)	1.0–08 (40–30)
Mechanical deburring	0.08–1.5 (0.003–0.060)		
Manual deburring	0.05–0.4 (0.002–0.015)[g]		

[a]Based on a burr 0.08 mm (0.003 in.) thick and 0.13 mm (0.005 in.) high in steel. Thinner burrs can generally be removed much more rapidly. Values shown are typical. Stock-loss values are for overall thickness or diameter. Location A implies that loss occurs over external surfaces, B that loss occurs over all surfaces, and C that loss occurs only near edge. (p) indicates best estimate.
[b]Values shown indicate typical before and after measurements in a deburring cycle.
[c]Abrasive is assumed to contact all surfaces.
[d]Stock loss occurs only at surfaces over which medium flows.
[e]Some additional stray etching occurs on some surfaces.
[f]Flat sanding produces a small burr and no radius.
[g]Chamfer is generally produced with a small burr.
Source: L. X. Gillespie, *American Machinist,* November 1985.

reach locations. Since the process does not use abrasive media, there is no change to any of the product dimensions. The product surfaces are rarely affected by the generated heat, and the cycle (including loading and unloading) can be repeated as many as 100 times an hour. Unfortunately, there is a thin recast layer and heat-affected zone that forms where the burrs were removed. This region is usually less than 0.001 in. thick but may be objectionable in hardened steels and highly stressed parts.

Of all of the burr removal methods, tumbling and vibratory finishing are usually the most economical, typically costing in the neighborhood of a few cents per part. Since most of the common methods also remove metal from exposed surfaces and produce a radius on all edges, it is important that the parts be designed for deburring. Table 35-5 provides a listing of the various deburring processes, as well as the edge radius, stock loss, and surface finish that would result from removal of a "typical burr" of 0.003 in. thickness.

DESIGN TO FACILITATE OR ELIMINATE BURR REMOVAL

By knowing how and where burrs are likely to *form*, the engineer may be able to design parts to make the burrs easy to remove or even eliminate them. As shown in Figure 35-19, extra recesses or grooves can eliminate the need for deburring, since the burr produced by a cutoff tool or slot milling cutter will now lie below the surface. In this approach, one must determine whether it is cheaper to perform another machining operation (*undercutting* or grooving) or to remove the resulting burr.

Chamfers on sharp corners can also eliminate the need to deburr. The chamfering tool removes the large burrs formed by facing, turning, or boring and produces a relief for mating parts. The small burr formed during chamfering may be allowable or can easily be removed. Often, it may be preferable to give the manufacturer the freedom to use either a chamfer (produced by machining) or an edge radius (formed during the deburring operation) on all exposed corners or edges.

■ 35.10 SURFACE INTEGRITY

Surface integrity has become the subject of intense interest because the traditional, non-traditional, and posttreatment methods used to manufacture hardware can change the material's properties. Although the consequence of these changes becomes a design problem, the preservation of properties is a manufacturing consideration. Designs that require a high degree of surface integrity are the ones that display the following qualities:

• Are highly stressed
• Employ low safety factors
• Operate in severe environments
• Must have prime reliability
• Have a high surface areas–to–volume ratio
• Are made with alloys that are sensitive to processing

FIGURE 35-19 Designing extra recesses and grooves into a part may eliminate the need to deburr. (*From L.X. Gillespie, American Machinist, November 1985.*)

Surface integrity should be a joint concern of manufacturing and engineering. Manufacturing must balance cost and producibility with design requirements. It bears repeating to say that engineering must design components with knowledge of manufacturing processes. A reduction in fatigue life resulting from processing can be reversed with a posttreatment. This is another example of design for manufacturing.

It is important to understand that the various manufacturing and surface-finishing processes each impart distinct properties to the materials that will influence the performance of the product. The achievement of satisfactory product performance obviously depends on a good design, high-quality manufacturing (including surface treatment), and proper assembly. The failure of parts in service, however, is usually the result of a combination of factors. A brief survey of features associated with surfaces and surface processing follows.

Each of the various machining processes produces characteristic surface textures (roughness, waviness, and lay) on the workpieces. In addition, the various processes tend to produce changes in the chemical, physical, mechanical, and metallurgical properties on or near the surfaces that are created. For the most part, these changes are limited to a depth of 0.005 to 0.050 in. below the surface. The effects can be beneficial or detrimental, depending on the process, material, and function of the product.

Machining processes (both chip forming and chipless) induce plastic deformation into the surface layer, as shown in Figure 35-20. The cut surfaces are generally left with tensile residual stresses, microcracks, and a hardness that is different from the bulk material. Processes such as EDM and laser machining leave a layer of hard, recast metal on the surface that usually contains microcracks. Ground surfaces can have either residual tension or residual compression, depending on the mix between chip formation and plowing or rubbing during the grinding operation. If sufficient heat is generated, phase transformations can occur in the surface and subsurface regions.

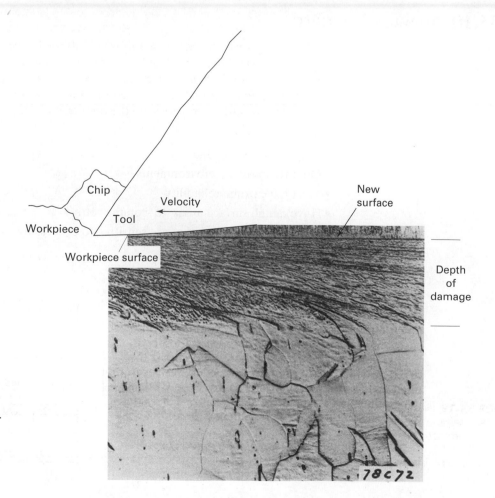

FIGURE 35-20 Plastic deformation in the surface layer after cutting. *(B. W. Kruszynski and C. W. Cuttervelt,* Advanced Manufacturing Engineering, *Vol. 1, 1989.)*

(a)   (b)

FIGURE 35-21   (a) Mechanism for formation of residual compressive stresses in surface by cold plastic deformation (shot peening). (b) Hardness increased in surface due to shot peening.

Processes such as *roller burnishing* (described in Chapter 18) produce a smooth surface with compressive residual stresses. Shot peening (and tumbling) can increase the hardness in the surface and introduce a residual compressive stress, as shown in Figure 35-21a and 35-21b. Welding processes produce tensile residual stresses as the deposited material shrinks upon cooling. Similar shrinkage occurs in castings, but the resulting stresses may be complex due to the variation of shrinkage or the lack of restraint. Tensile stresses on the surface can often be offset by a subsequent exposure to shot peening or tumbling.

In summary, the surface and subsurface regions of a material can be significantly altered due to (1) plastic strain or plastic deformation, (2) high temperatures, (3) differential expansions or contractions due to temperature changes or variations, and (4) chemical reactions.

To illustrate the complex nature of surface effects, consider Figure 35-22, which shows the depth of "surface damage" due to machining as a function of the rake angle of the tool. To increase the cutting speed (and thereby increase the rate of production), an engineer might change from a high-speed tool steel cutter with a large rake angle (such as 30°) to a carbide tool with a zero rake. While the resulting surface finish may be similar, the depth of "surface damage" is doubled. Failures may occur in service, whereas previous parts had performed quite admirably.

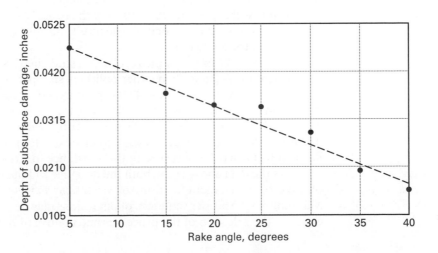

FIGURE 35-22   The depth of damage to the surface of a machined part increases with decreasing rake angle of the cutting tool.

FIGURE 35-23 *(Top)* A cantilever-loaded (bent) rotating beam, showing the normal distribution of surface stresses (i.e., tension at the top and compression at the bottom). *(Center)* The residual stresses induced by roller burnishing or shot peening. *(Bottom)* Net stress pattern obtained when loading a surface-treated beam. The reduced magnitude of the tensile stresses contributes to increased fatigue life.

INFLUENCE OF SURFACE FINISH ON FATIGUE

Fatigue failure occurs as the result of repeated loading at some point typically below the yield strength of the material. Fatigue failures have been shown to almost always nucleate on or near the surface of a component. Fine surface cracks begin at discontinuities (such as microcracks, grooves, ridges, cavities, machining marks, imbedded particles, etc.) at the surface, and the cracks propagate with repeated cyclic loads.

Tensile residual stresses in the altered surface layer have an additive effect on the applied stresses in the component. This means that tensile residual stresses in the material add to external stresses to the component, reducing its fatigue strength. Alternatively, as shown in Figure 35-23, compressive residual stresses subtract from tensile external stresses, and since tensile stresses are those ultimately responsible for fatigue failure, the fatigue strength of the material is increased.

Figure 35-23 shows how residual stresses couple with applied stresses to affect product performance. Suppose that a round beam has a load applied to it so that it is bent while rotating. At the top of the rotation, the surface is in tension, and at the bottom, it is in compression. The result is a condition of cyclic fatigue and the likelihood of a service life limited by fatigue failure. If the part is roller burnished or shot peened, the compressive residual stress pattern of the middle figure is added to the applied stresses, producing the net pattern shown at the bottom. The net effect is a lowering of the peak tensile stress experienced by the surface and a related extension in fatigue life. The specific results will depend on the details of the process. For *shot peening*, the key variables include shot size, shot velocity, exposure time, distance between the nozzle and the surface, and the angle of impact.

Figure 35-24 presents the results of a study in which specimens were prepared by milling and turning and then either polished, shot peened, or roller burnished. If an applied stress between 41,000 and 42,000 psi is experienced in a fatigue application, the difference in fatigue life between a milled specimen and one that has been milled and roller burnished is 610,000 cycles (90,000 cycles as opposed to 700,000 cycles). In essence, roller burnishing serves to induce a sevenfold extension to the fatigue life of the product. Similar results have been observed in the resistance to stress–corrosion cracking.

As the data above show, both the designer and the manufacturer need to be aware of the effects that manufacturing processes can have on the performance of a product. Maintaining the proper sequence of operations may be as important to the surface properties as the selection of the processes and control of the operating parameters.

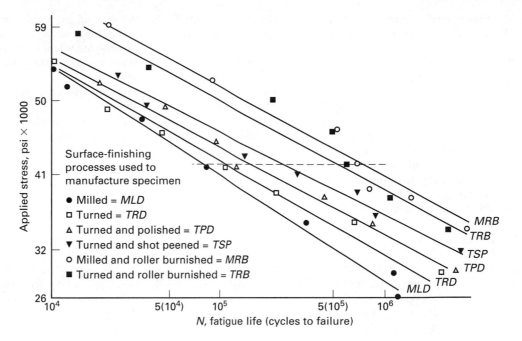

FIGURE 35-24 Fatigue life of rotating beam 2024-T4 aluminum specimens with a variety of surface-finishing operations. Note the enhanced performance that can be achieved by shot peening and roller burnishing.

(within figure)

Surface-finishing processes used to manufacture specimen

- ● Milled = *MLD*
- □ Turned = *TRD*
- △ Turned and polished = *TPD*
- ▼ Turned and shot peened = *TSP*
- ○ Milled and roller burnished = *MRB*
- ■ Turned and roller burnished = *TRB*

Applied stress, psi × 1000

N, fatigue life (cycles to failure)

■ Key Words

abrasive cleaning	chromate	hard chromium plate	shot peening
acid pickling	cladding	hot-dip coating	solvent cleaning
alkaline cleaning	coating	mechanical cleaning	spray painting
anodizing	coil-coated sheets	mechanical plating	surface integrity
barrel burnishing	color anodizing	media	surface roughness
barrel finishing	dipping	paint	textured surfaces
belt sanding	electrocoating	phosphate	tumbling
buffing	electroforming	porcelain enameling	ultrasonic cleaning
burr removal	electroless composite plating	powder coating	vacuum metallizing
case hardening	electroless plating	prime coat	vapor degreasing
centrifugal barrel tumbling	electroplating	residual stresses	vibratory finishing
chemical conversion coating	electropolishing	roller burnishing	waviness
chemical cleaning	electrostatic deposition	roughness	wire brushing
chemical vapor deposition	finishing compounds	sand blasting	

■ Review Questions

1. What are some possible objectives of surface engineering processes?
2. What are some of the factors that should be considered when selecting a surface-modification process?
3. How are the surface and its integrity altered by the process of metalcutting?
4. Two surfaces can have the same microinch roughness but be different in appearance. Explain!
5. What limits the resolution of a stylus-type surface-measuring device in finding profiles?
6. What is the general relationship between surface roughness and tolerance? Between tolerance and cost to produce the surface and/or tolerance?
7. What are some of the sources of foreign material on the surface of manufactured products?
8. What are some the common abrasive media used in blasting or abrasive cleaning operations?
9. What types of quantities and part sizes are most attractive for barrel finishing operations?
10. Describe why there might be an optimum fill level in barrel finishing. How might you find the optimum rotational speed?
11. Describe the primary differences between barrel finishing and vibratory finishing.
12. What are some of the possible functions of the compounds that are used in abrasive finishing operations?
13. How is electropolishing different from electroplating?
14. What are some of the mechanisms of alkaline cleaning, and what types of soils can be removed?
15. What types of surface contaminants cannot be removed by solvent cleaning?
16. In view of its many attractive features, why has vapor degreasing become an unattractive process?
17. What is the primary type of surface contaminant removed by acid pickling?

18. What is the difference between coating and cladding operations?
19. What are some of the reasons that paints may be specified for manufactured items?
20. What are some of the functions of a prime coat in a painting operation? What features are desired in the final coat?
21. What produces atomization and propulsion in airless spraying?
22. What features make industrial robots attractive for spray painting?
23. What are some of the attractive features of electrostatic spraying?
24. Why would it be difficult to apply electrostatic spray painting to products made from wood or plastic?
25. What are some of the metal coatings that can be applied by the hot-dip process?
26. What are the two most common types of chemical conversion coatings?
27. How can nonconductive materials such as plastic be coated by electroplating?
28. What are the attractive properties of hard chrome plate?
29. What are some of the common process variables in an electroplating cell?
30. Why is it difficult to mix parts of differing size and shape in an automated electroplating system?
31. How is electroforming different from electroplating?

32. When anodizing aluminum, what features determine the thickness of the resulting oxide when the oxide is not soluble in the electrolyte? When it is partially soluble?
33. What produces the various colors in the color anodizing process?
34. What are some of the attractive features of electroless plating?
35. What types of particulate composites can be deposited by electroless plating?
36. What is mechanical plating?
37. What are some of the attractive properties of a porcelain enamel coating?
38. How are burrs made by the milling process? See Figure 35-18.
39. What deburring processes are available that were not described in this chapter?
40. Do all machining processes leave a residual stress?
41. Why would the depth of damage (i.e., plastic deformation) increase with negative rake angle cutting tools?
42. What types of surface features or surface modifications result from machining-type processes?
43. Why might processes that produce residual compressive stresses on product surfaces be attractive for mechanically loaded products?

■ Problems

1. Fishermen are among the most superstitious people in the world, and their superstitions affect the type of equipment that they use. As a result, hook manufacturers generally offer their products in a wide range of colors and finishes. Your company manufactures a range of hooks from AISI 1080 carbon steel wire, forming them to precision shape (eye, bends, barbs, and point) and then heat treating them by a quench-and-temper treatment.

   Consider the size and shape of the product and the various properties that are required. The hooks must be strong enough to resist bending, but not so brittle that they might break. They must be corrosion-resistant to both fresh and salt water, and the desired appearance must be provided without fouling the point or the barbs. If the surface is applied before heat treatment, it must endure that process and maintain its appearance. If it is applied afterward, it cannot weaken or embrittle the hook.

   a. Of the various surface-modification processes, which ones might be attractive for such an application? (*Note:* Make sure that the process is appropriate! For example, barrel plating would probably produce a hopelessly snarled mass of wires!)
   b. For each of the possible processes, describe the advantages, limitations, possible colors or finishes, and relative cost.

2. Select one or more of the following products (as directed) and recommend a surface treatment or coating. Consider the appropriateness of the technique to the size, shape, quantity, and material. Cite the specific features that make the recommended treatment the most attractive. What, if any, are the primary limitations or production concerns?

   a. The exterior housing for the motor and drive unit of a chain saw that has been made as a magnesium-alloy die casting.

   b. Large quantities of steel bolts that are intended for use in outdoor construction. They have been fabricated from 4140 steel, have a shank diameter of $1/2$ in., and have been quenched and tempered to a final hardness of Rockwell C 45.
   c. The handle of a household utility knife (retractable-blade cutter) has been made as a two-part zinc die casting.
   d. The scoop portion of an inexpensive ice cream scoop that has been made as a zinc die casting.
   e. A decorative handle for a kitchen cabinet that is made as a zinc die casting.
   f. The exterior of an office filing cabinet that has been made from low-carbon steel sheet.
   g. A high-quality combination wrench (open-end and box-end) that has been forged from 4147 steel bar stock.
   h. The case of a moderately priced wristwatch that has been fabricated from yellow brass (to have a gold appearance).
   i. Tubular frame of a lightweight bicycle that has been made from age-hardened aluminum.
   j. The basket section of a grocery-store shopping cart that has been fabricated from welded steel mesh.
   k. The exterior of an automobile muffler to be fabricated from steel sheet. Describe how the coating treatment might best be integrated into the fabrication sequence.
   l. An inexpensive interior door knob that has been fabricated from deep-drawn cartridge brass sheet.
   m. High-quality steel sockets for a socket-wrench set. These have been forged from AISI 4145 bar stock and subsequently heat treated by a quench-and-temper process.
   n. Refrigerator door panels that have been fabricated from textured AISI 1010 steel sheet.
   o. The interior and exterior surfaces of a 1000-gallon water storage tank that is fabricated by welding 5000 series

aluminum plates. The water will be held at room temperature and is intended for human consumption.

p. A standard office paper clip.

q. A flashlight case that has been fabricated from deep-drawn yellow brass sheet.

r. High-speed drill bits that have been fabricated from Ml tool steel.

s. Injection-molded ABS plastic wheel covers for cars that are intended to look like chrome-plated metal.

t. Inexpensive household scissors that have been cast from gray cast iron.

u. The blade of a high-quality screwdriver that has been forged from AISI 1053 steel and quenched and tempered to Rockwell C 55.

v. The exterior of high-quality, thick-walled cast aluminum cookware.

w. A bathroom sink basin made from deep-drawn 1008 steel sheet.

x. The body section of a child's toy wagon that has been deep-drawn from 1008 steel sheet.

www.wiley.com/college/degarmo

Chapter 35 CASE STUDY

Dana Lynn's Fatigue Lesson

Dana Lynn has just come from her INSY 3000 lab in the manufacturing processes course.

The purpose of this lab was to introduce her to metal fatigue and its basic principles. Metal fatigue arises from the cyclic loading below the yield strength. It can be greatly influenced by the surface finish applied to the metal. Dr. Payton, her instructor, said fatigue most likely accounts for 90% of all mechanical failures, so it is important for an engineer to understand how materials respond to fatigue conditions.

The procedure of this lab was for each student to finish the aluminum specimen with the emery paper, then load the specimen into the fatigue machine (see Figure CS.35A). Students were required to record the number of cycles that was needed to fracture the specimen. It was important that the student wipe the specimen clean and inspect the specimen for burrs, ridges, or flats before inserting it into the drive spindle. The presence of stress raisers can decrease the cumulative number of cycles needed to start a fatigue crack and cause it to fail, hence reducing the fatigue life and static strength of the metal. Small surface cracks, surface flows, or machining marks are examples of stress raisers. Therefore, it is important that one strive to eliminate stress raiser or surface flaws in the specimen that will be exposed to cyclic loadings. The experimental factors for this experiment were the surface finish (grit size of energy paper), applied stress level, and the direction of the surface finish (parallel or perpendicular to the specimen axis). Dr. Payton explained that repeated applications of stress can cause metals to fracture, even if all of the stresses are less than the yield tensile strength and less than the ulti-

mate strength of the material. Surface conditions can heavily influence fatigue life because most fatigue cracks start at the surface of a metal. The fatigue data in Table CS 35A is to be analyzed using statistical experimental design techniques and summarized in the ANOVA table. Factor A was the applied load (50 and 55 lbs), factor B was the treatment (fine and coarse emery paper), and factor C was the surface finish direction (horizontal and vertical). The experiment was replicated so a total of 16 tests were run by the class (2 × 2 × 2 × 2). The specimen was tapered so that the stress acting along the test section (shaded) is constant. To make the specimen, you have to calculate θ and the diameter d (see Figure CS.35B).

Here are some of the questions that Dana Lynn had to answer.

1. Analyze the fatigue data collected by the class using statistical experiment (factorial experiment) design techniques. Include all calculations and summarize your results in an analysis of variance (ANOVA) table.

2. What are the effects of stress level, surface finish (grit size in our experiment) and surface finish direction. Are there any interactions between them on fatigue life? Discuss.

3. What effect would abrupt surface changes, such as tool marks or surface flaws, have on fatigue life?

4. The cylindrical specimen is designed such that the applied stress acting along the test section is constant. Why?

5. What is the endurance limit for a metal? Why is the endurance limit an important criteria in many design applications? Does aluminum have an EL?

Continued on next page

TABLE CS.35A	Data from 16 Fatigue Tests			
	Horizontal		**Vertical**	
	FINE	COARSE	FINE	COARSE
50 lb-in.	167000	126700	102200	88600
	145600	116800	78600	92600
55 lb-in.	89600	61300	56500	49400
	98800	59800	63200	41200

Figure CS 35A Rotary beam fatigue testing machine, used for cylindrical specimens.

$$\sigma_{max} = \frac{MC}{I} \qquad \theta = -----$$

Figure CS 35B Cylindrical specimen with constant stress section. Determine the angle θ and the diameter d.

CHAPTER 36

QUALITY ENGINEERING

■ 36.1 INTRODUCTION

All manufacturing processes display some level of variation. No two items coming from the process will be exactly the same. The primary objective of quality engineering is the systematic reduction of variability, as shown schematically in Figure 36-1. Variability is measured by sigma, σ, the standard deviation, which decreases with the reduction in variability. Early on, acceptance sampling techniques were used to screen incoming goods. This was followed by statistical process control (SPC) efforts, which gave way to companywide quality control (CWQC) and Total Quality Control (TQC) programs. Variation can be further reduced by the application of statistical techniques, like multiple variable analysis, designed experiments, and Taguchi methods, techniques that are a routine part of six sigma efforts many companies are implementing. The drive toward zero defects has been led by Toyota, which achieved exceptional levels of quality by redesigning the manufacturing system so that each step in the making of the car and all

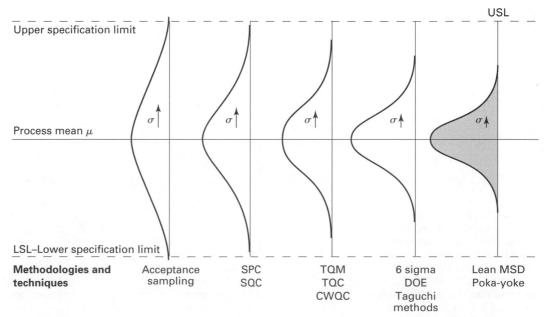

FIGURE 36-1 Over many years, many techniques have been used to reduce the variability in products and processes.

969

its components is checked before the part moves to the next step or stage in its manufacturing sequence. This system redesign is called lean manufacturing.

In a manufacturing process, the variation may be due to "chance causes" that produce random variations—these causes are said to be inherent and represent a stable source of variation. In addition, there are "assignable causes" of variation that can be detected and eliminated to help improve the process. For example, suppose that we view shooting at a metal target as a "process" for putting holes in a piece of metal. I hand you the gun and tell you to take nine shots at the bull's-eye. Figure 36-2 shows some possible results. You are the operator of the process. To measure the *process capability* (PC)—that is, your ability to consistently hit the bull's-eye you are aiming at—the target is inspected after you have finished shooting. So the capability of manufacturing processes is determined by measuring the output of the process. In *quality control* (QC), the product is examined to determine whether or not the processing accomplished was what was specified by the designer in the design, usually the nominal size and the tolerance. Of course, there are many other aspects of quality that quality engineers must address, such as performance, reliability, durability, aesthetics, and more. In this chapter, we will concentrate on the quality of conformance, meaning how well the product conforms to the specifications. And within that area, we will concentrate on PC studies that are directed at the machine tools used in the processing rather than the quality of the output or products from the processes. Going back to our example, a PC study would quantify the inherent accuracy and precision in the shooting process. Accuracy is reflected in your aim (the average of all your shots), whereas precision reflects the repeatability of the process. The objective is to root out problems that can cause defective products during production. Traditionally, the objective has been to find defects in the process. The more progressive point of view is to design the process to prevent the problems that can cause defective products from occurring during production.

■ 36.2 Determining Process Capability

The *nature of the process* refers to both the *variability* (or inherent uniformity) and the *accuracy* or the *aim* of the process. Thus in the target-shooting example, a perfect process would be capable of placing nine shots right in the middle of the bull's-eye, one right on

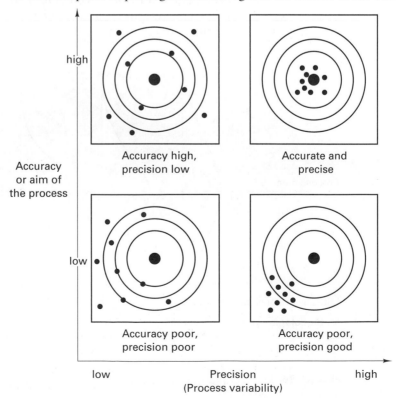

FIGURE 36-2   The concepts of accuracy (aim) and precision (repeatability) are shown in the four target outcomes. Accuracy refers to the ability of the process to hit the true value (nominal) on the average, while precision is a measure of the inherent variability of the process.

top of the other. The process would display no variability with perfect *accuracy*. Such performance would be very unusual in a real industrial process. The variability may have assignable causes and may be correctable if the cause can be found and eliminated. That variability to which no cause can be assigned and which cannot be eliminated is said to be inherent in the process and is therefore its nature.

Some examples of assignable causes of variation in processes include multiple machines for the same components, operator blunders, defective materials, or progressive wear in the tools during machining. Sources of inherent variability in the process include variation in material properties, operator variability, vibrations and chatter, and the wear of the sliding components in the machine, perhaps resulting in poorer operation of the machine. These kinds of variations, which occur naturally in processes, usually display a random nature and often cannot be eliminated. In quality control terms, these are referred to as *chance causes*. Sometimes the causes of assignable variation cannot be eliminated because of cost. Almost every process has multiple causes of variability occurring simultaneously, so it is extremely difficult to separate the effects of the different sources of variability during the analysis.

MAKING PC STUDIES BY THE TRADITIONAL METHODS

The object of the PC study is to determine the inherent nature of the process as compared to the desired specifications. The output of the process must be examined under normal conditions, or what is typically called *hands-off conditions*. The inputs (e.g., materials, setups, cycle times, temperature, pressure, and operator) are fixed or standardized. The process is allowed to run without tinkering or adjusting, while the output (i.e., the product or units or components) is documented with respect to (1) time, (2) source, and (3) order of production. A sufficient number of data have to be taken to ensure confidence in the statistical analysis of the data. The capability of the gage (its precision) used to measure the products must exceed the expected tolerance on the part by one order of magnitude. (See the discussion of the rule of 10 in Chapter 10.)

Prior to any data collection, these steps must be taken:

1. Design the PC experiment (standard method). Use normal or hands-off process conditions; specify machine settings for speed, feed, volume, pressure, material, temperature, operator, and so on.

2. Define the inspection method and the inspection means (the procedure and the instrumentation). In selecting the gage, consider these aspects:

   a. Features that the gage will be checking

   b. Speed or rate of operation

   c. Level of accuracy and precision

   d. Skill of the operator

   e. Portability of gages or part, or both

   f. Environment (clean and stable, cutting fluids)

   g. Workpiece (clean, lubricants present)

   h. Cost (initial, maintenance, daily)

3. Decide how many items (measurements) will be needed to perform the statistical analysis.

4. For a standard PC study, use homogeneous input material, and try to contrast it with normal (more variable) input material.

5. Data sheets must be designed to record date, time, source, order of production, and all the process parameters being used (or measured) while the data are being gathered.

6. Assuming that the standard PC study approach is being used, the process is run, and the parts are made and measured.

Now follow the steps outlined in Figure 36-3. Assume that the designer specified the part to be 1.000 ± 0.005 in. After manufacturing engineering has developed a process plan, some units are manufactured according to the process plan without any adjustment of the process. Each unit is measured, and the data are recorded on the data sheet.

FIGURE 36-3 The process capability study compares the part as made by the manufacturing process to the specifications called for by the designer. Measurements from the parts are collected for run charts and for histograms for analysis—see Figure 36-4.

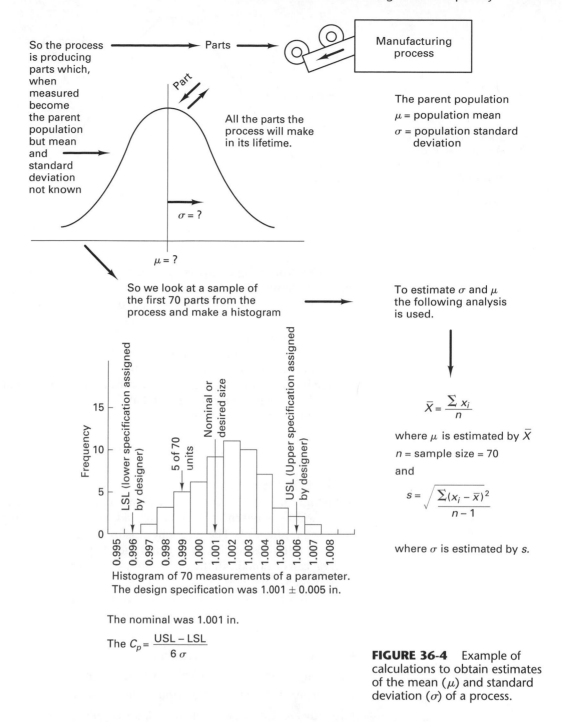

FIGURE 36-4 Example of calculations to obtain estimates of the mean (μ) and standard deviation (σ) of a process.

A frequency distribution, in the form of a *histogram*, or a *run chart* is developed. This histogram shows the raw data and the desired value, along with the upper and lower *specification limits*, where LSL represents lower specification limit and USL the upper specification limit. The statistical data are used to estimate the mean and the standard deviation of this distribution. The run chart shows the same data, but here the data are plotted against time.

The mechanics of this statistical analysis are outlined in Figure 36-4. The true mean of the distribution, designated μ, is to be compared with the nominal value specified by the designer. The estimate of the true *standard deviation*, designated σ (sigma), is used to determine how the process compares with the desired tolerance. *The purpose of the analysis is to obtain estimates of μ and σ values, the true process parameters, because they are not known.*

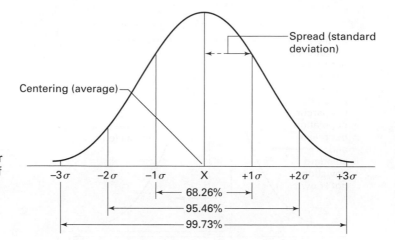

FIGURE 36-5 The normal or bell-shaped curve with the areas within $\pm 1\sigma$, $\pm 2\sigma$, and $\pm 3\sigma$ for a normal distribution; 68.26% of the observations will fall within $\pm1\sigma$ from the mean, and 99.73% will fall within $\pm 3\sigma$ from the mean.

The process capability is defined by $\pm 3\sigma$ or 6σ. Thus $\mu \pm 3\sigma$ defines the natural capability limits of the process, assuming the process is approximately normally distributed. Note that a distinction is made between a sample and a population. A sample is of a specified, limited size and is drawn from the population. The population is the large source of items, which can include all the items the process will ever produce under the specified conditions. Our calculations assume that this distribution was normal or bell-shaped. Figure 36-5 shows a typical normal curve and the areas under the curve as defined by the standard deviation. Other distributions, shown in Figure 36-6, are possible, but the histogram clearly suggested that this process can best be described by a normal probability distribution. Now it remains for the process engineer and the operator to combine their knowledge of the process with the results from the analysis in order to draw conclusions about the ability of this process to meet specifications.

HISTOGRAMS

A histogram is a representation of a frequency distribution that uses rectangles whose widths represent class intervals and whose heights are proportional to the corresponding frequencies. The frequency histogram is a type of diagram in which data are grouped into cells (or intervals), and the frequency of observations falling into each interval can be noted. All the observations within a cell are considered to have the same value, which is the midpoint of the cell. So, a histogram is a picture that describes the variation in a process. It is good to have this visual impression of the distribution of values, along with the mean and standard deviation. Histograms are used in many ways in QC, for example:

• To determine the process capability (central tendency and dispersion)
• To compare the process with the specifications
• To suggest the shape of the population (e.g., normality)
• To indicate discrepancies in data, such as gaps

There are several types of histograms. A histogram shows either absolute frequency (actual occurrence) or relative frequency (percentage). Cumulative histograms show cumulative frequency and reliability cumulative frequency. Each type has its own advantages and is used in different situations. Figure 36-7 shows frequency versus location for 150 measurements. The aim (accuracy) of the process is a bit low, but all the data are well within the tolerances. The disadvantage of the histogram is that it does not show trends and does not take time into account. We can take the data from the histogram and spread them out over time to create a run chart or run diagram.

RUN CHART OR DIAGRAM

A run diagram is a plot of a quality characteristic as a function of time. It provides some idea of general trends and degree of variability. Run charts reveal information that

Common probability distributions

Distribution	Form	Probability function	Comments
Normal		$y = \dfrac{1}{\sigma\sqrt{2\pi}}\,e-\dfrac{(X-\mu)/^2}{2\sigma^2}$ μ = Mean σ = Standard deviation	Applicable when there is a concentration of observations about the average and it is equally likely that observations will occur above and below the average. Variation in observations is usually the result of many small causes
Exponential		$y = \dfrac{1}{\mu}\,e^{-\frac{x}{\mu}}$	Applicable when it is likely that more observations will occur below the average than above
Weibull		$y = \alpha\beta(X-\gamma)^{\beta-1}e^{-\alpha(x-\gamma)^\beta}$ α = Scale parameter β = Shape parameter γ = Location parameter	Applicable in describing a wide variety of patterns of variation, including departures from the normal and exponential
Poisson*		$y = \dfrac{(np)^r e^{-np}}{r!}$ n = Number of trials r = Number of occurrences p = Probability of occurrence	Same as binomial but particularly applicable when there are many opportunities for occurence of an event but a low probability (less than 0.10) on each trial
Binomial*		$y = \dfrac{n!}{r!(n-r)!}\,p^r q^{n-r}$ n = Number of trials r = Number of occurrences p = Probability of occurrence $q = 1-p$	Applicable in defining the probability of r occurrences in n trials of an event that has a probability of occurrence of p on each trial

* = discrete distributions but shown as curves for ease of comparison

FIGURE 36-6 Common probability distributions that can be used to describe the outputs from manufacturing processes. (*Source:* Quality Control Handbook, *3rd ed.*)

histograms cannot, such as certain trends over time or at certain times of day. Individual measurements (not samples) are taken at regular time intervals, and the points are plotted on a connected line graph as a function of time. The graph can be used to find obvious trends in the process, as shown in Figure 36-8, a run diagram with measurements made every hour over four shifts.

Run diagrams are very important at startup to identify the basic nature of a process. Without this information, one may use an inappropriate tool in analyzing the data. For example, a control chart or histogram might hide tool wear if frequent tool changes and adjustments are made between groups of observations. As a result, run diagrams (with 100% inspection where feasible) should always precede the use of control charts for averages and ranges.

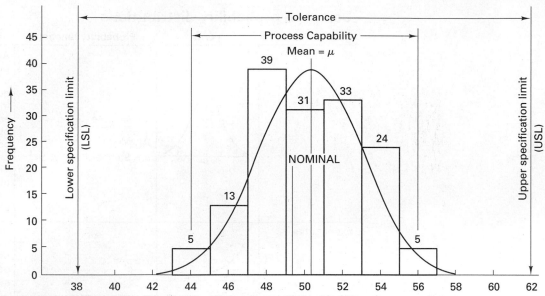

FIGURE 36-7   Histogram shows the output mean μ from the process versus nominal and the tolerance specified by the designer versus the spread as measured by the standard deviation σ. Here nominal = 49.2, USL = 60, LSL = 38, μ = 50.2, σ = 2.

The lengths of manufactured components are measured. A run diagram is constructed to determine how the process is behaving. During 34 hours, measurements are made every hour (60 minutes) and plotted in the order that rack bars are produced.

First shift	35	40	27	30	30	34	26	31
Second shift	24	23	28	15	23	17	16	21
Third shift	15	13	28	8	20	9	5	11
First shift	16	5	9	13	16	10	9	10

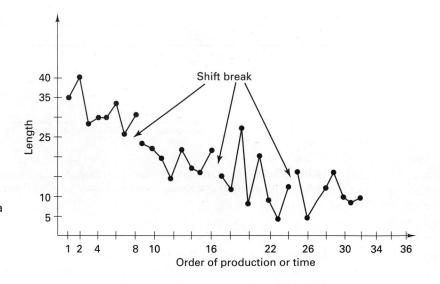

FIGURE 36-8   An example of a run chart or graph, which can reveal trends in the process behavior not shown by the histogram.

PROCESS CAPABILITY INDEXES

The most popular PC index tells you if the process has the ability to meet specifications. This process capability index, C_p, is often computed as follows:

$$C_p = \frac{\text{tolerance spread}}{6\sigma} = \frac{\text{USL} - \text{LSL}}{6\sigma} \qquad (36\text{-}1)$$

A value of $C_p \geq 1.33$ is considered good

The example in Figure 36-7 has

$$C_p = \frac{\text{USL} - \text{LSL}}{6\sigma} = \frac{24}{12} = 2 \tag{36-2}$$

The process capability ratio, C_p, does not, however, take into account the location of the process mean, μ, with respect to the nominal or the specifications. C_p mearly looks at the variability or spread of process (compared to specifications) in terms of sigmas. So another process capability ratio has been developed for off-center processes. This ratio is called C_{pk}, where

$$C_{pk} = \min C_{pu}, C_{pl} \tag{36-3}$$

$$= \min\left(C_{pu} = \frac{\text{USL} - \mu}{3\sigma}, C_{pl} = \frac{\mu - \text{LSL}}{3\sigma} \right)$$

C_{pk} is simply a one-sided ratio for the specification nearest to the process average, μ. To compare the two, look at Figure 36-9. All the histograms have the same standard deviation ($\sigma = 2$) and the same USL − LSL specifications (62−37). For Figure 36-9a, the two indexes are the same. For Figure 36-9b, the mean has shifted to $\mu = 54$ so $C_{pk} = (62-54)/3(2) = 1.5$.

Can you calculate C_{pk} for (c), (d), and (e)? Answers are on the figure.

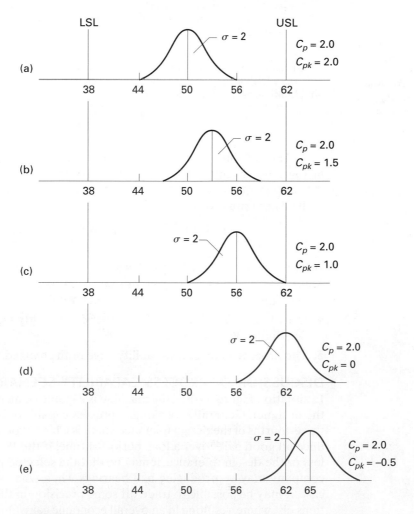

FIGURE 36-9   The output from the process is shifting toward the USL, which changes the C_{pk} ratio but not the C_p ratio.

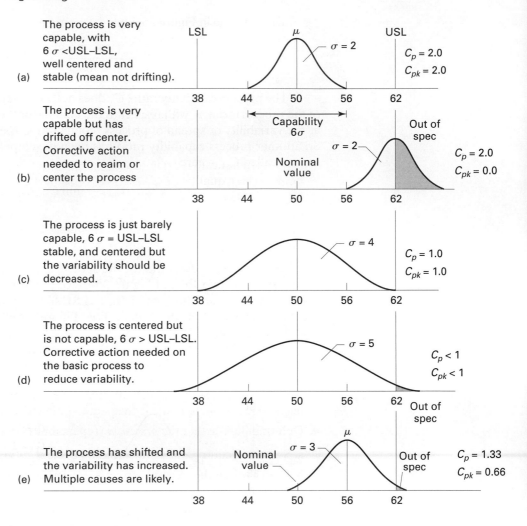

FIGURE 36-10 Five different scenarios for a process output versus the designer's specifications for the minimal (50) and upper and lower specifications of 65 and 38 respectively.

The capability indexes can tell you about the variance, where the width of the histogram is compared with the specifications (Figure 36-10). The natural spread of the process, 6σ, is computed and is then compared with the upper and lower tolerance limits. These situations can exist:

1. $6\sigma <$ USL $-$ LSL *or* $C_p > 1$, or process variability less than tolerance spread. See Figure 36-10a.

2. $6\sigma <$ USL $-$ LSL but process has shifted. See Figure 36-10b.

3. $6\sigma <$ USL $-$ LSL *or* $C_p = 1$, or process variability is just equal to tolerance spread. See Figure 36-10c.

4. $6\sigma <$ USL $-$ LSL *or* $C_p < 1$, or process variability is greater than tolerance spread. See Figure 36-10d.

5. The process mean and variability have both changed. See Figure 36-10d.

DISCUSSION OF PROCESS CAPABILITY SCENARIOS

In situation Figure 36-10a, the machine is capable of meeting the tolerances applied by the designer. Generally speaking, if process capability is on the order of two-thirds to three-fourths of the design tolerance, there is a high probability that the process will produce all good parts over a long period of time. If the PC is on the order of one-half or less of the design tolerance, it may be that the selected process is too good; that is, the company may be producing ball bearings when what is called for is marbles. In this case, it may be possible to trade off some precision in this process for looser specifications elsewhere, resulting in an overall economic gain. Quality in well-behaved processes can be maintained by checking the first, middle, and last part of a lot or production run. If these parts are good, then the lot is certain to be good. This is called $n = 3$.

Naturally, if the lot size is 3 or less, this is 100% inspection. Sampling and control charts are also used under these conditions to maintain the process aim and variability.

When the process is not capable of meeting the design specifications, there are a variety of alternatives, including the following:

1. Shifting this job to another machine with greater process capability.

2. Getting a review of the specifications to see if they may be relaxed.

3. Sorting the product to separate the good from the bad. This entails 100% inspection of the product, which may not be a feasible economic alternative unless it can be done automatically. Automatic sorting of the product on a 100% basis can ensure near-perfect quality of all the accepted parts. The automated station shown in Figure 36-11 checks parts for the proper diameter with the aid of a linear variable differential transformer (LVDT). As a part approaches the inspection station on a motor-driven conveyor system, a computer-based controller activates a clamping device. Embedded in the clamp is an LVDT position sensor with which the control computer can measure the diameter of the part. Once the measurement has been made, the computer releases the clamp, allowing the part to be carried away. If the diameter of the part is within a given tolerance, a solenoid-actuated gate operated by the computer lets the part pass. Otherwise, the part is ejected into a bin. With the fast-responding LVDT, 100% of manufactured parts can be automatically sorted quickly and economically.

   Sorting to find defects by automatic inspection is bad because you already paid to produce the defects. Also, automated sorting does not determine what caused the defects, so this example is an "automated defect finder." How would one change this inspection system to make it "inspect to prevent" the defect from occurring?

4. Determining whether the *precision* (repeatability) of the process can be improved by:

   a. Switching cutting tools, workholding devices, or materials

   b. Overhauling the existing process and/or developing a preventive maintenance program

   c. Finding and eliminating the causes of variability, using cause-and-effect diagrams

   d. Combinations of (a), (b), and (c)

   e. Using designed experiments and Taguchi methods to reduce the variability of the process

   In Figure 36-10c, the process capability is almost exactly equal to the assigned tolerance spread, so if the process is not perfectly centered, defective products will always result. Thus, this situation should be treated like the situation in Figure 36-10d unless the process can be perfectly centered and maintained. Tool wear, which causes the distribution to shift, must be negligible.

FIGURE 36-11   A linear variable differential transformer (LVDT) is a key element in an inspection station checking part diameters. Momentarily clamped into the sensor fixture, a part pushed the LVDT armature into the device winding. The LVDT output is proportional to the displacement of the armature. The transformer makes highly accurate measurements over a small displacement range.

PC studies can evaluate the ability of the process to maintain centering so that the average of the distribution comes as close as possible to the desired nominal value. Most processes can be re-aimed. Poor accuracy is often due to assignable causes, which can be eliminated.

In addition to direct information about the accuracy and the precision of the process, PC studies can also tell the manufacturing engineer how pilot processes compare with production processes, and vice versa. If the source and time of the manufacture of each product are carefully recorded, information about the instantaneous reproducibility can be found and compared with the repeatability of the process with respect to time (time-to-time variability). More important, since almost all processes are duplicated, PC studies generate information about machine-to-machine variability. Going back to our target-shooting example, suppose that nine different guns were used, all of the same make and type. The results would have been different, just as having nine marksmen use the same gun would have resulted in yet another outcome. Thus, PC studies generate information about the homogeneity and the differences in multiple machines and operators.

It is quite often the case in such studies that one variable dominates the process. Target shooting viewed as a process is "operator dominated" in that the outcome is highly dependent on the skill (the capability) of the "worker." Processes that are not well engineered or highly automated, or in which the worker is viewed as "highly skilled," are usually operator dominated. Processes that change or shift uniformly with time but that have good repeatability in the short run are often machine dominated. For example, the mean of a process (μ) will usually shift after a tool change, but the variability may decrease or remain unchanged. Machines tend to become more precise (to have less variability within a sample) after they have been "broken in" (i.e., the rough contact surfaces have smoothed out because of wear) but will later become less precise (will have less repeatability) due to poor fits between moving elements (called *backlash*) of the machine under varying loads. Other variables that can dominate processes are setup, input parameters, and even information.

In many machining processes in use today, the task of tool setting has been replaced by an automatic tool-positioning capability, which means that one source of variability in the process has been eliminated, making the process more repeatable. In the same light, it will be very important in the future for manufacturing engineers to know the process capability of robots they want to use in the workplace.

The discussion to this point has assumed that the *parent population* is normally distributed, that is, has the classic bell-shaped distribution in which the percentages (shown in Figure 36-5) are dictated by the number of standard deviations from the central value or mean. The shape of the histogram may reveal the nature of the process to be skewed to the left or the right (unsymmetrical), often indicating some natural limit in the process. Drilled holes exhibit such a trend, as the drill tends to make the hole oversize. Another possibility is a bimodal distribution (two distinct peaks), often caused by two processes being mixed together. Suppose you had a manufacturing cell as used in lean production and you were using a *check sheet* to gather data on one of the operations. See Figure 36-12. The time to perform an assembly task is being recorded. Why do you think the data are bimodal? In the cell design, shown in Chapter 1, Figure 1-9, operation 7 (caulking) was shared by two operators. The data would suggest that they do not do the assembly process the same way, since there appears to be a 5- or 6-second difference in their average time to complete this task. Clearly further study is needed to determine what the real problem was. For the next check sheet, use a different symbol for each worker.

The check sheet is an excellent way to view data while it is being collected. It can be constructed using predetermined parameters based on experience with the cell or system. The appropriate interval is checked as the data are being collected. This often allows the central tendency and the spread of the data to be seen. The check sheet can provide basically the same information as the histogram, but it is easier to build (once the check sheet is formatted). The possibilities are endless and require a careful recording of all the sources of the data to track down the factors that result in loss of precision and accuracy in the process. Rapid feedback on quality is perhaps the most important factor, so these data-gathering tasks are done right on the factory floor, by the operators.

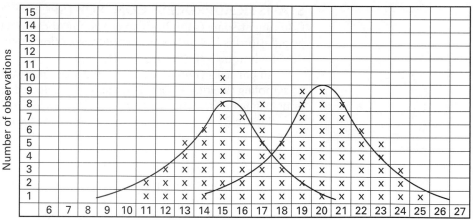

FIGURE 36-12 Example of a check sheet for gathering data on a process.

Time (in seconds) to perform an assembly task in a cell.

■ 36.3 INSPECTION TO CONTROL QUALITY

In virtually all manufacturing, it is extremely important that the dimensions and quality of individual parts be known and maintained. This is of particular importance where large quantities of parts, often made in widely separated plants, must be capable of interchangeable assembly. Otherwise, difficulty may be experienced in subsequent assembly or in service, and costly delays and failures may result. In recent years, defective products resulting in death or injury to the user have resulted in expensive litigation and damage awards against manufacturers. Inspection is the function that controls the quality (e.g., the dimensions, the performance, and the color) manually, by using operators or inspectors, or automatically, with machines, as discussed previously.

The economics-based question "How much should be inspected?" has three possible answers:

1. *Inspect every item being made.* 100% inspect every item being made; 100% checking with prompt execution of feedback and immediate corrective action can ensure perfect quality.

2. *Sample.* Inspect some of the product by sampling and make decisions about the quality of the process based on the sample.

3. *None.* Assume that everything made is acceptable or that the product is inspected by the consumer, who will exchange it if it is defective. (This is not a recommended procedure).

The reasons for not inspecting all of the product (i.e., for sampling) include the following:

1. Everything has not yet been manufactured—the process is continuing to make the item—so we have to look at some before we are done with all.

2. The test is destructive.

3. There is too much product for all of it to be inspected.

4. The testing takes too much time or is too complex or too expensive.

5. It is not economically feasible to inspect everything even though the test is simple, cheap, and quick.

Some characteristics are nondissectible; that is, they cannot be measured during the manufacturing process because they do not exist until after a whole series of operations have taken place. The final edge geometry of a razor blade is a good example, as is the yield strength of a rolled bar of steel.

Sampling (looking at some percentage of the whole) requires the use of statistical techniques that permit decisions about the acceptability of the whole based on the quality found in the sample. This is known as *statistical process control*.

STATISTICAL PROCESS CONTROL (SPC)

Looking at some (sampling) and deciding about the behavior of the whole (the parent population) is common in industrial inspection operations. The most widely used basic SPC technique is the control charts.

In particular, *control charts for variables* are used to monitor the output of a process by sampling (looking at some), by measuring selected quality characteristics, by plotting the sample data on the chart, and then by making decisions about the performance of the process.

Figure 36-13 shows the basic structure of two charts commonly used for variable types of measurements. The $\overline{X}$ chart tracks the aim (accuracy) of the process. The R chart (or σ chart) tracks the precision or variability of the process. Usually, only the $\overline{X}$ chart and the R chart are used unless the sample size is large, and then σ charts are used in place of R charts. For each sample, the following calculations are made:

$$(\text{mean})\overline{X} = \frac{\sum x_i}{n} \tag{36-4}$$

$$n = \text{sample size}$$

$$(\text{range})R = x_{\text{HIGH}} - x_{\text{LOW}} \tag{36-5}$$

Sometimes the standard deviation is calculated:

$$(\text{sigma})\sigma = \sqrt{\frac{\sum(x_i - \overline{x})^2}{n-1}} \tag{36-6}$$

$$n = \text{sample size}$$

After k samples, calculate the average range $\overline{R}$ and the process average $\overline{\overline{X}}$

$$\overline{R} = \frac{R_1 + R_2 + \dots R_k}{k} \qquad \overline{\overline{X}} = \frac{\overline{X}_1 + \overline{X}_2 + \dots \overline{X}_k}{k}$$

$$UCL_{\overline{X}} = \overline{\overline{X}} + A_2\overline{R} = \overline{\overline{X}} + 3\sigma_{\overline{X}}$$

$$LCL_{\overline{X}} = \overline{\overline{X}} - A_2\overline{R} = \overline{\overline{X}} - 3\sigma_{\overline{X}}$$

$$UCL_R = D_4\overline{R} \qquad LCL_R = D_3\overline{R}$$

$$3\sigma_{\overline{X}} = 3\frac{\sigma}{\sqrt{n}} = 3\frac{\overline{R}}{d_2\sqrt{n}} = A_2\overline{R}$$

FIGURE 36-13   Quality control chart calculations. On the charts, plot $\overline{X}$ and R values over time. The constants for calculating UCL and LCL values for the $\overline{X}$ and R charts are based on 3 standard deviations.

Sample Size	$\overline{X}$	R Chart		Est.	σ'
(n)	A_2	D_3	D_4	d_2	c_2
2	1.88	.00	3.27	1.13	.56
3	1.02	.00	2.57	1.69	.72
4	0.73	.00	2.28	2.06	.80
5	0.58	.00	2.11	2.33	.84
6	0.48	.00	2.00	2.53	.87
7	0.42	.08	1.92	2.70	.89
8	0.37	.14	1.86	2.85	.90
9	0.34	.18	1.82	2.97	.91
10	0.31	.22	1.78	3.08	.92

The samples are drawn over time.

Because some sample statistics tend to be normally distributed about their own mean, $\overline{X}$ values are normally distributed about $\overline{\overline{X}}$, R values are normally distributed about $\overline{R}$, and σ values are normally distributed about $\overline{\sigma}$.

Quality control charts are widely used as aids in maintaining quality and in achieving the objective of detecting trends in quality variation before defective parts are actually produced. These charts are based on the previously discussed concept that if only chance causes of variation are present, the deviation from the specified dimension or attribute will fall within predetermined limits.

When sampling inspection is used, the typical sample sizes are from 3 to about 12 units. The $\overline{X}$ chart tracks the sample averages ($\overline{X}$ values). The R chart plots the range values (R values). Figure 36-14 shows one example of $\overline{X}$ and R charts for measuring a dimension of a gap on a part called the retainers. Twenty-five samples of size 5 were taken over six days, and this sample data will be used to prepare the control charts.

The centerline of the $\overline{X}$ chart was computed prior to actual usage of the charts in control work:

$$\overline{\overline{X}} = \frac{\sum\limits_{i=1}^{k} \overline{X}}{k} \tag{36-7}$$

where $\overline{X}$ was a sample average and k was the number of sample averages. The horizontal axis for the charts is time, thus indicating *when* the sample was taken. $\overline{\overline{X}}$ serves as an estimate for μ, the true center of the process distribution. $\overline{\overline{X}}$ is also centerline of the $\overline{X}$ chart. The upper and lower control limits are commonly based on 3 standard error units, $3\sigma_{\overline{x}}$ ($\sigma_{\overline{x}}$ is the standard deviations for the distribution of $\overline{X}$'s about $\overline{\overline{X}}$).

Thus,

$$\text{UCL}_{\overline{X}} = \text{upper control limit on } \overline{X} \text{ chart} \tag{36-8}$$

$$= \mu + 3\sigma_{\overline{x}} \text{ or}$$

$$= \mu + A_2\overline{R}$$

$$\text{LCL}_{\overline{X}} = \text{lower control limit, } \overline{X} \text{ chart} = \mu - 3\sigma_{\overline{X}}$$

$$= \mu - A_2 R$$

(see Figure 36-13 for A_2 values). The upper and lower control limits are entered as dashed lines on the chart. The $\overline{X}$ chart is used to track the central tendency (aim) of the process. In this example, the samples were being taken 4 times a day. The R chart is used to track the variability or dispersion of the process. A σ chart could also be used. R is computed for each sample ($x_{\text{HIGH}} - x_{\text{LOW}}$). The value of $\overline{R}$ is calculated as:

$$\overline{R} = \frac{\sum\limits_{i=1}^{k} R}{k}$$

where $\overline{R}$ represents the average range of k range values. The range values are normally distributed about $\overline{R}$, with standard deviation σ_R. To determine the upper and lower control limits for the charts, the following relationships are used.

$$\text{UCL}_R = \text{upper control limit, } R \text{ chart} = \overline{R} + 3\sigma_R = D_4\overline{R} = 2.11\overline{R} \text{ for } n = 5$$

$$\text{LCL}_R = \text{lower control limit} = R = 3\sigma_R = D_3\overline{R} = 0$$

where D_4 and D_3 are constants and are given in Figure 36-13. For small values of n, the distance between centerline R and LCL$_R$ is more than $3\sigma_R$, but LCL$_R$ cannot be negative, as negative range values are not allowed, by definition. Hence, $D_3 = 0$ for values of n up to 6.

VARIABLES CONTROL CHART X&R
Averages & Ranges

Part/Asm. Name *Retainer*	Operation *Bend Clip*	Specification .50 – .90 mm	Nominal Size .70 mm
Part No. 1234567	Department 105	Gage *Depth Gage Micrometer*	
Parameter *Gap. Dim. "A"*	Machine 030	Sample Size/Frequency 5/2 *Hours*	

Date	6/8				6/9				6/10				6/11				6/12				6/15				6/16
Time of day	8	10	12	2	8	10	12	2	8	10	12	2	8	10	12	2	8	10	12	2	8	10	12	2	8
Operator																									
Sample Measurements — Value of X — 1	.65	.75	.75	.60	.70	.60	.75	.60	.65	.60	.80	.85	.70	.65	.90	.75	.75	.75	.65	.60	.50	.60	.80	.65	.65
2	.70	.85	.80	.70	.75	.75	.80	.70	.80	.70	.75	.75	.70	.70	.80	.80	.70	.70	.65	.60	.55	.80	.65	.60	.70
3	.65	.75	.80	.70	.65	.75	.65	.80	.85	.60	.90	.85	.75	.85	.80	.75	.75	.60	.85	.65	.65	.65	.75	.65	.70
4	.65	.85	.70	.75	.85	.85	.75	.75	.85	.80	.60	.65	.75	.75	.75	.80	.70	.70	.65	.60	.80	.65	.65	.60	.60
5	.85	.65	.75	.65	.80	.70	.70	.75	.75	.65	.80	.70	.70	.60	.85	.65	.80	.60	.70	.65	.80	.75	.65	.70	.65
Sum	3.50	3.85	3.80	3.40	3.75	3.65	3.65	3.60	3.90	3.35	3.75	3.80	3.60	3.55	4.10	3.75	3.80	3.35	3.50	3.10	3.30	3.45	3.50	3.20	3.30
Average $\bar{X}$	.70	.77	.76	.68	.75	.73	.73	.72	.78	.67	.75	.76	.72	.71	.82	.75	.76	.67	.70	.62	.66	.69	.70	.64	.66
Range R	.20	.20	.10	.15	.20	.25	.15	.20	.20	.20	.40	.20	.05	.25	.15	.15	.15	.15	.20	.05	.30	.20	.15	.10	.10

$\bar{\bar{X}} = (.70 + .77 + \ldots + .64 + .66)/25 = 17.90/25 = .716$ $UCL_{\bar{X}} = .716 + (.58 \times .178) = .819$

$\bar{R} = 4.45/25 = .176$ $UCL_R = 2.11 \times .178 = .376$ $LCL_R = 0$

FIGURE 36-14 Example of $\bar{X}$ and R charts and the data set of 25 samples [$k = 25$ of size 5 ($n = 5$)].
(*Source : Continuing Process Critical and Process Capability Improvement, Statistical Methods Office, Ford Motor Co., 1985.*)

After control charts have been established, and the average and range values have been plotted for each sample group, the charts act as a control indicator for the process. If the process is operating under chance cause conditions, the data will appear random (will have no trends or pattern). If $\bar{X}$, R, or σ values fall outside the control limits or if nonrandom trends occur (like 7 points on one side of the central line or 6 successive increasing or decreasing points appear), an assignable cause or change may have occurred, and some action should be taken to correct the problem.

Trends in the control charts often indicate the existence of an assignable cause factor before the process actually produces a point outside the control limit. In grinding operations, wheel wear (wheel undersize) results in the parts becoming oversized and corrective action should be taken. (Redress and reset the wheel or replace it with a new wheel.) Note that defective parts can be produced even if the points on the charts indicate the process is in control. That is, it is possible for something to change in the process, causing defective parts to be made, and the sample point still to be within the control limits. Since no corrective action was suggested by the charts, a type II error was made. Subsequent operations will then involve performing additional work on products already defective. Thus the effectiveness of the SPC approach in improving quality is often deterred by the time lag between the discovery of an abnormality and the corrective action.

With regard to control charts in general, it should be kept in mind that the charts are only capable of indicating that something has happened, not what happened, and that a certain amount of detective work will be necessary to find out what has occurred to cause a break from the random, normal pattern of sample points on the charts. Keeping careful track of when and where the sample was taken will be very helpful in such investigations, but the best procedure is to have the operator take the data and run the chart. In this way, quality feedback is very rapid and the causes of defects readily found.

■ 36.4 PROCESS CAPABILITY DETERMINATION FROM CONTROL CHART DATA

After the process is determined to be "under control," the data can be used to estimate the process capability parameters.

As an example, examine the data in Figure 36-14. These are measurements of the gaps in 125 retainers. They are supposed to have a nominal size of 0.70 mm. The population is assumed to be normal, and only chance variations are occurring. (Could you make a histogram of these 125 measurements?) The mean for all the data can be obtained by equation 36-4

$$\overline{X} = \frac{\sum\limits_{i=1}^{125} x_i}{n}$$

where x_i is an individual measurement and the number of items was 125. The mean is a measure of the central tendency around which the individual measurements tend to group. The variability of the individual measurements about the average may be indicated by the standard deviation, σ, where

$$\sigma = \frac{\sqrt{\sum\limits_{i=1}^{n} (x_i - \overline{X})^2}}{n} \quad \text{where } n = 125$$

But more commonly the sample data are used to make determinations of the process control.

A sample size of 5 was used in this example, so $n = 5$. Twenty-five groups or samples were drawn from the process, so $k = 25$. For each sample, the *sample mean* $\overline{X}$ and the *sample range R* are computed. [For large samples *(n > 12)*, the standard deviation of each sample should be computed rather than the range.] Next, the average of the sample averages, $\overline{\overline{X}}$, is computed as shown in Figure 33-14. This is sometimes called the *grand average*. This is used to estimate the mean of the process, μ. The standard deviation of the process, which is a measure of the spread or variability of the process, is estimated from either the average of the sample ranges, $\overline{R}$, or the average of the sample standard deviations, $\overline{\sigma}$, using either $\overline{R}/d_2$ or $\overline{\sigma}/c_2$. The factors d_2 and c_2 depend on the sample size n and are given in the table in Figure 36-13. So μ is estimated by $\overline{\overline{X}}$ and σ is estimated by $\overline{R}/d_2$ or $\overline{\sigma}/c_2$.

The standard deviation of the distribution of the $\overline{X}$ values about $\overline{\overline{X}}$, $\sigma_{\overline{X}}$, is related to the standard deviation of the parent population, σ, by $\sigma_{\overline{X}} = \sigma/\sqrt{n}$.

Now these estimates can be used to determine the process capability of the process in the same way the histogram was used.

■ 36.5 DETERMINING CAUSES FOR PROBLEMS IN QUALITY

The best way to quickly isolate quality problems is to make everyone an inspector. This means every worker, foreman, supervisor, engineer, manager, and so forth is responsible for making it right the first time and every time. One very helpful tool in this effort is the fishbone diagram. As shown in Figure 36-15, the fishbone diagram can be used in conjunction with the control chart to root out the causes of problems. The problem can have multiple causes, but in general, the cause will lie in the process, operators, materials, or method (i.e., the four main branches on the chart). Every time a quality problem is caused by one of these events, it is noted by the observer, and corrective action is taken. As before, experimental design procedures to be discussed later can help identify causes that affect performance.

Cause-and-effect (C&E) diagrams are also known as fishbone diagrams because of their structure. Initially developed by Kaorw Ishikowa in 1943, this diagram organizes theories about the probable cause of a problem. On the main line is a quality characteristic that is to be improved or the quality problem being investigated. Fishbone lines are drawn from the main line. These lines organize the main factors that could have caused the problem. Branching from each of these factors are even more detailed factors. Everyone taking part in making a diagram gains new knowledge of the process. When a diagram serves as a focus for the discussion, everyone knows the topic, and the conversation does not stray. The diagram is often structured around four branches: the machine tools (or processes), the operators (workers), the method, and the material being processed. Another version of the diagram is called the CEDAC, the cause-and-effect diagram with the addition of cards. The effect is often tracked with a control chart. The possible causes of the defect or problem are written on cards and inserted in slots in the charts.

The three main applications of C&E diagrams are as follows:

I. *Cause enumeration:* Every possible cause and subcause is listed.

    a. *Visual presentations* are one of the most widely used graphical techniques for QC.

    b. A better understanding of the relationships within the process yields a better understanding of the process as a whole.

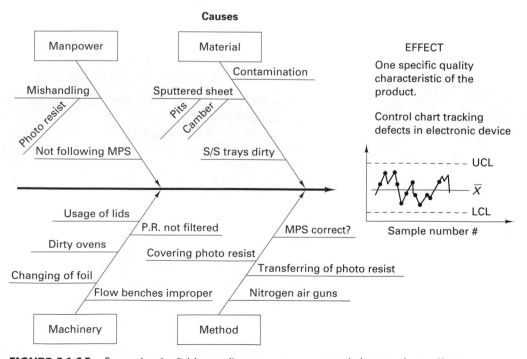

FIGURE 36-15   Example of a fishbone diagram using a control chart to show effects.

II. *Dispersion analysis* involves grouping causes under similar headings; the four M's are men, machines, materials, and methods.

    a. Each *major* cause is thoroughly analyzed.

    b. There is the possibility of not identifying the root cause (may not fall into main categories).

III. *Process analysis* is similar to creating a flow diagram.

    a. Each part of the process is listed in the sequence in which operations are performed.

In summary, data are gathered to develop the early PC studies and to prepare the initial control charts for the process. The removal of all assignable causes for variability and proper setting of the process average requires that operators and engineers work together to find the causes for variation. After the charts have been in place for some time and a large amount of data have been obtained from the process output, the PC study can be redone to obtain better estimates of the natural spread of the process during actual production. On-line or in-process methods such as SPC and off-line (Taguchi) methods are key elements in a total quality control programs.

SAMPLING ERRORS

It is important to understand that in sampling, two kinds of decision errors are always possible; see Figure 36-16. Suppose that the process is running perfectly, but the sample data indicate that something is wrong. You, the quality engineer, decide to stop the process to make adjustments. This is a type I error. Alternately, suppose that the process was not running perfectly and was making defective products. However, the sample data did not indicate that anything was wrong. You decided *not to stop* the process and set it right. This is a type II error. Both types of errors are possible in sampling. For a given sample size, reducing the chance of one type of error will increase the chance of the other. Increasing the sample size or the frequency of sampling reduces the probability of errors but increases the cost of the inspection. It is common practice in control chart work to set the upper and lower control limits at 3 standard deviations. This makes the probability of an alpha error very small and the probability of a beta error quite large. Many companies determine the size of the errors they are willing to accept according to the overall cost of making the errors plus the cost of inspection. If, for example, a type II error is very expensive in terms of product recalls or legal suits, the company may be willing to make more type I errors, to sample more, or even to go to 100% inspection on very critical items to ensure that the company is not accepting defective materials and passing them on to the customer. The inspection should take place immediately after the processing.

As mentioned earlier, in any continuing manufacturing process, variations from established standards are of two types: (1) *assignable-cause variations*, such as those due to malfunctioning equipment or personnel, to defective material, or to a worn or broken tool; and (2) *normal-chance variations*, resulting from the inherent nonuniformities that exist in materials and in machine motions and operations. Deviations due to assignable causes may vary greatly. Their magnitude and occurrence are unpredictable and thus should be prevented. However, if the assignable causes of variation are removed from a given opera-

	The process is running. It really has:	
	Changed (making defects)	Not changed
Based on the sample, you decide the process has changed	No error	Type I α error
Based on the sample, you decide the process has not changed	Type II β error	No error

Type I (α error) Saying the process has changed when it has not changed
Type II (β error) Saying the process has not changed when it has changed

FIGURE 36-16 When you look at some of the output from a process and decide about the whole (i.e., the quality of the process), you can make two kinds of errors.

tion, the magnitude and frequency of the chance variations can be predicted with great accuracy. Thus, if one can be assured that only chance variations will occur, the quality of the product will be better known, and manufacturing can proceed with assurance about the results. By using statistical process control procedures, one may detect the presence of an assignable-cause variation and after investigation to find the cause and remove it before it causes quality to become unacceptable. Also, the astute application of statistical experimental design methods (Taguchi experiments) can help to identify some assignable causes.

To sum up, PC analysis and process improvements help to get the process "under control." Control charts help to keep the process on center ($\overline{X}$ chart) with no increases in variability (R chart or σ chart).

GAGE CAPABILITY

The instrument (gage) used to measure the process will also have some inherent precision and accuracy, often referred to as gage capability. In other words, the observed variation in the component part being measured is really composed of the actual process variation plus the variation in the measured system; see Figure 36-17. The measuring system will display:

- *Bias*: poor accuracy or aim
- *Linearity*: accuracy changes over the span of measurements
- *Stability*: accuracy changes over time
- *Repeatability*: loss of precision in the gage (variability)
- *Reproducibility*: variation due to different operators

Measurements made by different operators will have different means and different variation about the mean when performing a measurement. Determining the capability of the gage is called an *R and R* study. In selecting a gage, the engineer tries to get the variation in the gage (as measured by the standard deviation) to be less than 10% of the total tolerance spread (USL − LSL). This is called the 10% rule, and the *R and R* study can determine the magnitude of the gage variability. Space does not permit a full discussion here of *R and R* studies, but detailed descriptions are found in Montgomery (2001), cited in the reference section at the end of the book. In particular, students involved in ISO9000 studies should examine the *Measurement System Analysis Reference Manual*, published by the Automotive Industry Action Group.

DESIGN OF EXPERIMENTS (DOE) AND TAGUCHI METHODS

Foreign competition has forced American manufacturers to take a second look at quality, as evidenced by the major emphasis (reemphasis) on SPC in American industry. This drive toward superior quality has led to the introduction of Taguchi methods for improvement in products, product design, and processes. Basically, SPC looks at processes and control, the latter loosely implying "improvement." DOE and Taguchi methods, however, span a much wider scope of functions and include the design aspects of products and processes, areas that were seldom, if ever, formally treated from the quality standpoint. Another threshold has been reached in quality control, witnessed by an expanding role of quality in the production of goods and services. The consumer is the central focus of attention on quality, and the methods of quality design and controls have been incorporated into all phases of production.

The Taguchi methods incorporate the following general features:

1. Quality is defined in relation to the total loss to the consumer (or society) from less-than-perfect quality of the product. The methods include placing a monetary value on quality loss. Anything less than perfect is waste.

FIGURE 36-17 Gage capability (variation) contributes to the total observed variation in the measurement of a part.

Variation in the process   +   Variation in the measurement system   =   Total observed variation

2. In a competitive society, continuous quality improvement and cost reduction are necessary for staying in business.

3. Continuous quality improvement requires continuous reduction in the variability of product performance characteristics with respect to their target values.

4. The quality and cost of a manufactured product are determined by the engineering designs of the product and its manufacturing system.

5. The variability in product and process performance characteristics can be reduced by exploiting the nonlinear (interactive) effects of the process or product parameters on the performance characteristics.

6. Statistically planned (Taguchi) experiments can be used to determine settings for processes and parameters that reduce the performance variation.

7. Design and improvement of products and processes can make them *robust*, or less sensitive to uncontrollable or difficult-to-control variations, called *noise* by Taguchi.

In the Taguchi approach, specified combinations of all of the input parameters, at various levels, that are believed to influence the quality characteristics being measured are used. These combinations should be run with the objective of selecting the best level for individual factors. For example, speed levels may be high, normal, and low; and operators may be fast or slow. For a Taguchi approach, material is often an input variable specified at different levels: normal, homogeneous, and highly variable. If a material is not controllable, it is considered a noise factor.

Designed experiments and Taguchi methods can be used as alternative approaches to making a PC study. The Taguchi approach uses a truncated experimental design (called an *orthogonal array*) to determine which process inputs have the greatest effect on process variability (i.e., precision) and which have the least. Those inputs that have the greatest influence are set at levels that minimize their effect on process variability. As shown in Figure 36-18, factors A, B, C, and D all have an effect on process variability V. By selecting a high level of A and low levels of B and C, the inherent variability of the process can be reduced. Those factors that have little effect on process variability,

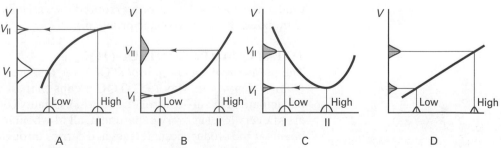

FIGURE 36-18 The use of Taguchi methods can reduce the inherent process variability, as shown in the upper figure. Factors A, B, C, and D versus process variable V are shown in the lower figure.

Selecting high and low values of input parameters changes the process output.

like factor D, are used to adjust or recenter the process aim. In other words, Taguchi methods seek to minimize or dampen the effect of the causes of variability and thus to reduce the total process variability. This is also the goal of the six sigma method, which is discussed next.

The methods are, however, more than just mechanical procedures. They infuse an overriding new philosophy into manufacturing management that basically makes quality the primary issue in manufacturing. The manufacturing world is rapidly becoming aware that the consumer is the ultimate judge of quality. Continuous quality improvement toward perfect quality is the ultimate goal. Finally, it is recognized that the ultimate quality and lowest cost of a manufactured product are determined to a large extent by the engineered designs of the following:

1. The product
2. The manufacturing process technology and the sequences of processes
3. The manufacturing system (integration of the product and the process)

So, a new understanding of quality has emerged. Process variability is not fixed. It can be improved! The noise level of a process can be reduced by exploring the nonlinear effects of the product (or process) parameters on the performance characteristics.

Process capability must be also addressed in the context of machining centers and [programmable numerical control (NC)] machines. In machine tools, accuracy and precision in processing are affected by machine alignment, the setup of the workholder, the design and rigidity (accuracy) of the workholder, the accuracy of the cutting tools, the design of the product, the temperature, and the operating parameters. DOE and Taguchi methods provide a means of determining which of the input parameters are most influential in product quality when the operator is no longer there to compensate for the variability.

MOTOROLA'S SIX SIGMA

In order to meet the quality challenge of the Japanese, an American company, Motorola, developed the six sigma concept. The concept is shown in Figure 36-19 in terms of four sigma and six sigma capability. Most people do not know how sigma is determined (σ is estimated from $\bar{R}/d_2$ or σ/C_2 sample data), that a sigma is a standard deviation, or what sigma measures (sigma measures the repeatability or variability or lack of precision in a process).

In essence, the six sigma concept calls for the process to be improved to the place where there are 12 standard deviations between USL and LSL. As the variability of a process changes, so does sigma. A reduction in sigma (a reduction in spread) reflects an improvement in process or an improvement in precision (better repeatability). As the process is improved, sigma decreases. So the question is: "How do I improve the process?" This is the essence of process capability work outlined in this chapter.

Here is an example. A foundry was having a problem with cores breaking in the molds during pouring. A PC study determined that the core strength was widely variable and that it was the low-strength cores that broke when the molten metal hit them during filling. Increasing the resin content in the cores and changing the gating in the mold did not eliminate the problem, so a Taguchi study was run that revealed that core strength was highly dependent on the grain size of the sand, which was also highly variable. The sand preparation process was revised to yield more uniform-sized grains, which, when used in the core-packing process, reduced the variability in the strength characteristic and eliminated the core breakage problem.

TOTAL QUALITY CONTROL (TQC)

The phrase *total quality control* (TQC) was first used by A. V. Feigenbaum in *Industrial Quality Control* in May 1957. TQC means that all departments of a company must participate in quality control (Table 36-1). Quality control is the responsibility of workers at every level in every department, all of whom have had quality control training. It begins at the product design stages and carries through the manufacturing system, where the emphasis is on making it right the first time. It is surprising how few companies have embarked on implementing Taguchi methods in order to reduce the inherent variability

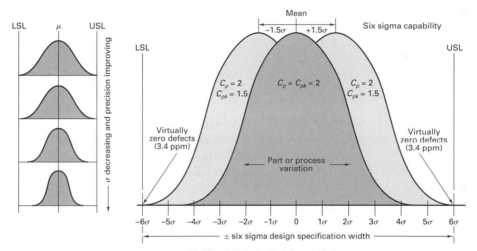

FIGURE 36-19   To move to six sigma capability from four sigma capability requires that the process capability (variability) be greatly improved (σ reduced). The curves in these figures represent histograms or curves fitted to histograms.

in their processes. This is probably because many manufacturing (mechanical) engineers have not had any coursework in this area.

Also, a change in company culture is often needed to give the responsibility for quality to the worker, along with the authority to stop the process when something goes wrong. An attitude of defect prevention and a habit of constant improvement of quality are fundamental to lean production, a system developed by Toyota. They have accomplished TQC (they call it company-wide quality control) by extensive education of the workers, giving them the analysis tools they need (control charts with cause-and-effect diagrams) to find and expose the problems. Workers are encouraged to correct their own errors, and 100% inspection (often done automatically) is the rule. Passing defective products on to the next process is not allowed. The goal is perfection. *Quality circles*, now popular in the United States, are just one of the methods used by Japanese industries to achieve perfection.

LINE STOP IN LEAN PRODUCTION

A pair of yellow and red lights hanging above the workers on the assembly line can be used to alert everyone in the area to the status of the processes. Many companies use Andon boards, which hang above the aisles. The number on the board reflects stations

TABLE 36-1	Total Quality Control: Concepts and Categories
TQC Category	**TQC Concept**
1. Organization	• Manufacturing engineering has responsibility for quality—quality circles
2. Goals	• Habit of improvement for everyone everywhere in the manufacturing
	• system
	• Perfection—zero defects—not a program, a goal
3. Basic principles	• Process control—defect prevention, not detection
	• Easy-to-see, quality—quality on display so customers can see and inspect processes—easy to understand quality
	• Insist on compliance with maintenance
	• Line stops when something goes wrong
	• Correct your own errors
	100% check in manufacturing and subassembly cells.
4. Facilitating concepts	• QC department acts as facilitator
	Audit suppliers
	Help in quality improvement projects
	Training workers, supervisors, suppliers
	• Small lot sizes through rapid changeover
	• Housekeeping
	• Less-than-full-capacity scheduling
	• Total preventive maintenance (TPM)
	• 8-4-8-4 two-shift scheduling
5. Techniques and aids	• Remove some inventory, expose problems, solve problems
	• Defect prevention, poka-yokes for checking 100% of parts
	• $n = 2$, for checking first and last item in lot (or $n = 3$, for large lots)
	• Analysis tools
	Cause-and-effect diagrams
	Histograms, and run charts, check sheets
	Control charts, $(\overline{X}, R, \sigma)$
	Scatter diagrams
	Pareto charts
	Process flow charts
	Taguchi and DOE methods

Source: Richard Schoenberger, 1983. Japanese Manufacturing Techniques.

on the line. A worker can turn on a yellow light when assistance is needed, and nearby workers will move to assist the worker having a problem. The line keeps moving, however, until the product reaches the end of the station. Only then is a red light turned on if the problem cannot be solved quickly and the line needs to be stopped. When the problem is solved and everyone is ready to go again, the red light goes off and everyone starts back to work, all in synch.

Every worker should be given the authority to stop the production line to correct quality problems. In systems using poka-yoke or automation, devices may stop the line automatically. The assembly line or manufacturing cell should be stopped immediately and started again only when the necessary corrections have been made. Although stopping the line takes time and money, it is advantageous in the long run. Problems can be found immediately, and the workers have more incentive to be attentive because they do not want to be responsible for stopping the line.

IMPLEMENTING QUALITY COMPANYWIDE

The basic idea of integration is to shift functions that were formerly done in the staff organization (called the production system) into the manufacturing system. What happens to the quality control department? The department serves as the facilitator and therefore acts to promote quality concepts throughout the plant. In addition, its staff educates and trains the workers in statistical and process control techniques and provides engineering assistance on visual and automatic inspection installations. Its most important functions will be training the entire company in quality control.

Another important function of the QC department will be to work with and audit the vendors. The vendor's quality must be raised to the level at which the buyer does not need to inspect incoming material, parts, or subassemblies. The vendor simply becomes an extension of the buyer's plant. Ultimately, each vendor will deliver to the plant perfect materials that need no incoming inspection. Note that this means the acceptable quality level (AQL) of incoming material is 0%. Perfection is the goal. For many years this country has lived with the unwritten rule that 2 or 3% defective was about as good as you could get: better quality just costs too much. For the mass-production systems, this was true. To achieve the kinds of quality that Toyota and many others have achieved, a company has to eliminate the job shop (a functional manufacturing system) and restructure the production system, integrating the quality function directly into the linked-cell manufacturing system, L-CMS.

The quality control department also performs complex or technical inspections, total performance checks (often called end item inspection), chemical analysis, nondestructive testing, X-ray analysis, destructive tests, and tests of long duration.

MAKING QUALITY VISIBLE

Visual display on quality should be placed throughout manufacturing facilities to make quality evident. These displays tell workers, managers, customers, and outside visitors what quality factors are being measured, what the current quality improvement projects are, and who has won awards for quality. Examples of visible quality are signs showing quality improvements, framed quality awards presented to or by the company, and displays of high-precision measuring equipment.

These displays have several benefits. When customers visit a plant to inspect processses, they want to see measurable standards of quality. Highly visible indicators of quality such as control charts and displays should be posted in every department. Everyone is informed on current quality goals and the progress being made. Displays and quality awards are also an effective way to show the workforce that the company is serious about quality.

SOURCE, SELF, AND SUCCESSFUL CHECKS AND POKA-YOKES

Many companies have developed an extensive QC program based on having many inspections. However, inspections can only find defects, not prevent them. Adding more inspectors and inspections merely uncovers more defects but does little to prevent them. Clearly, the least costly system is one that produces no defects. But is this possible? Yes, it can be accomplished through methods such as self and source inspection where quality control is in the hands of the operators.

Many people do not believe that the goal of zero defects is possible to reach, but many companies have achieved this goal or have reduced their defect level to virtually zero using techniques such as poka-yokes and source inspection. When you are inspecting to find defects, the components are compared with standards and defective items are removed. Sampling inspection is used when 100% inspection is not an option, but this assumes that defects are inevitable and that more rigorous inspection can reduce defects.

The truth is, to reduce the defects within a process, it must be recognized that defects are generated by the process itself and that most inspection techniques merely discover defects that already exist.

To achieve zero defects, a lean manufacturing concept must be developed where all operators are responsible for quality. You perform the following kinds of inspections:

• Successive checks, where the next operator checks the work of the previous worker

• Self-checks, where the operator checks his own work before passing it on

• Source inspection, where preventive action takes place at the error stage, to prevent errors from turning into defects

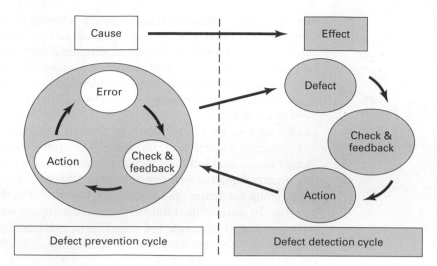

FIGURE 36-20 Source inspection involves defect prevention; that is, preventing errors from turning into defects. *(Source: Achieving Zero Defects Mistake-Proofing—The Zero Quality Control System, © 1999, Productivity, Inc. All rights reserved.)*

Source inspection (see Figure 36-20) involves rethinking the inspection part of the manufacturing process. First off, although it is necessary to have efficient inspection operations, they add little value in the product. Even the most efficient inspection operations are merely efficient forms of non-value-adding activity. Inspection plays a passive role in manufacturing and cannot by itself reduce defects.

Defects and *errors* are not the same thing in manufacturing. Errors can cause defects. For example, not setting the oven temperature correctly can burn the roast (too high) or cause it to be undercooked (too low). Incorrectly loading the original into the fax machine (the error) results in sending blank pages (the defect). When you discover the effect, you make corrections, but you have already spent the money to produce the defect. Table 36-2 outlines some common errors associated with manufacturing.

So *source inspection* looks for errors before they become defects. These techniques either stop the system to make corrections or automatically compensate or correct for the error condition to prevent a defective item from being made.

There are two ways to look at source inspections: vertically and horizontally. Vertical source inspections try to control upstream processes that can be the source or the cause of defects downstream. It is always necessary to examine source processes because they may have a much greater impact on quality than the processes being examined. Finding the source of a problem requires asking "Why?" at every opportunity. Here is an example.

Some steel bars were being cylindrically ground. After grinding, about 10% of the bars warped (bent longitudinally) and were rejected. The grinding process was studied extensively, and no sure solution was found. Looking upstream, a problem with the heat-treating process that preceded cylindrical grinding was detected. About 10% of the bars were not getting a complete, uniform heating prior to quench. Asking "Why?", it was found that these bars were always lying close to the door of the oven, and it was found that the door was not properly sealed, which resulted in a temperature gradient inside. Quenching of the bars induced a residual stress that was released by the grinding

---

TABLE 36-2   Common Errors That Can Produce Defects but Are Preventable

- Omissions in processing steps
- Errors in setting up the job in the machine
- Omission in the assembly process (missing parts)
- Inclusion of incorrect part
- Size errors due to wrong measurement
- Errors due to adjustments
- Errors in cutting-tool geometry or cutting-tool setting
- Errors in processing components (heat treating)

and caused the warping. Asking "Why?" at every opportunity uncovered the source of the problem.

Horizontal source inspections detect defect sources within the processes and then introduce corrections to keep from turning errors into defects. In metalcutting and forming, this is commonly called adaptive control (for preventing defects) or in-process quality control. One of the best ways to prevent errors from occurring is through the use of *poka-yokes*.

Poka-yoke is a Japanese word meaning "defect prevention." Poka-yoke devices and procedures are often devised mainly for preserving the safety of operations. The idea is to develop a method, mechanism, or device that will prevent the defect from occurring rather than to find the defect after it has occurred. Poka-yokes can be attached to machines to automatically check the products or parts in a process. Poka-yokes are source inspections that are usually attributes inspections. The production of a bad part is prevented by the device. Some devices may automatically shut down a machine if a defect is produced, preventing the production of an additional defective part. The poka-yoke system uses 100% inspection to guard against unavoidable human error.

Modern cars are equipped with many poka-yoke devices: you can't turn off the motor unless it is in park, you can't open a door while the car is moving, the headlights come on with the windshield wipers—all are devices to prevent you from making a mistake.

Such devices work very well when physical detection is needed, but many items can be checked only by sensory detection methods, such as the surface finish on a bearing race or the flatness of a glass plate. Variations in nonvisible conditions (air pressure, fluid velocity, temperature, electrical voltage, etc.) require detection devices where critical conditions are readily visible. For such problems a system of self-checks and successive checks can also be used.

TEAMS (AKA QUALITY CIRCLES)

Several popular programs are built upon the concept of participative management, such as quality circles, improvement teams, and task groups. These programs have been very successful in many companies but have failed miserably in others. The difference is often due to the way management implemented the program. Programs must be integrated and managed within the context of a lean manufacturing system design strategy. For example, asking an employee for a suggestion that management does not use (or cannot explain why it does not use) defeats a suggestion system. Management must learn to trust the employees' ideas and decisions and move the decision making to the factory floor.

A quality circle is usually a group of employees within the same department or factory floor area. Meetings are held to work on problems. An organization structure is usually composed of members, a team leader, a facilitator, a manufacturing engineer, and a steering committee.

Quality circles usually have the following main objectives: provide all workers with a chance to demonstrate their ideas; raise employee morale; and encourage and develop workers' knowledge, quality control techniques, and problem-solving methods. They also unify companywide QC activities, clarify managerial policies, and develop leadership and supervisory capabilities.

Quality circles have been implemented in U.S. companies with limited success when they are not part of a lean manufacturing strategy. It is possible for quality circles to work in the United States, but they must be encouraged and supported by management. Everyone must be taught the importance and benefits of integrated quality control.

SUPERIOR QUALITY IN MANUFACTURING/ASSEMBLY CELLS

In lean manufacturing subassembly and manufacturing cells, the cells are designed for a "make one–check one–move one on" (MO-CO-MOO) strategy. The part receives successive checks after each processing or assembly step. For *successive checks* to be successful, several rules should be followed. All the possible variables and attributes should not be measured, because this would eventually lead to errors and confusion in the inspection process. The part should be analyzed so that only one or two points are inspected after each step in the process. This is the heart of MO-CO-MOO. Only the most important elements just produced are inspected.

Another important rule is that the immediate feedback of a defect leads to immediate action. Since the parts are produced in an integrated manufacturing system, this will be very effective in preventing the production of more defective parts. Suppose the cell has only one or two workers and they are not in a position to directly check each other's work after each step. Here is where the decouplers can play an active role by providing automatic successive checking of the parts' critical features before proceeding to the next step. Only perfect parts are pulled from one process to the next through the decoupler.

In assembly lines, a worker may inspect each part immediately after producing it. This is called *self-checking*. There is an immediate feedback to the worker on quality. However, it would be difficult for many workers not to allow a certain degree of bias to creep into their inspection, whether they were aware of it or not, since they are inspecting their own work. Within cells operated by multiple workers, the operator of the downstream station or process can inspect the parts produced by the upstream operator. If there is a problem with the parts, the defective item is immediately passed back to the worker at the previous station. There the defect is verified and the problem corrected. Action is immediately taken to prevent any more defective parts. While this is going on, the line is shut down.

■ 36.6 SUMMARY

The designer of the product must have quality in mind during the quality design phase, seeking the least costly means to ensure the quality of the desired functional characteristics. Major factors that can be handled during the early stages of the product design cycle include temperature, humidity, power variations, and deterioration of materials and tools. Compensation for these factors is difficult or even impossible to implement after the product is in production. The distinction between superior- and poor-quality products can be seen in their variability in the face of internal and external causes. This is where Taguchi parameter design methods can be important.

The secret to successful process control is putting the control of quality in the hands of the workers. Many companies in this country are currently engaged in SQC (statistical quality control), but they are still inspecting to *find* defects. The number of defects will not be reduced merely by making the inspection stage better or faster or automated. You are simply more efficient at discovering defects. The trick is to *inspect* to *prevent* defects. How can this be done? Here are the basic ideas: Use source inspection techniques that control quality at the stage where defects originate. Use 100% inspection with immediate feedback rather than sampling. Make every worker an inspector. Minimize the time it takes to carry out corrective action. Remember that people are human and not infallible. Methods and devices can be developed to prevent them from making errors. Can you think of such a device? Does your car have a procedure that prevents you from locking the ignition keys inside the car? That is a poka-yoke.

Do not simply rely on inspection to control quality. Sorting to find defects by inspection is bad because you already paid to produce those defects.

Process improvement should drive toward defect prevention. In order to achieve the highest levels in quality, you have to implement a manufacturing system design (MSD) that has the highest objective (zero defects) built into (integrated into) the MSD.

Concentrate on making processes efficient, not simply on making the operators and operations more proficient. Continuous improvement requires that you redesign the manufacturing system continuously, reducing the time required for products to move through the system (i.e., the throughput time). This approach seems to be the American stumbling block. Industrial engineers can do operations improvement work such as buying a better machine or improving the ergonomics of a task. However, they need to do more systems improvement work. Too often fancy, complex, computerized solutions are devised to solve complex manufacturing process problems. Why not simplify the manufacturing system so that the need for complex solutions disappears?

■ Key Words

accuracy
control chart
control limits
fishbone diagram
histogram
nominal

parent population
poka-yoke
precision
process capability (PC)
quality control (QC)
range (R)

sample mean ($\overline{X}$)
sample size (n)
self-checking
source inspection
specification limit
standard deviation (σ)

statistical process control
  (SPC)
Taguchi methods
total quality control (TQC)
variability

■ Review Questions

1. Define a process capability study in terms of accuracy or precision.
2. What does the *nature of the process* refer to?
3. Suppose you have a "pistol-shooting" process that is accurate and precise. What might the target look like if, occasionally while shooting, a sharp gust of wind blew left to right?
4. What are the steps required to making a PC study of a process?
5. Why don't standard tables exist detailing the natural variability of a given process, such as rolling, extruding, or turning?
6. What are Taguchi or factorial experiments, and how might they be used to do a process capability study?
7. How does the Taguchi approach differ from the standard experimental method outlined in this chapter?
8. Why are Taguchi experiments so important compared to classical DOE-type experiments?
9. Here are some common, everyday processes with which you are familiar. What variable or aspect to the process might dominate the process in terms of quality, not output?

    a. Baking a cake (from scratch) or grilling a steak
    b. Mowing the lawn
    c. Washing dishes in a dishwasher

10. Explain why the diameter measurements for holes produced by the process of drilling could have a skewed rather than a normal distribution.
11. Name some common manufactured items that may receive the following:
    a. 100% inspection
    b. No final inspection
    c. Some final inspection, that is, sampling
12. What are common reasons for sampling inspection rather than 100% inspection?
13. Fill in this table with one of the four following statements: no error—the process is good; no error—the process is bad; type I or alpha error; type II or beta error.

		In reality, if we looked at everything the process made, we would know that it had:	
		changed	not changed
The sample suggested that the process had:	changed		
	not changed		

14. Explain why, when we sample, we cannot avoid making type I and type II errors?

15. Which type of error can lead to legal action from the consumer for a defective product that caused bodily injury?
16. Define and explain the difference between each of these:
    a. σ and $\sigma_{\bar{x}}$
    b. σ and $\overline{\sigma}$
    c. $\sigma_{\bar{x}}$, σ_R, and σ_σ
17. What is C_p, and why is a value of 0.80 not good? How about a value of 1.00? 1.3?
18. The designer of a component usually sets the nominal and tolerance values when designing the part. How do these decisions affect the decisions of the manufacturing engineer (MfE)?
19. What are some of the alternatives available to you when you have the situation where $6\sigma >$ USL $-$ LSL?
20. C_{pk} is also a process capability index. How does it differ from C_p?
21. In a sigma chart, are values for the samples normally distributed about $\overline{\sigma}$? Why or why not?
22. What is an assignable cause, and how is it different from a chance cause?
23. Why is the range used to measure variability when the standard deviation is really a better statistic?
24. How is the standard deviation of a distribution of sample means related to the standard deviation of the distribution from which the samples were drawn?
25. In the last two decades, the quality of automobiles has significantly improved. What do you think is the main cause for this marked quality improvement?
26. Figure 36-12 shows a bimodal check sheet indicating that the two operators performing an assembly task (in a cell) do the jobs at different rates. What would you recommend here?
27. Control charts use upper and lower control limits. Is a UCL the same as a USL?
28. In Figure 36-14, what are the USL and the LSL and why are they *not* shown on the charts?
29. What are four major branches (fishbones) on a cause-and-effect diagram?
30. How does variation in the measuring device (instrument) affect the measurements obtained on a component?
31. Explain what happed to improve the process in Figure 36-18.
32. In Table 36-1, explain these:

    a. MO-CO-MOO
    b. 8-4-8-4 scheduling
    c. $n = 2$ inspection
    d. Pareto chart

33. What is a quality circle, and how might you apply this concept to your college life?

■ Problems

1. For the items listed in the following chart, obtain a quantity of 48. Measure the indicated characteristics and determine the process mean and standard deviation. Use a sample size of 4, so that 12 samples are produced.

Item	Characteristic(s) You Can Measure
Flat washer	Weight, width, diameter of hole, outside diameter
Paper clip	Length, diameter of wire
Coin (penny, dime)	Diameter, thickness at point, weight
Your choice	Your choice

2. Perform a process capability study to determine the PC of the process that makes M&M candy. You will need to decide what characteristics you want to measure (weight, diameter, thickness, etc.), how you will measure it (use rule of 10), and what kind of M&Ms you want to inspect (how many bags of M&Ms you wish to sample). Take sample size of 4 ($n = 4$). Make a histogram of the individual data and estimate μ and σ as outlined in the chapter. If you decide to measure the weight characteristics, you can check your estimate of μ by weighing all the M&Ms together and dividing by the total number of M&Ms.

3. For the data given in Figure 36-4, compute the mean and standard deviation for the histogram and then C_p and C_{pk}, making any assumptions needed to perform the calculations.

4. For the data given in Figure 36-7, compute C_p and C_{pk}.

5. Calculate $\overline{X}$ and $\overline{R}$ and the control limits for the $\overline{X}$ and R control charts shown in Figure 36-A. The sample mean, $\overline{X}$, and range, R, for the first few subgroups and the data for each sample are given in the bottom of the figure. There are 25 samples of size 4. Therefore $k = 25$, $n = 4$. Complete the bottom part of the table and then compute the control limits for both charts. Construct the charts plotting $\overline{X}$ and $\overline{R}$ as solid lines and control limits as dashed lines, as shown in Figure 36-14. The first four data points have been plotted and the points

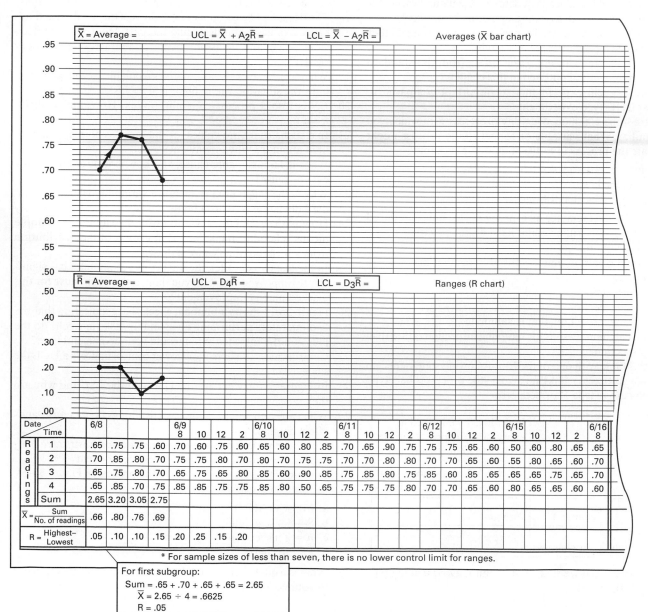

FIGURE 36-A

connected, but are they all correctly plotted? Replot any points that are incorrectly plotted. Plot the rest of the data on the charts and comment on your findings. (Use Figure 36-14 as a comparison.)

6. For the data given in Figure 36-A, estimate the mean and standard deviation for the process from which these samples were drawn (i.e., the parent population) and discuss the process capability in terms of C_p and C_{pk}. The USL and LSL for this dimension are 0.9 and 0.5, respectively, and the nominal is 0.7.

7. Figure 36-B contains data from a machining process that produces holes (drilling) with limits of 6.00 to 6.70 mm. The control charts for $\overline{X}$ and R using $n = 5$ and $k = 25$ are also shown in the figure. (*Note*: The numbers in the body of the table are 6.47, 6.19, 6.19, 6.29, etc.)

1. Recheck the calculation of the mean values and the range values for the 25 samples and then check the calculations for $\overline{\overline{X}}$ and $\overline{R}$ and the control limits for the charts.
2. Insert the centerlines for $\overline{X}$ and $\overline{R}$ on the charts.
3. Check the plotting of the points on the charts.
4. Discuss the charts.
5. Using the data to develop the process capability indexes and discuss the capability of this process.
6. Using the data $n = 5$ and $k = 25$, develop the σ control chart and use σ to estimate σ for the process capability indexes and C_p and C_{pk}.
7. Develop $\overline{X}$ and R charts for sample sizes of 4 (or 3) by ignoring X_5 (X_3 and X_5) or any combination of individual values. Use the charts to perform a process capability study. Did the findings change?

Product name	Cylinder		Sample size		3, 4 or 5		1996, 10.15	$\overline{X}$ Control chart	R Chart
Quality characteristic	Hole diameter	Samples	Timing of taking samples	Daily		Period	1975, 10.30	Center line $\overline{\overline{X}} = 6.299$	Center line $\overline{R} = 0.274$
Limits of allowable range	Max. 6.70 mm		Section		00 00	Person in charge	C. Black	$UCL_{\overline{x}} = \overline{\overline{X}} + A_2\overline{R}$	UCL $2.11 \times \overline{R} = 0.578$
	Min. 6.00 mm		Measuring instrument serial number		103037	Person in charge of inspection	Pogi Bear	$LCL_{\overline{x}} = \overline{\overline{X}} - A_2\overline{R}$	LCL = 0

	Lot No.	1	2	3	4	5	6	7	8	9	10	11	12	13	14	15	16	17	18	19	20	21	22	23	24	25	Total	Mean
Measured values	X_1	47	19	19	29	28	40	15	35	27	23	28	31	22	37	25	7	38	35	31	12	52	20	29	28	42		
	X_2	32	37	27	29	12	35	30	44	37	45	44	25	37	32	40	31	0	12	20	27	42	31	47	27	34		
	X_3	44	31	21	42	45	11	12	32	26	26	40	24	19	12	24	23	41	29	35	38	52	15	41	22	15		
	X_4	35	25	15	59	36	38	33	11	20	37	31	32	47	38	50	18	40	48	24	40	24	3	32	32	29		
	X_5	20	34	19	38	25	33	26	38	35	32	18	22	14	30	19	32	37	20	47	31	25	28	22	54	21		
	Total	178	146	101	197	146	157	116	160	145	163	161	134	139	149	158	111	156	144	157	148	195	97	171	163	141		
	Mean $\overline{x}$	35.6	29.2	20.2	39.4	29.2	31.4	23.2	32.0	29.0	32.6	32.2	26.8	27.8	29.8	31.6	22.2	31.2	28.8	31.4	29.6	39.0	19.4	34.2	32.6	28.2	746.6	29.86
	Range R	27	18	33	30	33	29	21	33	17	22	26	10	33	26	31	25	41	36	27	28	28	28	25	32	27	686	27.44

FIGURE 36-B

Chapter 36 CASE STUDY

Boring QC Chart Blunders

You have recently been hired by the Pippen Company as the 6 sigma Black Belt. Your new boss, Gabby Sorenson, has one of the leading textbooks in manufacturing engineering (NOT DeGarmo) and she shows you this problem. After reading the discussion and studying the figure, Gabby asked you the following questions.

1. What kind of control chart is this?
2. What do you estimate the standard deviation (a') of the parent population to be, assuming that the control limits on this chart are based on three sigma?
3. Is the dashed line in the figure labeled "mean" the mean or the average? (The mean is X and is equal to 0.00017.) What is the line really called?
4. From the information given in the discussion and in the figure, determine the values of C_p and C_{pk}.
5. What is the most glaring error in the figure (Hint: something one never does with control charts) and why is it so wrong?

Example: Maintaining accuracy in boring using control charts.

The workpiece shown in Figure CS 36 is made of gray cast iron and is bored to the tolerances indicated (5.5125/5.5115 in.). These parts were bored on a chucking machine. Each of the 19 points plotted on the vertical axis of the control chart represents the average of bore diameter measurements made on four parts (sample size). The horizontal broken lines at +0.0005 and 0.0005 represent upper and lower specified limits, respectively. The solid line X= 0.00017 in. is the estimate of the process capability based on a study of several samples bored on the machine. The upper and lower control limits are then calculated from X . We note that samples 4–9 show a definite trend toward undersized bored holes. If the operation had been continued without any changes, the successive bored holes very likely would have been out of tolerance. To avoid this situation (out of control), the boring tools were reset toward the upper control limit before parts in sample 10 and the rest were bored. *(Source: ASM International.)*

Figure CS 36

INDEX

SELECTED REFERENCES FOR ADDITIONAL STUDY

HANDBOOKS AND GENERAL REFERENCES

ASM Metals Reference Book. 3rd Ed., ASM, Materials Park, OH (1993).

ASM Engineered Materials Reference Book, 2nd Ed., ASM, Materials Park, OH (1993).

Tool and Manufacturing Engineers Handbook, 4th Ed., Society of Manufacturing Engineers, Dearborn, MI.
  Vol. 1, Machining (1983)
  Vol. 2, Forming (1984)
  Vol. 3, Materials, Finishing and Coating (1985)
  Vol. 4, Quality Control and Assembly (1987)
  Vol. 5, Manufacturing Management (1987)
  Vol. 6, Design for Manufacturability (1992)
  Vol. 7, Continuous Improvement (1994)
  Vol. 8, Plastic Part Manufacturing (1996)
  Vol. 9, Material and Part Handling in Manufacturing (1997)

ASME Handbook, McGraw-Hill, New York.
  Vol. 1, Metals Engineering: Design
  Vol. 2, Metals Properties
  Vol. 3, Engineering Tables
  Vol. 4, Metals Engineering: Processes

ASM Engineered Materials Handbook Series, ASM, Materials Park, OH.
  Vol. 1, Composites (1987)
  Vol. 2, Engineering Plastics (1988)
  Vol. 3, Adhesives and Sealants (1990)
  Vol. 4, Ceramics and Glasses (1991)

ASM Engineered Materials Handbook, Desk Ed., ASM, Materials Park, OH (1995).

ASTM Standards (multiple volumes, published annually), ASTM, Philadelphia, PA.

Manufacturing Engineering Handbook, Hinauyu Geng, McGraw-Hill (2004).

Manufacturing Processes Reference Guide, R.H. Todd, D. K. Allen, and F. Alting, Industrial Press (1994).

Materials Handbook, 13th Ed., George S. Brady and Henry R. Clauser, McGraw-Hill (1995).

Metals Handbook, Desk Ed., 2nd Ed., ASM, Materials Park, OH (1998).

Tool and Manufacturing Engineers Handbook – Desk Edition, Society of Manufacturing Engineers, Dearborn, MI (1988).

Machinery's Handbook, 27th Ed., Industrial Press (2004).

Metals Handbook, 9th Ed., ASM, Materials Park, OH.
  Vol. 1, Properties amid Selection: Irons and Steels (1978)
  Vol. 2, Properties and Selection: Nonferrous Alloys and Pure Metals (1979)
  Vol. 3, Properties and Selection: Tool Materials and Special-Purpose Metals (1980)
  Vol. 4, Heat Treating (1981)
  Vol. 5, Surface Cleaning, Finishing, and Coating (1982)
  Vol. 6, Welding, Brazing and Soldering (1983)
  Vol. 7, Powder Metallurgy (1984)
  Vol. 8, Mechanical Testing (1985)
  Vol. 9, Metallography and Microstructures (1985)
  Vol. 10, Materials Characterization (1986)
  Vol. 11, Failure Analysis and Prevention (1986)
  Vol. 12, Fractography (1987)
  Vol. 13, Corrosion (1987)
  Vol. 14, Forming and Forging (1988)
  Vol. 15, Casting (1988)
  Vol. 16, Machining (1989)
  Vol. 17, Nondestructive Evaluation & Quality Control (1989)

Metals Handbook, 10th Ed., ASM, Materials Park, OH.
  Vol. 1, Properties and Selection: Irons, Steels and High-Performance Alloys (1990)
  Vol. 2, Properties and Selection: Nonferrous Alloys and Special-Purpose Materials (1990)
  Vol. 3, Alloy Phase Diagrams (1992)
  Vol. 4, Heat Treating (1991)
  Vol. 5, Surface Engineering (1994)
  Vol. 6, Welding, Brazing and Soldering (1993)
  Vol. 7, Powder Metal Technologies and Applications (1998)
  Vol. 8, Mechanical Testing and Evaluation (2000)
  Vol. 9, Metallography and Microstructures (2004)
  Vol. 10, Materials Characterization (1986)
  Vol. 11, Failure Analysis and Prevention (2002)
  Vol. 12, Fractography (1987)
  Vol. 13A, Corrosion: Fundamentals, Testing, and Protection (2003)
  Vol. 13B, Corrosion: Materials (2005)
  Vol. 13C, Corrosion: Environments and Industries (2006)
  Vol. 14A, Metalworking: Bulk Forming (2005)
  Vol. 14B, Metalworking: Sheet Forming (2006)
  Vol. 15, Casting (1988)
  Vol. 16, Machining (1989)
  Vol. 17, Nondestructive Evaluation and Quality Control (1989)
  Vol. 18, Friction, Lubrication and Wear Technology (1992)
  Vol. 19, Fatigue & Fracture (1996)
  Vol. 20, Materials Selection and Design (1997)
  Vol. 21, Composites (2001)

MMPDS-02: Metallic Materials properties Development and Standardization, Battelle Memorial Institute (2005) [NOTE: This is the successor to MIL-HDBK-5].
>Vol. 1: Introduction and Guidelines
>Vol. 2: Steel and Magnesium Alloys
>Vol. 3: 2000-6000 Series Aluminum Alloys
>Vol. 4: 7000 Series and Cast Aluminum Alloys
>Vol. 5 Miscellaneous Alloys & Structural Joints

Production Handbook, 4th Ed., John A. White (ed.), Wiley (1987).

SAE Handbook, Part I—Materials (issued annually). Society of Automotive Engineers, Warrendale, PA.

Smithells Metals Reference Book, 8th Ed., W.F. Gale and T.C. Totemcier, Butterworths (2004).

Troubleshooting Manufacturing Processes, 4th Ed., SME (1988).

Woldman's Engineering Alloys, 9th Ed., J. Frick, ASM, Materials Park, OH (2000).

BASIC AND GENERAL TEXTBOOKS IN MATERIALS AND PROCESSES

21st Century Manufacturing, Paul Kenneth Wright, Prentice Hall (2001).

Engineering Materials, 8th Ed., Kenneth Budinski and Michael Budinski, Prentice Hall (2005).

Fundamentals of Modern Manufacturing, 3rd Ed. Mikell P. Groover, Wiley (2007).

Introduction to Materials Science for Engineers, 6th. Ed., James F. Shackelford, Prentice-Hall (2005).

Introduction of Manufacturing Processes, 3rd Ed., John A. Schey, McGraw-Hill (2000).

Introduction to Manufacturing Processes and Materials, Robert C. Creese, Marcel Dekker (1999).

Manufacturing Engineering and Technology, 5th Ed., Serope Kalpakjian and Steven R. Schmid, Prentice-Hall (2006).

Manufacturing Processes and Systems, 9th Ed., Phillip F. Ostwald and Jairo Muñoz, Wiley (1997).

Manufacturing Processes and Equipment, Jiri Tlusty, Prentice Hall, Upper Saddle River, NJ (1999).

Manufacturing Processes and Systems, 9th Ed., Phillip F. Ostwald and Jairo Muñoz, Wiley (1997).

Manufacturing Processes and Materials, 4th Ed., George F. Schrader and Ahmad K. Elshennawy, Society of Manufacturing Engineers (2000).

Materials Science and Engineering Materials: An Introduction, 6th Ed. W.D. Callister, Wiley (2003).

Mechanical Metallurgy, 3rd Ed., G. E.Dieter, McGraw-Hill, New York (1986).

Principles of Materials Science and Engineering, 4th Ed., W. F. Smith, McGraw-Hill (1996).

Processes and Design for Manufacturing, 2nd Ed., Sherif D. El Wakil, PWS Publishing Company (1998).

Processes and Materials of Manufacture, 4th Ed., Roy A. Lindberg, Allyn and Bacon (1990).

Structures and Properties of Engineering Alloys, 2nd Ed., W. F. Smith, McGraw-Hill New York (1993).

The Science and Engineering of Materials, 5th Ed., P.P. Puhle and D. R. Askeland, Thomson, Ontario, Canada (2006).

The Science and Design of Engineering Materials, 2nd Ed., J. P. Schaffer et al., McGraw-Hill (1999).

FERROUS METALS

ASM Specialty Handbook - Carbon and Alloy Steels, ASM (1996).

ASM Specialty Handbook - Cast Irons, ASM (1996).

ASM Specialty Handbook - Stainless Steels, ASM (1994).

Cast Iron: Physical and Engineering Properties, H.T. Angus, Butterworths (1976).

Design Guidelines for the Selection and Use of Stainless Steel, Designers Handbook Series, Specialty Steel Industry of the United States, Washington, DC (1992).

Engineering Properties of Steel, ASM, Materials Park, OH (1982).

Handbook of Stainless Steels, Donald Peckner and I. M. Bernstein (eds.), McGraw-Hill (1977).

Introduction to Stainless Steels, 3rd Ed., ASM (1999).

Metallurgy and Heat Treatment of Tool Steels, Robert Wilson, McGraw-Hill (1975).

Modern Steels and Their Properties, (various editions) Bethlehem Steel Corporation, Bethlehem, PA.

Properties and Selection of Tool Materials, ASM, Materials Park, OH (1975).

Stainless Steel, R.A. Lula, ASM, Metals Park, OH (1985).

Steel Selection: A Guide for Improving Performance and Profits, R. F. Kern and M. E. Suess, Wiley-Interscience (1979).

Steels: Metallurgy & Applications, D.T. Llewellyn, Butterworth Heinemann (1992).

Steels: Heat Treatment and Processing Principles, G. Krauss, ASM, Materials Park, OH (1990).

The Making, Shaping and Treating of Steel, 11th Ed., Association for Iron and Steel Technology (1998).

Tool Steels, 5th Ed., G.A. Roberts, G. Krauss and R. Kennedy, ASM, Materials Park, OH (1998).

Worldwide Guide to Equivalent Irons and Steels, 5th Ed., ASM, Materials Park, OH (2006).

Steels: Processing, Structure, and Performance, George Krauss, ASM, Materials Park, OH (2005).

NONFERROUS METALS

Aluminum Alloys: Structure and Properties, V. F. Mondolfo, Butterworths (1975).

Aluminum, ASM, Materials Park, OH (1967).
>Vol. 1, Properties, Physical Metallurgy and Phase Diagrams
>Vol. 2, Design and Application
>Vol. 3, Fabrication and Finishing

Aluminum Standards and Data, The Aluminum Association, Washington, DC (bi-annual updates).

Aluminum: Properties and Physical Metallurgy, John E. Hatch (ed.), ASM, Materials Park, OH (1984).

ASM Specialty Handbook - Aluminum and Aluminum Alloys, ASM (1993).

ASM Specialty Handbook - Copper and Copper Alloys, ASM (2001).

ASM Specialty Handbook - Magnesium and Magnesium Alloys, ASM (1999).

ASM Specialty Handbook - Nickel, Cobalt, and Their Alloys, ASM (2000).

Engineering Properties of Zinc Alloys, International Lead Zinc Research Organization, New York (1980).

Introduction to Aluminum Alloys and Tempers, J.G. Kaufman, ASM (2000).

Materials Properties Handbook : Titanium Alloys, R. Boyer, E.W. Collings, and G. Welsch, ASM, Materials Park, OH (1994).

Properties of Aluminum Alloys : Tensile, Creep and Fatigue Data at High and Low Temperatures, J.G. Kaufman (ed.), ASM and Aluminum Association (1999).

Standards Handbook: Copper, Brass, and Bronze (7 volumes) Copper Development Association.

Superalloys: A Technical Guide, 2nd Ed., ASM (2002).

Titanium: A Technical Guide, 2nd Ed., M. Donachie (ed.), ASM, Materials Park, OH (2000).

Titanium Alloys Handbook, Metals and Ceramics Information Center, Battelle Columbus Laboratories, Columbus, OH (1972).

Worldwide Guide to Equivalent Nonferrous Metals and Alloys, 4th Ed., ASM, Materials Park, OH (2001).

PLASTICS, COMPOSITES, AND CERAMICS

Advanced Polymer Composites: Principles and Applications, B. J. Jang. ASM, Materials Park, OH (1993).

Composite Materials Handbook, 2nd Ed., M.M. Schwartz, Mc-Graw-Hill, New York (1995).

Engineering Plastics and Composites, 2nd Ed., W.A. Woishnis (ed.), ASM, Materials Park, OH (1993).

Engineers Guide to Composite Materials, John W. Weeton (ed.), ASM, Materials Park, OH (1986).

Fabrication of Composite Materials—Source Book, M.M. Schwartz (ed.), ASM, Materials Park, OH (1985).

Fundamentals of Metal Matrix Composites, S. Suresh (ed.), Butterworth-Heinemann (1993).

Handbook of Composites, George Lubin (ed.), Van Nostrand Reinhold, New York (1982).

Handbook of Plastics, Elastomers and Composites, 2nd Ed., C.A. Harper (ed.), McGraw-Hill (1992).

Handbook of Thermoset Plastics, S.H. Goodman, Noyes Publications (1986).

Modern Plastics Encyclopedia, published annually by Modern Plastics Magazine, a McGraw-Hill publication.

Plastic Blow Molding Handbook, N. C. Lee (ed.), Van Nostrand Reinhold (1991).

Plastic Part Technology, E.A. Muccio, ASM, Materials Park, OH (1991).

Plastic Materials and Processes, S. Schwartz and S. H. Goldman, Van Nostrand Reinhold, New York (1982).

Plastics Product Design Engineering Handbook, 2nd Ed., Sidney Levy and J. H. DuBois, Chapman & Hall (1984).

Plastics Processing Technology, E.A. Muccio, ASM, Materials Park, OH (1994).

Plastics Process Engineering, James L. Throne, Marcel Dekker, New York (1979).

Plastics' Technology Handbook, 2nd Ed., M. Chanda and S. K. Roy (eds.), Marcel Dekker (1992).

Structural Ceramics, M. Schwartz, McGraw-Hill (1994).

Characterization and Failure Analysis of Plastics, ASM, Materials Park, OH (2003).

Fundamentals of Composite Manufacturing: Materials, Methods, and Applications, A. Brent Strong, Society of Manufacturing Engineers (1989).

Metal Matrix Composites, N. Chawla and K.K. Chawla, Springer, (2006).

Plastic Injection Molding – 4 volume set, Society of Manufacturing Engineers (1996-1999).
> Vol 1: Manufacturing Process Fundamentals
> Vol 2: Material Selection Fundamentals & Product Design
> Vol 3: Mold Design & Construction Fundamentals
> Vol 4: Manufacturing Startup and Management

ELEVATED TEMPERATURE APPLICATIONS

ASM Specialty Handbook - Heat-Resistant Materials, ASM (1997).

Engineer's Guide to High Temperature Materials, F. J. Clauss, Addison Wesley, Reading, MA (1969).

High Temperature Property Data: Ferrous Alloys, M. F. Rothman (ed.), ASM, Materials Park, OH (1987).

Properties of Refractory Metals, W.D. Wilkinson, Gordon and Breach (1969).

Source Book on Materials for Elevated Temperature Applications, ASM, Materials Park, OH (1979).

ELECTRONIC AND MAGNETIC APPLICATIONS

ASM Ready Reference: Electrical and Magnetic Properties of Materials, ASM (2000).

Electronic Materials Handbook, Vol. 1, Packaging, ASM (1989).

Electronic Materials & Processes Handbook, 2nd Ed., C.A. Harper and R. M. Sampson, McGraw-Hill (1994).

Electronic Packaging and Interconnection Handbook, C.A. Harper, McGraw-Hill, New York (2000).

Fundamentals of Semiconductor Processing Technology, B. El-Kareh Kluwer Academic Publishers, Boston (1995).

Integrated Circuit Fabrication Technology, D. J. Elliott, McGraw-Hill, New York (1989).

Soldering in Electronics Assembly, M. Judd and K. Brindley, Reed International Books, Boston (1992).

Springer Handbook of Electronic and Photonic Materials, S. Kasap and P. Capper (Eds.), Springer (2006).

The Magnetic Properties of Materials, J. E. Thompson, CRC Press, Cleveland, OH (1968).

The Science and Engineering of Microelectronic, S.A. Campbell, Oxford University Press (2001).

DESIGN

Application of Fracture Mechanics—for Selection of Metallic Structural Materials, J. E. Campbell, et al., ASM, Materials Park, OH (1982).

Atlas of Stress-Strain Curves, H. E. Boyer (ed.),ASM, Materials Park, OH (1986).

Atlas of Fatigue Curves, Howard E. Boyer (ed.),ASM, Materials Park, OH (1986).

Atlas of Stress Corrosion and Corrosion Fatigue Curves, A. J.McEvily,ASM,Materials Park, OH (1990).

Axiomatic Design, Nam P. Suh, Oxford University Press (2001).

Concurrent Design of Products and Processes,J. L. Nevins and Dan Whitney, McGraw-Hill (1989).

Engineering Design: A Materials and Processing Approach, 3rd Ed., George Dieter, McGraw-Hill (2000).

Fatigue Data Book: Light Structural Alloys, ASM, Materials Park, OH (1995).

Fracture Mechanics: Fundamentals and Applications,T. L. Anderson, CRC Press (1991).

Fundamentals of Tool Design, 5th Ed., D. Smith & J. Nee, Society of Manufacturing Engineers (2003).

Handbook of Product Design for Manufacturing—A Practical Guide for Low-Cost Production, James G. Bralla (ed.), McGraw-Hill, New York (1986).

Mechanical Assemblies, Daniel E. Whitney, Oxford University Press, (2004).

Introduction to Engineering Design, Andrew Samuel and John Weir, Buttorworth Heinemann, (1999).

Material Selection in Mechanical Design, Michael Ashby, Butterworth Heineman, (2000).

Product Design for Manufacture and Assembly, 2nd Edition, G. Boothroyd, P. Dewhurst, W. Knight, Marcel Decker, (2002).

The Principles of Design, Nam P. Suh, Oxford University Press (1990).

The Principles of Material Selection for Engineering Design, Pat L. Mangonon, Prentice Hall (1999).

Tool Design, 3rd Ed., C. Donaldson, G. LeCain, and V. C. Gould, Glencoe (1993).

CASTING

Cast Metals Handbook, 4th Ed., American Foundryman's Society (1957).

Iron Castings Handbook, Charles F.Walton, Iron Castings Society (1981).

Gray and Ductile Iron Casting Handbook,Gray and Ductile Iron Founder's Society, Cleveland, OH (1971).

Steel Castings Handbook, 6th Ed., Steel Founder's Society of America, OH (1995).

Investment Casting Handbook—1980, Investment Casting Institute, Chicago, IL (1979).

Principles of Metal Casting, 2nd Ed.,Heine, Loper and Rosenthal,McGraw-Hill (1967).

The NFFS (Non-Ferrous Founders Society) Guide to Aluminum Casting Design: Sand and Permanent Mold, NFFS (1994).

Copper, Brass and Bronze Castings: Their Structures, Properties, and Applications, Non-Ferrous Founder's Society, Cleveland, OH (1961)—2 volumes.

Copper Alloy Pressure Die Casting, A.A.Machonis (ed.), International Copper Research Assoc., New York (1975).

WELDING AND JOINING

Adhesives Technology Handbook, Arthur H. Landrock, Noyes Publications (1985).

Brazing Handbook, American Welding Society (1991).

Ceramic Joining, M.M. Schwartz,ASM, Materials Park, OH (1990).

Design of Welded Structures, James F. Lincoln Foundation.

Design of Weldments, James F. Lincoln Foundation.

Handbook of Adhesives, 2nd Ed., Irving Skiest (ed.),Van Nostrand Reinhold (1977).

The Basics of Soldering, A. Rahn,Wiley (1993).

Joining of Composite Matrix Materials, M.M. Schwartz,ASM, Materials Park, OH (1994).

Metals Joining Manual, M.M. Schwartz, McGraw-Hill (1979).

Modern Welding Technology, 2nd Ed.,H. B. Cary, Prentice Hall, NJ (1989).

Soldering Manual, 2nd Ed., American Welding Society (1978).

Solders and Soldering, 3rd Ed.,H.H.Manko,McGraw-Hill (1994).

Source Book on Electron Beam and Laser Welding, ASM,Materials Park, OH (1980).

Source Book on Brazing and Brazing Technology,ASM,Materials Park, OH (1980).

Standard Handbook of Fastening and Joining, 2nd Ed., R.O. Parmley, McGraw-Hill (1994).

Welding Encyclopedia, L.B.MacKenzie, 16th Ed., revised and re-edited by T.B. Jefferson, Monticello Books, Morton Grove, IL (1968).

Welding Handbook, 8th Ed., American Welding Society, New York (1987).

Brazing, 2nd Edition, M.M. Schwartz, ASM, Materials Park, OH (2003).

Friction Stir Welding and Processing, R.S. Mishra and M.W. Mahoney (Eds.), ASM, Materials Park, OH (2006).

Principles of Brazing, D.M. Jacobson and G. Humpston, ASM, Materials Park, OH (2005).

Principles of Soldering, G. Humpston and D.M. Jacobson, ASM, Materials Park, OH (2004).

Welding Technology for Engineers, B. Raj et. al., Narosa publishing House and ASM, Materials Park, OH (2006).

FABRICATION AND FORMING

Finite-Element Plasticity and Metalforming Analysts, G.W. Rowe, C. E. N. Sturgess, P. Hartley, and I. Pillingen, Cambridge University Press (1991).

Forging Industry Handbook, 3rd Ed.,T.G. Bryer (ed.), Forging Industry Association, Cleveland, OH and ASM (1981).

Forging Materials and Practices,A.M. Sabroff et al., Reinhold, New York (1968).

Handbook of Metalforming Processes, B.Avitzur,Wiley-Interscience,New York (1983).

Handbook of Metal Forming, Kurt Lange (ed.), McGraw-Hill, New York (1985).

Handbook of Fabrication Processes, O.D. Lascoe, ASM, Materials Park, OH (1988).

Mechanics of Plastic Deformation in Metal Processing, E.G.Thomsen, C.T.Yang, and S.Kobayashi, Macmillan (1965).

Mechanics of Sheet Metal Forming: Material Behavior and Deformation Analysis, D. Koistinen, Plenum (1977).

Metal Forming: Fundamentals and Applications, T.Altan, S-I Oh, and H. Gegal, ASM, Materials Park, OH (1983).

Metalforming and the Finite-Element Method, S. Kohavashi, S.Oh, and T. Altan, Oxford University Press (1989).

Metalworking Science and Engineering, Edward M. Mielnik, McGraw-Hill (1991).

Techniques of Pressworking Sheet Metal, 2nd Ed., D. F. Eary and E. A. Reed, Prentice Hall (1974).

Cold and Hot Forging: Fundamentals and Applications, T. Altan et. al., ASM, Materials Park, OH (2005).

Die Design Handbook, 3rd Edition, David A. Smith, Society of Manufacturing Engineers (1990).

Extrusion, 2nd Ed., M. Bauser et. al (Eds.), ASM, Materials Park, OH (2006).

Fundamentals of Hydroforming, Harjinder Singh, Society of Manufacturing Engineers (2003).

Fundamentals of Pressworking, David A. Smith Society of Manufacturing Engineers (1994).

Fundamentals of Tool Design, 5th Edition, David A. Smith and John Nee (Eds), Society of Manufacturing Engineers (2003).

Handbook of Workability and Process Design, G.E. Dieter et. al. (Eds.), ASM, Materials Park, OH (2003).

Press Brake Technology, S. Benson, Society of Manufacturing Engineers (1997).

Progressive Dies: Principles and Practices of Design and Construction, Don Peterson (Ed.), Society of Manufacturing Engineers (1994).

Roll Forming Handbook, George T. Halmos, Taylor and Francis (2005)

Sheet Metal Forming Processes and Die Design, Vukota Boljanovic, Industrial Press (2004).

Steel Forgings: Design, Production, Selection, Testing, and Application, E. G. Nisbett, ASTM, West Conshohocken, PA (2005).

Tube Forming Processes: A Comprehensive Guide, Greg Miller, Society of Manufacturing Engineers (2002).

POWDER METALLURGY

Handbook of Powder Metallurgy, H.H.Hausner, Chemical Publishing Co. (1973).

Metallurgy, J. S.Hirschorn, American Powder Metallurgy Institute, New York (1969).

Injection Molding of Metals and Ceramics, Randall M. German and Animesh Bose, Metal Powder Industries Federation (1997).

Powder Metallurgy Science, 2nd Ed., R.M.German, Metal Powder Industries Federation (1994).

Powder Metallurgy Processing: New Techniques and Analyses, H.A. Kuhn and A. Lawley, Academic Press (1978).

Powder Metallurgy Design Manual, 3rd Ed., Metal Powder Industries Federation (1998).

Powder Metallurgy, Principles and Applications, F.V. Lenel, Metal Powder Industries Federation, Princeton, NJ (1980).

Powder Metallurgy Equipment Manual—3, Samuel Bradbury (ed.), MPIF (1986).

Powder Injection Molding, R.M.German, Metal Powder Industries Federation (1990).

Fundamentals of Powder Metallurgy, Leander Pease and William West, Metal Powder Industries Federation (2002)

Introduction to Powder Metallurgy, F.Thummler and R. Oberacker, The Institute of Materials (1994).

Powder Injection Molding: Design & Applications, Randall M. German, Innovative Materials Solutions (2003).

Powder Metallurgy & Particulate Materials Processing, Randall M. German, Metal Powder Industries Federation (2005).

MACHINING

Analysis of Material Removal Processes, W. R.DeVries, Springer-Verlag (1992).

Fundamentals of Metal Machining and Machine Tools, 3rd Ed. G. Boothroyd and Winston Knight, Taylor & Francis (2006).

Machining of Plastics, Akira Kobayashi, Krieger Pub. Co. (1981).

Machining Data Handbook, 3rd Ed., (2 volumes) Machinability Data Center, Cincinnati, OH (1980).

Machining—Theory and Practice, ASM, Materials Park, OH (1950).

Machining and Metal Working Handbook, 3rd Ed. Bah. Walsh and D. Cormien, McGraw-Hill, (2006).

Mechanics of Machining, P. L. B. Oxley, Ellis Harwood Ltd. (1989).

Metal Cutting, 4th Ed., Edward M. Trent, Paul K. Wright, Butterworth Heineman (2000).

Metal Cutting Mechanics, N.N. Zorcv, Pergamon Press, (1966).

Metal Cutting Principles, 2nd Ed. M. C. Shaw, Oxford University Press (2005).

Metal Cutting Theory and Practice, 2nd Ed., David A. Stephenson and John S. Agapion, Taylor & Francis, (2006).

HEAT TREATMENT

Atlas of Isothermal and Cooling Transformation Diagrams, ASM, Materials Park, OH (1977).

Atlas of Isothermal and Cooling Transformation Diagrams for Engineering Steels, ASM, Materials Park, OH (1980).

Atlas of Time-Temperature Diagrams for Nonferrous Alloys, G.VanderVoort (ed.), ASM, Materials Park, OH (1991).

Atlas of Time-Temperature Diagrams for Irons and Steels, G.VanderVoort (ed.), ASM, Materials Park, OH (1991).

Handbook of Quenchants and Quenching Technology, G. E.Totten, C. E. Bates, and N. A. Clinton, ASM, Materials Park, OH (1992).

Heat Treater's Guide: Practices and Procedures for Irons and Steels, 2nd Ed., ASM, Materials Park, OH (1995).

Heat Treater's Guide: Practices and Procedures for Nonferrous Alloys, ASM (1996).

Heat Treatment, Structure and Properties of Nonferrous Alloys, Charles R. Brooks, ASM, Materials Park, OH (1982).

Heat Treating Data Book, 9th Ed., Seco/Warwick Corporation (2006).

Practical Heat Treating, 2nd Ed. J.L. Dossett and H.E. Boyer, ASM, Materials Park, OH (2006).

Practical Induction Heat Treating, R. E. Haimbaugh, ASM (2001).

Principles of Heat Treatment of Steel, G. Krauss, ASM, Materials Park, OH (1980).

Principles of Heat Treatment, M. A. Grossman and E. C. Bain, ASM, Materials Park, OH (1964).

Principles of the Heat Treatment Plain Carbon and Low Alloy Steels, C. R. Brooks, ASM, Materials Park, OH (1996).

Steel and Its Treatment, Bofors Handbook, K. E. Thelning, Butterworths (1975).

Steel Heat Treatment Handbook, 2nd Ed. G. E. Totten and M. A. H. Howes, Marcel Dekker (2006).

Surface Hardening of Steels: Understanding the Basics, J. R. Davis (ed.), ASM (2002).

SURFACES AND FINISHES

Advanced Machining Technology Handbook, James Brown, McGraw-Hill, (1998).

Metal Finishing: Guidebook Directory, Metals and Plastics Publications, Inc., Westwood, NJ (published annually).

Surface Finishing Systems, George Rudski, ASM, Materials Park, OH (1983).

Electroplating Engineering Handbook, 4th Ed., Lawrence Durney (ed.), Van Nostrand-Reinhold (1984).

Properties of Electroplated Metals and Alloys, 2nd Ed., Wm. H. Safranek, Amer. Electroplaters and Finishers Soc. (1986).

Metal Cutting and High Speed Machining, Ed. By D. Dudzinski, et.al, Kluwer Academic (2002).

Metallizing of Plastics: Handbook of Theory and Practice, R. Suchentruck, Finishing Publications Ltd and ASM (1993).

The Surface Treatment and Finishing of Aluminum and Its Alloys, 6th Ed., S. Wernick, R. Pinner and P. B. Sheasby, Finishing Publications Ltd. and ASM (2001).

Surface Engineering for Corrosion and Wear Resistance, J. R. Davis (ed.), IOM Communications and ASM (2001).

Coatings and Coating Processes for Metals, J.H. Lindsay (ed.), ASM (1998).

Guide to High Performance Powder Coating, Bob Utech, Society of Manufacturing Engineers (2002).

Handbook of Thermal Spray Technology, J.R. Davis (ed.), Thermal Spray Society and ASM, Metals park, OH (2004).

User's Guide to Powder Coating, 4th Edition, Nicholas Liberto, Society of Manufacturing Engineers (2003).

CORROSION

Handbook of Corrosion Data, 2nd Ed., B. Craig and D. Anderson (eds.), ASM, Materials Park, OH (1995).

NACE Corrosion Engineer's Reference Book, 2nd Ed., R. S. Treseder (ed.), National Association of Corrosion Engineers (NACE) (1991).

Corrosion Resistance Tables, 4th Ed., P.A. Schweitzer, Marcel Dekker (1995).

An Introduction to Metallic Corrosion, 3rd Ed., Ulick R. Evans, Edward Arnold Ltd. and ASM (1981).

Corrosion and Its Control—An Introduction to the Subject, Atkinson and Van Droffelaar, National Association of Corrosion Engineers, Houston, TX (1982).

Corrosion Engineering, 3rd Ed., M.G. Fontana, McGraw-Hill, New York (1986).

Corrosion and Corrosion Control, 3rd Ed., H.H. Uhlig and R. Winston Revie, Wiley-Interscience, New York (1985).

Corrosion, 2nd Ed.—Vol. 1: Corrosion of Metals and Alloys and Vol. 2: Corrosion Control, L. L. Shrier (ed.).

Material Selection for Corrosion Control, S. I. Chawla and R. K. Gupta, ASM, Materials Park, OH (1993).

Corrosion Prevention by Protective Coatings, Charles G. Munger, National Association of Corrosion Engineers (1984).

Fundamentals of Electrochemical Corrosion, E. E. Stansbury and R. A. Buchanan, ASM (2000).

Corrosion: Understanding the Basics, J. R. Davis (ed.), ASM (2000).

NDT AND METROLOGY

Hardness Testing, 2nd Ed., H. Chandler (ed.), ASM (1999).

ISO System of Limits and Fits, General Tolerances and Deviations, American National Standards Institute.

Metallography: Principles and Practice, George VanderVoort, McGraw-Hill (1984).

Nondestructive Testing Handbook, 2nd Ed. (multiple volumes), American Society for Nondestructive Testing (1982–1985).

Nondestructive Testing, Louis Cartz, ASM, Materials Park, OH (1995).

Practical Non-Destructive Testing, 2nd Ed., B. Raj, T. Jaykumar, and M. Thavasimuthu, Narosa Publishing and ASM (2002).

The Testing of Engineering Materials, 4th Ed., H. E Davis, G. E. Troxell, and G. F.W. Hauck, McGraw-Hill, New York (1982).

Springer handbook of Materials Measurement Methods, H. Czichos et. al. (Eds.), Springer (2006).

Tensile Testing, 2nd Ed., J.R. Davis (Ed.), ASM, Materials Park, OH (2004).

QUALITY CONTROL, TAGUCHI METHODS AND SIX SIGMA

Creating Quality, William J. Kolarik, McGraw-Hill (1995).

Guide to Quality Control, Kaoru Ishikawa, Asian Prod. Organization (1984).

Introduction to Statistical Quality Control, 4th Ed., D. C. Montgomery, Wiley (2001).

Introduction to Quality Engineering, Genichi Taguchi, Kraus Int. Pub. (1986).

Juran's Quality Planning & Analysis, 5th Ed. By F.M. Gryna, R.C.H. Chua and J.A. DeFeo, McGraw-Hill (2002).

Process Quality Control, 2nd Ed., E. R.Ott and E.G. Schilling, McGraw-Hill (1990).

Quality Control and Industrial Statistics, 5th Ed, Acheson J Duncan, Irwin, (1986).

Quality Engineering in Production Systems, G. Taguchi, E.A. ElSayed, and T. Hsiang, McGraw-Hill (1989).

Quality Control Handbook, 4th Ed., J.M. Juran and F. M. Gigna, McGraw-Hill (1988).

Quality, Productivity, and Competitive Position, W. Edwards Deming, MIT Press, Boston (1982).

Quality Planning and Analysis, 2nd Ed., Juran and Gryan, McGraw-Hill (1988).

Statistical Quality Design and Control, 2nd Ed. R. E.De Vor,T. Chang, and J.W. Sutherland, Macmillan (2007).

Statistical Quality Control, 6th Ed., Eugene L. Grant and Richard S. Leavenworth, McGraw-Hill (1988).

Taguchi Techniques for Quality Engineering, 2nd Ed. P. J. Ross, McGraw-Hill (1996).

The Six Sigma Handbook, Ed. By Thomas Pyzdek, McGraw-Hill, (2003).

Total Quality Control, A.V. Feigenbaumn, McGraw-Hill (1961).

Quality Engineering in Production Systems, Genichi Taguchi, Elsayed A. Edsayed, and Thomas C. Hsiang,McGraw-Hill (1989).

Reliability and Maintainability Management, Balbir S. Dhillon and Hans Reiche, Von Nostrand Reinhold (1985).

Quality Control for Managers and Engineers, Elwood G. Kirkpatrick, Wiley (1970).

Total Quality Management—Text, Cases and Readings, Joel E. Ross, St. Lucie Press (1993).

Zero Quality Control: Source Inspection and the Poka-Yoke System, Shigeo Shingo, Productivity Press (1986).

Lean Six Sigma, M.L. George, McGraw-Hill (2002).

The Six Sigma Black Belt Handbook, McCarty et.al. (2005).

The Six Sigma Performance Handbook, P. Gupta, McGraw-Hill (2005).

FAILURE ANALYSIS AND PRODUCT LIABILITY

Analysis of Metallurgical Failures, V. J. Conangelo and F. A. Heiser, Wiley-Interscience (1974).

Case Histories in Failure Analysis, ASM, Materials Park, OH (1979).

Engineering Aspects of Product Liability, V. J. Conangelo and P. A.Thornton, ASM, Materials Park, OH (1981).

Failure Analysis of Engineering Structures: Methodology and Case Histories, V. Ramachandran et. al, ASM, Materials Park, OH (2005).

Failure Analysis: Case Histories, F. K. Naumann, ASM, Materials Park, OH (1983).

Failure Analysis: The British Engine Technical Reports, F. K. Hutchings and P. M. Unterweiser, ASM, Materials Park, OH (1981).

Failure of Materials in Mechanical Design, J.A. Collins, Wiley-Interscience (1981).

Fractography in Failure Analysis, American Society for Testing and Materials, Philadelphia, PA (1978).

Handbook of Case Histories in Failure Analysis: Vols. 1 and 2, K.A. Esaklul (ed.),ASM, Materials Park, OH (1992–93).

Metal Failures: Mechanisms, Analysis, Prevention, A. J. McEvily (ed.), John Wiley & Sons, New York (2002).

Metallography in Failure Analysis, J. L. McCall and P. M. French, Plenum (1977).

Metallurgical Failure Analysis, C. R. Brooks and A. Choudhury, McGraw-Hill (1993).

Product Liability and the Reasonably Safe Product, A. S.Weinstein et al., Wiley-Interscience, New York (1978).

Residual Stresses and Fatigue in Metals, J.O. Almen and P. H. Black, McGraw-Hill, New York (1963).

Stress Corrosion Cracking: Materials Performance and Evaluation, Russell Jones (ed.),ASM, Materials Park, OH (1992).

Tool and Die Failures Source Book, ASM, Materials Park, OH (1982).

Understanding How Components Fail, 2nd Ed., Donald J. Wulpi, ASM, Materials Park, OH (1999).

Why Metals Fail, R.D. Barer and B. F. Peters, Gordon and Breach, New York (1970).

MANUFACTURING SYSTEMS ANALYSIS

Computational Intelligence in Design and Manufacturing, Andrew Kusiak, Wiley (2000).

Design and Analysis of Lean Production Systems, R.G. Askin and J.B. Goldberg, Wiley (2002).

Fundamentals of Systems Engineering, C. Jotin Kristy and Jamshid Mohammadi, Prentice-Hall (2001).

Manufacturing Systems Design and Analysis, 2nd Ed., B.Wu, Chapman and Hall (1994).

Manufacturing Systems Engineering, S. B.Gershwin, PTR, Prentice Hall (1994).

Manufacturing Systems Engineering, 2nd Ed., K.Hitomi, Taylor and Francis Ltd. (1996).

Modeling and Analysis of Manufacturing Systems, R.G.Askin and C. R. Standridge,Wiley (1993).

Performance Modeling of Automated Manufacturing Systems, N.Viswanadham and Y. Narahari, Prentice Hall (1992).

Stochastic Models of Manufacturing Systems, J.A. Buzacott and J. C. Shanthikumar, Prentice Hall (1993).

Systems Analysis and Modeling, onald W. Boyd, Academic Press (2001).

MANUFACTURING ENGINEERING

21st Century Manufacturing, P.K. Wright, Prentice-Hall, (2001).

Concurrent Design of Products and Processes, J. L. Nevins and D. E.Whitney, McGraw-Hill (1989).

Intelligent Manufacturing Systems, . Kusiak, Prentice Hall (1990).

Manufacturing Engineering Handbook, Ed. Hwaiyu Geng, McGraw-Hill, (2004),

Manufacturing Engineering and Technology, 4th Ed., S. Kalpakjan and S. R. Schmid, Prentice-Hall, (2001).

Manufacturing Intelligence, P. K. Wright and D. A. Bourne, Addison Wesley (1988).

ROBOTICS

Robot Analysis, L-W Tsai, Wiley (1999).

Handbook of Industrial Robotics, Nof (ed.), Wiley (1999).

Introduction to Robots in CIM Systems, J.A. Rehg, Prentice-Hall (2000).

Industrial Robots: Computer Interfacing and Control, W. E. Snyder, Prentice Hall (1985).

Introduction to Robot Technology, P. Coiffet and M. Chizouze, McGraw-Hill (1993).

Robotics and Manufacturing Automation, 2nd Ed., C. R. Asfahl, Wiley (1992).

AUTOMATION, CNC, FMS AND CIM

An Introduction to CNC Machining and Programming, .Gibbs and T. M. Crandall, Industrial Press (1991).

Anatomy of Automation, G.H. Amber and P. S. Amber, Prentice Hall (1962).

Automatic Assembly, C. Boothroyd, C. Poli, and L. F. March, Marcel Dekker (1982).

Automation, Production Systems, and Computer-Integrated Manufacturing, 2nd Ed., Mikell P. Groover, Prentice Hall (2001).

Computer-Aided Manufacturing, 3rd Ed. T.C. Chang, R.A. Wysk and Hsu-Pin Wang, Prentice Hall (2006).

Computer-Integrated Design and Manufacturing, D.D. Bedworth, M. R. Henderson, and P. M. Wolfe, McGraw-Hill (1991).

Engineering Robust Designs, J X. Wang, Prentice-Hall (2005).

Introduction to Computer Numerical Control, J.V. Valentino and J. Goldenberg, Regents/Prentice Hall (1993).

Numerical Control and Computer Aided Manufacturing, R. S. Pressman and I. J. E. Williams, Wiley (1977).

COST ESTIMATING

Cost Estimator's Reference Manual, R.D. Steward and R.W. Wyskida, Wiley (1987).

Engineering Cost Estimating, 3rd Ed., P. E. Ostwald, Prentice Hall (1992).

Realistic Cost Estimating for Manufacturing, 2nd Ed., W. Winchell, Society of Manufacturing Engineers (1989).

NTM/LASER MACHINING SOURCES

Advance Machining Technology Handbook, James Brown, McGraw-Hill, New York (1998).

Advanced Methods of Machining, J.A. McGeough, Chapman and Hall, New York (1988).

Advanced Machining Processes, H. El-Hofy, McGraw-Hill, (2005).

Laser Material Processing, W.M. Steen, Springer, London, England (1998).

Laser Materials Processing, Ed. L. Migliore, (1996).

Laser Machining, G. Chryssolouris, Springer-Verlag (1991).

Non-Traditional Machining Handbook, Carl Sommer, Advance Publishing, Inc., Houston, TX (2000).

The EDM Handbook, E. B.Guitrau and Hanser Gardner, Cincinnati, OH (1997).

RAPID PROTOTYPING

Automated Fabrication: Improving Productivity in Manufacturing, M. Burns, Prentice Hall, Englewood Cliffs, NJ (1993).

Rapid Prototyping and Other RP&M Technologies: From Rapid Prototyping to Rapid Tooling, P. F. Jacobs, SME, Dearborn, MI (1996).

Rapid Manufacturing: The Technologies and Applications of Rapid Prototyping and Rapid Tooling, D.T. Pham and S. S. Dimov, Springer-Verlag, London, UK (2001).

Rapid Tooling: Technologies and Industrial Applications, Peter D. Hilton and P. F. Jacobs, Marcel Dekker, New York (2000).

User's Guide to Rapid Prototyping, Todd Grimm, Society of Manufacturing Engineers, (2004).

Wohler's Report 2005: RP, RT and RM State of the Industry, Terry Wohlers, Timecompress.com

LEAN MANUFACTURING

A Revolution in Manufacturing: The SMED System, Shigeo Shingo, Productivity Press (1985).

A Study of The Toyota Production System, Shigeo Shingo, Productivity Press (1989).

Becoming Lean, Ed. J. Liker, Productivity Press, (2004).

Design, Analysis and Control and Manufacturing Cells, PED-Vol. 53, J.T. Black, B. C. Jiang, and G. J. Wiens, ASME (1991).

Handbook of Cellular Manufacturing Systems, Shahrukh A. Irani (ed.), Wiley (1999).

Introduction to TPM: Total Productive Maintenance, Seiichi Nakajima, Productivity Press (1988).

JIT Factory Revolution, H. Hirano and J T. Black, Productivity Press (1988).

Japanese Manufacturing Techniques: Nine Hidden Lessons in Simplicity, Richard J. Schonberger (1982). Free Press.

Kaizen for Quick Changeover, Kenichi Sekine and Keisuke Arai (translated by Bruce Talbot), Productivity Press (1992).

Kanban: Just-In-Time at Toyota, translation by D. J. Lu, Japan Management Association, Productivity Press (1986).

Kawasaki U.S.A.: A Case Study, Robert Hall, APICS (1982).

Lean Production Simplified, Pascal Dennis, (2002).

One-Piece Flow: Cell Design for Transforming the Production Process, Kenichi Sekine, Productivity Press (1990).

The Design of the Factory with A Future, J T. Black, McGraw-Hill (1991).

The Machine That Changed The World, James P. Womack, Daniel T. Jones, and Daniel Roos, Harper Perennial (1991).

Just-in-Time in American Manufacturing, William Duncan, SME, (1988).

Kaikaku, The Power and Magic of Lean, N. Bodek, PCS Press, (2004).

Lean Assembly, Michael Baudin, Productivity Press (2002).

Lean Enterprise Systems, Steve Bell, Wiley (2006).

Lean Manufacturing Implementation, D.P. Hobbs, APICS (2004).

The Toyota Product Development System, J.M. Morgan and J.K. Liker, Productivity Press, (2006).

Toyota Production System: Beyond Large-Scale Production, Taiichi Ohno, Productivity Press (1988).

Toyota Production System—An Integrated Approach to Just-In-Time, 3rd Ed., Yasuhiro Monden (1998).

World Class Manufacturing, Richard J. Schonberger, The Free Press (1986).

World Class Manufacturing Casebook: Implementing JIT and TOC, Richard J. Schonberger, The Free Press (1987).

Zero Inventories, Robert W. Hall, Dow Jones-Irwin (1983).

Inside the Mind of Toyota, Hino, Satuski, Productivity Press (2006).

Factory Physics, W.J. Hopp, and M.L. Spearman, McGraw-Hill (2001).

Lean Thinking, James Womack, Jones, Danie, Simon & Schuster, New York, (1996).

Competitive Manufacturing Management, John Nicholas, McGraw-Hill, (1998).

World Class Manufacturing, The Next Decade, Richard Schonberger, The Free Press, (1996).

Acronyms

AC	Adaptive Control
AFM	Abrasive Flow Machining
AGVS	Automated Guided Vehicle System
AI	Artificial Intelligence
APT	Automatic Programming of Tools
AQL	Acceptable Quality Limit (or Level)
ASCII	American Standard Code
AS/RS	Automatic Storage/Retrieval System
ATE	Automatic Test Equipment
AWJM	Abrasive Water Jet Machining
BASIC	Beginner's All-Purpose Symbolic Instruction Code
BTRI	Behind the Tape Reader Interface
CAD	Computer-Aided Design
CAD/CAM	Computer-Aided Design/Computer-Aided Manufacturing
CAD/D	Computer-Aided Drafting and Design
CAE	Computer-Aided Engineering
CAM	Computer-Aided Manufacturing
CAPP	Computer-Aided Process Planning
CATI	Computer-Aided Testing and Inspection
CDC	Cutter Diameter Compensation
CHM	Chemical Machining
CIM	Computer-Integrated Manufacturing
CL	Center Line
CMM	Coordinate Measuring Machine
CMS	Cellular Manufacturing System
CNC	Computer Numerical Control
COBOL	Common Business Oriented Language
CPR	Capacity Resources Planning
CPU	Central Processing Unit (Computer)
CRT	Cathode Ray Tube
CVD	Chemical Vapor Deposition
DBM	Data-Base Management
DDAS	Direct Data Acquisition System
DDC	Direct Digital Control
DNC	Digital (or Direct or Distributed) Numerical Control
DOS	Disk Operating System
DP	Diametrical Pitch
DPRO	Digital Position Readout
DRO	Digital Readout
EAROM	Electrically-Alterable Read-Only Memory
EBCDIC	Extended Binary Coded Decimal Interchange Code
EBM	Electron Beam Machining *(EBW = Welding) (EBC = Cutting)*
ECM	Electrochemical Machining
EDM	Electrodischarge Machining *(EDG = Grinding)*
EMI	Electromagnetic Interface
EOB	End of Block
EOP	End of Program (workpiece)
EOT	End of Tape
EROM	Eraseable Read-Only Memory
ESW	Electroslag Welding
FCAW	Flux Cored Arc Welding
FEM	Finite-Element Method
FMC	Flexible Manufacturing Cell
FMS	Flexible Manufacturing System
FORTRAN	Formula Translation
FRN	Feed Rate Number
GMAW	Gas Metal Arc Welding
GT	Group Technology
GTAW	Gas Tungsten Arc Welding
HAZ	Heat Affected Zone
HERF	High Energy Rate Forming